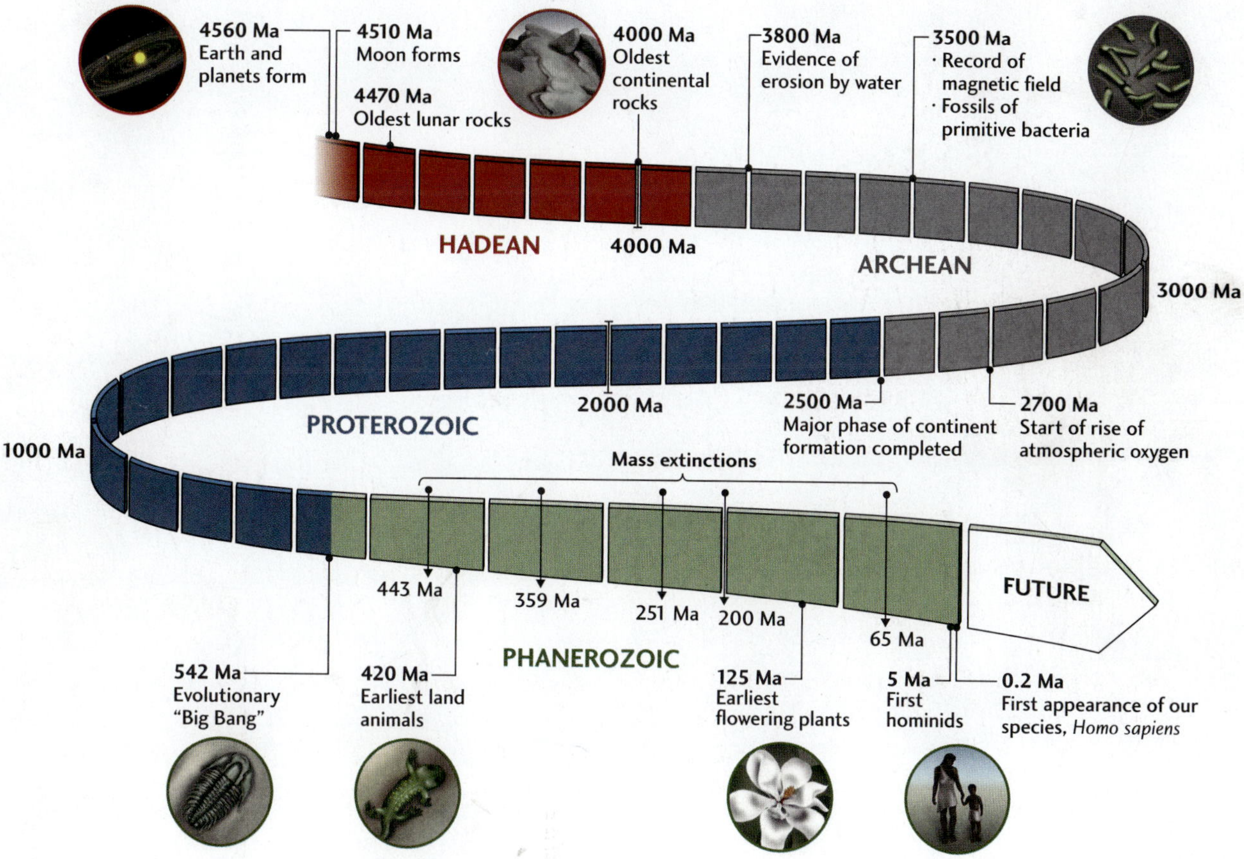

- **4560 Ma** Earth and planets form
- **4510 Ma** Moon forms
- **4470 Ma** Oldest lunar rocks
- **4000 Ma** Oldest continental rocks
- **3800 Ma** Evidence of erosion by water
- **3500 Ma**
  · Record of magnetic field
  · Fossils of primitive bacteria

HADEAN 4000 Ma **ARCHEAN** 3000 Ma

PROTEROZOIC 2000 Ma

- **2500 Ma** Major phase of continent formation completed
- **2700 Ma** Start of rise of atmospheric oxygen

1000 Ma

Mass extinctions

PHANEROZOIC

- **542 Ma** Evolutionary "Big Bang"
- **443 Ma**
- **420 Ma** Earliest land animals
- **359 Ma**
- **251 Ma**
- **200 Ma**
- **125 Ma** Earliest flowering plants
- **65 Ma**
- **5 Ma** First hominids
- **0.2 Ma** First appearance of our species, *Homo sapiens*

FUTURE

UNDERSTANDING EARTH

Eighth Edition

John Grotzinger
California Institute of Technology

Thomas H. Jordan
University of Southern California

Austin • Boston • New York • Plymouth

Vice President, STEM: Daryl Fox
Program Director: Andrew Dunaway
Director of Development, STEM: Lisa Samols
Senior Developmental Editor: Randi Blatt Rossignol
Marketing Manager: Leah Christians
Marketing Assistant: Maddie Inskeep
Executive Media Editor: Amy Thorne
Senior Media Editor: Alexandra Gordon
Senior Director of Digital Production: Keri deManigold
Director, Content Management Enhancement: Tracey Kuehn
Senior Managing Editor: Lisa Kinne
Senior Content Project Manager: Kerry O'Shaughnessy
Senior Workflow Project Supervisor: Susan Wein
Permissions Manager: Jennifer MacMillan
Photo Editor: Jennifer Atkins
Director of Design, Content Management: Diana Blume
Design Services Manager: Natasha Wolfe
Interior Designer: Gary Hespenheide
Cover Designer: John Callahan
Illustrations: Emily Cooper
Illustration Coordinator: Janice Donnola
Production Supervisor: Lawrence Guerra
Composition: Lumina Datamatics, Inc.
Printing and Binding: LSC Communications
Cover Photo: © Jimmy Chin

Library of Congress Control Number: 2019948814

ISBN-13: 978-1-319-05532-5
ISBN-10: 1-319-05532-X

© 2020, 2014, 2010, 2007 by W. H. Freeman and Company
All rights reserved.

Printed in the United States of America
1 2 3 4 5 6   24 23 22 21 20 19

Macmillan Learning
One New York Plaza
Suite 4600
New York, NY 10004-1562
www.macmillanlearning.com

In 1946, William Freeman founded W. H. Freeman and Company and published Linus Pauling's *General Chemistry*, which revolutionized the chemistry curriculum and established the prototype for a Freeman text. W. H. Freeman quickly became a publishing house where leading researchers can make significant contributions to mathematics and science. In 1996, W. H. Freeman joined Macmillan and we have since proudly continued the legacy of providing revolutionary, quality educational tools for teaching and learning in STEM.

Dedication

We dedicate this book to Frank Press and Ray Siever, pioneering educators in the era of modern geology. This book was possible only because they led the way.

About the Authors

John Grotzinger is a field geologist interested in the evolution of Earth's surface environments and biosphere. He also works on the early environmental evolution of Mars and on assessing its potential habitability. His research addresses the chemical development of the early oceans and atmosphere, the environmental context of early animal evolution, and the geologic factors that regulate sedimentary basins. His fieldwork has taken him to northwestern Canada, northern Siberia, southern Africa, the western United States, and via robot, to Mars. He received a B.S. in geoscience from Hobart College in 1979, an M.S. in geology from the University of Montana in 1981, and a Ph.D. in geology from Virginia Polytechnic Institute and State University in 1985. He spent three years as a research scientist at the Lamont-Doherty Geological Observatory before joining the MIT faculty in 1988. From 1979 to 1990, he was engaged in regional mapping for the Geological Survey of Canada. From 2008 to 2015, he served as the Chief Scientist for the Mars Curiosity Rover team, the first mission to assess the habitability of the ancient environments of another planet.

In 1998, Dr. Grotzinger was named the Waldemar Lindgren Distinguished Scholar at MIT, and in 2000, he became the Robert R. Shrock Professor of Earth and Planetary Sciences. In 2005, he moved from MIT to Caltech, where he is the Fletcher Jones Professor of Geology. He received the Presidential Young Investigator Award of the National Science Foundation in 1990, the Donath Medal of the Geological Society of America in 1992, the Charles Doolittle Walcott Medal of the National Academy of Sciences in 2007, and NASA's Outstanding Public Leadership Medal in 2013. He is a member of the American Academy of Arts and Sciences and the U.S. National Academy of Sciences.

Tom Jordan is a geophysicist interested in the composition, dynamics, and evolution of the solid Earth. He has conducted research into the nature of deep subduction, the formation of thickened keels beneath ancient continental cratons, and the question of mantle stratification. He has developed a number of seismological techniques for investigating Earth's interior that bear on geodynamic problems. He has also worked on modeling plate movements, measuring tectonic deformation, quantifying seafloor morphology, and characterizing large earthquakes. He received his Ph.D. in geophysics and applied mathematics at the California Institute of Technology (Caltech) in 1972 and taught at Princeton University and the Scripps Institution of Oceanography before joining the Massachusetts Institute of Technology (MIT) faculty as the Robert R. Shrock Professor of Earth and Planetary Sciences in 1984. He served as the head of MIT's Department of Earth, Atmospheric and Planetary Sciences for the decade 1988–1998. He moved from MIT to the University of Southern California (USC) in 2000, where he is University Professor and W. M. Keck Professor of Earth Sciences. From 2002 to 2017, he was the director of the Southern California Earthquake Center, where he coordinated an international research program in earthquake system science that involved over 600 scientists at more than 60 universities and research organizations.

Dr. Jordan received the Macelwane Medal of the American Geophysical Union in 1983, the Woollard Award of the Geological Society of America in 1998, and the Lehmann Medal of the American Geophysical Union in 2005. He is a member of the American Academy of Arts and Sciences, the U.S. National Academy of Sciences, and the American Philosophical Society.

Brief Contents

About the Authors   vi

Preface   xvii

Chapter 1
The Earth System   1

Chapter 2
Plate Tectonics: The Unifying Theory   23

Chapter 3
Earth Materials: Minerals and Rocks   53

Chapter 4
Igneous Rocks: Solids from Melts   91

Chapter 5
Volcanoes   117

Chapter 6
Sedimentation: Rocks Formed by Surface Processes   153

Chapter 7
Metamorphism: Alteration of Rocks by Temperature and Pressure   189

Chapter 8
Deformation: Modification of Rocks by Folding and Fracturing   213

Chapter 9
Clocks in Rocks: Timing the Geologic Record   239

Chapter 10
Earthquakes   267

Chapter 11
Exploring Earth's Interior   307

Chapter 12
The Climate System   337

Chapter 13
Civilization as a Global Geosystem   365

Chapter 14
Anthropogenic Global Change   393

Chapter 15
Glaciers: The Work of Ice   423

Chapter 16
Earth Surface Processes and Landscape Development   453

Chapter 17
The Hydrologic Cycle and Groundwater   495

Chapter 18
Stream Transport: From Mountains to Oceans   527

Chapter 19
Coastlines and Deserts 559

Chapter 20
Early History of the Terrestrial Planets 599

Chapter 21
History of the Continents 633

Chapter 22
Geobiology: Life Interacts with Earth 665

APPENDIX 1 Conversion Factors AP-1

APPENDIX 2 Numerical Data Pertaining to Earth AP-2

APPENDIX 3 Chemical Reactions AP-3

APPENDIX 4 Properties of the Most Common Minerals of Earth's Crust AP-5

APPENDIX 5 Practicing Geology Exercises: Answers to Bonus Problems AP-9

APPENDIX 6 Answers to Visual Literacy Exercises AP-11

Glossary GL-1

Index I-1

Contents

About the Authors   vi

Preface   xvii

Chapter 1
The Earth System   1

The Scientific Method   2
- Hypothesis and Theory   2
- Scientific Models   3
- Importance of Scientific Collaboration   3

Geology as a Science   3

Earth's Shape and Surface   6

Peeling the Onion: Discovery of a Layered Earth   6
- Earth's Density   7
- The Mantle and Core   8
- The Crust   10
- The Inner Core   10
- Chemical Composition of Earth's Major Layers   10

Earth as a System of Interacting Components   12
- The Climate System   14
- The Plate Tectonic System   14
- The Geodynamo   15
- Interactions Among Geosystems Support Life   16

An Overview of Geologic Time   16
- The Origin of Earth and Its Global Geosystems   16
- The Evolution of Life   17

Chapter 2
Plate Tectonics: The Unifying Theory   23

The Discovery of Plate Tectonics   24
- Continental Drift   24
- Seafloor Spreading   25
- The Great Synthesis: 1963–1968   26

The Plates and Their Boundaries   27
- Divergent Boundaries   32
- Convergent Boundaries   32
- Transform Faults   34
- Combinations of Plate Boundaries   34

Rates and History of Plate Movements   35
- The Seafloor as a Magnetic Tape Recorder   35
- Deep-Sea Drilling   37
- Measurements of Plate Movements by Geodesy   37

The Grand Reconstruction   38
- Seafloor Isochrons   38
- Reconstructing the History of Plate Movements   39
- The Breakup of Pangaea   41
- The Assembly of Pangaea by Continental Drift   41
- Implications of the Grand Reconstruction   44

Mantle Convection: The Engine of Plate Tectonics   44
- Where Do the Plate-Driving Forces Originate?   44
- How Deep Does Plate Recycling Occur?   45
- What Is the Nature of Rising Convection Currents?   46

Chapter 3
Earth Materials: Minerals and Rocks   53

What Are Minerals?   54

The Structure of Matter   54
- The Structure of Atoms   55
- Atomic Number and Atomic Mass   55
- Chemical Reactions   56
- Chemical Bonds   56

The Formation of Minerals   57
- The Atomic Structure of Minerals   58
- The Crystallization of Minerals   58
- How Do Minerals Form?   60

Classes of Rock-Forming Minerals   60
- Silicates   61
- Carbonates   63
- Oxides   64
- Sulfides   64
- Sulfates   65

Physical Properties of Minerals   65
- Hardness   66
- Cleavage   66
- Fracture   68
- Luster   68
- Color   68
- Density   69
- Crystal Habit   69

What Are Rocks?   70
- Properties of Rocks   70
- Igneous Rocks   71
- Sedimentary Rocks   73
- Metamorphic Rocks   74

The Rock Cycle: Interactions Between the Plate Tectonic and Climate Systems   75

CONTENTS

Concentrations of Valuable Mineral Resources   77
   Hydrothermal Deposits   79
   Igneous Deposits   82
   Sedimentary Deposits   82
   Mineral Evolution   83

Chapter 4
Igneous Rocks: Solids from Melts   91

How Do Igneous Rocks Differ from One Another?   92
   Texture   92
   Chemical and Mineral Composition   95

How Do Magmas Form?   97
   How Do Rocks Melt?   97
   The Formation of Magma Chambers   99
   Where Do Magmas Form?   99

Magmatic Differentiation   99
   Fractional Crystallization: Laboratory and Field Observations   100
   Granite from Basalt: Complexities of Magmatic Differentiation   101

Forms of Igneous Intrusions   104
   Plutons   105
   Sills and Dikes   105
   Veins   107

Igneous Processes and Plate Tectonics   107
   Spreading Centers as Magma Factories   108
   Subduction Zones as Magma Factories   111
   Mantle Plumes as Magma Factories   113

Chapter 5
Volcanoes   117

Volcanoes as Geosystems   118

Lavas and Other Volcanic Deposits   119
   Types of Lava   119
   Textures of Volcanic Rocks   123
   Pyroclastic Deposits   123

Eruptive Styles and Landforms   124
   Central Eruptions   125
   Fissure Eruptions   128

Interactions of Volcanoes with Other Geosystems   130
   Volcanism and the Hydrosphere   131
   Volcanism and the Atmosphere   132

The Global Pattern of Volcanism   133
   Volcanism at Spreading Centers   133
   Volcanism in Subduction Zones   134
   Intraplate Volcanism: The Mantle Plume Hypothesis   136
   2018 Eruption of Kilauea Volcano, Hawaii   137

Volcanism and Human Affairs   140
   Volcanic Hazards   141
   Reducing the Risks of Volcanic Hazards   144
   Natural Resources from Volcanoes   145

Chapter 6
Sedimentation: Rocks Formed by Surface Processes   153

Surface Processes of the Rock Cycle   154
   Weathering and Erosion: The Source of Sediments   155
   Transportation and Deposition: The Downhill Journey to Sedimentary Basins   157
   Oceans as Chemical Mixing Vats   160

Sedimentary Basins: The Sinks for Sediments   160
   Rift Basins and Thermal Subsidence Basins   160
   Flexural Basins   162

Sedimentary Environments   162
   Continental Sedimentary Environments   162
   Shoreline Sedimentary Environments   162
   Marine Sedimentary Environments   164
   Siliciclastic versus Chemical and Biological Sedimentary Environments   164

Sedimentary Structures   165
   Cross-Bedding   165
   Graded Bedding   165
   Ripples   166
   Bioturbation Structures   166
   Bedding Sequences   167

Burial and Diagenesis: From Sediment to Rock   168
   Burial   168
   Diagenesis   168

Classification of Siliciclastic Sediments and Sedimentary Rocks   172
   Coarse-Grained Siliciclastics: Gravel and Conglomerate   172
   Medium-Grained Siliciclastics: Sand and Sandstone   172
   Fine-Grained Siliciclastics   174

Classification of Chemical and Biological Sediments and Sedimentary Rocks   175
   Carbonate Sediments and Rocks   176
   Evaporite Sediments and Rocks: Products of Evaporation   178
   Other Biological and Chemical Sediments   181

Chapter 7
Metamorphism: Alteration of Rocks by Temperature and Pressure   189

Causes of Metamorphism   190
   The Role of Temperature   191
   The Role of Pressure   192
   The Role of Fluids   193

CONTENTS xi

Types of Metamorphism 193
    Regional Metamorphism 193
    Contact Metamorphism 193
    Seafloor Metamorphism 194
    Other Types of Metamorphism 194

Metamorphic Textures 195
    Foliation and Cleavage 195
    Foliated Rocks 195
    Granoblastic Rocks 197
    Porphyroblasts 198

Regional Metamorphism and Metamorphic Grade 199
    Mineral Isograds: Mapping Zones of Change 199
    Metamorphic Grade and Parent Rock Composition 200
    Metamorphic Facies 201

Plate Tectonics and Metamorphism 202
    Metamorphic Pressure-Temperature Paths 202
    Ocean-Continent Convergence 204
    Continent-Continent Collision 205
    Exhumation: A Link Between the Plate Tectonic and Climate Systems 206

Chapter 8
Deformation: Modification of Rocks by Folding and Fracturing 213

Plate Tectonic Forces 214

Mapping Geologic Structure 214
    Measuring Strike and Dip 214
    Geologic Maps 216
    Geologic Cross Sections 216

How Rocks Deform 217
    Brittle and Ductile Behavior of Rocks in the Laboratory 217
    Brittle and Ductile Behavior of Rocks in Earth's Crust 219

Basic Deformation Structures 219
    Faults 219
    Folds 222
    Circular Structures 224
    Joints 226
    Deformation Textures 227

Styles of Continental Deformation 228
    Tensional Tectonics 229
    Compressive Tectonics 229
    Shearing Tectonics 230

Unraveling Geologic History 231

Chapter 9
Clocks in Rocks: Timing the Geologic Record 239

Reconstructing Geologic History From the Stratigraphic Record 241
    Principles of Stratigraphy 242
    Fossils as Recorders of Geologic Time 242
    Unconformities: Gaps in the Geologic Record 245
    Cross-Cutting Relationships 246

The Geologic Time Scale: Relative Ages 247
    Intervals of Geologic Time 247
    Interval Boundaries Mark Mass Extinctions 247
    Ages of Petroleum Source Rocks 249

Measuring Absolute Time with Isotopic Clocks 249
    Discovery of Radioactivity 250
    Radioactive Isotopes: The Clocks in Rocks 253
    Isotopic Dating Methods 256

The Geologic Time Scale: Absolute Ages 256
    Eons: The Longest Intervals of Geologic Time 256
    Perspectives on Geologic Time 258

Recent Advances in Timing the Earth System 259
    Sequence Stratigraphy 259
    Chemical Stratigraphy 260
    Paleomagnetic Stratigraphy 260
    Clocking the Climate System 260

Chapter 10
Earthquakes 267

What Is an Earthquake? 269
    The Elastic Rebound Theory 269
    Fault Rupture During Earthquakes 271
    Foreshocks and Aftershocks 271

How Do We Study Earthquakes? 274
    Seismographs 274
    Seismic Waves 275
    Locating the Focus 277
    Measuring the Size of an Earthquake 277
    Determining Fault Mechanisms 281
    GPS Measurements and "Silent" Earthquakes 283

Earthquakes and Patterns of Faulting 284
    The Big Picture: Earthquakes and Plate Tectonics 284
    Regional Fault Systems 286

Earthquake Hazards and Risks 287
    How Earthquakes Cause Damage 288
    Reducing Earthquake Risk 293

xii CONTENTS

Can Earthquakes Be Predicted? 297
    Long-Term Forecasting 298
    Short-Term Prediction 300
    Medium-Term Forecasting 300

Chapter 11
Exploring Earth's Interior 307

Exploring Earth's Interior with Seismic Waves 308
    Basic Types of Waves 308
    Paths of Seismic Waves Through Earth 308
    Seismic Exploration of Near-Surface Layering 311
Layering and Composition of Earth's Interior 312
    The Crust 312
    The Mantle 312
    The Core-Mantle Boundary 314
    The Core 315
Earth's Internal Temperature 316
    Heat Flow Through Earth's Interior 316
    Temperatures Inside Earth 317
Visualizing Earth's Three-Dimensional Structure 319
    Seismic Tomography 320
    Earth's Gravitational Field 320
Earth's Magnetic Field and the Geodynamo 320
    The Dipole Field 323
    Complexity of the Magnetic Field 323
    Paleomagnetism 326
    The Magnetic Field and the Biosphere 328

Chapter 12
The Climate System 337

What Is Climate? 338
Components of the Climate System 339
    The Atmosphere 339
    The Hydrosphere 341
    The Cryosphere 342
    The Lithosphere 343
    The Biosphere 344
The Greenhouse Effect 345
    A Planet Without Greenhouse Gases 346
    Earth's Greenhouse Atmosphere 346
    Balancing the Climate System Through Feedbacks 347
    Climate Models and Their Limitations 348
Climate Variation 348
    Short-Term Regional Variations: El Niño and the Southern Oscillation 349
    Long-Term Global Variations: The Pleistocene Ice Ages 350
    Milankovitch Cycles 353
    Long-Term Global Variations: Ancient Ice Ages and Warm Periods 354
The Carbon Cycle 355
    Geochemical Cycles and How They Work 355
    The Cycling of Carbon 355
    Human Perturbations of the Carbon Cycle 357

Chapter 13
Civilization as a Global Geosystem 365

Growth and Impact of Civilization 366
    Human Population Growth 366
    Energy Resources 367
    Rise of the Hydrocarbon Economy 368
    Energy Consumption 369
    Carbon Flux from Energy Production 370
Fossil-Fuel Resources 370
    The Geologic Formation of Hydrocarbons 371
    Hydrocarbon Reservoirs 372
    Producing Oil and Gas from Tight Formations 373
    Distribution of Oil Reserves 373
    Oil Production 375
    Natural Gas 375
    Coal 376
    Other Hydrocarbon Resources 377
    Environmental Costs of Extracting Fossil Fuels 378
Alternative Energy Resources 381
    Nuclear Energy 381
    Biofuels 383
    Hydroelectric Energy 384
    Wind Energy 384
    Solar Energy 385
    Geothermal Energy 386
Our Energy Future 386

Chapter 14
Anthropogenic Global Change 393

Rise of Carbon Dioxide in the Atmosphere: The Keeling Curve 394
Types of Anthropogenic Global Change 395
    Chemical Change 396
    Physical Change 397
    Biological Change 400
Climate Change 400
    Projecting Future Climate Change 401
    Human Population and Global Change 402
    Consequences of Climate Change 403
Ocean Acidification 408
Loss of Biodiversity 410
    Dawning of the Anthropocene 410

Managing the Carbon Crisis   412
   Energy Policy   412
   Use of Alternative Energy Resources   413
   Engineering the Carbon Cycle   413
   Stabilizing Carbon Emissions   414

Chapter 15
Glaciers: The Work of Ice   423

Types of Glaciers   424
   Ice Accumulations as Rock Formations   424
   Valley Glaciers   425
   Continental Glaciers   425

How Glaciers Form   427
   Basic Ingredients: Freezing Cold and Lots of Snow   427
   Glacial Growth: Accumulation   428
   Glacial Shrinkage: Ablation   428
   The Glacial Budget: Accumulation Minus Ablation   429

How Glaciers Move   429
   Mechanisms of Glacial Flow   429
   Flow in Valley Glaciers   431
   Antarctica in Motion   431

Isostasy and Sea-Level Change   433

Glacial Landscapes   434
   Glacial Erosion and Erosional Landforms   435
   Glacial Sedimentation and Sedimentary Landforms   437
   Permafrost   441

Glacial Cycles and Climate Change   442
   The Wisconsin Glaciation   443
   Glaciation and Sea-Level Change   443
   The Geologic Record of Pleistocene Glaciations   444
   Variations During the Most Recent Glacial Cycle   444
   The Geologic Record of Ancient Glaciations   445

Chapter 16
Earth Surface Processes and Landscape Development   453

Controls on Weathering   454
   The Properties of Parent Rock   455
   Climate: Rainfall and Temperature   455
   The Presence or Absence of Soil   455
   The Length of Exposure   455

Chemical Weathering   456
   The Role of Water: Feldspar and Other Silicates   456
   Carbon Dioxide, Weathering, and the Climate System   457
   The Role of Oxygen: From Iron Silicates to Iron Oxides   459

   Chemical Stability   460

Physical Weathering   461
   How Do Rocks Break?   461

Soils: The Residue of Weathering   462
   Soils as Geosystems   463
   Paleosols: Working Backward from Soil to Climate   464

Erosion and Formation of Stream Valleys   464
   Interactions Between Weathering and Erosion   467

Mass Wasting   467
   Slope Materials   469
   Water Content   471
   Slope Steepness   471
   Triggers of Mass Movements   472

Classification of Mass Movements   472
   Mass Movements of Rock   475
   Mass Movements of Unconsolidated Material   478

Geomorphology and Landscape Development   482
   Types of Landforms   484
   Interactions Between Climate, Tectonics, and Topography Control Landscapes   486

Chapter 17
The Hydrologic Cycle and Groundwater   495

The Geologic Cycling of Water   496
   Flow and Reservoirs   497
   How Much Water Is There?   497
   The Hydrologic Cycle   497
   How Much Water Can We Use?   498

Hydrology and Climate   498
   Humidity, Rainfall, and Landscape   498
   Droughts   499
   The Hydrology of Runoff   501

The Hydrology of Groundwater   503
   Porosity and Permeability   504
   The Groundwater Table   506
   Aquifers   508
   Balancing Recharge and Discharge   508
   The Speed of Groundwater Flows   511
   Groundwater Resources and Their Management   513

Erosion by Groundwater   515

Water Quality   516
   Contamination of the Water Supply   517
   Reversing Contamination   518
   Is the Water Drinkable?   519

Water Deep in the Crust   520
   Hydrothermal Waters   520
   Ancient Microorganisms in Deep Aquifers   522

Chapter 18
Stream Transport: From Mountains to Oceans 527

The Form of Streams 528
Stream Valleys 528
Channel Patterns 529
Stream Floodplains 531
Drainage Basins 531
Drainage Networks 533
Drainage Patterns and Geologic History 533

Where Do Channels Begin? How Running Water Erodes Soil and Rock 535
Abrasion 536
Chemical and Physical Weathering 536
The Undercutting Action of Waterfalls 536

How Currents Flow and Transport Sediment 537
Erosion and Sediment Transport 538
Sediment Bed Forms: Dunes and Ripples 539

Deltas: The Mouths of Rivers 541
Delta Sedimentation 541
The Growth of Deltas 541
Human Effects on Deltas 543
The Effects of Ocean Currents, Tides, and Plate Tectonic Processes 543

Streams as Geosystems 543
Discharge 544
Floods 546
Longitudinal Profiles 549
Lakes 552

Chapter 19
Coastlines and Deserts 559

Coastal Processes 560
Wave Motion: The Key to Shoreline Dynamics 560
The Surf Zone 562
Wave Refraction 562
Tides 564

The Shaping of Shorelines 565
Beaches 565
Erosion and Deposition at Shorelines 568
Effects of Sea-Level Change 572

Hurricanes and Coastal Storm Surges 573
Hurricane Formation 574
Storm Surges 576
Hurricane Landfall 578

Desert Processes 579
Wind Strength 580
Particle Size 580

Windblown Sand and Dust 581
Sandblasting 581
Deflation 581
Sand Dunes 583
Windblown Dust 585
Dust Falls and Loess 587

The Desert Environment 588
Desert Weathering and Erosion 588
Desert Sediments and Sedimentation 589
Desert Landscapes 590

Tectonic, Climatic, and Human Controls on Deserts 591
The Role of Plate Tectonics 591
The Impact of Climate Change 592
The Influence of Humans 593

Chapter 20
Early History of the Terrestrial Planets 599

Origin of the Solar System 600
The Nebular Hypothesis 600
The Sun Forms 601
The Planets Form 601
Small Bodies of the Solar System 602

Early Earth: Formation of a Layered Planet 603
Earth Heats Up and Melts 603
Differentiation of Earth's Core, Mantle, and Crust 604
Earth's Oceans and Atmosphere Form 605

Diversity of the Planets 606

What's in a Face? The Age and Complexion of Planetary Surfaces 608
The Man in the Moon: A Planetary Time Scale 608
Mercury: The Ancient Planet 609
Pluto: The Dwarf Planet 610
Venus: The Volcanic Planet 611
Mars: The Red Planet 613
Earth: No Place Like Home 615

Mars Rocks! 616
Missions to Mars: Flybys, Orbiters, Landers, and Rovers 617
Mars Exploration Rovers (MER): *Spirit* and *Opportunity* 619
Mars Science Laboratory (MSL): *Curiosity* 619
InSight Lander Mission 622
Recent Missions: *Mars Reconnaissance Orbiter* (2006–) and *Phoenix* (May–November 2008) 622
Recent Discoveries: The Environmental Evolution of Mars 623

Exploring the Solar System and Beyond 626
Space Missions 626
The Cassini-Huygens Mission to Saturn 627
Other Solar Systems 628

Chapter 21
History of the Continents 633

The Structure of North America 634
The Stable Interior 635
The Appalachian Fold Belt 636

The Coastal Plain and Continental Shelf   637
The North American Cordillera   638

Tectonic Provinces Around the World   640
Types of Tectonic Provinces   640
Tectonic Ages   642
A Global Puzzle   642

How Continents Grow   643
Magmatic Addition   643
Accretion   643

How Continents Are Modified   646
Orogeny: Modification by Plate Collision   646
The Wilson Cycle   651
Epeirogeny: Modification by Vertical Movements   651

The Origin of Cratons   655

Deep Structure of Continents   657
Cratonic Keels   657
Composition of the Keels   657
Age of the Keels   658

Chapter 22
Geobiology: Life Interacts with Earth   665

The Biosphere as a System   666
Ecosystems   666
Inputs: The Stuff Life Is Made Of   668
Processes and Outputs: How Organisms Live and Grow   669
Biogeochemical Cycles   670

Microorganisms: Nature's Tiny Chemists   671
Abundance and Diversity of Microorganisms   671
Microorganism-Mineral Interactions   674

Geobiologic Events in Earth's History   679
Origin of Life and the Oldest Fossils   679
Prebiotic Soup: The Original Experiment on the Origin of Life   680
The Oldest Fossils and Early Life   680
Origin of Earth's Oxygenated Atmosphere   682

Evolutionary Radiations and Mass Extinctions   684
Radiation of Life: The Cambrian Explosion   685
Tail of the Devil: The Demise of Dinosaurs   687
Global Warming Disaster: The Paleocene-Eocene Mass Extinction   689

Astrobiology: The Search for Extraterrestrial Life   690
Habitable Zones Around Stars   691
Habitable Environments on Mars   692

APPENDIX 1   Conversion Factors   AP-1

APPENDIX 2   Numerical Data Pertaining to Earth   AP-2

APPENDIX 3   Chemical Reactions   AP-3

APPENDIX 4   Properties of the Most Common Minerals of Earth's Crust   AP-5

APPENDIX 5   Practicing Geology Exercises: Answers to Bonus Problems   AP-9

APPENDIX 6   Answers to Visual Literacy Exercises   AP-11

Glossary   GL-1

Index   I-1

Preface

Understanding Earth as a Changing Planet

Geology is everywhere in our daily lives. We are surrounded by materials and resources extracted from the Earth, from jewelry to the gasoline we use to fuel our cars, to the water we drink, and our understanding of Earth's past and future climate. Geological science routinely informs the decisions of public policy leaders in government, industry, and community organizations. Understanding our Earth has never been more important.

Because Earth science is so intertwined with our daily lives, our discipline evolves as the years go by, responding to the needs of what society compels us to understand. Decades ago, most geologists worked in oil and mining companies, but today there is an exploding need for environmental specialists and scientists who understand the oceans, atmosphere, and Earth's climate. As our world population grows, we see the increased impact of hurricanes, drought, tornadoes, and other environmental forces such as earthquakes and landslides. Even in the search for life on other planets, we increasingly see the need for geological expertise in helping to reconstruct the environments on planets like Mars. There, geologists are exploring for traces of past life in rocks that are billions of years old, with robots that are hundreds of millions of miles away.

These diverse needs require a strong understanding of the basic concepts and principles of Earth science. Although the times change and the applications vary, understanding the basic composition of geologic materials, their origins, and how the planet acts as a system is imperative to understanding Earth. Everything from climate change, to the abundance of groundwater, to the frequency of large storms and volcanic eruptions, to the location and cost of extracting rare elements from Earth is relevant. It is a simple fact that as the complexity of these challenges increases, the need for well-educated geologists to make wise decisions will increase as well. We bring that conviction to this book.

As Earth changes around us, so does society's need to understand the relationships among energy, climate, and the environment. *Understanding Earth*—while reinforcing the core concepts of physical geology—seeks to engage students in serious discussions of how fossil-fuel burning affects the environment and contributes to global change.

FIGURE 14.19 Bubbles of methane gas frozen into clear ice in Lake Baikal near the Mongolian border in Russia. [Streluk/Getty Images.]

FIGURE 5.23 Satellite image of the huge ash cloud spewing from the erupting Cordón Caulle volcano in central Chile on June 13, 2011. The ash plume extends 800 km from the snow-covered Andes mountains (on left side of photo) to the Argentine city of Buenos Aires (center right of photo). This ash cloud encircled the planet, closing airports in Australia and New Zealand. [NASA image courtesy Jeff Schmaltz, MODIS Rapid Response Team at NASA GSFC.]

NEW Study Tools to Help Students Learn and Engage

The eighth edition incorporates **Learning Objectives** at the beginning of each chapter to promote more purposeful reading and to help students navigate the chapter.

The chapter-ending **Review of Learning Objectives** reinforces the material with study assignments, exercises, and thought questions.

Visual Literacy Exercises have been added to each chapter to build reasoning skills as students interpret an illustration from the chapter. These exercises support student success in geology while developing a valuable skill for lifelong critical thinking.

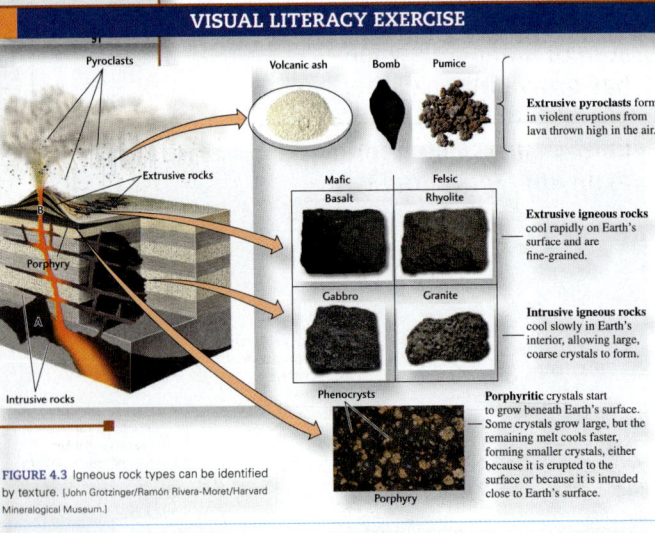

NEW Focus on Global Change

As global change becomes a more prominent topic of physical geology courses, a single chapter on climate is no longer sufficient to teach students what they need to know. In this edition of *Understanding Earth*, we have increased the coverage of climate and global change to three consecutive chapters. This comprehensive treatment is placed near the center of the book, after we introduce the key geological concepts, but before we discuss surface and hydrological processes.

FIGURE 12.13 Before and after images from December 2014 (left) and February 2015 (right) that show coral bleaching in the waters of the central Pacific. [The Ocean Agency / XL Catlin Seaview Survey.]

Chapter 12, The Climate System, discusses the components of the climate system, the greenhouse effect, climate variations, and the carbon cycle, laying the foundation for Chapters 13 and 14.

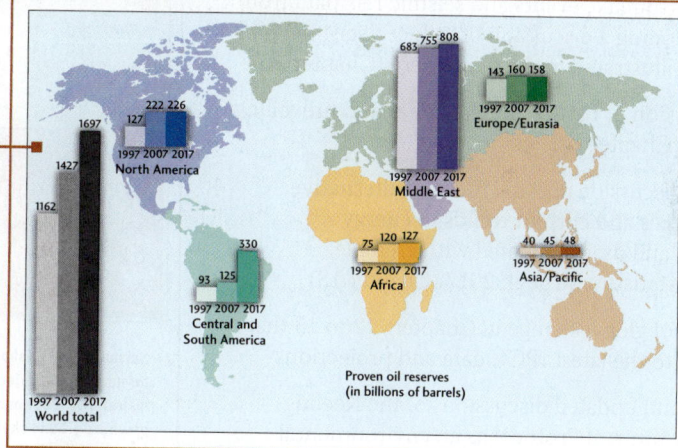

Chapter 13, Civilization as a Global Geosystem, demonstrates how civilization—the collective sum of human activities—has grown into a global geosystem that is changing Earth's surface environment at an unprecedented rate.

FIGURE 13.12 Regional estimates of world oil reserves in 1997 (left bar), 2007 (middle bar), and 2017 (right bar) in billions of barrels (bbl). The total world oil reserves in 2017 were 1.7 trillion barrels. [Data from the *British Petroleum Statistical Review of World Energy 2017*]

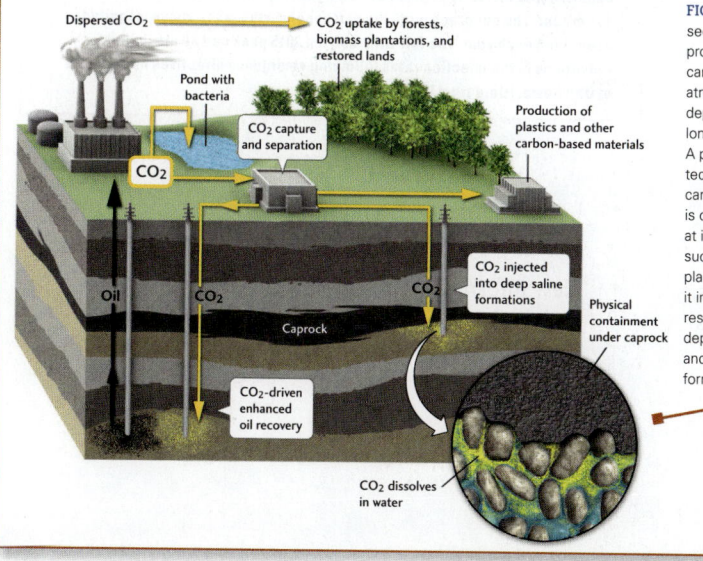

FIGURE 14.27 Carbon sequestration is the process of removing carbon from the atmosphere and depositing it into long-term reservoirs. A promising technology for carbon sequestration is capturing CO_2 at its source, such as a power plant, and storing it in underground reservoirs, such as depleted oil reservoirs and deep saline formations.

Chapter 14, Anthropogenic Global Change, discusses three of the most serious forms of human-caused global change: climate change, ocean acidification, and loss of species diversity.

NEW Changes to the Organization

- The chapter on Volcanoes (Chapter 5) has been moved forward to follow Igneous Rocks (Chapter 4).

- The chapter on Earthquakes (Chapter 9) has been moved forward to follow Deformation (Chapter 8).

- The chapter on Glaciers (Chapter 15) has been moved forward as a logical extension of the global change Chapters 12–14.

- Chapter 16, Earth Surface Processes and Landscape Development, now covers both weathering and erosion and landscape development.

- Chapter 19, Coastlines and Deserts, now covers both winds and deserts and coastlines and ocean basins. With the eighth edition's emphasis on interactions with the climate system and global change, it made sense to put these two topics together.

- Geobiology, a capstone topic, is now Chapter 22.

NEW Updated Science, Research, and Stories

In an effort to present current science and stories throughout the text, the following changes have been made in the eighth edition:

- New section on mineral evolution (Chapter 3)

- Updated earthquake events and seismic risk data from FEMA, including a discussion of induced seismicity in Oklahoma, illustrated in the figure here (Chapter 10)

- New discussion of Earth's climate zones and prevailing wind belts (Chapter 12)

- Expanded discussion of fossil fuel and alternative energy sources and current trends in energy production and usage, updated with credible, referenced statistics through 2017 (Chapter 13)

- Discussion of global change in Chapters 12 to 15 that are faithful to the latest IPCC data and projections

- Expanded and updated discussions of the Keeling curve, the Montreal Protocol as an environmental success story, ocean acidification, and loss of biodiversity (Chapter 14)

- Significantly revised chapter on glacial processes (Chapter 15) that puts more emphasis on the thickness and flow of continental glaciers that expands on the discussions of anthropogenic effects in Chapter 14; new section on isostasy and sea level change; new maps of Greenland and Antarctic ice thickness and Antarctic ice flow

- New data on annual precipitation in the United States, as well as groundwater contaminants and withdrawals (Chapter 17)

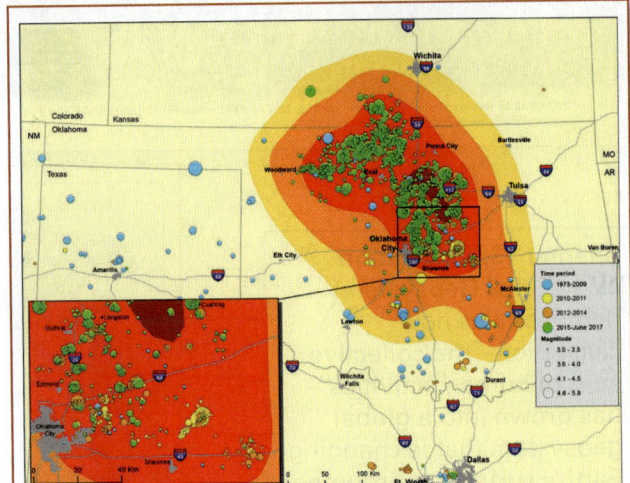

Seismicity in Oklahoma has increased dramatically since 2009, induced primarily by the injection of wastewater into deep geologic formations. The map shows locations of past earthquakes (colored dots), as well as projections of where ground shaking of Modified Mercalli Intensity VI is likely to occur (colored areas). The deepest red outlines the zone where the chances of experiencing Intensity VI ground shaking during the next year are about 10 to 14 percent. The bar graph plots the number of earthquakes above magnitude 3 recorded each year. The fall-off since the 2015 peak can be attributed to reductions in the injection rates following emergency directives by the State of Oklahoma. [Data from USGS.]

- Updated information on the expansion of deserts (Chapter 19)

- New information on the Mars Exploration Rover and the Mars Science Laboratory rover missions (Chapter 22)

NEW Student Engagement Tools

Achieve READ & PRACTICE Available with the eighth edition of *Understanding Earth*, Achieve Read & Practice is the marriage of Macmillan's popular LearningCurve adaptive quizzing and a mobile, accessible e-book all in one affordable product.

Achieve Read & Practice is easy to use. As an instructor, simply assign the materials you want students to read. When students log in, they will immediately see what reading is due. Comprehensive analytics allow instructors to monitor students' progress and identify topic areas that require additional instruction.

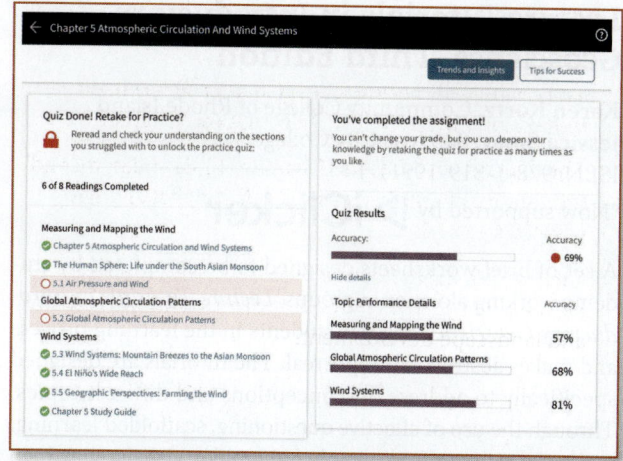

Students come to class prepared. After students complete the assigned reading, they begin a quiz that presents a variety of questions and challenges to each student on their own level of understanding. When the quiz is complete, the system prompts students to review topics in the e-book where they performed poorly and asks them additional questions to check their understanding. Students can retake the LearningCurve quiz to improve their performance and exam readiness.

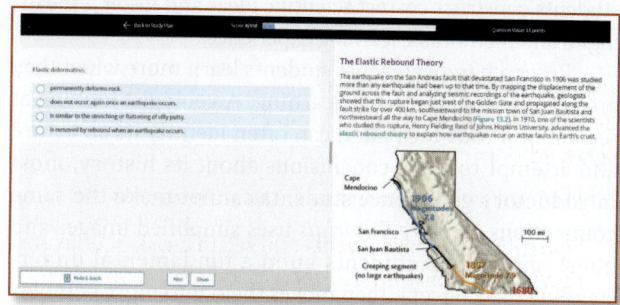

Instructors gain valuable insight into student comprehension. The performance analytics dashboard allows instructors to monitor their students' progress. This makes it easy to identify difficult topics that require additional instruction for the class or for an individual student.

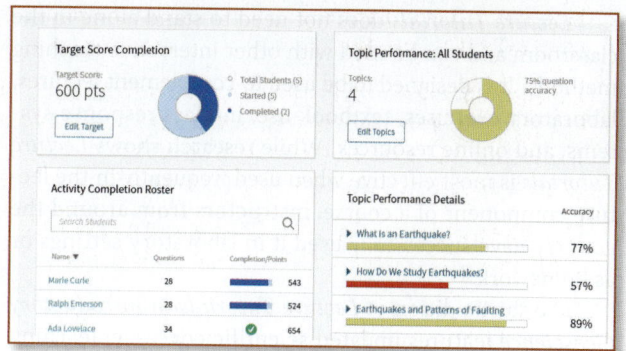

NEW Understanding Earth by Experimenting

Understanding Earth can also support your lab course! Macmillan offers a suite of geology lab exercises available in our KickStart Geology collection. KickStart Geology includes labs designed to cover common lab exercises in physical geology aligned to the organization of *Understanding Earth*, Eighth Edition.

You can choose the entire KickStart Geology collection for a ready-to-go lab manual, or customize it to fit your learning objectives by omitting, editing, and adding your own lab exercises.

NEW Understanding Earth by Engaging

Lecture Tutorials in Introductory Geoscience, Third Edition

Karen Kortz, Community College of Rhode Island
Jessica Smay, San Jose City College
ISBN: 978-1-319-19911-1
*Now supported by iClicker

A set of brief worksheets designed to be completed by students working alone or in groups, *Lecture Tutorials in Introductory Geoscience* engages students in the learning process and makes abstract concepts real. The tutorials are designed specifically to address misconceptions and difficult topics. Through the use of effective questioning, scaffolded learning, and a progression from simple to complex visuals, they help students construct correct scientific ideas and foster a meaningful and memorable learning experience.

Research indicates that students learn more when they are actively engaged while learning. A geologist looking at terrain or a rock formation can often identify its structure and attempt to draw conclusions about its history; most introductory geoscience students cannot make the same connections. *Lecture Tutorials* uses simplified images and questions to help students build a fundamental understanding of a concept, then moves them into more complex interpretations of that concept.

Lecture Tutorials does not need to stand alone in the classroom and can be used with other interactive teaching methods. It is designed to be used to complement lectures, laboratory exercises, textbook use, in-class response systems, and online resources. While research shows *Lecture Tutorials* is most effective when used frequently in the lecture component of a course, instructors from around the country have successfully used it in laboratory settings or as homework.

The third edition of *Lecture Tutorials in Introductory Geoscience* features updated scientific coverage, learning

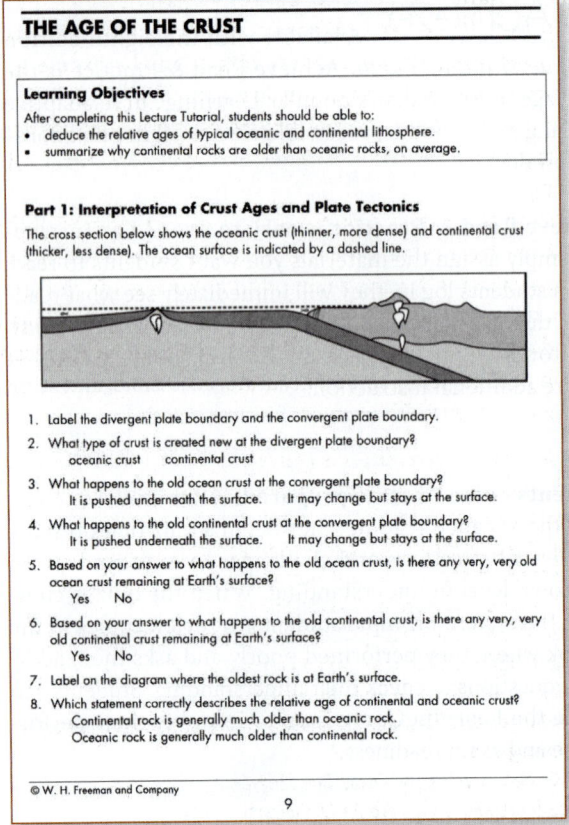

objectives aligned to every lesson, and new iClicker questions that give instructors real-time insights into students' understanding in courses of any size. *Lecture Tutorials in Introductory Geoscience* can be ordered as a stand-alone item, packaged with *Understanding Earth*, Eighth Edition, or bundled with student access to the iClicker app for only an additional $5. Visit www.macmillanlearning.com/geology for more information.

Instructor Resources

To support instructors who adopt *Understanding Earth*, Macmillan Learning offers instructor resources to support your teaching.

- PowerPoint lecture slides
- iClicker questions
- Image slides from the text
- Test bank

Understanding Earth and the Kortz/Smay *Lecture Tutorials* are also available in custom versions, so you can pick only the chapters and exercises that your class needs, and leave out the ones you don't cover. For more information, visit www.macmillanlearning.com/geology.

Acknowledgments

It is a challenge both to geology instructors and to authors of geology textbooks to compress the many important aspects of geology into a single course and to inspire interest and enthusiasm in their students. To meet this challenge, we have called on the advice of many colleagues who teach in all kinds of college and university settings.

From the earliest planning stages of each edition of this book, we have relied on a consensus of views in designing an organization for the text and in choosing which topics to include. As we wrote and rewrote the chapters, we again relied on our colleagues to guide us in making the presentation pedagogically sound, accurate, accessible, and stimulating to students. To each one we are grateful.

Dennis Ballero, *Georgia State University*
Mark Baskaran, *Wayne State University*
Margaret Benoit, *The College of New Jersey*
David Bradley, *Georgia Southern University*
Ian Brown, *Wake Technical Community College*
Tim Callahan, *College of Charleston*
Meredith Denton-Hedrick, *Austin Community College*
Glenn Dolphin, *University of Calgary*
Mark Everett, *Texas A&M University*
Mark Feigenson, *Rutgers, The State University of New Jersey–Livingston*
Lisa Greer, *Washington and Lee University*
Bruce Herbert, *Texas A&M University*
Joann Hochstein, *Lone Star College*
Qinhong Hu, *University of Texas at Arlington*
Ellery Ingall, *Georgia Institute of Technology*
Eric Jerde, *Morehead State University*
Amanda Julson, *Blinn College*
Haraldur Karlsson, *Texas Tech University*
David Kratzmann, *Santa Rosa Junior College*
Lawrence Lemke, *Wayne State University*
Pamela Nelson, *Glendale Community College*
Cassandra Runyon, *College of Charleston*
Joshua Schwartz, *California State University, Northridge*
Joanna Scheffler, *Mesa Community College*
Thomas Smith, *Kalamazoo College*
Barry Weaver, *University of Oklahoma*

We remain grateful to the instructors who were involved or assisted in the planning or reviewing stages of the seventh edition:

Marianne Caldwell, *Hillsborough Community College*
Courtney Clamons, *Austin Community College*
Ellen Cowan, *Appalachian State University*
Meredith Denton-Hedrick, *Austin Community College*
Mark Feigenson, *Rutgers, The State University of New Jersey–Livingston*
Edward Garnero, *Arizona State University*
Richard Gibson, *Texas A&M University*
Bruce Herbert, *Texas A&M University*
Bernard Housen, *Western Washington University*
Qinhong Hu, *University of Texas at Arlington*
Maureen McCurdy Hillard, *Louisiana Tech University*
Daniel Kelley, *Louisiana State University*
Steppen Murphy, *Central Piedmont Community College*
Alycia Stigall, *Ohio University*
John Tacinelli, *Rochester Community and Technical College*
J. M. Wampler, *Georgia State University*

Others have worked with us more directly in writing and preparing the manuscript for publication. At our side always was our Science Editor at W. H. Freeman and Company, Randi Rossignol. Randi has worked with us for the past 18 years, and through five revisions of the text. It has been a pleasure for us to work with her every time! Kerry O'Shaughnessy managed the production process, keeping us on schedule and with a keen eye to detail. Amy Thorne coordinated the media supplements, Natasha Wolfe designed the text, and Jennifer Atkins obtained many beautiful photographs. Crissy Dudonis coordinated the transmittal process. Andrew Dunaway, Program Director for STEM, was an enthusiastic and knowledgeable participant. And our heartfelt thanks go to Emily Cooper, who created so many beautiful illustrations.

The Earth System 1

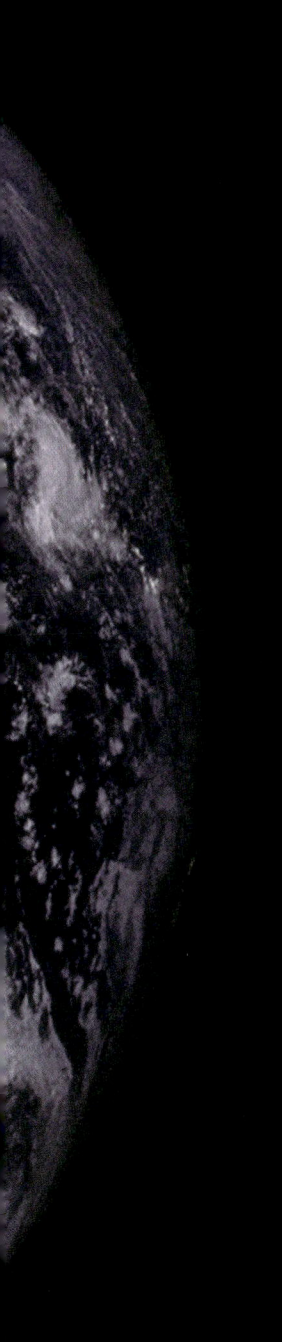

The Scientific Method	2
Geology as a Science	3
Earth's Shape and Surface	6
Peeling the Onion: Discovery of a Layered Earth	6
Earth as a System of Interacting Components	12
An Overview of Geologic Time	16

Learning Objectives

Each chapter of this textbook will begin with a set of "learning objectives" that will help you study the chapter. We will provide you with a brief summary of the concepts related to each objective at the end of the chapter. Chapter 1 describes how geologists investigate Earth as a system of interacting components. After you have studied the chapter, you should be able to

1.1 Describe key aspects of geology that relate it to, and distinguish it from, other sciences such as physics, chemistry, and biology.

1.2 Illustrate how the scientific method has been used to develop theories of Earth's shape and structure.

1.3 Compare the chemical compositions of Earth's crust, mantle, and core in terms of their most abundant elements.

1.4 Locate within the layered Earth system the global geosystems that explain climate, plate tectonics, and the geomagnetic field.

1.5 Recall Earth's age and some key events in the evolution of life that stand out in the geologic record.

First image of the whole Earth showing the Antarctic and African continents, taken by the *Apollo* astronauts on December 7, 1972. [NASA.]

CHAPTER 1 The Earth System

Earth is a unique place, home to a multitude of organisms, including ourselves. No other planet we have yet discovered has the same delicate balance of conditions necessary to sustain life. Geology is the science that studies Earth: how it was born, how it evolved, how it works, and how we can help preserve its habitats for life. Geologists seek answers to many basic questions: Of what material is the planet composed? Why are there continents and oceans? How did the Himalaya, Alps, and Rocky Mountains rise to their great heights? Why are some regions subject to earthquakes and volcanic eruptions while others are not? How did Earth's surface environment and the life it contains evolve over billions of years? What changes are likely in the future? We think you will find the answers to such questions fascinating. Welcome to the science of geology!

We have organized the discussion of geology in this textbook around three basic concepts that will appear in almost every chapter: (1) Earth as a system of interacting components, (2) plate tectonics as a unifying theory of geology, and (3) changes in the Earth system through geologic time.

This chapter gives a broad picture of how geologists think. It starts with the scientific method, the observational approach to the physical universe on which all scientific inquiry is based. Throughout this textbook, you will see the scientific method in action as we describe how Earth scientists gather and interpret information about our planet. In this first chapter, we will illustrate how the scientific method has been applied to discover some of Earth's basic features—its shape and its internal layering.

To explain features that are millions or even billions of years old, Earth scientists look at what is happening on Earth today. We will introduce the study of our complex natural world as an *Earth system* involving many interrelated components. Some of these components, such as the atmosphere and oceans, are clearly visible above Earth's solid surface; others lie hidden deep within its interior. By observing the ways in which these components interact, scientists have built up an understanding of how the Earth system has changed through geologic time.

We will also introduce you to a geologist's view of time. You will think about time differently as you begin to comprehend the immense span of geologic history. Earth and the other planets in our solar system formed about 4.5 billion years ago. More than 3 billion years ago, living cells developed on Earth's surface, and life has been evolving ever since. Yet our human origins date back only a few million years—less than a tenth of a percent of Earth's existence.

The Scientific Method

The term *geology* (from the Greek words for "Earth" and "knowledge") was coined by scientific philosophers more than 200 years ago to describe the study of rock formations and fossils. Through careful observations and reasoning, their successors developed the theories of biological evolution, continental drift, and plate tectonics—major topics of this textbook. Today, **geology** identifies the branch of Earth science that studies all aspects of the planet: its history, its composition and internal structure, and its surface features.

The goal of geology—and of science in general—is to explain the physical universe. Scientists believe that physical events have physical explanations, even if they may be beyond our present capacity to understand them. The **scientific method,** on which all scientists rely, is the general procedure for discovering how the universe works through systematic observations and experiments. Using the scientific method to make new discoveries and to confirm old ones is the process of *scientific research*.

Hypothesis and Theory

When scientists propose a *hypothesis*—a tentative explanation based on data collected through observations and experiments—they present it to the community of scientists for criticism and repeated testing. A hypothesis is supported if it explains new data or predicts the outcome of new experiments. Hypotheses confirmed by new observations gain credibility; those contradicted by the data can be rejected.

Here are four interesting scientific hypotheses we will encounter in this textbook:

- Earth is billions of years old.
- Coal is a rock formed from dead plants.
- Earthquakes are caused by the breaking of rocks along geologic faults.
- The burning of fossil fuels is causing global warming.

The first hypothesis agrees with the ages of thousands of ancient rocks as measured by precise laboratory techniques, and the next two hypotheses have also been confirmed by many independent observations. The fourth hypothesis has been more controversial, though so many new data support it that most scientists now accept it as true (see Chapters 12 and 14).

A coherent set of hypotheses proposed to explain some aspect of nature constitutes a scientific *theory*. Good theories are supported by substantial bodies of data and have survived repeated challenges. They usually obey *physical laws,* general principles about how the universe works that can be applied in almost every situation, such as Newton's law of gravity.

Some hypotheses and theories have been so extensively tested that all scientists accept them as true, at least to a good approximation. For instance, the theory that Earth is nearly spherical, which follows from Newton's law of gravity, is supported by so much experience and direct evidence (ask any astronaut) that we take it to be a fact. The longer a theory holds up to all scientific challenges, the more confidently it is held.

Yet theories can never be considered completely proved. The essence of science is that no explanation, no matter how believable or appealing, is closed to questioning. If convincing new evidence indicates that a theory is wrong, scientists will discard it or modify it to account for the data. A theory, like a hypothesis, must always be testable; any proposal about the universe that cannot be evaluated by observing the natural world should not be called a scientific theory.

For scientists engaged in research, the most interesting hypotheses are often the most controversial, rather than the most widely accepted. The hypothesis that fossil fuel burning is causing global warming has been repeatedly challenged in the decades since it was first proposed. Because the long-term predictions of climate change are so important, many Earth scientists have been vigorously testing this hypothesis and the climate theories based on it.

Scientific Models

Knowledge based on successful hypotheses and theories can be used to create a *scientific model*—a formal representation of how a natural process operates or how a natural system behaves. Scientists combine related ideas in a model to test the consistency of their knowledge and to make predictions. Like a good hypothesis or theory, a good model makes predictions that agree with observations.

A scientific model is often formulated as a computer program that simulates the behavior of a natural system through numerical calculations. The forecast of rain or sunshine you may see on TV tonight comes from a computer model of the weather. A computer can be programmed to simulate geologic phenomena that are too big to replicate in a laboratory or that operate over periods of time that are too long for humans to observe. For example, models used for predicting weather have been extended to predict climate changes decades into the future.

Importance of Scientific Collaboration

To encourage discussion of their ideas, scientists share them and the data on which they are based. They present their findings at professional meetings, publish them in professional journals, and explain them in informal conversations with colleagues. Scientists learn from one another's work and through teamwork, as well as by studying the discoveries of their predecessors. Most of the great concepts of science, whether they emerge as a flash of insight or in the course of painstaking analysis, result from untold numbers of such interactions. Albert Einstein put it this way: "In science . . . the work of the individual is so bound up with that of his scientific predecessors and contemporaries that it appears almost as an impersonal product of his generation."

Because such free intellectual exchange can be subject to abuses, a code of ethics has evolved among scientists. Scientists must not falsify data, use the work of others without recognizing them, or be otherwise deceitful in their work. They must also accept responsibility for training the next generation of researchers and teachers. These principles are supported by the basic values of scientific cooperation, which a former president of the National Academy of Sciences, Bruce Alberts, has aptly described as "honesty, generosity, a respect for evidence, openness to all ideas and opinions."

Geology as a Science

In the popular media, scientists are often portrayed as people who do experiments while wearing white coats. That stereotype is not inappropriate: Many scientific problems are best investigated in the laboratory. What forces keep atoms together? How do chemicals react with one another? Can viruses cause cancer? The phenomena that scientists observe to answer such questions are sufficiently small and happen quickly enough to be studied in the controlled environment of the laboratory.

The major questions of geology, however, involve processes that operate on much larger and longer scales. Controlled laboratory measurements yield critical data for testing geologic hypotheses and theories—the ages and properties of rocks, for instance—but they are usually insufficient to solve major geologic problems. Almost all of the great discoveries described in this textbook were made by observing Earth processes in their uncontrolled, natural environment.

For this reason, geology is an outdoor science with its own particular style and outlook. Geologists "go into the field" to observe nature directly (**Figure 1.1**). They learn how mountains were formed by climbing up steep slopes and examining the exposed rocks, and they deploy sensitive instruments to collect data on earthquakes, volcanic eruptions, and other activity within the solid Earth. They discover how ocean basins have evolved by sailing rough seas to map the ocean floor (**Figure 1.2**).

Geology is closely related to other areas of Earth science, including *oceanography*, the study of the oceans; *meteorology*, the study of the atmosphere; and *ecology*, which concerns the abundance and distribution of life. *Geophysics*, *geochemistry*, and *geobiology* are subfields of geology that apply the methods of physics, chemistry, and biology to geologic problems (**Figure 1.3**).

Geology is a *planetary science* that uses remote sensing devices, such as instruments mounted on Earth-orbiting spacecraft, to scan the entire globe (**Figure 1.4**). Geologists

FIGURE 1.1 Geology is principally an outdoor science. Here, Peter Gray welds one of the five Global Positioning System stations placed on the flanks of Mount St. Helens. The stations will monitor the changing shape of the land surface as molten rock moves upward within the volcano. [USGS photo by Lyn Topinka.]

FIGURE 1.3 A number of subfields contribute to the study of geology. Geobiologists investigate underground life inside Spider Cave at Carlsbad Caverns, New Mexico. [AP Photo/Val Hildreth-Werker.]

develop computer models that can analyze the huge quantities of data amassed by satellites to map the continents, chart the motions of the atmosphere and oceans, and monitor how our environment is changing.

A special aspect of geology is its ability to probe Earth's long history by reading what has been "written in stone."

The **geologic record** is the information preserved in the rocks that have been formed at various times throughout Earth's history (Figure 1.5). Geologists decipher the geologic record by combining information from many kinds of work: examination of rocks in the field; careful mapping of their positions relative to older and younger rock formations; collection of representative samples; and determination of their ages using sensitive laboratory instruments.

FIGURE 1.2 The research crew from the icebreaker Louis S. St-Laurent lowers a corer that will gather mud and sediment from the ocean floor. [AP Photo/The Canadian Press, Jonathan Hayward.]

FIGURE 1.4 An astronaut checks out instrumentation for monitoring Earth's surface. [StockTrek/SuperStock.]

FIGURE 1.5 The geologic record preserves evidence of Earth's long history. These multicolored layers of sand at Colorado National Monument were deposited more than 200 million years ago, when this part of the western United States was a vast Sahara-like desert. They were subsequently overlain by other rocks, welded by pressure into sandstone, uplifted by mountain-building events, and eroded by wind and water into today's stunning landforms.
[Mark Newman/Lonely Planet Images/Getty Images, Inc.]

In *Annals of the Former World*, a compendium of colorful stories about geologists, the popular writer John McPhee offers his view of how geologists bring field and laboratory observations together to visualize the big picture:

They look at mud and see mountains, in mountains oceans, in oceans mountains to be. They go up to some rock and figure out a story, another rock, another story,

and as the stories compile through time they connect—and long case histories are constructed and written from interpreted patterns of clues. This is detective work on a scale unimaginable to most detectives, with the notable exception of Sherlock Holmes.

The geologic record tells us that, for the most part, the processes we see in action on Earth today have worked in much the same way throughout the geologic past. This important concept is known as the **principle of uniformitarianism**. It was stated as a scientific hypothesis in the eighteenth century by a Scottish physician and geologist, James Hutton. In 1830, the British geologist Charles Lyell summarized the concept in a memorable line: "The present is the key to the past." For instance, by understanding the processes that build mountains today, we can infer how mountains were built a billion years ago.

The principle of uniformitarianism does not mean that all geologic phenomena proceed at the same gradual pace. Some of the most important geologic processes happen as sudden events. A large meteorite that impacts Earth can gouge out a vast crater in a matter of seconds. A volcano can blow its top and a fault can rupture the ground in an earthquake almost as quickly. Other processes do occur much more slowly. Millions of years are required for continents to drift apart, for mountains to be raised and eroded, and for river systems to deposit thick layers of sediments. Geologic processes take place over a tremendous range of scales in both space and time (**Figure 1.6**).

Nor does the principle of uniformitarianism mean that we have to observe a geologic event to know that it is important in the current Earth system. Humans have not witnessed a large meteorite impact in recorded history, but we know these impacts have occurred many times in the geologic past and will certainly happen again. The same can be said of the vast volcanic outpourings that have covered areas bigger than Texas with lava and poisoned the global atmosphere with volcanic gases. The long history of Earth is punctuated by many such extreme, though infrequent, events that result in rapid changes in the Earth system. Geology is the study of *extreme events* as well as gradual change.

From Hutton's day onward, geologists have observed nature at work and used the principle of uniformitarianism to interpret features found in rock formations. This approach has been very successful. However, Hutton's principle is too confining for geologic science as it is now practiced. Modern geology must deal with the entire range of Earth's history, which began more than 4.5 billion years ago. As we will see in Chapter 9, the violent processes that shaped Earth's early history were distinctly different from those that operate today. To understand that history, we will need some information about Earth's shape and surface, as well as its deep interior.

FIGURE 1.6 Some geologic processes take place over thousands of centuries, while others occur with dazzling speed. (a) The Grand Canyon, Arizona. (b) Meteor Crater, Arizona. [(a) John Wang/PhotoDisc/Getty Images; (b) David Parker/Science Photo Library/Getty Images.]

Earth's Shape and Surface

The scientific method has its roots in **geodesy,** a very old branch of Earth science that studies Earth's shape and surface. The concept that Earth is spherical rather than flat was advanced by Greek and Indian philosophers around the sixth century B.C., and it was the basis of Aristotle's theory of Earth put forward in his famous treatise, *Meteorologica*, published around 330 B.C. (the first Earth science textbook!). In the third century B.C., Eratosthenes used a clever experiment to measure Earth's radius, which turned out to be 6371 km (see the Practicing Geology Exercise *How Big Is Our Planet?*).

Much more precise measurements have shown that Earth is not a perfect sphere. Because of its daily rotation, it bulges out slightly at its equator and is slightly squashed at its poles. In addition, the smooth curvature of Earth's surface is broken by mountains and valleys and other ups and downs. This **topography** is measured with respect to *sea level,* a smooth surface set at the average level of ocean water that conforms closely to the squashed spherical shape expected for the rotating Earth.

Many features of geologic significance stand out in Earth's topography (**Figure 1.7**). Its two largest features are continents, which have typical elevations of 0 to 1 km above sea level, and ocean basins, which have typical depths of 4 to 5 km below sea level. The elevation of Earth's surface varies by nearly 20 km from its highest point (Mount Everest in the Himalaya at 8850 m above sea level) to its lowest point (Challenger Deep in the Marianas Trench in the Pacific Ocean at 11,030 m below sea level). Although the Himalaya may loom large to us, their elevation is a small fraction of Earth's radius, only about one part in a thousand, which is why the globe looks like a nearly smooth sphere when seen from outer space.

Peeling the Onion: Discovery of a Layered Earth

Ancient thinkers divided the universe into two parts, the heavens above and Hades below. The sky was transparent and full of light, and they could directly observe its stars and track its wandering planets. But Earth's interior was dark and closed to human view. In some places, the ground quaked and erupted hot lava. Surely something terrible was going on down there!

So it remained until about a century ago, when geologists began to peer downward into Earth's interior, not with waves of light (which cannot penetrate rock), but with waves produced by earthquakes. An earthquake occurs

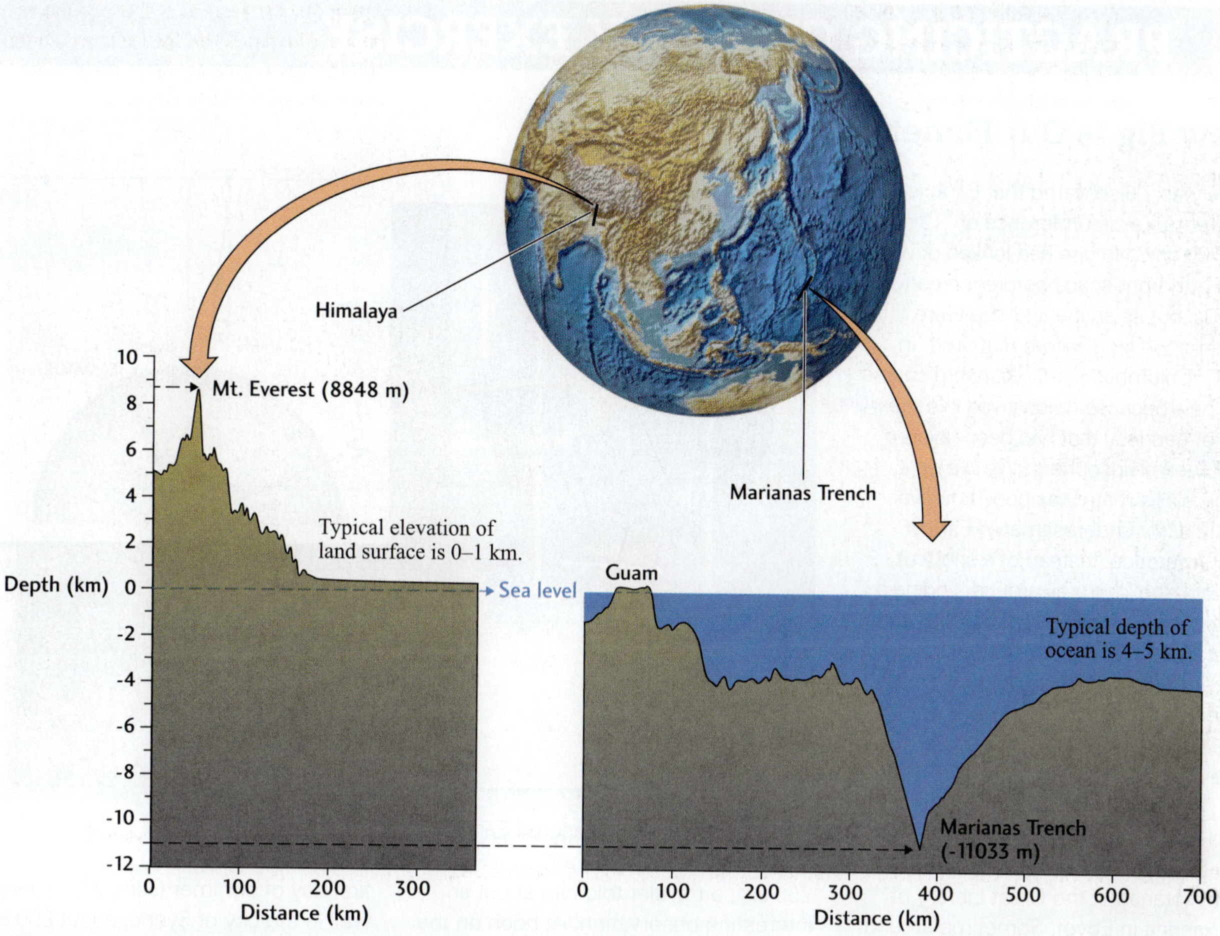

Peeling the Onion: Discovery of a Layered Earth

FIGURE 1.7 Earth's topography is measured with respect to sea level. The elevation scale in the diagram is greatly exaggerated.

when geologic forces cause brittle rocks to fracture, sending out vibrations like the cracking of ice on a river. These **seismic waves** (from the Greek word for earthquake, *seismos*), when recorded on sensitive instruments called *seismometers*, allow geologists to locate earthquakes and also to make pictures of Earth's inner workings, much as doctors use ultrasound and CAT scans to image the inside of your body. When the first networks of seismographs were installed around the world at the end of the nineteenth century, geologists began to discover that Earth's interior was divided into concentric layers of different compositions, separated by sharp, nearly spherical boundaries (**Figure 1.8**).

Earth's Density

Layering of Earth's deep interior was first proposed by the German physicist Emil Wiechert at the end of the nineteenth century, before much seismic data had become available. He wanted to understand why our planet is so heavy, or more precisely, so *dense*. The density of a substance is easy to calculate: just measure its mass on a scale and divide by its volume. A typical rock, such as the granite used for tombstones, has a density of about 2.7 grams per cubic centimeter (g/cm³). Estimating the density of the entire

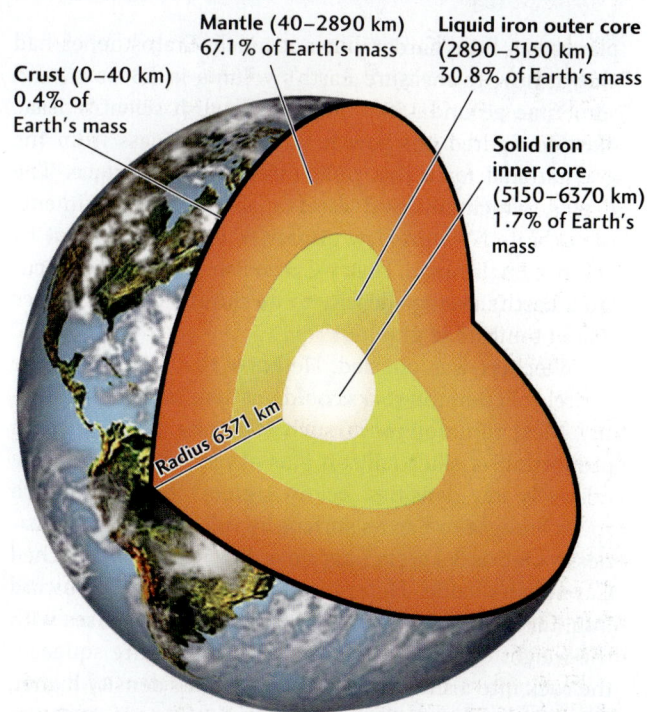

FIGURE 1.8 Earth's major layers, showing their depths and their masses expressed as a percentage of Earth's total mass.

PRACTICING GEOLOGY EXERCISE

How Big Is Our Planet?

How was it discovered that Earth is round with a circumference of 40,000 km? No one had looked down on Earth from space before the early 1960s, but its shape and size were understood long before that time. In 1492, Columbus set a westward course for India because he believed in a theory of geodesy that had been favored by Greek philosophers: *We live on a sphere.* His math was poor, however, so he badly underestimated Earth's circumference. Instead of a shortcut, he took the long way around, finding a New World instead of the Spice Islands! Had Columbus properly understood the ancient Greeks, he might not have made this fortuitous mistake, because they had accurately measured Earth's size more than 17 centuries earlier.

The credit for determining Earth's size goes to Eratosthenes (pronounced ěr′ə-tŏs′thə-nēz′), a Greek who was chief librarian at the Great Library of Alexandria in Egypt. Sometime around 250 B.C., a traveler told him about an interesting observation. At noon on the first day of summer (June 21), a deep well in the city of Syene, about 800 km

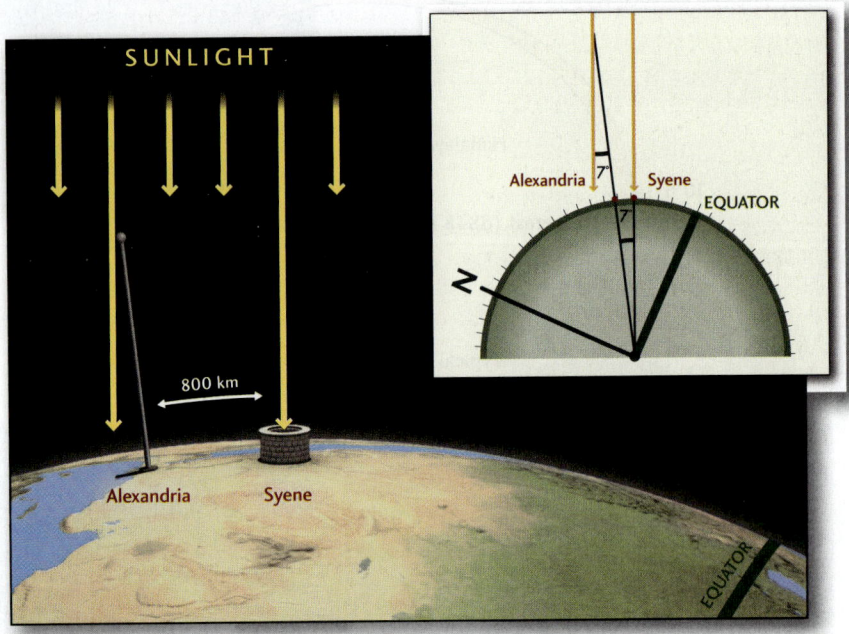

How Eratosthenes measured Earth's circumference.

planet is a little harder, but not much. Eratosthenes had shown how to measure Earth's volume in 250 B.C., and sometime around 1680, the great English scientist Isaac Newton figured out how to calculate its mass from the gravitational force that pulls objects to its surface. The details, which involved careful laboratory experiments to calibrate Newton's law of gravity, were worked out by another Englishman, Henry Cavendish. In 1798, he calculated Earth's average density to be about 5.5 g/cm³, twice that of tombstone granite.

Wiechert was puzzled. He knew that a planet made entirely of common rocks could not have such a high density. Most common rocks, such as granite, contain a high proportion of silica (silicon plus oxygen; SiO_2) and have relatively low densities, below 3 g/cm³. Some iron-rich rocks brought to Earth's surface by volcanoes have densities as high as 3.5 g/cm³, but no ordinary rock approached Cavendish's value. He also knew that, going downward into Earth's interior, the pressure on rock increases with the weight of the overlying mass. The pressure squeezes the rock into a smaller volume, making its density higher. But Wiechert found that even the effect of pressure was too small to account for the density Cavendish had calculated.

The Mantle and Core

In thinking about what lay beneath his feet, Wiechert turned outward to the solar system and, in particular, to meteorites, which are pieces of the solar system that have fallen to Earth. He knew that some meteorites are made of an *alloy* (a mixture) of two heavy metals, iron and nickel, and thus have densities as high as 8 g/cm³ (**Figure 1.9**). He also knew that these two elements are relatively abundant throughout our solar system. So, in 1896, he proposed a grand hypothesis: Sometime in Earth's past, most of the iron and nickel in its interior had dropped inward to its center under the force of gravity. This movement created a dense **core,** which was surrounded by a shell of silicate-rich rock that he called the **mantle** (using the German word for "coat"). With this hypothesis, he could come up with a two-layered Earth model that agreed with Cavendish's value for Earth's average density. He could also explain the existence of iron-nickel meteorites: They were chunks from the core of an Earthlike planet (or planets) that had broken apart, most likely by collision with other planets.

Wiechert got busy testing his hypothesis using seismic waves recorded by seismographs located around the globe (he designed one himself). The first results showed south of Alexandria, was completely lit up by sunlight because the Sun was directly overhead. Acting on a hunch, Eratosthenes did an experiment. He set up a vertical pole in his own city, and at high noon on the first day of summer, the pole cast a shadow.

Eratosthenes assumed that the Sun was very far away, so that the light rays falling on the two cities were parallel. Knowing that the Sun cast a shadow in Alexandria but was directly overhead at the same time in Syene, Eratosthenes could demonstrate with simple geometry that the ground surface must be curved. He knew that the most perfect curved surface is a sphere, so he hypothesized that Earth had a spherical shape (the Greeks admired geometric perfection). By measuring the length of the pole's shadow in Alexandria, he calculated that if vertical lines through the two cities could be extended to Earth's center, they would intersect at an angle of about 7°, which is about 1/50 of a full circle (360°). The distance between the two cities was known to be about 800 km in today's measurements. From these figures, Eratosthenes calculated a circumference for Earth that is very close to the modern value:

Earth's circumference
= 50 × distance from Syene to Alexandria = 50 × 800 km = 40,000 km

With this figure for Earth's circumference, it was a simple matter to calculate its radius. Eratosthenes knew that, for any circle, the circumference is equal to 2π (pi) times the radius, where π is about 3.14. Therefore, he divided his estimate of Earth's circumference by 2π to find its radius:

$$\text{radius} = \frac{\text{circumference}}{2\pi}$$

$$= \frac{40{,}000 \text{ km}}{6.28} = 6370 \text{ km}$$

By these calculations, Eratosthenes arrived at a simple and elegant scientific model: *Earth is a sphere with a radius of about 6370 km.*

In this powerful demonstration of the scientific method, Eratosthenes made observations (the length of the shadow), formed a hypothesis (spherical shape), and applied some mathematical theory (spherical geometry) to propose a remarkably accurate model of Earth's physical form. His model correctly predicted other types of measurements, such as the distance at which a ship's tall mast would disappear over the horizon. Moreover, knowing Earth's shape and size allowed Greek astronomers to calculate the sizes of the Moon and Sun and the distances of these bodies from Earth. This story makes clear why well-designed experiments and good measurements are central to the scientific method: They give us new information about the natural world.

BONUS PROBLEM: The volume of a sphere is given by

$$\text{volume} = \frac{4\pi}{3}(\text{radius})^3$$

From this formula, calculate Earth's volume in cubic kilometers.

a shadowy inner mass that he took to be the core, but he had problems identifying some of the seismic waves. These waves come in two basic types: *compressional waves*, which expand and compress the material they move through as they travel through a solid, liquid, or gas; and *shear waves*, which move the material from side to side. Shear waves can

(a) (b)

FIGURE 1.9 Two common types of meteorites. (a) This stony meteorite, which is similar in composition to Earth's silicate mantle, has a density of about 3 g/cm³. (b) This iron-nickel meteorite, which is similar in composition to Earth's core, has a density of about 8 g/cm³. [John Grotzinger/Ramón Rivera-Moret/Harvard Mineralogical Museum.]

propagate only through solids, which resist shearing, and not through fluids (liquids or gases) such as air and water, which have no resistance to this type of motion.

In 1906, a British seismologist, Robert Oldham, was able to sort out the paths traveled by these two types of seismic waves and show that shear waves did not propagate through the core. The core, at least in its outer part, was liquid! This finding turns out to be not too surprising. Iron melts at a lower temperature than silicates, which is why metallurgists can use containers made of ceramics (which are silicate materials) to hold molten iron. Earth's deep interior is hot enough to melt an iron-nickel alloy, but not silicate rock. Beno Gutenberg, one of Wiechert's students, confirmed Oldham's observations and, in 1914, determined that the depth of the *core-mantle boundary* was about 2890 km (see Figure 1.9).

The Crust

Five years earlier, a Croatian scientist had detected another boundary at the relatively shallow depth of 40 km beneath the European continent. This boundary, named the *Mohorovičić discontinuity* (Moho for short) after its discoverer, separates a **crust** composed of low-density silicates, which are rich in aluminum and potassium, from the higher-density silicates of the mantle, which contain more magnesium and iron.

Like the core-mantle boundary, the Moho is a global feature. However, it was found to be substantially shallower beneath the oceans than beneath the continents. On average, the thickness of oceanic crust is only about 7 km, compared with almost 40 km for continental crust. Moreover, continental rocks are richer in the light elements oxygen and silicon, making continental crust less dense than oceanic crust. Because continental crust is thicker but less dense than oceanic crust, the continents ride higher by floating like buoyant rafts on the denser mantle (**Figure 1.10**), much as icebergs float on the ocean. Continental buoyancy explains the most striking feature of Earth's surface topography: why the elevations fall into two main groups,

0 to 1 km above sea level for much of the land surface and 4 to 5 km below sea level for much of the deep sea.

Shear waves travel well through the mantle and crust, so we know that both are solid rock. How can continents float on solid rock? Rock can be solid and strong over the short term (seconds to years), but weak over the long term (thousands to millions of years). The mantle below a depth of about 100 km has little strength, and over very long periods, it flows as it adjusts to support the weight of continents and mountains.

The Inner Core

Because the mantle is solid and the outer part of the core is liquid, the core-mantle boundary reflects seismic waves, just as a mirror reflects light waves. In 1936, Danish seismologist Inge Lehmann discovered another sharp spherical boundary at the much greater depth of 5150 km, indicating a central mass with a higher density than the liquid core. Studies following her pioneering research showed that the inner core can transmit both shear waves and compressional waves. The **inner core** is therefore a solid metallic sphere suspended within the liquid **outer core**—a "planet within a planet." The radius of the inner core is 1220 km, about two-thirds the size of the Moon.

Geologists were puzzled by the existence of this "frozen" inner core. They knew that temperatures inside Earth should increase with depth. According to the best current estimates, Earth's temperature rises from about 3500°C at the core-mantle boundary to almost 5000°C at its center. If the inner core is hotter, how could it be solid while the outer core is molten? The mystery was eventually solved by laboratory experiments on iron-nickel alloys, which showed that the "freezing" was due to higher pressures, rather than lower temperatures, at Earth's center.

Chemical Composition of Earth's Major Layers

By the mid-twentieth century, geologists had discovered all of Earth's major layers—crust, mantle, outer core, and inner core—plus a number of more subtle features in its

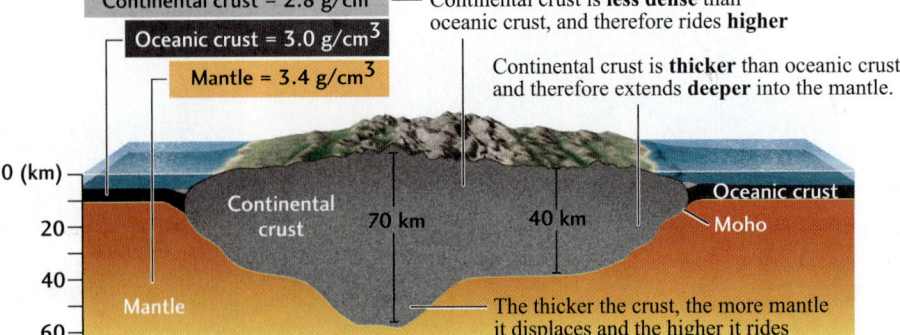

FIGURE 1.10 Because crustal rocks are less dense than mantle rocks, Earth's crust floats on the mantle. Continental crust is thicker and has a lower density than oceanic crust, which causes it to ride higher, explaining the difference in elevation between continents and the deep seafloor.

interior. They found, for example, that the mantle itself is layered into an *upper mantle* and a *lower mantle*, separated by a *transition zone* where the rock density increases in a series of steps. These density steps are not caused by changes in the rock's chemical composition, but rather by changes in the compactness of its constituent minerals due to the increasing pressure with depth. The two largest density jumps in the transition zone are located at depths of about 410 km and 660 km, but they are smaller than the density increases across the Moho and core-mantle boundaries, which *are* due to changes in chemical composition (**Figure 1.11**).

Geologists were also able to show that Earth's outer core cannot be made of a pure iron-nickel alloy, because the densities of these metals are higher than the observed density of the outer core. About 10 percent of the outer core's mass must be made of lighter elements, such as oxygen and sulfur. On the other hand, the density of the solid inner core is slightly higher than that of the outer core and is consistent with a nearly pure iron-nickel alloy.

By bringing together many lines of evidence, geologists have put together a model of the composition of Earth and its various layers. In addition to the seismic data, that evidence includes the compositions of crustal and mantle rocks as well as the compositions of meteorites, thought to be samples of the cosmic material from which planets like Earth were originally made.

Only eight elements, out of more than a hundred, make up 99 percent of Earth's mass (see Figure 1.11). In fact, about 90 percent of Earth consists of only four elements: iron, oxygen, silicon, and magnesium. The first two are the most abundant elements, each accounting for nearly a third of the planet's overall mass, but they are distributed very differently. Iron, the densest of these common elements, is concentrated in the core, whereas oxygen, the least dense, is concentrated in the crust and mantle. The crust contains more silica than the mantle, and

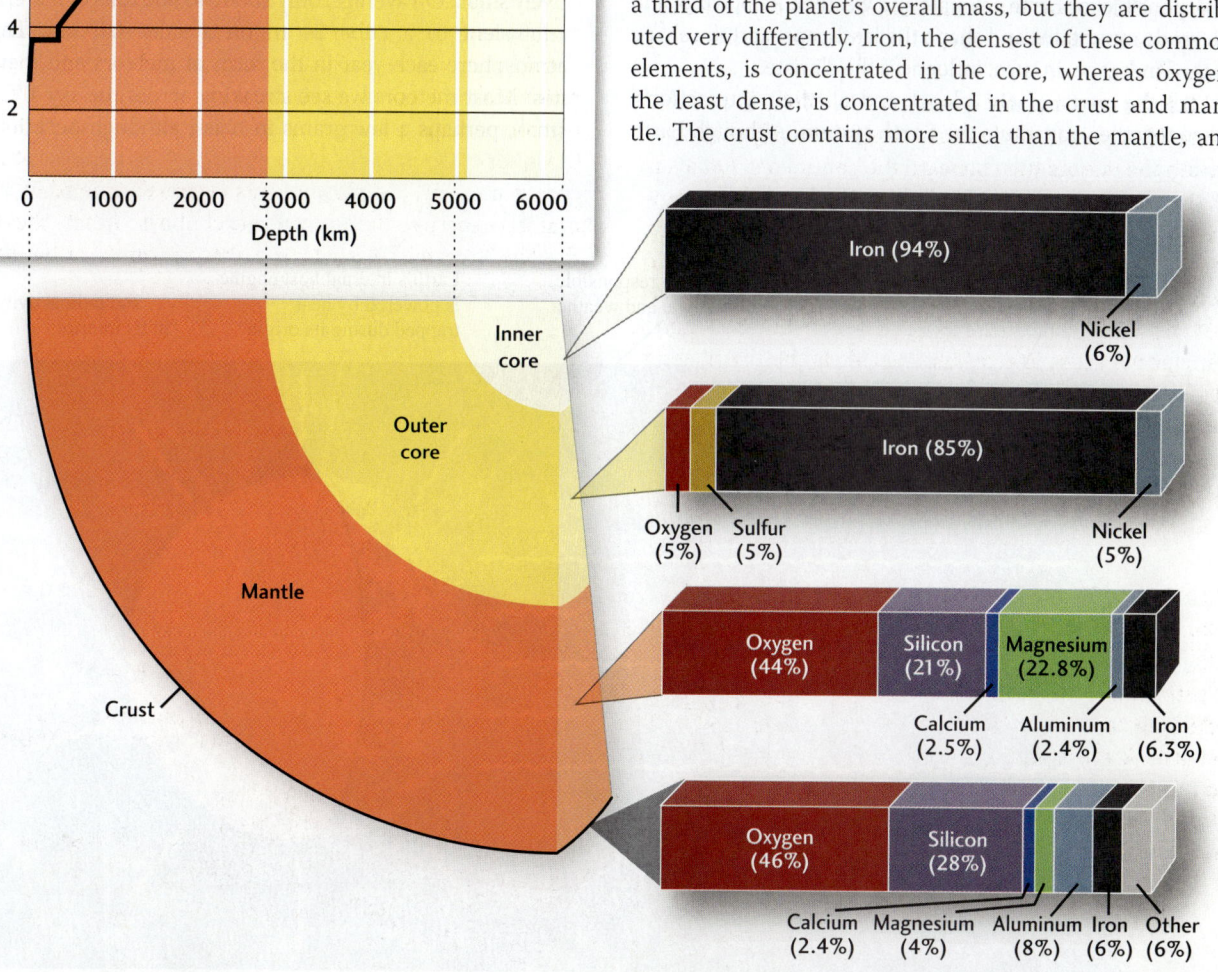

FIGURE 1.11 Jumps in density between Earth's major layers, shown in different colors, are caused primarily by differences in their chemical composition. The relative amounts of the main elements are depicted in the bars on the right.

the core almost none. These relationships confirm Wiechert's hypothesis that the different compositions of Earth's layers are primarily the work of gravity. As you can see in Figure 1.11, the crustal rocks on which we stand are almost 50 percent oxygen!

Earth as a System of Interacting Components

Earth is a restless planet, continually changing through geologic activity such as earthquakes, volcanoes, and glaciation. This activity is powered by two heat engines, one internal and one external (**Figure 1.12**). A *heat engine*—for example, the gasoline engine of an automobile—transforms heat into mechanical motion or work. Earth's *internal heat engine* is powered by the heat energy trapped in its deep interior during its violent origin and released inside the planet by radioactivity. This internal heat drives movement in the mantle and core, supplying the energy that melts rock, moves continents, and lifts up mountains. Earth's *external heat engine* is driven by solar energy: heat supplied to Earth's surface by the Sun. Heat from the Sun energizes the atmosphere and oceans and is responsible for Earth's climate and weather. Rain, wind, and ice erode mountains and shape the landscape, and the shape of the landscape, in turn, influences the climate.

All the parts of our planet and all their interactions, taken together, constitute the **Earth system.** Although Earth scientists have long thought in terms of natural systems, it was not until the late twentieth century that they had the tools to investigate how the Earth system actually works. Networks of instruments and Earth-orbiting satellites now collect information about the Earth system on a global scale, and computers are powerful enough to calculate the mass and energy transfers within the system. The major components of the Earth system can be represented as a set of domains, or "spheres" (**Figure 1.13**). We have discussed some of these components already; we will define the others shortly.

We will talk about the Earth system throughout this textbook. Let's get started by looking at some of its basic features. The Earth system is an *open system* that exchanges energy and mass with its surroundings (see Figure 1.13). Radiant energy from the Sun energizes the weathering and erosion of Earth's surface, as well as the growth of plants, which feed almost all living things. Earth's climate is controlled by the balance between the solar energy coming into the Earth system and the heat energy Earth radiates back into space.

Early in the life of the solar system, collisions between Earth and other solid bodies were a very important process, growing the planet's mass and forming the Moon. These days, the exchange of mass between Earth and space is relatively small: On average, only about 40,000 tons of material—equivalent to a cube 24 m on a side—fall into Earth's atmosphere each year in the form of meteors and meteorites. Most meteors we see streaking across the sky are very small, perhaps a few grams in mass, although occasionally

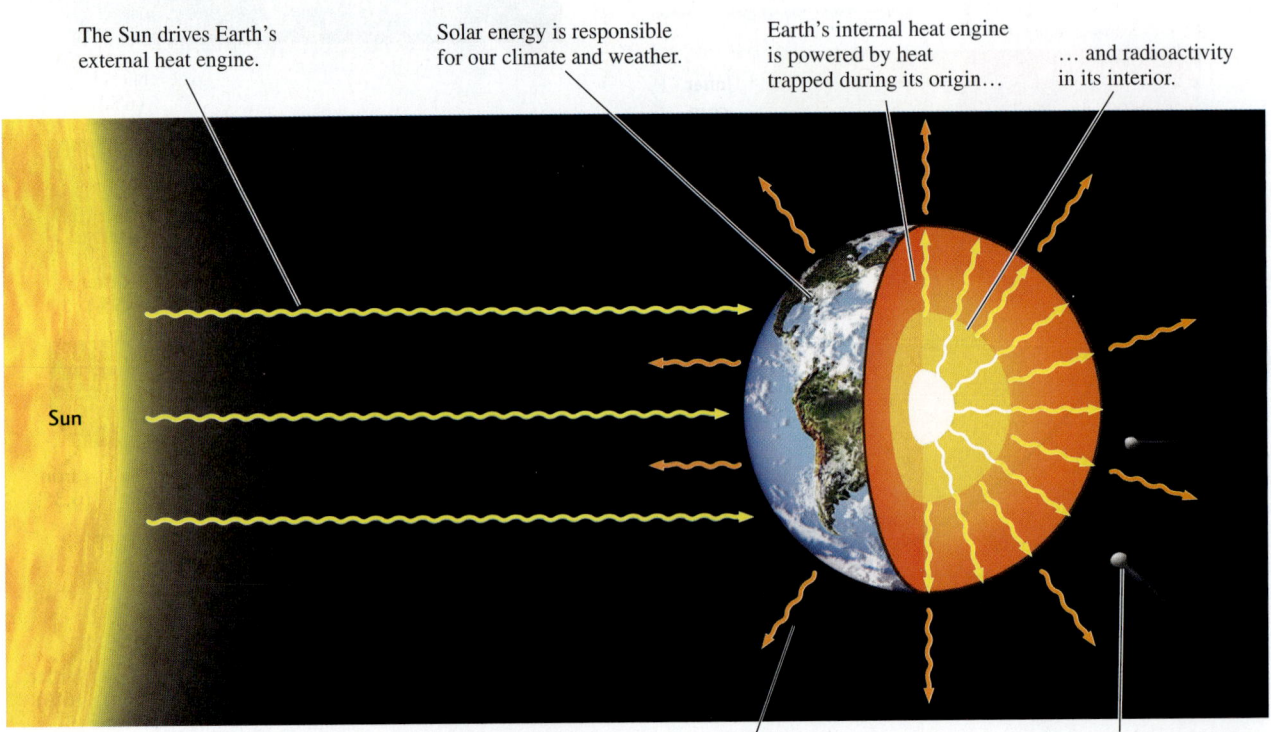

FIGURE 1.12 The Earth system is an open system that exchanges energy and mass with its surroundings.

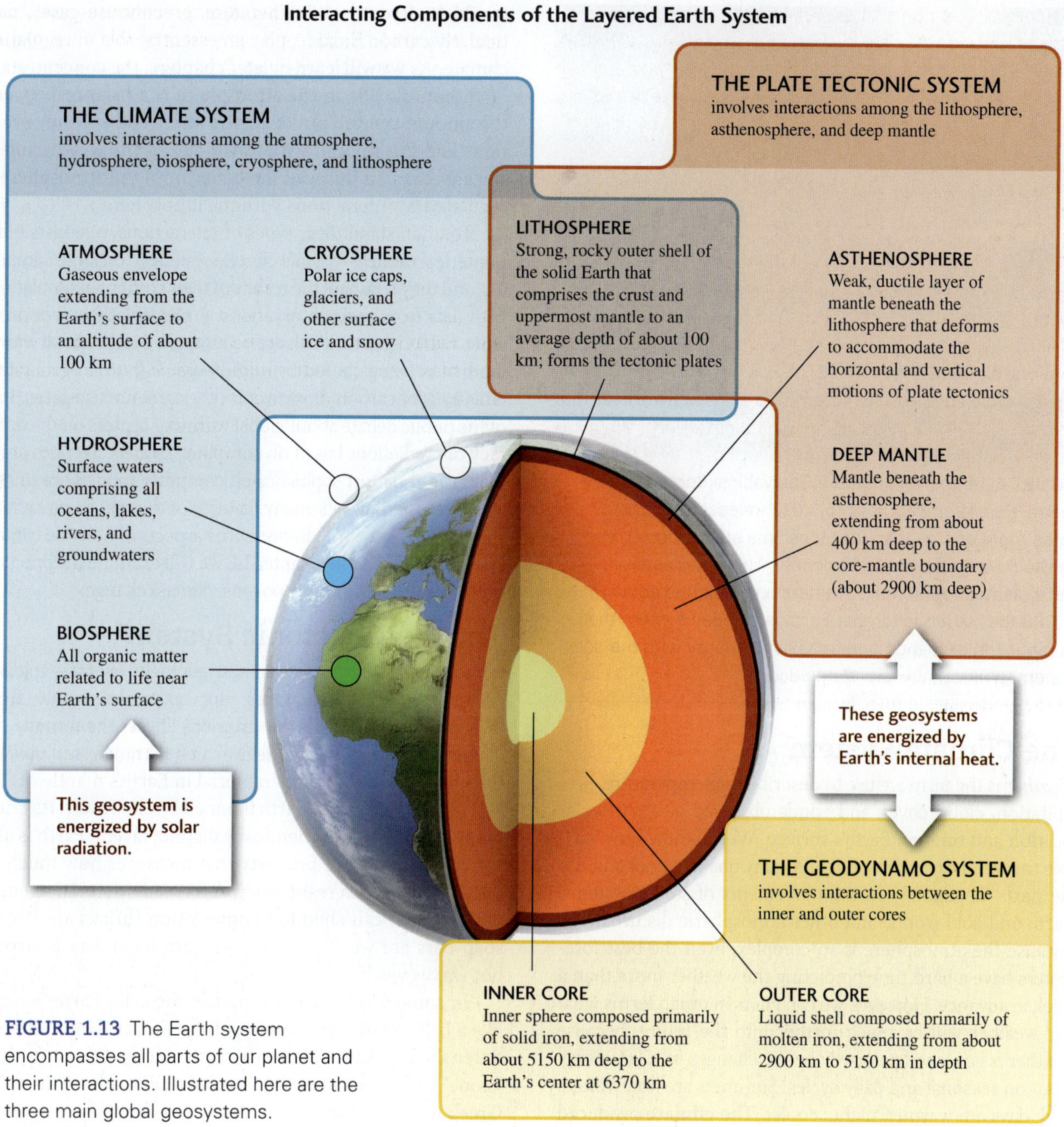

Interacting Components of the Layered Earth System

THE CLIMATE SYSTEM involves interactions among the atmosphere, hydrosphere, biosphere, cryosphere, and lithosphere

THE PLATE TECTONIC SYSTEM involves interactions among the lithosphere, asthenosphere, and deep mantle

ATMOSPHERE Gaseous envelope extending from the Earth's surface to an altitude of about 100 km

CRYOSPHERE Polar ice caps, glaciers, and other surface ice and snow

LITHOSPHERE Strong, rocky outer shell of the solid Earth that comprises the crust and uppermost mantle to an average depth of about 100 km; forms the tectonic plates

ASTHENOSPHERE Weak, ductile layer of mantle beneath the lithosphere that deforms to accommodate the horizontal and vertical motions of plate tectonics

HYDROSPHERE Surface waters comprising all oceans, lakes, rivers, and groundwaters

DEEP MANTLE Mantle beneath the asthenosphere, extending from about 400 km deep to the core-mantle boundary (about 2900 km deep)

BIOSPHERE All organic matter related to life near Earth's surface

This geosystem is energized by solar radiation.

These geosystems are energized by Earth's internal heat.

THE GEODYNAMO SYSTEM involves interactions between the inner and outer cores

INNER CORE Inner sphere composed primarily of solid iron, extending from about 5150 km deep to the Earth's center at 6370 km

OUTER CORE Liquid shell composed primarily of molten iron, extending from about 2900 km to 5150 km in depth

FIGURE 1.13 The Earth system encompasses all parts of our planet and their interactions. Illustrated here are the three main global geosystems.

Earth encounters a larger chunk, such as the Chelyabinsk meteor of 2013, with dangerous results (**Figure 1.14**).

Although we think of Earth as a single system, it is a challenge to study the whole thing all at once. Instead, we will focus our attention on the particular parts of the Earth system (subsystems) that we are trying to understand. For instance, in our discussion of global climate change, we will primarily consider interactions between the atmosphere and several other components that are driven by solar energy: the *hydrosphere* (Earth's surface waters and groundwaters), the *cryosphere* (Earth's ice caps, glaciers, and snowfields), and the *biosphere* (Earth's living organisms). Our coverage of how the continents are deformed to raise mountains will focus on interactions between the crust and the mantle that are driven by Earth's internal heat engine. Specialized subsystems that produce specific types of activity, such as climate change or mountain building, are called **geosystems**. The Earth system can be thought of as a collection of many open, interacting (and often overlapping) geosystems.

In this section, we will introduce three important geosystems that operate on a global scale: the climate system, the plate tectonic system, and the geodynamo. Later in this textbook, we will discuss a number of smaller geosystems, such as volcanoes that erupt hot lava (Chapter 5), hydrologic systems that give us our drinking water (Chapter 17), and petroleum reservoirs that produce oil and gas (Chapter 13).

FIGURE 1.14 Explosion of the Chelyabinsk meteor above central Russia on February 15, 2013, released 20–30 times more energy than the Hiroshima atomic bomb, and its shock wave injured 1500 people. This small asteroid had a diameter of about 20 meters and weighed about 11,000 tons, a reminder that Earth is an open system that continues to exchange both mass and energy with the solar system. [The Asahi Shimbun/Getty Images]

The Climate System

Weather is the term we use to describe the temperature, precipitation, cloud cover, and winds observed at a particular location and time on Earth's surface. We all know how variable the weather can be—hot and rainy one day, cool and dry the next—depending on the movements of storm systems, warm and cold fronts, and other atmospheric disturbances. Because the atmosphere is so complex, even the best forecasters have a hard time predicting the weather more than a week in advance. However, we can guess in rough terms what our weather will be much further into the future, because weather is governed primarily by the changes in solar energy input on seasonal and daily cycles: Summers are hot, winters cold; days are warmer, nights cooler. The **climate** produced by these weather cycles can be described by averaging temperatures and other variables over many years of observation. A complete description of climate also includes measures of how variable the weather has been, such as the highest and lowest temperatures ever recorded on a given day of the year.

The **climate system** includes all the Earth system components that determine climate on a global scale and how climate changes with time. In other words, the climate system involves not only the behavior of the atmosphere, but also its interactions with the hydrosphere, cryosphere, biosphere, and lithosphere (see Figure 1.13).

When the Sun warms Earth's surface, some of the heat is trapped by water vapor, carbon dioxide, and other gases in the atmosphere, similar to the way heat is trapped by frosted glass in a greenhouse. This *greenhouse effect* explains why Earth has a climate that makes life possible. If its atmosphere contained no greenhouse gases, Earth's surface would be frozen solid! Therefore, greenhouse gases, particularly carbon dioxide, play an essential role in regulating climate. As we will learn in later chapters, the concentration of carbon dioxide in the atmosphere is a balance between the amount spewed out of Earth's interior in volcanic eruptions and the amount withdrawn during the weathering of silicate rocks. In this way, the behavior of the atmosphere is regulated by interactions with the lithosphere.

To understand these types of interactions, scientists build numerical models—virtual climate systems—on large computers, and they compare the results of their computer simulations with data from their observations. A particularly urgent problem to which these models are being applied is the global warming that is being caused by *anthropogenic* (human-generated) emissions of carbon dioxide and other greenhouse gases. Part of the public debate about global warming centers on the accuracy of predictions based on computer models. Skeptics argue that even the most sophisticated computer models are unreliable because they lack many features of the real Earth system. In Chapter 12, we will discuss some aspects of how the climate system works, and in Chapter 14, we will examine the practical problems posed by anthropogenic climate change.

The Plate Tectonic System

Some of Earth's most dramatic geologic events—volcanic eruptions and earthquakes, for example—result from interactions within Earth's interior. These phenomena are driven by Earth's internal heat, which is transferred upward through the circulation of material in Earth's mantle.

We have seen that Earth is zoned by chemistry: Its crust, mantle, and core are chemically distinct layers. Earth is also zoned by *strength*, a property that measures how much an Earth material can resist being deformed. Material strength depends on both chemical composition (bricks are strong, soap bars are weak) and temperature (cold wax is strong, hot wax is weak).

In some ways, the outer part of the solid Earth behaves like a ball of hot wax. Cooling of the surface forms a strong outer shell, or **lithosphere** (from the Greek *lithos*, meaning "stone"), which encases a hot, weak **asthenosphere** (from the Greek *asthenes*, meaning "weak"). The lithosphere includes the crust and the top part of the mantle down to an average depth of about 100 km. The asthenosphere is the portion of mantle, perhaps 300 km thick, immediately below the lithosphere. When subjected to force, the lithosphere tends to behave like a nearly rigid and brittle shell, whereas the underlying asthenosphere flows like a moldable, or *ductile*, solid.

According to the remarkable theory of *plate tectonics*, the lithosphere is not a continuous shell; it is broken into about a dozen large plates that move over Earth's surface at rates of a few centimeters per year. Each lithospheric plate is a rigid unit that rides on the asthenosphere, which is also in motion. The lithosphere that forms a plate may vary from just a few kilometers thick in volcanically active areas to more than 200 km thick beneath the older, colder parts of continents. The discovery of plate tectonics in the 1960s led to the first unified theory that explained the worldwide distribution of earthquakes and volcanoes, continental drift,

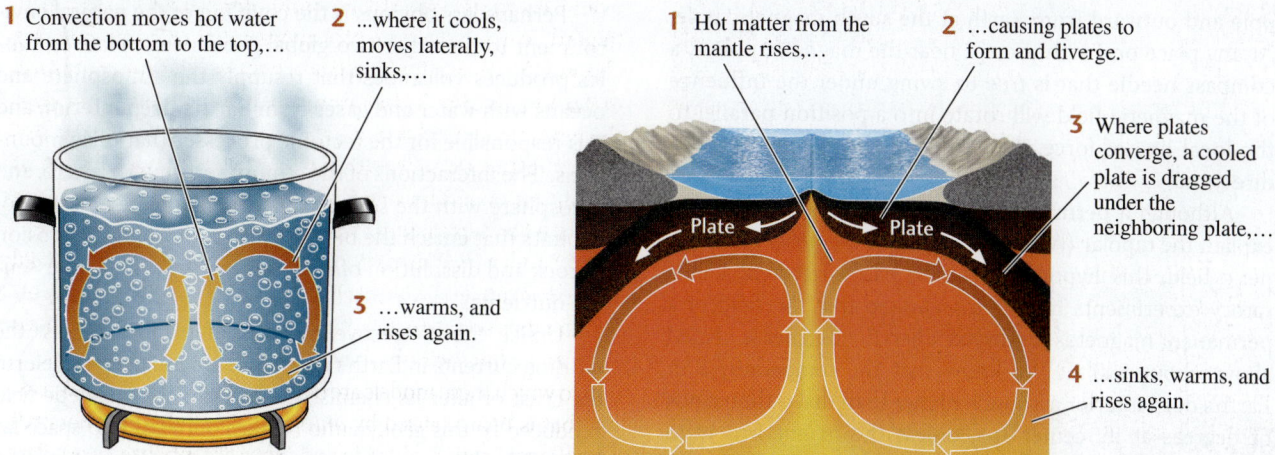

FIGURE 1.15 Convection in Earth's mantle can be compared to the pattern of movement in a pot of boiling water. Both processes carry heat upward through the movement of matter.

mountain building, and many other geologic phenomena. Chapter 2 describes the basic concepts of plate tectonics.

Why do the plates move across Earth's surface instead of locking up into a completely rigid shell? The forces that push and pull the plates come from the mantle. Driven by Earth's internal heat engine, hot mantle material rises at boundaries where plates separate, forming new lithosphere. The lithosphere cools and becomes more rigid as it moves away from these boundaries, eventually sinking back into the mantle under the pull of gravity at other boundaries where plates converge. This general process, in which hotter material rises and cooler material sinks, is called **convection** (Figure 1.15). Convection in the mantle can be compared to the pattern of movement in a pot of boiling water. Both processes transfer energy by the movement of mass, but mantle convection is much slower because the solid mantle rocks are much more resistant to deformation than ordinary fluids such as water.

The convecting mantle and its overlying mosaic of lithospheric plates constitute the **plate tectonic system.** As with the climate system (which involves a wide range of convective processes in the atmosphere and oceans), scientists use computer simulations to study the plate tectonic system and test the agreement of their models against observations.

The Geodynamo

The third global geosystem involves interactions that produce a **magnetic field** deep inside Earth in its liquid outer core. This magnetic field reaches far into outer space, causing compass needles to point north and shielding the biosphere from harmful solar radiation. When rocks form, they become slightly magnetized by this magnetic field, so geologists can study how the field behaved in the past and use it to help them decipher the geologic record.

Earth rotates about an axis that goes through its north and south poles. Earth's magnetic field behaves as if a powerful bar magnet were located at Earth's center and inclined about 11° from this rotational axis (Figure 1.16). The magnetic force points into Earth at the north magnetic

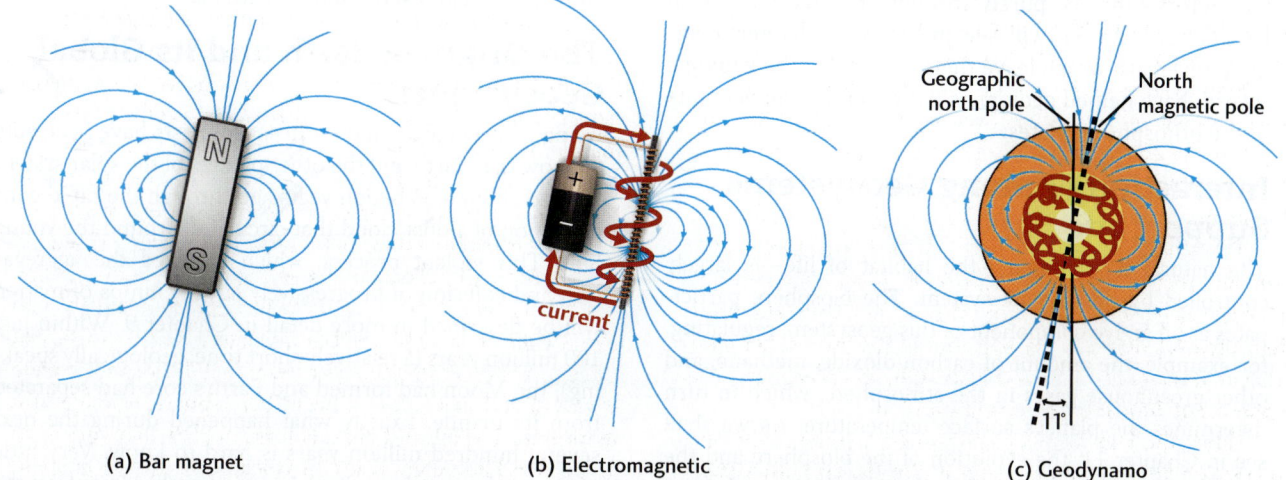

FIGURE 1.16 (a) A bar magnet creates a dipolar field with north and south poles. (b) A dipolar field can also be produced by electric currents flowing through a coil of metallic wire, as shown for this battery-powered electromagnet. (c) Earth's magnetic field, which is approximately dipolar above Earth's surface, is produced by electric currents flowing in the liquid-metal outer core, which are powered by convection.

pole and outward from Earth at the south magnetic pole. At any place on Earth (except near the magnetic poles), a compass needle that is free to swing under the influence of the magnetic field will rotate into a position parallel to the local line of force, approximately in the north-south direction.

Although a permanent magnet at Earth's center could explain the dipolar (two-pole) nature of the observed magnetic field, this hypothesis can be easily rejected. Laboratory experiments have demonstrated that the field of a permanent magnet is destroyed when the magnet is heated above about 500°C. We know that the temperatures in Earth's deep interior are much higher than that—thousands of degrees at its center—so, unless the magnetism were constantly regenerated, it could not be maintained.

Scientists theorize that heat flowing out of Earth's core causes convection that generates and maintains the magnetic field. Why is a magnetic field created by convection in the outer core, but not by convection in the mantle? First, the outer core is made primarily of iron, which is a very good electrical conductor, whereas the silicate rocks of the mantle are poor electrical conductors. Second, the convective flow is a million times more rapid in the liquid outer core than in the solid mantle. The rapid flow stirs up electric currents in the liquid iron-nickel alloy to produce the magnetic field. Thus, this **geodynamo** is more like an electromagnet than a bar magnet (see Figure 1.16).

For some 400 years, scientists have known that a compass needle points north because of Earth's magnetic field. Imagine how stunned they were half a century ago when they found geologic evidence that the direction of the magnetic force can be reversed. Over about half of geologic time, a compass needle would have pointed south! These *magnetic reversals* occur at irregular intervals ranging from tens of thousands to millions of years. The processes that cause them are not well understood, but computer models of the geodynamo show sporadic reversals occurring in the absence of any external factors—that is, purely through interactions within Earth's core. As we will see in the next chapter, magnetic reversals, which leave their imprint on the geologic record, have helped geologists figure out the movements of the lithospheric plates.

Interactions Among Geosystems Support Life

The natural environment—the habitat of life—is largely controlled by the climate system. The biosphere participates as an active component of this geosystem, regulating, for example, the amount of carbon dioxide, methane, and other greenhouse gases in the atmosphere, which in turn determines the planet's surface temperature. As we shall see in Chapter 11, the evolution of the biosphere and the atmosphere have gone hand-in-hand throughout the last 3.5 billion years of climate-system history.

Perhaps less obvious is the coupling of the natural environment to the other two global geosystems. Plate tectonics produces volcanoes that resupply the atmosphere and oceans with water and gases from Earth's deep interior, and it is responsible for the tectonic processes that raise mountains. The interactions of the atmosphere, hydrosphere, and cryosphere with the surface topography create a variety of habitats that enrich the biosphere and, through the erosion of rock and dissolution of minerals, provide life with essential nutrients.

Unlike the convective motions of plate tectonics, the swirling currents in Earth's outer core are too deep to deform the crust or alter its chemistry. However, the magnetic field produced by this geodynamo reaches outward into space far beyond Earth's atmosphere (see Figure 1.16). There it forms a barrier to highly energetic particles that stream outward from the Sun at speeds of more than 400 km/s—the *solar wind* (Figure 1.17). Without this shield, Earth's surface would be bombarded by harmful solar radiation, which would kill many forms of life that now prosper in its biosphere.

An Overview of Geologic Time

So far, we have discussed Earth's size and shape, its internal layering and composition, and the operation of its three major geosystems. How did Earth get its layered structure in the first place? How have the global geosystems evolved through geologic time? To begin to answer these questions, we present a brief overview of geologic time from the birth of the planet to the present. Later chapters will fill in the details.

Comprehending the immensity of geologic time is a challenge. John McPhee notes that geologists look into the "deep time" of Earth's early history (measured in billions of years), just as astronomers look into the "deep space" of the outer universe (measured in billions of light-years). Figure 1.18 presents the "ribbon of geologic time," marked with some major events and transitions.

The Origin of Earth and Its Global Geosystems

Using evidence from meteorites, geologists have been able to show that Earth and the other planets of the solar system formed about 4.56 billion years ago through the rapid condensation of a dust cloud that circulated around the young Sun. This violent process, which involved the aggregation and collision of progressively larger clumps of matter, will be described in more detail in Chapter 9. Within just 100 million years (a relatively short time, geologically speaking), the Moon had formed and Earth's core had separated from its mantle. Exactly what happened during the next several hundred million years is hard to know. Very little of the rock record survived intense bombardment by the large meteorites that were constantly smashing into Earth.

This early period of Earth's history is appropriately called the geologic "dark ages."

The oldest rocks now found on Earth's surface are over 4 billion years old. Rocks as ancient as 3.8 billion years show evidence of erosion by water, indicating the existence of a hydrosphere and the operation of a climate system not too different from that of the present. Rocks only slightly younger, 3.5 billion years old, record a magnetic field about as strong as the one we see today, showing that the geodynamo was operating by that time. By about 3 billion years ago, enough low-density crust had collected at Earth's surface to form large continental masses. The geologic processes that modified the continents after that time were very similar to those we see operating in plate tectonics today.

The Evolution of Life

Life also began very early in Earth's history, as we can tell from the study of **fossils,** traces of organisms preserved in the geologic record. Fossils of primitive bacteria have been found in rocks dated at 3.5 billion years ago. A key event was the evolution of organisms that release oxygen into the atmosphere and oceans. The buildup of oxygen in the atmosphere was under way by 2.7 billion years ago. Atmospheric oxygen concentrations probably rose to modern levels in a series of steps over a period as long as 2 billion years.

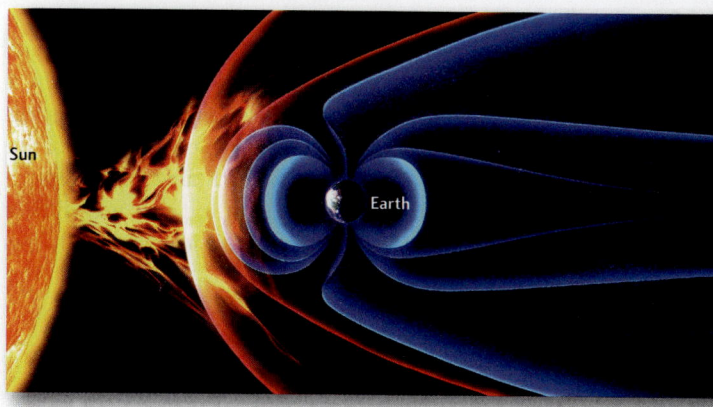

FIGURE 1.17 Earth's magnetic field protects life by shielding Earth's surface from harmful solar radiation. This solar wind contains highly energetic charged particles ejected from the Sun, which distorts Earth's magnetic field lines, shown here in light blue. The distances in this picture are not to scale. [Ikon Images/Getty Images.]

Life on early Earth was simple, consisting mostly of small, single-celled organisms that floated near the surface of the oceans or lived on the seafloor. Between 1 billion and

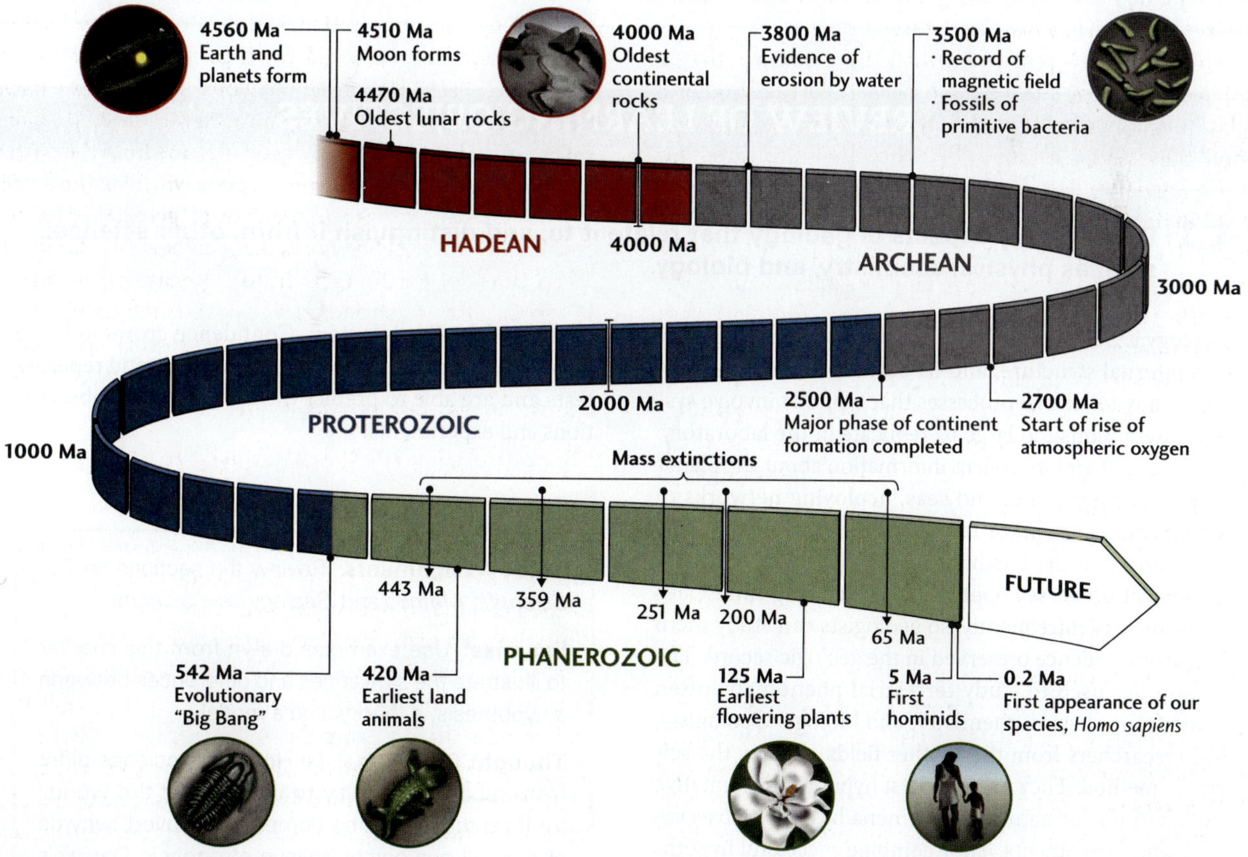

FIGURE 1.18 This geologic time line shows some of the major events observed in the geologic record, beginning with the formation of the planets. (Ma, million years ago.)

2 billion years ago, more complex life forms such as algae and seaweeds evolved. The first animals appeared about 600 million years ago, evolving in a series of waves. In a period starting 542 million years ago and probably lasting less than 10 million years, eight entirely new branches of the animal kingdom were established, including the ancestors of nearly all animals inhabiting Earth today. It was during this evolutionary explosion, sometimes called biology's "Big Bang," that animals with shells first left their fossils in the geologic record.

Although biological evolution is often viewed as a very slow process, it is punctuated by brief periods of rapid change. Spectacular examples are *mass extinctions*, during which many kinds of organisms suddenly disappeared from the geologic record. Five of these huge turnovers are marked on the geologic time line in Figure 1.18. The most recent one was caused by a large meteorite impact 65 million years ago. The meteorite, not much larger than 10 km in diameter, caused the extinction of half of Earth's species, including the giant animals of that geologic era, the dinosaurs.

KEY TERMS AND CONCEPTS

asthenosphere (p. 14)	geodesy (p. 6)	mantle (p. 8)
climate (p. 14)	geodynamo (p. 16)	outer core (p. 10)
climate system (p. 14)	geologic record (p. 4)	plate tectonic system (p. 15)
convection (p. 15)	geology (p. 2)	principle of uniformitarianism (p. 5)
core (p. 8)	geosystem (p. 13)	scientific method (p. 2)
crust (p. 10)	inner core (p. 10)	seismic wave (p. 7)
Earth system (p. 12)	lithosphere (p. 14)	topography (p. 6)
fossil (p. 17)	magnetic field (p. 15)	

REVIEW OF LEARNING OBJECTIVES

1.1 Describe key aspects of geology that relate it to, and distinguish it from, other sciences such as physics, chemistry, and biology.

Geology is the study of Earth—its history, its composition and internal structure, and its surface features. Geologists study terrestrial processes that typically involve system interactions too large to replicate in the laboratory. They must therefore collect information about the planet by surveying the land and seas, deploying networks of instruments to monitor terrestrial activity, and measuring changes in the Earth system using satellites. Many geological processes operate over very long timescales or occur very infrequently, so geologists can study them only from evidence preserved in the geologic record. The techniques used to study terrestrial phenomena often come from physics, chemistry, and biology. Geologists, like researchers from these other fields, employ the scientific method. They pose and test hypotheses—tentative explanations for natural phenomena based on observations and experiments. They combine successful hypotheses into theories, and they develop scientific models, often in the form of computer codes that can simulate aspects of the Earth system. Confidence grows in those hypotheses, theories, and models that withstand repeated tests and are able to predict the results of new observations and experiments.

Study Assignments: Review the sections on *The Scientific Method* and *Geology as a Science*.

Exercise: Use examples drawn from this chapter to illustrate the similarities and differences between a hypothesis, a theory, and a model.

Thought Questions: (a) How does science differ from religion as a way to understand the world? (b) If no theory can be completely proved, why do almost all geologists believe strongly in Darwin's theory of evolution?

1.2 Illustrate how the scientific method has been used to develop theories of Earth's shape and structure.

The Greek philosopher Eratosthenes obtained an accurate estimate of Earth's radius using two simple hypotheses (Earth is a sphere; Sun is far away), a mathematical theory (spherical geometry), and results from a clever experiment (measuring the shadow of a vertical pole). To explain Earth's high density, the German physicist Emil Wiechert hypothesized that Earth has an iron core encased in a silicate mantle. He supported his hypothesis with observations of stony and iron-nickel meteorites. The existence of a mantle and core has since been verified by many observations, including the direct imaging of Earth's internal structure using seismic waves from earthquakes. Geologists have built a layered Earth model that features a solid inner core and liquid outer core, and low-density silicate crust overlying a thick, higher-density silicate mantle.

Study Assignments: Complete the Practicing Geology Exercise and review Figure 1.8.

Exercises: (a) Give an example of how the model of Earth's spherical shape developed by Eratosthenes could be experimentally tested using observations made at Earth's surface. (b) Give two reasons why Earth's shape is not a perfect sphere. (c) If you made a model of Earth that was 10 cm in radius, how high would Mount Everest rise above sea level?

Thought Question: Why does the density of the material generally increase as you go deeper into Earth's interior?

1.3 Compare the chemical compositions of Earth's crust, mantle, and core in terms of their most abundant elements.

Ninety percent of Earth's mass consists of four elements. The core is primarily iron, whereas the crust and mantle are primarily oxygen, silicon, and magnesium. Oxygen and silicon combine with metals like magnesium and iron to form silicate minerals. The crust contains more oxygen and silicon than the mantle and therefore has a lower density. Variations in crustal composition and thickness explain the topography of continents and oceans.

Study Assignment: Review Figure 1.11.

Exercises: (a) From Earth's average radius and mean density given in this chapter, calculate its total mass in kilograms, and compare your results with the value given in Appendix 2. (b) Explain how Earth's outer core can be liquid while the mantle is solid.

Thought Questions: (a) Imagine you are a tour guide on a journey from Earth's surface to its center. How would you describe the material that your tour group encounters on the way down?

1.4 Locate within the layered Earth system the global geosystems that explain climate, plate tectonics, and the geomagnetic field.

To understand the Earth system, we focus on its subsystems (geosystems). The three major global geosystems are the climate system, which involves interactions among the atmosphere, hydrosphere, cryosphere, biosphere, and lithosphere; the plate tectonic system, which involves interactions among the lithosphere, asthenosphere, and deep mantle; and the geodynamo, which involves interactions within Earth's core. The climate system is driven by heat from the Sun, whereas the plate tectonic system and the geodynamo are driven by Earth's internal heat.

Study Assignment: Complete the Visual Literacy Exercise.

Exercises: (a) How do the terms *weather* and *climate* differ? Express the relationship between climate and weather using examples from your experience. (b) Earth's mantle is solid, but it undergoes convection as part of the plate tectonic system. Explain why these statements are not contradictory.

Thought Questions: (a) In what general ways are the climate system, the plate tectonic system, and the geodynamo similar? How do they differ? (b) Based on the material presented in this chapter, what can we say about how long ago the three major global geosystems began to operate? (c) Not every planet has a geodynamo. Why not? If Earth did not have a magnetic field, what might be different about our planet?

1.5 Recall Earth's age and some key events in the evolution of life that stand out in the geologic record.

Earth and the other planets in our solar system formed 4.56 billion years ago. Rocks more than 4 billion years old have survived in Earth's crust. Liquid water existed on Earth's surface by 3.8 billion years ago. Rocks about 3.5 billion years old show the earliest evidence of life. By 2.7 billion years ago, the oxygen content of the atmosphere was rising because of oxygen production by early organisms. Animals appeared suddenly about 600 million years ago, diversifying rapidly in a great evolutionary explosion. The subsequent evolution of life was marked by a series of extreme events that killed off many species, allowing new species to evolve. Our species, *Homo sapiens*, first appeared about 200,000 years ago.

Study Assignment: Review Figure 1.18.

Exercise: Imagine all of Earth history compressed into a single year, so that Earth formed on January 1st, and it is now midnight on December 31st. When would our human species first appear? At what time on New Year's Eve would you have been born?

Thought Question: It is thought that a large meteorite impacted Earth 65 million years ago causing the extinction of half of Earth's living species, including all the dinosaurs. Does this event disprove the principle of uniformitarianism? Explain your answer.

VISUAL LITERACY EXERCISE

Interacting Components of the Layered Earth System

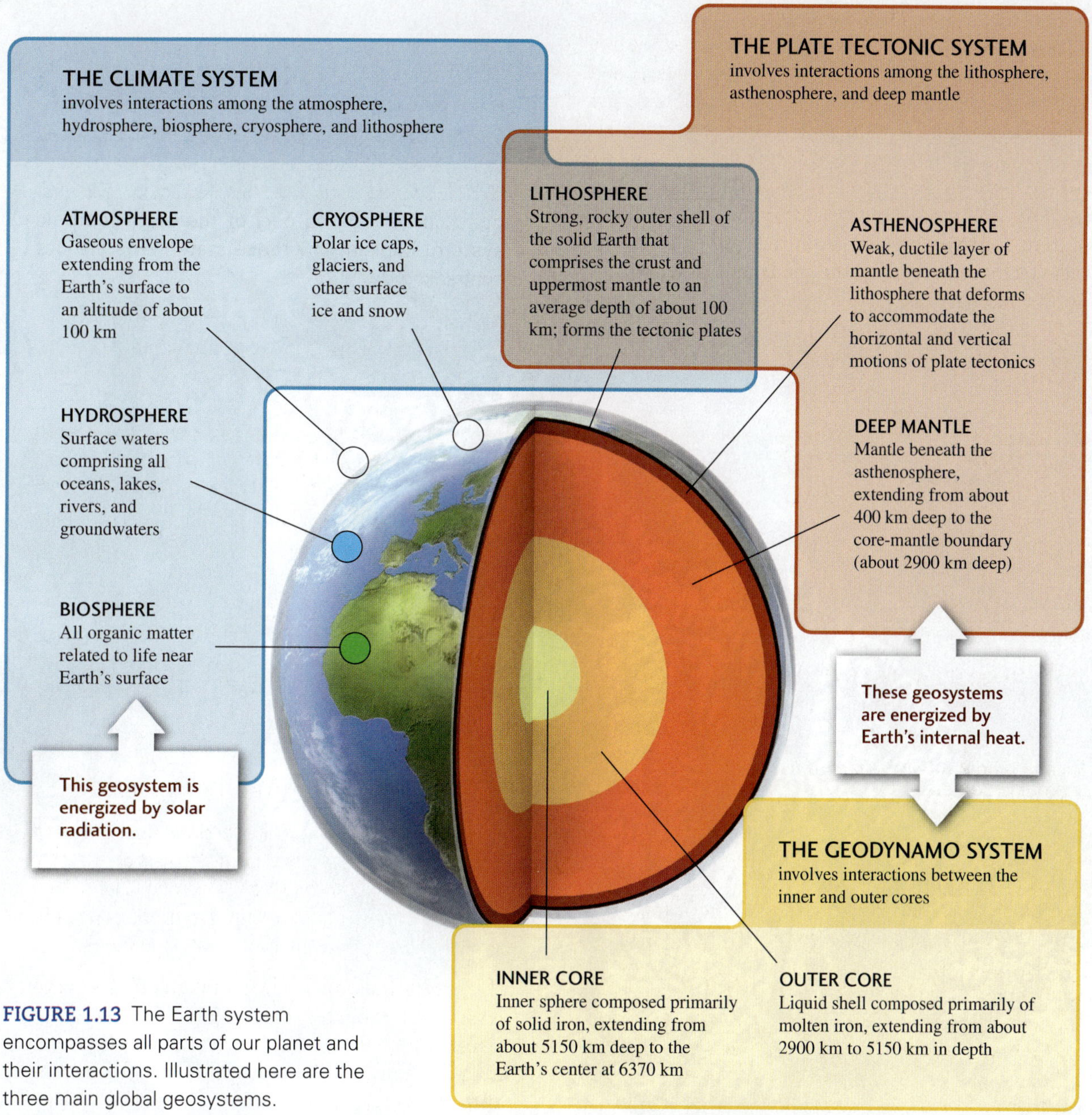

THE CLIMATE SYSTEM
involves interactions among the atmosphere, hydrosphere, biosphere, cryosphere, and lithosphere

THE PLATE TECTONIC SYSTEM
involves interactions among the lithosphere, asthenosphere, and deep mantle

ATMOSPHERE
Gaseous envelope extending from the Earth's surface to an altitude of about 100 km

CRYOSPHERE
Polar ice caps, glaciers, and other surface ice and snow

LITHOSPHERE
Strong, rocky outer shell of the solid Earth that comprises the crust and uppermost mantle to an average depth of about 100 km; forms the tectonic plates

ASTHENOSPHERE
Weak, ductile layer of mantle beneath the lithosphere that deforms to accommodate the horizontal and vertical motions of plate tectonics

HYDROSPHERE
Surface waters comprising all oceans, lakes, rivers, and groundwaters

DEEP MANTLE
Mantle beneath the asthenosphere, extending from about 400 km deep to the core-mantle boundary (about 2900 km deep)

BIOSPHERE
All organic matter related to life near Earth's surface

This geosystem is energized by solar radiation.

These geosystems are energized by Earth's internal heat.

THE GEODYNAMO SYSTEM
involves interactions between the inner and outer cores

INNER CORE
Inner sphere composed primarily of solid iron, extending from about 5150 km deep to the Earth's center at 6370 km

OUTER CORE
Liquid shell composed primarily of molten iron, extending from about 2900 km to 5150 km in depth

FIGURE 1.13 The Earth system encompasses all parts of our planet and their interactions. Illustrated here are the three main global geosystems.

1. Which of the three global geosystems are energized by Earth's internal heat?
2. Which of the three global geosystems primarily involves interactions within Earth's core?
3. With which global geosystem are interactions with the biosphere the strongest?
4. Which global geosystems are necessary for sustaining life over the long term?
5. Why is the lithosphere included in both the brown box and the blue box?

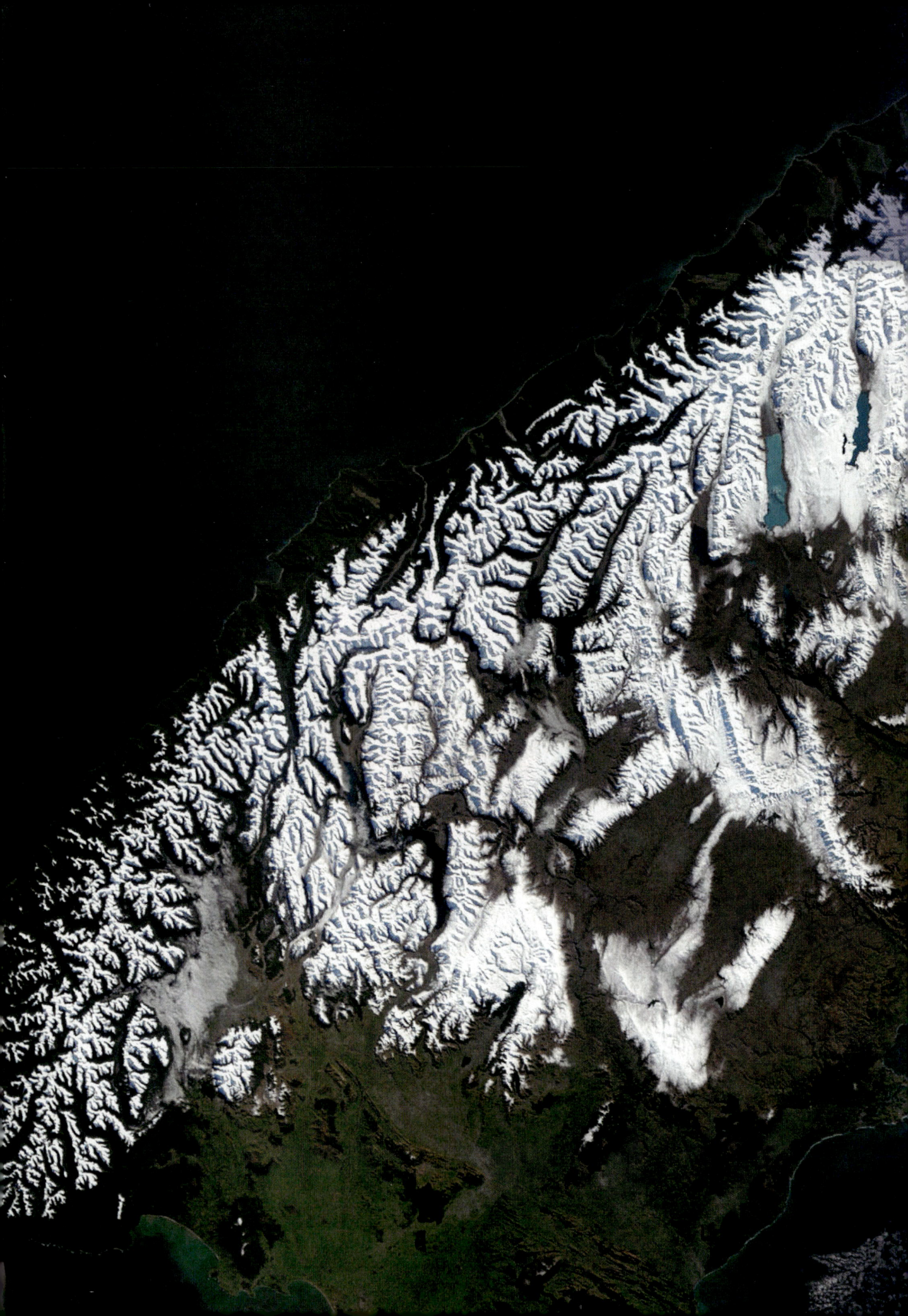

Plate Tectonics: The Unifying Theory

2

The Discovery of Plate Tectonics	24
The Plates and Their Boundaries	27
Rates and History of Plate Movements	35
The Grand Reconstruction	38
Mantle Convection: The Engine of Plate Tectonics	44

Learning Objectives

Plate tectonics theory explains geological processes on a global scale. From what you learn about plate tectonics in Chapter 2, you should be able to

2.1 Identify the largest tectonic plates and delineate their boundaries on a global map.

2.2 Summarize how geologists of the 1960s used seafloor spreading to explain Wegener's continental drift.

2.3 Based on observations of relative motions and geological activity, determine whether the edge of a tectonic plate is acting as part of a divergent, convergent, or transform-fault boundary.

2.4 Explain how magnetic anomalies recorded by ships at sea are used to estimate the age of the seafloor and the rate of seafloor spreading.

2.5 Reconstruct the supercontinent of Pangaea by running continental drift backwards over the past 200 million years.

2.6 Describe how the oceanic plates participate in mantle convection.

The snow line at the top of this satellite photo marks the Alpine Fault, an earthquake-prone plate boundary between the Australian and Pacific plates that crosses the South Island of New Zealand. [Jeff Schmaltz, MODIS Rapid Response Team, NASA/GSFC.]

The lithosphere—Earth's strong, rigid outer shell of rock—is broken into about a dozen plates, which slide past, converge with, or separate from each other as they move over the weaker, ductile asthenosphere. Plates are formed where they separate and recycled where they converge in a continuous process of creation and destruction. Continents, embedded in the lithosphere, drift along with the moving plates.

The theory of plate tectonics describes the movements of plates and the forces acting on them. It also explains volcanoes, earthquakes, and the distribution of mountain chains, rock assemblages, and structures on the seafloor—all of which result from events at plate boundaries. Plate tectonics provides a conceptual framework for a large part of this textbook and, indeed, for much of geology.

This chapter lays out the theory of plate tectonics and how it was discovered, describes plate movements today and in the geologic past, and examines how the forces that drive these movements arise from the mantle convection system.

The Discovery of Plate Tectonics

In the 1960s, a great revolution in thinking shook the world of geology. For almost 200 years, geologists had been developing various theories of *tectonics* (from the Greek *tekton,* meaning "builder")—the general term used to describe mountain building, volcanism, earthquakes, and other processes that construct geologic features on Earth's surface. It was not until the discovery of plate tectonics, however, that a single theory could satisfactorily explain the whole range of geologic processes. Physics had a comparable revolution at the beginning of the twentieth century, when the theory of relativity revised the physical laws that govern space, time, mass, and motion. Biology had a similar revolution in the middle of the twentieth century, when the discovery of DNA allowed biologists to explain how organisms transmit the information that controls their growth and functioning from generation to generation.

The basic ideas of plate tectonics were put together as a unified theory of geology about 50 years ago. The scientific synthesis that led to the theory of plate tectonics, however, really began much earlier in the twentieth century with the recognition of evidence for continental drift.

Continental Drift

Such changes in the superficial parts of the globe seemed to me unlikely to happen if the earth were solid to the center. I therefore imagined that the internal parts might be a fluid more dense, and of greater specific gravity than any of the solids we are acquainted with, which therefore might swim in or upon that fluid. Thus the surface of the earth would be a shell, capable of being broken and disordered by the violent movements of the fluid on which it rested.

(Benjamin Franklin, 1782, in a letter to French geologist Abbé J. L. Giraud-Soulavie)

The concept of **continental drift**—large-scale movements of continents—has been around for a long time. In the late sixteenth century and in the seventeenth century, European scientists noticed the jigsaw-puzzle fit of the coasts on both sides of the Atlantic Ocean, as if the Americas, Europe, and Africa had once been part of a single continent and had subsequently drifted apart. By the close of the nineteenth century, the Austrian geologist Eduard Suess had put together some of the pieces of the puzzle. He postulated that the present-day southern continents had once formed a single giant continent called *Gondwana* (or *Gondwanaland*). In 1915, Alfred Wegener (**Figure 2.1**), a German meteorologist who was recovering from wounds suffered in World War I, wrote a book on the breakup and drift of continents, in which he laid out the remarkable similarity of geologic features on opposite sides of the Atlantic (**Figure 2.2**). In the years that followed, Wegener postulated a supercontinent, which he called **Pangaea** (Greek for "all lands"), that broke up into the continents as we know them today.

FIGURE 2.1 Alfred Lothar Wegener (1880-1930) crossing a glacier during his final and fatal expedition to Greenland, November 1930. [Ullstein Bild/Getty Images.]

The Discovery of Plate Tectonics

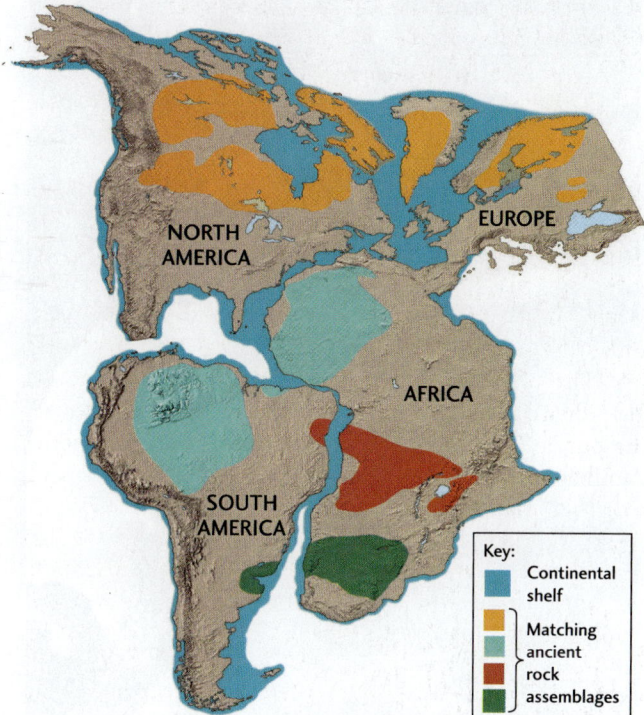

FIGURE 2.2 The jigsaw-puzzle fit of the continents bordering the Atlantic Ocean formed the basis of Alfred Wegener's theory of continental drift. In his book, titled *The Origin of Continents and Oceans,* Wegener cited as additional evidence the similarity of geologic features on opposite sides of the Atlantic. The matches shown here, between ancient crystalline rocks in adjacent regions of South America and Africa and of North America and Europe, are from maps refined by geologists in the mid-1960s.

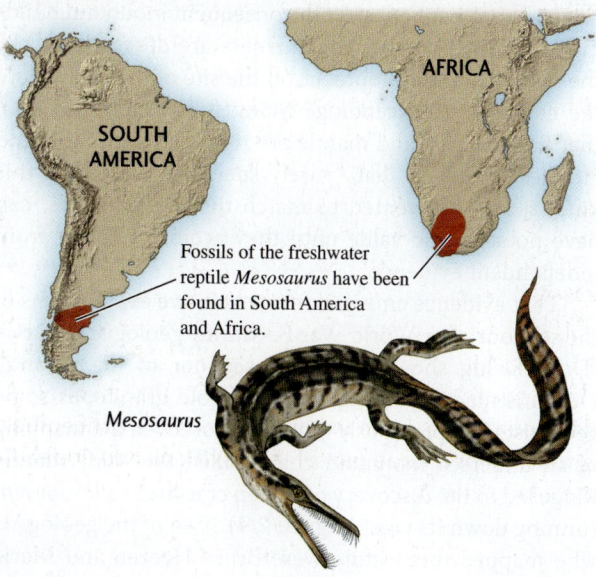

FIGURE 2.3 Fossils of the freshwater reptile *Mesosaurus,* 300 million years old, are found in South America and Africa and nowhere else in the world. If *Mesosaurus* were able to swim across the South Atlantic Ocean, it should have been able to cross other oceans and should have spread more widely. The observation that it did not suggests that South America and Africa must have been joined about 300 million years ago. [Information from A. Hallam, "Continental Drift and the Fossil Record," *Scientific American* (November 1972): 57–66.]

Although Wegener was correct in asserting that the continents had drifted apart, his hypotheses about how fast they were moving and what forces were pushing them across Earth's surface turned out to be wrong, as we will see, and those errors reduced his credibility among other scientists. After about a decade of spirited debate, physicists convinced geologists that Earth's outer layers were too rigid for continental drift to occur, and Wegener's ideas were rejected by all but a few geologists.

Wegener and other advocates of the drift hypothesis pointed not only to the geographic matching of geologic features, but also to similarities in rock ages and trends in geologic structures on opposite sides of the Atlantic. They also offered arguments, accepted now as good evidence of drift, based on fossil and climate data. Identical 300-million-year-old fossils of the reptile *Mesosaurus,* for example, have been found in Africa and in South America, but nowhere else, suggesting that the two continents were joined when *Mesosaurus* lived (**Figure 2.3**). The animals and plants on the different continents showed similarities in their evolution until the postulated breakup time. After that, they followed different evolutionary paths because of their isolation and changing environments on the separating continents. In addition, rocks deposited by glaciers that existed 300 million years ago were found distributed across South America, Africa, India, and Australia. If these southern continents had once been part of Gondwana near the South Pole, a single continental glacier could account for all of these glacial deposits.

Seafloor Spreading

The geologic evidence did not convince the skeptics, who maintained that continental drift was physically impossible. No one had yet come up with a plausible driving force that could have split Pangaea and moved the continents apart. Wegener, for example, thought the continents floated like boats across the solid oceanic crust, dragged along by the tidal forces of the Sun and Moon. His hypothesis was quickly rejected, however, because it could be shown that tidal forces are much too weak to move continents.

The breakthrough came when scientists realized that convection in Earth's mantle (described in Chapter 1) could push and pull continents apart, creating new oceanic crust through the process of **seafloor spreading.** In 1928, the British geologist Arthur Holmes proposed that convection currents "dragged the two halves of the original continent apart, with consequent mountain building in the front where the currents are descending, and the ocean floor development on the site of the gap, where the currents are ascending." Many still argued, however, that Earth's crust and mantle are rigid and immobile, and Holmes conceded that "purely speculative ideas of this kind, specially invented to match the requirements, can have no scientific value until they acquire support from independent evidence."

That evidence emerged from extensive explorations of the seafloor after World War II. Marine geologist Maurice "Doc" Ewing showed that the seafloor of the Atlantic Ocean is made of young basalt, not old granite, as some geologists had previously thought. Moreover, the mapping of an undersea mountain chain called the Mid-Atlantic Ridge led to the discovery of a deep cracklike valley, or *rift*, running down its crest (Figure 2.4). Two of the geologists who mapped this feature were Bruce Heezen and Marie Tharp, colleagues of Doc Ewing at Columbia University (Figure 2.5). "I thought it might be a rift valley," Tharp said years later. Heezen initially dismissed the idea as "girl talk," but they soon found that almost all earthquakes in the Atlantic Ocean occurred near the rift, confirming Tharp's hunch. Because most earthquakes are generated by faulting, their results indicated that the rift was a tectonically active feature. Other mid-ocean ridges with similar rifts and earthquake activity were found in the Pacific and Indian oceans.

FIGURE 2.5 Marie Tharp and Bruce Heezen inspecting a map of the seafloor. Their discovery of tectonically active rifts on mid-ocean ridges provided important evidence for seafloor spreading. [Estate of Marie Tharp.]

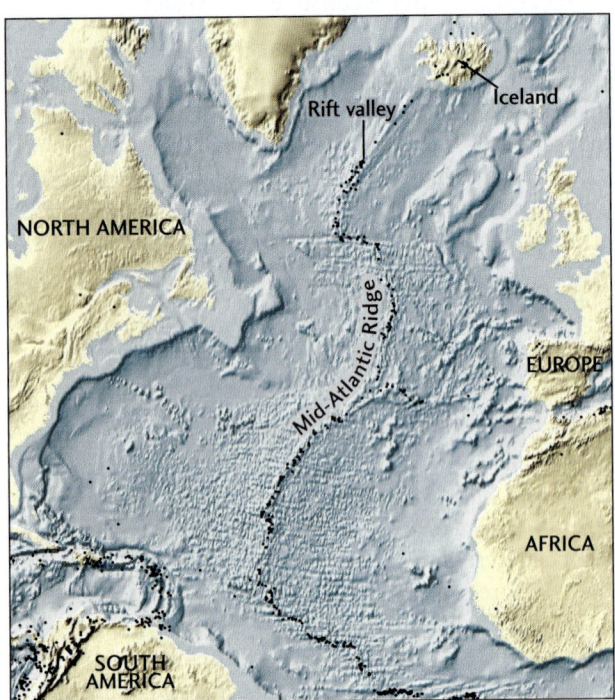

FIGURE 2.4 The North Atlantic seafloor, showing the cracklike rift valley running down the center of the Mid-Atlantic Ridge and the locations of earthquakes (black dots).

In the early 1960s, Harry Hess of Princeton University and Robert Dietz of the Scripps Institution of Oceanography proposed that Earth's crust separates along the rifts in mid-ocean ridges, and that new crust is formed by the upwelling of hot molten rock into these cracks. The new seafloor—actually the surface of newly created lithosphere—spreads laterally away from the rifts and is replaced by even newer crust in a continuing process of plate creation.

The Great Synthesis: 1963–1968

The seafloor spreading hypothesis put forward by Hess and Dietz explained how the continents could move apart through the creation of new lithosphere at mid-ocean ridges. But it raised another question: Could the seafloor and its underlying lithosphere be destroyed by recycling back into Earth's interior? If not, Earth's surface area would have to increase over time. For a while in the early 1960s, some physicists and geologists, including Heezen, believed in this idea of an expanding Earth. Other geologists recognized that the seafloor was indeed being recycled. They were convinced this was happening in several regions of intense volcanic and earthquake activity around the margins of the Pacific Ocean basin, known collectively as the Ring of Fire (**Figure 2.6**). The details of the process, however, remained unclear.

In 1965, the Canadian geologist J. Tuzo Wilson first described tectonics around the globe in terms of rigid plates moving over Earth's surface. He characterized three basic types of boundaries where plates move apart, come together, or slide past each other. Soon after, other scientists showed that almost all contemporary tectonic deformation—the process by which rocks are folded, faulted, sheared, or compressed by tectonic forces—is concentrated at these boundaries. They measured the rates and directions of crustal movements and demonstrated that these movements are mathematically consistent with a system of rigid plates moving over the planet's spherical surface.

The basic elements of the new theory of **plate tectonics** were established by the end of 1968. By 1970, the evidence for plate tectonics had become so persuasive that almost all Earth scientists embraced the theory. Textbooks were revised and specialists began to consider the implications of the new concept for their own fields.

The Plates and Their Boundaries

According to the theory of plate tectonics, the lithosphere is not a continuous shell, but is broken into a mosaic of rigid plates that move over Earth's surface (**Figure 2.7**). Each plate travels as a distinct unit, riding on the asthenosphere, which is also in motion. The largest is the Pacific Plate, which comprises much (though not all) of the Pacific Ocean basin. Some of the plates are named after the continents they include, but in no case is a plate identical with a continent. The North American Plate, for instance, extends from the Pacific coast of North America to the middle of the Atlantic Ocean, where it meets the Eurasian and African plates.

In addition to the 13 major plates, there are a number of smaller ones. An example is the Juan de Fuca Plate, a small piece of oceanic lithosphere trapped between the giant Pacific and North American plates just offshore of the northwestern United States. Others, not shown in Figure 2.7, are small continental fragments, such as the Anatolian Plate, which includes much of Turkey.

To see plate tectonics in action, go to a plate boundary! Depending on which boundary you visit, you may find earthquakes; volcanoes; rising mountains; long, narrow rifts; folding; or faulting. Many geologic features develop through the interactions of plates at their boundaries.

There are three basic types of plate boundaries (**Figure 2.8**), all defined by the direction of movement of the plates relative to each other:

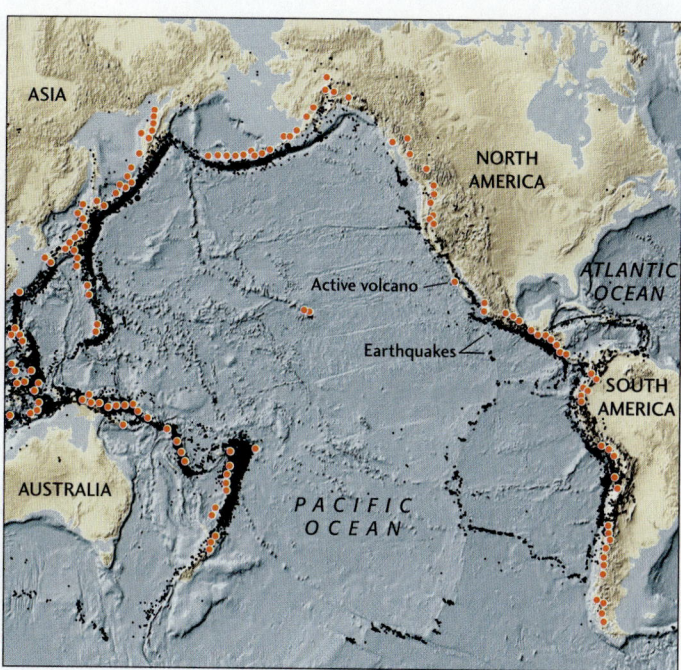

FIGURE 2.6 The Pacific Ring of Fire, with its active volcanoes (large red circles) and frequent earthquakes (small black dots), marks convergent plate boundaries where oceanic lithosphere is being recycled.

- At **divergent boundaries,** plates move apart and new lithosphere is created (plate area increases).
- At **convergent boundaries,** plates come together and one plate is recycled into the mantle (plate area decreases).
- At **transform faults,** plates slide horizontally past each other (plate area does not change).

Like many models of nature, these three plate boundary types are idealized. There are also "oblique" boundaries that combine divergence or convergence with some amount of transform faulting. Moreover, what actually goes on at a plate boundary depends on the type of lithosphere involved, because continental and oceanic lithosphere behave differently. The continental crust is made of rocks that are both lighter and weaker than those of either the oceanic crust or the mantle beneath the crust (see Figure 1.10). Later chapters will examine these differences in more detail; for now, you need to keep in mind only two of their consequences:

1. Because continental crust is lighter, it is not as easily recycled back into the mantle as oceanic crust.
2. Because continental crust is weaker, plate boundaries that involve continental crust tend to be more spread out and more complicated than those that involve oceanic crust.

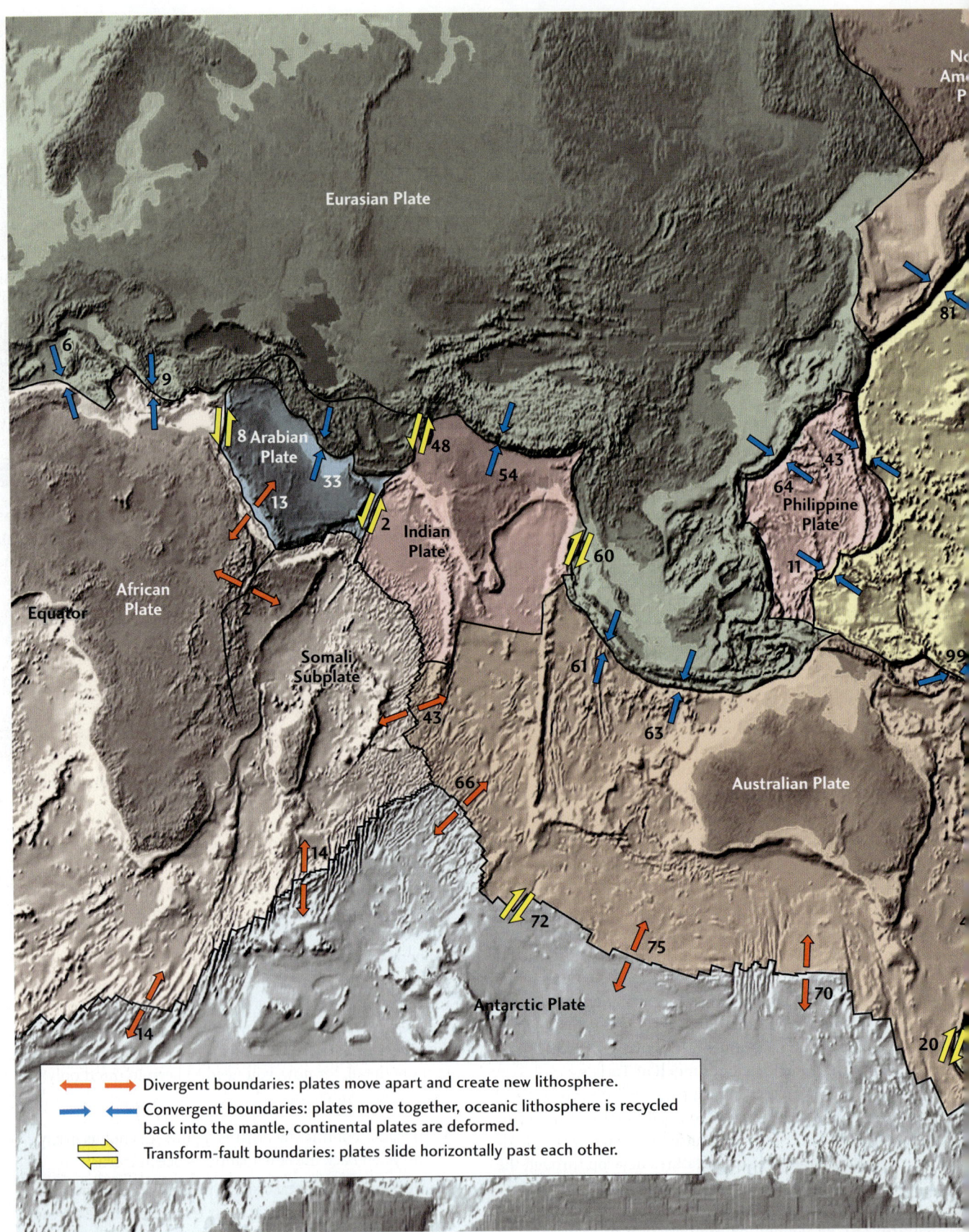

FIGURE 2.7 Earth's surface is a mosaic of 13 major plates, as well as a number of smaller plates, of rigid lithosphere that move slowly over the ductile asthenosphere. Only one of the smaller plates—the Juan de Fuca Plate, off the west coast of North America—is shown on this map. The arrows show the relative movement of two plates at a point on their boundary. The numbers next to the arrows give the relative plate velocities in millimeters per year. [Plate boundary data by Peter Bird, UCLA.]

The Plates and Their Boundaries

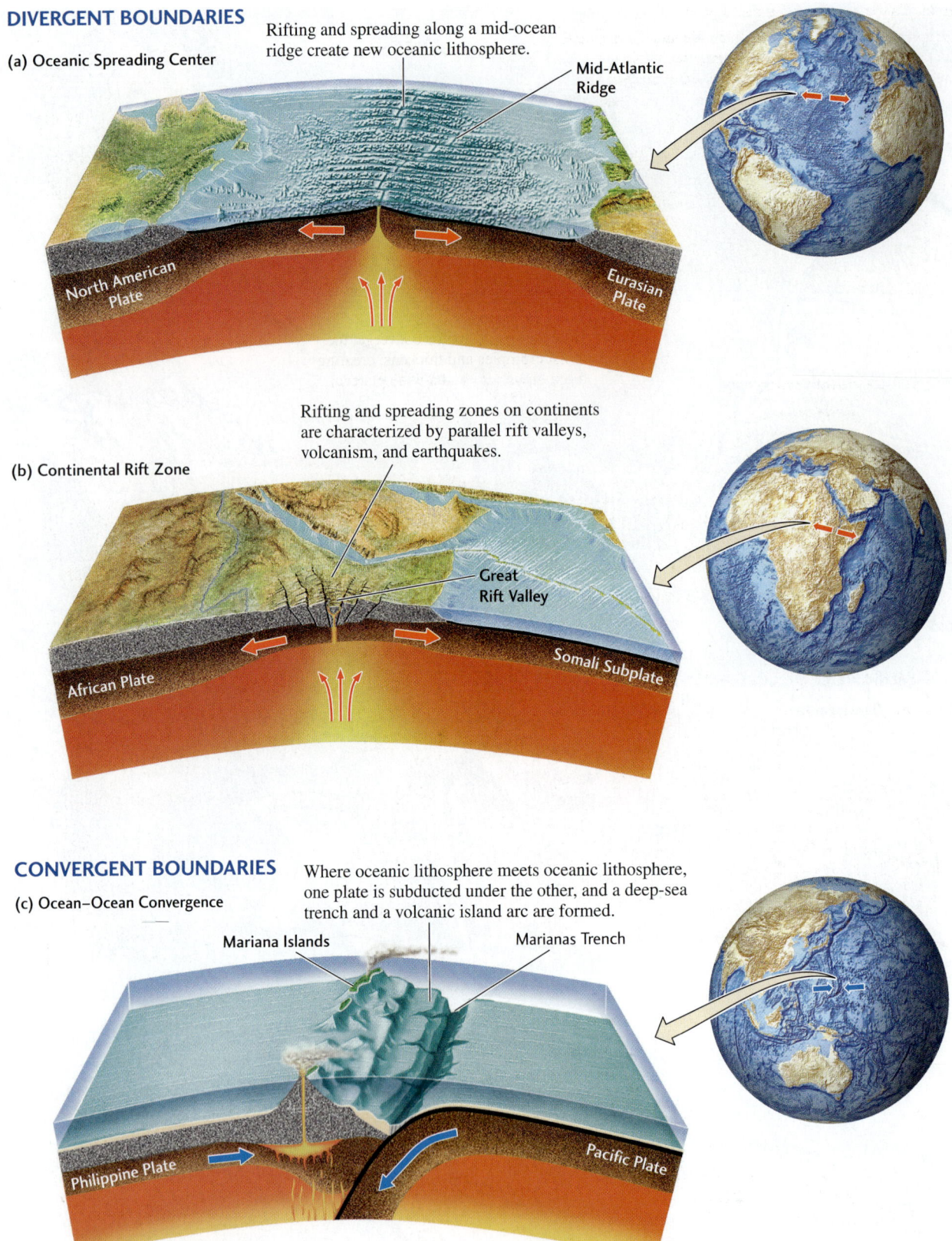

FIGURE 2.8 The interactions of lithospheric plates at their boundaries depend on the relative direction of plate movement and the type of lithosphere involved.

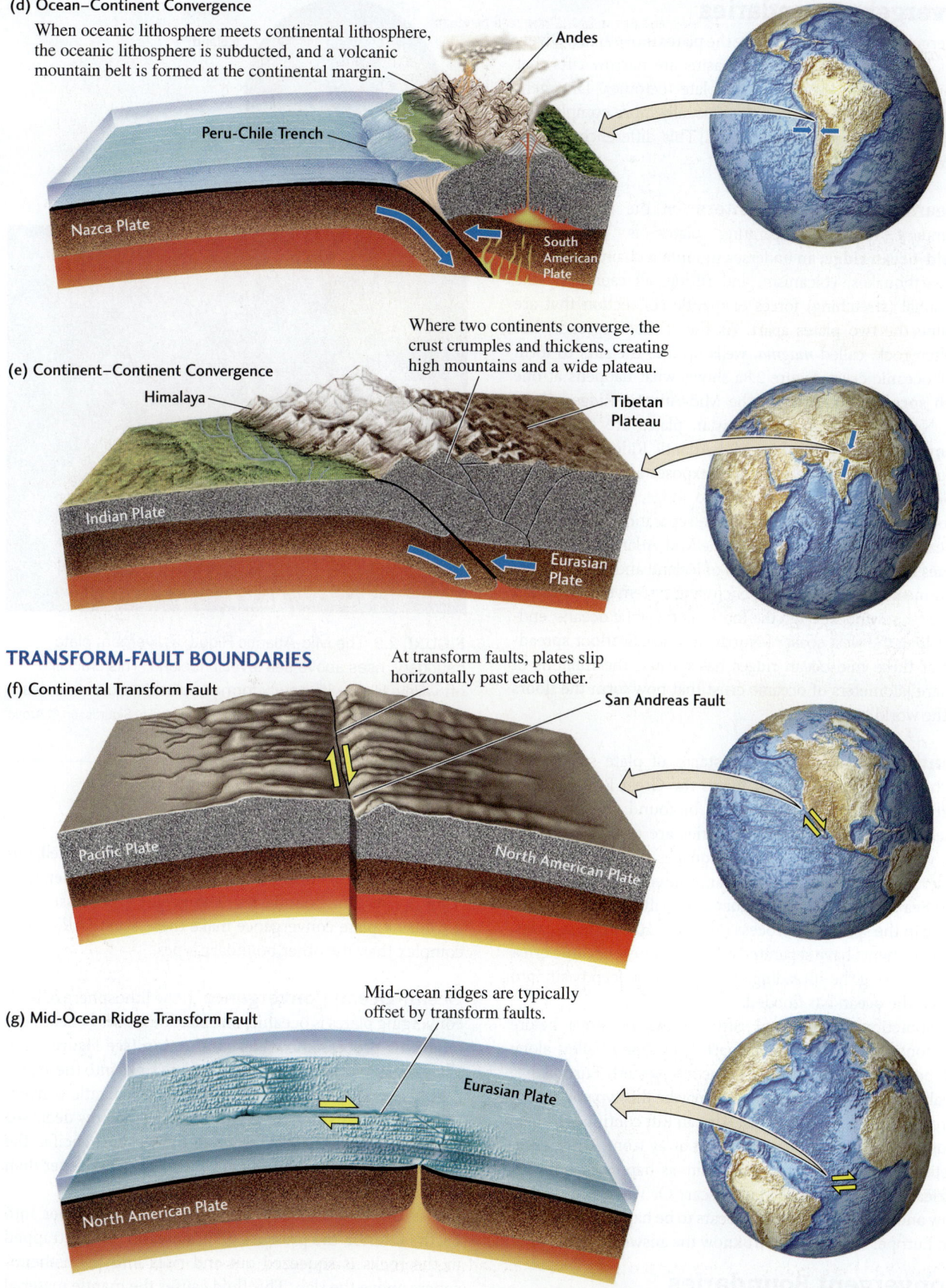

Divergent Boundaries

Divergent boundaries are where the plates move apart. Divergent boundaries within ocean basins are narrow rifts that approximate the idealization of plate tectonics. Divergent boundaries within continents are usually more complicated and distributed over a wider area. This difference is illustrated in Figures 2.8a and 2.8b.

Oceanic Spreading Centers On the seafloor, the boundary between separating plates is marked by a **mid-ocean ridge,** an undersea mountain chain that exhibits earthquakes, volcanism, and rifting, all caused by the tensional (stretching) forces of mantle convection that are pulling the two plates apart. As the seafloor spreads, hot molten rock, called *magma,* wells up into the rifts to form new oceanic crust. Figure 2.8a shows what happens at one such **spreading center** on the Mid-Atlantic Ridge, where the North American and Eurasian plates are separating. (A more detailed map of the Mid-Atlantic Ridge is shown in Figure 2.4.) The island of Iceland exposes a segment of the otherwise submerged Mid-Atlantic Ridge, allowing geologists to view the processes of plate separation and seafloor spreading directly (**Figure 2.9**). The Mid-Atlantic Ridge continues in the Arctic Ocean north of Iceland and, to the south, it connects to a nearly globe-encircling system of mid-ocean ridges that wind through the Indian and Pacific oceans, ending along the west coast of North America. Seafloor spreading at these mid-ocean ridges has created the millions of square kilometers of oceanic crust that now form the floors of the world's oceans.

Continental Rifting Early stages of plate separation, such as the divergence that forms the Great Rift Valley of east Africa (see Figure 2.8b), can be found on some continents. These divergent boundaries are characterized by rift valleys, volcanism, and earthquakes distributed over a wider zone than is found at oceanic spreading centers. The Red Sea and the Gulf of California are rifts that are further along in the spreading process (**Figure 2.10**). In these cases, the continents have separated enough for new oceanic crust to form along the spreading axis, creating a deep basin into which the ocean has flooded.

Sometimes continental rifting slows or stops before the continent actually splits apart. The Rhine Valley, along the border of Germany and France in western Europe, is a weakly active continental rift that may be this type of "failed" spreading center. Will the East African Rift continue to open, causing the Somali Subplate to split away from Africa completely and form a new ocean basin, as happened between Africa and the island of Madagascar? Or will the spreading slow and eventually stop, as appears to be happening in western Europe? Geologists don't know the answer.

Convergent Boundaries

Lithospheric plates cover the globe, so if they separate in one place, they must converge somewhere else if Earth's

FIGURE 2.9 The Mid-Atlantic Ridge, a divergent plate boundary, rises above sea level in Iceland. This cracklike rift valley, filled with newly formed volcanic rock, indicates that plates are being pulled apart. [Ragnar Th. Sigurdsson © Arctic Images/Alamy.]

surface area is to remain the same. (As far as we can tell, our planet is not expanding!) Where plates come together, they form convergent boundaries. The geologic processes that act during plate convergence make these boundaries more complex than the other boundary types.

Ocean-Ocean Convergence If the lithosphere of both converging plates is oceanic, one plate descends beneath the other in a process known as **subduction** (see Figure 2.8c). The lithosphere of the subducting plate sinks into the asthenosphere and is eventually recycled by the mantle convection system. This sinking produces a long, narrow deep-sea trench. In the Marianas Trench of the western Pacific, the ocean reaches its greatest depth, about 11 km—deeper than the height of Mount Everest (see Figure 1.7).

As the cold slab of lithosphere descends deeper into Earth's interior, the pressure on it increases. Water trapped in the rocks is squeezed out and rises into the asthenosphere above the slab. This fluid causes the mantle material above it to melt. The resulting magma produces a chain of volcanoes, called an **island arc,** behind the trench.

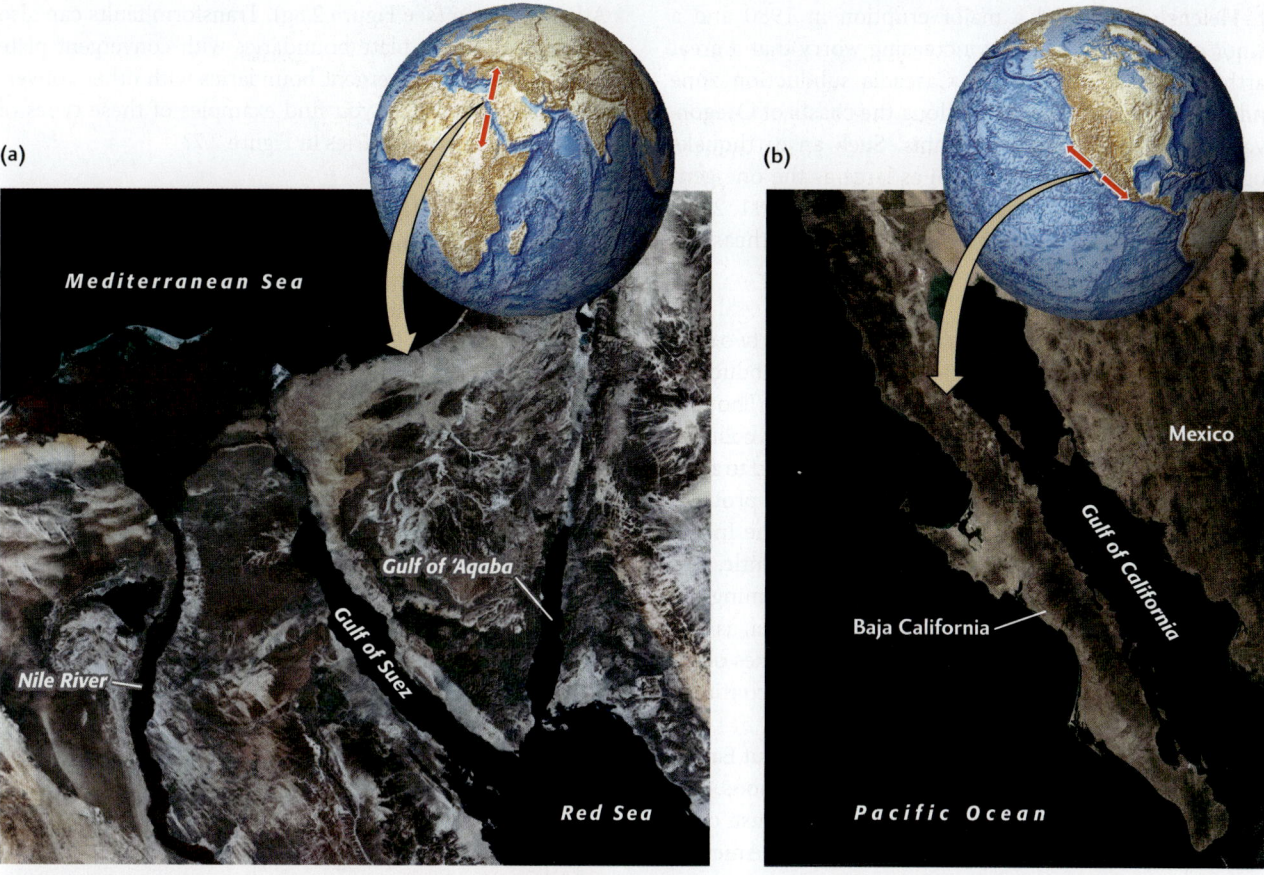

FIGURE 2.10 Rifting of continental crust. (a) The Arabian Plate, on the right, is moving northeastward relative to the African Plate, on the left, opening the Red Sea (lower right). The Gulf of Suez is a failed rift that became inactive about 5 million years ago. North of the Red Sea, most of the plate motion is now taken up by rifting and transform faulting along the Gulf of 'Aqaba and its northward extension. (b) Baja California, on the Pacific Plate, is moving northwestward relative to the North American Plate, opening the Gulf of California between Baja and the Mexican mainland. [(a) Courtesy MDA Information Systems LLC; (b) Jeff Schmaltz, MODIS Rapid Response Team, NASA/GSFC.]

The subduction of the Pacific Plate has formed the volcanically active Aleutian Islands west of Alaska as well as the Mariana Islands and other island arcs in the western Pacific. The lithospheric slabs descending into the mantle cause earthquakes as deep as 690 km beneath some island arcs.

Ocean-Continent Convergence If one plate has a continental edge, it overrides the oceanic lithosphere of the other plate because continental lithosphere is less dense and therefore less easily subducted (see Figure 2.8d). The submerged margin of the continent is crumpled by the convergence, deforming the continental crust and uplifting rocks into a mountain belt roughly parallel to the deep-sea trench. The enormous compressive (squeezing) forces of convergence and subduction produce great earthquakes along the subduction zone. Over time, materials are scraped off the descending slab and incorporated into the adjacent mountain belt, leaving geologists with a complex (and often confusing) record of the subduction process. As in the case of ocean-ocean convergence, the water carried downward by the subducting oceanic lithosphere causes mantle material to melt; the resulting magma rises and forms volcanoes in the mountain belt behind the trench.

The west coast of South America, where the South American Plate converges with the Nazca Plate, is a subduction zone of this type. A great chain of high mountains, the Andes, rises on the continental side of this convergent boundary, and a deep-sea trench lies just off the coast. The volcanoes here are active and deadly. One of them, Nevado del Ruiz in Colombia, killed 25,000 people when it erupted in 1985. Some of the world's largest earthquakes have been recorded along this boundary.

Another example is the Cascadia subduction zone, where the small Juan de Fuca Plate converges with the North American Plate off the coast of western North America. This convergent boundary gives rise to the dangerous volcanoes of the Cascade Range, such as Mount

St. Helens, which had a major eruption in 1980 and a minor one in 2004. There is increasing worry that a great earthquake will occur in the Cascadia subduction zone and cause devastating damage along the coasts of Oregon, Washington, and British Columbia. Such an earthquake could cause a disastrous tsunami as large as the one generated by the great Tohoku earthquake of March 11, 2011, which occurred in a subduction zone off the northeastern coast of Honshu, Japan.

Continent-Continent Convergence Where two continents converge (see Figure 2.8e), the kind of subduction seen at other convergent boundaries cannot occur. The geologic consequences of such a continent-continent collision are impressive. The collision of the Indian and Eurasian plates, both with continents at their leading edges, provides the best example. The Eurasian Plate overrides the Indian Plate, but India and Asia remain afloat on the mantle. The collision creates a double thickness of crust, forming the highest mountain range in the world, the Himalaya, as well as the vast high Tibetan Plateau. Severe earthquakes occur in the crumpling crust of this and other continent-continent collision zones.

Many episodes of mountain building throughout Earth's history were caused by continent-continent collisions. The Appalachian Mountains, which run along the east coast of North America, were uplifted when North America, Eurasia, and Africa collided to form the supercontinent Pangaea about 300 million years ago.

Transform Faults

At boundaries where plates slide past each other, lithosphere is neither created nor destroyed. Such boundaries are transform faults: fractures along which the plates slip horizontally past each other (see Figure 2.8f, g).

The San Andreas Fault in California, where the Pacific Plate slides past the North American Plate, is a prime example of a continental transform fault (see Figure 2.8f). Another is the Alpine Fault, which slices through the South Island of New Zealand, highlighted in the photo at the beginning of this chapter. Because the plates have been sliding past each other for millions of years, the rocks facing each other on the two sides of the fault are of different types and ages (**Figure 2.11**). Large earthquakes, such as the one that destroyed San Francisco in 1906, can occur on transform faults, which are often vertical and delineated by their seismic activity. There is much concern that, within the next several decades, a sudden rupture of the San Andreas Fault, or on related faults near Los Angeles and San Francisco, will result in an extremely destructive earthquake.

Transform-fault boundaries are typically found along mid-ocean ridges where the continuity of a spreading zone is broken and the boundary is offset in a steplike pattern. An example can be found along the boundary between the African and the South American plates in the central Atlantic Ocean (see Figure 2.8g). Transform faults can also connect divergent plate boundaries with convergent plate boundaries and convergent boundaries with other convergent boundaries. Can you find examples of these types of transform-fault boundaries in Figure 2.7?

Combinations of Plate Boundaries

Each plate is bordered by some combination of divergent, convergent, and transform-fault boundaries. For example, the Nazca Plate is bounded on three sides by spreading centers, offset in a steplike pattern by transform faults, and on one side by the Peru-Chile subduction zone (see Figure 2.7). The North American Plate is bounded on the east by the Mid-Atlantic Ridge, a spreading center, and on the west by subduction zones and transform-fault boundaries.

FIGURE 2.11 The San Andreas Fault runs through California for 970 km from Imperial Valley to Point Arena. This view is southeast along the fault in the Carrizo Plain of central California. The San Andreas is a transform-fault boundary between the Pacific Plate (right) and the North American Plate (left). [James Balog/Getty Images.]

Rates and History of Plate Movements

How fast do plates move? Do some plates move faster than others, and if so, why? Is the velocity of plate movements today the same as it was in the geologic past? Geologists have developed ingenious methods to answer these questions and thereby to gain a better understanding of plate tectonics. In this section, we will examine three of these methods.

The Seafloor as a Magnetic Tape Recorder

During World War II, extremely sensitive magnetometers were developed to detect submarines by the magnetic fields emanating from their steel hulls. Geologists modified these instruments slightly and towed them behind research ships to measure the local magnetic field created by magnetized rocks on the seafloor. Steaming back and forth across the ocean, seagoing scientists discovered regular patterns in the strength of the local magnetic field that completely surprised them. In many areas, the intensity of the magnetic field alternated between high and low values in long, narrow parallel bands, called **magnetic anomalies**, that were almost perfectly symmetrical with respect to the crest of a mid-ocean ridge (Figure 2.12). The detection of these patterns was one of the great discoveries that confirmed the seafloor spreading hypothesis and led to the theory of plate tectonics. It also allowed geologists to trace plate movements far back in geologic time. To understand these advances, we need to look more closely at how rocks become magnetized.

The Rock Record of Magnetic Reversals On Land
Magnetic anomalies are evidence that Earth's magnetic field does not remain constant over time. At present, the north magnetic pole is closely aligned with the geographic north pole (see Figure 1.16), but small changes in the geodynamo can flip the orientation of the north and south magnetic poles by 180°, causing a magnetic reversal.

In the early 1960s, geologists discovered that a precise record of this peculiar behavior can be obtained from layered flows of volcanic lava. When iron-rich lavas cool, they become slightly but permanently magnetized in the direction of Earth's magnetic field. This phenomenon is called *thermoremanent magnetization* because the cold lava "remembers" the magnetization long after the magnetic field has changed.

In layered lava flows, such as those in a volcanic cone, the rocks at the top represent the most recent layer, while layers deeper in the cone are older. The age of each layer can be determined by various dating methods (described in Chapter 8). The direction of magnetization of rock samples from each layer then reveals the direction of Earth's magnetic field at the time when that layer cooled (Figure 2.12b). By repeating these measurements at hundreds of places around the world, geologists have worked out the **magnetic time scale** of the past 200 million years. Figure 2.12c shows this time scale for the most recent 5 million years. About half of all volcanic rocks studied have been found to be magnetized in a direction opposite to that of Earth's present magnetic field. Apparently, the field has flipped frequently over geologic time, so normal fields (same as now) and reversed fields (opposite to now) are equally likely. Major periods during which the field is normal or reversed are called *magnetic chrons* (from the Greek for "time"); they last about half a million years, on average, although the pattern of reversals becomes highly irregular as we move back in geologic time. Within the major chrons are short-lived reversals of the field, known as *magnetic subchrons*, which may last anywhere from several thousand to 200,000 years.

Magnetic Anomaly Patterns on The Seafloor
The banded patterns of magnetism found on the seafloor puzzled scientists until 1963, when two Englishmen, F. J. Vine and D. H. Mathews—and, independently, two Canadians, L. Morley and A. Larochelle—made a startling proposal. Based on evidence of magnetic reversals that geologists had collected from lava flows on land, they reasoned that the bands of high and low magnetic intensity on the seafloor corresponded to bands of rock that were magnetized during ancient episodes of normal and reversed magnetism. That is, when a research ship was above rocks magnetized in the normal direction, it would record a locally stronger field, or a *positive magnetic anomaly*. When it was above rocks magnetized in the reversed direction, it would record a locally weaker field, or a *negative magnetic anomaly*.

The seafloor spreading hypothesis explained these observations: When the plates move apart at a mid-ocean ridge, magma rises from Earth's interior and flows into the rift, where it cools, solidifies, and becomes magnetized in the direction of Earth's magnetic field at the time. As the seafloor spreads away from the ridge, approximately half of the newly magnetized material moves to one side and half to the other, forming two symmetrical magnetized bands. Newer material fills the rift, continuing the process. In this way, the seafloor acts like a magnetic tape recorder that encodes the history of the opening of the oceans.

Inferring Seafloor Ages and Relative Plate Velocity
By using the ages of magnetic reversals that had been worked out from magnetized lavas on land, geologists were able to assign ages to the bands of magnetized rocks on the seafloor. They could then calculate how fast the seafloor was spreading by using the formula *speed = distance ÷ time*, where distance is measured from the mid-ocean ridge axis and time equals seafloor

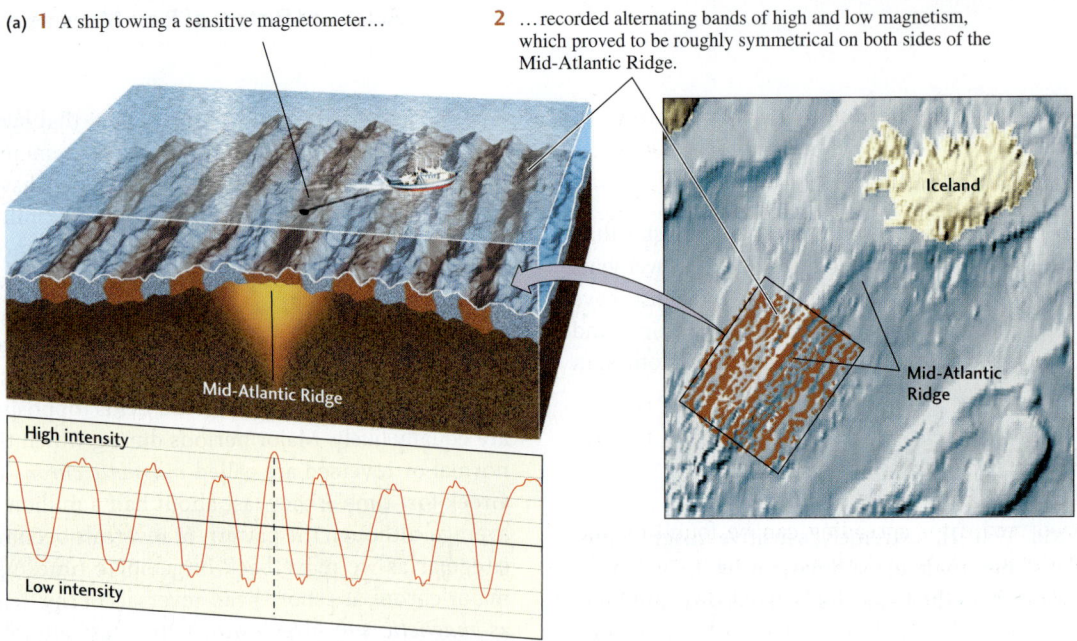

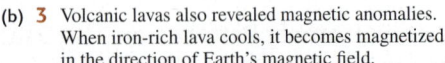

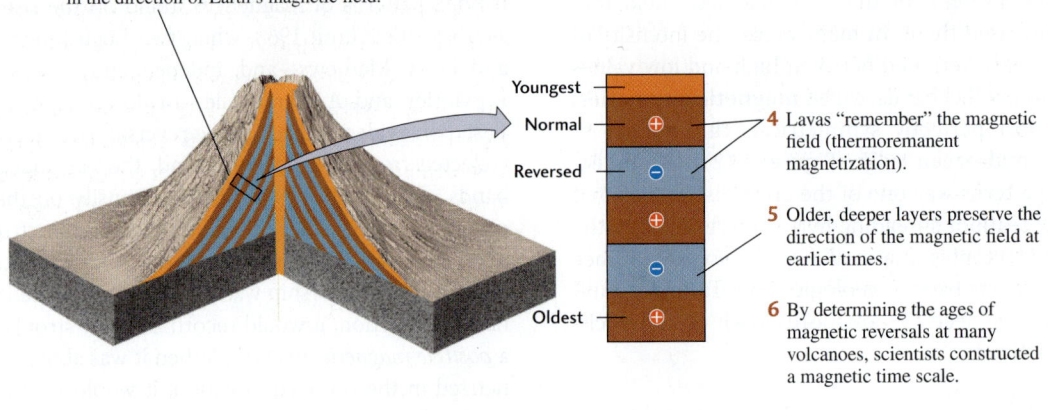

FIGURE 2.12 Magnetic anomalies allow geologists to measure the rate of seafloor spreading. (a) An oceanographic survey over the Mid-Atlantic Ridge just southwest of Iceland revealed a banded pattern of magnetic field intensity. (b) Geologists found and dated similar magnetic anomalies in volcanic lavas on land to construct a magnetic time scale. (c) That magnetic time scale was used to date the magnetic anomalies on the seafloor worldwide.

age. For instance, the magnetic anomaly pattern in Figure 2.12c shows that the boundary between the Gauss normal chron and the Gilbert reversed chron, which was dated from lava flows at 3.3 million years of age, is located about 30 km away from the crest of the Mid-Atlantic Ridge just southwest of Iceland. Thus, seafloor spreading has moved the North American and Eurasian plates apart by about 60 km in 3.3 million years, giving a spreading rate of 18 km per million years or, equivalently, 18 mm/year. On a divergent plate boundary, the combination of the spreading rate and the spreading direction gives the **relative plate velocity,** the velocity at which one plate moves relative to the other.

The speed record for spreading can be found on the East Pacific Rise just south of the equator, where the Pacific and Nazca plates are separating at a rate of about 150 mm/year—much faster than the rate in the North Atlantic. A rough average spreading rate for mid-ocean ridges around the world is 50 mm/year. This is approximately the rate at which your fingernails grow—so, geologically speaking, the plates move very fast indeed!

Mapping magnetic anomalies on the seafloor has been an amazingly effective and very expedient method for working out the history of ocean basins. Simply by steaming a ship back and forth over the ocean surface, measuring the magnetic fields of the seafloor rocks, and correlating the pattern of magnetic anomalies with the magnetic time scale, geologists have been able to determine the ages of various regions of the seafloor by remote sensing—that is, without directly sampling the oceanic crust far below the surface. In effect, they have learned how to "replay the tape."

Deep-Sea Drilling

In 1968, a seafloor drilling program was launched as a joint project of several major U.S. oceanographic institutions and the National Science Foundation. Later, many nations joined the effort (**Figure 2.13**). Using hollow drills, scientists brought up cores containing sections of seafloor rock from many locations in the oceans.

Small particles falling through the ocean waters—dust from the atmosphere, organic material from marine plants and animals—begin to accumulate as seafloor sediments on new oceanic crust as soon as it forms. Therefore, the age of the oldest sediments in a core—those immediately on top of the crust—tells the geologist how old the crust is at that spot. The ages of sediments can be calculated from the fossil skeletons of tiny single-celled planktonic organisms that live at the surface of the open ocean and sink to the bottom when they die. Geologists found that the ages of the samples in the drill cores increased with distance from mid-ocean ridges, and that the age of the samples at any one place agreed almost perfectly with the age of the seafloor determined from magnetic reversal data. This agreement validated the magnetic time scale and provided strong evidence for seafloor spreading.

FIGURE 2.13 The deep-sea drilling vessel *JOIDES Resolution* is 143 m long and carries a drilling derrick 61 m high that is capable of drilling into the seafloor beneath the deepest ocean. Rock samples recovered from the seafloor have confirmed the ages of seafloor rocks deduced from magnetic anomalies. Such samples have also shed new light on the history of ocean basins and ancient climate conditions. [Photo by Arito Sakaguchi; courtesy International Ocean Discovery Program (IODP).]

Measurements of Plate Movements by Geodesy

Astronomical Positioning Astronomical positioning—measuring the positions of points on Earth's surface in relation to the fixed stars in the night sky—is a technique of **geodesy,** the ancient science of measuring the shape of Earth and locating points on its surface. Surveyors have used astronomical positioning for centuries to determine geographic boundaries on land and sailors have used it to locate their ships at sea. Four thousand years ago, Egyptian builders used astronomical positioning to aim the Great Pyramid due north.

Because of the high accuracy that would have been required to observe plate movements directly, geodetic techniques did not play a significant role in the discovery of plate tectonics. Geologists had to rely on evidence of seafloor spreading from the geologic record—the magnetic anomalies and ages of fossils described earlier. Beginning in the late 1970s, however, an astronomical positioning method was developed that used signals from distant "quasi-stellar radio sources" (quasars) recorded by huge radio telescopes. This method can measure intercontinental distances to

Today, the Great Pyramid of Egypt is not aimed directly north, but slightly east of north. Did the ancient Egyptian astronomers make a mistake in orienting the pyramid 40 centuries ago? Archaeologists think not. Over this period, Africa drifted enough to rotate the pyramid out of alignment with true north.

The Global Positioning System Doing geodesy with big radio telescopes is expensive and is not a practical means of investigating plate movements in remote areas of the world where no radio telescopes exist. Since the mid-1980s, geologists have used a constellation of 24 Earth-orbiting satellites, called the Global Positioning System (GPS), to make the same types of measurements with the same astounding accuracy. The satellite constellation serves as an outside frame of reference, just as the fixed stars and quasars do in astronomical positioning. The satellites emit high-frequency radio waves keyed to precise atomic clocks on board. Those signals can be picked up by inexpensive, portable radio receivers not much bigger than this textbook (**Figure 2.14**). These devices are similar to the GPS receivers used in automobiles and smartphones, though much more precise.

Geologists use GPS to measure plate movements on a regular basis at many locations around the globe. Changes in distance between land-based GPS receivers placed on different plates, recorded over several years, agree in both magnitude and direction with those calculated from magnetic anomalies on the seafloor, indicating that plate movements are remarkably steady over periods ranging from just a few years to millions of years, as demonstrated in the Practicing Geology Exercise.

The Grand Reconstruction

The supercontinent Pangaea was the only major landmass that existed 250 million years ago. One of the great triumphs of modern geology was the reconstruction of the events that led to the assembly of Pangaea and to its later fragmentation into the continents we know today. Let's use what we have learned about plate tectonics to see how this feat was accomplished.

Seafloor Isochrons

The color map in **Figure 2.15** shows the ages of the rocks on the seafloor as determined from magnetic anomaly data and deep-sea drilling. Each colored band represents the span of time when the rocks within that band formed. Notice that the seafloor becomes progressively older on both sides of the mid-ocean ridges. The boundaries between bands are contours of equal seafloor age, or **isochrons.**

Isochrons tell us the time that has elapsed since the rocks were injected as magma into a spreading zone and, therefore, the amount of spreading that has occurred since they formed. For example, the distance from a ridge axis

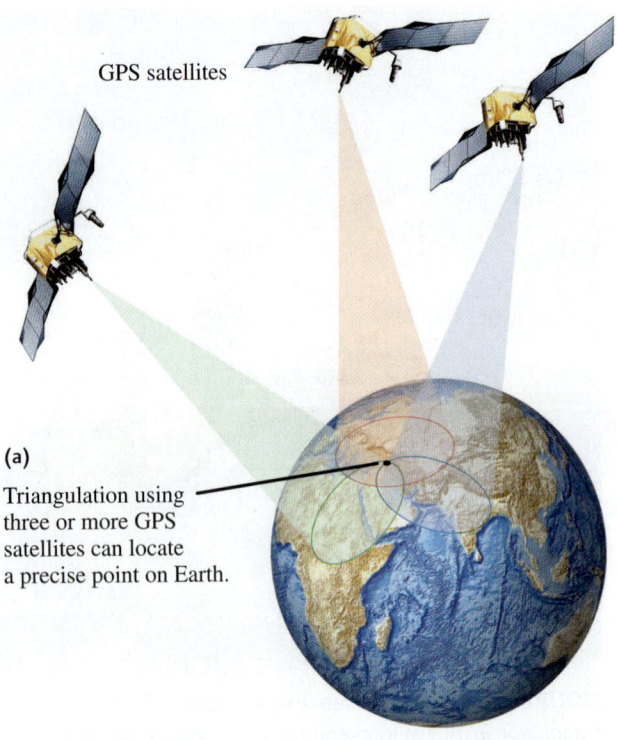

(a) Triangulation using three or more GPS satellites can locate a precise point on Earth.

(b) A GPS station

FIGURE 2.14 The Global Positioning System allows geologists to monitor plate movements. (a) GPS satellites provide a fixed frame of reference outside Earth. (b) Small GPS receivers, such as this one in Los Angeles, can be easily placed anywhere on Earth. Displacements of receiver locations over a period of years can be used to measure plate movements. [Photo courtesy of Southern California Earthquake Center.]

an amazing accuracy of 1 mm. In 1986, a team of scientists using this method showed that the distance between radio telescopes in Europe (Sweden) and North America (Massachusetts) had increased 19 mm/year over a period of 5 years, very close to the rate predicted by geologic models of plate tectonics.

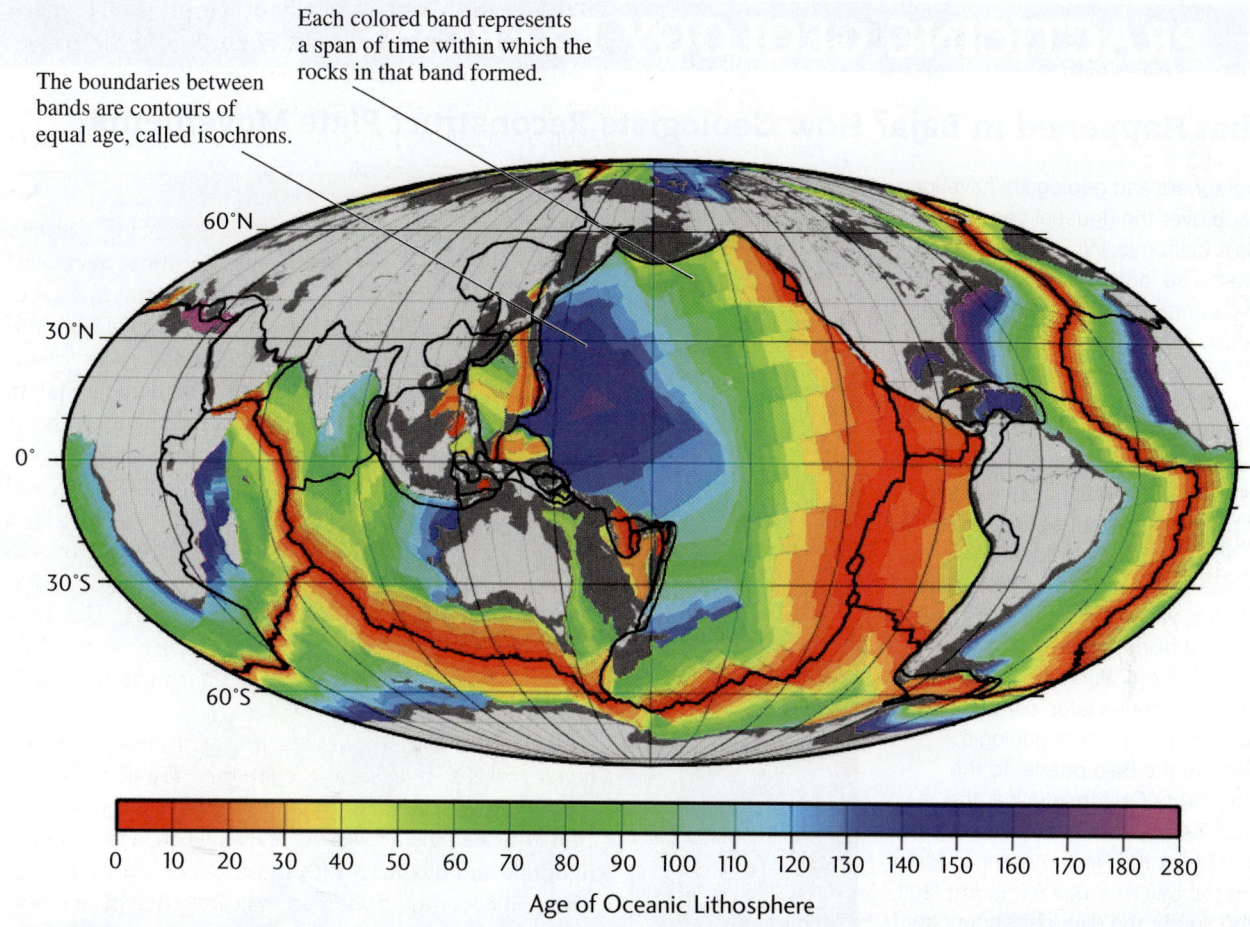

FIGURE 2.15 This global isochron map shows the ages of rocks on the seafloor. The time scale at the bottom gives the age of the seafloor in millions of years since its creation at mid-ocean ridges. Light gray indicates land; dark gray indicates shallow water over continental shelves. Mid-ocean ridges, along which new seafloor is extruded, coincide with the youngest rocks (red). [Dietmar Müller, University of Sydney.]

to a 100-million-year isochron (boundary between green and blue bands) indicates the extent of new seafloor created over that time span. The more widely spaced isochrons of the eastern Pacific signify faster spreading rates there than in the Atlantic.

In 1990, after a 20-year search, geologists found the oldest oceanic rocks by drilling into the seafloor of the western Pacific. These rocks turned out to be about 200 million years old, only 4 percent of Earth's age. In contrast, the oldest known continental rocks are over 4 billion years in age. All existing seafloor is geologically young compared with the continents.

Reconstructing the History of Plate Movements

Earth's plates behave as rigid bodies. That is, the distances between three points on the same rigid plate—say, New York, Miami, and Bermuda on the North American Plate—do not change very much, no matter how far the plate moves. But the distance between, say, New York and Lisbon increases over time because those two cities are on two different plates that are separating along the Mid-Atlantic Ridge. The direction of movement of one plate in relation to another depends on two geometric principles that govern the behavior of rigid plates on a sphere:

- *Transform-fault boundaries indicate the directions of relative plate movement.* With few exceptions, no overlap, buckling, or separation occurs along typical transform-fault boundaries in the oceans. The two plates merely slide past each other without creating or destroying plate material. Therefore, the orientation of the fault measures the direction in which one plate is sliding with respect to the other (see Figure 2.8f, g).

- *Seafloor isochrons reveal the positions of divergent boundaries in earlier times.* Isochrons on the seafloor are roughly parallel and symmetrical to the ridge axis along which they were created (see Figure 2.15). Because each isochron was at the divergent boundary

PRACTICING GEOLOGY EXERCISE

What Happened in Baja? How Geologists Reconstruct Plate Movements

Geographers and geologists have long puzzled over the unusual geography of Baja California. Why is the Gulf of California so long and thin? Why is the Baja California peninsula parallel to the Mexican coastline?

When the Spanish conquistador Hernando Cortés landed on the shores of California in 1535, he thought he had discovered an island. Decades passed before the Spanish realized that the northern half of *Isla California* was actually the west coast of North America and that its lower half, Baja California, was a long, thin peninsula, separated from the continent by the narrow Gulf of California.

Four centuries later, plate tectonic theory provided a neat geologic answer to the Baja puzzle. To the north, in Alta California (a.k.a. the Golden State), the Pacific Plate is moving past the North American Plate along the San Andreas transform fault. To the south, the divergent boundary between the Pacific Plate and the small Rivera Plate forms part of the East Pacific Rise, a mid-ocean ridge that produces new oceanic crust as the two plates spread apart.

By mapping earthquake locations and undersea volcanoes, marine geologists were able to show that the San Andreas Fault is connected to the East Pacific Rise by a dozen small spreading centers offset by transform faults—a plate boundary that steps like stairs up the entire length of the Gulf of California. The relative movement of the Pacific and North American plates is thus shifting Baja California away from the mainland

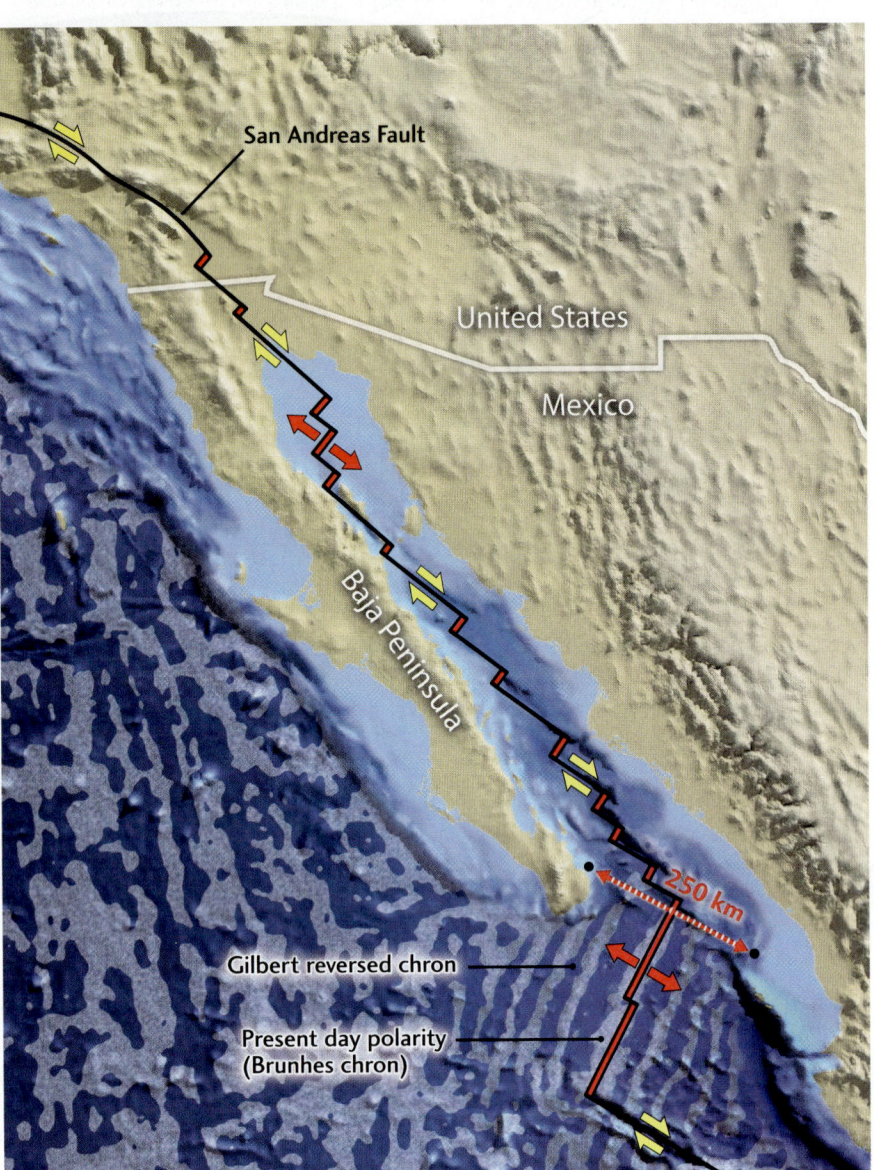

The Pacific Plate, on the left, is moving northwestward relative to the North American Plate, on the right, at a speed of about 50 mm/year, rifting the Baja California peninsula away from the Mexican mainland and opening the Gulf of California.

at an earlier time, isochrons that are of the same age but on opposite sides of a mid-ocean ridge can be brought together to show the positions of the plates and the configuration of the continents embedded in them, as they were in that earlier time.

Using these principles, geologists have reconstructed the history of continental drift. They have shown, for example, how the skinny peninsula of Baja California was rifted away from the Mexican mainland during the last 5 million years (see the Practicing Geology Exercise).

in a northwesterly direction, parallel to the transform faults, and the Gulf of California is being progressively widened by seafloor spreading.

How fast is this happening? An estimate can be made by using the equation

$$speed = distance \div time$$

We need two types of data to apply this equation:

- We can measure the *distance* by which Baja California has separated from Mexico directly from a seafloor map: about 250 km.

- We can estimate the *time* since the separation began from the pattern of magnetic anomalies across the East Pacific Rise. On both sides of that spreading center, the magnetic anomaly closest to the continental margin (and therefore the oldest) is the Gilbert reversed chron. Using the magnetic time scale in Figure 2.12c, we obtain a separation age of about 5 million years (My).

With this information, we can calculate the approximate speed of seafloor spreading in the Gulf of California:

$$speed = distance \div time$$
$$= 250 \text{ km}/5 \text{ My}$$
$$= 50 \text{ km/My}$$

or 50 mm/year.

Of course, this is only an average speed. How steady has it been? The plate separation could have started slowly and gradually picked up speed, or started fast and then slowed down. If the former is true, then the present-day separation rate should be greater than the average rate; if the latter, it should be less.

Using GPS, geologists were able to test these hypotheses using a totally different type of measurement. In the decade from 1990 to 2000, they repeatedly surveyed the locations of points on both sides of the Gulf of California oriented parallel to the plate movements. They found that the distances between these points increased by half a meter; that is, by 500 millimeters in 10 years, or 50 mm/year. Thus, the present-day speed of movement is approximately the same as the average speed; no speedup or slowdown of plate movements is necessary to account for it.

Based on the agreement between these two measurements as well as other data, geologists came up with a simple story. Before 5 million years ago, when Baja California was part of the mainland, the boundary between the Pacific and North American plates lay somewhere west of the North American continent. About 5 million years ago, this boundary jumped inland, initiating seafloor spreading in the Gulf of California. The plate movement has been nearly steady at 50 mm/year ever since.

This theory has survived various tests. For example, it predicts that the current slipping along the San Andreas Fault should also have begun about 5 million years ago, and that prediction agrees with the ages of rocks that have been displaced by the modern San Andreas Fault.

The Breakup of Pangaea

On a much grander scale, geologists have reconstructed the opening of the Atlantic Ocean and the breakup of Pangaea (**Figure 2.16**). Figure 2.16e shows the supercontinent Pangaea as it existed about 240 million years ago. It began to break apart when North America rifted away from Europe about 200 million years ago (Figure 2.16f). The opening of the North Atlantic was accompanied by the separation of the northern continents (referred to as Laurasia) from the southern continents (Gondwana) and the rifting of Gondwana along what is now the east coast of Africa (Figure 2.16g). The breakup of Gondwana separated South America, Africa, India, and Antarctica, creating the South Atlantic and Southern oceans and narrowing the Tethys Ocean (Figure 2.16h). The separation of Australia from Antarctica and the ramming of India into Eurasia closed the Tethys Ocean, giving us the world we see today (Figure 2.16i).

The plate movements have not ceased, of course, so the configuration of the continents will continue to evolve. A plausible scenario for the distribution of continents and plate boundaries 50 million years in the future is shown in Figure 2.16j.

The Assembly of Pangaea by Continental Drift

The isochron map in Figure 2.15 tells us that all of the seafloor on Earth's surface today has been created since the breakup of Pangaea. We know from the geologic record in older continental mountain belts, however, that plate tectonics had been operating for billions of years before this breakup. Evidently, seafloor spreading took place just as it does today, and there were previous episodes of continental drift and collision. Subduction into the mantle has destroyed the seafloor created in those earlier times, however, so we must rely on the older evidence preserved on continents to identify and chart the movements of ancient continents (*paleocontinents*).

Old mountain belts, such as the Appalachians of North America and the Urals, which separate Europe from Asia, help us locate ancient collisions of the paleocontinents. In

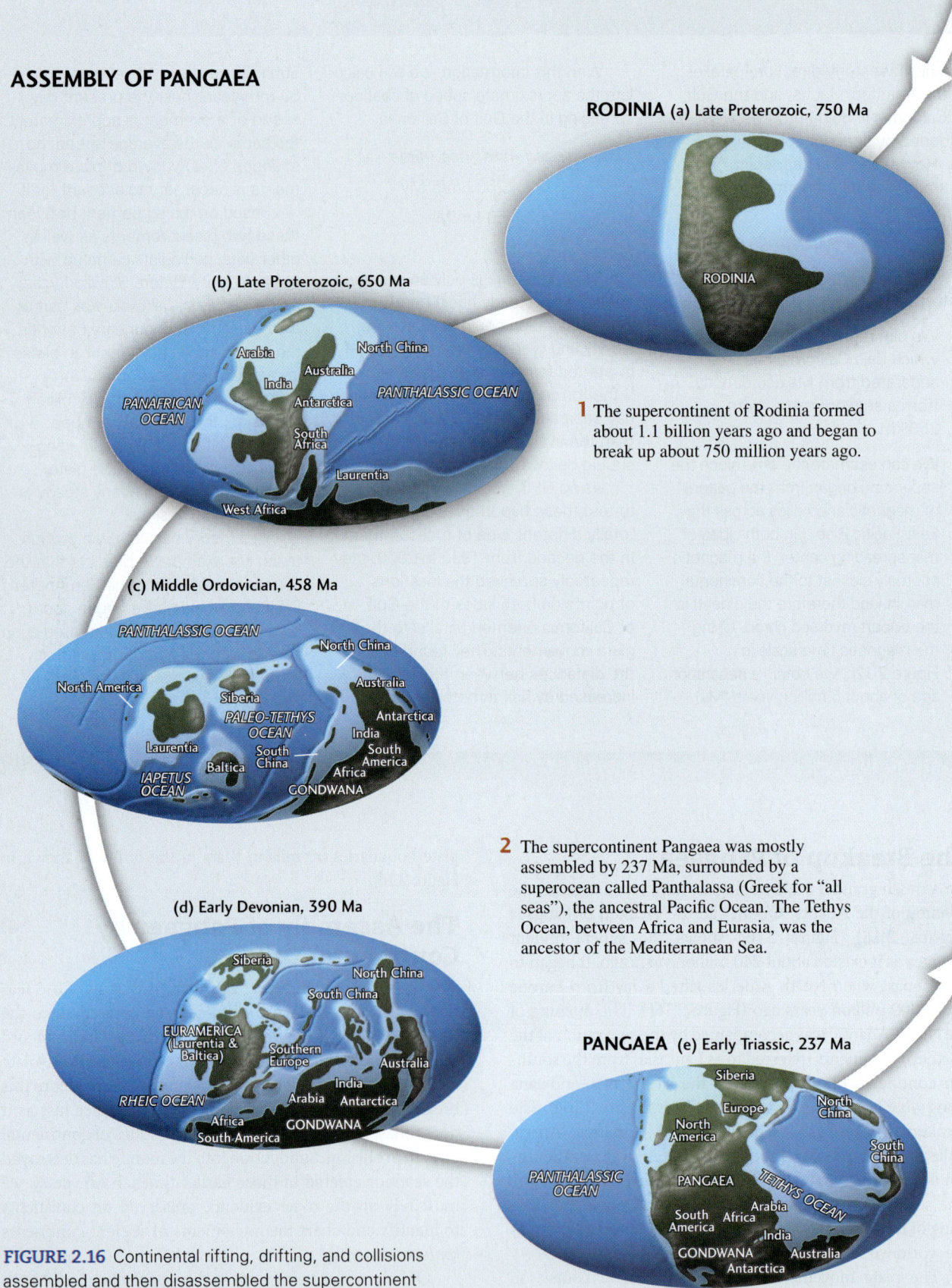

FIGURE 2.16 Continental rifting, drifting, and collisions assembled and then disassembled the supercontinent Pangaea. [Paleogeographic maps by Christopher R. Scotese, 2003 PALEOMAP Project (www.scotese.com).]

BREAKUP OF PANGAEA

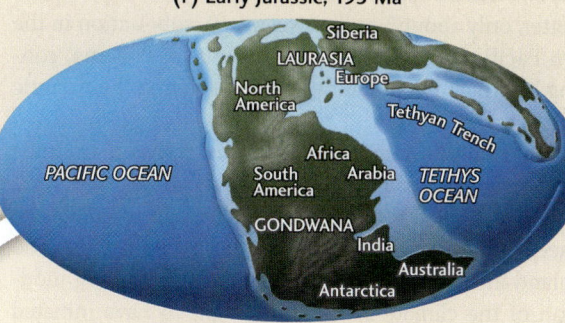
(f) Early Jurassic, 195 Ma

3 The breakup of Pangaea was signaled by the opening of rifts from which lava poured. Rock assemblages that are relics of this great event can be found today in 200-million-year-old volcanic rocks from Nova Scotia to North Carolina.

(g) Late Jurassic, 152 Ma

4 By about 150 million years ago, Pangaea was in the early stages of breakup. The Atlantic Ocean had partially opened, the Tethys Ocean had contracted, and the northern continents (Laurasia) had all but split away from the southern continents (Gondwana). India, Antarctica, and Australia began to split away from Africa.

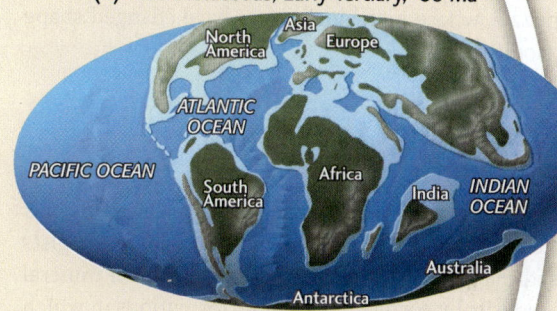
(h) Late Cretaceous, Early Tertiary, 66 Ma

5 By 66 million years ago, the South Atlantic had opened and widened. India was well on its way northward toward Asia, and the Tethys Ocean was closing to form the Mediterranean.

THE PRESENT-DAY AND FUTURE WORLD

6 The modern world has been produced over the past 65 million years. India collided with Asia, ending its trip across the ocean, and is still pushing northward into Asia. Australia has separated from Antarctica.

(i) PRESENT-DAY WORLD

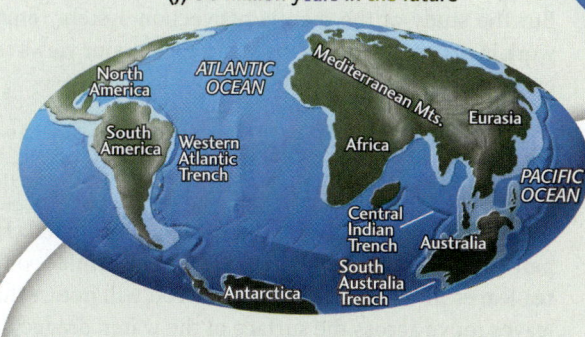
(j) 50 million years in the future

many places, the rocks reveal ancient episodes of rifting and subduction. Rock types and fossils also indicate the distribution of ancient seas, glaciers, lowlands, mountains, and climates. Knowledge of ancient climates enables geologists to locate the latitudes at which continental rocks formed, which in turn helps them to assemble the jigsaw puzzle of paleocontinents. When volcanism or mountain building produces new continental rocks, these rocks also record the direction of Earth's magnetic field, just as oceanic crust does when it is created by seafloor spreading. Like a compass frozen in time, the thermoremanent magnetization of a continental fragment records its ancient orientation and magnetic latitude.

The left side of Figure 2.16 shows one of the latest efforts to depict the pre-Pangaean configuration of continents. It is truly impressive that modern science can recover the geography of this strange world of hundreds of millions of years ago. The evidence from rock types, fossils, and magnetization has allowed scientists to reconstruct an earlier supercontinent, called **Rodinia,** that formed about 1.1 billion years ago and began to break up about 750 million years ago (Figure 2.16a). They have been able to chart its fragments over the subsequent 500 million years as those fragments drifted and reassembled into the supercontinent Pangaea. Geologists continue to sort out the details of this complex jigsaw puzzle, whose individual pieces have changed shape over geologic time.

Implications of the Grand Reconstruction

Hardly any branch of geology remains untouched by this grand reconstruction of the continents. Economic geologists have used the former fit of the continents to find mineral and oil deposits by correlating the rock formations in which these resources exist on one continent with their predrift continuations on another continent. Paleontologists have rethought some aspects of evolution in light of continental drift. Geologists have broadened their focus from the geology of a particular region to a world-encompassing picture. The concept of plate tectonics provides a way to interpret, in global terms, such geologic processes as rock formation, mountain building, and climate change.

Past Climate Changes Over millions of years, movements of the tectonic plates have rearranged the continents and oceans, affecting the climate system in profound ways. In the present arrangement, the waters of the Southern Ocean are able to circulate all the way around Antarctica, forming a circumpolar seaway or "Southern Ocean" that isolates the continent from the warmer water and air of tropical latitudes. This isolation keeps the southern polar regions colder than they might otherwise be, maintaining a massive ice sheet across the entire Antarctic continent.

The situation was rather different during the breakup of Pangaea (Figure 2.16). As shown in Figure 2.16h, 66 million years ago, Australia was still connected to Antarctica, allowing currents of warmer water to flow southward and heat the polar continent. Also at this time, the North and South American continents were separated, so that water could flow between the Atlantic and Pacific oceans. The circumpolar seaway did not form until Australia broke away from Antarctica around 40 million years ago. Somewhat later, only about 5 million years ago, subduction in the eastern Pacific Ocean formed the isthmus of Panama, connecting North and South America and isolating the Atlantic from the Pacific.

These changes, combined with the collision of India with Asia, which formed the high plateau of Tibet (see Figure 2.16i), cooled the entire planet enough to create the ice sheets of Antarctica in the southern hemisphere and Greenland in the northern hemisphere. The resulting modification of the climate system is thought to have initiated oscillations of climate between very cold periods (ice ages, described in Chapter 15) and somewhat warmer periods, such as the one we now enjoy.

Mantle Convection: The Engine of Plate Tectonics

Everything discussed in this chapter so far has been a description of *how* plates move. The theory of mantle convection provides an explanation of *why* plates move.

As Arthur Holmes and other early advocates of continental drift realized, mantle convection is the "engine" that drives the large-scale tectonic processes operating on Earth's surface. In Chapter 1, we described hot mantle as a ductile solid capable of flowing like a sticky fluid. Heat escaping from Earth's deep interior causes this material to undergo convection (circulation upward and downward) at speeds of a few tens of millimeters per year.

Almost all scientists now accept that the lithospheric plates somehow participate in the flow of this mantle convection system. As is often the case, however, "the devil is in the details." Many different hypotheses have been advanced on the basis of one piece of evidence or another, but no one has yet come up with a satisfactory, comprehensive theory that ties everything together. In what follows, we will pose three questions that get at the heart of the matter and give you our current understanding of their answers. But the study of the mantle convection system remains a work in progress, and we may have to alter our views as new evidence becomes available.

Where Do the Plate-Driving Forces Originate?

Here's an experiment you can do in your kitchen: Heat a pan of water until it is about to boil, then sprinkle some dry tea leaves in the center of the pan. You will notice that the leaves move across the surface of the water, dragged along by the convection currents in the hot water. Is this the way

Mantle Convection: The Engine of Plate Tectonics

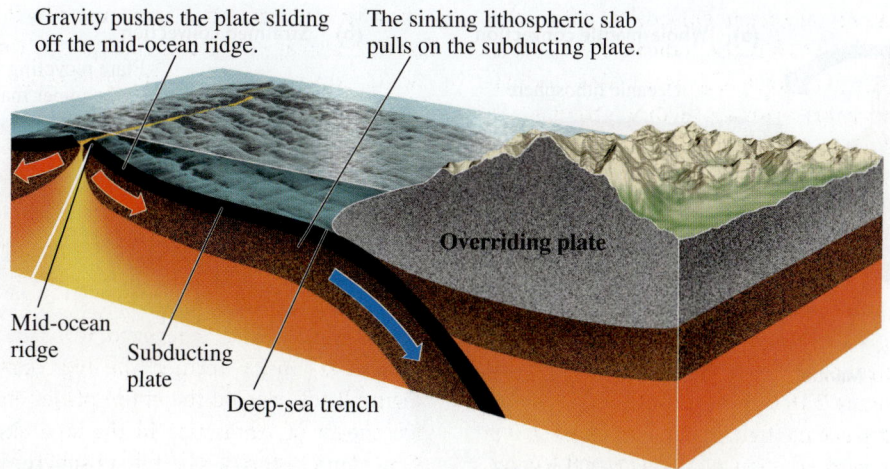

FIGURE 2.17 A schematic cross section through the outer part of Earth, illustrating two of the forces thought to be important in driving plate tectonics: the pulling force of a sinking lithospheric slab and the pushing force of plates sliding off a mid-ocean ridge.

plates move about, passively dragged to and fro on the backs of convection currents rising up from the mantle?

The answer appears to be no. The main evidence comes from the rates of plate movement we discussed earlier in this chapter. In Figure 2.7, we can see that the faster-moving plates (the Pacific, Nazca, Cocos, Indian, and Australian plates) are being subducted along a large fraction of their boundaries. In contrast, the slower-moving plates (the North American, South American, African, Eurasian, and Antarctic plates) do not have significant attachments of descending lithospheric slabs. These observations suggest that the gravitational pull exerted by the cold (and thus dense) slabs of subducting lithosphere pulls the plates downward into the mantle. In other words, the plates are not dragged along by convection currents rising from the mantle, but rather "fall back" into the mantle under their own weight. According to this hypothesis, seafloor spreading is the passive upwelling of mantle material where the plates have been pulled apart by subduction forces.

But if the only important force in plate tectonics is the gravitational pull of subducting slabs, why did Pangaea break apart and the Atlantic Ocean open up? The only subducting slabs of lithosphere currently attached to the North and South American plates are found in the small island arcs that bound the Caribbean and Scotia seas, which are thought to be too small to drag the Atlantic apart. One possibility is that the overriding plates, as well as the subducting plates, are pulled toward their convergent boundaries. For example, as the Nazca Plate subducts beneath South America, it may cause the plate boundary at the Peru-Chile Trench to retreat toward the Pacific, "sucking" the South American Plate to the west.

Other forces are evident from the history of plate movements. When the continents came together to form Pangaea, they acted as an insulating blanket, preventing heat from getting out of Earth's mantle (as it otherwise would through the process of seafloor spreading). That heat built up over time, forming hot bulges in the mantle beneath the supercontinent. These bulges raised Pangaea slightly and caused it to rift apart in a kind of "landslide" off the tops of the bulges. Gravitational forces continue to drive subsequent seafloor spreading as the plates "slide downhill" off the crest of the Mid-Atlantic Ridge. Earthquakes that sometimes occur in plate interiors provide direct evidence of the compression of plates by these ridge-related gravitational forces.

Convection in the mantle—the rising of hot matter in one place and the sinking of cold matter in another—is the driving force of plate tectonics. Although many questions remain, we can be reasonably sure of two things: first, that the plates themselves play an active role in this system, and second, that the forces associated with the sinking slabs and elevated ridges are probably the most important in governing the rates of plate movement (Figure 2.17). Scientists are attempting to resolve the other issues raised in this discussion by comparing their observations with detailed computer models of the mantle convection system. Some of their results will be discussed in Chapter 11.

How Deep Does Plate Recycling Occur?

For plate tectonics to work, the lithospheric material that descends into the mantle at subduction zones must be recycled through the mantle and eventually return to the

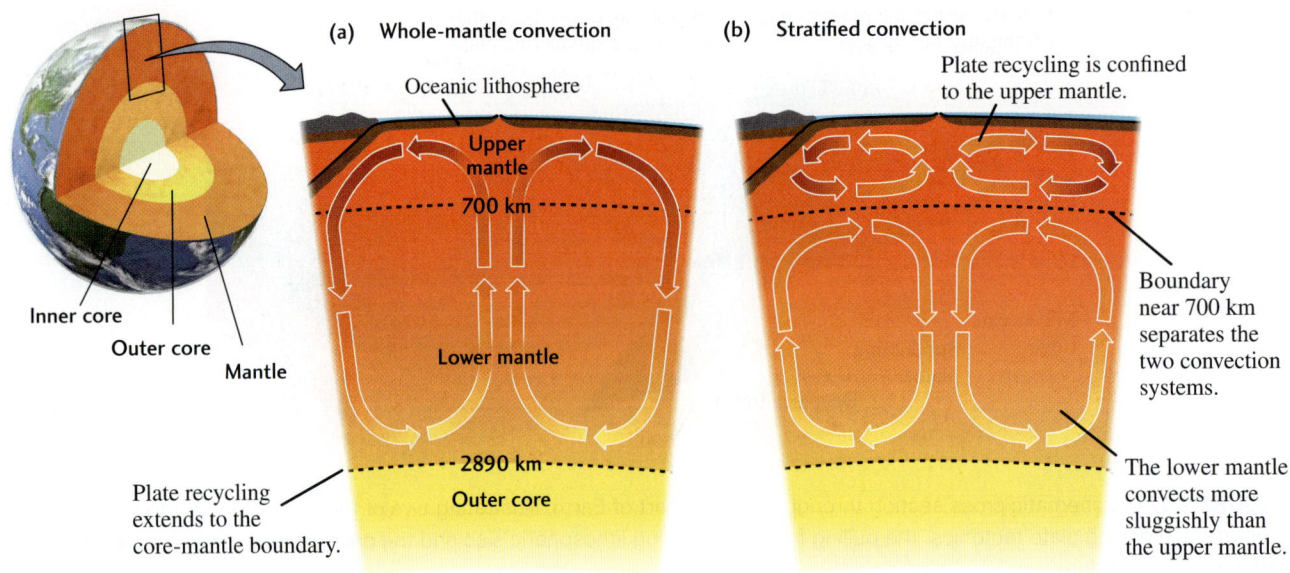

FIGURE 2.18 Two competing hypotheses on the depth extent of the mantle convection system that recycles the lithosphere.

crust as new lithosphere created at spreading centers. How deep into the mantle does this recycling process extend? That is, where is the lower boundary of the mantle convection system?

The lower boundary could be as deep as 2890 km below Earth's surface, where a sharp compositional boundary separates the mantle from the core (**Figure 2.18**). As we saw in Chapter 1, the iron-rich liquid below this core-mantle boundary is much denser than the solid rock of the mantle, preventing any significant exchange of material between the two layers. We can thus imagine a system of *whole-mantle convection* in which the material from the plates circulates all the way through the mantle, down as far as the core-mantle boundary (see Figure 2.18a).

However, some scientists think that the mantle might be divided into two layers: an upper mantle system above about 700 km, where the recycling of lithosphere takes place, and a lower mantle system from a depth of about 700 km to the core-mantle boundary, where convection is much more sluggish. According to this hypothesis, called *stratified convection*, the separation of the two systems is maintained because the upper system consists of lighter rock than the lower system and thus floats on top, in the same way the mantle floats on the core (see Figure 2.18b).

To test these two competing hypotheses, scientists have looked for "lithospheric graveyards" below convergent boundaries where plates have been subducted. Old subducted lithosphere is colder than the surrounding mantle and can therefore be "seen" using seismic waves. Moreover, there should be lots of it down there. From our knowledge of past plate movements, we can estimate that, just since the breakup of Pangaea, lithosphere equivalent to the surface area of Earth has been recycled into the mantle. Sure enough, scientists have found regions of colder material in the deep mantle under North and South America, eastern Asia, and other sites adjacent to convergent boundaries. These zones occur as extensions of descending lithospheric slabs, and some appear to go down as far as the core-mantle boundary. From this evidence, most scientists have concluded that plate recycling takes place through whole-mantle convection rather than stratified convection.

What Is the Nature of Rising Convection Currents?

What about the rising currents of hot mantle material needed to balance subduction? Are there concentrated, sheetlike upwellings directly beneath the mid-ocean ridges? Most scientists who study the problem think no; they instead believe that the rising currents are slower and spread out over broader regions. This view is consistent with the idea that seafloor spreading is a rather passive process: Pull the plates apart almost anywhere and you will generate a spreading center.

There is one exception, however: a type of narrow jetlike upwelling called a **mantle plume** (**Figure 2.19**). The best evidence for mantle plumes comes from regions of intense, localized volcanism (called *hot spots*), such as Hawaii, where huge volcanoes form in the middle of a plate, far from any spreading center. Mantle plumes are thought to be slender cylinders of fast-rising material, less than 100 km across,

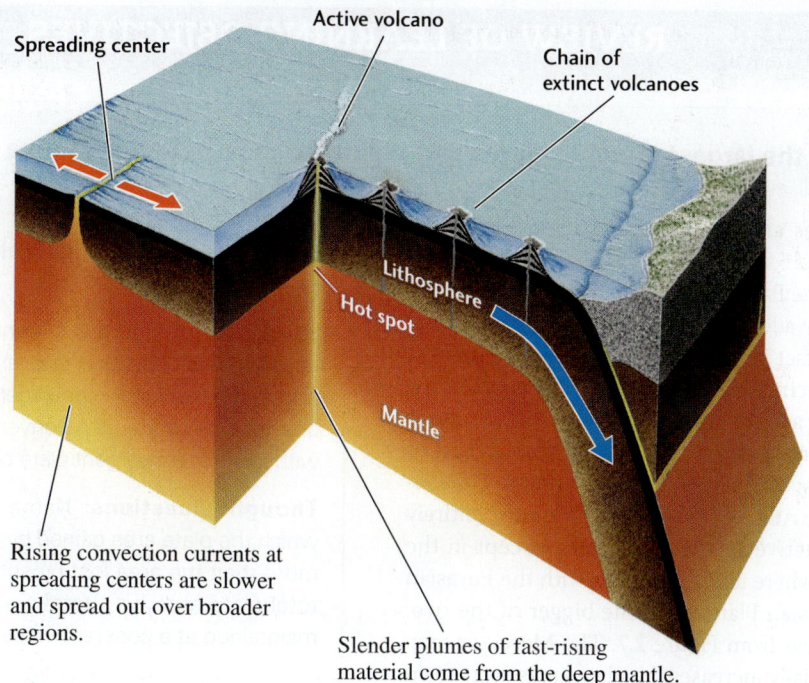

FIGURE 2.19 An illustration of the mantle plume hypothesis.

that come from the deep mantle (below the asthenosphere). They are intense enough to literally burn holes in the plates and produce tremendous volumes of lava. Mantle plumes may be responsible for outpourings of lava so massive that they may have changed Earth's climate and caused mass extinctions (see Chapter 1). We will describe mantle plumes in more detail in Chapter 11.

The mantle plume hypothesis was first put forward in 1970, soon after the theory of plate tectonics had been established, by one of its founders, W. Jason Morgan of Princeton University. Like other aspects of the mantle convection system, however, the observations that bear on rising convection currents are indirect, and the mantle plume hypothesis remains controversial.

KEY TERMS AND CONCEPTS

continental drift (p. 24)
convergent boundary (p. 27)
divergent boundary (p. 27)
geodesy (p. 37)
island arc (p. 32)
isochron (p. 38)

magnetic anomaly (p. 35)
magnetic time scale (p. 35)
mantle plume (p. 46)
mid-ocean ridge (p. 32)
Pangaea (p. 24)
plate tectonics (p. 27)

relative plate velocity (p. 37)
Rodinia (p. 44)
seafloor spreading (p. 25)
spreading center (p. 32)
subduction (p. 32)
transform fault (p. 27)

REVIEW OF LEARNING OBJECTIVES

2.1 Identify the largest tectonic plates and delineate their boundaries on a global map.

The major plates and their boundaries are mapped Figures 2.7 and 2.15. The largest, in yellow at the center of Figure 2.7, is the Pacific Plate, which is bounded on its east and south by a long chain of spreading centers that are irregularly offset by transform faults, and on its west by island arcs facing deep-sea trenches. The rainbow swath of seafloor ages in Figure 2.15 shows how seafloor spreading has continuously created the oceanic crust of the Pacific Plate over the past 200 million years. The second largest is the African Plate, which is almost entirely surrounded by active spreading centers, except in the Mediterranean, where it is converging with the Eurasian Plate. (The Eurasian Plate is *not* the bigger of the two, as you might guess from Figure 2.7. The Mercator map projection artificially increases the plate areas away from the equator. The Mollweide projection of Figure 2.15 preserves plate area, though it distorts the plate shapes. For an undistorted view of plate areas, look on a globe!)

Study Assignment: Complete the Visual Literacy Exercise.

Exercises: In Figure 2.7, identify an example of a transform-fault boundary that (a) connects a divergent plate boundary with a convergent plate boundary and one that (b) connects a convergent plate boundary with another convergent plate boundary.

Thought Questions: Name four large plates for which the plate area gained by seafloor spreading is more than the area lost by subduction. How is the total plate area (i.e., total area of Earth's surface) maintained at a constant value?

2.2 Summarize how geologists of the 1960s used seafloor spreading to explain Wegener's continental drift.

Wegener thought the continents drifted like ships, with continental crust pulled through oceanic crust by the tidal forces of the Moon and Sun—a theory quickly discounted as physically impossible. Marine geologists mapping magnetic anomalies at sea discovered that the seafloor was being created by divergence along the mid-ocean ridges, rafting the continents apart in a process driven by mantle convection. According to the theory of plate tectonics, developed in the 1960s, the lithosphere comprising the crust and uppermost mantle is broken into about a dozen large, rigid plates, and a number of smaller ones, that move over Earth's surface. Both continental crust and oceanic crust can be embedded within the same lithospheric plate, and the two types of crust can slide together across the asthenosphere. Beginning about 200 million years ago, the seafloor spreading that created the Atlantic Ocean basin caused North and South America to drift away from Eurasia and Africa as part of the breakup of the Pangaea supercontinent.

Study Assignment: Review the section *The Discovery of Plate Tectonics*.

Exercises: (a) Use a globe or Google Earth and the isochron map in Figure 2.15 to estimate the average speed of continental drift between North America and Africa. (b) How well does this speed compare with the present-day value of 23 mm/year determined using GPS?

Thought Questions: (a) What mistakes did Wegener make in formulating his theory of continental drift? (b) Do you think the geologists of his era were justified in rejecting his theory? (c) Would you classify plate tectonics as a hypothesis, a theory, or a fact? Why?

2.3 Based on observations of relative motions and geological activity, determine whether the edge of a tectonic plate is acting as part of a divergent, convergent, or transform-fault boundary.

Three types of plate boundaries are defined by the direction of the relative plate motions: divergent (moving apart), convergent (moving together), and transform-fault (moving sideways) boundaries. The geologic features that develop at plate boundaries are diagnostic of the boundary type. Divergent boundaries are typically marked by volcanism and shallow earthquakes at the crest of a mid-ocean ridge or within a continental rift zone. Convergent boundaries are marked by deep-sea trenches, deep earthquakes, mountain building, and volcanism. Transform faults, along which plates slide horizontally past each other, can be recognized by their lineated earthquake activity and horizontal offsets of geologic features.

Study Assignment: Review the plate-boundary types shown in Figure 2.8.

Exercises: (a) Using Figure 2.7, trace the boundaries of the South American Plate on a sheet of paper and identify segments that are divergent, convergent, and transform-fault boundaries. (b) Approximately what fraction of the plate area is occupied by the South American continent? (c) Is the fraction of the South American Plate occupied by oceanic crust increasing or decreasing over time? Explain your answer using the principles of plate tectonics.

Thought Questions: (a) Why are there active volcanoes along the Pacific coast in Washington and Oregon, but not along the east coast of the United States? (b) How do the differences between continental and oceanic crust affect the way lithospheric plates interact?

2.4 Explain how magnetic anomalies recorded by ships at sea are used to estimate the age of the seafloor and the rate of seafloor spreading.

Geologists have estimated seafloor ages by measuring the thermoremanent magnetization of basaltic lavas frozen at spreading centers. Magnetic anomaly patterns mapped on the seafloor have been compared with a magnetic time scale established using the magnetic anomalies observed in lavas of known ages. The average rate of seafloor spreading between two plates is computed by measuring the distance from a magnetic anomaly on one plate to the same anomaly on the other plate, and then dividing by the anomaly age given by the magnetic time scale. Seafloor ages have been verified through the dating of sediment and rock samples obtained by deep-sea drilling.

Study Assignment: Review the section *Seafloor as a Magnetic Tape Recorder*, paying special attention to Figure 2.12.

Exercise: On the Mid-Atlantic Ridge south of Iceland, the Gauss normal chron on the North American Plate is separated from this same chron on the Eurasian Plate by about 60 kilometers. From this fact and information about the magnetic time scale in Figure 2.12, estimate the rate of seafloor spreading at this location.

Thought Question: In Figure 2.15, the isochrons are symmetrically distributed in the Atlantic Ocean, but not in the Pacific Ocean. For example, seafloor as much as 180 million years old (in darkest blue) is found in the western Pacific, but not in the eastern Pacific. Why?

2.5 Reconstruct the supercontinent of Pangaea by running continental drift backwards over the past 200 million years.

Geologists have developed isochron maps for most of the world's oceans, which allow them to reconstruct the history of seafloor spreading over the past 200 million years (Figure 2.15). Using this method and other geologic data, geologists have developed a detailed model of how Pangaea broke apart and the continents drifted into their present configuration. "Rewinding the tape" yields the history of continental drift shown in the upper right of Figure 2.16.

Study Assignment: Complete the Practicing Geology Exercise.

Exercises: (a) Using the isochron map in Figure 2.15, estimate how long ago the continents of Australia and Antarctica were separated by seafloor spreading. Did this happen before or after South America separated from Africa? (b) Name three mountain belts formed by continental collisions that are occurring now or have occurred in the past.

Thought Question: The theory of plate tectonics was not widely accepted until the banded patterns of magnetism on the ocean floor were discovered. Why were these magnetic anomalies more convincing than earlier observations, such as the jigsaw-puzzle fit of the continents, the occurrence of the same fossils on both sides of the Atlantic, and the reconstruction of ancient climate conditions?

2.6 Describe how the oceanic plates participate in mantle convection.

The mantle convection that drives the plate tectonic system is by fueled by heat energy coming from Earth's deep interior. As part of this convection system, gravitational forces cause cooling oceanic lithosphere to slide downhill from spreading centers and sink into the mantle at subduction zones. Using seismic waves, we can observe that subducted lithosphere extends downward into the lower mantle and, in some places, as deep as the core-mantle boundary, which indicates that the whole mantle is involved in the convection system that recycles the plates. Rising convection currents may include mantle plumes, intense jets of material from the deep mantle that cause localized volcanism at hot spots in the middle of both oceanic and continental plates.

Study Assignment: Review the section *Mantle Convection: The Engine of Plate Tectonics*.

Exercises: Most active volcanoes are located on or near plate boundaries. (a) Give examples of volcanoes in ocean basins and on continents that are not on a plate boundary. (b) Describe a hypothesis consistent with plate tectonics that can explain this "intraplate" volcanic activity.

Thought Question: What observations favor whole-mantle convection over stratified convection?

VISUAL LITERACY EXERCISE

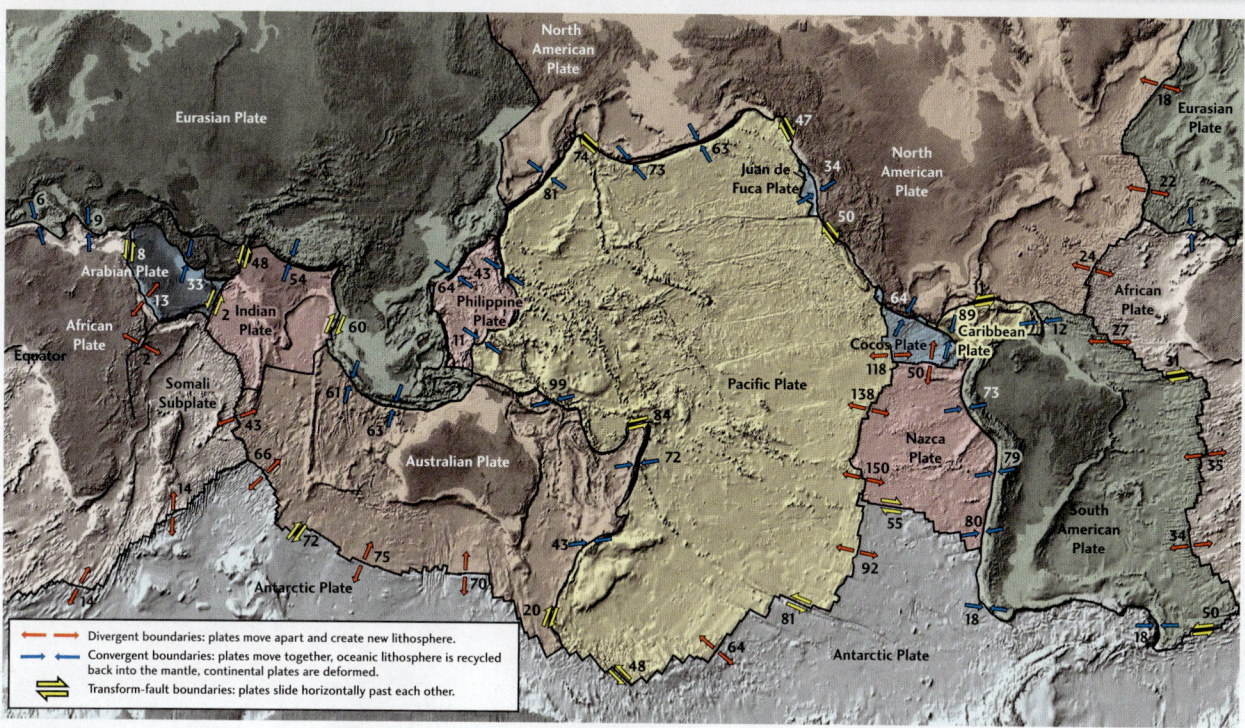

FIGURE 2.7 Earth's surface is a mosaic of 13 major plates, as well as a number of smaller plates, of rigid lithosphere that move slowly over the ductile asthenosphere. Only one of the smaller plates—the Juan de Fuca Plate, off the west coast of North America—is shown on this map. The arrows show the relative movement of two plates at a point on their boundary. The numbers next to the arrows give the relative plate velocities in millimeters per year. [Plate boundary data by Peter Bird, UCLA.]

1. What do the areas shaded in different pastel colors, such as yellow and green, represent?
   a. Continents
   b. Oceans
   c. Tectonic plates
   d. Plate boundaries

2. What do the numbers next to the arrows represent?
   a. Plate age in millions of years
   b. Length of a plate boundary segment in kilometers
   c. Amount of plate area created each year in square kilometers
   d. Relative plate velocities in millimeters per year

3. In what approximate direction is the Nazca Plate moving relative to the Antarctic Plate?
   a. Northward
   b. Eastward
   c. Southward
   d. Westward

4. Which of the following does the North American Plate include?
   a. North American continent
   b. North American continent and all of the Atlantic Ocean
   c. North American continent, Greenland, and part of the North Atlantic Ocean
   d. North American continent, part of the Atlantic Ocean, and the Caribbean Sea

5. Which plate boundary shows the fastest rates of seafloor spreading?
   a. Boundary between the Nazca and Pacific plates
   b. Boundary between the Antarctic and Australian plates
   c. Boundary between the North American and African plates
   d. Boundary between the Caribbean and South American plates

Earth Materials: Minerals and Rocks

What Are Minerals?	54
The Structure of Matter	54
The Formation of Minerals	57
Classes of Rock-Forming Minerals	60
Physical Properties of Minerals	65
What Are Rocks?	70
The Rock Cycle: Interactions Between the Plate Tectonic and Climate Systems	75
Concentrations of Valuable Mineral Resources	77

Learning Objectives

Chapter 3 describes how geologists break the Earth down into its fundamental building blocks: minerals and rocks. After you have studied the chapter, you should be able to

3.1 Understand the structure of matter and how atoms react through chemical bonds to form minerals, which are crystalline substances.

3.2 Recognize that all minerals are grouped into seven classes based on their chemical composition, and that all minerals also have distinct physical properties.

3.3 Understand that geologists' primary aim is to study the three basic classes of rocks in order to deduce their geologic origins.

3.4 Know that economic mineral resources occur in specific locations, and that the evolution of minerals over geologic time teaches us the history of our planet, including the emergence and evolution of life.

Crystals of amethyst and quartz, growing on top of epidote crystals (green). The planar surfaces are crystal faces, whose geometries are determined by the underlying arrangement of the atoms that make up the crystals. [John Grotzinger/Ramón Rivera-Moret/Harvard Mineralogical Museum.]

CHAPTER 3 Earth Materials: Minerals and Rocks

In Chapter 2, we saw how the plate tectonic system gives rise to Earth's large-scale structure and dynamics, but we touched only briefly on the wide variety of materials that appear in different plate tectonic settings. In this chapter, we focus on those materials: minerals and rocks. Minerals are the building blocks of rocks, which are, in turn, the records of geologic history. Rocks and minerals help determine the structure of Earth, much as concrete, steel, and plastic determine the structure, design, and architecture of large buildings.

To tell Earth's story, geologists often adopt a "Sherlock Holmes" approach: They use current evidence to deduce the processes and events that occurred in the past at some particular place. The kinds of minerals found in volcanic rocks, for example, provide evidence of eruptions that brought molten rock to Earth's surface, while the minerals in granite reveal that it crystallized deep in the crust under the very high temperatures and pressures produced when two continents collide. Understanding the geology of a region also allows us to make informed guesses about where undiscovered deposits of economically important mineral resources might lie.

This chapter begins with a description of minerals—what they are, how they form, and how they can be identified. We then turn our attention to the major groups of rocks formed from these minerals and the geologic environments in which they form.

What Are Minerals?

Minerals are the building blocks of rocks. **Mineralogy** is the branch of geology that studies the composition, structure, appearance, stability, occurrence, and associations of minerals. With the proper tools, most rocks can be separated into their constituent minerals. A few kinds of rocks, such as limestone, are made up primarily of a single mineral (in this case, calcite). Other rocks, such as granite, are made up of several different minerals. To identify and classify the many kinds of rocks that compose Earth and understand how they are formed, we must know how minerals are formed.

Geologists define a **mineral** as a naturally occurring, solid crystalline substance, usually inorganic, with a specific chemical composition. Minerals are homogeneous: They cannot be divided mechanically into smaller components.

Let's examine each part of our definition of a mineral in a little more detail.

Naturally occurring: To qualify as a mineral, a substance must be found in nature. The diamonds mined in South Africa, for example, are minerals. The synthetic versions produced in industrial laboratories are not minerals, nor are the thousands of laboratory products invented by chemists.

Solid crystalline substance: Minerals are solid substances—they are neither liquids nor gases. When we say that a mineral is crystalline, we mean that the tiny particles of matter, or atoms, that compose it are arranged in an orderly, repeating, three-dimensional array. Solid materials that have no such orderly arrangement are referred to as glassy or amorphous (without form) and are not conventionally called minerals. Windowpane glass is amorphous, as are some natural glasses formed during volcanic eruptions. Later in this chapter, we will explore in detail the process by which crystalline materials form.

Usually inorganic: Minerals are defined as inorganic substances and so exclude the organic materials that make up plant and animal bodies. Organic matter is composed of organic carbon, the form of carbon found in all organisms, living or dead. Decaying vegetation in a wetland may be geologically transformed into coal, which is also made of organic carbon, but although it is found in naturally occurring deposits, coal is not considered a mineral. Many minerals, however, are secreted by organisms. One such mineral, calcite (**Figure 3.1**), which forms the shells of oysters and many other marine organisms, contains inorganic carbon. These shells accumulate on the seafloor, where they may be geologically transformed into limestone. The calcite of these shells fits the definition of a mineral because it is inorganic and crystalline.

With a specific chemical composition: The key to understanding the composition of Earth materials lies in knowing how the chemical elements are organized into minerals. What makes each mineral unique is its chemical composition and the arrangement of its atoms in an internal structure. A mineral's chemical composition either is fixed or varies within defined limits. The mineral quartz, for example, has a fixed ratio of two atoms of oxygen to one atom of silicon. This ratio never varies, even though quartz is found in many different kinds of rocks. Similarly, the chemical elements that make up the mineral olivine—iron, magnesium, oxygen, and silicon—always have a fixed ratio. Although the numbers of iron and magnesium atoms may vary, the sum of those two atoms in relation to the number of silicon atoms always forms a fixed ratio.

The Structure of Matter

In 1805, the English chemist John Dalton hypothesized that each of the various chemical elements consists of a different kind of atom, that all atoms of any given element are identical, and that chemical compounds are formed by various combinations of atoms of different elements in definite proportions. By the early twentieth century, physicists, chemists, and mineralogists, building on Dalton's ideas, had come to understand the structure of matter much as we do today. We now know that an *atom* is the smallest unit of an element that retains the physical and chemical properties of that element. We also know that atoms are the small units of matter that combine in chemical reactions, and that atoms themselves are divisible into even smaller units.

The Structure of Matter

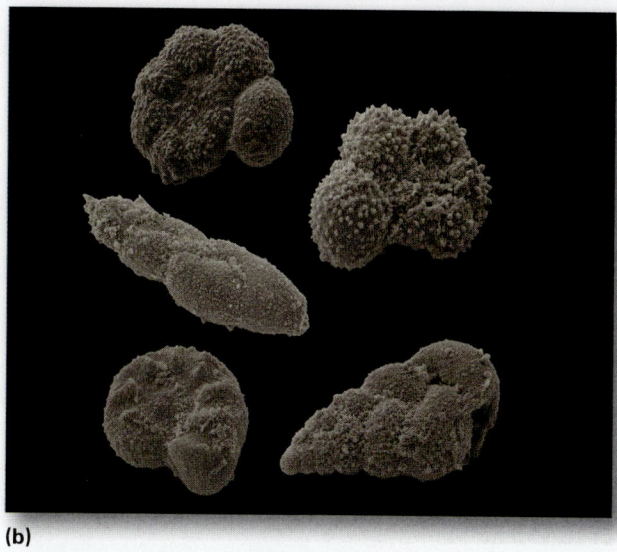

FIGURE 3.1 Many minerals are secreted by organisms. (a) The mineral calcite contains inorganic carbon. (b) Calcite is found in the shells of many marine organisms, such as these foraminifera. [(a) John Grotzinger/Ramón Rivera-Moret/Harvard Mineralogical Museum; (b) Andrew Syred/Science Source.]

The Structure of Atoms

Understanding the structure of atoms allows us to predict how chemical elements will react with one another and form new crystal structures. The structure of an atom is defined by a nucleus, which contains protons and neutrons, and which is surrounded by electrons. (For a more detailed review of the structure of atoms, see Appendix 3.)

The Nucleus: Protons and Neutrons At the center of every atom is a dense *nucleus* containing virtually all the mass of the atom in two kinds of particles: protons and neutrons (Figure 3.2). A *proton* has a positive electrical charge of 1. A *neutron* is electrically neutral—that is, uncharged. Atoms of the same chemical element may have different numbers of neutrons, but the number of protons does not vary. For instance, all carbon atoms have six protons.

Electrons Surrounding the nucleus is a cloud of moving particles called *electrons*, each with a mass so small that it is conventionally taken to be zero. Each electron carries a negative electrical charge of 21. The number of protons in the nucleus of any atom is balanced by the same number of electrons in the cloud surrounding the nucleus, so that the atom is electrically neutral. Thus, the nucleus of a carbon atom is surrounded by six electrons (see Figure 3.2).

Atomic Number and Atomic Mass

The number of protons in the nucleus of an atom is its **atomic number.** Because all atoms of the same element have the same number of protons, they also have the same atomic number. All atoms with six protons, for

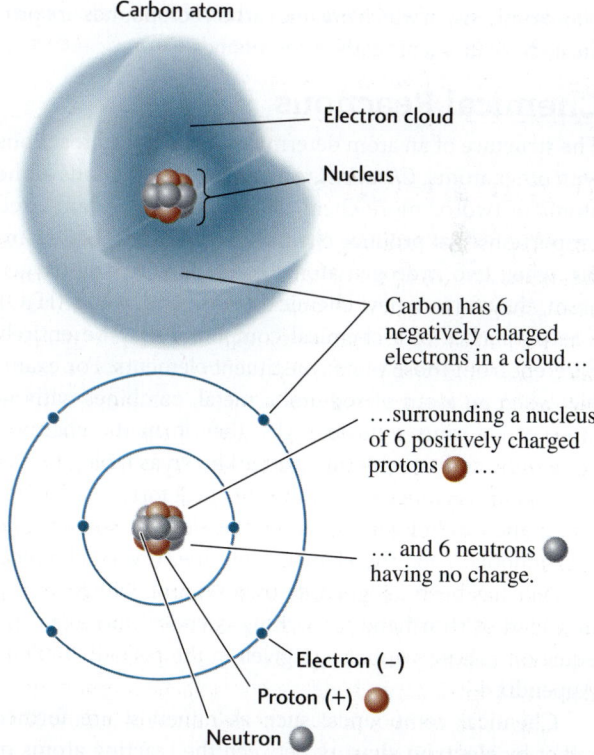

FIGURE 3.2 Structure of the carbon atom (carbon-12). The six electrons, each with a charge of −1, are represented as a negatively charged cloud surrounding the nucleus, which contains six protons, each with a charge of +1, and six neutrons, each with no charge. The size of the nucleus is greatly exaggerated in these drawings; it is much too small to show at a true scale.

example, are carbon atoms (atomic number 6). In fact, the atomic number of an element can tell us so much about that element's behavior that the periodic table organizes elements according to their atomic number (see Appendix 3). Elements in the same vertical column of the periodic table, such as carbon and silicon, tend to have similar chemical properties.

The **atomic mass** of an element is the sum of the masses of its protons and its neutrons. (Electrons, because they have so little mass, are not included in this sum.) Although atoms of the same element always have the same number of protons, they may have different numbers of neutrons, and therefore different atomic masses. Atoms of the same element with different numbers of neutrons are called **isotopes.** Isotopes of the element carbon, for example, all have six protons, but may have six, seven, or eight neutrons, giving atomic masses of 12, 13, and 14.

In nature, the chemical elements exist as mixtures of isotopes, so their average atomic masses are never whole numbers. The atomic mass of carbon, for example, is 12.011. It is close to 12 because the isotope carbon-12 is by far the most abundant. The relative abundances of the various isotopes of an element on Earth are determined by processes that enhance the abundances of some isotopes over others. Carbon-12, for example, is favored by some chemical reactions, such as photosynthesis, in which organic carbon compounds are produced from inorganic carbon compounds.

Chemical Reactions

The structure of an atom determines its chemical reactions with other atoms. *Chemical reactions* are interactions of the atoms of two or more chemical elements in certain fixed proportions that produce chemical compounds. For example, when two hydrogen atoms combine with one oxygen atom, they form a new chemical compound, water (H_2O). The properties of a chemical compound may be entirely different from those of its constituent elements. For example, when an atom of sodium, a metal, combines with an atom of chlorine, a noxious gas, they form the chemical compound sodium chloride, better known as table salt. We represent this compound by the chemical formula NaCl, in which the symbol Na stands for the element sodium and the symbol Cl for the element chlorine. (Every chemical element has been assigned its own symbol, which we use in a kind of shorthand for writing chemical formulas and equations; these symbols are given in the periodic table in Appendix 3.)

Chemical compounds, such as minerals, are formed either by **electron sharing** between the reacting atoms or by **electron transfer** between the reacting atoms. Carbon and silicon, two of the most abundant elements in Earth's crust, tend to form compounds by electron sharing. Diamond is a compound composed entirely of carbon atoms sharing electrons (**Figure 3.3**).

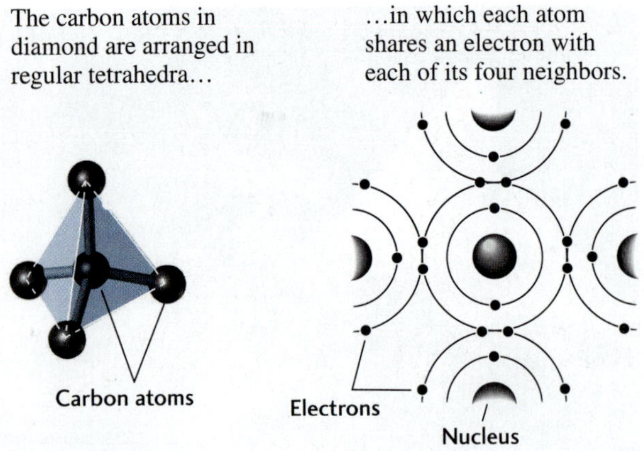

In the reaction between sodium (Na) and chlorine (Cl) atoms to form sodium chloride (NaCl), electrons are

The carbon atoms in diamond are arranged in regular tetrahedra…

…in which each atom shares an electron with each of its four neighbors.

FIGURE 3.3 Some atoms share electrons to form covalent bonds.

transferred. The sodium atom loses one electron, which the chlorine atom gains (**Figure 3.4a**). An atom or group of atoms that has an electrical charge, either positive or negative, because of the loss or gain of one or more electrons is called an **ion.** Because the chlorine atom has gained a negatively charged electron, it is now a negatively charged ion, Cl^-. Likewise, the loss of an electron gives the sodium atom a positive charge, making it a sodium ion, Na^+. The compound NaCl itself remains electrically neutral because the positive charge on Na^+ is exactly balanced by the negative charge on Cl^-. A positively charged ion is called a **cation,** and a negatively charged ion is called an **anion.**

Chemical Bonds

When a chemical compound is formed either by electron sharing or by electron transfer, the ions or atoms that make up the compound are held together by electrostatic attraction between negatively charged electrons and positively charged protons. The attractions, or *chemical bonds,* between shared electrons or between gained and lost electrons may be strong or weak. Strong bonds keep a substance from decomposing into its elements or into other compounds. They also make minerals hard and keep them from cracking or splitting. Two major types of bonds are found in most rock-forming minerals: ionic bonds and covalent bonds.

Ionic Bonds The simplest form of chemical bond is the **ionic bond.** Bonds of this type form by electrostatic attraction between ions of opposite charge, such as Na^+ and Cl^- in sodium chloride (see Figure 3.4a), when electrons are transferred. This attraction is of exactly the same nature as the static electricity that can make nylon or silk clothing cling to the body. The strength of an ionic bond decreases greatly as the distance between ions increases, and it increases as the electrical charges of the ions increase.

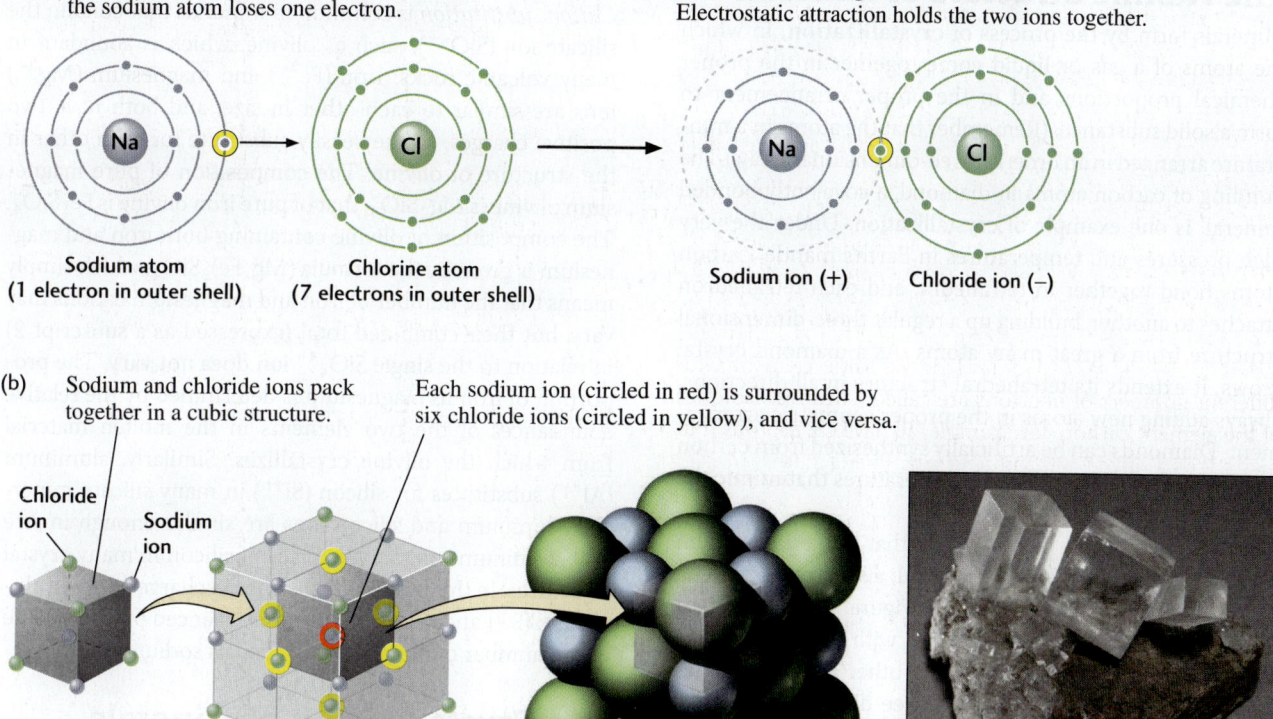

FIGURE 3.4 Some atoms transfer electrons to form ionic bonds. [Photo by John Grotzinger/ Ramón Rivera-Moret/ Harvard Mineralogical Museum.]

Ionic bonds are the dominant type of chemical bonds in mineral structures: About 90 percent of all minerals are essentially ionic compounds.

Covalent Bonds Elements that do not readily gain or lose electrons to form ions, and instead form compounds by sharing electrons, are held together by **covalent bonds.** These bonds are generally stronger than ionic bonds. One mineral with a covalently bonded crystal structure is diamond, which consists of a single element, carbon. Each carbon atom can share four of its own electrons with other carbon atoms and can acquire another four electrons by sharing with other carbon atoms. In diamond, every carbon atom is surrounded by four others arranged in a four-sided pyramidal form (a *tetrahedron*), each side of which is a triangle (see Figure 3.3). In this configuration, each carbon atom shares an electron with each of its four neighbors, resulting in a very stable configuration. (Figure 3.10 shows a network of carbon tetrahedra linked together.)

Metallic Bonds Atoms of metallic elements, which have strong tendencies to lose electrons, pack together as cations, and the freely mobile electrons are shared and dispersed among those cations. This free electron sharing results in a kind of covalent bond that we call a **metallic bond.** It is found in a small number of minerals, among them the metal copper and some sulfides.

The chemical bonds of some minerals are intermediate between pure ionic and pure covalent bonds because some electrons are exchanged and others are shared.

The Formation of Minerals

The orderly forms of minerals result from the chemical bonds we have just described. Minerals can be viewed in two complementary ways: as assemblages of submicroscopic atoms organized in an ordered three-dimensional array, and as crystals that we can see with the naked eye. In this section, we examine the crystal structures of minerals and the conditions under which minerals form. Later in this chapter, we will see how the crystal structures of minerals are manifested in their physical properties.

The Atomic Structure of Minerals

Minerals form by the process of **crystallization,** in which the atoms of a gas or liquid come together in the proper chemical proportions and in the proper arrangement to form a solid substance. (Remember that the atoms in a mineral are arranged in an orderly three-dimensional array.) The bonding of carbon atoms in diamond, a covalently bonded mineral, is one example of crystallization. Under the very high pressures and temperatures in Earth's mantle, carbon atoms bond together in tetrahedra, and each tetrahedron attaches to another, building up a regular three-dimensional structure from a great many atoms. As a diamond crystal grows, it extends its tetrahedral structure in all directions, always adding new atoms in the proper geometric arrangement. Diamonds can be artificially synthesized from carbon under very high pressures and temperatures that mimic the conditions in Earth's mantle.

The sodium and chloride ions that make up sodium chloride, an ionically bonded mineral, also crystallize in an orderly three-dimensional array. In Figure 3.4b, we can see the geometry of their arrangement, with each ion of one kind surrounded by six ions of the other kind in a series of cubic structures extending in three directions. We can think of ions as solid spheres, packed together in close-fitting structural units. Figure 3.4b also shows the relative sizes of the ions in NaCl. The relative sizes of the sodium and chloride ions allow them to fit together in a closely packed arrangement.

Many of the cations of abundant minerals are relatively small, while most anions—including the most common anion on Earth, oxygen (O^{2-})—are large (**Figure 3.5**). Because anions tend to be larger than cations, most of the space of a crystal is occupied by the anions, and the cations fit into the spaces between them. As a result, crystal structures are determined largely by how the anions are arranged and how the cations fit between them.

Cations of similar sizes and charges tend to substitute for one another and to form compounds having the same crystal structure but differing in chemical composition. *Cation substitution* is common in minerals that contain the silicate ion (SiO^{4+}), such as olivine, which is abundant in many volcanic rocks. Iron (Fe^{2+}) and magnesium (Mg^{2+}) ions are similar to each other in size, and both have two positive charges, so they easily substitute for each other in the structure of olivine. The composition of pure magnesium olivine is Mg_2SiO_4; that of pure iron olivine is Fe_2SiO_4. The composition of olivine containing both iron and magnesium is given by the formula $(Mg, Fe)_2SiO_4$, which simply means that the number of iron and magnesium cations may vary, but their combined total (expressed as a subscript 2) in relation to the single SiO_4^{4-} ion does not vary. The proportion of iron to magnesium is determined by the relative abundances of the two elements in the molten material from which the olivine crystallizes. Similarly, aluminum (Al^{3+}) substitutes for silicon (Si^{4+}) in many silicate minerals. Aluminum and silicon ions are similar enough in size that aluminum can take the place of silicon in many crystal structures. In this case, the difference in charge between aluminum (3+) and silicon (4+) ions is balanced by an increase in the number of other cations, such as sodium (1+).

The Crystallization of Minerals

Crystallization starts with the formation of microscopic single **crystals,** orderly three-dimensional arrays of atoms in which the basic arrangement is repeated in all directions. The boundaries of crystals are natural, flat (*planar*) surfaces called *crystal faces* (**Figure 3.6**). The crystal faces of a mineral are the external expression of the mineral's internal atomic structure. **Figure 3.7** pairs a drawing of a perfect quartz crystal with a photograph of the actual mineral. The six-sided (hexagonal) shape of the quartz crystal corresponds to its hexagonal internal atomic structure.

During crystallization, the initially microscopic crystals grow larger, maintaining their crystal faces as long as they are free to grow. Large crystals with well-defined faces form when growth is slow and steady, and when space is adequate

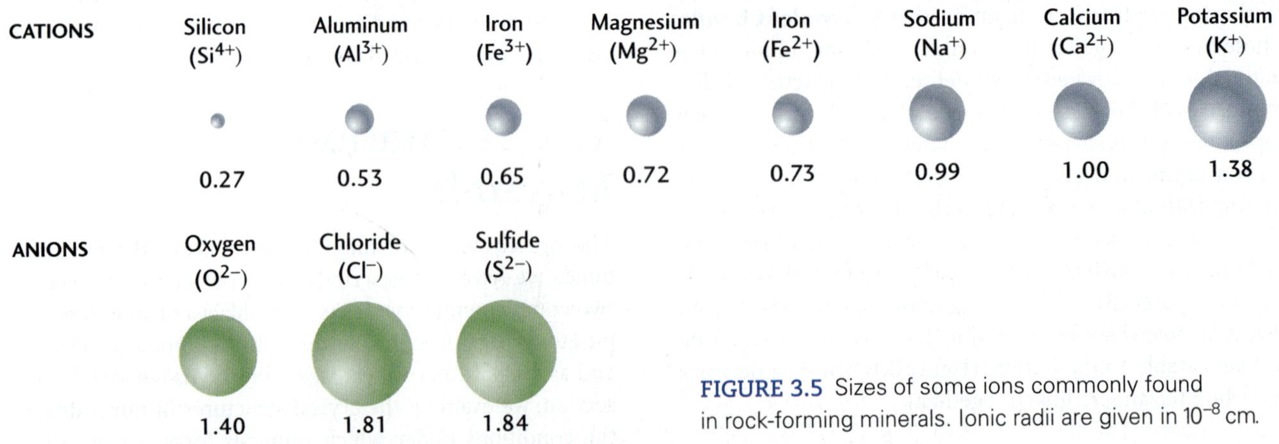

CATIONS	Silicon (Si^{4+})	Aluminum (Al^{3+})	Iron (Fe^{3+})	Magnesium (Mg^{2+})	Iron (Fe^{2+})	Sodium (Na^+)	Calcium (Ca^{2+})	Potassium (K^+)
	0.27	0.53	0.65	0.72	0.73	0.99	1.00	1.38
ANIONS	Oxygen (O^{2-})	Chloride (Cl^-)	Sulfide (S^{2-})					
	1.40	1.81	1.84					

FIGURE 3.5 Sizes of some ions commonly found in rock-forming minerals. Ionic radii are given in 10^{-8} cm.

The Formation of Minerals

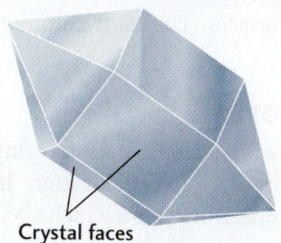
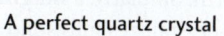
A perfect quartz crystal

A natural quartz crystal

FIGURE 3.7 A perfect crystal is rare in nature, but no matter how irregular the shapes of the crystal faces may be, the angles are always exactly the same. [Breck P. Kent.]

FIGURE 3.6 Crystals of amethyst and quartz growing on top of epidote crystals (green). The planar surfaces are crystal faces and reflect the mineral's internal atomic structure. [John Grotzinger/Ramón Rivera-Moret/Harvard Mineralogical Museum.]

to allow growth without interference from other crystals nearby (**Figure 3.8**). For this reason, most large mineral crystals form in open spaces in rocks, such as fractures or cavities.

More often, however, the spaces between growing crystals fill in, or crystallization proceeds rapidly. Crystals then grow over one another and coalesce to become a solid mass of crystalline particles, or **grains.** In this case, few or no grains show crystal faces. Large crystals that can be seen with the naked eye are relatively unusual, but many minerals in rocks display crystal faces that can be seen under a microscope.

Unlike minerals, glassy materials—which solidify from liquids so quickly that they lack any internal atomic order—do

FIGURE 3.8 Giant crystals are sometimes found in caves, where they have room to grow. These selenite crystals are a gem-quality form of gypsum (calcium sulfate). [Javier Trueba/ MSF/Science Source.]

not form crystals with planar faces. Instead, they are found as masses with curved, irregular surfaces. The most common natural glass is volcanic glass.

How Do Minerals Form?

Lowering the temperature of a liquid below its freezing point is one way to start the process of crystallization. In water, for example, 0°C is the temperature below which crystals of ice—a mineral—start to form. Similarly, a **magma**—a mass of hot, molten liquid rock—crystallizes into solid minerals when it cools. As a magma falls below its melting point, which may be higher than 1000°C depending on the elements it contains, crystals of silicate minerals such as olivine or feldspar begin to form. (Geologists usually refer to melting points of magmas rather than freezing points, because freezing implies cold.)

Crystallization can also occur as liquids evaporate from a solution. A *solution* is a homogeneous mixture of one chemical substance with another, such as salt and water. As the water evaporates from a salt solution, the concentration of salt eventually gets so high that the solution can hold no more salt and is said to be *saturated*. If evaporation continues, the salt starts to **precipitate,** or drop out of solution as crystals. Deposits of table salt, or halite, form under just these conditions when seawater evaporates to the point of saturation in some hot, arid bays or arms of the ocean (**Figure 3.9**).

Diamond and graphite (the material used as the "lead" in pencils) exemplify the dramatic effects that temperature and pressure can have on mineral formation. These two minerals are **polymorphs,** minerals with alternative structures formed from the same chemical element or compound (**Figure 3.10**). They are both formed from carbon, but have different crystal structures and very different appearances. From experimentation and geologic observation, we know that diamond forms and remains stable at the very high pressures and temperatures found in Earth's mantle. High pressures force the atoms in diamond into a closely packed structure. Diamond therefore has a higher **density** (mass per unit volume, usually expressed in grams per cubic centimeter, g/cm^3) than graphite, which is less closely packed: Diamond has a density of 3.5 g/cm^3, while that of graphite is only 2.1 g/cm^3. Graphite forms and is stable at moderate pressures and temperatures, such as those in Earth's crust.

Low temperatures can also produce close packing of atoms. Quartz and cristobalite, for example, are polymorphs of silica (SiO_2). Quartz forms at low temperatures and is relatively dense (2.7 g/cm^3). Cristobalite, which forms at higher temperatures, has a more open structure and is therefore less dense (2.3 g/cm^3).

Classes of Rock-Forming Minerals

All minerals on Earth have been grouped into seven classes according to their chemical composition (**Table 3.1**). Some minerals, such as copper, occur naturally as un-ionized pure elements; these minerals are classified as *native elements*. Most other minerals are classified by their anions. Olivine, for example, is classified as a silicate by its silicate anion, SiO_4^{4-}. Halite (sodium chloride, NaCl) is classified as a halide by its chloride anion, Cl^-. So is its close relative, sylvite (potassium chloride, KCl).

FIGURE 3.9 Halite crystals precipitating within a modern hypersaline lagoon on San Salvador Island in the Bahamas. Note the cubic shape of the crystals. [John Grotzinger.]

Classes of Rock-Forming Minerals

Natural **diamond** is formed at very high pressures and temperatures in Earth's mantle.

Strong bonds connect closely packed carbon atoms in a tetrahedral structure.

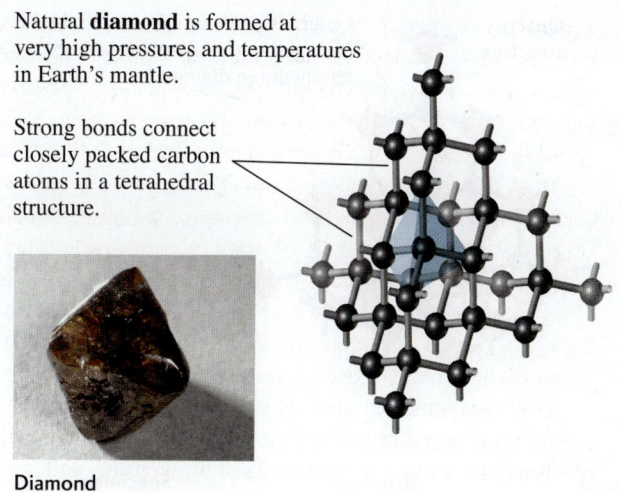

Diamond

Graphite is formed at lower pressures and temperatures than diamond. Strong bonds connect carbon atoms arranged in sheets.

Weak bonds connect carbon atoms between alternating sheets.

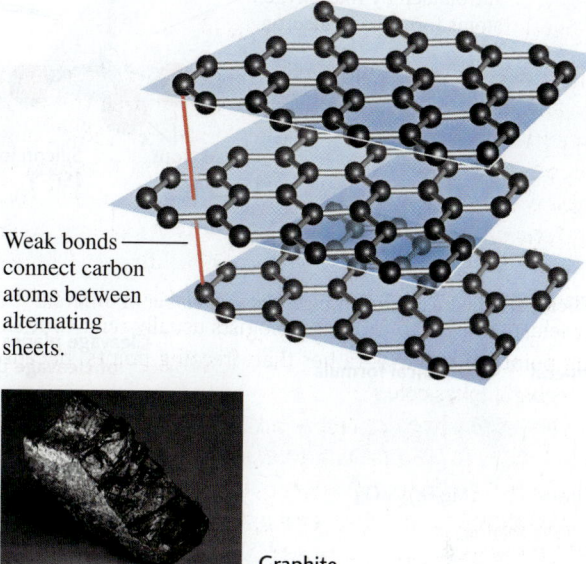

Graphite

FIGURE 3.10 Graphite and diamond are polymorphs, alternative structures formed from the same chemical compound, carbon. [John Grotzinger/Ramón Rivera-Moret/Harvard Mineralogical Museum.]

TABLE 3.1 Some Chemical Classes of Minerals

CLASS	DEFINING ANIONS	EXAMPLE
Native elements	None: no charged ions	Copper metal (Cu)
Oxides	Oxygen ion (O^{2-})	Hematite (Fe_2O_3)
Halides	Chloride (Cl^-), fluoride (F^-), bromide (Br^-), iodide (I^-)	Halite (NaCl)
Carbonates	Carbonate ion (CO_3^{2-})	Calcite ($CaCO_3$)
Sulfates	Sulfate ion (SO_4^{2-})	Anhydrite ($CaSO_4$)
Silicates	Silicate ion (SiO_4^{4-})	Olivine $(Mg,Fe)_2SiO_4$
Sulfides	Sulfide ion (S^{2-})	Pyrite (FeS^2)

Although many thousands of minerals are known, geologists commonly encounter only about 30 of them. These minerals are the building blocks of most crustal rocks and are called *rock-forming minerals*. Their relatively small number corresponds to the small number of elements that are abundant in Earth's crust.

In the following pages, we consider the five most common classes of rock-forming minerals:

- **Silicates,** the most abundant class of minerals in Earth's crust, are composed of oxygen (O) and silicon (Si)—the two most abundant elements in the crust—mostly in combination with cations of other elements.
- **Carbonates** are minerals composed of carbon and oxygen—in the form of the carbonate anion (CO_3^{2-})—in combination with calcium and magnesium. Calcite (calcium carbonate, $CaCO_3$) is one such mineral.
- **Oxides** are compounds of the oxygen anion (O^{2-}) and metallic cations; an example is the mineral hematite (iron oxide, Fe_2O_3).
- **Sulfides** are compounds of the sulfide anion (S^{2-}) and metallic cations; an example is the mineral pyrite (iron sulfide, FeS_2).
- **Sulfates** are compounds of the sulfate anion (SO_4^{2-}) and metallic cations; an example is the mineral anhydrite (calcium sulfate, $CaSO_4$).

The other two chemical classes of minerals—native elements, hydroxides, and halides—are less common as rock-forming minerals.

Silicates

The basic building block of all silicate mineral structures is the silicate ion. It is a tetrahedron composed of a central silicon ion (Si^{4+}) surrounded by four oxygen ions (O^{2-}), and thus has the formula SiO_4^{4-} (**Figure 3.11**).

CHAPTER 3 Earth Materials: Minerals and Rocks

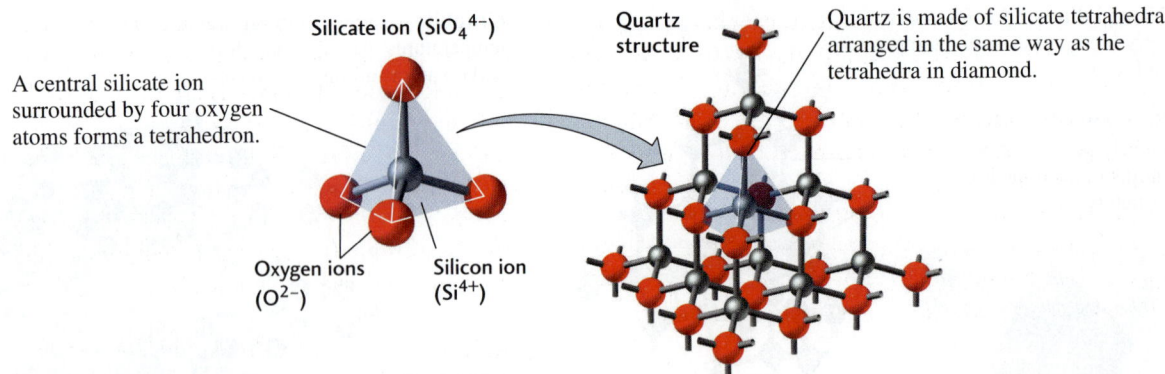

Silicate tetrahedra can be arranged into a number of different structures.

Mineral	Chemical formula	Cleavage planes and number of cleavage directions	Structure	Specimen
(a) Olivine	$(Mg,Fe)_2SiO_4$	1 plane	Isolated tetrahedra	
(b) Pyroxene	$(Mg,Fe)SiO_3$	2 planes at 90°	Single chains	
(c) Amphibole	$Ca_2(Mg,Fe)_5Si_8O_{22}(OH)_2$	2 planes at 60° and 120°	Double chains	
(d) Mica	Muscovite: $KAl_2(AlSi_3O_{10})(OH)_2$	1 plane	Sheets	
(e) Feldspar	Orthoclase feldspar: $KAlSi_3O_8$ Plagioclase feldspar: $(Ca,Na)AlSi_3O_8$	2 planes at 90°	Three-dimensional frameworks	

FIGURE 3.11 The silicate ion is the basic building block of silicate minerals. [John Grotzinger/Ramón Rivera-Moret/Harvard Mineralogical Museum.]

Because the silicate ion has a negative charge, it often bonds to cations to form minerals. The cations it typically bonds to include sodium (Na^+), potassium (K^+), calcium (Ca^{2+}), magnesium (Mg^{2+}), and iron (Fe^{2+}). Alternatively, the silicate ion can share oxygen ions with other silicate tetrahedra. Silicate tetrahedra can form a number of crystal structures: They may be isolated (linked only to cations), or they may be linked to other silicate tetrahedra in rings, single chains, double chains, sheets, or frameworks. Some of these structures are shown in Figure 3.11.

Isolated Tetrahedra
Isolated tetrahedra are linked by the bonding of each oxygen ion of the tetrahedron to a cation (Figure 3.11a). The cations, in turn, bond to the oxygen ions of other tetrahedra. The tetrahedra are thus isolated from one another by cations on all sides. Olivine is a rock-forming mineral with this structure.

Single-Chain Structures
Single chains are formed by the sharing of oxygen ions. Two oxygen ions of each silicate tetrahedron bond to adjacent tetrahedra in an open-ended chain (Figure 3.11b). These single chains are linked to other chains by cations. Minerals of the pyroxene group are single-chain silicate minerals. Enstatite, a pyroxene, contains iron or magnesium ions, or both; the two cations may substitute for each other, as in olivine. The formula $(Mg,Fe)SiO_3$ represents this structure.

Double-Chain Structures
Two single chains may combine to form double chains linked to each other by shared oxygen ions (Figure 3.11c). Adjacent double chains linked by cations form the structure of minerals in the amphibole group. Hornblende, a member of this group, is an extremely common mineral in both igneous and metamorphic rocks. It has a complex composition that includes calcium (Ca^{2+}), sodium (Na^+), magnesium (Mg^{2+}), iron (Fe^{2+}), and aluminum (Al^{3+}).

Sheet Structures
In sheet structures, each tetrahedron shares three of its oxygen ions with adjacent tetrahedra to build stacked sheets of tetrahedra (Figure 3.11d). Cations may be interlayered with the tetrahedral sheets. The micas and clay minerals are the most abundant sheet silicates. Muscovite, $KAl_2(AlSi_3O_{10})(OH)_2$, is one of the most common sheet silicates and is found in many types of rocks. It can be separated into extremely thin, transparent sheets. Kaolinite, $Al_2Si_2O_5(OH)_4$, which also has this structure, is a common clay mineral found in sediments and is the basic raw material for pottery.

Frameworks
Three-dimensional frameworks may form as each tetrahedron shares all its oxygen ions with other tetrahedra. Feldspars, the most abundant minerals in Earth's crust, are framework silicates (Figure 3.11e), as is another of the most common minerals, quartz (SiO_2).

Silicate Compositions
Chemically, the simplest silicate is silicon dioxide, also called silica (SiO_2), which is found most often as the mineral quartz. The silicate tetrahedra of quartz are linked, sharing two oxygen ions for each silicon ion, so the total formula adds up to SiO_2.

In other silicate minerals, as we have seen, the basic structural units—rings, chains, sheets, and frameworks—are bonded to cations such as sodium (Na^+), potassium (K^+), calcium (Ca^{2+}), magnesium (Mg^{2+}), and iron (Fe^{2+}). As noted in the discussion of cation substitution, aluminum (Al^{3+}) substitutes for silicon in many silicate minerals.

Carbonates

The basic building block of carbonate minerals is the carbonate ion (CO_3^{2-}), which consists of a carbon ion surrounded by, and covalently bonded to, three oxygen ions in a triangle (Figure 3.12a). Groups of carbonate ions are arranged in sheets somewhat like those of the sheet silicates, which are linked together by layers of cations. In calcite (calcium carbonate, $CaCO_3$), the sheets of carbonate ions are separated by layers of calcium ions (Figure 3.12b). Calcite is one of the most abundant minerals in Earth's crust and is the chief constituent of a group of rocks called limestones (Figure 3.12c). Dolomite ($CaMg(CO_3)_2$) is another

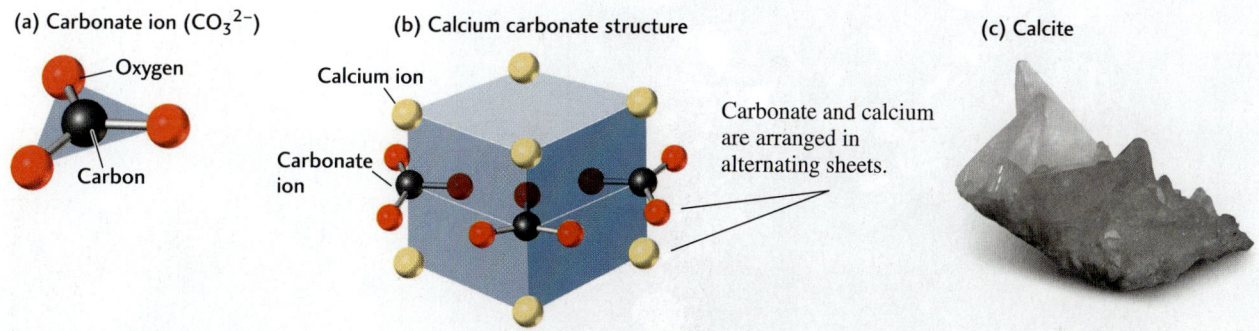

FIGURE 3.12 Carbonate minerals, such as calcite (calcium carbonate, $CaCO_3$), have a layered structure. (a) Top view of the carbonate ion, composed of a carbon ion surrounded by three oxygen ions in a triangle. (b) View of the alternating layers of calcium and carbonate ions in calcite. (c) Calcite. [John Grotzinger/Ramón Rivera-Moret/Harvard Mineralogical Museum.]

FIGURE 3.13 Oxides include many economically valuable minerals. (a) Hematite. (b) Spinel. [John Grotzinger/Ramón Rivera-Moret/Harvard Mineralogical Museum.]

major mineral of crustal rocks that is made up of the same carbonate sheets separated by alternating layers of calcium ions and magnesium ions.

Oxides

Oxide minerals are compounds in which oxygen is bonded to atoms or cations of other elements, usually metallic cations such as iron (Fe^{2+} or Fe^{3+}). Most oxide minerals are ionically bonded, and their structures vary with the size of the metallic cations. This class of minerals has great economic importance because it includes the ores containing many of the metals, such as chromium and titanium, used in the manufacture of metallic materials and devices. Hematite (Fe_2O_3) (Figure 3.13a) is a chief ore of iron.

Another group of abundant minerals in this class, the spinels (Figure 3.13b), are oxides of two metals, magnesium and aluminum ($MgAl_2O_4$). Spinels have a closely packed cubic structure and a high density (3.6 g/cm^3), reflecting the conditions of high pressure and temperature under which they form. Transparent gem-quality spinels may resemble ruby or sapphire and are found in the crown jewels of England and Russia.

Sulfides

The chief ores of other valuable minerals—such as copper, zinc, and nickel—are members of the sulfide class. The basic building block of this class is the sulfide ion (S^{2-}), a sulfur atom that has gained two electrons. In the sulfide minerals, the sulfide ion is bonded to metallic cations. Most sulfide minerals look like metals, and almost all are opaque. The most common sulfide mineral is pyrite (FeS_2), often called "fool's gold" because of its yellowish metallic appearance (Figure 3.14).

FIGURE 3.14 Pyrite, a sulfide mineral, is also known as "fool's gold." [John Grotzinger/Ramón Rivera-Moret/Harvard Mineralogical Museum.]

FIGURE 3.15 Gypsum is a sulfate formed when seawater evaporates. [John Grotzinger/Ramón Rivera-Moret/Harvard Mineralogical Museum.]

Sulfates

The basic building block of sulfates is the sulfate ion (SO_4^{2-}). It is a tetrahedron made up of a central sulfur atom surrounded by four oxygen ions (O^{2-}). One of the most abundant minerals of this class is gypsum (**Figure 3.15**), the primary component of plaster. Gypsum, a calcium sulfate, forms when seawater evaporates. During evaporation, Ca^{2+} and SO_4^{2-}, two ions that are abundant in seawater, combine and precipitate as layers of sediment, forming calcium sulfate ($CaSO_4 \cdot 2H_2O$). (The dot in this formula signifies that two water molecules are bonded to the calcium and sulfate ions.)

Another calcium sulfate, anhydrite ($CaSO_4$), differs from gypsum in that it contains no water. (Its name is derived from the word *anhydrous*, meaning "free from water.") Gypsum is stable at the low temperatures and pressures found at Earth's surface, whereas anhydrite is stable at the higher temperatures and pressures where sedimentary rocks are buried.

As scientists discovered in 2004, sulfate minerals precipitated from water and formed sedimentary layers early in the history of Mars. These minerals were precipitated by processes similar to those observed on Earth when lakes and shallow seas dried up. Many of these sulfate minerals, however, are quite different from the sulfate minerals commonly found on Earth and include strange iron-bearing sulfates that precipitated from very harsh, acidic waters (see Earth Issues 11.1).

Physical Properties of Minerals

Geologists use their knowledge of mineral composition and structure to understand the origins of rocks. First, they must identify the minerals that make up a rock. To do so, they rely greatly on chemical and physical properties that can be observed relatively easily. In the nineteenth and early twentieth centuries, geologists carried field kits for rough chemical analyses of minerals that would help in their identification. One such test is the origin of the phrase "the acid test." It consists of dropping diluted hydrochloric acid (HCl) on a mineral to see if it fizzes (**Figure 3.16**). Fizzing indicates that carbon dioxide (CO_2) is escaping, which means that the mineral is likely to be calcite, a carbonate.

In this section, we review the physical properties of minerals, many of which contribute to their practical and decorative value.

FIGURE 3.16 The acid test. One easy but effective way to identify certain minerals is to drop diluted hydrochloric acid (HCl) on the substance of interest. If it fizzes, indicating the escape of carbon dioxide, the mineral is likely to be calcite. [Chip Clark/Fundamental Photographs.]

Hardness

Hardness is a measure of the ease with which the surface of a mineral can be scratched. Just as diamond, the hardest mineral known, scratches glass, a quartz crystal, which is harder than feldspar, scratches a feldspar crystal. In 1822, Friedrich Mohs, an Austrian mineralogist, devised a scale (now known as the **Mohs scale of hardness**) based on the ability of one mineral to scratch another. At one extreme is the softest mineral (talc); at the other, the hardest (diamond) (Table 3.2). The Mohs scale is still one of the best practical tools for identifying an unknown mineral. With a knife blade and a few of the minerals on the hardness scale, a field geologist can gauge an unknown mineral's position on the scale. If the unknown mineral is scratched by a piece of quartz but not by the knife, for example, it lies between 5 and 7 on the scale.

Recall that covalent bonds are generally stronger than ionic bonds. The hardness of any mineral depends on the strength of its chemical bonds: the stronger the bonds, the harder the mineral. Within the silicate class of minerals, hardness varies with crystal structure, from 1 in talc, a sheet silicate, to 8 in topaz, a silicate with isolated tetrahedra. Most silicates fall in the 5 to 7 range on the Mohs scale. Only sheet silicates are relatively soft, with hardnesses between 1 and 3.

Within groups of minerals that have similar crystal structures, hardness is related to other factors that also affect bond strength:

- *Size:* The smaller the atoms or ions, the smaller the distance between them and the greater the electrostatic attraction—and thus the stronger the bond.
- *Charge:* The larger the charge of ions, the greater the attraction between them, and thus the stronger the bond.
- *Packing:* The closer the packing of atoms or ions, the smaller the distance between them, and thus the stronger the bond.

Size is an especially important factor for most metallic oxides and for most sulfides of metals with high atomic numbers, such as gold, silver, copper, and lead. Minerals of these groups are soft, with hardnesses of less than 3, because their metallic cations are so large. Carbonates and sulfates, whose structures are not closely packed, are also soft, with hardnesses of less than 5.

Cleavage

Cleavage is the tendency of a crystal to split along planar surfaces. The term *cleavage* is also used to describe the geometric pattern produced by such breakage. Cleavage varies inversely with bond strength: Strong bonds produce poor cleavage, while weak bonds produce good cleavage. Because of their strength, covalent bonds generally produce poor or no cleavage. Ionic bonds are relatively weak, so they produce good cleavage. Even within a mineral that is entirely covalently bonded or entirely ionically bonded, however, bond strength varies along the different planes. For example, all of the bonds in diamond are covalent bonds, which are very strong, but some planes are more weakly bonded than others. Thus, diamond, the hardest mineral of all, can be cleaved along these weaker planes to produce perfect planar surfaces. Muscovite, a mica sheet silicate, splits along smooth, lustrous, flat, parallel surfaces, forming transparent sheets less than a millimeter thick. The excellent cleavage of micas results from the relative weakness of the bonds between its layers of cations sandwiched within sheets of silicate tetrahedra (Figure 3.17).

Cleavage is classified according to two primary sets of characteristics: the number of planes and pattern of cleavage, and the quality of surfaces and ease of cleaving.

Number of Planes and Pattern of Cleavage The number of planes and pattern of cleavage are identifying hallmarks of many rock-forming minerals. Muscovite, for example, has only one plane of cleavage, whereas calcite and dolomite crystals have three cleavage planes that give them a rhomboidal shape (Figure 3.18).

A crystal's structure determines its cleavage planes and its crystal faces. Crystals have fewer cleavage planes than possible crystal faces. Faces may be formed along any of numerous planes defined by rows of atoms or ions. Cleavage occurs along any of those planes across which the bonding is weak. All crystals of a mineral exhibit its characteristic cleavage planes, whereas only some crystals display particular faces.

Distinctive angles of cleavage help identify two important groups of silicates, the pyroxenes and the amphiboles, that otherwise often look alike (Figure 3.19). Pyroxenes have a single-chain structure and are bonded so that their cleavage planes are almost at right angles (about 90°) to each

TABLE 3.2 Mohs Scale of Hardness

MINERAL	SCALE NUMBER	COMMON OBJECTS
Talc	1	
Gypsum	2	Fingernail
Calcite	3	Copper coin
Fluorite	4	
Apatite	5	Knife blade
Orthoclase	6	Window glass
Quartz	7	Steel file
Topaz	8	
Corundum	9	
Diamond	10	

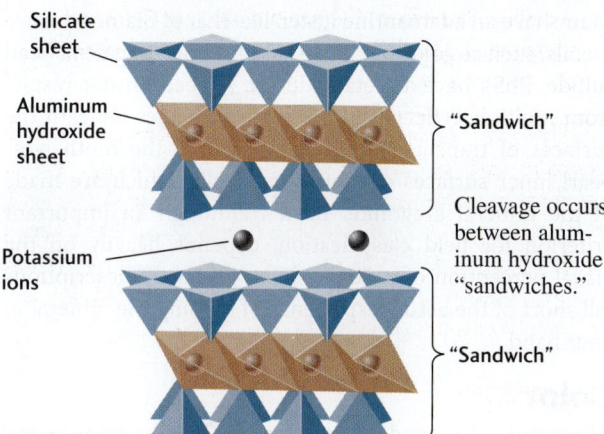

FIGURE 3.17 Cleavage of mica. The diagram shows the cleavage plane in the crystal structure, oriented perpendicular to the plane of the page. Horizontal lines mark the interfaces of silicate tetrahedral sheets and the sheets of aluminum hydroxide bonding the two tetrahedral sheets into a sandwich. Cleavage takes place between tetrahedral–aluminum hydroxide sandwiches. The photograph shows thin sheets of mica separating along the cleavage planes. [Chip Clark/Fundamental Photographs.]

other. In cross section, the cleavage pattern of pyroxenes is nearly a square. In contrast, amphiboles, which have a double-chain structure, are bonded so as to give two cleavage planes at about 60° and 120° to each other. They produce a diamond-shaped cross section.

Quality of Surfaces and Ease of Cleaving A mineral's cleavage is assessed as perfect, excellent, good, fair, poor, or none according to the quality of surfaces produced and the ease of cleaving. A few examples are described below.

Muscovite can be cleaved easily, and it produces extremely smooth surfaces; its cleavage is *perfect*. Single-chain and double-chain silicates (the pyroxenes and amphiboles, respectively) show *good* cleavage. Although these minerals

FIGURE 3.18 Example of rhomboidal cleavage in calcite. [Charles D. Winters/Science Source.]

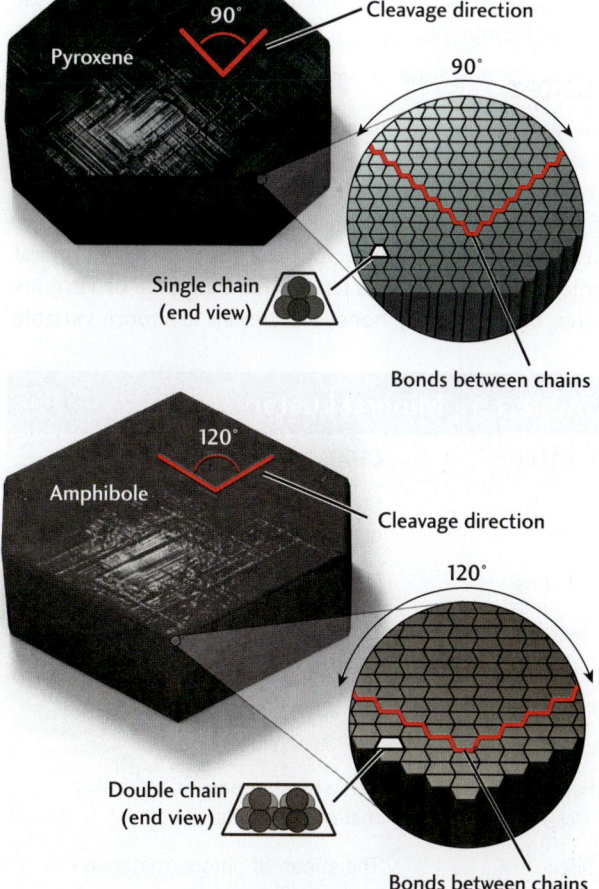

FIGURE 3.19 Pyroxene and amphibole often look very much alike, but their differing angles of cleavage can be used to identify and classify them.

split easily along the cleavage plane, they also break across it, producing cleavage surfaces that are not as smooth as those of micas. *Fair* cleavage is shown by the ring silicate beryl. Beryl's cleavage is irregular, and the mineral breaks relatively easily along directions other than cleavage planes.

Many minerals are so strongly bonded that they lack even fair cleavage. Quartz, a framework silicate, is so strongly bonded in all directions that it breaks only along irregular surfaces. Garnet, a silicate with isolated tetrahedra, is also bonded strongly in all directions and so shows no cleavage. This absence of a tendency to cleave is found in most framework and isolated tetrahedral silicates.

Fracture

Fracture is the tendency of a crystal to break along irregular surfaces other than cleavage planes. All minerals show fracture, either across cleavage planes or—in such minerals as quartz—with no cleavage in any direction. Fracture is related to how bond strengths are distributed in directions that cut across cleavage planes. Fractures may be *conchoidal*, showing smooth, curved surfaces like those of a thick piece of broken glass. Another common fracture surface with an appearance like split wood is described as *fibrous* or *splintery*. The shapes and appearances of fracture surfaces depend on the particular structure and composition of the mineral.

Luster

The way the surface of a mineral reflects light gives it a characteristic **luster.** Mineral lusters are described by the terms listed in Table 3.3. Luster is controlled by the kinds of atoms present and their bonding, both of which affect the way light passes through or is reflected by the mineral. Ionically bonded crystals tend to have a glassy, or vitreous, luster, but covalently bonded materials are more variable.

Many have an adamantine luster, like that of diamond. Pure metals, such as gold, and many sulfides, such as galena (lead sulfide, PbS), have a metallic luster. A pearly luster results from multiple reflections of light from planes beneath the surfaces of translucent minerals, such as the mother-of-pearl inner surfaces of many clamshells, which are made of the mineral aragonite. Luster, although an important criterion for field classification, depends heavily on the visual perception of reflected light. Textbook descriptions fall short of the actual experience of holding the mineral in your hand.

Color

The **color** of a mineral is imparted by light, either transmitted through or reflected by crystals or irregular masses of the mineral. The color of a mineral may be distinctive, but it is not the most reliable clue to its identity. Some minerals always show the same color; others may have a range of colors. Many minerals show a characteristic color only on freshly broken surfaces or only on weathered surfaces. Some—precious opals, for example—show a stunning display of colors on reflecting surfaces. Others change color slightly with a change in the angle of the light shining on their surfaces. Many ionically bonded crystals are colorless.

Streak refers to the color of the fine deposit of mineral powder left on an abrasive surface, such as a tile of unglazed porcelain, when a mineral is scraped across it. Such a tile, called a *streak plate* (Figure 3.20), is a good identification tool because the uniformly small grains of the mineral that are present in the powder are revealed on the plate. Hematite, for example, may look black, red, or brown, but this

TABLE 3.3	Mineral Lusters
LUSTER	CHARACTERISTICS
Metallic	Strong reflections produced by opaque substances
Vitreous	Bright, as in glass
Resinous	Characteristic of resins, such as amber
Greasy	The appearance of being coated with an oily substance
Pearly	The whitish iridescence of such materials as pearl
Silky	The sheen of fibrous materials such as silk
Adamantine	The brilliant luster of diamond and similar minerals

FIGURE 3.20 Hematite may be black, red, or brown, but it always leaves a reddish brown streak when scraped along a ceramic streak plate. [Breck P. Kent.]

mineral will always leave a trail of reddish brown powder on a streak plate.

Color is a complex and not yet fully understood property of minerals. It is determined both by the kinds of ions found in the pure mineral and by trace elements.

Ions and Mineral Color The color of pure minerals depends on the presence of certain ions, such as iron or chromium, that strongly absorb portions of the light spectrum. Olivine that contains iron, for example, absorbs all colors except green, which it reflects, so we see this type of olivine as green. We see pure magnesium olivine as white (transparent and colorless).

Trace Elements and Mineral Color All minerals contain impurities. Instruments can now measure even very small quantities of some elements—as little as a billionth of a gram in some cases. Elements that make up much less than 0.1 percent of a mineral are referred to as **trace elements.**

Some trace elements can be used to deduce the origins of the minerals in which they are found. Others, such as the traces of uranium in some granites, contribute to local natural radioactivity. Still others, such as small dispersed flakes of hematite that color a feldspar crystal brownish or reddish, are notable because they give a general color to an otherwise colorless mineral. Many of the gem varieties of minerals, such as emerald (green beryl) and sapphire (blue corundum), get their color from trace elements dissolved in the solid crystal (see **Figure 3.21**). Emerald derives its color from chromium; the sources of sapphire's blue color are iron and titanium.

FIGURE 3.21 Trace elements give gems their colors. Sapphire (*left*) and ruby (*center*) are formed of the same common mineral, corundum (aluminum oxide). Small amounts of impurities produce the intense colors that we value. Ruby, for example, is red because of small amounts of chromium, the same element that gives emerald (*right*) its green color. [John Grotzinger/Ramón Rivera-Moret/Harvard Mineralogical Museum.]

Density

You can easily feel the difference in weight between a piece of hematite and a piece of sulfur of the same size by lifting the two pieces. A great many common rock-forming minerals, however, are too similar in density for such a simple test. Scientists therefore need some easy method to measure this property of minerals. A standard measure of density is **specific gravity,** which is the weight of a mineral divided by the weight of an equal volume of pure water at 4°C.

Density depends on the atomic mass of a mineral's atoms or ions and how closely they are packed in its crystal structure. Consider magnetite, an iron oxide with a density of 5.2 g/cm^3. Its high density results partly from the high atomic mass of iron and partly from the closely packed structure that magnetite shares with the other members of the spinel group of oxides. The density of iron olivine, at 4.4 g/cm^3, is lower than that of magnetite for two reasons. First, the atomic mass of silicon, one of the elements that make up olivines, is lower than that of iron. Second, iron olivine has a more openly packed structure than minerals of the spinel group. The density of magnesium olivine is even lower, at 3.32 g/cm^3, because magnesium's atomic mass is much lower than that of iron.

Increases in density caused by increases in pressure affect the way minerals transmit light, heat, and seismic waves. Experiments at extremely high pressures have shown that the structure of olivine converts into the denser structure of spinel olivine at pressures corresponding to a depth in Earth's mantle of 410 km. At a greater depth, 660 km, mantle materials are further transformed into silicate minerals with the even more densely packed structure of perovskite olivine. Because of the huge volume of the lower mantle, perovskite olivine is probably the most abundant mineral in Earth as a whole. Temperature also affects density: the higher the temperature, the more open and expanded the structure of the mineral, and thus the lower its density.

Crystal Habit

A mineral's **crystal habit** is the shape in which individual crystals or aggregates of crystals grow. Some minerals have such a distinctive crystal habit that they are easily recognizable. An example is quartz, with its six-sided column topped by a pyramid-like set of faces (see Figure 3.7). Crystal habits are often named after common geometric shapes, such as blades, plates, and needles. These shapes indicate not only the planes of the mineral's crystal structure, but also the typical speed and direction of crystal growth. Thus, a needlelike crystal is one that grows very quickly in one direction and very slowly in all other directions. In contrast, a plate-shaped crystal (often referred to as *platy*) grows fast in all directions that are perpendicular to its single direction of slow growth. *Fibrous* crystals take shape as multiple long, narrow fibers, essentially aggregates of long needles. *Asbestos* is a generic name for a group of silicate minerals with a more or less

FIGURE 3.22 Chrysotile, a type of asbestos. Fibers are readily combed from the solid mineral. [Courtesy of Eurico Zimbres.]

fibrous habit that allows the crystals to become embedded in the lungs if they are inhaled (**Figure 3.22**).

Table 3.4 summarizes the physical properties of minerals that we have discussed in this section.

What Are Rocks?

A geologist's primary aim is to understand the properties of rocks and to deduce their geologic origins from those properties. Such deductions further our understanding of our planet, and they also provide important information about economically important resources. For example, knowing that oil forms in certain kinds of sedimentary rocks that are rich in organic matter allows us to explore for oil reserves more intelligently. Understanding how rocks form also guides us in solving environmental problems. For example, the underground storage of radioactive and other wastes depends on analysis of the rock to be used as a repository: Will this rock be prone to earthquake-triggered landslides? How might it transmit polluted waters in the ground?

Properties of Rocks

A **rock** is a naturally occurring solid aggregate of minerals or, in some cases, nonmineral solid matter. In an *aggregate*, minerals are joined in such a way that they retain their individual identity (**Figure 3.23**). A few rocks are composed of nonmineral matter. These rocks include the noncrystalline, glassy volcanic rocks obsidian and pumice as well as coal, which is made up of compacted plant remains.

What determines the physical appearance of a rock? Rocks vary in color, in the sizes of their crystals or grains, and in the kinds of minerals that compose them. Along a road cut, for example, we might find a rough white and pink speckled rock composed of interlocking crystals large enough to be seen with the naked eye. Nearby, we might see a grayish rock containing many large, glittering crystals of mica and some grains of quartz and feldspar. Overlying both the white and pink rock as well as the gray one, we might see horizontal layers of a striped white and mauve rock that appear to be made up of sand grains cemented together. And these rocks might all be overlain by a dark, fine-grained rock with tiny white dots in it.

The identity of a rock is determined partly by its mineralogy and partly by its texture. Here, the term *mineralogy* refers to the relative proportions of a rock's

TABLE 3.4	Physical Properties of Minerals
PROPERTY	**RELATION TO COMPOSITION AND CRYSTAL STRUCTURE**
Hardness	Strong chemical bonds result in hard minerals. Covalently bonded minerals are generally harder than ionically bonded minerals.
Cleavage	Cleavage is poor if bonds in crystal structure are strong, good if bonds are weak. Covalent bonds generally give poor or no cleavage; ionic bonds are weaker and so give good cleavage.
Fracture	Related to distribution of bond strengths across irregular surfaces other than cleavage planes.
Luster	Tends to be glassy for ionically bonded crystals, more variable for covalently bonded crystals.
Color	Determined by ions and trace elements. Many ionically bonded crystals are colorless. Iron tends to color strongly.
Streak	Color of fine mineral powder is more characteristic than that of massive mineral because of uniformly small size of grains.
Density	Depends on atomic weight of atoms or ions and their closeness of packing in crystal structure.
Crystal habit	Depends on the planes of a mineral's crystal structure and the typical speed and direction of crystal growth.

FIGURE 3.23 Rocks are naturally occurring aggregates of minerals. [John Grotzinger/Ramón Rivera-Moret/Harvard Mineralogical Museum.]

constituent minerals. **Texture** describes the sizes and shapes of a rock's mineral crystals or grains and the way they are put together. If the crystals or grains, which are only a few millimeters in diameter in most rocks, are large enough to be seen with the naked eye, the rock is categorized as *coarse-grained*. If they are not large enough to be seen, the rock is categorized as *fine-grained*. The mineralogy and texture that determine a rock's appearance are themselves determined by the rock's geologic origin—where and how it formed (**Figure 3.24**).

The dark rock that caps the sequence of rocks in our road cut, called basalt, was formed by a volcanic eruption. Its mineralogy and texture were determined by the chemical composition of rocks that were melted deep within Earth. All rocks formed by the solidification of molten rock, such as basalt and granite, are called **igneous rocks.**

The striped white and mauve layers in the road cut are sandstone, formed as sand particles accumulated, perhaps on an ancient beach, and eventually were covered over, buried, and cemented together. All rocks formed as the burial products of layers of sediments (such as sand, mud, or the calcium carbonate shells of marine organisms), whether they were laid down on land or under the sea, are called **sedimentary rocks.**

The grayish rock of our road cut, a gneiss, contains crystals of mica, quartz, and feldspar. It formed deep in Earth's crust as high temperatures and pressures transformed the mineralogy and texture of buried sedimentary rock. All rocks formed by the transformation of preexisting solid rock under the influence of high temperatures and pressures are called **metamorphic rocks.**

The four types of rocks seen in our road cut represent the three great families of rock: igneous, sedimentary, and metamorphic. Let's take a closer look at each of these families and at the geologic processes that form them.

Igneous Rocks

Igneous rocks (from the Latin *ignis*, meaning "fire") form by crystallization from magma. When a body of magma cools slowly in Earth's interior, the minerals it contains begin to

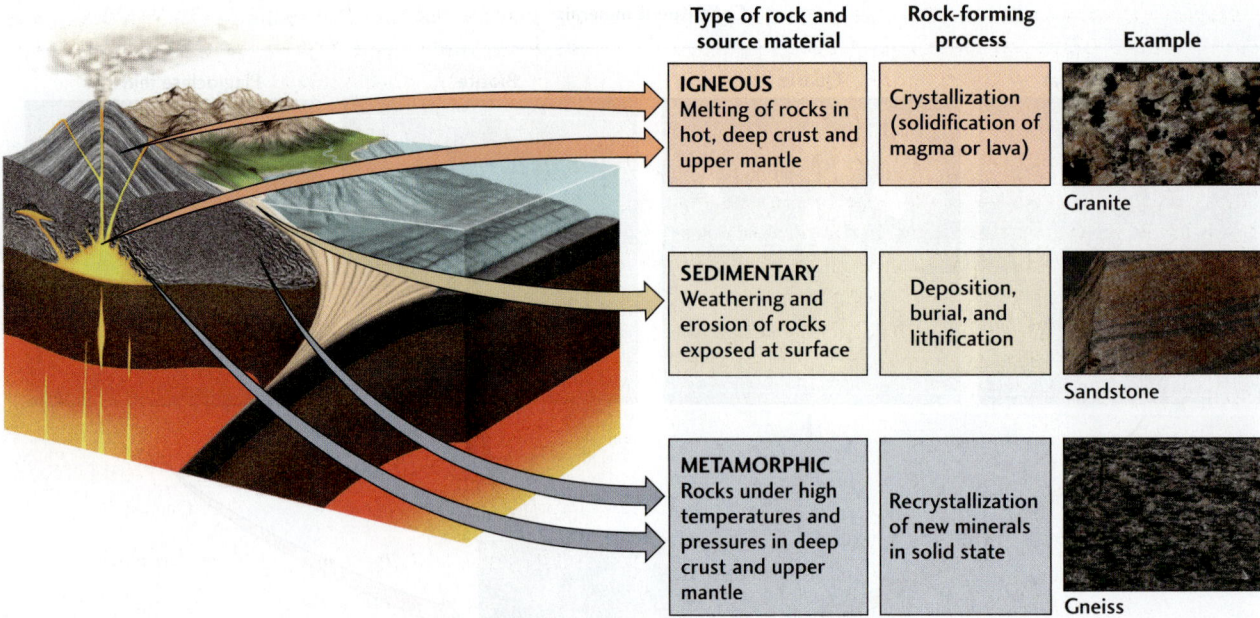

FIGURE 3.24 The three families of rocks are formed in different geologic environments by different geologic processes. [John Grotzinger/Ramón Rivera-Moret/Harvard Mineralogical Museum.]

form microscopic crystals. As the magma cools below its melting point, some of these crystals have time to grow to several millimeters in diameter or larger before the whole mass crystallizes as a coarse-grained igneous rock. But when magma erupts from a volcano onto Earth's surface as lava, it cools and solidifies so rapidly that individual crystals have no time to grow gradually. In that case, many tiny crystals form simultaneously, and the result is a fine-grained igneous rock. Geologists distinguish two major types of igneous rocks—intrusive and extrusive—on the basis of the sizes of their crystals.

Intrusive and Extrusive Igneous Rocks *Intrusive igneous rocks* crystallize when magma intrudes into unmelted rock masses deep in Earth's crust. Large crystals grow as the magma slowly cools, producing coarse-grained rocks. Intrusive igneous rocks can be recognized by their large, interlocking crystals (**Figure 3.25**). Granite is an intrusive igneous rock.

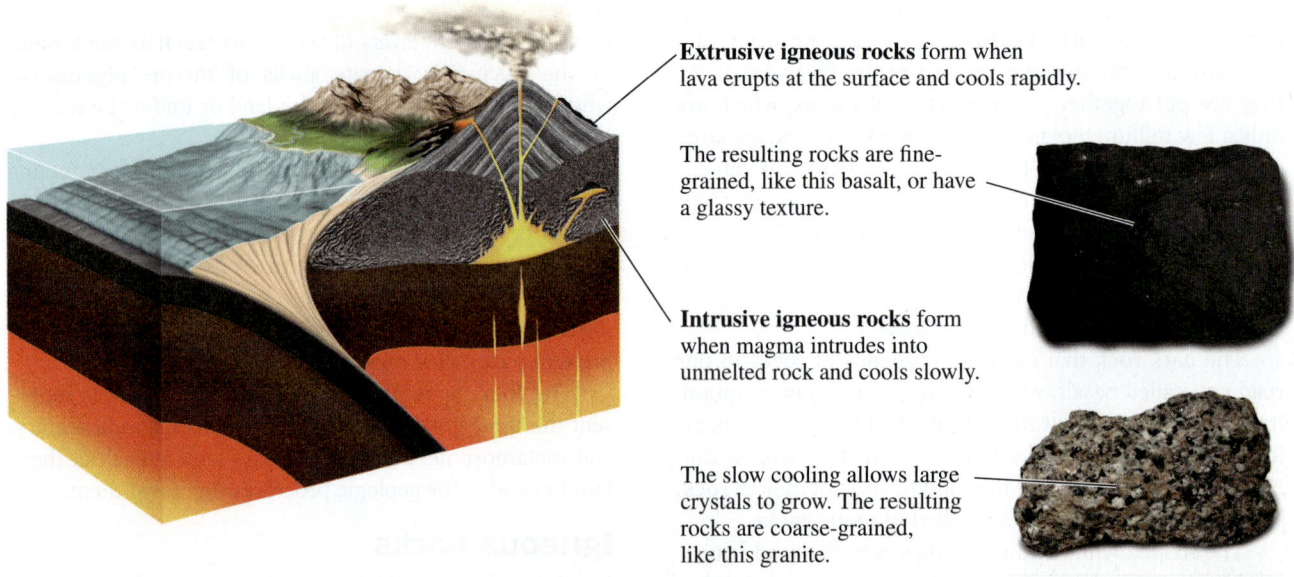

FIGURE 3.25 Igneous rocks are formed by the crystallization of magma. [John Grotzinger/Ramón Rivera-Moret/Harvard Mineralogical Museum.]

Extrusive igneous rocks form from magmas that erupt at Earth's surface as lava and cool rapidly. Extrusive igneous rocks, such as basalt, are easily recognized by their glassy or fine-grained texture.

Common Minerals of Igneous Rocks Most of the minerals of igneous rocks are silicates, partly because silicon is so abundant in Earth's crust and partly because many silicate minerals melt at the high temperatures and pressures reached in deeper parts of the crust and in the mantle. The silicate minerals most commonly found in igneous rocks include quartz, feldspars, micas, pyroxenes, amphiboles, and olivines (Table 3.5).

Sedimentary Rocks

Sediments, the precursors of sedimentary rocks, are found at Earth's surface as layers of loose particles, such as sand, silt, and the shells of organisms. These particles originate in the processes of weathering and erosion. **Weathering** refers to all of the chemical and physical processes that break up and decay rocks into fragments and dissolved substances of various sizes. These particles are then transported by **erosion,** the set of processes that loosen soil and rock and move them downhill or downstream to the spot where they are deposited as layers of sediment (Figure 3.26).

TABLE 3.5 Some Common Minerals of Igneous, Sedimentary, and Metamorphic Rocks

IGNEOUS ROCKS	SEDIMENTARY ROCKS	METAMORPHIC ROCKS
Quartz	Quartz	Quartz
Feldspar	Clay minerals	Feldspar
Mica	Feldspar	Mica
Pyroxene	*Calcite	Garnet
Amphibole	*Dolomite	Pyroxene
Olivine	*Gypsum	Staurolite
	*Halite	Kyanite

*Nonsilicate minerals.

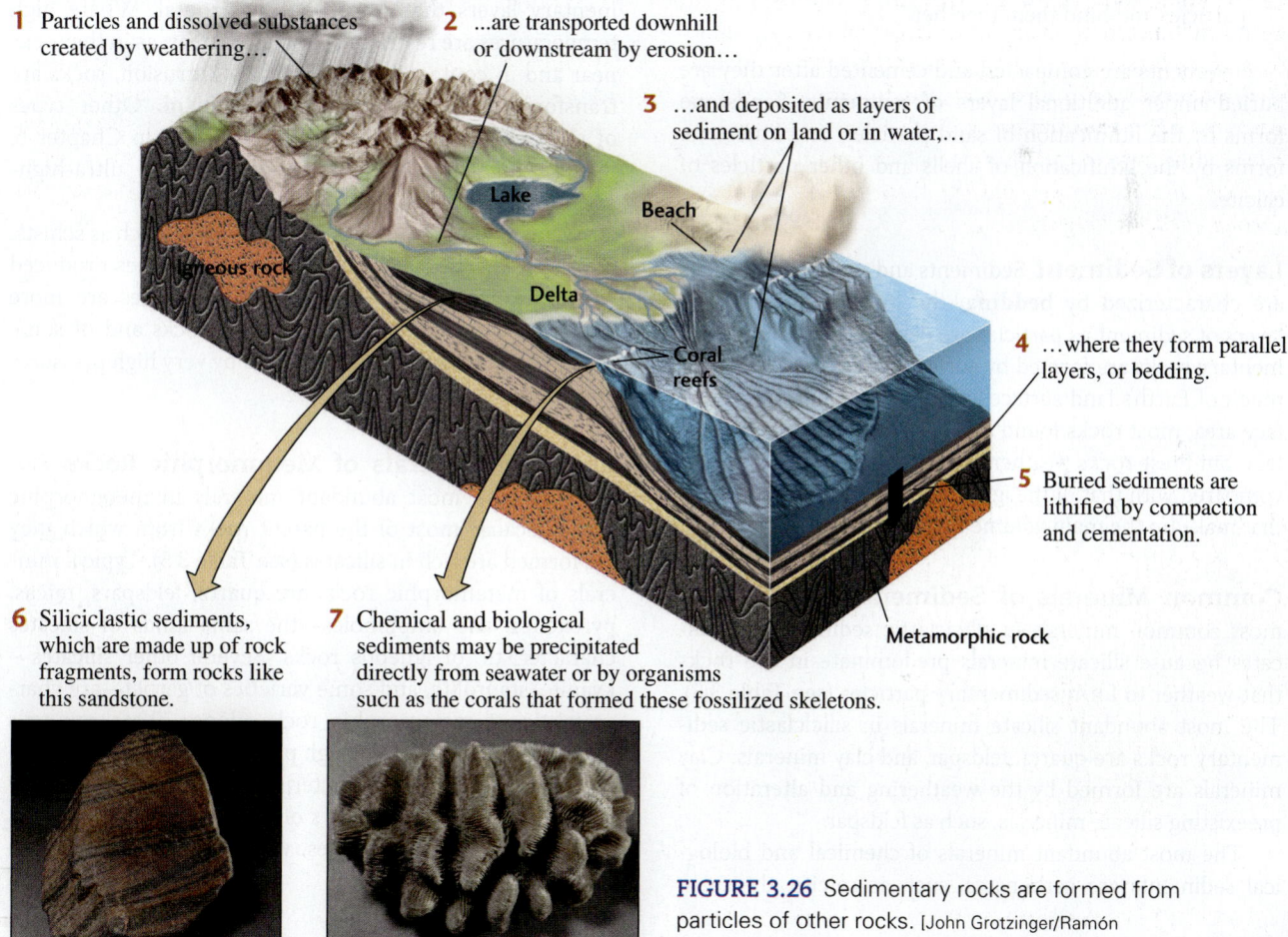

FIGURE 3.26 Sedimentary rocks are formed from particles of other rocks. [John Grotzinger/Ramón Rivera-Moret/MIT.]

Sediments are deposited in two ways:

- **Siliciclastic sediments** are made up of physically deposited particles, such as grains of quartz and feldspar derived from weathered granite. (*Clastic* is derived from the Greek word *klastos*, meaning "broken.") These sediments are laid down by running water, wind, and ice.

- **Chemical sediments** and **biological sediments** are new chemical substances that form by precipitation. Weathering dissolves some of a rock's components, which are carried in stream waters to the ocean. Halite is a chemical sediment that precipitates directly from evaporating seawater. Calcite is precipitated by marine organisms to form shells or skeletons, which form biological sediments when the organisms die.

From Sediment to Solid Rock Lithification is the process that converts sediments into solid rock. It occurs in two ways:

- In *compaction*, particles are squeezed together by the weight of overlying sediments into a mass denser than the original.
- In *cementation*, minerals precipitate around deposited particles and bind them together.

Sediments are compacted and cemented after they are buried under additional layers of sediments. Sandstone forms by the lithification of sand particles, and limestone forms by the lithification of shells and other particles of calcite.

Layers of Sediment Sediments and sedimentary rocks are characterized by **bedding,** the formation of parallel layers of sediment as particles are deposited. Because sedimentary rocks are formed by surface processes, they cover much of Earth's land surface and seafloor. In terms of surface area, most rocks found at Earth's surface are sedimentary, but these rocks weather easily, so their volume is small compared with that of the igneous and metamorphic rocks that make up the main volume of the crust.

Common Minerals of Sedimentary Rocks The most common minerals in siliciclastic sediments are silicates because silicate minerals predominate in the rocks that weather to form sedimentary particles (see Table 3.5). The most abundant silicate minerals in siliciclastic sedimentary rocks are quartz, feldspar, and clay minerals. Clay minerals are formed by the weathering and alteration of preexisting silicate minerals, such as feldspar.

The most abundant minerals of chemical and biological sediments are carbonates, such as calcite, the main constituent of limestone. Dolomite is a calcium-magnesium carbonate formed by precipitation during lithification. Two other chemical sediments—gypsum and halite—form by precipitation as seawater evaporates.

Metamorphic Rocks

Metamorphic rocks take their name from the Greek words for "change" (*meta*) and "form" (*morphe*). These rocks are produced when high temperatures and pressures deep within Earth cause changes in the mineralogy, texture, or chemical composition of any kind of preexisting rock—igneous, sedimentary, or other metamorphic rock—while maintaining its solid form. The temperatures of metamorphism are below the melting point of the rocks (about 700°C), but high enough (above 250°C) for the rocks to be changed by recrystallization and chemical reactions.

Regional and Contact Metamorphism Metamorphism may take place over a widespread area or a limited one (**Figure 3.27**). **Regional metamorphism** occurs where high pressures and temperatures extend over large regions, as happens where plates collide. Regional metamorphism accompanies plate collisions that result in mountain building and the folding and breaking of sedimentary layers that were once horizontal. Where high temperatures are restricted to smaller areas, as in the rocks near and in contact with a magmatic intrusion, rocks are transformed by **contact metamorphism.** Other types of metamorphism, which we will describe in Chapter 6, include high-pressure metamorphism and ultra-high-pressure metamorphism.

Many regionally metamorphosed rocks, such as schists, have characteristic *foliation,* wavy or flat planes produced when the rock was folded. Granular textures are more typical of most contact metamorphic rocks and of some regional metamorphic rocks formed by very high pressures and temperatures.

Common Minerals of Metamorphic Rocks Silicates are the most abundant minerals in metamorphic rocks because most of the parent rocks from which they are formed are rich in silicates (see Table 3.5). Typical minerals of metamorphic rocks are quartz, feldspars, micas, pyroxenes, and amphiboles—the same kinds of silicates characteristic of igneous rocks. Several other silicates—kyanite, staurolite, and some varieties of garnet—are characteristic of metamorphic rocks alone. These minerals form under conditions of high pressure and temperature in the crust and are not characteristic of igneous rocks. They are therefore good indicators of metamorphism. Calcite is the main mineral of marbles, which are metamorphosed limestones.

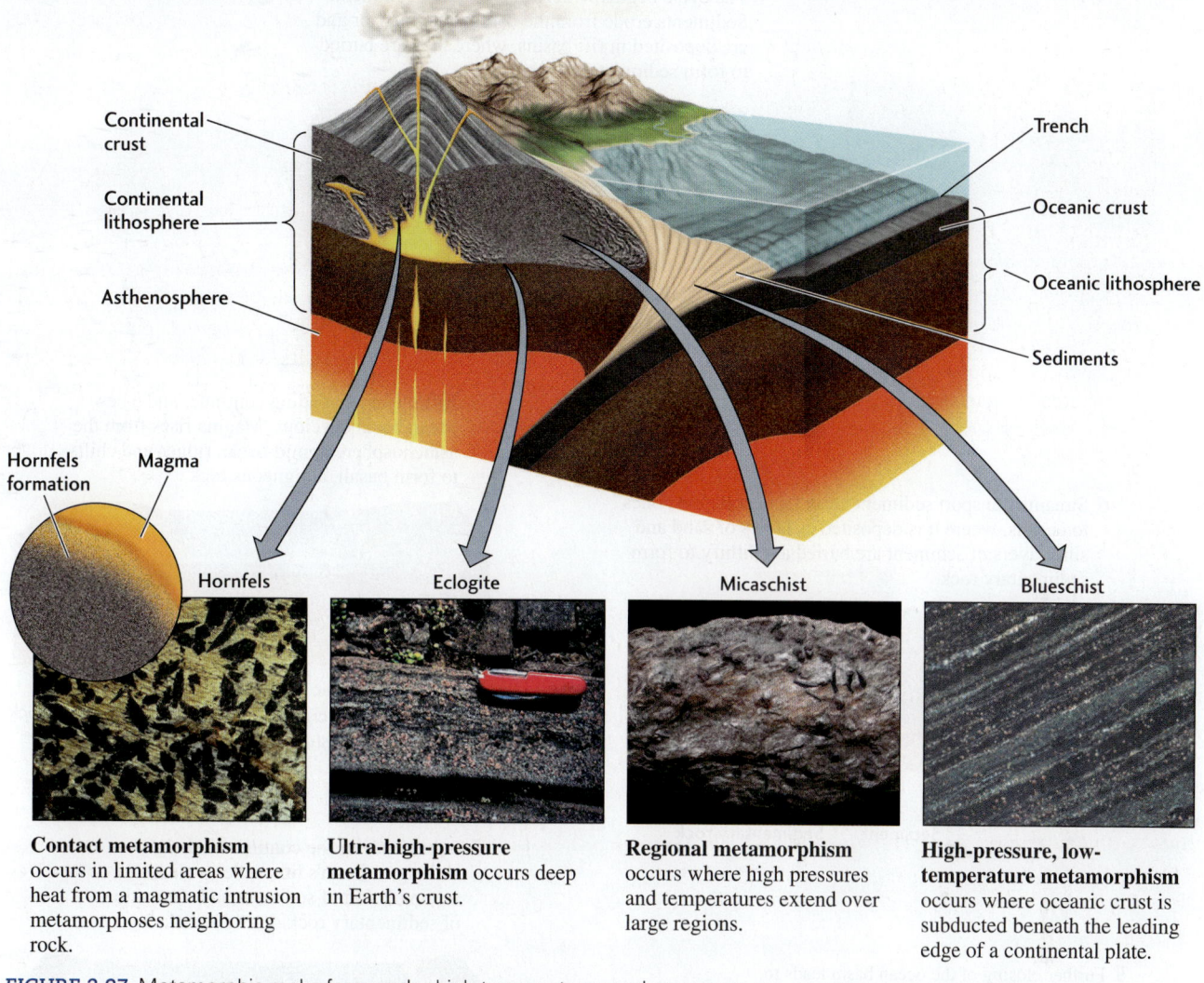

FIGURE 3.27 Metamorphic rocks form under high temperatures and pressures. [*hornfels*: Biophoto Associates/Science Source; *eclogite*: Courtesy of Julie Baldwin; *micaschist*: John Grotzinger; *blueschist*: Courtesy of Mark Cloos.]

The Rock Cycle: Interactions Between the Plate Tectonic and Climate Systems

Earth scientists have known for over 200 years that the three families of rocks—igneous, metamorphic, and sedimentary—all can evolve from one to another. Their observations gave rise to the concept of a **rock cycle,** which explains how each type of rock is transformed into one of the other two types. The rock cycle is now known to be the result of interactions between two of the three global geosystems: the plate tectonic system and the climate system. Interactions between these two geosystems drive transfers of materials and energy among Earth's interior, the land surface, the ocean, and the atmosphere. For example, the formation of magmas at subduction zones results from processes operating within the plate tectonic system. When these magmas erupt, materials and energy are transferred to the land surface, where the materials (newly formed rocks) are subject to weathering by the climate system. The eruption process also injects volcanic ash and carbon dioxide gas high into the atmosphere, where they may affect global climates. As global climates change, perhaps becoming warmer or cooler, the rate of weathering changes, which in turn influences the rate at which materials (sediments) are returned to Earth's interior.

Let's trace one turn of the rock cycle, beginning with the creation of new oceanic lithosphere at a mid-ocean ridge spreading center as two continents drift apart (**Figure 3.28**).

1 The cycle begins with rifting within a continent. Sediments erode from the continental interior and are deposited in rift basins, where they are buried to form sedimentary rocks.

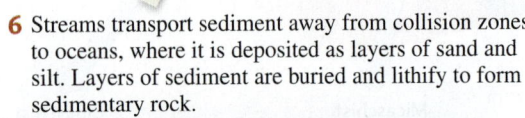

2 Rifting and spreading continue, and a new ocean basin develops. Magma rises from the asthenosphere at mid-ocean ridges and chills to form basalt, an igneous rock.

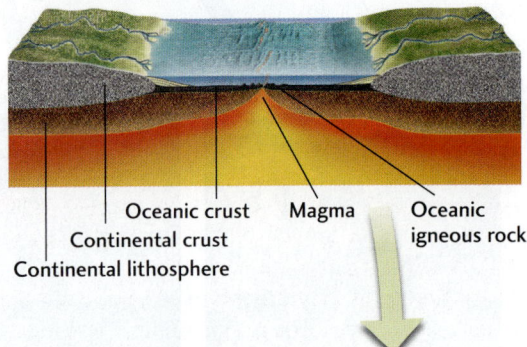

6 Streams transport sediment away from collision zones to oceans, where it is deposited as layers of sand and silt. Layers of sediment are buried and lithify to form sedimentary rock.

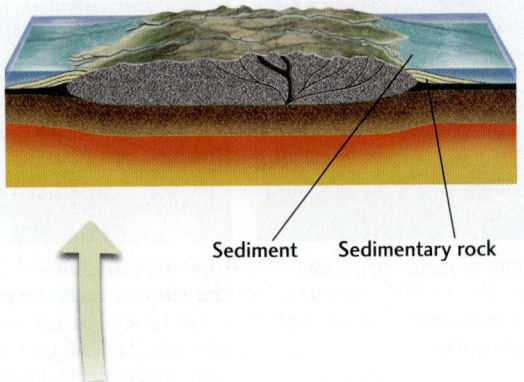

3 Subsidence of the continental margin—sinking of Earth's lithosphere—leads to accumulation of sediment and formation of sedimentary rock during burial.

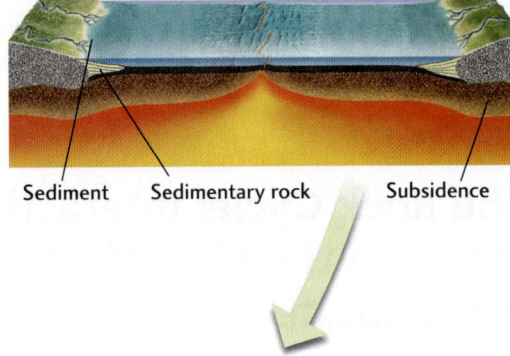

5 Further closing of the ocean basin leads to continental collision, forming high mountain ranges. Where continents collide, rocks are buried deeper or modified by heat and pressure, forming metamorphic rocks. Uplifted mountains force moisture-laden air to rise, cool, and release its moisture as precipitation. Weathering creates loose material—soils and sediment—that erosion strips away.

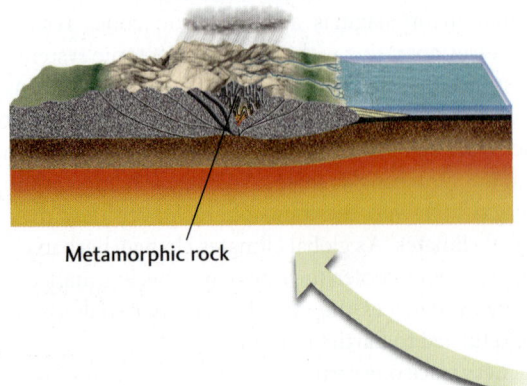

4 Oceanic crust subducts beneath a continent, building a volcanic mountain chain. The subducting plate melts as it descends. Magma rises from the melting plate and mantle and cools to make granitic igneous rocks.

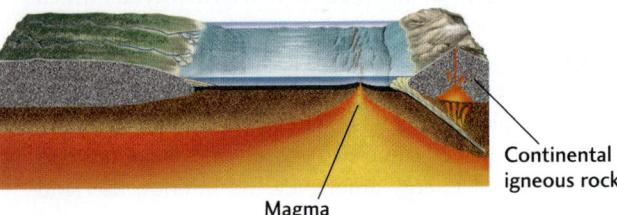

FIGURE 3.28 The rock cycle results from the interaction of the plate tectonic and climate systems.

The ocean gets wider and wider, until at some point the process reverses itself and the ocean closes. As the ocean basin closes, igneous rocks created at the mid-ocean ridge are eventually subducted beneath a continent. Sediments that were formed on the continent and deposited at its edge may also be dragged down into the subduction zone. Ultimately, the two continents, which were once drifting apart, may collide. As the igneous rocks and sediments that descend into the subduction zone go deeper and deeper into Earth's interior, they begin to melt to form a new generation of igneous rocks. The great heat associated with the intrusion of these igneous rocks, coupled with the heat and pressure that come with being pushed to levels deep within Earth, transforms these igneous rocks—and other surrounding rocks—into metamorphic rocks. When the continents collide, these igneous and metamorphic rocks are uplifted into a high mountain chain as a section of Earth's crust crumples, deforms, and undergoes further metamorphism.

The rocks of the uplifted mountains are exposed to the influences of the climate system, but they affect the climate system in turn, forcing moving air to rise, cool, and release precipitation. The rocks are slowly weathered, forming loose materials that erosion then strips away. Water and wind transport some of these materials across the continent and eventually to the edges of the continent, where they are deposited as sediments. The sediments laid down where the land meets the ocean are buried under successive layers of sediments, where they slowly lithify into sedimentary rock. These oceans, like those mentioned at the beginning of the cycle, were probably formed by seafloor spreading along mid-ocean ridges, thus completing the rock cycle.

The particular pathway illustrated here—that of a continent breaking apart, forming a new ocean basin, then closing back up again—is only one variation among many that may take place in the rock cycle. Any type of rock—igneous, sedimentary, or metamorphic—can be uplifted during a mountain-building event and then weathered and eroded to form new sediments. Some stages may be omitted: As a sedimentary rock is uplifted and eroded, for example, metamorphism and melting are skipped. In some cases, the rock cycle proceeds very slowly. For example, we know that some igneous and metamorphic rocks many kilometers deep in the crust may be uplifted or exposed to weathering and erosion only after billions of years have passed.

The rock cycle never ends. It is always operating at different stages in various parts of the world, forming and eroding mountains in one place and laying down and burying sediments in another. The rocks that make up the solid Earth are recycled continuously, but we can see only the surface parts of the cycle. We must deduce the recycling of the deep crust and the mantle from indirect evidence.

Concentrations of Valuable Mineral Resources

The rock cycle turns out to be crucial in creating economically important concentrations of the many valuable minerals found in Earth's crust. Minerals are not only sources of metals, which will be our focus here, but also provide us with stone for buildings and roads, phosphates for fertilizers, cement for construction, clays for ceramics, sand for silicon chips and fiber-optic cables, and many other items we use in our daily lives. Finding these minerals and extracting them is a vital job for Earth scientists, so we turn our attention next to how and where some of these geologic prizes are formed.

The chemical elements of Earth's crust are widely distributed in many kinds of minerals, and those minerals are found in a great variety of rocks. In most places, any given element will be found homogenized with other elements in amounts close to its average concentration in the crust. An ordinary granitic rock, for example, may contain a small percentage of iron, close to the average concentration of iron in Earth's crust.

When an element is present in concentrations higher than the average, it means that the rock underwent some geologic process that concentrated larger quantities of that element than normal. The *concentration factor* of an element in a mineral deposit is the ratio of the element's abundance in the deposit to its average abundance in the crust. High concentrations of elements are found in a limited number of specific geologic settings. These settings are of economic interest because the higher the concentration of a resource in a given deposit, the lower the cost to recover it.

Ores are rich deposits of minerals from which valuable metals can be recovered profitably (see the Practicing Geology Exercise). The minerals containing these metals are referred to as *ore minerals*. Ore minerals include sulfides (the largest group), oxides, and silicates. The ore minerals in each of these groups are compounds of metallic elements with sulfur, with oxygen, and with silicon and oxygen, respectively. The copper ore mineral covellite, for example, is a copper sulfide (CuS). The iron ore mineral hematite (Fe_2O_3) is an iron oxide. The nickel ore mineral garnierite is a nickel silicate ($Ni_3Si_2O_5(OH)_4$). In addition, some metals, such as gold, are found in their

PRACTICING GEOLOGY EXERCISE

Is It Worth Mining?

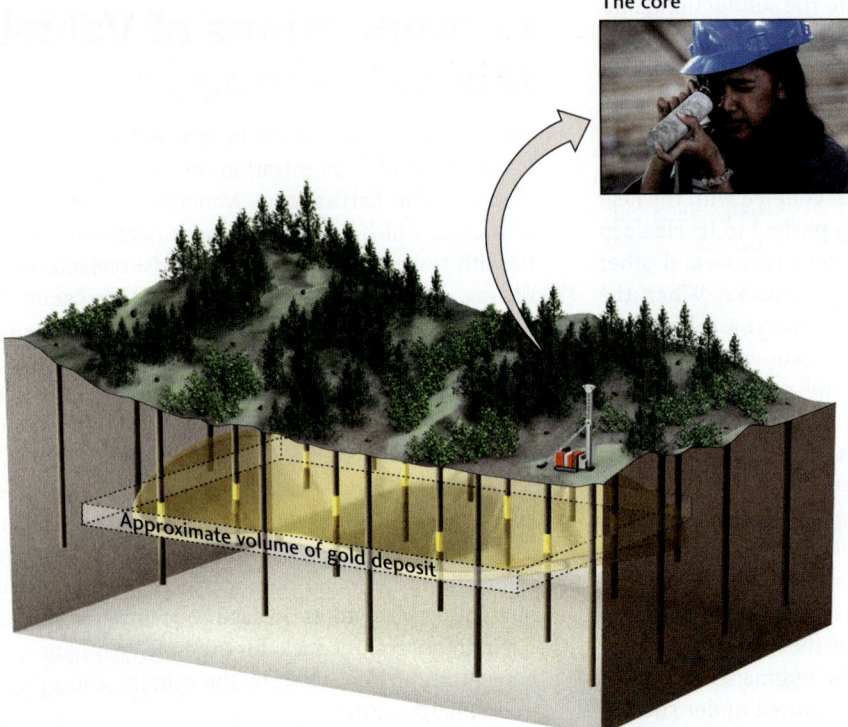

An ore deposit is drilled to provide core samples for geochemical and mineralogical analysis. A rotating metal tube, studded with diamond teeth, cuts into the deposit. The hollow space in the tube becomes filled with solid rock, which is extracted when the tube is pulled out of the rock. The core has the shape of a cylinder. [drill rig: SPL/Science Source; core: Dadang Tri/Bloomberg via Getty Images.]

Geologists employed by the Rocks-r-Us Corporation have discovered basaltic volcanic rocks laced with gold. The corporate executives ponder their figures and measurements and study their three-dimensional model of the ore mineral deposit, but in the end they have just one question: Should we open a mine?

Exploration for ore minerals is an important and challenging activity that employs many geologists. Finding a promising deposit is only the first step toward extracting useful materials, however. The shape of the deposit, and the distribution and concentration of the ore, must be estimated before mining begins. This is done by drilling closely spaced holes and obtaining continuous cores through the ore deposit and the surrounding rock. Information from the cores is used to create a three-dimensional model of the ore deposit. That model is then used to evaluate whether or not the deposit is large enough and has a high enough concentration of minerals to justify opening a mine. Geologists contribute key information of direct economic significance to this very practical decision-making process.

The planning of mining operations is typically based on chemical and mineralogical analyses of the extracted cores, from which two quantities are calculated:

▶ *Grade* refers to the concentration of ore minerals within economically valueless parent rock (referred to as *waste rock*).

▶ *Mass* refers to the amount of ore that could potentially be extracted from the deposit.

Both quantities are important because neither grade nor mass alone is sufficient to identify an economically valuable deposit. For example, the grade could be locally very high in veins, but the overall mass could be low because the veins are rare. In another case, the mass might be high, but the ore minerals might be so dispersed within the waste rock that the costs of processing it to extract the ore would become too high. Thus, the ideal ore deposit is one that has both high grade and high mass.

Grade is calculated by determining the percentage of ore minerals within a volume of rock. Laboratory analysis of core samples provides this measurement. Mass is calculated by assigning the grade value determined for individual cores to the unknown volume of rock between drill holes. Mass is the amount of ore that could be extracted if all of it could be extracted from the rock, but it is rare that all of it can be. In the parlance of the mining industry, mass is often calculated in tons and referred to as *tonnage* because of the enormous volumes of rock that are involved.

Drilling and analysis of core samples has shown that the gold in the Rocks-r-Us deposit has an average grade of 0.02 percent across all cores. The deposit has been determined to have a rectangular geometry extending laterally for 50 meters in one direction and 1500 meters in the other, with a thickness of 2 meters.

What is the volume of the ore deposit?

$$V_{deposit} = length \times width \times thickness$$
$$= 50 \text{ m} \times 1500 \text{ m} \times 2 \text{ m}$$
$$= 150,000 \text{ m}^3$$

What is the volume of gold in the ore deposit?

$$V_{gold} = V_{deposit} \times grade$$
$$= 150,000 \text{ m}^3 \times 0.02\%$$
$$= 30 \text{ m}^3$$

Given that gold has a density of 19 g/cm³ (about 6800 ounces/m³), what is the mass of the gold in ounces?

$$mass = V_{gold} \times density$$
$$= 30 \text{ m}^3 \times 6800 \text{ ounces/m}^3$$
$$= 204,000 \text{ ounces}$$

The price of gold is typically reported in dollars per ounce. At the time of this writing, the price of gold is roughly $800/ounce. What is the potential value of this ore deposit?

$$value = mass \times price$$
$$= 204,000 \text{ ounces} \times \$800/\text{ounce}$$
$$= \$163,200,000$$

BONUS PROBLEM: You bring this information to a meeting with the Rocks-r-Us corporate executives. They calculate that, over its lifetime, the mine will cost about $120,000,000 to operate, including restoration of the land after mining is completed. Is the value of the gold worth it? What simple calculation will give you the answer?

native state—that is, uncombined with other elements (**Figure 3.29**).

Hydrothermal Deposits

Many of the most valuable ores are formed in regions of volcanism by the interaction of igneous processes with the hydrosphere. Recall from our discussion of the rock cycle that subduction zones may be associated with the melting of oceanic lithosphere to form igneous rocks. Very large ore deposits can be formed in such plate tectonic settings when hot water solutions—also known as **hydrothermal solutions**—are formed around bodies of molten rock.

(a)

(b)

FIGURE 3.29 Some metals are found in their native state. (a) A geologist examines rock samples in an underground gold mine in Zimbabwe, in southern Africa. (b) Native gold on a quartz crystal. [(a) Peter Bowater/Science Source; (b) 97-35023 by Chip Clark, Smithsonian.]

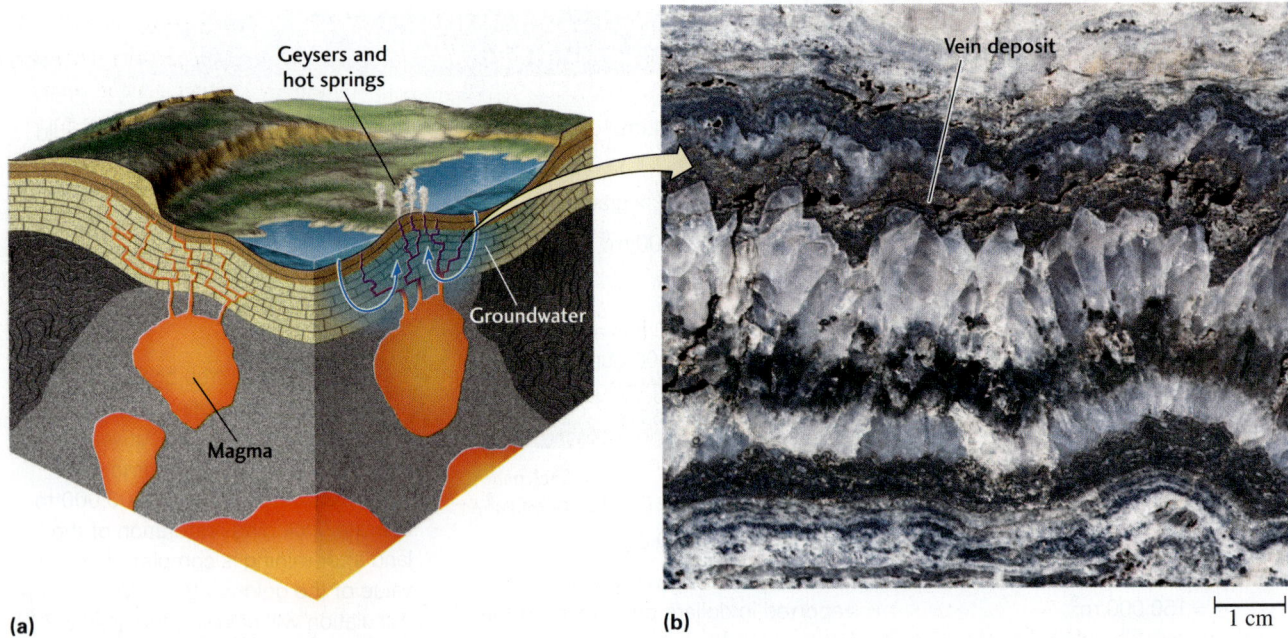

FIGURE 3.30 Many deposits of ore minerals are found in veins formed by hydrothermal solutions. (a) Groundwater percolating through fractured rock dissolves metal oxides and sulfides. When heated by a magmatic intrusion, it rises, precipitating metallic ores in the rock fractures. (b) This quartz vein deposit (about 1 cm thick) in Oatman, Arizona, which contains gold and silver ores, was formed by such a process. [Peter Kresan.]

This happens when circulating groundwater or seawater comes into contact with a magmatic intrusion, reacts with it, and carries off significant quantities of elements and ions released by the reaction. These elements and ions then interact with one another to form ore minerals, usually as the solution cools.

Veins Hydrothermal solutions moving through rocks often deposit ore minerals (**FIGURE 3.30**). These fluids flow easily through fractures in the rocks, cooling rapidly in the process. Quick cooling causes rapid precipitation of the ore minerals. The resulting *tabular* (sheetlike) deposits of precipitated minerals in the fractures are called **veins**. Some ore minerals are found in veins; others are found in the rocks surrounding the veins, which are altered when the hydrothermal solutions heat and infiltrate those rocks. As the solutions react with the surrounding rocks, they may precipitate ore minerals together with quartz, calcite, or other common vein-filling minerals. Vein deposits are a major source of gold.

Hydrothermal vein deposits are among the most important sources of metallic ores. Typically, metallic ores exist as sulfides, such as iron sulfide (pyrite), lead sulfide (galena), zinc sulfide (sphalerite), and mercury sulfide (cinnabar) (**Figure 3.31**). Hydrothermal solutions reach the surface as hot springs and geysers, many of which precipitate metallic ores—including ores of lead, zinc, and mercury—as they cool.

FIGURE 3.31 Some metallic sulfide ores. Sulfides are the most common types of metallic ores. [Chip Clark/Fundamental Photographs.]

Concentrations of Valuable Mineral Resources   81

FIGURE 3.32 Copper ores. Chalcopyrite and chalcocite are copper sulfide ores. Malachite is a carbonate of copper found in association with sulfides of copper. [Chip Clark/Fundamental Photographs.]

Chalcopyrite (a copper sulfide)   Malachite (a copper carbonate)   Chalcocite (a copper sulfide)

Disseminated Deposits Deposits of ore minerals that are scattered through volumes of rock much larger than veins are called **disseminated deposits.** In both igneous and sedimentary rocks, minerals are disseminated along abundant cracks and fractures. Among the economically important disseminated deposits are the copper deposits of Chile and the southwestern United States. These deposits develop in geologic regions with abundant igneous rocks, usually emplaced as large intrusive bodies. In Chile, these intrusive igneous rocks are related to the subduction of oceanic lithosphere beneath the Andes (an event very similar to what was described in our example of the rock cycle). The most common copper mineral in these deposits is chalcopyrite, a copper sulfide (**Figure 3.32**). The copper was deposited when ore minerals were introduced into a great number of tiny fractures in granitic intrusive rocks and in the rocks surrounding the upper parts of the igneous intrusions. Some unknown process associated with the magmatic intrusion or its aftermath broke these rocks into millions of pieces. Hydrothermal solutions penetrated and re-cemented the rocks by precipitating ore minerals throughout the extensive network of tiny fractures. This widespread dispersal produced a low-grade but very large resource of many millions of tons of ore, which can be mined economically by large-scale methods (**Figure 3.33**).

The lead-zinc deposits of the Upper Mississippi Valley, which extend from southwestern Wisconsin to Kansas and Oklahoma, are found in sedimentary rocks. The ores in this disseminated hydrothermal deposit are not associated with a known magmatic intrusion that could have been a source of

FIGURE 3.33 Kennecott Bingham Canyon Mine, Utah, an open-pit mine. Open-pit mining is typical of the large-scale methods used to exploit disseminated ore deposits. [Royce Bair/Getty Images.]

hydrothermal solutions, so their origin must be different. Some geologists speculate that the ores were deposited by groundwater that was driven out of the ancestral Appalachian Mountains when they were much higher. A continent-continent collision between North America and Africa may have created a continental-scale squeegee that pushed fluids from deep within the collision zone all the way into the continental interior of North America. Groundwater may have penetrated hot crustal rocks at great depths and dissolved soluble ore minerals, then moved upward into the overlying sedimentary rocks, where it precipitated the minerals as fillings in cavities. In some cases, it appears that these solutions infiltrated limestone formations and dissolved some carbonates, then replaced the carbonates with equal volumes of sulfide crystals. The major minerals of these deposits are lead sulfide (galena) and zinc sulfide (sphalerite).

Igneous Deposits

The most important deposits of ore minerals in igneous rocks are found as segregations of ore minerals near the bottoms of magmatic intrusions (see Chapter 5, Practicing Geology). These deposits form when minerals with relatively high melting temperatures crystallize from a body of cooling magma, settle, and accumulate at the base of the magma. Most of the chromium and platinum ores of the world, such as the deposits in South Africa and Montana, are found as layered accumulations of minerals that formed in this way (**Figure 3.34**). One of the richest ore deposits ever found, at Sudbury, Ontario, is a large igneous intrusion containing great quantities of layered nickel, copper, and iron sulfides near its base. Geologists believe that these sulfide deposits formed from the crystallization of a dense, sulfide-rich liquid that separated from the rest of a cooling magmatic intrusion and sank to the bottom before it congealed.

As the magma in a large granite-forming intrusion cools, the last material to crystallize forms *pegmatites*, extremely coarse-grained rocks in which minerals present in only trace amounts in the magma are concentrated. Pegmatites may contain rare ore minerals rich in elements such as beryllium, boron, fluorine, lithium, niobium, and uranium, as well as gem minerals such as tourmaline.

Sedimentary Deposits

Sedimentary deposits include some of the world's most valuable mineral sources. Many economically important minerals, such as copper, iron, and other metals, segregate as an ordinary result of sedimentary processes. These deposits are chemically precipitated in sedimentary environments to which large quantities of metals are transported in solution. Some of the important sedimentary copper ores, such as those of the Permian Kupferschiefer (German for "copper slate") beds of Germany, may have precipitated from hydrothermal solutions rich in metal sulfides that interacted with sediments on the seafloor. The plate tectonic setting of these deposits may have been something like the mid-ocean ridge described in our example of the rock cycle, except that it developed within a continent. Here, rifting of the continental crust led to development of a deep trough, where sediments and ore minerals were deposited in a very still, narrow sea.

Many rich deposits of gold, diamonds, and other heavy minerals such as magnetite and chromite are found in *placers*, sedimentary ore deposits that have been concentrated by the mechanical sorting action of river currents. These ore deposits originate where uplifted rocks

FIGURE 3.34 Chromite (chromium ore, visible as dark layers) in a layered igneous intrusion in the Bushveld Complex, South Africa.
[Spencer Titley.]

weather to form grains of sediment, which are then sorted by weight when currents of water flow over them. Because heavy minerals settle out of a current more quickly than lighter minerals such as quartz and feldspar, they tend to accumulate on streambeds and sandbars. Similarly, ocean waves preferentially deposit heavy minerals on beaches or shallow offshore bars. A gold panner accomplishes the same thing: The shaking of a water-filled pan allows the lighter minerals to be washed away, leaving the heavier gold in the bottom of the pan (Figure 3.35).

Some placers can be traced upstream to the location of the original mineral deposit, usually of igneous origin, from which the minerals were eroded. Erosion of the Mother Lode, an extensive gold-bearing vein system lying along the western flanks of the Sierra Nevada, produced the placers that were discovered in 1848 and led to the California gold rush. The placers were found before their source was discovered. Placers also led to the discovery of the Kimberley diamond mines of South Africa two decades later.

Mineral Evolution

The study of minerals also has an important historic dimension. Modern Earth is composed of over 4000 minerals that make up a very broad range of rock types and economically important mineral deposits. Yet the young Earth was vastly depleted in the diversity of minerals we see today. This difference records an evolution of minerals that mirrors the evolution of the Earth, and also our solar system (Figure 3.36). The history of Earth and our solar system is further explored in Chapters 20 (Early History of the Terrestrial Planets) and 21 (History of the Continents).

The story of mineral evolution goes all the way to the "Big Bang," when the universe was formed. In the very beginning there were no minerals, just lighter chemical elements, which gravitational forces pulled together to form stars. Some giant stars then exploded as supernovas that synthesized the rest of the chemical elements. The first minerals formed were likely microscopic crystals of graphite and diamond, joined soon after by a dozen or so other hardy minerals made of silicon, carbon, titanium, and nitrogen.

As our solar system formed, pulses of heat coming from the infant Sun melted and remixed elements, forming dozens of new minerals, including the first iron-nickel alloys, sulfides, and a broad range of silicates and oxides. Remnants of these earliest minerals are preserved in a class of primitive meteorites known as "chondrites." During the course of planet formation, when dust, sand, and pebble-sized masses of primitive rock aggregated to form larger masses, as many 250 new minerals were created. These minerals are the raw material from which all rocky planets are formed, and all of

(a)

(b)

FIGURE 3.35 (a) Panning for gold was popularized by "forty-niners" during the California gold rush and is still popular in the San Gabriel River today. (b) Gold is denser than the other materials from the streambed, so it sinks to the bottom of the pan. [(a) Bo Zaunders/Getty Images; (b) David Butow/Getty Images.]

CHAPTER 3   Earth Materials: Minerals and Rocks

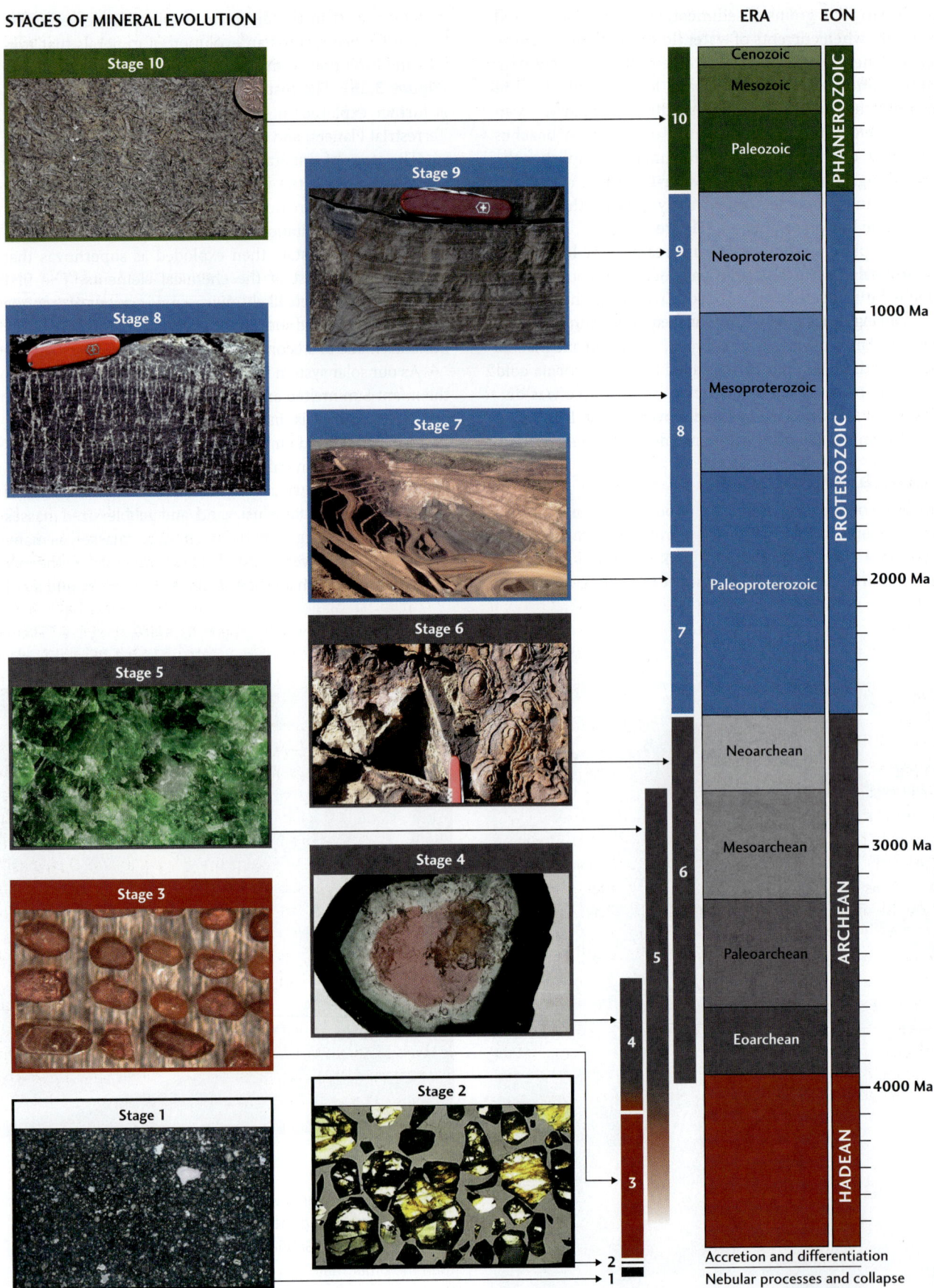

FIGURE 3.36 Each of the 10 stages of mineral evolution saw a change in the diversity and/or surface distribution of mineral species. This timeline is accompanied by photos of near-surface Earth materials illustrative of each stage. **Stage 1:** chondrite meteorite; **Stage 2:** pallasite meteorite; **Stage 3:** zircon grains; **Stage 4:** tourmaline; **Stage 5:** jadeite; **Stage 6:** the oldest stromatolites, simple domes; **Stage 7:** iron formation; **Stage 8:** columnar stromatolites widespread in younger oceans; **Stage 9:** granite clast dropped into soft marine sediments from melting iceberg; **Stage 10:** trilobite, the earliest animal fossils. [Data from Lafuente, B., Downs, R. T., Yang, H., and Stone, N. (2015). The power of databases: The RRUFF project. In: *Highlights in Mineralogical Crystallography*, T. Armbruster and R. M. Danisi, eds. Berlin, Germany: W. De Gruyter, pp. 1–30. Stages 1–2: Courtesy of the Smithsonian Institution; Stage 3: Courtesy Aaron J. Cavosie; Stages 4–5: Courtesy of Robert Downs; Stages 6–10: Courtesy John Grotzinger.]

them are represented in the diverse suites of meteorites that still fall to Earth today.

During Earth's first few hundred million years, cycles of melting and solidifying of its rocky crust increased mineral diversity. After some time granites were formed, along with another 500 minerals rich in lithium, beryllium, boron, cesium, uranium, and a dozen rare earth elements. Additionally, the formation of an ocean and atmosphere caused a new process to occur—rock weathering. We will learn about rock weathering in Chapter 6 (Sedimentation: Rocks Formed by Surface Processes), but for now, just think of it as the result of all the processes that affect rocks when the atmosphere and ocean interact with them, such as rain, which dissolves rocks, and freezing/thawing, which breaks rocks apart. These processes of crustal cycling and weathering, operating over countless cycles, gradually concentrated uncommon elements enough to form a broad diversity of exotic minerals. Not every planet has these minerals. Mercury and the Moon became frozen before they could extensively interact with water. Early on in its history, Mars had significant interactions with water before it too became desiccated; indeed, an important goal of current robotic missions to Mars is to assess its inventory of minerals that record formation in water to see how similar its environments were to that of early Earth. Probably 1500 minerals formed at or near the Earth's surface during its first 2 billion years. But what happened next was what really distinguishes Earth from every other planet in the solar system, and it tripled Earth's mineralogical diversity.

The answer is the origin and evolution of life. Earth's biosphere distinguishes it from all other planets and moons and it has irreversibly transformed the near surface of the planet. This is most obvious in studying the oceans and atmosphere; it also holds true for the minerals and rocks. Initially Earth's atmosphere lacked abundant oxygen, so vital to animal and human survival, but by 2.5–2.0 billion years, significant quantities began to build up due to the emergence of photosynthesis by microorganisms (see Chapter 22, Geobiology). One of the most important new rock types to form were "banded iron formations" in which the oceans rained out iron compounds that sequestered the newly formed oxygen.

The presence of oxygen and water paved the way for perhaps as many as 2500 new mineral species to form—directly the result of Earth's new biogeochemical processes. And on the surface of the planet, all the black basalt was transformed into red rock and soil. From space, Earth's continents 2 billion years ago might have looked like the red surface of Mars does today, albeit with blue oceans and white clouds. The red color of Mars is also caused by oxidation, but in this case, oxygen was produced by a nonbiologic process: Sunlight split water molecules apart high in the atmosphere, with the hydrogen escaping to space because it is so light. That process created enough oxygen to rust the planet's surface and turn it red, although not enough to create the great diversity (and abundance) that is observed on the more geologically and biologically active Earth.

In Chapter 20 (Early History of the Terrestrial Planets), we will learn that outside of our own solar system, there are thousands of other solar systems. And the search continues for any that might have a geologically and biologically active surface, like Earth's. If oxygen is one day discovered in the atmosphere of one of these "exoplanets," then it will be taken as a strong signal that life may also have originated on that planet, becoming advanced enough to produce oxygen by photosynthesis. The next question we ask will be "what is that planet's mineral diversity," because that would strengthen the case for extraterrestrial life. We live in a universe primed to increase its complexity: Hydrogen atoms form stars, stars form the elements, elements form planets, which in turn create minerals. And minerals catalyze the formation of life. Viewed in this way, it is understandable to imagine that we may not be alone in the universe.

CHAPTER 3 Earth Materials: Minerals and Rocks

KEY TERMS AND CONCEPTS

anion (p. 56)
atomic mass (p. 56)
atomic number (p. 55)
bedding (p. 74)
biological sediment (p. 74)
carbonate (p. 61)
cation (p. 56)
chemical sediment (p. 74)
cleavage (p. 66)
color (p. 68)
contact metamorphism (p. 74)
covalent bond (p. 57)
crystal (p. 58)
crystal habit (p. 69)
crystallization (p. 58)
density (p. 60)
disseminated deposit (p. 81)
electron sharing (p. 56)
electron transfer (p. 56)

erosion (p. 73)
fracture (p. 68)
grain (p. 59)
hardness (p. 66)
hydrothermal solution (p. 79)
igneous rock (p. 71)
ion (p. 56)
ionic bond (p. 56)
isotope (p. 56)
lithification (p. 74)
luster (p. 68)
magma (p. 60)
metallic bond (p. 57)
metamorphic rock (p. 71)
mineral (p. 54)
mineralogy (p. 54)
Mohs scale of hardness (p. 66)
ore (p. 77)
oxides (p. 61)

polymorph (p. 60)
precipitate (p. 60)
regional metamorphism (p. 74)
rock (p. 70)
rock cycle (p. 75)
sediment (p. 73)
sedimentary rock (p. 71)
silicate (p. 61)
siliciclastic sediment (p. 74)
specific gravity (p. 69)
streak (p. 68)
sulfate (p. 61)
sulfide (p. 61)
texture (p. 71)
trace element (p. 69)
vein (p. 80)
weathering (p. 73)

REVIEW OF LEARNING OBJECTIVES

3.1 Understand the structure of matter and how atoms react through chemical bonds to form minerals, which are crystalline substances.

Minerals, the building blocks of rocks, are naturally occurring, usually inorganic solids with specific crystal structures and chemical compositions. A mineral is constructed of atoms, the small units of matter that combine in chemical reactions. An atom is composed of a nucleus made up of protons and neutrons and surrounded by electrons. The atomic number of an element is the number of protons in its nucleus, and its atomic mass is the sum of the masses of its protons and neutrons.

Chemical elements react with one another to form compounds either by gaining or losing electrons to become ions, or by sharing electrons. Ionic bonds, which form by electrostatic attraction between positive ions (cations) and negative ions (anions), are the dominant type of chemical bond in mineral structures. Atoms that form compounds by sharing electrons are held together by covalent bonds. When a mineral crystallizes, atoms or ions come together in the proper proportions to form a crystal structure—an orderly three-dimensional array in which the basic arrangement of the atoms is repeated in all directions.

Study Assignments: Figures 3.3 and 3.4

Exercise: Draw a simple diagram to show how silicon and oxygen in silicate minerals share electrons.

Thought Question: What is the difference between an atom and an ion?

3.2 Recognize that all minerals are grouped into seven classes based on their chemical composition, and that all minerals also have distinct physical properties.

Silicate minerals, the most abundant minerals in Earth's crust, are built of silicate ions that are linked in various ways. Silicate tetrahedra may be isolated (linked together only by cations) or bonded together in structures such as single chains, double chains, sheets, or frameworks. Carbonate minerals are made up of carbonate ions bonded to calcium, magnesium, or both. Oxide minerals are compounds of oxygen and metallic elements. Sulfide and sulfate minerals are composed of sulfide and sulfate ions, respectively, in combination with metallic elements.

Mineralogy (the kinds and proportions of minerals that make up a rock) and texture (the sizes, shapes, and spatial arrangement of its crystals or grains) define a rock. Geologists use the physical properties of minerals to identify them. These physical properties include hardness—the ease with which a mineral's surface is scratched; cleavage—its tendency to split along planar surfaces; fracture—the way it breaks along irregular surfaces; luster—the way it reflects light; color—imparted by transmitted or reflected light to crystals or irregular masses or visible as streak (the color of a fine powder); density—mass per unit volume; and crystal habit—the shape in which individual crystals or aggregates of crystals grow.

Study Assignments: Table 3.2, Appendix 4

Exercise: Choose two minerals from Appendix 4 that you think might make good abrasive or grinding stones for sharpening steel, and describe the physical property that causes you to believe they would be suitable for that purpose.

Thought Question: How would a field geologist measure hardness?

3.3 Understand that geologists' primary aim is to study the three basic classes of rocks to deduce their geologic origins.

The mineralogy and texture of a rock are determined by the geologic processes by which it formed. Igneous rocks form by the crystallization of magmas as they cool. Intrusive igneous rocks cool slowly in Earth's interior and have large crystals. Extrusive igneous rocks, which cool rapidly at Earth's surface, have a glassy or fine-grained texture. Sedimentary rocks form by the lithification of sediments after burial. Sediments are derived from the weathering and erosion of rocks at Earth's surface. Metamorphic rocks form when igneous, sedimentary, or other metamorphic rocks are subjected to high temperatures and pressures in Earth's interior that change their mineralogy, texture, or chemical composition.

The rock cycle relates geologic processes driven by the plate tectonic system and the climate system to the formation of the three families of rocks. We can view these processes by starting at any point in the cycle, such as the creation of new oceanic lithosphere at a spreading center as two continents drift apart. The ocean basin gets wider until at some point the process reverses itself. As the basin closes and igneous rocks and sediments are subducted beneath a continent, they begin to melt to form a new generation of igneous rocks. The heat and pressure associated with subduction and with the intrusion of these igneous rocks transforms surrounding rocks into metamorphic rocks. Ultimately, the two continents collide, and these igneous and metamorphic rocks are uplifted into a high mountain chain. The uplifted rocks slowly weather, and their fragments are deposited as sediments.

Study Assignment: Figure 3.28

Exercise: Using the rock cycle, trace the path from a magma to a granitic intrusion to a metamorphic gneiss to a sandstone. Be sure to include the roles of the plate tectonic and climate systems and the specific processes that create the rocks.

Thought Question: What process is the "engine" that causes recycling of rocks on Earth?

CHAPTER 3 Earth Materials: Minerals and Rocks

3.4 Know that economic mineral resources occur in specific locations, and that the evolution of minerals over geologic time teaches us the history of our planet, including the emergence and evolution of life.

Ores are deposits of minerals from which valuable metals can be recovered profitably. Hydrothermal deposits of ore minerals are formed when groundwater or seawater reacts with a magmatic intrusion to form a hydrothermal solution. The heated water transports soluble minerals to cooler rocks, where they are precipitated in fractures. The resulting ores may be found in veins or in disseminated deposits. Igneous ore deposits typically form when minerals crystallize from cooling magma, settle, and accumulate at the base of the magma body. They are often found as layered accumulations of minerals. Other ore minerals are chemically precipitated in sedimentary environments to which metals are transported in solution.

Minerals are important in recording the history of the Earth. The appearance of new mineral species over time is a consequence of a variety of physical, chemical, and biological processes. In the beginning, there were only a handful of minerals, including oxides and silicates, but today there are more than 4000 types of minerals. This dramatic increase in abundance was due to critical events in Earth history, such as the recycling of rocks due to plate tectonics, the emergence of life, and the origin of photosynthesis by microorganisms that created an oxygen-rich atmosphere.

> **Study Assignment:** Practicing Geology Exercise
>
> **Exercise:** Work through the "bonus problem."
>
> **Thought Question:** Why is it important to collect three-dimensional data to determine the economic value of an ore deposit?

VISUAL LITERACY EXERCISE

1. **What causes rocks to be eroded into sediment particles?**
   a. Meteorite impact events
   b. Weathering due to exposure to the atmosphere
   c. Plate tectonic collisions
   d. Longshore transport along beaches

2. **How is sediment moved from mountains to the oceans?**
   a. During volcanic eruptions
   b. By regional windstorms
   c. By downslope transport in streams
   d. During magnetic field reversals

3. **When sediments arrive at the oceans, what kind of deposit is formed?**
   a. A submarine fan
   b. A dune field
   c. A desert pavement
   d. A delta

4. **How is sediment converted into rock?**
   a. By a process called "lithification," involving compaction and cementation
   b. During "metamorphism," a process that involves heating of rocks
   c. By a process called "subduction," when one tectonic plate overrides another
   d. Within debris flows, which form on steep slopes

5. **Where do reefs form?**
   a. In deserts where transport by wind is important
   b. Along the peaks of mountains where rainfall is high
   c. Along the coastline, but away from deltas where accumulating sediments might inhibit their growth
   d. In playa lakes where evaporation rates are high

FIGURE 3.36 Each of the 10 stages of mineral evolution saw a change in the diversity and/or surface distribution of mineral species. This timeline is accompanied by photos of near-surface Earth materials illustrative of each stage. **Stage 1:** chondrite meteorite; **Stage 2:** pallasite meteorite; **Stage 3:** zircon grains; **Stage 4:** tourmaline; **Stage 5:** jadeite; **Stage 6:** the oldest stromatolites, simple domes; **Stage 7:** iron formation; **Stage 8:** columnar stromatolites widespread in younger oceans; **Stage 9:** granite clast dropped into soft marine sediments from melting iceberg; **Stage 10:** trilobite, the earliest animal fossils. [Data from Lafuente, B., Downs, R. T., Yang, H., and Stone, N. (2015). The power of databases: The RRUFF project. In: *Highlights in Mineralogical Crystallography*, T. Armbruster and R. M. Danisi, eds. Berlin, Germany: W. De Gruyter, pp. 1–30. Stages 1–2: Courtesy of the Smithsonian Institution; Stage 3: Courtesy Aaron J. Cavosie; Stages 4–5: Courtesy of Robert Downs; Stages 6–10: Courtesy John Grotzinger.]

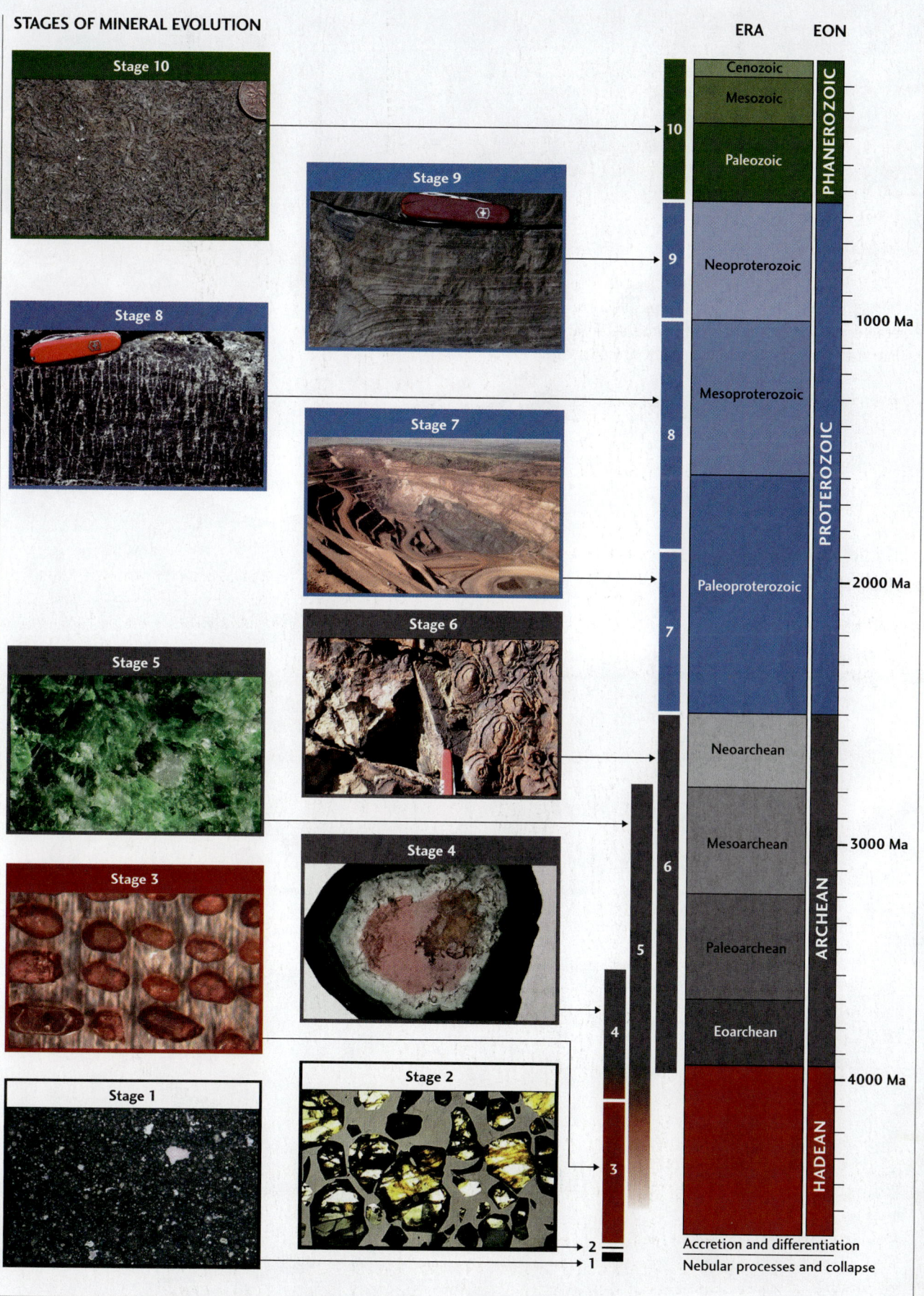

Igneous Rocks: Solids from Melts

4

How Do Igneous Rocks Differ from One Another?	92
How Do Magmas Form?	97
Magmatic Differentiation	99
Forms of Igneous Intrusions	104
Igneous Processes and Plate Tectonics	107

Learning Objectives

In this chapter, we will examine the wide range of igneous rock types, both intrusive and extrusive, and the processes by which they form. We will explore the forces that cause rock to melt and form magmas and the ways in which those magmas reach the locations at and below Earth's surface where they solidify. We will then take a more detailed look at the igneous processes associated with specific plate tectonic settings. After you have studied the chapter, you should be able to:

4.1 Explain how igneous rocks can be classified and differentiated from one another.

4.2 Describe how magmas form and what causes a rock to melt.

4.3 Demonstrate how magmas of different compositions can be made from a uniform parent material.

4.4 Summarize the types of igneous intrusions.

4.5 Discuss the link between igneous processes and plate tectonics.

Granite, such as that shown in this image of Mt. Whitney, the tallest peak in the continental United States, makes up nearly all of the Sierra Nevada mountain range. [Courtesy Jennifer Griffes.]

CHAPTER 4   Igneous Rocks: Solids from Melts

More than 2000 years ago, the Greek scientist and geographer Strabo traveled to Sicily to view the eruptions of Mount Etna. He observed that the hot liquid lava spilling down from the volcano onto Earth's surface cooled and hardened into solid rock within a few hours. By the eighteenth century, geologists began to understand that some sheets of rock that cut across other rock formations had also been formed by the cooling and solidification of molten rock. In these cases, the magma had cooled much more slowly because it had remained buried in Earth's crust.

Today we know that deep in Earth's crust and mantle, rock melts and rises toward Earth's surface. Some magmas solidify before they reach the surface, and some break through and solidify on the surface. Both processes produce igneous rocks.

Understanding the processes that melt and resolidify rock is a key to understanding how Earth's crust forms. Although we still have much to learn about the exact *mechanisms* of melting and solidification, we do have good answers to some fundamental questions: How do types of igneous rock differ from one another? Where and how do magmas form? How do rocks solidify from those magmas?

In answering these questions, Chapter 4 will focus on the central role of igneous processes in the Earth system. Observations of igneous rocks by geologists from Strabo until today make sense only in light of plate tectonic theory. Specifically, igneous rocks form at spreading centers where plates move apart, along convergent boundaries where one plate descends beneath another, and at "hot spots" where hot mantle material ascends to the crust.

How Do Igneous Rocks Differ from One Another?

Geologists today classify igneous rock samples in the same way some geologists did in the late nineteenth century: by their texture and by their mineral and chemical composition.

Texture

Two hundred years ago, the first division of igneous rocks was made on the basis of texture, which largely reflects differences in mineral grain size: Geologists classified rocks as either coarse-grained or fine-grained (see Chapter 3). Grain size is a simple characteristic that geologists can easily see in the field. A coarse-grained rock, such as granite, has distinct crystals that are easily visible to the naked eye. In contrast, the crystals of a fine-grained rock, such as basalt, are too small to be seen, even with a magnifying glass. **Figure 4.1** shows samples of granite and basalt, accompanied by photomicrographs of very thin, transparent slices of each rock. *Photomicrographs*, which are simply photographs taken through a microscope, give us an enlarged view of minerals and their textures. Textural differences were clear to early geologists, but several more clues were needed to unravel the meaning of those differences.

First Clue: Volcanic Rocks   Early geologists observed volcanic rocks forming from lava during volcanic eruptions. (**Lava** is the term that we apply to magma flowing out onto Earth's surface.) They noted that where lava cooled rapidly, it formed either a fine-grained rock or a glassy one in which

FIGURE 4.1 Igneous rocks were first classified by texture. Early geologists assessed rock texture with a small hand-held magnifying glass. Modern geologists have access to high-powered polarizing microscopes, which can produce photomicrographs of thin, transparent rock slices like those shown here. [John Grotzinger/Ramón Rivera-Moret/Harvard Mineralogical Museum; Courtesy Steven Chemtob.]

no crystals could be distinguished. Where lava cooled more slowly, as in the middle of a thick flow many meters high, somewhat larger crystals were formed.

Second Clue: Laboratory Studies of Crystallization

Just over a hundred years ago, experimental scientists began to understand the nature of crystallization. Anyone who has frozen a tray of ice cubes knows that water solidifies to ice in a few hours as its temperature drops below the freezing point. If you have ever attempted to retrieve your ice cubes before they were completely solid, you may have seen thin ice crystals forming at the surface and along the sides of the tray. During crystallization, the water molecules take up fixed positions in the solidifying crystal structure, and they are no longer able to move freely, as they did when the water was liquid. All other liquids, including magmas, crystallize in this way.

The first tiny crystals form a pattern. Other atoms or ions in the crystallizing liquid then attach themselves in such a way that the tiny crystals grow larger. It takes some time for the atoms or ions to "find" their correct places on a growing crystal, so crystals grow large only if they have time to grow slowly. If a liquid solidifies very quickly, as a magma does when it erupts onto the cool surface of Earth, the crystals have no time to grow. Instead, a large number of tiny crystals form simultaneously as the liquid cools and solidifies.

Third Clue: Granite as Evidence of Slow Cooling

By studying volcanoes, early geologists determined that fine-grained textures indicate quick cooling at Earth's surface and that fine-grained igneous rocks are evidence of former volcanism. But in the absence of direct observation, how could geologists deduce that coarse-grained rocks form by *slow* cooling deep in Earth's interior? Granite—one of the most common rocks of the continents—turned out to be the crucial clue (**Figure 4.2**). James Hutton, one of geology's founding fathers, saw granite cutting across and disrupting layers of sedimentary rock as he worked in the field in Scotland. He noticed that the granite had somehow fractured and invaded the sedimentary rock, as though the granite had been forced into the fractures as a liquid.

As Hutton looked at more and more granites, he began to focus on the sedimentary rocks bordering them. He observed that the minerals of the sedimentary rocks in contact with the granite were different from those found in sedimentary rocks at some distance from the granite. He concluded that the changes in the sedimentary rocks must have resulted from great heat, and that the heat must have come from the granite. Hutton also noted that the granite was composed of interlocked crystals (see Figure 4.1). By this time, chemists had established that a slow crystallization process produces this pattern.

FIGURE 4.2 Granite pegmatite sill or dike (the lighter-colored rock) in an outcrop of schist (darker colored rock) along the Harlem River, New York, suggests to geologists that the intruding rock had been forced into the fractures as a liquid.
[Catherine Ursillo/Science Source.]

With these three lines of evidence, Hutton proposed that granite forms from hot molten material that solidifies deep within Earth. The evidence was conclusive because no other explanation could accommodate all the facts. Other geologists, who saw the same characteristics of granites in widely separated places throughout the world, came to recognize that granite and many similar coarse-grained rocks were the products of magma that had crystallized slowly in Earth's interior.

Intrusive and Extrusive Textures The full significance of an igneous rock's texture is now clear: It is linked to the rate, and therefore the place, of cooling. An **intrusive igneous rock** is one that has forced its way into the surrounding rock, called **country rock**, and solidified without reaching Earth's surface. Slow cooling of magma in Earth's interior allows adequate time for the growth of the large, interlocking crystals that characterize intrusive igneous rocks (**Figure 4.3**).

Rapid cooling at Earth's surface produces the fine-grained texture or glassy appearance of **extrusive igneous rocks** (see Figure 4.3). These rocks, formed partly or largely of volcanic glass, are formed from material that erupts from volcanoes. For this reason, they are also known as *volcanic rocks*. They fall into two major categories based on the type of erupted material from which they are formed:

- *Lavas:* Volcanic rocks formed from flowing lavas range in appearance from smooth and ropy to sharp, spiky, and jagged, depending on the conditions under which they are formed.

- *Pyroclasts:* In more violent eruptions, **pyroclasts** form when fragments of lava are thrown high into the air. **Volcanic ash** is made up of extremely small fragments, usually of glass, that form when escaping gases force a fine spray of magma from a volcano. **Bombs** are larger particles hurled from the volcano and streamlined by the air as they hurtle through it. As they fall to the ground and cool, these fragments of volcanic debris may stick together to form rocks.

One volcanic rock type is **pumice**, a frothy mass of volcanic glass in which a great number of spaces remain after trapped gas has escaped from the solidifying melt. Another wholly glassy volcanic rock type is **obsidian**; unlike pumice, it contains only tiny vesicles and so is solid and dense. Chipped or fragmented obsidian produces very sharp edges, and Native Americans and many other hunting groups used it for arrowheads and a variety of cutting tools.

A **porphyry** is an igneous rock that has a mixed texture in which large crystals "float" in a predominantly fine-grained matrix (see Figure 4.3). The large crystals, called *phenocrysts*, form in magma while it is still below Earth's surface. Then, before other crystals can grow, a volcanic eruption brings the magma to the surface, where it cools quickly to a finely crystalline mass. In some cases, porphyries

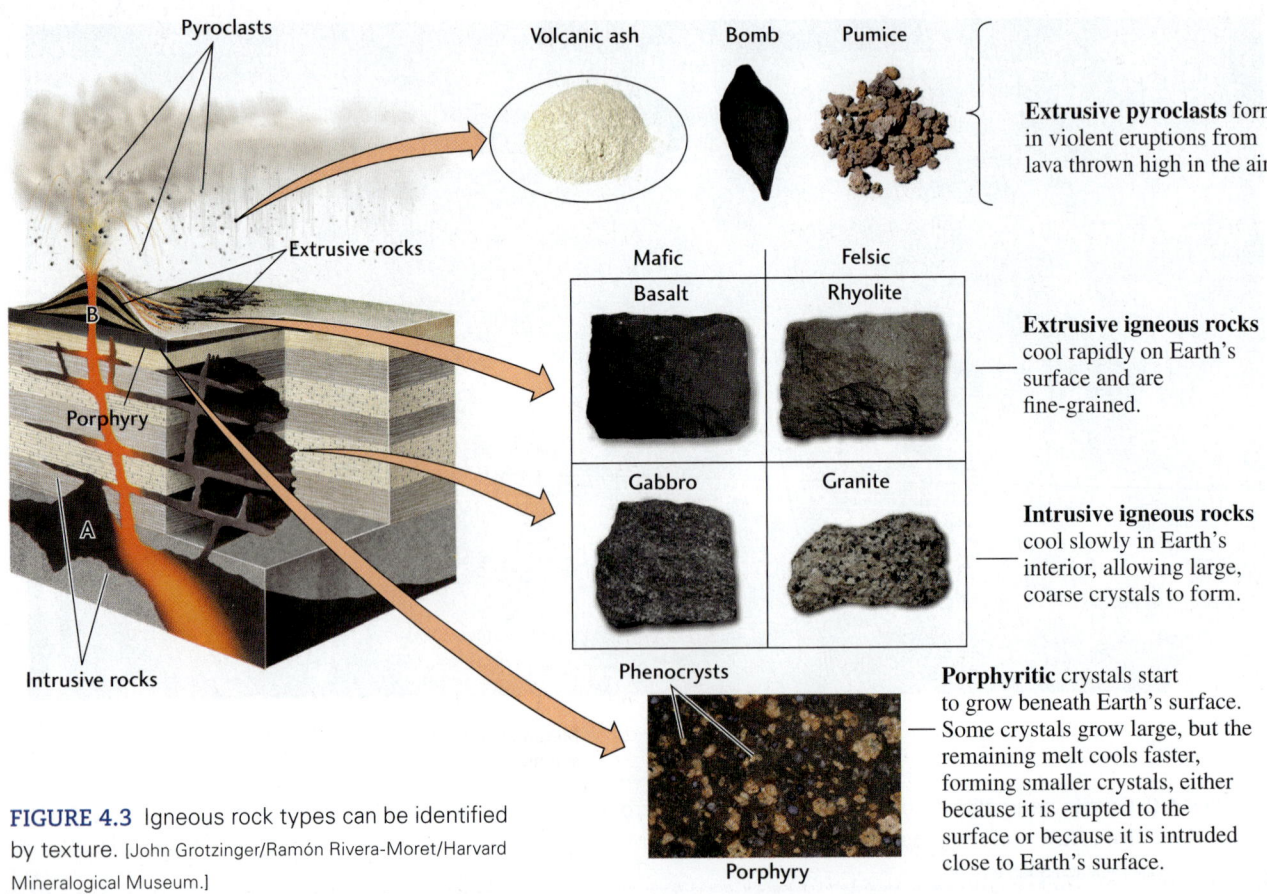

FIGURE 4.3 Igneous rock types can be identified by texture. [John Grotzinger/Ramón Rivera-Moret/Harvard Mineralogical Museum.]

form as intrusive igneous rocks; for example, they may form where magmas cool quickly at very shallow levels in the crust. Porphyry textures are important to geologists because they show that different minerals crystallize at different rates, a point that will be emphasized later in this chapter.

In Chapter 12, we will look more closely at how volcanic processes form extrusive igneous rocks. Now, however, we turn to the second way in which the family of igneous rocks is subdivided.

Chemical and Mineral Composition

We have just seen how igneous rocks can be subdivided according to their texture. They can also be classified on the basis of their chemical and mineral composition. Volcanic glass, which is formless even under a microscope, is often classified by chemical analysis alone. One of the earliest classifications of igneous rocks was based on a simple chemical analysis of their silica content. Silica (SiO_2) is abundant in most igneous rocks, accounting for 40 to 70 percent of their total weight.

Modern classifications group igneous rocks according to their relative proportions of silicate minerals (Table 4.1; see also Appendix 4).

The silicate minerals—quartz, feldspars, muscovite and biotite micas, amphiboles and pyroxenes, and olivine—form a systematic series. *Felsic* minerals are the highest in silica; *mafic* minerals are the lowest in silica. The adjectives *felsic* (from *fel*dspar and *si*lica) and *mafic* (from *ma*gnesium and *f*erric, from the Latin *ferrum*, "iron") are applied both to minerals and to rocks containing large proportions of those minerals. Mafic minerals crystallize at higher temperatures—that is, earlier in the cooling of a magma—than felsic minerals.

As the mineral and chemical compositions of igneous rocks became known, geologists soon noticed that some extrusive and intrusive rocks were identical in composition and differed only in texture. Basalt, for example, is an extrusive rock formed from lava. Gabbro has exactly the same mineral and chemical composition as basalt, but forms deep in Earth's crust (see Figure 4.3). Similarly, rhyolite and granite are identical in composition, but differ in texture. Thus, extrusive and intrusive rocks form two chemically and mineralogically parallel sets of igneous rocks. Conversely, most of the chemical and mineral compositions in the felsic-to-mafic series we have just described can appear in either extrusive or intrusive rocks. The only exceptions are very highly mafic rocks, which rarely appear as extrusive igneous rocks.

TABLE 4.1 Common Minerals of Igneous Rocks

COMPOSITIONAL GROUP	MINERAL	CHEMICAL COMPOSITION	SILICATE STRUCTURE
FELSIC	Quartz	SiO_2	Frameworks
	Orthoclase feldspar	$KAlSi_3O_8$	
	Plagioclase feldspar	$NaAlSi_3O_8$; $CaAl_2Si_2O_8$	
	Muscovite (mica)	$KAl_3Si_3O_{10}(OH)_2$	Sheets
MAFIC	Biotite (mica)	$\left.\begin{array}{l}K\\Mg\\Fe\\Al\end{array}\right\} Si_3O_{10}(OH)_2$	
	Amphibole group	$\left.\begin{array}{l}Mg\\Fe\\Ca\\Na\end{array}\right\} Si_8O_{22}(OH)_2$	Double chains
	Pyroxene group	$\left.\begin{array}{l}Mg\\Fe\\Ca\\Al\end{array}\right\} SiO_3$	Single chains
	Olivine	$(Mg,Fe)_2SiO_4$	Isolated tetrahedral

FIGURE 4.4 Classification model for igneous rocks. The vertical axis shows the minerals contained in a given rock as a percentage of its volume. The horizontal axis shows the silica content of a given rock as a percentage of its weight. Thus, if you knew by chemical analysis that a coarsely textured rock sample was about 70 percent silica, you could deduce that its composition was about 6 percent amphibole, 3 percent biotite, 5 percent muscovite, 14 percent plagioclase feldspar, 22 percent quartz, and 50 percent orthoclase feldspar. Your rock would be granite. Although rhyolite has the same mineral composition, its fine texture would eliminate it from consideration.

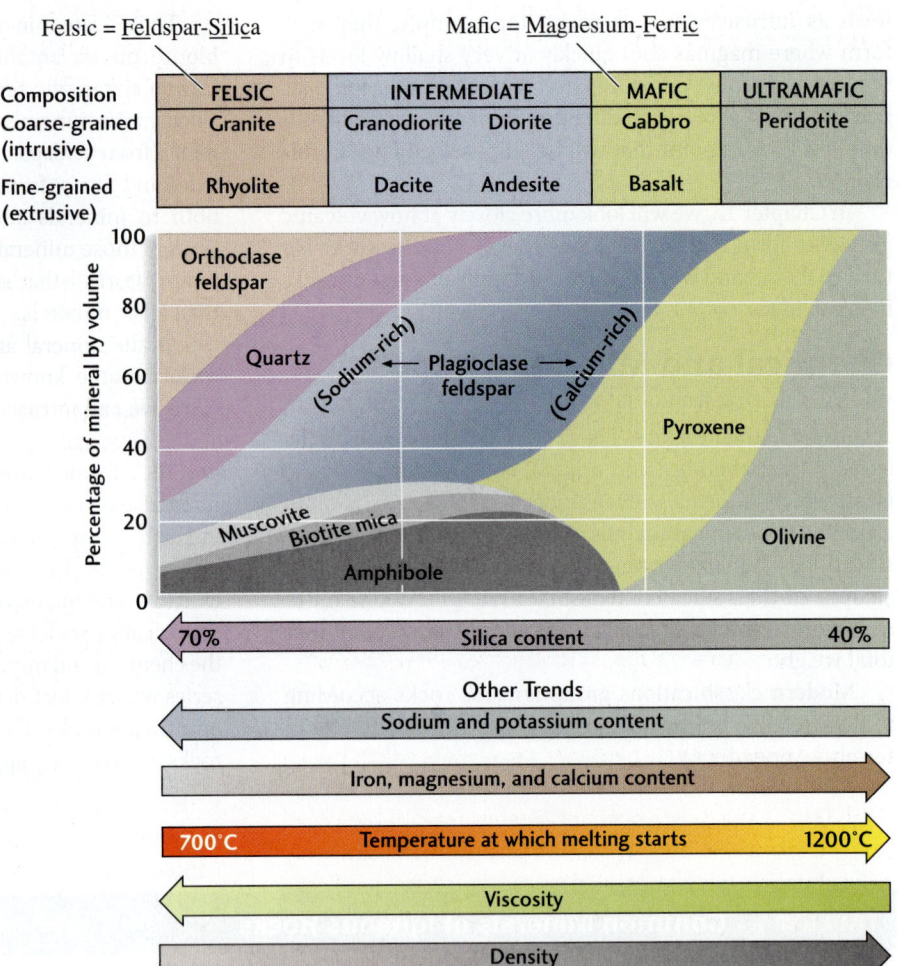

Figure 4.4 is a model that portrays these relationships. The horizontal axis plots silica content as a percentage of a given rock's weight. The percentages given—from high silica content at 70 percent to low silica content at 40 percent—cover the range found in igneous rocks. The vertical axis plots mineral content as a percentage of a given rock's volume. This model can be used to classify an unknown rock sample with a known silica content: By finding its silica content on the horizontal axis, you can determine its mineral composition and, from that, the type of rock it is.

We can use Figure 4.4 to guide our discussion of intrusive and extrusive igneous rocks. We begin with the felsic rocks at the far left of the model.

Felsic Rocks **Felsic rocks** are poor in iron and magnesium and rich in felsic minerals, which are high in silica. Such minerals include quartz, orthoclase feldspar, and plagioclase feldspar. Orthoclase feldspars, which contain potassium, are more abundant than plagioclase feldspars. Plagioclase feldspars contain varying amounts of calcium and sodium; as Figure 4.4 indicates, they are richer in sodium near the felsic end and richer in calcium near the mafic end of the scale. Thus, just as mafic minerals crystallize at higher temperatures than felsic minerals, calcium-rich plagioclases crystallize at higher temperatures than sodium-rich plagioclases.

Felsic rocks tend to be light in color. **Granite,** one of the most abundant intrusive igneous rocks, contains about 70 percent silica. Its mineral composition includes abundant quartz and orthoclase feldspar and a smaller amount of plagioclase feldspar (see the far left of Figure 4.4). These light-colored felsic minerals give granite its pink or gray color. Granite also contains small amounts of muscovite and biotite micas and amphibole. **Rhyolite** is the extrusive equivalent of granite. This light brown to gray rock has the same felsic composition and light coloration as granite, but it is much more fine-grained. Many rhyolites are formed largely or entirely of volcanic glass.

Intermediate Igneous Rocks Midway between the felsic and mafic ends of the scale are the **intermediate igneous rocks.** As their name indicates, these rocks are neither as rich in silica as the felsic rocks nor as poor in silica as the mafic rocks. We find the intermediate intrusive igneous rocks to the right of granite in Figure 4.4. The first is **granodiorite,** a light-colored rock that looks something like granite. It is also similar to granite in having abundant quartz, but its predominant feldspar is plagioclase, not orthoclase. To its right is **diorite,** which contains still less silica and is dominated by plagioclase feldspar, with little or no quartz. Diorites contain a moderate amount of the mafic minerals biotite, amphibole, and pyroxene. They tend to be darker than granite or granodiorite.

The volcanic equivalent of granodiorite is **dacite.** To its right in the extrusive series is **andesite,** the volcanic equivalent of diorite. Andesite derives its name from the Andes, the volcanic mountain belt in South America.

Mafic Rocks **Mafic rocks** contain large proportions of pyroxenes and olivines. These minerals are relatively poor in silica but are rich in magnesium and iron, from which they get their characteristic dark colors. **Gabbro** is a coarse-grained, dark gray intrusive igneous rock. Gabbro has an abundance of mafic minerals, especially pyroxenes. It contains no quartz and only moderate amounts of calcium-rich plagioclase feldspar.

Basalt is the most abundant igneous rock of the crust, and it underlies virtually the entire seafloor. This dark gray to black rock is the fine-grained extrusive equivalent of gabbro. In some places, extensive thick sheets of basalt, called *flood basalts,* form large plateaus. The Columbia River basalts of Washington State and the remarkable formation known as the Giant's Causeway in Northern Ireland are two examples. The Deccan flood basalts of India and the Siberian flood basalts of northern Russia were formed by enormous outpourings of basalt that appear to coincide closely with two of the greatest periods of mass extinction in the fossil record.

Ultramafic Rocks **Ultramafic rocks** consist primarily of mafic minerals and contain less than 10 percent feldspar. Here, at the far right of Figure 4.4, with a silica content of only about 45 percent, we find **peridotite,** a coarse-grained, dark greenish gray rock made up primarily of olivine with smaller amounts of pyroxene. Peridotites are the dominant rocks in Earth's mantle, and as we will see, they are the source of the basaltic magmas that form rocks at mid-ocean ridges. Ultramafic rocks are rarely found as extrusives. Because they solidify at such high temperatures, they are rarely liquid and hence do not form typical lavas.

Trends in the Felsic-to-Mafic Series The names and exact compositions of the various rocks in the felsic-to-mafic series are less important to remember than the trends shown in Figure 4.4. There is a strong correlation between a rock's mineralogy and its temperature of crystallization

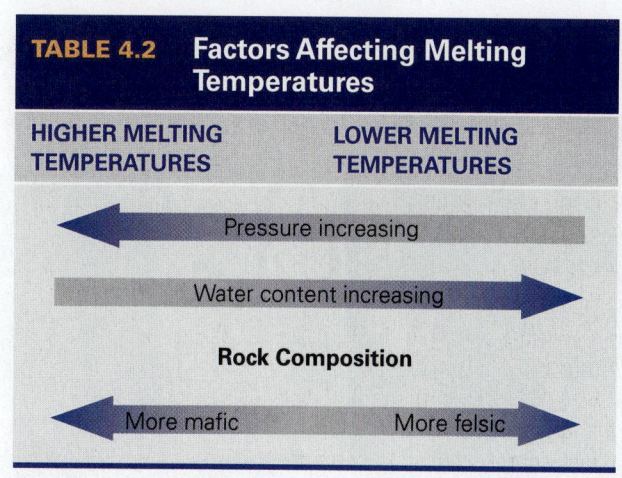

TABLE 4.2 Factors Affecting Melting Temperatures

HIGHER MELTING TEMPERATURES	LOWER MELTING TEMPERATURES
← Pressure increasing	
Water content increasing →	
Rock Composition	
← More mafic	More felsic →

or melting. As Table 4.2 indicates, mafic minerals melt at higher temperatures than felsic minerals. At temperatures below their melting point, minerals crystallize; therefore, mafic minerals also crystallize at higher temperatures than felsic minerals. We can also see that silica content increases as we move from the mafic end to the felsic end of the series. Increasing silica content results in increasingly complex silicate structures (see Table 4.1), which interfere with a melted rock's ability to flow. Thus, **viscosity**—the measure of a liquid's *resistance* to flow—increases as silica content increases. Viscosity is an important factor in the behavior of lavas, as we will see in Chapter 12. Increasing silica content also results in decreasing density, as we saw in Chapter 1.

It is clear that an igneous rock's mineralogy provides a great deal of information about the conditions under which the rock's parent magma formed and crystallized. To interpret this information accurately, however, we must understand more about igneous processes. We turn to that topic next.

How Do Magmas Form?

We know from the way Earth transmits seismic waves that the bulk of the planet is solid for thousands of kilometers down to the core-mantle boundary (see Chapter 1). The evidence of volcanic eruptions, however, tells us that there must also be liquid regions where magmas originate. How do we resolve this apparent contradiction? The answer lies in the processes that melt rocks and create magmas.

How Do Rocks Melt?

Although we do not yet understand the exact mechanisms of rock melting and solidification within Earth, we have learned a great deal from laboratory experiments using high-temperature furnaces (Figure 4.5). From these experiments, we know that a rock's melting point depends on its chemical and mineral composition, and on conditions of temperature and pressure (see Table 4.2).

FIGURE 4.5 Experimental device used to melt rocks in a laboratory. [Courtesy Jon Blundy.]

Temperature and Melting A hundred years ago, geologists discovered that rock does not melt completely at a given temperature. Instead, rocks undergo **partial melting** because the minerals that compose them melt at different temperatures. As temperatures rise, some minerals melt and others remain solid. If the same conditions are maintained at any given temperature, the same mixture of solid and melted rock is maintained. The fraction of rock that has melted at a given temperature is called a *partial melt*. To visualize a partial melt, think of how a chocolate chip cookie would look if you heated it to the point at which the chocolate chips melted while the main part of the cookie stayed solid. The chips represent the partial melt, or magma.

The ratio of solid to partial melt depends on the proportions and melting temperatures of the minerals that make up the original rock. It also depends on the temperature at the depth in the crust or mantle where melting takes place. At the lower end of a rock's melting range, a partial melt might be less than 1 percent of the volume of the original rock. Much of the hot rock would still be solid, but significant amounts of liquid would be present as small droplets in the tiny spaces between crystals throughout the mass. In the upper mantle, for example, some basaltic magmas are produced by only 1 to 2 percent melting of peridotite. However, 15 to 20 percent melting of mantle peridotite to form basaltic magmas is common beneath mid-ocean ridges. At the high end of a rock's melting range, much of the rock would be liquid, containing lesser amounts of unmelted crystals. An example would be the reservoir of basaltic magma and crystals just beneath a volcano such as the island of Hawaii. Geologists have used this knowledge of partial melts to determine how different kinds of magmas form at different temperatures and in different regions of Earth's interior. As you can imagine, the composition of a magma formed from completely melted rock may be very different from that of a magma formed from rock in which only the minerals with the lowest melting points have melted. Thus, basaltic magmas that form in different regions of the mantle may have somewhat different compositions.

Pressure and Melting To get the whole story on melting, we must consider pressure as well as temperature. Pressure increases with depth within Earth as a result of the increasing weight of overlying rock. Geologists found that as they melted rocks under various pressures in the laboratory, higher pressures led to higher melting temperatures. Thus, rocks that would melt at a given temperature at Earth's surface would remain solid at the same temperature in Earth's interior. For example, a rock that melts at 1000°C at Earth's surface might have a much higher melting temperature, perhaps 1300°C, deep in the interior, where pressures are many thousands of times greater than those at the surface. It is the effect of pressure that explains why the rocks in most of the crust and mantle do not melt. Rock can melt only when both temperature and pressure conditions are right.

Just as an increase in pressure can keep rock solid, a decrease in pressure can make rock melt, given a sufficiently high temperature. Because of convection currents in the mantle, mantle material rises to Earth's surface at mid-ocean ridges at a more or less constant temperature. As the material rises and the pressure on it decreases below a critical point, the solid rock melts spontaneously, without the introduction of any additional heat. This process, known as **decompression melting,** produces the greatest volume of magma anywhere on Earth. It is the process by which most basalts form on the seafloor.

Water and Melting The many experiments on melting temperatures and partial melting of rocks paid other dividends as well. One of them was a better understanding of the role of water in melting. Geologists studying natural lavas in the field determined that water was present in some magmas. This finding gave them the idea of adding water to their experimental melts back in the laboratory. By adding small but varying amounts of water, they discovered that the compositions of partial melts varied not only with temperature and pressure, but also with the amount of water present.

Consider, for example, the effect of dissolved water on pure albite, a sodium-rich plagioclase feldspar, at the low pressures at Earth's surface. If only a small amount of water is present in the rock, the rock will remain solid at temperatures just over 1000°C, hundreds of degrees above the boiling point of water. At these temperatures, the water in the albite is present as a vapor (gas). If large amounts of water are dissolved in the albite, however, its melting temperature will decrease, dropping to as low as 800°C. This behavior follows the general rule that dissolving one substance (in this case, water vapor) in another (in this case, albite) lowers the melting temperature of the solution. If you live in a cold climate, you are probably familiar with this principle because you know that salt sprinkled on icy roads lowers the melting temperature of the ice. By the same principle, the melting temperature of albite—and of all silicate minerals—drops considerably in the presence of large amounts of water. The melting points of these minerals decrease in proportion to the amount of water dissolved in the molten silicate.

Melting of rock induced by the presence of water that lowers its melting point is referred to as **fluid-induced melting.** Water content is a significant factor in the melting of sedimentary rocks, which contain an especially large volume of water in their pore spaces, more than is found in igneous or metamorphic rocks. As we will see later in this chapter, the water in sedimentary rocks plays an important role in the melting that gives rise to much of the volcanic activity at subduction zones.

The Formation of Magma Chambers

Most substances are less dense in their liquid form than in their solid form. The density of melted rock is lower than the density of solid rock of the same composition. With this knowledge, geologists reasoned that large bodies of magma could form in the following way: If the less dense melted rock were given a chance to move, it would move upward—just as oil, which is less dense than water, rises to the surface of a mixture of oil and water. Being liquid, a partial melt could move slowly upward through pores and along the boundaries between crystals of the surrounding solid rock. As the hot drops of melted rock moved upward, they would mix with other drops, gradually forming larger pools of magma within Earth's solid interior.

The rise of magmas through the mantle and crust may be slow or rapid. Magmas rise at rates from 0.3 m/year to almost 50 m/year, over periods of tens of thousands or even hundreds of thousands of years. As they ascend, magmas may mix with other melts and may also melt portions of the crust. We now know that large pools of molten rock, called **magma chambers,** form in the lithosphere as rising magmas melt and push aside surrounding solid rock. We know that they exist because seismic waves have shown us the depth, size, and general outlines of the magma chambers underlying some active volcanoes. A magma chamber may encompass a volume as large as several cubic kilometers. We cannot yet say exactly how magma chambers form, nor exactly what they look like in three dimensions. We can think of them as large, liquid-filled cavities in solid rock, which expand as more of the surrounding rock melts or as magma migrates through cracks and other small openings. Magma chambers contract as they expel magma to the surface in volcanic eruptions.

Where Do Magmas Form?

Our understanding of igneous processes stems from geologic inferences as well as laboratory experimentation. One important source of information is volcanoes, which give us information about where magmas are located. Another is the record of temperatures measured in deep drill holes and mine shafts. This record shows that the temperature of Earth's interior increases with depth. Using these measurements, scientists have been able to estimate the rate at which temperature rises as depth increases.

The temperatures recorded at a given depth in some locations are much higher than the temperatures recorded at the same depth in other locations. These results indicate that some parts of Earth's mantle and crust are hotter than others. For example, the Great Basin of the western United States is an area where the North American continent is being stretched and thinned, with the result that the temperature increases with depth at an exceptionally rapid rate, reaching 1000°C at 40 km, not far below the base of the crust. This temperature is almost high enough to melt basalt. By contrast, in tectonically stable regions, such as the central parts of continents, the temperature increases much more slowly, reaching only 500°C at the same depth.

Magmatic Differentiation

The processes we've discussed so far account for the melting of rocks to form magmas. But what accounts for the variety of igneous rocks? Are magmas of different chemical compositions made by the melting of different kinds of rock? Or do igneous processes produce a variety of rocks from an originally uniform parent material?

Again, the answers to these questions came from laboratory experiments. Geologists mixed chemical elements in proportions that simulated the compositions of natural igneous rocks, then melted those mixtures. As the melts cooled and solidified, the geologists observed and recorded the temperatures at which crystals formed, as well as the chemical compositions of those crystals. This research gave rise to the theory of **magmatic differentiation,** a process by which rocks of varying composition can arise from a uniform parent magma. Magmatic differentiation occurs because different minerals crystallize at different temperatures.

In a kind of mirror image of partial melting, the last minerals to melt are the first minerals to crystallize from a cooling magma. This initial crystallization withdraws chemical elements from the melt, changing the magma's composition. Continued cooling crystallizes the minerals that melted at the next lower temperature range. Again, the magma's chemical composition changes as various elements are withdrawn. Finally, as the magma solidifies completely, the last minerals to crystallize are the ones that melted first. Thus, the same parent magma, because of its changing chemical composition throughout the crystallization process, can give rise to different types of igneous rocks.

Fractional Crystallization: Laboratory and Field Observations

Fractional crystallization is the process by which the crystals formed in a cooling magma are segregated from the remaining liquid rock. This segregation happens in several ways, following a sequence commonly described as *Bowen's reaction series* (**Figure 4.6**). In the simplest scenario, crystals formed in a magma chamber settle to the chamber's floor and are therefore removed from further reaction with the remaining liquid. Thus, crystals that form early are segregated from the remaining magma, which continues to crystallize as it cools.

The effects of fractional crystallization can be seen in the Palisades, a line of imposing cliffs that faces the city of New York on the west bank of the Hudson River (**Figure 4.7**). This igneous formation is about 80 km long and, in places, more than 300 m high. It formed as a magma of basaltic composition intruded into nearly horizontal layers of sedimentary rock. It contains abundant olivine near the bottom, pyroxene and calcium-rich plagioclase feldspar in the middle, and mostly sodium-rich plagioclase feldspar near the top. This variation in mineral composition from bottom to top made the Palisades a perfect site for testing the theory of fractional crystallization.

Geologists melted rocks with about the same mineral compositions as those found in the Palisades intrusion and determined that the initial temperature of the magma from which it formed had to have been about 1200°C. The parts of the magma within a few meters of the relatively cold country rock above and below it cooled quickly. This quick cooling formed a fine-grained basalt and preserved the chemical composition of the original magma. The hot interior cooled more slowly, as evidenced by the slightly larger crystals found in the intrusion's interior.

The theory of fractional crystallization leads us to expect that the first mineral to crystallize from the slowly cooling interior of the Palisades intrusion would have been olivine, as this heavy mineral would sink through the melt to the bottom of the intrusion. It can be found today as a coarse-grained, olivine-rich layer just above the chilled, fine-grained basaltic layer along the bottom zone of contact with the underlying sedimentary rock. Plagioclase feldspar would have started to crystallize at about the same time; it has a lower density than olivine, however, and so would have settled to the bottom more slowly (see the Practicing Geology Exercise *How Do Valuable Metallic Ores Form?*). Continued cooling would have produced pyroxene crystals, which would have reached the bottom next, followed almost immediately by calcium-rich plagioclase feldspar. The abundance of plagioclase feldspar in the upper parts of the intrusion is evidence that the magma continued to change in composition until successive layers of settled crystals were topped off by a

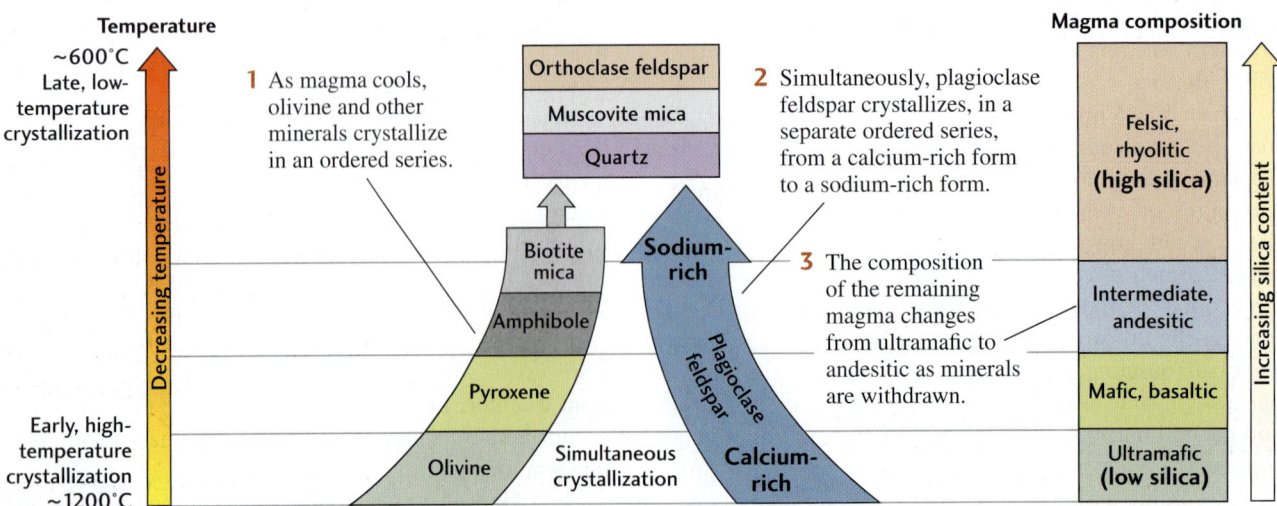

FIGURE 4.6 Bowen's reaction series provides a model of fractional crystallization.

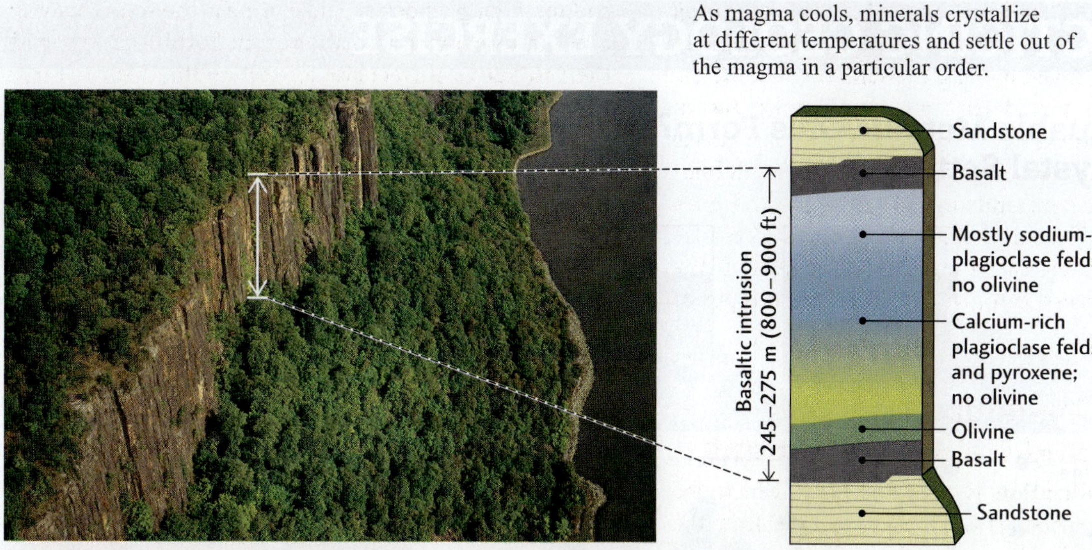

FIGURE 4.7 Fractional crystallization explains the composition of the basaltic intrusion that forms the Palisades. Minerals in the Palisades intrusion are ordered with olivine at the bottom, a gradient of pyroxene and calcium-rich plagioclase feldspar in the center, and sodium-rich plagioclase feldspar at the top. Layers of fine-grained basalt, which cooled quickly at the edges of the intrusion, surround the more slowly cooled interior. [Jim Wark/AirPhoto.]

layer of mostly sodium-rich plagioclase feldspar. In addition to crystallizing at a lower temperature, sodium-rich plagioclase feldspar is less dense than either olivine or pyroxene, so it would have settled out last.

Being able to explain the layering of the Palisades intrusion as the result of fractional crystallization was an early success in understanding magmatic differentiation. It firmly tied field observations to laboratory results and was solidly based on chemical knowledge. We now know that this intrusion actually has a more complex history that includes several injections of magma and a more complicated process of olivine settling. Nevertheless, the Palisades intrusion remains a valid example of fractional crystallization.

Granite from Basalt: Complexities of Magmatic Differentiation

Studies of volcanic lavas have shown that basaltic magmas are common—far more common than the rhyolitic magmas that correspond in composition to granites. How, then, could granites have become so abundant in Earth's crust? The answer is that the process of magmatic differentiation is much more complex than geologists first thought.

The original theory of magmatic differentiation suggested that a basaltic magma would gradually cool and differentiate into a more felsic magma by fractional crystallization. The early stages of this differentiation would produce an andesitic magma, which might erupt to form andesitic lavas or solidify by slow crystallization to form dioritic intrusions. Intermediate stages would result in magmas of granodioritic composition. If the process were carried far enough, its late stages would form rhyolitic lavas and granitic intrusions. One line of research has shown, however, that so much time would be needed for small crystals of olivine to settle through a dense, viscous magma that they might never reach the bottom of a magma chamber. Other researchers have demonstrated that many layered intrusions—similar to, but much larger than, the Palisades intrusion—do not show the simple progression of layers predicted by the original theory.

The greatest sticking point in the original theory, however, was the source of granite. The great volume of granite found on Earth could not have formed from basaltic magmas by magmatic differentiation, because large quantities of liquid volume are lost by crystallization during successive stages of differentiation. To produce a given amount of granite, an initial volume of basaltic magma 10 times that of the granitic intrusion would be required. Based on that observation, there should be huge quantities of basalt underlying granitic intrusions. But geologists could not find anything like that amount of basalt. Even where great volumes of basalt are found—at mid-ocean ridges—there is no wholesale conversion into granite through magmatic differentiation.

Most in question is the original idea that all granitic rocks are formed from the differentiation of a single type of magma, a basaltic melt. Instead, geologists now believe that the melting of varied rock types in the upper mantle and

PRACTICING GEOLOGY EXERCISE

How Do Valuable Metallic Ores Form? Magmatic Differentiation Through Crystal Settling

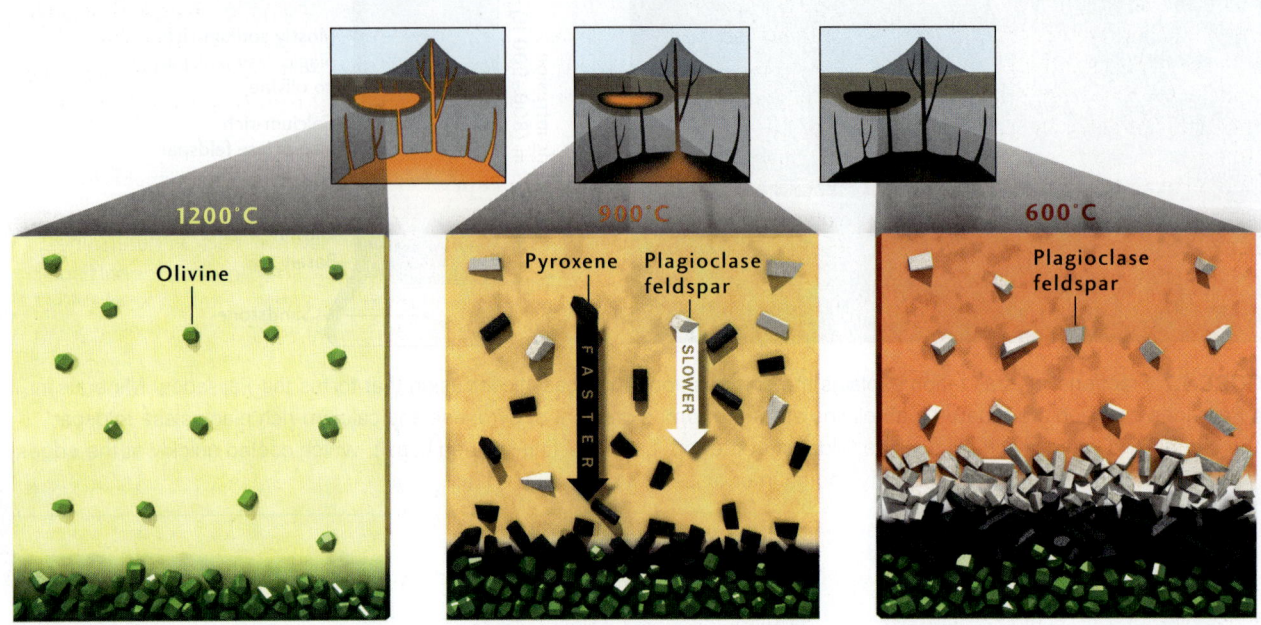

Fractional crystallization in the Palisades sill.

Some of the most economically important mineral deposits in the world are formed by differential settling of crystals in magma chambers. The Bushveld deposit in South Africa and the Stillwater deposit in Montana, just north of Yellowstone National Park, are two of the most famous examples. These deposits contain some of the world's largest reserves of metals in the platinum group—such as platinum and palladium—but vast quantities of iron, tin, chromium, and titanium are also found there. These deposits represent ancient magma chambers in which fractional crystallization led to the formation of different minerals over time, which settled to the bottom of the magma chamber in economically important concentrations. Geologists realized that the process of crystal settling was the key to understanding how the deposits formed.

Many geologic processes involve the movement of particles within crust is responsible for much of the observed variation in the composition of magmas:

1. Rocks in the upper mantle undergo partial melting to produce basaltic magmas.
2. Mixtures of sedimentary rock and basaltic oceanic crust, such as those found in subduction zones, melt to form andesitic magmas.
3. Mixtures of sedimentary, igneous, and metamorphic continental crustal rocks melt to produce granitic magmas.

Thus, the mechanisms of magmatic differentiation must be much more complex than first recognized in a number of ways:

- Magmatic differentiation can begin with the partial melting of mantle and crustal rocks with a range of water contents over a range of temperatures.
- Magmas do not cool uniformly; they may exist transiently at a range of temperatures within a magma chamber. Differences in temperature within and among

fluids. We see the same basic principles in the movement of sand grains in rivers, in the hurtling of debris through the atmosphere when a volcano erupts, and in the settling of crystals through magmas. In each case, the movement of the particles is regulated by a number of factors.

In the example of the Palisades sill, we saw that as a basaltic magma cools, olivine crystallizes first, followed by pyroxene and plagioclase feldspar. Once crystallized, each mineral sinks through the remaining liquid magma to settle out at the bottom of the magma chamber. Thus, the Palisades sill is layered with olivine at its base, followed by pyroxene and plagioclase feldspar overlying it (see Figure 4.7).

Olivine not only crystallizes before feldspar, but also settles more quickly than feldspar due to its higher density. Both fractional crystallization and crystal settling rates contribute to the segregation of minerals within magma chambers.

The rate at which crystals settle depends both on their density and size and on the viscosity of the remaining magma. That rate can be calculated using a mathematical relationship called *Stokes' law*:

$$V = \frac{gr^2(d_c - d_m)}{u}$$

where V is the velocity at which the crystals settle through the magma; g is the acceleration due to Earth's gravity (980 cm/s^2); r is the radius of the crystal; d_c is the density of the crystal; d_m is the density of the magma; and u is the viscosity of the magma.

Stokes' law states that three factors are important in regulating the velocity at which crystals will move through magma:

1. As crystals grow larger, their radius (r) increases. Because r is in the numerator of the equation, Stokes' law tells us that larger crystals will settle faster than smaller crystals. Furthermore, r is squared, which tells us that small increases in crystal size will result in much larger increases in their settling velocity.

2. Magma viscosity (u) is a measure of the resistance of the magma to flow or, in this case, to moving out of the way of a sinking crystal. Because u is in the denominator of the equation, it tells us that an increase in magma viscosity will result in a decrease in settling velocity.

3. Settling velocity (V) also depends on the difference between the density of the crystal (d_c) and the density of the magma (d_m). V will increase as the density of the crystal increases and as the density of the magma decreases. Thus, in a magma of constant density, crystals with higher density will settle faster than crystals with lower density.

If we now consider fractional crystallization in the Palisades sill, Stokes' law will help us determine the actual settling rates for particular minerals. Consider an olivine crystal with a radius of 0.1 cm and a density of 3.7 g/cm^3. The magma through which the crystal settles has a density of 2.6 g/cm^3 and a viscosity of 3000 poise (1 poise = 1 g/cm^3 s). How fast will this olivine crystal fall through the magma?

$$V = \frac{(980 \text{ cm/s}^2) \times (0.1)^2 \times (3.7 - 2.6 \text{ g/cm}^3)}{3000 \text{ g/cm}^3}$$

$$= 0.0036 \text{ cm/s}$$

$$= 12.9 \text{ cm/hr}$$

BONUS PROBLEM: Try the same calculation for a plagioclase feldspar crystal of the same size, which has a density of 2.7 g/cm^3. Which mineral settles through the magma at a faster rate? How much faster does it settle?

magma chambers may cause the chemical composition of magma to vary from one region to another.

- A few magma types are *immiscible*—they do not mix with one another, just as oil and water do not mix. When such magmas coexist in one magma chamber, each forms its own fractional crystallization series. Magmas that are *miscible*—that *do* mix—may follow a crystallization path different from that followed by any one magma type alone.

We now know more about the physical processes that interact with fractional crystallization within magma chambers (**Figure 4.8**). Magma at various temperatures in different parts of a magma chamber may flow turbulently, crystallizing as it circulates. Crystals may settle, then be caught up in currents again, and eventually be deposited on the chamber's walls. The margins of such a magma chamber may be a "mushy" zone of crystals and melt lying between the solid rock border of the chamber and the completely liquid magma within the heart of the chamber. And, at some mid-ocean ridges, such as the East Pacific Rise, a mushroom-shaped magma chamber may be surrounded by hot basaltic rock containing only small amounts (1 to 3 percent) of partial melt.

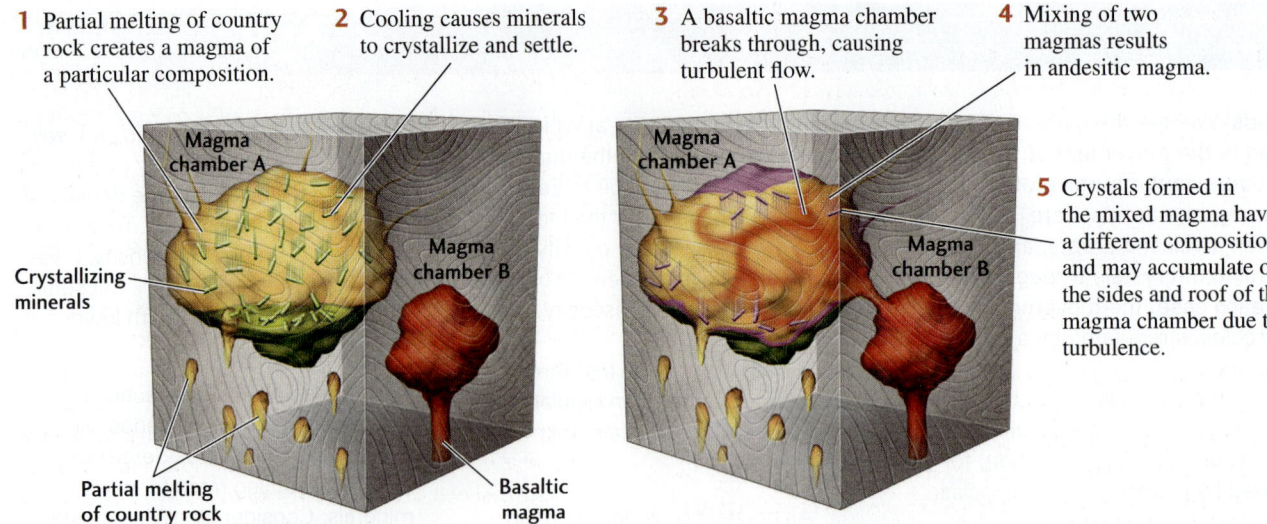

FIGURE 4.8 Magmatic differentiation is a more complex process than first recognized. Some magmas derived from rocks of varying compositions may mix, whereas others are immiscible. Crystals may be transported to various parts of a magma chamber by turbulent flow in the liquid magma.

Forms of Igneous Intrusions

As noted earlier, we cannot directly observe the shapes of igneous intrusions. We can deduce their shapes and distributions only by observing parts of them where they have been uplifted and exposed by erosion, millions of years after the magma that formed them intruded and cooled.

We do have indirect evidence of current magmatic activity. Seismic waves, for example, have shown us the general outlines of the magma chambers that underlie some active volcanoes. In some nonvolcanic but tectonically active regions, such as an area near the Salton Sea in Southern California, measurements in deep drill holes reveal crustal temperatures much hotter than normal, which may be evidence of a magmatic intrusion nearby. But these methods cannot reveal the detailed shapes or sizes of intrusions.

Most of what we know about igneous intrusions is based on the work of field geologists who have examined and compared a wide variety of outcrops and have reconstructed their histories. In the following pages, we consider some of these forms. **Figure 4.9** illustrates a variety of intrusive and extrusive structures.

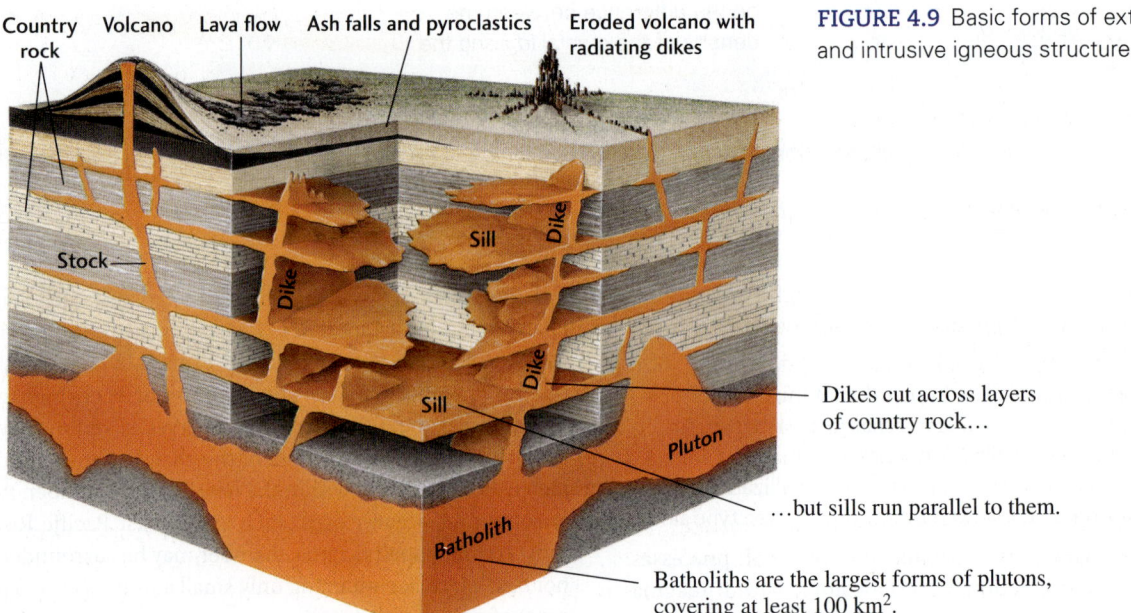

FIGURE 4.9 Basic forms of extrusive and intrusive igneous structures.

Plutons

Plutons are large igneous bodies formed deep in Earth's crust. They range in size from a cubic kilometer to hundreds of cubic kilometers. We can study these large bodies when uplift and erosion uncover them or when mines or drill holes cut into them. Plutons are highly variable, not only in size but also in shape and in their relationship to the country rock.

This wide variation is due in part to the different ways in which magma makes space for itself as it rises through the crust. Most plutons intrude at depths greater than 8 to 10 km. At these depths, there are few holes or openings in the country rock because the pressure of the overlying rock would close them. But the upwelling magma overcomes even that great pressure.

Magma rising through the crust makes space for itself in three ways (Figure 4.10) that may be referred to collectively as *magmatic stoping*:

1. *Wedging open the overlying rock.* As the rising magma lifts the great weight of the overlying rock, it fractures that rock, penetrates the cracks, wedges them open, and so flows into the rock. Overlying rocks may bow upward during this process.
2. *Melting surrounding rock.* Magma also makes its way by melting country rock.
3. *Breaking off large blocks of rock.* Magma can push its way upward by breaking off blocks of the invaded crust. These blocks, known as *xenoliths* (from the Greek for "foreign rocks"), sink into the magma, where they may melt and blend into the liquid, in some places changing the composition of the magma.

Most plutons show sharp zones of contact with country rock and other evidence of the intrusion of liquid magma into solid rock. Some plutons grade into country rock and contain structures vaguely resembling those of sedimentary rocks. The features of these plutons suggest that they formed by partial or complete melting of preexisting sedimentary rock.

Batholiths, the largest plutons, are great irregular masses of coarse-grained igneous rock that, by definition, cover at least 100 km² (see Figure 4.10). They are thick, horizontal, sheetlike or lobe-shaped bodies extending from a funnel-shaped central region. Their bottoms may extend 10 to 15 km deep, and a few are estimated to go even deeper. The coarse grain of batholiths results from slow cooling at great depths. Other, smaller plutons are called **stocks**. Both batholiths and stocks are **discordant intrusions**; that is, they cut across the layers of the country rock that they intrude.

Sills and Dikes

Sills and dikes are similar to plutons in many ways, but they are smaller and have a different relationship to the layering of the country rock (Figure 4.11). A **sill** is a sheetlike body formed by the injection of magma between parallel layers of bedded country rock. Sills are **concordant intrusions**; that is, their boundaries lie parallel to the country rock layers, whether or not those layers are horizontal. Sills range in thickness from a single centimeter to hundreds of meters, and they can extend over considerable areas. Figure 4.11a shows a large sill at Finger Mountain, Antarctica. The 300-m-thick Palisades intrusion (see Figure 4.7) is another large sill.

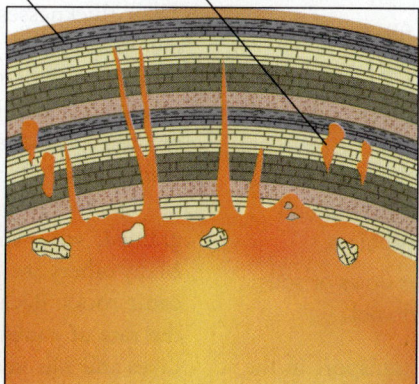
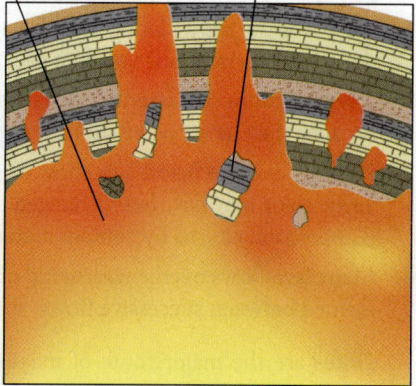

FIGURE 4.10 Magmas make their way into country rock in three basic ways: by invading cracks and wedging open overlying rock, by melting country rock, and by breaking off pieces of rock. Pieces of broken-off country rock, called xenoliths, can become completely dissolved in the magma. If the country rock differs in composition from the magma, the composition of the magma may change.

FIGURE 4.11 Sills and dikes. (a) Sills are concordant intrusions. In Glacier National Park, Montana, the diorite sill intrudes into a sequence of sedimentary rocks. (b) Dikes are discordant intrusions. This dike of igneous rock (dark) intrudes into sedimentary rock in Grand Canyon National Park, Arizona. [(a) Marli Bryant Miller; (b) Asa Thorsen/Photo Researchers/Getty Images.]

Sills may superficially resemble lava flows and pyroclastic deposits, but they differ from those layers in four ways:

1. They lack the ropy, blocky, and vesicle-filled structures that characterize many volcanic rocks (see Chapter 12).
2. They are more coarse-grained than volcanic rocks because they have cooled more slowly.
3. Rocks above and below sills show the effects of heating: Their color may have been changed or their mineral composition altered by contact metamorphism.
4. Many lava flows overlie weathered older flows or soils formed between successive flows; sills do not.

Dikes are the major route of magma transport in the crust. Dikes, like sills, are sheetlike igneous bodies, but dikes cut across the layers in bedded country rock (Figure 4.11b) and so are discordant intrusions. Dikes sometimes form by forcing open existing fractures in the country rock, but more often they create channels through new cracks opened by the pressure of rising magma. Some individual dikes can be followed in the field for tens of kilometers. Their thicknesses vary from many meters to a few centimeters. In some dikes, xenoliths provide evidence of disruption of the country rock during the intrusion process. Dikes rarely exist alone; more typically, swarms of hundreds or thousands of dikes are found in a region that has been deformed by a large igneous intrusion.

The textures of dikes and sills vary. Many are coarse-grained, with an appearance typical of intrusive rocks. Many others are fine-grained and look much more like volcanic rocks. Because we know that texture corresponds to the rate of cooling, we can conclude that the fine-grained dikes and sills invaded country rock nearer Earth's surface, where the country rock was cold compared with the intrusions. Their fine texture is the result of rapid cooling. The coarse-grained ones formed at depths of many kilometers and invaded warmer rocks whose temperatures were much closer to their own.

FIGURE 4.12 A granitic pegmatite vein. The center of the intrusion (*upper right*) cooled more slowly and developed coarser crystals. The margin of the intrusion (*lower left*) has finer crystals due to more rapid cooling. [John Grotzinger/Ramón Rivera-Moret/Harvard Mineralogical Museum.]

Veins

Veins are deposits of minerals found within a rock fracture that are foreign to the country rock. Irregular pencil-shaped or sheet-shaped veins branch off from the tops and sides of many igneous intrusions. They may be a few millimeters to several meters across, and they tend to be tens of meters to kilometers long or wide. The formation of veins is described in more detail in Chapter 3.

Veins of extremely coarse-grained granite cutting across much finer-grained country rock are called **pegmatites** (**Figure 4.12**). Pegmatites crystallize from a water-rich magma in the late stages of solidification.

Other veins are filled with hydrous minerals that are known to crystallize from hydrothermal solutions. From laboratory experiments, we know that these minerals typically crystallize at temperatures of 250°C to 350°C—high temperatures, but not nearly as high as the temperatures of magmas. The solubility and composition of the minerals in these hydrothermal veins indicate that abundant water was present as the veins formed. Some of the water may have come from the magma itself, but some may have been underground water in the cracks and pore spaces of the intruded rocks. On land, groundwaters originate as rainwater seeps into the soil and surface rocks. Hydrothermal veins are also abundant along mid-ocean ridges, where seawater infiltrates cracks in the newly formed seafloor, circulates down into hotter regions of the ridge, and reemerges at hydrothermal vents in the rift valley between the spreading plates. Hydrothermal processes at mid-ocean ridges are examined in more detail in Chapter 12.

Igneous Processes and Plate Tectonics

Geologists have observed that the facts and theories of igneous rock formation fit nicely into a framework based on plate tectonic theory. The geometry of plate movements is the link we need to tie tectonic activity and rock composition to igneous processes (**Figure 4.13**). Batholiths, for example, are found in the cores of many mountain ranges formed by the convergence of two plates. This observation implies a connection between pluton formation and the mountain-building process, and between both of those processes and plate movements. Similarly, our knowledge of the temperatures and pressures at which different kinds of rock melt gives us some idea of where melting takes place. For example, we know that mixtures of sedimentary rocks, because of their composition and water content, should melt at temperatures several hundred degrees below the melting point of basalt. This information leads us to predict that basalt will start to melt near the base of the crust in tectonically active regions of the upper mantle and that sedimentary rocks will melt at shallower depths.

Magma forms most abundantly in two plate tectonic settings: mid-ocean ridges, where two plates diverge and the seafloor spreads, and subduction zones, where one plate dives beneath another. Mantle plumes, though not associated with plate boundaries, also produce large amounts of magma.

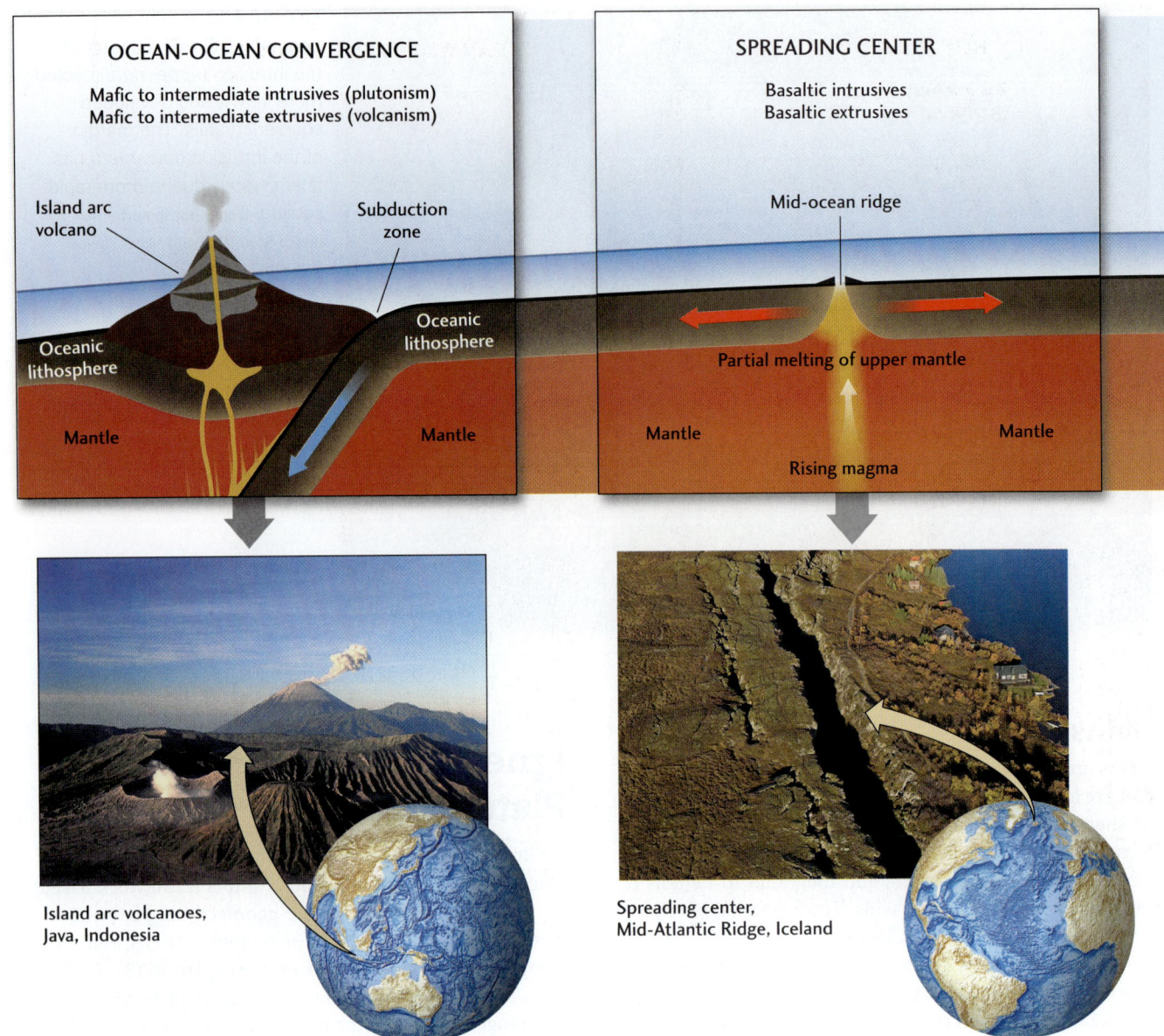

FIGURE 4.13 Magmatic activity is related to plate tectonic settings. [Photos (left to right): 24BY36/Alamy; Ragnar Th Sigurdsson © Arctic Images/Alamy; G. Brad Lewis/Stone/Getty Images; © Michael Sedam/Age Fotostock.]

Spreading Centers as Magma Factories

Most igneous rocks are formed at mid-ocean ridges by the process of seafloor spreading. Each year, approximately 19 km³ of basaltic magma is produced along the mid-ocean ridges in the process of seafloor spreading—a truly enormous volume. In comparison, all the active volcanoes along convergent plate boundaries (about 400) generate volcanic rock at a rate of less than 1 km³/year. Enough magma has erupted during seafloor spreading over the past 200 million years to create all of the present-day seafloor, which covers nearly two-thirds of Earth's surface. Throughout the mid-ocean ridge network, decompression melting of mantle material that wells up along rising convection currents creates magma chambers below the ridge axis. These magmas are extruded as lavas and form new seafloor. At the same time, intrusions of gabbro are emplaced at depth.

Before the advent of plate tectonic theory, geologists were puzzled by unusual assemblages of rocks that were characteristic of the seafloor but were found on land. Known as **ophiolite suites,** these assemblages consist of deep-sea sediments, submarine basaltic lavas, and mafic igneous

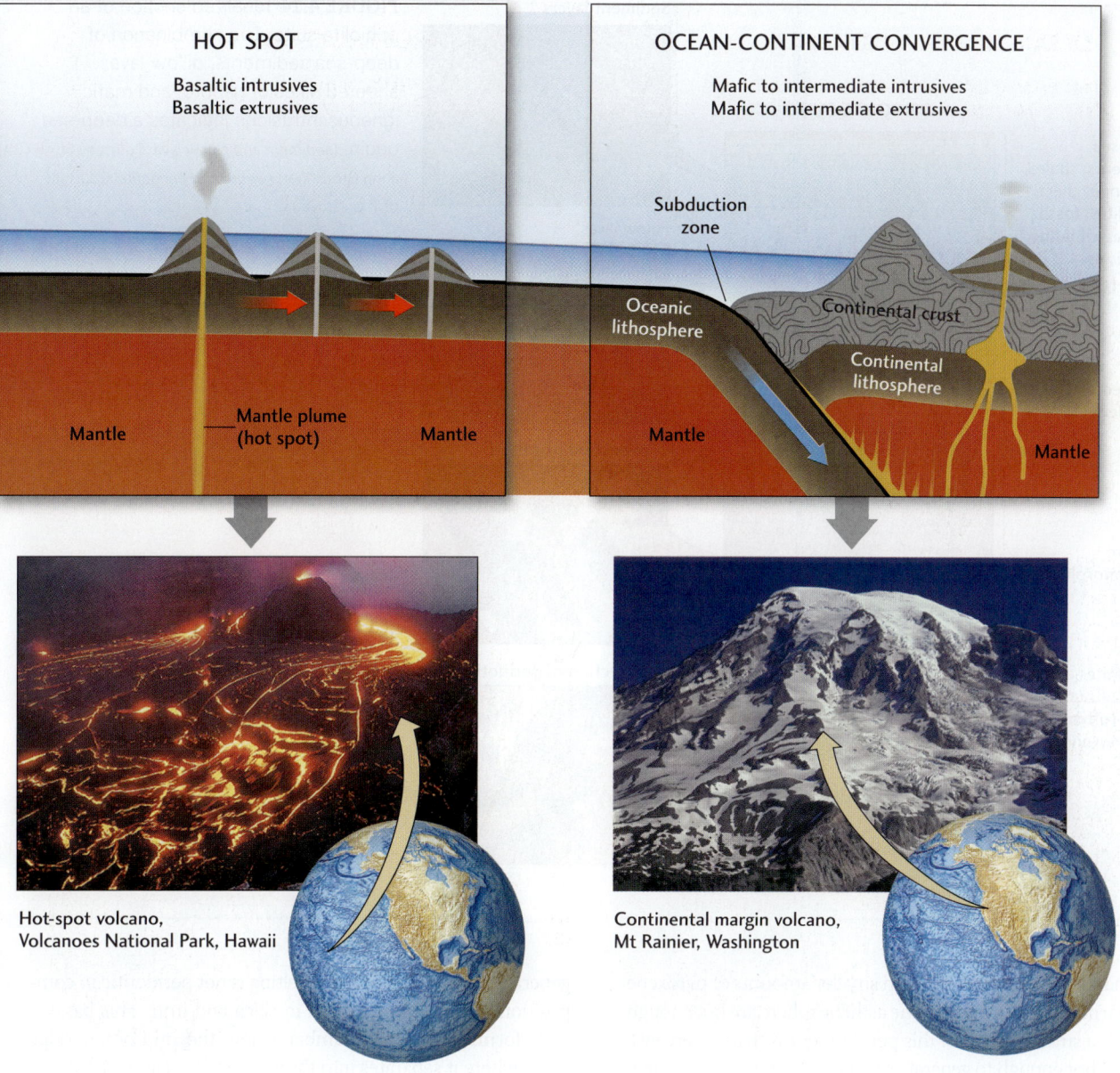

intrusions (**Figure 4.14**). Using data gathered from deep-diving submarines, dredging, deep-sea drilling, and seismic exploration, geologists now explain these rocks as fragments of oceanic lithosphere that were transported by seafloor spreading and then raised above sea level and thrust onto a continent in a later episode of plate collision. On some of the more complete ophiolite suites preserved on land, we can literally walk across rocks that used to lie along the boundary between Earth's oceanic crust and mantle.

How does seafloor spreading work? We can think of a spreading center as a huge factory that processes mantle material to produce oceanic crust. **Figure 4.15** is a highly schematic and simplified representation of what may be happening, based in part on studies of ophiolite suites found on land and on information gleaned from deep-sea drilling and seismic profiling. Deep-sea drilling has penetrated to the gabbro layer of the seafloor, but not to the crust-mantle boundary below. Seismic profiling has found several small magma chambers similar to the one shown in Figure 4.15.

Input Material: Peridotite in the Mantle The raw material fed into this magma factory comes from the convecting asthenosphere, in which the dominant rock type is peridotite. The mineral composition of the average peridotite in

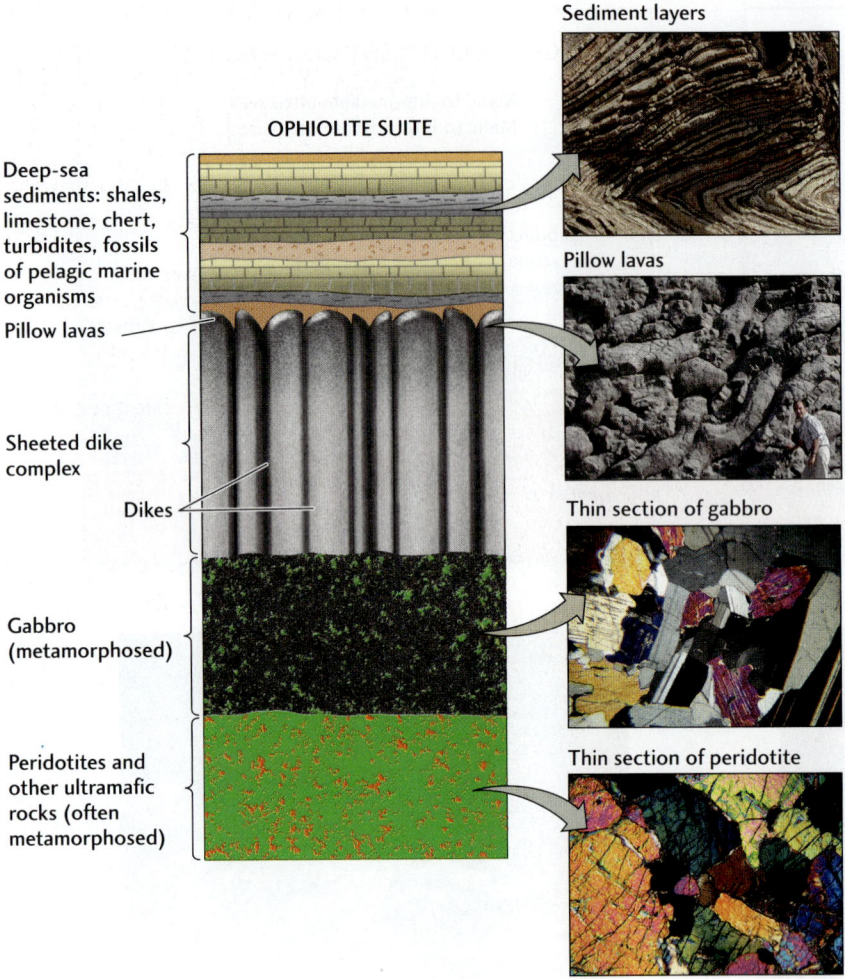

FIGURE 4.14 Idealized section of an ophiolite suite. The combination of deep-sea sediments, pillow lavas, sheeted dikes of gabbro, and mafic igneous intrusions indicates a deep-sea origin. [sediment and pillow lava: Courtesy of John Grotzinger; gabbro and peridotite: Courtesy of T. L. Grove.]

the mantle is chiefly olivine, with smaller amounts of pyroxene and garnet. Temperatures in the asthenosphere are hot enough to melt a small fraction of this peridotite (less than 1 percent), but not hot enough to generate substantial volumes of magma.

Process: Decompression Melting Decompression melting is the process that generates great volumes of magma from peridotite at spreading centers. Recall that a decrease in pressure generally lowers a mineral's melting temperature. As the plates pull apart, the partially molten peridotite is sucked inward and upward toward the spreading center. The decrease in pressure as the peridotite rises causes a large fraction of the rock (up to 15 percent) to melt. The buoyancy of the melt causes it to rise faster than the denser surrounding rock. This process separates the liquid rock from the remaining crystal mush to produce large volumes of magma.

Output Material: Oceanic Crust Plus Mantle Lithosphere The peridotites subjected to this process do not melt evenly: The garnet and pyroxenes they contain melt at lower temperatures than the olivine. For this reason, the magma generated by decompression melting is not peridotitic in composition; rather, it is enriched in silica and iron. This basaltic melt forms a magma chamber below the mid-ocean ridge crest, where it separates into three layers (see Figure 4.15):

1. Some magma rises through the narrow cracks that open where the plates are separating and erupts into the ocean, forming the basaltic *pillow lavas* that cover the seafloor.
2. Some magma solidifies in the cracks as vertical, sheeted dikes of gabbro.
3. The remaining magma solidifies as massive intrusions of gabbro as the underlying magma chamber is pulled apart by seafloor spreading.

These igneous units—pillow lavas, sheeted dikes, and massive gabbros—are the basic layers of the crust that geologists have found throughout the world's oceans.

Seafloor spreading results in another layer beneath this oceanic crust: the residual peridotite from which the basaltic magma was originally derived. Geologists consider this layer to be part of the mantle, but its composition is different

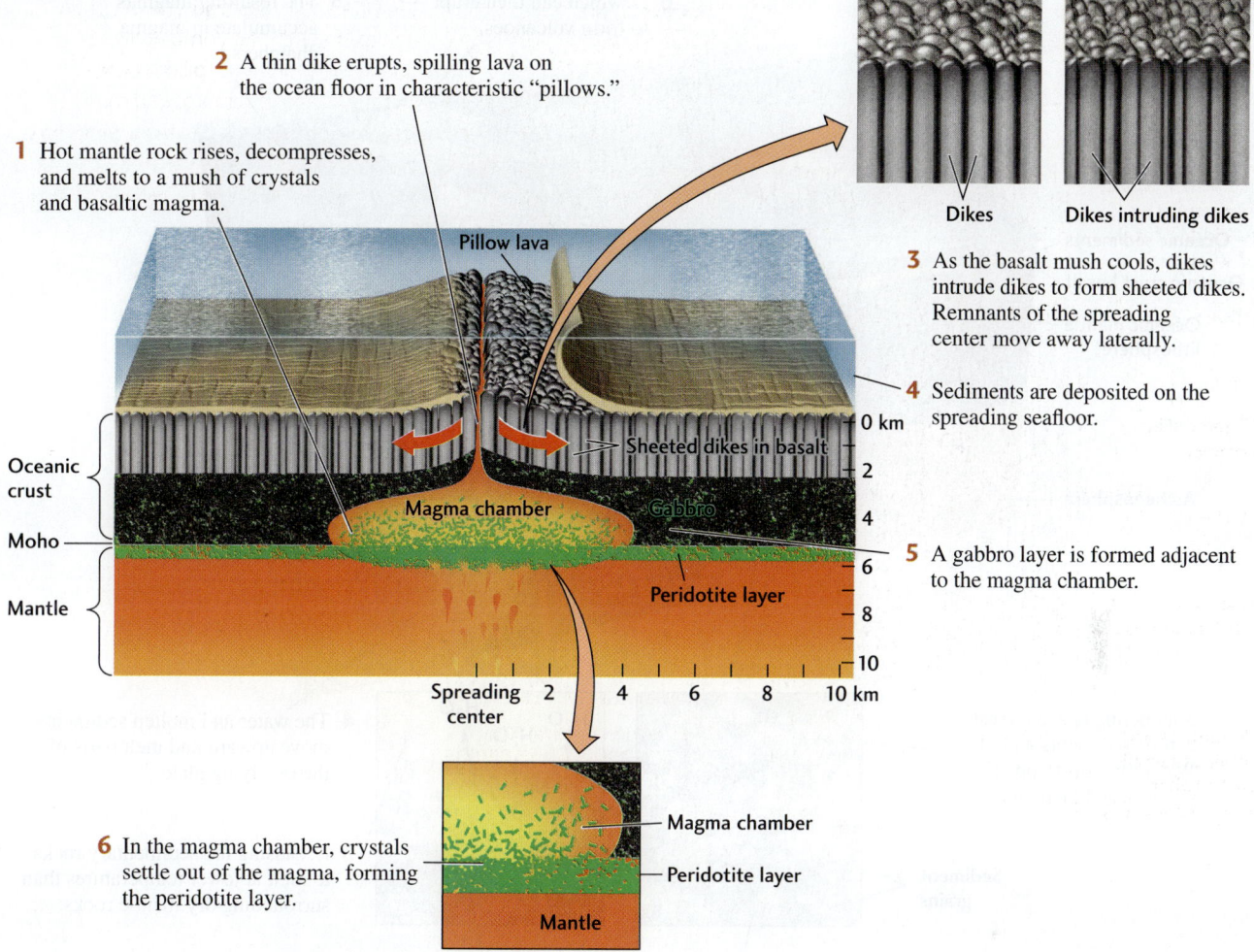

FIGURE 4.15 Decompression melting creates magma at seafloor spreading centers.

from that of the convecting asthenosphere. In particular, the extraction of the basaltic melt makes the residual peridotite richer in olivine and stronger than ordinary mantle material. Geologists now believe it is this olivine-rich layer at the top of the mantle that gives the oceanic lithosphere its great rigidity.

Above the pillow lavas, a blanket of deep-sea sediment begins to cover the newly formed oceanic crust. As the seafloor spreads, these layers of sediment, pillow lavas, dikes, and gabbro are transported away from the mid-ocean ridge where this characteristic sequence of rocks is assembled, almost as if they were moving along a production line.

Subduction Zones as Magma Factories

Other types of magmas underlie regions where volcanoes are highly concentrated, such as the Andes of South America and the Aleutian Islands of Alaska. Both of these regions lie over subduction zones, which are major magma factories (**Figure 4.16**). They generate magmas of varying composition, depending on how much and what kinds of materials are subducted.

Where oceanic lithosphere is subducted beneath a continent, the resulting volcanoes and volcanic rocks form a volcanic mountain belt on the continent. The Andes, which mark the subduction of the Nazca Plate beneath South America, are one such mountain belt. Similarly, subduction of the small Juan de Fuca Plate beneath western North America has generated the Cascade Range, with its active volcanoes, in northern California, Oregon, and Washington. Where oceanic lithosphere is subducted beneath oceanic lithosphere, a deep-sea trench and a volcanic island arc are formed.

Input Material: A Mixed Bag The variation in the chemical and mineral composition of magmas at subduction zones is a clue that the magma factories at convergent boundaries operate differently from those at spreading centers. The raw materials for these magma factories include mixtures of seafloor sediments, mixtures of basaltic oceanic crust and felsic continental crust, mantle peridotite, and water.

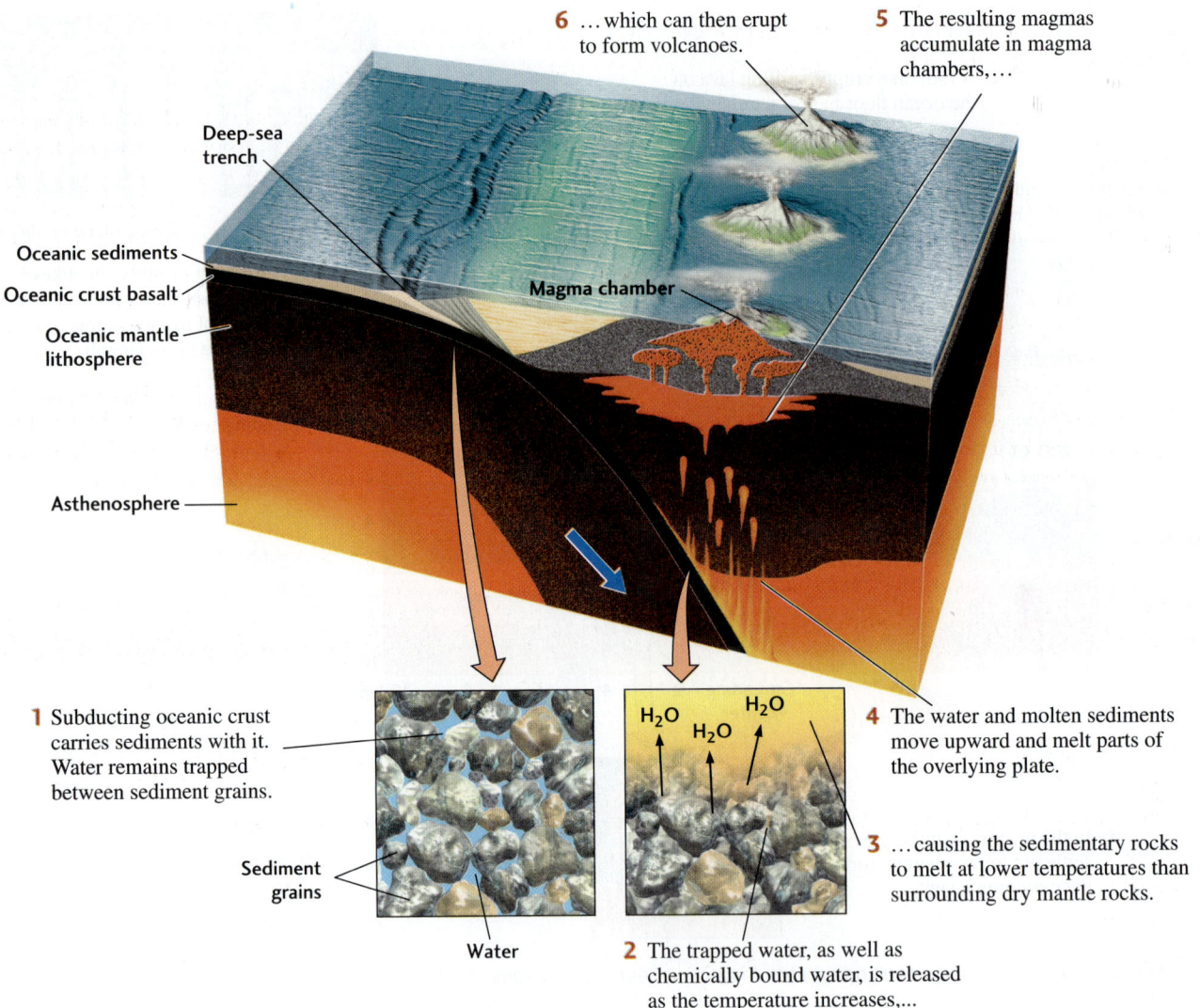

FIGURE 4.16 Fluid-induced melting creates magma in subduction zones.

Process: Fluid-Induced Melting The basic mechanism of magma formation at subduction zones is fluid-induced melting. The fluid involved is primarily water, which, as we have seen, lowers the melting temperature of rock. By the time oceanic lithosphere is subducted at a convergent plate boundary, a lot of water has been incorporated into its outer layers. We have already mentioned one of the processes responsible: hydrothermal activity during the formation of oceanic lithosphere. Some of the seawater that circulates through the crust near a spreading center reacts with basalt to form new hydrous minerals. In addition, as the lithosphere ages and is transported across the ocean basin, sediments containing water are deposited on its surface. The rocks formed from these sediments include shales, which contain large proportions of hydrous clay minerals. Some of these sediments get scraped off the subducting plate at the deep-sea trench, but much of this water-laden material is carried downward into the subduction zone.

As the lithospheric slab moves downward, it is subjected to increasing pressure. Water is squeezed out of the minerals in the outer layers of the descending crust and rises buoyantly into the mantle above the descending crust. At moderate depths of about 5 km, the temperature increases to about 150°C. Here, more water is released by metamorphic chemical reactions as basalt is converted to *amphibolite*, which is composed of amphibole and plagioclase feldspar (see Chapter 6). As other chemical reactions take place, additional water is released at depths ranging from 10 to 20 km. Finally, at depths greater than 100 km, the temperature increases to 1200°C–1500°C, and the subducted slab undergoes an additional metamorphic transition induced by the increased pressure. Amphibolite is converted to *eclogite*, which is composed of pyroxene and garnet (see Chapter 6). Here, the increase in both pressure and temperature in the subducting slab releases all of its remaining water in addition to other materials.

During subduction, the released water induces melting in the descending basalt-rich oceanic crust and in the overlying peridotite-rich mantle material. Most of the resulting mafic magma accumulates at the base of the crust of the overriding plate, and some of it intrudes into the crust to form magma chambers, resulting in the formation of volcanoes.

Output: Magmas Of Varying Composition The magmas produced by fluid-induced melting at subduction zones are essentially basaltic, although their composition is more variable than that of mid-ocean ridge basalts. The composition of these magmas is further altered during their residence in the crust. Within magma chambers, the process of fractional crystallization increases the magma's silica content, producing eruptions of andesitic lavas. Where the overlying plate is continental, the heat from the magmas can melt the felsic rocks in the crust, forming magmas with an even higher silica content, such as dacitic and rhyolitic magmas. The contribution of lithospheric fluids to these magmas is suggested by the presence in the magmas of trace elements known to be present in oceanic crust and sediments.

Mantle Plumes as Magma Factories

Mantle plumes, like spreading centers, are sites of decompression melting, but they differ from spreading centers by forming within lithospheric plates rather than along the margins of plates. These plumes of hot mantle material rise from deep within Earth, possibly as deep as the core-mantle boundary. Mantle plumes that reach the surface, most of them far from plate boundaries, form the "hot spots" of Earth. At these locations, basaltic magmas produced by decompression melting of mantle material may erupt in huge outpourings to form islands, such as the Hawaiian Islands, or basalt plateaus, such as the Columbia Plateau in the Pacific Northwest of North America. Mantle plumes and hot spots are discussed in more detail in Chapter 12.

KEY TERMS AND CONCEPTS

andesite (p. 97)
basalt (p. 97)
batholith (p. 105)
bomb (p. 94)
concordant intrusion (p. 105)
country rock (p. 94)
dacite (p. 97)
decompression melting (p. 98)
dike (p. 106)
diorite (p. 97)
discordant intrusion (p. 105)
extrusive igneous rock (p. 94)
felsic rock (p. 96)
fluid-induced melting (p. 99)

fractional crystallization (p. 100)
gabbro (p. 97)
granite (p. 96)
granodiorite (p. 97)
intermediate igneous rock (p. 96)
intrusive igneous rock (p. 94)
lava (p. 92)
mafic rock (p. 97)
magma chamber (p. 99)
magmatic differentiation (p. 99)
obsidian (p. 94)
ophiolite suite (p. 108)
partial melting (p. 98)
pegmatite (p. 107)

peridotite (p. 97)
pluton (p. 105)
porphyry (p. 94)
pumice (p. 94)
pyroclast (p. 94)
rhyolite (p. 96)
sill (p. 105)
stock (p. 105)
ultramafic rock (p. 97)
vein (p. 107)
viscosity (p. 97)
volcanic ash (p. 94)

REVIEW OF LEARNING OBJECTIVES

4.1 Explain how igneous rocks can be classified and differentiated from one another.

Igneous rocks can be divided into two broad textural classes: coarse-grained rocks, which are intrusive and therefore cooled slowly; and fine-grained rocks, which are extrusive and cooled rapidly. Igneous rocks can also be classified on the basis of their chemical and mineral composition. Modern classification schemes group igneous rocks on the basis of their silica content using a scale that runs from felsic (rich in silica) to ultramafic (poor in silica). Mafic minerals crystallize at higher temperatures (earlier in the cooling of the magma) than felsic minerals.

Study Assignments: Figures 4.3 and 4.4

Exercise: Why are intrusive igneous rocks coarse-grained and extrusive rocks fine-grained?

Thought Question: How would you classify a coarse-grained igneous rock that contains about 50 percent pyroxene and 50 percent olivine?

4.2 Describe how magmas form and what causes a rock to melt.

Magmas form at places in the lower crust and mantle where temperatures are high enough for partial melting of rock. Because the minerals within a rock melt at different temperatures, the composition of magmas varies with temperature. Pressure raises the melting temperature of rock, and the presence of water lowers it. Because melted rock is less dense than solid rock, magma rises through the surrounding rock, and drops of magma come together to form magma chambers. The density of melted rock is lower than the density of solid rock of the same composition. If the less-dense melted rock is given the opportunity to move, it will move upward. Because liquids can flow, a partial melt can move upward through pores and along the boundaries between crystals of the surrounding solid rock. The rise of magmas through the mantle and crust may be slow or rapid.

Study Assignments: Table 4.2, Figure 4.5

Exercise: Using Figure 4.5, design an experiment that would allow you to determine the silica content of an igneous rock.

Thought Question: What factors would affect the ratio of solid to partial melt?

4.3 Demonstrate how magmas of different compositions can be made from a uniform parent material.

Magmatic differentiation is a process by which rocks of varying composition can arise from a uniform parent magma. Because different minerals crystallize at different temperatures, the composition of magma changes as it cools and various minerals are lost through crystallization. Fractional crystallization is the process by which the crystals formed in a cooling magma are segregated from the remaining liquid rock. In a simple scenario, crystals formed in the magma chamber settle to the bottom, and are removed from further reaction with the remaining liquid.

Study Assignments: Figure 4.6, Figure 4.8

Exercise: Describe how a change in temperature of a cooling melt can change feldspar composition.

Thought Question: How could a rock that is composed almost entirely of olivine form?

4.4 Summarize the types of igneous intrusions.

Large intrusive igneous bodies are called plutons. The largest plutons are batholiths, which are thick horizontal masses extending from a funnel-shaped central region. Stocks are smaller plutons. Less massive than plutons are sills, which lie parallel to the layers of bedded country rock, and dikes, which cut across those layers. Veins form where water is abundant, either in the magma or in the surrounding country rock.

Study Assignment: Figure 4.9

Exercise: Describe the differences between a volcano and an igneous intrusion. What kinds of textures would support your arguments?

Thought Question: What observations might you make to show that a dike postdates a set of sedimentary "country rock"?

4.5 Discuss the link between igneous processes and plate tectonics.

Magmas are produced at two types of plate boundaries. At spreading centers, peridotite rises from the mantle and undergoes decompression melting to form basaltic magma. At subduction zones, subducting oceanic lithosphere undergoes fluid-induced melting to generate magmas of varying composition. Mantle plumes within lithospheric plates are also sites of decompression melting that produce basaltic magmas.

Study Assignment: Figure 4.13

Exercise: Outline how plate tectonic processes cause decompression melting.

Thought Question: Why might volcanic rocks of intermediate composition form above subduction zones, whereas magmas of mafic composition form above spreading centers?

VISUAL LITERACY EXERCISE

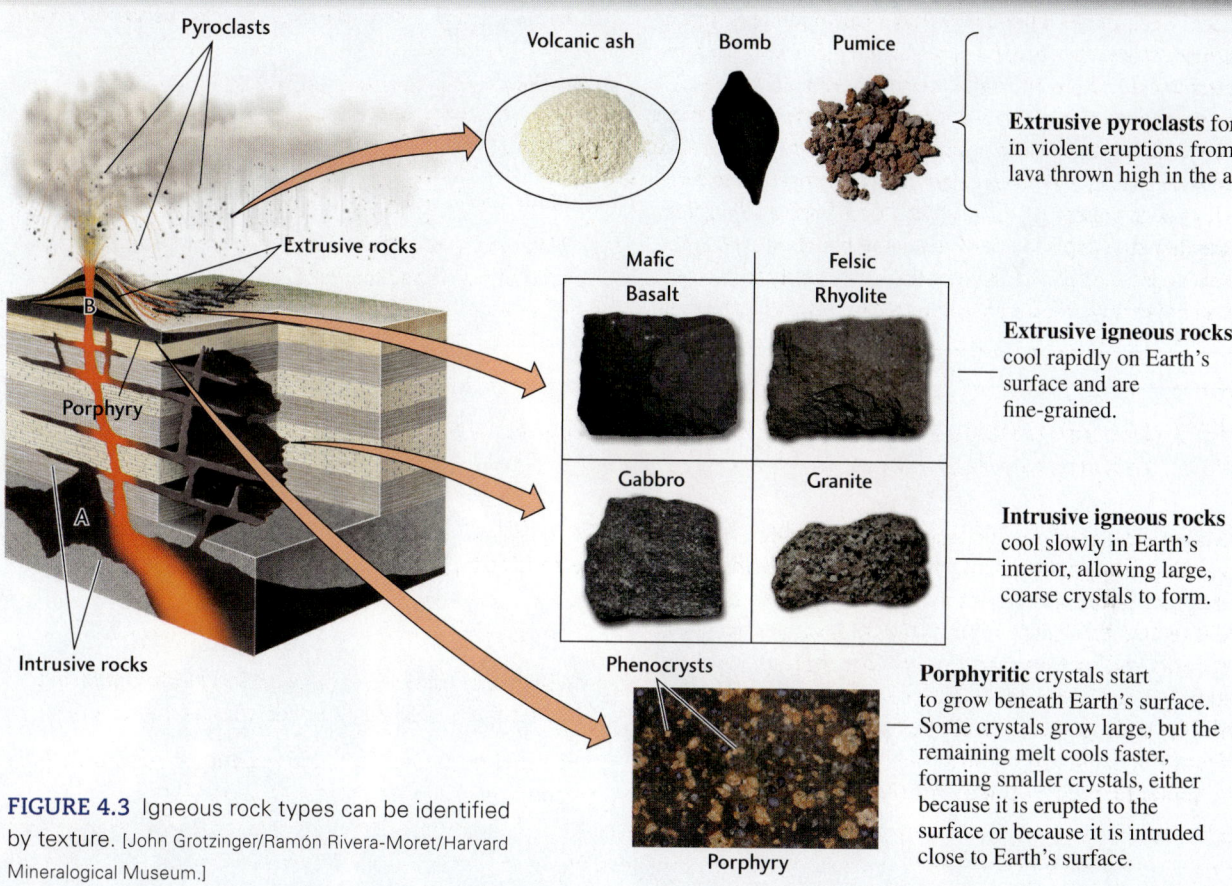

FIGURE 4.3 Igneous rock types can be identified by texture. [John Grotzinger/Ramón Rivera-Moret/Harvard Mineralogical Museum.]

1. What makes granite and gabbro different?
    a. Cooling rate (fast vs. slow)
    b. Size of minerals (large vs. small)
    c. Location of formation (at the surface vs. deep in the crust)
    d. Composition (mafic vs. felsic)

2. Which rocks could form at Location A and Location B, respectively?
    a. A: Basalt and B: Gabbro
    b. A: Gabbro and B: Granite
    c. A: Granite and B: Rhyolite
    d. A: Rhyolite and B: Basalt

3. The early part of the formation of a porphyry is similar to the formation of what type of rock?
    a. Extrusive igneous pyroclasts
    b. Extrusive igneous rocks
    c. Intrusive igneous rocks

4. Which is the finest-grained volcanic material?
    a. Ash
    b. Bombs
    c. Pumice
    d. Lava

5. Where do porphyritic rocks form?
    a. In Earth's mantle
    b. In Earth's core
    c. At shallow levels in the crust
    d. At deep levels in the crust

Volcanoes 5

Volcanoes as Geosystems	118
Lavas and Other Volcanic Deposits	119
Eruptive Styles and Landforms	124
Interactions of Volcanoes with Other Geosystems	130
The Global Pattern of Volcanism	133
Volcanism and Human Affairs	140

Learning Objectives

After you have studied this chapter, you should be able to:

5.1 Describe how volcanoes transport magma from the Earth's interior to its surface.

5.2 Differentiate between the major types of volcanic deposits and explain how the textures of volcanic rocks can reflect the conditions under which they solidified.

5.3 Summarize how volcanic landforms are shaped.

5.4 Discuss how volcanic gases can affect the hydrosphere and atmosphere.

5.5 Explain how the global pattern of volcanism is related to plate tectonics.

5.6 Illustrate the hazards and beneficial effects of volcanism.

Soufriere Hills is a stratovolcano composed of alternating layers of hardened lava, solidified ash, and rocks ejected by previous eruptions. The eruption of Soufriere Hills, Montserrat, Caribbean, began on Friday, January 8th, 2010. Residents said it was one of the largest eruptions they have witnessed at the volcano since its reawakening in 1995. Scientists don't believe there was a major collapse of the dome, but a significant amount of material was lost. After the seventeenth century, the volcano experienced no recorded eruptions until 1995, when a series of major eruptions eventually forced the evacuation of Montserrat's former capital, Plymouth. [Marco Fulle/Barcroft Media.]

The northwestern corner of Wyoming is a geologic wonderland of geysers, hot springs, and steam vents—the visible signs of a vast active volcano that stretches across the wilderness of Yellowstone National Park. Every day, this volcano expels more energy in the form of heat than is consumed as electric power in the three surrounding states of Wyoming, Idaho, and Montana combined. This energy is not released steadily; some of it builds up in hot magma chambers until the volcano blows its top. A cataclysmic eruption of the Yellowstone volcano 630,000 years ago ejected 1000 km^3 of rock into the air, covering regions as far away as Texas and California with a layer of volcanic ash.

The geologic record shows that volcanic explosions nearly this big, or even bigger, have occurred in the western United States at least six times during the last 2 million years, so we can be fairly certain that such an eruption will happen again. We can only imagine what it might do to human civilization. Hot ash would snuff out all life within 100 km or more, and cooler but choking ash would blanket the ground more than 1000 km away. Ash thrown high into the stratosphere would dim the Sun for several years, dropping temperatures and plunging the Northern Hemisphere into an extended volcanic winter.

The hazards volcanoes pose to human society, as well as the mineral resources and energy they provide, are certainly good enough reasons to study them. In addition, volcanoes are fascinating because they are windows through which we can look into Earth's deep interior to understand the igneous and plate tectonic processes that have generated its oceanic and continental crust.

In this chapter, we will examine how magma rises through Earth's crust, emerges onto its surface as lava, and cools into solid volcanic rock. We will see how plate tectonic processes and mantle convection explain volcanism at plate boundaries and at "hot spots" within plates. We will see how volcanoes interact with other components of the Earth system, particularly the hydrosphere and the atmosphere. Finally, we will consider their destructive power as well as the potential benefits they can provide for human society.

Volcanoes as Geosystems

The geologic processes that give rise to volcanoes and volcanic rocks are known collectively as *volcanism*. We had a glimpse of some of these processes when we examined the formation of igneous rocks in Chapter 4, but we will take a more detailed look at them here.

Ancient philosophers were awed by volcanoes and their fearsome eruptions of molten rock. In their efforts to explain volcanoes, they spun myths about a hot, hellish underworld below Earth's surface. Basically, they had the right idea. Modern researchers also see evidence of Earth's internal heat in volcanoes. Temperature readings of rocks as far down as humans have drilled (about 10 km) show that Earth does indeed get hotter with depth. We now believe that temperatures at depths of 100 km and more—within the asthenosphere—reach at least 1300°C, high enough for the rocks there to begin to melt. For this reason, we identify the asthenosphere as a main source of *magma*, the molten rock that we call *lava* after it rises to the surface and erupts. Portions of the solid lithosphere that ride above the asthenosphere may also melt to form magma.

Because magma is liquid, it is less dense than the rocks that produce it. Therefore, as magma accumulates, it begins to float upward through the lithosphere. In some places, the magma may find a path to the surface by fracturing the lithosphere along zones of weakness. In other places, the rising magma melts its way toward the surface. Most of the magma freezes at depth, but some fraction, probably only 10 to 30 percent, eventually reaches the surface and erupts as lava. A **volcano** is a hill or mountain constructed from the accumulation of lava and other erupted materials.

Taken together, the rocks, magmas, and processes needed to describe the entire sequence of events from melting to eruption constitute a **volcanic geosystem.** This type of geosystem can be viewed as a chemical factory that processes the input material (magmas from the asthenosphere) and transports the end product (lava) to the surface through an internal plumbing system.

Figure 5.1 is a simplified diagram of a volcano, showing the plumbing system through which magma travels to the surface. Magmas rising buoyantly through the lithosphere pool together in a magma chamber, usually at shallow depths in the crust. This chamber periodically empties through a pipelike feeder channel to a central vent on the surface in repeated cycles of *central eruptions.* Lava can also erupt from vertical cracks and other vents on the flanks of a volcano.

As we saw in Chapter 4, only a small fraction of the asthenosphere melts in the first place. The resulting magma gains chemical components as it melts the surrounding rocks while rising through the lithosphere. It loses other components as crystals settle out during transport or in shallow magma chambers. And its gaseous constituents escape to the atmosphere or ocean as it erupts at the surface. By accounting for these changes, we can extract clues to the chemical composition and physical state of the upper mantle where the lavas originated. We can also learn about eruptions that occurred millions or even billions of years ago by using isotopic dating (see Chapter 8) to determine the ages of lavas.

Lavas and Other Volcanic Deposits

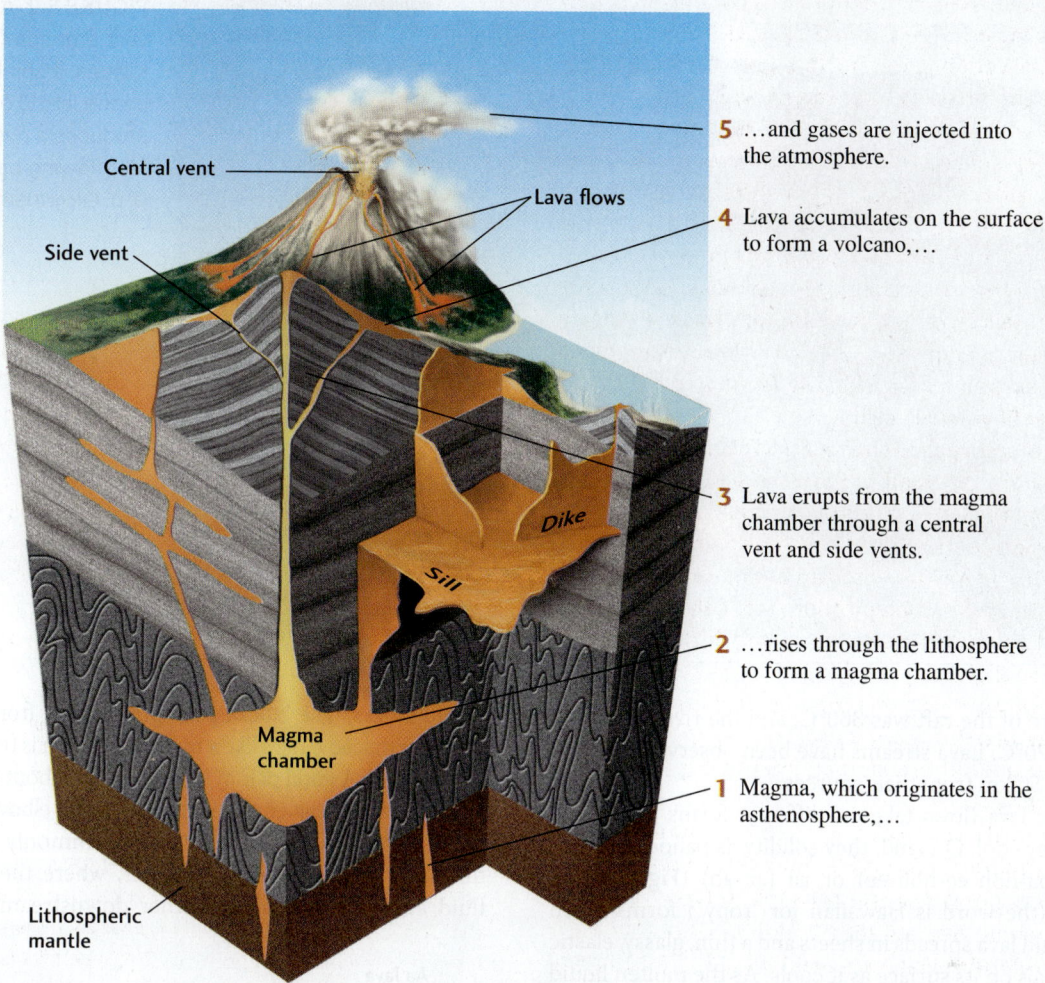

FIGURE 5.1 Volcanoes transport magma from Earth's interior to its surface, where rocks are formed and gases are injected into the atmosphere (or hydrosphere, in the case of an underwater eruption).

Lavas and Other Volcanic Deposits

Lavas of different types produce different landforms. The differences depend on the chemical composition, gas content, and temperature of the lavas. The higher the silica content and the lower the temperature, for example, the more viscous the lava is, and the more slowly it moves. The more gas a lava contains, the more violent its eruption is likely to be.

Types of Lava

Erupted lavas, the end products of volcanic geosystems, usually solidify into one of three major types of igneous rock (see Chapter 4): basalt, andesite, or rhyolite.

Basaltic Lavas Basalt is an extrusive igneous rock of mafic composition (high in magnesium, iron, and calcium) and has the lowest silica content of the three igneous rock types; its intrusive equivalent is gabbro. Basaltic magma, the product of mantle melting, is the most common magma type. It is produced along mid-ocean ridges and at hot spots within plates, as well as in continental rift valleys and other zones of extension. The volcanic island of Hawaii, which is made up primarily of basaltic lava, lies above a hot spot.

Basaltic lavas erupt when hot, fluid magmas fill up a volcano's plumbing system and overflow (**Figure 5.2**). Basaltic eruptions are rarely explosive. On land, a basaltic eruption sends lava down the flanks of the volcano in great streams that can engulf everything in their path (**Figure 5.3**). When cool, these lavas are black or dark gray, but at their high eruption temperatures (1000°C to 1200°C), they glow in reds and yellows. Because their temperatures are high and their silica content low, they are extremely fluid and can flow downhill fast and far. Lava streams flowing as fast as 100 km/hour have been observed, although velocities of a few kilometers per hour are more common. In 1938, two daring Russian volcanologists measured temperatures and collected gas samples while floating down a river of molten basalt on a raft of colder solidified lava. The surface

FIGURE 5.2 A central vent eruption from Kilauea, a shield volcano on the island of Hawaii, produces a river of hot, fast-flowing basaltic lava. [J. D. Griggs/USGS.]

temperature of the raft was 300°C, and the river temperature was 870°C. Lava streams have been observed to travel more than 50 km from their sources.

Basaltic lava flows take on different forms depending on how they cool. On land, they solidify as pahoehoe (pronounced pa-hoh-ee-hoh-ee) or aa (ah-ah) (**Figure 5.4**). *Pahoehoe* (the word is Hawaiian for "ropy") forms when a highly fluid lava spreads in sheets and a thin, glassy, elastic skin congeals on its surface as it cools. As the molten liquid continues to flow below the surface, the skin is dragged and twisted into coiled folds that resemble rope.

"*Aa*" is what the unwary exclaim after venturing barefoot onto lava that looks like clumps of moist, freshly plowed earth. Aa forms when lava loses its gases and consequently flows more slowly than pahoehoe, allowing a thick skin to form. As the flow continues to move, the thick skin breaks into rough, jagged blocks. The blocks pile up in a steep front of angular boulders that advances like a tractor tread. Aa is truly treacherous to cross. A good pair of boots may last about a week on it, and the traveler can count on cut knees and elbows.

A single downhill basaltic flow commonly has the features of pahoehoe near its source, where the lava is still fluid and hot, and of aa farther downstream, where the

FIGURE 5.3 A partly buried school bus in Kalapana, Hawaii. The village was buried by a basaltic lava flow from Kilauea. [J. D. Griggs/USGS.]

FIGURE 5.4 The two forms of basaltic lava are shown here: The jagged aa lava flow is moving over a pahoehoe lava flow on the island of Hawaii. [InterNetwork Media/Getty Images.]

Lavas and Other Volcanic Deposits   121

(a)

FIGURE 5.5 These bulbous pillow lavas, which were recently extruded on the Mid-Atlantic Ridge, were photographed from the deep-sea submersible *Alvin*. [OAR/National Undersea Research Program (NURP)/NOAA.]

flow's surface—having been exposed to cool air longer—has developed a thicker outer skin.

Basaltic lava that cools under water forms *pillow lavas*: piles of ellipsoidal, pillowlike blocks of basalt about a meter wide (**Figure 5.5**). Pillow lavas are an important indicator that a region on dry land was once under water. Scuba-diving geologists have actually observed pillow lavas forming on the ocean floor off Hawaii. Tongues of molten basaltic lava develop a tough, plastic skin on contact with the cold ocean water. Because the lava inside the skin cools more slowly, the pillow's interior develops a crystalline texture, whereas the quickly chilled skin solidifies to a crystal-less glass.

Andesitic Lavas Andesite is an extrusive igneous rock with an intermediate silica content; its intrusive equivalent is diorite. Andesitic magmas are produced mainly in the volcanic mountain belts above subduction zones. The name comes from a prime example: the Andes of South America.

The temperatures of **andesitic lavas** are lower than those of basalts, and because their silica content is higher, they flow more slowly and lump up in sticky masses. If one of these sticky masses plugs the central vent of a volcano, gases can build up beneath the plug and eventually blow off the top of the volcano. The explosive eruption of Mount St. Helens in 1980 (**Figure 5.6**) is a famous example.

(b)

(c)

FIGURE 5.6 Mount St. Helens, an andesitic volcano in southwestern Washington State, before, during, and after its cataclysmic eruption in May 1980, which ejected about 1 km^3 of pyroclastic material. The collapsed northern flank can be seen in the bottom photo. [(a) U.S. Forest Service/USGS; (b) USGS; (c) Lyn Topinka/USGS.]

FIGURE 5.7 A phreatic eruption of an island-arc volcano spews out plumes of steam into the atmosphere. The volcano, about 6 miles off the Tongan island of Tongatau, is one of about 36 in the area. [Dana Stephenson/Getty Images.]

Some of the most destructive volcanic eruptions in history have been *phreatic*, or steam, explosions, which occur when hot, gas-charged magma encounters groundwater or seawater, generating vast quantities of superheated steam (Figure 5.7). In 1883, the island of Krakatau, an andesitic volcano in Indonesia, was destroyed by a phreatic explosion. This legendary eruption was heard thousands of kilometers away, and it generated a tsunami that killed more than 40,000 people.

Rhyolitic Lavas Rhyolite is an extrusive igneous rock of felsic composition (high in sodium and potassium) with a silica content greater than 68 percent; its intrusive equivalent is granite. It is light in color, often a pretty pink. Rhyolitic magmas are produced in zones where heat from the mantle has melted large volumes of continental crust. Today, the Yellowstone volcano is producing huge amounts of rhyolitic magma that are building up in shallow chambers.

Rhyolite has a lower melting point than andesite, becoming liquid at temperatures of only 600°C to 800°C. Because **rhyolitic lavas** are richer in silica than any other lava type, they are the most viscous. A rhyolitic flow typically moves about 10 times more slowly than a basaltic flow, and it tends to pile up in thick, bulbous deposits (Figure 5.8). Gases are

FIGURE 5.8 Aerial view of a rhyolite dome that erupted about 1300 years ago in Newberry Caldera, Oregon. The light-colored rhyolite flow stands out against the trees with Paulina Peak in the background. Its dome shape indicates that the lava was very viscous. [William Scott/USGS.]

easily trapped beneath rhyolitic lavas, and large rhyolitic volcanoes such as Yellowstone produce the most explosive of all volcanic eruptions.

Textures of Volcanic Rocks

The textures of volcanic rocks, like the surfaces of solidified lava flows, reflect the conditions under which they solidified. Coarse-grained textures with visible crystals can result if lavas cool slowly. Lavas that cool quickly tend to have fine-grained textures. If they are silica-rich, rapidly cooled lavas can form *obsidian*, a volcanic glass.

Volcanic rock often contains little bubbles, created as gases are released during an eruption. As we have seen, magma is typically charged with gas, like soda in an unopened bottle. When magma rises toward Earth's surface, the pressure on it decreases, just as the pressure on the soda drops when the bottle cap is removed. And just as the carbon dioxide in the soda forms bubbles when the pressure is released, the water vapor and other dissolved gases escaping from lava as it erupts create gas cavities, or *vesicles* (**Figure 5.9**). *Pumice* is an extremely vesicular volcanic rock, usually rhyolitic in composition. Some pumice has so many vesicles that it is light enough to float on water.

Pyroclastic Deposits

Water and gases in magma can have even more dramatic effects than bubble formation. Before magma erupts, the confining pressure of the overlying rock keeps these volatiles from escaping. When the magma rises close to the surface and the pressure drops, the volatiles may be released with explosive force, shattering the lava and any overlying solidified rock and sending fragments of various sizes, shapes, and textures into the air (**Figure 5.10**). These fragments, known as *pyroclasts*, are classified according to their size.

Volcanic Ejecta The finest pyroclasts are fragments less than 2 mm in diameter, which are classified as *volcanic ash*. Volcanic eruptions can spray ash high into the atmosphere, where ash that is fine enough to stay aloft can be carried great distances. Within two weeks of the 1991 eruption of Mount Pinatubo in the Philippines, for example, its ash was traced all the way around the world by Earth-orbiting satellites.

Fragments ejected as blobs of lava that cool in flight and become rounded, or as chunks torn loose from previously solidified volcanic rock, can be much larger. These

FIGURE 5.9 A sample of vesicular basalt. [Courtesy John Grotzinger.]

FIGURE 5.10 An explosive eruption at Arenal volcano, Costa Rica, hurls pyroclasts into the air. [Gregory G. Dimijian/Science Source.]

FIGURE 5.11 Volcanic bombs within a stratified pyroclastic deposit at Hawaii Volcanoes National Park. These explosive eruptions eject volcanic materials up into the atmosphere, which then fall down and accumulate as poorly sorted deposits in pyroclastic flows. [Courtesy John Grotzinger.]

fragments are called *volcanic bombs* (**Figure 5.11**). Volcanic bombs as large as houses have been thrown more than 10 km by explosive eruptions.

Sooner or later, these pyroclasts fall to Earth, building the largest deposits near their source. As they cool, the hot, sticky fragments become welded together (lithified). Rocks created from smaller fragments are called **tuffs**; those formed from larger fragments are called **breccias** (**Figure 5.12**).

Pyroclastic Flows **Pyroclastic flows,** which are particularly spectacular and often deadly, occur when a volcano ejects hot ash and gases in a glowing cloud that rolls downhill at high speeds. The solid particles are buoyed up by the hot gases, so there is little frictional resistance to their movement.

In 1902, with very little warning, a pyroclastic flow with an internal temperature of 800°C exploded from the side of Mont Pelée, on the Caribbean island of Martinique. The avalanche of choking hot gas and glowing volcanic ash plunged down the slopes at a hurricane speed of 160 km/hour. In one minute and with hardly a sound, the searing emulsion of gas and ash enveloped the town of St. Pierre and killed 29,000 people. It is sobering to recall the statement of one Professor Landes, issued the day before the cataclysm: "The Montagne Pelée presents no more danger to the inhabitants of St. Pierre than does Vesuvius to those of Naples." Professor Landes perished with the others. In 1991, Mount Pinatubo erupted and created a major pyroclastic flow that was captured on camera in this impressive image (**Figure 5.13**).

Eruptive Styles and Landforms

The surface features produced by a volcano as it ejects material vary with the properties of the magma, especially its chemical composition and gas content, the type of material (lava versus pyroclasts) erupted, and the environmental conditions under which it erupts—on land or under the sea.

(a)

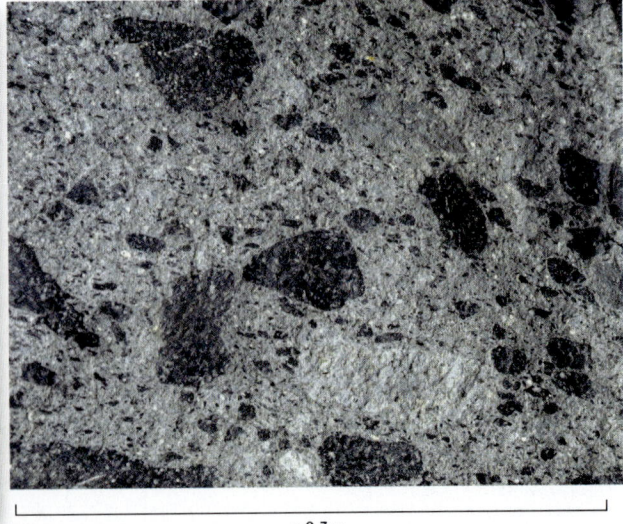
(b)
~0.3 m

FIGURE 5.12 Welded tuff is a volcanic igneous rock that forms when still warm ash-flow deposits weld together under the weight of overlying deposits. [(a) Courtesy John Grotzinger; (b) W. K. Fletcher/Science Source.]

FIGURE 5.13 Pyroclastic flow emanating from Mount Pinatubo, Philippines. After being dormant for 611 years, Mount Pinatubo erupted with massive violence, destroying everything in its path and killing 847 people. The Mount Pinatubo eruption is considered the world's most violent and destructive volcanic eruption of the 20th century. [Alberto Garcia/Redux.]

Volcanic landforms also depend on the rate at which lava is produced and the plumbing system that gets it to the surface (Figure 5.14).

Central Eruptions

Central eruptions discharge lava or pyroclasts from a *central vent*, an opening atop a pipelike feeder channel rising from the magma chamber. The magma ascends through this channel to erupt at Earth's surface. Central eruptions create the most familiar of all volcanic features: the volcanic mountain, shaped like a cone.

Shield Volcanoes A *lava cone* is built by successive flows of lava from a central vent. If the lava is basaltic, it flows easily and spreads widely. If flows are copious and frequent, they create a broad, shield-shaped volcano two or more kilometers high and many tens of kilometers in circumference, with relatively gentle slopes. Mauna Loa, on the island of Hawaii, is the classic example of such a **shield volcano** (Figure 5.14a). Although it rises only 4 km above sea level, it is actually the world's tallest mountain: Measured from its base on the seafloor, Mauna Loa is 10 km high, taller than Mount Everest! It grew to this enormous size by the accumulation of thousands of lava flows, each only a few meters thick, over a period of about a million years. The island of Hawaii actually consists of the overlapping tops of several active shield volcanoes emerging from the ocean.

Volcanic Domes In contrast to basaltic lavas, andesitic and rhyolitic lavas are so viscous they can barely flow. They often produce a *volcanic dome*, a bulbous, steep-sided mass of rock (see Figure 5.8). Domes look as though the lava has been squeezed out of a vent like toothpaste, with very little lateral spreading. Domes often plug vents, trapping gases beneath them (Figure 5.14b). Pressure can increase until an explosion occurs, blasting the dome into fragments.

Cinder Cones When volcanic vents discharge pyroclasts, these solid fragments can build up to create *cinder cones*. The profile of a cinder cone is determined by the *angle of repose* of the fragments: the maximum angle at which the fragments will remain stable rather than sliding downhill. The larger fragments, which fall near the vent, form very steep but stable slopes. Finer particles are carried farther from the vent and form gentler slopes at the base of the cone. The classic concave-shaped volcanic cone with its central vent at the summit develops in this way (Figure 5.14c).

Stratovolcanoes When a volcano emits lava as well as pyroclasts, alternating lava flows and beds of pyroclasts build a concave-shaped composite volcano, or **stratovolcano** (Figure 5.14d). Lava that solidifies in the central feeder channel and in radiating dikes strengthen the cone structure. Stratovolcanoes are common above subduction zones. Famous examples are Mount Fuji in Japan, Mount Vesuvius and Mount Etna in Italy, and Mount Rainier in Washington State. Mount St. Helens had a near-perfect stratovolcano shape until its eruption in 1980 destroyed its northern flank (see Figure 5.6c).

Craters A bowl-shaped pit, or **crater,** is found at the summit of most volcanic mountains, surrounding the central vent. During an eruption, the upwelling lava overflows

(a) Shield volcano

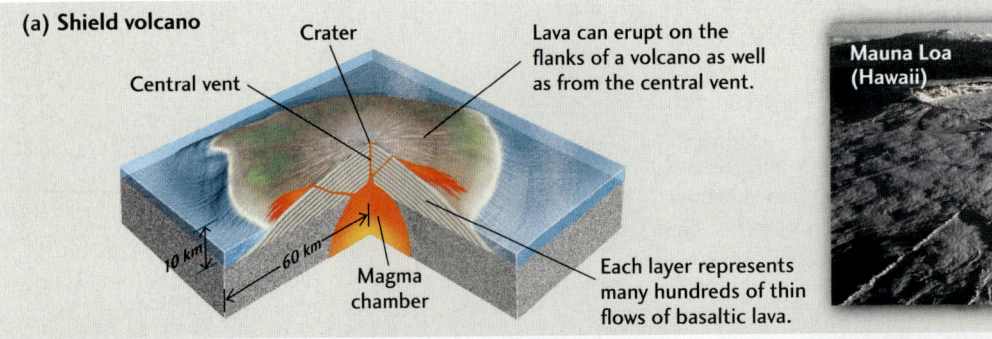 

- Crater
- Central vent
- Lava can erupt on the flanks of a volcano as well as from the central vent.
- 10 km
- 60 km
- Magma chamber
- Each layer represents many hundreds of thin flows of basaltic lava.

Mauna Loa (Hawaii)

(b) Volcanic dome

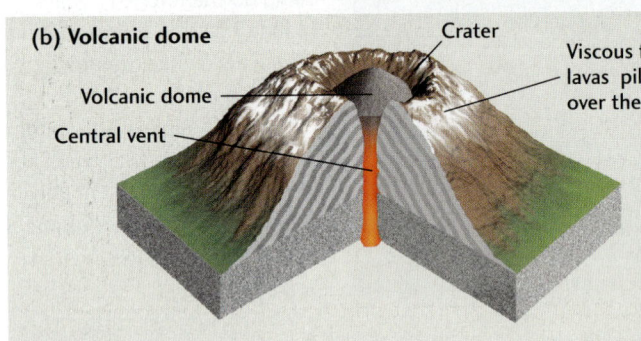 

- Crater
- Volcanic dome
- Central vent
- Viscous felsic lavas pile up over the vent.

A dome has been growing within the center of Mount St. Helens since its 1980 eruption.

Mount St. Helens (Washington)

(c) Cinder-cone volcano

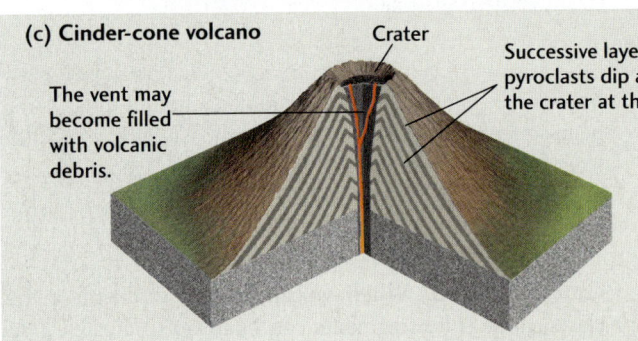 

- Crater
- The vent may become filled with volcanic debris.
- Successive layers of ejected pyroclasts dip away from the crater at the summit.

This eruption of Cerro Negro in 1968 built a cinder cone on an older terrain of lava flows.

Cerro Negro (Nicaragua)

(d) Stratovolcano

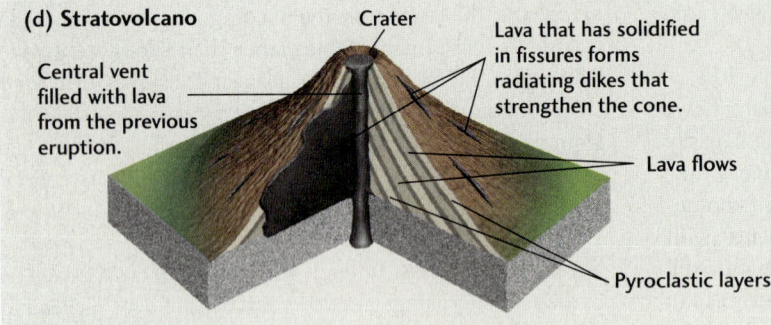 

- Crater
- Central vent filled with lava from the previous eruption.
- Lava that has solidified in fissures forms radiating dikes that strengthen the cone.
- Lava flows
- Pyroclastic layers

Mount Fuji (Japan)

(e) Caldera

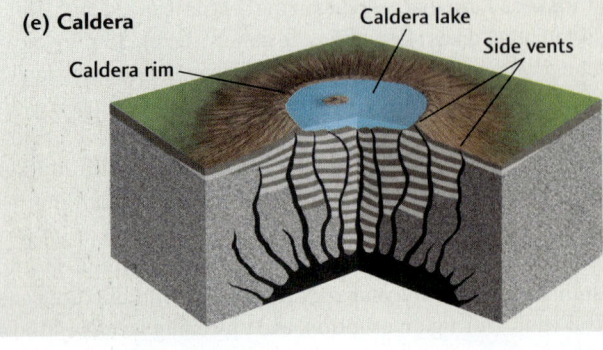 

- Caldera lake
- Caldera rim
- Side vents

Calderas result when a violent eruption empties a volcano's magma chamber, which then cannot support the overlying rock. It collapses, leaving a large, steep-walled basin.

Crater Lake (Oregon)

FIGURE 5.14 The eruptive styles of volcanoes and the landforms they create are determined principally by the composition of magma. [(a) USGS; (b) Lyn Topinka/USGS Cascades Volcano Observatory; (c) Smithsonian; (d) Corbis; (e) Bates Littlehales/Getty Images.]

the crater walls. When the eruption ceases, the lava that remains in the crater often sinks back into the vent and solidifies, and the crater may become partly filled by rock fragments that fall back into it. When the next eruption occurs, that material may be blasted out of the crater. Because a crater's walls are steep, they may cave in or become eroded over time. In this way, a crater can grow to several times the diameter of the vent and hundreds of meters deep. The crater of Mount Etna in Sicily, for example, is currently 300 m in diameter.

Calderas When great volumes of magma are discharged rapidly from a large magma chamber, the chamber can no longer support its roof. In such cases, the overlying volcanic structure can collapse catastrophically, leaving a large, steep-walled, basin-shaped depression much larger than a crater, called a **caldera** (Figure 5.14e). The development of the caldera that forms Crater Lake in Oregon is shown in **Figure 5.15**. Calderas can be impressive features, ranging in diameter from a few kilometers to 50 km or more. Owing to their size and high-volume eruptions, large caldera systems are sometimes called "supervolcanoes." The Yellowstone supervolcano, which is the largest active volcano in the United States, has a caldera with an area greater than Rhode Island.

After some hundreds of thousands of years, enough fresh magma may reenter the collapsed magma chamber to reinflate it, forcing the caldera floor to dome upward again to create a *resurgent caldera*. The cycle of eruption, collapse, and resurgence may occur repeatedly over geologic time. Three times over the last 2 million years, the Yellowstone supervolcano has erupted catastrophically, in each instance ejecting hundreds or thousands of times more material than the 1980 Mount St. Helens eruption and depositing ash over much of what is now the western United States. Other resurgent calderas are Valles Caldera in New Mexico and Long Valley Caldera in California, which last erupted about 1.2 million and 760,000 years ago, respectively.

Diatremes When magma from Earth's deep interior escapes explosively, the vent and the feeder channel below it are often left filled with volcanic breccia as the eruption wanes. The resulting structure is called a **diatreme.** Shiprock, a formation that towers over the surrounding plain in New Mexico, is a diatreme exposed by erosion of the sedimentary rocks through which it erupted. To transcontinental air travelers, Shiprock looks like a gigantic black skyscraper in the red desert (**Figure 5.16**).

The eruptive mechanism that produces diatremes has been pieced together from the geologic record. The kinds of minerals and rocks found in some diatremes could have formed only at great depths—100 km or so, well within the

FIGURE 5.15 Stages in the formation of the Crater Lake caldera. The Stage 3 collapse occurred about 7700 years ago.

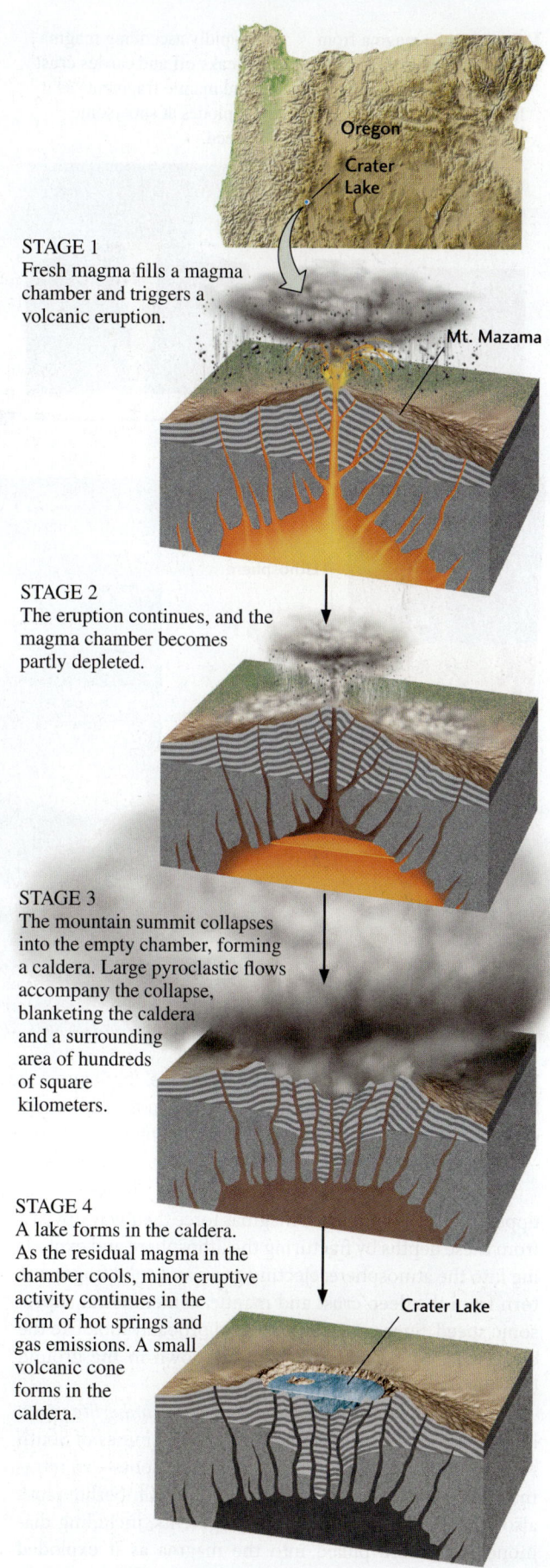

STAGE 1
Fresh magma fills a magma chamber and triggers a volcanic eruption.

STAGE 2
The eruption continues, and the magma chamber becomes partly depleted.

STAGE 3
The mountain summit collapses into the empty chamber, forming a caldera. Large pyroclastic flows accompany the collapse, blanketing the caldera and a surrounding area of hundreds of square kilometers.

STAGE 4
A lake forms in the caldera. As the residual magma in the chamber cools, minor eruptive activity continues in the form of hot springs and gas emissions. A small volcanic cone forms in the caldera.

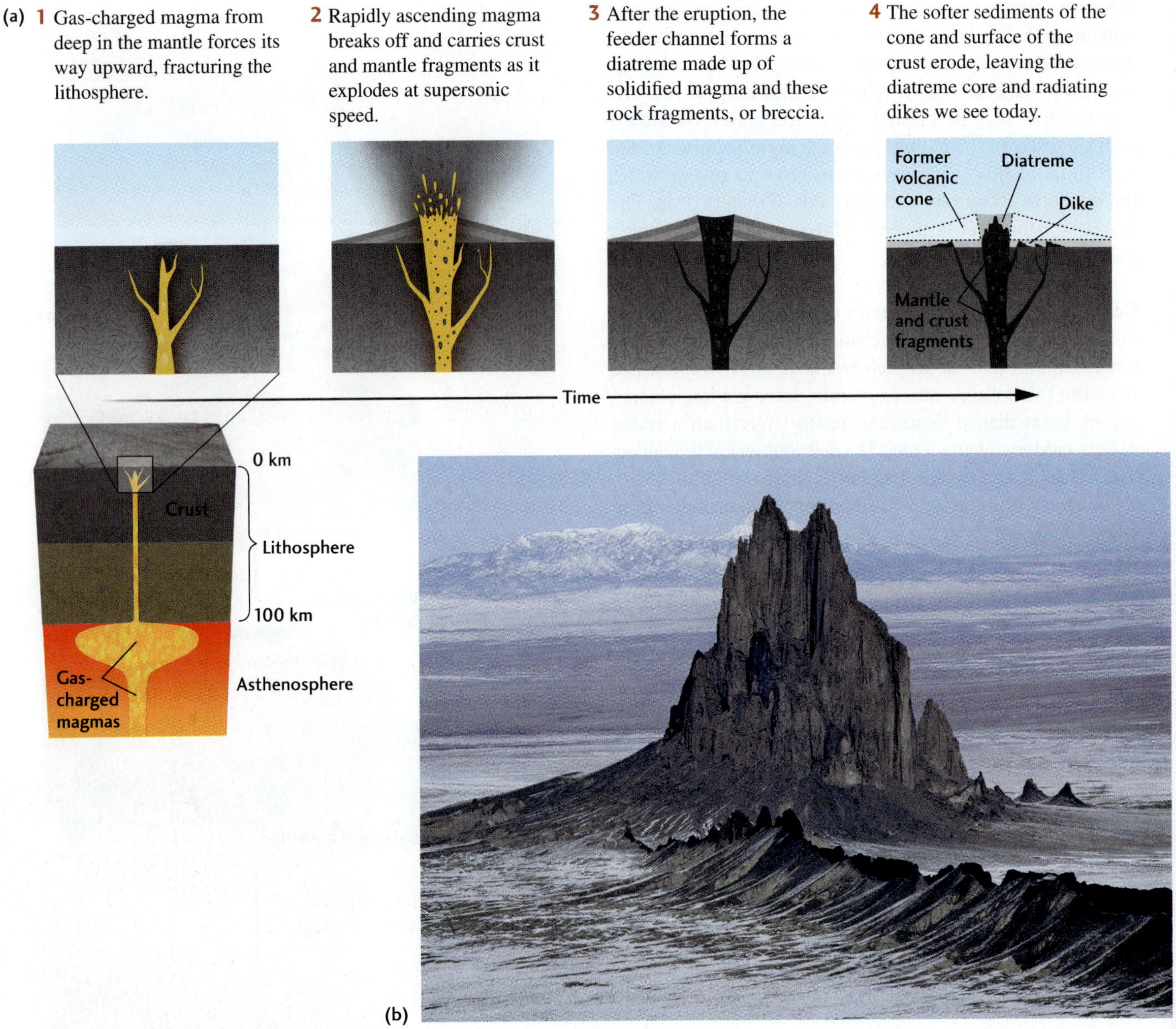

(a)

1 Gas-charged magma from deep in the mantle forces its way upward, fracturing the lithosphere.

2 Rapidly ascending magma breaks off and carries crust and mantle fragments as it explodes at supersonic speed.

3 After the eruption, the feeder channel forms a diatreme made up of solidified magma and these rock fragments, or breccia.

4 The softer sediments of the cone and surface of the crust erode, leaving the diatreme core and radiating dikes we see today.

← Time →

0 km — Crust
Lithosphere
100 km
Asthenosphere
Gas-charged magmas

Former volcanic cone — Diatreme — Dike
Mantle and crust fragments

(b)

FIGURE 5.16 (a) The formation of a diatreme. (b) Shiprock, towering 515 m above the surrounding flat-lying sediments of New Mexico, is a diatreme that has been exposed by erosion of the softer sedimentary rocks that once enclosed it. Note the vertical dike, one of six radiating from the central vent. [Airphoto—Jim Wark.]

---

upper mantle. Gas-charged magmas force their way upward from these depths by fracturing the lithosphere and exploding into the atmosphere, ejecting gases and solid fragments torn from the deep crust and mantle, sometimes at supersonic speed. Such an eruption would probably look like the exhaust jet of a giant rocket upside down in the ground blowing rocks and gases into the air.

Perhaps the most exotic diatremes are *kimberlite pipes*, named after the fabled Kimberley diamond mines of South Africa. Kimberlite is a volcanic type of peridotite—an ultramafic rock composed primarily of olivine. Kimberlite pipes also contain a variety of mantle fragments, including diamonds that were pulled into the magma as it exploded toward the surface (see Figure 21.25). The extremely high pressures needed to squeeze carbon into the mineral diamond are reached only at depths greater than 150 km. From careful studies of diamonds and other mantle fragments found in kimberlite pipes, geologists have been able to reconstruct sections of the mantle as if they had had been able to drill down to 200 km or more. These studies provide strong support for the theory that the upper mantle is made primarily of peridotite.

Fissure Eruptions

The largest volcanic eruptions do not come from a central vent, but through large, nearly vertical cracks in Earth's

FIGURE 5.17 A fissure eruption generates a "curtain of fire" on Kilauea, Hawaii, in 1992. [USGS.]

surface, sometimes tens of kilometers long (**Figure 5.17**). Such **fissure eruptions** are the main style of volcanism along mid-ocean ridges, where new oceanic crust is formed. A moderate-sized fissure eruption occurred in 1783 on a segment of the Mid-Atlantic Ridge that comes ashore in Iceland (**Figure 5.18**). A fissure 32 km long opened and, in six months, spewed out 12 km³ of basalt, enough to cover Manhattan to a height about halfway up the Empire State Building. The eruption also released more than 100 megatons of sulfur dioxide, creating a poisonous blue haze that

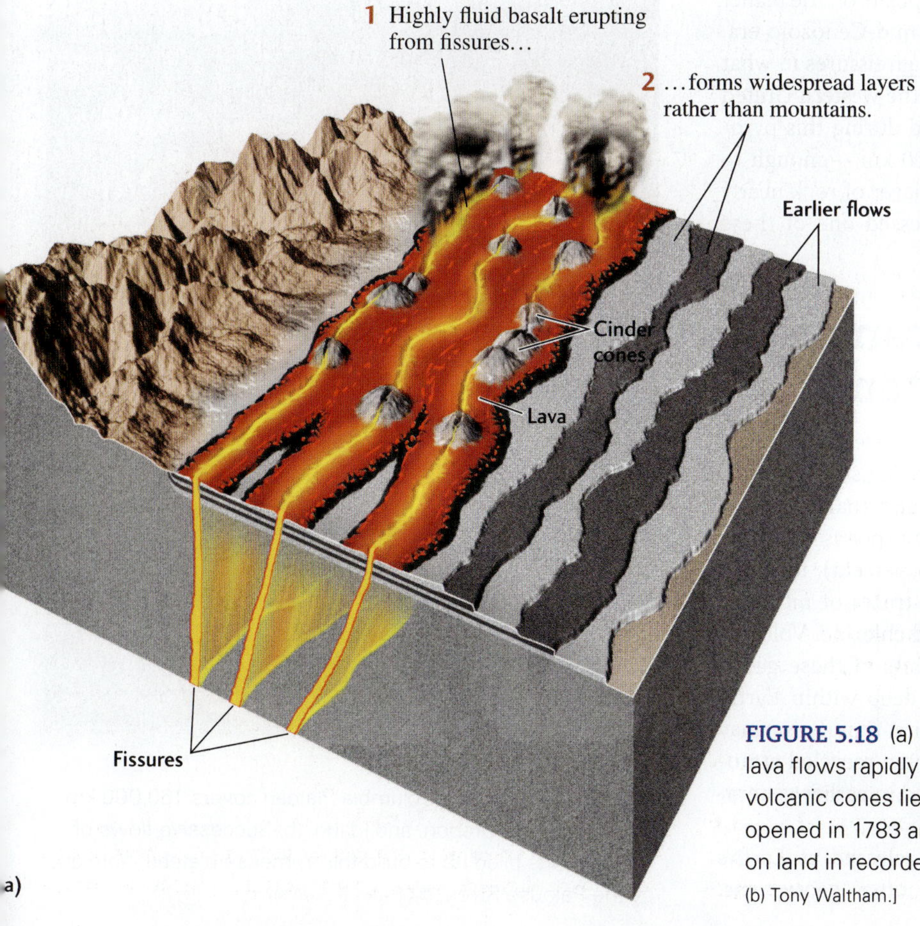

FIGURE 5.18 (a) In a fissure eruption, highly fluid basaltic lava flows rapidly away from the fissures. (b) These volcanic cones lie along the Laki fissure in Iceland, which opened in 1783 and erupted the largest flow of lava on land in recorded history. [(a) Data from R. S. Fiske/USGS; (b) Tony Waltham.]

hung over Iceland for more than a year. The resulting crop failures caused three-quarters of the island's livestock and one-fifth of its human population to die of starvation. Sulphuric aerosols from the Laki eruption were transported by the prevailing winds across Europe, causing crop damage and respiratory illnesses in many countries.

Flood Basalts Highly fluid basaltic lavas erupting from fissures on continents can spread out in sheets over flat terrain. Successive flows often pile up into immense basalt plateaus, called **flood basalts,** rather than forming a shield volcano as they do when the eruption is confined to a central vent. In North America, a huge eruption of flood basalts about 16 million years ago buried 160,000 km^2 of preexisting topography in what is now Washington, Oregon, and Idaho to form the Columbia Plateau (Figure 5.19). Individual flows were more than 100 m thick and some were so fluid that they traveled more than 500 km from their source. An entirely new landscape with new river valleys has since developed atop the lava that buried the old surface. Plateaus formed by flood basalts are found on every continent as well as on the seafloor.

Ash-Flow Deposits Eruptions of pyroclasts on continents have produced extensive sheets of hard volcanic tuffs called **ash-flow deposits.** A succession of forests in Yellowstone National Park has been buried under such ash flows. Some of the largest pyroclastic deposits on the planet are the ash flows that erupted in the mid-Cenozoic era, 45 million to 30 million years ago, through fissures in what is now the Basin and Range province of the western United States. The amount of material released during this pyroclastic flare-up was a staggering 500,000 km^3—enough to cover the entire state of Nevada with a layer of rock nearly 2 km thick! Humans have never witnessed one of these spectacular events.

Interactions of Volcanoes with Other Geosystems

Volcanoes are chemical factories that produce gases as well as solid materials. Courageous volcanologists have collected volcanic gases during eruptions and analyzed them to determine their composition. Water vapor is the main constituent of volcanic gases (70 to 95 percent), followed by carbon dioxide, sulfur dioxide, and traces of nitrogen, hydrogen, carbon monoxide, sulfur, and chlorine. Volcanic eruptions can release enormous amounts of these gases. Some volcanic gases may come from deep within Earth, making their way to the surface for the first time. Some may be recycled groundwater and ocean water, recycled atmospheric gases, or gases that were trapped in earlier generations of rocks.

As we have seen, volcanic gases released at Earth's surface have a number of effects on other geosystems.

(a)

(b)

FIGURE 5.19 (a) The Columbia Plateau covers 160,000 km^2 in Washington, Oregon, and Idaho. (b) Successive flows of flood basalts piled up to build this immense plateau, here cut by the Palouse River. [© Charles Bolin/Alamy.]

Interactions of Volcanoes with Other Geosystems   131

The emission of volcanic gases during Earth's early history is thought to have created the oceans and the atmosphere, and volcanic gas emissions continue to influence those components of the Earth system today. Periods of intense volcanic activity have affected Earth's climate repeatedly, and they may have been responsible for some of the mass extinctions documented in the geologic record.

Volcanism and the Hydrosphere

Volcanic activity does not stop when lava or pyroclastic materials cease to flow. For decades, or even centuries after a major eruption, volcanoes continue to emit steam and other gases through small vents called *fumaroles* (**Figure 5.20**). These emanations contain dissolved materials that precipitate onto surrounding surfaces as the water evaporates or cools, forming various encrusting deposits. Some of these precipitates contain valuable minerals.

Fumaroles are a surface manifestation of **hydrothermal activity**: the circulation of water through hot volcanic rocks and magmas. Circulating groundwater that comes into contact with buried magma (which may remain hot for hundreds of thousands of years) is heated and returned to the surface as hot springs and geysers. A *geyser* is a hot-water fountain that spouts intermittently with great force, frequently accompanied by a thunderous roar. The best-known geyser in the United States is Old Faithful in Yellowstone National Park, which erupts about every 65 or 90 minutes, sending a jet of hot water as high as 60 m into the air (**Figure 5.21**). We'll take a closer look at the mechanisms that drive hot springs and geysers in Chapter 17.

Hydrothermal activity is especially intense in the spreading centers at mid-ocean ridges, where huge volumes of water and magma come into contact. Fissures created by tensional forces allow seawater to circulate throughout the newly formed oceanic crust. Heat from the hot volcanic rocks and deeper magmas drives a vigorous convection current that pulls cold seawater into the crust, heats it, and expels the hot water back into the overlying ocean through vents on the rift valley floor (**Figure 5.22**).

FIGURE 5.21 Old Faithful geyser, in Yellowstone National Park, erupts regularly about every 65 or 90 minutes. [SPL/Science Source.]

FIGURE 5.20 A fumarole encrusted with sulfur deposits on the Merapi volcano in Indonesia. [R. L. Christiansen/USGS.]

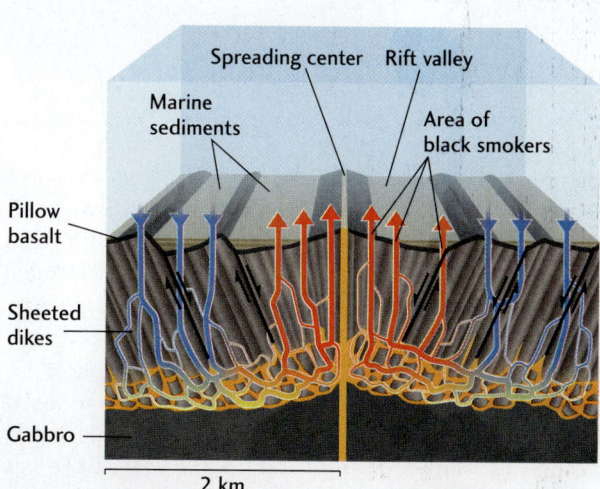

FIGURE 5.22 Near spreading centers, seawater circulates through the oceanic crust, is heated by magma, and is reinjected into the ocean, forming black smokers and depositing minerals on the seafloor.

Given the common occurrence of hot springs and geysers in volcanic geosystems on land, the evidence for pervasive hydrothermal activity at spreading centers immersed in deep water should come as no surprise. Nevertheless, geologists were amazed once they recognized the intensity of the convection and discovered some of its chemical and biological consequences. The most spectacular manifestations of this process were first found in the eastern Pacific Ocean in 1977. Plumes of hot, mineral-laden water with temperatures as high as 350°C were seen spouting through hydrothermal vents at the crest of the East Pacific Rise (see Chapter 22). The rates of fluid flow turned out to be very high. Marine geologists have estimated that the entire volume of the ocean's water is circulated through the cracks and vents of Earth's spreading centers in only 10 million years.

Scientists have come to realize that the interactions between the lithosphere and the hydrosphere at spreading centers profoundly affect the geology, chemistry, and biology of the oceans in a number of ways:

- The creation of new lithosphere accounts for almost 60 percent of the energy flowing out of Earth's interior. Circulating seawater cools the new lithosphere very efficiently and therefore plays a major role in the outward transport of Earth's internal heat.
- Hydrothermal activity leaches metals and other elements from the new crust, injecting them into the oceans. These elements contribute as much to seawater chemistry as the mineral components dumped into the oceans by all the world's rivers.
- Metal-rich minerals precipitate out of the circulating seawater and form ores of zinc, copper, and iron in shallow parts of the oceanic crust. These ores form when seawater sinks through porous volcanic rocks, is heated, and leaches these elements from the new crust. When the heated seawater, enriched with dissolved minerals, rises and reenters the cold ocean, the ore-forming minerals precipitate.

The energy and nutrients at hydrothermal vents feed unusual colonies of strange organisms whose energy comes from Earth's interior rather than from sunlight. Chemoautotrophic hyperthermophiles, similar to those that populate hot springs on land, form the base of complex ecosystems, providing food for giant clams and tube worms up to several meters long. Some scientists have speculated that life on Earth may have begun in the energetic, chemically rich environments of hydrothermal vents (see Chapter 22).

Volcanism and the Atmosphere

Volcanism in the lithosphere affects weather and climate by changing the composition and properties of the atmosphere. Large eruptions can inject sulfurous gases into the atmosphere tens of kilometers above Earth (**Figure 5.23**). Through various chemical reactions, these gases form an aerosol (a fine airborne mist) containing tens of millions of

FIGURE 5.23 Satellite image of the huge ash cloud spewing from the erupting Cordón Caulle volcano in central Chile on June 13, 2011. The ash plume extends 800 km from the snow-covered Andes mountains (on left side of photo) to the Argentine city of Buenos Aires (center right of photo). This ash cloud encircled the planet, closing airports in Australia and New Zealand. [NASA image courtesy Jeff Schmaltz, MODIS Rapid Response Team at NASA GSFC.]

metric tons of sulfuric acid. Such aerosols may block enough of the Sun's radiation from reaching Earth's surface to lower global temperatures for a year or two. The eruption of Mount Pinatubo, one of the largest explosive eruptions of the twentieth century, led to a global cooling of at least 0.5°C in 1992. (Chlorine emissions from Mount Pinatubo also hastened the loss of ozone in the atmosphere, nature's shield that protects the biosphere from the Sun's ultraviolet radiation.)

The debris lofted into the atmosphere during the 1815 eruption of Mount Tambora in Indonesia resulted in even greater cooling. The next year, the Northern Hemisphere suffered a very cold summer; according to a diarist in Vermont, "no month passed without a frost, nor one without a snow." The drop in temperature and the ash fall caused widespread crop failures. More than 90,000 people perished in that "year without a summer," which inspired Lord Byron's gloomy poem, "Darkness":

I had a dream, which was not all a dream.
The bright sun was extinguish'd, and the stars
Did wander darkling in the eternal space,
Rayless, and pathless, and the icy earth
Swung blind and blackening in the moonless air;
Morn came and went—and came, and brought no day.
And men forgot their passions in the dread
Of this their desolation; and all hearts
Were chill'd into a selfish prayer for light.

The Global Pattern of Volcanism

Before the advent of plate tectonic theory, geologists noted a concentration of volcanoes around the rim of the Pacific Ocean and nicknamed it the Ring of Fire (see Figure 2.6). The explanation of the Ring of Fire in terms of subduction zones was one of the great successes of the new theory. As we will see in this section, plate tectonics can explain essentially all major features in the global pattern of volcanism (**Figure 5.24**).

Figure 5.25 shows the locations of the world's active volcanoes that occur on land or above the ocean surface. About 80 percent are found at convergent plate boundaries, 15 percent at divergent plate boundaries, and the remaining few within plate interiors. There are many more active volcanoes than shown on this map, however. Most of the lava erupted on Earth's surface comes from vents beneath the oceans, located at spreading centers on mid-ocean ridges.

Volcanism at Spreading Centers

As we have seen, enormous volumes of basaltic lava erupt continually along the global network of mid-ocean ridges—enough to have created all of the present-day seafloor. This "crustal factory" lies beneath a rift valley a few kilometers

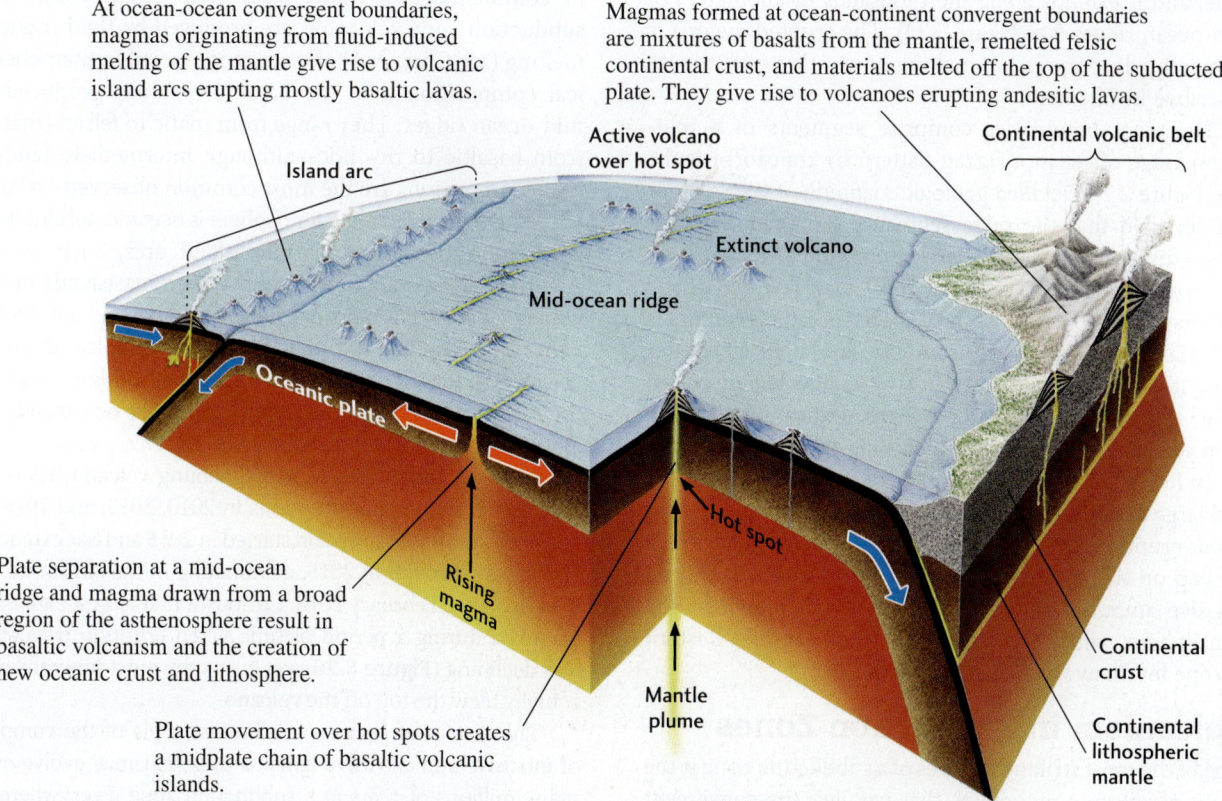

FIGURE 5.24 Plate tectonic processes explain the global pattern of volcanism.

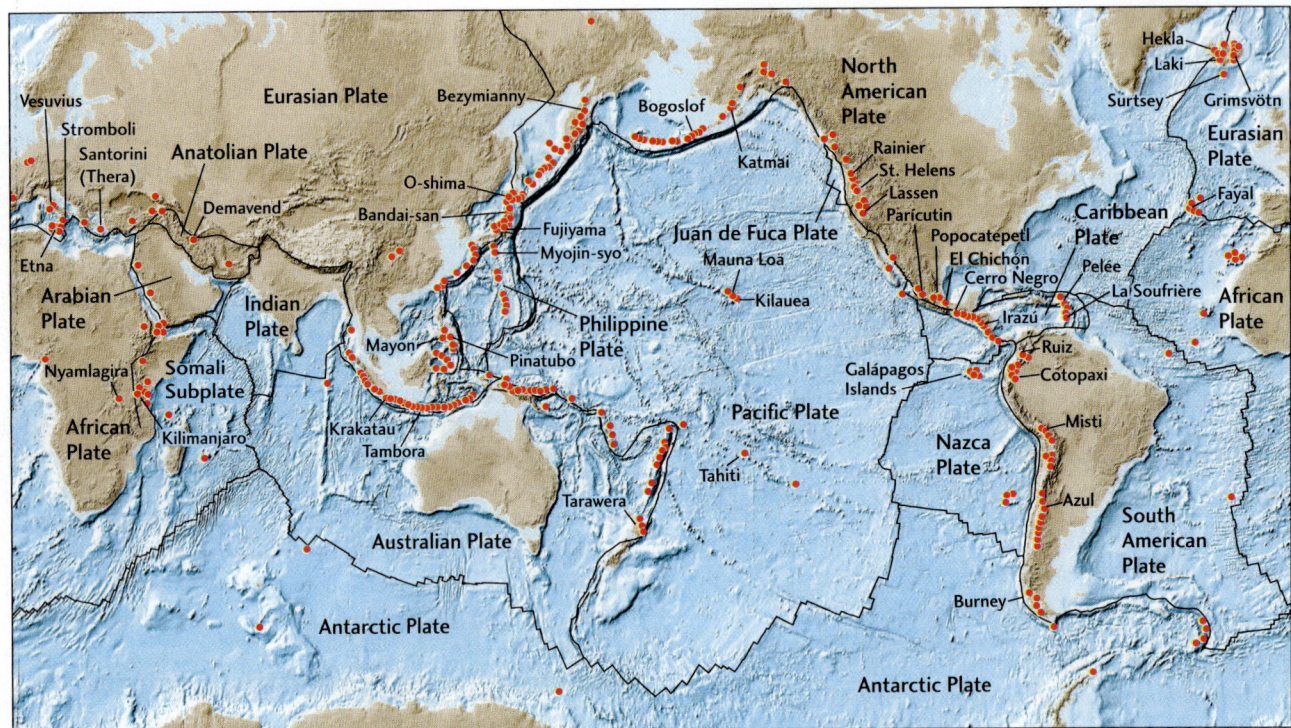

FIGURE 5.25 The active volcanoes of the world with vents on land or above the ocean surface are represented on this map by red dots. Black lines represent plate boundaries. Not shown on this map are the numerous vents of the mid-ocean ridge system below the ocean surface.

wide, and it extends along the thousands of kilometers of mid-ocean ridges (see Figure 5.24). The erupted magma is formed by decompression melting of mantle peridotite, as described in Chapter 4.

Divergent boundaries comprise segments of a mid-ocean ridge offset in a zigzag pattern by transform faults (see Figure 2.7). Detailed geologic mapping of the seafloor has revealed that the ridge segments can themselves be quite complex. They are often composed of shorter, parallel spreading centers that are offset by a few kilometers and may partly overlap. Each of these spreading centers is an "axial volcano" that erupts basaltic lava at variable rates along its length. Basalts from nearby axial volcanoes often show slight geochemical differences, indicating that the axial volcanoes have separate plumbing systems.

In Iceland, the Mid-Atlantic Ridge rises above the ocean and large basaltic eruptions are common. The most recent major eruption, from a volcano beneath the Eyjafjallajökull ice cap on the southern coast of Iceland in 2010, ejected massive amounts of very fine-grained ash high into the atmosphere, which disrupted air traffic across western Europe for many weeks (see Earth Issues 5.1).

Volcanism in Subduction Zones

One of the most striking features of a subduction zone is the chain of volcanic mountains that parallels the convergent boundary above the sinking slab of oceanic lithosphere, regardless of whether the overriding lithosphere is oceanic or continental (see Figure 5.24). The magmas that feed subduction-zone volcanoes are produced by fluid-induced melting (see Chapter 4) and are more varied in their chemical composition than the basaltic magmas produced at mid-ocean ridges. They range from mafic to felsic—that is, from basaltic to rhyolitic—although intermediate (andesitic) compositions are the most common observed on land.

Where the overriding lithosphere is oceanic, subduction-zone volcanoes form volcanic island arcs, such as the Aleutian Islands of Alaska and the Mariana Islands of the western Pacific. Where oceanic lithosphere is subducted beneath a continent, the volcanoes and volcanic rocks coalesce to form a volcanic mountain belt on land, such as the Andes, which mark the subduction of the oceanic Nazca Plate beneath continental South America.

Recently, Indonesia's Mount Sinabung volcano has been very active, with major eruptions in 2010, 2013, and 2018. A phase of continuous eruption started in 2013 and has extended all the way through to 2018, culminating in the largest explosion to date in February 2018. Curiously, that largest explosion occurred during a period of time when eruption frequency was declining (**Figure 5.26**), yet it was the most powerful and actually blew the top off the volcano.

The terrain of Japan is a prime example of the complex of intrusive and extrusive igneous rock that may evolve over many millions of years at a subduction zone. Everywhere in this small country are all kinds of extrusive igneous rocks of various ages, mixed in with mafic and intermediate intrusives,

Earth Issues 5.1 Volcanic Ash Clouds over Europe

On April 14, 2010, the Eyjafjallajökull volcano in Iceland began a series of eruptions that shut down air travel over western and northern Europe for a period of six days. [According to the joke, "Eyjafjallajökull" is Icelandic for "name that no one can pronounce." Actually, it's pronounced: aye-ya-fyah-dla-jow-kudl and means "island-mountain glacier."] These eruptions led to the closure of most of Europe's larger airports and caused the cancellation of many flights to and from Europe, resulting in the highest level of air traffic disruption since World War II. Many people were stranded for days with little comfort as flights were sequentially canceled, leaving people to struggle to find alternative means of transportation or accommodations. In the week following the eruption, it is estimated that 250,000 British, French, and Irish citizens were stranded abroad, the European economy may have lost almost 2 billion dollars, and the aviation industry lost up to 250 million dollars per day.

The eruptions were predicted well in advance. Seismic activity in and around Eyjafjallajökull began in late 2009 and increased in intensity and frequency until March 20, 2010, when a small eruption occurred. A second, much larger eruption occurred on April 14th, ejecting 250 million cubic meters of volcanic ash. The ash cloud rose to elevations of 9000 meters and, though not as large as the 1980 Mount St. Helens eruption in Oregon, it was high enough to enter the jet stream, which flowed directly over Iceland at the time. The eastward flow of the jet stream transported the ash to Europe, where it spread out over a large part of the continent.

Much of this volcanic ash was produced by the interaction of hot magma with glacial ice and water,

In this April 16, 2010, photo, the Eyjafjallajökull volcano in southern Iceland sends ash into the air just prior to sunset. [AP Photo/Brynjar Gauti.]

which made it very fine-grained, less than 2 mm in size. When ash of this size gets caught up in jet engines, the high temperatures of the jet (up to 2000°C) can remelt it, re-creating a sticky lava that can cause engine failure. In extreme cases, planes have had to literally glide their way out of the ash cloud before engines can be restarted.

The Eyjafjallajökull eruptions lasted for only one month; by June 2010, very little ash was being ejected. But future eruptions in Iceland are inevitable, and the agricultural and environmental consequences for Europe are potentially dire. Right now, geologists are carefully monitoring the nearby Katla volcano, whose historical eruptions have often followed those of Eyjafjallajökull.

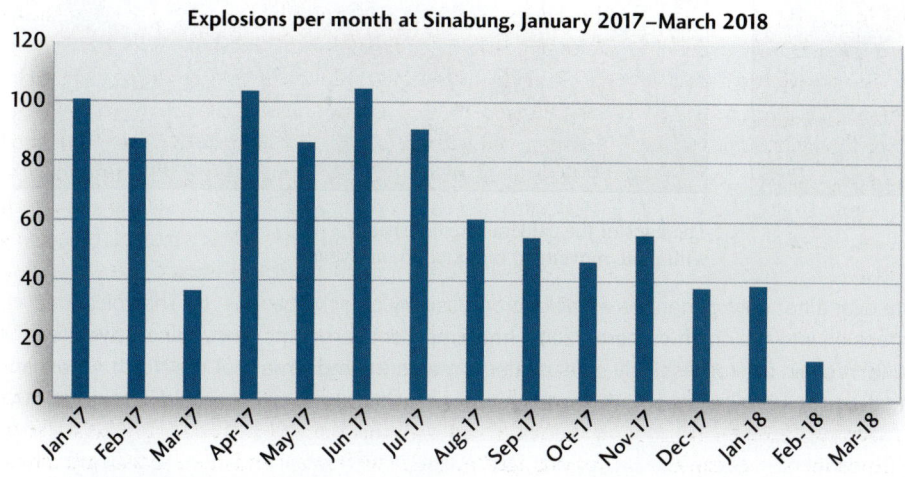

FIGURE 5.26 This graph shows the number of explosions per month at Sinabung reported from January 2017 to March 2018. Only partial data was reported for January 18 to 31 and no explosions were reported during March 2018. [Data from the Center for Volcanology and Geological Disaster Mitigation (PVMBG).]

metamorphosed volcanic rocks, and sedimentary rocks derived from erosion of the igneous rocks. The erosion of these various rocks has contributed to the distinctive landscapes portrayed in so many classic and modern Japanese paintings.

Intraplate Volcanism: The Mantle Plume Hypothesis

Decompression melting explains volcanism at spreading centers, and fluid-induced melting can account for the volcanism above subduction zones, but how can plate tectonic theory explain *intraplate volcanism*—that is, volcanoes far from plate boundaries? Geologists have found a clue in the ages of such volcanoes.

Hot Spots and Mantle Plumes Consider the Hawaiian Islands, which stretch across the middle of the Pacific Plate. This island chain begins with the active volcanoes on the island of Hawaii and continues to the northwest as a string of progressively older, extinct, eroded, and submerged volcanic mountains and ridges. In contrast to the seismically active mid-ocean ridges, the Hawaiian island chain is not marked by frequent large earthquakes (except near the active volcanoes). It is essentially aseismic (without earthquakes), and is therefore called an *aseismic ridge*. Active volcanoes at the beginnings of progressively older aseismic ridges can be found elsewhere in the Pacific and in other large ocean basins. Two examples are the active volcanoes of Tahiti, at the southeastern end of the Society

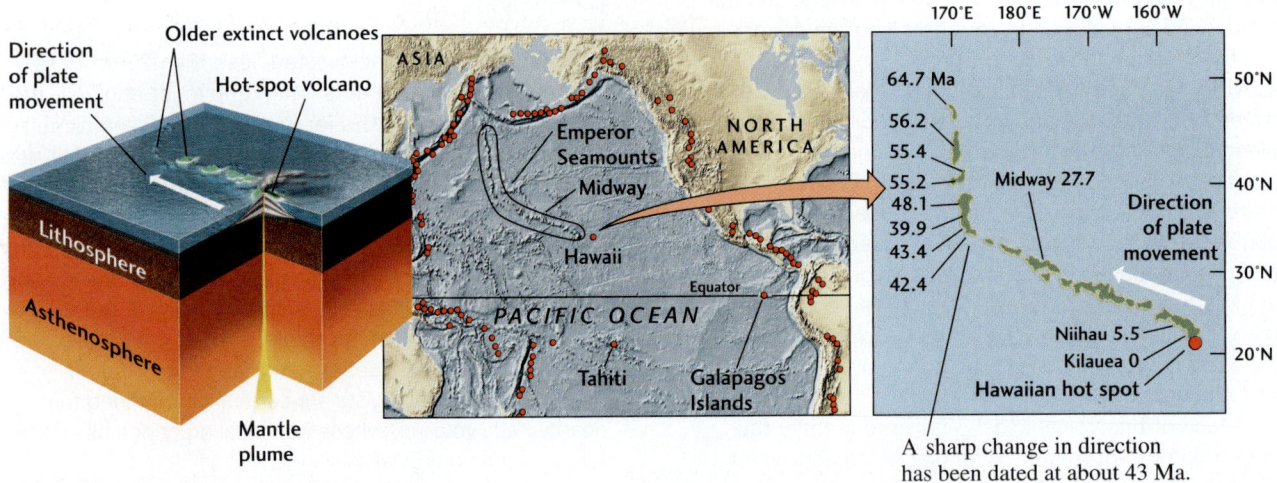

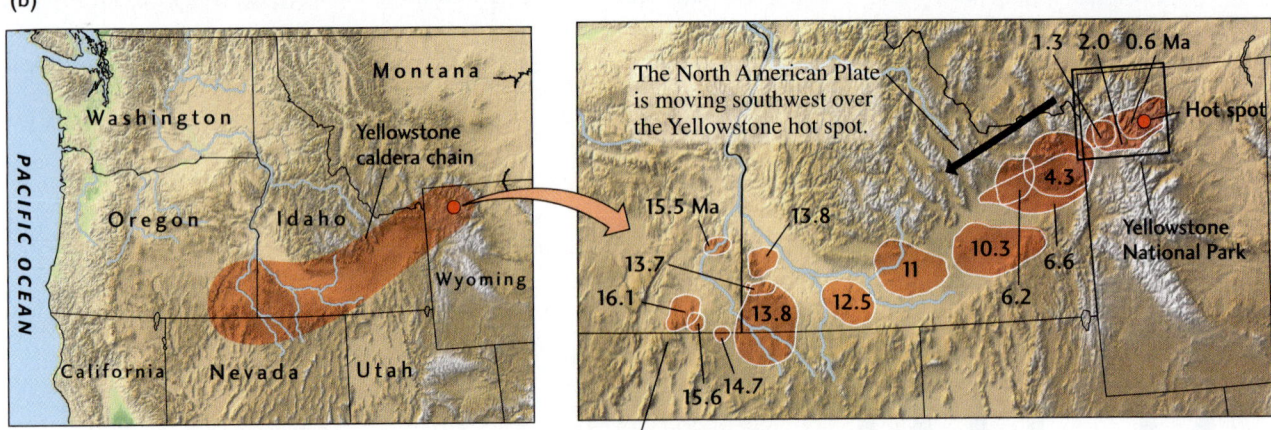

FIGURE 5.27 The movement of a plate over a hot spot generates a trail of progressively older volcanoes. (a) The volcanoes of the Hawaiian island chain and its extension into the northwestern Pacific (the Emperor seamounts) show a northwestward trend toward progressively older ages. (b) A chain of progressively older calderas marks the movement of the North American Plate over a continental hot spot during the past 16 million years. (Ma, million years ago.) [Data from Wheeling Jesuit University/NASA Classroom of the Future.]

Islands, and the Galápagos Islands, at the western end of the aseismic Nazca Ridge (see Figure 5.25).

Once the general pattern of plate movements had been worked out, geologists were able to show that these aseismic ridges approximated the paths that the plates would take over a set of volcanically active **hot spots** that were fixed relative to one another, as if they were blowtorches anchored in Earth's mantle (Figure 5.27). Based on this evidence, they hypothesized that hot spots were caused by hot, solid material rising in narrow, cylindrical jets from deep within the mantle (perhaps as deep as the core-mantle boundary), called **mantle plumes.** According to the mantle plume hypothesis, when peridotites transported upward in a mantle plume reach lower pressures at shallower depths, they begin to melt, producing basaltic magma. The magma penetrates the lithosphere and erupts at the surface. The current position of a plate over the hot spot is marked by an active volcano, which becomes inactive as plate movement carries it away from the hot spot. The movement of the plate thus generates a trail of extinct, progressively older volcanoes. As shown in Figure 5.27a, the Hawaiian Islands fit this pattern well. Dating of the volcanoes yields a rate of movement of the Pacific Plate over the Hawaiian hot spot of about 100 mm/year.

Some aspects of intraplate volcanism within continents can also been explained by the mantle plume hypothesis. The modern Yellowstone Caldera, only 630,000 years old, is still volcanically active, as evidenced by the geysers, hot springs, uplift, and earthquakes observed in the area. It is the youngest member of a chain of sequentially older and now-extinct calderas that supposedly mark the movement of the North American Plate over the Yellowstone hot spot (Figure 5.27b). The oldest member of the chain, a volcanic area in Oregon, erupted about 16 million years ago, producing some of the flood basalts of the Columbia Plateau. The North American Plate has moved over the Yellowstone hot spot to the southwest at a rate of about 25 mm/year during the past 16 million years. Accounting for the relative movement of the Pacific and North American plates, this rate and direction are consistent with the plate movements inferred from the Hawaiian Islands.

2018 Eruption of Kilauea Volcano, Hawaii

The 2018 eruption of the Kilauea volcano, on the island of Hawaii, is a reminder of the ongoing and nearly constant impact of mid-plate volcanism. The current phase of eruption began on May 3, 2018, and has continued through at least August of 2018 (Figure 5.28). Lava has been pouring through fissures associated with the volcano's East Rift Zone in Puna (Figure 5.29). The eruption closely coincided with a magnitude 6.9 earthquake and opening of dozens of fissures through which lava poured, causing the evacuation of approximately 2000 residents. By August of 2018, almost 15 square miles have been covered by fresh lava, destroying over 700 homes and a major geothermal energy facility that created about 25 percent of the island's electricity. This eruption is the most destructive in the United States since the eruption of Mount St. Helens in 1980.

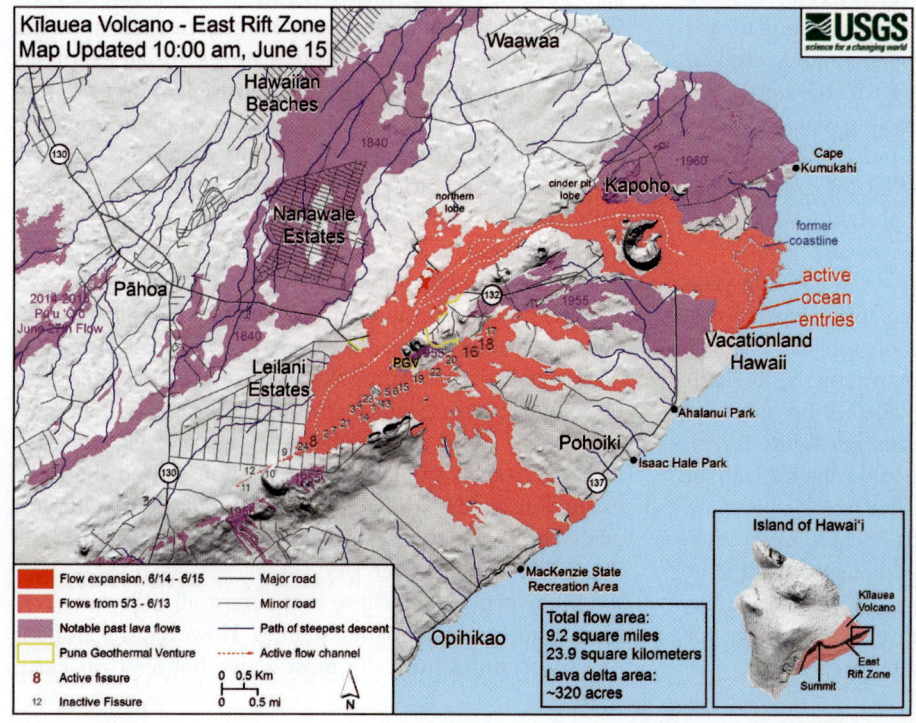

FIGURE 5.28 The 2018 eruption of the Kilauea volcano, on the island of Hawaii, is an example of mid-plate volcanism. The current phase of eruption began on May 3, 2018, and has continued through at least August of 2018. [USGS.]

FIGURE 5.29 Lava. This photo taken on March 6, 2011, shows lava pouring through fissures associated with the Kilauea volcano's East Rift Zone in Puna. [USGS.]

Measuring Plate Movements Using Hot-Spot Tracks Assuming that hot spots are anchored by plumes rising from the deep mantle, geologists can use the worldwide distribution of their volcanic tracks to compute how the global system of plates is moving with respect to the deep mantle. The results are sometimes called "absolute plate movements" to distinguish them from the movements of plates relative to each other. The absolute plate movements calculated from hot-spot tracks have helped geologists understand the forces driving the plates. Plates that are being subducted along large fractions of their boundaries—such as the Pacific, Nazca, Cocos, Indian, and Australian plates—are moving rapidly with respect to the hot spots, whereas plates without much subducting slab—such as the Eurasian and African plates—are moving slowly. This observation supports the hypothesis that the gravitational pull of the dense sinking slabs is an important force driving plate movements (see Chapter 2).

The use of hot-spot tracks to reconstruct absolute plate movements works fairly well for recent plate movements. Over longer periods, however, a number of problems arise. For instance, according to the fixed-hot-spot hypothesis, the sharp bend in the Hawaiian aseismic ridge (where it becomes the north-trending Emperor seamount chain; see Figure 5.27a), dated at about 43 million years ago, should coincide with an abrupt shift in the direction of the Pacific Plate. However, no sign of such a shift is evident in magnetic isochron maps, leading some geologists to question the fixed-hot-spot hypothesis. Others have pointed out that, in a convecting mantle, plumes would not necessarily remain fixed relative to one another, but might be moved about by shifting convection currents.

Large Igneous Provinces The origin of fissure eruptions on continents—such as those that formed the Columbia Plateau and even larger basalt plateaus in Brazil and Paraguay, India, and Siberia—is a major puzzle. The geologic record shows that these eruptions can release immense amounts of lava—up to several million cubic kilometers—in a period as short as a million years.

Flood basalts are not limited to continents; they also create large oceanic plateaus, such as the Ontong Java Plateau on the northern side of the island of New Guinea and major parts of the Kerguelen Plateau in the southern Indian Ocean. These features are all examples of what geologists call **large igneous provinces** (LIPs) (**Figure 5.30**). LIPs are large volumes of predominantly mafic extrusive and intrusive igneous rock whose origins lie in processes other than normal seafloor spreading. LIPs include continental flood basalts and associated intrusive rocks, oceanic basalt plateaus, and the aseismic ridges produced by hot spots.

The fissure eruption that covered much of Siberia with basaltic lava is of special interest to geobiologists because it happened at the same time as the greatest mass extinction in the geologic record, which occurred at the end of the Permian period, about 251 million years ago (see Chapter 22). Some geologists think that the eruption caused the mass extinction, perhaps by polluting the atmosphere with volcanic gases that triggered major climate changes (see the Practicing Geology Exercise).

Many geologists believe that almost all LIPs were created at hot spots by mantle plumes. However, the amount of lava erupting from the most active hot spot on Earth

The Global Pattern of Volcanism

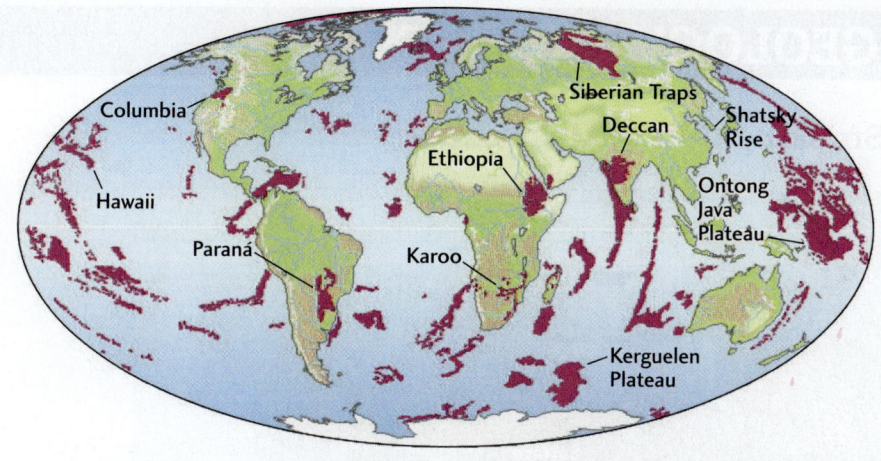

FIGURE 5.30 The global distribution of large igneous provinces on continents and in ocean basins. These provinces are marked by unusually large deposits of basaltic magma. [Data from M. Coffin and O. Eldholm, *Reviews of Geophysics, 32* (1994): 1–36, Figure 1.]

today, Hawaii, is paltry compared with the enormous outpourings of fissure eruptions. What explains these unusual bursts of basaltic magma from the mantle? Some geologists speculate that they result when a new plume rises from the core-mantle boundary. According to this hypothesis, a large, turbulent blob of hot material—a "plume head"—leads the way. When this plume head reaches the top of the mantle, it generates a huge quantity of magma by decompression melting, which erupts in massive flood basalts (**Figure 5.31**). Others dispute this hypothesis, pointing out that continental flood basalts often seem to be associated with preexisting zones of weakness in the continental crust and suggesting that the magmas are generated by convective processes localized in the upper mantle. Sorting out the origins of LIPs is one of the most exciting areas of current geologic research.

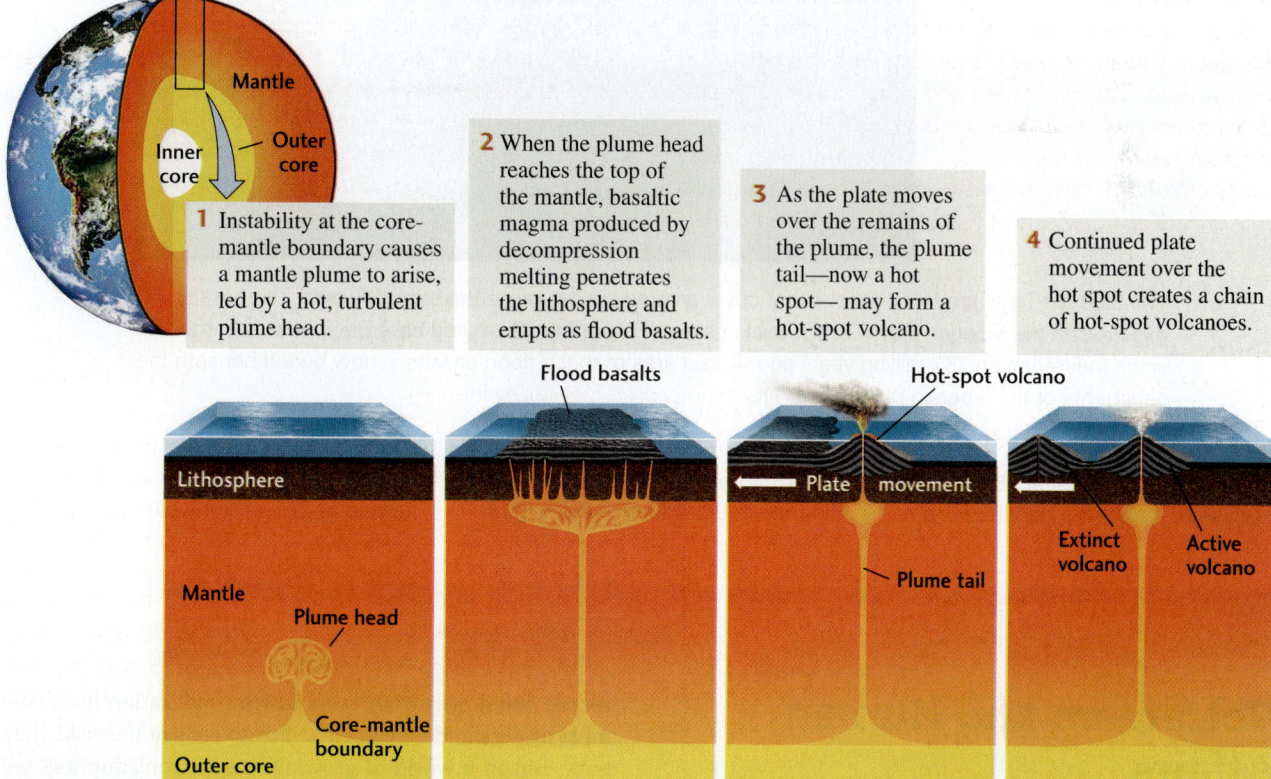

FIGURE 5.31 A speculative model for the formation of flood basalts and other large igneous provinces. A new mantle plume rises from the core-mantle boundary, led by a hot, turbulent plume head. When the plume head reaches the top of the mantle, it flattens, generating a huge volume of basaltic magma, which erupts as flood basalts.

PRACTICING GEOLOGY EXERCISE

Are the Siberian Traps a Smoking Gun of Mass Extinction?

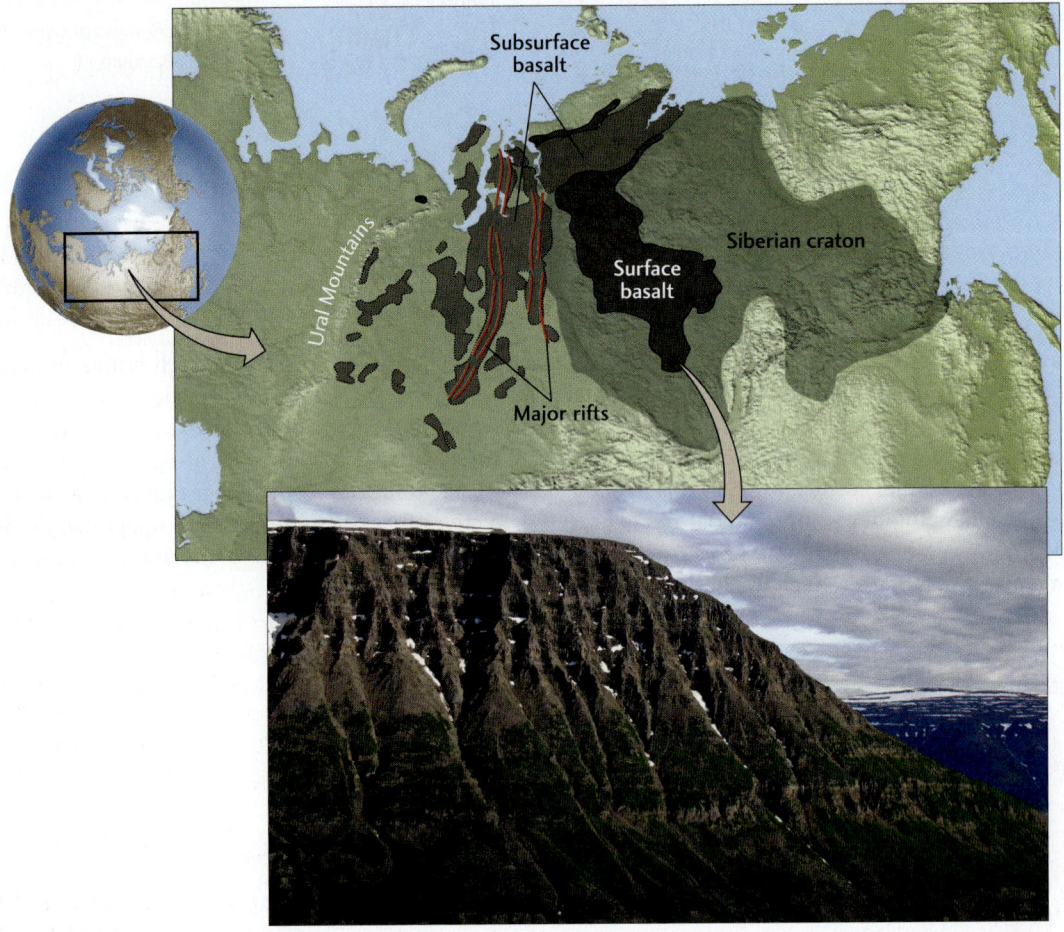

The Siberian Traps are flood basalts that cover an area almost twice the size of Alaska. The basalts exposed on the Siberian craton reach thicknesses of more than 6 km and have been heavily eroded since their eruption 251 million years ago. A vast area of these flood basalts is now buried beneath the sediments of the Siberian platform. [Sergey Anatolievich Pristyazhnyuk/123RF.com.]

The mass extinction at the end of the Permian period, dated at 251 million years ago, marks the transition from the Paleozoic era to the Mesozoic era (see Chapter 9). The flood basalts of Siberia—the product of the largest continental volcanic eruption in the Phanerozoic eon—have also been dated at 251 million years ago. Is this

Volcanism and Human Affairs

Large volcanic eruptions are not just of academic interest to geologists. More than 600 million people live close enough to active volcanoes to be directly affected by eruptions. A repeat of the largest eruptions observed in the geologic record could disrupt or even destroy civilization itself. We must understand volcanic hazards to reduce the risks they pose. But in a world of growing human consumption, we also need to understand and appreciate the benefits volcanism provides to society in the form of mineral resources, fertile soils, and thermal energy.

just a coincidence, or was the eruption of the flood basalts responsible for the end-Permian mass extinction?

Let's first consider the size and rate of the Siberian eruption. Geologic mapping of these flood basalts, called the Siberian Traps, shows that they once extended across much of the Siberian platform and craton, covering an area exceeding 4 million square kilometers. Although much has been eroded away or buried beneath younger sediments, the total volume of the basalts must have originally exceeded 2 million cubic kilometers and may have been as much as 4 million cubic kilometers. Isotopic dating indicates that the basalts were extruded over a period of about 1 million years, implying an average eruption rate of 2 to 4 km^3/year.

To appreciate how large this rate really is, we can compare it with the volcanism at rapidly diverging plate boundaries. Enough basalt is extruded along mid-ocean ridges to form the entire oceanic crust, so the production rate of seafloor spreading is given by the formula

production rate = spreading rate × crustal thickness × ridge length

The fastest spreading we see today is along the East Pacific Rise near the equator, where the Pacific Plate is separating from the Nazca Plate at an average rate of about 140 mm/year, or 1.4×10^{-4} km/year (see Figure 2.7), creating a basaltic crust with an average thickness of 7 km. The length of the Pacific-Nazca Plate boundary is about 3600 km, so the production rate along this spreading center is

1.4×10^{-4} km/year × 7 km × 3600 km
= 3.5 km^3/year

From this calculation, we see that the Siberian eruption produced basalt at a rate comparable to that of the entire Pacific-Nazca Plate boundary, the largest magma factory on Earth today!

You can sail on the tropical sea surface over the Pacific-Nazca Plate boundary and be completely unaware of the magmatic activity deep beneath you. Most of the magma generated by seafloor spreading solidifies as igneous intrusions to form the basaltic dikes and massive gabbros of the oceanic crust (see Figure 4.15). The basalts that are extruded onto the seafloor are quickly quenched by seawater to produce pillow lavas, and the gases that are emitted dissolve into the ocean.

But if you were visiting Siberia about 251 million years ago, you would probably not be so comfortable. The Siberian basalts were erupted directly onto the land surface through fissures in the continental crust, flooding millions of square kilometers. This exceptionally rapid extrusion of lavas would have generated huge pyroclastic deposits—much more than typical flood basalt eruptions, such as those of the Columbia Plateau—and it would also have discharged massive amounts of ash and gases, including carbon dioxide and methane, into the atmosphere. Such an eruption could have triggered changes in Earth's climate of a magnitude that might have led to the end-Permian mass extinction, in which 95 percent of the species living at the time were completely wiped out (see Chapter 9).

Some geologists have argued for years that the end-Permian mass extinction was the result of this intense Siberian volcanism, possibly caused by the sudden arrival of a "plume head" at Earth's surface (see Figure 5.31). Others have preferred alternative hypotheses, such as a meteorite impact or a sudden release of gases from the ocean. However, recent isotopic dating with improved techniques has shown that the Siberian volcanism occurred immediately before or during the end-Permian mass extinction. The finding that these extreme events so precisely coincide has convinced many more geologists that the Siberian Traps are the "smoking gun" behind the largest killing of species in Earth history.

PROBLEM: The Big Island of Hawaii, which has a total rock volume of about 100,000 km^3, has been formed by a series of basaltic eruptions over the last 1 million years. Calculate the production rate of the Hawaiian basalts and compare it with that of the Siberian Traps. What length of the Nazca-Pacific Plate boundary produces basalt at a rate equivalent to the Hawaiian hot spot?

Volcanic Hazards

Volcanic eruptions have a prominent place in human history and mythology. The myth of the lost continent of Atlantis may have its source in the explosion of Thera, a volcanic island in the Aegean Sea (also known as Santorini). The eruption, which has been dated at 1623 B.C., formed a caldera 7 km by 10 km in diameter, visible today as a lagoon up to 500 m deep with two small active volcanoes in the center. The eruption and the tsunami that followed it destroyed dozens of coastal settlements over a large part of the eastern Mediterranean. Some scientists have attributed the mysterious demise of the Minoan civilization to this ancient catastrophe.

Of the 500 to 600 active volcanoes that rise above sea level (see Figure 5.25), at least one in six is known to have claimed human lives. So far in this century, only about 600 people have died in volcanic eruptions, more than half of them in the 2010 eruptions of Mount Merapi in Indonesia. But history teaches us that this luck will not hold over the longer term. In the past 500 years alone, more than 250,000 people have been killed by volcanic eruptions (**Figure 5.32a**). Volcanoes can kill people and damage property in many ways, some of which are listed in Figure 5.32b and depicted in **Figure 5.33**. We have already mentioned some of these hazards, including pyroclastic flows and tsunamis. Several additional volcanic hazards are of special concern.

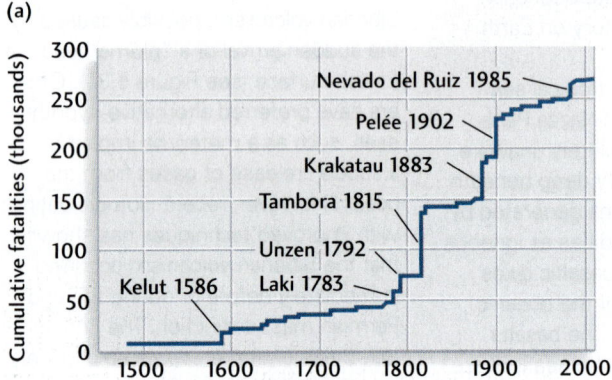

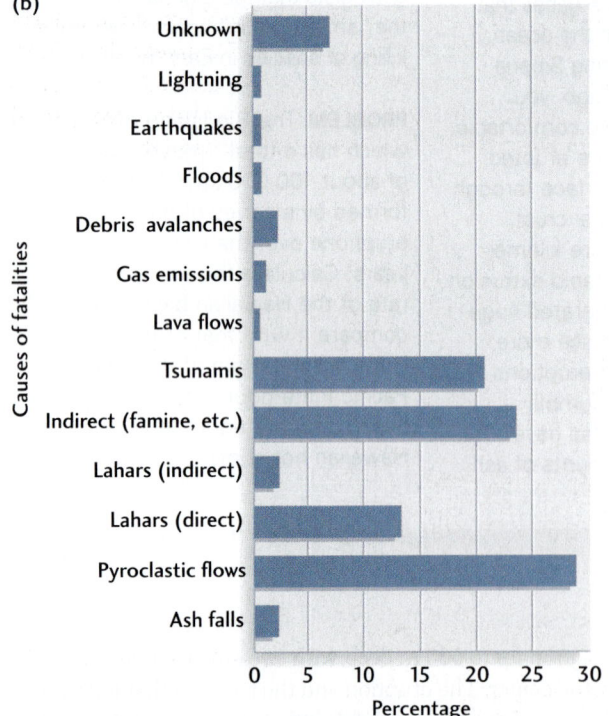

FIGURE 5.32 (a) The eight eruptions that dominate the fatality record, each of which claimed 10,000 or more victims. These eruptions account for two-thirds of the total deaths caused by volcanic eruptions. (b) Specific causes of volcano fatalities since A.D. 1500.

Lahars Among the most dangerous volcanic events are the torrential flows of wet volcanic debris called **lahars**. They can occur when a pyroclastic flow meets a river or a snowbank; when the wall of a water-filled crater breaks; when a lava flow melts glacial ice; or when heavy rainfall transforms new ash deposits into mud. One extensive layer of volcanic debris in the Sierra Nevada of California contains 8000 km^3 of material of lahar origin, enough to cover all of Delaware with a deposit more than a kilometer thick. Lahars have been known to carry huge boulders for tens of kilometers. When Nevado del Ruiz in the Colombian Andes erupted in 1985, lahars triggered by the melting of glacial ice near the summit plunged down the slopes and buried the town of Armero 50 km away, killing more than 25,000 people. In volcanic terrains beneath icecaps, a common danger is the torrential release of floodwater when magma melts large volumes of glacial ice; this very fluidized type of lahar is called a *jökulhlaup* (you-kyl-loop) in Icelandic.

Following the 1997 eruption of the Soufriere Hills volcano on Montserrat, a lahar was generated that blanketed part of the island, resulting in 19 deaths (**Figure 5.34**). Montserrat's tourist industry collapsed and it took 15 years to begin to recover.

Flank Collapse A volcanic mountain is constructed from thousands of deposits of lava or pyroclasts or both—not the best way to build a stable structure. The volcano's sides may become too steep and break or slip off. In recent years, volcanologists have discovered many prehistoric examples of catastrophic structural failures in which a big piece of a volcano broke off, perhaps because of an earthquake, and slid downhill in a massive, destructive landslide. On a worldwide basis, such *flank collapses* occur at an average rate of about four times per century. The collapse of one side of Mount St. Helens was the most damaging part of its 1980 eruption (see Figure 5.6).

Surveys of the seafloor off the Hawaiian Islands have revealed many giant landslides on the underwater flanks of the Hawaiian Ridge. When they occurred, these massive earth movements probably triggered huge tsunamis. In fact, coral-bearing marine sediments have been found some 300 m above sea level on one of the Hawaiian Islands. These sediments were probably deposited by a giant tsunami caused by a prehistoric flank collapse.

The southern flank of Kilauea, on the island of Hawaii, is advancing toward the ocean at a rate of 100 mm/year, which is relatively fast, geologically speaking. This advance became even more worrisome when it suddenly accelerated by a factor of several hundred on November 8, 2000. A network of motion sensors detected an ominous surge in velocity of about 50 mm/day. The surge lasted for 36 hours, after which the normal motion was reestablished. Since then, similar surge events, though variable in size, have been observed every two to three years. Someday—maybe thousands of years from now, but perhaps sooner—the southern flank of the volcano is likely to break off and slide into the ocean. This catastrophic event would trigger a tsunami

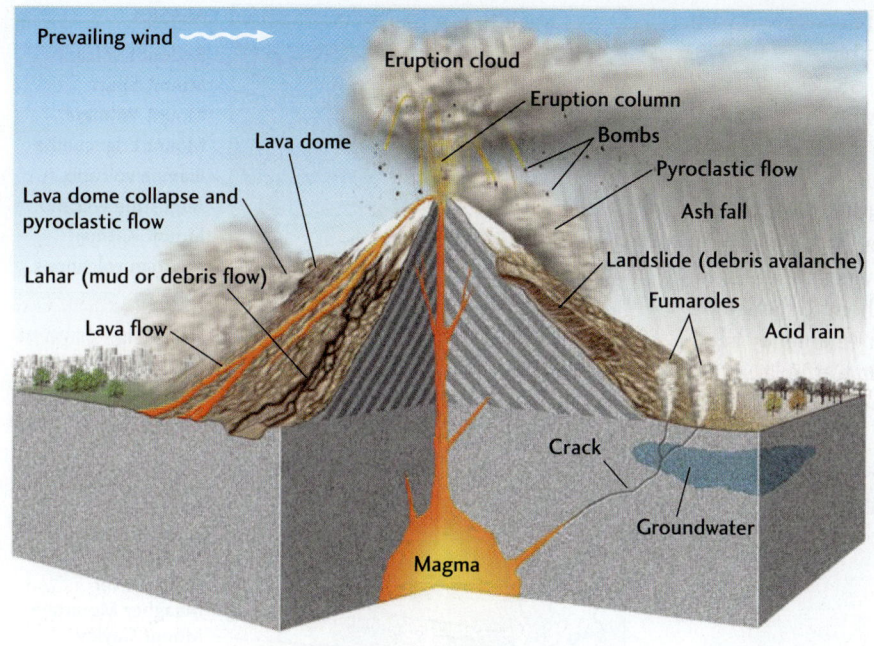

FIGURE 5.33 Some of the volcanic hazards that can kill people and destroy property.

that could prove disastrous for Hawaii, California, and other Pacific coastal areas.

Caldera Collapse Although infrequent, collapses of large calderas are some of the most destructive natural phenomena on Earth. Monitoring the activity of calderas is very important because of their long-term potential for widespread destruction. Fortunately, no catastrophic collapses have occurred in North America during recorded history, but geologists are concerned about an increase in small earthquakes in the Yellowstone and Long Valley calderas as well as other indications of activity in their underlying magma chambers. For example, carbon dioxide leaking into the soil from magma in the crust has been killing trees since 1992 on Mammoth Mountain, a volcano on the boundary of Long Valley Caldera. Regions of the Yellowstone Caldera have been rising at rates as high as 7 cm/year since 2004, and a swarm of more than a thousand small earthquakes occurred near the center of the caldera in a two-week period from December 2008 to January 2009. As in the case of the

FIGURE 5.34 Montserrat in the Caribbean was flooded by lahars after an eruption of the Soufriere Hills volcano in 1997. [Prisma Bildagentur AG/Alamy.]

Long Valley Caldera, these observations are consistent with the injection of magma at mid-crustal depths.

Eruption Clouds A less deadly but still costly hazard comes from the eruption of ash clouds that can damage the jet engines of airplanes that fly through them. More than 60 commercial jet passenger planes have been damaged by such clouds. One Boeing 747 temporarily lost all four engines when ash from an erupting volcano in Alaska was sucked into the engines and caused them to flame out. Fortunately, the pilot was able to make an emergency landing. Warnings of eruption clouds near air traffic lanes are now being issued by several countries. The eruptions of the Eyjafjallajökull, Iceland, volcano in April and May of 2010 disrupted air traffic across the North Atlantic, resulting in over a billion dollars of losses to commercial airlines (see Earth Issues 5.1).

Reducing the Risks of Volcanic Hazards

There are about 100 high-risk volcanoes in the world today, and some 50 volcanic eruptions occur each year. These volcanic eruptions cannot be prevented, but their catastrophic effects can be greatly reduced by a combination of science and enlightened public policy. Volcanology has progressed to the point that we can identify the world's dangerous volcanoes and characterize their potential hazards by studying deposits laid down in earlier eruptions. Some potentially dangerous volcanoes in the United States and Canada are identified in **Figure 5.35**. Assessments of their hazards can be used to guide zoning regulations to restrict land use—the most effective measure to reduce property losses and casualties.

Such studies indicate that Mount Rainier, because of its proximity to the heavily populated cities of Seattle and Tacoma, probably poses the greatest volcanic risk in the United States (**Figure 5.36**). At least 80,000 people and their homes are at risk in Mount Rainier's lahar-hazard zones. An eruption could kill thousands of people and cripple the economy of the Pacific Northwest.

Predicting Eruptions Can volcanic eruptions be predicted? In many cases, the answer is yes, as long as it's close to the actual time of eruption. The "grumblings" (seismic tremors) of volcanoes tend to increase just shortly before they erupt. Instrumented monitoring can detect signals such as earthquakes, magnetic pulses, changes in gravity, swelling of the volcano, and gas emissions that warn of impending eruptions. High-resolution GPS measurements can detect movements in the magma chamber beneath the surface of the volcano. People at risk can be evacuated if the authorities are organized and prepared. Scientists monitoring Mount St. Helens were able to warn people before its eruption in 1980 (see Earth Issues 5.2). Government infrastructure was in place to evaluate the warnings and to enforce evacuation orders, so very few people were killed.

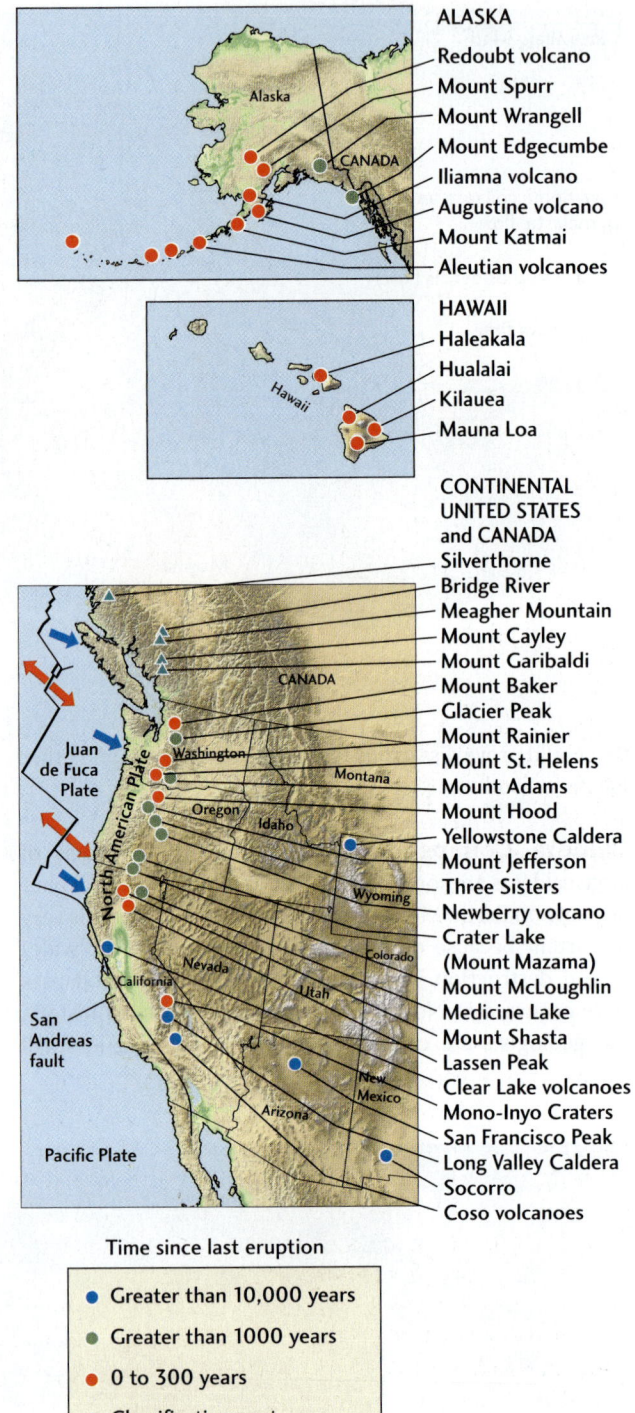

FIGURE 5.35 Locations of potentially hazardous volcanoes in the United States and Canada. Volcanoes within each U.S. group are color-coded by time since their last eruption; those that have erupted most recently are thought to present the greatest cause for concern. (These classifications are subject to revision as studies progress and are not available for Canadian volcanoes.) Note the relationship between the volcanoes extending from northern California to British Columbia and the convergent boundary between the North American Plate and the Juan de Fuca Plate. [Data from R. A. Bailey, P. R. Beauchemin, F. P. Kapinos, and D. W. Klick/USGS.]

Volcanism and Human Affairs

FIGURE 5.36 Mount Rainier, seen from Tacoma, Washington. [Patrick Lynch/Alamy.]

an accurate eruptive history over the many hundreds, or tens of thousands, of years necessary to establish a reliable recurrence interval; and 2) Few volcanoes maintain the same behavior for long. More often than not, as soon as a repetitive pattern becomes apparent, the volcano changes behavior.

Controlling Eruptions Can we go further by actually controlling volcanic eruptions? Not likely, because large volcanoes release energy on a scale that dwarfs our capabilities for control. Under special circumstances and on a small scale, however, the damage can be reduced. Perhaps the most successful attempt to manage volcanic activity was made on the Icelandic island of Heimaey in January 1973. By spraying advancing lava with seawater, residents cooled and slowed the flow, preventing the lava from blocking the entrance to their harbor and saving some homes from destruction. The best focus for our efforts, however, will be the establishment of more warning and evacuation systems and more rigorous restriction of settlements in potentially dangerous locations.

Natural Resources from Volcanoes

In this chapter, we have seen something of the beauty of volcanoes and something of their destructiveness. But it should be kept in mind that volcanoes contribute to our well-being in many, though often indirect, ways. Soils derived from volcanic materials are exceptionally fertile because of the mineral nutrients they contain. Volcanic rock, gases, and steam are also sources of important industrial materials and chemicals, such as pumice, boric acid, ammonia, sulfur, carbon dioxide, and some metals. Hydrothermal activity is responsible for the deposition of unusual minerals that concentrate relatively rare elements, particularly metals, into ore deposits of great economic value. Seawater circulating through mid-ocean ridges is a major factor in the formation of such ores and in the maintenance of the chemical balance of the oceans.

In some regions where geothermal gradients are steep, Earth's internal heat can be tapped to heat homes and drive electric generators. **Geothermal energy** depends on the heating of water as it passes through a region of hot rock (a *heat reservoir*) that may be hundreds or thousands of meters beneath Earth's surface. Hot water or steam can be brought to the surface through boreholes drilled for the purpose. Usually, the water is naturally occurring groundwater that seeps downward along fractures in rock. Less typically, the water is artificially introduced by pumping from the surface.

By far the most abundant source of geothermal energy is naturally occurring groundwater that has been heated to temperatures of 80°C to 180°C. Water at these relatively low temperatures is used for residential, commercial, and industrial heating. Warm groundwater drawn from a heat reservoir in the Paris sedimentary basin now heats more

Another successful warning was issued a few days before the cataclysmic eruption of Mount Pinatubo in the Philippines on June 15, 1991. A quarter of a million people were evacuated, including some 16,000 residents of the nearby U.S. Clark Air Force Base (which was heavily damaged by the eruption and has since been permanently abandoned). Tens of thousands of lives were saved from the lahars that destroyed everything in their paths. Casualties were limited to the few who disregarded the evacuation order. And in 1994, 30,000 residents of Rabaul, Papua New Guinea, were successfully evacuated by land and sea hours before the two volcanoes on either side of the town erupted, destroying or damaging most of it. Many owe their lives to the government, which conducted evacuation drills, and to scientists at the local volcano observatory, who issued a warning when their seismographs recorded the ground tremors that signaled magma moving toward the surface.

Over the long term, prediction of eruptions is much more difficult because they must be based on the history of volcanic eruptions, so-called "eruption recurrence intervals." These are notoriously unreliable for two reasons: 1) Few volcanoes are sufficiently well studied to provide

Earth Issues 5.2 Mount St. Helens: Dangerous but Predictable

Mount St. Helens, in the Cascade Range of the Pacific Northwest, is the most active and explosive volcano in the contiguous United States (see Figure 5.6). It has a documented 4500-year history of destructive lava flows, pyroclastic flows, lahars, and distant ash falls. Beginning on March 20, 1980, a series of small to moderate earthquakes under the volcano signaled the start of a new eruptive phase after 123 years of dormancy, motivating the U.S. Geological Survey to issue a formal hazard alert. The first outburst of ash and steam erupted from a newly opened crater on the summit one week later.

Throughout April, the seismic tremors increased, indicating that magma was moving beneath the summit, and instruments detected an ominous swelling of the northeastern flank of the mountain. The USGS issued a more serious warning and people were ordered out of the vicinity. On May 18, the main eruption began abruptly. A large earthquake apparently triggered the collapse of the north side of the mountain, loosening a massive landslide, the largest ever recorded anywhere. As this huge debris avalanche plummeted down the mountain, gas and steam under high pressure were released in a tremendous lateral blast that blew out the northern flank of the mountain.

USGS geologist David A. Johnston was monitoring the volcano from his observation post 8 km to the north. He must have seen the advancing blast wave before he radioed his last message: "Vancouver, Vancouver, this is it!" A northward-directed jet of superheated (500°C) ash, gas, and steam roared out of the breach with hurricane force, devastating a zone 20 km outward from the volcano and 30 km wide. A vertical eruption column sent an ash plume 25 km into the sky, twice as high as a commercial jet flies. The ash plume drifted to the east and northeast with the prevailing winds, bringing darkness at noon to an area 250 km to the east and depositing a layer of ash as deep as 10 cm over much of Washington, northern Idaho, and western Montana. The energy of the blast was equivalent to about 25 million tons of TNT. The volcano's summit was destroyed, its elevation was reduced by 400 m, and its northern flank disappeared. In effect, the mountain was hollowed out.

Earthquakes and magmatic activity have continued off and on since the 1980 eruption. After more than a decade of relative quiescence, the volcano reawoke in September 2004 with a series of minor steam and ash eruptions that continued into 2005. Growth of the central volcanic dome (see Figure 5.14b) suggests that the current phase of eruptive activity may persist for some time into the future.

The eruption of Mount St. Helens on May 18, 1980, sent an ash plume into the stratosphere and an avalanche and blast wave toward the north.

(May 17, 3 P.M.) View of Mount St. Helens the day before its eruption. The north side of the volcano has bulged outward from magma intruded at shallow levels during the previous two months. [Keith Ronnholm.]

(May 18, 8:33 A.M.) An earthquake and massive landslide "uncork" the volcano, releasing an ash plume and a powerful lateral blast wave. [Keith Ronnholm.]

FIGURE 5.37 The Geysers, one of the world's largest supplies of natural steam. The geothermal energy is converted into electricity for San Francisco, 120 km to the south. [Charles Rotkin/Getty Images.]

than 20,000 apartments in France. Reykjavik, the capital of Iceland, which sits atop the Mid-Atlantic Ridge, is almost entirely heated by geothermal energy.

Heat reservoirs with temperatures above 180°C are useful for generating electricity. They are present primarily in regions of recent volcanism as hot, dry rock, natural hot water, or natural steam. Naturally occurring water heated above the boiling point and naturally occurring steam are highly prized resources. The world's largest facility for producing electricity from natural steam, located at The Geysers, 120 km north of San Francisco, generates more than 600 megawatts of electricity (**Figure 5.37**). Some 70 geothermal electricity-generating plants operate in California, Utah, Nevada, and Hawaii, producing 2800 megawatts of power—enough to supply about a million people.

KEY TERMS AND CONCEPTS

andesitic lava (p. 121)
ash-flow deposit (p. 130)
basaltic lava (p. 119)
breccia (p. 124)
caldera (p. 127)
crater (p. 125)
diatreme (p. 127)
fissure eruption (p. 129)

flood basalt (p. 130)
geothermal energy (p. 145)
hot spot (p. 137)
hydrothermal activity (p. 131)
lahar (p. 142)
large igneous province (p. 138)
mantle plume (p. 137)
pyroclastic flow (p. 124)

rhyolitic lava (p. 122)
shield volcano (p. 125)
stratovolcano (p. 125)
tuff (p. 124)
volcanic geosystem (p. 118)
volcano (p. 118)

REVIEW OF LEARNING OBJECTIVES

5.1 Describe how volcanoes transport magma from the Earth's interior to its surface.

Temperatures within the asthenosphere can reach at least 1300°C, high enough to begin to melt rocks. The asthenosphere is the main source of magma, and once that molten rock reaches the surface and erupts it is called lava. Because magma is liquid, it is less dense than the rocks that produce it, and will begin to float upward through the lithosphere. In some places, the magma may fracture the lithosphere along zones of weakness. The rocks, magmas, and processes needed to describe the entire sequence of events from melting to eruption constitute a volcanic geosystem.

Study Assignment: Figure 5.1

Exercise: What is the difference between magma and lava? Describe a geologic situation in which a magma does not form a lava.

Thought Question: Give a few examples of what geologists have learned about Earth's interior by studying volcanoes and volcanic rocks.

5.2 Differentiate between the major types of volcanic deposits and explain how the textures of volcanic rocks can reflect the conditions under which they solidified.

Lavas of different types produce different landforms. These differences depend on chemical composition, gas content, and temperature of the lavas. The textures of volcanic rocks reflect the conditions under which they solidified. Erupted lavas usually solidify into one of three major types of rock: basaltic (mafic), andesitic (intermediate), or rhyolitic (felsic), which are classified on the basis of their content of silica and other minerals. Basaltic lavas are relatively fluid and flow freely; andesitic and rhyolitic lavas are more viscous. Lavas differ from pyroclasts, which are formed by explosive eruptions and vary in size from fine ash particles to house-sized bombs.

Study Assignment: Figures 5.4, 5.6, 5.8

Exercise: What are the three major types of volcanic rocks and their intrusive counterparts? Is kimberlite one of these three types?

Thought Question: While on a field trip, you come across a volcanic formation that resembles a field of sandbags. The individual ellipsoidal forms have a smooth, glassy surface texture. What type of lava is this, and what information does this give you about its history?

5.3 Summarize how volcanic landforms are shaped.

The chemical composition and gas content of magma are important factors in a volcano's eruptive style and in the shape of the landforms it creates. Volcanic landforms are also dependent on the rate at which lava is produced and the plumbing system that gets it to the surface. A shield volcano grows from repeated eruptions of basaltic lava from a central vent. Andesitic and rhyolitic lavas tend to erupt explosively. The erupted pyroclasts may pile up into a cinder cone. A stratovolcano is built of alternating layers of lava flows and pyroclastic deposits. The rapid ejection of magma from a large magma chamber, followed by collapse of the chamber's roof, results in a large depression, or caldera. Basaltic lavas can erupt from fissures along mid-ocean ridges as well as on continents, where they flow over the landscape in sheets to form flood basalts. Pyroclastic eruptions from fissures can cover an extensive area with ash-flow deposits.

Study Assignment: Figure 5.14

Exercise: What type of volcano is the Arenal volcano, shown in Figure 5.10?

Thought Question: Why are eruptions of stratovolcanoes generally more explosive than those of shield volcanoes?

5.4 Discuss how volcanic gases can affect the hydrosphere and atmosphere.

Volcanoes produce gases as well as solid materials. These gases may come from deep within the Earth, making their way to the surface for the first time. These gases have a number of effects on other geosystems, including the hydrosphere and atmosphere. Even when lavas and pyroclastics cease to flow, volcanoes continue to emit steam and gases for many years. Hydrothermal activity is the circulation of water through hot volcanic rocks and magmas, which is heated and returned to the surface as hot springs and geysers. Hydrothermal activity is especially intense at spreading centers and mid-ocean ridges, where large volumes of water and magma come into contact. Volcanism can also affect the water and climate by changing the composition and properties of the atmosphere. Large eruptions can inject sulfurous gases into the atmosphere, blocking the Sun's radiation and lowering global temperatures.

Study Assignment: Figure 5.22

Exercise: Describe how interactions between the lithosphere and hydrosphere at spreading centers affect the geology, chemistry, and biology of the oceans.

Thought Question: How could periods of intense volcanic activity be responsible for some of the mass extinctions documented in the geologic record?

5.5 Explain how the global pattern of volcanism is related to plate tectonics.

Of the world's active volcanoes that occur on land or above the ocean surface, 80 percent are found at convergent plate boundaries, 15 percent at divergent plate boundaries, and 5 percent within plate interiors. The huge volumes of basaltic magma that form oceanic crust are produced by decompression melting and erupted at spreading centers on mid-ocean ridges. Interactions between the lithosphere and hydrosphere at spreading centers affect the geology, chemistry, and biology of the oceans. Andesitic lavas are the most common lava type in the volcanic mountain belts of ocean-continent subduction zones. Rhyolitic lavas are produced by the melting of felsic continental crust. Within plates, basaltic volcanism occurs above hot spots, which are manifestations of rising plumes of hot mantle material.

Study Assignment: Figure 5.24

Exercise: On Earth's surface as a whole, what process generates the greater volume of volcanic rock, decompression melting or fluid-induced melting? Which of these processes creates the more dangerous volcanoes?

Thought Question: Why are the volcanoes on the northwestern side of the island of Hawaii dormant whereas those on the southeastern side are more active?

5.6 Illustrate the hazards and beneficial effects of volcanism.

Volcanic hazards that can kill people and damage property include pyroclastic flows, tsunamis, lahars, flank collapses, caldera collapses, eruption clouds, and ash falls. Volcanic eruptions have killed about 250,000 people in the past 500 years. On the positive side, volcanic materials produce nutrient-rich soils, and hydrothermal processes are important in the formation of many economically valuable mineral ores. Seawater circulating through mid-ocean ridges is a major factor in the formation of such ores and in the maintenance of the chemical balance of the oceans. Geothermal heat drawn from areas of hydrothermal activity is a useful source of energy in some regions.

Study Assignment: Figure 5.33

Exercise: How do scientists predict volcanic eruptions?

Thought Question: What might be the effects on civilization of a Yellowstone-type caldera eruption, such as the one described at the opening of this chapter?

VISUAL LITERACY EXERCISE

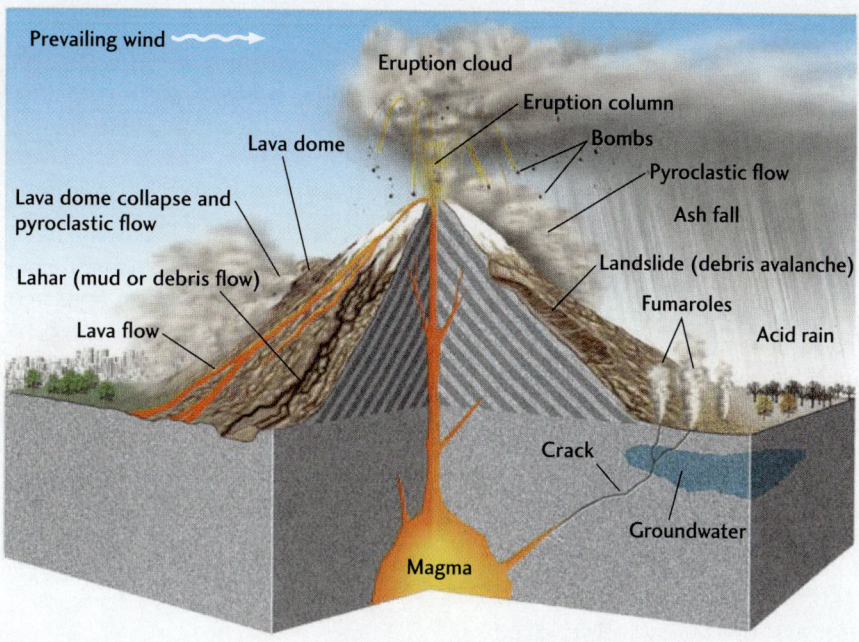

FIGURE 5.33 Some of the volcanic hazards that can kill people and destroy property.

1. What causes a pyroclastic flow?
   a. Lava dome collapse
   b. Acid rain
   c. Erupting groundwater
   d. Entrainment by prevailing winds

2. When do volcanic bombs move?
   a. During a landslide
   b. In flowing lava
   c. During a volcanic eruption
   d. During formation of fumaroles

3. How is water heated that feeds fumaroles?
   a. When groundwater comes in contact with rocks heated by magma
   b. During rainfall
   c. By friction generated in debris flows
   d. When lava flows down the sides of volcanoes

4. Which is correct?
   a. A pyroclastic flow is generated by a volcanic eruption.
   b. Volcanic ash falls from an eruption cloud.
   c. Prevailing winds push an eruption cloud in the direction of that wind.
   d. Ash can travel further than bombs.

5. What are lahars composed of? How many of these are correct?
   a. Acid rain
   b. Groundwater
   c. Bombs
   d. Mud

Sedimentation: Rocks Formed by Surface Processes

6

Surface Processes of the Rock Cycle	154
Sedimentary Basins: The Sinks for Sediments	160
Sedimentary Environments	162
Sedimentary Structures	165
Burial and Diagenesis: From Sediment to Rock	168
Classification of Siliciclastic Sediments and Sedimentary Rocks	172
Classification of Chemical and Biological Sediments and Sedimentary Rocks	175

Learning Objectives

Chapter 6 describes how the rock cycle produces sediments and sedimentary rocks. We will describe the compositions, textures, and structures of sediments and sedimentary rocks, and examine how they correlate with the kinds of environments in which the sediments and rocks are laid down. Throughout the chapter, we will apply our understanding of sediment origins to the study of human environmental problems and to the exploration for energy and mineral resources. After you have studied the chapter, you should be able to:

6.1 Describe the major processes that form sedimentary rocks.

6.2 Compare the various types of sedimentary basins.

6.3 Define the different types of sedimentary environments and the two main types of sediments and sedimentary rocks.

6.4 Summarize the types of sedimentary structures.

6.5 Discuss the process of burial and diagenesis of sediments.

6.6 Describe how the major kinds of siliciclastic, chemical, and biological sediments are classified.

6.7 Discuss how carbonate and evaporite rocks are formed.

The large-scale cross-bedding visible in this sandstone records the history of its formation in an ancient desert. [Courtesy John Grotzinger.]

Much of Earth's surface, including its seafloor, is covered with sediments. These layers of loose particles have diverse origins. Most sediments are created by weathering of the continental crust. Some are the remains of mineral shells secreted by organisms. Still others consist of inorganic crystals that precipitated when dissolved chemicals in oceans and lakes combined to form new minerals.

Sedimentary rocks were once sediments, so they are records of the conditions at Earth's surface when and where those sediments were deposited. Geologists can work backward to infer the sources of the sediments from which the rocks were formed and the kinds of places in which the sediments were originally deposited. The top of Mount Everest, for example, is composed of fossiliferous (fossil-containing) limestones. Because we know that such limestones are formed from carbonate minerals in seawater, we can conclude that Mount Everest must once have been part of an ocean floor! This kind of analysis can be applied just as well to ancient shorelines, mountains, plains, deserts, and wetlands. By reconstructing such environments, we can map the continents and oceans of long ago.

Sedimentary rocks may also reveal former plate tectonic events and processes by their presence within or adjacent to volcanic island arcs, continental rift valleys, or collisional or volcanic mountain belts. In cases in which sediments and sedimentary rocks are derived from the weathering of preexisting rocks, we can form hypotheses about ancient climates and environments. We can also use sedimentary rocks formed by precipitation from seawater to read the history of changes in Earth's climate and seawater chemistry.

The study of sediments and sedimentary rocks has great practical value as well. Oil, natural gas, and coal, our most valuable fossil-fuel resources, are found in these rocks. A number of other important mineral resources are also sedimentary, such as the phosphate rock used for fertilizer and much of the world's iron ore. Knowing how these kinds of sediments form helps us to find and use these limited resources.

Finally, because virtually all sedimentary processes take place at or near Earth's surface where we humans live, they provide a background for our understanding of environmental problems. We once studied sedimentary rocks primarily to understand how to exploit the natural resources just mentioned. Increasingly, however, we study these rocks to improve our understanding of Earth's environment.

Surface Processes of the Rock Cycle

Sediments, and the sedimentary rocks formed from them, are produced by the surface processes of the rock cycle. These processes act on rocks after they have been moved from Earth's interior to its surface during mountain building, and before they are returned to Earth's interior by subduction. They move materials from a *source area*, where sediment particles are created, to a *sink area*, where they are deposited in layers. The path that sediment particles follow from source to sink may be a very long one—one that involves several important processes resulting from interactions between the plate tectonic and climate systems.

Let's look at the role of the Mississippi River in a typical sedimentary process. Plate movement lifts up rocks in the Rocky Mountains. Rainfall in those mountains—a source area—causes weathering of the rocks there. If precipitation increases in the mountains, weathering also increases. Faster weathering produces more sediment to be released into the river and transported downhill and downstream. At the same time, if the flow in the river also increases because of the higher rainfall, transportation of sediment down the length of the river increases, and the volume of sediment delivered to sink areas—sites of deposition, also known as *sedimentary basins*—in the Mississippi delta and the Gulf of Mexico increases as well. In these sedimentary basins, the sediments pile up on top of one another—layer after layer—and are eventually buried deep in Earth's crust, to depths where they may become filled with valuable oil and natural gas.

The surface processes of the rock cycle that are important in the formation of sedimentary rocks are reviewed in **Figure 6.1** and summarized here:

- *Weathering* is the general process by which rocks are broken down at Earth's surface to produce sediment particles. There are two types of weathering. **Physical weathering** takes place when solid rock is fragmented by mechanical processes, such as freezing and thawing or wedging by tree roots (**Figure 6.2**), that do not change its chemical composition. The rubble of broken stone often seen at the tops of mountains and hills is primarily the result of physical weathering. **Chemical weathering** refers to processes by which the minerals in a rock are chemically altered or dissolved. The blurring or disappearance of lettering on old gravestones and monuments is caused mainly by chemical weathering.

- *Erosion* refers to processes that dislodge particles of rock produced by weathering and move them away from the source area. Erosion occurs most commonly when rainwater runs downhill.

- *Transportation* refers to processes by which sediment particles are moved to sink areas. Transportation occurs when water, wind, or the moving ice of glaciers transport particles to new locations downhill or downstream.

- *Deposition* (also called *sedimentation*) refers to processes by which sediment particles settle out as water currents slow, winds die down, or glacier edges melt to form layers of sediment in sink areas.

Surface Processes of the Rock Cycle

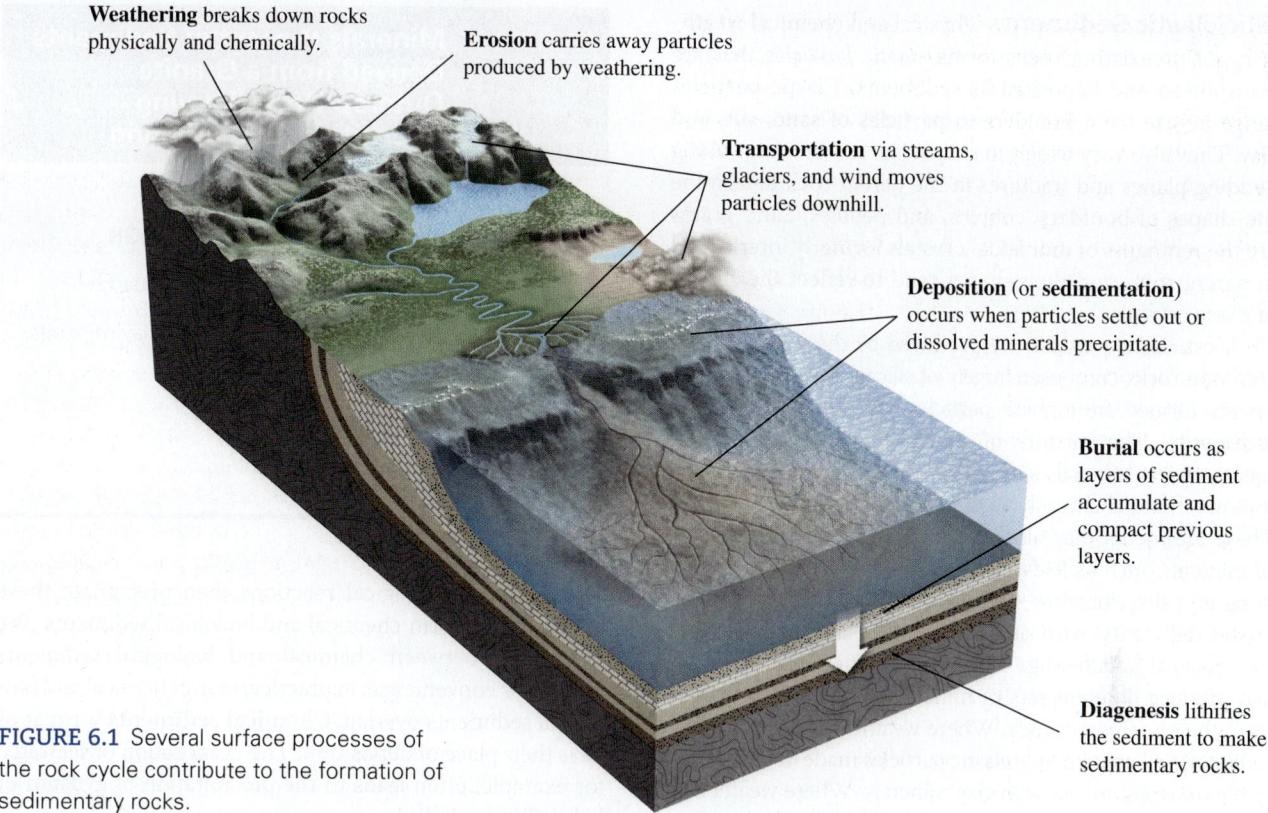

FIGURE 6.1 Several surface processes of the rock cycle contribute to the formation of sedimentary rocks.

In aquatic environments, particles settle out, chemical precipitates form and are deposited, and the bodies and shells of dead organisms are broken up and deposited.

- *Burial* occurs as layers of sediment accumulate in sink areas on top of older, previously deposited sediments, which are compacted and progressively buried deep within a sedimentary basin. These sediments will remain at depth, as part of Earth's crust, until they are either uplifted again or subducted by plate tectonic processes.

- *Diagenesis* refers to the physical and chemical changes—caused by pressure, heat, and chemical reactions—by which sediments buried within sedimentary basins are *lithified*, or converted into sedimentary rock.

Weathering and Erosion: The Source of Sediments

Chemical and physical weathering reinforce each other. Chemical weathering weakens rocks and makes them more susceptible to fragmentation. The smaller the fragments produced by physical weathering, the greater the surface area exposed to chemical weathering. Together, physical and chemical weathering of rock create both solid particles and dissolved products, and erosion carries them away. These end products can be classified as either siliciclastic sediments or chemical and biological sediments.

FIGURE 6.2 Plant roots contribute to physical weathering by penetrating fractures and wedging rocks apart. [David R. Frazier/Science Source.]

Siliciclastic Sediments Physical and chemical weathering of preexisting rocks forms *clastic particles* that are transported and deposited as sediments. Clastic particles range in size from boulders to particles of sand, silt, and clay. They also vary widely in shape. Natural breakage along bedding planes and fractures in the parent rock determine the shapes of boulders, cobbles, and pebbles. Sand grains are the remnants of individual crystals formerly interlocked in parent rock, and their shapes tend to reflect the shapes of those crystals.

Most clastic particles are produced by the weathering of common rocks composed largely of silicate minerals, so sediments formed from these particles are called **siliciclastic sediments.** The mixture of minerals in siliciclastic sediments varies. Minerals such as quartz resist weathering and thus are found chemically unaltered in siliciclastic sediments. These sediments may also contain partly altered fragments of minerals, such as feldspar, that are less resistant to weathering and therefore less stable. Still other minerals in siliciclastic sediments, such as clay minerals, are newly formed by chemical weathering. Varying intensities of weathering can produce different sets of minerals in sediments derived from the same parent rock. Where weathering is intense, the sediment will contain only clastic particles made of chemically stable minerals, mixed with clay minerals. Where weathering is slight, many minerals that are unstable under land surface conditions will survive as clastic particles in the sediment. Table 6.1 shows three possible sets of minerals in sediments derived from a typical granite outcrop.

Chemical and Biological Sediments Chemical weathering produces dissolved ions and molecules that accumulate in the waters of soils, rivers, lakes, and oceans.

TABLE 6.1	Minerals Present in Sediments Derived from a Granite Outcrop Under Varying Intensities of Weathering	
INTENSITY OF WEATHERING		
LOW	**MEDIUM**	**HIGH**
Quartz	Quartz	Quartz
Feldspar	Feldspar	Clay minerals
Mica	Mica	
Pyroxene	Clay minerals	
Amphibole		

Chemical and biological reactions then precipitate these substances to form chemical and biological sediments. We distinguish between chemical and biological sediments mainly for convenience; in practice, many chemical and biological sediments overlap. **Chemical sediments** form at or near their place of deposition. The evaporation of seawater, for example, often leads to the precipitation of gypsum or halite (**Figure 6.3**).

Biological sediments also form near their place of deposition, but they are the result of mineral precipitation by organisms. Some organisms, such as mollusks and corals, precipitate minerals as they grow. After the organisms die, their shells or skeletons accumulate on the seafloor as sediments. In these cases, the organism *directly* controls mineral precipitation. However, in a second but equally important

FIGURE 6.3 Salts precipitate when water containing dissolved minerals evaporates, which has occurred here in Death Valley, California. [John G. Wilbanks/ Age fotostock.]

FIGURE 6.4 One kind of sedimentary rock of biological origin is formed entirely of shell fragments. [Courtesy John Grotzinger.]

process, organisms control mineral precipitation only *indirectly*. Instead of taking up minerals from the water to form a shell, these organisms change the surrounding environment so that mineral precipitation occurs on the outside of the organism, or even away from the organism. Certain microorganisms are thought to enable the precipitation of pyrite (an iron sulfide mineral) in this way (see Chapter 22).

In shallow marine environments, directly precipitated biological sediments consist of layers of particles, such as whole or fragmented shells of marine organisms (**Figure 6.4**). Many different types of organisms, ranging from corals to clams to algae, may contribute their shells. Sometimes the shells are transported, further broken up, and deposited as **bioclastic sediments.** These shallow-water sediments consist predominantly of two calcium carbonate minerals, calcite and aragonite, in variable proportions. Other minerals, such as phosphates and sulfates, are only locally abundant in bioclastic sediments.

In the deep sea, biological sediments are made up of the shells of only a few kinds of planktonic organisms. Most of these organisms secrete shells composed primarily of calcite and aragonite, but a few species form silica shells, which are precipitated broadly over some parts of the deep seafloor. Because these biological particles accumulate in very deep water, where agitation by sediment-transporting currents is uncommon, they rarely form bioclastic sediments.

Transportation and Deposition: The Downhill Journey to Sedimentary Basins

After clastic particles and dissolved ions have been formed by weathering and dislodged by erosion, they start their journey to a sedimentary basin. This journey may be very long; for example, as we have seen, it might span the thousands of kilometers from the tributaries of the Mississippi River in the Rocky Mountains to the wetlands of the Mississippi delta.

Most agents of sediment transport carry sediments on a one-way trip downhill. Rocks falling from a cliff, a river flowing to the ocean carrying a load of sand, and glacial ice slowly dragging boulders downhill are all responses to gravity. Although wind may blow material from a low elevation to a higher one, in the long run the effects of gravity prevail. When a windblown particle drops into the ocean and settles through the water, it is trapped. It can be picked up again only by an ocean current, which can transport it to and deposit it in another site on the seafloor. Ocean currents transport sediments over shorter distances than do big rivers on land, and the short transportation distances of chemical and biological sediments contrast with the much greater distances over which siliciclastic sediments are transported. But eventually, all sediment transportation paths, as simple or complicated as they may be, lead downhill into a sedimentary basin.

Currents as Transport Agents Most sediments are transported by currents of air or water. The enormous quantities of all kinds of sediments found in the oceans result primarily from the transport capacities of rivers, which annually carry a solid and dissolved sediment load of about 25 billion tons (25×10^{15} g) (**Figure 6.5**). Air currents—winds—move sediments, too, but in far smaller quantities than rivers or ocean currents. As particles are lifted into the air or water, the current carries them downwind or downstream. The stronger the current—that is, the faster it flows—the larger the particles it can transport.

FIGURE 6.5 Sediments are easily transported by flowing water. In this photo, small ripples of sand in the channel are evidence of sediment transportation. [Courtesy John Grotzinger.]

Current Strength, Particle Size, and Sorting

Deposition starts where transportation stops. For clastic particles, gravity is the driving force of deposition. The tendency of particles to settle under the pull of gravity works against a current's ability to carry them. A particle's settling velocity is proportional to its density and its size (see the Practicing Geology Exercise in Chapter 4). Because all clastic particles have roughly the same density, we use particle size as the best indicator of how quickly a particle will settle. (We will take a more specific look at particle size categories later in this chapter.) In water, large particles settle faster than small ones. This is also true in air, but the difference is much smaller.

Current strength, which is directly related to current velocity, determines the size of the particles deposited in a particular place. As a wind or water current begins to slow, it can no longer keep the largest particles suspended, and those particles settle. As the current slows even more, smaller particles settle. When the current stops completely, even the smallest particles settle. Currents segregate particles in the following ways:

- *Strong currents* (faster than 50 cm/s) carry gravel (which includes boulders, cobbles, and pebbles), along with an abundant supply of smaller particles. Such currents are common in swiftly flowing streams in mountainous terrain, where erosion is rapid. Beach gravels are deposited where ocean waves erode rocky shores.

- *Moderately strong currents* (20–50 cm/s) lay down sand beds. Currents of moderate strength are common in most rivers, which carry and deposit sand in their channels. Rapidly flowing floodwaters may spread sand over the width of a river valley. Waves and currents deposit sand on beaches and in the ocean. Winds also blow and deposit sand, especially in deserts. However, because air is much less dense than water, much higher current velocities are required for it to move sediments of the same size and density.

- *Weak currents* (slower than 20 cm/s) carry muds composed of the finest clastic particles (silt and clay). Weak currents are found on the floor of a river valley when floodwaters recede slowly or stop flowing entirely. In the ocean, muds are generally deposited some distance from shore, where currents are too slow to keep even fine particles in suspension. Much of the floor of the open ocean is covered with mud particles originally transported by surface

FIGURE 6.6 As the strength of a current changes, sediments are sorted according to particle size. The relatively homogeneous group of sand grains on the left is well sorted; the group on the right is poorly sorted. [Courtesy John Grotzinger.]

waves and currents or by wind. These particles slowly settle to depths where currents and waves are stilled and, ultimately, all the way to the bottom of the ocean.

As you can see, currents may begin by carrying particles of widely varying sizes, which then become separated as the strength of the current changes. A strong, fast current may lay down a bed of gravel while keeping sand and mud in suspension. If the current weakens and slows, it will lay down a bed of sand on top of the gravel. If the current then stops altogether, it will deposit a layer of mud on top of the sand bed. This tendency for variations in current velocity to segregate sediments according to size is called **sorting**. A well-sorted sediment consists mostly of particles of a uniform size. A poorly sorted sediment contains particles of many sizes (**Figure 6.6**).

As cobbles, pebbles, and large sand grains are being transported by water or air currents, they tumble and strike one another or rub against the underlying rock. The resulting *abrasion* affects the particles in two ways: It reduces their size and it rounds off their rough edges (**Figure 6.7**). These effects apply mostly to the larger particles; smaller sand grains and silt undergo little abrasion.

Particles are generally transported intermittently rather than steadily. A river may transport large quantities of sand and gravel when it floods, then drop them as the flood recedes, only to pick them up again and carry them even farther in the next flood. Similarly, strong winds may carry large amounts of dust for a few days,

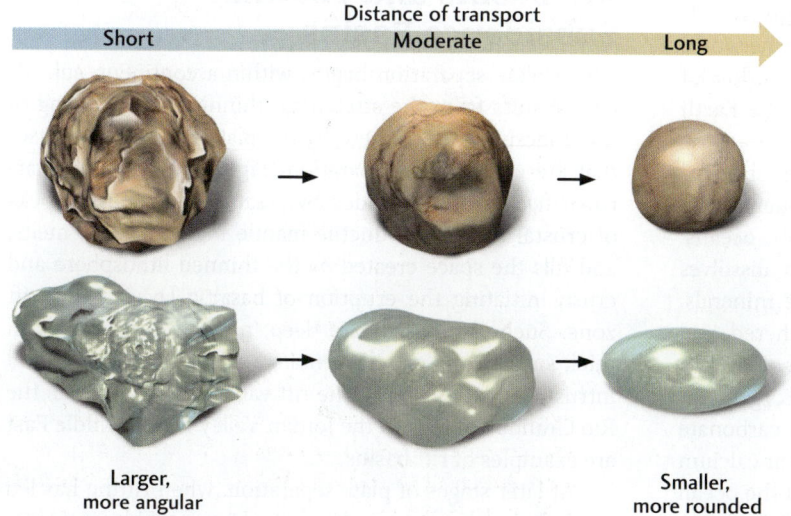

FIGURE 6.7 Abrasion during transportation reduces the size and angularity of clastic particles. Particles become more rounded and slightly smaller as they are transported, although the general shape of the particle may not change significantly.

then die down and deposit the dust as a layer of sediment. The strong tidal currents along some shorelines may transport broken shell fragments to places farther offshore and drop them there.

The total time it takes for clastic particles to be transported may be many hundreds or thousands of years, depending on the distance to the final sedimentary basin and the number of stops along the way. Clastic particles eroded in the mountains of western Montana, for example, take hundreds of years to travel the 3200 km down to the Mississippi River and out into the Gulf of Mexico.

Oceans as Chemical Mixing Vats

The driving force of chemical and biological sedimentation is precipitation rather than gravity. Substances dissolved in water by chemical weathering are carried along with the water. These materials are part of the aqueous solution itself, so gravity cannot cause them to settle out. As the dissolved materials are carried down rivers, they ultimately enter the ocean.

Oceans may be thought of as huge chemical mixing vats. Rivers, rain, wind, and glaciers constantly bring in dissolved materials. Smaller quantities of dissolved materials enter the oceans through hydrothermal chemical reactions between seawater and hot basalt at mid-ocean ridges. The oceans lose water continuously by evaporation at the surface. The inflow and outflow of water is so exactly balanced that the amount of water in the oceans remains constant over such geologically short times as years, decades, or even centuries. Over a time scale of thousands to millions of years, however, the balance may shift. During the most recent ice age, for example, significant quantities of seawater were converted into glacial ice, and sea level was drawn down by more than 100 m.

The entry and exit of dissolved materials, too, is balanced. Each of the many dissolved components of seawater participates in some chemical or biological reaction that eventually precipitates it out of the water and onto the seafloor. As a result, the ocean's **salinity**—the total amount of dissolved material in a given volume of water—remains constant. Totaled over the world ocean, mineral precipitation balances the total inflow of dissolved material—yet another way in which the Earth system maintains balance.

We can better understand this chemical balance by considering the element calcium. Calcium is a component of the most abundant biological precipitate formed in the oceans: calcium carbonate ($CaCO_3$). On land, calcium dissolves when limestone and calcium-containing silicate minerals, such as some feldspars and pyroxenes, are weathered, and that calcium is transported to the ocean as dissolved calcium ions (Ca^{2+}). There, a wide variety of marine organisms take up calcium ions and combine them with carbonate ions (CO_3^{2-}), also present in seawater, to form their calcium carbonate shells. Thus, the calcium that entered the ocean as dissolved ions leaves it as solid sediment particles when the organisms die and their shells settle to the seafloor and accumulate there as calcium carbonate sediments. Ultimately, the calcium carbonate sediments will be buried and transformed into limestone. The chemical balance that keeps the concentrations of calcium dissolved in the ocean constant is thus controlled in part by the activities of organisms.

Nonbiological mechanisms also maintain chemical balance in the oceans. For example, sodium ions (Na^+) transported to the oceans react chemically with chloride ions (Cl^-) to form the precipitate sodium chloride (NaCl). This happens when evaporation raises the concentrations of sodium and chloride ions past the point of saturation. As we saw in Chapter 3, minerals precipitate when solutions become so saturated with dissolved materials that they can hold no more. The intense evaporation required to crystallize sodium chloride may take place in warm, shallow arms of the sea or in saline lakes.

Sedimentary Basins: The Sinks for Sediments

As we have seen, the currents that move sediments across Earth's surface generally flow downhill. Therefore, sediments tend to accumulate in depressions in Earth's crust. Such depressions are formed by **subsidence,** in which a broad area of the crust sinks (subsides) relative to the surrounding crust. Subsidence is induced partly by the weight of sediments on the crust, but is caused mainly by plate tectonic processes.

Sedimentary basins are regions of variable size where the combination of sedimentation and subsidence has formed thick accumulations of sediments and sedimentary rocks. Sedimentary basins are Earth's primary sources of oil and natural gas. Commercial exploration for these resources has helped us better understand the deep structure of sedimentary basins and of the continental lithosphere.

Rift Basins and Thermal Subsidence Basins

When plate separation begins within a continent, subsidence results from the stretching, thinning, and heating of the underlying lithosphere by the plate tectonic processes that are causing the separation (**Figure 6.8**). A long, narrow rift develops, bounded by great downdropped blocks of crustal rock. Hot, ductile mantle material rises, melts, and fills the space created by the thinned lithosphere and crust, initiating the eruption of basaltic lavas in the rift zone. Such **rift basins** are deep, narrow, and long, with thick successions of sedimentary rocks and extrusive and intrusive igneous rocks. The rift valleys of East Africa, the Rio Grande Valley, and the Jordan Valley in the Middle East are examples of rift basins.

At later stages of plate separation, when rifting has led to seafloor spreading and the newly separated continents

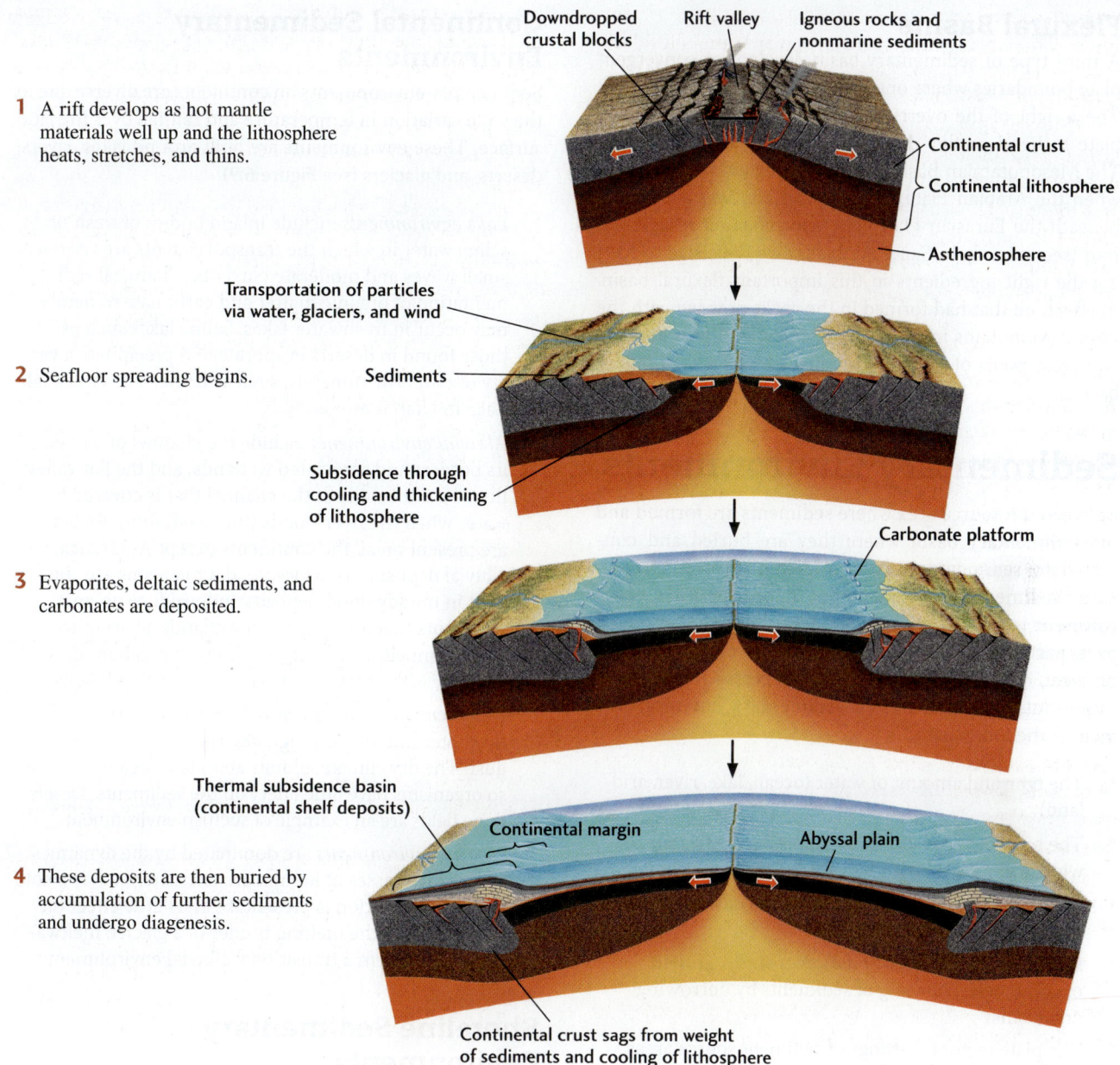

FIGURE 6.8 Sedimentary basins are formed by plate separation.

are drifting away from each other, subsidence continues through the cooling of the lithosphere that was thinned and heated during the earlier rifting stage (see Figure 6.8). Cooling leads to an increase in the density of the lithosphere, which in turn leads to its subsidence below sea level, where sediments can accumulate. Because cooling of the lithosphere is the main process creating the sedimentary basins at this stage, they are called **thermal subsidence basins**. Sediments from erosion of the adjacent land fill the basin nearly to sea level along the edge of the continent, creating a **continental shelf**.

The continental shelf continues to receive sediments for a long time because the trailing edge of the drifting continent subsides slowly and because the continent provides a tremendous land area from which sediments can be derived. The load of the growing mass of sediment further depresses the crust, so that the basin can receive still more material from the land. As a result of this continuous subsidence and sediment transportation, continental shelf deposits can accumulate in an orderly fashion to thicknesses of 10 km or more. The continental shelves off the Atlantic coasts of North and South America, Europe, and Africa are good examples of thermal subsidence basins. These basins began to form when the supercontinent Pangaea split apart about 200 million years ago and the North American and South American plates separated from the Eurasian and African plates.

Flexural Basins

A third type of sedimentary basin develops at convergent plate boundaries where one plate pushes up over the other. The weight of the overriding plate causes the underlying plate to bend or flex down, producing a **flexural basin.** The Mesopotamian Basin in Iraq is a flexural basin formed when the Arabian Plate collided with and was subducted beneath the Eurasian Plate. The enormous oil reserves in Iraq (second only to Saudi Arabia's) owe their size to having the right ingredients in this important flexural basin. In effect, oil that had formed in the rocks now beneath the Zagros Mountains in Iran was squeezed out, forming several great pools of oil with volumes larger than 10 billion barrels.

Sedimentary Environments

Between the source area where sediments are formed and the sedimentary basin where they are buried and converted to sedimentary rocks, sediments travel through many sedimentary environments. A **sedimentary environment** is an area of sediment deposition characterized by a particular combination of climate conditions and physical, chemical, and biological processes (**Figure 6.9**). Important characteristics of sedimentary environments include the following:

- The type and amount of water (ocean, lake, river, arid land)
- The type and strength of transport agents (water, wind, ice)
- The topography (lowland, mountain, coastal plain, shallow sea, deep sea)
- Biological activity (precipitation of shells, growth of coral reefs, churning of sediments by burrowing organisms)
- The plate tectonic settings of sediment source areas (volcanic mountain belt, continent-continent collision zone) and sedimentary basins (rift, thermal subsidence, flexural)
- The climate (cold climates may form glaciers; arid climates may form deserts where minerals precipitate by evaporation)

Consider the beaches of Hawaii, famous for their unusual green sands, which are a result of their distinctive sedimentary environment. The volcanic island of Hawaii is composed of olivine-bearing basalt, from which the olivine is released during weathering. Rivers transport the olivine to the beach, where waves and wave-produced currents concentrate the olivine and remove fragments of basalt to form olivine-rich sand deposits.

Sedimentary environments are often grouped by location: on continents, near shorelines, or in the ocean. This very general subdivision highlights the processes that give sedimentary environments their distinct identities.

Continental Sedimentary Environments

Sedimentary environments on continents are diverse due to the wide variation in temperature and rainfall over the land surface. These environments are built around lakes, rivers, deserts, and glaciers (see Figure 6.9).

- *Lake environments* include inland bodies of fresh or saline water in which the transport agents are relatively small waves and moderate currents. Chemical sedimentation of organic matter and carbonate minerals may occur in freshwater lakes. Saline lakes such as those found in deserts evaporate and precipitate a variety of *evaporite* minerals, such as halite. The Great Salt Lake in Utah is an example.
- *Alluvial environments* include the channel of a river, its borders and associated wetlands, and the flat valley floor on either side of the channel that is covered by water when the river floods (the *floodplain*). Rivers are present on all the continents except Antarctica, so alluvial deposits are widespread. Organisms are abundant in muddy flood deposits and produce organic sediments that accumulate in wetlands adjacent to river channels. Climates vary from arid to humid. An example is the Mississippi River and its floodplains.
- *Desert environments* are arid. Wind and the rivers that flow intermittently through deserts transport sand and dust. The dry climate inhibits abundant organic growth, so organisms have little effect on the sediments. Desert dune fields are an example of such an environment.
- *Glacial environments* are dominated by the dynamics of moving masses of ice and are characterized by a cold climate. Vegetation is present, but has little effect on sediments. At the melting border of a glacier, meltwater streams form a transitional alluvial environment.

Shoreline Sedimentary Environments

The dynamics of waves, tides, and river currents on sandy shores dominate shoreline sedimentary environments (see Figure 6.9):

- *Deltas*, where rivers enter lakes or oceans
- *Tidal flats*, where extensive areas exposed at low tide are dominated by tidal currents
- *Beaches*, where the strong waves approaching and breaking on the shore distribute sediments on the beach, depositing strips of sand or gravel

In most cases, the sediments that accumulate in shoreline environments are siliciclastic. Organisms affect these sediments mainly by burrowing into and mixing them. However, in some tropical and subtropical settings, sediment particles, particularly carbonate sediments, may be of biological origin. These biological sediments are also subject to transportation by waves and tidal currents.

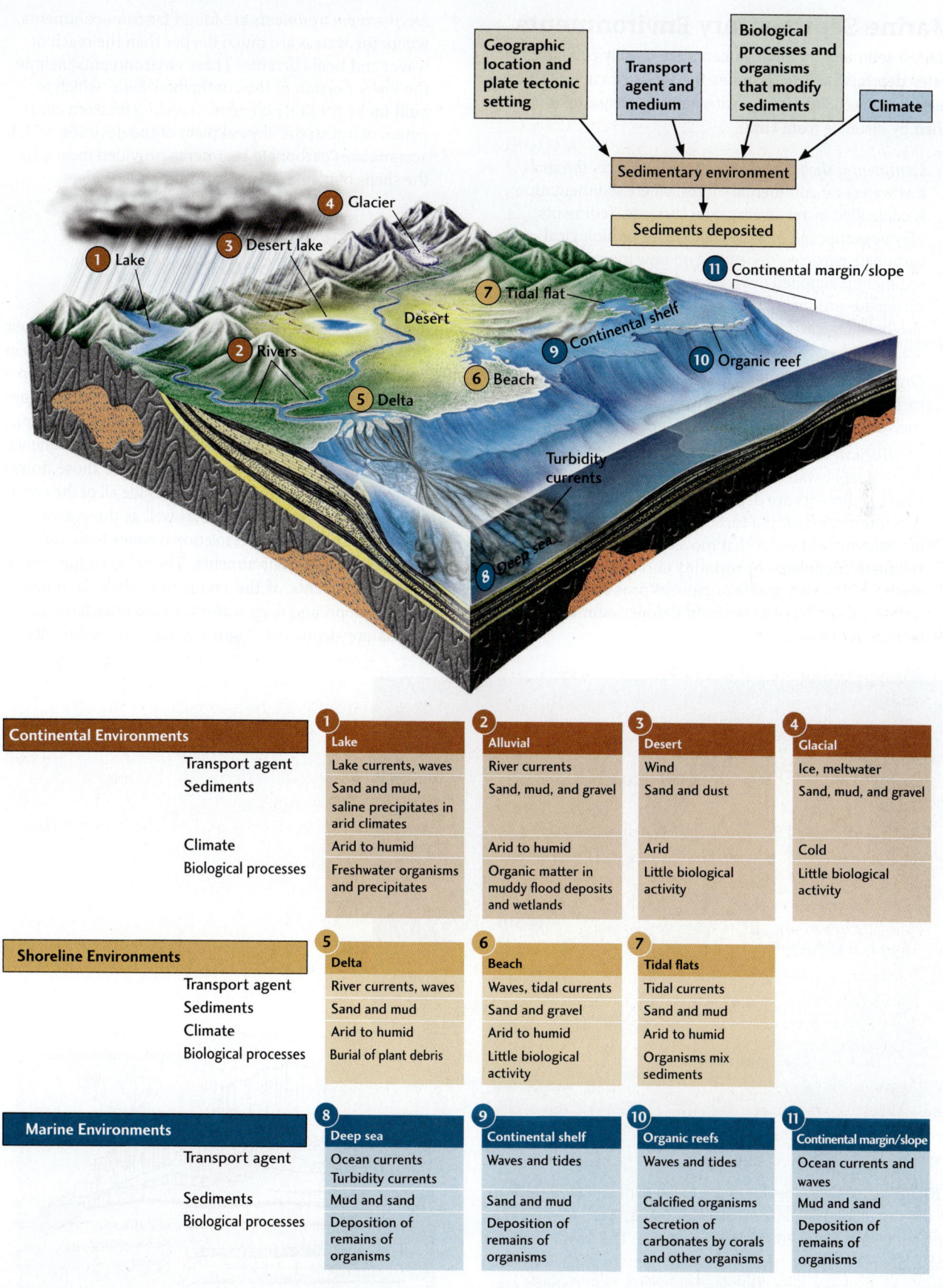

FIGURE 6.9 Multiple factors interact to create sedimentary environments.

Marine Sedimentary Environments

Marine sedimentary environments are usually classified by water depth, which determines the kinds of currents that are present (see Figure 6.9). Alternatively, they can be classified by distance from land.

- *Continental shelf environments* are located in the shallow waters off continental shores, where sedimentation is controlled by relatively gentle currents. Sediments may be composed of either siliciclastic or biological carbonate particles, depending on how much siliciclastic sediment is supplied by rivers and on the abundance of carbonate-producing organisms. Sedimentation may also be chemical if the climate is arid and an arm of the sea becomes isolated and evaporates.
- *Organic reefs* are carbonate structures, formed by carbonate-secreting organisms, built up on continental shelves or on oceanic volcanic islands.
- *Continental margin and slope environments* are found in the deeper waters at and off the edges of continents, where sediments are deposited by turbidity currents. A *turbidity current* is a turbulent submarine avalanche of sediment and water that moves downslope. Most sediments deposited by turbidity currents are siliciclastic, but at sites where organisms produce abundant carbonate sediments, continental slope sediments may be rich in carbonates.
- *Deep-sea environments* are found far from continents, where the waters are much deeper than the reach of waves and tidal currents. These environments include the lower portion of the continental slope, which is built up by turbidity currents traveling far from continental margins; the abyssal plain of the deep sea, which accumulates carbonate sediments provided mostly by the shells of plankton; and the mid-ocean ridges.

Siliciclastic versus Chemical and Biological Sedimentary Environments

Sedimentary environments can be grouped not only by their location, but also according to the kinds of sediments found in them or according to the dominant sediment formation process. Grouping of sedimentary environments in this manner produces two broad classes: siliciclastic sedimentary environments and chemical and biological sedimentary environments.

Siliciclastic sedimentary environments are those dominated by siliciclastic sediments. They include all of the continental sedimentary environments as well as those shoreline environments that serve as transitional zones between continental and marine environments. They also include those marine environments of the continental shelf, continental margin and slope, and deep seafloor where siliciclastic sands and muds are deposited (**Figure 6.10**). The sediments of

FIGURE 6.10 These sedimentary rocks exposed at El Capitan, in the Guadalupe Mountains of West Texas, were formed in an ancient ocean about 260 million years ago. The lower slopes of the mountains contain siliciclastic sedimentary rocks formed in deep-sea environments. The overlying cliffs of El Capitan are limestone and dolostone, formed from sediments deposited in a shallow sea when carbonate-secreting organisms died, leaving their shells in the form of a reef. [Courtesy John Grotzinger.]

TABLE 6.2 Major Chemical and Biological Sedimentary Environments

ENVIRONMENT	AGENT OF PRECIPITATION	SEDIMENTS
Shoreline and Marine		
Carbonate (reefs, platforms, deep sea, etc.)	Shelled organisms, some algae; inorganic precipitation from seawater	Carbonate sands and muds, reefs
Evaporite	Evaporation of seawater	Gypsum, halite, other salts
Siliceous (deep sea)	Shelled organisms	Silica
Continental		
Evaporite	Evaporation of lake water	Halite, borates, nitrates, carbonates, other salts
Wetland	Vegetation	Peat

these siliciclastic environments are often called **terrigenous sediments** to indicate their origin on land.

Chemical and biological sedimentary environments are characterized principally by chemical and biological precipitation (Table 6.2).

Carbonate environments are marine settings where calcium carbonate, mostly secreted by organisms, is the main sediment. They are by far the most abundant chemical and biological sedimentary environments. Hundreds of species of mollusks and other invertebrate animals, as well as calcareous (calcium-containing) algae and microorganisms, secrete carbonate shells or skeletons. Various populations of these organisms live at different depths, both in quiet areas and in places where waves and currents are strong. As they die, their shells and skeletons accumulate to form carbonate sediments.

Except for those of the deep sea, carbonate environments are found mostly in the warmer tropical and subtropical regions of the oceans, where carbonate-secreting organisms flourish. These environments include organic reefs, carbonate sand beaches, tidal flats, and shallow carbonate platforms. In a few places, carbonate sediments form in cooler waters that are supersaturated with carbonate ions—waters that are generally below 20°C, such as in some regions of the Southern Ocean south of Australia. These carbonate sediments are formed by a very limited group of organisms, most of which secrete calcite shells.

Siliceous environments are unique deep-sea sedimentary environments named for the silica shells deposited in them. The planktonic organisms that secrete these silica shells grow in surface waters where nutrients are abundant. When they die, their shells settle to the deep seafloor and accumulate as layers of siliceous sediments.

An *evaporite environment* is created when the warm seawater of an arid inlet or arm of the sea evaporates more rapidly than it can mix with seawater from the open ocean. The degree of evaporation and the length of time it has proceeded controls the salinity of the evaporating seawater and thus the kinds of chemical sediments formed. Evaporite environments also form in lakes lacking river outlets. Such lakes may produce sediments of halite, borate, nitrates, and other salts.

Sedimentary Structures

Sedimentary structures include all kinds of features formed at the time of deposition. Sediments and sedimentary rocks are characterized by *bedding*, or *stratification*, which occurs when layers of sediment, or *beds*, with different particle sizes or compositions, are deposited on top of one another. These beds range from only millimeters or centimeters thick to meters or even many meters thick. Most bedding is horizontal, or nearly so, at the time of deposition. Some types of bedding, however, form at a high angle relative to the horizontal.

Cross-Bedding

Cross-bedding consists of beds deposited by wind or water and inclined at angles as much as 35° from the horizontal (Figure 6.11). Cross-beds form when sediment particles are deposited on the steeper, downcurrent (leeward) slopes of sand dunes on land or sandbars in rivers and on the seafloor. Cross-bedding patterns in wind-deposited sand dunes may be complex as a result of rapidly changing wind directions (as in the photograph at the opening of this chapter). Cross-bedding is common in sandstones and is also found in gravels and some carbonate sediments. It is easier to see in sandstones than in sands, which must be excavated to see a cross section.

Graded Bedding

Graded bedding is most abundant in continental slope and deep-sea sediments deposited by dense, muddy turbidity

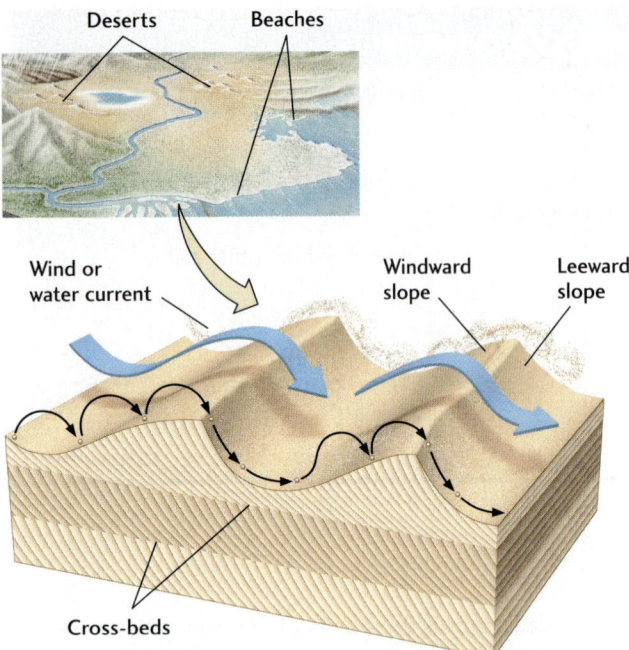

FIGURE 6.11 Sediment particles transported down the steeper, downcurrent slope of a sand dune, sandbar, or ripple form cross-bedding.

currents, which hug the bottom of the ocean as they move downhill. Each bed progresses from large particles at the bottom to small particles at the top. As the current progressively slows, it drops progressively smaller particles. The grading indicates a weakening of the current that deposited the particles. A graded bed comprises one set of sediment particles, normally ranging from a few centimeters to several meters thick, that formed a horizontal or nearly horizontal layer at the time of deposition. Accumulations of many individual graded beds can reach a total thickness of hundreds of meters. A graded bed formed as a result of deposition by a turbidity current is called a *turbidite*.

Ripples

Ripples are very small ridges of sand or silt whose long dimensions are at right angles to the current. They form low, narrow ridges, usually only a centimeter or two high, separated by wider troughs. These sedimentary structures are common in both modern sands and ancient sandstones (**Figure 6.12**). Ripples can be seen on the surfaces of wind-swept dunes, on underwater sandbars in shallow streams, and under the waves at beaches. Geologists can distinguish the symmetrical ripples made by waves moving back and forth on a beach from the asymmetrical ripples formed by currents moving in a single direction over river sandbars or windswept dunes (**Figure 6.13**).

Bioturbation Structures

In many sedimentary rocks, the bedding is broken or disrupted by roughly cylindrical tubes a few centimeters in diameter that extend vertically through several beds. These sedimentary structures are remnants of burrows and tunnels excavated by clams, worms, and other marine organisms that live on the ocean bottom. These organisms churn and burrow through muds and sands—a process called **bioturbation.** They ingest the sediment, digest the bits of organic matter it contains, and leave behind the reworked sediment, which fills the burrow (**Figure 6.14**). From bioturbation structures, geologists can determine the behavior of the organisms that burrowed in the sediment. Since the behavior of burrowing organisms is controlled partly by environmental factors, such as the strength of currents or the availability of nutrients, bioturbation structures can help us reconstruct past sedimentary environments.

(a)

(b)

FIGURE 6.12 Ripples. (a) Ripples in modern sand on a beach [John Grotzinger.] (b) Ancient ripple-marked sandstone. [John Grotzinger/ Ramón Rivera-Moret/ MIT]

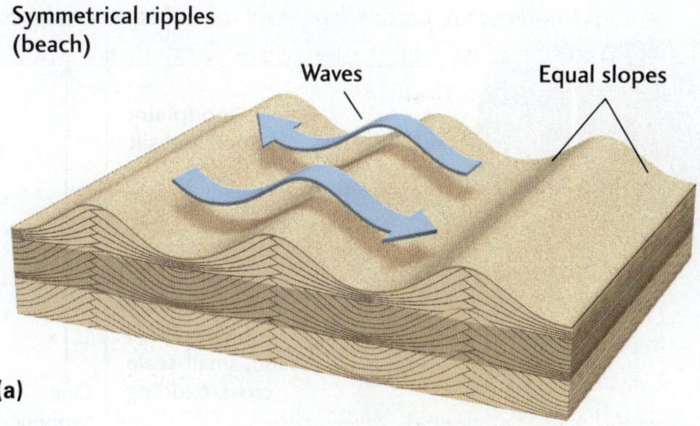

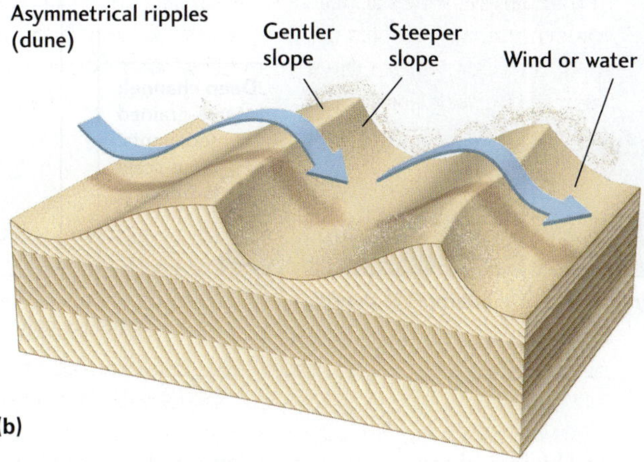

FIGURE 6.13 Geologists can distinguish ripples formed by waves from ripples formed by currents. (a) The shapes of ripples on beach sand, produced by the back-and-forth movements of waves, are symmetrical. [Courtesy John Grotzinger.] (b) Ripples on dunes and river sandbars, produced by the movement of a current in one direction, are asymmetrical. [Courtesy John Grotzinger.]

Bedding Sequences

Bedding sequences are built of interbedded and vertically stacked layers of different sedimentary rock types. A bedding sequence might consist of cross-bedded sandstone, overlain by bioturbated siltstone, overlain in turn by rippled sandstone—in any combination of thicknesses for each rock type in the sequence.

Bedding sequences help geologists reconstruct the ways in which sediments were deposited and so provide insight into the history of geologic processes and events

FIGURE 6.14 Bioturbation structures. This rock is crisscrossed with fossilized tunnels originally made by organisms burrowing through the mud. [John Grotzinger/ Ramón Rivera-Moret/ Harvard Mineralogical Museum]

FIGURE 6.15 A typical bedding sequence formed by a meandering river. [USDA-NRCS photo by Jim R. Fortner]

that occurred at Earth's surface long ago. **Figure 6.15** shows a bedding sequence typically formed in alluvial sedimentary environments. A river lays down sediments as its channel meanders back and forth across the valley floor. Thus, the lower part of the sequence contains the beds deposited in the deepest part of the river channel, where the current was strongest. The middle part contains the beds deposited in the shallower parts of the channel, where the current was weaker, and the upper part contains the beds deposited on the floodplain. Typically, a bedding sequence formed in this manner consists of sediment particles that grade upward from large to small. This sequence may be repeated a number of times if the river meanders back and forth.

Most bedding sequences consist of a number of small-scale subdivisions. In the example shown in Figure 6.15, the basal layers contain cross-bedding. These layers are overlain by more cross-bedded layers, but the cross-beds are smaller in scale. Horizontal bedding occurs at the top of the bedding sequence. Today, computer models are used to analyze how bedding sequences of sands were deposited in alluvial environments.

Burial and Diagenesis: From Sediment to Rock

Most of the clastic particles produced by weathering on land end up deposited in various sedimentary basins in the oceans. A smaller amount of siliciclastic sediment is deposited in sedimentary environments on land. Most chemical and biological sediments are also deposited in ocean basins, although some are deposited in lakes and wetlands.

Burial

Once sediments reach the ocean floor, they are trapped there. The deep seafloor is the ultimate sedimentary basin and, for most sediments, their final resting place. As the sediments are buried under new layers of sediments, they are subjected to increasingly high temperatures and pressures as well as chemical changes.

Diagenesis

After sediments are deposited and buried, they are subject to **diagenesis**—the many physical and chemical changes that result from the increasing temperatures and pressures as they are buried ever deeper in Earth's crust. These changes continue until the sediment or sedimentary rock is either exposed to weathering or metamorphosed by more extreme heat and pressure (**Figure 6.16**).

Temperature increases with depth in Earth's crust at an average rate of 30°C for each kilometer of depth, although that rate varies somewhat among sedimentary basins. Thus, at a depth of 4 km, buried sediments may reach 120°C or more, the temperature at which certain types of organic matter may be converted to oil and natural gas (see the Practicing Geology Exercise). Pressure also increases with depth—on average, about 1 atmosphere for each 4.4 m of depth. This increased pressure is responsible for the compaction of buried sediments.

Buried sediments are also continuously bathed in groundwater full of dissolved minerals. These minerals can

FIGURE 6.16 Diagenesis is the set of physical and chemical changes that convert sediments into sedimentary rocks. [*mud, sand, gravel:* Courtesy John Grotzinger; *shale:* John Grotzinger/Ramón Rivera-Moret/Harvard Mineralogical Museum; *sandstone, conglomerate, coal:* Courtesy John Grotzinger/Ramón Rivera-Moret/MIT; *diatoms:* Mark B. Edlund, Ph.D./Science Museum of Minnesota; *plant material:* Roman Gorielov/Shutterstock; *oil and gas:* Wasabi/Alamy.]

precipitate in the pores between the sediment particles and bind them together—a chemical change called **cementation**. Cementation decreases **porosity**, the percentage of a rock's volume consisting of open pores between particles. In some sands, for example, calcium carbonate is precipitated as calcite, which acts as a cement that binds the grains and hardens the resulting mass into sandstone (**Figure 6.17**). Other minerals, such as quartz, may cement sands, muds, and gravels into sandstone, mudstone, or conglomerate.

The major physical diagenetic change is **compaction**, a decrease in the volume and porosity of a sediment. Compaction occurs as sediment particles are squeezed

PRACTICING GEOLOGY EXERCISE

Where Do We Look for Oil and Gas?

The search for new deposits of oil and natural gas is taking on ever-greater urgency as fuel supplies dwindle and geopolitical issues make nations eager to produce their own energy supplies. The search for these deposits must be guided by an understanding of how and where oil and gas form.

The first step in exploring for oil and gas is a search for sedimentary rocks formed from sediments that are likely to have been rich in organic matter. Once such rocks have been located, the next step is to determine how deeply they have been buried and the maximum temperature they might have achieved. These factors determine the prospectivity of the rocks—their likelihood of containing oil or gas.

Many fine-grained sediments and sedimentary rocks, such as shale, contain organic matter. Subsidence of sedimentary basins, coupled with deposition of overlying sedimentary layers, may result in deep burial of these organic-rich sediments. As they are buried progressively deeper, the sediments become increasingly hotter. The rate at which temperatures

closer together by the weight of overlying sediments. Sands are fairly well packed during deposition, so they do not compact much. However, newly deposited muds, including carbonate-containing muds, are highly porous. In many of these sediments, more than 60 percent of the volume consists of water in pore spaces. As a result, muds compact greatly after burial, losing more than half their water.

Both cementation and compaction result in **lithification,** the hardening of soft sediment into rock.

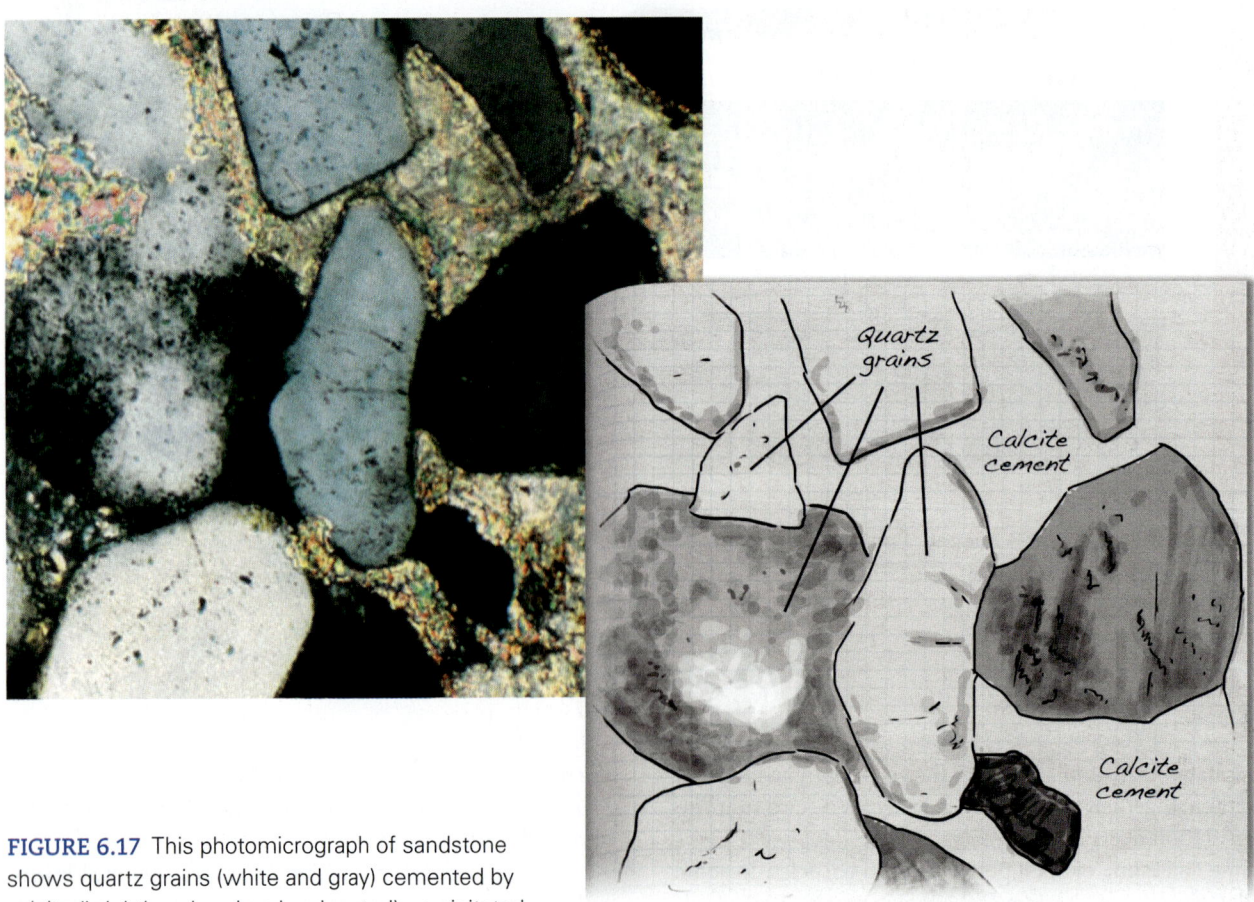

FIGURE 6.17 This photomicrograph of sandstone shows quartz grains (white and gray) cemented by calcite (brightly colored and variegated) precipitated after deposition. [Peter Kresan.]

Burial and Diagenesis: From Sediment to Rock

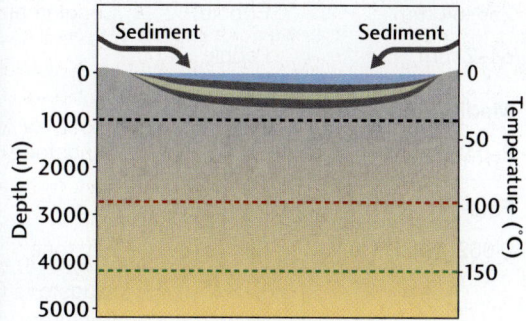
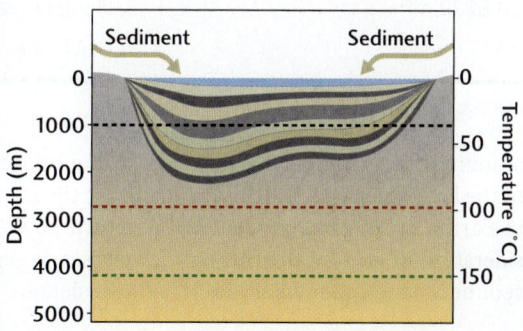
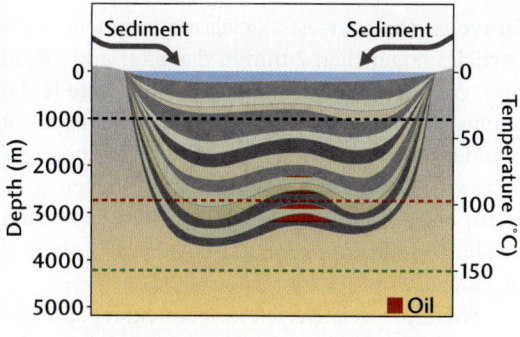
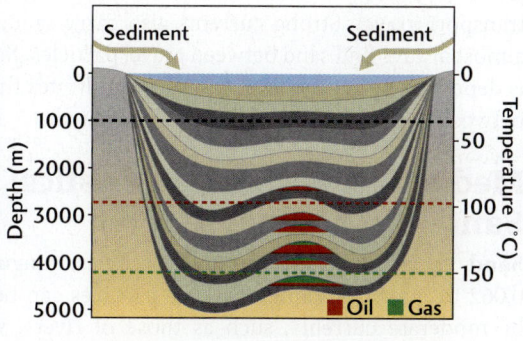

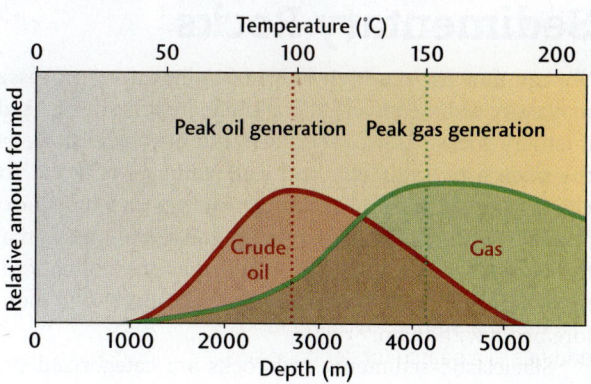

increase with depth is called the geothermal gradient (see Chapter 7).

Depending on the geothermal gradient in the sedimentary basin, organic-rich sedimentary rocks may eventually become hot enough that the organic matter they contain is transformed into oil or gas. That process of transformation (described in more detail in Chapter 13) is known as maturation. Maturation begins shortly after the sediments are deposited, but increases dramatically above 50°C. Oil is generated as the sediments are heated to temperatures between 60°C and 150°C. At higher temperatures, the oil becomes unstable and breaks down, or "cracks," to form natural gas.

Geologists have discovered organic-rich shales in the Rocknest Basin, which has a geothermal gradient of 35°C/km. The accompanying diagram shows the relationship between depth of burial, temperature, and the relative amounts of oil and gas formed in shales in this sedimentary basin. Assuming that peak oil generation occurs at about 100°C, calculate the depth at which peak oil generation would occur in the Rocknest Basin.

Depth of peak oil generation = Temperature of peak oil generation ÷ Geothermal gradient
= 100°C ÷ 35°C/km
= 2.85 km (2850 m)

If the organic-rich shales in the Rocknest Basin were buried to depths of 2850 m or greater, then one might expect to find oil in the basin. However, if the depth of burial were shallower than 2850 m, then the prospectivity of the basin would be downgraded.

BONUS PROBLEM: The depth of peak gas generation in the Rocknest Basin is 3575 m. Rearrange the equation above and solve for the temperature at which gas generation would peak.

Classification of Siliciclastic Sediments and Sedimentary Rocks

We can now use our knowledge of sedimentary processes to classify sediments and their lithified counterparts, sedimentary rocks. As we have seen, the major divisions are the siliciclastic sediments and sedimentary rocks and the chemical and biological sediments and sedimentary rocks. Siliciclastic sediments and rocks constitute more than three-fourths of the total mass of all types of sediments and sedimentary rocks in Earth's crust (Figure 6.18). We therefore begin with them.

Siliciclastic sediments and rocks are categorized primarily by particle size (Table 6.3):

- *Coarse-grained:* gravel and conglomerate
- *Medium-grained:* sand and sandstone
- *Fine-grained:* silt and siltstone; mud, mudstone, and shale; clay and claystone

We classify siliciclastic sediments and rocks on the basis of their particle size because it distinguishes them by one of the most important conditions of sedimentation: current strength. As we have seen, the larger the particle, the stronger the current needed to transport and deposit it. This relationship between current strength and particle size is the reason like-sized particles tend to accumulate in sorted beds. In other words, most sand beds do not contain pebbles or mud, and most muds consist only of particles finer than sand.

Of the various types of siliciclastic sediments and sedimentary rocks, the fine-grained siliciclastics are by far the most abundant—about three times more common than the coarser-grained siliciclastics (see Figure 6.18). The abundance of the fine-grained siliciclastics, which contain large amounts of clay minerals, is due to the chemical weathering of the large quantities of feldspar and other silicate minerals in Earth's crust into clay minerals. We turn now to a consideration of each of the three major classes of siliciclastic sediments and sedimentary rocks in more detail.

TABLE 6.3 Major Classes of Siliciclastic Sediments and Sedimentary Rocks

PARTICLE SIZE	SEDIMENT	ROCK
Coarse-Grained	Gravel	
Larger than 256 mm	Boulder	
256–64 mm	Cobble	Conglomerate
64–2 mm	Pebble	
Medium-Grained		
2–0.062 mm	Sand	Sandstone
Fine-Grained	Mud	
0.062–0.0039 mm	Silt	Siltstone
		Mudstone (blocky fracture)
Finer than 0.0039 mm	Clay	Shale (breaks along bedding)
		Claystone

Coarse-Grained Siliciclastics: Gravel and Conglomerate

Gravel is the coarsest siliciclastic sediment, consisting of particles larger than 2 mm in diameter and including pebbles, cobbles, and boulders. **Conglomerate** is the lithified equivalent of gravel (Figure 6.19a). Pebbles, cobbles, and boulders are easy to study and identify because of their large size, which tells us the strength of the currents that transported them. In addition, their composition can tell us about the nature of the distant terrain where they were produced.

There are relatively few sedimentary environments—mountain streams, rocky beaches with high waves, and glacier meltwaters—in which currents are strong enough to transport gravel. Strong currents also carry sand, and we almost always find sand between gravel particles. Some of it is deposited with the gravel and some infiltrates the spaces between particles after the gravel is deposited.

Medium-Grained Siliciclastics: Sand and Sandstone

Sand consists of medium-sized particles, ranging from 0.062 to 2 mm in diameter. These particles can be moved by moderate currents, such as those of rivers, waves at

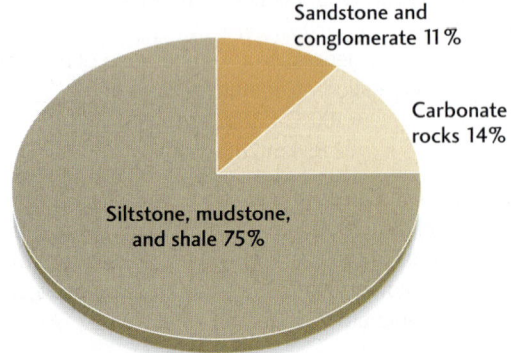

FIGURE 6.18 The relative abundances of the major sedimentary rock types. In comparison with these three types, all other sedimentary rock types—including evaporites, cherts, and other chemical sedimentary rocks—exist in only minor amounts.

(a) Conglomerate

(b) Sandstone

(c) Shale

FIGURE 6.19 Examples of the three major classes of siliciclastic sedimentary rocks. [*conglomerate and sandstone*: John Grotzinger/Ramón Rivera-Moret/MIT; *shale*: John Grotzinger/Ramón Rivera-Moret/Harvard Mineralogical Museum.]

shorelines, and the winds that blow sand into dunes. Sand grains are large enough to be seen with the naked eye, and many of their features are easily discerned with a low-power magnifying glass. The lithified equivalent of sand is **sandstone** (Figure 6.19b).

Both groundwater geologists and petroleum geologists have a special interest in sandstones. Groundwater geologists study the origin of sandstones to predict possible supplies of water in areas of porous sandstone, such as those found in the western plains of North America. Petroleum geologists must understand the porosity and cementation of sandstones because much of the oil and natural gas discovered in the past 150 years has been found in buried sandstones. In addition, much of the uranium used for nuclear power plants and weapons has come from uranium deposits precipitated in sandstones.

Sizes and Shapes of Sand Grains Medium-sized siliciclastic particles—sand grains—are subdivided into fine, medium, and coarse grains. The average size of the grains in any one sandstone can be an important clue to both the strength of the current that carried them and the sizes of the crystals eroded from the parent rock. The range of grain sizes and their relative abundances are also significant. If all the grains are close to the average size, the sand is well sorted. If many grains are much larger or smaller than the average, the sand is poorly sorted (see Figure 6.6). The degree of sorting can help us distinguish, for example, between sands deposited on beaches (which tend to be well sorted) and sands deposited by glaciers (which tend to be muddy and poorly sorted). The shapes of sand grains can also be important clues to their origin. Sand grains, like pebbles and cobbles, are abraded and rounded during transportation. Angular grains imply short transport distances; rounded ones indicate long journeys down a large river system (see Figure 6.7).

Mineralogy of Sands and Sandstones Siliciclastics can be further subdivided by their mineralogy, which can help identify the parent rocks. Thus, there are quartz-rich sandstones and feldspar-rich sandstones. Some sands are bioclastics, rather than siliciclastics; they are formed from materials such as carbonate minerals that were originally precipitated as shells, but then broken up and transported by currents. Thus, the mineralogy of sands and sandstones indicates the source areas and materials that were eroded to produce the sand grains. Sodium- and potassium-rich feldspars with abundant quartz, for example, might indicate that the sediments were eroded from a granitic terrain. Other minerals, as we will see in Chapter 7, might indicate metamorphic parent rocks.

The mineral content of sands and sandstones also indicates the plate tectonic setting of the parent rock. Sandstones containing abundant fragments of mafic volcanic rock, for example, might indicate that the sand grains were derived from a volcanic mountain belt at a subduction zone.

Major Kinds of Sandstones Sandstones can be divided into four major groups on the basis of their mineralogy and texture (Figure 6.20):

- **Quartz arenites** are made up almost entirely of quartz grains, usually well sorted and rounded. Pure quartz sands result from extensive weathering before and during transportation that removed everything but quartz, the most stable silicate mineral.

- **Arkoses** are more than 25 percent feldspar. Their grains tend to be more angular and less well sorted than those of quartz arenites. These feldspar-rich sandstones come from rapidly eroding granitic and metamorphic terrains where chemical weathering is subordinate to physical weathering.

- **Lithic sandstones** contain many particles derived from fine-grained rocks, mostly shales, volcanic rocks, and fine-grained metamorphic rocks.

- **Graywacke** is a heterogeneous mixture of rock fragments and angular grains of quartz and feldspar in

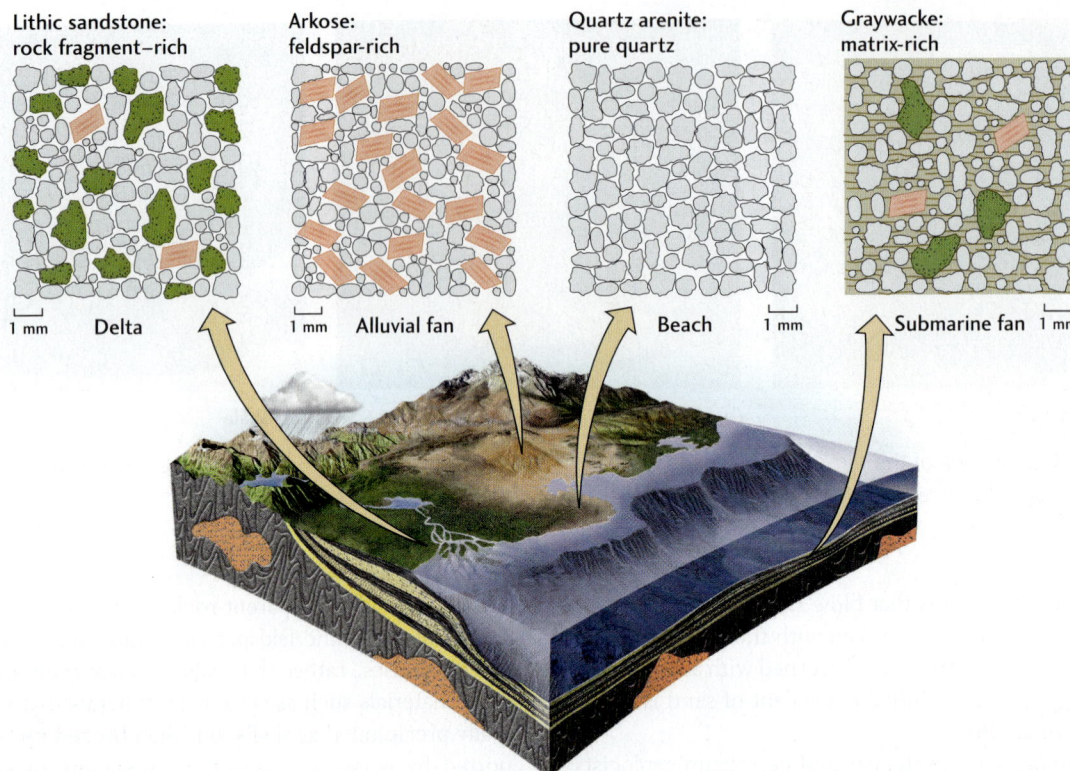

FIGURE 6.20 The mineralogy of four major groups of sandstones and the sedimentary environments where they are most likely to be found.

which the sand grains are surrounded by a fine-grained clay matrix. Much of this matrix is formed from fragments of relatively soft rock, such as shale and some volcanic rocks, that are chemically altered and physically compacted after deep burial of the sandstone formation.

Fine-Grained Siliciclastics

The finest-grained siliciclastic sediments and sedimentary rocks are the silts and siltstones; the muds, mudstones, and shales; and the clays and claystones. All of them consist of particles that are less than 0.062 mm in diameter, but they vary widely in their ranges of grain sizes and in their mineral compositions. Fine-grained sediments are deposited by the gentlest currents, which allow the finest sediment particles to settle slowly to the bottom in quiet waves.

Silt and Siltstone Siltstone is the lithified equivalent of **silt**, a siliciclastic sediment in which most of the grains are between 0.0039 and 0.062 mm in diameter. Siltstone looks similar to mudstone or very fine-grained sandstone.

Mud, Mudstone, and Shale Mud is a siliciclastic sediment containing water in which most of the particles are less than 0.062 mm in diameter. Thus, mud can be made of silt- or clay-sized sediment particles, or varying quantities of both. The general term "mud" is very useful in fieldwork because it is often difficult to distinguish between silt- and clay-sized particles without a microscope.

Muds are deposited by rivers and tides. As a river recedes after flooding, the current slows, and mud, some of it containing abundant organic matter, settles on the floodplain. This mud contributes to the fertility of river floodplains. Muds are also left behind by ebbing tides along many tidal flats where wave action is mild. Much of the deep seafloor, where currents are weak or absent, is blanketed by muds.

The fine-grained rock equivalents of muds are mudstones and shales. **Mudstones** are blocky and show poor or no bedding. Distinct beds may have been present when the sediments were first deposited, but then lost through bioturbation. **Shales** (Figure 6.19c) are composed of silt plus a significant component of clay, which causes them to break readily along bedding planes. Many muds contain more than 10 percent calcium carbonate sediments, forming calcareous mudstones and shales. Black, or organic, shales contain abundant organic matter. Some, called oil shales, contain large quantities of oily organic material, which makes them a potentially important source of oil.

Hydraulic fracturing, also known as "fracking," is caused by the injection of highly pressurized fluids into shale. This creates new channels (fractures) in the rock, which link together tiny pores filled with oil and natural gas to create a flow of larger scale that is economically viable. The Marcellus Formation, found in the northeastern United States (see Figure 6.21), was named for Marcellus, New York. It

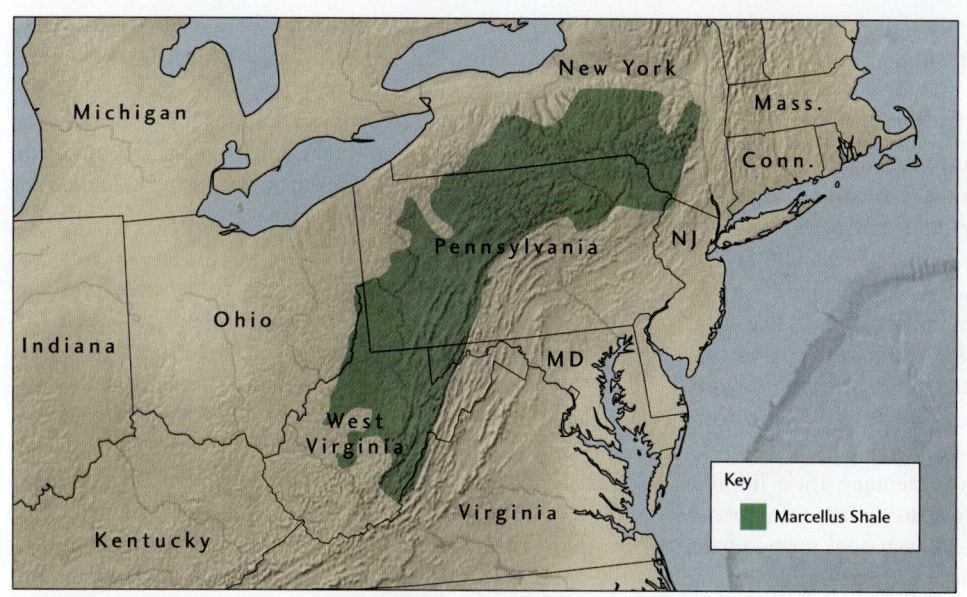

FIGURE 6.21 The Marcellus Formation, found in the northeastern United States, has previously untapped natural gas reserves. The shaded area of the map indicates the most economically promising parts of the Marcellus Shale.

is a unit of shale that had previously untapped natural gas reserves. In 2007, the Marcellus Shale was first drilled into and, using fracking methods, the extraction of natural gas became economically viable. The environmental impacts of fracking are debated, though, due to the effects of the chemicals used, the water supply, and the safety of drilling.

Clay and Claystone Clay is the most abundant component of fine-grained sediments and sedimentary rocks, and consists largely of clay minerals. Clay-sized particles are less than 0.0039 mm in diameter. Rocks made up exclusively of clay-sized particles are called **claystones**.

Classification of Chemical and Biological Sediments and Sedimentary Rocks

Chemical and biological sediments and sedimentary rocks can be classified by their chemical composition (Table 6.4). Geologists distinguish between chemical sediments and biological sediments not only for convenience, but also to emphasize the importance of organisms as the chief mediators of biological sedimentation. Both kinds of sediments

TABLE 6.4 Classification of Biological and Chemical Sediments and Sedimentary Rocks

SEDIMENT	ROCK	CHEMICAL COMPOSITION	MINERALS
Biological			
Sand and mud (primarily bioclastic)	Limestone	Calcium carbonate ($CaCO_3$)	Calcite, aragonite
Siliceous sediment	Chert	Silica (SiO_2)	Opal, chalcedony, quartz
Peat, organic matter	Organics	Carbon compounds; carbon compounded with oxygen and hydrogen	(Coal, oil, natural gas)
No primary sediment (formed by diagenesis)	Phosphorite	Calcium phosphate ($Ca_3(PO_4)_2$)	Apatite
Chemical			
No primary sediment (formed by diagenesis)	Dolostone	Calcium-magnesium carbonate ($CaMg(CO_3)_2$)	Dolomite
Iron oxide sediment	Iron formation	Iron silicate; oxide (Fe_2O_3); limonite, carbonate	Hematite, siderite
Evaporite sediment	Evaporite	Calcium sulfate ($CaSO_4$); sodium chloride ($NaCl$)	Gypsum, anhydrite, halite, other salts

can tell us about chemical conditions in the ocean, their predominant environment of deposition.

Carbonate Sediments and Rocks

Most **carbonate sediments** and **carbonate rocks** are formed by the accumulation and lithification of carbonate minerals that are directly or indirectly precipitated by organisms. The most abundant of these carbonate minerals is calcite (calcium carbonate, $CaCO_3$); in addition, most carbonate sediments contain aragonite, a less stable form of calcium carbonate. Some organisms precipitate calcite, some precipitate aragonite, and some precipitate both. During burial and diagenesis, carbonate sediments react with water to form a new suite of carbonate minerals.

The dominant biological sedimentary rock lithified from carbonate sediments is **limestone,** which is composed mainly of calcite (**Figure 6.22a**). Limestone is formed from carbonate sands and muds and, in some cases, ancient reefs (see Figure 5.10).

Another abundant carbonate rock is **dolostone,** made up of the mineral dolomite, which is composed of calcium-magnesium carbonate. Dolostones are diagenetically altered carbonate sediments and limestones. Dolomite does not form as a primary precipitate from ordinary seawater, and no organisms secrete shells of dolomite. Instead, some calcium ions in the calcite or aragonite of a carbonate sediment are exchanged for magnesium ions from seawater (or magnesium-rich groundwater) slowly passing through the pores of the sediment. This exchange converts calcium carbonate ($CaCO_3$) into dolomite ($CaMg(CO_3)_2$).

Direct Biological Precipitation of Carbonate Sediments

Carbonate rocks are abundant because of the large amounts of calcium and carbonate minerals dissolved in seawater, which organisms can convert directly into shells. Calcium is supplied by the weathering of feldspars and other minerals in igneous and metamorphic rocks. Carbonate minerals are derived from the carbon dioxide in the atmosphere. Calcium and carbonate minerals also come from the easily weathered limestone on the continents.

Most carbonate sediments of shallow marine environments are bioclastic sediments originally secreted as shells by organisms living near the surface or on the bottom. After the organisms die, they break apart, producing shells or fragments of shells that constitute individual particles, or *clasts*, of carbonate sediment. These sediments are found in

(a) Limestone (b) Gypsum (c) Halite (d) Chert

FIGURE 6.22 Chemical and biological sedimentary rocks: (a) limestone, lithified from carbonate sediments; (b) gypsum and (c) halite, marine evaporites that precipitate in shallow seawater basins; (d) chert, made up of siliceous sediments. [John Grotzinger/Ramón Rivera-Moret/Harvard Mineralogical Museum.]

Classification of Chemical and Biological Sediments and Sedimentary Rocks

tropical and subtropical environments from Pacific islands to the Caribbean and the Bahamas. Carbonate sediments are most accessible for study in these spectacular vacation spots, but the deep sea is where most carbonate sediments are deposited today.

Most of the carbonate sediments deposited on the abyssal plain of the deep sea are derived from the calcite shells of **foraminifera** (see Figure 3.1b) and other planktonic organisms that live in the surface waters and secrete calcium carbonate. When the organisms die, their shells settle to the seafloor and accumulate there as sediments.

Reefs are moundlike or ridgelike organic structures composed of the carbonate skeletons and shells of millions of organisms (**Figure 6.23**). In the warm seas of the present,

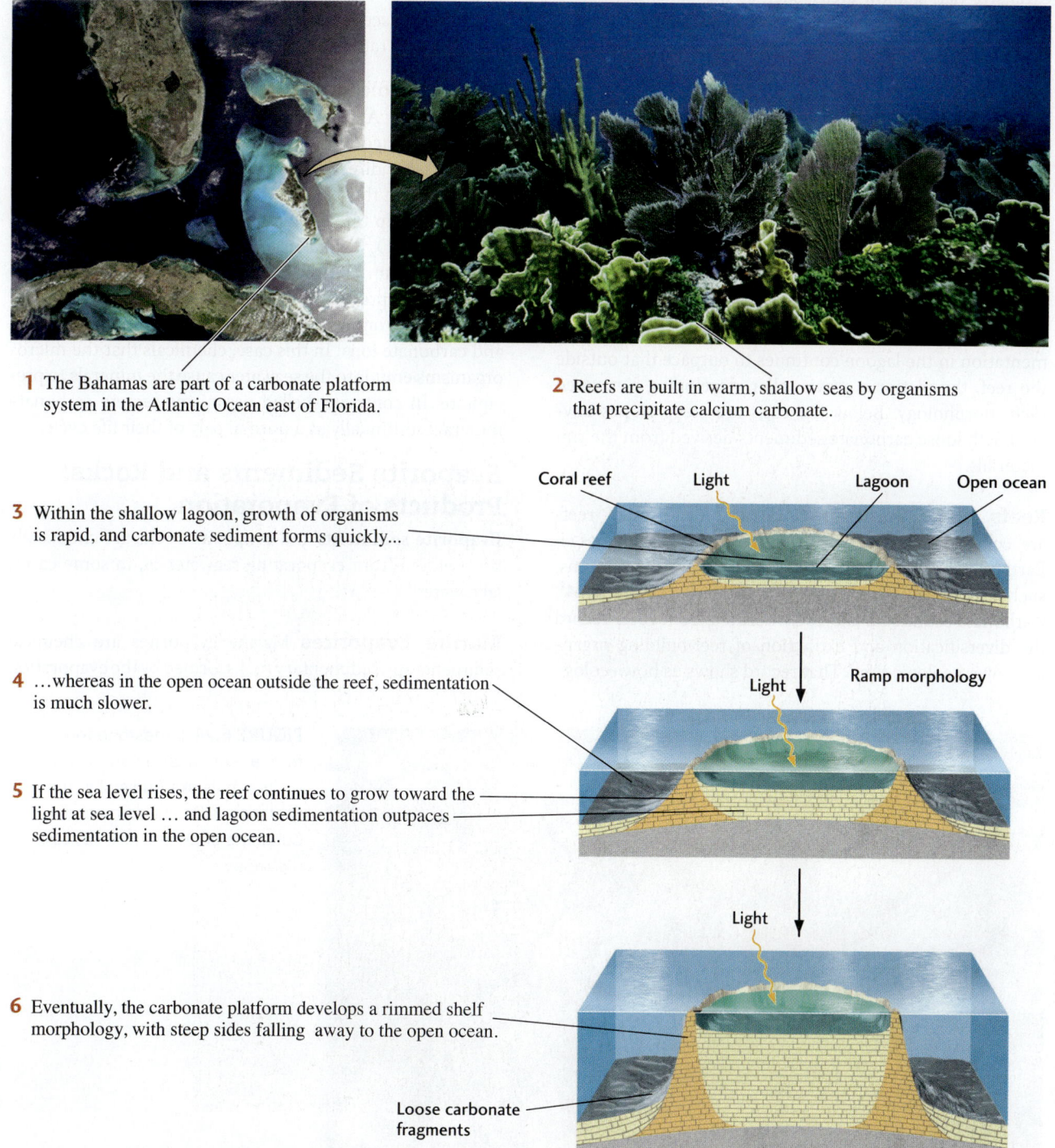

1. The Bahamas are part of a carbonate platform system in the Atlantic Ocean east of Florida.

2. Reefs are built in warm, shallow seas by organisms that precipitate calcium carbonate.

3. Within the shallow lagoon, growth of organisms is rapid, and carbonate sediment forms quickly...

4. ...whereas in the open ocean outside the reef, sedimentation is much slower.

5. If the sea level rises, the reef continues to grow toward the light at sea level ... and lagoon sedimentation outpaces sedimentation in the open ocean.

6. Eventually, the carbonate platform develops a rimmed shelf morphology, with steep sides falling away to the open ocean.

FIGURE 6.23 Marine organisms create carbonate platforms. [*left*: Jacques Descloitres, MODIS Land Rapid Response Team, NASA/GSFC; *right*: © Stephen Frink/Getty Images.]

reefs are built mainly by corals, but hundreds of other organisms, such as algae, clams, and snails, also contribute. In contrast to the soft, loose sediments produced in other carbonate environments, the reef forms a rigid, wave-resistant structure of solid calcite and aragonite that is built up to and slightly above sea level. The solid calcite and aragonite of the reef is produced directly by the carbonate-cementing action of the organisms; there is no loose sediment stage.

Coral reefs may give rise to *carbonate platforms*: extensive flat, shallow areas, such as the Bahamas, where both biological and nonbiological carbonate sediments are deposited (see Figure 6.23). Carbonate platforms are among the most important carbonate environments, both in past geologic ages and at present. The building of a carbonate platform results from interactions between the biosphere, hydrosphere, and lithosphere (see Earth Issues 6.1). The process begins with a reef that encloses and shelters an area of shallow ocean water known as a *lagoon*. Carbonate-secreting organisms proliferate in and around the lagoon, and carbonate sediments accumulate rapidly, while in the open ocean outside the reef, sedimentation is much slower. At this point, the carbonate platform has a *ramp* morphology, with gentle slopes leading to deeper water. As sedimentation in the lagoon continues to outpace that outside the reef, the platform grows taller, developing a *rimmed shelf* morphology. Below the rims are steep slopes covered with loose carbonate sediments derived from the rim materials.

Reefs and Evolutionary Processes Today, reefs are constructed mainly by corals, but at earlier times in Earth's history, they were constructed by other organisms, such as a now-extinct variety of mollusk (**Figure 6.24**). Carbonate sediments and rocks formed from reefs record the diversification and extinction of reef-building organisms over geologic time. That record shows us how ecology and environmental change help to regulate the process of evolution.

Today, natural and anthropogenic changes threaten the growth of coral reefs, which are very sensitive to environmental change. In 1998, an El Niño event (described in Chapter 12) raised sea surface temperatures so much that many reefs in the western Indian Ocean were killed. The reefs of the Florida Keys are dying off for a completely different reason: They are getting too much of a good thing. Groundwaters originating in the farmlands of the Florida Peninsula are seeping out to the reefs and exposing them to lethal concentrations of nutrients.

Indirect Biological Precipitation of Carbonate Sediments A significant fraction of the carbonate mud deposited in lagoons and on shallow carbonate platforms is precipitated indirectly from seawater. Microorganisms may be involved in this process, but their role is still uncertain. They may help to shift the balance of calcium (Ca^{2+}) and carbonate (CO_3^{2-}) ions in the seawater surrounding them so that calcium carbonate ($CaCO_3$) is formed. Microorganisms can precipitate carbonate minerals only if their external environment already contains abundant calcium and carbonate ions. In this case, chemicals that the microorganisms emit into the seawater cause the minerals to precipitate. In contrast, shelled organisms secrete carbonate minerals continually as a normal part of their life cycle.

Evaporite Sediments and Rocks: Products of Evaporation

Evaporite sediments and **evaporite rocks** are chemically precipitated from evaporating seawater or, in some cases, lake water.

Marine Evaporites Marine evaporites are chemical sediments and sedimentary rocks formed by the evaporation

FIGURE 6.24 Limestone formed from a reef constructed by now-extinct mollusks (rudists) in the Cretaceous Shuiba formation, Sultanate of Oman. [Courtesy John Grotzinger.]

Earth Issues 6.1 Darwin's Coral Reefs and Atolls

For more than 200 years, coral reefs have attracted explorers and travel writers. Ever since Charles Darwin sailed the oceans on the *Beagle* from 1831 to 1836, these reefs have been a matter of scientific discussion as well. Darwin was one of the first scientists to analyze the geology of coral reefs and his theory of the origin of one type of coral reef is still accepted today.

The coral reefs that Darwin studied were atolls, coral islands in the open ocean surrounding circular lagoons. The outermost part of an atoll is a slightly submerged, wave-resistant reef front: a steep slope facing the ocean. The reef front is composed of the interlaced skeletons of corals and calcareous algae, which form a tough, hard limestone. Behind the reef front is a flat platform extending into a shallow lagoon. An island may lie at the center of the lagoon. Parts of the reef, as well as the central island, are above sea level and may become forested. A great many plant and animal species inhabit the reef and the lagoon.

Coral reefs are generally limited to waters less than 20 m deep because, below that depth, seawater does not transmit enough light to enable reef-building organisms to grow. How, then, could an atoll be built up from the bottom of the deep, dark ocean? Darwin proposed that the process starts with a volcano building up to the sea surface from the seafloor and forming an island. As the volcano becomes dormant, temporarily or permanently, corals and algae colonize the shore of the island and build fringing reefs. Erosion may then lower the volcanic island almost to sea level.

Darwin reasoned that if such a volcanic island were to subside slowly beneath the waves, actively growing corals and algae might keep pace with its subsidence, continuously building up the reef over geologic time. In this way, the volcanic island would disappear, leaving an atoll in its place. More than 100 years after Darwin proposed his theory, deep drilling on several atolls found volcanic rock below their coralline limestone. And some decades later, the theory of plate tectonics explained both volcanism and the subsidence that resulted from plate cooling and contraction.

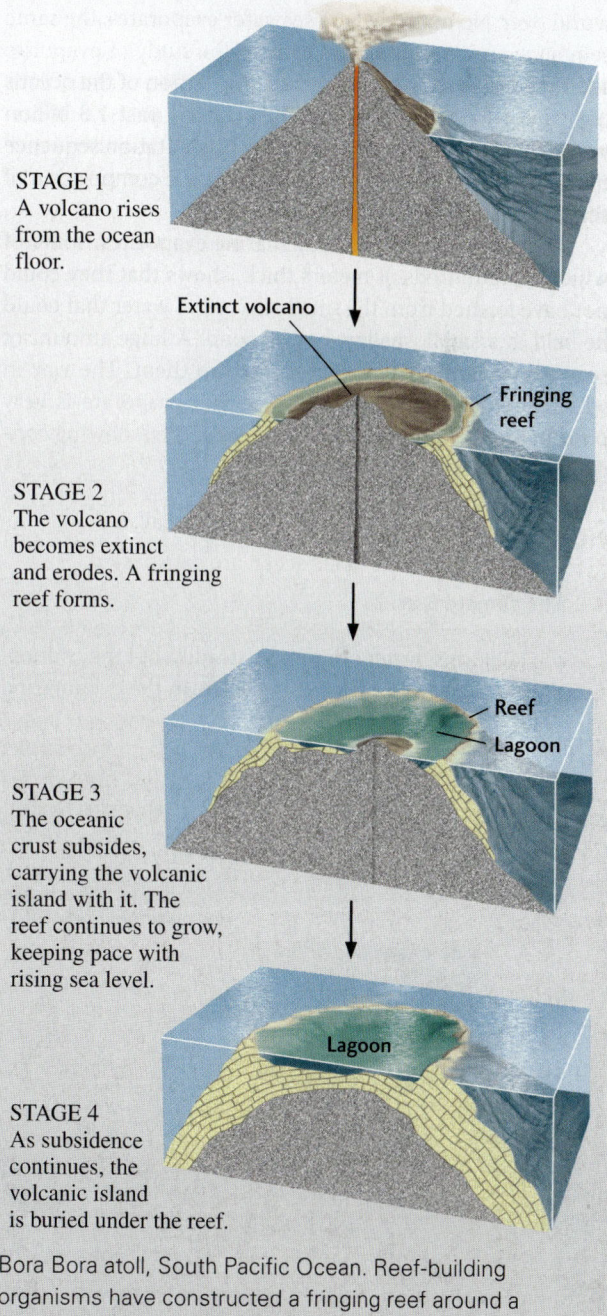

STAGE 1 A volcano rises from the ocean floor.

STAGE 2 The volcano becomes extinct and erodes. A fringing reef forms.

STAGE 3 The oceanic crust subsides, carrying the volcanic island with it. The reef continues to grow, keeping pace with rising sea level.

STAGE 4 As subsidence continues, the volcanic island is buried under the reef.

Bora Bora atoll, South Pacific Ocean. Reef-building organisms have constructed a fringing reef around a volcanic island, forming a protected lagoon. [Jean-Marc Truchet/Stone/Getty Images.]

of seawater. These sediments and rocks contain minerals formed by the crystallization of sodium chloride (halite), calcium sulfate (gypsum and anhydrite), and other combinations of ions commonly found in seawater. As evaporation proceeds and the ions in the seawater become more concentrated, those minerals crystallize in a set sequence. As dissolved ions precipitate to form each mineral, the composition of the evaporating seawater changes.

Seawater has the same composition in all the oceans, which explains why marine evaporites are so similar the world over. No matter where seawater evaporates, the same sequence of minerals always forms. The study of evaporite sediments also shows us that the composition of the oceans has stayed more or less constant over the past 1.8 billion years. Before that time, however, the precipitation sequence may have been different, indicating that the composition of seawater may also have been different.

The great volume of many marine evaporites, some of which are hundreds of meters thick, shows that they could not have formed from the small amount of water that could be held in a small, shallow bay or pond. A huge amount of seawater must have evaporated to form them. The way in which such large quantities of seawater evaporate is very clear in bays or arms of the sea that meet the following conditions (**Figure 6.25**):

* The freshwater supply from rivers is small.
* Connections to the open ocean are constricted.
* The climate is arid.

In such locations, water evaporates steadily, but the connections allow seawater to flow in to replenish the evaporating waters of the bay. As a result, those waters stay at a constant volume, but become more saline than the open ocean. The evaporating bay waters remain more or less constantly supersaturated and steadily deposit evaporite minerals on the floor of the bay.

As seawater evaporates, the first precipitates to form are the carbonates. Continued evaporation leads to the precipitation of gypsum, or calcium sulfate ($CaSO_4 \cdot 2H_2O$) (Figure 6.22b). By the time gypsum precipitates, almost no carbonate ions are left in the water. Gypsum is the principal component of plaster of Paris and is used in the manufacture of wallboard, which lines the walls of most new houses.

After still further evaporation, the mineral halite, or sodium chloride (NaCl)—one of the most common chemical sediments precipitated from evaporating seawater—starts to form (Figure 6.22c). Halite, as you may remember from Chapter 3, is table salt. Deep under the city of Detroit, Michigan, beds of salt laid down by an evaporating arm of an ancient ocean are commercially mined.

In the final stages of evaporation, after the sodium chloride is gone, magnesium and potassium chlorides and sulfates precipitate from the water. The salt mines near Carlsbad, New Mexico, contain commercial quantities of potassium chloride. Potassium chloride is often used as a substitute for table salt by people with certain dietary restrictions.

This sequence of mineral precipitation from seawater has been studied in the laboratory and is matched by the bedding sequences found in certain natural evaporite formations. Most of the world's evaporites consist of thick sequences of dolomite, gypsum, and halite, and do not

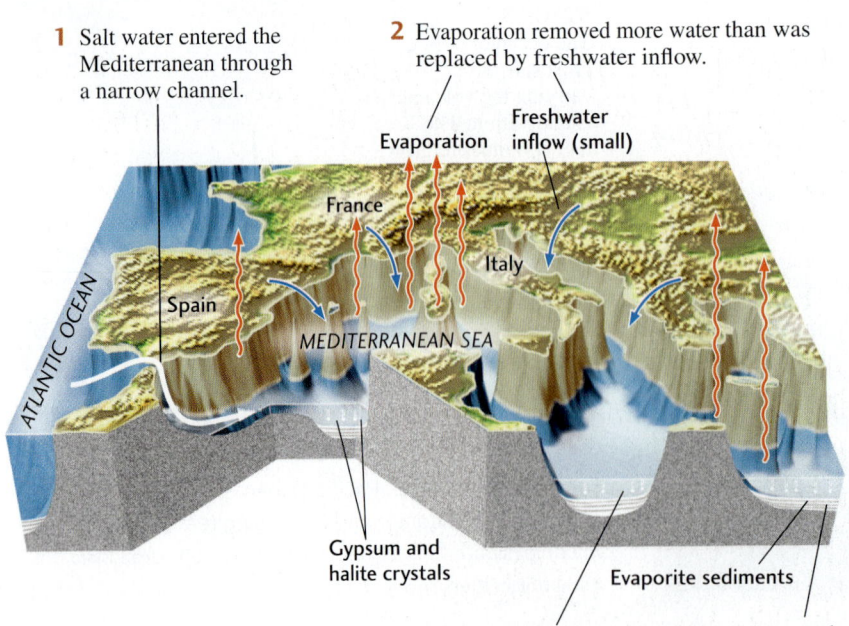

1 Salt water entered the Mediterranean through a narrow channel.

2 Evaporation removed more water than was replaced by freshwater inflow.

3 As the basin became more saline, gypsum and halite precipitated, forming evaporite sediments.

FIGURE 6.25 A marine evaporite environment of the past. The drier climate of the Miocene epoch made the Mediterranean Sea shallower than it is today, and its restricted connection to the open ocean created conditions suitable for evaporite formation. As seawater evaporated, gypsum precipitated to form evaporite sediments. A further increase in salinity led to the crystallization of halite. (The basin depth is greatly exaggerated in this diagram.)

contain the final-stage precipitates. Many do not go even as far as halite. The absence of the final stages indicates that the water did not evaporate completely, but was replenished by normal seawater as evaporation continued.

Nonmarine Evaporites Evaporite sediments also form in arid-region lakes that typically have few or no river outlets. In such lakes, evaporation controls the lake level, and incoming minerals derived from chemical weathering accumulate as sediments. The Great Salt Lake is one of the best known of these lakes (**Figure 6.26**). In the dry climate of Utah, evaporation has more than balanced the inflow of fresh water from rivers and rain. As a result, the concentrations of dissolved ions in the lake make it one of the saltiest bodies of water in the world—eight times more saline than seawater. Sediments form when these ions precipitate.

Small lakes in arid regions may precipitate unusual salts, such as borates (compounds of the element boron), and some become alkaline. The water in this kind of lake is poisonous. Economically valuable deposits of borates and nitrates (minerals containing the element nitrogen) are found in the sediments beneath some of these lakes.

Other Biological and Chemical Sediments

Carbonate minerals secreted by organisms are the principal source of biological sediments, and minerals precipitated from evaporating seawater are the principal source of chemical sediments. However, there are several less common biological and chemical sediments that are locally abundant. They include chert, coal, phosphorite, iron ore, and the organic carbon-rich sediments that produce oil and natural gas. The role of biological versus chemical processes in forming these sediments is variable.

Siliceous Sediments: Sources of Chert One of the first sedimentary rocks to be used for practical purposes by our prehistoric ancestors was **chert**, which is composed of silica (SiO_2) (Figure 6.22d). Early hunters used it for arrowheads and other tools because it could be chipped and shaped to form hard, sharp implements. A common name for chert is *flint;* the two terms are virtually interchangeable. The silica in most cherts is in the form of extremely fine grained quartz. Some geologically young cherts consist of opal, a less fine-grained form of silica.

Like calcium carbonate sediments, many siliceous sediments are precipitated biologically as silica shells secreted by planktonic organisms that settle to the deep seafloor and accumulate as layers of sediment. After these sediments are buried by later sediments, they are cemented into chert. Chert may also form as nodules and irregular masses, replacing carbonate in limestones and dolostones.

Phosphorite Sediments Among the many other kinds of chemical and biological sediments deposited in the ocean is **phosphorite.** Sometimes called *phosphate rock*, phosphorite is composed of calcium phosphate precipitated from phosphate-rich seawater in places where currents of deep, cold water containing phosphate and other nutrients rise along continental margins. Organisms play an important role in creating phosphate-rich water, and bacteria that live on sulfur may play a key role in precipitating phosphate minerals. The phosphorite forms diagenetically by the interaction of calcium phosphate with muddy or carbonate sediments.

FIGURE 6.26 The high concentrations of dissolved ions in the Great Salt Lake make it one of the saltiest bodies of water in the world—eight times more saline than seawater. Evaporite sediments form when these ions precipitate.
[© Jon Mclean/Alamy.]

Iron Oxide Sediments: Source of Iron Formations Iron formations are sedimentary rocks that usually contain more than 15 percent iron in the form of iron oxides and some iron silicates and iron carbonates. Iron oxides were once thought to be of chemical origin, but there is now some evidence that they may have been precipitated indirectly by microorganisms. Most of these rocks formed early in Earth's history, when there was less oxygen in the atmosphere and, as a result, iron dissolved more easily. Iron was transported to the ocean in soluble form and, where microorganisms were producing oxygen, it reacted with that oxygen and precipitated from solution as iron oxides (see Chapter 22).

Organic Sediments: Sources of Coal, Oil, and Natural Gas Coal is a biological sedimentary rock composed almost entirely of organic carbon and formed by the diagenesis of wetland vegetation. In wetland environments, vegetation may be preserved from decay and accumulate as a rich organic material called **peat,** which contains more than 50 percent carbon. If peat is ultimately buried, it may be transformed into coal. Coal is classified as an **organic sedimentary rock,** a class that consists entirely or partly of organic carbon-rich deposits formed by the diagenesis of once-living material that has been buried.

In both lake and ocean waters, the remains of algae, bacteria, and other microscopic organisms may accumulate in fine-grained sediments as organic matter that can be transformed by diagenesis into oil and natural gas. **Crude oil** (petroleum) and **natural gas** are fluids that are not normally classed with sedimentary rocks. They can be considered organic sediments, however, because they form by the diagenesis of organic material in the pores of sedimentary rocks. Deep burial changes the organic matter originally deposited along with inorganic sediments into a fluid that then escapes to porous rock formations and becomes trapped there. Oil and natural gas are found mainly in sandstones and limestones.

As supplies of oil and natural gas begin to diminish, the challenges for geologists increase. These challenges include finding new oil fields as well as squeezing out what is left behind in existing fields. Ultimately, it is the availability of organic sediments that limits how much oil and gas can be found. These sediments were more abundant in some periods of Earth's history and they were formed more easily in certain parts of the world, so there are geologic constraints that we must learn to accept. But we can learn to be smarter about how we explore for what little oil is left, and the need for well-trained geologists has never been greater.

KEY TERMS AND CONCEPTS

arkose (p. 173)	dolostone (p. 178)	quartz arenite (p. 173)
bedding sequence (p. 167)	evaporite rock (p. 178)	reef (p. 177)
bioclastic sediment (p. 157)	evaporite sediment (p. 178)	rift basin (p. 160)
biological sediment (p. 156)	flexural basin (p. 162)	ripple (p. 166)
bioturbation (p. 166)	foraminifera (p. 177)	salinity (p. 160)
carbonate rock (p. 178)	graded bedding (p. 165)	sand (p. 172)
carbonate sediment (p. 178)	gravel (p. 172)	sandstone (p. 173)
cementation (p. 169)	graywacke (p. 173)	sedimentary basin (p. 160)
chemical sediment (p. 156)	iron formation (p. 182)	sedimentary environment (p. 162)
chemical weathering (p. 154)	limestone (p. 178)	sedimentary structure (p. 165)
chert (p. 181)	lithic sandstone (p. 173)	shale (p. 174)
clay (p. 175)	lithification (p. 170)	siliciclastic sediments (p. 156)
claystone (p. 175)	mud (p. 174)	silt (p. 174)
coal (p. 182)	mudstone (p. 174)	siltstone (p. 174)
compaction (p. 169)	natural gas (p. 182)	sorting (p. 159)
conglomerate (p. 172)	organic sedimentary rock (p. 182)	subsidence (p. 160)
continental shelf (p. 161)	peat (p. 182)	terrigenous sediment (p. 165)
cross-bedding (p. 165)	phosphorite (p. 181)	thermal subsidence basin (p. 161)
crude oil (p. 182)	physical weathering (p. 154)	
diagenesis (p. 168)	porosity (p. 169)	

REVIEW OF LEARNING OBJECTIVES

6.1 Describe the major processes that form sedimentary rocks.

Sediments, and the sedimentary rocks that form them, are produced by the surface processes of the rock cycle. These processes move material from a source area to a sink area, where they are deposited. The path from source to sink involves many processes, including weathering, erosion, transport, deposition, burial, and diagenesis. Weathering breaks down rock into the particles that compose siliciclastic sediments and the dissolved ions and molecules that are precipitated to form chemical and biological sediments. Erosion mobilizes the particles produced by weathering. Currents of water and air and the movement of glaciers transport the sediments to their ultimate resting place in a sedimentary basin. Deposition (also called sedimentation) is the settling out of particles or precipitation of minerals to form layers of sediments. Burial and diagenesis compress and harden the sediments into sedimentary rock.

Study Assignment: Figure 6.1

Exercise: What processes change sediments into sedimentary rock?

Thought Question: Weathering of the continents has been much more widespread and intense in the past 10 million years than it was in earlier times. How might this observation be borne out in the sediments that now cover the Earth's surface?

6.2 Compare the various types of sedimentary basins.

The currents that move sediments across the surface generally flow downhill. Therefore, these sediments tend to accumulate in depressions formed by subsidence, in which a broad area of crust sinks relative to the surrounding crust. Sedimentary basins are regions where the combination of sedimentation and subsidence has formed thick accumulations of sediments and sedimentary rock. Some of these basin types include rift basins, thermal subsidence basins, and flexural basins. Rift basins are caused when plate separation begins within a continent; subsidence results from the stretching, thinning, and heating of the underlying lithosphere, and the rift develops bounded by down-dropped blocks of crustal rock. Once that rift has led to seafloor spreading, subsidence continues and the thinned lithosphere cools. Since cooling of the lithosphere is the main process creating the sedimentary basins at this stage, they are called thermal subsidence basins. In areas where one plate pushes up over the other, the weight of the overriding plates causes the underlying plate to flex down, producing a flexural basin.

Study Assignment: Figure 6.8

Exercise: Explain how plate tectonic processes control the development of sedimentary basins.

Thought Question: How did the continental shelves off the Atlantic coasts of North and South America, Europe, and Africa form and what type of basin are they?

6.3 Define the different types of sedimentary environments and the two main types of sediments and sedimentary rocks.

A sedimentary environment is an area of sediment deposition characterized by a particular combination of climate conditions and physical, chemical, and biological processes. Characteristics of a sedimentary environment include the type and amount of water, type and strength of transport agents, the topography, biological activity, plate tectonic setting, and the climate. Sedimentary environments can be grouped not only by their location, but also by the kinds of sediments found in them. Sediments and the sedimentary rocks that form from them can be classified as one of two types: siliciclastic sediments or chemical and biological sediments. Siliciclastic sediments form from the fragmentation of parent rock by physical and chemical weathering and are transported to sedimentary basins by water, wind, or ice. Chemical and biological sediments originate from minerals dissolved in and transported by water. Through chemical and biological reactions, these minerals are precipitated from solution to form sediments.

Study Assignment: Figure 6.9

Exercises: Define a sedimentary environment and name three siliciclastic sedimentary environments. How and on what basis are the siliciclastic sedimentary rocks classified?

Thought Question: In what sedimentary environments would you expect to find carbonate muds?

6.4 Summarize the types of sedimentary structures.

Sedimentary structures include features formed at the time of deposition. Sediments and sedimentary rocks are characterized by bedding, or stratification, which occurs when layers of sediment, or beds, are deposited on top of one another. This bedding can be horizontal or at angles relative to the horizontal. Examples of this include cross-bedding, graded bedding, ripples, bioturbation structures, and bedding sequences. Cross-bedding is formed when beds are deposited at angles by wind or water and deposited on the steeper, downcurrent slopes of sand dunes or sandbars. Patterns of cross-bedding can be complex as a result of rapidly changing wind direction. Graded bedding progresses from large particles at the bottom to small particles at the top. Grading indicates a weakening of the current that deposited the particles. Ripples are small ridges of sand or silt whose long dimensions are at right angles to the current. Ripples can be found on the surface of dunes, on underwater sandbars in shallow streams, or under the waves at beaches. Bioturbation structures are remnants of burrows and tunnels excavated by clams, worms, and other marine organisms that live on the ocean floor. They ingest the sediment, digest the organic matter it contains, and leave behind the reworked sediment, which fills the burrow. Bedding sequences are built of interbedded and vertically stacked layers of different sedimentary types. All of these structures can help geologists reconstruct ways in which sediments were deposited to give insight into events that happened long ago.

Study Assignments: Figure 6.11, Figure 6.15

Exercise: How do organisms produce or modify sediments?

Thought Question: You are looking at a cross section of a rippled sandstone. What sedimentary structure would tell you the direction of the current that deposited the sand?

6.5 Discuss the process of burial and diagenesis of sediments.

Most clastic particles produced by weathering on land end up in the oceans, although a smaller amount of siliciclastic sediment is deposited in sedimentary environments on land. Once sediments reach the ocean floor, they are trapped there. As the sediments are buried under new layers of sediments, they are subjected to increasingly high pressures and temperatures, as well as chemical changes. After they are buried, they are subject to diagenesis, which results in physical and chemical changes until they are either exposed to weathering or metamorphosed. Buried sediments are also in contact with groundwater rich in dissolved minerals. These minerals can precipitate in the pores between the sediment particles and bind them together, which is called cementation. Cementation and the compaction of the sediments due to the weight of the overlying sediments result in lithification, the hardening of soft sediment into rock.

Study Assignment: Figure 6.16

Exercise: Describe how diagenesis converts sediments into sedimentary rocks.

Thought Questions: If you drilled an oil well into the bottom of a sedimentary basin that is 1 km deep and another that is 5 km deep, which would have the higher pressures and temperatures? Oil turns into natural gas at high basin temperatures. In which well would you expect to find more natural gas?

6.6 Describe how the major kinds of siliciclastic, chemical, and biological sediments are classified.

Siliciclastic sediments and sedimentary rocks are classified by particle size. The three major classes, in order of descending particle size, are coarse-grained siliciclastics (gravels and conglomerates); medium-grained siliciclastics (sands and sandstones); and fine-grained siliciclastics (silts and siltstones; muds, mudstones, and shales; and clays and claystones). This classification method emphasizes the importance of the strength of the current that transported the sediments. Chemical and biological sediments and sedimentary rocks are classified on the basis of their chemical composition. The most abundant of these rocks are the carbonate rocks: limestone and dolostone. Limestone is made up largely of biologically precipitated calcite. Dolostone is formed by the diagenetic alteration of limestone. Other chemical and biological sediments include evaporites; siliceous sediments such as chert; phosphorite; iron formations; and peat and other organic matter that is transformed into coal, oil, and natural gas.

Study Assignment: Table 6.3

Exercise: In what kinds of sedimentary rocks are oil and natural gas found?

Thought Question: How can you use the size and sorting of sediment particles to distinguish between sediments deposited in a glacial environment and those deposited in a desert?

6.7 Discuss how carbonate and evaporite rocks are formed.

Most carbonate rocks are formed by the accumulation and lithification of carbonate minerals that are directly or indirectly precipitated by organisms. The most abundant of these carbonate minerals is calcite, and most contain aragonite, a less stable form of calcium carbonate. During burial and diagenesis, carbonate sediments react with water to form a new suite of carbonate minerals. The dominant biological rocks lithified from carbonate sediments are limestone and dolostone. Carbonate rocks are abundant due to large amounts of calcium and carbonate minerals dissolved in seawater, which organisms can convert to shells. Reefs are mound-like organic structures composed of the carbonate skeletons and shells of millions of organisms, which can give rise to carbonate platforms.

Evaporite sediments are chemically precipitated from evaporating seawater (marine evaporites) or lake water (nonmarine evaporites). As evaporation proceeds and the ions in water become more concentrated, minerals such a halite, gypsum, and anhydrite will crystallize.

Study Assignments: Figure 6.23, Figure 6.25

Exercise: Name the two ions that take part in the precipitation of calcium carbonate.

Thought Question: From the base upward, a bedding sequence begins with a bioclastic limestone, passes upward into a dense carbonate rock made of carbonate-cementing organisms, and ends with beds of dolostone. What are the possible sedimentary environments represented by this sequence?

VISUAL LITERACY EXERCISE

1. Which is an example of a shoreline environment?
   a. Desert
   b. Continental shelf
   c. Tidal flat
   d. Glacier

2. Where do turbidity currents form?
   a. Desert
   b. Tidal flat
   c. Continental slope/deep sea
   d. Reefs

3. Which geologic features lead into deltas?
   a. Glaciers
   b. Rivers
   c. Reefs
   d. Tidal flats

4. Where do glaciers form?
   a. Beach
   b. Deep sea
   c. Lakes
   d. Mountains

5. What kinds of sediment form in reefs?
   a. Mud and sand
   b. Gravel
   c. Salts
   d. Calcified organisms

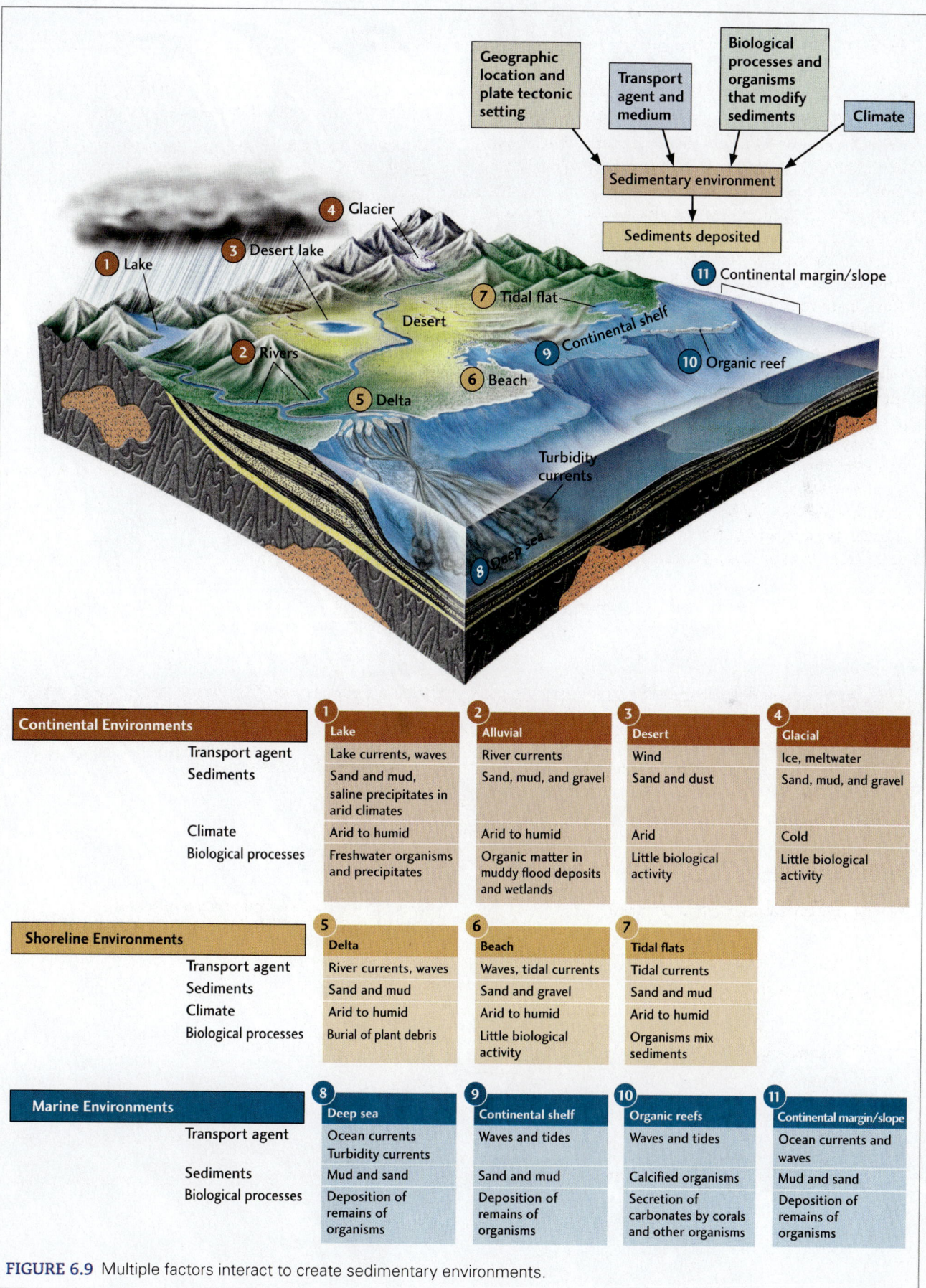

FIGURE 6.9 Multiple factors interact to create sedimentary environments.

Metamorphism: Alteration of Rocks by Temperature and Pressure

7

Causes of Metamorphism	**190**
Types of Metamorphism	**193**
Metamorphic Textures	**195**
Regional Metamorphism and Metamorphic Grade	**199**
Plate Tectonics and Metamorphism	**202**

Learning Objectives

Chapter 7 examines the causes of metamorphism, the types of metamorphism that take place in certain geologic settings, and the origins of the various textures that characterize metamorphic rocks. It shows how geologists use characteristics of metamorphic rocks to understand how and where they were transformed, and it looks at what their journey through the rock cycle tells us about the processes that shape Earth's crust. After you have studied this chapter, you should be able to:

7.1 Explain the causes of metamorphism.

7.2 Describe the various types of metamorphism.

7.3 Summarize the various types of textural features displayed in metamorphic rocks.

7.4 Discuss the ways metamorphic rocks reveal the conditions under which they were formed.

7.5 Illustrate how metamorphic rocks are related to plate tectonic processes.

Connemara marble, found in western Ireland, shows strong deformation by folding during metamorphism. [Courtesy Jennifer Griffes.]

During the rock cycle, rocks may be subjected to temperatures and pressures great enough to cause changes in their mineralogy, texture, or chemical composition. We are all familiar with some of the ways in which heat and pressure can transform materials. Cooking batter in a waffle iron not only heats up the batter but also puts pressure on it, transforming it into a rigid solid. In similar ways, rocks are transformed as they encounter high temperatures and pressures deep in Earth's crust.

Tens of kilometers below the surface, temperatures and pressures are high enough to cause chemical reactions and recrystallization that transform rock without being high enough to melt it. Increases in temperature and pressure and changes in the chemical environment can alter the mineral composition and crystalline texture of igneous and sedimentary rock, *even though it remains solid all the while.* The result is the third large class of rocks: metamorphic rocks, which have undergone changes in mineralogy, texture, chemical composition, or all three.

It is important to understand that most metamorphism is a dynamic process, not a static event. Earth's internal heat engine drives the plate tectonic processes that push rocks formed at Earth's surface down to great depths, thereby subjecting them to high pressures as well as high temperatures. But the transformed rocks return to Earth's surface eventually, and that process is largely powered by weathering and erosion—in other words, by the climate system.

Causes of Metamorphism

Sediments and sedimentary rocks are products of Earth's surface environments, whereas igneous rocks are products of the magmas that originate in the lower crust and mantle. Metamorphic rocks are the products of processes acting on rocks at depths ranging from the upper to the lower crust.

When a rock is subjected to significant changes in temperature or pressure, it will, given enough time—short by geologic standards, but usually a million years or more—undergo changes in its chemical composition, mineralogy, and texture, or all three, until it is in equilibrium with the new temperature and pressure. A limestone filled with fossils, for example, may be transformed into a white marble in which no trace of fossils remains. The mineral and chemical composition of the rock may be unaltered, but its texture may have changed drastically, from small calcite crystals to large, interlocked calcite crystals that overprint and distort former features such as fossils. Shale, a well-bedded sedimentary rock so fine-grained that no individual crystal can be seen with the naked eye, may become schist, in which the original bedding is obscured and the texture is dominated by large crystals of mica. In this case, both mineralogy and texture have changed, but the overall chemical composition of the rock has remained the same.

Most metamorphic rocks are formed at depths of 10 to 30 km, in the middle to lower half of the crust. Only later are those rocks *exhumed*, or transported back to Earth's surface, where they may be exposed as outcrops. But metamorphism can also occur at Earth's surface. We can see metamorphic changes, for example, in the baked surfaces of soils and sediments just beneath volcanic lava flows.

The heat and pressure in Earth's interior and its fluid composition are the three principal factors that drive metamorphism. In much of Earth's crust, the temperature increases at a rate of 30°C per kilometer of depth, although that rate varies considerably among different regions, as we will see shortly. Thus, at a depth of 15 km, the temperature will be about 450°C—much higher than the average temperature at Earth's surface, which ranges from 10°C to 20°C in most regions. The contribution of pressure is the result of vertically oriented forces exerted by the weight of overlying rocks as well as horizontally oriented forces developed as the rocks are deformed by plate tectonic processes. The average pressure at a depth of 15 km amounts to about 4000 times the pressure at the surface.

As high as these temperatures and pressures may seem, they are only in the middle range of conditions for metamorphism, as **Figure 7.1** shows. A rock's *metamorphic grade* reflects the temperatures and pressures it was subjected to during metamorphism. We refer to metamorphic rocks formed under the lower temperatures and pressures of shallower crustal regions as *low-grade* metamorphic rocks and those formed under the higher temperatures and pressures at greater depths as *high-grade* metamorphic rocks.

As the grade of metamorphism changes, the assemblages of minerals within metamorphic rocks also change. Some silicate minerals are found mostly in metamorphic rocks: These minerals include kyanite, andalusite, sillimanite,

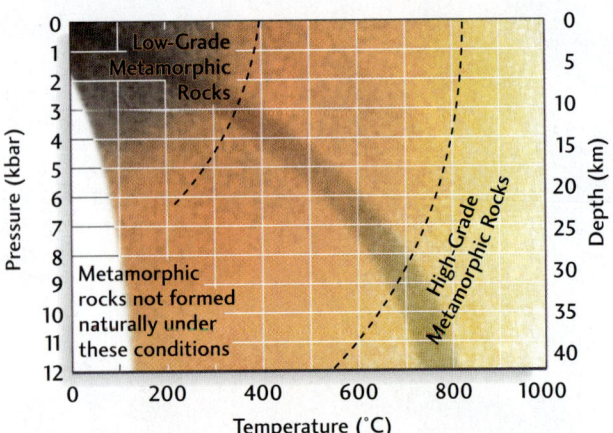

FIGURE 7.1 Temperatures, pressures, and depths at which low-grade and high-grade metamorphic rocks form. The dark band shows the rates at which temperature and pressure increase with depth over much of the continental lithosphere.

staurolite, garnet, and epidote. Geologists use distinctive textures as well as mineral composition to help guide their studies of metamorphic rocks.

The Role of Temperature

Heat can transform a rock's chemical composition, mineralogy, and texture by breaking chemical bonds and altering the existing crystal structures of the rock. When rock is moved from Earth's surface to its interior, where temperatures are higher, the rock adjusts to the new temperature. Its atoms and ions recrystallize, linking up in new arrangements and creating new mineral assemblages. Many new crystals grow larger than the crystals in the original rock.

The increase in temperature with increasing depth in Earth's interior is called the *geothermal gradient*. The geothermal gradient varies among plate tectonic settings, but on average it is about 30°C per kilometer of depth. In areas where the continental lithosphere has been stretched and thinned, such as Nevada's Great Basin, the geothermal gradient is *steep* (for example, 50°C per kilometer of depth). In areas where the continental lithosphere is old and thick, such as central North America, the geothermal gradient is *shallow* (for example, 20°C per kilometer of depth) (**Figure 7.2**).

Because different minerals crystallize and remain stable at different temperatures, we can use a rock's mineral composition as a kind of *geothermometer* to gauge the

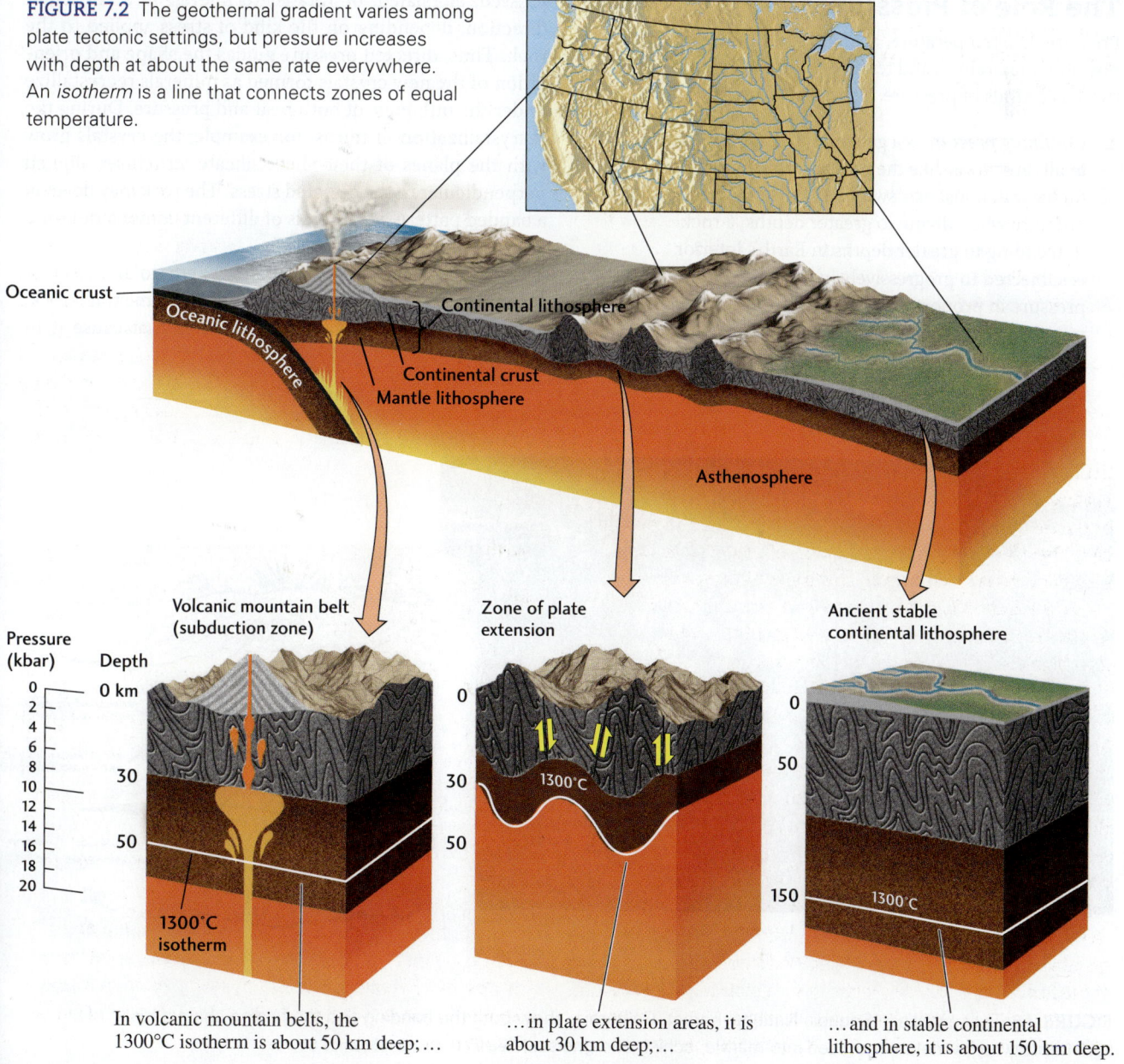

FIGURE 7.2 The geothermal gradient varies among plate tectonic settings, but pressure increases with depth at about the same rate everywhere. An *isotherm* is a line that connects zones of equal temperature.

In volcanic mountain belts, the 1300°C isotherm is about 50 km deep;…

…in plate extension areas, it is about 30 km deep;…

…and in stable continental lithosphere, it is about 150 km deep.

temperature at which it formed. For example, as sedimentary rocks containing clay minerals are buried deeper and deeper, the clay minerals begin to recrystallize and form new minerals, such as micas. With additional burial at greater depths and temperatures, the micas become unstable and begin to recrystallize into new minerals, such as garnet.

Plate tectonic processes such as subduction and continent-continent collision, which transport rocks and sediments into the hot depths of the crust, are the mechanisms that form most metamorphic rocks. In addition, limited metamorphism may occur where rocks are subjected to elevated temperatures near igneous intrusions. The heat is locally intense, but does not penetrate deeply; thus, the intrusions can metamorphose the surrounding country rock, but the effect is local in extent.

The Role of Pressure

Pressure, like temperature, changes a rock's chemical composition, mineralogy, and texture. Solid rock is subjected to two basic kinds of pressure, also called **stress:**

1. *Confining pressure* is a general force applied equally in all directions, like the pressure a swimmer feels under water. Just as a swimmer feels greater confining pressure when diving to greater depths, a rock descending to greater depths in Earth's interior is subjected to progressively increasing confining pressure in proportion to the weight of the overlying mass.

2. *Directed pressure*, or *differential stress*, is force exerted in a particular direction, as when you squeeze a ball of clay between your thumb and forefinger. Directed pressure is usually concentrated within particular zones or along discrete planes.

The compressive force exerted where lithospheric plates converge is a form of directed pressure, and it results in deformation of the rocks near the plate boundary. Heat reduces the strength of a rock, so directed pressure is likely to cause severe folding and other forms of ductile deformation, as well as metamorphism, where temperatures are high. Rocks subjected to differential stress may be severely distorted, becoming flattened in the direction the force is applied and elongated in the direction perpendicular to the force (**Figure 7.3**).

The minerals in a rock under pressure may be compressed, elongated, or rotated to line up in a particular direction, depending on the kind of stress applied to the rock. Thus, directed pressure guides the shape and orientation of the new crystals formed as minerals recrystallize under the influence of both heat and pressure. During the recrystallization of micas, for example, the crystals grow with the planes of their sheet silicate structures aligned perpendicular to the directed stress. The rock may develop a banded pattern as minerals of different compositions are segregated into separate planes.

Marble owes its remarkable strength to this recrystallization process. When limestone, a sedimentary rock, is heated to the very high temperatures that cause it to

FIGURE 7.3 These rocks in Sequoia National Forest, California, show both the banding and the folding characteristic of sedimentary rocks metamorphosed into marble, schist, and gneiss. [Gregory G. Dimijian/Science Source.]

recrystallize, the original minerals and crystals become reoriented and tightly interlocked to form a very strong structure with no planes of weakness.

The pressure to which rock is subjected deep in Earth's crust is related to both the thickness and the density of the overlying rocks. Pressure, which is usually recorded in *kilobars* (1000 bars, abbreviated kbar), increases at a rate of 0.3 to 0.4 kbar per kilometer of depth (see Figure 7.1). One bar is approximately equivalent to the pressure of air at Earth's surface. A diver touring the deeper part of a coral reef at a depth of 10 m would experience an additional bar of pressure.

Minerals that are stable at the lower pressures near Earth's surface become unstable and recrystallize into new minerals under the increased pressures deep in Earth's crust. As we will see in Chapter 8, geologists have subjected rocks to extremely high pressures in the laboratory and recorded the pressures required to cause these changes. With these laboratory data in hand, we can examine the mineralogy and texture of metamorphic rock samples and infer what the pressures were in the area where they formed. Thus, metamorphic mineral assemblages can be used as pressure gauges, or *geobarometers*. Given a specific assemblage of minerals in a metamorphic rock, we can determine the range of pressures, and therefore the depth, at which the rock must have formed.

The Role of Fluids

Metamorphic processes can alter a rock's mineralogy by introducing or removing chemical components that are soluble in heated water. Hydrothermal fluids accelerate metamorphic chemical reactions because they carry dissolved carbon dioxide as well as other chemical substances—such as sodium, potassium, silica, copper, and zinc—that are soluble in hot water under pressure. As hydrothermal solutions percolate up to the shallower parts of the crust, they react with the rocks they penetrate, changing their chemical and mineral compositions and sometimes completely replacing one mineral with another without changing the rock's texture. This kind of change in a rock's composition by fluid transport of chemical substances into or out of it is called **metasomatism.** Many valuable deposits of copper, zinc, lead, and other metallic ores are formed by this kind of chemical substitution, as we saw in Chapter 3.

Where do these chemically reactive fluids originate? Although most rocks appear to be completely dry and to have extremely low porosity, they characteristically contain water in minute pores (the spaces between grains). This water comes not from the pores of sedimentary rocks—from which most of the water is expelled during diagenesis—but rather from chemically bound water in clays. Water forms part of the crystal structure of metamorphic minerals such as micas and amphiboles. The carbon dioxide dissolved in hydrothermal fluids is derived largely from sedimentary carbonates: limestones and dolostones.

Types of Metamorphism

Geologists can duplicate metamorphic conditions in the laboratory and determine the precise combinations of pressure, temperature, and parent rock composition under which particular transformations might take place. But to understand when, where, and how such conditions came about in Earth's interior, we must categorize metamorphic rocks on the basis of their geologic settings (Figure 7.4).

Regional Metamorphism

Regional metamorphism, the most widespread type of metamorphism, takes place where both high temperatures and high pressures are imposed over large parts of the crust. We use this term to distinguish this type of metamorphism from more localized transformations near igneous intrusions or faults. Regional metamorphism is a characteristic feature of convergent plate boundaries. It occurs in volcanic mountain belts, such as the Andes of South America, and in the cores of mountain chains produced by continent-continent collisions, such as the Himalaya of central Asia. These mountain chains are often linear features, so zones of regional metamorphism are often linear in their distribution. In fact, geologists usually interpret regionally extensive belts of metamorphic rocks as representing sites of former mountain chains that were eroded over millions of years, exposing the rocks at their core.

Some regional metamorphic belts are created by the high temperatures and moderate to high pressures near volcanic mountain belts formed where subducted plates sink deep into the mantle. Others are formed under the very high pressures and temperatures found deeper in the crust along boundaries where colliding continents deform rock and raise high mountain chains. In both cases, the metamorphosed rocks are typically transported to great depths in Earth's crust, then eventually uplifted, exposed, and eroded at Earth's surface. A full understanding of the patterns of regional metamorphism, including how rocks respond to systematic changes in temperature and pressure over time, depends on an understanding of the specific plate tectonic settings in which metamorphic rocks form. We will discuss that topic later in this chapter.

Contact Metamorphism

In **contact metamorphism,** the heat from an igneous intrusion metamorphoses the rock immediately surrounding it. This type of localized transformation normally affects only a thin zone of country rock along the zone of contact. In many contact metamorphic rocks, especially at the margins of shallow intrusions, the mineral and chemical transformations are largely related to the high temperature of the intruding magma. Pressure effects are important only where the magma is intruded at great depths. Here, the pressure results not from the intrusion forcing its way into the country rock, but from the presence of regional confining pressure. Contact metamorphism by volcanic deposits is limited to very

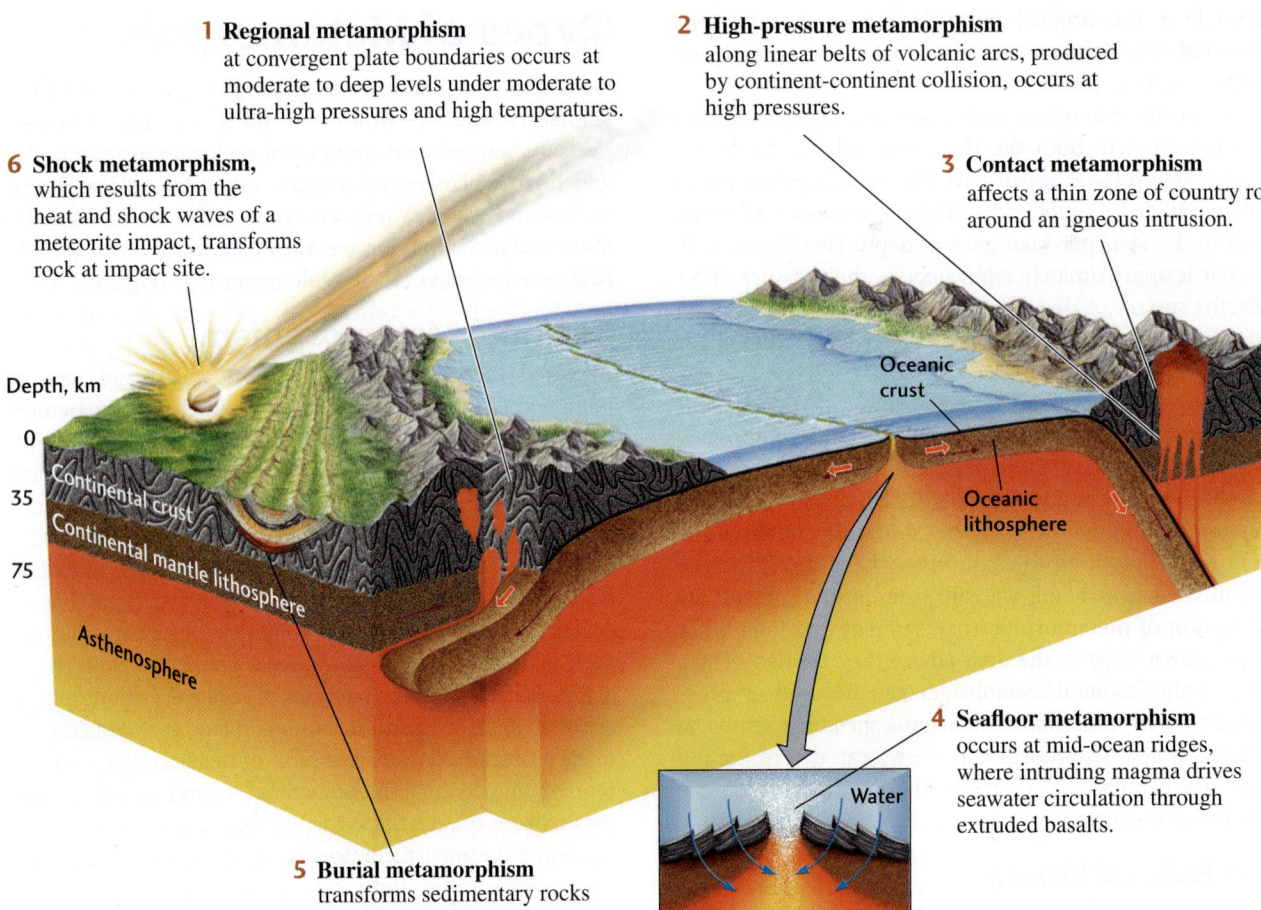

FIGURE 7.4 Different types of metamorphism occur in different geologic settings.

thin zones because lavas cool quickly at Earth's surface and their heat has little time to penetrate the surrounding rocks deeply and cause metamorphic changes. Contact metamorphism may also affect xenoliths that are not completely melted. Blocks of rock up to several meters wide may be torn off the sides of magma chambers and completely surrounded by hot magma. Heat projects into these xenoliths from all directions, and they may become completely metamorphosed.

Seafloor Metamorphism

Another type of metamorphism, a form of metasomatism called **seafloor metamorphism,** is often associated with mid-ocean ridges (see Chapter 4). Hot basaltic lava at a seafloor spreading center heats infiltrating seawater, which starts to circulate through the newly forming oceanic crust by convection. The increase in temperature promotes chemical reactions between the seawater and the rock, forming altered basalts whose chemical compositions differ from that of the original basalt. Metasomatism resulting from percolation of high-temperature fluids also takes place on continents when hydrothermal solutions circulating near igneous intrusions metamorphose the rocks they intrude.

Other Types of Metamorphism

There are several other types of metamorphism that produce smaller amounts of metamorphic rock. Some of these types are extremely important in helping geologists understand conditions deep within Earth's crust.

Burial Metamorphism Recall from Chapter 6 that sedimentary rocks are transformed by diagenesis as they are gradually buried. Diagenesis grades into **burial metamorphism,** low-grade metamorphism that is caused by the progressive increase in pressure exerted by the growing layers of overlying sediments and sedimentary rocks and by the increase in heat associated with increased depth of burial.

Depending on the local geothermal gradient, burial metamorphism typically begins at depths of 6 to 10 km, where temperatures range between 100°C and 200°C and pressures are less than 3 kbar. This fact is of great importance to the oil and gas industry, which defines its "economic basement" as the depth where low-grade metamorphism begins. Oil and gas wells are rarely drilled below this depth because temperatures above 150°C convert organic matter trapped in sedimentary rocks into carbon dioxide rather than crude oil and natural gas.

High-Pressure and Ultra-High-Pressure Metamorphism

Metamorphic rocks formed by **high-pressure metamorphism** (at 8 to 12 kbar) and **ultra-high-pressure metamorphism** (at pressures greater than 28 kbar) are rarely exposed at Earth's surface for geologists to study. These rocks are rare because they form at such great depths that it takes a very long time for them to be recycled to the surface. Most high-pressure metamorphic rocks form in subduction zones as sediments scraped from subducting oceanic crust are plunged to depths of over 30 km, where they experience pressures of up to 12 kbar.

Unusual metamorphic rocks once located at the base of Earth's crust can sometimes be found at Earth's surface. These rocks, called **eclogites** (see Figure 3.27), may contain minerals such as *coesite* (a very dense, high-pressure form of quartz) that indicate pressures of greater than 28 kbar, suggesting depths of over 80 km. Such rocks form at moderate to high temperatures, ranging from 800°C to 1000°C. In a few cases, these rocks contain *microscopic diamonds*, indicative of pressures greater than 40 kbar and depths greater than 120 km! Surprisingly, outcrop exposures of these ultra-high-pressure metamorphic rocks may cover areas greater than 400 km by 200 km. The only other two rocks known to come from these depths are diatremes and kimberlites (see Chapter 5), igneous rocks that form narrow pipes just a few hundred meters wide. Geologists agree that these latter rock types form by volcanic eruption, albeit from very unusual depths. In contrast, the mechanisms required to bring eclogites to the surface are hotly debated. It appears that these rocks represent pieces of the leading edges of continents that were subducted during continent-continent collisions and subsequently rebounded (via some unknown mechanism) to the surface before they had time to recrystallize at lower pressures.

Shock Metamorphism

Shock metamorphism occurs when a meteorite collides with Earth. Upon impact, the energy represented by the meteorite's mass and velocity is transformed into heat and shock waves that pass through the impacted country rock. The country rock can be shattered and even partially melted to produce *tektites*. The smallest tektites look like droplets of glass. In some cases, quartz is transformed into coesite and *stishovite*, two of its high-pressure forms.

Most large impacts on Earth have left no trace of a meteorite because these bodies are usually destroyed in the collision with Earth. The occurrence of coesite and craters with distinctive fringing fractures, however, provides evidence of these collisions. Earth's dense atmosphere causes most meteorites to burn up before they strike its surface, so shock metamorphism is rare on Earth. On the surface of the Moon, however, shock metamorphism is pervasive. It is characterized by extremely high pressures of many tens to hundreds of kilobars.

Metamorphic Textures

Metamorphism imprints new textures on the rocks it alters. The texture of a metamorphic rock is determined by the sizes, shapes, and arrangement of its constituent crystals. Some metamorphic rock textures depend on the particular kinds of minerals formed under metamorphic conditions. Variation in grain size is also important. In general, grain size increases as metamorphic grade increases. Each textural variety of metamorphic rock tells us something about the metamorphic process that created it. In this section, we examine those processes, then describe two major textural classes of metamorphic rocks: foliated rocks and granoblastic rocks.

Foliation and Cleavage

The most prominent textural feature of regionally metamorphosed rocks is **foliation**, a set of flat or wavy parallel cleavage planes produced by deformation of igneous and sedimentary rocks under directed pressure (**Figure 7.5**). These foliation planes may cut through the bedding of the original sedimentary rock at any angle or be parallel to the bedding (Figure 7.5). In general, as the grade of regional metamorphism increases, foliation becomes more pronounced.

A major cause of foliation is the formation of minerals with a platy crystal habit, chiefly the micas and chlorite. The planes of all the platy crystals are aligned parallel to the foliation, an alignment called the *preferred orientation* of the minerals (Figure 7.5). As platy minerals crystallize, their preferred orientation is usually perpendicular to the main direction of the forces squeezing the rock during metamorphism. Crystals of preexisting minerals may contribute to the foliation by rotating until they also lie parallel to the developing foliation plane.

The most familiar form of foliation is seen in slate, a common metamorphic rock, which is easily split into thin sheets along smooth, parallel surfaces. This *slaty cleavage* (not to be confused with the perfect cleavage of sheet silicates such as micas) develops at small, regular intervals in the rock.

Minerals with an elongate, needlelike crystal habit also tend to assume a preferred orientation during metamorphism: These crystals, too, normally line up parallel to the foliation plane. Rocks that contain abundant amphiboles (typically, metamorphosed mafic volcanic rocks) have this kind of texture.

Foliated Rocks

The **foliated rocks** are classified according to four main criteria:

1. Metamorphic grade
2. Grain (crystal) size
3. Type of foliation
4. Banding

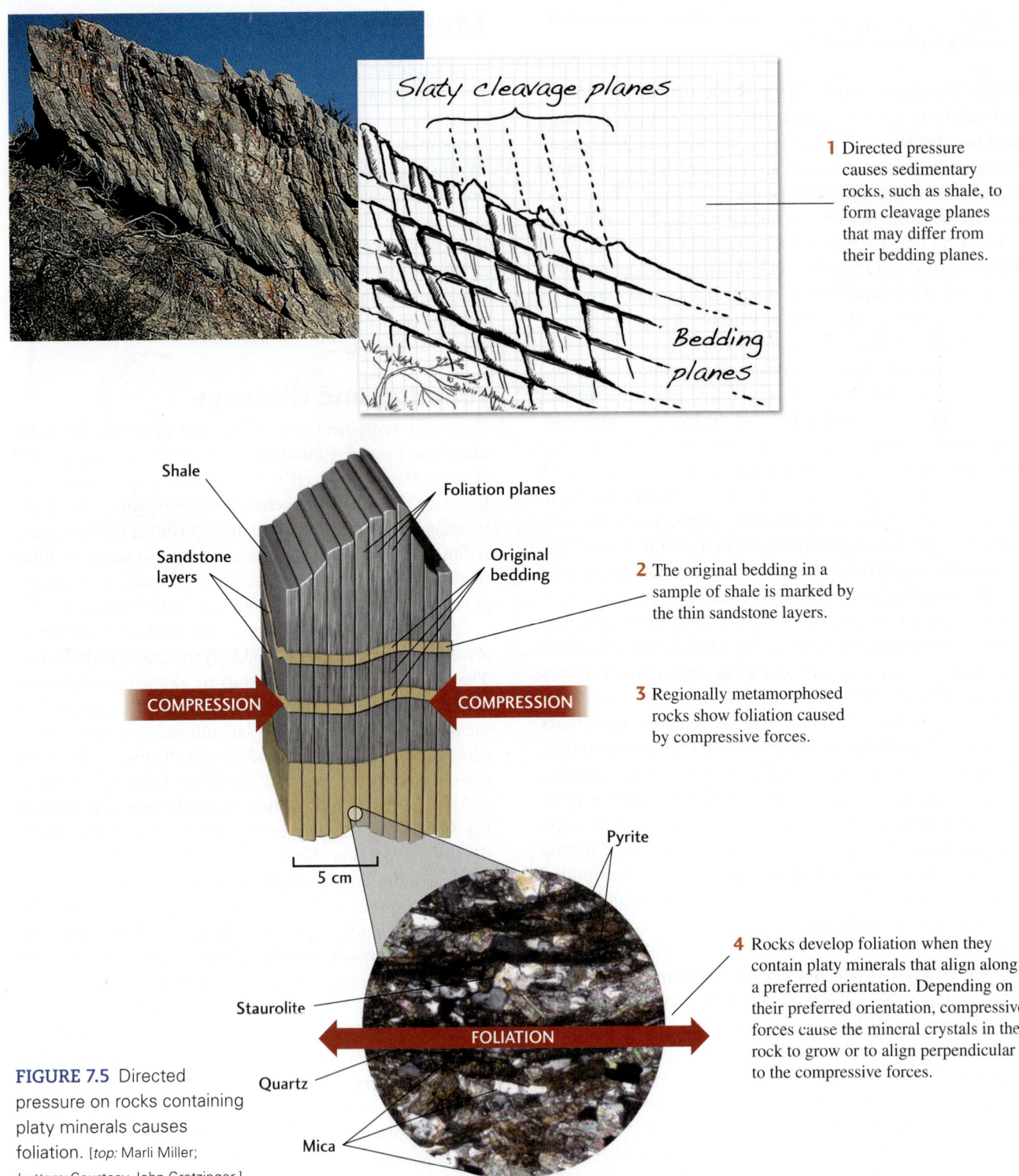

FIGURE 7.5 Directed pressure on rocks containing platy minerals causes foliation. [top: Marli Miller; bottom: Courtesy John Grotzinger.]

1 Directed pressure causes sedimentary rocks, such as shale, to form cleavage planes that may differ from their bedding planes.

2 The original bedding in a sample of shale is marked by the thin sandstone layers.

3 Regionally metamorphosed rocks show foliation caused by compressive forces.

4 Rocks develop foliation when they contain platy minerals that align along a preferred orientation. Depending on their preferred orientation, compressive forces cause the mineral crystals in the rock to grow or to align perpendicular to the compressive forces.

Figure 7.6 shows examples of the major types of foliated rocks. In general, foliation progresses from one texture to another with increasing metamorphic grade. In this progression, as temperature and pressure increase, a shale may metamorphose first to a slate, then to a phyllite, then to a schist, then to a gneiss, and finally to a migmatite.

Slate Slates are the lowest grade of foliated rocks. These rocks are so fine-grained that their individual crystals cannot be seen easily without a microscope. They are commonly produced by the metamorphism of shales or, less frequently, of volcanic ash deposits. Slates usually range from dark gray to black, colored by small amounts of organic material originally present in the parent shale. Slate splitters learned long ago to recognize foliation planes and use them to make thick or thin slabs for roofing tiles and blackboards. Flat slabs of slate are still used for flagstone walks in places where slate is abundant.

FIGURE 7.6 Foliated rocks are classified by metamorphic grade, grain size, type of foliation, and banding. [*slate, phyllite, schist, gneiss:* John Grotzinger/Ramón Rivera-Moret/Harvard Mineralogical Museum; *migmatite:* Courtesy of Kip Hodges.]

Phyllite Phyllites are rocks of a slightly higher grade than the slates, but are similar to them in character and origin. They tend to have a more or less glossy sheen resulting from crystals of mica and chlorite that have grown a little larger than those of slates. Phyllites, like slates, tend to split into thin sheets, but less perfectly than slates.

Schist At low grades of metamorphism, the crystals of platy minerals are generally too small to be seen, and foliation planes are closely spaced. As rocks are subjected to higher temperatures and pressures, however, the platy crystals grow large enough to be visible to the naked eye, and the minerals tend to segregate into lighter and darker bands. This parallel arrangement of platy minerals produces the coarse, wavy foliation known as *schistosity,* which characterizes **schists**. Schists, which are intermediate-grade rocks, are among the most abundant metamorphic rock types. They contain more than 50 percent platy minerals, mainly the micas muscovite and biotite. Schists may contain thin layers of quartz, feldspar, or both, depending on the quartz content of the parent shale.

Gneiss Even coarser foliation is shown by **gneisses**, light-colored rocks with coarse bands of light and dark minerals throughout the rock. This *gneissic foliation* results from the segregation of lighter-colored quartz and feldspar from darker-colored amphiboles and other mafic minerals. Gneisses are high-grade, coarse-grained metamorphic rocks in which the ratio of granular to platy minerals is higher than that in slate or schist. The result is poor foliation and thus little tendency to split. Under high pressures and temperatures, the mineral assemblages of lower-grade rocks containing micas and chlorite are transformed into new assemblages dominated by quartz and feldspars, with lesser amounts of micas and amphiboles.

Migmatite Temperatures higher than those necessary to produce gneiss may begin to melt the country rock. In this case, as with igneous rocks (see Chapter 4), the first minerals to melt will be those with the lowest melting temperatures. Therefore, only part of the country rock melts, and the melt migrates only a short distance before solidifying again. Rocks produced in this way are badly deformed and contorted, and they are penetrated by many veins, small pods, and lenses of melted rock. The result is a mixture of igneous and metamorphic rock called **migmatite**. Some migmatites are mainly metamorphic, with only a small proportion of igneous material. Others have been so affected by melting that they are considered almost entirely igneous.

Granoblastic Rocks

Granoblastic rocks are nonfoliated metamorphic rocks composed mainly of crystals that grow in equant (equidimensional) shapes, such as cubes and spheres, rather than in platy or elongate shapes. These rocks result from metamorphic processes, such as contact metamorphism, in which directed pressure is absent, so foliation does not occur. Granoblastic rocks include hornfels, quartzite, marble, greenstone, amphibolite, and granulite (Figure 7.7). All granoblastic rocks except hornfels are defined by their mineralogy rather than their texture because all of them have a homogeneous granular texture.

Hornfels is a high-temperature contact metamorphic rock of uniform grain size that has undergone little or no deformation (see Figure 3.27). It is formed from fine-grained sedimentary rock and other types of rock containing an abundance of silicate minerals. Hornfels has a granular texture overall, even though it commonly contains pyroxene, which makes elongate crystals, and some micas. It is not foliated, and its platy or elongate crystals are oriented randomly.

Quartzites are very hard, white rocks derived from quartz-rich sandstones. Some quartzites are homogeneous, unbroken by preserved bedding or foliation (Figure 7.7a). Others contain thin bands of slate or schist, relics of former interbedded layers of clay or shale.

Marbles are the metamorphic products of heat and pressure acting on limestones and dolomites. Some white, pure marbles, such as the famous Italian Carrara marbles prized by sculptors, show a smooth, even texture of interlocked calcite crystals of uniform size. Other marbles show irregular banding or mottling from silicate and other mineral impurities in the original limestone (Figure 7.7b).

Greenstones are metamorphosed mafic volcanic rocks. Many of these low-grade metamorphic rocks form by seafloor metamorphism. Large areas of the seafloor are covered with basalts that have been slightly or extensively altered in this way at mid-ocean ridges. An abundance of chlorite gives these rocks their greenish cast.

Amphibolites are made up of amphibole and plagioclase feldspar. They are typically the product of medium- to high-grade metamorphism of mafic volcanic rocks. Foliated amphibolites can be produced by directed pressure.

Granulite, a high-grade metamorphic rock that is also referred to as *granofels*, has a homogeneous granular texture. It is a medium- to coarse-grained rock in which the crystals are equant and show only faint foliation at most. It is formed by the metamorphism of shale, impure sandstone, and many kinds of igneous rock.

Porphyroblasts

Newly formed metamorphic minerals may grow into large crystals surrounded by a much finer-grained matrix of other minerals (Figure 7.8). These large crystals, called **porphyroblasts,** are found in rocks formed both by contact and by regional metamorphism. Porphyroblasts form from minerals that are stable over a broad range of pressures and temperatures. Crystals of these minerals grow large while the minerals of the matrix are being continuously recrystallized as pressures and temperatures change, so they replace

(a) Quartzite

(b) Marble

FIGURE 7.7 Granoblastic (nonfoliated) metamorphic rocks: (a) quartzite; (b) marble. [*quartzite:* Breck P. Kent; *marble:* Courtesy John Grotzinger.]

FIGURE 7.8 Garnet porphyroblasts in a schist matrix. The minerals in the matrix are continuously recrystallized as pressures and temperatures change and therefore grow to only a small size. In contrast, porphyroblasts grow to a large size because they are stable over a broad range of pressures and temperatures. [MSA 260 by Chip Clark, Smithsonian.]

TABLE 7.1	Classification of Metamorphic Rocks by Texture		
CLASSIFICATION	CHARACTERISTICS	ROCK NAME	TYPICAL PARENT ROCK
Foliated	Distinguished by slaty cleavage, schistosity, or gneissic foliation; mineral grains show preferred orientation	Slate Phyllite Schist Gneiss	Shale, sandstone
Granoblastic (nonfoliated)	Granular, characterized by coarse or fine interlocking grains; little or no preferred orientation	Hornfels Quartzite Marble Argillite Greenstone Amphibolite Granulite	Shale, volcanic Quartz-rich sandstone Limestone, dolomite Shale Basalt Shale, basalt Shale, basalt
Porphyroblastic	Large crystals set in fine-grained matrix	Slate to gneiss	Shale

parts of the matrix. Porphyroblasts vary in size, ranging from a few millimeters to several centimeters in diameter. Garnet and staurolite are two common minerals that form porphyroblasts, although many others are also found. The precise composition and distribution of porphyroblasts of these two minerals can be used to infer the pressures and temperatures that occurred during metamorphism, as we will see later in this chapter.

Table 7.1 summarizes the textural classes of metamorphic rocks and their main characteristics.

Regional Metamorphism and Metamorphic Grade

As we have seen, metamorphic rocks form under a wide range of conditions, and their mineralogies and textures are clues to the pressures and temperatures in the crust where and when they formed. Geologists who study the formation of metamorphic rocks constantly seek to determine the intensity and character of metamorphism more precisely than is indicated by a designation of "low grade" or "high grade." To make these finer distinctions, geologists "read" minerals as though they were pressure gauges and thermometers. These techniques are best illustrated by their application to regional metamorphism.

Mineral Isograds: Mapping Zones of Change

When we study a broad belt of regional metamorphism, we can see many outcrops, some showing one set of minerals, some showing others. Different zones within the belt may be distinguished by *index minerals*: abundant minerals that each form under a limited range of temperatures and pressures (**Figure 7.9**). For example, a zone of unmetamorphosed shales may lie next to a zone of weakly metamorphosed slates (Figure 7.9a). As we move from the shale zone into the slate zone, a new mineral—chlorite—appears. Chlorite is an index mineral marking the point at which we move into a new zone with a higher metamorphic grade. If laboratory studies have determined the temperature and pressure at which the index mineral forms, we can draw conclusions about the conditions that existed when the rocks in the zone were formed.

We can use the occurrences of index minerals to make a map of the boundaries between metamorphic zones. Geologists define these boundaries by drawing lines called *isograds* that plot the transitions from one zone to the next. Isograds are used in Figure 7.9a to show a series of mineral assemblages produced by the regional metamorphism of shale in New England. A pattern of isograds tends to follow the deformation features (folds and faults) of a region. An isograd based on a single index mineral, such as the chlorite isograd in Figure 7.9a, provides a good approximate measure of metamorphic pressure and temperature.

To determine metamorphic pressure and temperature more precisely, geologists can examine a group of two or three minerals that have crystallized together. For example, based on laboratory data, a geologist knows that a sillimanite zone that contains orthoclase feldspar and sillimanite must have formed by the reaction of muscovite and quartz at temperatures of about 600°C and pressures of about 5 kbar, liberating water (as water vapor) in the process. The sillimanite isograd records the following reaction:

$$\text{Muscovite} + \text{quartz} \rightarrow$$
$$\text{KAl}_3\text{Si}_3\text{O}_{10}(\text{OH}) \quad \text{SiO}_2$$

$$\text{orthoclase feldspar} + \text{sillimanite} + \text{water}$$
$$\text{KAlSi}_3\text{O}_8 \quad\quad \text{Al}_2\text{SiO}_5 \quad\quad \text{H}_2\text{O}$$

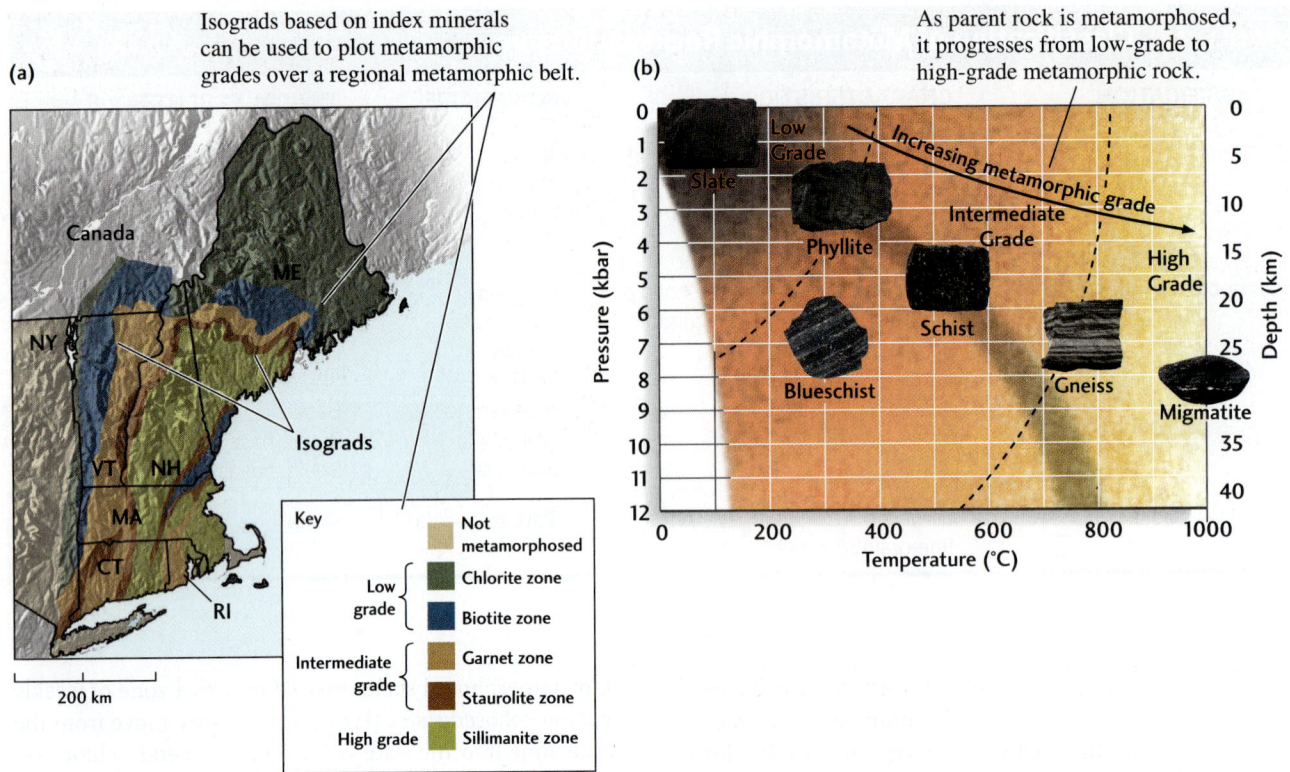

FIGURE 7.9 Index minerals define the different metamorphic zones within a belt of regional metamorphism. (a) Map of New England, showing metamorphic zones based on index minerals found in rocks metamorphosed from shale. (b) Rocks produced by the metamorphism of shale at different temperatures and pressures. [*slate, phyllite, schist, gneiss:* John Grotzinger/Ramón Rivera-Moret/Harvard Mineralogical Museum; *blueschist:* courtesy of Mark Cloos; *migmatite:* courtesy of Kip Hodges.]

Because isograds reflect the pressures and temperatures at which the minerals in a regional metamorphic belt formed, the isograd sequence in one belt may differ from that in another. The reason for this variation is that, as we have seen, pressures and temperatures do not increase at the same rate in all plate tectonic settings.

Metamorphic Grade and Parent Rock Composition

The kind of metamorphic rock that results from a given grade of metamorphism depends partly on the mineralogy of the parent rock (**Figure 7.10**). The metamorphism of shale, as shown in Figure 7.10a, reveals the effects of pressure and temperature on parent rock that is rich in clay minerals, quartz, and perhaps some carbonate minerals. The metamorphism of mafic volcanic rock, composed predominantly of feldspars and pyroxene, follows a different course (Figure 7.10b).

In the regional metamorphism of a basalt, for example, the lowest-grade rocks characteristically contain various **zeolite** minerals. The silicate minerals in the zeolite class contain water within their crystal structure. Zeolite minerals form at very low temperatures and pressures. Rocks that include this group of minerals are thus placed in the zeolite grade.

Overlapping with the zeolite grade is a higher grade of metamorphosed mafic volcanic rocks, the **greenschists**, whose abundant minerals include chlorite. Next are the amphibolites, which contain large amounts of amphiboles. The granulites, coarse-grained rocks containing pyroxene and calcium-rich plagioclase, constitute the highest grade of metamorphosed mafic volcanic rocks.

Rocks of the greenschist, amphibolite, and granulite grades are also formed during metamorphism of sedimentary rocks such as shale, as shown in Figure 7.10b. The pyroxene-bearing granulites are the products of high-grade metamorphism in which the temperature is high and the pressure is moderate. The opposite conditions, in which the pressure is high and the temperature moderate, produce rocks of the **blueschist** grade from parent rock of various starting compositions, from mafic volcanic rocks to shaley sedimentary rocks (see Figure 7.9b).

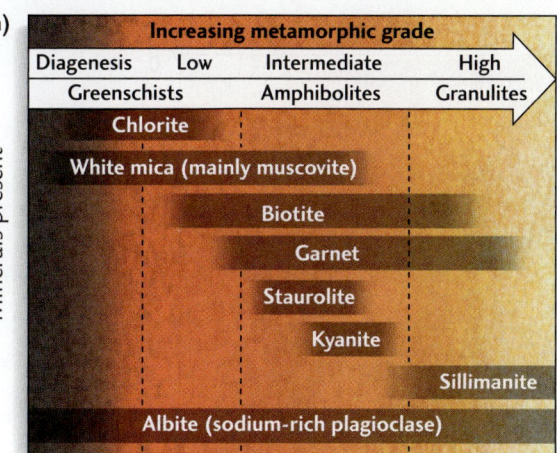

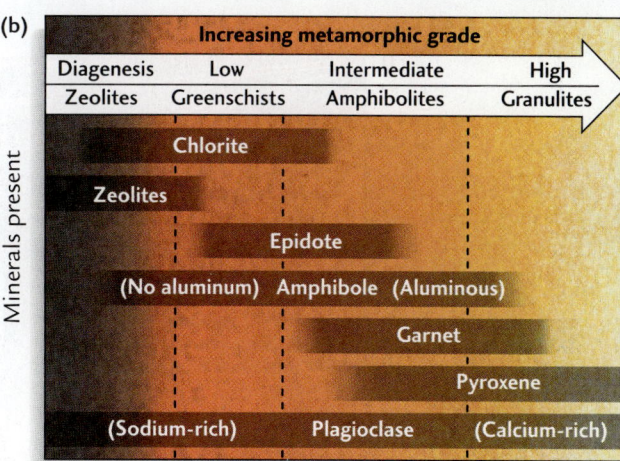

Metamorphic Facies	Minerals Produced from Shale Parent	Minerals Produced from Basalt Parent
Greenschist	Muscovite, chlorite, quartz, albite	Albite, epidote, chlorite
Amphibolite	Muscovite, biotite, garnet, quartz, albite, staurolite, kyanite, sillimanite	Amphibole, plagioclase feldspar
Granulite	Garnet, sillimanite, albite, orthoclase, quartz, biotite	Calcium-rich pyroxene, calcium-rich plagioclase feldspar
Eclogite	Garnet, sodium-rich pyroxene, quartz/coesite, kyanite	Sodium-rich pyroxene, garnet

FIGURE 7.10 The kind of metamorphic rock that results from a given grade of metamorphism depends partly on the mineralogy of the parent rock. (a) Changes in the mineral composition of shale (a mafic volcanic rock) with increasing metamorphic grade. (b) Changes in the mineral composition of basalt (a sedimentary rock) with increasing metamorphic grade. (c) Major minerals of metamorphic facies produced from shale and basalt.

The name comes from the abundance of glaucophane, a blue amphibole, in these rocks. Still another metamorphic rock, formed at extremely high pressures and moderate to high temperatures, is eclogite, which is rich in garnet and pyroxene.

Metamorphic Facies

We can plot this information about the grades of the metamorphic rocks in a regional metamorphic belt—derived from parent rocks of many different chemical compositions—on a graph of temperature and pressure (**Figure 7.11**). **Metamorphic facies** are groupings of rocks of various mineral compositions formed under particular conditions of temperature and pressure from different parent rocks. By delineating metamorphic facies, we can be more specific about the grades of metamorphism observed in rocks. Two essential points characterize the concept of metamorphic facies:

1. Different kinds of metamorphic rocks of the same metamorphic grade form from parent rocks of different composition.

2. Different kinds of metamorphic rocks of different metamorphic grades form from parent rocks of the same composition.

Figure 7.10c shows the major minerals of the metamorphic facies produced from shale and basalt. Because parent rocks vary so greatly in composition, there are no sharp boundaries between metamorphic facies (see Figure 7.11). Perhaps the most important reason for analyzing metamorphic facies is that they give us clues to the plate tectonic processes responsible for metamorphism, as we shall see next.

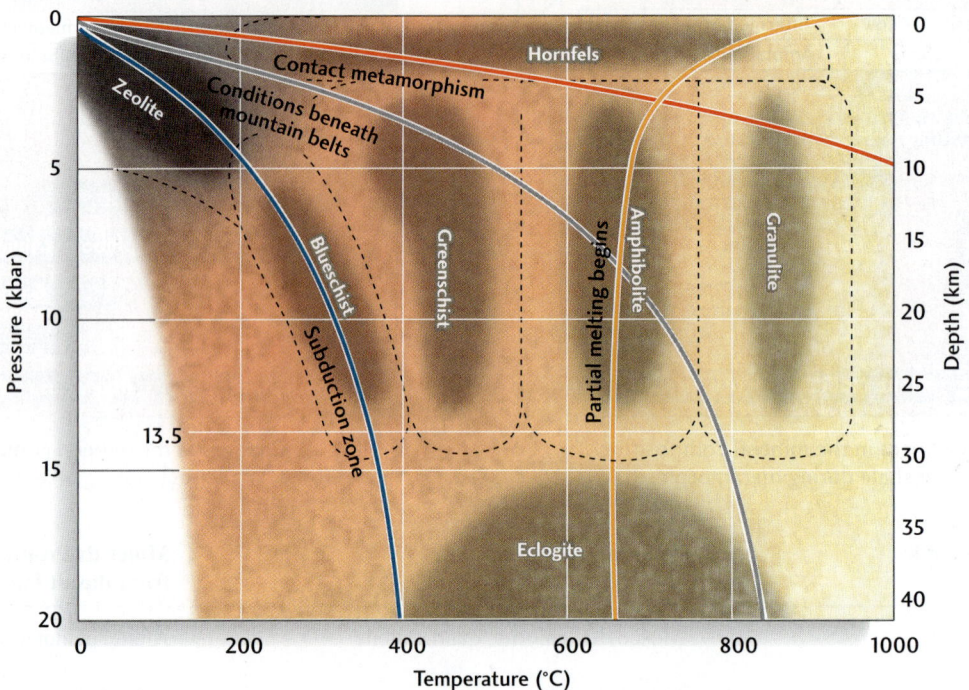

FIGURE 7.11 Metamorphic facies correspond to particular combinations of temperature and pressure, which correspond to particular plate tectonic settings. The dashed lines indicate the overlapping nature of the boundaries between metamorphic facies.

Plate Tectonics and Metamorphism

Soon after the theory of plate tectonics was proposed, geologists started to see how patterns of metamorphism fit into the larger framework of plate movements. Different types of metamorphism are likely to occur in different plate tectonic settings (see Figure 7.4):

- *Continental interiors.* Contact metamorphism, burial metamorphism, and perhaps regional metamorphism occur at different levels in the crust. Shock metamorphism is likely to be best preserved in continental interiors because their large areal extent provides a large target area to record rare meteorite impact events.
- *Divergent plate boundaries.* Seafloor metamorphism and contact metamorphism around plutons intruding into the oceanic crust occur at divergent plate boundaries.
- *Convergent plate boundaries.* Regional metamorphism, high-pressure and ultra-high-pressure metamorphism, and contact metamorphism.
- *Transform faults.* In oceanic settings, seafloor metamorphism may occur. In both oceanic and continental settings, we find extensive metamorphism caused by shearing forces along transform faults.

Metamorphic Pressure-Temperature Paths

As we have seen, the concept of metamorphic grade can inform us of the maximum pressure or temperature to which a metamorphic rock has been subjected, but it tells us nothing about where the rock encountered those conditions. Nor does it tell us anything about how the rock was **exhumed,** or transported back to Earth's surface.

Each metamorphic rock has a distinctive history of changing temperature and pressure that is reflected in its texture and mineralogy. This history is called a metamorphic pressure-temperature path, or **P-T path.** The P-T path can be a sensitive recorder of many important factors that influence metamorphism—such as sources of heat, which changes temperatures, and rates of tectonic transport (burial and exhumation), which changes pressures. Thus, P-T paths are characteristic of particular plate tectonic settings.

To obtain a P-T path, geologists must analyze specific minerals from metamorphic rock samples in the laboratory. One of the minerals most widely used for this purpose is garnet, a common porphyroblast that serves as a sort of P-T path recording device (**Figure 7.12**). Garnet crystals grow steadily during metamorphism, and as the pressure and temperature of the environment change, the

Plate Tectonics and Metamorphism

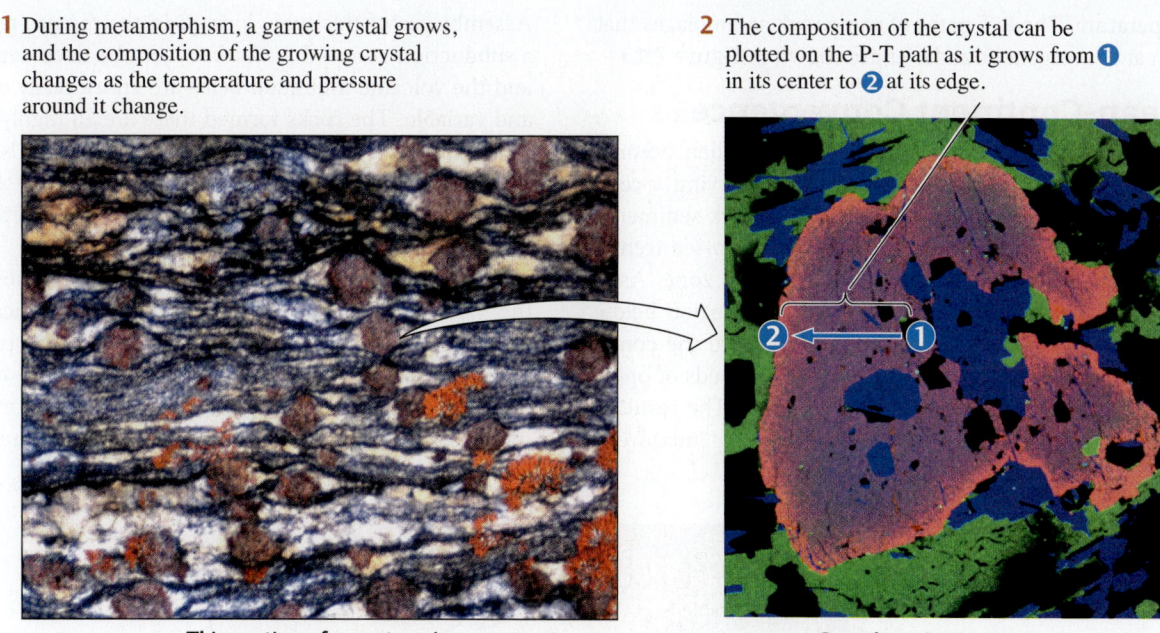

1 During metamorphism, a garnet crystal grows, and the composition of the growing crystal changes as the temperature and pressure around it change.

Thin section of garnet gneiss

2 The composition of the crystal can be plotted on the P-T path as it grows from ❶ in its center to ❷ at its edge.

Growth zoning in garnet

3 As rock is carried deeper in Earth's crust and is subjected to higher temperatures and pressures (the **prograde** path), the garnet crystal initially grows in a schist but ends up growing in a gneiss as metamorphism progresses.

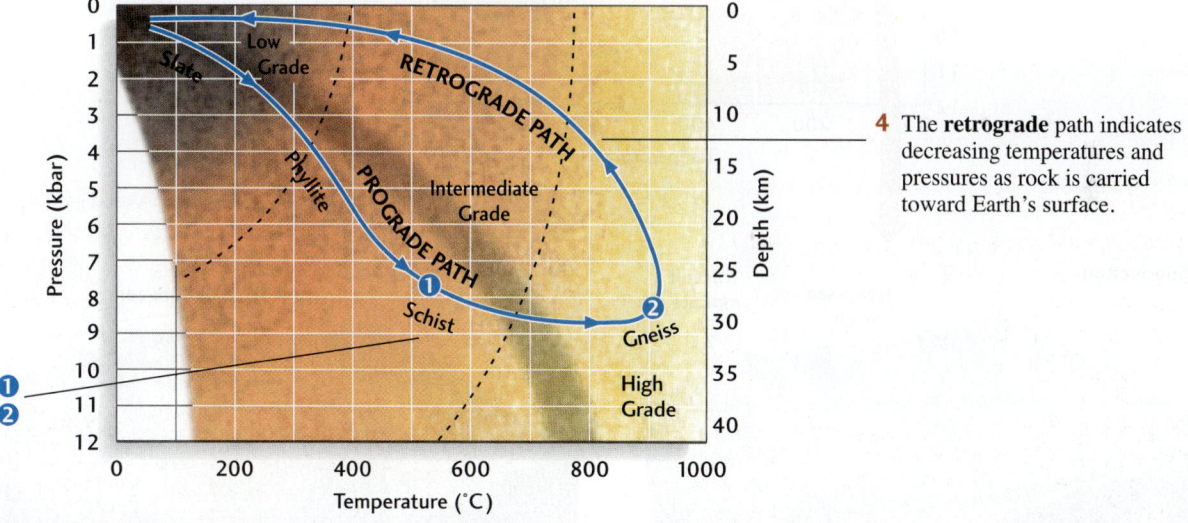

4 The **retrograde** path indicates decreasing temperatures and pressures as rock is carried toward Earth's surface.

FIGURE 7.12 Porphyroblasts such as garnet can be used to plot the P-T paths of metamorphic rocks. The P-T path that a metamorphic rock typically follows begins with an increase in pressure and temperature (the prograde path), followed by a decrease in pressure and temperature (the retrograde path). [Photos courtesy of Kip Hodges.]

chemical composition of the garnet changes. The oldest part of a garnet crystal is its core, and the youngest is its outer edge, so the variation in its composition from core to edge will yield the history of the metamorphic conditions under which it formed. Geologists can measure the chemical composition of a garnet porphyroblast in the laboratory and plot the corresponding pressure and temperature values as a P-T path (see Practicing Geology Exercise).

P-T paths have two segments: a *prograde* segment that indicates increasing pressure and temperature, and a *retrograde* segment that indicates decreasing pressure and temperature. The P-T paths of some rock assemblages that form at convergent boundaries are shown in Figure 7.13.

Ocean-Continent Convergence

A distinct metamorphic assemblage forms when oceanic lithosphere is subducted beneath a plate carrying a continent on its leading edge (Figure 7.13a). Thick sediments eroded from the continent rapidly fill the deep-sea trench that forms a flexural basin at the subduction zone. As it descends, the oceanic lithosphere stuffs the region below the inner wall of the trench (the wall closer to the continent) with these sediments, as well as with shreds of ophiolite suites scraped off the descending plate. The result is a chaotic mix known as a **mélange** (French for "mixture").

Assemblages of this type, located in the *forearc* region of a subduction zone—the area between the deep-sea trench and the volcanic mountain belt—are enormously complex and variable. The rocks formed there are all highly folded, intricately faulted, and metamorphosed (Figure 7.14). They are difficult to map in detail, but are recognizable by their distinctive combination of minerals and structural features.

Subduction-Related Metamorphism Blueschist—the metamorphic rock type whose minerals indicate that they were produced under high pressures but at relatively low temperatures (see Figure 7.9b)—forms from mélange in the forearc region of a subduction zone. Here, materials may be carried rapidly into the subduction zone to depths

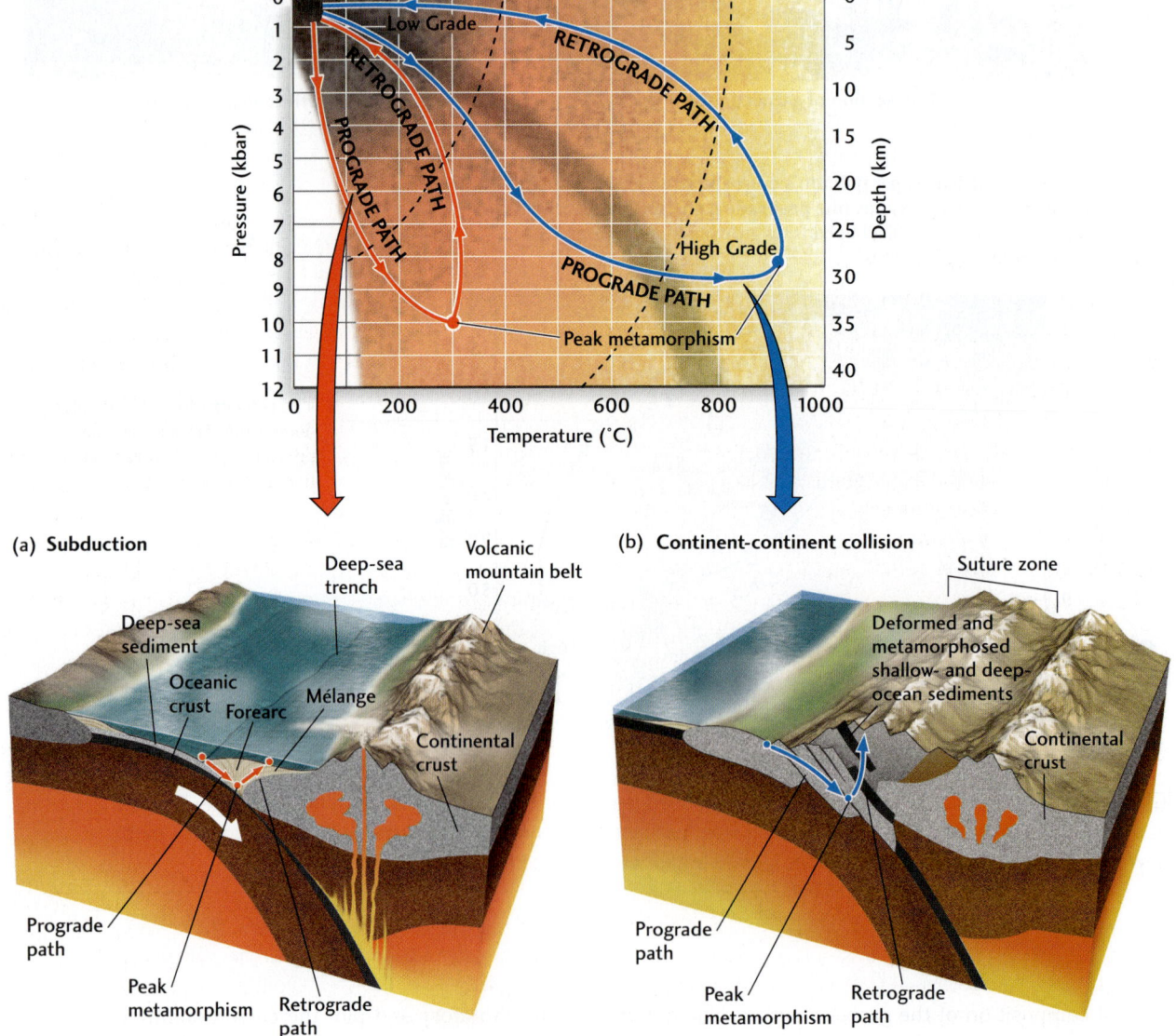

FIGURE 7.13 P-T paths indicate the trajectory of rocks during metamorphism. (a) Metamorphism of mélange at an ocean-continent convergence zone. (b) Metamorphism at a continent-continent convergence zone. The different P-T paths of rocks formed in different plate tectonic settings indicate differences in geothermal gradients. Rocks transported to similar depths—and pressures—beneath mountain belts become much hotter than rocks transported to an equivalent depth by subduction.

FIGURE 7.14 Mélange is a kind of breccia composed of rock fragments formed by churning in subduction zones. [John Platt.]

as great as 30 km. The cool subducting slab moves downward so quickly that there is little time for it to heat up, but pressure on the slab increases rapidly.

Eventually, as part of the subduction process, the material rises back to the surface. This exhumation results from two forces: buoyancy and circulation. Imagine trying to push a basketball below the surface of a swimming pool. The air-filled basketball has a lower density than the surrounding water, so it tends to rise back to the surface. In a similar way, the subducted metamorphic rocks are driven upward by their inherent buoyancy relative to the denser crust that surrounds them. But what "pushes" the material down to begin with? A natural circulation pattern is set up in the subduction zone. You can think of a subduction zone as an eggbeater. As the eggbeater rotates, it moves the froth in a circular direction. What moves in one direction eventually moves in the opposite direction because of the circular motion. In an analogous way, the sinking slab in a subduction zone sets up a circular motion of material above the slab, first pulling it down to great depths, then returning it to the surface.

Figure 7.13a shows the typical P-T path of rocks subjected to blueschist-grade metamorphism during subduction and exhumation. Note that the P-T path forms a loop in this diagram. If we compare the graph in Figure 7.13a with the metamorphic facies diagram in Figure 7.11, we can see that the prograde segment of the path represents subduction, as shown by a rapid increase in pressure and only a relatively small increase in temperature. During exhumation, the path loops back around because temperature is still slowly increasing, but now pressure is rapidly decreasing. The retrograde segment of the P-T path represents the exhumation process described above.

Evidence of Ancient Ocean-Continent Convergence The essential elements of these subduction-related rock assemblages have been found at many places in the geologic record, particularly around the Pacific Ocean basin. One can see mélange in the Franciscan formation of the California Coast Ranges and in the parallel volcanic mountain belt in the Sierra Nevada to the east. These rocks mark the Mesozoic collision between the North American Plate and the Farallon Plate, which has nearly disappeared by seduction. The location of mélange to the west and a volcanic mountain belt to the east shows that the Farallon Plate to the west was the subducted one. Analysis of the P-T paths of metamorphic minerals in the blueschist-grade Franciscan mélange reveals a loop similar to that shown in Figure 7.13a, indicating a rapid increase in pressure, which is characteristic of subduction.

Continent-Continent Collision

Because continental crust is buoyant, when a continent collides with another continent, both continents resist subduction and stay afloat on the mantle. As a result, a wide zone of intense deformation develops at the convergent boundary where the continents grind together. The remnant of such a boundary left behind in the geologic record is called a **suture**. The intense deformation results in a much-thickened continental crust in the collision zone, often producing high mountains. Ophiolite suites are often found near the suture.

As the lithosphere thickens, the deep parts of the continental crust heat up and undergo varying grades of metamorphism. In still deeper zones, melting may begin at the same time, forming magma chambers deep within the core of the mountain range. In this way, a complex mixture

PRACTICING GEOLOGY EXERCISE

How Do We Read Geologic History in Crystals?

What can a tiny crystal of garnet tell us about the history of the place where it was found? Knowing the plate tectonic setting in which a rock sample was formed tells us what other kinds of minerals might be found there. Geologists use variations in the chemical composition of garnet porphyroblasts to deduce the relative rates at which the rocks containing them were buried and then exhumed. These rates, in turn, reflect particular plate tectonic settings, as shown in Figure 7.13.

The chemical composition of a garnet porphyroblast generally varies progressively from its center to its edges (see Figure 7.12). This progression gives us a sense of change in pressure or temperature as a function of time: The center of the crystal records earlier conditions, and the edge of the crystal records later conditions. Changes in the calcium content of garnet track changes in pressure, whereas its iron content is more sensitive to temperature changes. We have noted that the increase in pressure over a given range of temperature values is much higher during subduction at ocean-continent convergence zones than it is during mountain building at continent-continent convergence zones. Conversely, a rock heated by an igneous intrusion experiences an increase in temperature, but little change in pressure.

By analyzing the chemical composition of garnet crystals, we can distinguish between these different metamorphic processes. To do so, we can compare changes in the abundance of the element of interest (calcium or iron) with the sum of changes in the abundances of all the elements that can vary in garnet: calcium (Ca), iron (Fe), magnesium (Mg), and manganese (Mn). Calculating the actual changes in pressure and temperature that a rock experienced during metamorphism requires

---

of metamorphic and igneous rock forms the core of the mountain belt. Millions of years afterward, when erosion has stripped off the surface layers of the mountains, their cores, containing schists, gneisses, and other metamorphic rocks, are exposed, providing a rock record of the metamorphic processes that formed them.

P-T paths of metamorphic rocks produced by continent-continent collision have a different shape from those of rocks produced by subduction alone. Continent-continent collision generates higher temperatures than subduction; therefore, as rock is pushed to greater depths, the temperature that corresponds to a given pressure will be higher (Figure 7.13b). The P-T path begins at the same place as the path for subduction, but shows a more rapid increase in temperature as greater pressures and depths are reached. Geologists generally interpret the prograde segment of a P-T path with this shape as indicating the burial of rocks beneath high mountains. The retrograde segment represents uplift and exhumation of the buried rocks during the collapse of mountains, either by erosion or by post-collision stretching and thinning of the continental crust.

The prime example of a continent-continent collision zone is the Himalaya, which began to form some 50 million years ago when the Indian continent collided with the Asian continent. That collision continues today: India is moving into Asia at a rate of a few centimeters per year, and the mountain building is still going on, together with faulting and very rapid erosion by rivers and glaciers.

Exhumation: A Link Between the Plate Tectonic and Climate Systems

Forty years ago, plate tectonic theory provided a ready explanation for how metamorphic rocks could be produced by seafloor spreading, subduction, and continent-continent collision. By the mid-1980s, the study of P-T paths provided a clearer picture of the specific tectonic mechanisms involved in the deep burial and metamorphism of rocks. At the same time, it surprised geologists by providing an equally clear picture of the subsequent, and often very rapid, uplift and exhumation of these deeply buried rocks. Since the time of this discovery, geologists have been searching for exclusively tectonic mechanisms that could bring these rocks back to Earth's surface so quickly.

One popular idea is that mountains, having been built to great elevations during collisional crustal thickening, suddenly fail by gravitational collapse. The old saying "what goes up must come down" applies here, but with surprisingly fast results—so fast, in fact, that some geologists don't believe gravity is the only important mechanism involved. Other forces must also be at work.

As we will learn in Chapter 16, geologists who study landscapes have discovered that glaciers and streams in tectonically active mountainous regions can produce extremely high erosion rates. Over the past decade, these geologists have presented a new hypothesis that links rapid rates of uplift and exhumation to rapid erosion rates.

additional data, including the composition of the complete mineral assemblage. Even without those details, however, it is possible to make some rough estimates.

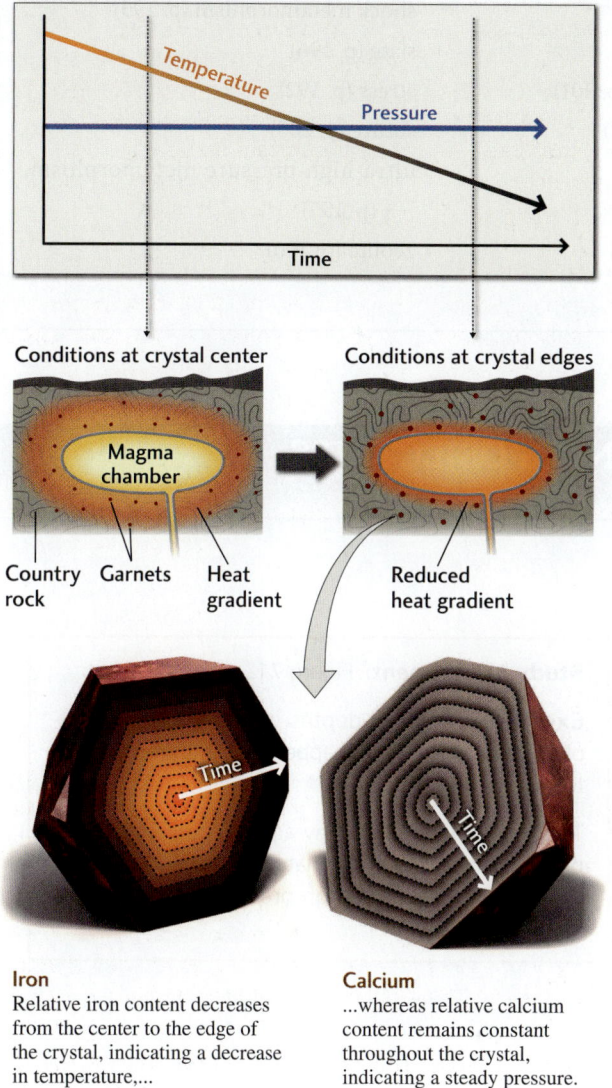

Iron
Relative iron content decreases from the center to the edge of the crystal, indicating a decrease in temperature,...

Calcium
...whereas relative calcium content remains constant throughout the crystal, indicating a steady pressure.

The following data were obtained by measuring the number of atoms of four elements in the center and at the edge of a garnet porphyroblast:

Element	Abundance at center	Abundance at edge
Ca	0.30	0.30
Fe	2.25	1.98
Mg	0.20	0.52
Mn	0.25	0.20

We first calculate the relative abundances of calcium and iron at the center of the crystal using the following ratios:

$$\left(\frac{Ca}{Ca+Fe+Mg+Mn}\right)_{center} = \frac{0.30}{0.30+1.98+0.52+0.20} = 0.10$$

$$\left(\frac{Fe}{Ca+Fe+Mg+Mn}\right)_{center} = \frac{2.25}{0.30+1.98+0.52+0.20} = 0.75$$

Then we do the same for the relative abundances of calcium and iron at the edge:

$$\left(\frac{Ca}{Ca+Fe+Mg+Mn}\right)_{edge} = \frac{0.30}{0.30+1.98+0.52+0.20} = 0.10$$

$$\left(\frac{Fe}{Ca+Fe+Mg+Mn}\right)_{edge} = \frac{1.98}{0.30+1.98+0.52+0.20} = 0.66$$

Based on these data, what can you say about the metamorphic event that caused this garnet crystal to grow? Was the rock carried down into a subduction zone, or was it sitting next to an igneous intrusion?

The decrease in iron content from 0.75 to 0.66 from center to edge is not associated with any change in calcium content. This observation indicates that metamorphism resulted mainly from a change in temperature with no change in pressure. These conditions are more consistent with metamorphism near an igneous intrusion than they are with subduction.

BONUS PROBLEM: Assume that the same calculations showed that iron content was constant, but calcium content changed significantly from center to edge. Would this pattern be consistent with transport of the rocks into a subduction zone?

The idea is that the climate system, not tectonic processes alone, drives the movement of rocks from the deep crust to the shallow crust through the process of rapid erosion. Thus, plate tectonic processes—which act through mountain building—and climate processes—which act through weathering and erosion—interact to control the flow of metamorphic rocks to Earth's surface. After decades of emphasis solely on plate tectonic explanations for regional and global geologic processes, it now seems that two apparently unrelated processes—metamorphism and erosion—are linked in an elegant way. As one geologist exclaimed: "Savor the irony should the metamorphic muscles that push mountains to the sky be driven by the pitter-patter of tiny raindrops."

KEY TERMS AND CONCEPTS

amphibolite (p. 198)
blueschist (p. 200)
burial metamorphism (p. 194)
contact metamorphism (p. 193)
eclogite (p. 195)
exhumation (p. 202)
foliated rock (p. 195)
foliation (p. 195)
gneiss (p. 197)
granoblastic rock (p. 197)
granulite (p. 198)
greenschist (p. 200)

greenstone (p. 198)
high-pressure metamorphism (p. 195)
hornfels (p. 198)
marble (p. 198)
mélange (p. 204)
metamorphic facies (p. 201)
metasomatism (p. 193)
migmatite (p. 197)
phyllite (p. 197)
porphyroblast (p. 198)
P-T path (p. 202)

quartzite (p. 198)
regional metamorphism (p. 193)
schist (p. 197)
seafloor metamorphism (p. 194)
shock metamorphism (p. 195)
slate (p. 196)
stress (p. 192)
suture (p. 205)
ultra-high-pressure metamorphism (p. 195)
zeolite (p. 200)

REVIEW OF LEARNING OBJECTIVES

7.1 Explain the causes of metamorphism.

Metamorphism is alteration in the mineralogy, texture, or chemical composition of solid rock. It is caused by increases in pressure and temperature and by reactions with chemical components introduced by hydrothermal solutions. As rocks are pushed deep within the crust by plate tectonic processes and exposed to increasing temperatures and pressures, the chemical components of the parent rock rearrange themselves into a new set of minerals that are stable under the new conditions. Metamorphic rocks that form at relatively low temperatures and pressures are referred to as low-grade metamorphic rocks; those that form at high temperatures and pressures are high-grade metamorphic rocks. Fluids also play an important role in metamorphism by altering a rock's mineralogy. Chemical components may be added to or removed from a rock during metamorphism, usually by hydrothermal solutions.

Study Assignment: Figure 7.1

Exercises: At what depths in Earth do metamorphic rocks form? What happens if temperatures get too high?

Thought Question: Why are there no metamorphic rocks formed under natural conditions of very low temperature and high pressure, as shown in Figure 7.1?

7.2 Describe the various types of metamorphism.

The three most common types of metamorphism are regional metamorphism, during which rocks over large areas are metamorphosed by high pressures and temperatures generated during mountain building; contact metamorphism, during which country rock close to an igneous intrusion is transformed by the heat of the intruding magma; and seafloor metamorphism, during which hot fluids percolate through and metamorphose oceanic crust. Less common types are burial metamorphism, during which deeply buried sedimentary rocks are altered by pressures and temperatures higher than those that result in diagenesis; high-pressure and ultra-high-pressure metamorphism, which occur at great depths, as when sediments are subducted; and shock metamorphism, which results from meteorite impacts.

> **Study Assignment:** Figure 7.4
>
> **Exercise:** Draw a sketch showing how seafloor metamorphism might take place.
>
> **Thought Question:** What type of metamorphism is related to igneous intrusions?

7.3 Summarize the various types of textural features displayed in metamorphic rocks.

The textural diversity of metamorphic rocks tells us something about the metamorphic processes that created them. Metamorphic rocks fall into two major textural classes: foliated rocks (displaying foliation, a pattern of parallel cleavage planes resulting from a preferred orientation of crystals) and granoblastic, or nonfoliated, rocks. The kinds of rocks produced depend on the composition of the parent rock and the grade of metamorphism. Regional metamorphism of shale leads to zones of foliated rock of progressively higher grade, from slate to phyllite, schist, gneiss, and finally migmatite. Among granoblastic rocks, marble is derived from the metamorphism of limestone, quartzite from quartz-rich sandstone, and greenstone from basalt. Hornfels is the product of contact metamorphism of fine-grained sedimentary rocks and other types of rock containing an abundance of silicate minerals. Regional metamorphism of mafic volcanic rocks progresses from zeolite facies to greenschist facies and then to amphibolite and granulite facies.

> **Study Assignments:** Figure 7.6, Table 7.1
>
> **Exercise:** What does preferred orientation refer to in a metamorphic rock? Think about how the alignment of minerals relates to metamorphic processes.
>
> **Thought Question:** How do the minerals that define foliation establish its metamorphic grade?

7.4 Discuss the ways metamorphic rocks reveal the conditions under which they were formed.

Zones of metamorphism can be mapped with isograds defined by the first appearance of an index mineral. The presence of an index mineral can indicate the temperature and pressure under which the rocks in the zone were formed. According to the concept of metamorphic facies, rocks of the same metamorphic grade may differ because of variations in the chemical composition of the parent rock, whereas rocks metamorphosed from the same parent rock may vary because they were subjected to different grades of metamorphism.

> **Study Assignment:** Figure 7.10
>
> **Exercise:** You have mapped an area of regional metamorphism, such as the region in Figure 7.9a, and have observed a series of metamorphic zones, marked by north-south isograds, running from sillimanite in the east to chlorite in the west. Were metamorphic temperatures higher in the east or in the west?
>
> **Thought Question:** How are isograds related to metamorphic facies?

7.5 Illustrate how metamorphic rocks are related to plate tectonic processes.

Different types of metamorphism are likely to occur in different plate tectonic settings, including continental interiors, divergent plate boundaries, convergent plate boundaries, and transform faults. During subduction and continent-continent collision at convergent plate boundaries, rocks and sediments are pushed to great depths in Earth's crust, where they are subjected to increasing pressures and temperatures that result in metamorphism. The shapes of metamorphic pressure-temperature ("P-T") paths provide insight into the plate tectonic settings where these rocks were metamorphosed. In the case of ocean-continent convergence, P-T paths indicate rapid subduction of rocks and sediments to environments with high pressures and relatively low temperatures. In continent-continent collision zones, rocks are pushed down to depths where pressures and temperatures are both high. In both settings, the P-T paths show that after the rocks experience the maximum pressures and temperatures, they are returned to shallow depths. This process of exhumation may be driven by weathering and erosion at Earth's surface as well as by plate tectonic processes.

Study Assignment: Figure 7.13

Exercise: Draw a P-T path for an amphibolite-grade metamorphic rock exposed near the top of Mount Everest.

Thought Question: What controls exhumation of metamorphic rocks?

VISUAL LITERACY EXERCISE

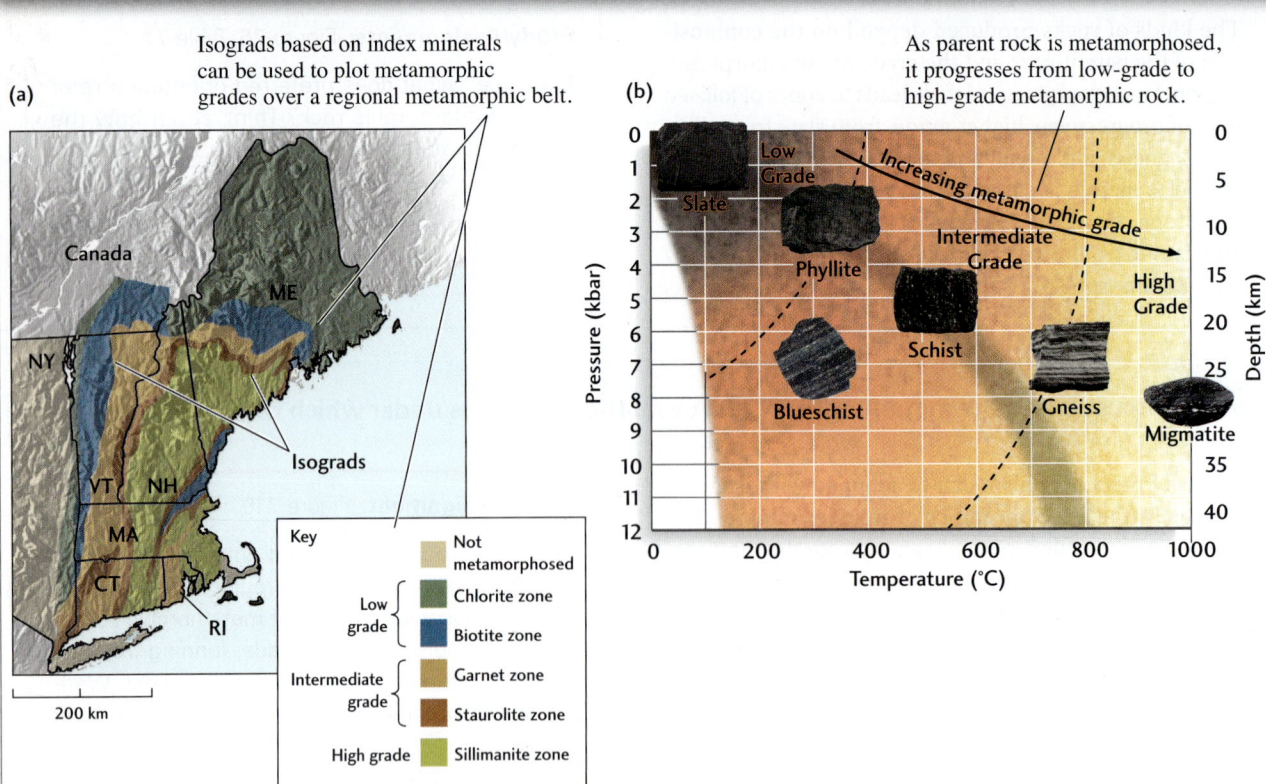

FIGURE 7.9 Index minerals define the different metamorphic zones within a belt of regional metamorphism. (a) Map of New England, showing metamorphic zones based on index minerals found in rocks metamorphosed from shale. (b) Rocks produced by the metamorphism of shale at different temperatures and pressures. [*slate, phyllite, schist, gneiss:* John Grotzinger/Ramón Rivera-Moret/Harvard Mineralogical Museum; *blueschist:* courtesy of Mark Cloos; *migmatite:* courtesy of Kip Hodges.]

1. **What sequence of rock types defines increasing metamorphic grade?**
   a. Migmatite → Gneiss → Schist → Phyllite → Slate
   b. Slate → Phyllite → Schist → Gneiss → Migmatite
   c. Phyllite → Gneiss → Slate → Migmatite → Slate
   d. Migmatite → Gneiss → Schist → Phyllite → Slate

2. **What minerals define intermediate metamorphic grade?**
   a. Sillimanite
   b. Garnet/Staurolite
   c. Chlorite/Biotite
   d. Not metamorphosed

3. **What New England state has the largest amount of low-grade metamorphic rock?**
   a. New Hampshire
   b. Maine
   c. Massachusetts
   d. New York

4. **At what temperature does gneiss form?**
   a. 100 degrees
   b. 400 degrees
   c. 1000 degrees
   d. 800 degrees

5. **Why do geologists map isograds?**
   a. To plot metamorphic grade over a regional metamorphic belt
   b. To trace the path of volcanic eruptions
   c. To search for ancient reefs
   d. To delineate magnetic reversals

Deformation: Modification of Rocks by Folding and Fracturing

Plate Tectonic Forces	214
Mapping Geologic Structure	214
How Rocks Deform	217
Basic Deformation Structures	219
Styles of Continental Deformation	228
Unraveling Geologic History	231

Learning Objectives

Chapter 8 describes how rocks can be tilted, bent, and fractured into the patterns we see at the land surface. The chapter concentrates on the processes of folding and faulting that deform continental rocks near plate boundaries. You will learn how geologists collect and interpret field observations to make geologic maps, and then explore what those maps can tell us about the history of deformation, as well as the tectonic forces that caused it. After you have studied this chapter, you should be able to:

8.1 Compare the different types of tectonic forces at plate boundaries that deform rocks.

8.2 Describe how maps and diagrams are used to represent geologic structures.

8.3 Explain how laboratory experiments can help us understand the way rocks deform.

8.4 Summarize the basic deformation structures observed in the field and the types of forces that produce these deformation structures.

8.5 Define the main styles of continental deformation.

8.6 Describe how we reconstruct the geologic history of a region.

Aerial view of the Tree River Folds in northwestern Canada. These large-scale folds have a wavelength of about 1 km.
[Courtesy John Grotzinger.]

When rocks are caught up in plate boundaries, their textures and mineralogy can be transformed by metamorphism, as we saw in Chapter 7. Among the processes that cause regional metamorphism in continents, the most important is *deformation:* the modification of rocks by squeezing, stretching, folding, and faulting. On the scale of individual rocks, deformation can change granites into gneisses and sediments into schists. On a large scale, deformation can distort layers of sediments, which were deposited almost horizontally, into crazy-looking patterns.

Early geologists understood that most sedimentary rocks had originally been deposited as soft horizontal layers at the bottom of the ocean and had hardened over time. But what forces could have acted on these rocks, which seemed so strong and rigid, to produce the patterns they observed? Why were particular patterns of deformation repeated time and time again throughout geologic history? The discovery of plate tectonics in the 1960s provided the answers.

Plate Tectonic Forces

Deformation is a general term that encompasses the folding, faulting, shearing, compression, and extension of rock by plate tectonic forces. The types of deformation we see exposed at Earth's surface are caused mainly by the horizontal movements of the lithospheric plates relative to one another. For this reason, the tectonic forces that deform rocks at plate boundaries are predominantly *horizontally directed* and depend on the direction of relative plate movement:

- **Tensional forces,** which stretch and pull rock formations apart, dominate at divergent boundaries, where plates move away from each other.
- **Compressive forces,** which squeeze and shorten rock formations, dominate at convergent boundaries, where plates move toward each other.
- **Shearing forces,** which shear two parts of a rock formation in opposite directions, dominate at transform-fault boundaries, where plates slide past each other.

If plates were perfectly rigid, the plate boundaries would be sharp lineations, and points on either side of those boundaries would move at the relative plate velocity. This idealization is often a good approximation in the oceans, where rift valleys at mid-ocean ridges, deep-sea trenches, and nearly vertical transform faults form narrow plate boundary zones, often just a few kilometers wide.

Within continents, however, the deformation caused by plate movements can be "smeared out" across a plate boundary zone hundreds or even thousands of kilometers wide. The continental crust does not behave rigidly within these broad zones, so rocks at the surface are deformed by folding and faulting. Folds in rocks are like folds in clothing. Just as cloth pushed together from opposite sides bunches up in folds, layers of rock slowly compressed by tectonic forces in the crust can be pushed into folds (**Figure 8.1a**). Tectonic forces can also cause a rock formation to break and slip on both sides of a fracture, forming a fault (Figure 8.1b). When such a break occurs suddenly, the result is an earthquake. Active zones of continental deformation are marked by frequent earthquakes.

Geologic folds and faults can range in size from centimeters to meters (as in Figure 8.1) to tens of kilometers or more. Many mountain ranges are actually a series of large folds and faults that have been weathered and eroded. From the geologic record of deformation laid out on Earth's surface, geologists can deduce the directions of movement at ancient plate boundaries and reconstruct the tectonic history of the continental crust.

Mapping Geologic Structure

Faults and folds are examples of the basic features geologists observe and map to reconstruct crustal deformation. To better understand this process, we need information about the geometry of faults and folds. The best place to find this information is at an *outcrop,* where the solid rock that underlies the ground surface—the *bedrock*—is exposed (not obscured by vegetation, soil, or loose boulders). At an outcrop, geologists can identify distinct **formations:** groups of rock layers that can be identified throughout a region by their physical properties. Some formations consist of a single rock type, such as limestone. Others are made up of thin, interlayered beds of different kinds of rock, such as sandstone and shale. However they vary, each formation comprises a distinct set of rock layers that can be recognized and mapped as a unit.

Figure 8.1a shows an outcrop in which the folding of sedimentary beds is clearly visible. Often, however, folded rocks are only partly exposed in an outcrop and can be seen only as inclined layers (**Figure 8.2**). The orientation of those layers is an important clue we can use to piece together a picture of the overall geologic structure. Two measurements describe the orientation of a rock layer exposed at an outcrop: the strike and the dip of the layer's surface.

Measuring Strike and Dip

The **strike** is the compass direction of a rock layer where it intersects with a horizontal surface. The **dip,** which is measured at right angles to the strike, is simply the amount of tilting—the angle at which the rock layer inclines from the

(a)

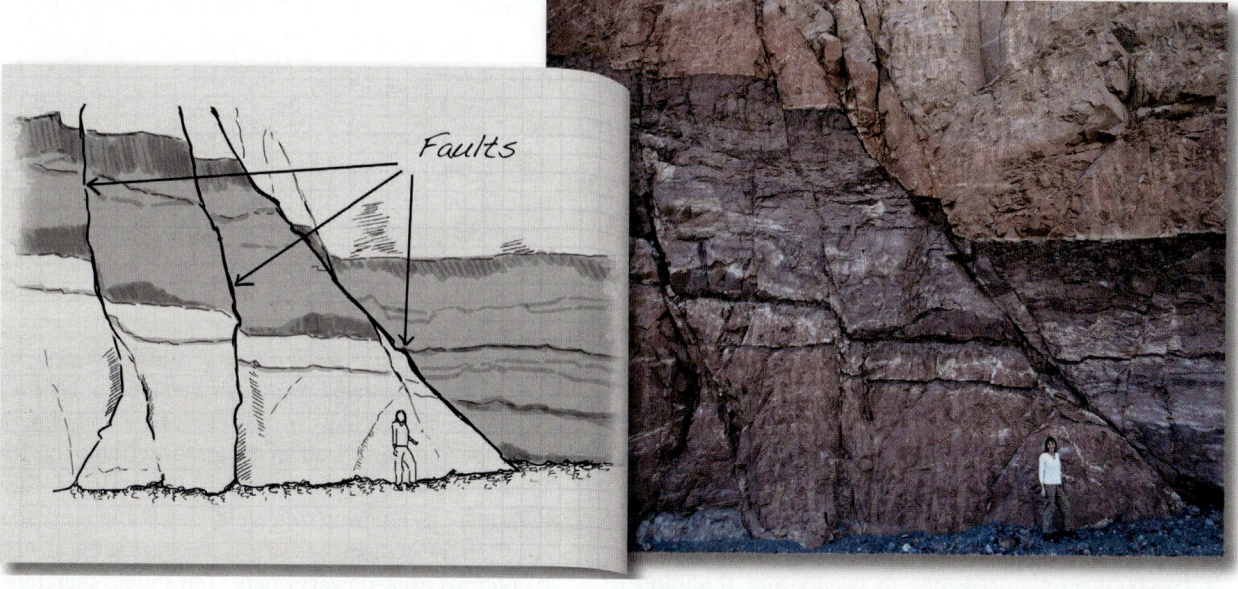

(b)

FIGURE 8.1 Rocks subjected to tectonic forces are deformed by folding and faulting. (a) An outcrop of originally horizontal rock layers has been bent into folds by compressive tectonic forces. (b) An outcrop of once-continuous rock layers has been displaced along small faults by tensional tectonic forces. [(a) Tony Waltham; (b) Marli Miller.]

FIGURE 8.2 Dipping limestone and shale beds on the coast of Somerset County, England. The children are walking along the strike of the beds. The beds dip to the left at an angle of about 15°. [Chris Pellant.]

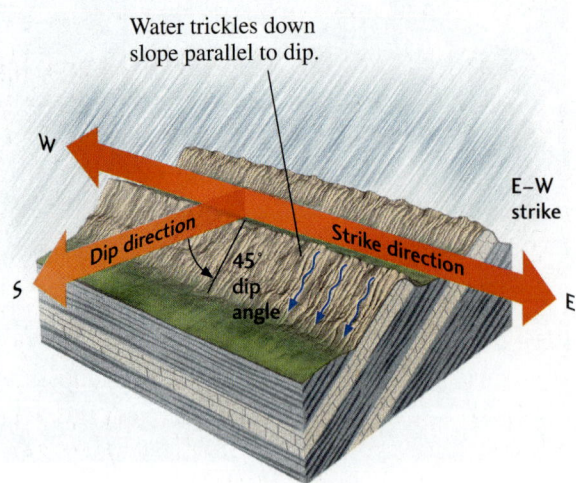

FIGURE 8.3 The strike and dip of a rock layer define its orientation at a particular place. The strike is the compass direction of a rock layer along the line of its intersection with a horizontal surface. The dip is the angle and direction of the steepest descent of a rock layer from the horizontal, measured at right angles to the strike. Here, the strike is east-west and the dip is 45° to the south.

horizontal. **Figure 8.3** shows how strike and dip are measured in the field. A geologist might describe the outcrop in this figure as "a bed of coarse-grained sandstone striking east-west and dipping 45° south."

Geologic Maps

Geologic maps are two-dimensional representations of the rock formations exposed at Earth's surface (**Figure 8.4**). To create a geologic map, a geologist must choose an appropriate *scale*—the ratio of distance on the map to true surface distance. A common scale for geologic field mapping is 1:24,000 (pronounced "one to twenty-four thousand"), which means that 1 inch on the map corresponds to 24,000 inches (2000 feet) on Earth's surface. To depict the geology of an entire state, a geologist would choose a smaller scale: say, 1:1,000,000, where 1 cm represents 10 km and 1 inch almost 16 miles. The smaller the scale, the less detail can be depicted on the map.

Geologists keep track of different rock formations by assigning each formation a particular color on the map, usually keyed to the rock's type and age (see Figure 8.4). Many different rock formations may be exposed in highly deformed regions, so geologic maps can be very colorful!

Softer rocks, such as mudstones and other poorly consolidated sediments, are more easily eroded than harder rocks, such as limestones, sandstones, or metamorphic rocks. Consequently, rock types can exert a strong influence on the topography of the land surface and the exposure of rock formations (see Figure 8.4). The important relationships between geology and topography can be made clear by plotting the contours of the land surface on a geologic map.

Because geologic maps can represent such a huge amount of information, they have been called "textbooks on a piece of paper." To convey this information more concisely, geologic maps are annotated with special symbols that indicate the local strike and dip of rock formations and with special types of lines that mark faults and other significant features. For instance, the strike and dip of rock formations are indicated on a geologic map by T-like symbols:

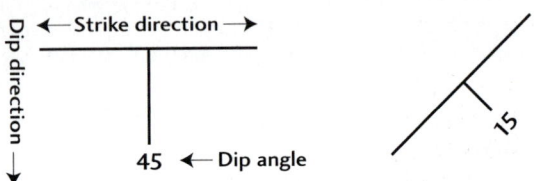

The top of the T indicates the strike direction, the shank of the T indicates the dip direction, and the number gives the dip angle in degrees. On a map where north is up, the symbol on the left thus describes the sandstone bed in Figure 8.3, which has an east-west strike and a dip of 45° to the south. The one on the right describes a formation that strikes northeast-southwest and dips to the southeast at an angle of 15°, such as the beds in Figure 8.2.

Of course, not every detail of surface geology can be represented on a map, so geologists must simplify the structures they see, perhaps by representing a complex zone of faulting as a single fault or by ignoring folds too small to show at the scale they have chosen. They may also "dust off" their maps by ignoring thin layers of soil and loose rock that cover up the geologic structure, portraying the structure as if outcrops existed everywhere. You should therefore think of a geologic map as a simplified *scientific model* of the surface geology.

Geologic Cross Sections

Once a region has been mapped, the two-dimensional geologic map must be interpreted in terms of the underlying three-dimensional geologic structure. How can the shapes of the rock layers be reconstructed, even when erosion has removed parts of a formation? The process is like putting together a three-dimensional jigsaw puzzle with some of the pieces missing. Common sense and intuition play important roles, as do basic geologic principles.

To piece together the puzzle, geologists construct **geologic cross sections**—diagrams showing the features that would be visible if vertical slices were made through part of the crust. Some small cross sections can actually be seen in the vertical faces of cliffs, quarries, and

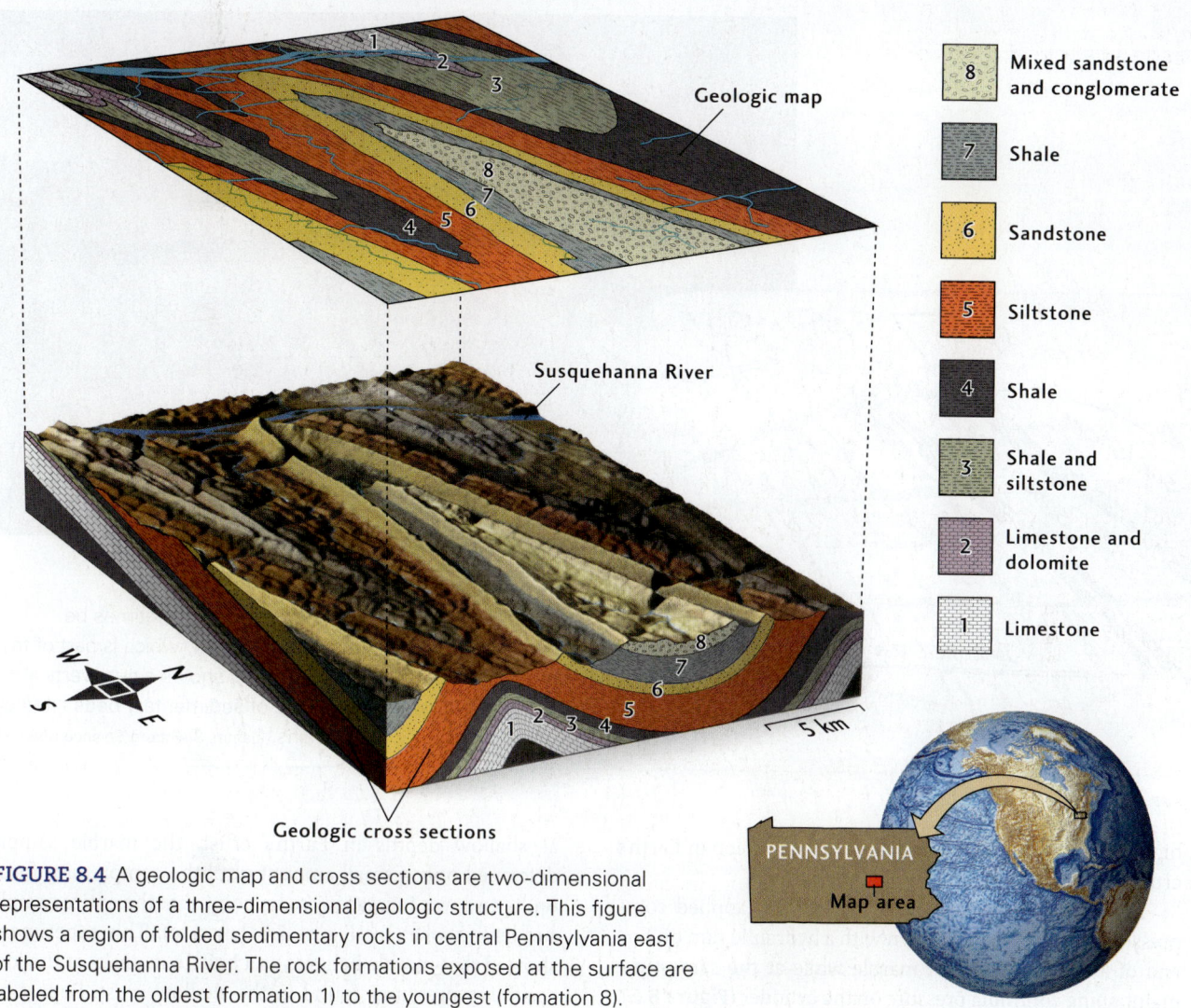

FIGURE 8.4 A geologic map and cross sections are two-dimensional representations of a three-dimensional geologic structure. This figure shows a region of folded sedimentary rocks in central Pennsylvania east of the Susquehanna River. The rock formations exposed at the surface are labeled from the oldest (formation 1) to the youngest (formation 8).

road cuts (Figure 8.5). Cross sections spanning much larger areas can be constructed from the information on a geologic map, including the strikes and dips observed at outcrops. The accuracy of cross sections based on surface mapping can be improved by drilling boreholes to collect rock samples as well as by seismic imaging. But drilling and seismic imaging are expensive, so data collected by these methods are usually available only for areas that have been explored for oil, water, or other valuable natural resources.

Figure 8.4 shows a geologic map of an area where originally horizontal beds of sedimentary rock were bent into a series of folds and eroded into a set of zigzagging ridges and valleys. We will explore some of the geologic relationships seen in this map later in this chapter. But first we will investigate the basic processes by which rocks deform.

How Rocks Deform

Rocks deform in response to the tectonic forces acting on them. Whether they will respond to tectonic forces by folding, faulting, or some combination of the two depends on the orientation of the forces, the rock type, and the physical conditions (such as temperature and pressure) during deformation.

Brittle and Ductile Behavior of Rocks in the Laboratory

In the mid-1900s, geologists began to explore the forces of deformation by using powerful hydraulic rams to bend and break small samples of rock. Engineers had invented such machines to measure the strength of concrete and other building materials, but geologists modified them to discover how rocks deform at pressures and temperatures

CHAPTER 8 Deformation: Modification of Rocks by Folding and Fracturing

FIGURE 8.5 Geologic cross sections can sometimes be observed directly in the field. This road cut, which is part of the Lariat Loop Scenic Byway in Colorado, shows a near-vertical cross section through a sequence of sedimentary beds tilted by the uplift of the Rocky Mountains. [James Steinberg/Science Source.]

high enough to simulate physical conditions deep in Earth's crust.

In one such experiment, the researchers applied compressive force by pushing down with a hydraulic ram on one end of a small cylinder of marble while at the same time maintaining confining pressure on the cylinder (**Figure 8.6**). Under low confining pressures, equivalent to those found at shallow depths in Earth's crust, the marble sample deformed only a small amount until the compressive force on its end was increased to the point that the entire sample suddenly broke (see Figure 8.6, left side). This experiment showed that marble behaves as a **brittle** material at the low confining pressures found in the shallow crust. Repeating the experiment under high confining pressures, equivalent

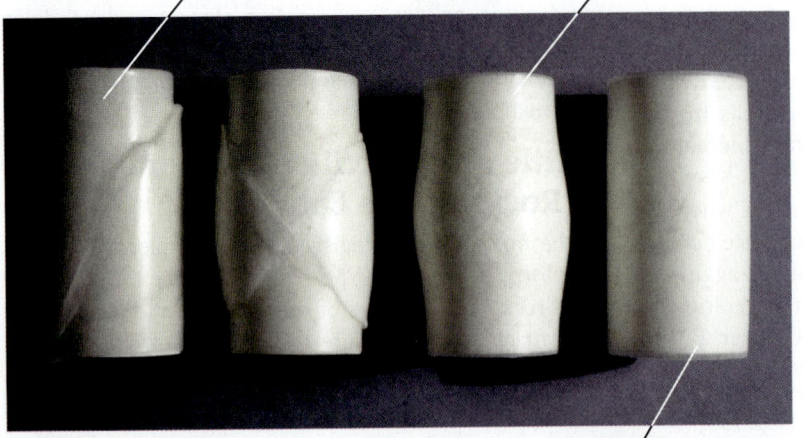

FIGURE 8.6 Results of laboratory experiments conducted to discover how rock—in this case, marble—is deformed by compressive forces. The marble samples are encased in translucent plastic jackets, which explains their shiny appearance. [Fred and Judith Chester/John Handin Rock Deformation Laboratory of the Center for Tectonophysics.]

to those that often accompany metamorphism, produced a different result: The marble sample slowly and steadily deformed into a shortened, bulging shape without fracturing (see Figure 8.6, right side). Marble thus behaves as a pliable, or **ductile,** material at the high confining pressures found deep in the crust.

Other experiments showed that when marble was heated to temperatures as high as those that accompany metamorphism, it acted as a ductile material at a lower confining pressure—just as heating wax changes it from a hard material that can break into a soft material that flows. The researchers concluded that the particular marble they were working with would deform by faulting at depths shallower than a few kilometers, but would deform by folding at the greater crustal depths where metamorphism generally occurs.

Brittle and Ductile Behavior of Rocks in Earth's Crust

Natural conditions in Earth's crust cannot be reproduced exactly in the laboratory. Tectonic forces are applied over millions of years, whereas laboratory experiments are rarely conducted over more than a few hours, or at most a few weeks. Nevertheless, the results of laboratory experiments can help us interpret what we see in the field. Geologists keep the following points in mind when mapping crustal folds and faults:

- The same rock can be brittle at shallow depths (where temperatures and pressures are relatively low) and ductile deep in the crust (where temperatures and pressures are higher). Metamorphism is usually accompanied by ductile deformation.

- Rock type affects deformation. In particular, the hard igneous and metamorphic rocks that form the crystalline *basement* of a continent (the crust beneath layers of sediments) often behave as brittle materials, fracturing along faults during deformation, while softer sedimentary rocks that overlie them often behave as ductile materials, folding gradually.

- A rock formation that would behave as a ductile material if deformed slowly may behave as a brittle material if deformed more rapidly. (Think of Silly Putty, which flows as a ductile clay when you squeeze it slowly, but breaks into pieces when you pull it apart very quickly.)

- Rocks break more easily when subjected to tensional (pulling and stretching) forces than when subjected to compressive forces. Sedimentary rock formations that will deform by folding during compression will often break along faults when subjected to tension.

Basic Deformation Structures

Geologists use the simple geometric concepts and measurements described earlier in this chapter to classify features such as faults and folds into different types of deformation structures.

Faults

A **fault** is a fracture that displaces the rock on either side of it. We can measure the orientation of the fracture surface, or *fault surface,* by its strike and dip, just as we do for other geologic surfaces (see Figure 8.3). The movement of the block of rock on one side of the fault with respect to that on the other side can be described by a *slip direction* and by the total displacement, or offset. For small faults, such as the ones pictured in Figure 8.1b, the offset might be only a couple of meters, whereas the offset along a major transform fault, such as the San Andreas fault, can amount to hundreds of kilometers (Figure 8.7).

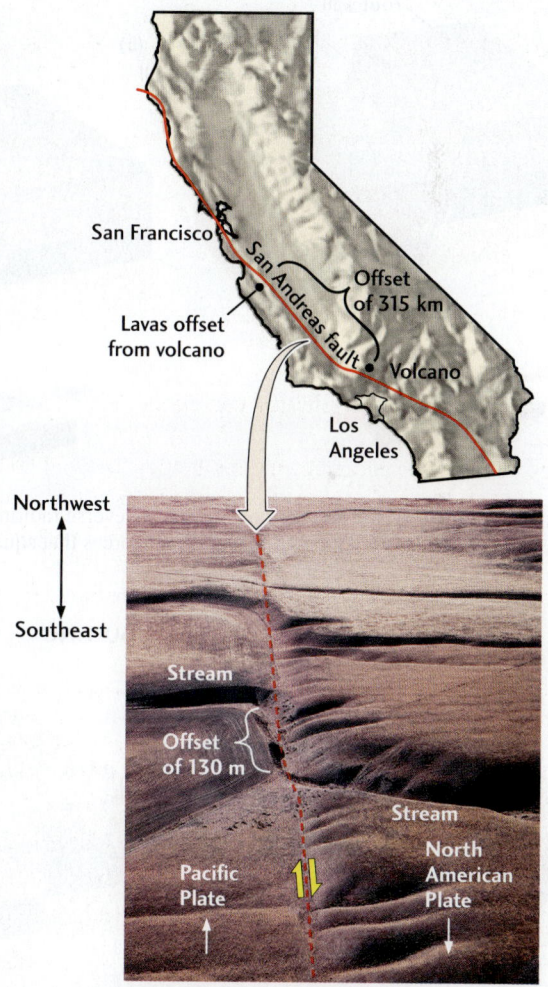

FIGURE 8.7 View of the San Andreas fault, showing the northwestward movement of the Pacific Plate with respect to the North American Plate. The map shows a formation of volcanic rocks 23 million years old that has been displaced by 315 km. The fault runs from top to bottom (dashed line) near the middle of the photograph. Note the offset of the stream (Wallace Creek) by 130 m as it crosses the fault.
[University of Washington Libraries, Special Collections, John Shelton Collection, KCN7-23.]

Rocks on either side of a fault cannot interpenetrate one another, and at high pressures below the surface, they cannot open up, so the slip direction during faulting must be parallel to the fault surface. Faults can therefore be classified by their slip direction along this surface (Figure 8.8). A **dip-slip fault** is one on which there has been relative movement of blocks of rock up or down the dip of the fault plane. A **strike-slip fault** is a fault on which the movement of blocks has been horizontal, parallel to the strike of the fault plane. When blocks of rock move along the strike and simultaneously up or down the dip, the result is an *oblique-slip fault*. Dip-slip faults are caused by compressive or tensional forces, whereas strike-slip faults are the work of shearing forces. An oblique-slip fault results from shearing in combination with either compression or tension.

These fault types require further classification because the movement of rock can be up or down, or right or left. To describe these movements, geologists borrow some terminology used by miners, calling the block of rock above a dipping fault plane the **hanging wall** and the block of rock below it the **foot wall**. A dip-slip fault is called a **normal fault** if the hanging wall moves downward relative

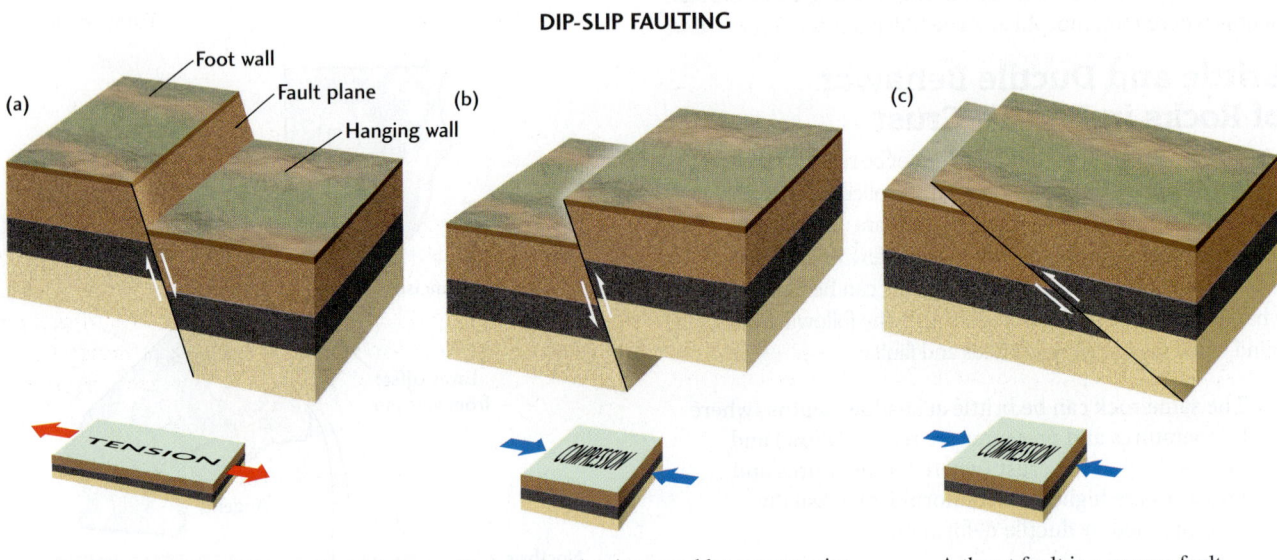

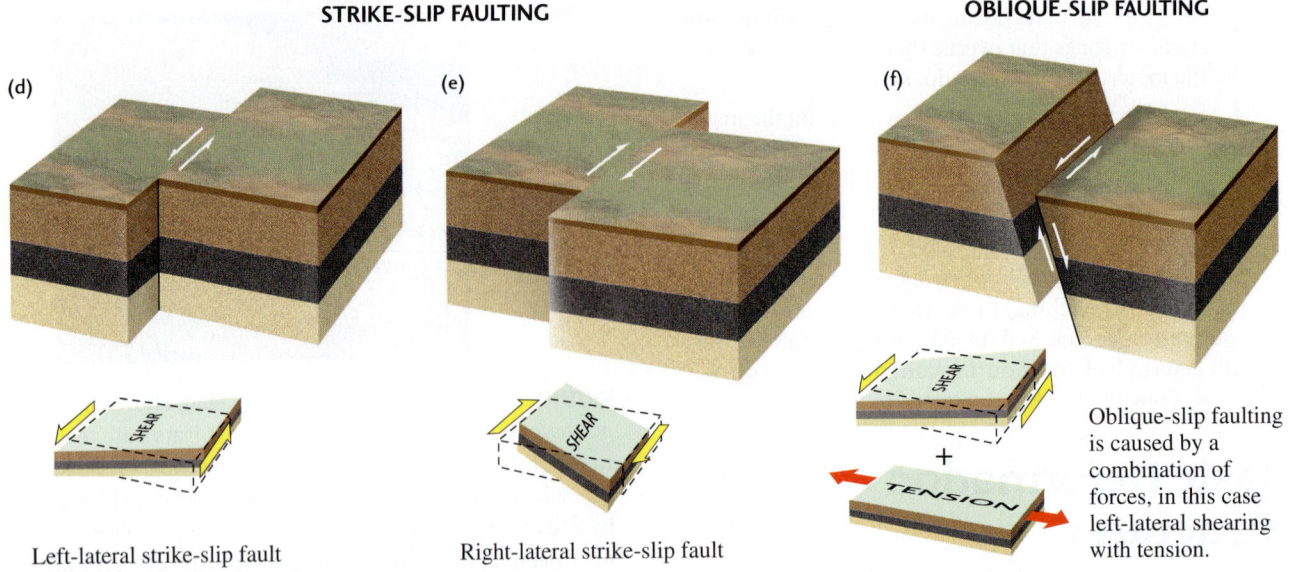

FIGURE 8.8 The orientation of tectonic forces determines the style of faulting. Dip-slip faulting (a–c) is caused by compressive or tensional forces. Strike-slip faulting (d, e) is caused by shearing forces. Oblique-slip faulting (f) is caused by a combination of shearing and compressive or tensional forces.

to the foot wall, extending the structure horizontally (Figure 8.8a). A dip-slip fault is called a *reverse fault* if the hanging wall moves upward relative to the foot wall, causing a shortening of the structure (Figure 8.8b)—the reverse of what geologists have (somewhat arbitrarily) chosen as "normal." A **thrust fault** is a low-angled reverse fault—that is, one with a dip of less than 45°, so that the movement is more horizontal than vertical (Figure 8.8c). When subjected to horizontal compression, brittle rocks of the continental crust usually break along thrust faults with dips of 30° or less, rather than along more steeply dipping reverse faults.

A strike-slip fault is a *left-lateral fault* if an observer on one side of the fault sees that the block on the opposite side has moved to the left (Figure 8.8d). It is a *right-lateral fault* if the block on the opposite side appears to have moved to the right (Figure 8.8e). As you can tell from the stream offset in Figure 8.7, the San Andreas fault is a right-lateral transform fault. Other faults show both strike-slip and dip-slip motions. These faults are known as **oblique-slip faults** (Figure 8.8f).

Geologists can recognize faults in the field in several ways. A fault may form a *scarp* (a cliff) that marks where the fault intersects the ground surface (**Figure 8.9**). If the offset has been large, as it is for transform faults such as the San Andreas, the rock formations facing each other across the fault may differ in type and age. When movements are smaller, offset features can be observed and measured. (As an exercise, try to match up the beds offset by the small faults in Figure 8.1b.) In establishing the time of faulting, geologists apply a simple rule: A fault must be younger than the youngest rocks it cuts (the rocks had to be there before they could break!) and older than the oldest undeformed formation that covers it.

On geologic maps, faults are represented by *fault traces:* lines that indicate the point where a fault intersects the ground surface. Normal faults are distinguished from thrust faults by the different types of "teeth" that annotate the fault trace:

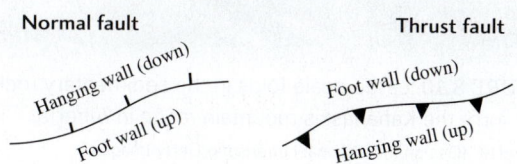

For both types of dip-slip faults, the teeth point toward the hanging wall. Examples of normal faults represented this way are shown in Figure 8.20; examples of thrust faults are shown in Figure 8.22. For strike-slip faults, the direction of movement, right-lateral or left-lateral, is indicated by a pair of arrows (yellow) that bracket the fault trace (see Figure 8.7).

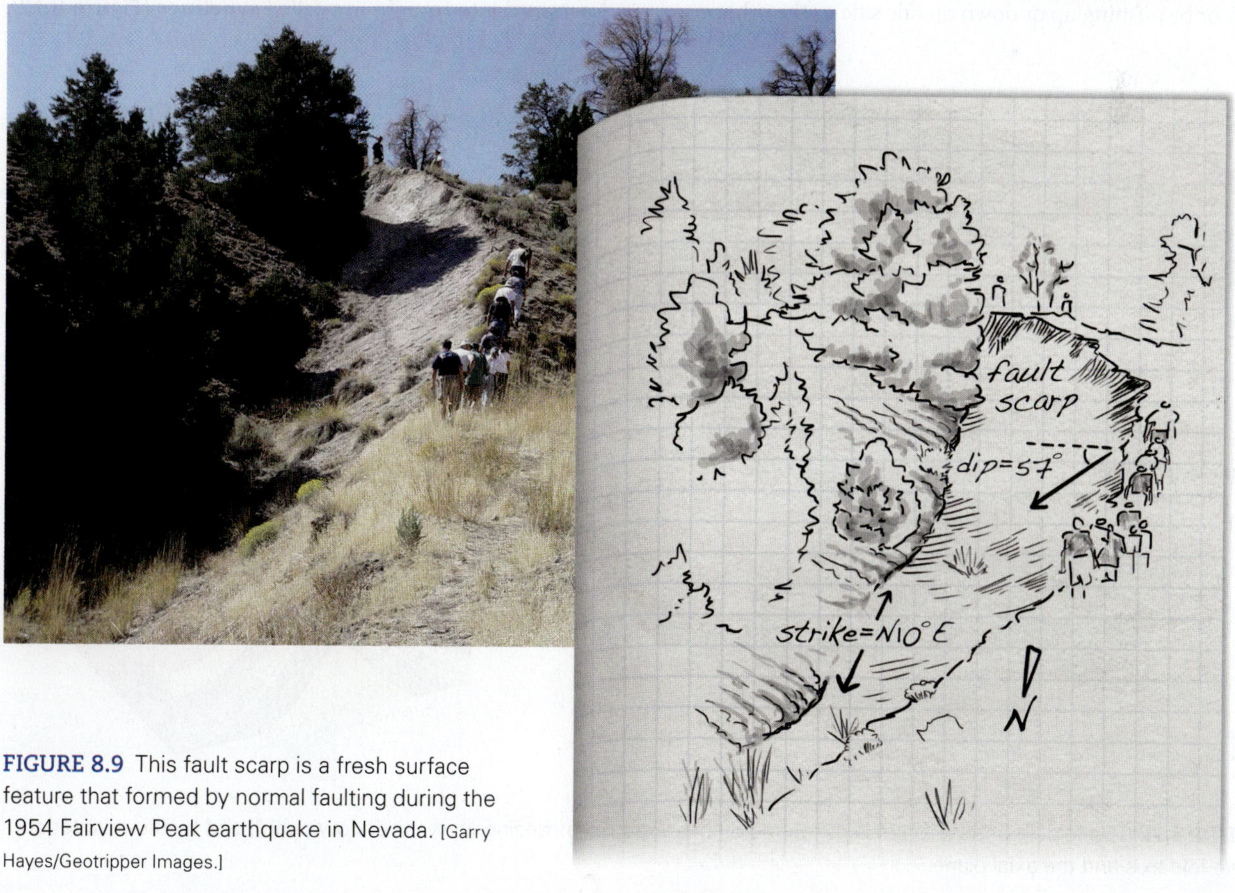

FIGURE 8.9 This fault scarp is a fresh surface feature that formed by normal faulting during the 1954 Fairview Peak earthquake in Nevada. [Garry Hayes/Geotripper Images.]

222　CHAPTER 8　Deformation: Modification of Rocks by Folding and Fracturing

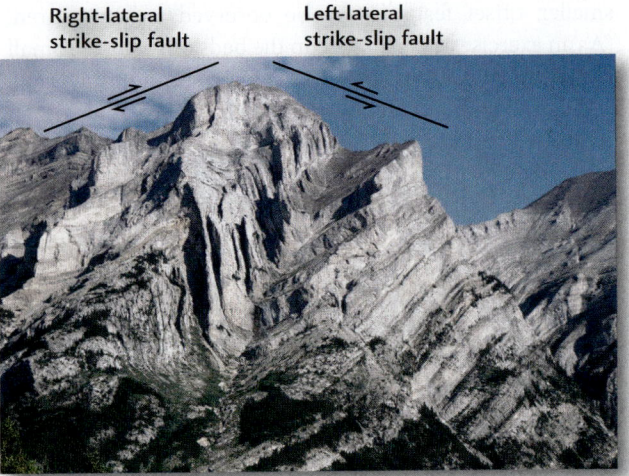

FIGURE 8.10 Large-scale folds in the sedimentary rocks that form the Kananaskis mountain range in Alberta, Canada. [Design Pics/Michael Interisano/Getty Images.]

FIGURE 8.11 Small-scale folds in sedimentary beds of anhydrite (light) and shale (dark) in West Texas. [John Grotzinger/Ramón Rivera-Moret/Harvard Mineralogical Museum.]

Folds

Folding is a common form of deformation observed in layered rocks (as in Figure 8.1a). **Folds** occur when an originally planar structure, such as a sedimentary bed, is bent into a curved structure. The bending can be produced by either horizontally or vertically directed forces in the crust, just as a piece of paper can be folded by pushing together its opposite edges or by pushing up or down on one side or the other.

Like faults, folds come in all sizes. In many mountain belts, majestic, sweeping folds can be traced over many kilometers (Figure 8.10). On a much smaller scale, very thin sedimentary beds can be crumpled into folds a few centimeters long (Figure 8.11). The bending can be gentle or severe, depending on the magnitude of the applied forces, the length of time over which they were applied, and the resistance of the rocks to deformation.

Folds in which layered rocks are bent upward into arches are called **anticlines;** those in which rocks are bent downward into troughs are called **synclines** (Figure 8.12).

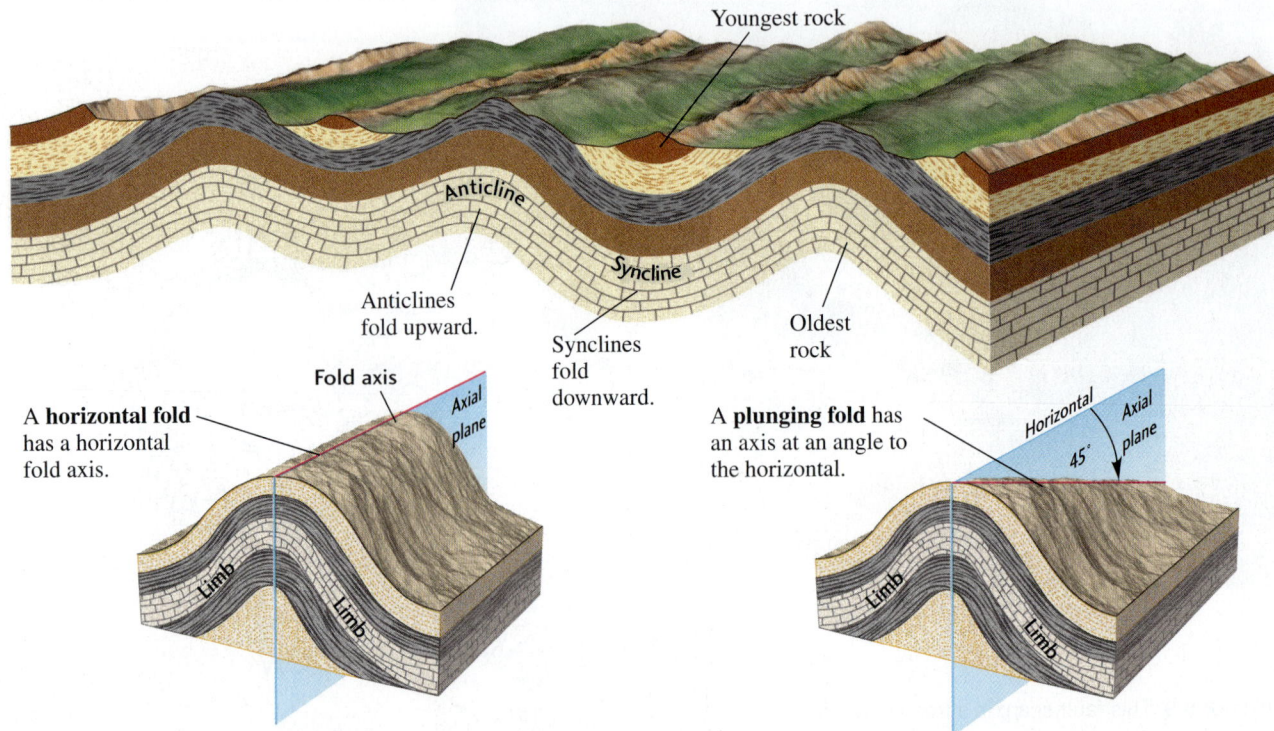

FIGURE 8.12 The folding of rock layers is described by the direction of folding (upward or downward) and by the orientation of the fold axis and the axial plane.

Basic Deformation Structures 223

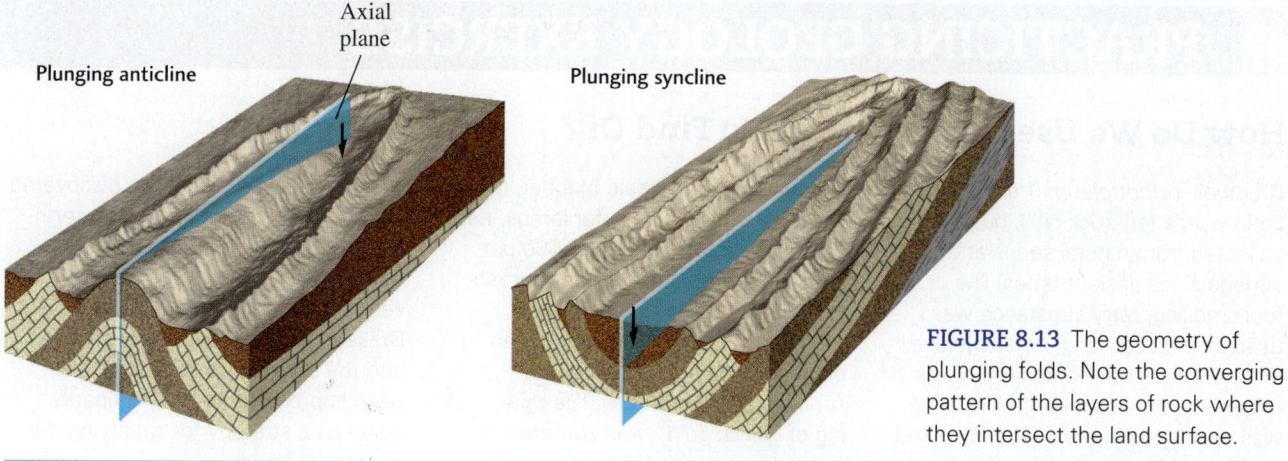

FIGURE 8.13 The geometry of plunging folds. Note the converging pattern of the layers of rock where they intersect the land surface.

The two sides of a fold are its *limbs*. The *axial plane* of a fold is an imaginary surface that divides the fold as symmetrically as possible, with one limb on either side of the plane. The line made by the lengthwise intersection of the axial plane with the rock layers is the *fold axis*. A symmetrical horizontal fold has a horizontal fold axis and a vertical axial plane with limbs dipping symmetrically away from the axis.

Folds rarely stay horizontal, however. Follow the axis of any fold in the field, and sooner or later the fold dies out or appears to plunge into the ground. If a fold's axis is not horizontal, it is called a *plunging fold*. **Figure 8.13** diagrams the geometry of plunging anticlines and plunging synclines. In eroded mountain belts, a zigzag pattern of outcrops may appear in the field after erosion has removed much of the surface rock from the folds. The geologic map in Figure 8.4 shows this characteristic pattern.

Nor do folds usually remain symmetrical. With increasing amounts of deformation, folds can be pushed into asymmetrical shapes, with one limb dipping more steeply than the other (**Figure 8.14**). Such *asymmetrical folds* are

Symmetrical folds have limbs that dip symmetrically from the axial plane.

Asymmetrical folds have one limb that dips more steeply than the other.

Overturned folds have limbs that dip in the same direction, but one limb has been tilted beyond the vertical.

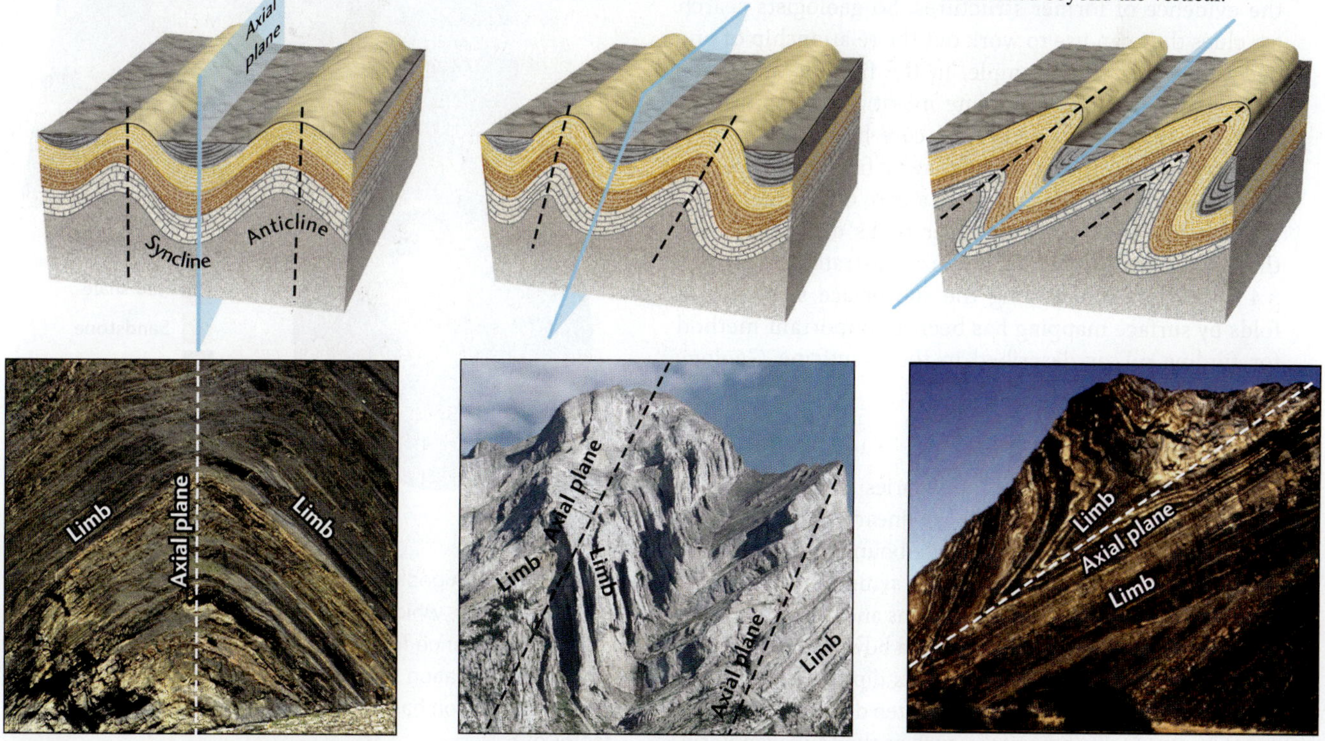

FIGURE 8.14 With increasing deformation, folds are pushed into asymmetrical shapes. [(left) Tony Waltham; (center) Design Pics/Michael Interisano/Getty Images; (right) Courtesy John Grotzinger.]

PRACTICING GEOLOGY EXERCISE

How Do We Use Geologic Maps to Find Oil?

Crude oil, or *petroleum* (from the Latin words for "rock oil"), has been collected from natural seeps at Earth's surface since ancient times. The foul-smelling, tarry substance was used as boat caulking, wheel grease, and medicine, but not commonly as a fuel until the process of oil refining was developed in the 1850s. Demand skyrocketed at that time, primarily because oil from whale blubber, the best fuel then available for lamps, had become terribly expensive ($60 per gallon in today's dollars!) as overfishing decimated whale populations.

The ability to refine clean lamp oil from petroleum set off North America's first oil boom. The mining of "black gold" was centered in areas around Lake Erie, where major petroleum seeps had been discovered in northwestern Pennsylvania, northeastern Ohio, and southern Ontario. Early petroleum explorers, such as self-proclaimed "Colonel" Edwin Drake of Pennsylvania, simply drilled into the seeps, but this straightforward approach soon proved inadequate as a strategy for satisfying the new thirst for oil.

---

common. When the deformation is so intense that one limb has been tilted beyond the vertical, the fold is called an *overturned fold*. Both limbs of an overturned fold dip in the same direction, but the order of the layers in the bottom limb is precisely the reverse of their original sequence—that is, older rocks are on top of younger rocks.

Observations in the field seldom provide complete information about folds. Bedrock may be obscured by overlying soils, or erosion may have removed much of the evidence of former structures. So geologists search for clues they can use to work out the relationship of one bed to another. For example, in the field or on a geologic map, an eroded anticline might be recognized as a strip of older rocks forming a core bordered on both sides by younger rocks dipping away from the core. An eroded syncline might appear as a core of younger rocks bordered on both sides by older rocks dipping toward the core. These relationships are illustrated in Figures 8.4 and 8.13. Determining the subsurface structure of folds by surface mapping has been an important method for finding oil, as described in the Practicing Geology Exercise.

Circular Structures

Deformation along plate boundaries by horizontally directed forces usually results in linear faults and folds oriented nearly parallel to the plate boundary. Some types of deformation, however, are more symmetrical, forming nearly circular structures called basins and domes.

A **basin** is a synclinal structure, a bowl-shaped depression of rock layers in which the beds dip toward a central point (Figure 8.15). Sediments are often deposited in basins (see Chapter 6). In some cases, such as the Michigan Basin,

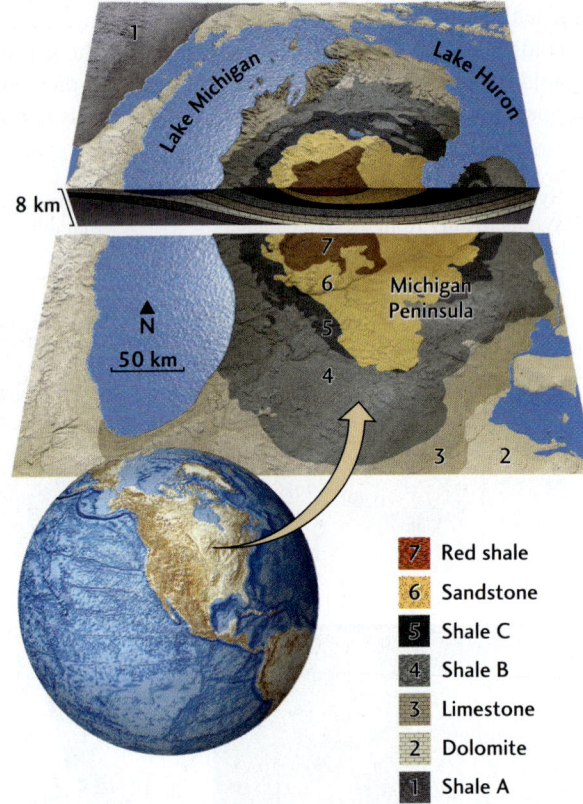

FIGURE 8.15 A geologic map and cross section of the Michigan Basin, which shows sedimentary layers deposited in a thick sequence from the oldest (formation 1) to the youngest (formation 7) during subsidence of the basin. The cross section has been vertically exaggerated by a factor of 5:1.

Could geologic knowledge be used to locate large petroleum reservoirs hidden underground—that is, in regions where no oil seeped to the surface? An affirmative answer was provided in 1861 by T. Sterry Hunt, a Connecticut-born geochemist. As a member of the Geological Survey of Canada, Hunt had been active in the new science of mapping natural resources. He documented the petroleum seeps of southern Ontario in 1850. As the oil production of the region increased, he noticed that seeps and successful wells tended to be aligned along the crests of geologic folds.

Hunt had also studied the physical and chemical properties of petroleum in the laboratory, and he knew it was formed when sedimentary rocks rich in organic material were subjected to heat and pressure (see Chapter 6). Petroleum is lighter than water; because of this buoyancy, it tends to rise toward the surface. Hunt hypothesized that the rising petroleum could accumulate in porous "reservoir rocks," such as sandstones, if such rocks were overlain by impermeable "cap rocks," such as shales, that prevented the petroleum from rising farther. Moreover, the most likely place to find large reservoirs would be along the fold axes of anticlines, where substantial amounts of petroleum could be trapped without escaping to the surface.

The accompanying figure illustrates a typical anticlinal trap, for which we can imagine the following narrative of geologic discovery. Erosion of the fold has exposed a sequence of sandstones, limestones, and shales. Mapping by an enterprising geologist shows that the axis of the anticline strikes to the northeast. Drilling at point A on the axis of the anticline first penetrates a thick sandstone layer exposed at the surface and then a thinner shale layer. Immediately below the shale, the drilling crew encounters another sandstone layer containing gas and, below the gas, significant quantities of oil. The geologist infers that the shale is capping a major petroleum reservoir in the deeper sandstone layer, so he instructs his crew to move along the strike of the anticline and drill at point B. Bingo—another successful oil well!

Hunt's "anticlinal theory" allowed geologists to discover oil (and some to get rich) by mapping fold structures at the surface and, later, by the three-dimensional imaging of such structures using seismic techniques. The results have been impressive: Most of the one trillion barrels of crude oil produced since 1861 have come from anticlinal oil traps of the type that Hunt first described.

BONUS PROBLEM: The company that manages the petroleum claim in the figure would like to expand its operations, and they propose to drill a new well further along the axis of the anticline at point C. As a consulting geologist, how would you rate their chances of bringing in another successful well? Illustrate your answer by sketching a geologic cross section.

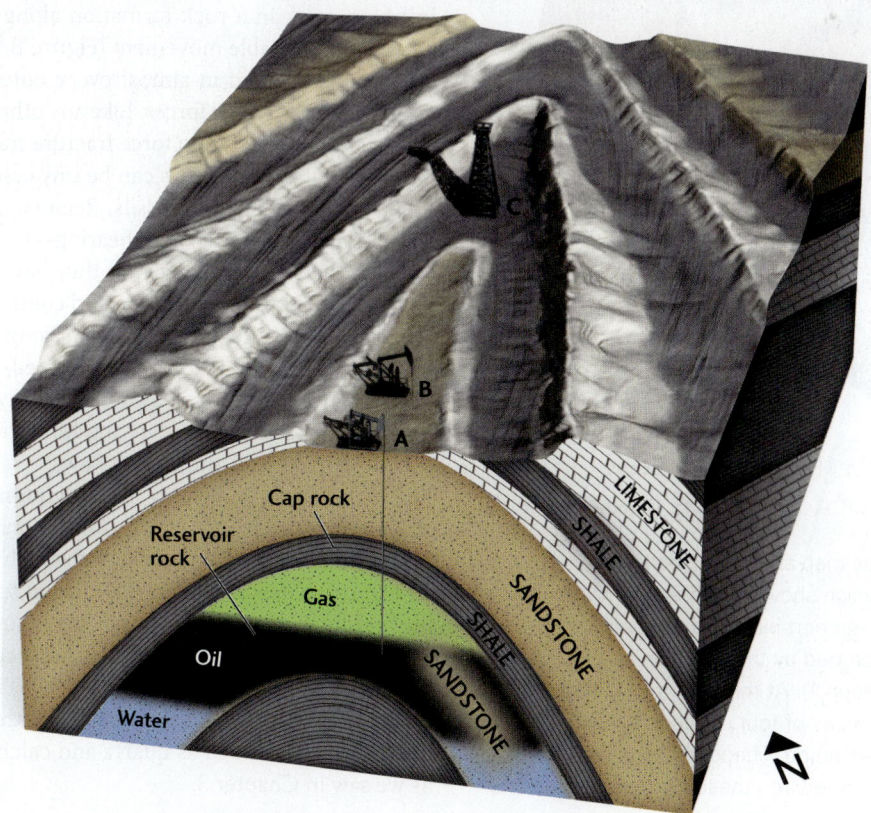

shown in Figure 8.15, this deposition can produce sedimentary sequences many kilometers in thickness. A **dome** is an anticlinal structure, a broad circular or oval upward bulge of rock layers. The flanking beds of a dome encircle a central point and dip radially away from it (**Figure 8.16**). Domes, like other anticlines, are important in petroleum geology because oil is buoyant and tends to migrate upward through permeable rocks (see the Practicing Geology Exercise). If the rocks at the high point of a dome are impermeable to oil, the oil becomes trapped beneath them.

Domes and basins are typically many kilometers in diameter, and some extend for hundreds of kilometers. They are recognized in the field by outcrops that outline their characteristic circular or oval shapes. At these outcrops, the rock layers dip downward toward the center of the basin or upward toward the top of the dome (see Figures 8.15 and 8.16).

Some circular structures are formed by multiple episodes of deformation—for instance, when rocks are compressed in one direction and then again in a direction nearly perpendicular to the original direction. In many other cases, however, these structures result from the upward force of rising material or the downward force of sinking material, rather than the horizontally directed forces of plate tectonics. Not surprisingly, such circular structures tend to be more common in the interiors of plates, far away from active plate boundaries. There are many domes and basins, for example, in the central portion of the United States. Almost all of the Lower Peninsula of Michigan is a large sedimentary basin (see Figure 8.15); the Black Hills of South Dakota are an eroded dome (see Figure 8.16).

Several types of deformation can produce domes and basins. Some domes are formed by rising bodies of buoyant material—magma, hot igneous rock, or salt—that push the overlying sediments upward. As we saw in Chapter 6, some sedimentary basins form when a heated portion of the crust cools and contracts, causing the overlying sediments to subside (thermal subsidence basins). Others result when tectonic forces stretch and thin the crust (rift basins) or compress it downward (flexural basins). The weight of sediments deposited by a river delta can depress the crust into a sedimentary basin, such as the very large basin now forming at the mouth of the Mississippi River in the Gulf of Mexico.

Joints

As we have seen, a fracture that has displaced the rock on either side is called a fault. A second type of fracture is a **joint**—a crack in a rock formation along which there has been no appreciable movement (**Figure 8.17a**).

Joints are found in almost every outcrop. Some joints are caused by tectonic forces. Like any other brittle material, brittle rocks subjected to force fracture most easily at flaws or weak spots. These flaws can be tiny cracks, fragments of other materials, or even fossils. Regional tectonic forces—compressive, tensional, or shearing—may leave a set of joints as their imprint long after they have vanished.

The nontectonic expansion and contraction of rock can also form joints. Regular patterns of joints are often found in plutons and lavas that have cooled, contracted, and cracked. Erosion can strip away surface layers, releasing the confining pressure on underlying formations and allowing the rocks to expand and split at flaws.

Joints are usually only the beginning of a series of changes that greatly alter rock formations as they age. For example, joints provide channels through which water and air can reach deep into a formation and speed the weathering and internal weakening of its structure. If two or more sets of joints intersect, weathering may cause the formation to break into large columns or blocks (Figure 8.17b). The circulation of hydrothermal solutions through joints can deposit minerals such as quartz and calcite, forming veins, as we saw in Chapter 3.

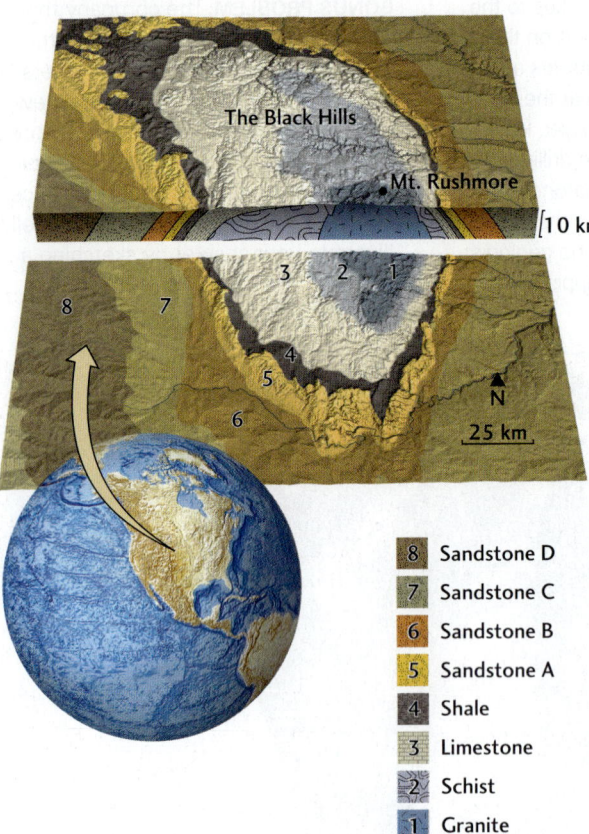

FIGURE 8.16 A geologic map and cross section of the Black Hills dome, which shows sedimentary rocks (formations 3–8) and metamorphic rocks (formation 2) that were uplifted and eroded by the intrusion of a granitic batholith (formation 1). At the Mount Rushmore National Memorial, the faces of four presidents—George Washington, Thomas Jefferson, Theodore Roosevelt, and Abraham Lincoln—are carved into these granitic rocks.

Basic Deformation Structures 227

(a)

(b)

FIGURE 8.17 Joint patterns. (a) Intersecting joints in a massive granite outcrop, Joshua Tree National Park, California. (b) Columnar joints in basalt, Giants Causeway, Northern Ireland. [(a) Sean Russell/Getty Images; (b) Michael Brooke/Photolibrary/Getty Images.]

Deformation Textures

Joints are examples of small features in rock formations that are best observed up close at an outcrop. Another type of small-scale deformation structure is the texture of a rock mass in areas of localized shearing, such as fault zones.

As we have seen, tectonic forces cause the brittle parts of Earth's crust to crack and slip. As the rocks along a fault plane shear past each other, they grind and mechanically fragment solid rock. Where rocks behave as brittle materials (usually in the upper crust), shearing produces rocks with *cataclastic textures*, in which the grains are broken, angular fragments. One such rock type, called *fault breccia*, is shown in **Figure 8.18a**.

Deeper in the crust, where temperatures and pressures are high enough to allow ductile deformation, shearing forces can

(a)

(b)

FIGURE 8.18 (a) Fault breccia developed on a fault in eastern Nevada. The rust-colored breccia shows a cataclastic texture. The gray rocks on either side are limestones. (b) Mylonites developed in the Great Slave Lake shear zone, Northwest Territories, Canada. The rock was originally a granite. As a result of intense shearing forces, the large, originally angular potassium feldspar crystals have been rolled and transformed into smooth balls. [(a) Marli Miller, University of Oregon/Earth Science World Image Bank; (b) Courtesy John Grotzinger.]

produce metamorphic rocks called *mylonites* (Figure 8.18b). The movement of rock surfaces against one another recrystallizes minerals and strings them out in bands or streaks. Development of mylonites typically occurs at the greenschist to amphibolite grades of metamorphism (see Chapter 7). The textural effects of deformation are most obvious in mylonites, but they are also prominent in cataclastic rocks.

The San Andreas fault of Southern California makes a good case study of how deformation textures might relate to changes in temperature and pressure with depth. This fault, which marks the boundary between the Pacific Plate and the North American Plate (see Figure 8.7), extends through the crust and probably down into the mantle. At depths up to about 20 km, the fault zone is thought to be very narrow and characterized by cataclastic textures, indicating brittle deformation. Earthquakes are generated in this zone. Below 20 km, however, earthquakes do not occur, and the fault is thought to be characterized by a broad zone of ductile deformation that produces mylonites.

Styles of Continental Deformation

If we look closely enough, we can find all the basic deformation structures—faults, folds, domes, basins, joints—in any zone of continental deformation. But when we view continental deformation at a regional scale, we see distinctive patterns of faulting and folding that relate directly to the tectonic forces causing the deformation. **Figure 8.19**

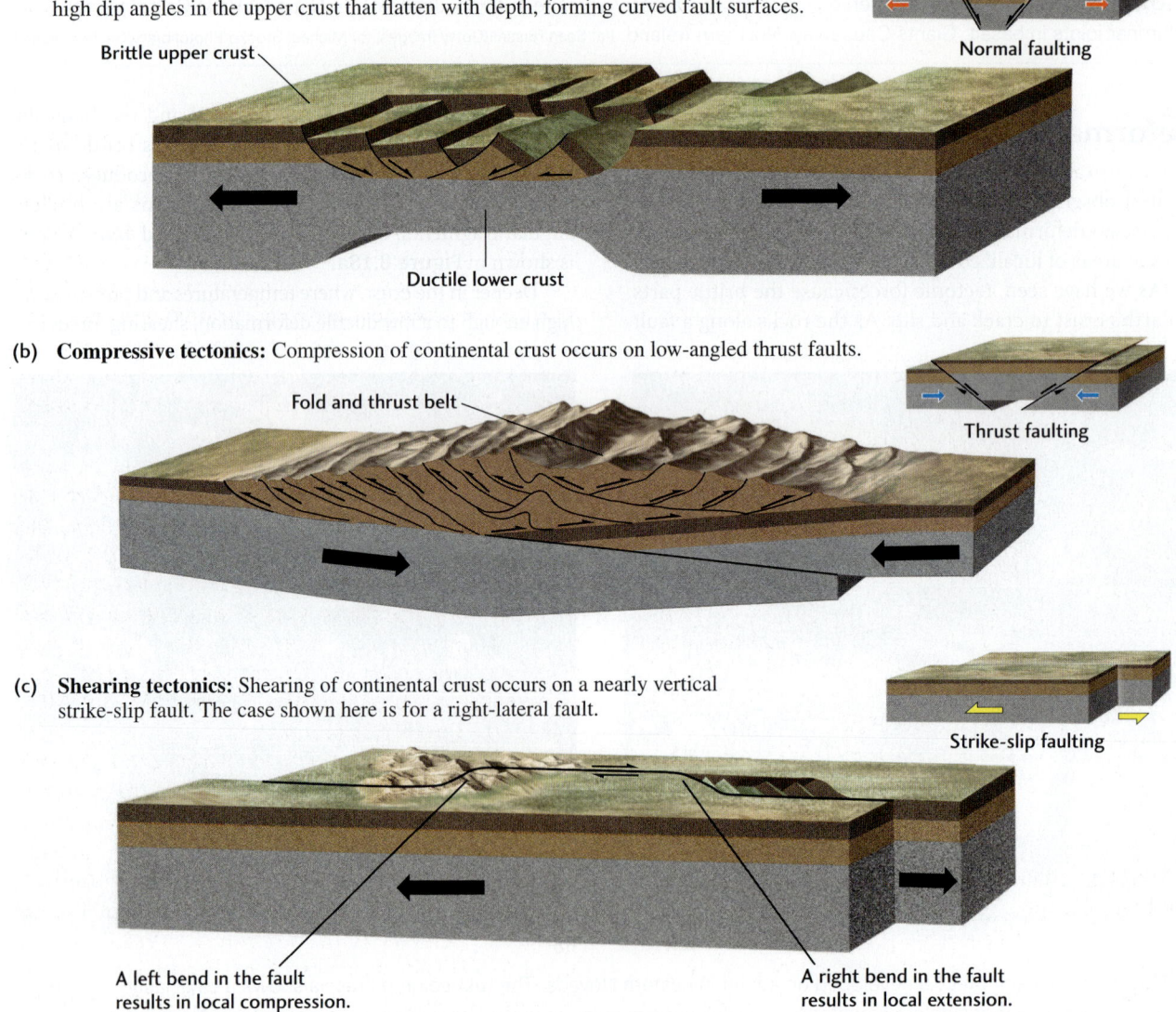

FIGURE 8.19 The orientation of tectonic forces—(a) tensional, (b) compressive, and (c) shearing—determines the style of continental deformation. On a regional scale, the basic types of faulting shown in the inset figures can lead to distinctive, complex patterns of deformation.

depicts the deformation styles typical of the three main types of tectonic forces.

Tensional Tectonics

In brittle crust, the tensional forces that produce normal faulting may split a plate apart, resulting in a *rift valley*—a long, narrow trough formed when a block of rock drops downward relative to its two flanking blocks along nearly parallel, steeply dipping normal faults (Figure 8.19a). Well-known examples include the Rhine Valley, the rift valleys of East Africa (**Figure 8.20**), and the Red Sea, as well as the rift valleys of mid-ocean ridges. As we saw in Chapter 6, these structures form basins that fill with sediments eroded from the rift walls as well as with volcanic rocks extruded from tensional cracks in the crust.

Tension on the shallow continental crust usually produces normal faults with high dip angles, typically 60° or more. Below a depth of about 20 km, however, crustal rocks are hot enough to behave as ductile materials, and deformation occurs by stretching rather than by fracturing. This change in rock behavior causes the dip of the faults to flatten with increasing depth, which results in normal faults with curved fault surfaces (called *listric* faults), as shown in Figure 8.19a. The crustal blocks moving along these curved faults are tilted backward as the stretching continues.

The Basin and Range province, which is centered on the Great Basin of Nevada and Utah, is a good example of a region defined by many adjacent rift valleys. This region, which is now more than 800 km wide, has been stretched and extended in a northwest-southeast direction by a factor of 2 during the last 15 million years. Here, normal faulting has created an immense landscape of eroded, rugged fault-block mountains and smooth, sediment-filled valleys, some covered with recent volcanic rocks (see Figure 10.5). This tensional deformation, which appears to be caused by upwelling convection currents beneath the Basin and Range province, continues today.

Compressive Tectonics

In subduction zones, oceanic lithosphere slips beneath an overriding plate along a huge thrust fault, or *megathrust*. The world's largest earthquakes, such as the great Tohoku, Japan, earthquake of March 11, 2011, which generated a disastrous tsunami that killed more than 19,000 people, are caused by sudden slips on megathrusts. Thrust faulting is also the most common type of faulting within continents undergoing tectonic compression. Sheets of crust may glide over one another for tens of kilometers along nearly horizontal thrust faults, forming *overthrust* structures (**Figure 8.21**).

When two continents collide, the crust can be compressed across a wide zone, resulting in spectacular episodes of mountain building. During such collisions, the brittle basement rocks ride over one another by thrust faulting, while the more ductile overlying sedimentary rocks are compressed into a series of great folds, forming a *fold and thrust belt* (Figure 8.19b). Large earthquakes are common in fold and thrust belts; a recent example is the great Wenchuan earthquake that hit Sichuan, China, on May 12, 2008, killing more than 80,000 people.

The ongoing collisions of Africa, Arabia, and India with the southern margin of the Eurasian continent have created fold and thrust belts from the Alps to the Himalaya, many of which are still active. The great oil reservoirs of the Middle East are trapped in anticlines formed by this deformation. Compression across western North America, caused by that continent's westward movement during the opening of the Atlantic Ocean, created the fold and thrust belt of the Canadian Rockies. The Valley and Ridge province of the Appalachian Mountains is an ancient fold and thrust belt that dates back to the collisions that created the supercontinent Pangaea.

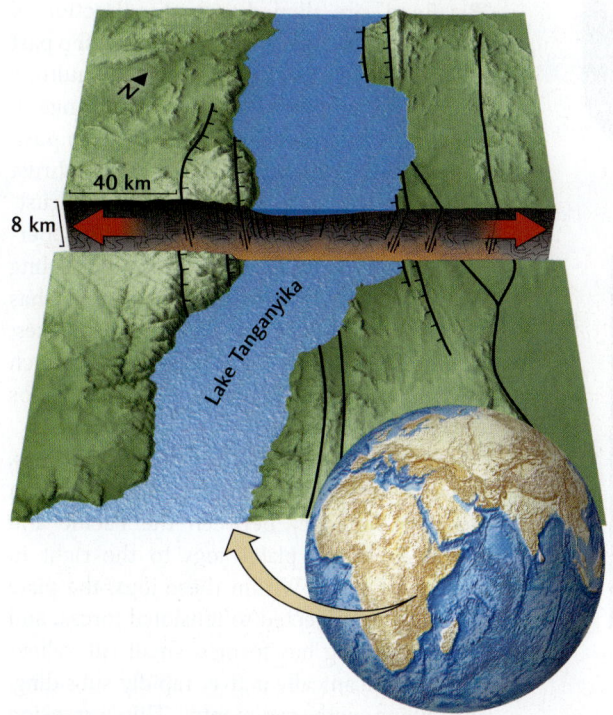

FIGURE 8.20 In East Africa, tensional forces are pulling the Somali subplate away from the African Plate, creating rift valleys bounded by normal faults (see Figure 2.8b). The rift valley shown here is filled by sediments and Lake Tanganyika, on the boundary between Tanzania and the Democratic Republic of Congo. The cross section has been vertically exaggerated by a factor of 2.5:1, which exaggerates the fault dips; the actual dips of the normal faults are about 60°.

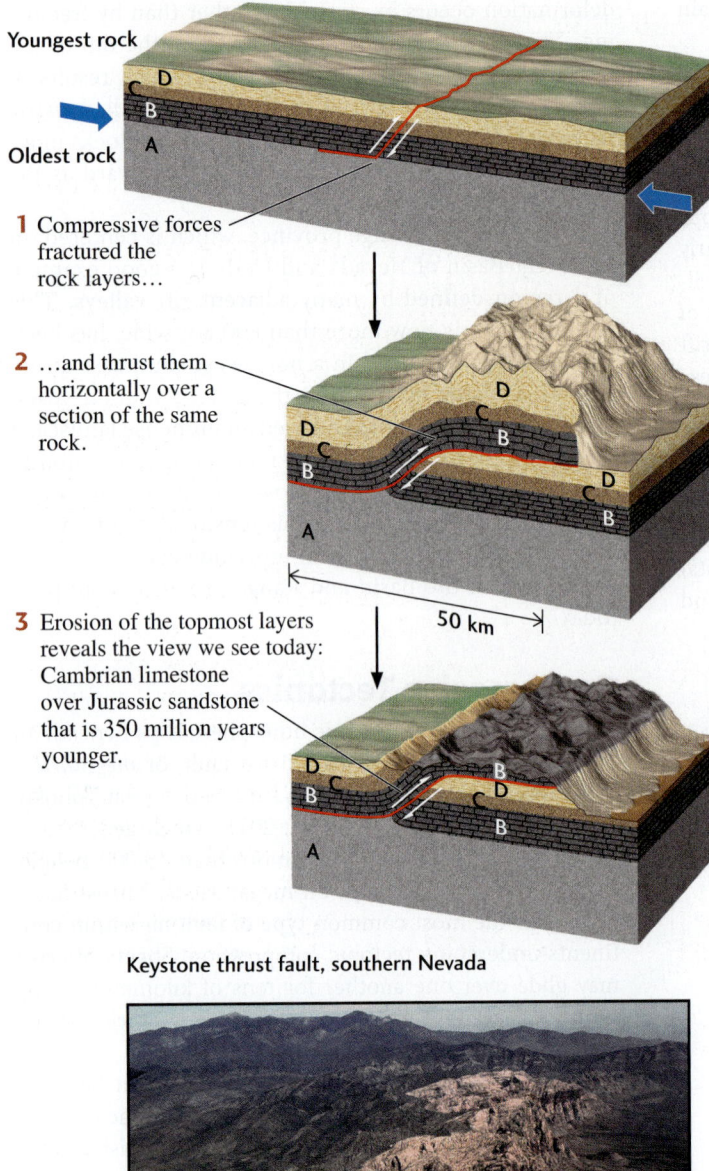

FIGURE 8.21 The Keystone thrust fault of southern Nevada is a large-scale overthrust structure of a kind formed during episodes of continental compression. Compressive forces have detached a sheet of rock layers (D, C, B) and thrust it a great distance horizontally over a section of the same rock layers (D, C, B, A). [Courtesy Vince Matthews.]

Shearing Tectonics

A transform fault is a strike-slip fault that forms a plate boundary. Transform faults such as the San Andreas can offset geologic formations by long distances (see Figure 8.7), but as long as they stay aligned with the direction of relative plate movement, the blocks on either side can slide past each other without much internal deformation. Long transform faults are rarely straight, however, so deformation patterns along these faults can be much more complicated. The faults may have bends and jogs that change the tectonic forces acting across portions of the plate boundary from shearing forces to compressive or tensional forces. Those forces, in turn, cause secondary faulting and folding (Figure 8.19c).

A good example of these complications can be found in Southern California, where the right-lateral San Andreas fault bends first to the left and then to the right as one moves along its trace from south to north (**Figure 8.22**). The segments of the fault on both sides of this "Big Bend" are aligned with the direction of relative plate movement, so the blocks slip past each other there by simple strike-slip faulting. Within the Big Bend, however, the change in the fault orientation causes the plates to push against each other, which produces thrust faulting to the south of the fault. This thrusting has raised the San Gabriel and San Bernardino mountains to elevations exceeding 3000 m and, during the last half century, has produced a series of destructive earthquakes, including the 1994 Northridge quake, which caused more than $40 billion in damage to Los Angeles (see Chapter 9).

At the southern end of the San Andreas, between the Gulf of California and the Salton Sea, the boundary between the Pacific and North American plates jogs to the right in a series of steps. Within these jogs, the plate boundary is subjected to tensional forces, and normal faulting has formed small rift valleys that are volcanically active, rapidly subsiding, and filling with sediments. This extension occurs within 200 km of the Big Bend compression, demonstrating how variable the tectonics along continental transform faults can be!

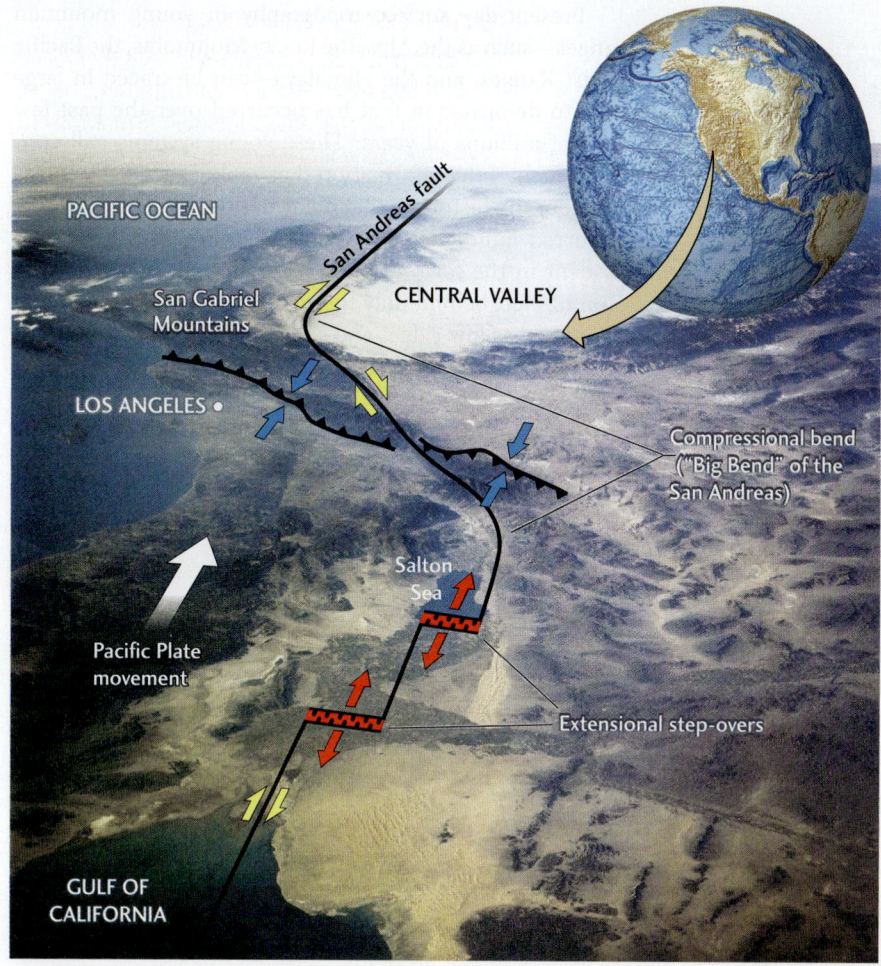

FIGURE 8.22 Space shuttle photograph showing an oblique view of the San Andreas fault system. The annotations illustrate how the deviations of a transform fault's strike from the direction of plate motion can cause local extension and compression. Between the Gulf of California and the Salton Sea (near bottom of figure), the fault system jogs to the right in two major steps; the right-lateral fault segments (black lines), which are parallel to the Pacific–North America plate motion, are separated by rift valleys (in red) that are volcanically active, subsiding, and filling with sediments. As we look northward, the fault trace first bends to the left, away from the direction of plate motion, and then to the right, realigning with the plate motion in central California (near top of figure). This "Big Bend" in the San Andreas fault causes compression, which is taken up by reverse faulting in the Los Angeles region (middle of figure). [Image Science & Analysis Laboratory, NASA Johnson Space Center.]

Unraveling Geologic History

The geologic history of a region is a succession of episodes of deformation and other geologic processes. Let's see how some of the concepts and methods introduced in this chapter can be used to reconstruct that history.

The cross sections in **Figure 8.23** represent a few tens of kilometers of a geologic region that underwent a succession of tectonic events. First, horizontal layers of sediment were deposited on the seafloor. Those layers were tilted and folded, and eventually uplifted above sea level, by horizontal compressive forces. There, erosion gave them a new horizontal surface. That surface was covered by lava when forces deep in Earth's interior caused a volcanic eruption. In the most recent stage, tensional forces resulted in normal faulting, which broke the crust into blocks.

Geologists see only the last stage, but visualize the entire sequence. They begin by identifying and determining the ages of the rock layers and recording the orientation of layers, folds, and faults on geologic maps. Then they use those maps to construct cross sections of the subsurface features. Once the geologists have identified sedimentary beds, they can start with the knowledge that the beds must originally have been horizontal and undeformed at the bottom of an ancient ocean. The succeeding events can then be reconstructed.

CHAPTER 8 Deformation: Modification of Rocks by Folding and Fracturing

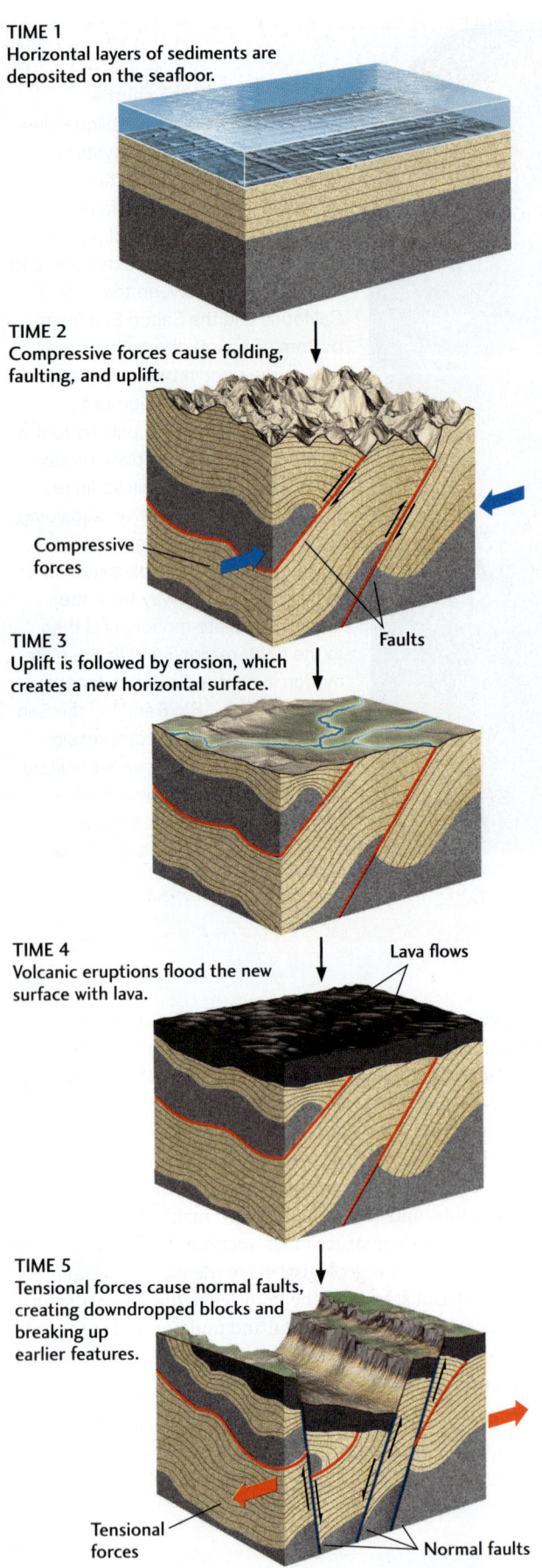

TIME 1
Horizontal layers of sediments are deposited on the seafloor.

TIME 2
Compressive forces cause folding, faulting, and uplift.

Compressive forces

Faults

TIME 3
Uplift is followed by erosion, which creates a new horizontal surface.

TIME 4
Volcanic eruptions flood the new surface with lava.

Lava flows

TIME 5
Tensional forces cause normal faults, creating downdropped blocks and breaking up earlier features.

Tensional forces

Normal faults

Present-day surface topography in young mountain ranges—such as the Alps, the Rocky Mountains, the Pacific Coast Ranges, and the Himalaya—can be traced in large part to deformation that has occurred over the past few tens of millions of years. These young systems still contain much of the information that geologists need to piece together the history of that deformation. Deformation that occurred hundreds of millions of years ago is no longer evident in the form of rugged mountains, however. Erosion has left behind only the remnants of folds and faults, expressed as low ridges and shallow valleys. As we will see in Chapter 21, even older episodes of mountain building are evident only in the twisted, highly metamorphosed formations that constitute the basement rocks of the interiors of continents.

FIGURE 8.23 Stages in the development of a geologic region. A geologist sees only the last stage and attempts to reconstruct all the earlier stages from the observable structural features.

KEY TERMS AND CONCEPTS

anticline (p. 222)	fault (p. 219)	oblique-slip fault (p. 221)
basin (p. 224)	fold (p. 222)	shearing force (p. 214)
brittle (p. 218)	foot wall (p. 220)	strike (p. 214)
compressive force (p. 214)	formation (p. 214)	strike-slip fault (p. 220)
deformation (p. 214)	geologic cross section (p. 216)	syncline (p. 222)
dip (p. 214)	geologic map (p. 216)	tensional force (p. 214)
dip-slip fault (p. 220)	hanging wall (p. 220)	thrust fault (p. 221)
dome (p. 226)	joint (p. 226)	
ductile (p. 219)	normal fault (p. 220)	

REVIEW OF LEARNING OBJECTIVES

8.1 Compare the different types of tectonic forces at plate boundaries that deform rocks.

Deformation includes the processes of folding, faulting, shearing, compression, and extension of rock by plate tectonic forces. At plate boundaries, these tectonic forces are tensional, compressive, or shearing. Tensional forces stretch and pull the rock formations apart, and are dominant at divergent boundaries where plates move away from each other. Compressive forces squeeze and shorten the rock, and are dominant at convergent boundaries where plates move toward each other. Shearing forces shear two parts of a rock in different directions, and are dominant at transform-fault boundaries, where plates slide past each other.

Tectonic forces can also cause a rock formation to break and slip on both sides of a fracture, forming a fault.

Study Assignment: Figure 8.1

Exercise: Describe how geologic patterns of deformation can be used to deduce directions of movement along ancient plate boundaries.

Thought Question: What types of force results in folding?

8.2 Describe how maps and diagrams are used to represent geologic structures.

Two important measures in geologic maps and diagrams are strike and dip. Strike is the compass direction of a rock layer along its intersection with a horizontal surface. Dip is the angle at which the rock layer inclines from the horizontal, measured at right angles to the strike. A geologic map is a two-dimensional model of the geologic features exposed at Earth's surface, showing various rock formations as well as other features such as faults. A geologic cross section is a diagram representing the geologic features that would be visible if a vertical slice were made through part of the crust. Geologic cross sections can be constructed from the information on a geologic map, although they can often be improved with subsurface data collected by drilling or seismic imaging.

Study Assignments: Figures 8.3 and 8.4

Exercise: On a geologic map of 1:250,000 scale, how many centimeters would represent an actual distance of 2.5 km? What is the actual distance in miles of 1 inch on the same map?

Thought Question: Why is it correct to say "large-scale geologic structures should be represented on small-scale geologic maps"? How big a piece of paper would be required to make a map of the entire U.S. Rocky Mountains at 1:24,000 scale?

8.3 Explain how laboratory experiments can help us understand the way rocks deform.

Rocks deform in response to tectonic forces acting upon them, whether by folding, faulting, or some combination of the two. Laboratory studies show that rocks subjected to tectonic forces may behave as brittle materials or as ductile materials. These behaviors depend on temperature and pressure, the type of rock, the speed of deformation, and the orientation of tectonic forces. Rocks that are brittle at shallow depths can act as ductile material deep in the crust. Hard igneous and metamorphic rocks often behave as brittle materials, while softer sedimentary rocks often behave as ductile materials, folding gradually.

Study Assignment: Figure 8.6

Exercise: Describe why a rock formation would behave as a ductile material if it deformed slowly, but may behave as a brittle material if deformed more rapidly.

Thought Question: The submerged margin of a continent has a thick layer of sediments overlying metamorphic basement rocks. That continental margin collides with another continental mass, and the compressive forces deform it into a fold and thrust belt. During the deformation, which of the following geologic formations would be likely to behave as brittle materials and which as ductile materials: (a) the sedimentary formations in the upper few kilometers; (b) the metamorphic basement rocks at depths of 5 to 15 km; (c) lower crustal rocks at depths below 20 km? In which of these layers would you expect earthquakes?

8.4 Summarize the basic deformation structures that are observed in the field and the types of forces that produce these deformation structures.

Among the geologic structures that result from deformation are folds, faults, circular structures, joints, and deformation textures caused by shearing. Fractures are known as faults if rocks are displaced across the fracture surface, and as joints if no displacement is observed. Faults and folds are produced primarily by horizontally directed tectonic forces at plate boundaries. Horizontal tensional forces at divergent boundaries produce normal faults, horizontal compressive forces at convergent boundaries produce thrust faults, and horizontal shearing forces at transform-fault boundaries produce strike-slip faults. Folds are usually formed in layered rock by compressive forces, especially in regions where continents collide. Circular structures, such as domes and basins, can be produced by vertically directed forces far from plate boundaries. Some domes are caused by the rise of buoyant materials. Basins can form when tensional forces stretch the crust or when a heated portion of the crust cools, contracts, and subsides. Joints can be caused by tectonic stresses or by the cooling and contraction of rock formations.

Study Assignments: Figures 8.8 and 8.12

Exercise: Show that a left jog in a right-lateral strike-slip fault will produce compression, whereas a right jog in a right-lateral strike-slip fault will produce extension. Write a similar rule for left-lateral strike-slip faults.

Thought Question: Dip-slip faulting causes offset of blocks of rock on either side of the fault. How would you distinguish between blocks that are moved during extension versus those that are moved during compression?

8.5 Define the main styles of continental deformation.

There are three main styles of continental deformation. Tensional tectonics produce rift valleys with normal faulting; in continental regions undergoing extension, the dip angles of the normal faults flatten with depth, causing the fault blocks to tilt away from the rift as the faulting continues. Compressive tectonics produce thrust faulting; in the case of continent-continent collisions, compression may produce fold and thrust belts. Shearing tectonics produce strike-slip faulting, but bends and jogs in the fault may cause local thrust faulting and normal faulting.

Study Assignment: Figure 8.21

Exercise: Describe how a repeated sequence of stratigraphic layers would require the presence of an overthrust structure.

Thought Question: How does compressive tectonics create overthrust structures?

8.6 Describe how we reconstruct the geologic history of a region.

We can observe only the end result of a succession of events: deposition, deformation, erosion, volcanism, and so forth. The deformational history of a region is deduced by identifying and determining the ages of rock layers, recording the geometric orientation of rock layers on geologic maps, mapping folds and faults, and constructing cross sections of subsurface structure consistent with their surface observations.

Study Assignment: Figure 8.23

Exercise: Construct a geologic cross section that tells the following story: A series of marine sediments was deposited and subsequently deformed by compressive forces into a fold and thrust belt. The mountains of the fold and thrust belt eroded to sea level, and new sediments were deposited. The region then began to be extended, and lava intruded the new sediments to create a sill. In the latest stage, tensional forces broke the crust to form a rift valley bounded by steeply dipping normal faults.

Thought Question: Discuss how unraveling a series of geologic events leads to understanding geologic history.

CHAPTER 8 Deformation: Modification of Rocks by Folding and Fracturing

VISUAL LITERACY EXERCISE

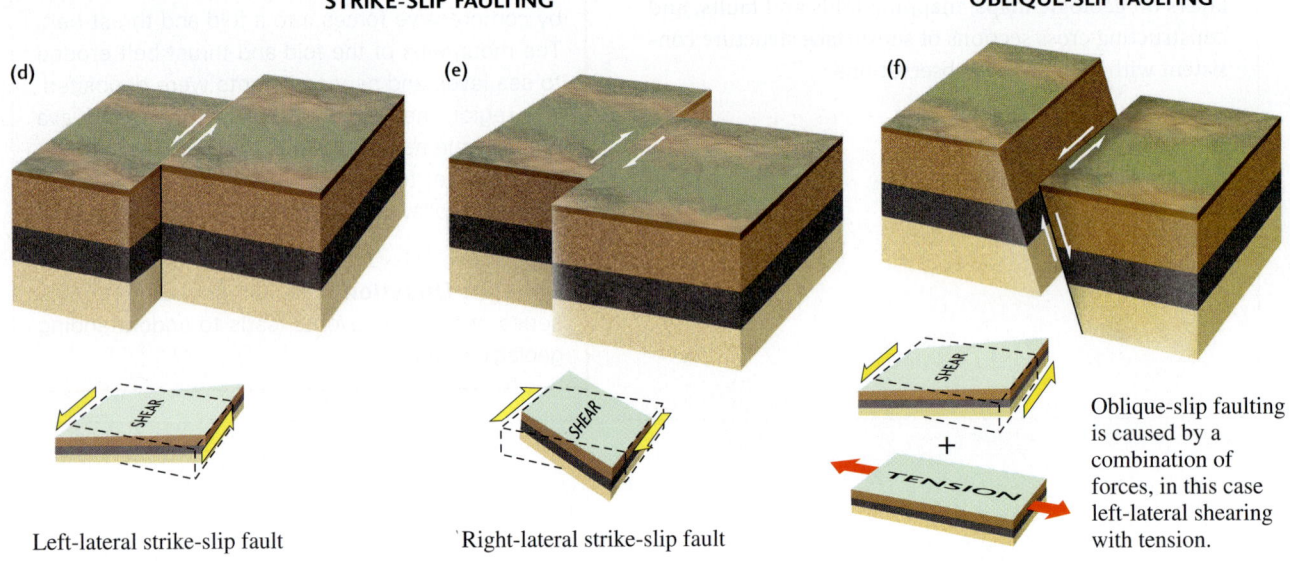

FIGURE 8.8 The orientation of tectonic forces determines the style of faulting. Dip-slip faulting (a–c) is caused by compressive or tensional forces. Strike-slip faulting (d, e) is caused by shearing forces. Oblique-slip faulting (f) is caused by a combination of shearing and compressive or tensional forces.

1. What kind of forces create a normal fault?
   a. Tensional
   b. Compressional
   c. Shear
   d. Forces are not involved in faulting.

2. What kind of forces create an oblique-slip fault?
   a. Tensional + Shear
   b. Tensional + Compressional
   c. Shear
   d. Compressional

3. What is a thrust fault?
   a. A right-lateral strike-slip fault
   b. A left-lateral strike-slip fault
   c. A shallowly dipping reverse fault
   d. A high-angle dip-slip fault

4. What do compressive forces do to a rock?
   a. Stretch and pull it apart
   b. Shear it
   c. Squeeze and shorten it
   d. Bend it

5. If a geologist faces a strike-slip fault and observes the rocks on the other side of the fault to have moved to the right, how is the fault classified?
   a. Normal fault
   b. Thrust fault
   c. Right-lateral strike-slip fault
   d. Left-lateral strike-slip fault

Clocks in Rocks: Timing the Geologic Record

Reconstructing Geologic History From the Stratigraphic Record	241
The Geologic Time Scale: Relative Ages	247
Measuring Absolute Time with Isotopic Clocks	249
The Geologic Time Scale: Absolute Ages	256
Recent Advances in Timing the Earth System	259

Learning Objectives

From the information in this chapter, you will be able to:

9.1 Apply stratigraphic principles to geological field observations to infer whether one rock is older than another.

9.2 Classify unconformities in stratigraphic sequences according to the relationships between the layers above and below them.

9.3 Understand how the absolute ages of rocks can be determined by the decay of radioactive isotopes, and know which isotopic systems are suitable for dating rocks of various ages.

9.4 Recall the major divisions and subdivisions of the geological time scale from Earth's formation to the present day.

9.5 Summarize how geologists created and refined the geological time scale.

These trilobites are preserved as fossils in rocks about 365 million years old. [Science Photo Library/Science Source.]

CHAPTER 9 Clocks in Rocks: Timing the Geologic Record

Philosophers have struggled with the notion of time throughout human history, but until fairly recently, they have had very little data to constrain their speculations. The immensity of time—"deep time" measured in billions of years—was a great geologic discovery that changed our thinking about how Earth operates as a system.

Pioneering geologists such as James Hutton and Charles Lyell understood that the planet was not shaped by a series of catastrophic events over a mere few thousand years, as many people had believed. Rather, what we see today is the product of ordinary geologic processes operating over much longer time intervals. Hutton stated this understanding as the *principle of uniformitarianism,* described in Chapter 1. Knowledge of geologic time helped Charles Darwin formulate his theory of evolution, and it has led to many other insights about the workings of the Earth system, the solar system, and the universe as a whole.

Geologic processes occur on time scales that range from seconds (meteorite impacts, volcanic explosions, earthquakes) to tens of millions of years (the recycling of oceanic lithosphere) and even billions of years (the tectonic evolution of continents). If we are careful enough, we can measure the rates of short-term processes, such as beach erosion or seasonal variations in the transport of sediments by rivers, in a few years. Precise surveying can monitor the slow movements of glaciers (meters per year), and with the Global Positioning System, we can track the even slower movements of the lithospheric plates (centimeters per year). Historical documents can provide certain types of geologic data, such as the dates of major earthquakes or volcanic eruptions, from hundreds or, in some cases, thousands of years ago.

However, the record of human observation is far too short for the study of many slow geologic processes (Figure 9.1). In fact, it's not even long enough to capture some types of rapid but infrequent events; for example, we have never witnessed a meteorite impact as big as the one that left the crater shown in Figure 1.6. We must rely instead on the geologic record: the information preserved in rocks that have survived erosion and subduction. Almost all oceanic crust older than 200 million years has been subducted back into the mantle, so most of Earth's history is documented only in the older rocks of the continents. Geologists can reconstruct subsidence from the record of sedimentation; uplift from the erosion of rock layers; and deformation from faults, folds, and metamorphic rocks. To measure the pace of these processes and understand their common causes, we must be able to assign ages to events observed in the geologic record.

In this chapter, we will learn how geologists first plumbed the abyss of time by finding chronological order in the geologic record. Then we will see how they used the discovery of radioactive "clocks in rocks" to develop a precise and detailed geologic time scale and to date the events that have occurred throughout Earth's 4.56-billion-year history.

FIGURE 9.1 Two photographs of Bowknot Bend on the Green River in Utah, taken nearly 100 years apart, show that the configuration of rocks and geologic structures has changed very little in that time interval. [(left) E. O. Beaman/USGS; (right) H. G. Stevens/USGS.]

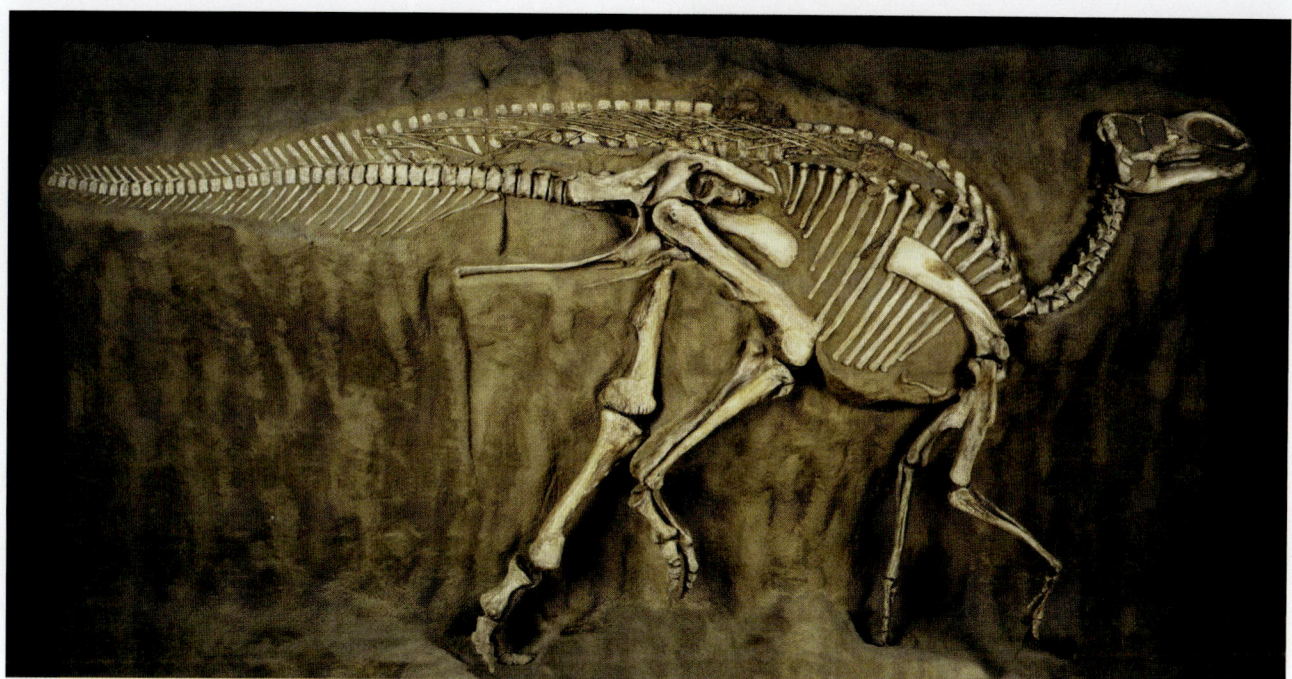

(a)

FIGURE 9.2 Fossils are traces of living organisms preserved in the geologic record. (a) Fossilized skeleton found at Dinosaur National Monument near the Colorado-Utah line. This individual is a Gryposaurus, a hook-nosed duckbill dinosaur that lived during the Late Cretaceous. (b) Petrified Forest, Arizona. These ancient logs are more than two hundred million years old. Their substance has been completely replaced by silica, which preserved the details of their original forms. [(a) Dorling Kindersley/Science Source. (b) Thinkstock.]

(b)

Reconstructing Geologic History From the Stratigraphic Record

Geologists speak carefully about time. To them, *dating* refers not to a popular social activity, but to measuring the **absolute age** of an event in the geologic record: the number of years elapsed from that event until now. Before the twentieth century, no one knew much about absolute ages; geologists could determine only whether one event was earlier or later than another—their **relative age.** They could say, for instance, that fish bones were first deposited in marine sediments before mammal bones first appeared in sediments on land, but they couldn't tell how many millions of years ago the first fish or mammals appeared.

The first geologic observations pertaining to the question of deep time came in the mid-seventeenth century from the study of fossils. A *fossil* is an artifact of life preserved in the geologic record (**Figure 9.2**). However, few people living in seventeenth-century Europe would have understood this definition. Most thought that the seashells and other lifelike forms they found in rocks dated from Earth's beginnings about 6000 years earlier, or grew there spontaneously.

In 1667, the Danish scientist Nicolaus Steno, who was working for the royal court in Florence, Italy, demonstrated that the peculiar "tongue stones" found in certain Mediterranean sedimentary rocks were essentially identical to the teeth of modern sharks (**Figure 9.3**). He concluded that tongue stones *really were* ancient shark teeth preserved in the rocks and, more generally, that fossils were the remains of ancient life deposited with sediments. To convince people of his ideas, Steno wrote a short but brilliant book about the geology of Tuscany, in which he laid the foundation for the modern science of **stratigraphy**—the study of *strata* (layers) in rocks.

Principles of Stratigraphy

Geologists still use the principles set forth by Steno to interpret sedimentary strata. Two of his basic rules are so simple they seem obvious to us today:

1. The **principle of original horizontality** states that sediments are deposited under the influence of gravity as nearly horizontal beds. Observations in a wide variety of sedimentary environments support this principle. If we find folded or faulted strata, we know that the beds were deformed by tectonic forces after the sediments were deposited.

2. The **principle of superposition** states that each layer of an undeformed sedimentary sequence is younger than the one beneath it and older than the one above it. A new layer cannot be deposited beneath an existing layer. Thus, strata can be vertically ordered in time from the lowest (oldest) bed to the uppermost (youngest) bed (**Figure 9.4**). A chronologically ordered set of strata is known as a **stratigraphic succession.**

We can apply Steno's principles in the field to determine whether one sedimentary formation is older than another. Then, by piecing together the formations exposed in different outcrops, we can sort them into chronological order and thus construct the stratigraphic succession of a region—at least in principle.

In practice, there were two problems with this strategy. First, geologists almost always found gaps in a region's stratigraphic succession, indicating time intervals that had gone entirely unrecorded. Some of these intervals were short, such as periods of drought between floods. Others lasted for millions of years—for example, periods of regional tectonic uplift when thick sequences of sedimentary rocks were removed by erosion. Second, it was difficult to determine the relative ages of two formations that were widely separated in space. Stratigraphy alone couldn't determine whether a sequence of mudstones in, say, Tuscany was older, younger, or the same age as a similar sequence in England. It was necessary to expand Steno's ideas about the biological origin of fossils to solve these problems.

Fossils as Recorders of Geologic Time

In 1793, William Smith, a surveyor working on the construction of canals in southern England, recognized that fossils could help geologists determine the relative ages of sedimentary rocks. Smith was fascinated by the variety of fossils he collected from the strata exposed along canal cuts. He observed that different layers contained different sets of fossils, and he was able to tell one layer from another by the characteristic fossils in each. He

(a)

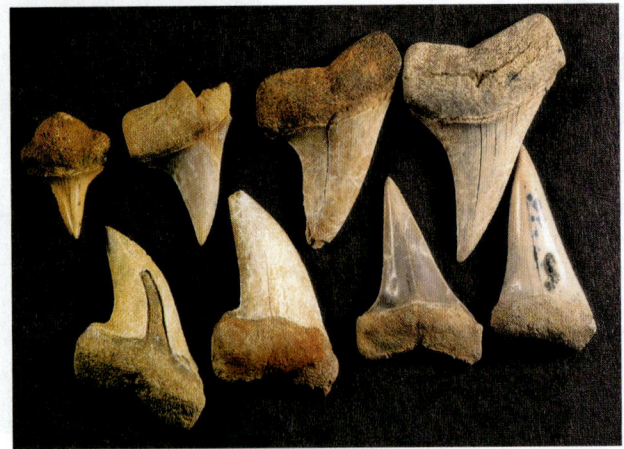
(b)

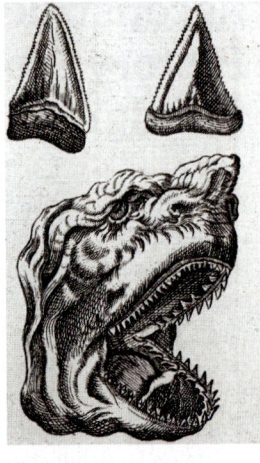
(c)

FIGURE 9.3 Nicolaus Steno was the first to demonstrate that fossils are the remains of ancient life. (a) A portrait of Nicolaus Steno (1638–1686). (b) "Tongue stones" of the type found in sedimentary rocks in the Mediterranean region, where Steno worked. (c) This diagram is from Steno's 1667 book, which demonstrated that tongue stones are the fossilized teeth of ancient sharks. [(left) SPL/Science Source; (center) Corbis RF/Alamy; (right) Paul D. Stewart/SPL/Science Source.]

FIGURE 9.4 Steno's principles guide the study of sedimentary strata. (a) Sediments are deposited in horizontal layers before being slowly transformed into sedimentary rock. If left undisturbed by tectonic processes, the youngest layers remain on the top and the oldest on the bottom. (b) Marble Canyon, part of the Grand Canyon, was cut by the Colorado River through what is now northern Arizona, revealing these undisturbed strata, which record millions of years of geologic history. [W. K. Fletcher/Science Source.]

established a general order for the sequence of fossil assemblages and strata, from lowest (oldest) to uppermost (youngest). Regardless of its location, Smith could predict the stratigraphic position of any particular layer or formation in any outcrop in southern England based on its fossil assemblages. This stratigraphic ordering of the fossils of animal species (fauna) produces a sequence known as a *faunal succession*. Smith's **principle of faunal succession** states that the sedimentary strata in an outcrop contain fossils in a definite sequence. The same sequence can be found in outcrops at other locations, so that strata in one location can be matched to strata in another location.

Using faunal successions, Smith was able to identify formations of the same age in different outcrops. By noting the vertical order in which the formations were found in each place, he compiled a composite stratigraphic succession for the entire region. His compilation showed how the complete succession would have looked if the formations at different levels in all the various outcrops could have been brought together at a single spot. Figure 9.5 shows such a composite stratigraphic succession for two outcrops. Smith

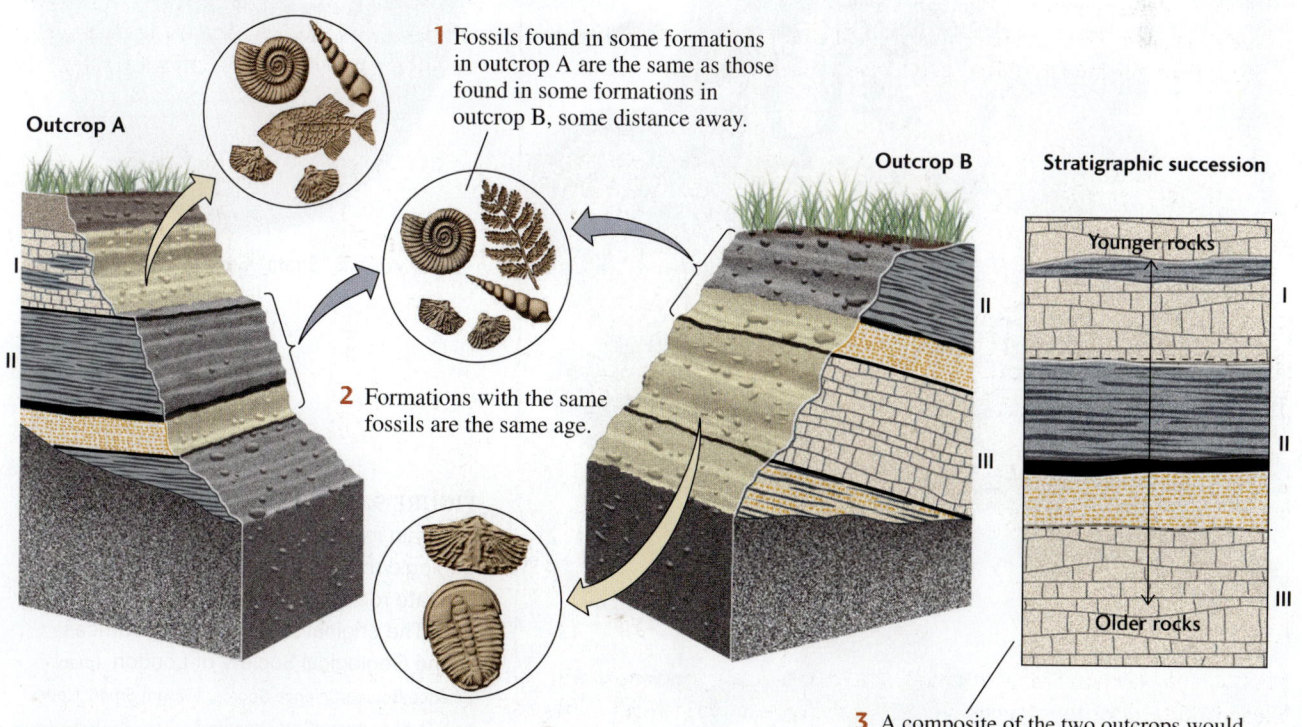

FIGURE 9.5 The principle of faunal succession can be used to correlate rock formations in different outcrops.

kept track of his work by mapping outcrops using colors assigned to specific formations, thus inventing the geologic map. In 1815, he summarized his lifelong research by publishing his *General Map of Strata in England and Wales*, a hand-colored masterpiece 8 feet tall and 6 feet wide—the first geologic map of an entire country (Figure 9.6).

The geologists who followed in Steno's and Smith's footsteps described and catalogued hundreds of fossils and their relationships to modern organisms, establishing the new science of *paleontology*: the historical study of ancient lifeforms. The most common fossils they found were the shells of invertebrate animals. Some were similar to clams, oysters, and other living shellfish; others represented strange species with no living examples, such as the trilobites shown in the chapter opening photo. Less common were the bones of vertebrates, such as mammals, birds, and the huge extinct reptiles they called dinosaurs. Plant fossils were found to be abundant in some rocks, particularly in coal beds, where leaves, twigs, branches, and even whole tree trunks could be recognized. Fossils were not found in intrusive igneous rocks—no surprise, since any biological material would have been destroyed when the rocks melted—nor in high-grade metamorphic rocks—where any remains of organisms would have been distorted beyond recognition.

By the beginning of the nineteenth century, paleontology had become the single most important source of information about geologic history. The systematic study of fossils affected science far beyond geology, however.

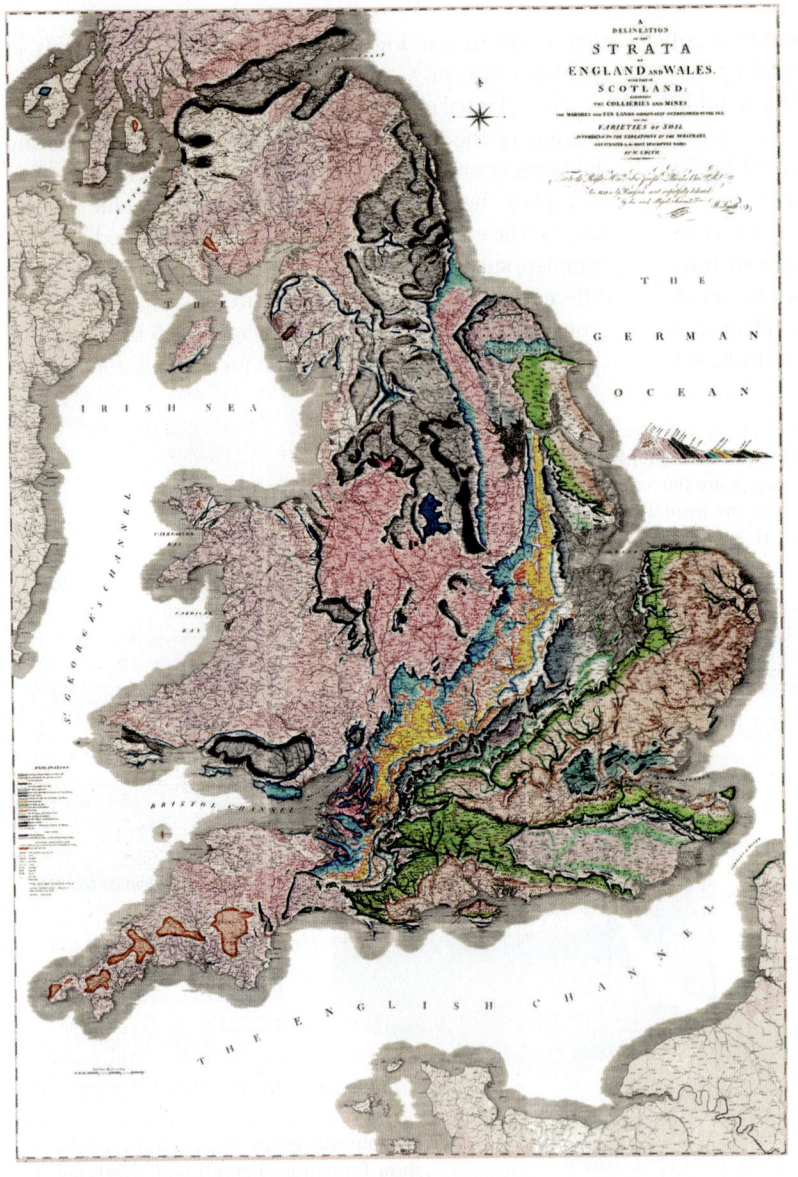

William "Strata" Smith, 1769–1839

FIGURE 9.6 William Smith's *General Map of Strata in England and Wales*, the first geologic map of an entire country. The colors indicate rock formations of the same relative age. The original still hangs in the offices of the Geological Society of London. [*map:* Science Source/Science Source; *William Smith:* New York Public Library/Getty Images.]

Charles Darwin studied paleontology as a young scientist, and he collected many unusual fossils on his famous voyage aboard the *Beagle* (1831–1836). During this world-circling tour, he also studied many unfamiliar animal and plant species in their native habitats. Darwin pondered what he had seen until 1859, when he proposed his theory of evolution by natural selection. His theory revolutionized the science of biology and provided a sound theoretical framework for paleontology: If organisms evolve progressively over time, then the fossils in each sedimentary bed must represent the organisms living when that bed was deposited.

Unconformities: Gaps in the Geologic Record

In compiling the stratigraphic succession of a region, geologists often find places in the geologic record where a formation is missing. Either no rock was ever deposited, or it was eroded away before the next strata were laid down. The surface between two beds that were laid down with a time gap between them—the boundary representing the missing time—is called an **unconformity** (Figure 9.7). A series of beds bounded above and below by unconformities is referred to as a *sedimentary sequence*. An unconformity, like a sedimentary sequence, represents the passage of time.

An unconformity may imply that tectonic forces raised the rock above sea level, where erosion removed some rock layers. Alternatively, the unconformity may have been produced by the erosion of newly exposed rock as sea level fell. As we will see in Chapter 12, global sea level can be lowered by hundreds of meters during ice ages, when water is withdrawn from the oceans to form continental ice sheets.

Unconformities are classified according to the relationships between the layers above and below them:

- A *disconformity* is an unconformity at which an upper sedimentary sequence overlies an erosional surface developed on an undeformed, still-horizontal lower sedimentary sequence (see Figure 9.7). Disconformities are often created when sea level drops or during broad tectonic uplifts.

- A *nonconformity* is an unconformity at which the upper sedimentary beds overlie metamorphic or igneous rocks (see Earth Issues 9.1 for an example).

- An *angular unconformity* is an unconformity at which the upper beds overlie lower beds that have been folded by tectonic processes and then eroded to a more or less even plane. In an angular unconformity, the two sequences have bedding planes that are not parallel (Figure 9.8). The formation of an angular unconformity by tectonic processes is illustrated in Figure 9.9.

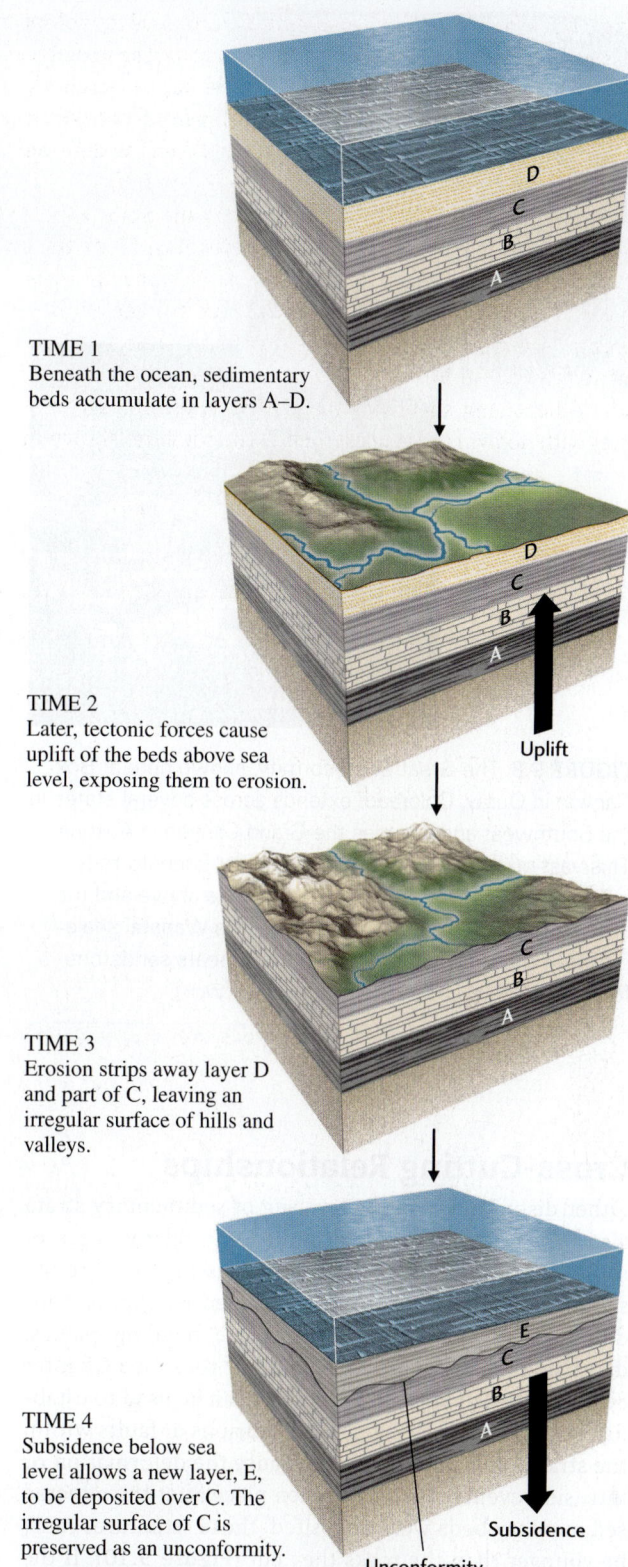

TIME 1 Beneath the ocean, sedimentary beds accumulate in layers A–D.

TIME 2 Later, tectonic forces cause uplift of the beds above sea level, exposing them to erosion.

TIME 3 Erosion strips away layer D and part of C, leaving an irregular surface of hills and valleys.

TIME 4 Subsidence below sea level allows a new layer, E, to be deposited over C. The irregular surface of C is preserved as an unconformity.

FIGURE 9.7 An unconformity is a surface between two rock layers representing a layer that never formed or was eroded away. The type of unconformity shown here is a disconformity, created through uplift and erosion followed by subsidence and another round of sedimentation on top of an undeformed surface.

FIGURE 9.8 The Great Unconformity, shown here at Box Canyon in Ouray, Colorado, extends across several states in the Southwest and includes the Grand Canyon in Arizona. This vast geologic formation is an angular unconformity between the horizontal Tapeats sandstone above and the steeply dipping Wapatai shale below. The Wapatai shale is part of the Canyon beds, while the Tapeats sandstone formed during the Cambrian period. [Ron Wolf.]

Cross-Cutting Relationships

Other disturbances of the layering of sedimentary strata also provide clues for determining the relative ages of rocks. Recall from Chapter 4 that dikes can cut through sedimentary beds; sills can be intruded parallel to bedding planes; and faults can displace bedding planes, dikes, and sills as they shift blocks of rock (see Chapter 8). These *cross-cutting relationships* can be used to establish the relative ages of igneous intrusions or faults within the stratigraphic succession. Because the deformation or intrusion events must have taken place after the affected sedimentary beds were deposited, those structures must be younger than the rocks they cut (**Figure 9.10**). If the intrusions or fault displacements are eroded and planed off and then overlaid by younger sedimentary beds, we know that those structures are older than the younger strata.

Geologists can combine field observations of cross-cutting relationships, unconformities, and stratigraphic successions to decipher the history of geologically

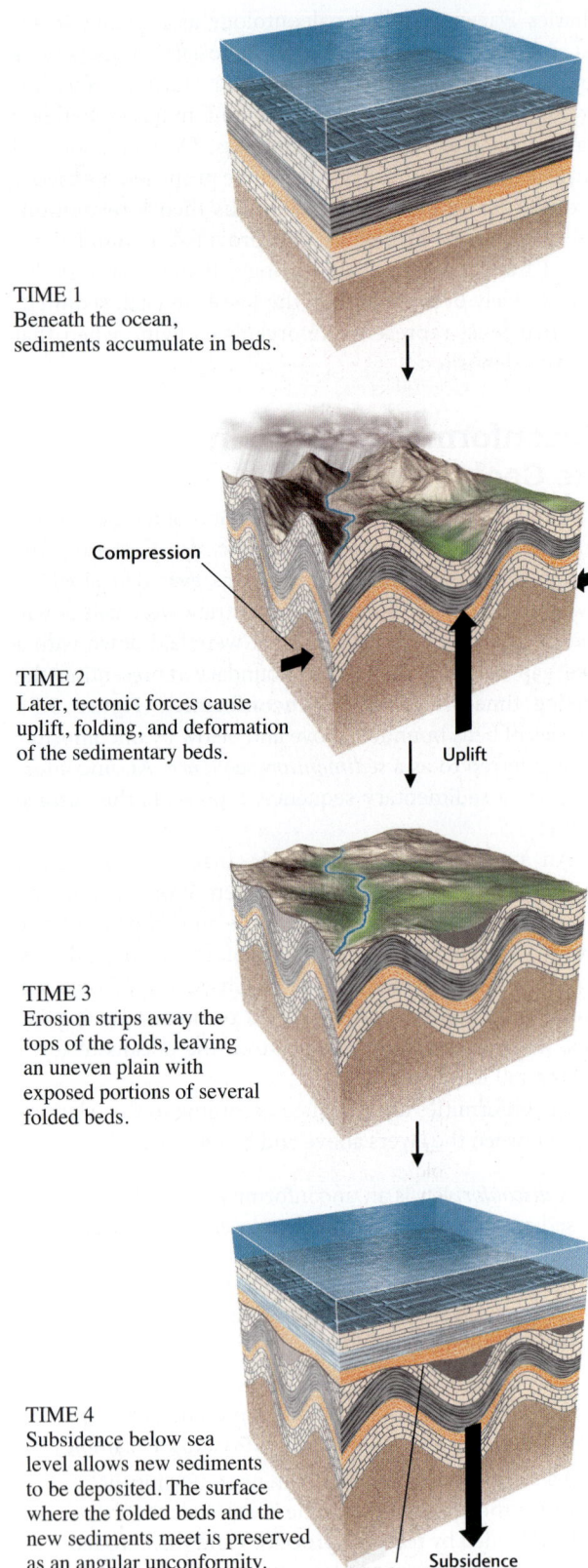

TIME 1
Beneath the ocean, sediments accumulate in beds.

TIME 2
Later, tectonic forces cause uplift, folding, and deformation of the sedimentary beds.

TIME 3
Erosion strips away the tops of the folds, leaving an uneven plain with exposed portions of several folded beds.

TIME 4
Subsidence below sea level allows new sediments to be deposited. The surface where the folded beds and the new sediments meet is preserved as an angular unconformity.

FIGURE 9.9 An angular unconformity is a surface that separates two sedimentary sequences whose bedding planes are not parallel. This series of drawings shows how an angular unconformity can form.

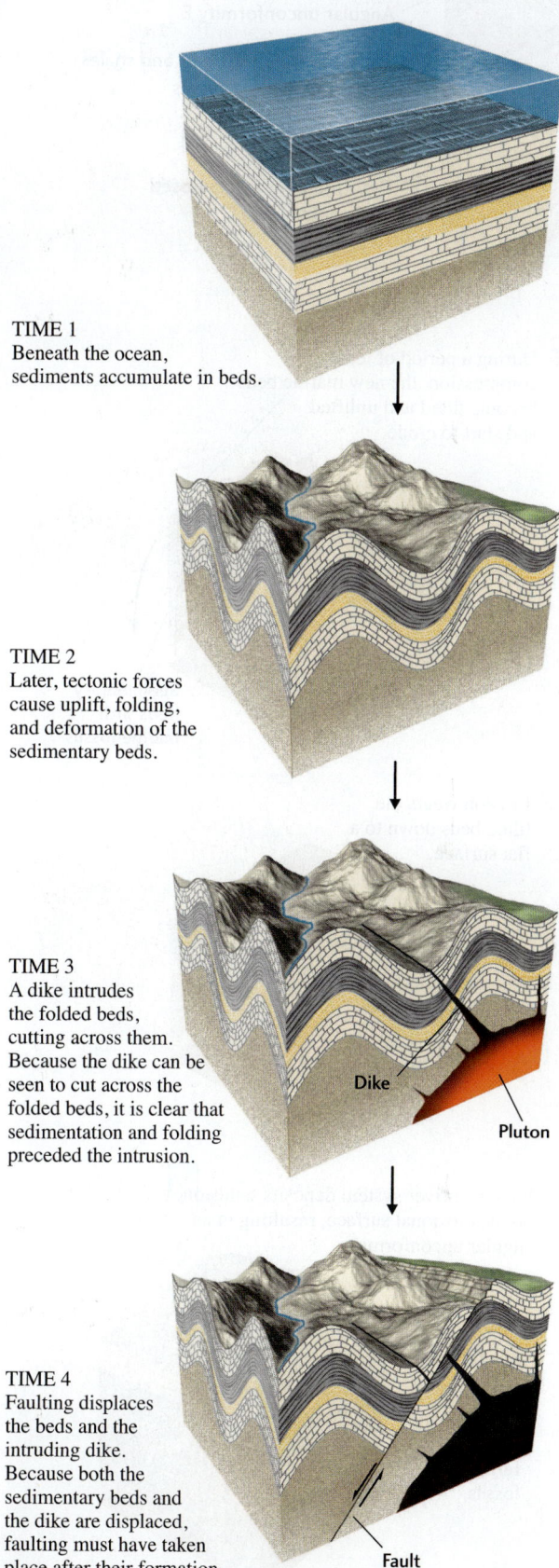

TIME 1
Beneath the ocean, sediments accumulate in beds.

TIME 2
Later, tectonic forces cause uplift, folding, and deformation of the sedimentary beds.

TIME 3
A dike intrudes the folded beds, cutting across them. Because the dike can be seen to cut across the folded beds, it is clear that sedimentation and folding preceded the intrusion.

TIME 4
Faulting displaces the beds and the intruding dike. Because both the sedimentary beds and the dike are displaced, faulting must have taken place after their formation.

FIGURE 9.10 Cross-cutting relationships allow geologists to establish the relative ages of igneous intrusions and faults within a stratigraphic succession.

complicated regions (**Figure 9.11**). Earth Issues 9.1 gives a more detailed example of how geologists work backward in time to determine the relative ages of the rocks in a region.

The Geologic Time Scale: Relative Ages

Early in the nineteenth century, geologists began to apply Steno's and Smith's stratigraphic principles to outcrops all over the world. They discovered the same distinctive fossils in similar sedimentary formations on many continents. They also found that faunal successions from different continents often displayed the same changes in fossil assemblages. By matching up faunal successions and using cross-cutting relationships, geologists could determine the relative ages of rock formations on a global basis. By the end of the century, they had pieced together a worldwide history of geologic events—a **geologic time scale.**

Intervals of Geologic Time

The geologic time scale divides Earth's history into intervals marked by distinct sets of fossils, and it places the boundaries of those intervals at times when those sets of fossils changed abruptly (**Figure 9.12**). The basic divisions of this time scale are the **eras:** the Paleozoic (from the Greek *paleo,* meaning "old," and *zoi,* meaning "life"), the Mesozoic ("middle life"), and the Cenozoic ("new life").

The eras are subdivided into **periods,** usually named for the locality in which the formations representing them were first or best described, or for some distinguishing characteristic of the formations. The Jurassic period, for example, is named for the Jura mountain range of France and Switzerland, and the Carboniferous period is named for the coal-bearing sedimentary rocks of Europe and North America. The Paleogene and Neogene periods of the Cenozoic are two exceptions: These Greek names mean "old origin" and "new origin," respectively.

Some periods are further subdivided into **epochs,** such as the Miocene, Pliocene, and Pleistocene epochs of the Neogene period (see Figure 9.12). The most recent is the Holocene ("completely new") epoch of the Neogene period in the Cenozoic era.

Interval Boundaries Mark Mass Extinctions

Many of the major boundaries in the geologic time scale represent **mass extinctions:** short intervals during which a large proportion of the species living at the time simply disappeared from the fossil record, followed by the blossoming of many new species. These abrupt changes in faunal successions were a great mystery to the geologists who discovered them. Darwin's theory of evolution explained how new species could evolve, but what had caused the mass extinctions?

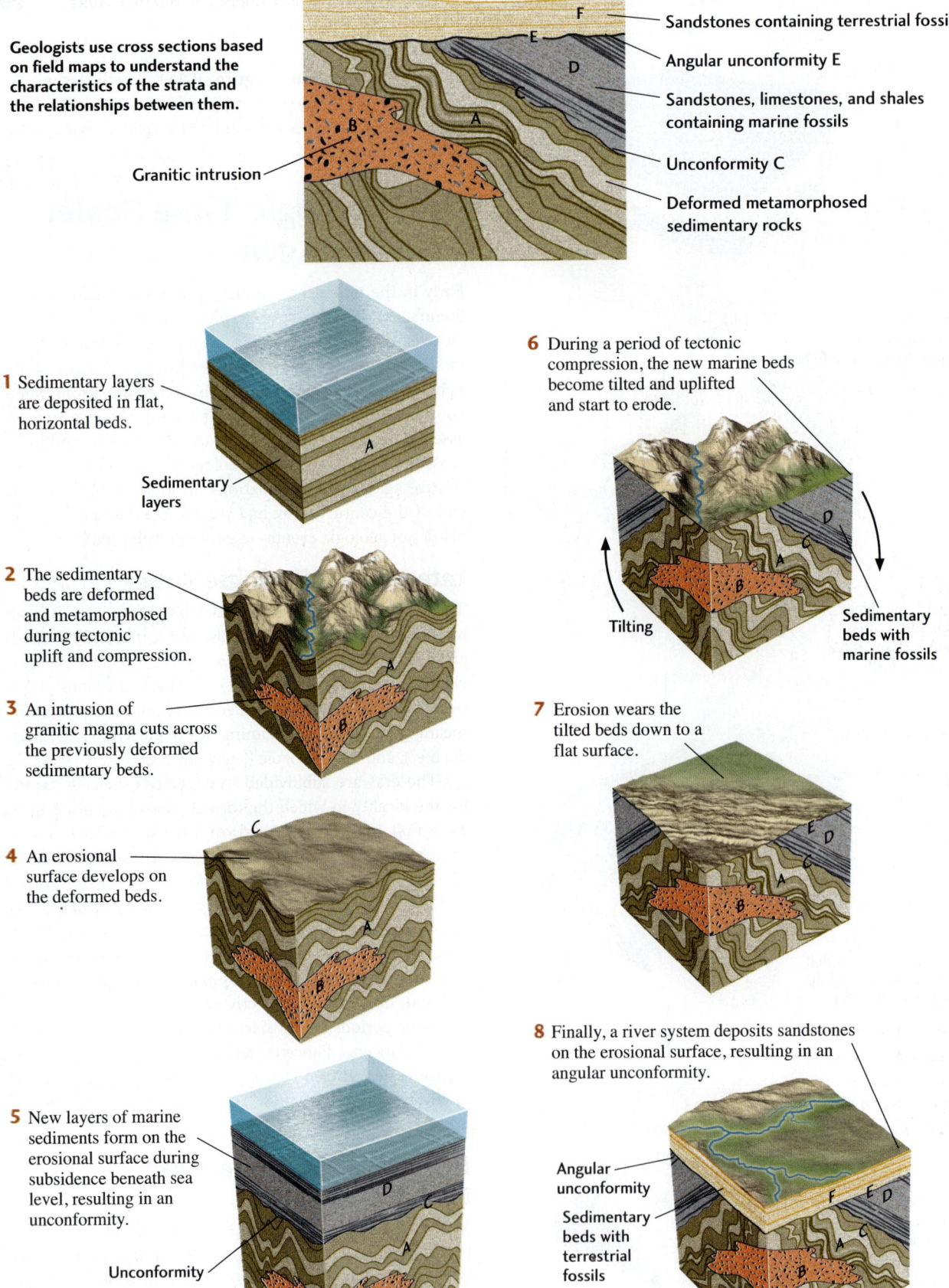

FIGURE 9.11 Geologists use stratigraphic principles and cross-cutting relationships to establish a relative chronology of geologic events.

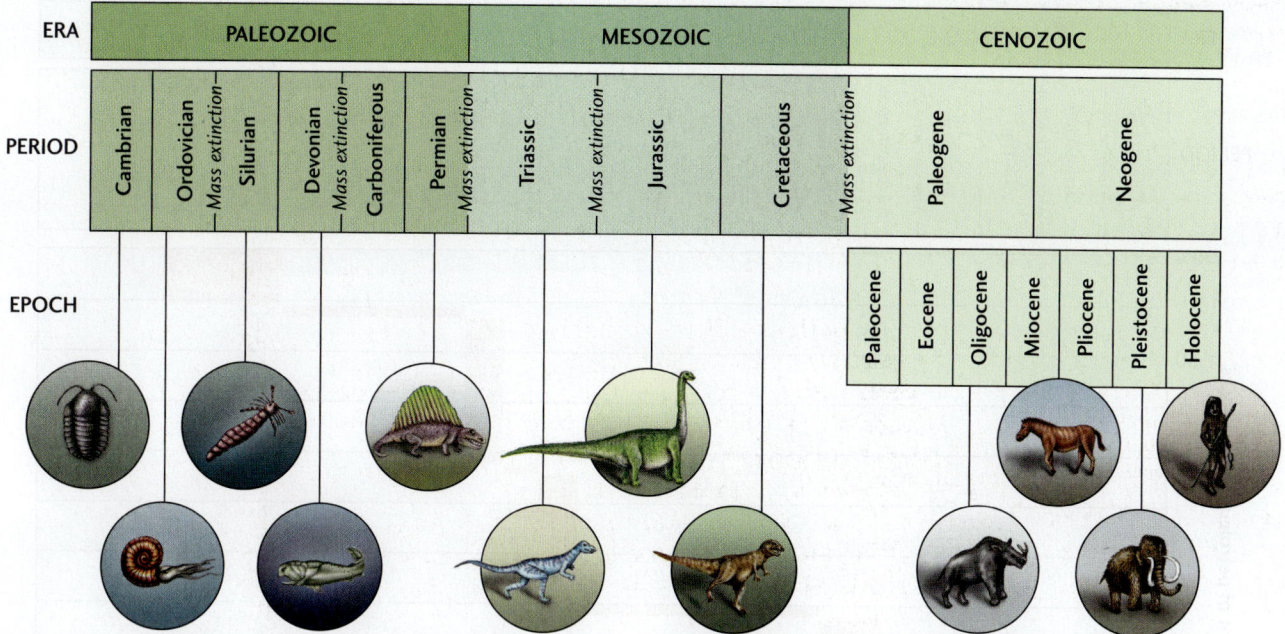

FIGURE 9.12 The geologic time scale, showing the relative ages of the eras, periods, and epochs of the Phanerozoic eon distinguished by assemblages of fossils. The boundaries of these intervals are marked by the abrupt disappearance of some life-forms and the appearance of new ones. The five most dramatic mass extinctions are indicated. Note that this diagram shows only the relative ages of the intervals.

In some cases, we think we know. The mass extinction at the end of the Cretaceous period, which killed off 75 percent of the living species, including essentially all of the dinosaurs, was almost certainly the result of a large meteorite impact that darkened and poisoned the atmosphere and plunged Earth's climate into many years of bitter cold. This disaster marks the end of the Mesozoic era and the beginning of the Cenozoic. In other cases, we are still not sure. The largest mass extinction, at the end of the Permian period, which defines the Paleozoic-Mesozoic boundary, eliminated nearly 95 percent of all living species, but the cause of this event is still debated. The extreme events that separate intervals of geologic time are the subject of very active research, as we will see in Chapter 22.

Ages of Petroleum Source Rocks

Oil and natural gas come from organic matter that was buried in sedimentary rock formations at some time in the geologic past. The relative ages of these "petroleum source rocks" provide important clues about where to look for new oil and gas resources. Global surveys have shown that very little petroleum has come from Precambrian rocks, which makes sense, because the primitive organisms that existed before the Cambrian period generated little organic matter.

Petroleum source rocks were deposited during all three of the geologic eras following the Cambrian, although certain periods of geologic time have produced much more of this resource than others (**Figure 9.13**). The clear winners are the Jurassic and Cretaceous periods of the Mesozoic era, which together have accounted for almost 60 percent of the world's petroleum production. Sedimentary formations of Jurassic and Cretaceous age were the source rocks for the great oil fields of the Middle East, the Gulf of Mexico, Venezuela, and the North Slope of Alaska.

If you examine Figure 2.16, you can see that during these periods of geologic time, the supercontinent of Pangaea was breaking up into the modern continents. This tectonic activity formed many marine sedimentary basins and increased the rate at which sediments were deposited into these basins. During the Jurassic and Cretaceous periods, which include the Age of Dinosaurs, marine life was abundant, providing much of the organic matter that was buried in the sediments. This carbon-rich material has since been "cooked" and transported into the oil reservoirs, where we find it today.

Measuring Absolute Time with Isotopic Clocks

The geologic time scale based on stratigraphy and faunal successions is a relative time scale. It tells us whether one formation or fossil assemblage is older than another, but not how long the eras, periods, and epochs were in actual years. Estimates of how long it takes mountains to erode and sediments to accumulate suggested that most geologic periods had lasted for millions of years, but nineteenth-century geologists did not know whether the duration of

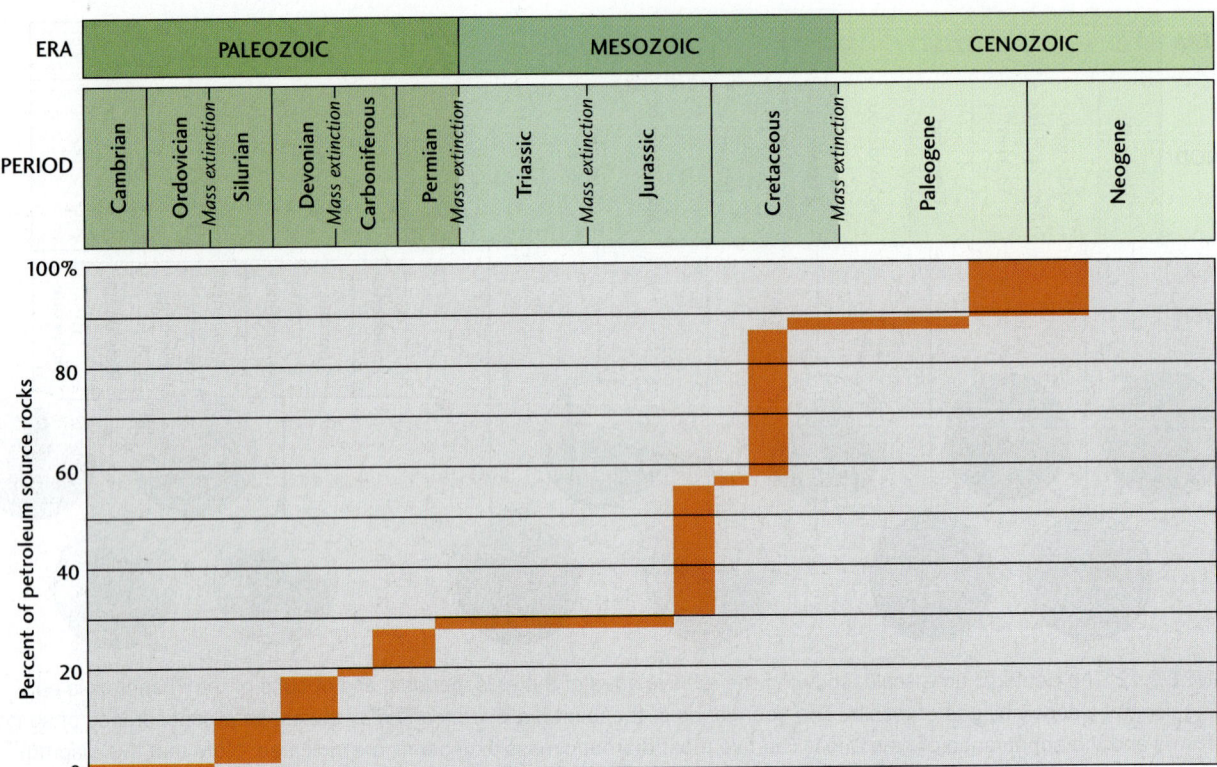

FIGURE 9.13 Relative ages and amounts of sedimentary rocks that contained the organic matter now found as oil and natural gas. Bars in the lower graph show the percentage of these petroleum source rocks found worldwide (height of the bar) within a given age range (width of the bar). Almost 60 percent of the total inventory was deposited during the Jurassic and Cretaceous periods of the Mesozoic era. [Data from H. D. Klemme & G. F. Ulmishek, *AAPG Bulletin*, 1999.]

any specific period was 10 million years, 100 million years, or even longer.

They did know that the geologic time scale was incomplete. The earliest period of geologic history recorded by faunal successions was the Cambrian, when animal life, in the form of shelly fossils, suddenly appeared in the geologic record. Many rock formations were clearly older, however, because they occurred below Cambrian rocks in stratigraphic successions. But these formations contained no recognizable fossils, so there was no way to determine their relative ages. All such rocks were lumped into the general category **Precambrian.** What fraction of Earth's history was locked up in these cryptic rocks? How old was the oldest Precambrian rock? How old was Earth itself?

These questions sparked a huge debate in the latter half of the nineteenth century. Physicists and astronomers argued for a maximum age of less than 100 million years, but most geologists regarded this age as much too young, even though they had no precise data to back them up.

Discovery of Radioactivity

In 1896, a major advance in physics paved the way for reliable and accurate measurements of absolute ages. Henri Becquerel, a French physicist, discovered radioactivity in uranium. Within three years, the French chemist Marie Sklodowska-Curie discovered and isolated a new and highly radioactive element, radium.

In 1905, the physicist Ernest Rutherford suggested that the absolute age of a rock could be determined by measuring the decay of radioactive elements found in it. He calculated the age of one rock from measurements of its uranium content. This was the start of **isotopic dating,** the use of naturally occurring radioactive elements to determine the ages of rocks. Isotopic dating methods were refined over the next few years as more radioactive elements were discovered and the processes of radioactive decay became better understood. Within a decade of Rutherford's first attempt, geologists were able to show that some Precambrian rocks were billions of years old.

In 1956, the geochemist Clair Patterson measured the decay of uranium in meteorites and terrestrial rocks to determine that the solar system—and, by implication, Earth—was formed 4.56 billion years ago. That age has been modified by less than 10 million years since Patterson's original measurement, so we might say that he completed the discovery of geologic time.

Earth Issues

9.1 Stratigraphy of the Colorado Plateau: An Exercise in Relative Dating

We can use the strata exposed in the Grand Canyon and other parts of the Colorado Plateau to illustrate how relative dating works (see the figure on the next page). These beds record a long history of sedimentation in a variety of environments, sometimes on land and sometimes under water. By matching the rock formations exposed at different localities, geologists have constructed a stratigraphic succession over a billion years long that spans both the Paleozoic and Mesozoic eras.

The lowest—and therefore oldest—rocks exposed in the Grand Canyon are dark igneous and metamorphic rocks that make up the Vishnu Schist, a group of formations that have been shown to be about 1.8 billion years old.

Above the Vishnu Schist are the younger Grand Canyon Beds. Although these sedimentary rocks contain fossils of single-celled microorganisms that provide evidence of early life, they do not contain the shelly fossils distinctive of the Cambrian and later periods, so they are categorized as Precambrian rocks.

A nonconformity separates the Vishnu Schist and the Grand Canyon Beds. This structure indicates a period of deformation accompanying metamorphism of the Vishnu Schist, and then a period of erosion, before the deposition of the Grand Canyon Beds. The tilting of the Grand Canyon Beds from their originally horizontal position shows that they, too, were folded after deposition and burial.

A major angular unconformity known as the Great Unconformity divides the Grand Canyon Beds from the overlying horizontal Tapeats Sandstone (see Figure 9.8). This unconformity indicates a long period of erosion after the lower rocks had been tilted. The Tapeats Sandstone and Bright Angel Shale can be dated as Cambrian by their fossils, many of which are trilobites.

Above the Bright Angel Shale is a group of horizontal limestone and shale formations (Muav Limestone, Temple Butte Limestone, Redwall Limestone) that represent a roughly 200-million-year history of marine sedimentation from the late Cambrian to the Carboniferous. There are so many time gaps represented by disconformities in these rocks that less than 40 percent of the Paleozoic is actually represented by rock strata.

The next set of strata, high up on the canyon wall, is the Supai Formation (Carboniferous and Permian), which contains fossils of land plants like those found in the coal beds of North America and other continents. Overlying the Supai Formation is the Hermit Shale, a sandy red shale deposited in a terrestrial environment.

Continuing up the canyon wall, we find another continental deposit, the Coconino Sandstone. This formation contains vertebrate animal tracks, which indicates that the Coconino was formed in a terrestrial environment during the Permian period. At the top of the cliffs at the canyon rim are two more formations of Permian age: The Toroweap, made mostly of limestone, is overlain by the Kaibab, a massive layer of sandy and cherty limestone. These two formations record subsidence below sea level and the deposition of marine sediments.

Above the Kaibab Limestone and the canyon rim itself, but exposed within Grand Canyon National Park, is the Moenkopi Formation, a red sandstone of Triassic age—the first appearance of rocks from the Mesozoic era in this stratigraphic succession.

The stratigraphic succession at the Grand Canyon, though picturesque and informative, represents an incomplete picture of Earth's history. Younger intervals of geologic time are not preserved there and we must travel to nearby locations in Utah, such as Zion Canyon and Bryce Canyon, to fill in this more recent history. At Zion Canyon, we find equivalents of the Kaibab Limestone and Moenkopi Formation, which allow us to correlate this stratigraphic succession with the one at the Grand Canyon and establish a link. Unlike the Grand Canyon strata, however, the Zion strata extend upward to Jurassic time, including ancient sand dunes represented by the Navajo Sandstone. In Bryce Canyon, to the east of Zion, we again find the Navajo Sandstone, as well as strata that extend still farther upward, to the Wasatch Formation of the Paleogene period.

The correlation of strata among these three areas of the Colorado Plateau shows how widely separated localities—each with an incomplete record of geologic time—can be pieced together to build a composite record of Earth's history.

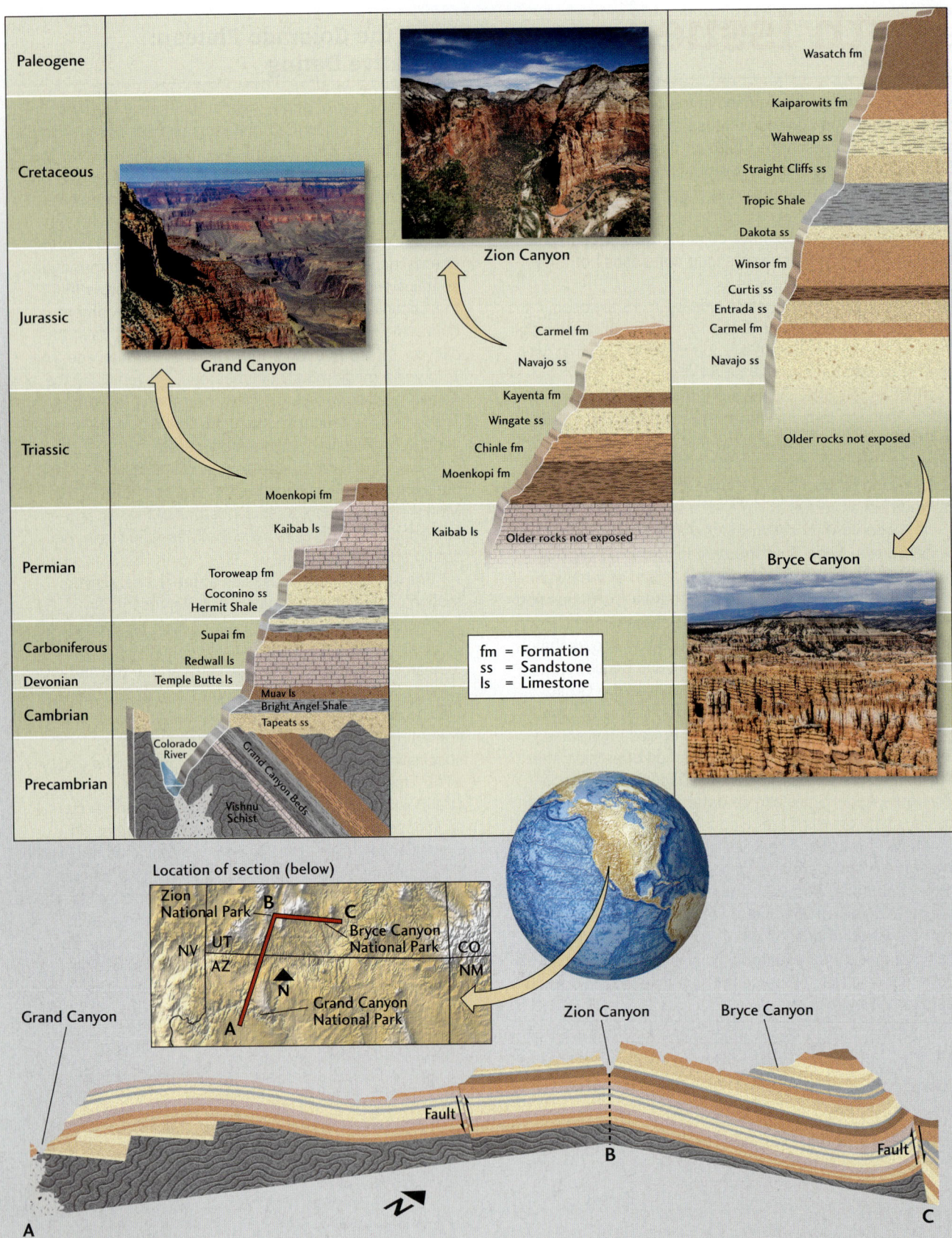

Stratigraphic succession of the Colorado Plateau, reconstructed from strata exposed in Grand Canyon, Zion Canyon, and Bryce Canyon National Parks. [*Grand Canyon:* Richard A. McMillin/Shutterstock; *Zion Canyon:* Bryan Brazil/Shutterstock; *Bryce Canyon:* Filip Fuxa/Shutterstock.]

Radioactive Isotopes: The Clocks in Rocks

How is radioactivity used to determine the age of a rock? Recall that the nucleus of an atom consists of protons and neutrons. For a given element, the number of protons is constant, but the number of neutrons can vary among different *isotopes* of the same element (see Chapter 3). Most isotopes are stable, but the nucleus of a *radioactive* isotope can spontaneously disintegrate, or *decay*, emitting particles and transforming the atom into an atom of a different element. The original atom is called the *parent* and the product of decay, its *daughter*.

One useful element for isotopic dating is rubidium, which has 37 protons and two naturally occurring isotopes: rubidium-85, which has 48 neutrons and is stable, and rubidium-87, which has 50 neutrons and is radioactive. A neutron in the nucleus of a rubidium-87 atom can spontaneously emit an electron, thus changing into a proton, which remains in the nucleus. The parent rubidium atom therefore forms a daughter strontium-87 atom, with 38 protons and 49 neutrons (**Figure 9.14**). Thus, isotopes that decay slowly over billions of years, such as rubidium-87, are most useful in measuring the ages of older rocks, whereas those that decay rapidly, such as carbon-14, can only be used to date younger rocks (see **Table 9.1** and the Practicing Geology Exercise).

A parent isotope decays into a daughter isotope at a constant rate. The rate of radioactive decay is measured by the isotope's **half-life**: the time required for one-half of the original number of parent atoms to be transformed into daughter atoms. At the end of the first half-life, the number of parent atoms has decreased by a factor of two; at the end of the second half-life, by a factor of four; at the end of the third half-life, by a factor of eight; and so forth. As the parent decays, the amount of the daughter isotope increases, preserving the total number of atoms (**Figure 9.15**). The half-lives of radioactive elements commonly used for isotopic dating are given in Table 9.1.

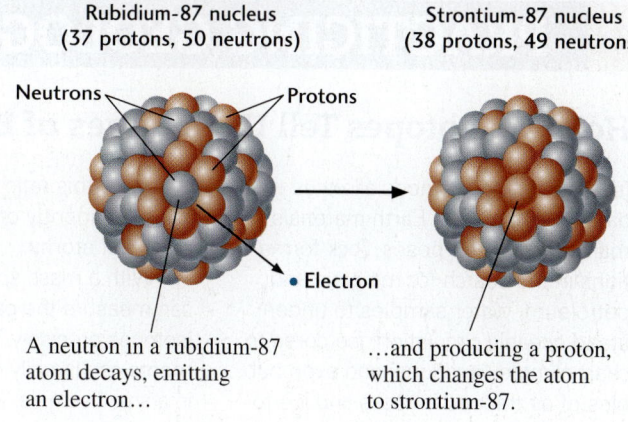

FIGURE 9.14 The radioactive decay of rubidium to strontium.

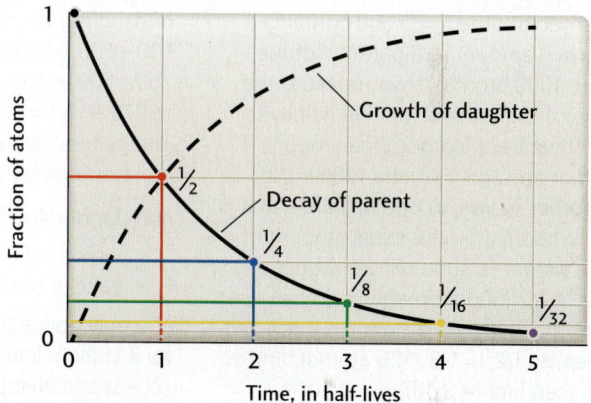

FIGURE 9.15 The fraction of atoms of the parent isotope declines at a constant rate over time. This rate of decay is measured by the half-life of the isotope. As the parent decays, the amount of the daughter isotope grows, preserving the total number of atoms.

TABLE 9-1 Major Radioactive Elements Used in Isotopic Dating

ISOTOPES		HALF-LIFE OF PARENT (YEARS)	EFFECTIVE DATING RANGE (YEARS)	EXAMPLES OF MINERALS AND MATERIALS THAT CAN BE DATED
PARENT	DAUGHTER			
Rubidium-87	Strontium-87	49 billion	10 million–4.6 billion	Muscovite, biotite, orthoclase feldspar
Uranium-238	Lead-206	4.5 billion	10 million–4.6 billion	Zircon, apatite
Potassium-40	Argon-40	1.3 billion	50,000–4.6 billion	Muscovite, biotite, hornblende
Uranium-235	Lead-207	0.7 billion	10 million–4.6 billion	Zircon, apatite
Carbon-14	Nitrogen-14	5730	100–70,000	Wood, charcoal, peat; bone and tissue; shells and other calcium carbonates

PRACTICING GEOLOGY EXERCISE

How Do Isotopes Tell Us the Ages of Earth Materials?

Isotopic dating methods allow us to date many types of Earth materials for many practical purposes: rock formations in the search for minerals and petroleum; water samples to understand oceanic circulation; ice cores to chart climate variations; and even bubbles of air trapped in rocks and ice to measure changes in the composition of the atmosphere. So it's worthwhile to understand in more detail how geologists actually determine the ages of materials using isotopes.

Consider a mineral grain that was formed at time $T = 0$ and contains a certain amount of a parent isotope—say, 1000 atoms. If we measure the age of the mineral grain in half-lives of the parent isotope, the amount left at any age T will be $1000 \times 1/2^T$. In other words, in one half-life—that is, when $T = 1$—the initial amount of the parent isotope will be reduced to $1/2^1 = 1/2$ (500 atoms); in two half-lives, to $1/2^2 = 1/4$ (250 atoms); in three half-lives, to $1/2^3 = 1/8$ (125 atoms); and so on (see Figure 9.14).

The radioactive decay of each atom of the parent isotope generates one new atom of the daughter isotope. If the mineral grain remains a closed system (that is, if no isotopes are transferred into or out of the grain), the number of new daughter atoms produced from the parent atoms by age T must equal $1000 \times (1 - 1/2^T)$, because the new daughters and remaining parents must add up to the initial amount of the parent isotope (1000 atoms). The ratio of new daughters to remaining parents thus depends only on the age of the mineral grain:

$$\left(\frac{\text{number of new daughters}}{\text{number of remaining parents}}\right) =$$

$$\frac{1 - 1/2^T}{1/2^T} = 2^T - 1$$

As the age of the mineral grain increases from 0 to 3 half-lives, for instance, this ratio increases from 0 to 7, independently of the initial number of parent atoms.

With a mass spectrometer, we can measure the parent and daughter isotopes precisely. Today, such instruments can literally count the atoms in a small sample. However, to determine the age of a mineral grain, we must account for any daughter isotope incorporated into the mineral grain at the time it crystallized. In our example, if there were 100 daughter atoms in the grain at $T = 0$, then the number of daughter atoms would increase to $500 + 100 = 600$ after one half-life; to $750 + 100 = 850$ after two half-lives; and to $875 + 100 = 975$ after three half-lives. The general expression for the total number of daughter atoms is therefore:

number of daughters = $(2^T - 1)$
× number of remaining parents
+ number of initial daughters

You may notice that this is an equation for a straight line with a slope of $(2^T - 1)$ and an intercept equal to the initial number of daughter atoms, as illustrated in part (a) of the accompanying illustration.

Although we can measure only the total amount of the daughter isotope, we can often infer the initial amount from another isotope of the same element. For example, strontium-87 is created by the decay of rubidium-87 (see Figure 9.14), but another isotope, strontium-86, is not produced by radioactive decay and is not itself radioactive. Thus, if a mineral grain remains a closed system after crystallization, the number of strontium-86 atoms will not change with age. The trick is to divide the daughter-parent relationship by the amount of strontium-86:

$$\left(\frac{\text{number of strontium-87}}{\text{number of strontium-86}}\right) = (2^T - 1)$$

$$\times \left(\frac{\text{number of rubidium-87}}{\text{number of strontium-86}}\right)$$

$$+ \left(\frac{\text{number of initial strontium-87}}{\text{number of strontium-86}}\right)$$

Different mineral grains in a rock will crystallize with differing initial amounts of strontium and rubidium. However, because the two strontium isotopes behave similarly in the chemical reactions that take place before crystallization, the strontium-87/strontium-86 ratio at crystallization will be the same for all the grains in the same rock. Therefore, by fitting a line to the data from several mineral grains, we can determine the initial strontium-87/strontium-86 ratio as well as the age T.

In part (b) of the accompanying figure, we apply this method to strontium and rubidium measurements from a famous stony meteorite, called Juvinas, that fell in southern France in 1821. The Juvinas meteorite, which is similar to the one shown in Figure 1.9a, is thought to have come from a planetary body that formed at the same time as Earth, but was subsequently destroyed by planetary collisions (see Chapter 20). Using mass spectrometer measurements of four samples from this meteorite, we can plot the strontium-87/strontium-86 and rubidium-87/strontium-86 ratios along a line whose intercept gives an initial strontium-87/strontium-86 ratio of 0.699. That line is an isochron (a locus of equal time) with a slope of 0.067.

To solve for T, we begin with

$$(2^T - 1) = 0.067$$

Adding 1 to both sides of this equation and taking base-10 logarithms of both sides yields

$$T \log(2) = \log(1.067)$$

or

$$T = \log(1.067) / \log(2).$$

Using a scientific calculator (there's probably one on your smartphone), we find that $\log(1.067) = 0.0282$ and $\log(2) = 0.301$, which gives

$$T = \frac{0.0282}{0.301} = 0.094 \text{ half-lives.}$$

Multiplying the number of half-lives by the half-life of rubidium-87, 49 billion years (see Table 9.1), yields a meteorite age of

0.094 × 49 billion years = 4.59 billion years

The uncertainty of this estimate is about 0.07 billion years, so it's consistent with the age of 4.56 billion years first obtained for Earth by Patterson in 1956.

BONUS PROBLEM: When plotted on a diagram like part (b) of the accompanying figure, the rubidium and strontium isotope ratios from several mineral grains collected from the same rock lie on a line with a slope of 0.0143. Assuming these mineral grains have been closed systems since they crystallized, calculate the age of the rock. *Hint:* log(1.0143) = 0.00617.

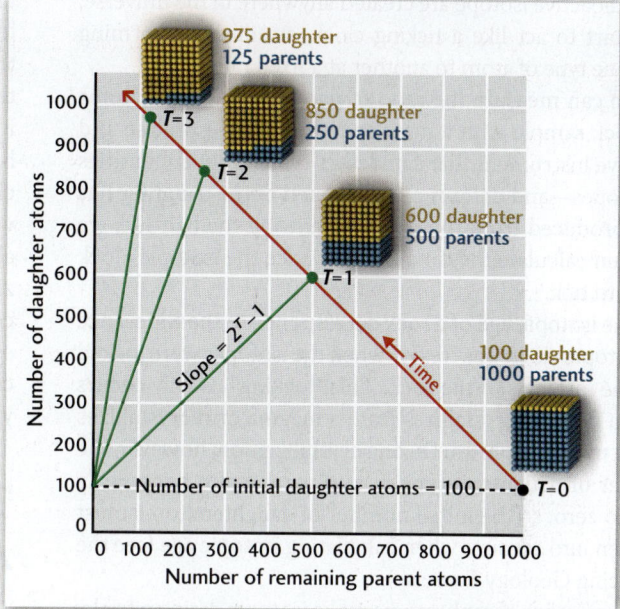

(a)

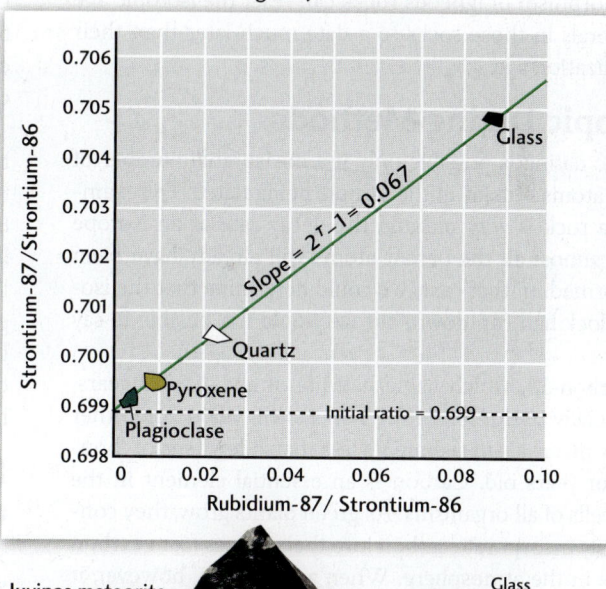

(a) The number of parent atoms in a mineral grain decreases and the number of daughter atoms increases during radioactive decay. As a mineral grain ages, its representation on this plot moves continuously upward and to the left along the red line. The labeled points represent 0, 1, 2, and 3 half-lives. (b) A graph of the strontium-87/strontium-86 versus rubidium-87/strontium-86 ratios for the Juvinas meteorite. The data are obtained from mass spectrometer measurements of different minerals from the meteorite. [Martin Prinz/American Museum of Natural History.]

(b)

Radioactive isotopes make good clocks because their half-lives do not vary with the changes in temperature, pressure, chemical environment, or other factors that can accompany geologic processes on Earth or other planets. So when atoms of a radioactive isotope are created anywhere in the universe, they start to act like a ticking clock, steadily transforming from one type of atom to another at a fixed rate.

We can measure the ratio of parent to daughter atoms in a rock sample with a mass spectrometer—a precise and sensitive instrument that can detect even minute quantities of isotopes—and determine how much of the daughter has been produced from the parent. Knowing the half-life, we can then calculate the time elapsed since the isotopic clock began to tick.

The isotopic age of a rock corresponds to the time since the isotopic clock was "reset," when the isotopes were locked into the minerals of the rock. This "locking" usually occurs when a mineral crystallizes from a magma or recrystallizes during metamorphism. During crystallization, however, the number of daughter atoms in a mineral is not necessarily reset to zero, so the initial number of daughter atoms must be taken into account when calculating isotopic age (see the Practicing Geology Exercise).

Many other complications make isotopic dating a tricky business. A mineral can lose daughter isotopes by weathering or be contaminated by fluids circulating in the rock. Metamorphism of igneous rocks can reset the isotopic age of minerals in those rocks to a date much later than their crystallization age.

Isotopic Dating Methods

Isotopic dating is possible only if a measurable number of parent atoms remain in the sample being dated. For example, if a rock is very old and the decay rate of an isotope is fast, almost all the parent atoms will already have been transformed. In that case, we could determine that the isotopic clock had run down, but we would not be able to say when.

Carbon-14, which has a half-life of about 5700 years, is especially useful for dating bone, shell, wood, and other organic materials in sediments less than a few tens of thousands of years old. Carbon is an essential element in the living cells of all organisms. As green plants grow, they continuously incorporate carbon into their tissues from carbon dioxide in the atmosphere. When a plant dies, however, it stops absorbing carbon dioxide. At the moment of death, the ratio of carbon-14 to the stable isotope carbon-12 in the plant is the same as that in the atmosphere. Thereafter, the ratio decreases as the carbon-14 in the dead tissue decays. Nitrogen-14, the daughter isotope of carbon-14, is a gas and thus leaks from the material, so it cannot be measured to determine the time that has elapsed since the plant died. We can, however, estimate the absolute age of the plant material by comparing the carbon-14 to the carbon-12 left in the plant material with the ratio in the atmosphere at the time the plant died. The latter can be estimated from carbon-14 ages calibrated using other measures of absolute time, such as counting tree rings (dendrochronology).

One of the most precise dating methods for old rocks is based on the decay of two related isotopes: the decay of uranium-238 to lead-206 and the decay of uranium-235 to lead-207. Isotopes of the same element behave similarly in the chemical reactions that alter rocks because the chemistry of an element depends mainly on its atomic number, not its atomic mass. The two uranium isotopes have different half-lives, however, so together they provide a consistency check that helps geologists compensate for the problems of weathering, contamination, and metamorphism discussed above. The lead isotopes from single crystals of zircon—zirconium silicate, a crustal mineral with a relatively high concentration of uranium—can be used to date the oldest rocks on Earth with an uncertainty of less than 1 percent. These formations turn out to be more than 4 billion years old.

The Geologic Time Scale: Absolute Ages

Armed with isotopic dating techniques, geologists of the twentieth century were able to nail down the absolute ages of the key events on which their predecessors had based the geologic time scale. More important, they were able to explore the early history of the planet recorded in Precambrian rocks. Figure 9.16 presents the results of this century-long effort.

The assignment of absolute ages to the geologic time scale revealed great differences in the lengths of the geologic periods. The Cretaceous period (spanning 80 million years) turned out to be more than three times longer than the Neogene period (only 23 million years), and the Paleozoic era (291 million years) was found to be longer than the Mesozoic and Cenozoic eras combined. The biggest surprise was the Precambrian, which had a duration of over 4000 million years—almost nine-tenths of Earth's history!

Eons: The Longest Intervals of Geologic Time

To represent the rich history of the Precambrian, a division of the geologic time scale longer than the era, called the **eon**, was introduced. Four eons, based on the isotopic ages of terrestrial rocks and meteorites, are now recognized.

Hadean Eon The earliest eon, whose name comes from *Hades* (the Greek word for "hell"), began with the formation of Earth 4.56 billion years ago and ended about 3.9 billion years ago. During its first 650 million years, Earth was bombarded by chunks of material from the early solar system. Although very few rock formations survived this violent period, individual zircon grains with ages as great as 4.4 billion

The Geologic Time Scale: Absolute Ages   257

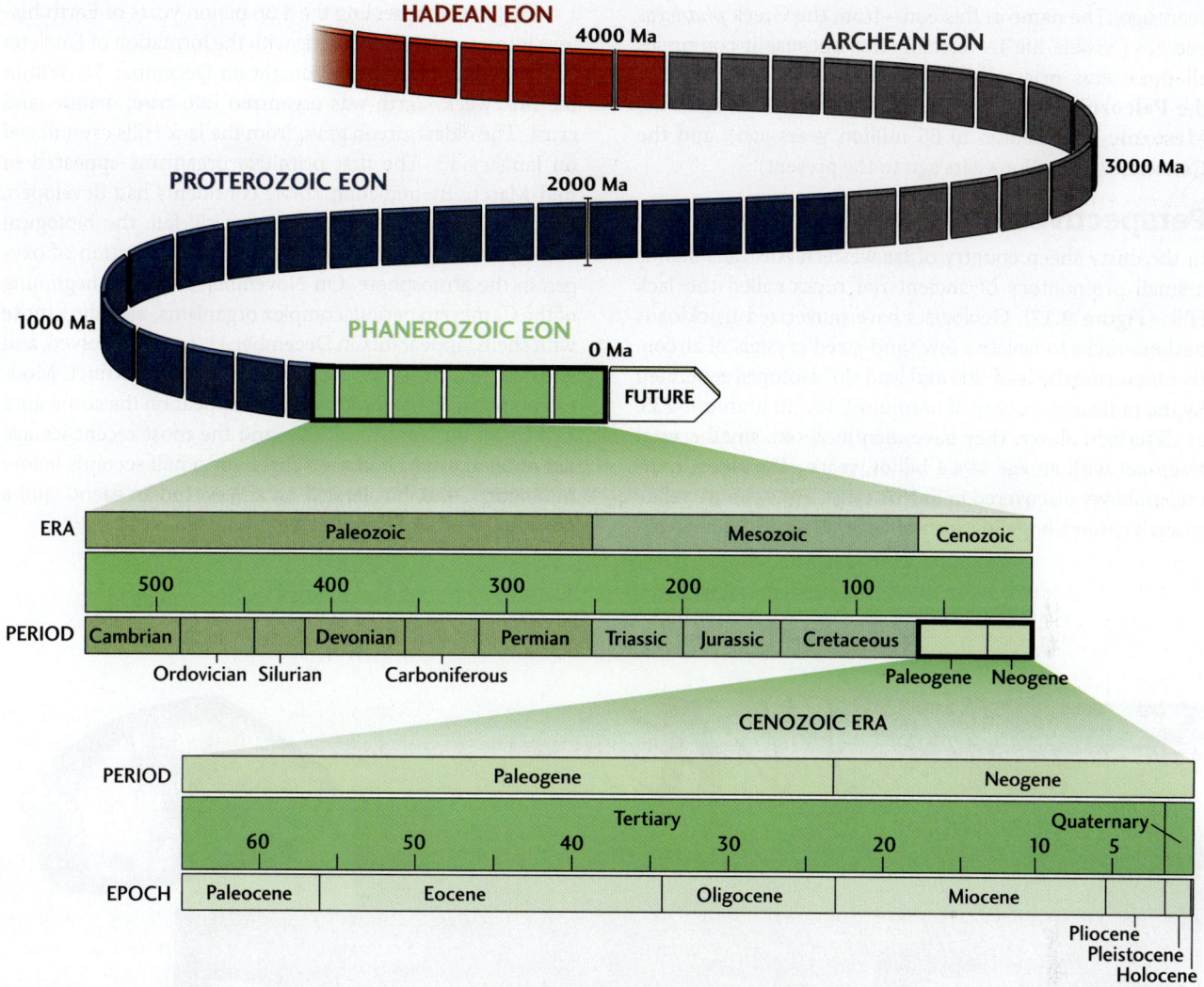

FIGURE 9.16 The complete geologic time scale. (Ma, million years ago.) The intervals labeled "Tertiary" and "Quaternary" are older divisions that have been largely replaced by the Paleogene and Neogene periods, but are still sometimes used by geologists.

years have been found, indicating that Earth had a felsic crust within 200 million years of its formation. There is also evidence that some liquid water existed on Earth's surface at about this time, suggesting that the planet cooled rapidly. In Chapter 20, we will explore this early phase of Earth's history in more detail.

Archean Eon The name of the next eon comes from *archaios* (the Greek word for "ancient"). Rocks of Archean age range from 3.9 billion to 2.5 billion years old. The geodynamo and the climate systems were established during the **Archean eon,** and felsic crust accumulated to form the first stable continental masses, as we will see in Chapter 21. The processes of plate tectonics were probably operating by the end of the Archean, although perhaps substantially differently from later in Earth's history. Life, in the form of primitive single-celled microorganisms, became established, as indicated by the fossils found in sedimentary rocks of this age.

Proterozoic Eon The last part of the Precambrian is the **Proterozoic eon** (from the Greek words *proteros* and *zoi*, meaning "earlier life"), which spans the time interval from 2.5 billion to 542 million years ago. By the beginning of this eon, the plate tectonic and climate systems were working much as they do today. Throughout the Proterozoic, organisms that produced oxygen as a waste product (as plants do today) increased the amount of oxygen in Earth's atmosphere. We will explore the early evolution of life and its effects on the Earth system in Chapter 22.

Phanerozoic Eon The start of the **Phanerozoic eon** is marked by the first appearance of shelly fossils at the beginning of the Cambrian period, now dated at 542 million years ago. The name of this eon—from the Greek *phaneros* and *zoi* ("visible life")—certainly fits, because it comprises all three eras originally recognized in the fossil record: the **Paleozoic** (542 million to 251 million years ago), the **Mesozoic** (251 million to 65 million years ago), and the **Cenozoic** (65 million years ago to the present).

Perspectives on Geologic Time

In the dusty sheep country of far western Australia stands a small promontory of ancient red rocks called the Jack Hills (**Figure 9.17**). Geologists have pulverized truckloads of these rocks to isolate a few sand-sized crystals of zircon. By measuring the lead-206 and lead-207 isotopes generated by the radioactive decay of uranium-238 and uranium-235, as described above, they have identified one small crystal fragment with an age of 4.4 billion years—the oldest mineral grain yet discovered in Earth's crust. How can we relate to such a mind-boggling span of time?

Imagine compressing the 4.56 billion years of Earth history into a single year, starting with the formation of Earth on January 1 and ending at midnight on December 31. Within the first week, Earth was organized into core, mantle, and crust. The oldest zircon grain from the Jack Hills crystallized on January 13. The first primitive organisms appeared in mid-March. By mid-June, stable continents had developed, and throughout the summer and early fall, the biological activity of evolving life increased the concentration of oxygen in the atmosphere. On November 18, at the beginning of the Cambrian period, complex organisms, including those with shells, appeared. On December 11, reptiles evolved, and late on Christmas Day, the dinosaurs became extinct. Modern humans, *Homo sapiens*, did not appear on the scene until 11:42 P.M. on New Year's Eve, and the most recent ice age did not end until 11:58 P.M. Three and a half seconds before midnight, Columbus landed on a West Indian island, and a couple of tenths of a second ago, you were born!

FIGURE 9.17 The outcrop in the Jack Hills in Western Australia, in which geologists have found zircon grains as old as 4.4 billion years. *Inset:* A zircon crystal ($ZrSiO_4$) from the Hadean eon extracted from the Jack Hills. [Bruce Watson, Rensselaer Polytechnic Institute; *inset:* Dr. Martina Menneken.]

Recent Advances in Timing the Earth System

We have seen that the time scales of geologic processes are not uniform, but vary from seconds to billions of years. We must therefore use a variety of methods for timing the Earth system: some to determine the ages of very old rocks, others to measure rapid changes. New methods for determining the relative and absolute ages of Earth materials have steadily improved our understanding of how the Earth system works. To conclude our story of the geologic time scale, we will describe a few of these recent advances.

Sequence Stratigraphy

Until a few decades ago, geologists had to rely on rocks exposed at outcrops, in mines, and by drilling to map stratigraphic successions. As mentioned in Chapter 1 (and further described in Chapter 11), technological innovations in the field of seismic imaging now allow us to see below Earth's surface without actually digging holes. From recordings of seismic waves generated by controlled explosions or by natural earthquakes, we can construct three-dimensional images of deeply buried structures (**Figure 9.18**). Seismic imaging of sedimentary rocks allows geologists to identify sedimentary sequences and map their distribution in three dimensions, a type of geologic mapping called *sequence stratigraphy*.

Sedimentary sequences commonly form on the edges of continents; here, sediment deposition by rivers is modified by fluctuations in sea level. In the example shown in Figure 9.18, sediments were laid down in a delta where a river entered the ocean. As the sediments accumulated, the delta advanced seaward. When sea level fell because of continental glaciation, the deltaic deposits were exposed and eroded. Once the glaciers melted and sea level rose, the shoreline shifted inland, and a new deltaic sequence began to cover the old one, creating an unconformity.

(a) Seismic profile

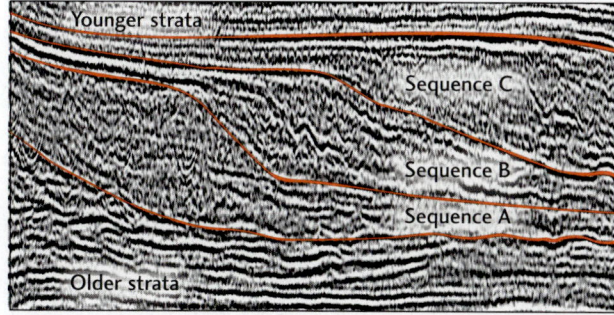

(b) Sedimentary sequences

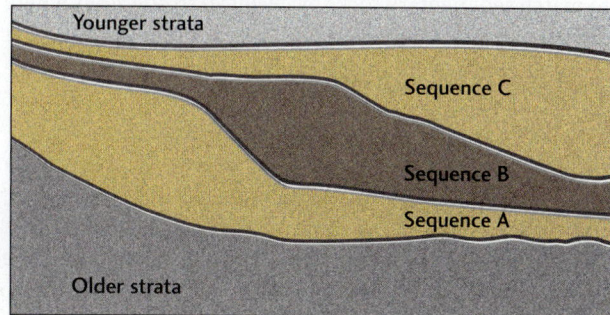

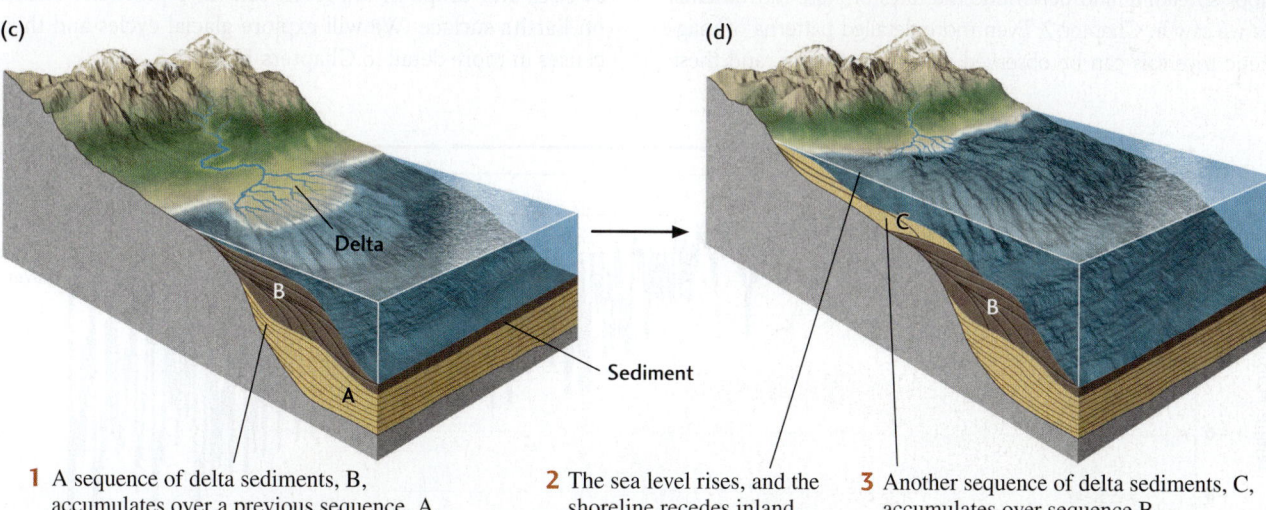

1 A sequence of delta sediments, B, accumulates over a previous sequence, A.

2 The sea level rises, and the shoreline recedes inland.

3 Another sequence of delta sediments, C, accumulates over sequence B.

FIGURE 9.18 Sequence stratigraphy can be used to understand how sedimentary bedding patterns were created. (a) A seismic profile reveals individual sedimentary beds. (b) Geologists can group these beds into sedimentary sequences. (c) and (d) In this case, seismic imaging reveals a stratigraphic succession that is characteristic of a series of deltaic sequences.

Over millions of years, cycles such as this one can be repeated many times, producing a complex set of sedimentary sequences. Because sea level fluctuations are global, geologists can match sedimentary sequences of the same age over large areas. The relative ages of these sequences can then be used to reconstruct the geologic history of a region, including any regional tectonic uplift or subsidence that contributed to sea level changes. Sequence stratigraphy has been especially effective in finding deeply buried oil and gas on continental margins, such as the Gulf of Mexico and the Atlantic margin of North America.

Chemical Stratigraphy

Many sedimentary beds contain minerals and chemicals that identify them as distinct units. For example, the amount of iron or manganese in carbonate sediments may vary from bed to bed if the composition of seawater changed during precipitation of the carbonate minerals. When the sediments are buried and transformed into sedimentary rocks, these chemical variations may be preserved, "fingerprinting" the formations. These chemical fingerprints may extend regionally or even globally, allowing us to match sedimentary rocks by *chemical stratigraphy* where no other features, such as fossils, are available.

Paleomagnetic Stratigraphy

Another technique for fingerprinting rock formations is *paleomagnetic stratigraphy*. As we saw in Chapter 1, Earth's magnetic field reverses itself at irregular intervals. These magnetic reversals are recorded by thermoremanent magnetization in volcanic rocks, which can be dated by isotopic methods. The resulting chronology of magnetic reversals—the magnetic time scale—allows us to "replay the magnetic tape" of seafloor spreading and determine the rates of plate movements, as we saw in Chapter 2. Even more detailed patterns of magnetic reversals can be observed in sediment cores, and these magnetic fingerprints can be dated using faunal successions. Paleomagnetic stratigraphy has recently become one of the main methods for measuring sedimentation rates along the continental margins and in the deep sea. We will discuss paleomagnetic stratigraphy in more detail in Chapter 11.

Clocking the Climate System

The Pliocene and Pleistocene epochs were times of rapid and dramatic global climate change. We can chart these climate changes from the isotopes contained in shelly fossils buried in deep-sea sediments. Deep-sea drilling vessels such as the *JOIDES Resolution* (see Figure 2.13) have taken cores from sedimentary beds around the world's oceans. Geologists can use the carbon-14 dating method to estimate when the shells recovered from these sediment cores were formed, and they can measure the stable isotopes of oxygen to estimate temperature of the seawater in which the shell-producing organisms lived.

The carefully tabulation of both temperature and age estimates for many sedimentary layers has provided us with a precise record of global climate during the last 5 million years (**Figure 9.19**). The record shows a general cooling trend beginning about 3.5 million years ago and the subsequent development of rapid climate cycles that became especially large during the Pleistocene epoch. The low temperatures during these cycles, which were as much as 8° C below the average present-day temperature of Earth's surface, correspond to the Pleistocene "ice ages," when glaciers covered large areas of North America, Europe, and Asia.

Repeated cycles of glaciation have occurred with dominant periods ranging from 40,000 to 100,000 years. Shorter-term cycles lasting a few thousand years or less are also evident. The effects of these climate cycles, such as rises and drops in sea level, can have profound effects on Earth's surface. We will explore glacial cycles and their causes in more detail in Chapters 12 and 15.

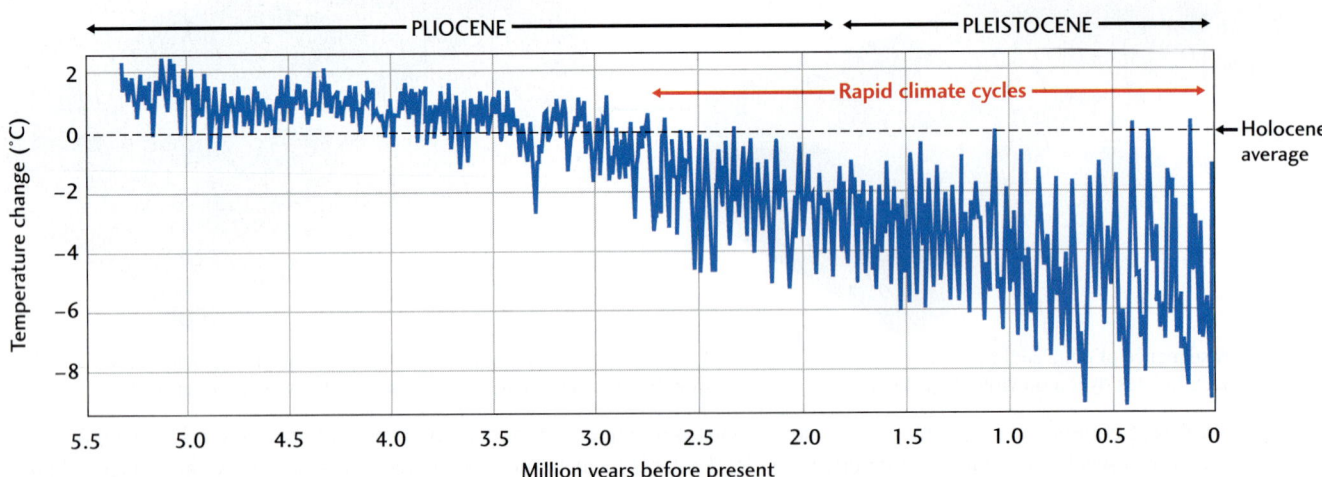

FIGURE 9.19 Changes in Earth's average surface temperature (jagged blue line) during the Pliocene and Pleistocene epochs, measured from temperature indicators in well-dated oceanic sediments. Zero change (dashed black line) corresponds to the average temperature during the Holocene epoch of the last 11,000 years. Note the rapid climate cycles since about 2.7 million years ago. The low temperatures during these cycles correspond to "ice ages." [Data from L. E. Lisiecki and M. E. Raymo.]

KEY TERMS AND CONCEPTS

absolute age (p. 241)
Archean eon (p. 257)
Cenozoic era (p. 258)
eon (p. 256)
epoch (p. 247)
era (p. 247)
geologic time scale (p. 247)
Hadean eon (p. 256)
half-life (p. 253)

isotopic dating (p. 250)
mass extinction (p. 247)
Mesozoic era (p. 258)
period (p. 247)
Paleozoic era (p. 258)
Phanerozoic eon (p. 257)
Precambrian (p. 250)
principle of faunal succession (p. 243)

principle of original horizontality (p. 242)
principle of superposition (p. 242)
Proterozoic eon (p. 257)
relative age (p. 241)
stratigraphic succession (p. 242)
stratigraphy (p. 242)
unconformity (p. 245)

REVIEW OF LEARNING OBJECTIVES

9.1 Apply stratigraphic principles to geological field observations to infer whether one rock is older than another.

The relative ages of rocks can be determined by studying the stratigraphy, fossils, and cross-cutting relationships of rock formations observed at outcrops. According to Steno's principles, an undeformed sequence of sedimentary beds will be horizontal, with each bed younger than the beds beneath it and older than the beds above it. Sedimentary strata contain fossils in a definite sequence, which allows strata in one location to be matched with strata in another location. Cross-cutting relationships establish the relative ages of igneous intrusions and faults within the stratigraphic succession.

Study Assignment: Figure 9.11

Exercise: Construct a cross section similar to the one at the top of Figure 9.11 to show the following sequence of geologic events: (a) deposition of a limestone formation; (b) uplift and folding of the limestone; (c) erosion of the folded rock; (d) subsidence and deposition of a sandstone formation.

Thought Questions: (a) The theory of evolution suggests a "principle of floral (plant) succession" to complement Smith's principle of faunal succession. Why do you think Smith relied primarily on animal fossils rather than plant fossils in his stratigraphic mapping? (b) You encounter an isolated outcrop of a sedimentary sequence where the strata have been tilted into a vertical position. You observe that the sediments are turbidites with graded bedding (see Chapter 6). How might you determine which strata are the oldest and which are the youngest?

9.2 Classify unconformities in stratigraphic sequences according to the relationships between the layers above and below them.

An unconformity is a surface between two rock layers in a stratigraphic succession that have a time gap between them, indicating that strata were not deposited during the missing interval or were eroded away. Three types of unconformity are (1) a disconformity, at which an upper sedimentary sequence overlies an erosional surface developed on a lower, undeformed sedimentary sequence, (2) a nonconformity, at which the upper sedimentary beds overlie metamorphic or igneous rocks, and (3) an angular unconformity, at which the upper beds overlie lower beds that have been folded by tectonic processes and then eroded.

Study Assignment: Earth Issues 9.1

Exercises: (a) How many formations can you count in the geologic cross section of the Grand Canyon in Earth Issues 9.1? How many are the same formations observed in Zion Canyon? Are any of the formations observed in both the Grand Canyon and the Bryce Canyon cross sections? (b) By comparing the sequence of formations illustrated in Earth Issues 9.1 with the relative time scale in Figure 9.12, identify a major disconformity in the Grand Canyon stratigraphic succession. (i) Which periods of geologic time are missing? (ii) Consult Figure 9.16 to estimate the minimum amount of geologic time, measured in millions of years, that is represented by this disconformity.

Thought Question: In studying an area of tectonic compression, a geologist discovers a sequence of older, more deformed sedimentary rocks on top of a younger, less deformed sequence, separated by an angular unconformity. What sequence of plate tectonic processes might have created the angular unconformity?

9.3 Understand how the absolute ages of rocks can be determined by the decay of radioactive isotopes, and know which isotopic systems are suitable for dating rocks of various ages.

Isotopic dating is based on the decay of radioactive isotopes, in which unstable parent atoms are transformed into stable daughter atoms at a constant rate given by the half-life of the parent. The half-life does not vary with the changes in temperature, pressure, and chemical environment. The isotopic clock starts ticking when radioactive isotopes are locked into minerals as igneous rocks crystallize or metamorphic rocks recrystallize. By measuring the amounts of parent and daughter atoms in a sample, geologists can estimate the absolute ages of rocks. Isotopes that decay slowly over billions of years, such as rubidium-87, are most useful in measuring the ages of older rocks, whereas those that decay rapidly, such as carbon-14, can only be used to date younger rocks.

Study Assignment: Practicing Geology Exercise

Exercises: (a) If an isolated mineral initially contains 1000 atoms of a radioactive parent isotope and no atoms of its stable daughter isotope, how much of the daughter isotope will exist after two half-lives of the parent? (b) If the parent isotope in this same mineral has a half-life of 700 million years, and the number of parent atoms is measured to be 125 atoms, how old is the rock? (c) A geologist discovers a distinctive set of fish fossils that dates from the Devonian period within a low-grade metamorphic rock. The rubidium-strontium isotopic age of the rock is determined to be only 70 million years. Give a possible explanation for the discrepancy.

Thought Questions: (a) How does determining the ages of igneous rocks by isotopic methods help to date fossils? (b) Is carbon-14 a suitable isotope for dating geologic events in the Pliocene epoch? Why?

Review of Learning Objectives

9.4 Recall the major divisions and subdivisions of the geological time scale from Earth's formation to the present day.

The geologic time scale is divided into four eons: the Hadean (4.56 billion to 3.9 billion years ago), Archean (3.9 billion to 2.5 billion years ago), Proterozoic (2.5 billion to 542 million years ago), and Phanerozoic (542 million years ago to the present). The Phanerozoic eon is divided into three eras, the Paleozoic, Mesozoic, and Cenozoic, each of which is subdivided into shorter periods. The boundaries of the eras and periods are marked by abrupt changes in the fossil record; many correspond to mass extinctions.

Study Assignment: Visual Literacy Exercise

Exercises: (a) Explain why the last eon of geologic time is named the Phanerozoic. (b) At the present rate of seafloor spreading, the entire seafloor is recycled every 200 million years. Assuming that the past rate of seafloor generation has been this fast or faster, calculate the minimum number of times the seafloor has been recycled since the end of the Archean eon. (c) If the 4.56 billion years of Earth history were compressed into a single year, what dates would correspond to the Paleozoic-Mesozoic and Mesozoic-Cenozoic boundaries?

Thought Question: Why did geologists choose the boundary between the Permian and Triassic periods to separate the Paleozoic era from the Mesozoic era and not, say, the Carboniferous-Permian boundary?

9.5 Summarize how geologists created and refined the geological time scale.

By using faunal successions to match rocks in outcrops around the world, geologists of the nineteenth century compiled composite stratigraphic successions, from which they developed a relative time scale. They divided this time scale into eras, periods, and epochs based on abrupt changes in the fossil record, many of which we now know to be associated with mass extinctions due to rapid environmental changes. Isotopic dating methods developed in the twentieth century allowed them to assign absolute ages to the time scale and to determine that the Precambrian spanned almost nine-tenths of Earth history. Recent refinements of the geological time scale have come from the seismic imaging and fossil dating of sedimentary sequences on continental margins produced by the cyclical rise and fall of sea level during glacial cycles. Glacial cycles recorded in sediments have also been dated using ice cores taken from the Antarctic and Greenland ice caps. Chemical fingerprints and magnetic reversals provide additional information about the ages of sedimentary sequences.

Study Assignment: This entire chapter

Exercises: (a) Mass extinctions have been dated at 444 million years ago, 416 million years ago, and 359 million years ago. How are these events expressed in the geologic time scale of Figure 9.16? (b) True or false: No rocks survived the hellish conditions at Earth's surface during the Hadean eon.

Thought Questions: (a) Why did the nineteenth-century geologists constructing the geologic time scale find sedimentary strata deposited in oceans and shallow seas more useful than strata deposited on land? (b) A geologist documents a distinctive chemical signature caused by organisms of the Proterozoic eon that has been preserved in sedimentary rock. Would you consider this chemical signature to be a fossil?

VISUAL LITERACY EXERCISE

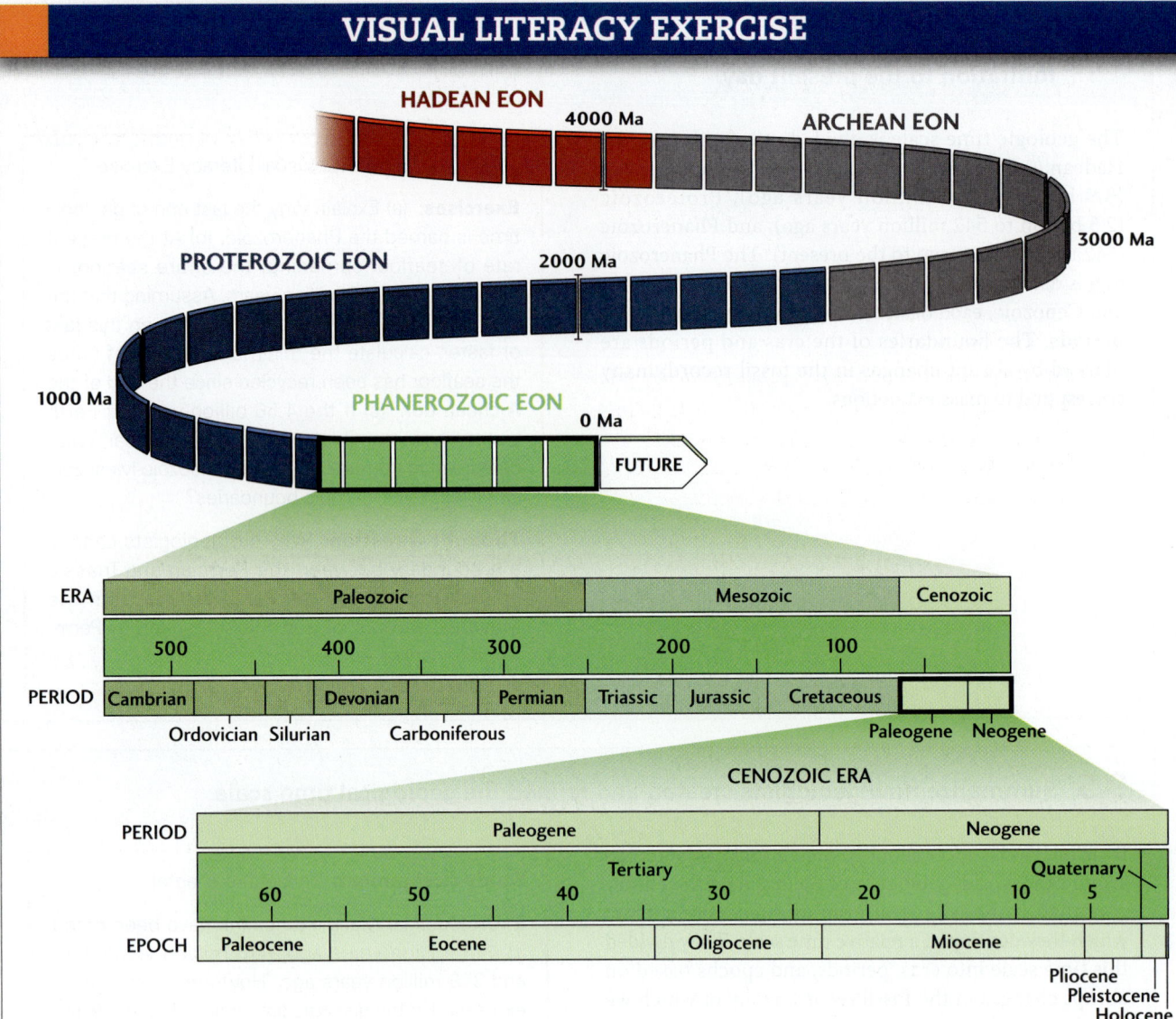

FIGURE 9.16 The complete geologic time scale. (Ma, million years ago.) The intervals labeled "Tertiary" and "Quaternary" are older divisions that have been largely replaced by the Paleogene and Neogene periods, but are still sometimes used by geologists.

1. The ribbon at the top of Figure 9.16 partitions geologic time into eons.
   a. How many eons are there, and what are their names?
   b. Which eon is the longest, and when did it begin and end?
   c. Which eon is the oldest, and when did it begin and end?
2. The bars below the ribbon of geologic time show the main divisions of the Phanerozoic eon on an expanded time scale.
   a. How many periods does the Paleozoic era include?
   b. How many periods does the Mesozoic era include?
   c. The Mesozoic-Cenozoic boundary corresponds to the boundary between which two periods?
3. What interval of geologic time is shown by the lower set of bars, and what is the name of this interval?
4. What is the time span of the Neogene period, and how many epochs does it include?

Earthquakes 10

What Is an Earthquake?	269
How Do We Study Earthquakes?	274
Earthquakes and Patterns of Faulting	284
Earthquake Hazards and Risks	287
Can Earthquakes Be Predicted?	297

Learning Objectives

Earthquakes are a key process of plate tectonics, the violent consequence of active faulting. From material learned in this chapter, you will be able to:

10.1 Explain how stress builds up on faults and is released in earthquakes through the process of elastic rebound.

10.2 Identify the three main types of seismic waves and describe the ground motions they produce.

10.3 Distinguish shaking intensity from earthquake magnitude and understand how they are related.

10.4 Classify earthquakes according to their plate-tectonic setting and patterns of faulting.

10.5 Characterize seismic hazards and the risks they pose to society and its built environment.

10.6 Defend the statement that "earthquakes can be forecast in the long term but not predicted in the short term."

The tsunami from the 2011 Tohoku earthquake crashes over a seawall designed to protect the city of Miyako from destructive sea waves. [AP Photo/Mainichi Shimbun, Tomohiko Kano.]

Earthquakes rival all other natural disasters in the threat they pose to human life and property. Our fragile "built environment" is necessarily anchored in Earth's active crust, which makes it extremely vulnerable to earthquakes and their secondary effects, such as ground ruptures, landslides, and tsunamis. Some events of the past century are sobering illustrations of this fact.

On a fine April morning in 1906, the citizens of northern California were awakened by the roar and violent shaking caused by the breaking of the San Andreas transform fault, which created the most destructive earthquake the United States has yet experienced. The ensuing fires destroyed the city of San Francisco; by the time the flames died away, nearly 3000 of its inhabitants were dead (**Figure 10.1**).

Almost a century later, on December 26, 2004, a much larger fault—a subduction-zone megathrust—broke west of the Indonesian island of Sumatra, lifting up the seafloor and sending a huge tsunami across the Indian Ocean. The monster waves drowned more than 220,000 people living on coastlines from Thailand to Africa. Another megathrust ruptured off Japan on March 11, 2011, creating an even larger tsunami, pictured here, that drowned almost 20,000. Between the 1906 earthquake and now, earthquakes have killed more than 2 million people around the world.

To cope with the all-too-frequent death and destruction caused by earthquakes, we have long sought to improve our ability to predict where and when these events might occur and our understanding of what happens when they do. Science has shown that seismic activity can be understood in terms of the basic machinery of plate tectonics. As a result, attempts to reduce earthquake risk have become increasingly fused with the quest for a more fundamental understanding of the geologically active Earth.

This chapter will examine what happens during earthquakes, how scientists use seismic waves to locate and measure earthquakes, and what can be done to reduce the casualties and economic damage caused by earthquakes. We cannot yet reliably predict when large earthquakes will occur, but we can take measures to reduce their destructiveness. We can use our geologic knowledge to identify places where large earthquakes are likely; work with engineers to design buildings, dams, bridges, and other structures that can withstand seismic shaking; and help endangered communities prepare for and respond to seismic events.

FIGURE 10.1 This photograph, taken from a balloon by George Lawrence 5 weeks after the San Francisco earthquake of April 18, 1906, shows the devastation of the city caused by the quake and subsequent fires. The view is looking over Nob Hill toward the business district. [Library of Congress/Getty Images.]

What Is an Earthquake?

The movement of lithospheric plates generates enormous forces at the boundaries between those plates, as we saw in Chapter 8. These forces deform brittle crustal rocks in ways that can be described by the concepts of stress, strain, and strength. *Stress* is the local force per unit area that causes rocks to deform. *Strain* is the relative amount of deformation, expressed as the percentage of distortion (for example, compression of a rock by 1 percent of its length). Rocks *fail*—that is, they lose cohesion and break into two or more parts—when they are stressed beyond a critical value, called their *strength*.

An **earthquake** occurs when rocks under stress suddenly fail along a geologic fault. Most large earthquakes are caused by ruptures of preexisting faults, where past earthquakes have already weakened the rocks on the fault surface. The two blocks of rock on either side of the fault slip suddenly, releasing energy in the form of seismic waves, which we feel as ground shaking. When the fault slips, the stress is reduced, dropping to a value below the rock strength. After the earthquake, the stress begins to increase again, eventually leading to another large earthquake. The faults involved in this repeated earthquake cycle are called *active faults*, and they are concentrated in the zones that form plate boundaries, where most of the stress and strain caused by plate movements is concentrated.

The Elastic Rebound Theory

The earthquake on the San Andreas Fault that devastated San Francisco in 1906 was studied more than any earthquake had been up to that time. By mapping the displacement of the ground across the fault and analyzing seismic recordings of the earthquake, geologists showed that this rupture began just west of the Golden Gate and propagated along the fault strike for almost 500 km, southeastward to the mission town of San Juan Bautista and northwestward all the way to Cape Mendocino (**Figure 10.2**). In 1910, one of the scientists who studied this rupture, Henry Fielding Reid of Johns Hopkins University, advanced the **elastic rebound theory** to explain how earthquakes recur on active faults in Earth's crust.

Picture a strike-slip fault between two crustal blocks, and imagine that surveyors have painted straight lines on the ground, running perpendicular to the fault and extending from one block to the other, as in **Figure 10.3a**. The two blocks are being pushed in opposite directions by plate movements. The weight of the overlying rock presses them together, however, so friction locks them in place along the fault. They do not move, just as a pushed car does not move when the emergency brake is engaged. Instead of slipping along the fault as stress builds up, the blocks are strained elastically near the fault, as shown by the bent lines in Figure 10.3b. By *elastically*, we mean that the blocks would spring

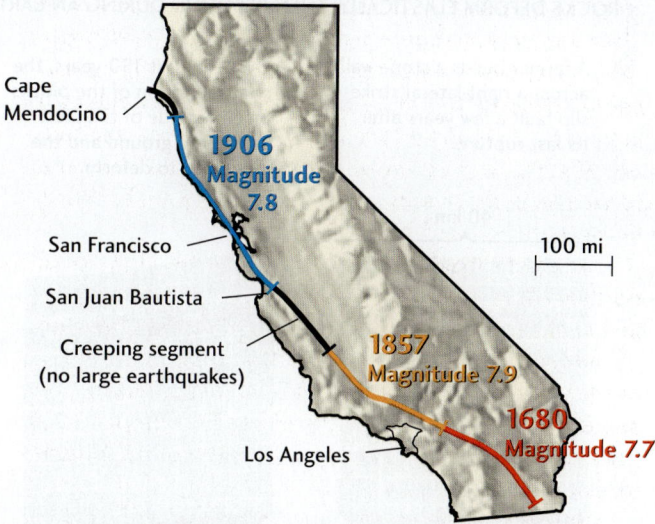

FIGURE 10.2 Map of California, showing the segments of the San Andreas Fault that ruptured in 1680, 1857, and 1906. [Southern California Earthquake Center.]

back and return to their undeformed, stress-free shape if the fault were suddenly to unlock.

As the slow plate movements continue to push the blocks in opposite directions, the elastic strain in the rocks—measured by the bending of the survey lines—continues to build up over decades, centuries, or even millennia (Figure 10.3c). At some point, the strength of the rocks is exceeded. Somewhere along the fault surface, the frictional bond that locks the fault can no longer hold, and it breaks. The blocks slip suddenly in a rupture that starts at a point but quickly extends over a section of the fault.

Figure 10.3d shows how the two blocks have rebounded—sprung back to their undeformed state—after the earthquake. The bent survey lines have straightened, and the two blocks have been displaced. (Note that a fence built just before the rupture has been bent during the rebound.) The distance of the displacement is called the **fault slip**. In the inset photograph of Figure 10.3, you can see that the fault slip during the 1906 earthquake was about 4 m. The peak velocity of the slipping at any point on the fault is about 1 m/s, so the entire episode of fault slipping at a given point takes only a few seconds. After the fault has slipped, it locks up again. The steady movement of the blocks on either side of the fault then causes stress to rise again, and the earthquake cycle repeats.

The energy that is slowly built up by elastic strain as the two blocks are pushed in opposite directions is like the elastic energy stored in a rubber band when it is slowly stretched. The sudden release of energy when a fault slips is like the violent backlash, or *rebound*, that occurs when the rubber band breaks. Some of this elastic energy is radiated

ROCKS DEFORM ELASTICALLY, THEN REBOUND DURING AN EARTHQUAKE RUPTURE

A A farmer builds a stone wall across a right-lateral strike-slip fault a few years after its last rupture.

B Over the next 150 years, the relative motion of the blocks on either side of the fault causes the ground and the stone wall to deform.

C Just before the next rupture, a new fence is built across the already deformed land.

D The fault slips, lowering the stress, and the elastic rebound restores the blocks to their prestressed state. Both the rock wall and the fence are shifted equal amounts along the fault.

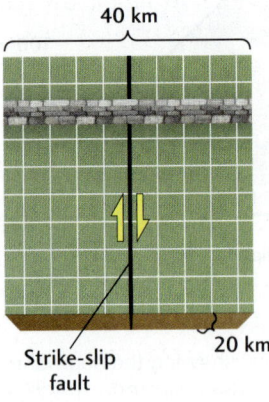
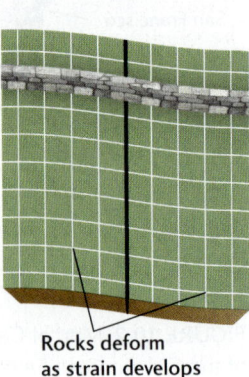
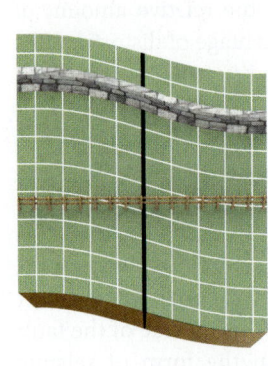
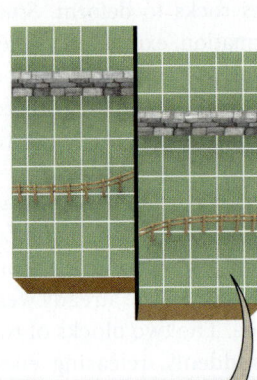

STRESS BUILDS UNTIL IT EXCEEDS ROCK STRENGTH

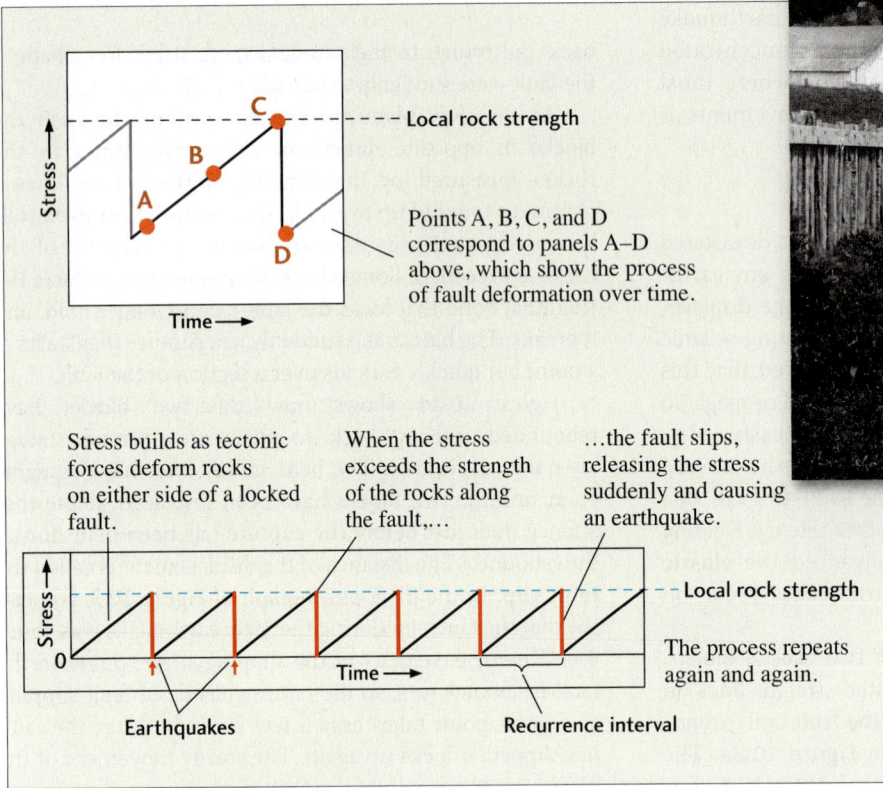

A fence built across the San Andreas Fault near Bolinas, California, is offset by nearly 4 m after the great San Francisco earthquake of 1906.

FIGURE 10.3 The elastic rebound theory explains the earthquake cycle. According to the theory, stress on rocks builds up over time as a result of plate movements. Earthquakes occur when that stress exceeds rock strength. Rocks under stress deform elastically, then rebound when an earthquake occurs. Panels A–D show deformation at the points labeled A–D in the bottom panel. [Photo: G. K. Gilbert/USGS.]

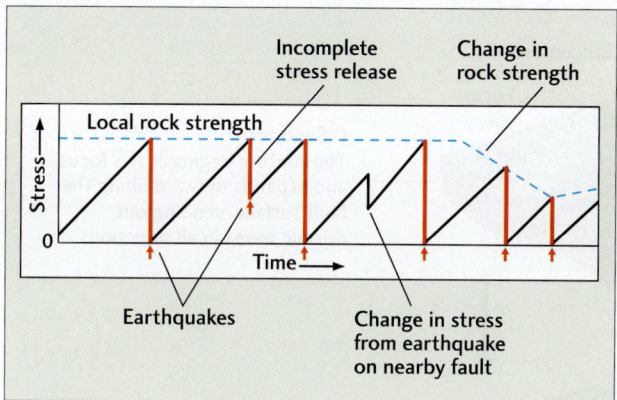

FIGURE 10.4 Irregularities in the earthquake cycle can be caused by incomplete stress release, changes in stress caused by earthquakes on nearby faults, and local variations in rock strength.

in seismic waves, which can cause violent shaking many kilometers away from the fault.

The elastic rebound theory implies that there should be a periodic buildup and release of elastic energy at faults, and that the time between ruptures, called the **recurrence interval,** should be a constant (as shown in the bottom panel of Figure 10.3). The recurrence interval can be calculated by dividing the fault slip in each rupture by the long-term slip rate. The long-term slip rate on the San Andreas Fault is about 30 mm/year, for example, so earthquakes with 4 m of slip should occur with a recurrence interval of about once every 130 years.

However, most active faults, including the San Andreas, do not conform to this simple theory. For instance, all of the strain accumulated since the last earthquake may not be released in the next—that is, the rebound may be incomplete—or the stress on one fault may change because of earthquakes on nearby faults (**Figure 10.4**). Over the long term, the strength of the fault rocks themselves may change. These irregularities are some of the reasons why earthquakes are so difficult to predict.

Fault Rupture During Earthquakes

The point at which fault slipping begins is the **focus** of an earthquake (**Figure 10.5**). The **epicenter** is the geographic point on Earth's surface directly above the focus. For example, you might hear in a news report, "The U.S. Geological Survey reports that the epicenter of last night's destructive earthquake in California was located 6 kilometers east of Los Angeles City Hall. The depth of the focus was 10 kilometers."

The focal depths of most earthquakes in continental crust range from 2 to 20 km. Continental earthquakes below 20 km are rare because under the high temperatures and pressures found at those greater depths, continental crust behaves as a ductile rather than a brittle material (just as hot wax flows when stressed, whereas cold wax breaks; see Chapter 8). In subduction zones, however, where cold oceanic lithosphere plunges into the mantle, earthquakes can originate at depths of almost 700 km.

The fault rupture does not happen all at once. It begins at the focus and expands outward along the fault surface, typically at 2 to 3 km/s (see Figure 10.5). The rupture stops where the stresses become insufficient to continue breaking the fault (where the rocks are stronger) or where the rupture enters ductile material in which it can no longer propagate. As we will see later in this chapter, the magnitude of an earthquake is related to the total area of fault rupture. Most earthquakes are very small, with rupture dimensions much less than the depth of the focus, so that the rupture never breaks the ground surface.

In large, destructive earthquakes, however, surface breaks are common. The 1906 San Francisco earthquake caused surface displacements averaging about 4 m along the 470 km section of the San Andreas Fault that ruptured in that event (see the inset of Figure 10.3). Fault ruptures in the largest earthquakes can extend for more than 1000 km, and the fault slip can be tens of meters. Generally, the longer the fault rupture, the greater the fault slip.

As we have seen, the sudden slipping of the blocks at the time of the earthquake reduces the stress on the fault and releases much of the stored elastic energy. Most of this stored energy is converted to frictional heat in the fault zone or dissipated by rock fracturing, but part of it is released as seismic waves that travel outward from the rupture, much as waves ripple outward from the spot where a stone is dropped into a still pond. The focus of the earthquake generates the first seismic waves, but slipping parts of the fault continue to generate waves until the rupture stops. In a large event, the propagating rupture continues to produce waves for many tens of seconds. These waves can cause destruction in regions all along the fault rupture, even far from the epicenter. Towns along the San Andreas Fault far north of San Francisco were badly damaged in the 1906 earthquake.

Foreshocks and Aftershocks

Almost all large earthquakes trigger smaller earthquakes called **aftershocks.** Aftershocks follow the triggering event, or *mainshock,* in sequences, and their foci are distributed in and around the rupture surface of the

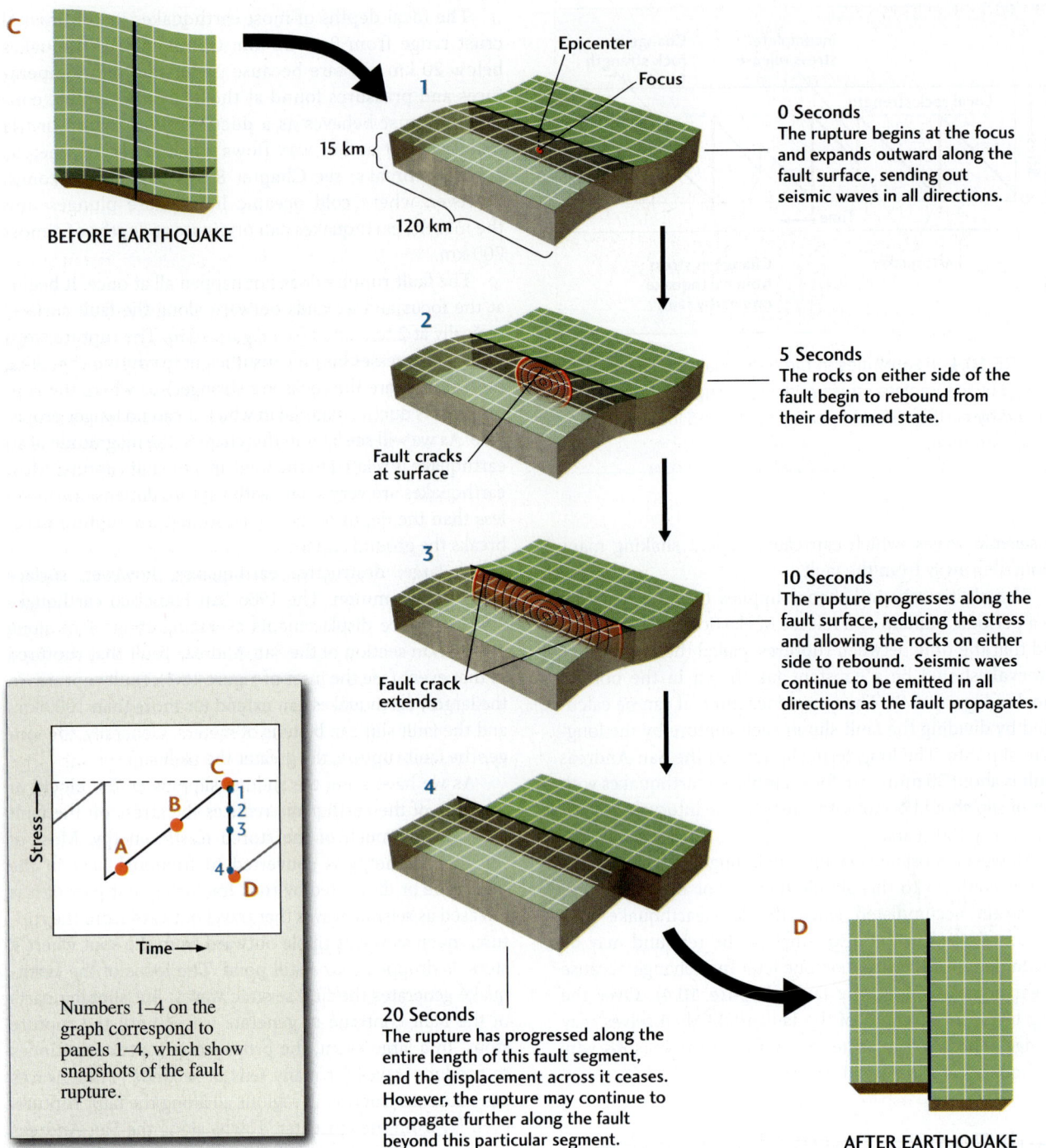

FIGURE 10.5 During an earthquake, fault slipping begins at the focus and spreads out along the fault surface. Panels 1–4 are snapshots of the fault rupture corresponding to the numbered points on the graph.

mainshock (**Figure 10.6**). Aftershock sequences exemplify the complexities of earthquakes that cannot be described by simple elastic rebound theory. Although fault slipping during the mainshock decreases the stress along most of the rupture surface, it can increase the stress on parts of the fault surface that did not slip, or where the slip was incomplete, as well as in surrounding regions. Aftershocks happen where that stress exceeds the rock strength.

The number and sizes of aftershocks depend on the magnitude of the mainshock, and their frequencies decrease with time after the mainshock. Noticeable aftershocks of a magnitude-5 earthquake might last for only a few weeks, whereas those of a magnitude-7 earthquake occur over a larger region and can continue for several years. The size of the largest aftershock is generally about one magnitude unit smaller than the mainshock. According to this rough rule of

What Is an Earthquake? 273

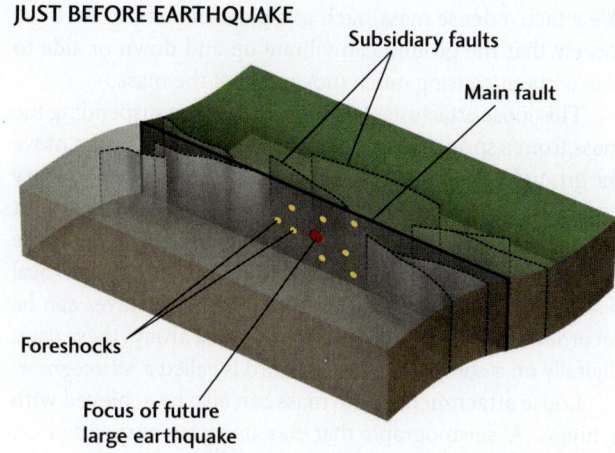

JUST BEFORE EARTHQUAKE
- Subsidiary faults
- Main fault
- Foreshocks
- Focus of future large earthquake

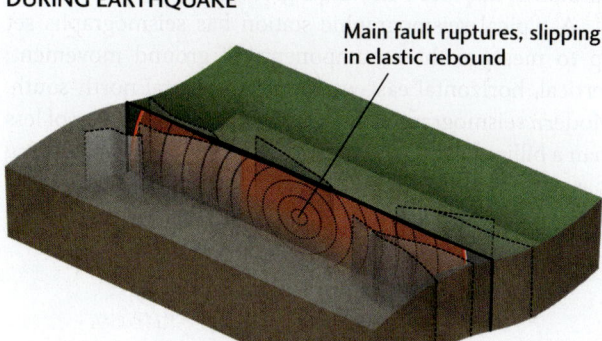

DURING EARTHQUAKE
- Main fault ruptures, slipping in elastic rebound

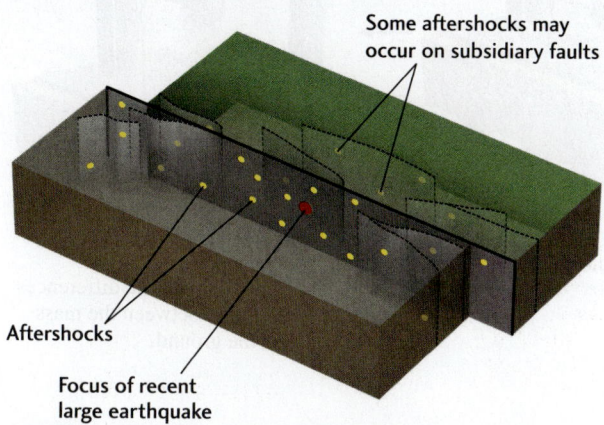

JUST AFTER EARTHQUAKE
- Some aftershocks may occur on subsidiary faults
- Aftershocks
- Focus of recent large earthquake

FIGURE 10.6 Aftershocks are smaller earthquakes that follow a large earthquake (the mainshock). Foreshocks occur shortly before the mainshock, near its focus.

in New Zealand, caused extensive damage, but nobody was killed and only a few people were injured. However, a magnitude-6.3 aftershock struck right beneath the center of Christchurch on February 22, 2011, collapsing buildings and killing 185 people (**Figure 10.7**). The economic losses from this aftershock, estimated at $15 billion, were several times greater than the losses caused by the mainshock five months before. Other strong aftershocks hit the city on June 13 and in December of 2011, injuring dozens and causing $4 billion in additional damages. More aftershocks can be expected in the years to come.

A **foreshock** is a small earthquake that occurs shortly before a mainshock, near its focus (see Figure 10.6). One or more foreshocks have preceded many large earthquakes, so scientists have tried to use foreshocks to predict when and where large earthquakes might happen. Unfortunately, it is usually very hard to distinguish foreshocks from other small earthquakes that occur randomly and frequently on active faults, so this method has only rarely proved successful. The magnitude-9 Tohoku earthquake that caused a great tsunami to hit Honshu, Japan on March 11, 2011 (see Earth Issues 10.1), was preceded by a foreshock of magnitude 7.2 about 50 hours before the mainshock. In some sense, the mainshock was an anomalously big "aftershock" associated with the first event. But what turned out to be a foreshock was considered at the time to be only an ordinary magnitude-7.2 earthquake in the subduction zone.

As this example illustrates, foreshocks, mainshocks, and aftershocks can be classified in a definitive way only after the earthquake sequence has ended; during the sequence we cannot be sure whether the mainshock—the biggest event in the sequence—is yet to come.

FIGURE 10.7 Ruins of the Christchurch Basilica, one of many buildings in downtown Christchurch, New Zealand, destroyed by an earthquake on February 22, 2011. This event was an aftershock of a larger but less damaging earthquake that occurred west of Christchurch on September 4, 2010. [Mark Longley/Alamy.]

thumb, a magnitude-7 earthquake might have an aftershock as large as magnitude 6.

In populated regions, the shaking from large aftershocks can be very dangerous, compounding the damage caused by the mainshock. On September 4, 2010, a magnitude-7.1 earthquake west of Christchurch, the second largest city

How Do We Study Earthquakes?

As in any experimental science, instruments and field observations provide the basic data used to study earthquakes. These data enable investigators to locate earthquakes, determine their sizes and numbers, and understand their relationships to faults.

Seismographs

The **seismograph,** an instrument that records the seismic waves, is to the Earth scientist what the telescope is to the astronomer: a tool for peering into inaccessible regions (Figure 10.8). The ideal seismograph would be a device affixed to a stationary frame that is not attached to Earth. When the ground shook, the seismograph would measure the changing distance between the frame, which did not move, and the vibrating ground, which did. As yet, we have no way to position a seismograph that is not attached to Earth—although Global Positioning System (GPS) technology is beginning to remove this limitation. So we compromise.

We attach a dense mass, such as a piece of steel, to Earth so loosely that the ground can vibrate up and down or side to side without causing much movement of the mass.

This loose attachment can be achieved by suspending the mass from a spring (Figure 10.8a). When seismic waves move the ground up and down, the mass tends to remain stationary because of its inertia (an object at rest tends to stay at rest), but the mass and the ground move relative to each other because the spring can compress or stretch. In this way, the vertical displacement of the ground caused by seismic waves can be recorded by a pen on chart paper or, almost always these days, digitally on a computer. Such a record is called a *seismogram.*

Loose attachment of the mass can also be achieved with a hinge. A seismograph that has its mass suspended on hinges, like a swinging gate (Figure 10.8b), can record the horizontal displacement of the ground.

A typical seismographic station has seismographs set up to measure three components of ground movement: vertical, horizontal east-west, and horizontal north-south. Modern seismographs can detect ground oscillations of less than a billionth of a meter—an astounding feat, considering that such small displacements are of atomic size!

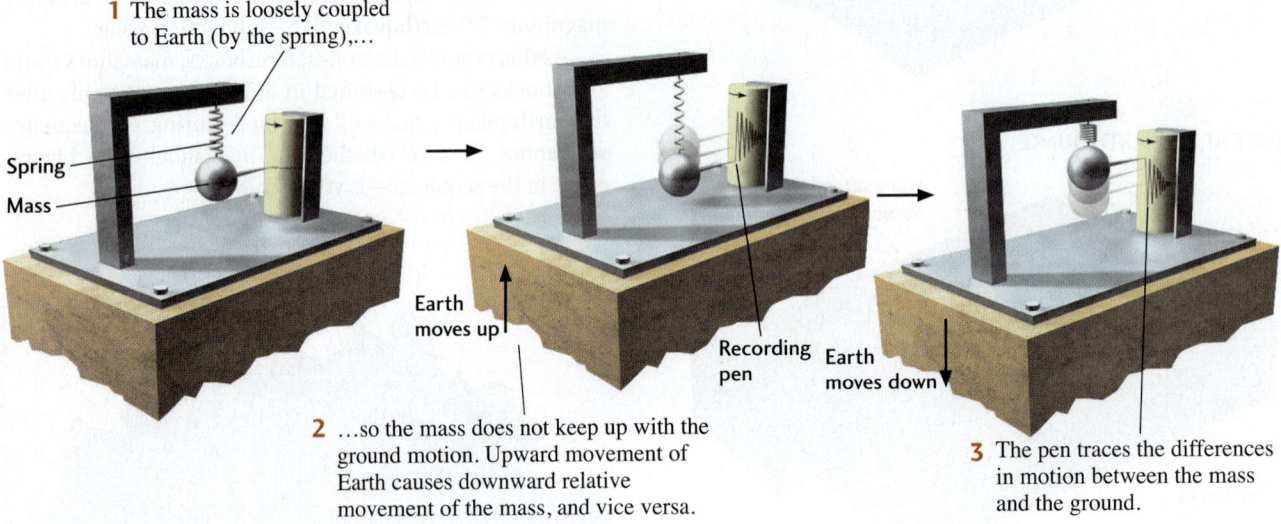

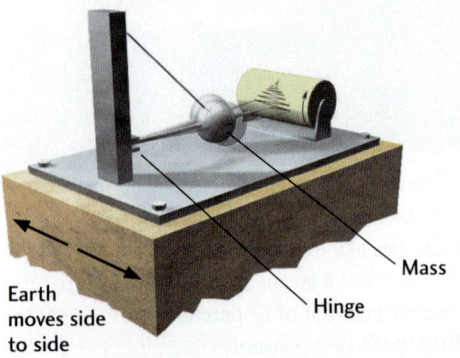

FIGURE 10.8 A seismograph consists of a dense mass (such as a steel ball) attached to a recording device. Because of its inertia and its loose coupling to Earth through (a) a spring or (b) a hinge, the mass does not keep up with the motion of the ground.

Earth Issues 10.1 The Tsunami Stone of Aneyoshi

On a hill along the Tohoku coastline of northeastern Honshu, in the fishing hamlet of Aneyoshi, sits a stone monument of uncertain age, inscribed with Japanese characters that read, "High dwellings are the peace and harmony of our descendants. Remember the calamity of the great tsunamis. Do not build any homes below this point." Aneyoshi, now part of the city of Miyako, was once more conveniently located down by the sea where the fishermen tied their boats, but only four of its residents survived the tsunami of 1896 and only two survived the tsunami of 1933. The stone reminds people why they now live on higher ground.

History became prophecy at 2:46 P.M. on March 11, 2011, when the offshore thrust fault that separates Japan from the Pacific Plate began to slip. The rupture started on a small patch of the fault surface 30 kilometers beneath the ocean, about 100 km southeast of Aneyoshi, and accelerated outward like a crack through glass, reaching speeds of nearly 3 km/s (more than 6000 miles/hour). By the time it stopped several minutes later, the Pacific Plate had moved under Japan by as much as 50 m along a fault surface the size of South Carolina. Seismic waves from this Tohoku megaquake, which measured magnitude 9, propagated over Earth's surface and through its deep interior, causing the planet to ring like a bell for many days.

The thrusting of Honshu eastward and upward over the Pacific Plate raised the seafloor as much as 10 m almost instantly, displacing several hundred billion tons of water, which flowed away from the fault in a huge tsunami. In less than an hour, the water waves, slower than the seismic waves but much more deadly, passed into the bays and inlets of the Japanese coastline like an undulating monster, gaining height as they approached the shore (see chapter-opening photograph). Funneled into harbors, the waves created immense walls of water—tsunami is Japanese for "harbor wave"—which inundated the near-shore communities, sweeping up boats, cars, and buildings, in some places traveling several kilometers inland.

The fast-moving swath of devastation was captured on horrific videos from helicopters overhead and by survivors who made it to high ground and the tops of buildings. The tsunami overran the seawalls designed to protect the city center of Miyako, destroying all but 30 of the 1000 boats in its famous fishing fleet and killing many hundreds who could not, or did not, escape in time. Though the exact number remains uncertain, the death toll along the Tohoku coastline was almost 20,000. One of the highest levels reached by the enormous wave—the greatest in recent Japanese history—was 39 m (128 ft) above the shoreline, just below the Aneyoshi tsunami stone. The residents in their houses above the stone were safe.

The tsunami stone at Aneyoshi. [Ko Sasaki/*The New York Times*/Redux]

Seismic Waves

Install a seismograph almost anywhere, and within a few hours it will record the passage of seismic waves generated by an earthquake somewhere on Earth. The waves travel from the earthquake focus through Earth and arrive at the seismograph in three distinct groups (**Figure 10.9a**). The first waves to arrive are called primary waves, or **P waves.** The secondary waves, or **S waves,** follow. Both P waves and S waves travel through Earth's interior. Afterwards come the slower **surface waves,** which travel around Earth's surface.

P waves in rock are similar to sound waves in air, except that P waves travel through Earth's crust at about 6 km/s, which is about 20 times faster than the speed of sound. Like sound waves, P waves are *compressional waves,* so called because they move through solid, liquid, or gaseous materials as a succession of compressions and expansions

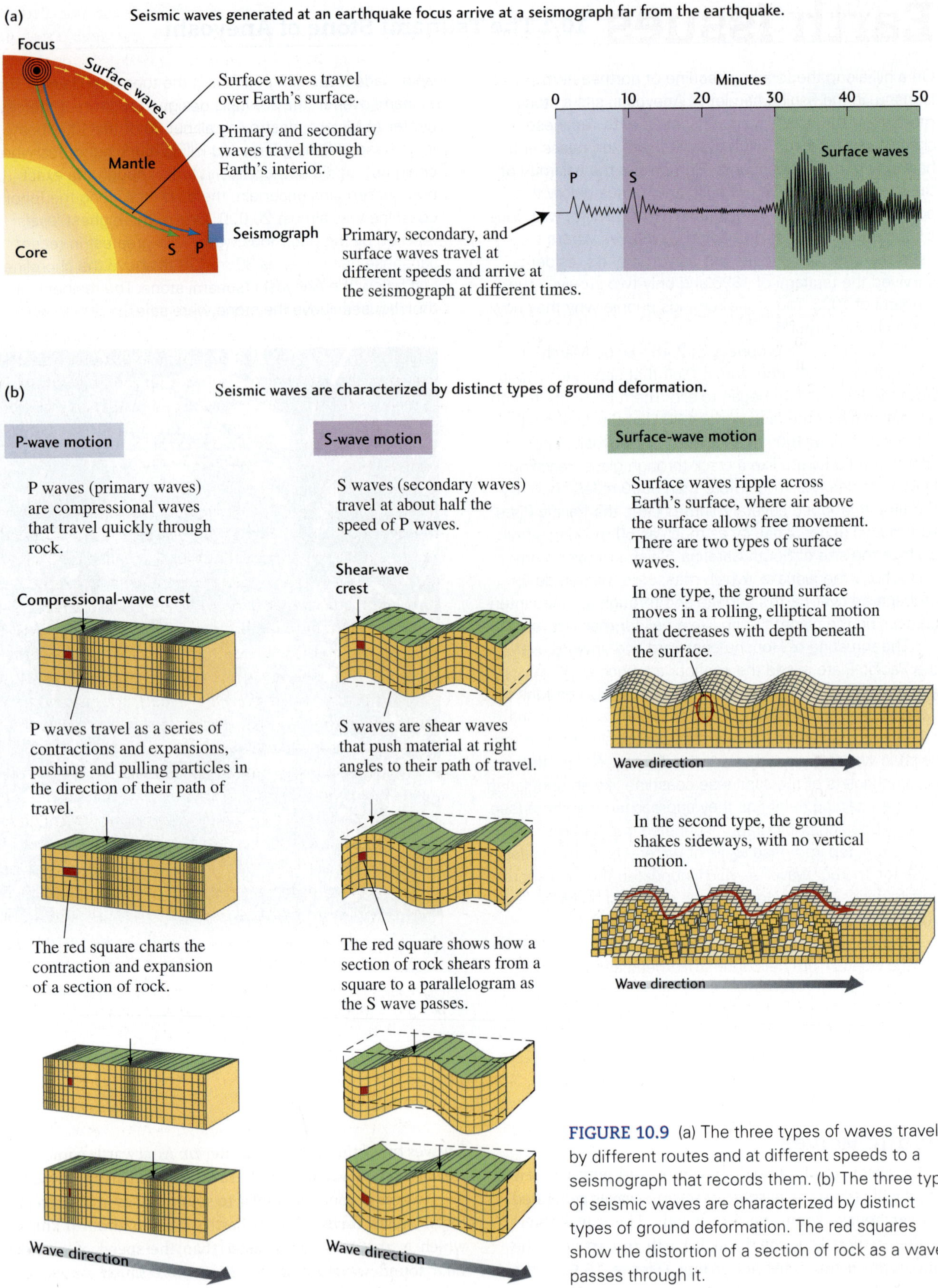

(a)
Seismic waves generated at an earthquake focus arrive at a seismograph far from the earthquake.
Focus
Surface waves
Surface waves travel over Earth's surface.
Primary and secondary waves travel through Earth's interior.
Mantle
Core
S
P
Seismograph
Primary, secondary, and surface waves travel at different speeds and arrive at the seismograph at different times.
Minutes
0
10
20
30
40
50
P
S
Surface waves
(b)
Seismic waves are characterized by distinct types of ground deformation.
P-wave motion
P waves (primary waves) are compressional waves that travel quickly through rock.
Compressional-wave crest
P waves travel as a series of contractions and expansions, pushing and pulling particles in the direction of their path of travel.
The red square charts the contraction and expansion of a section of rock.
Wave direction
S-wave motion
S waves (secondary waves) travel at about half the speed of P waves.
Shear-wave crest
S waves are shear waves that push material at right angles to their path of travel.
The red square shows how a section of rock shears from a square to a parallelogram as the S wave passes.
Wave direction
Surface-wave motion
Surface waves ripple across Earth's surface, where air above the surface allows free movement. There are two types of surface waves.
In one type, the ground surface moves in a rolling, elliptical motion that decreases with depth beneath the surface.
Wave direction
In the second type, the ground shakes sideways, with no vertical motion.
Wave direction

FIGURE 10.9 (a) The three types of waves travel by different routes and at different speeds to a seismograph that records them. (b) The three types of seismic waves are characterized by distinct types of ground deformation. The red squares show the distortion of a section of rock as a wave passes through it.

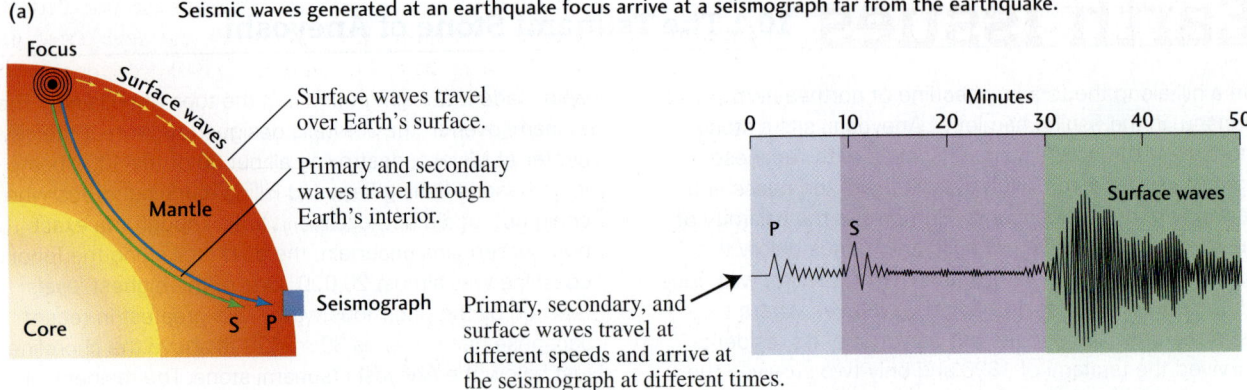

(Figure 10.9b). P waves can be thought of as push-pull waves: They push or pull particles of matter in the direction of their path of travel.

S waves travel through solid rock at a little more than half the velocity of P waves. They are *shear waves* that displace material at right angles to their path of travel (see Figure 10.9b). Shear waves cannot travel through liquids or gases.

The velocities of P and S waves are higher when the resistance to their movement is greater. It takes more force to compress solids than to shear them, so P waves always travel faster than S waves through a solid, which is why the P waves from an earthquake arrive at a seismograph before the S waves. This physical principle also explains why S waves cannot travel through air, water, or Earth's liquid outer core: Gases and liquids put up no resistance to shear.

Surface waves are confined to Earth's surface and outer layers, like waves on the ocean. Their velocity is slightly less than that of S waves. One type of surface wave sets up a rolling motion in the ground; another type shakes the ground sideways (see Figure 10.9b). Surface waves are usually the most destructive waves in a large, shallow-focus earthquake, especially in sedimentary basins, where reverberations in the soft, near-surface sediments can increase their amplitudes, causing much stronger shaking than in the basement rocks.

People have felt seismic waves and witnessed their destructiveness throughout history, but not until the close of the nineteenth century were scientists able to devise seismographs to record them accurately. Seismic waves enable us to study earthquakes and also provide our most important means of probing Earth's deep interior, as we will see in Chapter 11.

Locating the Focus

Locating an earthquake's focus is like deducing the distance of a lightning strike from the time interval between the flash of light and the sound of thunder: the greater the distance to the lightning bolt, the longer the time interval. Light travels faster than sound, so the lightning flash may be likened to the P waves of an earthquake and the thunder to the slower S waves.

The time interval between the arrival of P waves and S waves depends on the distance the waves have traveled from the focus: the longer the interval, the longer the distance the waves have traveled. Seismologists have used networks of sensitive seismographs around the world and highly accurate clocks to time the arrival of seismic waves from earthquakes as well as from underground nuclear explosions at known locations. From the results, they have constructed *travel-time curves*, which show how long it takes seismic waves of each type to travel a certain distance.

To estimate the distance to a new quake's focus, seismologists read from a seismogram the amount of time between the arrival of the first P waves and the arrival of the first S waves. Then they use travel-time curves, like the ones shown in Figure 10.9, to determine the distance from the seismograph to the focus. If they can determine the distances from three or more seismographic stations, they can locate the focus (see Figure 10.10). They can also deduce the time of the quake at the focus—the earthquake's *origin time*—because the arrival time of the first P waves at each station is known, and it is possible to determine from the travel-time curves how long those waves took to reach the station. Today, this entire process is done automatically by computers, which use data from a large network of seismographs to determine each earthquake's epicenter, depth of focus, and origin time.

Measuring the Size of an Earthquake

Locating earthquakes is only one step on the way to understanding them. We must also determine their sizes, or *magnitudes*. Other things being equal (such as distance from the focus and regional geology), an earthquake's magnitude is the main factor that determines the intensity and duration of the seismic waves it produces, and thus the earthquake's potential destructiveness.

Richter Magnitude In 1935, Charles Richter, a California seismologist, devised a simple procedure that assigned a numerical size to each earthquake, now called the *Richter magnitude* (Figure 10.11). Richter studied astronomy as a young man and learned how astronomers use a logarithmic scale to measure the brightness of stars, which varies over a huge range of values. Adapting this idea to earthquakes, Richter took as his measure of earthquake size the logarithm of the largest ground movement registered by a standard type of seismograph at a standard distance, thus defining a **magnitude scale.**

On Richter's magnitude scale, two earthquakes at the same distance from a seismograph differ by one magnitude unit if the peak amplitude of the ground movement they produce differs by a factor of 10. The ground movement of an earthquake of magnitude 3, therefore, is 10 times that of an earthquake of magnitude 2. Similarly, a magnitude-6 earthquake produces ground movements that are 100 times greater than those of a magnitude-4 earthquake. The energy released as seismic waves increases even more with earthquake magnitude, by a factor of about 32 for each magnitude unit. A magnitude-7 earthquake releases 32×32, or about 1000, times the energy of a magnitude-5 earthquake. According to this energy scale, the Tohoku megaquake was a million times more powerful than a magnitude-5 event!

Seismic waves gradually weaken as they move away from the focus, so to make his procedure work for any seismograph, Richter had to find a way to correct the measurement of ground movement for the distance between the seismograph and the focus. He devised a simple graph that allowed seismologists at different locations to quickly come up with nearly the same value for the magnitude of an earthquake no matter how far their instruments were from the focus (see Figure 10.11). His procedure came to be used throughout the world.

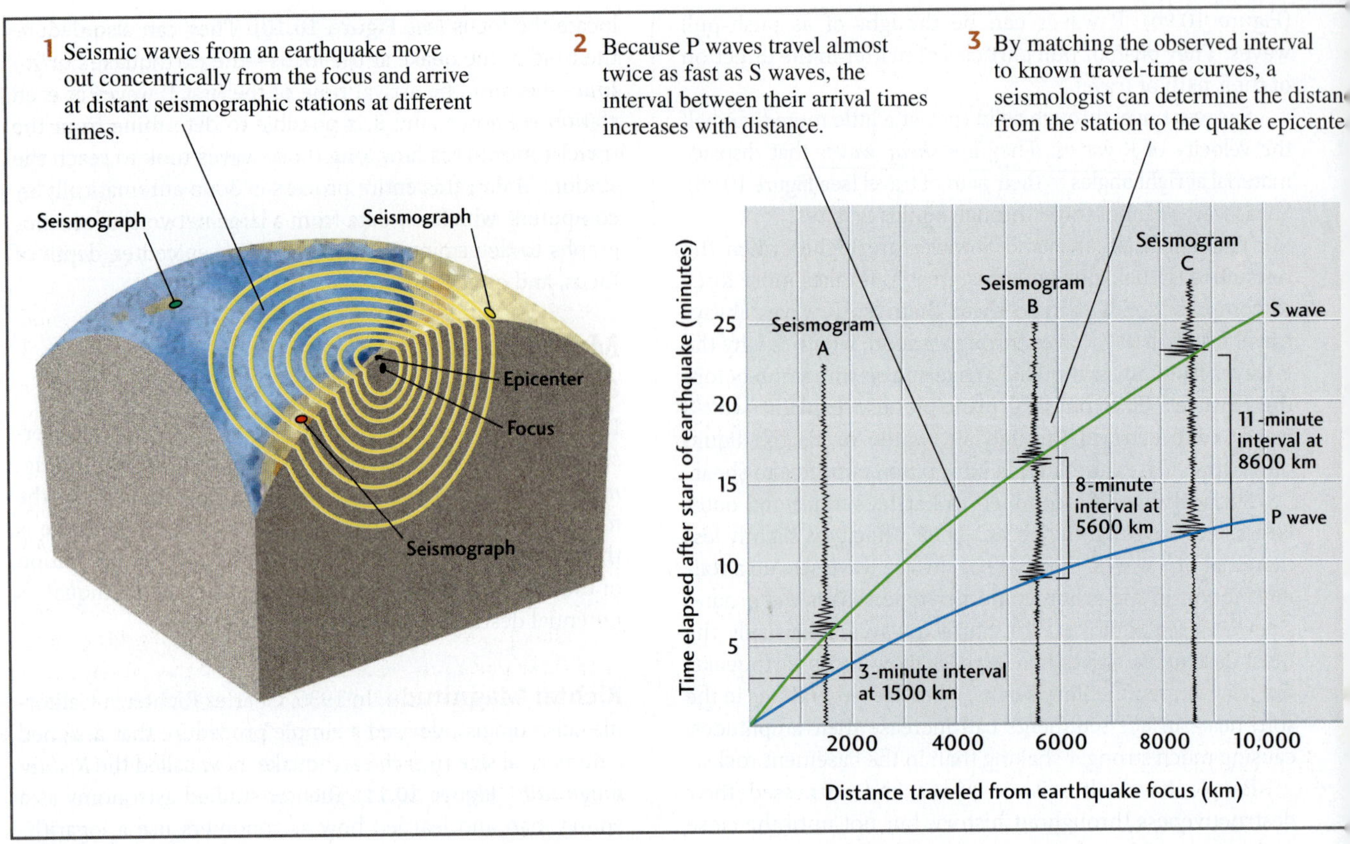

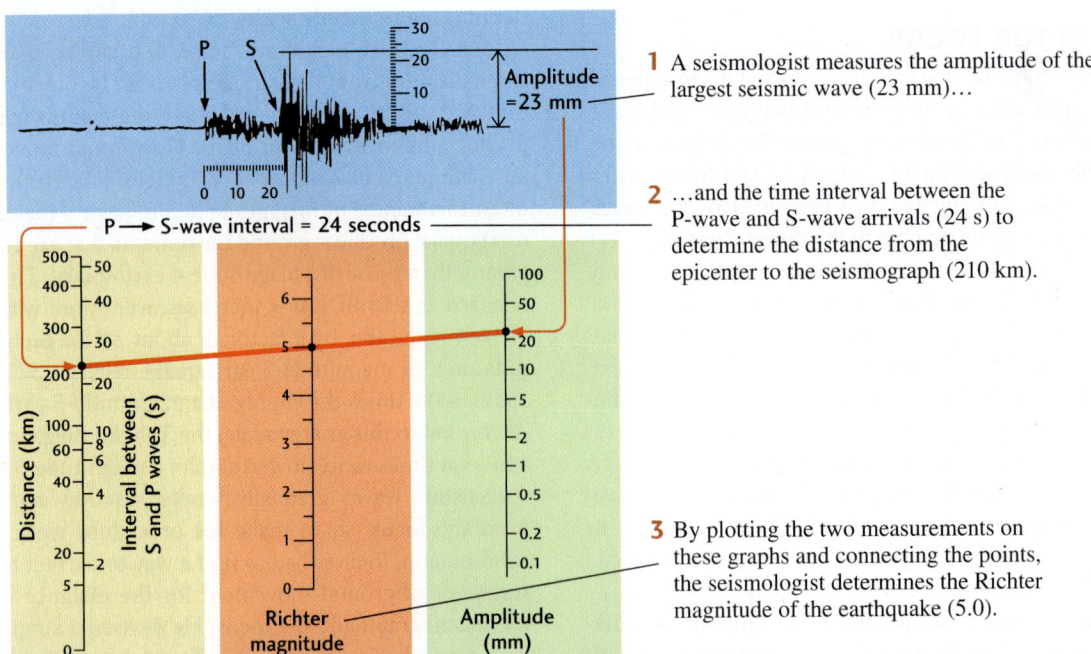

FIGURE 10.10 Readings from three or more seismographic stations can be used to determine the location of an earthquake's focus.

1 A seismologist measures the amplitude of the largest seismic wave (23 mm)...

2 ...and the time interval between the P-wave and S-wave arrivals (24 s) to determine the distance from the epicenter to the seismograph (210 km).

3 By plotting the two measurements on these graphs and connecting the points, the seismologist determines the Richter magnitude of the earthquake (5.0).

FIGURE 10.11 The maximum amplitude of ground movement, corrected by the P-S wave interval, is used to assign a Richter magnitude to an earthquake. [Data from California Institute of Technology.]

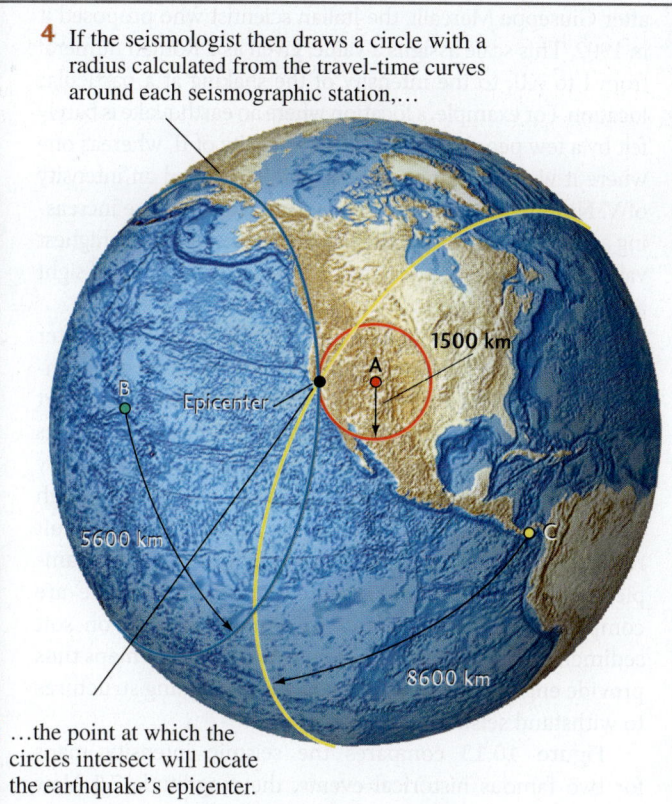

4 If the seismologist then draws a circle with a radius calculated from the travel-time curves around each seismographic station,...

5 ...the point at which the circles intersect will locate the earthquake's epicenter.

Moment Magnitude Although "Richter scale" is a household term, seismologists prefer a measure of earthquake size more directly related to the physical properties of the faulting that causes an earthquake. The *seismic moment* of an earthquake is defined as a number proportional to the product of the area of faulting and the average fault slip. The corresponding *moment magnitude* increases by about one unit for every tenfold increase in the area of faulting (see the Practicing Geology Exercise in the next section).

Although Richter's method and the moment method produce roughly the same numerical values, the moment magnitude can be measured more accurately from seismographs, and it can sometimes be determined directly from field measurements of the faulting.

Magnitude and Frequency Large earthquakes occur much less often than small ones. This observation can be expressed by a simple relationship between earthquake frequency and magnitude (**Figure 10.12**). Worldwide, approximately 1,000,000 earthquakes with magnitudes greater than 2 take place each year. This number decreases by a factor of 10 for each magnitude unit. Hence, there are about 100,000 earthquakes with magnitudes greater than 3, about 1000 with magnitudes greater than 5, and about 10 with magnitudes greater than 7 each year.

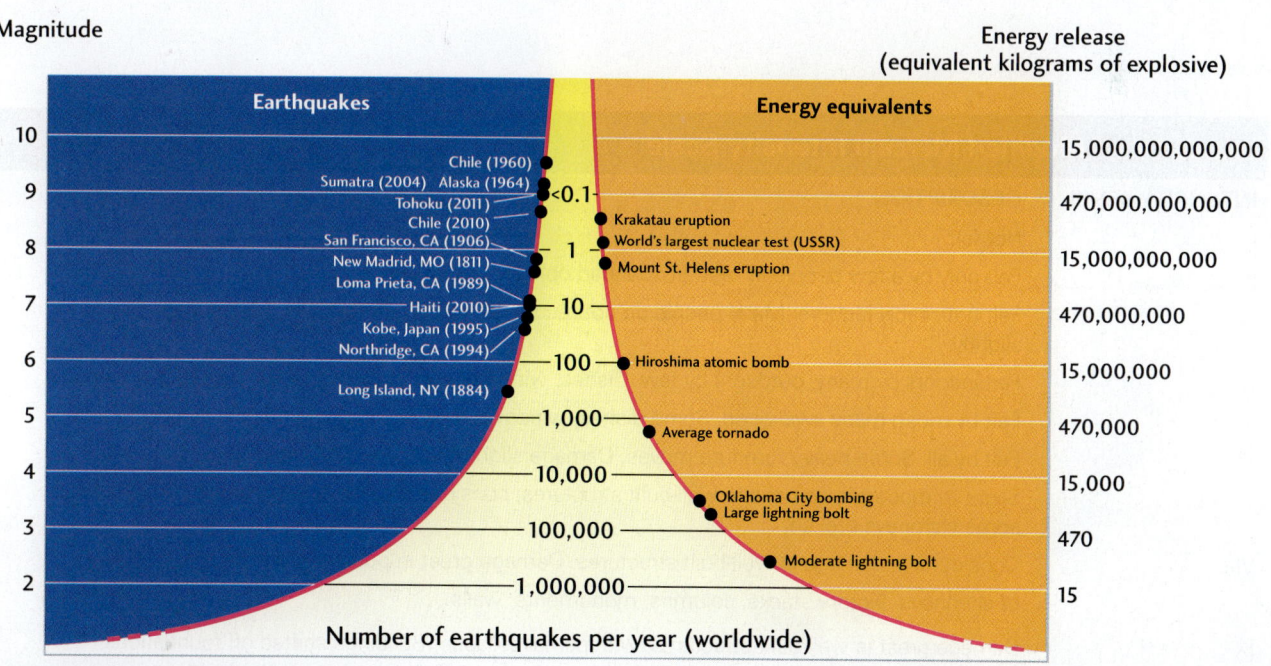

FIGURE 10.12 Relationship between moment magnitude, seismic energy release, and number of earthquakes per year worldwide. Examples of earthquakes of various magnitudes and of other large sources of sudden energy release are included for comparison. [Data from IRIS Consortium, http://www.iris.edu.]

According to these statistics, there should be, on average, about 1 earthquake with a magnitude greater than 8 per year and 1 earthquake with a magnitude greater than 9 every 10 years. In fact, the very largest earthquakes, such as the ones that occurred on thrust faults in the subduction zones off Japan in 2011 (moment magnitude 9.0), Sumatra in 2004 (moment magnitude 9.2), Alaska in 1964 (moment magnitude 9.1), and Chile in 1960 (moment magnitude 9.5) and 2010 (moment magnitude 8.8), are almost this common when averaged over many decades. However, even the largest subduction zone megathrusts are too small to create magnitude 10 earthquakes, so seismologists believe that events of such extreme size do not follow this rule; that is, they are less frequent than once per century.

Shaking Intensity Earthquake magnitude by itself does not describe seismic hazard because the shaking that causes destruction generally weakens with distance from the fault rupture. A magnitude-8 earthquake in a remote area far from the nearest city might cause no human or economic losses, whereas a magnitude-6 quake immediately beneath a city would likely cause serious damage. The destruction in Christchurch by the earthquakes of February 22 and June 13, 2011, illustrates this important point (see Figure 10.7).

In the late nineteenth century, before Richter invented his magnitude scale, seismologists and earthquake engineers developed **intensity scales** for estimating the intensity of shaking directly from an earthquake's destructive effects. Table 10.1 shows one intensity scale that remains in common use today, called the *modified Mercalli intensity scale*

after Giuseppe Mercalli, the Italian scientist who proposed it in 1902. This scale assigns a value, given as a Roman numeral from I to XII, to the intensity of the shaking at a particular location. For example, a location where an earthquake is barely felt by a few people is assigned an intensity of II, whereas one where it was felt by nearly everyone is assigned an intensity of V. Numbers at the upper end of the scale indicate increasing amounts of damage. The narrative attached to the highest value, XII, is tersely apocalyptic: "Damage total. Lines of sight and level are distorted. Objects thrown into the air."

Observations of damage and felt effects collected after an earthquake can be compiled into maps showing contours of equal shaking intensity. Many people now report such data directly to the U.S. Geological Survey through its popular "Did You Feel It?" website, which posts intensity maps immediately following significant events. Although earthquake intensities are generally highest near the fault rupture, they also depend on the local geology. For example, when sites at equal distances from the rupture are compared, the shaking tends to be more intense on soft sediments than on hard basement rock. Intensity maps thus provide engineers with crucial data for designing structures to withstand seismic shaking.

Figure 10.13 compares the seismic intensity maps for two famous historical events, the magnitude-7.5 New Madrid earthquake, which occurred near the southern tip of Missouri on December 16, 1811, and the magnitude-7.8 1906 San Francisco earthquake. The New Madrid quake rattled Boston, over 1700 km from the epicenter, whereas the San Francisco quake was not felt even in Los Angeles, less

TABLE 10.1	Modified Mercalli Intensity Scale
INTENSITY LEVEL	DESCRIPTION
I	Not felt.
II	Felt only by a few people at rest. Suspended objects may swing.
III	Felt noticeably indoors. Many people do not recognize it as an earthquake. Parked cars may rock slightly.
IV	Felt indoors by many, outdoors by few. Dishes, windows, doors rattle. Parked cars rock noticeably.
V	Felt by most; many awakened. Some dishes, windows broken. Unstable objects overturned.
VI	Felt by all. Some heavy furniture moves. Damage slight.
VII	Slight to moderate damage in well-built structures; considerable damage in poorly built structures; some chimneys broken.
VIII	Considerable damage in well-built structures. Damage great in poorly built structures. Fall of chimneys, factory stacks, columns, monuments, walls.
IX	Damage great in well-built structures, with partial collapse. Buildings shifted off foundations.
X	Some well-built wooden structures destroyed; most masonry and frame structures destroyed. Rails bent.
XI	Few if any masonry structures remain standing. Bridges destroyed. Rails bent greatly.
XII	Damage total. Lines of sight and level are distorted. Objects thrown into the air.

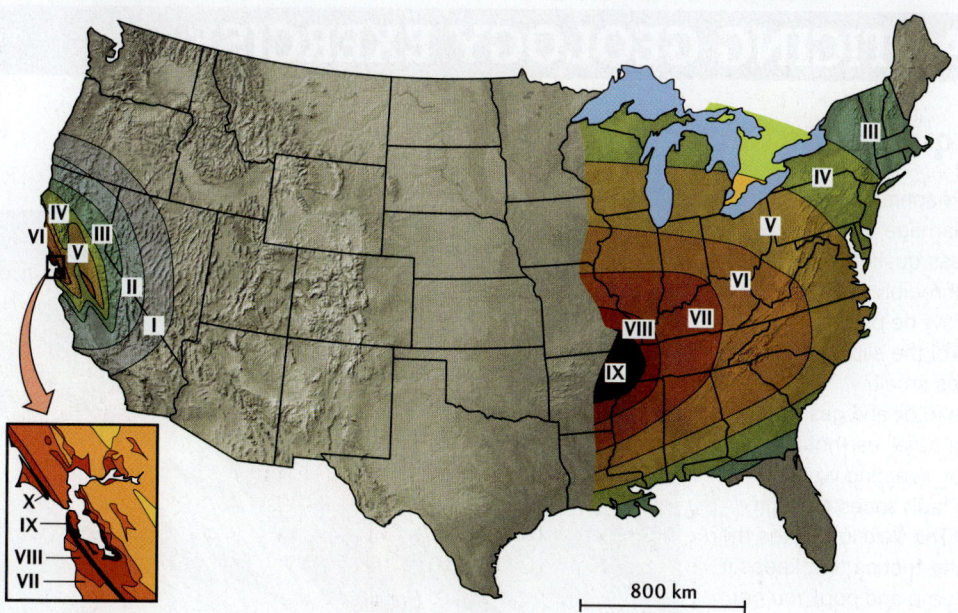

FIGURE 10.13 Measurements of modified Mercalli intensities for two famous historical events, the magnitude-7.5 New Madrid earthquake, which occurred near the southern tip of Missouri on December 16, 1811, and the magnitude-7.8 1906 San Francisco earthquake. During the New Madrid earthquake, intensities as high as VI were observed 200 km from the epicenter, whereas the San Francisco quake was not felt even in Los Angeles, less than 500 km away (see Table 10.1). Owing to the lack of observations, the intensities of the New Madrid earthquake are not shown west of the Mississippi River.

than 500 km away. Why was the area of intense shaking so much larger in the former case, even though the earthquake magnitude was smaller? The crust and uppermost mantle in the continental interior of the eastern United States comprise stiffer rocks at lower temperatures than in the tectonically active regions of the west; therefore, seismic waves propagate much more efficiently, and the shaking from an earthquake of given magnitude can be much more intense out to much greater distances. Engineers must take these regional differences into account when assessing seismic risk.

Determining Fault Mechanisms

The pattern of ground shaking also depends on the orientation of the fault rupture and the direction of slipping, which together specify the **fault mechanism** of an earthquake. The fault mechanism tells us whether the rupture was on a *normal*, a *reverse*, or a *strike-slip* fault. If the rupture was on a strike-slip fault, the fault mechanism also tells us whether the movement was *right-lateral* or *left-lateral* (see Figure 8.8 for the definitions of these terms). We can then use this information to infer the regional pattern of tectonic forces (**Figure 10.14**).

For shallow ruptures that break the surface, we can sometimes determine the fault mechanism from field observations of the fault scarp. As we have seen, however, most ruptures are too deep to break the surface, so we must deduce the fault mechanism from seismograms.

For large earthquakes at any depth, this task turns out to be easy, because there are enough seismographic

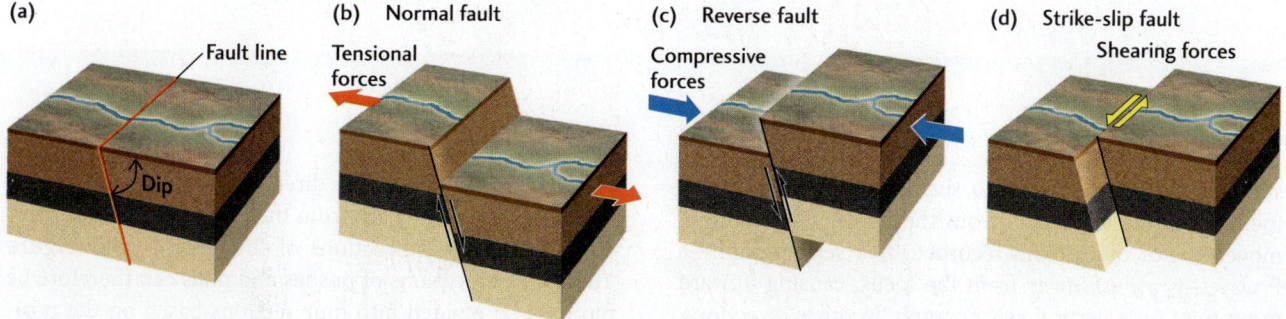

FIGURE 10.14 The three main types of fault mechanisms that initiate earthquakes and the stresses that cause them. (a) A fault before movement takes place. (b) Normal faulting due to tensional stress. (c) Reverse faulting due to compressive stress. (d) Strike-slip faulting due to shearing stress (in this case, left-lateral).

PRACTICING GEOLOGY EXERCISE

Can Earthquakes Be Controlled?

Earthquakes of magnitude 4 rarely result in much damage to nearby communities, whereas quakes of magnitude 8 can be incredibly destructive. Would it somehow be possible for humans to control the slip on a fault to keep earthquakes small?

Experiments in oil and gas fields have shown that small earthquakes can be caused by injecting water or other fluids into fault zones through deep drill holes. The fluid lubricates the fault, reducing the friction that keeps it from slipping. Pump and pop! You get an earthquake. Why not control the sizes of earthquakes by using this fluid injection technique to release energy on a fault only in ruptures smaller than magnitude 4?

The feasibility of this method depends on how many events of magnitude 4 would produce the same fault slip over the same area as one event of magnitude 8. From observations of many earthquakes, seismologists have discovered two simple rules about moment magnitude that can guide this calculation:

1. *Area rule:* The area of faulting increases by a factor of 10 for each unit of moment magnitude. Therefore, a magnitude-8 rupture has 10,000 times the area of a magnitude-4 rupture (because $10^{(8-4)} = 10^4$).

2. *Slip rule:* The average slip of a fault rupture increases by a factor of 10

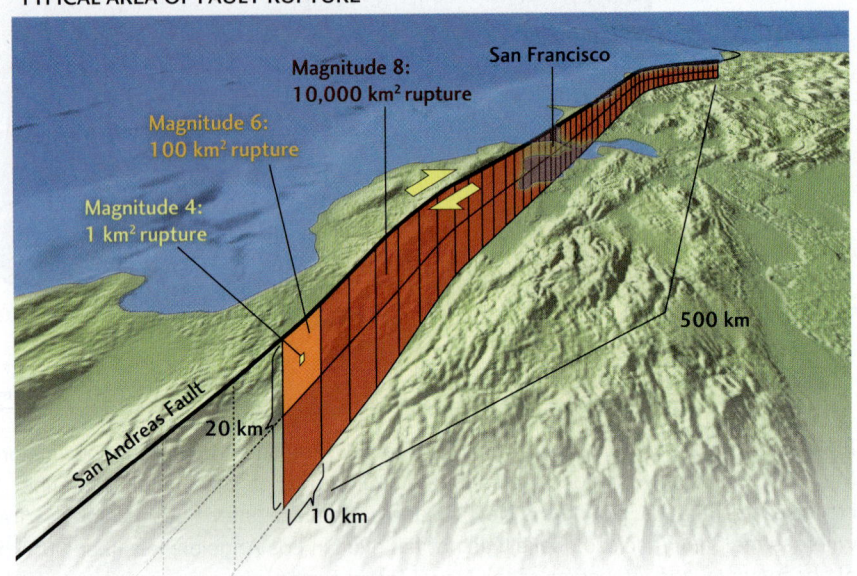

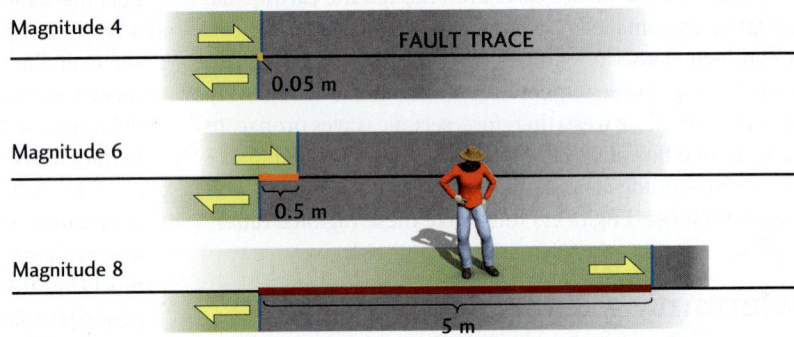

The top panel shows how the rupture area of the San Andreas Fault increases with earthquake magnitude. The lower panel shows how the slip distance of the fault rupture increases with magnitude

stations around the world to surround any earthquake's focus. In some directions from the focus, the very first movement of the ground recorded by a seismograph—a P wave—is a *push away* from the focus, causing upward movement on a vertical seismograph. In other directions, the initial ground movement is a *pull toward* the focus, causing downward movement on a vertical seismograph.

For strike-slip ruptures, the directions of the largest pushes lie on an axis rotated 45° from the fault plane and are perpendicular to the directions of the largest pulls (**Figure 10.15**). The locations of pushes and pulls can therefore be plotted and divided into four sections based on the positions of the seismographic stations. One of the two boundaries between those sections will be the fault orientation;

for each two units of moment magnitude. Therefore, the slip of a magnitude-8 rupture is 100 times that of a magnitude-4 rupture (because $10^{(8-4)/2} = 10^2$).

The area of a magnitude-8 rupture is typically about 10,000 km², and the average slip is about 5 meters per event.

▶ The area rule implies that the area of a magnitude-4 rupture will be 10,000 times smaller than that of a magnitude-8 rupture, or 1 km².
▶ The slip rule implies that the slip of a magnitude-4 rupture will be 100 times smaller than the slip of a magnitude-8 rupture, or 0.05 m (5 cm).

Therefore, the number of magnitude-4 events needed to equal a single magnitude-8 event is

$$10{,}000 \times 100 = 1{,}000{,}000$$

This calculation shows that small earthquakes don't add up to much of the displacement that occurs across a fault; the big ones are what really count. On a fault like the San Andreas, which has earthquakes of nearly magnitude 8 every 100 years or so, we would have to generate magnitude-4 earthquakes at a rate of almost 10,000 per year to take up the same amount of fault movement.

Injecting faults with fluids to increase the rate of small earthquakes would be a bad idea for at least two reasons. It would be prohibitively expensive: Drilling thousands of holes along the fault and pumping all that water down to earthquake focal depths would cost billions of dollars. It would also be dangerous: One of the ruptures induced by the fluid injection could become a much larger earthquake than intended. An effort to control earthquakes could end up causing a big one!

BONUS PROBLEM: How many magnitude-4 earthquakes would provide the same slip over the same area as one magnitude-6 earthquake?

---

the other will be a plane perpendicular to the fault. The slip direction on the fault plane can also be determined from the arrangement of pushes and pulls. In this manner, seismologists can deduce whether the horizontal crustal forces that triggered an earthquake were tensional, compressive, or shearing forces.

GPS Measurements and "Silent" Earthquakes

As we saw in Chapter 2, GPS receivers can record the slow movements of lithospheric plates. These instruments can also measure the strain that builds up from such

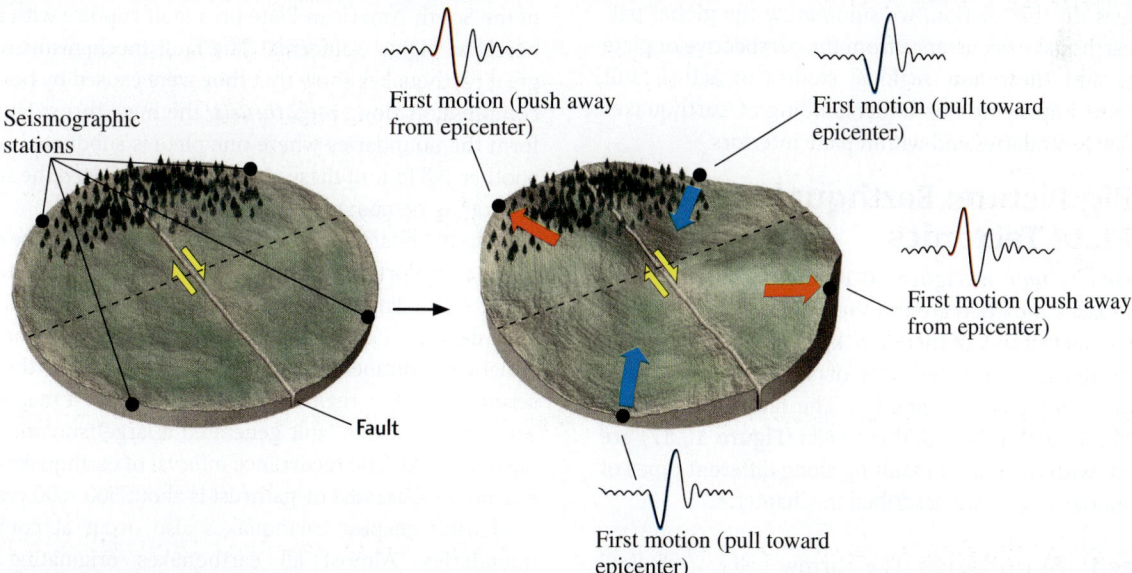

FIGURE 10.15 The movement marking the first P wave arriving from a fault rupture at each of several seismographic stations is used to determine the orientation of the fault and the direction of fault slipping. The case shown here is the rupture of a right-lateral strike-slip fault. Note that the alternating pattern of pushes and pulls would remain the same if a plane perpendicular to the fault ruptured with left-lateral displacement. Seismologists can usually choose between the two possibilities using additional information, such as field mapping of the fault scarp or the alignment of aftershocks along the fault.

movements, as well as the sudden slipping on a fault when it ruptures in an earthquake.

Seismologists use GPS observations to study another kind of movement along active faults. It has been known for many years that a section of the San Andreas Fault in central California creeps continuously, rather than rupturing suddenly. This creep slowly deforms structures and cracks pavements that cross the fault trace. More recently, GPS receivers have found surface movements at convergent plate boundaries that reflect short-lived creep events, which commonly last days to weeks at a time. They have been named *silent earthquakes* because these gradual movements do not trigger destructive seismic waves. Nevertheless, they can release large amounts of stored strain energy.

These observations raise many questions that geologists are now trying to answer: What causes faults to stick and slip catastrophically in some places and creep in others? Will the release of strain energy by silent earthquakes make destructive earthquakes in those regions less frequent or less severe? Can silent earthquakes be used to predict potentially destructive earthquakes?

Earthquakes and Patterns of Faulting

As we have seen, seismologists are using networks of sensitive seismographs to locate earthquakes around the world, measure their magnitudes, and deduce their fault mechanisms. These methods are revealing new information about tectonic processes on scales much smaller than the plates themselves. In this section, we summarize the global pattern of earthquake occurrence from the perspective of plate tectonics and show how regional studies of active fault systems are improving our understanding of earthquakes along plate boundaries and within plate interiors.

The Big Picture: Earthquakes and Plate Tectonics

The *seismicity map* in **Figure 10.16** shows the epicenters of earthquakes recorded around the world since 1976. The most obvious features of this map, known to seismologists for many decades, are the belts of seismic activity that mark the major plate boundaries. The fault mechanisms observed for earthquakes in these belts (**Figure 10.17**) are consistent with the types of faulting along different types of plate boundaries that we described in Chapter 8.

Divergent Boundaries The narrow belts of shallow earthquakes that run through ocean basins coincide with mid-ocean ridge crests and their offsets on transform faults. The P waves recorded from the ridge-crest quakes indicate that they are caused by normal faulting. The faults strike parallel to the ridge and dip toward the rift valley at the ridge crest. Normal faulting implies that tensional forces are at work as the plates are pulled apart during seafloor spreading. Earthquakes also have normal fault mechanisms in zones where continental crust is being pulled apart, such as in the East African rift valleys and in the Basin and Range Province of western North America.

Transform-Fault Boundaries Earthquake activity is even greater along the transform-fault boundaries that offset mid-ocean ridge segments, where plates create horizontal shearing forces as they slide past each other in opposite directions. As expected, earthquakes along these transform faults have strike-slip fault mechanisms. You will notice that the slip direction indicated by the fault mechanisms is left-lateral where the ridge crest steps right and right-lateral where it steps left. These directions are the opposite of what would be needed to create the offsets of the ridge crest, but are consistent with the direction of slip predicted by seafloor spreading. In the mid-1960s, seismologists used this property of the transform faults to support the hypothesis of seafloor spreading. Slip directions on transform faults that run through continental crust, such as California's San Andreas Fault and New Zealand's Alpine Fault (both right-lateral), also agree with the predictions of plate tectonics.

Convergent Boundaries The world's largest earthquakes occur at convergent plate boundaries. The four greatest earthquakes of the last hundred years were of this type: the 2011 Tohoku earthquake; the Sumatra earthquake of 2004; the Alaska earthquake of 1964; and the largest of all, the 1960 earthquake in the subduction zone west of Chile (magnitude 9.5). During the Chile earthquake, the crust of the Nazca Plate slipped an average of 20 m beneath the crust of the South American Plate on a fault rupture with an area nearly as big as California! The fault mechanisms of these great earthquakes show that they were caused by horizontal compression along *megathrusts*, the huge thrust faults that form the boundaries where one plate is subducted beneath another. All four of these earthquakes displaced the seafloor, generating tsunamis that devastated coastlines.

The boundary of the North American Plate off the coast of the Pacific Northwest is a 1000-km-long megathrust known as the Cascadia Fault, where the oceanic lithosphere of the Juan de Fuca Plate is being subducted beneath the North American continent (see Figure 2.7). Although the recent seismicity in this region has been fairly low, a magnitude-9 earthquake on this fault generated a large tsunami that hit Japan in 1700. The recurrence interval of earthquakes of this size on the Cascadia megathrust is about 500–600 years.

Earth's deepest earthquakes also occur at convergent boundaries. Almost all earthquakes originating below 100 km rupture the descending plate in a subduction zone. The fault mechanisms of these deeper earthquakes show a variety of orientations, but they are consistent with the deformation expected within the descending plate as gravity pulls it back into the convecting mantle. The deepest earthquakes take place in the oldest—and therefore coldest—descending plates, such as those beneath South America, Japan, and

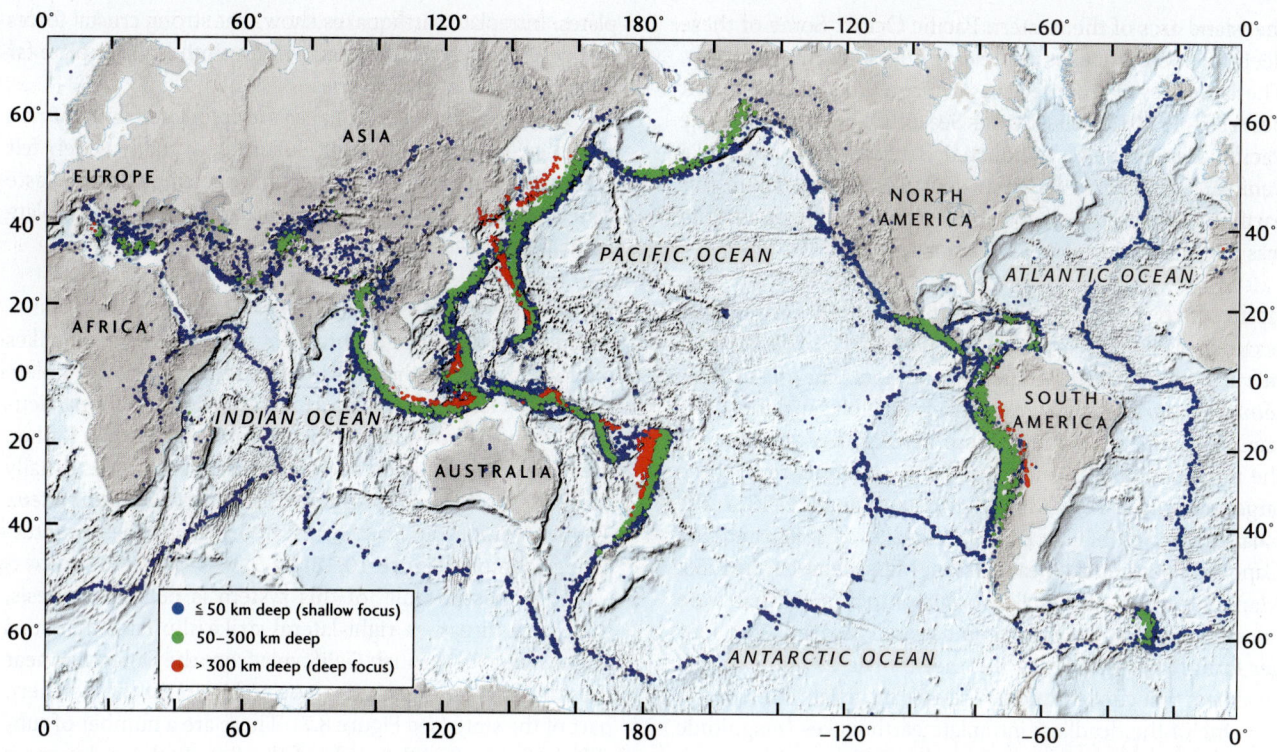

FIGURE 10.16 Global map of seismic activity from January 1976 through October 2013. Each dot represents the epicenter of an earthquake larger than magnitude 5. Colors indicate the focal depth. Note the concentration of earthquakes along the boundaries between major lithospheric plates. [Map based on data from USGS.]

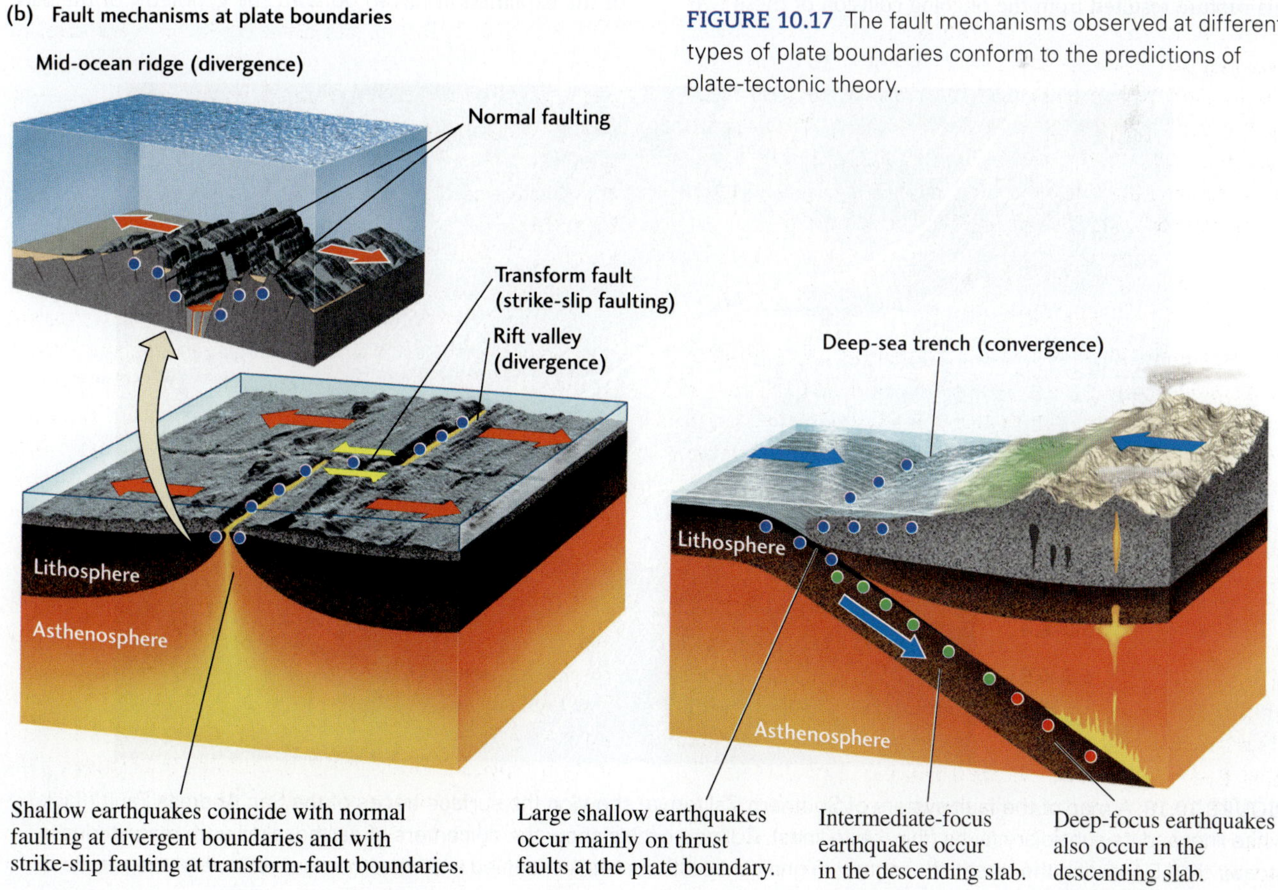

(b) Fault mechanisms at plate boundaries

FIGURE 10.17 The fault mechanisms observed at different types of plate boundaries conform to the predictions of plate-tectonic theory.

Shallow earthquakes coincide with normal faulting at divergent boundaries and with strike-slip faulting at transform-fault boundaries.

Large shallow earthquakes occur mainly on thrust faults at the plate boundary.

Intermediate-focus earthquakes occur in the descending slab.

Deep-focus earthquakes also occur in the descending slab.

the island arcs of the western Pacific Ocean. Some of these deep-focus earthquakes can release huge amounts of energy. The largest ever recorded was a magnitude-8.3 event that occurred in 2013 beneath the Sea of Okhotsk, where the Pacific Plate is being subducted beneath the Kamchatka Peninsula. Owing to its great depth, around 600 km, this earthquake caused no substantial damage, though shaking was felt in Moscow over 7000 km away.

Intraplate Earthquakes Although most earthquakes occur at plate boundaries, a small percentage of global seismic activity originates within plate interiors. The foci of these *intraplate earthquakes* are relatively shallow, and most occur on continents. Among these earthquakes are some of the most famous in American history: a sequence of three large events near New Madrid, Missouri, in 1811–1812; the Charleston, South Carolina, earthquake of 1886; and the Cape Ann earthquake near Boston, Massachusetts, in 1755. Many intraplate earthquakes occur on old faults that were once parts of ancient plate boundaries. These faults no longer form plate boundaries, but they remain zones of crustal weakness that concentrate and release intraplate stresses.

One of the deadliest intraplate earthquakes (magnitude 7.6) occurred near Bhuj, in the state of Gujarat, in western India, in 2001. It is estimated that some 20,000 lives were lost. The Bhuj earthquake occurred on a previously unknown thrust fault 1000 km south of the boundary between the Indian and Eurasian plates, but the compressive stresses responsible for this rupture resulted from the ongoing collision of these two plates. Intraplate earthquakes show that strong crustal forces can develop and cause faulting within a lithospheric plate far from modern plate boundaries.

Some intraplate earthquakes are associated with human activities, such as the recent clustering of many widely felt earthquakes in Oklahoma, where the deep injection of waste water into basement rocks has lubricated ancient intraplate faults (Earth Issues 10.2).

Regional Fault Systems

Although the fault mechanisms of most major earthquakes conform to the predictions of plate tectonic theory, a plate boundary can rarely be described as a single fault, particularly when the boundary involves continental crust. Rather, the zone of deformation between two moving plates usually comprises a network of interacting faults—a *fault system*. The fault system in Southern California provides an interesting example (Figure 10.18).

The "master fault" of this system is our old nemesis, the San Andreas—a right-lateral strike-slip fault that runs northwestward through California from the Salton Sea near the Mexican border until it goes offshore in the northern part of the state (see Figure 8.7). There are a number of subsidiary faults on either side of the San Andreas, however, that generate large earthquakes. In fact, most of the damaging earthquakes in Southern California during the past century have occurred on those subsidiary faults.

Why is the San Andreas Fault system so complex? Part of the explanation has to do with the geometry of the San

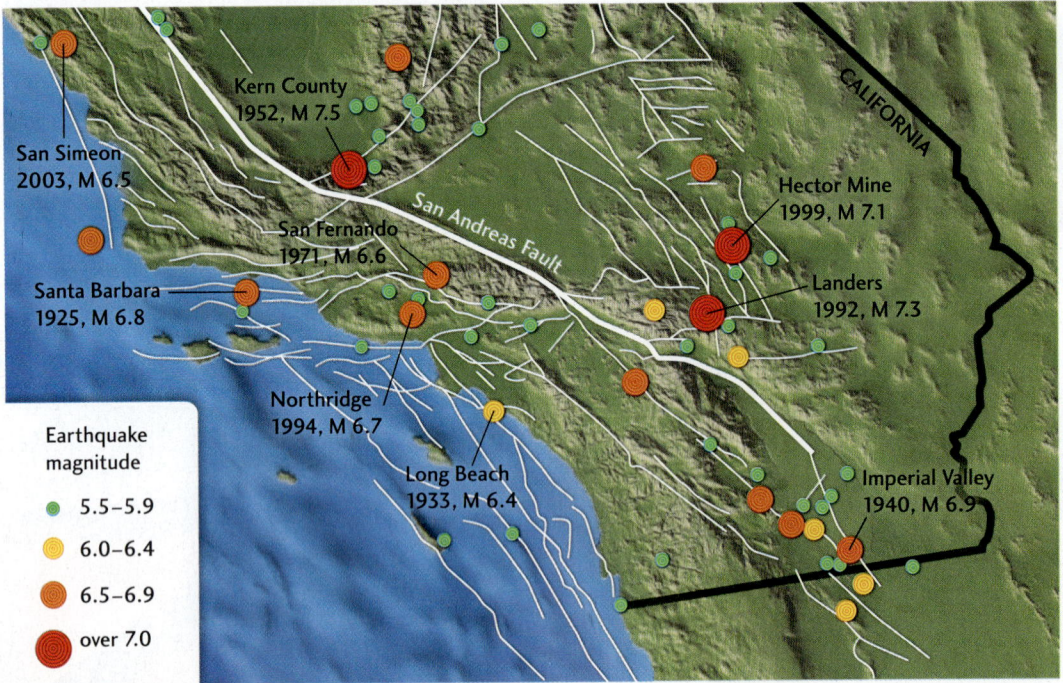

FIGURE 10.18 A map of the fault system of Southern California, showing the surface traces of the San Andreas Fault (thick white line) and its subsidiary faults (thin white lines). Colored circles show the epicenters of earthquakes with magnitudes greater than 5.5 during the twentieth century. Significant earthquakes are labeled with their names, years, and magnitudes.
[Courtesy of Southern California Earthquake Center.]

FIGURE 10.19 Sixteen people died in the Northridge Meadows apartment building in Los Angeles during the 1994 Northridge earthquake. The victims lived on the first floor and were crushed when the upper levels collapsed. Many more buildings would have collapsed if the newer buildings in the area had not been constructed according to stringent building codes. [Nick Ut/AP Photo.]

Andreas Fault itself. A bend in the fault creates compressive forces that cause thrust faulting in the area north of Los Angeles (see Figure 8.22). Thrust faulting in this "Big Bend" was responsible for two recent deadly earthquakes, the San Fernando earthquake of 1971 (magnitude 6.6, 65 people killed) and the Northridge earthquake of 1994 (magnitude 6.7, 58 people killed) (see Figure 10.18). Over the past several million years, this thrust faulting has raised the San Gabriel Mountains to elevations of 1800 to 3000 m.

Another complication is the plate extension taking place east of California in the Basin and Range Province, which spans the state of Nevada and much of Utah and Arizona (see Chapters 8 and 21). This broad zone of extensional deformation is connected with the San Andreas Fault system through a series of faults that run along the eastern side of the Sierra Nevada and through the Mojave Desert. Faults of this system were responsible for the 1992 Landers earthquake (magnitude 7.3) and the 1999 Hector Mine earthquake (magnitude 7.1), as well as the 1872 Owens Valley earthquake (magnitude 7.6).

Earthquake Hazards and Risks

Since 2000, earthquakes worldwide have killed more than 700,000 people and disrupted the economies of entire regions. The United States has been incredibly lucky; during this period, earthquakes claimed only three fatalities. However, two California temblors—the 1989 Loma Prieta earthquake (magnitude 7.1), which occurred 80 km south of San Francisco, and the 1994 Northridge earthquake in Los Angeles—were among the costliest disasters in the nation's history. Damage amounted to more than $10 billion in the Loma Prieta quake and $40 billion in the Northridge quake because of their proximity to urban centers. About 60 people died in each event, but the death toll would have been many times higher if stringent building codes had not been in place (**Figure 10.19**).

Destructive earthquakes are even more frequent in Japan than in California. The recorded history of destructive earthquakes in Japan, going back 2000 years, has left an indelible impression on the Japanese people, which is why Japan is the best prepared nation in the world to deal with earthquakes. It has impressive public education campaigns, building codes, and warning systems. Despite this preparedness, more than 5600 people were killed in a devastating (magnitude 6.9) earthquake that struck the city of Kobe on January 16, 1995 (**Figure 10.20**).

FIGURE 10.20 This elevated expressway in Hyogo, Japan, was overturned during the 1995 Kobe earthquake. [Pacific Press Service/Alamy.]

Earth Issues 10.2 Oklahoma, Earthquake Country?

Oklahoma is situated on the interior platform of North America, far away from the seismically active plate boundaries. Until recently, earthquakes strong enough to be felt at Earth's surface, magnitude 3 or larger, struck that prairie state at a rate fewer than 2 per year. But seismic activity began to rise rapidly in 2009 (see figure). By 2015, earthquakes of this size were occurring at a rate of more than 2 per *day*, making Oklahoma the nation's most seismically active state. Some of these events, such as the magnitude-5.6 earthquake that struck the town of Prague in 2011, have produced shaking as high as level VII on the Modified Mercalli Intensity scale (Table 10.1)—strong enough to damage buildings and injure people.

This remarkable rise in seismicity can be attributed to a rapidly increasing human activity—the injection of wastewater deep into Earth's crust. Huge amounts of oil and gas lie beneath the Oklahoma hills, but much of it is trapped in tight formations of sedimentary rocks. Such impermeable formations were difficult to tap by conventional drilling methods, but enhanced production techniques, which include horizontal drilling and hydraulic fracturing ("fracking"), have overcome those limitations, setting off an oil boom in the central and eastern United States. Unfortunately, what comes out of these new oil wells is more water than oil—typically a gallon of water for every cup of oil. Much of this "produced water" is from salty brines entrapped in the same sediments as the oil and gas, and it is frequently laden with poisonous salts and minerals that make it unsuitable for other uses. The cheapest way to prevent this wastewater from contaminating the environment is to inject it back into the crust. The injection must be well below our groundwater supplies and thus into deep rock formations that are the breeding ground of earthquakes.

Experiments in oil fields going back to the 1960s have shown that deep injection can generate earthquakes by changing the crustal stresses. In particular, injecting wastewater can increase the fluid pressure in the rocks, which reduces the friction on preexisting faults, allowing them to rupture in earthquakes. Many factors must be favorable for this mechanism to work; for example, the faults must be large enough to produce felt earthquakes, and they must be properly oriented for the crustal stresses to cause the faults to rupture. Understanding these preconditions helps to explain why, among the tens of thousands of deep injection wells in the United States, only a small fraction are known to cause significant seismicity.

Earthquakes generated by human activities that alter stresses in Earth's crust are called **induced seismicity**. Wastewater injection is the primary cause of induced seismicity in Oklahoma, but earthquakes can also be induced by the filling of large reservoirs, which locally increases the load on the crust, and by the production of geothermal energy, which can alter the temperatures and fluid pressures in the crust. The hazard from natural earthquakes cannot be avoided, but the hazard from induced seismicity can. Knowing just how injection causes earthquakes is allowing society to make better plans for reducing the risk of damaging earthquakes induced by future industrial activity.

Oklahoma is not alone. Many parts of the central and eastern United States have seen substantial increases in induced seismicity. In March of 2016, the U.S. Geological Survey issued, for the first time, a one-year forecast of both natural and induced seismicity for the central and eastern United States. Several of the affected states are increasing their regulation of deep wastewater injection. A regulatory commission in Oklahoma, for example, has asked well operators in central Oklahoma to reduce the volume of deep injections by 40 percent, and these actions appear to be lowering the induced seismicity rate from the peak in 2015.

The large numbers of casualties and structural failures (50,000 buildings destroyed) resulted partly from the less stringent building codes that were in effect before 1980, when much of the city was built, and partly from the location of the earthquake rupture, which was very close to the city. The tsunami of the 2011 Tohoku earthquake caused an even greater loss of life (almost 20,000). The disaster was compounded by meltdowns and explosions at the Fukushima-Daiichi power plant, one of the world's largest nuclear facilities, discussed later in the chapter. Although the economic costs are still being counted, the Tohoku earthquake is already the most expensive natural disaster in recorded history.

How Earthquakes Cause Damage

Earthquakes proceed as chain reactions in which the primary effects of earthquakes—faulting and ground shaking—trigger secondary effects, which include landslides and tsunamis as well as destructive processes within the built environment, such as collapsing structures and fires.

Earthquake Hazards and Risks

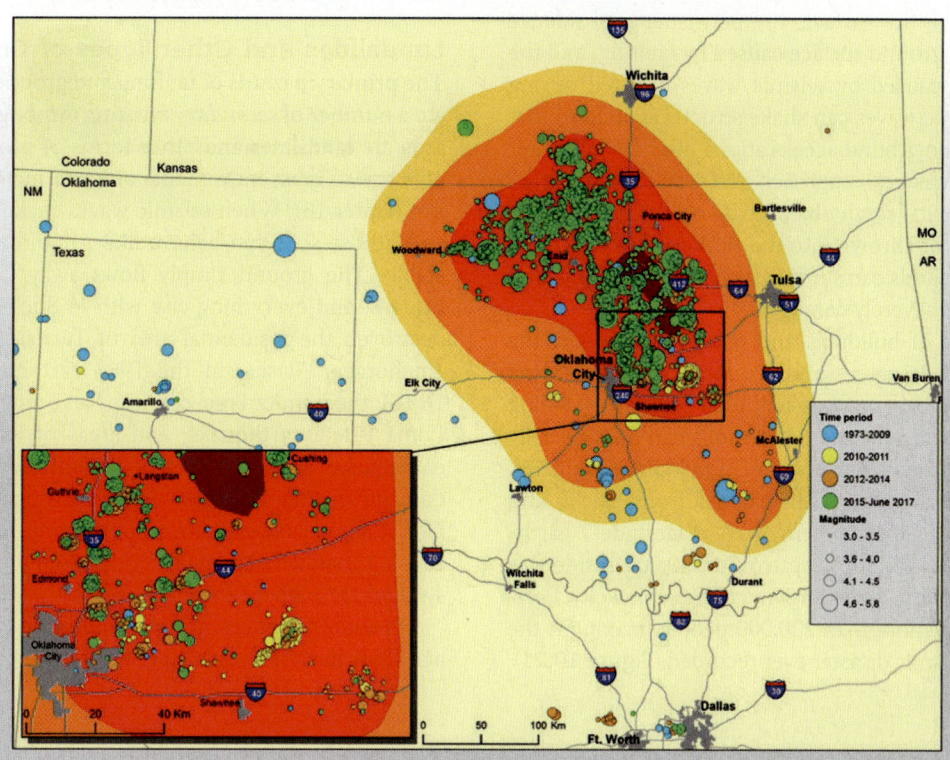

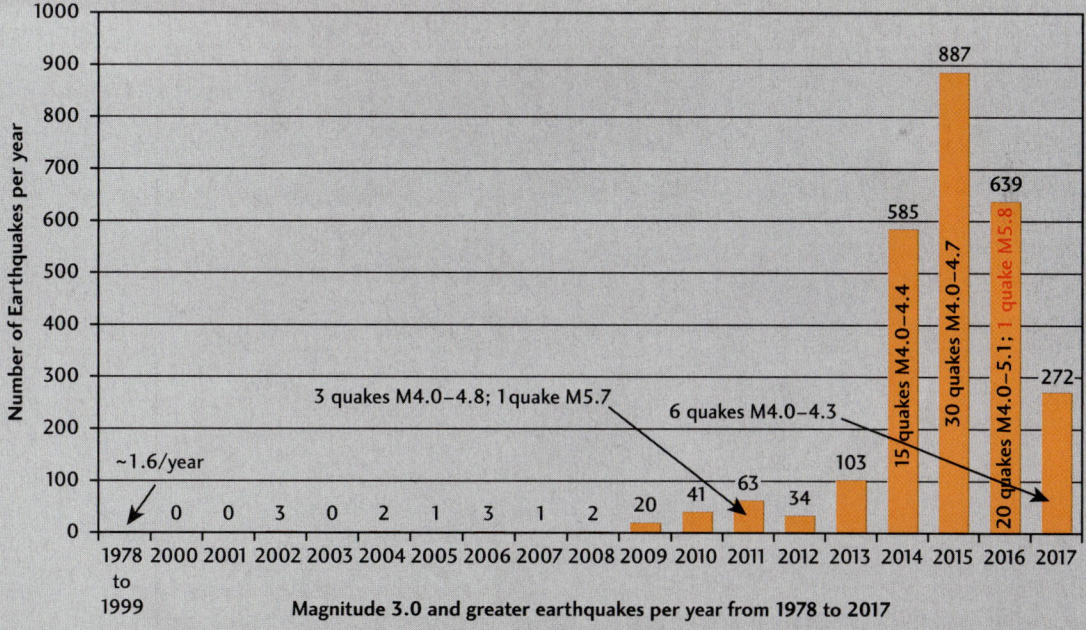

Seismicity in Oklahoma has increased dramatically since 2009, induced primarily by the injection of wastewater into deep geologic formations. The map shows locations of past earthquakes (colored dots), as well as projections of where ground shaking of Modified Mercalli Intensity VI is likely to occur (colored areas). The deepest red outlines the zone where the chances of experiencing Intensity VI ground shaking during the next year are about 10 to 14 percent. The bar graph plots the number of earthquakes above magnitude 3 recorded each year. The fall-off since the 2015 peak can be attributed to reductions in the injection rates following emergency directives by the State of Oklahoma. [Data from USGS.]

Faulting and Shaking The *primary hazards* of earthquakes are the ruptures in the ground surface that occur when faults break the surface, the permanent subsidence and uplift of the ground surface caused by faulting, and the ground shaking caused by seismic waves radiated during the quake. Seismic waves can shake structures so hard that they collapse. The ground accelerations near the epicenter of a large earthquake can approach and even exceed the acceleration of gravity, so an object lying on the ground surface can literally be thrown into the air. Very few structures built by human hands can survive such severe shaking, and those that do are severely damaged.

The collapse of buildings and other structures is the leading cause of casualties and economic damage during earthquakes. In cities, most casualties are caused by falling buildings and their contents. Death tolls can be especially high in densely populated areas of developing countries, where buildings are often constructed from bricks and mortar without steel reinforcement. A magnitude-7 earthquake on January 12, 2010, destroyed 250,000 residences and 30,000 commercial buildings in Haiti's capital city, Port-au-Prince, killing more than 230,000 people, making it the fifth deadliest seismic disaster ever recorded (**Figure 10.21**). Improving construction practices so that buildings are able to withstand shaking is the key to avoiding such tragedies.

Landslides and Other Types of Ground Failure
The primary hazards of faulting and ground shaking generate a number of *secondary hazards.* Among secondary hazards are landslides and other forms of ground failure that give rise to mass movements of Earth materials (described in Chapter 16). When seismic waves shake water-saturated soils, those soils can behave like a liquid and become unstable. The ground simply flows away, taking buildings, bridges, and everything else with it. Such soil *liquefaction* destroyed the residential area of Turnagain Heights near Anchorage, Alaska, in the 1964 earthquake (see Figure 16.18); the Nimitz Freeway near San Francisco in the 1989 Loma Prieta earthquake; and areas of Kobe in the 1995 earthquake. Liquefaction was responsible for much of the damage during the 2010–2011 earthquake sequence in Christchurch, New Zealand, destroying many homes and causing massive damage to underground water and sewer systems throughout the city.

In some instances, ground failure can cause more damage than the ground shaking itself. A 1970 earthquake in

FIGURE 10.21 Homes and buildings in Port-au-Prince destroyed by the Haiti earthquake on January 12, 2010. [Daniel Aguilar/Reuters/Newscom.]

Peru triggered an immense avalanche of rock and snow (up to 50 million cubic meters) that destroyed the mountain towns of Yungay and Ranrahirca (see Figure 16.27). Of the more than 66,000 people killed in the earthquake, about 18,000 of them died in the avalanche.

Tsunamis A large earthquake that occurs beneath the ocean can generate a destructive sea wave, sometimes called a "tidal wave," but more accurately named a **tsunami** (Japanese for "harbor wave"), since it has nothing to do with tides. Tsunamis are by far the deadliest and most destructive hazards associated with the world's largest earthquakes, the megathrust events that occur in subduction zones.

When a megathrust ruptures, it can push the seafloor landward of the deep-sea trench upward by tens of meters, displacing a large mass of the overlying ocean water. This disturbance flows outward in waves that travel across the ocean at speeds of up to 800 km/hour (about as fast as a commercial jetliner). In the deep sea, a tsunami is hardly noticeable, but when it approaches shallow coastal waters, the waves slow down and pile up, inundating the shoreline in walls of water that can reach heights of tens of meters (**Figure 10.22**). This "run-up" can propagate inland for hundreds of meters or even kilometers, depending on the slope of the land surface.

Tsunamis caused by megathrust events are most common in the Pacific Ocean, which is ringed with very active subduction zones. The destructive power of a great tsunami was brought home by the terrifying video images captured on March 11, 2011, as the Tohoku tsunami swept over the shoreline of northeastern Japan. In the coastal city of Miyako, the height of the water mass reached an astounding 38 m (123 ft!) above normal sea level, destroying nearly everything in its path (see Earth Issues 10.1). In low-lying regions near the port city of Sendai, the tsunami traveled up to 10 km inland, transporting huge floating debris fields of buildings, boats, cars, and trucks (**Figure 10.23**). The waves propagated across the entire Pacific Ocean, attaining heights of more than 2 m along the coast of Chile, 16,000 km away.

Tsunami warning systems in Japan and the circum-Pacific region worked according to design. The warning times along the Japanese coast nearest the earthquake were too short for complete evacuation. Nevertheless, the system is credited with saving many thousands of lives.

No tsunami warning system was in place when the magnitude 9.2 Sumatra earthquake of September 26, 2004 unleashed an ocean-wide tsunami that swept over low-lying coastal areas from Indonesia and Thailand to Sri Lanka,

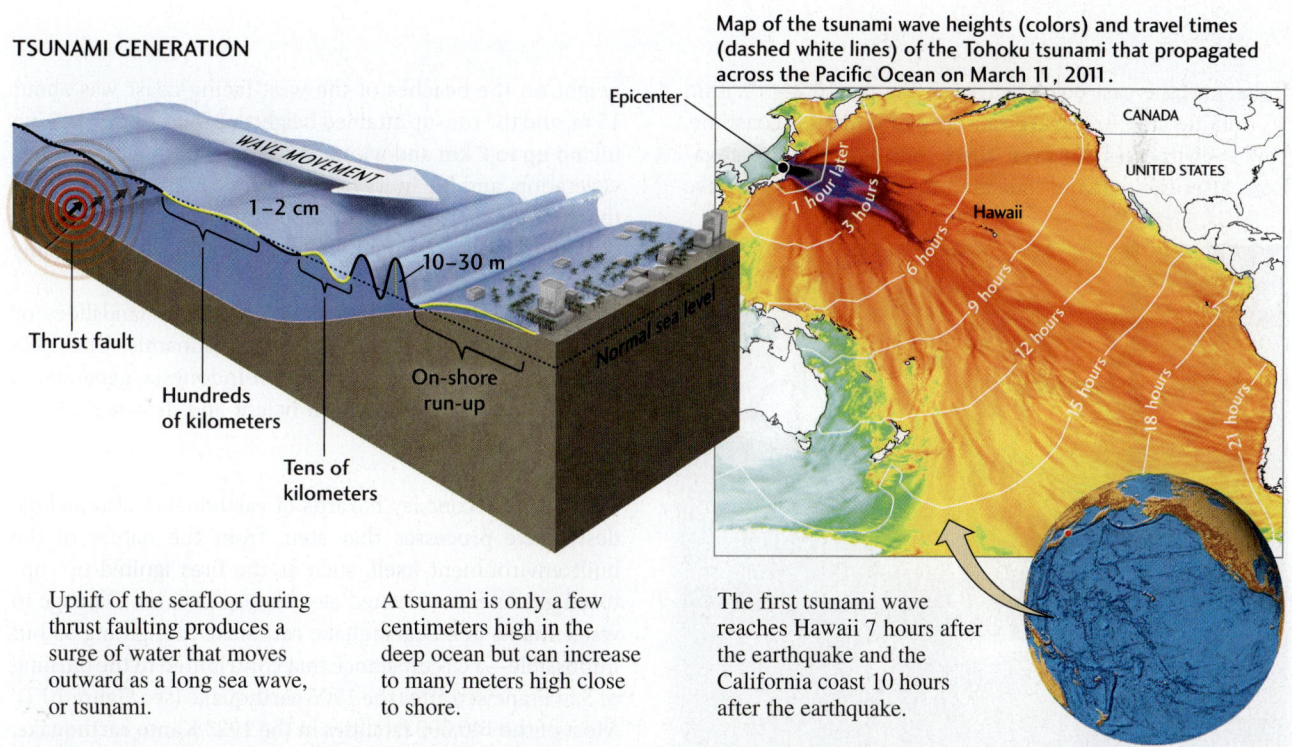

FIGURE 10.22 Earthquakes on megathrusts may generate tsunamis that can propagate across ocean basins. [NOAA, Pacific Marine Environmental Laboratory.]

FIGURE 10.23 Video image taken from a helicopter, showing the tsunami surge carrying debris across farmland near Sendai, Japan, following the Tohoku earthquake on March 11, 2011. [Xinhua/Xinhua News Agency/Newscom.]

India, and the east coast of Africa (Figure 10.24). Within 15 minutes, the first wave ran up the Sumatran coastline. Few eyewitnesses survived there, but geologic investigations after the tsunami indicated that the maximum wave height on the beaches of the west-facing coast was about 15 m, and the run-up attained heights of 25 to 35 m, reaching inland up to 2 km and wiping out most building structures, vegetation, and life in its path (Figure 10.25). It is believed that more than 150,000 people perished along the Sumatran coastline, though no one will ever be sure because many bodies were washed out to sea.

Disturbances of the seafloor caused by landslides or volcanic eruptions can also produce tsunamis. The 1883 explosion of Krakatau, a volcano in Indonesia, generated a tsunami that reached 40 m in height and drowned 36,000 people on nearby coasts.

Fires The secondary hazards of earthquakes also include destructive processes that stem from the nature of the built environment itself, such as the fires ignited by ruptured gas lines or downed electric power lines. Damage to water mains in an earthquake can make firefighting all but impossible—a circumstance that contributed to the burning of San Francisco after the 1906 earthquake (see Figure 10.1). Most of the 140,000 fatalities in the 1923 Kanto earthquake, one of Japan's greatest disasters, resulted from fires in the cities of Tokyo and Yokohama. Scenario studies predict that half of the damage from a large California earthquake, perhaps more, could come from quake-ignited firestorms.

FIGURE 10.24 The tsunami caused by the 2004 Sumatra earthquake struck without warning on a beach in Phuket, Thailand. [Courtesy David Rydevik.]

FIGURE 10.25 This small headland near Banda Aceh, on the west coast of Sumatra, was previously covered by dense vegetation to the waterline, but was stripped clean to a height of about 15 m by the 2004 tsunami. [Courtesy Jose Borrero, University of Southern California Tsunami Research Group.]

Reducing Earthquake Risk

In assessing the possibility of damage from earthquakes, or from any type of natural disaster, it is important to distinguish between *hazard* and *risk*. In the case of earthquakes, **seismic hazard** describes the frequency and intensity of earthquake shaking and ground disruption that can be expected over the long term at some specified location. Seismic hazard, which depends on the proximity of the site to active faults that might generate earthquakes, can be expressed in the form of a seismic hazard map. **Figure 10.26** displays the national seismic hazard map produced by the U.S. Geological Survey.

In contrast, **seismic risk** describes the *damage* that can be expected over the long term in a specified region, such as a county or state, usually measured in terms of human casualties and dollar losses per year. A region's risk depends not only on its seismic hazard, but also on its exposure to seismic damage (its population and density of buildings and other infrastructure) and its fragility (the vulnerability of its built environment to seismic shaking). Because so many geologic and economic variables must be considered, estimating seismic risk is a complex job. A comprehensive study of seismic risk in the United States, published by the Federal Emergency Management Agency in 2017, is presented in **Figure 10.27**.

The differences between seismic hazard and seismic risk can be appreciated by comparing the national maps. For instance, although the seismic hazard levels in Alaska and California are both high, California's exposure to seismic damage is much greater, yielding a much larger total risk. California leads the nation in seismic risk, with about 61 percent of the national total; in fact, a single county, Los Angeles, accounts for 22 percent. Nonetheless, the problem is truly national: 46 million people in several metropolitan areas outside of California face substantial earthquake risks. Those areas include Hilo, Honolulu, Anchorage, Seattle, Tacoma, Portland, Salt Lake City, Reno, Las Vegas, Albuquerque, Charleston, Memphis, Atlanta, St. Louis, New York, Boston, and Philadelphia.

Not much can be done about seismic hazard because we have no way to prevent or control earthquakes (see the Practicing Geology Exercise). However, there are many important steps that society can take to reduce seismic risk.

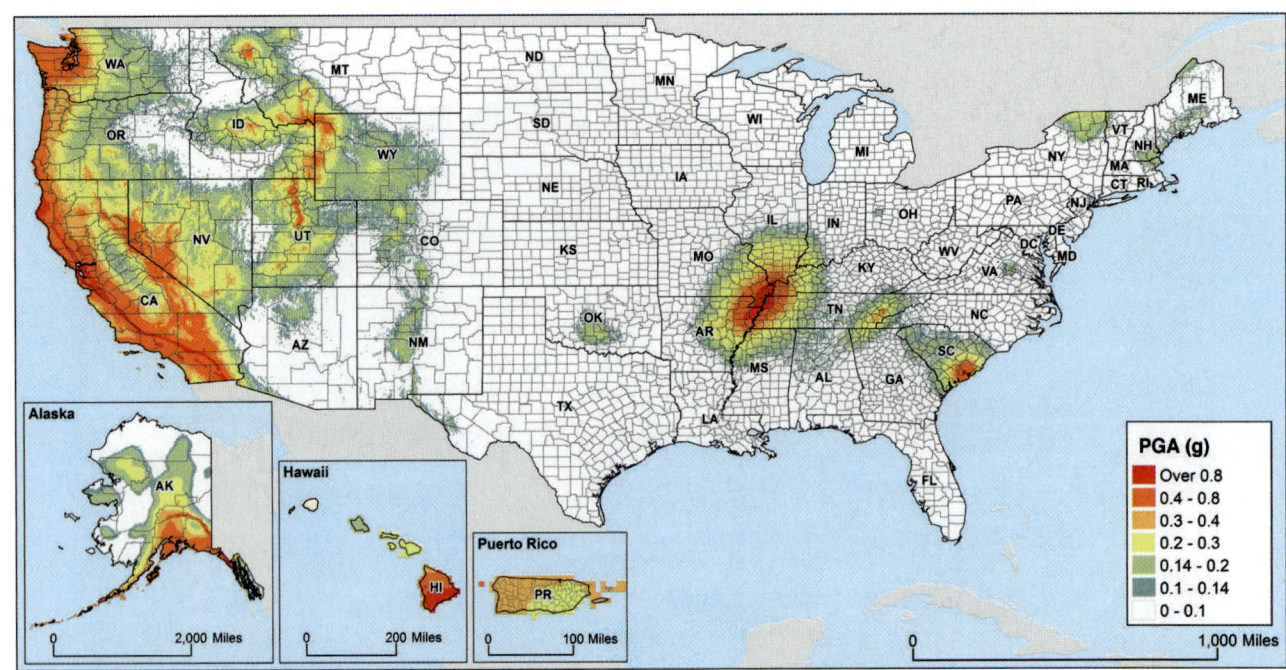

FIGURE 10.26 Seismic hazard map for the United States, measured as the peak ground acceleration (PGA) in gravity units (g) expected with 5 percent probability in the next 50 years. The regions of highest hazard lie along the plate boundaries of the West Coast and Alaska and on the southern side of the Big Island of Hawaii. In the central and eastern United States, the areas of highest hazard are near New Madrid, Missouri, and Charleston, South Carolina; in eastern Tennessee; and in portions of the Northeast. [U.S. Geological Survey, http://geohazards.cr.usgs.gov.]

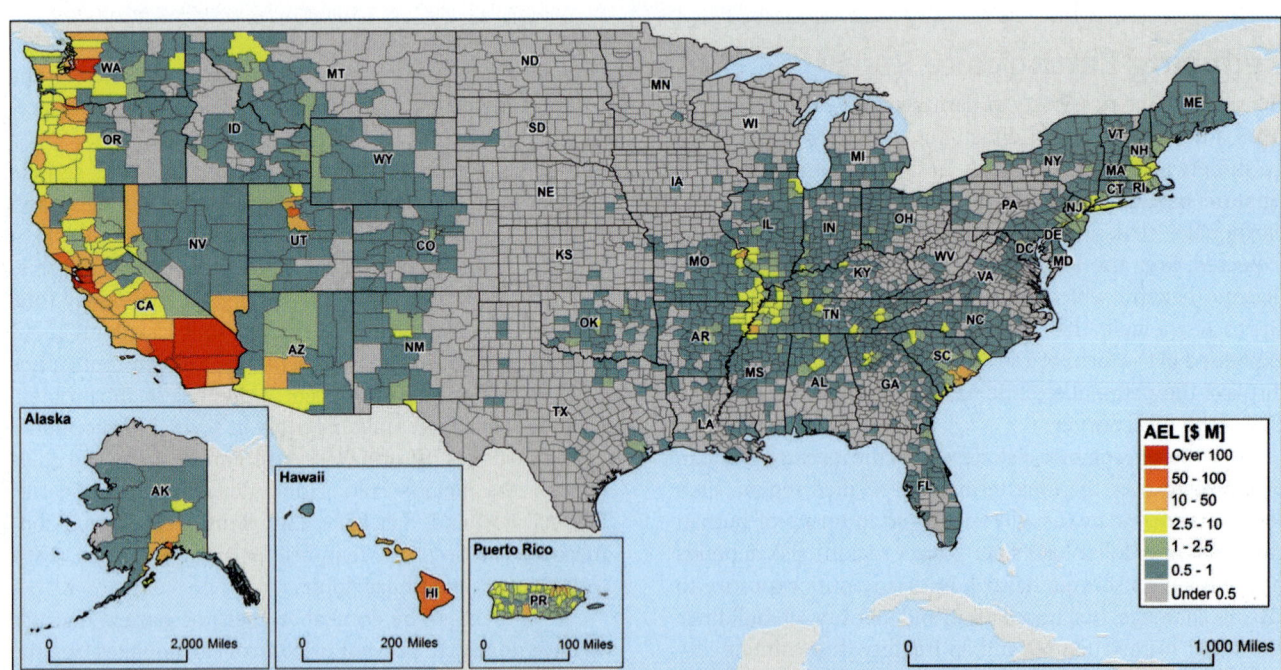

FIGURE 10.27 Seismic risk map for the United States. The map shows current annualized earthquake losses (AEL) in millions of dollars on a county-by-county basis. Measured this way, about two-thirds of the national earthquake risk is concentrated in California. [USGS.]

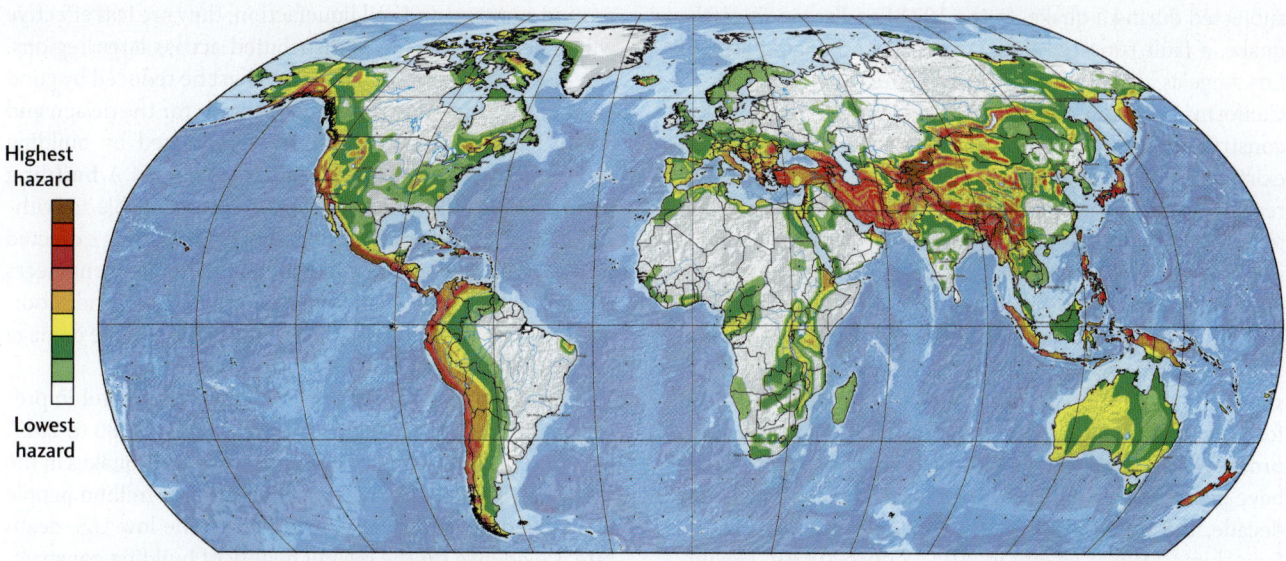

FIGURE 10.28 Global seismic hazard map. [Data from K. M. Shedlock et al., *Seismological Research Letters* 71(2000): 679–686.]

Hazard Characterization The first step is to follow the advice of the old proverb, "Know thy enemy." We still have much to learn about the sizes and frequencies of ruptures on active faults. For example, it is only in the past couple of decades that we have come to appreciate that an earthquake in the Cascadia subduction zone, which stretches from northern California through Oregon and Washington to British Columbia, could produce a tsunami as large as the one that devastated the Indian Ocean region in 2004 and Japan in 2011. These dangers became apparent when geologists found evidence of a magnitude-9 earthquake that occurred in 1700, before any written historical accounts of the area existed. This monstrous rupture caused major ground subsidence along the Cascadia coastline and left a record of flooded, dead coastal forests. A tsunami at least 5 m high hit Japan, where historical records pin down its exact date (January 26, 1700). Geologists know that the Juan de Fuca Plate is being subducted under the North American Plate at a rate of about 40 mm/year. They have debated whether this movement occurs seismically, by continuous creep, or perhaps by silent earthquakes, but current thinking pegs the average time between magnitude-9 earthquakes in the Cascadia subduction zone at 500 to 600 years.

Although we have a good understanding of seismic hazard in some parts of the world—the United States and Japan in particular—we know much less about other regions. During the 1990s, the United Nations sponsored an effort to map seismic hazard worldwide as part of the International Decade of Natural Disaster Reduction. This effort resulted in the first global seismic hazard map, shown in **Figure 10.28**. The map is based primarily on historical earthquakes, so it may underestimate the hazard in some regions where the historical record is short. Scientists and engineers from around the world are now working together through the Global Earthquake Model Foundation, based in Italy, to improve knowledge of seismic hazards and risk on a global scale.

Land-Use Policies The exposure of buildings and other structures to seismic hazard can be reduced by policies that restrict land uses in high-hazard areas. This approach works well where the hazard is localized, as in the case of known faults. Erecting buildings on active faults, as was done in the case of the residential developments pictured in **Figure 10.29**, is clearly unwise, because few buildings can withstand the deformation to which they might be

FIGURE 10.29 Housing tracts constructed within the San Andreas Fault Zone, on the San Francisco Peninsula, before the state of California passed legislation restricting this practice. The red line indicates the approximate fault trace, along which the ground ruptured and slipped about 2 m during the earthquake of 1906. [Copyright 2013 TerraMetrics, Inc., www.terrametrics.com.]

subjected during a quake. In the 1971 San Fernando earthquake, a fault ruptured under a densely populated area of Los Angeles, destroying almost 100 structures. The state of California responded in 1972 with a law that restricts the construction of new buildings across an active fault. If an existing residence is on or very near a fault, the owners and real estate agents are required to disclose that information to potential buyers. A notable omission is that the act does not cover publicly owned or industrial facilities.

Siting of nuclear power plants and other critical industrial facilities to avoid seismic and tsunamic hazards would seem to be an obvious priority, but the experience in Japan shows how additional considerations, such as the need for water to cool the reactors, can lead to unwise compromises. Two nuclear facilities along the Japanese coast have been severely damaged by the earthquakes in the last decade, the Kashiwazaki-Kariwa facility in 2007 and the Fukushima-Daiichi facility in 2011 (Figure 10.30). Heightened public concern after the Fukushima disaster led the Japanese government to shut down most of the country's nuclear plants, raising concerns about the future costs of electrical power, both economic and environmental.

Earthquake Engineering Although land-use policies help to reduce the risk from localized hazards such as ground ruptures and soil liquefaction, they are less effective where seismic shaking is distributed across large regions. The risks from seismic shaking can best be reduced by good engineering and construction. Standards for the design and construction of new buildings are regulated by building codes enacted by state and local governments. A **building code** specifies the forces a structure must be able to withstand, based on the maximum intensity of shaking expected in the area. In the aftermath of an earthquake, engineers study buildings that were damaged and recommend modifications to building codes that could reduce future damage from similar earthquakes.

U.S. building codes have been largely successful in preventing loss of life during earthquakes. From 1980 to 2017, for example, 146 people died in 11 severe earthquakes in the western United States, whereas more than 1 million people were killed by earthquakes worldwide. The low U.S. death rate is evidence for the overall quality of building construction promoted by strong building codes. Nevertheless, more can be done. Damage from inevitable earthquakes could be reduced by retrofitting older, more vulnerable structures to be seismically safe, as well as by using specialized construction materials and advanced engineering methods in new construction, such as putting entire buildings on movable supports to isolate them from seismic shaking.

FIGURE 10.30 Aerial photo taken by a small, unmanned drone on March 24, 2011, showing the reactor containment buildings of the Fukushima-Daiichi nuclear power plant that were damaged by explosion after the Tohoku tsunami crippled the plant. [Air Photo Service/ABACAUSA.COM/Newscom.]

Emergency Preparedness and Response Public authorities must plan ahead and be prepared with emergency supplies, rescue teams, evacuation procedures, firefighting plans, and other steps to minimize the consequences of a severe earthquake. For individuals, earthquake preparedness begins at home. Earth Issues 10.3 summarizes some of the steps you can take to protect yourself and your family from earthquakes.

Once an earthquake happens, networks of seismographs can transmit signals automatically to central processing facilities. In a fraction of a minute, computers can pinpoint the earthquake's focus, measure its magnitude, and determine its fault mechanism. If equipped with strong-motion sensors that accurately record the most violent shaking, these automated systems can also deliver accurate maps in nearly real time, showing where the ground shaking was strong enough to cause significant damage. Such "ShakeMaps" can help emergency managers and other officials deploy equipment and personnel as quickly as possible to save people trapped in rubble and to reduce further economic losses from fires and other secondary hazards. Bulletins about the magnitude and area of the shaking can also be channeled through the mass media to reduce public confusion during disasters or to allay the fears aroused by minor tremors.

Earthquake Early Warning Systems With the technology just described, it is possible to detect earthquakes in the early stages of fault rupture, rapidly predict the intensity of the future ground motions, and warn people before they experience the intense shaking that might be damaging. Earthquake early warning systems detect strong shaking near an earthquake's epicenter and transmit alerts at the speed of light, ahead of the seismic waves. Potential warning times depend primarily on the distance between the user and the earthquake epicenter. There is a "blind zone" near an earthquake epicenter where early warning is too slow to beat the seismic waves, but at more distant sites, warnings can be issued from a few seconds up to about one minute prior to strong ground shaking.

Earthquake early warning systems have already been deployed in at least four countries—Japan, Romania, Taiwan, and Turkey—and an operational system is being developed in the United States. Japan is the only country with a nationwide system that provides public alerts. A national seismic network of nearly 1000 seismographs is used to detect earthquakes and issue warnings, which are transmitted via the Internet, satellite, cell phones, as well as automated control systems that do such things as stop trains and place sensitive equipment in a safe mode. An earthquake early warning system called "ShakeAlert" is being deployed in California and the Pacific Northwest by a public-private partnership with financial support from both federal and state governments.

Tsunami Warning Systems The great tsunamis generated by the Sumatra earthquake of 2004 and the Tohoku earthquake of 2011 illustrate the issues associated with tsunami warning. Because tsunami waves travel 10 times more slowly than seismic waves, there is enough time after a large suboceanic quake occurs, sometimes many hours, to warn people on distant shorelines of an impending disaster. Warnings broadcast by the Pacific Tsunami Warning Center, based in Hawaii, after the Tohoku earthquake allowed islands such as Hawaii and the western coastlines of the Americas to be evacuated prior to the tsunami arrival (see Figure 10.22). Unfortunately, no such system had been installed in the Indian Ocean, so the 2004 tsunami struck with essentially no warning, killing tens of thousands.

The most difficult situations arise in areas located close to active offshore faults, where tsunamis arrive so quickly that there is no time for a warning. One such place is Papua New Guinea, where in 1998, a tsunami killed as many as 3000 people in coastal villages near the epicenter of the quake that caused it. Such communities could be protected by building barrier walls to block inundation by ocean water, but this type of construction is expensive and has been tried in Japan with mixed results (**Figure 10.31**). In these places, the best warning system is a very simple one: If you feel a strong earthquake, move quickly away from the coastal lowlands to higher ground!

Can Earthquakes Be Predicted?

If we could predict earthquakes reliably, communities could be prepared, people could be evacuated from dangerous locations, and many aspects of the impending disaster might be averted. How well can we predict earthquakes?

Predicting an earthquake means specifying its time, location, and size. By combining plate tectonic theory with

FIGURE 10.31 This tsunami barrier was designed to protect the town of Taro, Japan, but it was breached by the tsunami from the great Tohoku earthquake. [Carlos Barria/Reuters/Newscom.]

Earth Issues 10.3 Seven Steps to Earthquake Safety

Individuals living in seismically active areas need to prepare for earthquakes and know how to respond when one strikes. Here are seven steps to earthquake safety, recommended by the Southern California Earthquake Center, that you can use to protect yourself and your family.

Before an earthquake occurs:

1. *Identify potential hazards in your home and begin to fix them.* As buildings are becoming better designed to withstand seismic shaking, more of the damage and injuries that occur are resulting from falling objects. You should secure items in your home that are heavy enough to cause damage or injury if they fall or valuable enough to be a significant loss if they break.

2. *Create your disaster plan.* With your family or housemates, plan now what you will do before, during, and after an earthquake. The plan should include safe spots you can go to during the shaking, such as under sturdy desks and tables; a safe spot outside your home where you can meet after the shaking stops; and contact phone numbers, including someone outside the area who can be called to relay information in case local communications are disrupted.

3. *Create your disaster kit.* Stock your disaster kit with essential items. Your personal kit should include medications, a first-aid kit, a whistle, sturdy shoes, high-energy snacks, a flashlight with extra batteries, and personal hygiene supplies. Your home kit should include a fire extinguisher, wrenches to turn off gas and water mains, a portable radio, drinking water, food supplies, and extra clothing.

4. *Identify your building's potential weaknesses and begin to fix them.* Consult a building inspector or contractor to identify potential safety problems. Common problems include inadequate foundations, unbraced cripple walls, weak first stories, unreinforced masonry, and vulnerable pipes.

During an earthquake:

5. *Drop, cover, and hold on.* During an earthquake or severe aftershock, drop to the floor, take cover under a sturdy desk or table, and hold on to it so that it doesn't move away from you. Wait there until the shaking stops. Stay away from danger zones, such as those near the exterior walls of buildings, near windows, and under architectural façades.

After an earthquake:

6. *After the shaking stops, check for damage and injuries needing immediate attention.* Take care of your own situation first; get to a safe location and remember your disaster plan. If you are trapped, protect your mouth, nose, and eyes from dust; signal for help using a cell phone or whistle or by knocking loudly on a solid part of the building three times every few minutes (rescuers will be listening for such knocks). Check for injuries and treat people needing assistance. Check for fires, gas leaks, damaged electrical systems, and spills. Stay away from damaged structures.

7. *When safe, follow your disaster plan.* Be in communication by turning on your radio and listening for advisories; check your phones, call your out-of-area contact to report your status, and then stay off the phone except for emergencies. Check your food and water supplies and check on your neighbors.

For further information, see *Putting Down Roots in Earthquake Country,* Southern California Earthquake Center, available online at http://www.earthquakecountry.org.

---

detailed geologic mapping of regional fault systems, geologists can reliably predict which faults are likely to produce large earthquakes over the long term. However, specifying precisely *when* a particular fault will rupture in a large earthquake has turned out to be very difficult.

Long-Term Forecasting

Ask a seismologist to predict the time of the next large earthquake at a particular location and the response is likely to be, "The longer the time since the last big quake, the sooner the next one will be." As we have seen, the recurrence interval—the time required to accumulate the strain that will be released by fault slipping in a future earthquake—can be calculated from the rate of relative plate movement and the expected fault slip, as estimated from the displacements observed in past earthquakes. Geologists can also estimate the intervals between large earthquakes up to several thousand years in the past by finding and dating soil layers that were offset by fault displacements (**Figure 10.32**).

Although these two methods usually give similar results, the uncertainty of the predictions turns out to be large—as much as 100 percent of the recurrence interval. In Southern California, for example, the recurrence interval for the San Andreas Fault is estimated to be 110 to 180 years, but the observed intervals between individual earthquakes can be appreciably shorter or longer than this average value. One part of this fault experienced a large earthquake in 1857, whereas another part (the southernmost) appears to have remained locked since a large earthquake that occurred around 1680 (see Figure 10.2). Therefore, an earthquake can be expected there at any time—tomorrow, or decades from now.

Prepare

Before the next big earthquake we recommend these four steps that will make you, your family, or your workplace better prepared to survive and recover quickly:

Step 1:
Secure your space by identifying hazards and securing moveable items.

Step 2:
Plan to be safe by creating a disaster plan and deciding how you will communicate in an emergency.

Step 3:
Organize disaster supplies in convenient locations.

Step 4:
Minimize financial hardship by organizing important documents, strengthening your property, and considering insurance.

Survive And Recover

During the next big earthquake, and immediately after, is when your level of preparedness will make a difference in how you and others survive and can respond to emergencies:

Step 5:
Drop, Cover, and Hold On when the earth shakes.

Step 6:
Improve safety after earthquakes by evacuating if necessary, helping the injured, and preventing further injuries or damage.

After the immediate threat of the earthquake has passed, your level of preparedness will determine your quality of life in the weeks and months that follow:

Step 7:
Reconnect and Restore
Restore daily life by reconnecting with others, repairing damage, and rebuilding community.

The 7 steps for earthquake preparedness and survival. [Courtesy of Southern California Earthquake Center.]

FIGURE 10.32 Geologist Gordon Seitz examines layers of rock and peat that have been disturbed by prehistoric earthquakes in a trench crossing the San Jacinto Fault, a major strand of the San Andreas Fault system in Southern California. By dating the peat layers using the carbon-14 method, geologists can reconstruct the history of large earthquakes on this fault. Such information helps scientists to forecast future events. [Courtesy of Tom Rockwell, San Diego State University.]

Because the prediction intervals are decades to centuries, these methods of earthquake prediction are referred to as *long-term forecasting* to distinguish them from what most people would really want: a *short-term prediction* of a large rupture on a specific fault, accurate to within days or even hours of the actual event.

Short-Term Prediction

There have been a few successful short-term earthquake predictions. In 1975, an earthquake was predicted only hours before it occurred near Haicheng, in northeastern China. Chinese seismologists used what they considered to be *precursors* to make their predictions: swarms of tiny earthquakes and a rapid deformation of the ground several hours before the mainshock. Almost a million people, prepared in advance by a public education campaign, evacuated their homes and factories in the hours before the quake. Although many towns and villages were destroyed and several hundred people were killed, it appears that many were saved. The very next year, however, an unpredicted earthquake struck the Chinese city of Tangshan, killing more than 240,000 people. Obvious precursors such as those seen in Haicheng have not been repeated in subsequent large events.

Although many schemes have been proposed, we have not yet found a reliable method of predicting earthquakes minutes to weeks ahead of time. We cannot say that short-term earthquake prediction is impossible, but seismologists do not expect that it will be feasible in the near future.

We do have some useful guidelines about how the earthquake probabilities change over time. We know that earthquakes tend to cluster together in both space and time—for example, large earthquakes have nearby aftershocks—and seismologists have shown how the chances of a potentially damaging earthquake tend to go up during periods of increased seismic activity. Interpreting this type of information can be tricky, however, because, even when the seismic activity is high, accurate predictions of large earthquakes are still not possible. During seismic crises, it is easy for the public to become confused about how the hazard is changing. For example, a miscommunication of short-term earthquake probabilities before the damaging L'Aquila earthquake of April 6, 2009, led to the criminal prosecution of scientific advisors to the Italian government on charges of manslaughter. Forecasts based on earthquakes clustering are now being deployed to help Italians assess how the seismic hazards are changing. These short-term forecasting methods are also being developed in other regions, including California.

Medium-Term Forecasting

Uncertainties in long-term forecasting can be reduced by studying the behavior of regional fault systems. One strategy is to generalize the elastic rebound theory. The simple version of the theory depicted in Figure 10.3 describes how the tectonic stress that builds steadily on an isolated fault segment is released in a periodic sequence of fault ruptures. However, as we have seen in the case of Southern California (see Figure 10.18), faults are rarely isolated. Instead, they are connected to one another in complex networks. Thus, a rupture on one fault segment changes the stresses throughout the surrounding region (see Figure 10.4). Depending on the geometry of the fault system, these changes can either increase or decrease the likelihood of earthquakes on nearby fault segments. In other words, when and where earthquakes happen in one part of a fault system influences when and where they happen elsewhere in the system.

If Earth scientists can understand how variations in stress raise or lower the frequency of small seismic events, they might be able to predict earthquakes over time intervals as short as a few years, or maybe even a few months, although still with substantial uncertainties. Monitoring of such events on networks of seismographs could then provide a regional "stress gauge." Someday you might hear a news report that says, "The National Earthquake Prediction Evaluation Council estimates that, during the next year, there is a 50 percent probability of a magnitude 7 or larger earthquake on the southern segment of the San Andreas Fault."

The ability to issue such *medium-term forecasts* would raise some difficult questions, however. How should society respond to a threat that is neither imminent nor long-term? A medium-term forecast would give the probability of an earthquake only on time scales of months to years—not precisely enough to evacuate regions that might be damaged. False alarms would be common. What effect would such predictions have on property values and other investments in the threatened region? These questions will have to be addressed by policy makers and economists, as well as by scientists.

KEY TERMS AND CONCEPTS

aftershock (p. 271)
building code (p. 296)
earthquake (p. 269)
elastic rebound theory (p. 269)
epicenter (p. 271)
fault mechanism (p. 281)
fault slip (p. 269)

focus (p. 271)
foreshock (p. 273)
induced seismicity (p. 288)
intensity scale (p. 280)
magnitude scale (p. 277)
P wave (p. 275)
recurrence interval (p. 271)

S wave (p. 275)
seismic hazard (p. 293)
seismic risk (p. 293)
seismograph (p. 274)
surface wave (p. 275)
tsunami (p. 291)

REVIEW OF LEARNING OBJECTIVES

10.1 Explain how stress builds up on faults and is released in earthquakes through the process of elastic rebound.

An earthquake is a shaking of the ground that occurs when brittle rocks being stressed by tectonic forces break suddenly along a fault. Slow plate movements displace the blocks on either side of plate-boundary faults in opposite directions, building up elastic strain over decades, centuries, or even millennia (Figure 10.3). At some point, the strength of the rocks is exceeded. Somewhere along the fault surface, the frictional bond that locks the fault can no longer hold, and it breaks. When faults rupture, the elastic energy is released rapidly, and some of this energy is radiated as seismic waves. The focus of an earthquake is the point at which the fault first breaks; the epicenter is the point on Earth's surface directly above the focus. The foci of most continental earthquakes are shallow. In subduction zones, however, earthquakes can occur in cold, descending lithospheric slabs at depths as great as 690 km.

Study Assignment: Visual Literacy Exercise

Exercises: The last big earthquake to occur on the San Andreas Fault, north of Los Angeles, was in 1857 (magnitude 7.8). Geological analysis of soils in a trench across this fault shows that prehistoric earthquakes with magnitudes greater than about 7 occurred in 1812, 1547, 1360, 1084, 1067, and 956. (a) Estimate the recurrence interval for earthquakes of magnitude 7 and greater on this section of the fault. (b) How does this estimate compare with the recurrence interval obtained by assuming that the long-term fault slip rate of 25 mm/year is accommodated by earthquakes with average displacements of 4 m?

Thought Question: The recurrence interval is the time between earthquakes averaged over many events. Why does the time between individual events differ from this average? (*Hint:* Consult Figure 10.4.)

10.2 Identify the three main types of seismic waves and describe the ground motions they produce.

Earthquakes generate three types of seismic waves that can be distinguished on seismograms. Two types of waves travel through Earth's interior: P (primary) waves, which are transmitted by all forms of matter and move fastest, and S (secondary) waves, which are transmitted only by solids and move at a little more than half the velocity of P waves. P waves are compressional waves that travel as a succession of compressions and expansions. S waves are shear waves that displace material at right angles to their path of travel. Surface waves are confined to Earth's surface and outer layers. They travel more slowly than S waves and come in two basic types: those that involve vertical ground motions, and those where the ground motions are side-to-side perpendicular to their direction of propagation.

Study Assignment: Review the seismic wave types in Figure 10.9.

Exercise: Seismographic stations report the following S-wave–P-wave arrival time intervals for an earthquake: Dallas, S-P = 3 minutes; Los Angeles, S-P = 2 minutes; San Francisco, S-P = 2 minutes. Use a map of the United States (or better yet, a globe) and the travel-time curves in Figure 10.9 to obtain a rough location for the epicenter of the earthquake.

Thought Question: Which of the three types of seismic waves cause the most damage during large earthquakes?

10.3 Distinguish shaking intensity from earthquake magnitude and understand how they are related.

Earthquake magnitude is a measure of the size of an earthquake, whereas shaking intensity is a measure of the ground motions at a particular site. Seismologists now measure earthquake size by moment magnitude, which is derived from the physical properties of the faulting that causes the earthquake: the area of faulting and the average fault slip. The area of faulting increases by a factor of 10 for each unit of moment magnitude, and the slip increases by a factor of 10 for each two units of moment magnitude. Shaking intensity is measured on the Modified Mercalli Intensity scale (Table 10.1) with unit values I–XII that describe the effects of shaking. Shaking intensities are generally higher nearer the faulting, and they decay with distance owing to the spreading out and dissipation of the seismic energy. The area of strongest shaking increases with magnitude. For a given magnitude, the area of strongest shaking tends to be larger in the central and eastern United States, where the rocks are old, cool, and strong, than on the West Coast, where the rocks are younger, hotter, and weaker.

Study Assignment: Practicing Geology Exercise

Exercises: (a) Describe the two types of scales for measuring the size of an earthquake. (b) Which is the more appropriate scale for measuring the amount of faulting that caused the earthquake? (c) Which is more appropriate for measuring the amount of shaking experienced by a particular observer?

Thought Questions: (a) Why do the largest earthquakes occur on megathrusts at subduction zones and not, say, on continental strike-slip faults? (b) How big would a fault have to be to produce a magnitude-10 earthquake? Do you think such a large earthquake could occur by the rupturing of subduction-zone megathrusts?

10.4 Classify earthquakes according to their plate-tectonic setting and patterns of faulting.

The fault mechanism of an earthquake is determined by the type of plate boundary at which it occurs. Normal faulting, caused by tensional forces, occurs at divergent boundaries. Strike-slip faulting, caused by shearing forces, occurs along transform-fault boundaries. A small number of earthquakes happen far from plate boundaries, mostly on continents. Worldwide, about 1,000,000 earthquakes with magnitudes greater than 2 take place each year. This number decreases by a factor of 10 for each magnitude unit. Hence, there are about 100,000 earthquakes with magnitudes greater than 3, about 1000 with magnitudes greater than 5, and about 10 with magnitudes greater than 7. The largest earthquakes, caused by compressive forces, occur on megathrusts at convergent boundaries, such as the devastating 2011 Tohoku, Japan, earthquake (magnitude 9.0).

Study Assignment: Compare the global seismicity map in Figure 10.16 with the plate-tectonic map of Figure 2.7.

Exercises: (a) What are the dominant fault mechanisms of earthquakes at the three types of plate boundaries? (b) In Southern California, a magnitude-5 earthquake occurs about once per year. Approximately how many magnitude-4 earthquakes would you expect each year? How many magnitude-2 earthquakes?

Thought Questions: (a) Destructive earthquakes occasionally occur within lithospheric plates, far from plate boundaries. Why? (b) The belts of shallow-focus earthquakes shown by the blue dots in Figure 10.16 are wider and more diffuse on the continents than in the oceans. Why? (*Hint:* Review Chapter 8.)

10.5 Characterize seismic hazards and the risks they pose to society and its built environment.

Faulting and ground shaking, the primary hazards during an earthquake, can damage or destroy buildings and other infrastructure. They can also trigger secondary hazards, such as landslides and fires. Earthquakes that lift up the seafloor can generate tsunamis, which may cause widespread destruction when they pile up in shallow coastal waters, forcing a wall of water inland. Risk can be reduced in several ways. Land-use regulations can restrict new building near active fault zones, and construction in high-hazard areas can be regulated by building codes so that buildings and other structures will be strong enough to withstand the expected intensity of seismic shaking. Systems using networks of seismographs and other sensors have been developed to provide early warnings of earthquakes and tsunamis. Public authorities can plan ahead, be prepared, and put early warning systems in place. People living in earthquake-prone areas can be informed about how to prepare and what to do when an earthquake occurs.

Study Assignment: Use the maps in Figures 10.26 and 10.27 to assess the seismic hazard and risk in your home state.

Exercise: How much more energy is released by a magnitude-7.5 earthquake than by a magnitude-6.5 earthquake?

Thought Question: Would you support legislation to prevent landowners from building structures close to active faults?

10.6 Defend the statement that "earthquakes can be forecast in the long term but not predicted in the short term."

Scientists can characterize the long-term level of seismic hazard in a region by knowing the potential magnitude of faulting (from the fault area), the rate of fault slip (from geodetic and geologic observations), and the frequency of large earthquakes (from historic and geologic records). The results can be viewed as seismic hazard maps (Figure 10.26), which, for example, guide the seismic safety provisions in building codes. But scientists cannot consistently predict earthquakes with the accuracy that would be needed to alert a population hours to weeks in advance. In particular, no precursory signals have been identified that can reliably predict large earthquakes in the short term. The best hope of making such predictions in the future may lie in a better understanding of how variations in stress raise or lower the frequency of seismic events in a regional fault system.

Study Assignment: Review the section *Can Earthquakes Be Predicted?*

Exercise: At a location along the boundary between the Nazca Plate and the South American Plate, the relative plate movement is 80 mm/year. The last large earthquake there, in 1880, showed a fault slip of 12 m. If the next large earthquake has a similar displacement, estimate when it might be expected to occur. What makes this estimate uncertain?

Thought Question: Taking into account the possibility of false alarms, mass hysteria, economic depression, and other possible negative consequences of earthquake prediction, do you think the objective of predicting earthquakes should have a high priority?

VISUAL LITERACY EXERCISE

ROCKS DEFORM ELASTICALLY, THEN REBOUND DURING AN EARTHQUAKE RUPTURE

A A farmer builds a stone wall across a right-lateral strike-slip fault a few years after its last rupture.

B Over the next 150 years, the relative motion of the blocks on either side of the fault causes the ground and the stone wall to deform.

C Just before the next rupture, a new fence is built across the already deformed land.

D The fault slips, lowering the stress, and the elastic rebound restores the blocks to their prestressed state. Both the rock wall and the fence are shifted equal amounts along the fault.

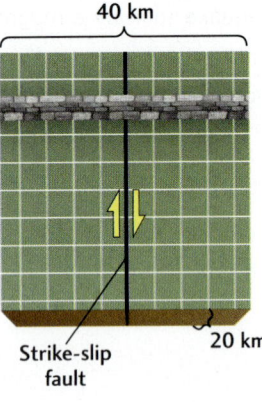
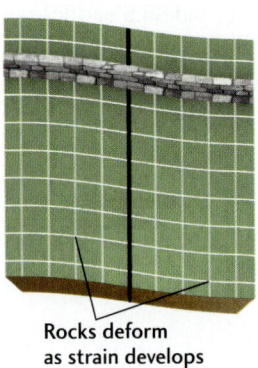
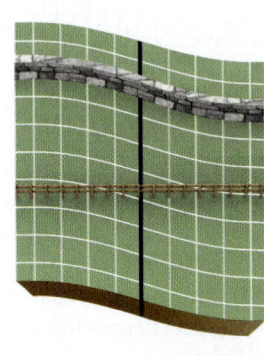
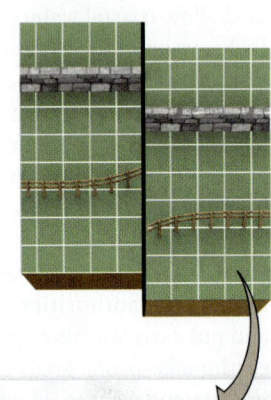

STRESS BUILDS UNTIL IT EXCEEDS ROCK STRENGTH

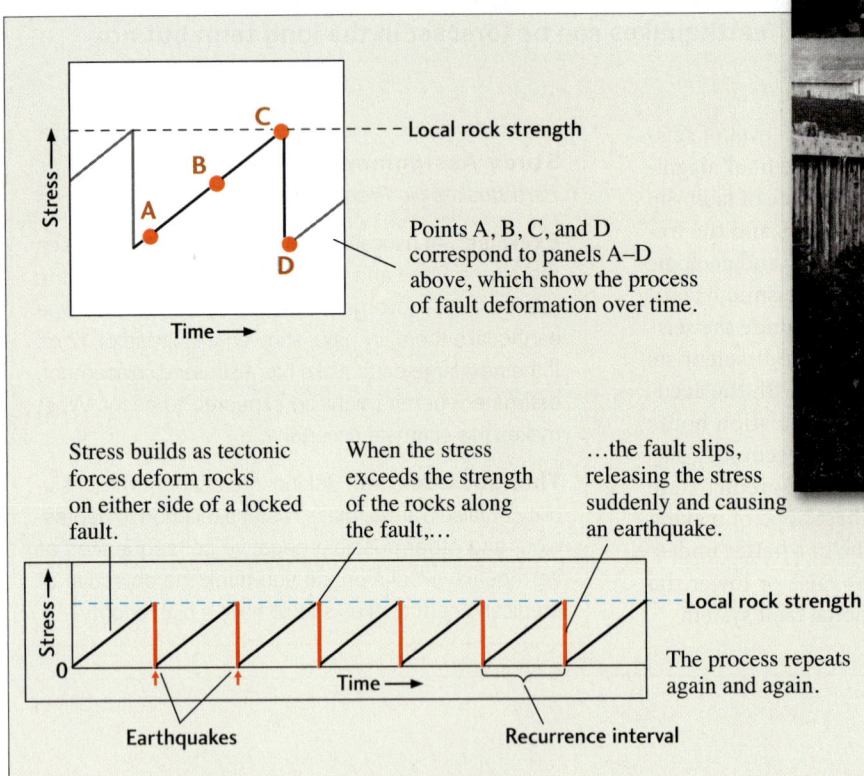

A fence built across the San Andreas Fault near Bolinas, California, is offset by nearly 4 m after the great San Francisco earthquake of 1906.

FIGURE 10.3 The elastic rebound theory explains the earthquake cycle. According to the theory, stress on rocks builds up over time as a result of plate movements. Earthquakes occur when that stress exceeds rock strength. Rocks under stress deform elastically, then rebound when an earthquake occurs. Panels A–D show deformation at the points labeled A–D in the bottom panel. [Photo: G. K. Gilbert/USGS.]

Figure 10.3 illustrates four stages of the earthquake cycle in panels A–D, and it relates those stages to stress on the fault.

1. What type of faulting is shown in this figure?
   a. Normal
   b. Thrust
   c. Left-lateral strike-slip
   d. Right-lateral strike-slip

2. The earthquake rupture occurs between which two panels?
   a. A and B
   b. B and C
   c. C and D
   d. D and A

3. In which panel is the stress on the fault the highest?
   a. A
   b. B
   c. C
   d. D

4. Which gives the best measure of the earthquake recurrence interval?
   a. Time between A and B
   b. Time between A and D
   c. Time between B and C
   d. Time between C and D

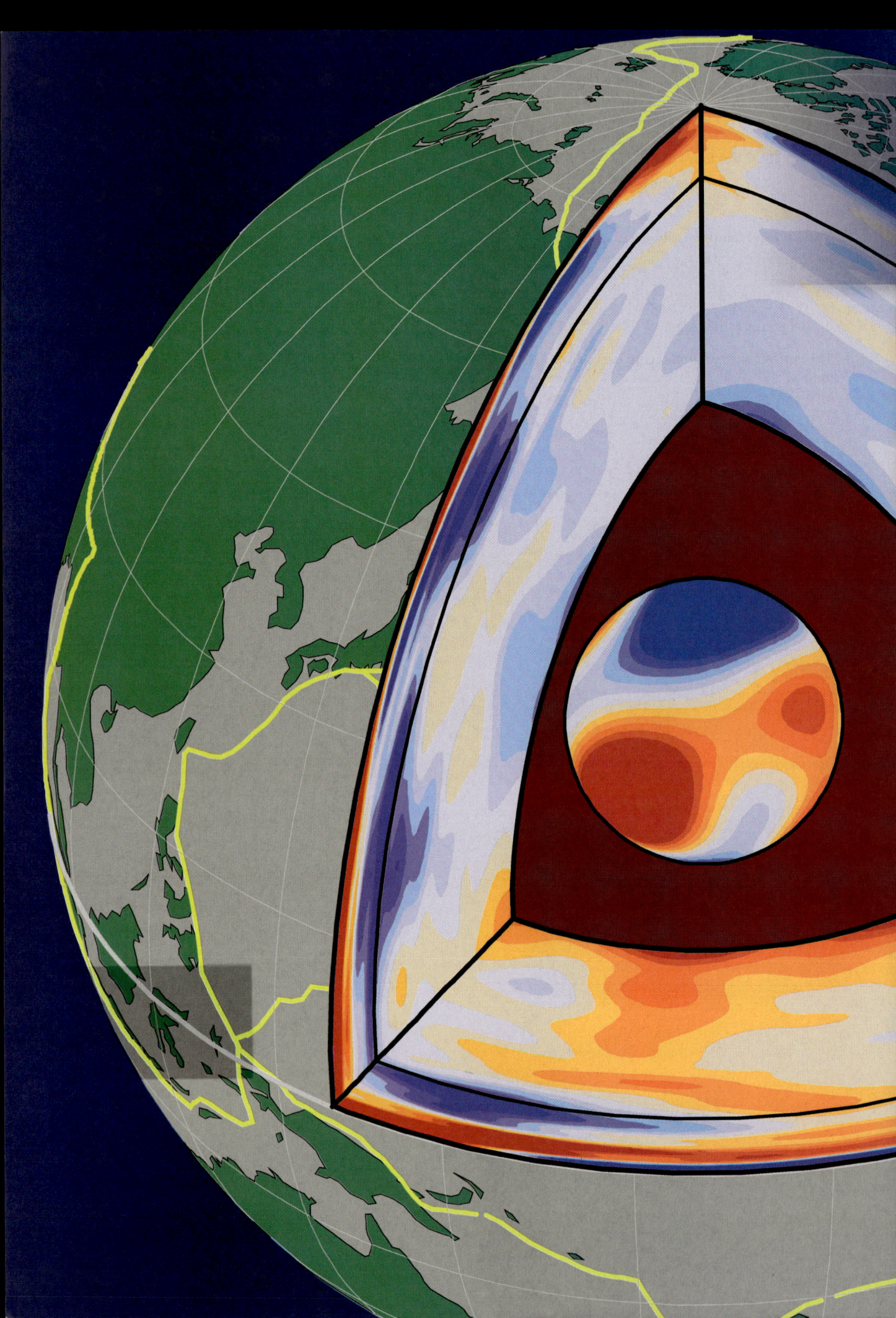

Exploring Earth's Interior 11

Seismic waves can be used to map features generated by dynamic processes in Earth's interior. This image shows variations in shear wave speed on cross sections through the mantle and on the surface of the inner core. Yellow lines on the surface of the globe are the plate boundaries. [Courtesy of J. H. Woodhouse, Oxford University.]

Exploring Earth's Interior with Seismic Waves	308
Layering and Composition of Earth's Interior	312
Earth's Internal Temperature	316
Visualizing Earth's Three-Dimensional Structure	319
Earth's Magnetic Field and the Geodynamo	320

Learning Objectives

Because Earth's deep interior is opaque to direct observation, you will have to use your imagination in thinking about what might be going on in this planetary "underworld." From what you learn in this chapter, you should be able to:

11.1 Trace the paths of the major seismic waves through Earth's interior and know what they reveal about the layering of Earth's crust, mantle, and core.

11.2 Apply the principle of isostasy to explain why continents ride higher than oceans and why the oceans deepen as the geologic age of the seafloor increases.

11.3 Recognize where heat in Earth's interior is transported upward primarily by conduction and where primarily by convection.

11.4 Describe how temperature increases from Earth's surface to its center, and identify the depths where the geotherm crosses the melting curve.

11.5 Visualize Earth's three-dimensional structure from the images provided by seismic tomography.

11.6 Summarize what Earth's magnetic field tells us about the liquid outer core.

11.7 Explain how magnetic field reversals are used to date rock sequences.

Humans have burrowed in mines to depths as great as 4 km to extract gold and other minerals, and they have drilled down to more than 10 km in search of petroleum. But these heroic efforts have barely scratched the surface of our massive planet. The crushing pressures and red-hot temperatures of Earth's deeper layers make the planet's interior inaccessible to us for the foreseeable future. Nevertheless, we can learn much about the structure and composition of Earth's interior from our position on its surface.

Some of the best information comes from seismology. Chapter 10 described the terrible shaking and destruction that can be wrought by seismic waves. Yet this same seismic energy can be harnessed to illuminate Earth's deepest regions, allowing us to construct three-dimensional images of geologic features in the lower crust, the rising and falling of convection currents in the mantle, and even the workings of the outer and inner core. Our understanding of Earth's interior has been further enriched by material erupted from volcanoes, by the behavior of Earth materials under high temperatures and pressures in the laboratory, and by the information contained in Earth's gravitational and magnetic fields.

In this chapter, we will explore Earth's interior down to its center, nearly 6400 km beneath our feet. We will see how seismic waves have been used to image the structure of Earth's crust, mantle, and core. We will investigate temperatures deep inside Earth and the machinery of two great geosystems powered by its internal heat engine: the plate tectonic system, which is driven by convection in the mantle, and the geodynamo in the outer core, which generates Earth's magnetic field.

Exploring Earth's Interior with Seismic Waves

Different types of waves—light, sound, and seismic—have a common characteristic: The velocity at which they travel depends on the material through which they are passing. Light waves travel fastest through a vacuum, more slowly through air, and even more slowly through water. Sound waves, on the other hand, travel faster through water than through air and not at all through a vacuum. Why?

Sound waves are simply propagating variations in pressure. Without something to compress, such as air or water, they cannot exist. The more force it takes to compress a material, the faster sound will travel through it. The speed of sound in air—Mach 1, in the jargon of jet pilots—is typically 0.34 km/s, or about 760 miles per hour, at Earth's surface. Water resists compression much more than air, so the speed of sound waves in water is correspondingly higher, about 1.5 km/s. Solid materials are even more resistant to compression, so sound waves travel through them at even higher speeds. Sound travels through granite at about 6 km/s—nearly 13,500 miles per hour!

Basic Types of Waves

As we will see in Chapter 13, some of the seismic waves created by earthquakes are **compressional waves** (like sound waves), which travel with a push-pull motion, while others are **shear waves,** which travel with a side-to-side motion, displacing material at right angles to their path of travel (see Figure 10.10). Solids are more resistant to compression than to shearing, so compressional waves always travel through solids faster than shear waves do. This physical principle explains a relationship we discussed in Chapter 13: Compressional waves are always the first arrivals at a seismographic station (and hence are called primary, or P waves), and shear waves are the secondary arrivals (S waves). It also explains why the speed of shear waves in gases and liquids is zero: Those materials have no resistance to shearing. Shear waves cannot propagate through any fluid—air, water, or the liquid iron in Earth's outer core.

Geologists can calculate the speed of a P or an S wave by dividing the distance traveled by the travel time. These wave speeds can then be used to infer which materials the waves encountered along their paths.

The concepts of travel times and wave paths sound simple enough, but complications arise when waves pass through more than one type of material. At the boundary between two different materials, some of the waves bounce off—that is, they are *reflected*—and others are transmitted into the second material, just as light is partly reflected and partly transmitted when it strikes a windowpane. The waves that cross the boundary between two materials are bent, or *refracted*, as their velocity changes from that in the first material to that in the second. **Figure 11.1** shows a laser light beam whose path bends as it goes from water into air, much as a P or an S wave bends as it travels from one material to another. By studying the speeds at which seismic waves travel and how they are refracted and reflected at Earth's internal boundaries, seismologists have been able to model the layering of Earth's crust, mantle, and core with great precision (see Figure 1.11).

Paths of Seismic Waves Through Earth

If Earth were made of a single material with constant properties from the surface to the center, P and S waves would travel from the focus of an earthquake to a distant seismographic station along straight lines through Earth's interior, just as the Sun's rays travel in straight lines through outer space. When the first global networks of seismographs were installed about a century ago, however, seismologists discovered that **seismic ray paths** are not straight lines; the waves are refracted and reflected by Earth's layered structure.

Exploring Earth's Interior with Seismic Waves

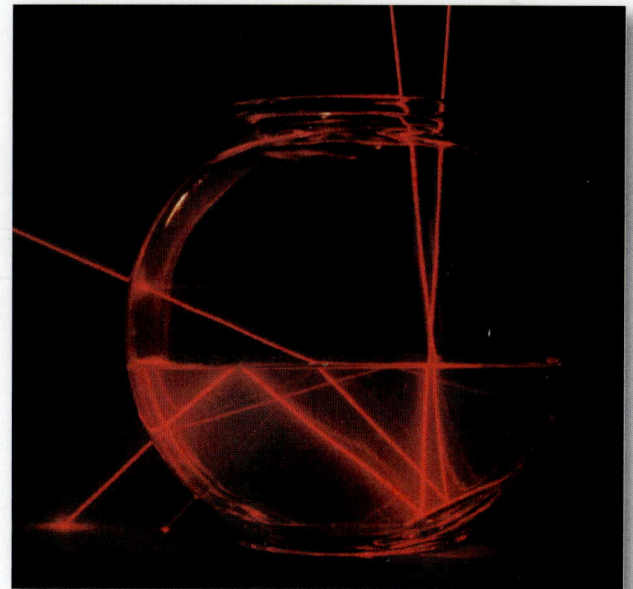

FIGURE 11.1 In this experiment, two beams of laser light enter a bowl of water from the top at slightly different angles. Both beams are reflected from a mirror on the bottom of the bowl. One is then reflected at the water-air boundary and passes through the bowl to make a bright spot on the table. Most of the light in the other beam is bent (refracted) as it passes from the water to the air, although a small amount is reflected to form a second spot on the table. [Susan Schwartzenberg/The Exploratorium.]

Waves Refracted Through Earth's Interior From their observations of travel times and the amount of upward refraction of the ray paths, seismologists were able to conclude that P waves travel much faster through rock deep within Earth than they do through rock at Earth's surface. This was hardly surprising, because rock subjected to the great pressures in Earth's interior would be squeezed into tighter crystal structures. The atoms in these tighter structures would be more resistant to further compression, which would cause P waves to travel through them more quickly.

Seismologists were very surprised, however, by what they found at progressively greater distances from an earthquake focus (Figure 11.2). After the P and S waves had traveled beyond about 11,600 km, they suddenly disappeared! Like airplane pilots and ship captains, seismologists prefer to measure distances traveled over Earth's surface in angular degrees—from 0° at the earthquake focus to 180° at a point on the opposite side of Earth. Each degree measures 111 km at the surface, so 11,600 km corresponds to an angular distance of 105°, as shown in Figure 11.2. When they looked at seismograms recorded beyond 105° from the focus, they did not see the distinct P- and S-wave arrivals that were so clear on seismograms recorded at shorter distances. Beyond about 15,800 km from the focus (142°), the P waves suddenly reappeared, although they were much delayed compared with their expected travel times. The S waves never reappeared.

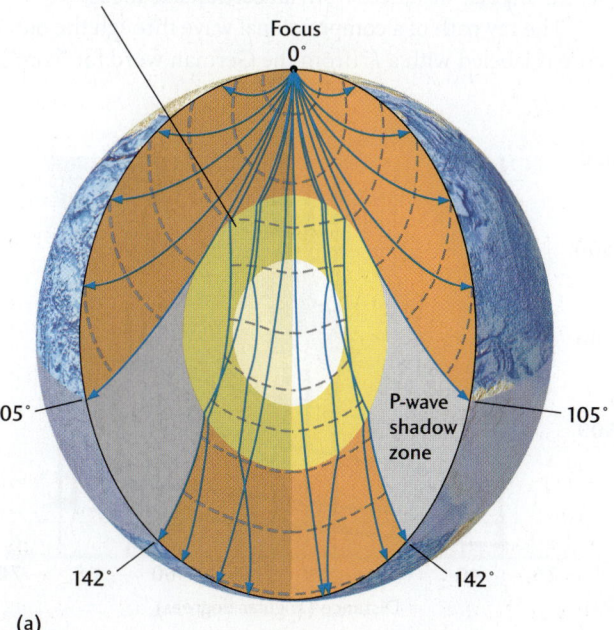

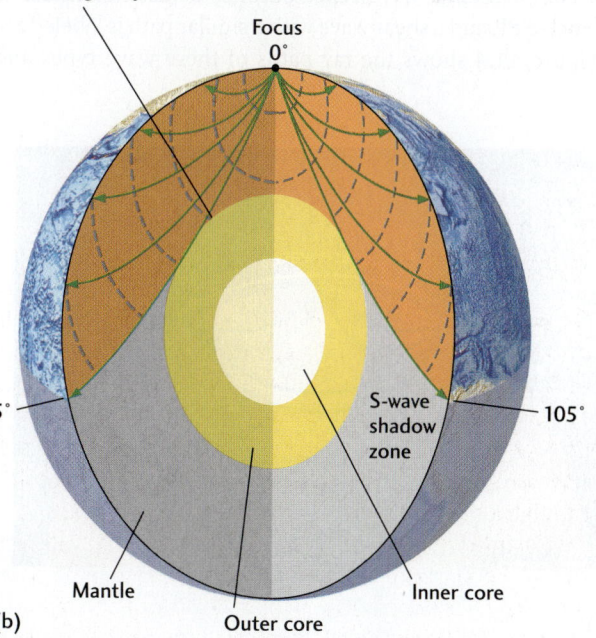

FIGURE 11.2 Earth's core creates P-wave and S-wave shadow zones. The ray paths of the seismic waves from an earthquake focus through Earth's interior are shown by solid lines (blue for P waves, green for S waves). The dashed lines show the progress of the waves at 2-minute intervals. Distances are measured in angular degrees from the earthquake focus. (a) The P-wave shadow zone extends from 105° to 142°. (b) The larger S-wave shadow zone extends from 105° to 180°.

In 1906, the British seismologist R. D. Oldham put these observations together to provide the first evidence that Earth has a liquid outer core. S waves cannot travel through the outer core, he argued, because it is liquid, and liquids have no resistance to shearing. Thus, there is an S-wave **shadow zone** beyond 105° from the earthquake focus, where S-wave ray paths encounter the core-mantle boundary (see Figure 11.2b). The propagation of P waves is more complicated (see Figure 11.2a). At 105°, their ray paths also encounter the core-mantle boundary. At that boundary, P-wave velocity drops by almost a factor of two. Therefore, the waves are refracted downward into the core and emerge at greater angular distances after the delay caused by their detour through the core. This refraction effect forms a P-wave shadow zone at angular distances between 105° and 142°.

Waves Reflected by Earth's Internal Boundaries

When seismologists looked at records of seismic waves made at angular distances of less than 105° from an earthquake focus, they found waves that must have been reflected from the core-mantle boundary. They labeled a compressional wave reflected from the top of the outer core PcP and a shear wave ScS. (The lowercase c indicates a core reflection.) In 1914, a German seismologist, Beno Gutenberg, used the travel times of these reflected waves to determine the depth of the core-mantle boundary, which modern estimates put at about 2890 km. **Figure 11.3** shows examples of the ray paths taken by these core-reflected waves.

Figure 11.3 also shows the ray paths of some other prominent wave types seen on seismograms, along with the labels seismologists have attached to them. For example, a compressional wave reflected once at Earth's surface is labeled PP, and a shear wave with a similar path is labeled SS. **Figure 11.4** shows the ray paths of these wave types and

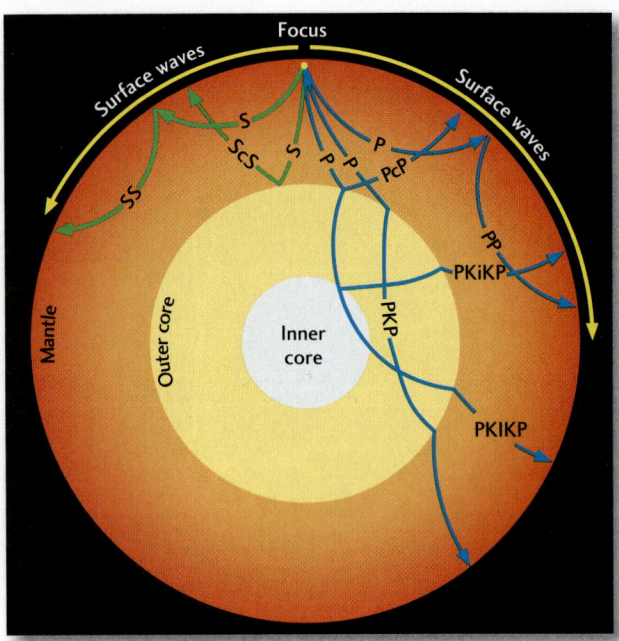

FIGURE 11.3 Seismologists use a simple labeling scheme to describe the various ray paths taken by seismic waves. PcP and ScS are compressional and shear waves, respectively, that are reflected by the core. PP and SS waves are internally reflected from Earth's surface. A PKP wave travels through the liquid outer core, a PKIKP wave travels through the solid inner core, and a PKiKP wave is reflected by the inner core. Surface waves propagate along Earth's outer surface, like waves on the surface of a pond.

their internal reflections on seismograms recorded at different angular distances from an earthquake focus.

The ray path of a compressional wave through the outer core is labeled with a K (from the German word for "core"),

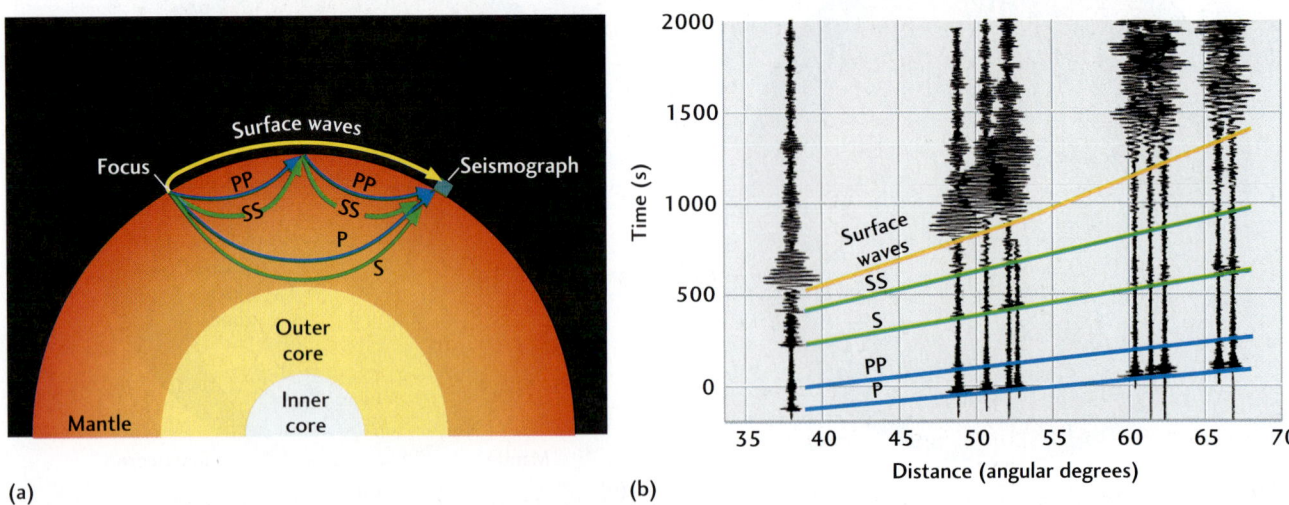

FIGURE 11.4 (a) P and S waves are refracted upward in the mantle and also can be reflected from Earth's surface. A seismic wave that has been reflected once from Earth's surface is labeled with a double letter (PP or SS). (b) Seismograms recorded at various distances from an earthquake focus in the Aleutian Islands, Alaska. The colored lines identify the arrival times of the P and S waves, the surface waves, and the PP and SS waves reflected from Earth's surface.

so PKP designates a compressional wave that propagates through the crust and mantle, into the outer core, and back through the mantle and crust to a seismograph at Earth's surface. In 1936, Danish seismologist Inge Lehmann (Figure 11.5) discovered Earth's inner core by observing compressional waves refracted by its outer boundary, which she determined to be at a depth of about 5150 km. Ray paths through the inner core are labeled with an *I*, so the waves Lehmann observed are labeled PKIKP. Other seismologists have since observed compressional waves (labeled PKiKP) reflected from the top side of the inner core–outer core boundary (the lowercase *i* indicates a reflection rather than a refraction; see Figure 11.3).

Seismic Exploration of Near-Surface Layering

Seismic waves can also be used to probe the shallow parts of Earth's crust. One technique, called *seismic profiling*, has a number of practical applications. Seismic waves generated by artificial sources, such as dynamite explosions on land and compressed-air explosions at sea, are reflected by geologic structures at shallow depths in the crust (Figure 11.6). Recording of these reflections has proved to be the most successful method for finding deeply buried oil and gas reservoirs. This type of seismic

FIGURE 11.5 Danish seismologist Inge Lehmann discovered Earth's inner core in 1936. [SPL/Science Source.]

(a)

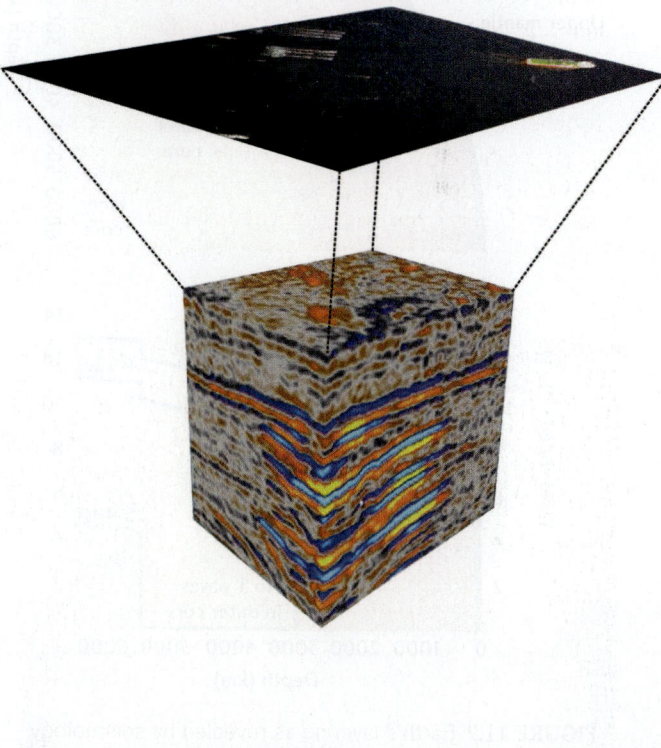

(b)

FIGURE 11.6 (a) The *Geco Topaz*, a vessel operated by WesternGeco Inc., towing hydrophones, conducting a three-dimensional seismic survey. The bubbles behind the ship are compressed-air explosions that send out compressional waves; the reflections of those waves from the rocks below are recorded by seismographs pulled on cables behind the ship to produce an image of the subsurface structure. (b) A three-dimensional image produced by a seismic survey. The colors represent layers of sediment beneath the seafloor, some of which may trap oil and natural gas. [a: John Lawrence Photography/Alamy; b: Courtesy of BP.]

exploration is now a multibillion-dollar industry. Reflected seismic waves are also used to measure the depth of water tables and the thickness of glaciers. At sea, compressional waves can be generated by mechanical devices similar to loudspeakers, and oceanographic ships routinely use the underwater sound they produce to measure the depth of the ocean and the thickness of sediments on the seafloor.

Layering and Composition of Earth's Interior

By measuring the travel times of compressional and shear waves from earthquakes around the world, geologists have developed a detailed model of how the wave speeds change with depth from Earth's surface to its center. We will explore this Earth model, which is shown in Figure 11.7, by taking an imaginary downward journey through Earth's interior, from its outer crust to its inner core.

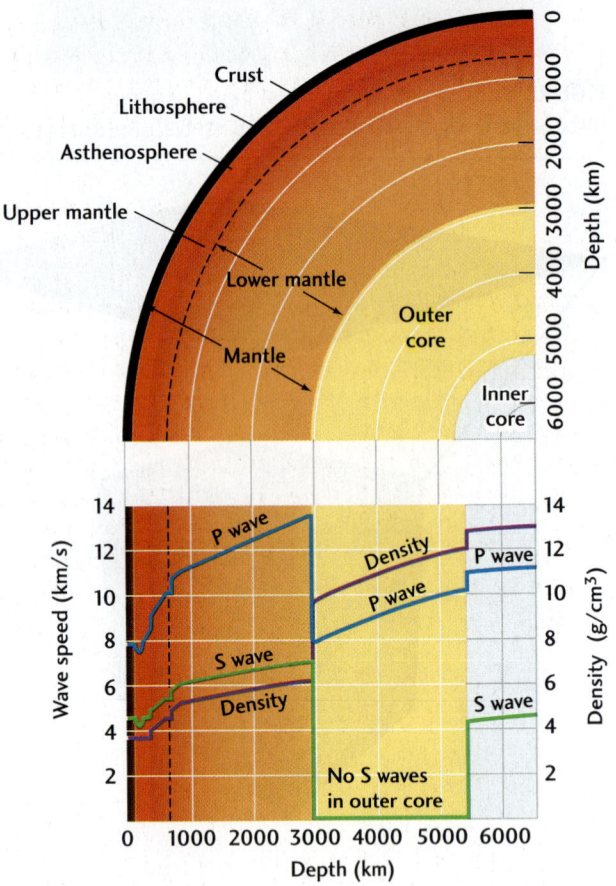

FIGURE 11.7 Earth's layering as revealed by seismology. The lower diagram shows changes in P-wave and S-wave velocities and rock densities with depth. The upper diagram is a cross section through Earth on the same depth scale, showing how those changes are related to the major layers (see also Figure 1.11).

The Crust

By measuring the velocities of seismic waves passing through samples of various materials in the laboratory, seismologists have compiled a library of seismic wave velocities through different rock types. Rough values for P-wave velocities in igneous rocks, for example, are as follows:

- Felsic rocks typical of the upper continental crust (granite): 6 km/s
- Mafic rocks typical of oceanic crust or the lower continental crust (gabbro): 7 km/s
- Ultramafic rocks typical of the upper mantle (peridotite): 8 km/s

These velocities vary because they depend on a rock's density and its resistance to compression and shear, which depend on chemical composition and crystal structure. In general, higher densities correspond to higher P-wave velocities; typical densities for granite, gabbro, and peridotite are 2.6 g/cm^3, 2.9 g/cm^3, and 3.3 g/cm^3, respectively.

We know from measurements of P-wave velocities that the upper part of the continental crust is made mostly of low-density granitic rocks. The measurements also show that no granite exists on the deep seafloor; the crust there consists entirely of basalt and gabbro overlain by sediments. The velocity of P waves increases abruptly to 8 km/s at the **Mohorovičić discontinuity,** or *Moho,* which marks the base of the crust (see Chapter 1). That velocity indicates that the mantle below the Moho is made primarily of dense peridotite.

Seismic data show that Earth's crust is thin (about 7 km) under the oceans, thicker (about 33 km) under the stable, flat-lying continents, and thickest (as much as 70 km) under the high mountains of orogenic zones. The elevations of continents relative to the deep seafloor can be explained by the **principle of isostasy,** which states that the total weight of lithospheric columns should be same for continents and oceans. This balance is called *isostatic equilibrium* (see the Practicing Geology Exercise following this section).

The Mantle

The **upper mantle,** which extends from the Moho to 410 km, is made primarily of peridotite, a dense ultramafic rock composed primarily of olivine and pyroxene. These minerals contain less silica and more magnesium and iron than those in typical crustal rocks (see Chapter 4). S-wave velocities have been used to explore the layering of the mantle (Figure 11.8). The layering of the upper mantle is caused primarily by the effects of increasing temperature and pressure on peridotite. Olivine and pyroxene undergo partial melting under the conditions found in the uppermost part of the mantle. At greater depths, increasing pressure forces the atoms of these minerals closer together into more compact crystal structures.

The mantle just below the Moho is relatively cold. Like the crust, it is part of the lithosphere, the rigid layer that through it. Because partial melting also allows the rock to flow more easily, geologists identify the low-velocity zone with the top part of the *asthenosphere*: the weak, ductile layer on which the rigid lithospheric plates slide. This idea fits nicely with evidence that the asthenosphere is the source of most basaltic magmas (see Chapters 4 and 5).

The base of the low-velocity zone lies at about 200 to 250 km below oceanic crust, where S-wave velocity increases to a value consistent with solid peridotite. The low-velocity zone is not as well defined under stable continental cratons, where colder lithospheric mantle extends to these depths.

At depths of about 200 to 400 km, the S-wave velocity again increases with depth. Within this zone, pressure continues to increase with depth, but the temperature does not rise as rapidly as it does near the surface, due to the effects of convection within the asthenosphere. (We will discuss why this is so in the next section.) The combined effects of pressure and temperature decrease the amount of melting with depth and cause the rock rigidity—and thus S-wave velocity—to rise.

About 400 km below the surface, S-wave velocity increases by about 10 percent within a narrow zone less than 20 km thick. This jump in S-wave velocity can be explained by a **phase change** in olivine, the major mineral constituent of the upper mantle. When olivine is squeezed with high pressures in the laboratory, the atoms that form its crystal structure collapse into a more compact arrangement at the temperatures and pressures corresponding to depths of about 410 km. The jumps in the P- and S-wave velocities measured for this phase change in the laboratory match the increase observed from seismic waves passing through this depth.

At a depth of about 660 km, the S-wave velocity abruptly increases again, indicating another major phase change in olivine to an even more closely packed crystal structure. Laboratory experiments have confirmed the existence of a second major mineralogical phase change at this depth.

Because it contains at least two major phase changes, the layer between 410 km and 670 km in depth is called the **transition zone**. Phase changes are transitions in a rock's mineralogy, but not in its chemical composition. Some geologists have argued, however, that the increase in seismic wave velocities at 660 km comes in part from a change in the chemical composition of mantle rock. This debate was critical to understanding the plate tectonic system, because a chemical change would imply that the convection that drives plate tectonics does not penetrate much beyond this depth—in other words, that convection in the mantle is stratified (as shown in Figure 2.19b). Evidence from detailed studies of mantle structure now indicates very little, if any, chemical change in this region of the mantle.

Below the transition zone at 660 km, seismic wave velocities increase gradually and do not show any more

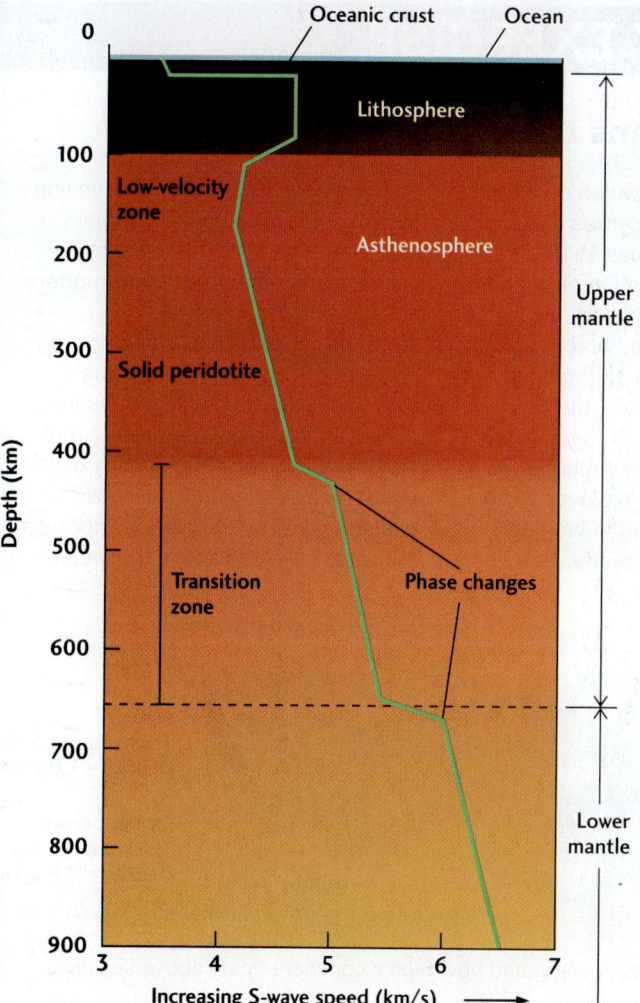

FIGURE 11.8 The structure of the mantle beneath old oceanic lithosphere, showing S-wave velocities to a depth of 900 km. Changes in S-wave velocity mark the strong, brittle lithosphere, the weak, ductile asthenosphere, and a transition zone, in which increasing pressure forces rearrangements of atoms into denser and more compact crystal structures (phase changes).

forms the plates (see Chapter 1). On average, the thickness of the lithosphere is about 100 km, but it is highly variable geographically, ranging from almost no thickness near spreading centers, where new oceanic lithosphere is forming from hot, rising mantle material, to over 200 km beneath the cold, stable continental cratons.

Near the base of the oceanic lithosphere, S-wave velocity abruptly decreases, marking the start of a **low-velocity zone**. At about 100 km, the temperature approaches the melting temperature of peridotite, partially melting some of its minerals. Although the amount of melting is small (in most places, less than 1 percent), it is sufficient to decrease the rigidity of the rock, which slows the S waves passing

PRACTICING GEOLOGY EXERCISE

The Principle of Isostasy: Why Are Oceans Deep and Mountains High?

Earth's topography is dominated by continents, which typically have elevations of 0 to 1 km above sea level, and ocean basins, which typically have elevations of 4 to 5 km below sea level. Why the difference? The answer comes from the principle of isostasy, which relates the elevations of continents and oceans to the densities of crustal and mantle rocks. This amazingly useful principle not only explains much of Earth's topography, but also allows scientists to use changes in crustal elevation over time to investigate the properties of the mantle (see Earth Issues 11.1).

Isostasy (from the Greek for "equal standing") is based on Archimedes' principle, which states that the weight of a floating solid is equal to the weight of the fluid it displaces. (According to legend, the Greek philosopher Archimedes discovered this principle over 2200 years ago while sitting in his bath; boggled by its implications, he rushed naked into the street, yelling "Eureka, I have found it!" Major discoveries rarely provoke such enthusiastic responses from modern scientists.)

Consider a block of wood floating in water. In each unit of area, the block's mass is its density times its thickness, whereas the mass of the water it displaces is the density of water times the water thickness, which is given by the block's thickness minus its elevation above the water.

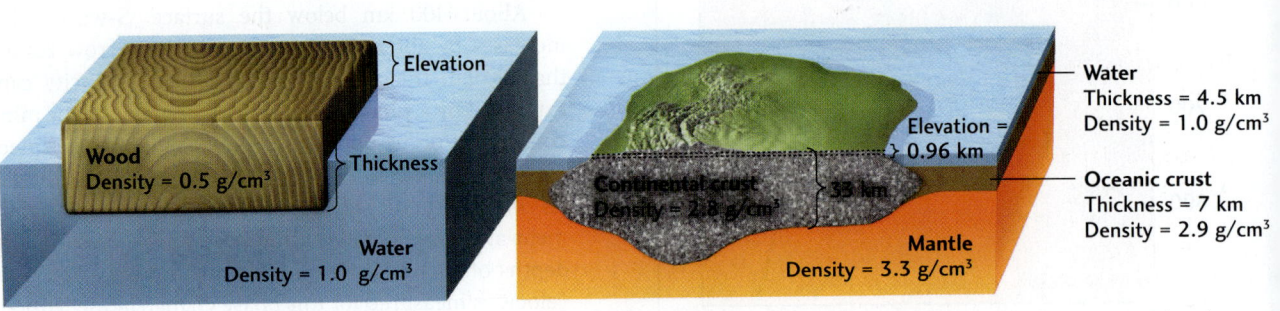

The principle of isostasy explains how high a wood block floats in water and how high a continent floats above sea level.

---

unusual features until close to the core-mantle boundary. This relatively homogeneous region, more than 2000 km thick, is called the **lower mantle.** The lower mantle is convecting and exchanges mass with the upper mantle, driven in part by slabs of oceanic lithosphere subducting through the upper mantle into the lower mantle.

The Core-Mantle Boundary

At the **core-mantle boundary,** about 2890 km below the surface, we encounter the most extreme change in properties found anywhere in Earth's interior. From the way seismic waves are reflected by this boundary, seismologists can tell that it is a very sharp interface. The material changes abruptly from solid silicate rock to a liquid iron alloy. Because of the complete loss of rigidity, the S-wave velocity drops from about 7.5 km/s to zero, and the P-wave velocity drops from more than 13 km/s to about 8 km/s, resulting in the core shadow zones. Density, on the other hand, increases by about 4 g/cm³ (see Figure 11.7). This large density difference, which is even greater than the increase in density from atmosphere to lithosphere at Earth's solid surface, keeps the core-mantle boundary very flat (you could probably skateboard on it!) and prevents any large-scale mixing of the mantle and core.

The core-mantle boundary appears to be a very active place. Heat conducted out of the core increases the temperatures at the base of the mantle by as much as 1000°C (see Figure 11.10). Indeed, the paths of seismic waves that pass near the base of the mantle show peculiar complications, suggesting a region of exceptional geologic activity. In a thin layer above the core-mantle boundary, there is a steep (10 percent or more) decrease in seismic wave velocities, which may be an indication that the mantle in contact with the core is partially molten, at least in some places. As we noted in Chapter 5, some geologists believe this hot region to be the source of mantle plumes that rise all the way to Earth's surface, creating volcanic hot spots such as Hawaii and Yellowstone.

The lowest boundary layer of the mantle, a region about 300 km thick, may be the ultimate graveyard of some subducted lithospheric material, such as the dense, iron-rich parts of the oceanic crust. It is possible that this zone experiences an upside-down version of the tectonics we see at Earth's surface. For example, accumulations of heavy,

Archimedes' principle states that these quantities must be equal:

$$\text{wood density} \times \text{wood thickness} =$$
$$\text{water density} \times \text{water thickness} =$$
$$\text{water density} \times (\text{wood thickness} - \text{wood elevation})$$

We can solve this equation to find the elevation:

$$\text{wood elevation} = \left(1 - \frac{\text{wood density}}{\text{water density}}\right) \times \text{wood thickness}$$

The expression in brackets is called the "buoyancy factor" because it tells us what fraction of the wood will rise above the water surface. A light wood, such as young pine, has only half the density of water, so its buoyancy factor is

$$1 - \frac{0.5 \text{ g/cm}^3}{1.0 \text{ g/cm}^3} = 0.5$$

The pine block will float high, with half of its volume out of the water. However, in the case of old oak, which has a density of 0.9 g/cm³, the buoyancy factor is only 0.1, so the block will float low, with only one-tenth of its thickness above the water.

If continental crust (density = 2.8 g/cm³) floated alone on top of mantle material (3.3 g/cm³), the previous equation could be modified to give continental elevation by simply replacing "wood" with "continent" and "water" with "mantle." However, we must account for the oceanic crust (2.9 g/cm³) and the ocean water (1.0 g/cm³) that also float on the mantle. Since these two layers fill up the basins around the continents, we must subtract from the continental elevation the height that each of those layers alone would float above the mantle, given by its buoyancy factor times its thickness. The isostatic equation for continents therefore has three terms, one positive and two negative:

$$\text{continent elevation} = \left(1 - \frac{\text{continental crust density}}{\text{mantle density}}\right) \times \text{continental crust thickness}$$

$$- \left(1 - \frac{\text{oceanic crust density}}{\text{mantle density}}\right) \times \text{oceanic crust thickness}$$

$$- \left(1 - \frac{\text{oceanic water density}}{\text{mantle density}}\right) \times \text{oceanic water thickness}$$

Using thicknesses of 33 km and 7 km for continental and oceanic crust, respectively, and a water depth of 4.5 km, we obtain

$$\text{continent elevation} =$$
$$(0.15 \times 33 \text{ km}) - (0.12 \times 7.0 \text{ km}) - (0.7 \times 4.5 \text{ km})$$
$$= 0.96 \text{ km above sea level}$$

This result is consistent with the overall distribution of Earth's topography (see Figure 1.7).

Because of isostasy, elevation is a sensitive indicator of crustal thickness, so regions of lower elevation must have thinner crust (or higher average density), whereas regions of higher elevation, such as the Tibetan Plateau, must have thicker crust (or lower average density).

The principle of isostasy explains how high a wood block floats in water and how high a continent floats above sea level.

BONUS PROBLEM: The average elevation of the Tibetan Plateau is about 5 km above sea level. Use the isostatic equation for continents to compute the average thickness of the crust in this region, assuming that its average density is 2.8 g/cm³.

---

iron-rich material might form chemically distinct "anticontinents" that are constantly pushed to and fro across the core-mantle boundary by convection currents. We still have a lot to learn about the geologic processes that might be active in this strange place.

The Core

Many lines of evidence support the hypothesis that Earth's core is made of iron and nickel. These metals are abundant in the universe (see Chapter 1); in addition, they are dense enough to explain the mass of the core (about one-third of Earth's total mass) and to be consistent with the theory that the core formed by gravitational differentiation (see Chapter 20). This hypothesis, first proposed by Emil Wiechert in the late nineteenth century, was buttressed by discoveries of meteorites made almost entirely of iron and nickel, which presumably came from the breakup of a planetary body that also had an iron-nickel core (see Figure 1.9).

Laboratory measurements at appropriately high pressures and temperatures have led to a slight revision of this hypothesis. A pure iron-nickel alloy turns out to be about 10 percent too dense to match the data for the outer core. Therefore, it has been proposed that the core includes minor amounts of some lighter elements. Oxygen and sulfur are leading candidates, although the precise composition remains the subject of research and debate.

Seismology tells us that the core below the mantle is liquid, but the core is not liquid to the very center of Earth.

As Lehmann first discovered, P waves that penetrate to depths of 5150 km suddenly speed up, indicating the presence of an *inner core*, a metallic sphere two-thirds the size of the Moon. Seismologists have shown that the inner core transmits shear waves, confirming early speculations that it is solid. In fact, some calculations suggest that the inner core spins at a slightly faster rate than the mantle, acting like a "planet within a planet."

The very center of Earth is not a place you would like to be. The pressures are immense, over 4 million times the atmospheric pressure at Earth's surface. And it's also very hot, as we are about to see.

Earth's Internal Temperature

The evidence of Earth's internal heat is everywhere: volcanoes, hot springs, and the elevated temperatures measured in mines and boreholes. Earth's internal heat fuels convection in the mantle, which drives the plate-tectonic system, as well as the geodynamo in the core, which produces Earth's magnetic field.

Earth's internal heat engines are powered by several sources. During the planet's violent origin, kinetic energy released by impacts with planetesimals heated its outer regions, while gravitational energy released by differentiation of the core heated its deep interior (see Chapter 20). The decay of radioactive isotopes in Earth's interior continues to generate heat.

After Earth formed, it began to cool, and it is cooling to this day as heat flows from the hot interior to the cool surface. The temperatures in the planet's interior result from a balance between heat gained and heat lost.

Heat Flow Through Earth's Interior

Earth cools in two main ways: through the transport of heat by conduction and through the more efficient transport of heat by convection. Conduction dominates in the lithosphere, whereas convection is more important throughout most of Earth's interior.

Conduction Through the Lithosphere Heat energy exists in a material as vibrations of atoms; the higher the temperature, the more intense the vibrations. The **conduction** of heat occurs when thermally agitated atoms and molecules jostle one another, mechanically transferring kinetic energy from a hot region to a cool one. Heat is transferred from regions of high temperature to regions of low temperature by this process.

Materials vary in their ability to conduct heat. Metal is a better conductor than plastic (think of how rapidly a metal handle on a frying pan heats up compared with one made of plastic). Rock and soil are very poor heat conductors, which is why underground pipes are less susceptible to freezing than those above ground. Rock conducts heat so poorly that a lava flow 100 m thick takes about 300 years to cool from 1000°C to ground surface temperatures. Moreover, the cooling time of a layer increases with the square of its thickness, so a lava flow twice as thick (200 m) would take four times as long to cool (about 1200 years).

The conduction of heat through the outer surface of the lithosphere causes the lithosphere to cool slowly over time. As it cools, its thickness increases, just as the cold crust on a bowl of hot wax thickens over time. Rock, like wax, contracts and becomes denser with decreasing temperature, so the average density of the lithosphere must increase over time, and therefore, according to the principle of isostasy, its surface must sink to lower levels. Thus, the mid-ocean ridges stand high because the lithosphere there is young, hot, and thin, whereas the abyssal plains are deep because the lithosphere there is old, cold, thick, and dense.

From these principles, geologists have constructed a simple but precise theory of seafloor topography that uses conductive cooling to explain the large-scale features of ocean basins. The theory predicts that ocean depth should depend primarily on the age of the seafloor. Because the cooling depth goes as the square root of cooling time, ocean depth should increase as the square root of seafloor age. In other words, seafloor that is 40 million years old should have subsided twice as much as seafloor that is only 10 million years old (because $\sqrt{40/10} = \sqrt{4} = 2$). This simple mathematical relationship matches seafloor topography near the mid-ocean ridge crests amazingly well, as demonstrated in **Figure 11.9**.

Conductive cooling of the lithosphere accounts for a wide variety of other geologic phenomena, including the subsidence of passive continental margins and thermal subsidence basins (see Chapter 6). It explains why the amount of heat flowing out of oceanic lithosphere is high near spreading centers and decreases as the oceanic lithosphere gets older, and it tells us why the average thickness of the oceanic lithosphere is about 100 km. The establishment of this theory was one of the great successes of plate tectonics.

Conductive cooling does not explain all aspects of heat flow through Earth's outer surface, however. Marine geologists have found that seafloor older than about 100 million years does not continue to subside as the simple theory would predict. Moreover, simple conductive cooling is far too inefficient to account for the cooling of Earth over its entire history. It can be shown that if the 4.5-billion-year-old Earth cooled by conduction alone, very little of the heat from depths greater than about 500 km would have reached the surface. The mantle, which was molten in Earth's early history, would be far hotter than it is now. To understand these observations, we must consider the second mode of heat transport, convection, which is more efficient than conduction in getting heat out of Earth's interior.

Convection in the Mantle and Core When a fluid—either liquid or gas—is heated, it expands and rises because it has become less dense than the surrounding material. The upward movement of the heated fluid displaces cooler fluid downward, where it is heated and then rises to continue the cycle. This process, called **convection,** transfers heat more efficiently than conduction because the heated

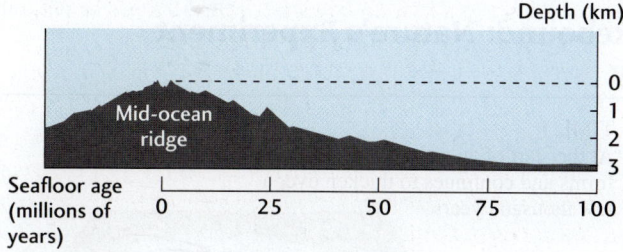
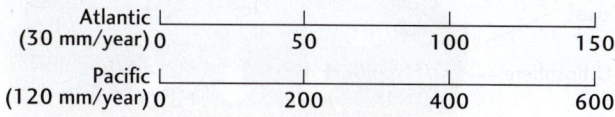
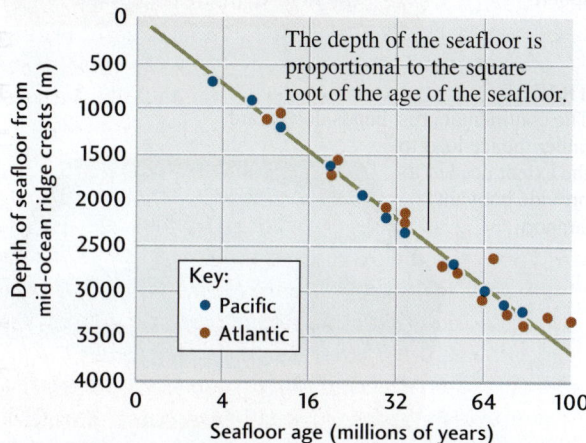

FIGURE 11.9 Topography of mid-ocean ridges in the Atlantic and Pacific oceans, showing how ocean depth increases in proportion to the square root of lithosphere age as plates move away from spreading centers. The same theoretical curve, derived by assuming that the lithosphere cools by conduction, matches the data for both ocean basins, even though seafloor spreading is much faster in the Pacific than in the Atlantic.

lithosphere behaves as a ductile material; over long periods, it can flow like a very viscous fluid (see Earth Issues 11.1). As we saw in Chapters 1 and 2, seafloor spreading and plate movements are direct evidence of this solid-state convection at work. The hot mantle material that rises under mid-ocean ridges builds new lithosphere, which cools as it spreads away. In time, it sinks back into the mantle at subduction zones, where it is eventually resorbed and reheated. Through this process, heat is carried from Earth's interior to its surface.

Temperatures Inside Earth

Geologists have many reasons for wanting to understand the *geothermal gradient*—the increase in temperature with depth—in Earth's interior. Temperature and pressure determine the state of matter (solid or molten), its viscosity (resistance to flow), and how its atoms are packed together in crystals. The curve that describes how temperature increases with depth in Earth's interior is called the **geotherm**. In Figure 11.10, we compare one possible geotherm (in yellow) with the melting curves for mantle and core materials (in red).

material itself moves, carrying its heat with it. Convection is the same process by which water is heated in a kettle on the stove (see Figure 1.15). Liquids conduct heat poorly, so a kettle of water would take a long time to boil if convection did not distribute the heat rapidly. Convection is what moves heat when a chimney draws, when warm tobacco smoke rises, or when thunderclouds form on a hot day.

We have already seen how seismic waves revealed that Earth's outer core is liquid. Other types of data demonstrate that the iron-rich material in the outer core has a low viscosity and can therefore convect very easily. Convective movement in the outer core distributes heat through the core very efficiently, and it generates Earth's magnetic field, a phenomenon we will examine in more detail later in this chapter. At the core-mantle boundary, heat from the core flows into the mantle.

The existence of convection in the solid mantle is more surprising, but we now know that mantle rock below the

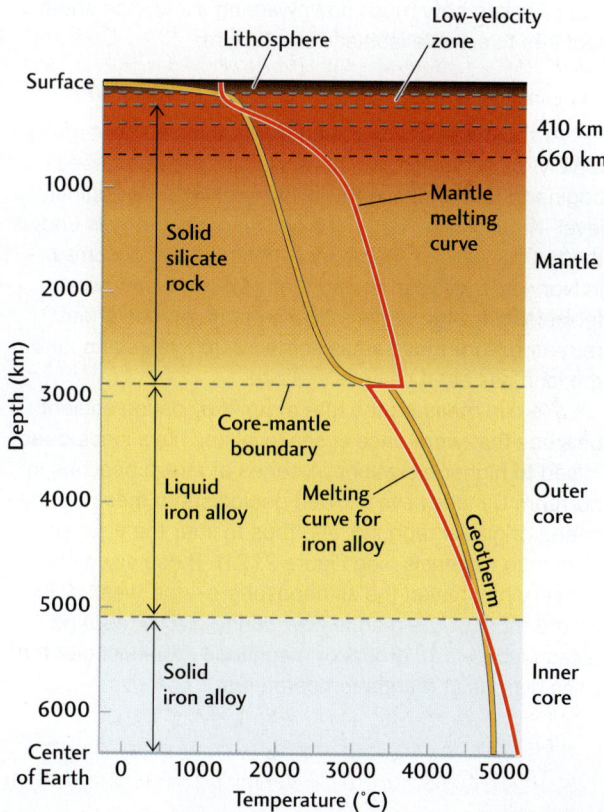

FIGURE 11.10 An estimate of Earth's geotherm, which describes the increase in temperature with depth (yellow line). The geotherm first rises above the melting curve—the temperature at which peridotite begins to melt (red line)—in the upper mantle, forming the partially molten low-velocity zone. It does so again in the outer core, where the iron-nickel alloy is in a liquid state. The geotherm falls below the melting curve throughout most of the mantle and in the solid inner core.

Earth Issues 11.1 Glacial Rebound: Nature's Experiment with Isostasy

If you depress a cork floating in water with your finger and then release it, the cork pops up almost instantly. A cork floating in molasses would rise more slowly because the drag of the viscous fluid would slow the process. How convenient it would be if we could push Earth's crust down somewhere, remove the depressive force, and then sit back and watch the depressed area rise. From its response, we could observe how isostatic equilibrium is achieved—in particular, how the viscosity of the mantle determines the rate of uplift and subsidence.

Nature has actually done this experiment for us. The depressive force is the weight of a continental glacier—an ice sheet 2 to 3 km thick. During the onset of an ice age, ice sheets can form in only a few thousand years. The immense ice load depresses the crust and a downward bulge develops under the ice sheet, displacing enough mantle to provide buoyant support. Using the information in the Practicing Geology Exercise and the densities of ice (0.92 g/cm^3) and mantle material (3.3 g/cm^3), we can calculate how much downwarping a 3-km ice sheet requires to achieve isostatic equilibrium:

$$(0.92 \text{ g/cm}^3 \div 3.3 \text{ g/cm}^3) \times 3.0 \text{ km} = 0.84 \text{ km}$$

At the onset of a warming trend, the ice sheet melts rapidly. With the removal of its weight, the depressed crust begins to rebound, eventually rising back to its original level—in this case, 840 m higher than when it was under the full glacial load. Such *glacial rebound* has occurred in Norway, Sweden, Finland, Canada, and elsewhere in formerly glaciated regions. The most recent ice sheet retreated from those areas some 12,000 years ago, and the land has been rising ever since.

We can measure the rate of uplift by dating ancient beaches that were once at sea level and have since been raised to higher elevations. A series of raised beaches in northern Canada have allowed geologists to measure the speed of glacial rebound, and thus to infer the viscosities of mantle materials (see Figure 21.21). Those viscosities are very high. Even the asthenosphere—the weak layer where most of the mantle flow during glacial rebound takes place—is 10 orders of magnitude more viscous than silica glass is at mantle temperatures.

TIME 1
At the start of an ice age, a continental ice sheet forms and continues to thicken over a few thousand years.

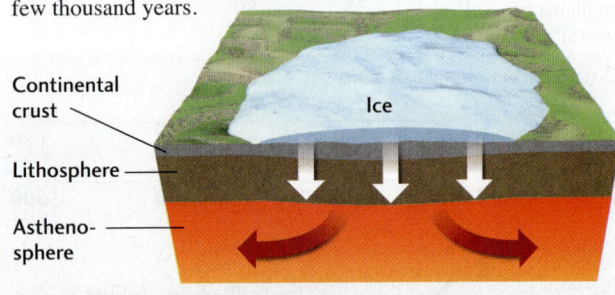

TIME 2
The continental crust bends downward under the ice load to the extent needed to provide buoyant support.

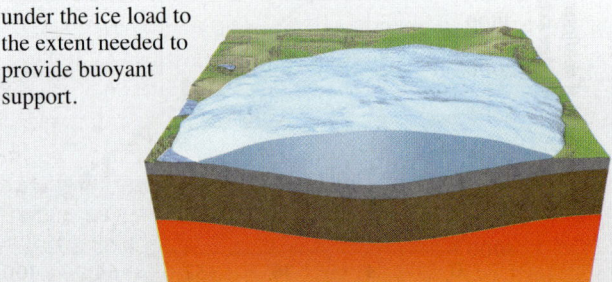

TIME 3
At the end of the ice age, rapid warming melts the glacier. The depressed crust begins to rebound.

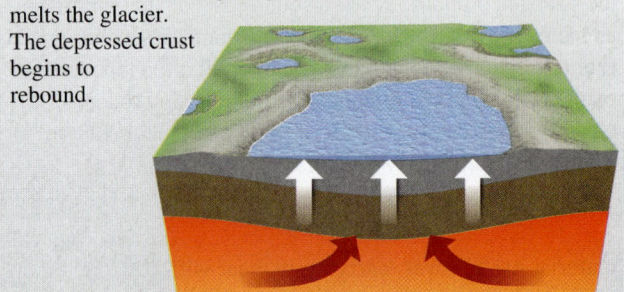

TIME 4
Rebound continues long after the glacier has melted, slowly returning the crust to its pre-ice age elevation.

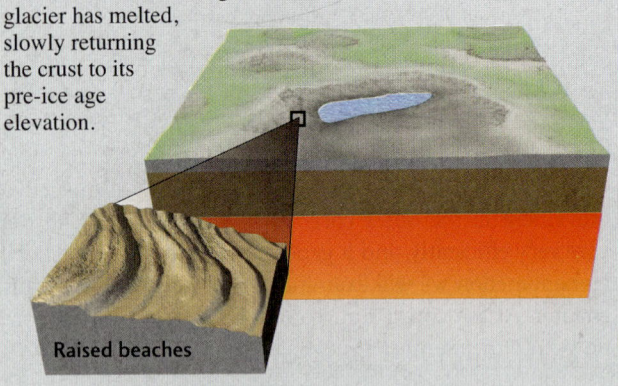

The principle of isostasy explains glacial rebound. Rates of uplift measured from raised beaches allow geologists to infer mantle viscosities, including that of the asthenosphere, where most of the mantle flow takes place during uplift.

The melting curves show how the onset of melting depends on pressure, which increases with depth.

Geologists at Earth's surface can directly measure temperatures at depths of up to 4 km in mines and more than 10 km in boreholes. They have found that the geothermal gradient is 20°C to 30°C per kilometer in most continental crust. Conditions below the crust can be inferred from the properties of lavas and rocks erupted by volcanoes. These data indicate that temperatures near the base of the lithosphere range from 1300°C to 1400°C. As Figure 11.10 shows, it is at these temperatures that the geotherm rises above the melting point of mantle rock. The geotherm intersects the melting curve at about 100 km beneath most oceanic crust, and somewhat deeper (150 to 200 km) beneath most continental crust. From this depth to where the geotherm drops below the melting curve, at depths of 200 to 250 km, mantle material is partially molten. These observations are consistent with the existence of a shear-wave low-velocity zone (see Figure 11.8), as well as with widespread evidence suggesting that basaltic magmas are produced by partial melting in the upper part of the asthenosphere.

The steep geothermal gradient near Earth's surface tells us that heat is transported through the lithosphere by conduction. Below the lithosphere, the temperature does not rise as rapidly. If it did, temperatures in the deeper parts of the mantle would be so high (tens of thousands of degrees) that the lower mantle would be entirely molten, which is inconsistent with seismological observations. Instead, the change in temperature with depth drops to about 0.5°C per kilometer, which is the geothermal gradient expected in a convecting mantle. This drop occurs because convection mixes cooler material near the top of the mantle with warmer material at greater depths, averaging out the temperature differences (just as temperatures are evened out when you stir your bathwater).

Phase changes—observed as steep increases in seismic wave velocities—occur in the transition zone at depths of 410 km and 660 km (see Figure 11.8). Seismology can accurately determine the depths (and thus the pressures) of these phase changes, so the temperatures at which the phase changes take place can be calibrated using high-pressure laboratory experiments. The values obtained from the laboratory data are consistent with the geotherm shown in Figure 11.10.

We have more limited information about the temperatures at greater depths. Most geologists agree that convection extends throughout the mantle, vertically mixing material and keeping the geothermal gradient low. Near the base of the mantle, however, we would expect temperatures to increase more rapidly, because the core-mantle boundary restricts vertical mixing. Convective movements near the core-mantle boundary, like those near the crust, are primarily horizontal rather than vertical. Heat is transported from the core into the mantle through this boundary layer mainly by conduction. The rate at which heat flows is proportional to the geothermal gradient. To transport the heat fast enough, this gradient must be high, as it is in the lithosphere.

Seismology tells us that the outer core is liquid, which means that its temperature is high enough to melt the iron alloy that constitutes it. Laboratory data indicate that this temperature is probably greater than 3000°C, and that estimate is consistent with the high geothermal gradient at the base of the mantle predicted by convection models. The inner core, on the other hand, is solid. Because its iron-nickel composition is nearly the same as that of the outer core, the boundary between the inner core and outer core should correspond to the depth where the geotherm crosses the melting curve for the core material. This hypothesis implies that the temperature at Earth's center is slightly less than 5000°C.

Many aspects of this story can be debated, however, especially in regard to the deeper parts of the geotherm. For example, some geologists believe that the temperature at Earth's center may be as high as 6000°C to 7000°C. More laboratory experiments and better calculations are required to reconcile these differences.

Visualizing Earth's Three-Dimensional Structure

So far, we have investigated how the properties of Earth's materials vary with depth. Such a one-dimensional description would suffice if our planet were a perfectly symmetrical sphere, but of course it is not. At the surface, we can see *lateral variations* (geographic differences) in Earth's structure associated with oceans and continents and with the basic features of plate tectonics: mid-ocean ridges at spreading centers, deep-sea trenches at subduction zones, and mountain belts uplifted by continent-continent collisions.

Below the crust, we can expect that convection will cause variations in temperature from one part of the mantle to another. Downwelling currents, such as those associated with subducted lithospheric plates, will be relatively cold, whereas upwelling currents, such as those associated with mantle plumes, will be relatively hot. Computer models tell us that lateral variations in temperature due to mantle convection should be on the order of several hundred degrees. From laboratory experiments on rocks, we know that such temperature differences should cause small variations in seismic wave velocities from place to place. For example, a temperature increase of 100°C reduces the speed of an S wave traveling through mantle peridotite by about 1 percent (or even more if the rock is close to its melting temperature). If the mantle is indeed convecting, then seismic wave velocities should vary by several percentage points from place to place. Seismologists can make three-dimensional maps of these small lateral variations in wave velocities using the techniques of seismic tomography.

Seismic Tomography

Seismic tomography is an adaptation of a medical technique commonly used to map the human body, called computerized axial tomography (CAT). CAT scanners construct three-dimensional images of organs by measuring small differences in X rays that sweep the body in many directions. Similarly, we can use the travel times of seismic waves from earthquakes, as recorded on thousands of seismographs all over the world, to sweep Earth's interior in many different directions and construct a three-dimensional image of what's inside. A reasonable hypothesis, consistent with the results of laboratory experiments, is that regions where seismic waves speed up are composed of relatively cool, dense rock (for example, subducted lithosphere), whereas regions where seismic waves slow down are composed of relatively hot, buoyant rock (for example, rising plumes).

Seismic tomography has revealed features in the mantle that are clearly associated with mantle convection. In the 1990s, researchers at Harvard University constructed a tomographic model of the mantle. Their model is displayed in Figure 11.11 as a cross section of Earth and as a series of global maps at depths ranging from just below the crust down to the core-mantle boundary. Near Earth's surface (Figure 11.11b), you can clearly see the structures of plate tectonics. The upwelling of hot mantle material along the mid-ocean ridges is visible in warm colors; the cold lithosphere in old ocean basins and beneath the continental cratons is visible in cool colors.

At greater depths, the features become more variable and less coherent with surface tectonic features, reflecting what is probably a complex pattern of mantle convection. Some large-scale features stand out particularly well. You will notice that just above the core-mantle boundary (Figure 11.11e), there is a red region of relatively low S-wave velocities beneath the central Pacific Ocean, surrounded by a broad blue ring of higher S-wave velocities. Seismologists have speculated that the high velocities represent a "graveyard" of cold oceanic lithosphere subducted beneath the Pacific's volcanic island arcs and mountain belts—the Ring of Fire—during the last 100 million years or so.

The cross section through the mantle (Figure 11.11a) clearly reveals material from the once-large Farallon Plate, which has been almost completely subducted under North America (see Chapter 10). The obliquely sinking slab material (in blue) appears to have penetrated the entire mantle. The image also indicates sinking colder rock beneath Indonesia, another subduction zone. In addition, a large yellow blob of hotter rock, thought to be a "superplume," can be seen rising at an angle from the core-mantle boundary to a position beneath southern Africa. This hot, buoyant mass pushing up the cooler material above it may explain the uplifted, mile-high plateaus of South Africa. The other visible blobs of hotter and cooler material may be evidence of material exchanges among the lithosphere, the mantle, and the layer of hotter material at the core-mantle boundary.

Earth's Gravitational Field

The same temperature variations that speed up and slow down seismic waves also influence the densities of mantle rocks. Laboratory experiments have shown that the expansion of rock caused by a 300°C increase in temperature reduces its density by about 1 percent. This might seem to be a small effect, but the mass of Earth's mantle is enormous (about 4 billion trillion tons!), so even small changes in the distribution of its mass can lead to observable variations in the pull of Earth's gravity.

Geologists can determine features of Earth's mass distribution by observing variations in the gravitational field above its surface, as well as from bulges and dimples in the shape of the planet. They have been able to show that the shape measured by Earth-orbiting satellites matches the pattern of mantle convection imaged by seismic tomography (see Earth Issues 11.2). This agreement has allowed us to refine our models of the mantle convection system.

Earth's Magnetic Field and the Geodynamo

Like the mantle, Earth's outer core transports most of its heat by convection. But the same techniques that have revealed so much about mantle convection—seismic tomography and the study of Earth's gravitational field—have provided almost no information about convection in the core. Why not?

The problem has to do with the fluidity of the outer core. The mantle is a viscous solid that flows very slowly. As a result, convection creates regions where temperatures are significantly higher or lower than the average mantle geotherm. We can see these regions in Figure 11.11 as places where seismic wave velocities are lower or higher than the average at that depth. The outer core, in contrast, has a very low viscosity: Its molten material can flow as easily as water or liquid mercury. Even small density variations caused by convection are quickly smoothed out by the rapid flow under the force of gravity. Any lateral variations in seismic wave velocity caused by convection will be much too small for us to see using seismic tomography, and they will not cause measurable distortions in the shape of the planet.

We can, however, investigate convection in the outer core through observations of Earth's magnetic field. In Chapter 1, we briefly described the magnetic field and its generation by the geodynamo in the outer core. In Chapter 2, we discussed magnetic reversals and the use of magnetic anomalies in volcanic rocks to measure seafloor spreading. In this section, we will further explore the nature of Earth's magnetic field and its origin in the geodynamo.

(a) Tomographic cross section

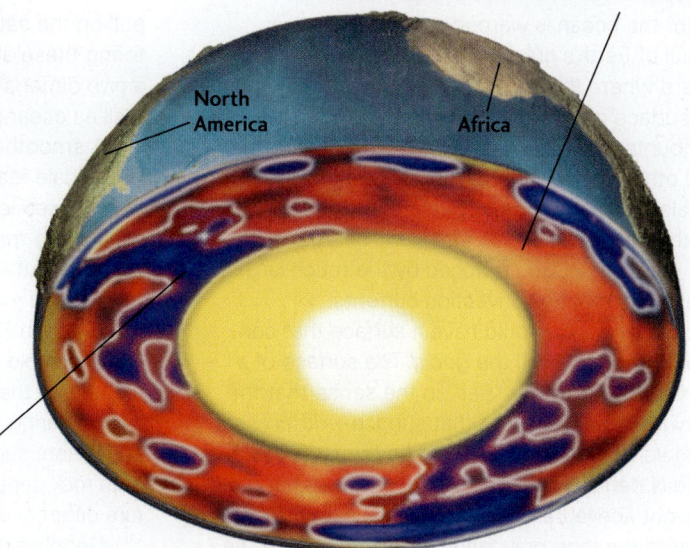

A tomographic cross section through Earth reveals hot regions, such as a superplume rising from Earth's core beneath South Africa,...

...and cooler regions, such as the descending remnants of the Farallon Plate under the North American Plate.

(b–e) Global maps at four different depths

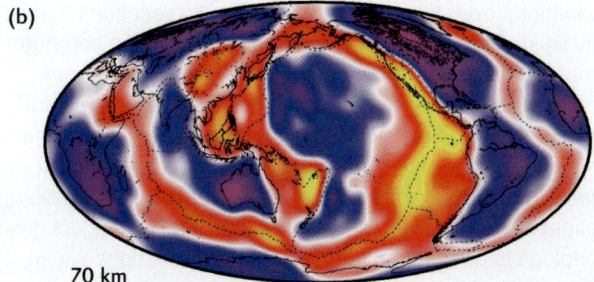

70 km

Near Earth's surface, hot rocks in the asthenosphere lie beneath oceanic spreading centers.

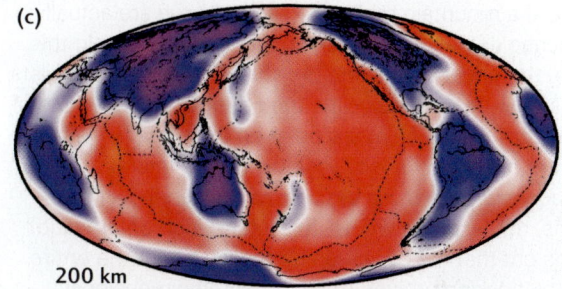

200 km

Moving deeper, we see the cold lithosphere of stable continental cratons and the warmer asthenosphere beneath ocean basins.

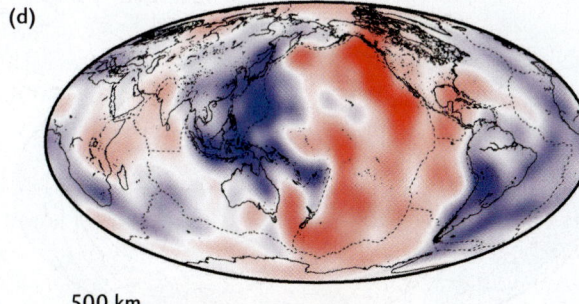

500 km

Deeper in the mantle, the features no longer match the continental positions.

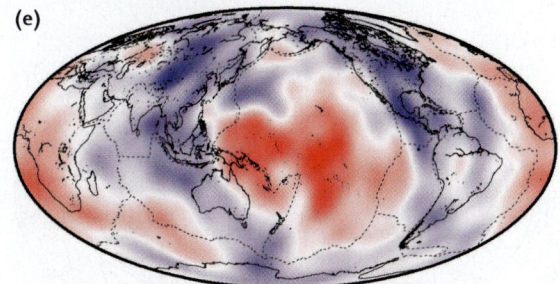

2800 km (near core-mantle boundary)

Near the core-mantle boundary, the colder regions around the Pacific may be the "graveyards" of sinking lithospheric slabs.

FIGURE 11.11 A three-dimensional image of Earth's mantle created by seismic tomography. Regions with faster S-wave velocities (blue and purple) indicate colder, denser rock; regions with slower S-wave velocities (red and yellow) indicate hotter, less dense rock. (a) Cross section of Earth. (b–e) Global maps at four different depths. The S-wave velocity variations are measured relative to the average value at a given depth; that is, velocities near this average are shown in white. [cross section (a) From M. Gurnis, *Scientific American* (March 2001): 40; maps (b–e) by L. Chen and T. Jordan, University of Southern California.]

Earth Issues 11.2 The Geoid: The Shape of Planet Earth

The surface of the ocean is warped upward in places where the pull of Earth's gravitational field is stronger and downward where the pull is weaker. The shape of the ocean's surface can be precisely measured by radar altimeters mounted on satellites. By averaging out wave motions and other fluctuations, geologists can map the small-scale variations in gravity caused by geologic features on the seafloor, such as faults and seamounts. Variations in gravity are also produced by the much larger features caused by mantle convection currents.

A perfectly still ocean would have a surface that conforms to what geologists call the *geoid*. The surface of a still body of water is perfectly "flat" in the sense that the pull of gravity is perpendicular to that surface—otherwise, the water would flow "downhill" to make the surface flatter. The geoid is defined as an imaginary surface at some reference height above Earth adjusted to be everywhere perpendicular to the local gravitational force. Because the ocean surface approximates the geoid, we usually take the reference height to be sea level. When we measure the height of a mountain relative to sea level, we are actually measuring its height above the geoid at that point. In this sense, the geoid represents the shape of Earth. Geologists can use the geoid to calculate the strength and direction of the gravitational force at any point on the planet's surface and infer how rock density varies in Earth's interior.

Radar altimeters can easily map the geoid over the oceans, but how can we get this information on dry land? It turns out that the geoid can be measured for the entire Earth by tracking orbiting satellites. Three-dimensional mass variations in the mantle exert a small gravitational pull on the satellites, shifting their orbits slightly. By monitoring these shifts over long periods, scientists can create a two-dimensional map of the geoid over continents as well as oceans.

A smoothed version of the observed geoid reveals the large-scale features of Earth's gravitational field. Relative to what sea level would be on an Earth without any lateral variation in mass, the elevation of the geoid varies from a low of about −110 m at a point near the coast of Antarctica to a high of just over 100 m on the island of New Guinea in the western Pacific.

The geoid shows some similarities to the large-scale features of the deeper parts of the mantle, as you can see by comparing the geoid map with Figures 11.11d and e. This agreement suggests that the three-dimensional variations in both rock density and S-wave velocity are related to temperature differences arising from mantle convection.

Geophysicists Brad Hager and Mark Richards tested this hypothesis in the mid-1980s. Using laboratory data for calibration, they first calculated three-dimensional density variations from the seismic wave velocity variations mapped by seismic tomography. They then constructed a computer model of convective flow by assuming that the heavier parts of the mantle are sinking while the lighter parts are rising. Finally, they calculated what the geoid shape should be according to this convection model. You can see that their model results match the observed geoid quite well, especially for the largest features. This agreement has given geologists confidence that temperature variations within the mantle convection system can explain what we see both in seismic images and in the gravitational field.

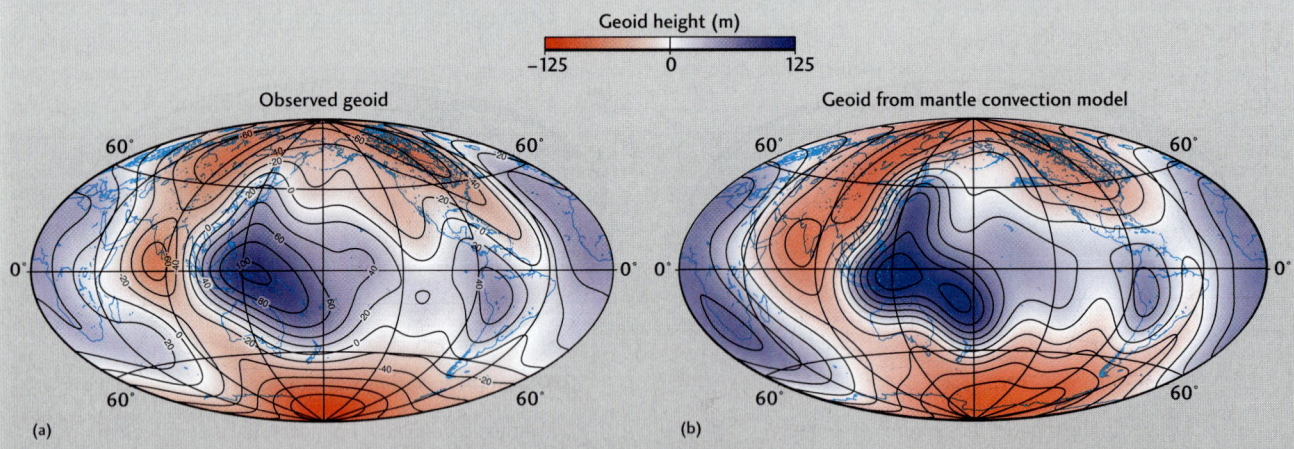

A smoothed map of the geoid, or "shape of Earth," derived from satellite observations. The contours, given here in meters, show how the observed sea level deviates from that on an ideal Earth without any lateral variation in rock density. (b) A map of the geoid computed from a model of mantle convection that is consistent with the temperature structure of the mantle derived from seismic tomography. By matching the observed geoid with such theoretical models, geologists have improved their understanding of the mantle convection system. [a: NASA; map by L. Chen and T. Jordan; b: model by B. Hager, Massachusetts Institute of Technology; map by L. Chen and T. Jordan.]

The Dipole Field

The most basic instrument used for sensing Earth's magnetic field is the magnetic compass, invented by the Chinese more than 22 centuries ago. For hundreds of years, explorers and ship captains used magnetic compasses to navigate, but they had little understanding of how these ancient devices actually worked. In 1600, William Gilbert, physician to Queen Elizabeth I, provided a scientific explanation. He proposed that "the whole Earth is itself a great magnet" whose field acts on the small magnet of a compass needle to align it in the direction of the north magnetic pole.

Scientists of Gilbert's day had begun to visualize a magnetic field as lines of force, such as those revealed by the alignment of iron filings on a piece of paper above a bar magnet. Gilbert showed that Earth's magnetic lines of force point into the ground at the north magnetic pole and outward at the south magnetic pole, as if a powerful bar magnet were located at Earth's center and oriented along an axis inclined about 11° from Earth's axis of rotation (see Figure 1.16). In other words, the lines of force revealed a **dipole** (two-pole) magnetic field.

Complexity of the Magnetic Field

Gilbert solved an important problem for a seafaring nation dependent on the compass for navigation, but his explanation was only partly correct. We now know that the source of the magnetic field is a geodynamo powered by core convection, not a permanent magnet at Earth's center (which would be quickly destroyed by the high temperatures in the core).

The geodynamo is formed by rapid convective movements in the liquid, iron-rich, electrically conducting outer core. The magnetic field produced by the geodynamo is considerably more complex than a simple dipole field, and it is constantly changing over time due to these fluid movements.

Within a few decades after Gilbert's famous pronouncement, careful observers had realized that the magnetic field varies over time. Not surprisingly, some of the best evidence for these changes came from the compass measurements systematically recorded by the British navy. Navigators had to correct their compass bearings to account for the displacement of the north magnetic pole (magnetic north) from the north rotational pole (true north), and these corrections showed that the north magnetic pole was moving at rates of 5° to 10° per century (**Figure 11.12**). Little did the British sailors know that these changes were caused by convective movements deep in Earth's core!

The Nondipole Field Measurements at Earth's surface have revealed that only about 90 percent of Earth's magnetic field can be described by the simple dipole field illustrated in Figure 1.16. The remaining 10 percent, which geologists refer to as the *nondipole field,* has a more complex structure. This structure can be seen by comparing the magnetic field strengths calculated for a simple dipole field (**Figure 11.13a**) with those of the observed field (Figure 11.13b). If we extrapolate the observed lines of force down to the core-mantle boundary using a computer model, the size of the nondipole field actually increases

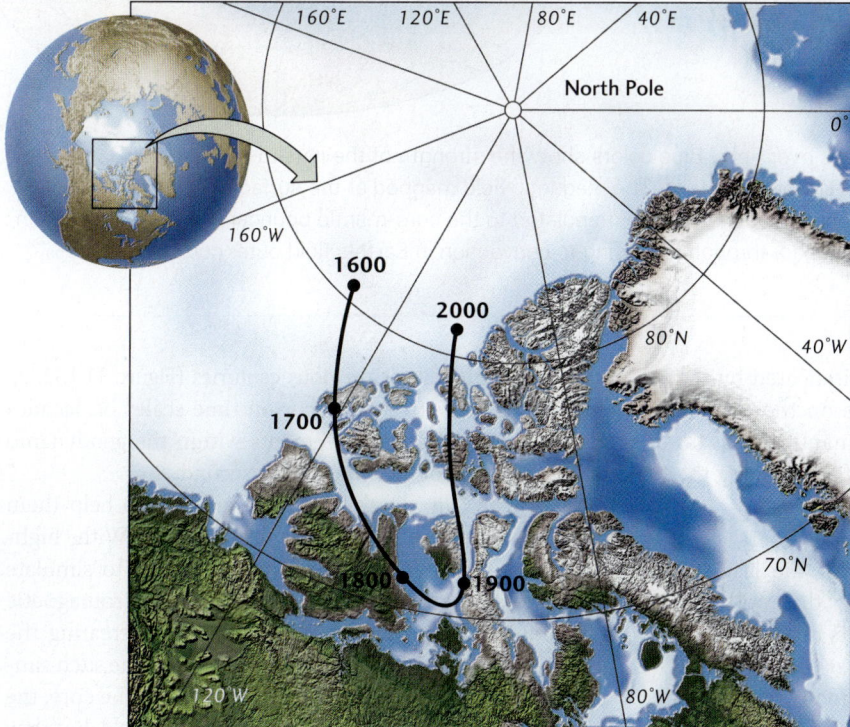

FIGURE 11.12 Path of the north magnetic pole, as mapped by compass readings and other measurements of Earth's magnetic field since 1600. Changes in the pole location are caused by convective movements within Earth's fluid outer core.

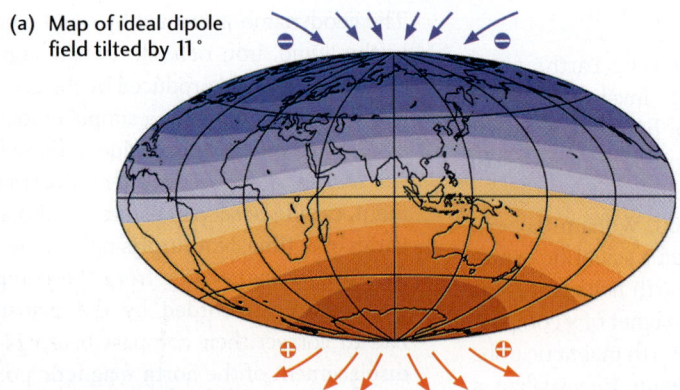
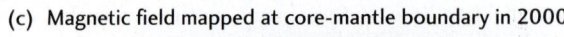
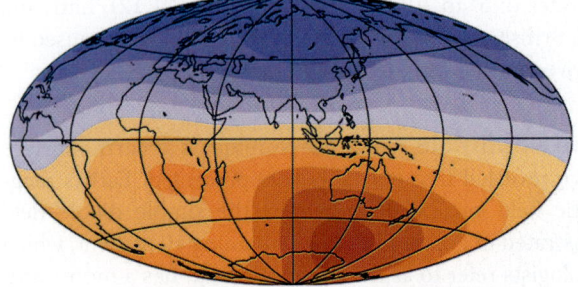
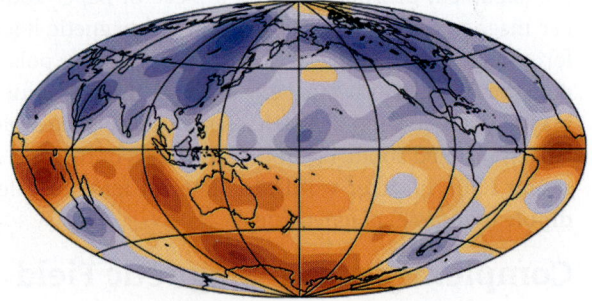
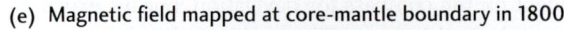
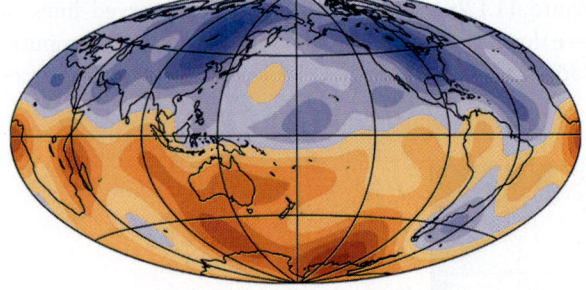
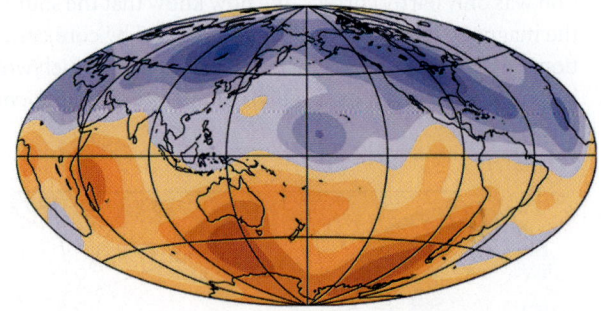

FIGURE 11.13 Earth's magnetic field changes over time. Blue colors show the strength of the inward-pointing field, and orange colors show the strength of the outward-pointing field. The magnetic field mapped at the surface (b) is more complex than a simple dipole (a), and it shows more complications when extrapolated to the core-mantle boundary (c). The features on the nondipole field change over time, as seen in (c) through (e), owing to convection in Earth's fluid outer core. [Data for maps from J. Bloxham, Harvard University.]

relative to the size of the dipole field, as indicated by the bumpiness of the orange and blue colors on the map in Figure 11.13c. The poorly conducting mantle tends to smooth out complexities in the magnetic field, making the dipole field seem bigger than it really is.

Secular Variation Magnetic records for the last 300 years (many from the British navy) show that both the dipole and nondipole components of Earth's magnetic field are changing over time, but that this *secular* (time-related) *variation* is fastest for the nondipole component. Secular variation is evident when we compare a map of today's magnetic field at the core-mantle boundary (Figure 11.13c) with maps reconstructed for previous centuries (Figure 11.13d, e). Changes in field strength occur on time scales of decades and indicate that fluid movements within the geodynamo are on the order of millimeters per second.

Scientists can use this secular variation to help them understand convection in the outer core. With high-performance computers, they have been able to simulate the complex convective movements and electromagnetic interactions in the outer core that might be creating the geodynamo. The magnetic lines of force from one such simulation are shown in **Figure 11.14**. Away from the core, the lines of force can be approximated by a dipole field, but they become more complicated near the core-mantle boundary.

Within the core itself, they are hopelessly entangled by the strong convective movements.

Magnetic Reversals These same computer simulations also allow us to understand a remarkable behavior of the geodynamo: spontaneous reversals of the magnetic field. As discussed in Chapter 2, the magnetic field reverses its direction at irregular intervals (ranging from tens of thousands to millions of years), exchanging the north and south magnetic poles as if the magnet depicted in Figure 1.16 were flipped 180°. Computer simulations of the geodynamo have been able to reproduce these sporadic reversals in the absence of any external triggers (Figure 11.14). In other words, it is possible for Earth's magnetic field to reverse

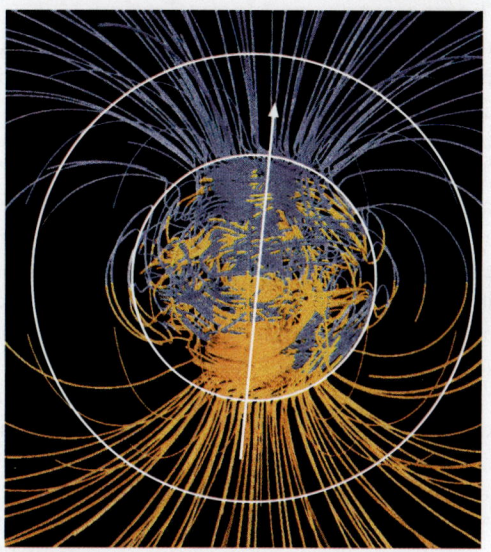

Time 1

Magnetic lines of force with normal orientation prior to reversal. The magnetic lines of force in the mantle approximate those of a dipole field.

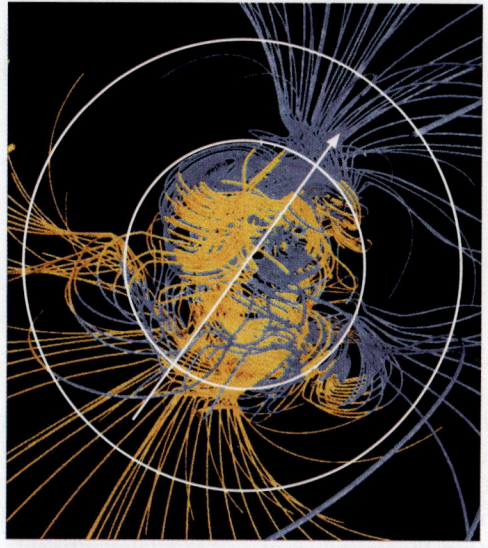

Time 2

Beginning of magnetic reversal. Geodynamo spontaneously begins to reorganize, increasing the complexity of the lines of force within the outer core and decreasing the strength of the dipole component of the magnetic field.

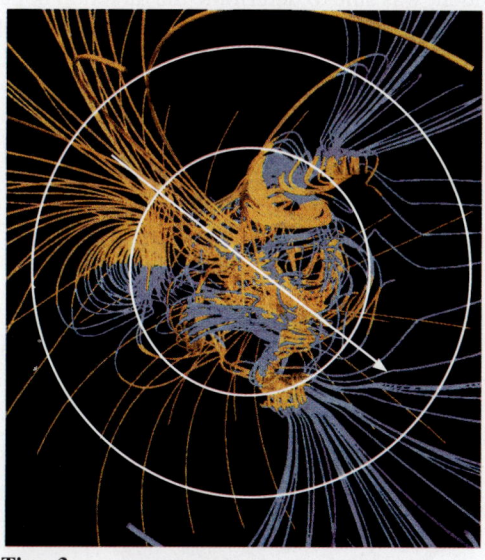

Time 3

Reversal continues with rapid changes in the structure of the magnetic field, which continues to have a weak dipole component.

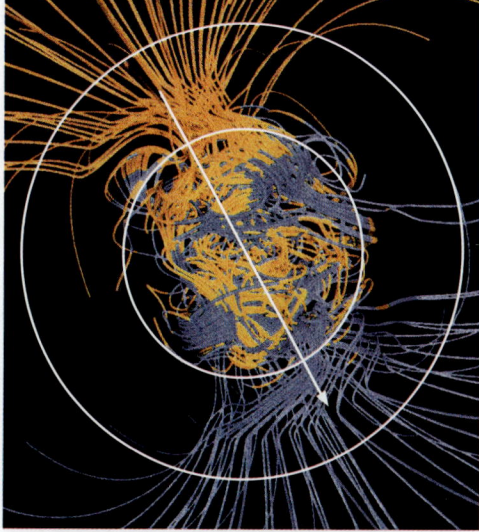

Time 4

Reversal nearly complete. Dipole field restrengthens with its north pole now pointing south.

FIGURE 11.14 Computer models have shown that spontaneous changes in the geodynamo could cause magnetic reversals. Large and small circles show the positions of Earth's surface and core-mantle boundary, respectively. [Courtesy G. Glatzmaier, University of California, Santa Cruz.]

itself spontaneously, purely through internal interactions within Earth's core.

This behavior illustrates a fundamental difference between the geodynamo and the dynamos used in power plants. A steam-powered dynamo is an artificial system engineered by humans to do a particular job. The geodynamo, in contrast, exemplifies a *self-organized natural system*—one whose behavior is not predetermined by external constraints, but emerges from interactions within the system. Two other global geosystems, the plate tectonic and climate systems, also display a wide variety of self-organized behaviors. Understanding how these natural systems organize themselves is one of the greatest challenges to geoscience. We will return to this subject when we discuss the climate system in Chapter 12.

Paleomagnetism

We have seen repeatedly how the geologic record of ancient magnetism, or **paleomagnetism,** has provided crucial information for understanding Earth's history. Magnetic anomalies mapped on oceanic crust confirmed the existence of seafloor spreading, and they still provide the best data for tracking plate movements since the breakup of Pangaea 200 million years ago (see Chapter 2). Paleomagnetic data from old continental rocks have been essential for establishing the existence of earlier supercontinents, such as Rodinia (see Chapter 21).

Scientists have also used paleomagnetic data to reconstruct the history of Earth's magnetic field. The oldest magnetized rocks found so far, which formed about 3.5 billion years ago, indicate that Earth had a magnetic field similar to the present one at that time. The presence of magnetization in the most ancient rocks is consistent with the ideas about Earth's differentiation discussed in Chapter 1, which imply that a convecting liquid core must have been established very early in Earth's 4.5-billion-year history.

Let's delve a little more deeply into the rock-forming processes that have allowed geologists to draw these remarkable conclusions. You may find it helpful to consult the material in Figure 2.12 and its accompanying text as you read this section.

Thermoremanent Magnetization In the early 1960s, an Australian graduate student found a fireplace in an ancient campsite where Aborigines had cooked their meals. He carefully removed several stones that had been baked by the fires, first noting their physical orientation. Then he measured the direction of the stones' magnetization and found that it was exactly the reverse of Earth's present magnetic field. He proposed to his disbelieving professor that, as recently as 30,000 years ago, when the campsite was occupied, the direction of the magnetic field was the reverse of the present one—that is, a compass needle would have pointed south rather than north.

Recall that high temperatures destroy magnetization. An important property of many magnetizable materials is that, as they cool below about 500°C, they become magnetized in the direction of the surrounding magnetic field. This happens because groups of atoms of the material align themselves in the direction of the magnetic field when the material is hot. When the material has cooled, these atoms are locked in place. This process is called **thermoremanent magnetization,** because the magnetization caused by heating and cooling is "remembered" by the rock long after the magnetizing field has disappeared. Thus, the Australian student was able to determine the direction of Earth's magnetic field at the time the stones cooled after the last campfire (**Figure 11.15**).

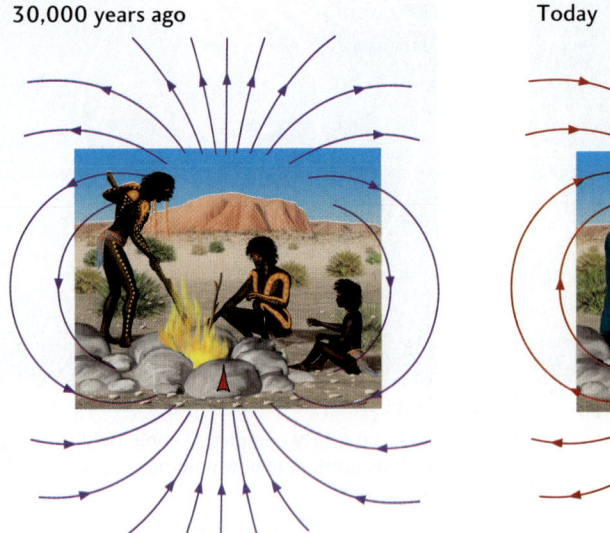

FIGURE 11.15 The orientation of Earth's magnetic field 30,000 years ago was the reverse of today's, as evidenced by magnetized rocks found in the fireplace of an ancient campsite. The rocks, when cooling after the last fire, became magnetized in the direction of the ancient magnetic field, leaving a permanent record of its orientation.

Thermoremanent magnetization is the same process that magnetizes lava flows and newly formed oceanic crust, as described in Chapter 2. The discovery of magnetic reversals in these igneous rock types was a key ingredient in formulating the theory of plate tectonics.

Depositional Remanent Magnetization Some sedimentary rocks can take on a different type of remanent magnetization. Marine sedimentary rocks form when particles of sediment that have settled to the seafloor become lithified. Magnetic grains among the particles—chips of the mineral magnetite (Fe_3O_4), for example—become aligned in the direction of Earth's magnetic field as they fall through the water, and this orientation may be incorporated into the rock when the sediments become lithified. The **depositional remanent magnetization** found in some sedimentary rocks results from the parallel alignment of all these tiny magnets, as if they were compasses pointing in the direction of the magnetic field prevailing at the time of deposition (Figure 11.16).

Paleomagnetic Stratigraphy Geologists have used paleomagnetism in combination with isotopic dating methods to work out the time sequence of magnetic reversals over the last 170 million years (Figure 11.17).

FIGURE 11.16 Newly formed sediments can become magnetized in the direction of the magnetic field at the time of their deposition.

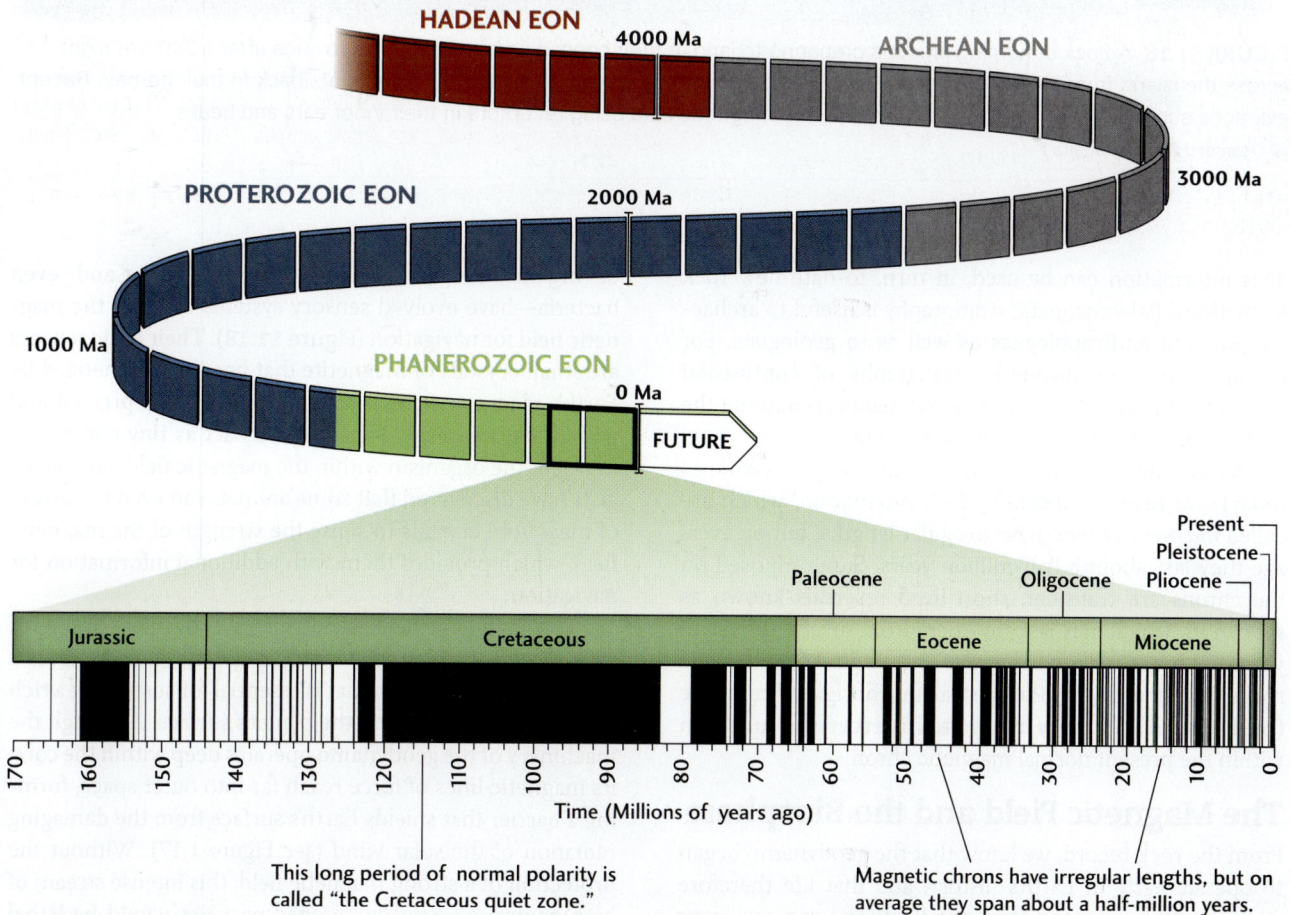

FIGURE 11.17 The paleomagnetic time scale from 170 million years ago to the present, showing normal chrons (black bands) and reversed chrons (white bands).

FIGURE 11.18 A flock of homing pigeons preparing to land at their coop in Cuba's Villaclara province after a 240-km flight across the island from Havana. These birds use Earth's magnetic field to navigate the long flights back to their homes. Recent evidence suggests that homing pigeons sense the magnetic field using receptors in their inner ears and beaks. [© Desmond Boylan/Reuters.]

This information can be used, in turn, to date new rock formations. Paleomagnetic stratigraphy is useful to archaeologists and anthropologists as well as to geologists. For example, the paleomagnetic stratigraphy of continental sediments has been used to date sediments containing the remains of predecessors of our own species.

As we saw in Chapter 2, periods of "normal" (same as today) and reversed magnetic field orientation, which are called *magnetic chrons*, have irregular lengths, but on average they last about a half-million years. Superimposed on the chrons are transient, short-lived reversals known as *subchrons*, which may last anywhere from several thousand years to tens of millions of years. The reversal found in the rocks remagnetized by the Australian Aborigines' campfire (see Figure 11.15) can be interpreted as a reversed subchron within the present normal magnetic chron.

The Magnetic Field and the Biosphere

From the rock record, we know that the geodynamo began to operate early in Earth's history, and that life therefore evolved within a strong magnetic field. The consequences turn out to be rather surprising. For example, many types of organisms—pigeons, sea turtles, whales, and even bacteria—have evolved sensory systems that use the magnetic field for navigation (Figure 11.18). Their basic sensors are small crystals of magnetite that become magnetized by Earth's magnetic field as they are biologically precipitated within the organism. These crystals act as tiny compasses to orient the organism within the magnetic field. Geobiologists have discovered that some animals can even use arrays of magnetite crystals to sense the strength of the magnetic field, which provides them with additional information for navigation.

The magnetic field is not just a convenient frame of reference for flying and swimming species. It constitutes a part of the Earth system that is essential for sustaining a rich and delicate biosphere at the planet's surface. Although the machinery of the geodynamo operates deep within the core, its magnetic lines of force reach far into outer space, forming a barrier that shields Earth's surface from the damaging radiation of the solar wind (see Figure 1.17). Without the protection of a strong magnetic field, this intense stream of high-energy, electrically charged particles would be lethal to many organisms.

Moreover, if the geodynamo were to stop producing a magnetic field, bombardment by the solar wind would gradually strip away Earth's atmosphere, further degrading the terrestrial environment. This appears to have happened in the case of Mars. Paleomagnetism in the ancient Martian crust has been detected by orbiting spacecraft, so we know the Red Planet once had an active geodynamo that generated a strong magnetic field. Sometime early in the planet's history, however, its geodynamo ceased to operate, perhaps because the Martian core cooled enough to freeze. Exposure to the solar wind subsequently eroded its atmosphere to the tenuous state we observe today.

KEY TERMS AND CONCEPTS

- compressional wave (p. 308)
- conduction (p. 316)
- convection (p. 316)
- core-mantle boundary (p. 314)
- depositional remanent magnetization (p. 327)
- dipole (p. 323)
- geotherm (p. 317)
- low-velocity zone (p. 313)
- lower mantle (p. 314)
- Mohorovičić discontinuity (p. 312)
- paleomagnetism (p. 326)
- phase change (p. 313)
- principle of isostasy (p. 312)
- seismic ray path (p. 308)
- seismic tomography (p. 320)
- shadow zone (p. 310)
- shear wave (p. 308)
- thermoremanent magnetization (p. 326)
- transition zone (p. 313)
- upper mantle (p. 312)

REVIEW OF LEARNING OBJECTIVES

11.1 Trace the paths of the major seismic waves through Earth's interior and know what they reveal about the layering of Earth's crust, mantle, and core.

Seismic waves are refracted and reflected by Earth's layered structure. By observing how the travel times of various wave types change with distance, seismologists can determine how the P-wave and S-wave velocities vary with depth. From correlations of these velocities with rock types, they can infer the composition of Earth's interior. The continental crust is made mostly of low-density granitic rock and that the deep seafloor is composed of basalt and gabbro. The crust and outer part of the mantle make up the rigid lithosphere, which moves across the surface in plate tectonics. Beneath the lithosphere lies the asthenosphere, the weak ductile layer of the mantle on which the lithospheric plates slide. At the top of the asthenosphere, the temperature is high enough to partially melt peridotite, forming an S-wave low-velocity zone. Below 250 km, S-wave velocities again increase with depth. At 410 km and 660 km below the surface, the seismic wave velocities show jumps caused by phase changes in mantle minerals. Below 660 km lies the lower mantle, a relatively homogeneous layer about 2000 km thick, in which seismic wave velocities increase gradually.

Seismic waves reflected from the core-mantle boundary show that it is a sharp interface located at a depth of 2890 km. At the core-mantle boundary, the density abruptly increases and the seismic velocities abruptly decrease. The failure of S waves to penetrate below the core-mantle boundary indicates that the outer core is liquid. At a depth of 5150 km, a jump in P-wave velocity marks the boundary between the liquid outer core and the solid inner core. Several lines of evidence suggest that the core is composed mostly of iron and nickel, with minor amounts of some lighter element, such as oxygen or sulfur.

Study Assignment: Identify the main seismic wave paths in Figures 11.1, 11.2, and 11.3. Review the section *Layering and Composition of Earth's Interior*.

Exercises: (a) The velocities of compressional waves in the lower part of the oceanic crust average about 7 km/s. What rock type is most consistent with this observation as well as with your knowledge of the oceanic crust? (b) Where in the mantle would you look to find regions of anomalously high S-wave velocities?

Thought Question: How would you use seismic waves to find a chamber of molten magma in the crust?

11.2 Apply the principle of isostasy to explain why continents ride higher than oceans and why the oceans deepen as the geologic age of the seafloor increases.

The principle of isostasy states that, when measured to the same depth near the base of the lithosphere (say, at 100 km), the total weight of two lithospheric columns will be the same. Therefore, the mean densities of high-standing regions such as continents must be lower than the mean densities of low-standing regions such as oceans, which is consistent with the low-density, felsic composition of continental crust. The cooling of oceanic lithosphere by conduction increases the mean lithospheric density by an amount proportional to the square root of lithospheric age. According to the principle of isostasy, the depth of the seafloor must therefore increase as the square root of lithospheric age, which is consistent with observed seafloor depths in all of the major ocean basins.

Study Assignment: Practicing Geology Exercise and Figure 11.9

Exercises: (a) The average elevation of the Tibetan Plateau is about 5 km above sea level. Use the isostatic equation for continents to compute the average thickness of the crust in this region, assuming that its average density is 2.8 g/cm^3 (Practicing Geology bonus problem). (b) If oceanic crust had the same density as mantle rocks (3.3 gm/cm^3), what would be the average elevation of the continents above sea level?

Thought Question: How does Earth maintain isostatic equilibrium?

11.3 Recognize where heat in Earth's interior is transported upward primarily by conduction and where primarily by convection.

Earth's interior is cooled by the upward flow of heat. This heat is efficiently transported by convection in the solid mantle and liquid outer core. Throughout most of the interior, the stirring motions of convection maintain a low geothermal gradient (less than 1°C/km). The exceptions are boundary layers where the mass motions are primarily horizontal and heat is transported upward primarily by conduction. One of these layers is the lithosphere, which forms horizontally moving plates at Earth's surface. Another is in the lowermost mantle just above the core-mantle boundary, where vertical motions are inhibited by the large density increase from the silicate mantle to the iron-rich core. Because conduction is less efficient than convection, the geothermal gradients in these boundary layers have to be much larger (more than 20°C/km) in order to transport the same amount of heat.

Study Assignment: Review the section *Heat Flow Through Earth's Interior.*

Exercises: (a) Is the temperature at the Moho beneath a continental craton hotter or cooler than the temperature at the Moho beneath an ocean basin? (b) Seafloor that is 4 million years old is observed to have subsided by about 700 m below the depth where it was formed at a spreading center. How much more will this seafloor have subsided when it reaches 16 million years in age?

Thought Question: The Moon shows no evidence of plate-tectonic processes, nor has it been volcanically active for billions of years. What does this observation imply about the state and temperature of the interior of that planetary body?

11.4 Describe how temperature increases from Earth's surface to its center, and identify the depths where the geotherm crosses the melting curve.

Earth's interior is hot because it still retains much of the heat accumulated during its violent formation as well as heat currently being generated by the decay of radioactive isotopes. The geotherm is a curve showing how temperature increases with depth. Within most continental crust, the geotherm increases at a rate of 20°C to 30°C per kilometer. Temperatures rise even more rapidly with depth in the oceanic lithosphere, where they reach 1300°C to 1400°C near its base (~100 km). Here the conditions are hot enough for mantle peridotites to begin to melt, forming a partially molten low-velocity zone. Mantle convection maintains a low geothermal gradient below the lithosphere, whereas the melting curve increases with pressure, keeping it above the geotherm in deeper regions of the mantle. The core-mantle boundary is a sharp transition from silicate rock with a high melting temperature to the iron-nickel alloy with a low melting temperature. Here the melting curve drops below the geotherm, which explains why the outer core is in a liquid state. However, the geothermal gradient in the convecting outer core is lower than the gradient of the melting curve; therefore, the two curves cross again at the inner core–outer core boundary, which explains why the inner core is solid. The temperature at Earth's center is about 5000°C.

Study Assignment: Figure 11.10

Exercises: (a) What evidence suggests that the asthenosphere is in some places partially molten? (b) Given the mantle melting curve in Figure 11.10, would you expect to find an S-wave low velocity zone beneath stable continental interiors, where the geothermal gradients are observed to be less than 30°C/km?

Thought Question: As Earth cools off, does the inner core grow or shrink in size?

11.5 Visualize Earth's three-dimensional structure from the images provided by seismic tomography.

Seismologists can use seismic tomography to create three-dimensional images of Earth's interior. Regions where seismic wave velocities increase indicate relatively cool, dense rock; regions where they decrease indicate relatively hot, less dense rock. Tomographic images reveal the structures of plate tectonics close to Earth's surface, from the upwelling of hot mantle material under mid-ocean ridges to the cold lithosphere that extends deep beneath continental cratons. They also reveal many features of mantle convection, such as the sinking of lithospheric slabs into the lower mantle and the rising of plumes from deep within the mantle.

Study Assignment: Visual Literacy Exercise

Exercises: (a) How deep do subducted slabs go before they are recycled by mantle convection? (b) From Figure 11.11b, identify the geographic region where the S-wave velocities of the uppermost mantle are the lowest, and provide a plate-tectonic explanation of these low velocities.

Thought Question: Seismic tomography of Earth's outer core shows that the lateral variations in P-wave velocity are undetectably small. Why is the outer core so homogeneous?

Review of Learning Objectives

11.6 Summarize what the Earth's magnetic field tells us about the liquid outer core.

Convective movements in the outer core stir its electrically conducting iron-rich liquid, forming a geodynamo that produces the magnetic field. At Earth's surface, the magnetic field from the geodynamo is primarily a dipole field, but it has a small nondipole component. Maps of the magnetic field derived from compass readings show that the pattern of magnetic field strengths at Earth's surface has changed over the last several centuries. These observations tell us that convective movements driving the geodynamo are much more rapid than those of mantle convection, consistent with predictions that the liquid outer core flows as easily as water or liquid mercury.

Study Assignment: Review the section *Earth's Magnetic Field and the Geodynamo*.

Exercises: (a) What evidence supports the hypothesis that Earth's magnetic field is generated by a geodynamo in its outer core? (b) What evidence do we have that the magnetic field changes by an observable amount over the span of a human lifetime?

Thought Question: How do the existence of Earth's magnetic field, iron meteorites, and the abundance of iron in the cosmos support the ideas that Earth's core is mostly iron and that the outer core is liquid?

11.7 Explain how magnetic field reversals are used to date rock sequences.

Geologists have discovered that minerals in some rock types align themselves in the direction of Earth's magnetic field at the time the rocks form. This remanent magnetization can be preserved in rocks for millions of years. Paleomagnetic stratigraphy tells us that Earth's magnetic field has reversed (flipped back and forth) over geologic time. The chronology of reversals over the last several hundred million years has been worked out. Changes in the directions of remanent magnetization measured in rock sequences can thus be compared with this paleomagnetic time scale to determine the ages of rocks in the sequence.

Study Assignment: Paleomagnetic time scale in Figure 11.17

Exercises: (a) How do igneous rocks become magnetized when they form? (b) How does the magnetization of sedimentary rocks differ from this process?

Thought Question: How is the paleomagnetic time scale used to determine the age of oceanic lithosphere?

VISUAL LITERACY EXERCISE

(a) Tomographic cross section

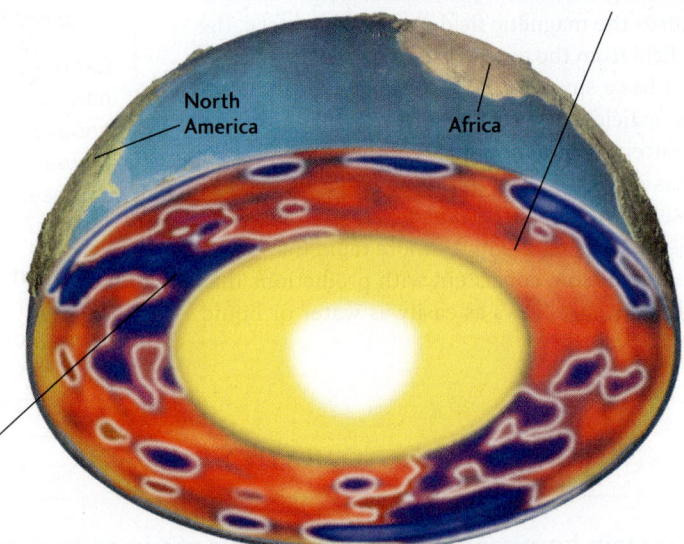

A tomographic cross section through Earth reveals hot regions, such as a superplume rising from Earth's core beneath South Africa,...

...and cooler regions, such as the descending remnants of the Farallon Plate under the North American Plate.

(b–e) Global maps at four different depths

(b)
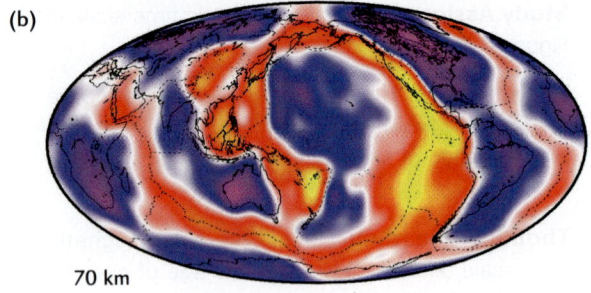
70 km

Near Earth's surface, hot rocks in the asthenosphere lie beneath oceanic spreading centers.

(c)
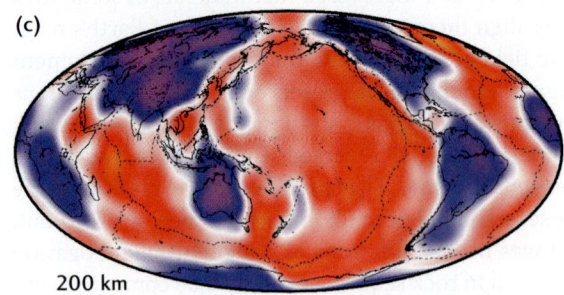
200 km

Moving deeper, we see the cold lithosphere of stable continental cratons and the warmer asthenosphere beneath ocean basins.

(d)
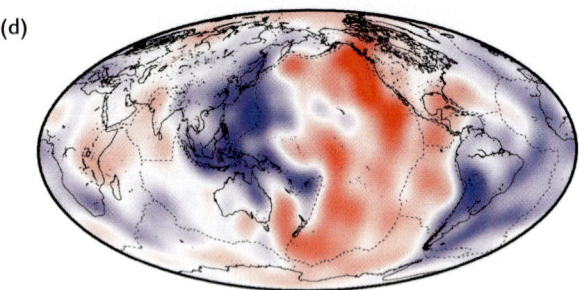
500 km

Deeper in the mantle, the features no longer match the continental positions.

(e)
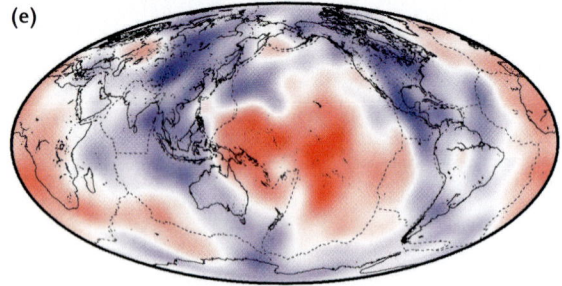
2800 km (near core-mantle boundary)

Near the core-mantle boundary, the colder regions around the Pacific may be the "graveyards" of sinking lithospheric slabs.

FIGURE 11.11 A three-dimensional image of Earth's mantle created by seismic tomography. Regions with faster S-wave velocities (blue and purple) indicate colder, denser rock; regions with slower S-wave velocities (red and yellow) indicate hotter, less dense rock. (a) Cross section of Earth. (b–e) Global maps at four different depths. The S-wave velocity variations are measured relative to the average value at a given depth; that is, velocities near this average are shown in white. [Cross section (a) data from M. Gurnis, *Scientific American* (March 2001): 40; maps (b–e) by L. Chen and T. Jordan, University of Southern California.]

1. How are the S-wave variations color-coded?
   a. Which color shows regions where the velocity is near the average for a given depth?
   b. Which color shows the lowest relative velocities?
   c. Which color shows the highest relative velocities?
2. Panel (a) is a tomographic cross section.
   a. What is the circular yellow region in this cross section?
   b. What is the central white region?
   c. Parts of the mantle that are likely to be sinking downward in mantle convection are colored blue and purple. Why would this material be sinking?
3. Panels (b)–(e) are maps at various depths.
   a. Which map shows the lateral S-wave velocity variations in the lithosphere?
   b. What are the dominant high-velocity features in map (c), and what does this map imply about their temperatures at 200 km depth?
   c. What plate-tectonic process might explain the bluish regions that surround the Pacific Ocean in map (e)?

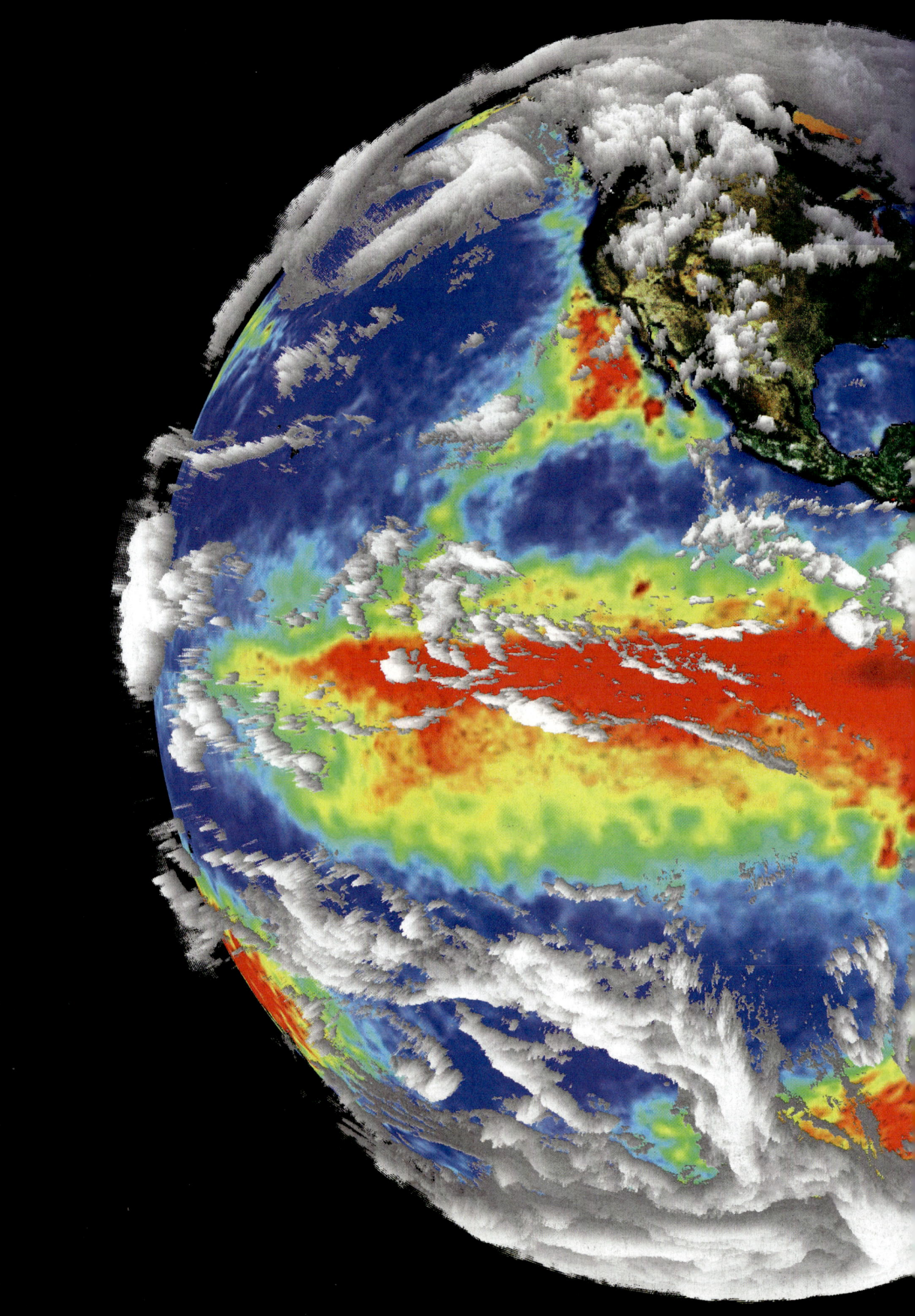

The Climate System

12

What Is Climate?	338
Components of the Climate System	339
The Greenhouse Effect	345
Climate Variation	348
The Carbon Cycle	355

Learning Objectives

12.1 Recognize the major climate zones of today's world from annual variations in temperature and precipitation.

12.2 Consider each major component of the climate system and describe how the transport of energy and mass within and among these components affects climate zonation.

12.3 Explain the greenhouse effect in terms of radiation balance and feedbacks within the climate system.

12.4 Describe how the climate changes over the short term during the El Niño–Southern Oscillation and over the long term during glaciations.

12.5 Identify the signatures of Milankovitch cycles in the Pleistocene glacial record.

12.6 Quantify the carbon cycle in terms of the flux of carbon between the atmosphere and the other principal carbon reservoirs.

A snapshot of the climate system taken by sensors on several spacecraft, showing cloud cover (in white), variations in sea surface temperature (from the warmest in red to the coolest in dark blue), and land surface properties, including density of vegetation (from the lowest density in brown to the highest in green). [R.B. Husar, Washington University/NASA Visible Earth.]

In the last chapter, we descended into Earth's deep interior to explore the internal heat engine that drives the plate-tectonic system and the geodynamo. In this chapter, we return to Earth's surface to examine a global geosystem powered not by Earth's internal heat, but by external heat from the Sun: the climate system.

What we learn about the climate system in this chapter will equip us to study the wide range of geologic processes that shape the face of our planet—weathering, erosion, sediment transport, and the interaction of the plate tectonic and climate systems. In particular, it will set the stage for the next two chapters, in which we examine more closely how our civilization, as a global carbon engine, is driving changes in climate, ocean chemistry, and the biosphere, and what we might do to mitigate those changes.

No aspect of Earth science is more important to our continued well-being than the study of the climate system. Throughout geologic time, evolutionary radiations and extinctions of organisms have been closely connected to changes in climate. Even the short history of our own species is deeply imprinted by climate change: Agricultural societies began to flourish only about 10,000 years ago, when the harsh climate of the most recent ice age was rapidly transformed into the mild and steady climate of the Holocene epoch. Now, a globalized human society energized by fossil fuels is injecting greenhouse gases into the atmosphere at an ever-increasing rate, with potentially dire consequences: global warming, ocean acidification, and unfavorable changes to the biosphere. The climate system is a huge, incredibly complex machine, and, like it or not, our hands are on the controls. We're in the driver's seat, with the pedal to the metal, so we had better understand how the darn thing works!

In this chapter, we will examine the main components of the climate system and the ways in which those components interact to produce the climate zones that we live in today. We will investigate the geologic record of climate change and discuss the key role of the carbon cycle in regulating climate.

What Is Climate?

At any point on Earth's surface, the amount of energy received from the Sun changes on daily, yearly, and longer-term cycles owing to Earth's movement through the solar system. This cyclical variation in the total input of solar energy, known as **solar forcing,** causes changes in the surface environment: Temperatures rise during the day and fall at night, and they rise in summer and fall in winter. The term *climate* refers to the average conditions at a point on Earth's surface and their variation during these cycles of solar forcing.

Climate is described by daily and seasonal statistics on the atmospheric temperature near Earth's surface (the *surface temperature*) as well as atmospheric pressure, surface humidity, cloud cover, rate of rainfall, wind speed, and other weather conditions. Table 12.1 gives an example of seasonal temperature statistics for New York City, which include measures of temperature variation (record highs and lows) as well as average values. In addition to these common weather statistics, a full scientific description of climate includes the non-atmospheric components of the surface environment, such as soil moisture on land and the surface temperature of the oceans.

Weather records at many points on the land surface allow us to identify **climate zones** that have similar seasonal variations in temperature and precipitation (Figure 12.1). *Tropical zones* are defined to be where the average temperature of the coldest month is greater than 18°C, and *polar zones* where the average temperature of the warmest month is less than 10°C. Between these extremes are *temperate zones* and *boreal zones*, differentiated by whether the coldest monthly average is greater than or less than freezing (0°C). *Arid zones* are dry, characterized by low precipitation thresholds. Within each of these major climate zones, there are subzones defined by seasonal variations in temperature and precipitation, often correlated with dominant types of vegetation. Some tropical regions are wet (rain forests), whereas others are dry (savannas). Some polar regions are expanses of treeless frozen ground (tundra), while others are covered with permanent snow and ice. Climate at intermediate latitudes is highly variable

TABLE 12.1	Seasonal Temperatures (°F) in Central Park, New York City			
DATA TYPE*	JANUARY 1	APRIL 1	JULY 1	OCTOBER 1
Record high	62	83	100	88
Average high	39	56	83	69
Average low	28	40	68	55
Record low	24	12	52	36

*Temperatures are averages for the date shown for the period 1981–2010; record temperatures are those for the period 1869–2017.

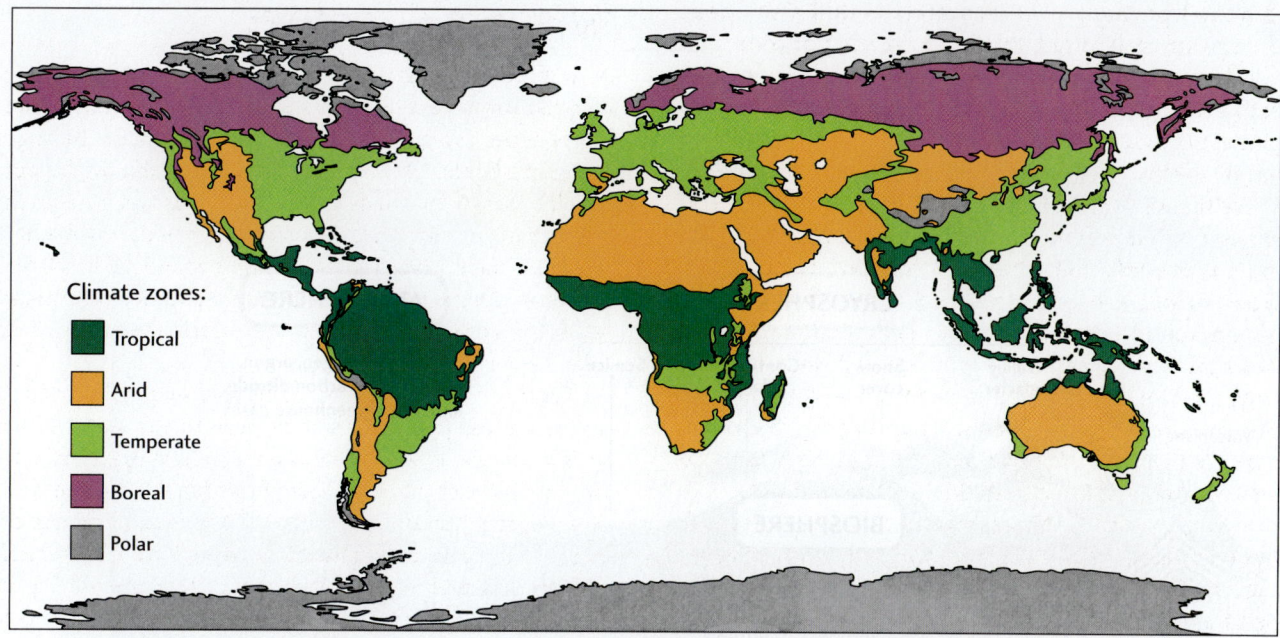

FIGURE 12.1 Earth's climate zones: tropical, arid, temperate, boreal, and polar. [Data from M. C. Peel, B. L. Finlayson, T. A. McMahon.]

with subzones defined by the length and timing of dry and rainy periods. A Mediterranean climate, for example, is a temperate climate distinguished by hot, dry summers and cool, wet winters.

Components of the Climate System

Climate zones are the product the *climate system*. This global geosystem is defined to include all the components of the Earth system and all the interactions among those components that determine how climate varies in space and time (Figure 12.2). The main components of the climate system are the atmosphere, hydrosphere, cryosphere, lithosphere, and biosphere. Each component plays a different role in the climate system, and that role depends on its ability to store and transport mass and energy.

The Atmosphere

Earth's atmosphere is the most mobile and rapidly changing part of the climate system. Like Earth's interior, the atmosphere is layered (Figure 12.3). About three-fourths of its mass is concentrated in the layer closest to Earth's surface, the **troposphere,** which is variable in its thickness, extending up to 20 km near the equator but only up to 7 km near the poles. Above the troposphere is the **stratosphere,** a dryer layer that extends to an altitude of about 50 km. The outer atmosphere, above the stratosphere, has no abrupt cutoff; it slowly becomes thinner and fades away into outer space.

The dry atmosphere is a mixture of gases, mainly nitrogen (78 percent by volume) and oxygen (21 percent by volume). The remaining 1 percent consists of argon (0.92 percent), carbon dioxide (0.04 percent), and other minor gases, including methane and ozone. Water vapor is concentrated in the troposphere in highly variable amounts, up to 3 percent, but typically about 1 percent. Water vapor and carbon dioxide, though minor constituents of the atmosphere, play a large role in the climate system because they are the principal *greenhouse gases*.

Ozone (O_3^+) is a highly reactive greenhouse gas produced primarily by the ionization of molecular oxygen by ultraviolet radiation from the Sun. In the lower part of the atmosphere, ozone exists in only tiny amounts, although it is a strong enough greenhouse gas to play a role in regulating Earth's surface temperature. Most atmospheric ozone is found in the stratosphere, where its concentration reaches a maximum at an altitude of 25 to 30 km (see Figure 12.3). This stratospheric ozone layer filters out incoming ultraviolet radiation, protecting the biosphere at Earth's surface from its potentially damaging effects.

The troposphere convects vigorously due to the uneven heating of Earth's surface by the Sun (*tropos* is the Greek word for "turn" or "mix"). When air is warmed, it expands, becomes less dense than cooler air, and tends to rise; conversely, cool air tends to sink. The resulting convection patterns in the troposphere, combined with Earth's rotation,

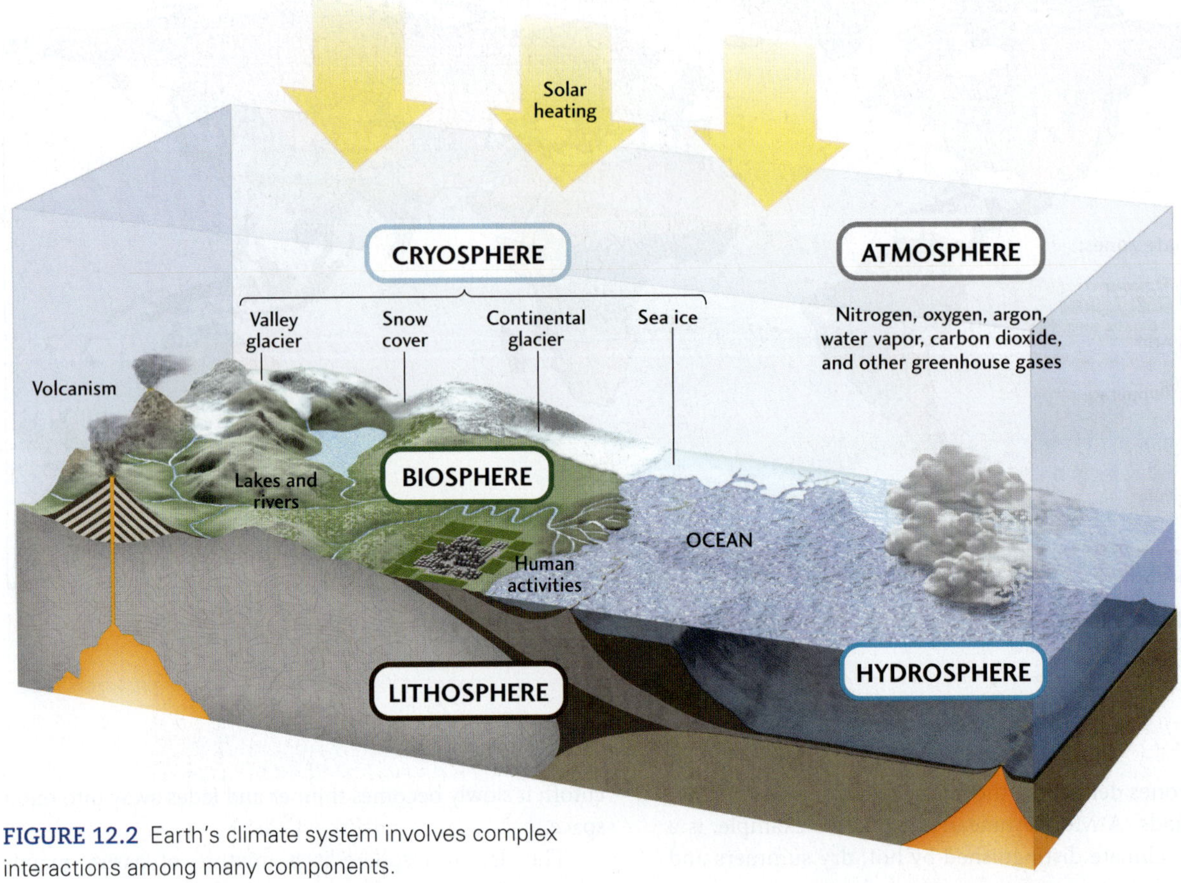

FIGURE 12.2 Earth's climate system involves complex interactions among many components.

create a banded structure of prevailing wind belts, each belt subtending about 30° in latitude (**Figure 12.4**). These wind belts arise because the Sun warms Earth's surface most intensely at the equator, where the Sun's rays are almost perpendicular. The intensity of sunlight is less at high latitudes where the rays strike the surface at an angle, spreading the same amount of solar energy over a larger surface area.

As hot air rises near the equator, it cools and releases its moisture, causing the cloudiness and abundant rain of the tropics. The most intense atmospheric upwelling occurs in the *intertropical convergence zone*, a band of clouds and thunderstorms that shifts northward during Northern Hemisphere summer and southward during the winter, regulating the tropical wet and dry seasons. The upwelling air dries and cools as it moves away from the equator, sinking back to ground level at latitudes near 30°N and 30°S. The belts of increased atmospheric pressures associated with this sinking are known as the *subtropical highs*. Many of the world's great deserts, such as the Sahara in Africa and the Great Victoria Desert of Australia, lie in these regions of dry atmospheric downwelling.

Within temperate latitudes, the air at Earth's surface flows poleward from the subtropical highs, gaining moisture. The warm, moist air converges with the cold, dry air of the polar regions at latitudes near 60°N and 60°S, forming stormy belts of decreased atmospheric pressure called the *subpolar lows*. Atmospheric circulation is thus organized into six major "convection cells" that, acting together, transport

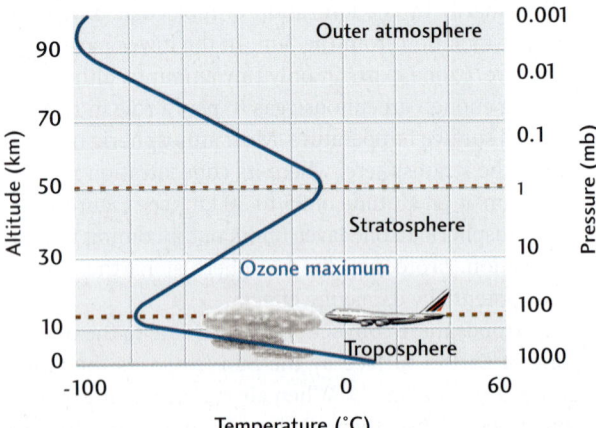

FIGURE 12.3 Layers of the atmosphere, showing variations in temperature (indicated by the blue line) and in pressure, which decreases rapidly with altitude.

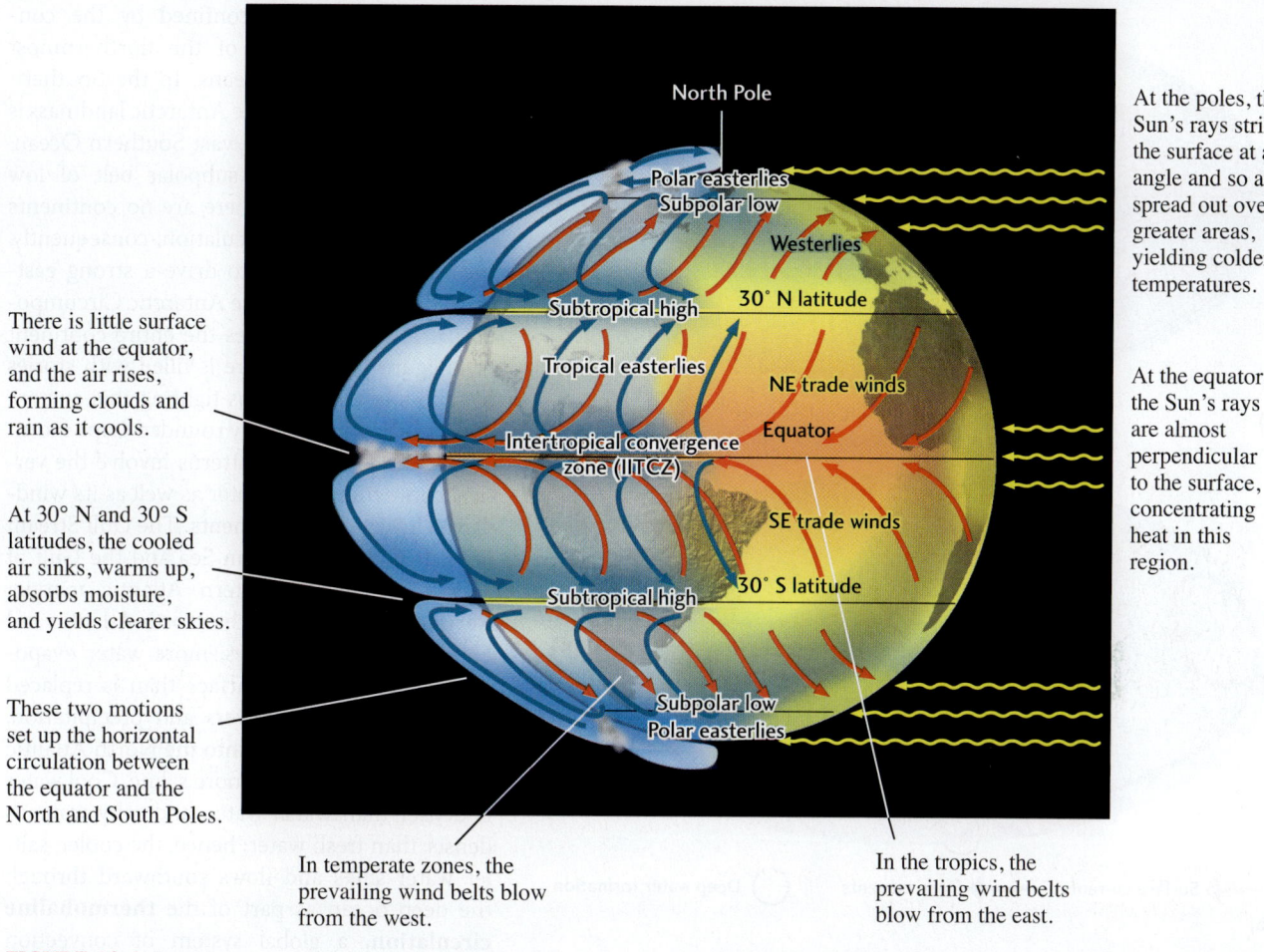

FIGURE 12.4 Earth's atmosphere circulates in prevailing wind belts created by variable solar heating and by Earth's rotation.

heat energy from lower latitudes to higher latitudes (see Figure 12.4). The tropical and temperate cells are bounded by an upwelling of air on one side (the intertropical convergence zone or a subpolar low) and a downwelling on the other (a subtropical high). In a polar cell, the downwelling occurs in a high-pressure region centered on the pole, around which swirls a huge low-pressure windbelt. This *polar vortex* circulates counterclockwise in the Northern Hemisphere and clockwise in the Southern Hemisphere.

These simple north-south circulatory patterns are complicated by Earth's rotation, which acts to bend any current of air (or water) to the right in the Northern Hemisphere and to the left in the Southern Hemisphere. This phenomenon, named the *Coriolis effect* after its discoverer, profoundly alters the pattern of atmospheric circulation. For example, as surface air in the Northern Hemisphere flows southward from the subtropical high towards the intertropical convergence zone, it is deflected to the right (westward), resulting in *trade winds* that blow from the northeast rather than from the north. Similarly, the northward movement of air at the surface in temperate regions is also deflected to the right, causing the prevailing winds to come from the southwest. These *westerlies* transport parcels of air eastward around the globe in about a month, which is why it takes a few days for storms to blow from west to east across the continental United States. The strongest westerly winds are *jet streams*, currents of air flowing at high velocity (100–400 km/hour) in the upper troposphere above the subtropical highs and the subpolar lows. The paths of these jet streams meander north and south in wavelike motions that guide the tracks of storms, causing rapid changes in the weather at midlatitudes.

The Hydrosphere

The *hydrosphere* comprises all the liquid water on, over, and under Earth's surface, including oceans, lakes, streams, and groundwater. Almost all of that liquid water resides in the oceans (1400 million cubic kilometers); lakes, streams, and groundwater constitute a mere 1 percent (15 million cubic kilometers) of the hydrosphere. Though small, these continental components of the hydrosphere play a vital role in the climate system. They are reservoirs for moisture on land, and they transport water, salt, and organic matter to the oceans.

Although seawater circulates more slowly than air, it can store much more heat energy. Owing to this high heat

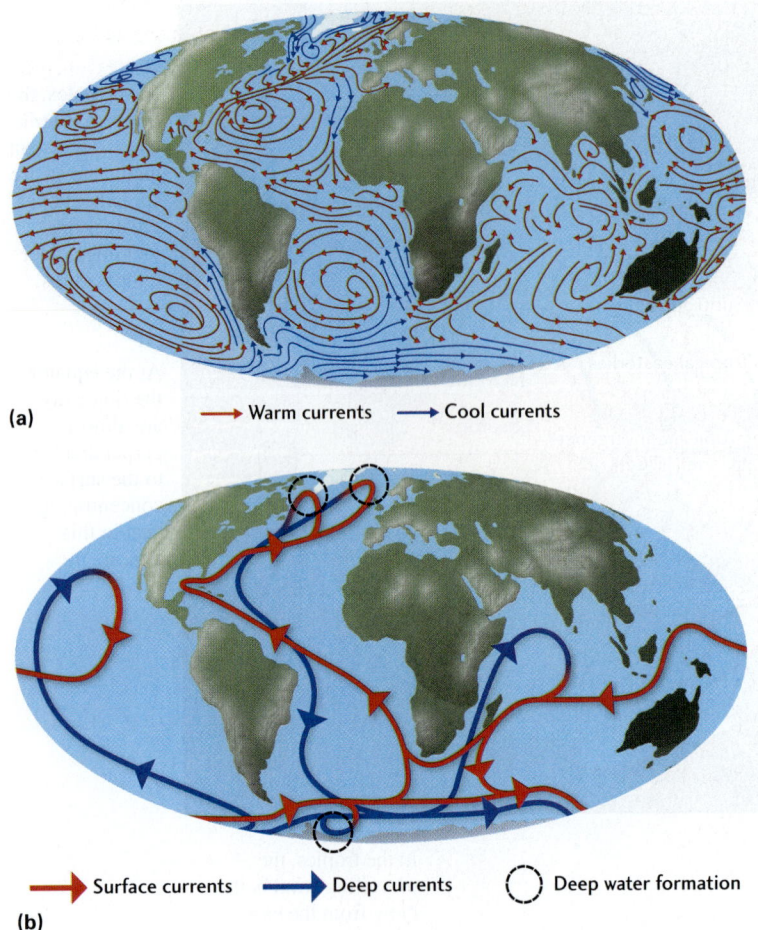

FIGURE 12.5 Two major circulation systems in the oceans. (a) Currents at the surface of the oceans are generated by winds. [Data from sercarleton.edu.] (b) A schematic representation of thermohaline circulation, which acts like a conveyor belt to transport heat from warm equatorial regions to cool polar regions. [Data from NASA.]

capacity, ocean currents are very efficient in transporting heat from the equatorial regions to the polar regions. Prevailing winds blowing across the oceans generate surface currents, which organize into large-scale circulations called *gyres* that span the ocean basins (**Figure 12.5a**). Ocean water is driven westward by the trade winds at low latitudes and eastward by the westerlies at high latitudes, creating the subtropical gyres, which rotate clockwise in the Northern Hemisphere and counterclockwise in the Southern Hemisphere. Owing to the Coriolis effect, the most intense surface currents are the poleward flows of warm water on the western sides of the subtropical gyres. Well-known examples include the Gulf Stream on the western side of the North Atlantic Gyre and the Kuroshio Current of the western side of the North Pacific Gyre.

Surface winds at higher latitudes drive the subpolar gyres. The northern subpolar gyres, which circulate counterclockwise, are confined by the continents to small areas of the northernmost Atlantic and Pacific Oceans. In the Southern Hemisphere, however, the Antarctic landmass is entirely surrounded by a vast Southern Ocean. Near 60°S, within the subpolar belt of low atmospheric pressure, there are no continents to impede the ocean circulation; consequently, the westerlies are able to drive a strong eastward flow of seawater, the Antarctic Circumpolar Current, that encircles the entire continent (see Figure 12.5). Folklore is filled with stories of sailors in clipper ships fighting this current and its high winds as they rounded Cape Horn.

Ocean circulation patterns involve the vertical convection of seawater as well as its wind-driven horizontal movements. The Gulf Stream flows from the Caribbean Sea and the Gulf of Mexico along the western Atlantic margin, warming the climate of the North Atlantic and Europe. At high latitudes, more water evaporates from the ocean surface than is replaced by fresh water from rivers and precipitation, so the seawater flowing into the North Atlantic cools and also becomes more saline. Cool water is denser than warm water, and salty water is denser than fresh water; hence, the cooler, saltier water sinks and flows southward through the deep ocean as part of the **thermohaline circulation,** a global system of convection driven by differences in both temperature and salinity. Thermohaline circulation acts like a set of enormous conveyor belts running through the oceans and moving heat from the equatorial regions toward the poles (Figure 12.5b). Changes in this circulation pattern can strongly influence global climate.

The Cryosphere

The icy component of the climate system is called the *cryosphere*. It comprises 33 million cubic kilometers of ice, primarily in ice caps near the poles; regions of permafrost, where the ground is permanently frozen; and land covered by seasonal snows (**Figure 12.6**). Today, ice caps and continental glaciers cover about 10 percent of the land surface (15 million square kilometers), storing about 75 percent of the world's fresh water. Floating ice includes *sea ice* in the open ocean, as well as frozen lake and river water. The role of the cryosphere in the climate system differs from that of the liquid hydrosphere because ice is relatively immobile, conducts heat poorly, and reflects back into space almost all of the solar energy that falls on it.

The seasonal exchange of water between the cryosphere and the hydrosphere is an important process of the climate system. During winter, sea ice typically covers

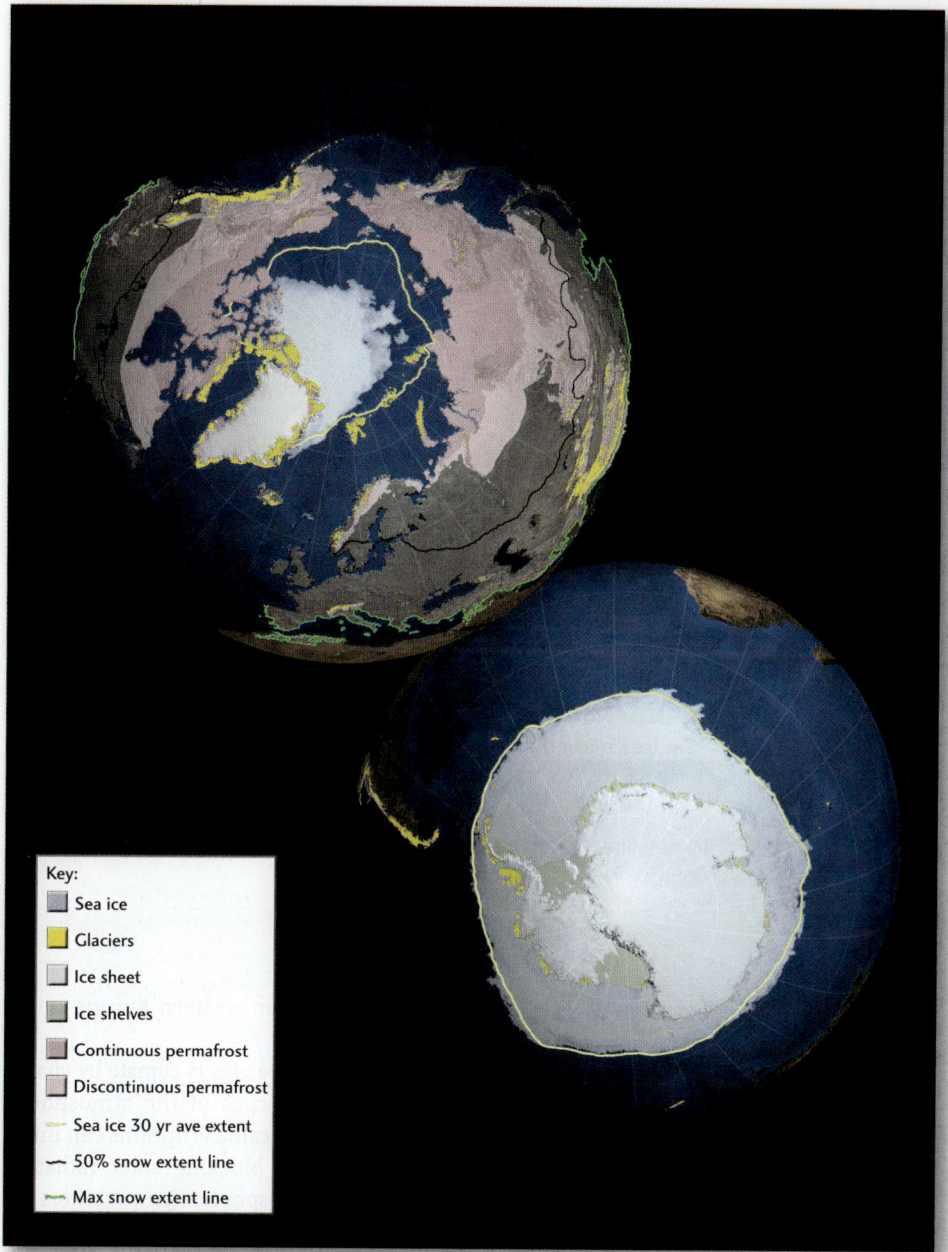

FIGURE 12.6 Elements of the Earth's cryosphere. [NASA/Goddard Space Flight Center Scientific Visualization Studio.]

14 million to 16 million square kilometers of the Arctic Ocean (**Figure 12.7**) and 17 million to 20 million square kilometers of the Southern Ocean, shrinking to about one-third of that area in summer. About one-third of the land surface is covered by seasonal snows, almost entirely (all but 2 percent) in the Northern Hemisphere. Melting snow is the source of much of the fresh water in the hydrosphere. In the U.S. Sierra Nevada and Rocky Mountains, for example, 60 to 70 percent of annual precipitation is snowfall, which is later released as water during spring snowmelt and stream runoff. Much larger amounts of water are exchanged between the cryosphere and the hydrosphere during glacial cycles. At the peak of the most recent ice age, 20,000 years ago, sea level was about 130 m lower than it is today, and the volume of the cryosphere was three times larger.

The Lithosphere

The part of the lithosphere that is most important to the climate system is the land surface, which accounts for about 30 percent of Earth's total surface area. The composition of the surface affects the way land absorbs and releases solar energy. As the temperature of the land surface rises, more heat energy is radiated back into the atmosphere, and more water evaporates into the atmosphere. Because evaporation requires considerable energy, it causes the land surface to cool. Consequently, soil moisture and other factors that

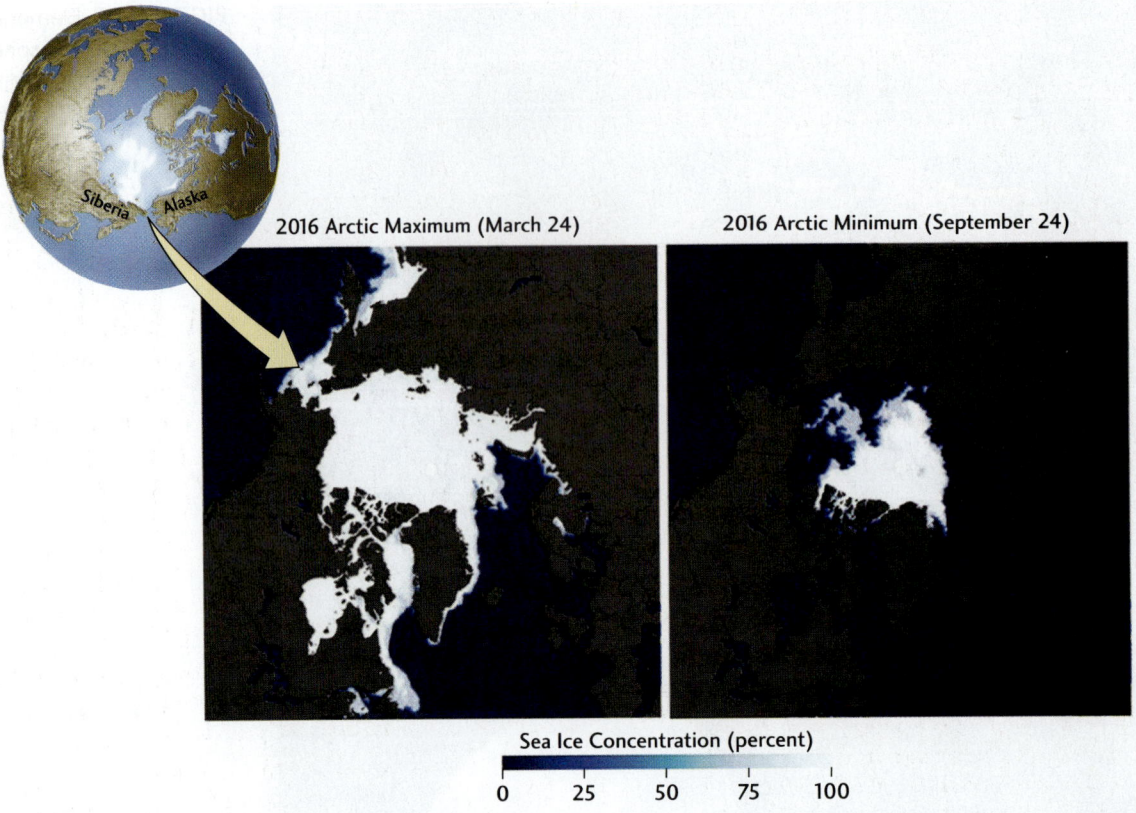

FIGURE 12.7 The volume of sea ice varies seasonally. This satellite image shows Arctic sea ice maximum and minimum in 2016. [NASA Earth Observatory maps by Joshua Stevens, based on AMSR2-E data from NSIDC.]

influence rates of evaporation—such as vegetation cover and the subsurface flow of water—are important in controlling atmospheric temperatures.

The topography created by plate tectonics has a direct effect on climate through its influence on atmospheric circulation. Air masses that flow over mountain ranges dump rain on the windward side, creating a rain shadow on the leeward side of the mountains (see Figure 17.3). Tectonic plate movement affects the climate system in more profound ways through rearrangements of the continents and ocean basins. At the present time, the land area of the Northern Hemisphere is about twice that of the Southern Hemisphere, which causes hemispheric asymmetries in the distribution of climate zones; for example, the average position of the intertropical convergence zone is shifted several hundred kilometers north of the equator.

By changing the average depth of the ocean basins, seafloor spreading can cause sea level to rise and fall, and the drift of continents over the poles can lead to the growth of continental glaciers. The movements of continents can also block ocean currents or open gateways through which seawater can flow, inhibiting or facilitating the poleward transfer of heat. For example, if future tectonic activity were to close the narrow channel between the Bahamas and Florida through which the Gulf Stream flows, the average temperature in western Europe might see a substantial drop.

Volcanism from the lithosphere affects climate by changing the composition and properties of the atmosphere. As we saw in Chapter 5, large volcanic eruptions can inject aerosols into the stratosphere, blocking solar radiation and temporarily lowering atmospheric temperatures on a global scale. After the massive April 1815 eruption of Mount Tambora in Indonesia, New England suffered through a "year without a summer" in 1816, which caused widespread crop failures. Recent large volcanic eruptions—including those of Krakatau (1883), El Chichón (1982), and Mount Pinatubo (1991)—each produced an average dip of 0.3°C in global surface temperatures about 14 months after the eruption. Temperatures returned to normal in about 4 years. Of course, the most extreme volcanic eruptions, such as a Yellowstone-type caldera collapse or a flood basalt eruption on the scale of the Siberian Traps, are capable of perturbing the climate system much more severely for much longer durations (see the Chapter 5 Practicing Geology Exercise).

The Biosphere

The *biosphere* comprises all the organisms living on and beneath Earth's surface, in its atmosphere, and in its waters.

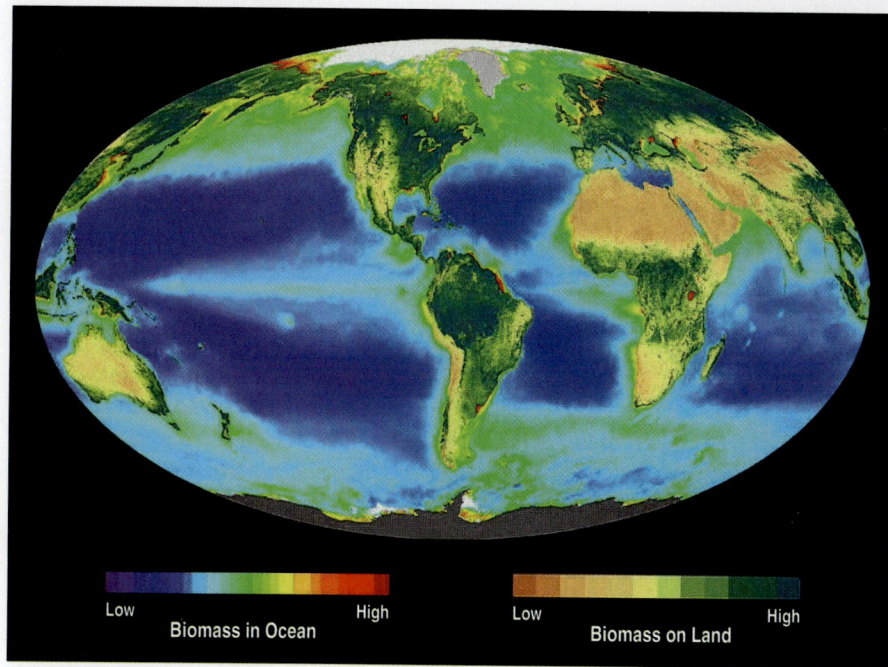

FIGURE 12.8 The biosphere, represented by the global distribution of algal and plant biomass in the oceans and on land, as mapped by NASA's SeaWiFS satellite. [NASA/Goddard Space Flight Center.]

Life is found almost everywhere on Earth, but the amount of life at any location depends on local climate conditions, as we can see from the satellite image of plant and algal biomass in **Figure 12.8**.

The total energy contained and transported by living organisms is relatively small on a global scale: Less than 0.1 percent of incoming solar energy is used by plants in photosynthesis. The biosphere, however, is strongly coupled to the other components of the climate system by the metabolic processes described in Chapter 22. For example, terrestrial vegetation can affect atmospheric temperature, because plants absorb solar radiation for photosynthesis and release it as heat during respiration, as well as atmospheric moisture, because they take up groundwater and release it as water vapor. Organisms also regulate the composition of the atmosphere by taking up or releasing greenhouse gases such as carbon dioxide (CO_2) and methane (CH_4). Through photosynthesis, plants and algae transfer CO_2 from the atmosphere to the biosphere. As noted above, some of that carbon moves from the biosphere to the lithosphere when it is buried as organic matter in marine sediments or precipitated as calcium carbonate shells. The biosphere thus plays a central role in the carbon cycle.

Humans (*anthropoi* in Greek) are part of the biosphere, of course, though hardly an ordinary part. Our influence over the biosphere is growing rapidly, and we have become the planet's most active agents of environmental change. Modification of the global environment by humans is called **anthropogenic global change**. As an organized society with a highly developed technology, we behave in fundamentally different ways from other species.

For example, we can study anthropogenic global change scientifically and modify our actions according to what we have learned (see Chapter 14).

The anthropogenic global change of greatest concern is the increase in atmospheric concentrations of CO_2 and other greenhouse gases due to fossil-fuel burning, which is causing **global warming**—an increase over time in the average surface temperature of the planet. To understand this problem, we need to study the role that greenhouse gases play in regulating Earth's surface temperature.

The Greenhouse Effect

The Sun is a yellow star that emits about half its radiant energy as visible light. The other half is split between infrared waves, which have longer wavelengths and lower energy intensities than visible light (which we perceive as heat), and ultraviolet waves with shorter wavelengths and higher energy intensities than visible light (which can give you a sunburn). The average amount of solar radiation Earth's surface receives throughout the year is 340 watts per square meter of surface area (340 W/m^2; 1 *watt* = 1 joule per second, where *joule* is a unit of energy or heat). In comparison, the average amount of heat flowing out of Earth's deep interior by mantle convection is minuscule, only 0.06 W/m^2. Essentially all the energy driving the climate system ultimately comes from the Sun (**Figure 12.9**).

We know that the global surface temperature, averaged over daily and seasonal cycles, remains approximately constant. Therefore, Earth's surface must be radiating energy back into space at a rate of precisely 340 W/m^2. Any less

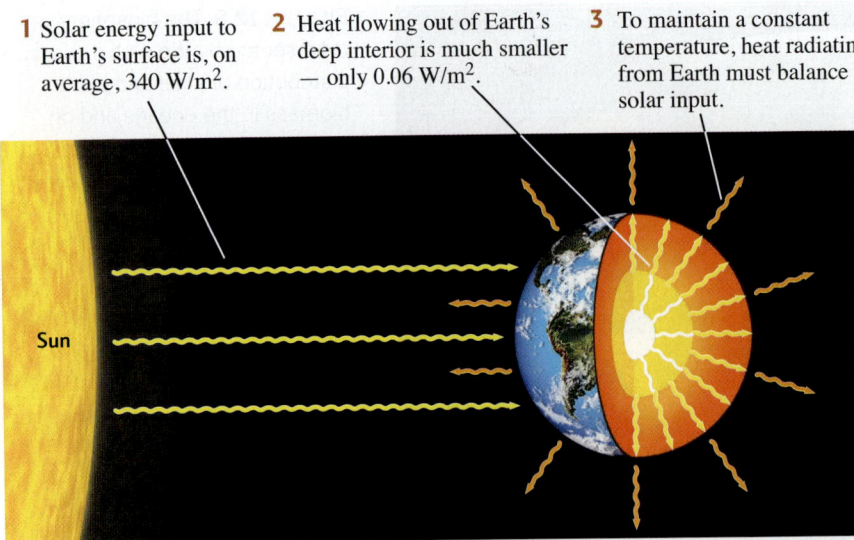

1 Solar energy input to Earth's surface is, on average, 340 W/m².

2 Heat flowing out of Earth's deep interior is much smaller — only 0.06 W/m².

3 To maintain a constant temperature, heat radiating from Earth must balance solar input.

FIGURE 12.9 Earth's energy balance is achieved by the radiation of incoming solar energy back into space. Heat gain from Earth's interior is negligible in comparison to that from solar energy.

would cause the surface to heat up; any more would cause it to cool down. In other words, Earth maintains a *radiation balance*: an equilibrium between incoming and outgoing radiant energy. How is this equilibrium achieved?

A Planet Without Greenhouse Gases

Suppose Earth were a rocky sphere like the Moon, with no atmosphere at all. Some of the sunlight falling on the surface would be reflected back into space, and some would be absorbed by the rocks, the amount depending on the color of the rock surface. A perfectly white planet would reflect all the solar energy falling on it, whereas a perfectly black planet would absorb it all. The fraction of the solar energy reflected by a surface is called its **albedo** (from the Latin word *albus*, meaning "white"). Although the full Moon looks bright to us, the rocks on its surface are mainly dark basalts, so its albedo is only about 7 percent. In other words, the Moon is dark gray—very nearly black.

The energy radiated by a black body increases rapidly as its temperature increases. A cold bar of iron is black and gives off little heat. If you heat the bar to 100°C, it gives off warmth in the form of infrared radiation (like a steam radiator). If you heat the bar to 1000°C, it becomes bright orange, radiating heat at visible wavelengths (like the burner on an electric stove).

A black body exposed to the Sun heats up until its temperature is at just the right value for it to radiate the incoming solar energy back into space. The same principle applies to a "gray body" like the Moon, except that the reflected energy must be excluded from the radiation balance. And, in the case of rotating bodies like the Moon and Earth, day-and-night cycles must be taken into account. The Moon's daytime temperatures rise to 130°C, and its nighttime temperatures drop to −170°C. Not a pleasant environment!

Earth rotates much faster than the Moon (once per day rather than once per month), which evens out the day-and-night extremes of temperature. Earth's albedo, at about 29 percent, is much higher than the Moon's because Earth's blue oceans, white clouds, and ice caps are more reflective than dark lunar basalts. If our atmosphere did not contain greenhouse gases, the average surface temperature required to balance the absorbed solar radiation would be about −19°C (−2°F), cold enough to freeze all the water on the planet. Instead, Earth's average surface temperature remains a balmy 14°C (57°F). The difference of 33°C is a result of the greenhouse effect.

Earth's Greenhouse Atmosphere

Greenhouse gases, such as water vapor, carbon dioxide, methane, and ozone, absorb energy—coming directly from the Sun as well as radiated by Earth's surface—and reradiate it as infrared energy in all directions, including downward to the surface. In this way, they act like the glass in a greenhouse, allowing light energy to pass through, but trapping heat in the atmosphere. This trapping of heat, which increases the temperature at the surface relative to the temperature higher in the atmosphere, is known as the **greenhouse effect.**

How Earth's atmosphere balances incoming and outgoing radiation is illustrated in Figure 12.10. Incoming solar radiation that is not directly reflected is absorbed by Earth's atmosphere and surface. To achieve radiation balance, Earth radiates this same amount of energy back into space as infrared energy. Because of the heat trapped by greenhouse gases, the amount of energy transported away from Earth's surface, both by radiation and by the flow of warm air and moisture from the surface, is significantly larger than the amount Earth receives as direct solar radiation. The excess is exactly the energy received as Earthward infrared radiation from the greenhouse gases. It is this "back radiation" that causes Earth's

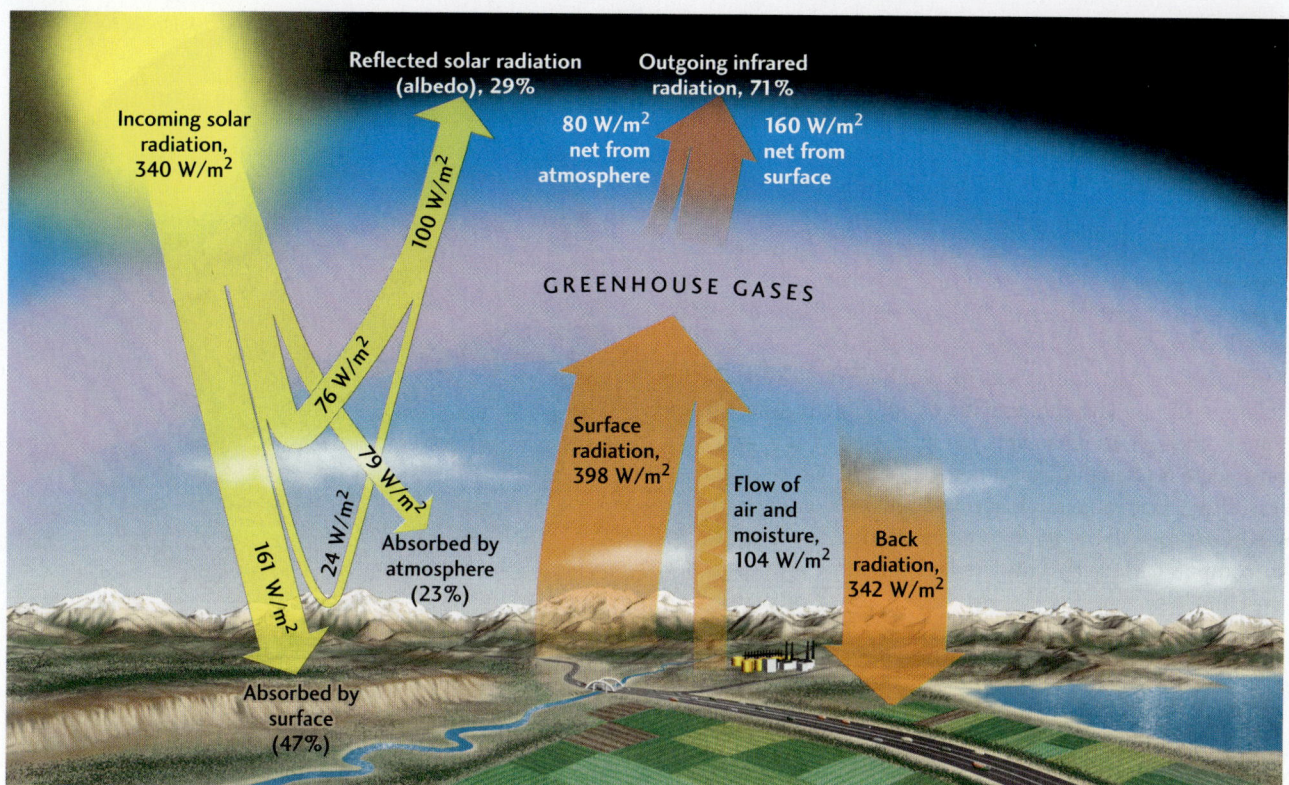

FIGURE 12.10 To maintain radiation balance, Earth radiates as much energy into outer space, on average, as it receives from the Sun (340 W/m²). Of the incoming radiation, 100 W/m² (29 percent) is reflected, 161 W/m² is absorbed by Earth's surface, and 79 W/m² is absorbed by Earth's atmosphere. Radiation and flows of warm air and moisture transport more energy away from Earth's surface (502 W/m²) than it receives. The greenhouse gases in the atmosphere reflect most of this energy (342 W/m²) back to Earth's surface as infrared radiation. [Data from IPCC, *Climate Change 2013: The Physical Science Basis*.]

surface to be 33°C warmer than it would be if the atmosphere contained no greenhouse gases.

Balancing the Climate System Through Feedbacks

How does the climate system actually achieve the radiation balance illustrated in Figure 12.10? Why does the greenhouse effect yield an overall warming of 33°C and not some larger or smaller amount? The answers to these questions are not simple because they depend on interactions among the many components of the climate system. The most important of those interactions involve *feedbacks*.

Feedbacks come in two basic types: **positive feedbacks**, in which a change in one component is *enhanced* by the changes it induces in other components, and **negative feedbacks**, in which a change in one component is *reduced* by the changes it induces in other components. Positive feedbacks tend to amplify changes in a system, whereas negative feedbacks tend to stabilize the system against change.

Here are some of the feedbacks within the climate system that affect the surface temperature achieved by radiation balance:

- *Water vapor feedback.* A rise in temperature increases the amount of water vapor that moves from Earth's surface into the atmosphere through evaporation. Water vapor is a greenhouse gas, so this increase enhances the greenhouse effect, and the temperature rises further—a positive feedback.

- *Albedo feedback.* A rise in temperature reduces the accumulation of ice and snow in the cryosphere, which decreases Earth's albedo and increases the energy its surface absorbs. This increased warming enhances the temperature rise—another positive feedback.

- *Radiative damping.* A rise in atmospheric temperature increases the amount of infrared energy radiated back into space, which moderates the temperature rise—a negative feedback. This "radiative damping" stabilizes Earth's climate against major temperature changes, keeping the oceans from freezing up or boiling off and thus maintaining an equable habitat for aquatic life.

- *Plant growth feedback.* Increasing atmospheric CO_2 concentrations stimulate plant growth. Growing plants remove CO_2 from the atmosphere by converting it into carbon-rich organic matter, thus reducing the greenhouse effect—another negative feedback.

Feedbacks can involve much more complex interactions among components of the climate system. For example, an increase in atmospheric water vapor produces more clouds. Because clouds reflect solar energy, they increase the planetary albedo, which sets up a negative feedback between atmospheric water vapor and temperature. On the other hand, clouds absorb infrared radiation efficiently, so increasing the cloud cover enhances the greenhouse effect, thus providing a positive feedback between atmospheric water vapor and temperature. Does the net effect of clouds produce a positive or a negative feedback?

Scientists have found it surprisingly difficult to answer such questions. The components of our climate system are joined through an amazingly complex web of interactions on a scale far beyond experimental control. Consequently, it is often impossible to gather data that isolate one type of feedback from all the others. Scientists must therefore turn to computer models to understand the inner workings of the climate system.

Climate Models and Their Limitations

Generally speaking, a **climate model** is any representation of the climate system that can reproduce one or more aspects of its behavior. Some models are designed to study local or regional climate processes, such as the relationships between water vapor and clouds, but the most interesting representations are global models that describe how climate has changed in the past or predict how it might change in the future.

At the heart of such global climate models are schemes for computing movements within the atmosphere and oceans based on the fundamental laws of physics. These *general circulation models* represent the currents of air and water driven by solar energy on scales ranging from small disturbances (storms in the atmosphere, eddies in the oceans) to global circulation patterns (wind belts in the atmosphere, thermohaline circulation in the oceans). Scientists represent the basic physical variables (temperature, pressure, density, velocity, and so forth) on three-dimensional grids comprising millions, or even billions, of geographic points. They use supercomputers to solve mathematical equations that describe how the variables change over time at each of these points (Figure 12.11). You see the results of these calculations whenever you check the weather report. These days, most weather predictions are made by entering the current conditions observed at thousands of weather stations into a general circulation model and running it forward in time. Weather predictions thus use the same basic computer programs that are used for climate modeling.

Climate modeling is more difficult than weather prediction, however. In forecasting weather a few days from now, scientists can ignore such slow processes as changes in atmospheric greenhouse gas concentrations or in oceanic circulation. Climate predictions, on the other hand, require that we properly model these processes that occur more slowly, including all important feedbacks. Moreover, the simulation must be extended years or decades into the future.

FIGURE 12.11 Numerical climate models are used to predict future climate change. This global climate model, developed with support from the U.S. Department of Energy, portrays interactions among the atmosphere, hydrosphere, cryosphere, and land surface. The colors in the oceans represent sea surface temperatures, and the arrows represent surface wind velocities. [Warren Washington and Gary Strand/National Center for Atmospheric Research.]

Because climate models are so complex and subject to errors in our understanding of the climate system, their predictions must be viewed with some skepticism. Many questions remain about how the climate system works—for instance, how clouds affect atmospheric temperatures. Though most climate scientists now think the cloud feedback is positive (destabilizing), substantial uncertainties in these and other key processes remain. The predictions of climate models have been a topic of much debate among experts and government authorities who must understand and deal with the consequences of human-induced climate change. We will take a closer look at those predictions in Chapter 14.

Climate Variation

Earth's climate varies considerably from place to place: Its poles are frigid and arid, its tropics sweltering and humid. Comparable variations in climate can also occur over time. The geologic record shows us that periods of global warmth have alternated with periods of glacial cold many times in the past. This climate variation is erratic; dramatic changes can happen in just a few decades or evolve over time scales of many millions of years.

Some climate variation can be attributed to factors outside the climate system, such as solar forcing and changes in the distribution of land and sea surfaces caused by continental drift. Others result from changes within the climate system itself, such as the growth of continental glaciers that increase Earth's albedo. Both types of variations, external and internal, can be amplified or suppressed by various types of feedbacks. In this section, we will examine several types of climate variation and discuss their causes, beginning with short-term variation on a regional scale.

Short-Term Regional Variations: El Niño and the Southern Oscillation

Local and regional climates are much more variable than the average global climate: Averaging over large surface areas, like averaging over time, tends to smooth out small-scale fluctuations. Over periods of years to decades, the predominant regional variations result from interactions between atmospheric circulation and the sea and land surfaces. They generally occur in distinct geographic patterns, although their timing and amplitudes can be highly irregular.

One of the best-known examples is a warming of the eastern Pacific Ocean that occurs every 3 to 7 years and lasts for a year or so. Peruvian fishermen named this event **El Niño** ("the boy child" in Spanish) because the warming typically reaches the surface waters off the coast of South America around Christmastime. El Niño events can decimate fish populations, which depend on the upwelling of cold water for their nutrient supply, and can thus be disastrous for populations that depend on fishing for their livelihood.

El Niño and a complementary cooling event, known as *La Niña* ("the girl child"), are part of a natural variation in the exchange of heat between the atmosphere and the tropical Pacific Ocean. Normally, atmospheric pressure gradients cause prevailing winds, the trade winds, to blow from east to west, pushing the warm tropical waters westward. This movement of water causes colder water to well up from the ocean depths off Peru. Sporadically, the trade winds weaken and occasionally reverse direction, cutting off the upwelling and equalizing water temperatures across the tropical Pacific (an El Niño event). At other times, the trade winds strengthen, enhancing the temperature difference between the eastern and western Pacific (a La Niña event). This recurring swing in air pressure is called the Southern Oscillation, and the overall variation in the climate of the equatorial Pacific associated with these phenomena is known as the El Niño–Southern Oscillation, or **ENSO** (Figure 12.12).

Good instrumental records of ENSO did not become available until around 1950. The three largest El Niño

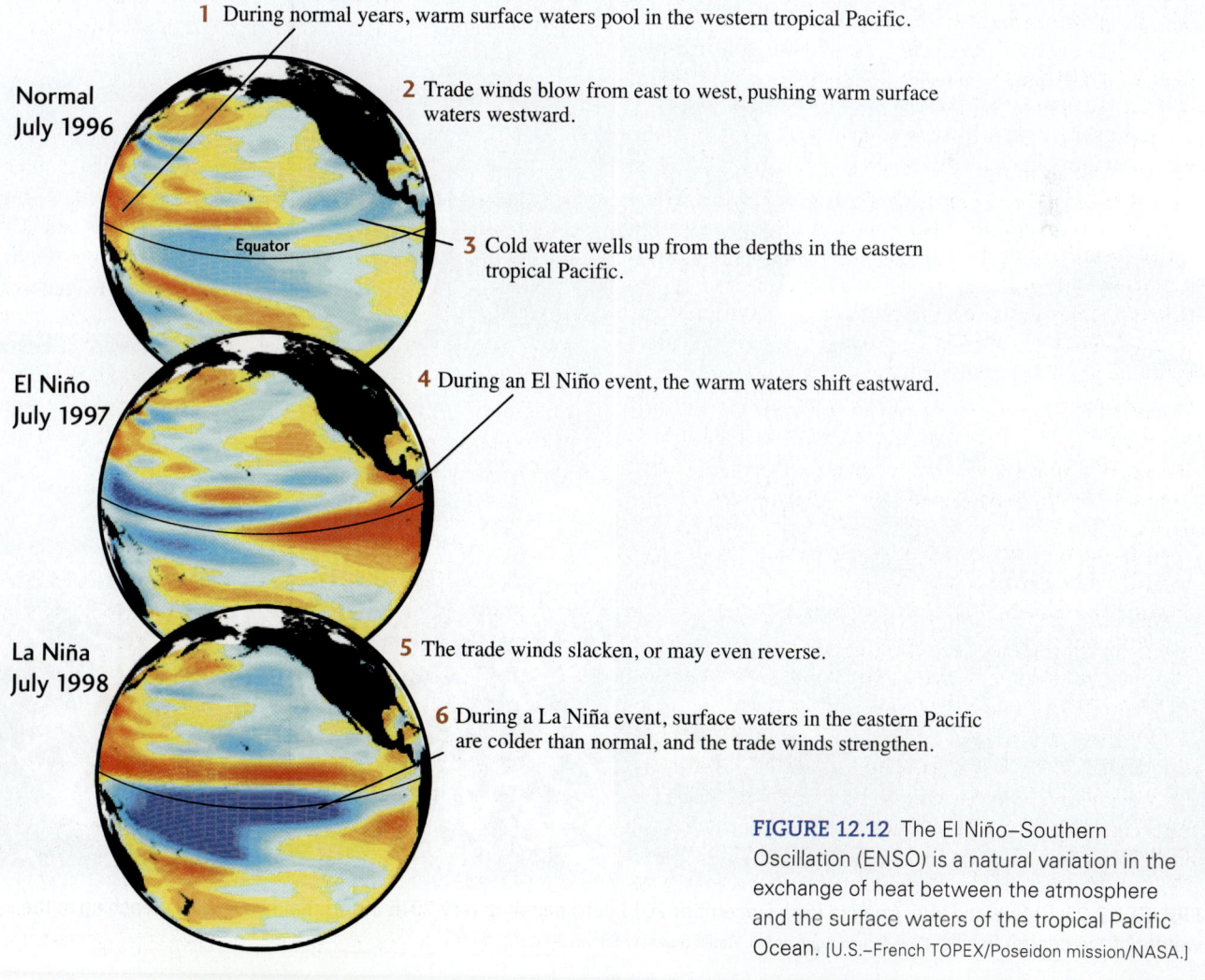

FIGURE 12.12 The El Niño–Southern Oscillation (ENSO) is a natural variation in the exchange of heat between the atmosphere and the surface waters of the tropical Pacific Ocean. [U.S.–French TOPEX/Poseidon mission/NASA.]

events since that time occurred in 1982–83, 1997–1998, and 2015–2016, when the temperatures in the central tropical Pacific rose by as much as 3°C above average. The latest El Niño, nicknamed "Godzilla" by meteorologists because of its size and intensity, began in March 2015 and ended in May 2016. It affected weather around the world, bringing heavy rains to southern South America, central Africa, southeastern China, and the southeastern United States and droughts to northern South America, Indonesia, and northern Australia. The above-normal water temperatures in the central Pacific Ocean were blamed for anomalous cyclone activity and massive coral bleaching (Figure 12.13). The warmer waters contributed to an increase in Earth's average temperature, making 2015 and 2016 the warmest years ever recorded.

Oscillatory patterns of weather and climate variation occur in other regions. One example is the North Atlantic Oscillation, a highly irregular, multiyear fluctuation in the balance of atmospheric pressures between Iceland and the Azores; this oscillation has a strong influence on the movement of storms across the North Atlantic and thus affects weather conditions throughout Europe and parts of Asia. A better understanding of these short-term climate variations is improving long-range weather forecasting and may provide important information about the regional effects of anthropogenic climate change.

Long-Term Global Variations: The Pleistocene Ice Ages

Some of the most dramatic climate variations that can be seen in the geologic record are the glacial cycles of the Pleistocene epoch, which began 1.8 million years ago. A **glacial cycle** starts with a gradual decline in temperature of about 6°C to 8°C from a warm **interglacial period** to a cold *glacial period*, or **ice age.** As the climate cools, water is transferred from the hydrosphere to the cryosphere. The amount of sea ice increases, and more snow falls on the continents in winter than melts in summer, increasing

FIGURE 12.13 Before and after images from December 2014 (left) and February 2015 (right) that show coral bleaching in the waters of the central Pacific. [The Ocean Agency / XL Catlin Seaview Survey.]

FIGURE 12.14 At the last glacial maximum around 20,000 years ago, continental glaciers covered most of North America. The continental shelves were exposed by the lowering of the sea level, illustrated here by the expanded coastline of Florida. [Wm. Robert Johnston.]

the volume and area of polar ice caps and decreasing the volume of the oceans. As the ice caps expand into lower latitudes, they reflect more solar energy back into space, and Earth's surface temperatures fall further—an example of albedo feedback. Sea level falls, exposing areas of the continental shelves that are normally under water. At the peak of the ice age—the *glacial maximum*—great continental glaciers up to 2 or 3 m thick cover vast land areas (Figure 12.14). The ice age ends abruptly with a rapid rise in temperature. Water is transferred from the cryosphere to the hydrosphere as the ice caps melt and sea level rises.

Timing the Pleistocene Ice Ages A precise record of Pleistocene temperature variations can be obtained by measuring oxygen isotopes preserved in marine sediments and in glacial ice. Pleistocene marine sediments contain numerous fossils of foraminifera: small, single-celled marine organisms that secrete shells of calcite ($CaCO_3$). The proportions of oxygen isotopes incorporated into these shells depend on the oxygen isotope ratio of the seawater in which the organisms lived. Water (H_2O) containing the lighter and more common isotope, oxygen-16 (^{16}O), has a greater tendency to evaporate than water containing the heavier oxygen-18 (^{18}O). Therefore, during ice ages, ^{18}O is preferentially left behind in the oceans as water-containing ^{16}O evaporates from the ocean surface and is trapped in glacial ice, and the $^{18}O/^{16}O$ ratio in the oceans rises. Paleoclimatologists can use $^{18}O/^{16}O$ ratios in marine sediment beds to estimate sea surface temperature and ice volume at the time the beds were deposited. Figure 12.15

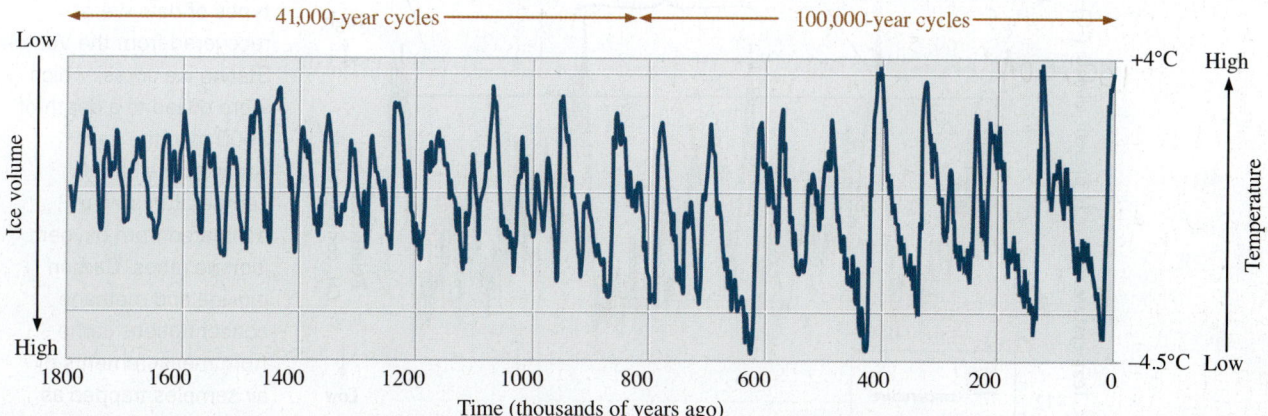

FIGURE 12.15 Changes in global climate over the last 1.8 million years, as inferred from oxygen isotope ratios in marine sediments. The peaks indicate interglacial periods (high temperatures, low ice volumes, high sea level), and the valleys indicate ice ages (low temperatures, high ice volumes, low sea level). [Data from L. E. Lisiecki and M. E. Raymo, *Paleoceanography* 20 (2005): 1003.]

Earth Issues 12.1 Ice-Core Drilling in Antarctica and Greenland

At the Vostok Station in the frozen Antarctic, Russian and French scientists have worked for decades to uncover the history of Earth's climate hidden in glacial ice. By 1998, they had drilled the ice to a depth of 3600 m, recovering ice cores as old as 400,000 years. By carefully counting the annual cycles of ice formation, working from the top down, the researchers were able to determine the age of the ice with depth, much as tree rings reveal the age of a tree. They measured oxygen isotope ratios in the ice as well as the gas composition of small bubbles trapped in the ice. From this stratigraphic record, they produced a detailed history of the last four glacial cycles.

The data supported other evidence suggesting that variations in Earth's orbit—Milankovitch cycles—control the alternation of ice ages and interglacial periods, and they showed that surface temperatures were correlated with concentrations of greenhouse gases in the atmosphere (see Figure 12.16). The Vostok results have been confirmed by ice core drilling at a number of other locations on the Antarctic and Greenland ice sheets. A core drilled at a site about 600 km from Vostok by a European project provides the longest ice record, yielding temperature and greenhouse gas data for the last eight glacial cycles. During this 800,000-year period, the concentration of CO_2 in the atmosphere varied from a low of about 170 parts per million (ppm) during glacial maxima to about 300 ppm during interglacial periods. In comparison, the CO_2 concentration has risen from about 280 ppm in the mid-1800s, at the start of the Industrial Revolution, to over 410 ppm today, and it continues to increase at a rate of 2 ppm/year.

These scientific discoveries have not been won easily. The Vostok Station, located at an elevation of 3500 m near the center of Antarctica, is an especially grueling place to do research. Its average annual temperature is only about −55°C, and the lowest reliably measured temperature on Earth's surface, −89.2°C, was recorded there in 1983. The scientists not only had to endure these extreme conditions, but they also had to be careful not to melt and contaminate the ice cores while drilling them, transporting them to laboratories, and storing them. It is a tribute to the patience and ingenuity of these hardy bands of researchers that glacial ice cores have contributed so much to our understanding of the history of global climate change.

French, Russian, and American scientists in the Vostok team photo holding freshly drilled ice cores. The cores were then cut into one-meter sections and analyzed in an on-site laboratory. [NOAA.]

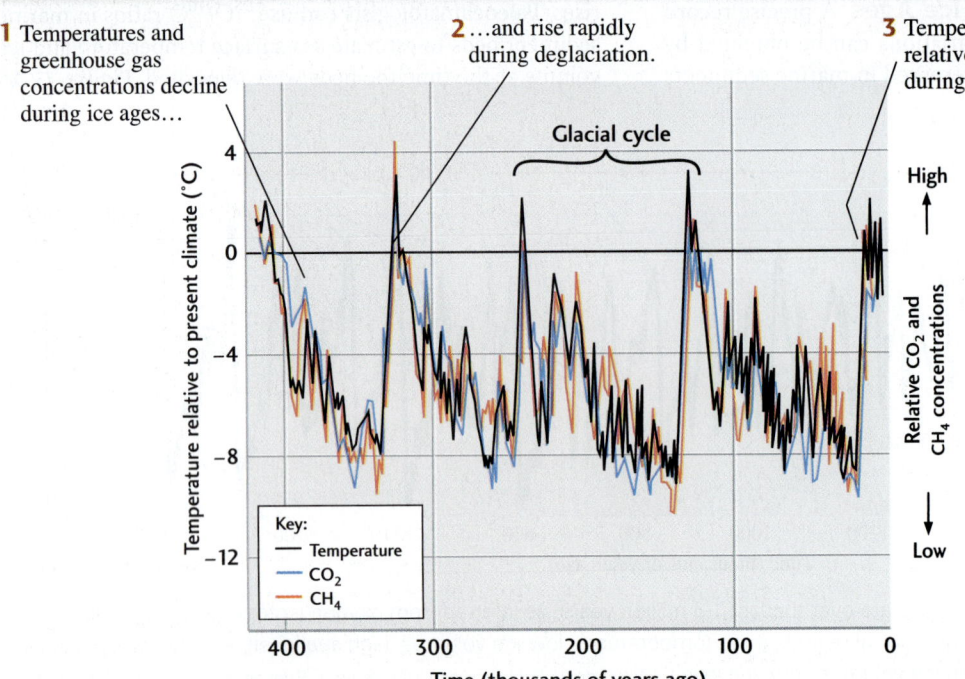

1 Temperatures and greenhouse gas concentrations decline during ice ages…

2 …and rise rapidly during deglaciation.

3 Temperatures have been relatively warm and stable during the Holocene.

FIGURE 12.16 Three types of data were recovered from the Vostok Station ice cores, which were drilled to a depth of 3600 m in the East Antarctic ice sheet. Temperatures were estimated from oxygen isotope ratios. Carbon dioxide and methane concentrations came from measurements of air samples trapped as tiny bubbles within the Antarctic ice. [Data from IPCC, *Climate Change 2001: The Scientific Basis*.]

shows changes in global climate over the past 1.8 million years inferred from these estimates.

As $^{18}O/^{16}O$ ratios in the oceans increase during ice ages, those in the layers of ice that form the growing glaciers decrease. The best records of climate variation during the last half-million years come from ice cores drilled in the East Antarctic and Greenland ice sheets (see Earth Issues 12.1). The oxygen isotope ratios of the ice layers in the cores can be used to estimate atmospheric temperatures at the time the ice formed. The composition of the atmosphere, including the concentrations of carbon dioxide and methane, can also be measured in tiny bubbles of air trapped at the time the ice formed. Figure 12.16 displays these three types of measurements from ice core drilled at the Vostok Station on the East Antarctic ice sheet.

Milankovitch Cycles

The major ups and downs in the marine sediment record (see Figure 12.15) during the Pleistocene epoch match the glacial cycles in the ice core record (see Figure 12.16). What causes these climate fluctuations? The answer is small periodic variations in the amount of radiation Earth receives from the Sun—*solar forcing*. These variations arise from complexities in Earth's movement around the Sun, called **Milankovitch cycles** after the Serbian geophysicist who first calculated them in the early twentieth century. Three kinds of Milankovitch cycles can be correlated with global climate variation:

- First, the shape of Earth's orbit around the Sun changes periodically, becoming more circular at some times and more elliptical at others. The degree of ellipticity of Earth's orbit around the Sun is known as its *eccentricity*. A nearly circular orbit has low eccentricity, and a more elliptical orbit has high eccentricity (Figure 12.17a). The amount of solar radiation Earth receives each year, averaged over its surface, decreases as the eccentricity increases. The length of one cycle of variation in eccentricity is about 100,000 years.

- Second, the angle or *tilt* of Earth's axis of rotation changes periodically. Today this angle is 23.4°, but it cycles between 22.1° and 24.5° with a period of about 41,000 years. These variations also slightly change the amount of radiation Earth receives from the Sun (Figure 12.17b).

- Third, Earth's axis of rotation wobbles like a top, giving rise to a pattern of variation called *precession* with a period of about 23,000 years (Figure 12.17c). Precession, too, modifies the amount of radiation Earth receives from the Sun, though by less than variations in eccentricity and tilt.

Correlations With Glacial Cycles You can see lots of small ups and downs in the oxygen-isotope record of Figure 12.15, but in the last half-million years, the record

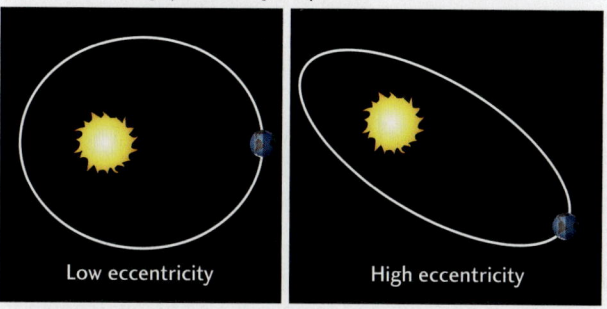

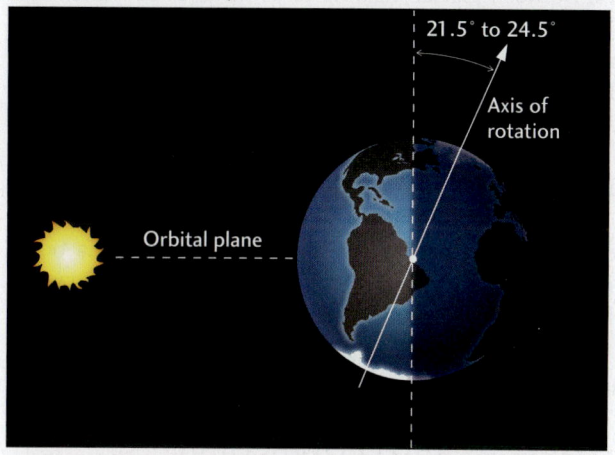

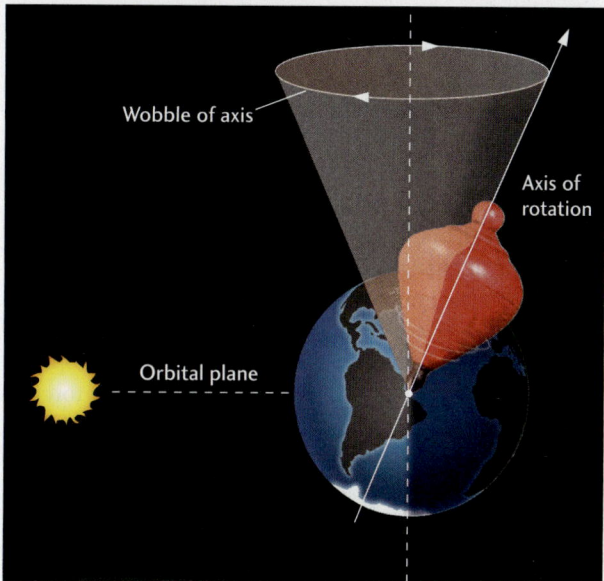

FIGURE 12.17 Three kinds of Milankovitch cycles (much exaggerated in these diagrams) affect the amount of solar radiation Earth receives. (a) Eccentricity is the degree of ellipticity of Earth's orbit. (b) Tilt is the angle between Earth's axis of rotation and the angle perpendicular to the orbital plane. (c) Precession is the wobble of the axis of rotation. One can imagine this motion by thinking of the wobble of a spinning top.

reveals a sawtooth pattern of major glacial cycles that looks roughly like this:

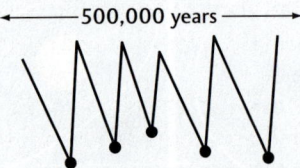

In particular, you can count five glacial maxima, in which ice volumes are high and temperatures are low (shown in the sketch as black dots), revealing an average time interval between glacial maxima of about 100,000 years. This 100,000-year spacing of minimum temperatures closely matches the times of high orbital eccentricity, when Earth received slightly less radiation from the Sun—a Milankovitch cycle.

Now let's move backward in time to examine the first half-million years of the record in Figure 12.15, from 1.8 million to 1.3 million years ago. Again, we see many small fluctuations, but major maxima and minima occur more frequently than they do in the later record, as approximated in the following sketch:

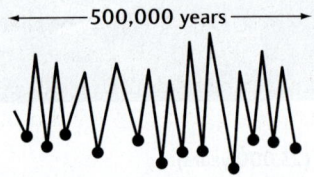

During this period, we find 12 glacial maxima, with an average spacing between them of 500,000 years/12 cycles, or 41,667 years. This shorter interval is very close to the 41,000-year cycle of variation in the tilt of Earth's axis—another Milankovitch cycle! Like the variation in eccentricity, the variation in tilt is very small, only about 3° (see Figure 15.14b), but it's evidently enough to trigger ice ages.

The small changes in solar radiation caused by Milankovitch cycles cannot by themselves explain the large drops in Earth's surface temperature from interglacial periods to ice ages. Some type of positive feedback must be operating within the climate system to amplify the solar forcing. The data in Figure 12.16 strongly suggest that this feedback involves greenhouse gases. Atmospheric concentrations of carbon dioxide and methane precisely track the temperature variations throughout the glacial cycles: Warm interglacial periods are marked by high concentrations, cold glacial periods by low concentrations. Exactly how this feedback works has not yet been fully explained, but it demonstrates the importance of the greenhouse effect in long-term climate variations.

Many other aspects of this story are not yet understood. For example, you will notice in Figure 12.15 that the 41,000-year periodicity continued to dominate the climate record up to about 1 million years ago. Then, the highs and lows became more variable, eventually shifting to the 100,000-year periodicity after about 700,000 years ago. What caused this transition? Climate scientists are still scratching their heads.

In fact, we don't really know what triggered the Pleistocene ice ages. The climate record shows that the 41,000-year glacial cycles were not confined to the Pleistocene, but extended back at least into the Pliocene epoch (5.3 million to 1.8 million years ago), when Antarctica became covered in ice. The global cooling of Earth's climate that preceded these glaciations began during the Miocene epoch (23 million to 5.3 million years ago). Its cause continues to be debated, although most geologists believe it is somehow related to continental drift. According to one hypothesis, the collision of the Indian subcontinent with Eurasia and the resulting Himalayan orogeny led to an increase in the weathering of silicate rocks, and the chemical processes of weathering decreased the amount of CO_2 in the atmosphere. Other hypotheses are based on changes in oceanic circulation associated with the opening of the Drake Passage between South America and Antarctica (25 million to 20 million years ago) or the closing of the Isthmus of Panama between North and South America (about 5 million years ago). Perhaps the cooling resulted from a combination of these events.

Long-Term Global Variations: Ancient Ice Ages and Warm Periods

In addition to the Pleistocene ice ages, there is good evidence in the geologic record for earlier episodes of continental glaciation during the Permian-Carboniferous and Ordovician periods and at least twice in the Proterozoic eon. In most cases, these events can be explained by plate-tectonic processes, coupled with the albedo effect and other feedbacks in the climate system.

For much of Earth's history, there were no extensive land areas in the polar regions and hence no permanent ice caps. Oceanic circulation extended from equatorial regions into polar regions, transporting heat fairly evenly over Earth's surface. During the ice-free Paleocene and Eocene epochs of the early Cenozoic Era, for instance, the global climate was warmer than it is today by about 8°C. The boundary between these two epochs, 56 million years ago, was an exceptionally warm period, evidently caused by massive injections of carbon into the atmosphere from reservoirs on land and in oceans. Changes in the fossil record at this spike-like "Paleocene-Eocene Thermal Maximum" show mass extinctions of benthic foraminifera and the rise of modern mammals, including primates.

When large land areas drifted to positions that obstructed the efficient transport of heat by the oceans, the differences in temperature between the poles and the equator increased. As the poles cooled, ice caps formed. For example, when the first Antarctic glaciation occurred at the end of the Eocene, about 34 million years ago, the global temperature abruptly dropped by several degrees. Some geologists believe that at one time in the late Proterozoic, Earth was very cold and completely covered by ice, and that the atmosphere warmed only after volcanic eruptions increased the concentration of greenhouse gases. We'll take a closer look at this "Snowball Earth hypothesis" in Chapter 15.

The Carbon Cycle

In the past 200 years, atmospheric CO_2 concentrations have risen by nearly 50 percent, from about 280 ppm to over 410 ppm (reached in mid-2017). Earth's atmosphere has not contained this much CO_2 for at least the last million years, and probably not for the last 14 million years. Atmospheric CO_2 concentrations are now increasing at an unprecedented rate of 0.5 percent per year, faster than at any time in recent geologic history.

Yet the situation could be worse. Over the decade 2000–2009, human activities injected an average of 8.9 gigatons (Gt) of carbon into the atmosphere each year. (A gigaton, or 1 billion tons, is 10^{12} kg, the mass of 1 km^3 of water.) Fossil-fuel burning and other industrial activities emitted about 7.8 Gt/year of carbon, and the burning of forests and other changes in land use emitted an additional 1.1 Gt/year. If all of that carbon had stayed in the air, the atmospheric CO_2 increase would have been over 1 percent per year, more than twice the observed rate. Instead, 4.9 Gt of carbon was removed from the atmosphere each year by natural processes. Where did all that carbon go?

We will address this question by examining the **carbon cycle**: the continual movement of carbon between different components of the Earth system. Let's begin with a broader look at geochemical cycles.

Geochemical Cycles and How They Work

Geochemical cycles describe the flow, or *flux,* of chemicals from one component of the Earth system to another. In discussing geochemical cycles, we view components of the Earth system—atmosphere, hydrosphere, cryosphere, lithosphere, and biosphere—as **geochemical reservoirs** for the storage of carbon and other chemicals, linked by processes that transport chemicals among them. By quantifying the amounts of the chemicals that are stored in and moved among the various reservoirs, we can gain new insights into the workings of the Earth system.

Residence Time Reservoirs gain chemicals from inflows and lose chemicals from outflows. If inflow equals outflow, the amount of the chemical in the reservoir stays the same, even though the chemical is constantly entering and leaving it. On average, a molecule of the chemical spends a certain amount of time, called the **residence time,** in a reservoir. For example, the residence time of sodium in the ocean is extremely long—about 48 million years—because sodium is highly soluble in seawater (that is, the capacity of the reservoir to store sodium is high) and because rivers contain relatively small amounts of sodium (its inflow into the reservoir is low). In contrast, iron has a residence time in the ocean of only 100 years because its solubility in seawater is very low and the inflow from rivers is relatively high.

Residence times of chemicals in the atmosphere are usually shorter than those in the oceans because, in terms of total mass, the atmosphere is a smaller reservoir than the oceans, and fluxes into and out of the atmosphere can be larger. Sulfur dioxide, for example, has a residence time in the atmosphere of hours to weeks, and oxygen, which makes up about 21 percent of the atmosphere, has a residence time of 6000 years. Atmospheric nitrogen gas is abundant (about 78 percent of the atmosphere) and very stable, so its residence time is almost 400 million years. A molecule of nitrogen that entered the atmosphere in the late Paleozoic era is still likely to be there!

Chemical Reactions In many cases, reactions with other chemicals govern a chemical's residence time in a reservoir. For example, as we learned in Chapter 6, a calcium ion (Ca^{2+}) can be removed from solution in seawater by reacting with a carbonate ion (CO_3^{2-}) to form calcium carbonate ($CaCO_3$), which can precipitate as carbonate sediment. The amount of calcium that remains dissolved in seawater thus depends on the availability of carbonate ions, which in turn depends on the influx of carbon dioxide (CO_2) into the ocean.

When carbon dioxide dissolves in water, most of it reacts with the water to form carbonic acid (H_2CO_3), which can dissociate into hydrogen (H^+) and bicarbonate ions (HCO_3^-). Some of the hydrogen ions then react with carbonate ions to form more bicarbonate ions (**Figure 12.18**). The net effect is to increase the acid content of seawater and decrease the concentration of carbonate ions. The decrease in carbonate affects the ability of marine organisms such as corals, clams, and foraminifera to build their shells and skeletons by precipitating calcium carbonate. As we shall see, this process of *ocean acidification* is one of the most threatening aspects of anthropogenic global change.

The Cycling of Carbon

Carbon cycles among four main reservoirs: the atmosphere; the oceans, including marine organisms; the land surface, including plants and soils; and the deeper lithosphere (**Figure 12.19**). We can describe the flux of carbon among these reservoirs in terms of several basic subcycles. During times when Earth's climate is stable, each subcycle can be characterized by a constant flux.

Atmosphere-Ocean Gas Exchange The exchange of CO_2 directly across the interface between the oceans and the atmosphere amounts to an average carbon flux of about 80 Gt per year. The flux through this subcycle depends on many factors, including air and sea temperatures and the

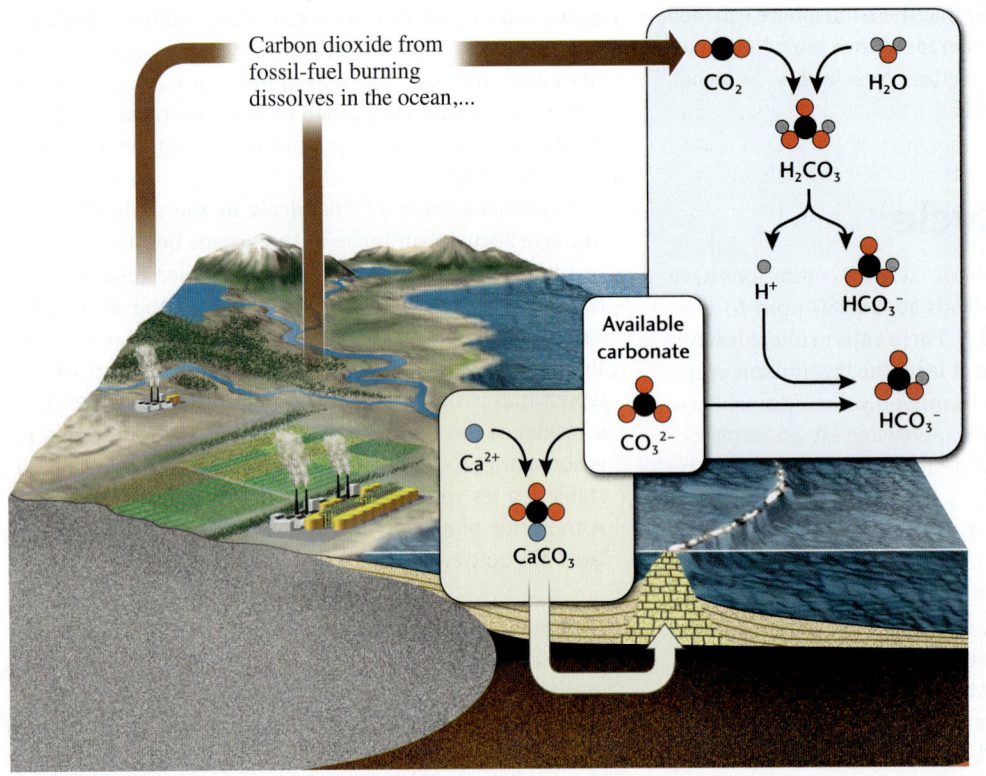

FIGURE 12.18 Increasing CO₂ concentrations in the atmosphere drive a series of chemical reactions in seawater, causing ocean acidification and reducing the ability of marine organisms to form shells and skeletons of calcium carbonate.

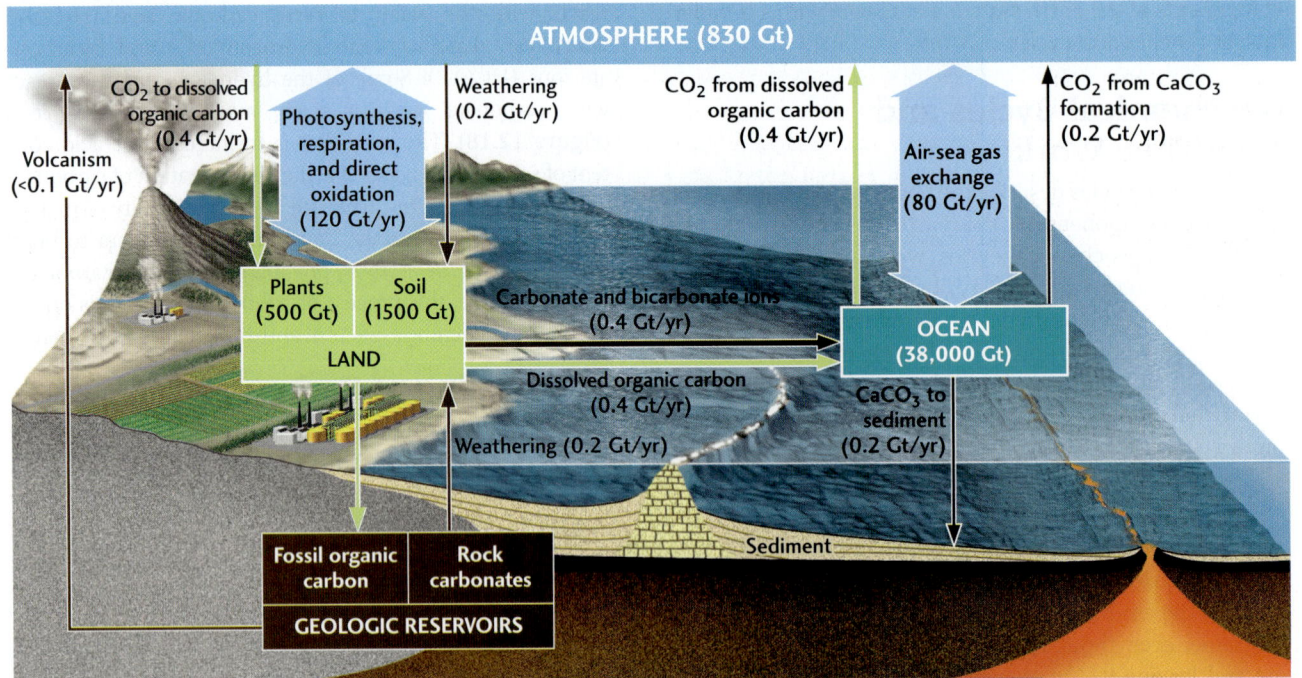

FIGURE 12.19 The carbon cycle describes the fluxes of carbon between the atmosphere and its other principal reservoirs. Amounts of carbon stored in each reservoir are given in gigatons; fluxes are given in gigatons per year. [Data from IPCC, *Climate Change 2001: The Scientific Basis*, updated according to IPCC, *Climate Change 2013: The Physical Science Basis*.]

composition of the seawater, but it is particularly sensitive to wind velocity, which increases the transfer of CO_2 and other gases by stirring up the surface water and generating sea spray. Carbon dioxide dissolved in seawater escapes from solution and enters the atmosphere by evaporating from sea spray, while atmospheric CO_2 enters the ocean by dissolving in sea spray and rain or directly across the sea surface.

Atmosphere-Biosphere Gas Exchange The subcycle with the greatest carbon flux, 120 Gt per year, is the exchange of CO_2 between the terrestrial biosphere and the atmosphere by photosynthesis, respiration, and decomposition. Plants take in this entire amount of CO_2 during photosynthesis and respire about half of it back into the atmosphere. The other half is incorporated into plant tissues—leaves, wood, and roots—as organic carbon. Animals eat the plants, and microorganisms decompose them; both processes result in the breakdown of plant tissues and the respiration of CO_2. Much of the organic carbon released by these processes—about three times the total plant mass—is stored in soils. A significant fraction (about 4 Gt/year) reenters the atmosphere through direct oxidation by forest fires and other combustion of plant material.

A small fraction of the CO_2 incorporated into plant tissues (0.4 Gt/year) is dissolved in surface waters and transported by rivers to the ocean, where it is respired back into the atmosphere by marine organisms and eventually taken up again by plants through photosynthesis.

Lithosphere-Atmosphere Gas Exchange The lithosphere is a long-term reservoir for carbon, storing it as calcium carbonate ($CaCO_3$) in limestones and other sedimentary rocks or as buried organic matter such as coal and petroleum. The weathering of carbonate rock removes about 0.2 Gt of carbon per year from the lithosphere and an equal amount from the atmosphere. The CO_2 dissolved in rainwater forms carbonic acid, which reacts with carbonates in the rock, releasing dissolved calcium ions (Ca^{2+}), carbonate ions (CO_3^-), and bicarbonate ions (HCO_3^-), which are transported by rivers to the ocean. There marine organisms reverse the weathering reaction, combining the dissolved calcium with carbonate to form shells of calcium carbonate. The reversed reaction releases an equal amount of carbon (0.2 Gt) back into the atmosphere as CO_2. The calcium carbonate shells fall to the seafloor and are buried in sediments, returning 0.2 Gt of carbon to the lithosphere.

On geological time scales, atmospheric CO_2 is regulated by a second chemical weathering process in which carbon dioxide dissolved in rain water reacts with calcium silicate ($CaSiO_3$), releasing dissolved calcium ions (Ca^{2+}), bicarbonate ions (HCO_3^-), and silica (SiO_2). These raw materials wash into the oceans, where some species of marine organisms, such as corals and coccolithophores (a type of phytoplankton), make shells of calcium carbonate, while others, such as diatoms and radiolarians, make shells of silica (see Chapter 16). Note the difference. The geochemical cycle involving limestone weathering does not change the amount of carbon in the atmosphere, because each unit of $CaCO_3$ weathered out of the lithosphere is returned as a unit of $CaCO_3$ in marine shells. In contrast, silicate weathering replaces each unit of $CaSiO_3$ with one of $CaCO_3$, thus removing a CO_2 molecule from the atmosphere.

Silicate weathering is a slow process, so the net flux of carbon from silicate weathering is relatively small (less than 0.1 Gt/year). Therefore, like volcanism (which releases minor amounts of CO_2 into the atmosphere), it is usually neglected in short-term climate modeling. Acting over millions of years, however, this reaction pumps CO_2 from the atmosphere into the lithosphere, reducing the amount of this greenhouse gas and thereby cooling the global climate. For example, the uplifting of the Himalaya and the Tibetan Plateau, which began about 40 million years ago, may have increased weathering rates enough to reduce the concentration of CO_2 in the atmosphere, contributing to the subsequent climate cooling that led to the Pleistocene glaciations.

Human Perturbations of the Carbon Cycle

With this background, let's return to the fate of anthropogenic carbon emissions. **Figure 12.20** shows what happened to the carbon that was added to the atmosphere by human activities in the decade 2000–2009. Out of a total of 8.9 Gt/year injected into the atmosphere by human activities, only 45 percent (4.0 Gt/year) remained in the atmosphere as CO_2. The rest was absorbed in nearly equal amounts by the oceans (2.3 Gt/year) and the land surface (2.6 Gt/year). Through the carbon cycle, the hydrosphere and lithosphere have clearly been doing their fair share of absorbing our increasing carbon emissions!

Although this removal of carbon from the atmosphere acts to reduce the rate of global warming—a good thing, no doubt—its effects on marine life can be deadly. Anthropogenic carbon emissions are being absorbed by the oceans, making seawater more acidic, and this ocean acidification is increasing the solubility of calcium in seawater, making it more difficult for key marine organisms to form their calcium carbonate shells and skeletons (see Figure 12.18). Coral reefs are already in trouble (see Figure 12.13), and if the present trends continue, ocean acidification could cause population declines in common marine organisms such as starfish and mollusks within the next few decades. In fact, some biologists believe that this type of global change has already contributed to massive die-offs of starfish recently reported on both the east and west coasts of North America.

What will happen on land is less clear. In fact, exactly what is happening to the huge amount of carbon dioxide being pulled out of the atmosphere by terrestrial plants has been a real puzzle (see the Practicing Geology Exercise).

PRACTICING GEOLOGY EXERCISE

Balancing Carbon Emission with Carbon Accumulation: The Case of the Missing Sink

Understanding how humans are changing the carbon cycle is one of the most pressing issues in Earth science today because it holds the key to learning to manage anthropogenic global change. We can see from Figure 12.20 that, of the 8.9 Gt/year of carbon emitted by humans in 2000–2009, 2.6 Gt/year—almost a third—was absorbed by the land surface. Plant photosynthesis and respiration dominate the exchange of CO_2 between the atmosphere and the land surface, so an increase in the rate of photosynthesis by land plants must clearly be the cause. But where on Earth is this happening? This question was so hard to answer that scientists for years called it the "missing sink" problem. The answer turns out to be important, because future treaties in which nations agree to regulate their carbon emissions will need to take into account all carbon sources and sinks within each nation's boundaries.

As shown in the accompanying figure, the total anthropogenic carbon emissions rose from an average of 6.9 Gt/year in the 1980s, to 8.0 Gt/year in the 1990s, to 8.9 Gt/year in the 2000s. The rate at which these atmospheric emissions were absorbed into the ocean also increased, so that the

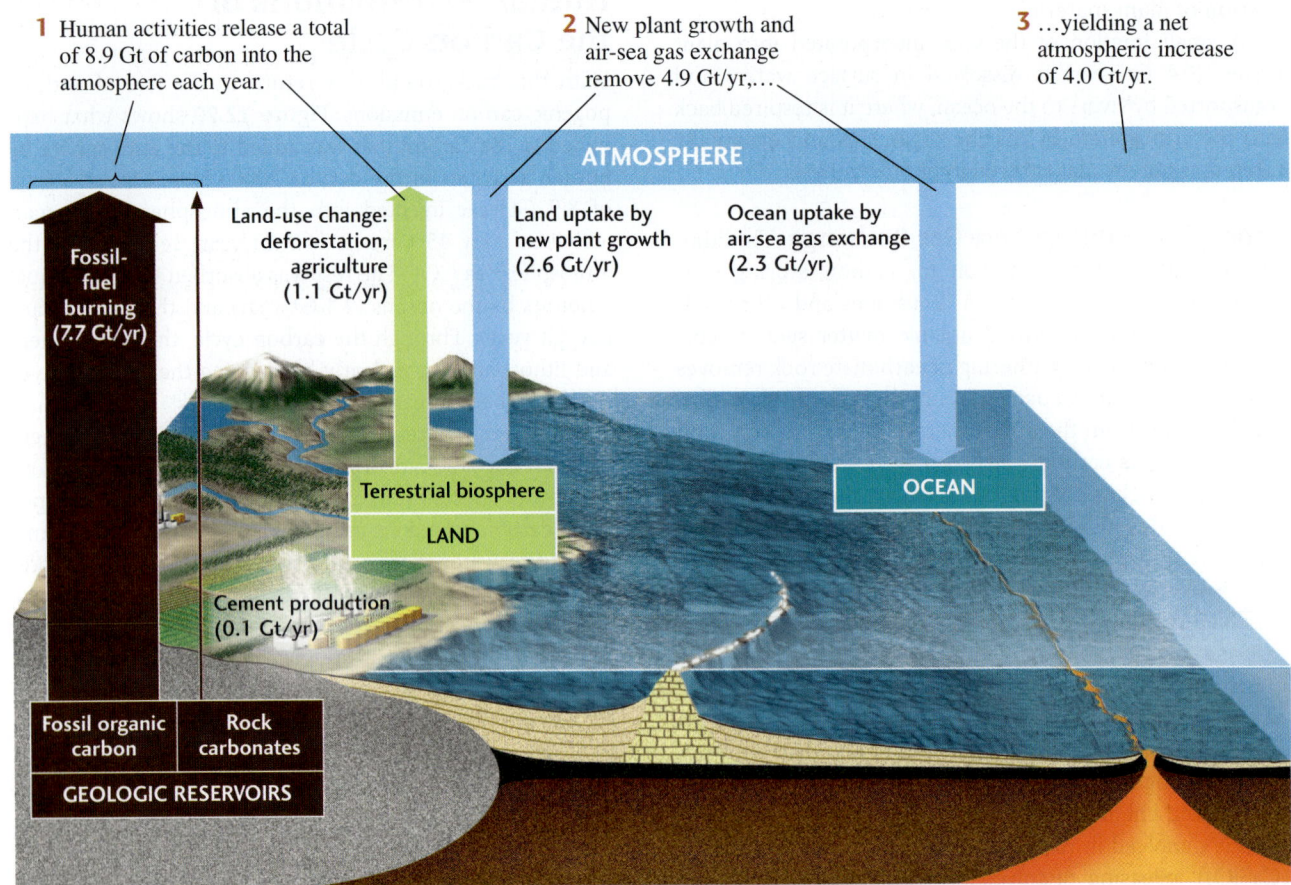

FIGURE 12.20 Much of the CO_2 emitted into the atmosphere by human activities is absorbed by the oceans and by plant growth on land. The remainder stays in the atmosphere, increasing the concentration of CO_2. The fluxes shown in this figure (given in gigatons of carbon per year) are for the decade 2000–2009. [Data from IPCC, *Climate Change 2013: The Physical Science Basis*.]

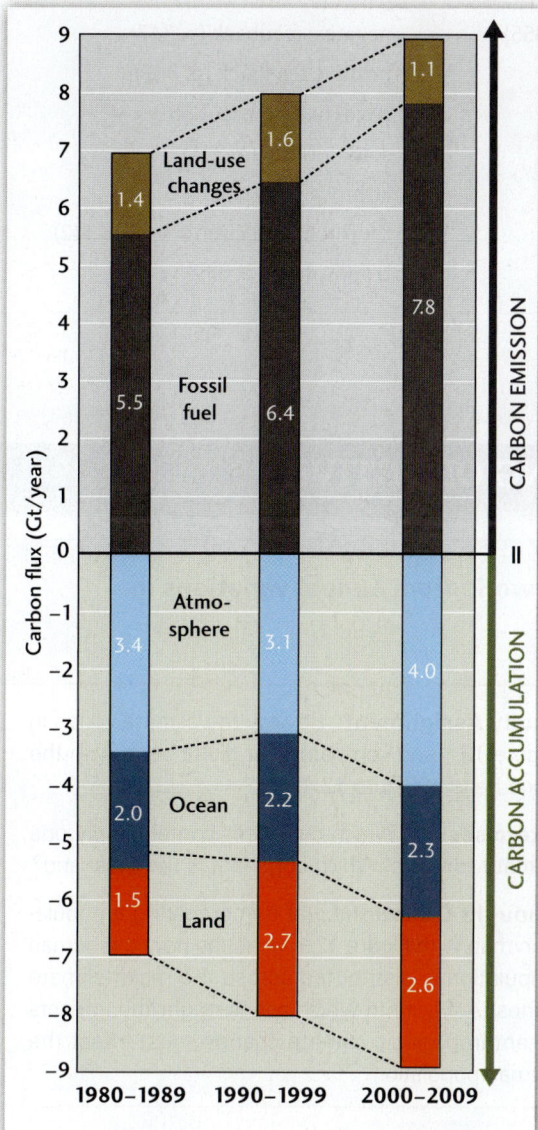

Comparison of the global carbon balance for the last three decades. [Data from IPCC, *Climate Change 2013: The Physical Science Basis*.]

percentage captured by the oceans remained nearly constant. However, the fraction of anthropogenic carbon accumulating in the atmosphere was not nearly so steady; in fact, the rate actually went down between the 1980s and 1990s from 3.4 Gt/year to 3.1 Gt/year, and then jumped to 4.0 Gt/year in the 2000s.

Atmospheric accumulation varies inversely with the carbon absorbed by the missing sink. The total amount of carbon is conserved; therefore, when summed over all geochemical reservoirs, carbon accumulation must balance carbon emission, as shown graphically by the bar charts in the figure. This balance allows us to calculate the amount of carbon absorbed by the missing sink:

$$\text{missing sink} = \text{total carbon emissions} - \text{atmospheric accumulation} - \text{ocean accumulation}$$

The values are given by the red bars in the figure. Carbon absorption by the missing sink increased by 80 percent from 1.5 Gt/year in the 1980s to 2.7 Gt/year in the 1990s, and then decreased slightly to 2.6 Gt/year in the 2000s.

An obvious place to look for the missing carbon is in the world's forests, which account for about half the annual terrestrial uptake of CO_2 by photosynthesis. Forests are classified according to the climate zones in which they grow: boreal, temperate, or tropical (see Figure 12.1). Early models suggested that the growth of temperate forests might account for most of the missing carbon. However, data in the IPCC's *Fifth Assessment Report* indicate that the current carbon accumulation is almost as large in the boreal regions and actually greater in the tropical regions.

$$\underset{(2.6\text{ Gt/year})}{\text{missing sink}} = \underset{(0.5\text{ Gt/year})}{\text{boreal accumulation}} + \underset{(0.8\text{ Gt/year})}{\text{temperate forest growth}} + \underset{(1.3\text{ Gt/year})}{\text{tropical forest growth}}$$

This newer model suggests that tropical forest growth (1.3 Gt/year) is more than offsetting tropical deforestation, which accounts for most of the carbon emissions from land-use changes (1.1 Gt/year).

These estimates not only demonstrate the importance of our forests as carbon sinks, but also raise major policy issues about how they should be managed. For example, how much "carbon credit" should nations such as the United States and Brazil receive for the carbon taken up by their forests? Such issues will figure prominently in the negotiation of international treaties to deal with anthropogenic global change.

BONUS PROBLEM: From the data shown in the figure, calculate the net carbon flux from the land surface by balancing the carbon emitted by land-use changes with the carbon absorbed by the missing sink. In the three decades shown, was this net carbon flux from the land surface positive (net emission) or negative (net absorption)? What factors might explain why the magnitude of this net flux increased steadily from the 1980s to the 2000s?

KEY TERMS AND CONCEPTS

albedo (p. 346)	geochemical cycle (p. 355)	Milankovitch cycle (p. 353)
anthropogenic global change (p. 345)	geochemical reservoir (p. 355)	negative feedback (p. 347)
	glacial cycle (p. 350)	positive feedback (p. 347)
carbon cycle (p. 355)	global warming (p. 345)	residence time (p. 355)
climate model (p. 348)	greenhouse effect (p. 346)	solar forcing (p. 338)
climate zones (p. 338)	greenhouse gas (p. 346)	stratosphere (p. 339)
El Niño (p. 349)	ice age (p. 350)	thermohaline circulation (p. 342)
ENSO (p. 349)	interglacial period (p. 350)	troposphere (p. 339)

REVIEW OF LEARNING OBJECTIVES

12.1 Recognize the major climate zones of today's world from annual variations in temperature and precipitation.

In tropical climate zones, the average temperature of the coldest month is greater than 18°C; in polar zones, the average temperature of the warmest month is less than 10°C. Temperate and boreal zones, which lie between these extremes, are differentiated by whether the coldest monthly average is greater than or less than 0°C. Arid zones are areas of low precipitation. Climate subzones are distinguished by seasonal variations in temperature and precipitation, often correlated with dominant types of vegetation; for example, tropical subzones characterized by wet rainforests and dry savannas.

Study Assignment: Review the climate zones in Figure 12.1 and compare their distribution with the wind belts of Figure 12.4.

Exercises: (a) Which continent contains only one climate zone? (b) Which continent is the most arid?

Thought Questions: (a) By comparing a population map with Figure 12.1, describe how the human population is distributed across the major climate zones (A–E). (b) In which zones might the impacts of anthropogenic climate change most affect the human population?

12.2 Separate the climate system into its major components and describe how the transport of energy and mass within and among these components affects climate zonation.

The climate system includes all of the components of the Earth system, and all of the interactions among those components, that determine how climate varies in space and time. The main components of the climate system are the atmosphere, hydrosphere, cryosphere, lithosphere, and biosphere. Each component plays a role in the climate system that depends on its ability to store and transport mass and energy. Tropospheric convection is organized of a banded structure of prevailing wind belts, each belt subtending about 30° in latitude, that transport heat energy from lower latitudes to higher latitudes. Seawater has a high heat capacity, so the much slower ocean currents are also important in transporting heat from the equatorial regions to the polar regions. The exchange of water between the cryosphere and hydrosphere plays a major role in the climate system, in particular by changing Earth's albedo. The land surface absorbs and releases solar energy, and its topography governs the flow of air and water. Volcanism from the lithosphere affects climate by changing the composition and properties of the atmosphere.

Study Assignment: Review the section *Components of the Climate System*.

Exercises: (a) Why are most of the world's arid zones concentrated near latitudes of 30°N and 30°S? (b) Why is the climate of Great Britain much milder than regions of Canada at the same latitude?

Thought Question: Why do the subtropical ocean gyres circulate water clockwise in the Northern Hemisphere and counterclockwise in the Southern Hemisphere?

12.3 Explain the greenhouse effect in terms of radiation balance and feedbacks within the climate system.

When Earth's surface is warmed by the Sun, it radiates heat back into the atmosphere. Carbon dioxide and other greenhouse gases absorb some of this infrared radiation and reradiate it in all directions, including downward to Earth's surface. This radiation maintains the atmosphere at a warmer temperature than it would be if there were no greenhouse gases, similar to the warmer air temperature maintained in a greenhouse. Positive feedbacks associated with variations in water vapor and albedo can enhance the greenhouse effect, while radiative damping, a negative feedback, stabilizes Earth's climate against major temperature changes.

Study Assignment: Review the radiation balance shown in Figure 12.10.

Exercises: (a) Name three atmospheric gases that contribute to the greenhouse effect. (b) The energy fluxes in Figure 12.10 are average values. What are they averaged over?

Thought Questions: (a) Why is it incorrect to assert that greenhouse gases prevent heat energy from escaping to outer space? (b) Give an example not discussed in this chapter of a positive feedback and a negative feedback in the climate system.

12.4 Describe how the climate changes on the short term during the El Niño–Southern Oscillation and over the long term during glaciations.

Natural variations in climate occur on a wide range of scales in both time and space. Some variations result from factors outside the climate system, such as solar forcing and changes in the distribution of land and sea surfaces caused by continental drift. Others result from variations within the climate system itself. Regional climate fluctuations with durations of a few years include the El Niño–Southern Oscillation. ENSO is characterized by two extreme phases, El Niño, when the trade winds weaken, cutting off the Peruvian upwelling and equalizing water temperatures across the tropical Pacific, and La Niña, when the trade winds strengthen, enhancing the temperature difference between the eastern and western Pacific. Long-term global climate variations are exemplified by the Pleistocene glaciations, continental ice sheets advanced and retreated many times, during which average surface temperatures changed by as much as 6°C to 8°C. Each ice age involved a massive transfer of water from the hydrosphere to the cryosphere, resulting in expansion of glaciers and a lowering of sea level.

Study Assignment: Observe which features the maps in Figure 12.12 correspond to the ENSO phenomena discussed in the text.

Exercises: (a) What do the data from ice cores say about the role of greenhouse gases in glacial cycles? (b) List three causes of climate change that result from plate-tectonic processes.

Thought Questions: (a) In what ways does an El Niño event change global weather patterns? (b) Evaluate the hypothesis that Pleistocene glacial cycles may have resulted from increased weathering rates due to the Cenozoic uplift of the Tibetan Plateau. In particular, discuss why increased weathering rates might cause global cooling.

12.5 Identify the signatures of Milankovitch cycles in the Pleistocene glacial record.

Ice advance and retreat during the Pleistocene glaciations has been driven by Milankovitch cycles, small periodic variations in Earth's movement through the solar system that alter the amount of solar radiation received at Earth's surface. Milankovitch cycles are caused by changes in the eccentricity of Earth's orbit (100,000-year period), the tilting of Earth's rotation axis (41,000-year period), and the precession of its rotation axis (25,000-year period). These variations have been amplified by positive feedbacks involving atmospheric concentrations of greenhouse gases.

Study Assignment: Review the section *Milankovitch Cycles*, paying special attention to Figure 12.17.

Exercises: (a) Which Milankovitch cycle has dominated glacial cycles during the last 600,000 years? (b) Which dominated during the period from 1.3–1.8 million years ago?

Thought Question: Do Milankovitch cycles fully explain the warming and cooling of the global climate during the Pleistocene glacial cycles?

12.6 Quantify the carbon cycle in terms of the flux of carbon between the atmosphere and the other principal carbon reservoirs.

The carbon cycle is the flux of carbon among its four principal reservoirs: the atmosphere, the lithosphere, the oceans, and the land surface. Major fluxes of carbon between these reservoirs include gas exchange between the atmosphere and the ocean surface (80 Gt/year); the movement of carbon dioxide between the biosphere and the atmosphere through photosynthesis, respiration, and direct oxidation (120 Gt/year); the transport of dissolved organic carbon in surface waters to the ocean and then into the atmosphere (0.4 Gt/year); and the weathering and precipitation of calcium carbonate (0.2 Gt/year).

Study Assignment: Practicing Geology Exercise

Exercises: (a) Between 2000 and 2010, emissions of carbon into the atmosphere from fossil-fuel burning and other industrial activities averaged about 7.8 Gt/year, almost all of it in the form of carbon dioxide. What was the mass of carbon dioxide emitted? (b) From the information given in Figure 15.18, estimate the residence time of carbon dioxide (i) in the ocean and (ii) in the atmosphere.

Thought Questions: How is the carbon cycle affected by (a) the weathering of silicate rocks and (b) human activity?

VISUAL LITERACY EXERCISE

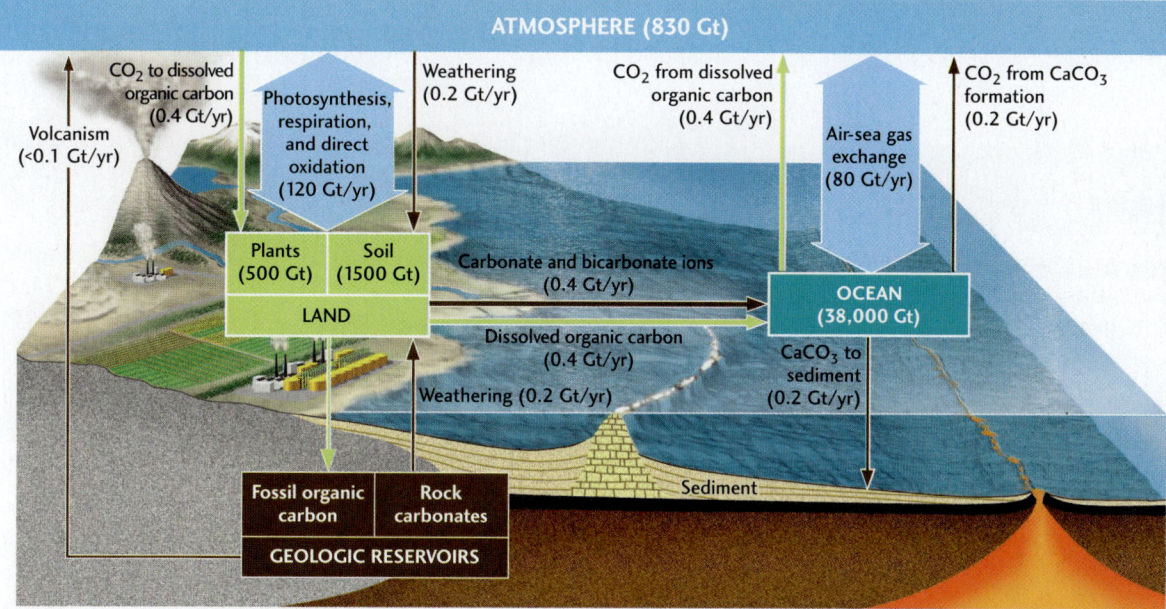

FIGURE 12.19 The carbon cycle describes the fluxes of carbon between the atmosphere and its other principal reservoirs. Amounts of carbon stored in each reservoir are given in gigatons; fluxes are given in gigatons per year. [Data from IPCC, *Climate Change 2001: The Scientific Basis*, updated according to IPCC, *Climate Change 2013: The Physical Science Basis*.]

1. What is the main topic of this figure?
   a. Calcium cycle
   b. Flux of carbonate and bicarbonate from the land to the oceans
   c. Carbon cycle
   d. Air-sea gas exchange

2. What is the annual flux of carbon between the atmosphere and ocean?

   a. 120 Gt/year
   b. 80 Gt/year
   c. 0.4 Gt/year
   d. This flux is not given in the figure.

3. Why do the blue arrows point both up and down?

4. True or false: The amount of carbon in the oceans is more than 40 times greater than the amount of carbon in the atmosphere.

Civilization as a Global Geosystem 13

Growth and Impact of Civilization	366
Fossil-Fuel Resources	370
Alternative Energy Resources	381
Our Energy Future	386

Learning Objectives

Human population and the energy needed to sustain this population are multiplying at phenomenal rates. This chapter contains the essential facts and figures that you will need to:

13.1 Explain how the impact of civilization on the Earth system qualifies it as a global geosystem.

13.2 Categorize our natural resources as renewable and nonrenewable, and differentiate energy reserves from energy resources.

13.3 Understand the geological processes that form fossil fuels and the energy available from their reserves.

13.4 Answer the question: When will civilization run out of oil?

13.5 Compute the carbon intensities of fossil fuels from the energy they produce and the carbon dioxide they emit, and use carbon intensities to compute the changes in carbon flux from changes in energy production.

13.6 Quantify the relative contributions of alternative energy resources to energy production, and estimate their potential to satisfy future energy needs.

13.7 Project the worldwide growth of energy consumption by geographic region and fuel type.

North and South America at night, showing the lights of our globalized, energy-intensive civilization. [Image by NOAA's National Geophysical Data Center from data collected by US Air Force Weather Agency.]

A better understanding of the Earth system has helped us find natural resources, sustain the natural environment, and reduce the risks from natural hazards. But the upward progress of our civilization cannot be taken for granted. The human population is growing at a phenomenal rate, and Earth's natural resources are limited. Environmental conditions and overall prosperity are not improving in some parts of the world, and the prospects for detrimental changes to the global environment loom large. Balancing the benefits we reap from our use of natural resources against the costs of that use—particularly harmful long-term changes to our environment—raises new challenges for Earth science and society.

In this chapter, we show how civilization—the collective sum of human activities—has grown into a global geosystem that is changing Earth's surface environment at a phenomenal rate. We survey the energy resources that fuel our economy and examine how our future use of those resources will affect our environment. We focus on two of civilization's most pressing problems: the need for more energy resources to power economic development and the potential for energy usage to cause detrimental global change.

Our economy depends on the burning of non-renewable energy resources (fossil fuels) that produce a dangerous greenhouse gas (carbon dioxide). This stark reality poses some difficult questions: How long will our fossil-fuel resources last? How quickly is fossil-fuel burning increasing greenhouse gases in the atmosphere, and how will those increases change the climate system? How quickly will we need to replace fossil fuels with alternative energy sources to sustain our economy and environment? These questions have political and economic dimensions that extend far beyond Earth science, so they do not have strictly scientific answers. Nevertheless, the decisions we will make as a society must be informed by our best scientific predictions about how the Earth system will change over the next decades and centuries. Realistic predictions about the future global environment can be made only if we include civilization as part of the Earth system.

Growth and Impact of Civilization

The human habitat is the thin, water-rich interface where Earth meets sky, where three global geosystems—the climate system, plate tectonics, and the geodynamo—interact to provide a life-sustaining environment. We have increased our standard of living by developing technologies that can ever more efficiently exploit this environment: grow more food, extract more minerals, transport more materials, build more structures, and manufacture more goods of all kinds. One result has been an explosion in the human population.

Human Population Growth

Early in the Holocene, about 10,000 years ago, when the climate was warming and agriculture first began to flourish, roughly 100 million people were living on the planet. Population grew slowly (**Figure 13.1**). The first doubling, to 200 million, was achieved early in the Bronze Age, about 5000 years ago, when humans first learned how to mine ores and refine them into metals such as copper and tin (of which bronze is an alloy). The second doubling, to 400 million, was not achieved until the Middle Ages, about 700 years ago. But once industrialization began in the late eighteenth century, the global population really took off, climbing to 1 billion in about 1800, 2 billion in 1927, and 4 billion in 1974. By the mid-twentieth century, the doubling time for the human population had dropped to only 47 years—less than a human lifetime. The world population exceeded 7 billion in early 2012, and, although its growth rate is expected to decline, the total will almost surely top 8 billion by 2030.

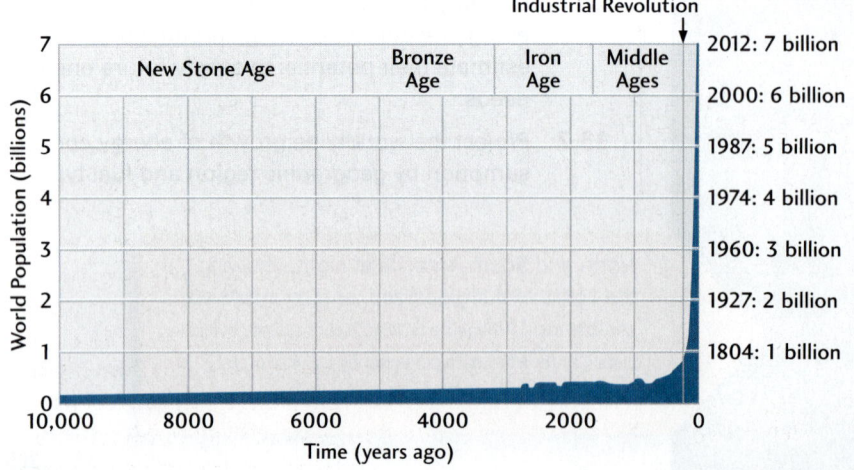

FIGURE 13.1 Human population growth over the last 10,000 years. The global population is expected to reach 8 billion by 2028. [Data from the U.S. Census Bureau.]

As our population has exploded, the demand for natural resources has skyrocketed. Our energy usage has risen 1000 percent over the last 70 years and is now increasing twice as fast as the human population. The view of Earth from space in this chapter's opening photograph shows a glowing lattice of highly energized urbanization spreading rapidly across the planet's surface.

Human activities have altered the environment by deforestation, agriculture, and other land-use changes since civilization began. But in earlier times, the effects of our species, *Homo sapiens*, were usually restricted to local or regional habitats. Today, industrialized energy production makes it possible for civilization to compete with the climate system and plate tectonics in modifying Earth's surface environment. The global scale of these modifications can be illustrated by some startling observations:

- Human activities are currently eroding the land surface 10 times faster than all natural processes combined.
- Since the beginning of the Industrial Revolution, humans have increased the sediment load of the world's rivers by almost 30 percent. Dams and reservoirs built by humans now trap almost 40 percent of this global sediment load before it reaches the oceans.
- In the last century, humans have converted about one-third of the world's forested area to other land uses, primarily agriculture.
- Within 50 years after the invention of the artificial coolant freon, enough of it had leaked out of refrigerators and air conditioners and floated into the upper atmosphere to damage Earth's protective ozone layer.
- Burning of fossil fuels has increased the concentration of carbon dioxide in the atmosphere by almost 50 percent over preindustrial levels.

We are not just part of the Earth system; we are transforming how the Earth system works, perhaps in fundamental ways. In the blink of a geologic eye, civilization has developed into a full-fledged global geosystem.

Energy Resources

Energy is required to do work; hence, access to energy is fundamental to all aspects of civilization, including population growth. **Energy resources** refer to the total energy that civilization could potentially produce from the natural environment. A century and a half ago, most of civilization's available energy was produced by burning wood (**Figure 13.2**). A wood fire, in chemical terms, is the combustion of *biomass*, organic matter consisting of carbon and hydrogen compounds, or **hydrocarbons.** Biomass is produced by plants and animals in a food web based on photosynthesis. Thus, the ultimate source of the energy in wood is the sunlight plants use to convert carbon dioxide and water into hydrocarbons. Combustion of wood or other biomass produces heat energy and returns carbon dioxide and water to the environment. In this capacity, the biomass acts as a short-term reservoir for storing solar energy. Biomass is a **renewable resource** because the biosphere is constantly producing new hydrocarbons. Before the mid-nineteenth century, the burning of wood and other biomass derived from plants and animals (e.g., whale oil, dried buffalo dung) satisfied most of society's need for fuel. Today, the energy derived from biomass is only a small share of our total

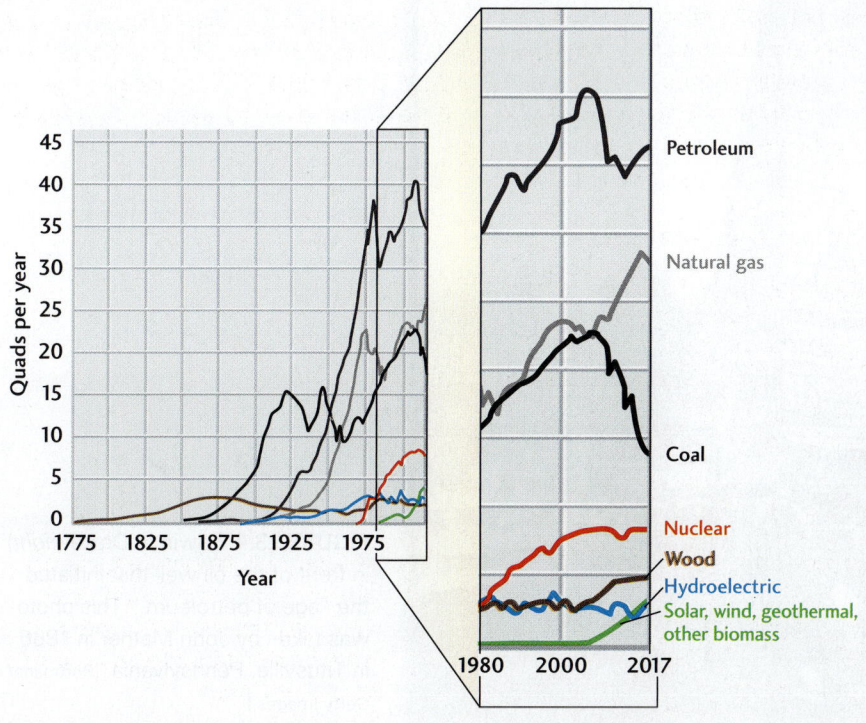

FIGURE 13.2 U.S. energy consumption, shown here for the period 1775–2017, has successively been dominated by the burning of wood, coal, and oil. In just the past few years, the energy from coal has dropped while that from natural gas and renewable sources such as biomass, solar, and wind has increased. Here and elsewhere in this chapter, the energy units are quads per year (1 quad = 10^{15} Btu = 1.054×10^{18} J).
[Data from U.S. Energy Information Agency.]

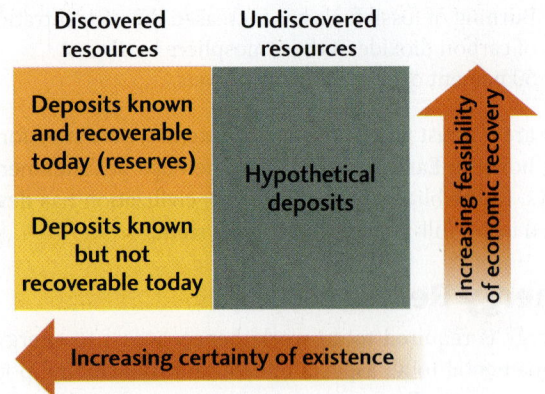

FIGURE 13.3 Fossil-fuel resources include reserves, plus known but currently unrecoverable deposits, plus undiscovered deposits that geologists think may eventually be found.

energy usage, though that fraction is now increasing owing to the industrial production of ethanol and other biofuels.

Coal, a combustible rock, was formed from biomass that was buried in sedimentary rock formations millions of years ago, particularly during the Carboniferous period. When we burn coal, we are using energy stored from Paleozoic sunlight. Hence, the primary source of this fossilized energy is the same solar power that drives the climate system. Our other major fuels, crude oil (petroleum) and natural gas (methane), are also created from dead organic matter. Coal, oil, and natural gas are hydrocarbon-rich substances known collectively as **fossil fuels.** Fossil-fuel resources are the entire supply of fossil fuels in the lithosphere that may become available for human use in the future. Geologists use a more restrictive term, **reserves,** to describe proven deposits that can be exploited economically (and legally) at the present time. Resources are more uncertain "guesstimates" that include reserves plus known but currently unrecoverable supplies plus undiscovered supplies that are likely be found sometime in the future (Figure 13.3). At current rates of use, our fossil-fuel reserves will be exhausted long before geologic processes can replenish them; therefore, they are **nonrenewable resources.**

Rise of the Hydrocarbon Economy

Civilization has used a variety of renewable energy sources to power mills and other machinery for thousands of years, including wind, falling water, and the work of horses, oxen, and elephants. By the late eighteenth century, however, industrialization was increasing the demand for energy beyond what these traditional renewable sources could supply. About 1780, James Watt and others developed coal-fired steam engines that could do the work of hundreds of horses, sparking the Industrial Revolution. Steam technology lowered the price of energy dramatically, in part because it made coal mining possible on an industrial scale. By the end of the nineteenth century, coal accounted for more than 60 percent of the U.S. energy supply (see Figure 13.2).

The first oil well was drilled in Pennsylvania by Colonel Edwin L. Drake in 1859. The idea that petroleum could be profitably mined like coal provoked skeptics to call the project "Drake's Folly" (Figure 13.4). They were wrong, of course. By the early twentieth century, oil and natural gas

FIGURE 13.4 Edwin L. Drake (right) in front of the oil well that initiated the "age of petroleum." This photo was taken by John Mather in 1866 in Titusville, Pennsylvania. [Bettmann/Getty Images.]

were beginning to displace coal as the fuels of choice. Not only did they burn more cleanly than coal, producing no ash, but they could be transported by pipeline as well as by rail and ship. Moreover, gasoline and diesel fuels refined from crude oil were suitable for burning in the newly invented internal combustion engine.

Today, the engine of civilization runs primarily on fossil fuels. Taken together, oil, natural gas, and coal account for 81 percent of global energy consumption. We can fairly call the civilization fed by this energy system a **hydrocarbon economy**: one that takes hydrocarbons stored in the lithosphere and burns them in the atmosphere to produce the energy that humans require. This burning also produces waste materials that humans must either bury in the lithosphere or release into the environment. Essentially all of the carbon dioxide produced by fossil-fuel burning is released into the atmosphere.

To understand the implications, we will undertake an accounting of our energy usage in terms of the carbon dioxide it produces, comparing fossil fuels with alternative sources of energy. This accounting requires a careful estimation of how much carbon dioxide is produced by each unit of energy captured during the burning of a fuel, a ratio called the **carbon intensity** (see the Practicing Geology Exercise at the end of the next section).

Energy Consumption

Energy use is often measured in units appropriate to the fuel—for example, barrels of oil, cubic feet of natural gas, and tons of coal. But comparisons are much easier if we express fuel consumption in terms of its *equivalent energy*, the amount of energy the fuel produces on burning. A popular choice for measuring equivalent energy is the British thermal unit (Btu). One Btu is the amount of energy needed to raise the temperature of 1 pound of water by 1°F (1055 joules). When we measure large quantities, such as a nation's annual energy use, we use units of 1 quadrillion (10^{15}) Btu, or **quads.** One quad (1.055×10^{18} joules) is a huge amount of energy, equivalent to burning 170 million barrels of oil, 970 billion cubic feet of natural gas, or 36 million tons of coal.

Energy production of the United States in 2015 was about 97.7 quads of energy (**Figure 13.5**), compared with a global total of 536 quads. Thus, the United States, with 4.4 percent of the world's population, consumes about four times more energy per person (or *per capita*) than the

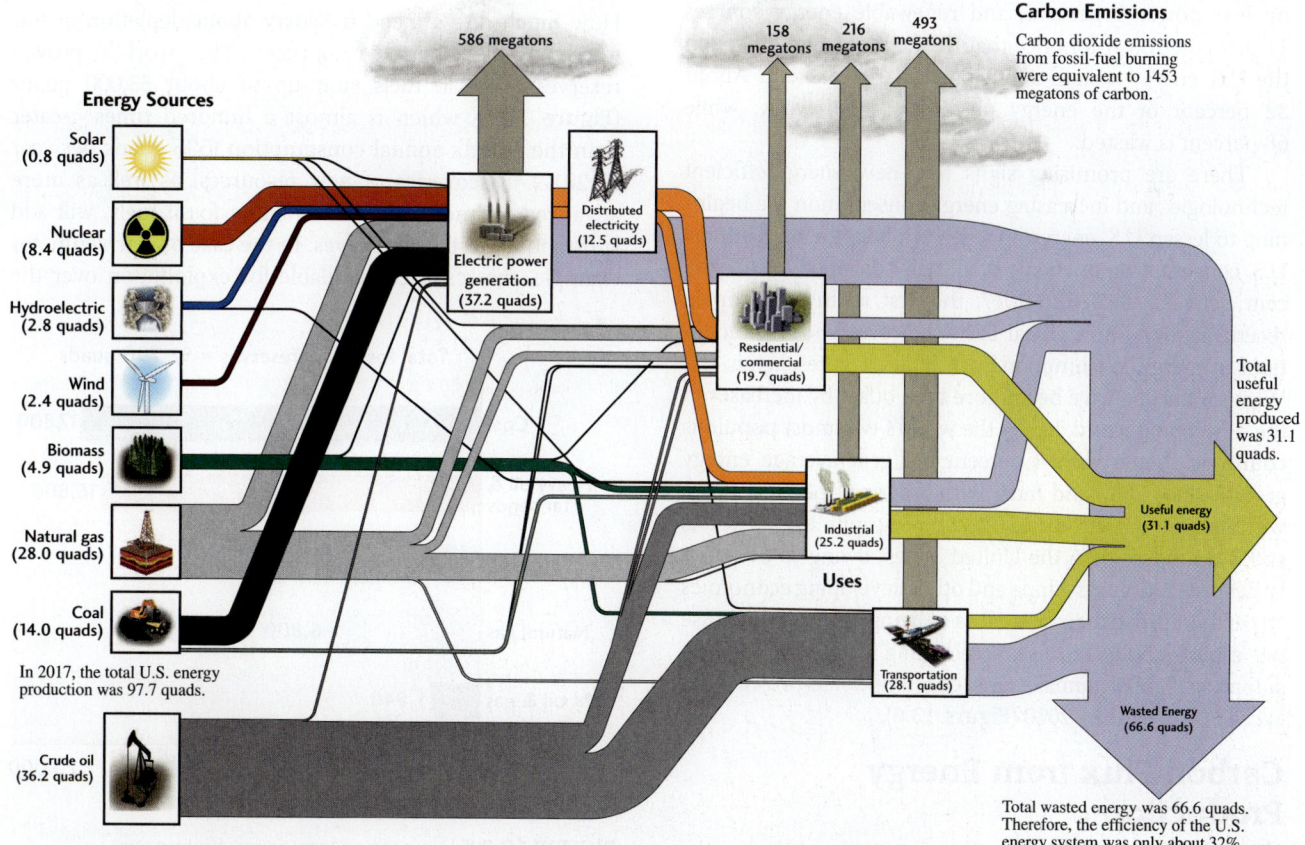

FIGURE 13.5 Energy consumption in the United States in 2017 (in quads). Energy from primary fuel sources (boxes on left side) is delivered to the residential, commercial, industrial, and transportation sectors (boxes in middle to right side). Not represented are small contributions to electric power generation from geothermal energy (0.2 quad). [Information from Lawrence Livermore National Laboratory, based on data from the Energy Information Administration.]

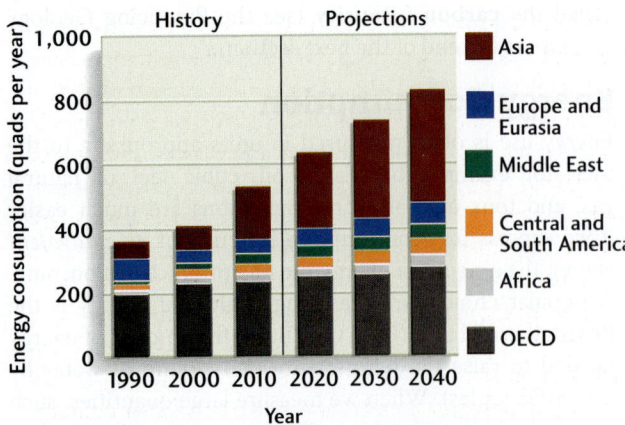

FIGURE 13.6 Historical and projected energy consumption in quads per year by world regional groupings, 1990–2040. The Organization for Economic Cooperation and Development (OECD) includes countries in Western Europe, North America, and Australia. About 70 percent of the growth in energy consumption will come from developing nations outside the OECD. [Data from the U.S. Energy Information Agency, 2015.]

global average. Fossil fuels provided 80 percent of that total, nuclear power 9 percent, and renewable energy sources 11 percent. You will notice that the flow of energy through the U.S. energy system is not particularly efficient: About 32 percent of the energy performs useful work, while 68 percent is wasted.

There are promising signs that new energy-efficient technologies and increasing energy conservation are beginning to lessen U.S. energy appetites. In fact, the total annual U.S. consumption of energy of all types dropped by 3.7 percent between 2007 and 2017, the first multiyear drop in recent history. On a global basis, however, modest reductions in energy consumption by the United States, Japan, and Western Europe have been more than offset by increases in the developing world, led by the world's two most populous countries, China with 4 percent per year average energy growth since 2007 and India with 5 percent per year. Even with its high growth rates, China's per capita energy use is still 3 times less than the United States, owing to its much larger population. As China and other developing economies strive to improve their standards of living, global energy use per capita is bound to rise, accelerating overall energy consumption. Global annual energy consumption is projected to exceed 600 quads by 2020 (Figure 13.6).

Carbon Flux from Energy Production

In the prehuman world, the exchange of carbon between the lithosphere and the other components of the Earth system was regulated by the slow rates at which geologic processes buried and unearthed organic matter. This natural carbon cycle has been disrupted by the rise of the hydrocarbon economy, which is now pumping huge amounts of carbon from the lithosphere directly into the atmosphere. In Figure 13.5, you can see that the U.S. energy system released about 1.5 gigatons (Gt) of carbon into the atmosphere in 2017, primarily as CO_2 (1 Gt = 1 billion metric tons = 10^{12} kg). Over the decade 2000–2009, the total global emissions from fossil-fuel burning averaged 7.8 Gt per year, but this mass has increased to about 10 Gt per year in 2018.

As we saw in Chapter 12, the climate system is tightly coupled to the global carbon cycle because carbon dioxide is a greenhouse gas. The concentration of this gas has risen rapidly from its preindustrial level of about 280 ppm to over 410 ppm today. If the burning of fossil fuels continues unabated, the amount of CO_2 in the atmosphere will double its preindustrial level by mid-century.

Anthropogenic increases in carbon dioxide and other greenhouse gases have already led to enhancement of the greenhouse effect and global climate warming. It is clear that the future of the climate system and its living component, the biosphere, depends on how our society manages its energy resources, which we will now consider in more detail.

Fossil-Fuel Resources

How much do we need to worry about depletion of our nonrenewable energy resources? The world's proven reserves of fossil fuels sum up to about 53,000 quads (Figure 13.7), which is almost a hundred times greater than the world's annual consumption (536 quads per year in 2017). Discoveries of new resources, as well as more advanced technologies for extracting fossil fuels, will add substantially to these reserves. In the case of petroleum, for example, the resources available for exploitation over the

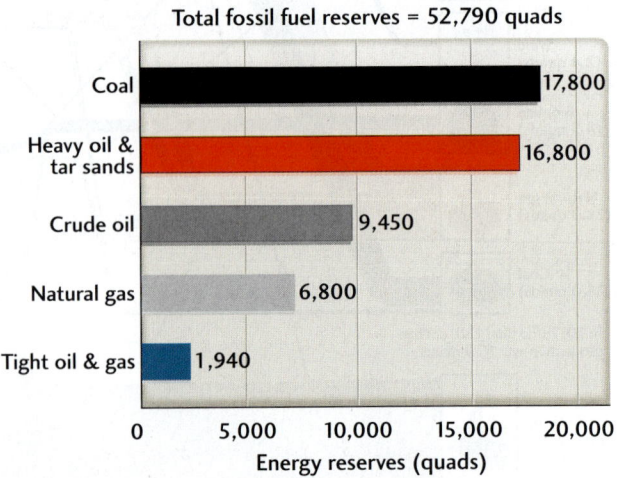

FIGURE 13.7 Estimates of total fossil-fuel energy reserves amount to about 53,000 quads. The total fossil-fuel resources that could be recovered with improved technologies and lower costs are at least four times this amount. [Data from the World Energy Council.]

next century are estimated to be two to three times the current reserves. Hence, hydrocarbon resources will no doubt be available to power civilization for many decades to come. A number of factors complicate the economics of energy production, including the environmental costs of fossil-fuel burning and the increasing availability of renewable energy resources. In this section, we describe how fossil fuels are formed by geological processes, and we take stock of our current reserves.

The Geologic Formation of Hydrocarbons

Economically valuable deposits of hydrocarbons develop under special geologic conditions. Fossil fuels come from the organic debris of former life: plants, algae, bacteria, and other organisms that have been buried, transformed, and preserved in sediments.

Coal is a biological sediment formed from large accumulations of plant material in wetlands where plant growth is very rapid (Figure 13.8). Burial in waterlogged soil preserves the plant material by cutting it off from the oxygen needed for bacterial decay. As the plant material accumulates, it turns into *peat*, a porous brown mass of organic matter in which leaves, twigs, roots, and other plant parts can still be recognized. The accumulation of peat in oxygen-poor environments can be seen in modern swamps and peat bogs. When dried, peat burns readily because it contains up to 50 percent carbon.

Chemical reactions during compression and heating of the organic-rich sediments (diagenesis) can convert peat first to *lignite*, a very soft, brown coal containing 60–70 percent carbon and then to *bituminous* coal that contains up to 80 percent carbon. At still higher temperatures and pressures, low-grade metamorphic reactions transform soft coal to hard coal, or *anthracite*, which contains more than 90 percent carbon, further increasing its energy content.

Liquid and gaseous hydrocarbon fuels also begin to form in sedimentary basins where the production of organic

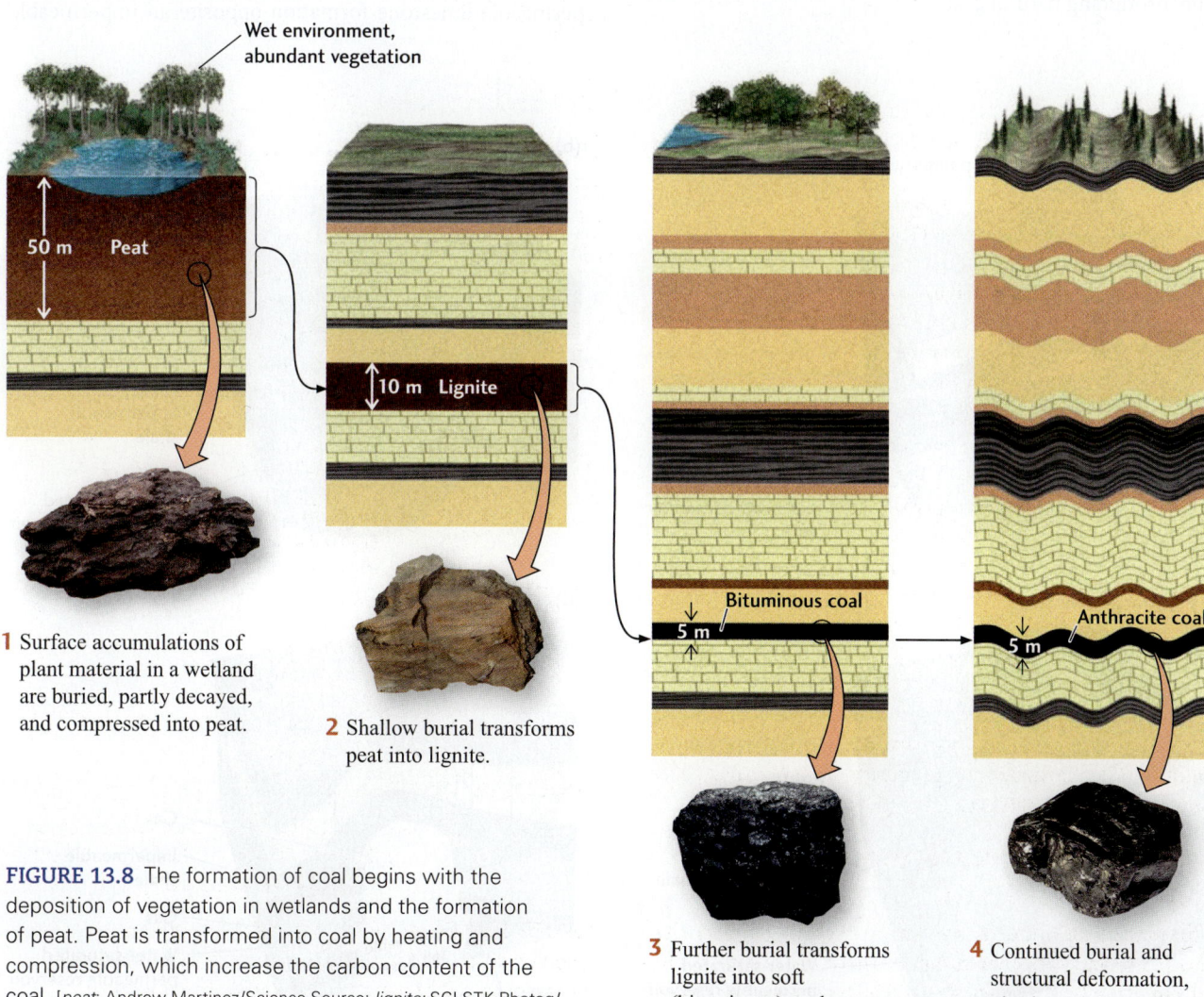

FIGURE 13.8 The formation of coal begins with the deposition of vegetation in wetlands and the formation of peat. Peat is transformed into coal by heating and compression, which increase the carbon content of the coal. [*peat*: Andrew Martinez/Science Source; *lignite*: SCI STK Photog/Science Source; *bituminous coal*: Mark A. Schneider/Science Source; *anthracite*: Gary Ombler/Getty Images.]

1 Surface accumulations of plant material in a wetland are buried, partly decayed, and compressed into peat.

2 Shallow burial transforms peat into lignite.

3 Further burial transforms lignite into soft (bituminous) coal.

4 Continued burial and structural deformation, plus heat, metamorphose soft coal into hard coal (anthracite).

matter is high and the supply of oxygen in the sediments is inadequate to decompose much of the organic matter they contain. Many offshore thermal subsidence basins on continental margins satisfy these two conditions, as do some river deltas and inland seas. During millions of years of burial, diagenetic reactions at elevated temperatures and pressures slowly transform the organic material in these *source beds* into combustible hydrocarbons. The simplest hydrocarbon is methane (CH_4), the primary constituent of **natural gas.** Raw petroleum, or **crude oil,** includes a diverse class of liquids composed of more complex hydrocarbons.

Crude oil forms at a limited range of pressures and temperatures, known as the **oil window,** usually found at depths between about 2 and 5 km (see the Chapter 6 Practicing Geology Exercise, *Where Do We Look for Oil and Gas?*). Above the oil window, temperatures are too low (generally below 50°C) for the maturation of organic material into hydrocarbons, whereas below the oil window, temperatures are so high (greater than 150°C) that the hydrocarbons that form are broken down into methane, producing natural gas.

Hydrocarbon Reservoirs

As sediment burial progresses, compaction of the source beds forces crude oil and natural gas into adjacent beds of permeable rock, such as sandstones or fractured limestones, which act as *hydrocarbon reservoirs*. The relatively low densities of oil and gas cause them to rise, so that they float atop the water that almost always occupies the pores of permeable rock formations. The conditions that favor large-scale accumulation of oil and natural gas are where permeable geologic formations are capped by layers of impermeable rock, such as shales or evaporites. This barrier to upward migration forms an **oil trap** (Figure 13.9).

Some oil traps, called *structural traps*, are created by rock deformation. One type of structural trap is formed by an anticline in which an impermeable layer of shale overlies a permeable sandstone formation (Figure 13.9a). The oil and gas accumulate at the crest of the anticline—the gas highest, the oil next—both floating on the groundwater that saturates the sandstone. Similarly, an angular unconformity or displacement across a fault may place a dipping permeable limestone formation opposite an impermeable

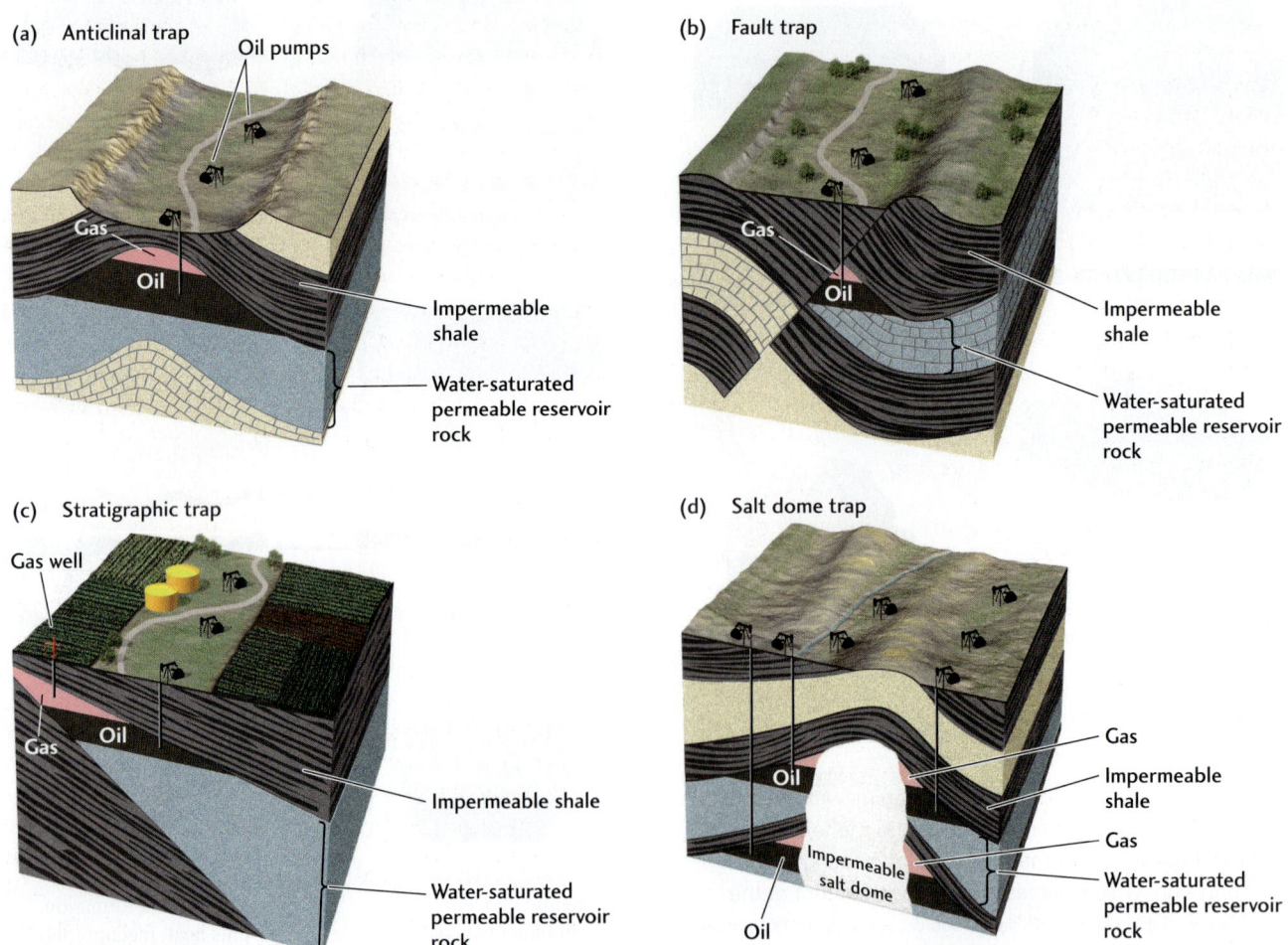

FIGURE 13.9 Oil and gas accumulate in traps formed by geologic structures. Four types of oil traps are illustrated here.

FIGURE 13.10 New technologies used aboard offshore platforms in the Gulf of Mexico can recover oil and gas from rock reservoirs below very deep waters. Drilling from a single platform like this one can cost over $100 million. [Larry Lee Photography/Getty Images.]

shale, creating another type of structural trap (Figure 13.9b). Other types of oil traps are created by the original pattern of sedimentation, such as when a dipping permeable sandstone formation thins out against an impermeable shale (Figure 13.9c). These structures are called *stratigraphic traps*. Oil can also be confined against an impermeable mass of salt in a *salt dome trap* (Figure 13.9d).

Extracting oil and gas by drilling deep into Earth's crust has become a very sophisticated and expensive business. Offshore platforms can drill into hydrocarbon reservoirs deep beneath the seafloor (**Figure 13.10**), which has opened vast areas of the continental margins to oil and gas production. Seismic imaging can map the reservoir rocks in three dimensions (see Figure 11.6), showing where the bulk of the oil and gas is located and mapping how it will flow from holes drilled into the reservoir. Water and carbon dioxide can be pumped down strategically positioned drill holes to push the oil into areas where it can be more efficiently recovered through other drill holes. These methods have increased the amount of oil and gas that can be extracted from hydrocarbon reservoirs and have thus increased our fossil-fuel reserves.

In their search for petroleum, geologists have seismically mapped thousands of oil traps throughout the world. Only a fraction of them have proven to contain economically valuable amounts of oil or gas, because traps alone are not enough to create a hydrocarbon reservoir. A trap will contain oil only if source beds were present, the necessary chemical reactions took place, and the oil migrated into the trap and stayed there without being disturbed by subsequent heating or deformation. Most of the large hydrocarbon reservoirs have already been discovered, and finding major new reservoirs is becoming more difficult.

Producing Oil and Gas from Tight Formations

Hydrocarbons are very common in sedimentary rocks, but only a small fraction has been concentrated as oil and gas in easily drilled reservoirs. The bulk of our petroleum resources are widely distributed, sealed within impermeable source beds called **tight formations.** Examples include extensive gas-rich shale deposits throughout North America, such as the Marcellus Formation that underlies the northern Appalachian Mountains and the Allegheny Plateau of the eastern United States (see Figure 6.21). Petroleum engineers have developed more efficient ways to extract oil and natural gas from tight formations. They use three-dimensional models of the sedimentary structures and sophisticated navigation systems to steer drill bits on horizontal paths through flat-lying sedimentary rocks, and they inject large amounts of water and sand through holes in the drill pipe to create tiny fissures in the rock, which allows gas to flow more readily back into the pipe (**Figure 13.11**). This combination of **horizontal drilling** and **hydraulic fracturing** (or "**fracking**") has revolutionized the oil and gas industry. The proven reserves of gas from shales and other tight formations now exceeds conventional gas reserves.

Distribution of Oil Reserves

In the decade 2007–2017, the world consumed about 0.33 trillion barrels of oil (1 barrel = 42 gallons). Yet, the worldwide reserves of oil did not decline; they *increased* by about the same amount, from about 1.43 trillion barrels in 2007 to 1.70 trillion barrels in 2017. Oil exploration is an immensely successful geologic activity!

Oil reserves and their decadal changes are displayed by region in **Figure 13.12**. The oil fields of the Middle East—including Iran, Kuwait, Saudi Arabia, Iraq, and the Baku region of Azerbaijan—account for 48 percent of the world's total. Here, sediments rich in organic material have been folded and faulted by the closure of the ancient Tethys Ocean, forming a nearly ideal environment for oil accumulation. The extensive reservoirs discovered in this vast convergence zone include the world's largest,

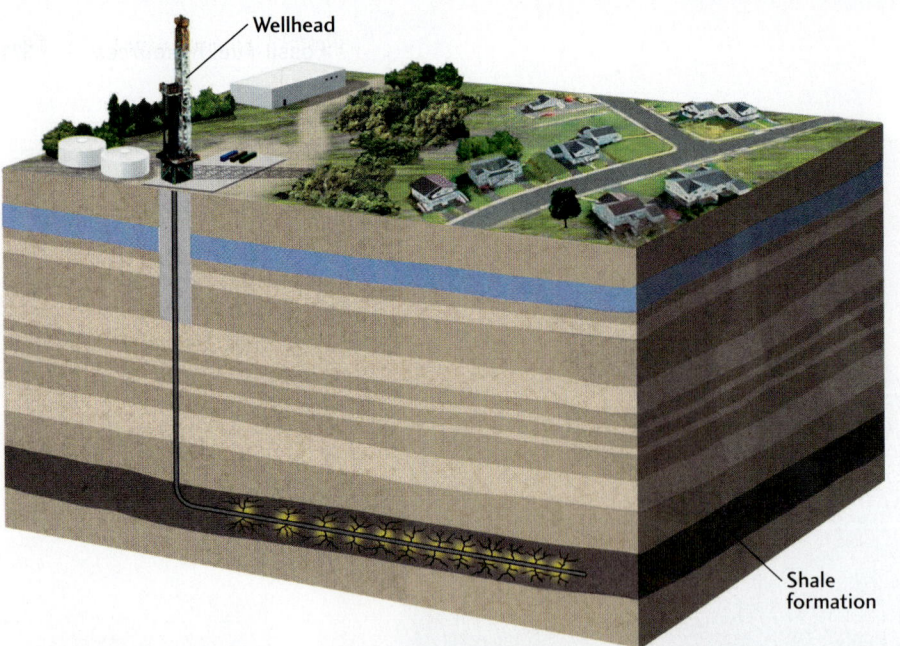

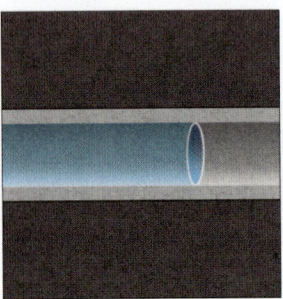
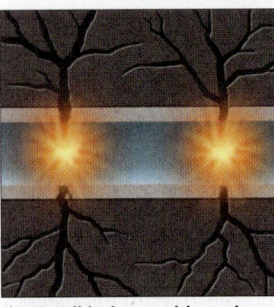
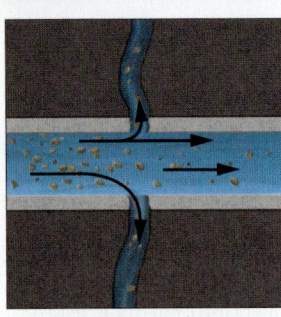
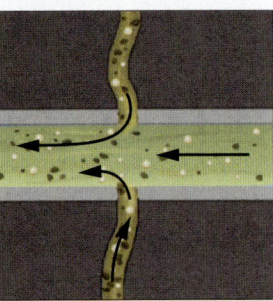

FIGURE 13.11 Hydraulic fracturing or "fracking" is a technique for withdrawing oil and gas from shale and other tight formations by first pumping water and sand into a borehole at high pressures to create fractures through which the oil and gas can more readily flow. The boreholes are commonly drilled horizontally through nearly flat-lying shale formations.

(a) Borehole is cased and surrounded by cement.

(b) Small holes are blasted through casing and cement.

(c) Surrounding rocks are hydraulically fractured by pumping water and sand into borehole at high pressure.

(d) Fracking generates small fissures that are kept open by the sand, allowing oil and gas to flow up the borehole to the wellhead.

the Ghawar field in Saudi Arabia. Ghawar has produced about 70 billion barrels of oil since its opening in 1948 and may produce another 70 billion barrels over its remaining lifetime.

Most of the oil reserves in the Western Hemisphere are located in the highly productive Gulf Coast–Caribbean area, which includes the Louisiana–Texas region, Mexico, Colombia, and Venezuela. The threefold increase in South

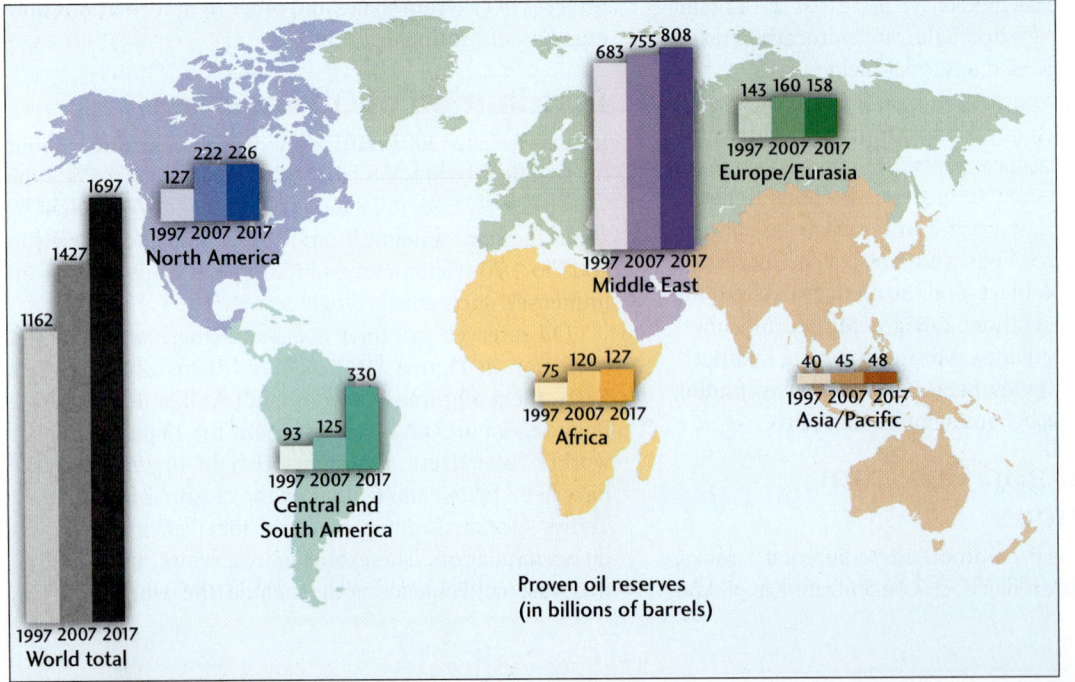

FIGURE 13.12 Regional estimates of world oil reserves in 1997 (left bar), 2007 (middle bar), and 2017 (right bar) in billions of barrels (bbl). The total world oil reserves in 2017 were 1.7 trillion barrels.
[Data from the *British Petroleum Statistical Review of World Energy 2017*.]

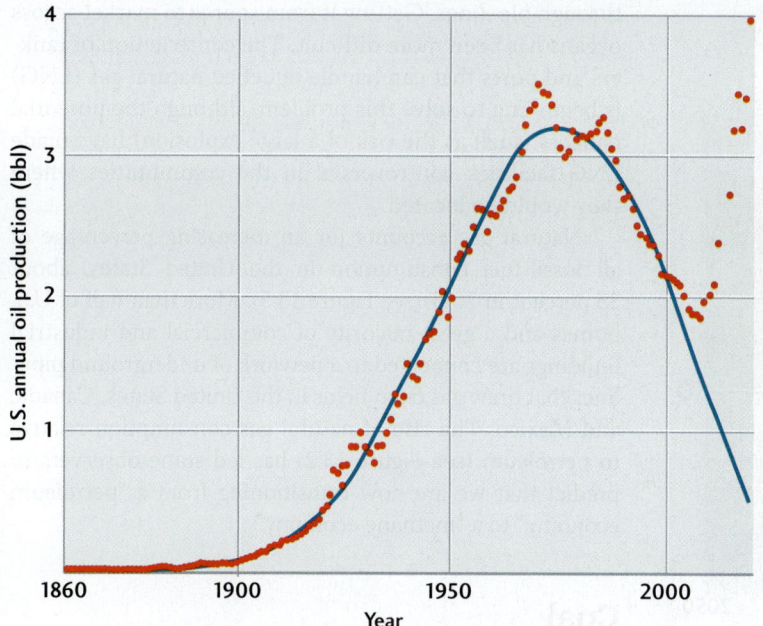

FIGURE 13.13 U.S. annual crude oil production in billions of barrels from 1860 to 2018. The red points show production figures for each year. The solid blue line is similar to Hubbert's 1956 projection, which predicted the peak in the 1970s and the subsequent decline. However, production reached a minimum in 2008 and has since been rapidly increasing, reaching nearly 4 billion barrels in 2018. [Data from the U.S. Energy Information Administration and from K. Deffeyes, *Hubbert's Peak*. Princeton, NJ: Princeton University Press, 2001.]

American oil reserves from 2007 to 2017 came mainly from improvements in oil-recovery technology, which will allow the heavy oil of Venezuela's Orinoco Basin to be exploited economically, as well as the discovery of huge new oil fields in the Atlantic Ocean off Brazil.

The United States' oil reserves also increased, from 30 billion barrels in 2007 to 50 billion barrels in 2017, placing it ninth worldwide. Thirty-one U.S. states have commercial oil reserves, and small, noncommercial resources can be found in most of the others.

Oil Production

Global oil production in 2018 was about 30 billion barrels per year worldwide. The United States produced 4 billion barrels, more than any other nation. The United States is a mature oil producer, whose history of production provides insights into the future of the global fossil-fuel supply. Production reached a maximum in 1970 and declined, approximately following a bell-shaped curve (Figure 13.13).

The high point is referred to as **Hubbert's peak,** named for petroleum geologist M. King Hubbert. In the mid-1950s, Hubbert used a simple mathematical relationship between the production rate and the rate at which new reserves were being discovered to predict that U.S. oil production would begin to decline sometime in the early 1970s. His arguments were roundly dismissed as overly pessimistic because, at the time, oil production was still growing rapidly. But history proved him right. Production did indeed peak in 1970, beginning a decline that continued much as Hubbert predicted throughout the late twentieth century.

In 2009, however, U.S. oil production suddenly began to increase again, and that trend has accelerated: 2018 production was almost 4 billion barrels, more than twice the 2008 minimum and surpassing the 1970 value of 3.5 billion barrels. This increase signals a new U.S. oil boom, fed by the rapid development of offshore oil fields and improved technology for recovering oil on land, including the controversial technique of fracking. These same technologies, which are now being applied in other regions, have created an oil glut that has caused the price of oil to drop from over $100 per barrel in 2008 to about $50 in 2017. We are not about to run out of oil anytime soon! (See Earth Issues 13.1.)

Natural Gas

The world's reserves of natural gas are comparable to its crude oil reserves (see Figure 13.7) and will likely exceed them in the decades ahead. Estimates of natural gas resources have been rising in recent years because exploration for natural gas has increased, and gas reservoirs have been identified in new settings, such as very deep rock formations, overthrust structures, coal beds, tight formations of sandstones and shales. Fracking technology has created a boom in the extraction of natural gas from shale formations that are extensive throughout (see Chapter 6). The production of "tight gas" from shale and other tight formations has increased more than a factor of three since 2000 and now accounts for about two-thirds of U.S. natural gas production (Figure 13.14).

Natural gas is a premium fuel for a number of reasons. In combustion, methane combines with atmospheric oxygen, releasing energy in the form of heat and producing only carbon dioxide and water. Natural gas therefore burns much more cleanly than oil or coal, which also produces sulfur dioxide (the major cause of acid rain). Moreover, natural gas emits 30 percent less CO_2 per unit of energy than oil and more than 60 percent less than coal. Therefore,

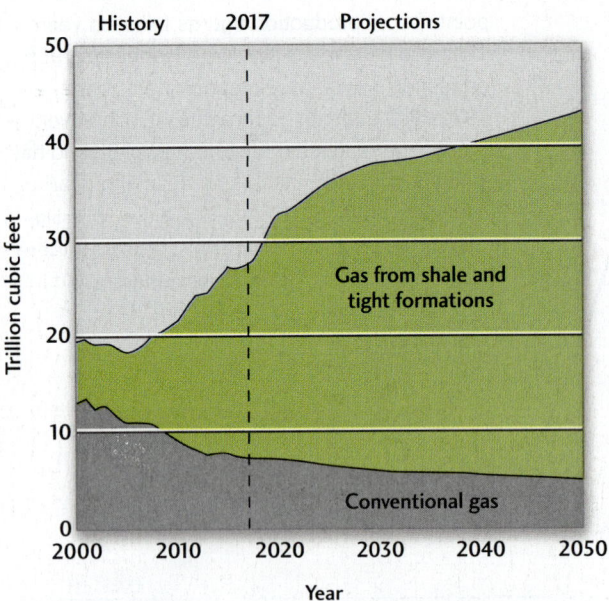

FIGURE 13.14 Gas produced from shales and other tight formations now accounts for almost two-thirds of natural gas production in the United States, and it is expected to constitute almost three-fourths by 2040. [Data from the U.S. Energy Information Administration, 2017.]

through pipelines. Getting it from source to market across oceans has been more difficult. The construction of tankers and ports that can handle liquefied natural gas (LNG) is beginning to solve this problem, although the potential dangers (such as the risk of a large explosion) have made LNG facilities controversial in the communities where they would be located.

Natural gas accounts for an increasing percentage of all fossil-fuel consumption in the United States, about 35 percent in 2017 (see Figure 13.5). More than half of U.S. homes and a great majority of commercial and industrial buildings are connected to a network of underground pipelines that draw gas from fields in the United States, Canada, and Mexico. The rise of natural gas consumption relative to petroleum (see Figure 13.2) has led some observers to predict that we are now transitioning from a "petroleum economy" to a "methane economy."

substituting natural gas for coal as, say, the fuel for power plants, lowers the carbon intensity of energy production (see the Practicing Geology Exercise at the end of this section). Natural gas is easily transported across continents

Coal

There are huge resources of coal in sedimentary rocks. Although coal has been a major energy source since the late nineteenth century, only about a few percent of the world's coal reserves have been consumed. According to the best estimates, these reserves amount to 860 billion metric tons, which are capable of producing 17,800 quads of energy, more than any other fossil fuel (see Figure 13.7). About 85 percent of the world's coal resources are concentrated in the Russian Federation, China, and the United States; these countries are also the world's largest coal producers. The United States has the largest reserves (**Figure 13.15**)—enough to last for a few hundred years at the nation's

FIGURE 13.15 Coal resources of the United States. In 2017, U.S. coal reserves were 251 billion metric tons, about one-quarter of the global total. [U.S. Bureau of Mines.]

Earth Issues 13.1 When Will We Run Out of Oil?

At the current production rate, the world would consume all of today's oil reserves in about 55 years. Does that mean we will run out of oil before the end of this century? No, because oil resources are much greater than oil reserves.

In fact, we will never really "run out" of oil. As resources diminish, prices will eventually rise so high that we cannot afford to waste oil by burning it as a fuel. Its main use will then be as a raw material for producing plastics, fertilizers, and a host of other petrochemical products. The petrochemical industry is already a very big business, consuming 7 percent of global oil production. As oil geologist Ken Deffeyes has noted, future generations will probably look back on the Petroleum Age with a certain amount of disbelief: "They burned it? All those lovely organic molecules, and they just burned it?"

The key question is not when oil will run out, but when oil production will stop rising and begin to decline. This milestone—Hubbert's peak for world oil production—would be the real tipping point; once it is reached, the gap between supply and demand would grow rapidly, driving oil prices sky-high.

So how close are we to Hubbert's peak? The answer to this question has been the subject of considerable debate. Hubbert himself predicted that world oil production would peak around the year 2000. Production using conventional recovery techniques on land did indeed reach a maximum about that time (2003) and have since declined. But Hubbert and other "oil pessimists" did not account for new technologies such as fracking and offshore drilling, which have substantially increased global oil reserves (see Figure 13.12) and spurred the remarkable resurgence of U.S. oil production (see Figure 13.13). Reaching Hubbert's peak now appears to be much less of an economic danger than the costly environmental effects of burning fossil fuels, especially the tremendous potential costs of polluting the atmosphere with carbon dioxide and other greenhouse gases.

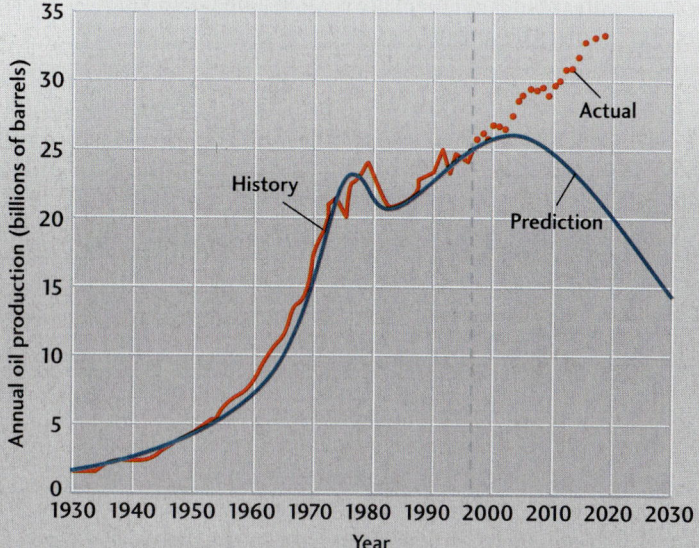

Based on data up to 1996 (red line), the oil geologist Colin Campbell published a prediction (blue line) that Hubbert's peak in global oil production would occur in the early 2000s. Actual production (red dots) has instead grown rapidly, owing almost entirely to increases from unconventional oil resources such as fracking and offshore drilling. [Prediction from C. J. Campbell & J. H. Laherrère, *Scientific American*, March 1998; data from *British Petroleum Statistical Review of World Energy 2018*.]

current rate of use (about 800 million tons per year). From 1975, when the price of oil began its precipitous rise, until 2005, coal supplied an increasing proportion of U.S. energy needs, primarily as fuel for electrical power generation. But coal usage has since declined precipitously as natural gas production has increased (see Figure 13.2). Coal currently accounts for about 14 percent of U.S. energy consumption.

Other Hydrocarbon Resources

Extensive deposits of hydrocarbons occur in two other forms: (1) source beds that are rich in organic material but never reached the oil window, and (2) formations that once contained oil but have since "dried out," losing many of their volatile components, to form a tarlike substance called *natural bitumen* (not to be confused with bituminous coal).

A hydrocarbon resource of the first type is *oil shale*, a fine-grained, clay-rich sedimentary rock containing large amounts of organic matter. In the 1970s, oil producers began trying to commercialize the extensive oil shales of western Colorado and eastern Utah, but those efforts were largely abandoned by the 1980s as petroleum prices fell, concerns over environmental damage increased, and technical problems persisted. The methods in use require the mining of oil shale, which is then crushed and heated to extract "shale oil" (not to be confused with "tight oil" derived by the fracking of shale beds and other tight formations). This process is expensive, consuming so much energy and producing so much environmentally dangerous waste material that the commercial production of shale oil is not currently feasible. However, new techniques, such as the production of shale oil by heating the rock formations in place, could change the economics. The resource implications would be significant, because the total amount of petroleum potential

PRACTICING GEOLOGY EXERCISE

Carbon Intensity of Fossil-Fuel Burning

The carbon intensity of a fuel is defined to be the mass of carbon emitted as CO_2 per unit of useful energy produced by burning the fuel. The amount of carbon emitted for a particular fossil-fuel type is a fixed fraction of its mass. For example, the burning of methane is represented by the following chemical equation:

$$CH_4 + 2O_2 = CO_2 + 2H_2O + \text{heat energy}$$

The amount of carbon (atomic mass of 12) emitted by the burning of 1 gigaton (10^{12} kg) of methane (atomic mass of 16) is $1 \text{ Gt} \times 12/16 = 0.75$ Gt of carbon. We can also measure the heat energy released during the reaction, which is 52 quads per gigaton of methane. The carbon intensity of methane burning is the ratio of these two quantities:

$$\text{carbon intensity} = \frac{\text{carbon emitted}}{\text{energy produced}}$$
$$= \frac{0.75 \text{ Gt}}{52 \text{ quads}} = 0.014 \text{ Gt/quad}$$

The carbon intensities of other fossil fuels are 0.020 Gt/quad for crude oil and 0.025 for coal, summarized in the following graph:

recoverable from known oil shale deposits is about 3 trillion barrels, twice the reserves of conventional petroleum.

A deposit of the second type, the *tar sands* of Alberta, Canada, is estimated to contain a hydrocarbon reserve equivalent to 170 billion barrels of oil and a total resource perhaps 10 times that amount—comparable to the worldwide reserves of conventional petroleum (**Figure 13.16**). More than 900 million barrels of oil are now extracted from the Alberta tar sands each year, but further development of the tar sands, like oil shales, raises important environmental concerns. It takes 2 tons of mined sand and 3 barrels of water to produce 1 barrel of oil, leaving lots of waste sand and water, which are environmental pollutants. Moreover, production of oil from the tar sands is an inefficient process that sucks up about two-thirds of the energy they ultimately render, and its carbon intensity is significantly higher than conventional oil production.

Environmental Costs of Extracting Fossil Fuels

Although the pollution of the atmosphere with greenhouse gases is the most serious environmental problem associated with the use of fossil fuels, the process of extracting a fossil fuel from the lithosphere can itself have serious detrimental effects on the environment. We illustrate the problems with a few examples.

Coal The extraction and combustion of coal present serious problems that make it a less desirable fuel than oil or natural gas. Underground coal mining is a dangerous profession; more than 2000 miners are killed each year in China alone. Many more coal miners suffer from black lung, a debilitating inflammation of the lungs caused by the inhalation of coal particles. Surface or "strip"

FIGURE 13.16 Surface mining of the Athabasca tar sands in Alberta, Canada, now produces more than 900 million barrels of oil each year. [dan_prat/Getty Images.]

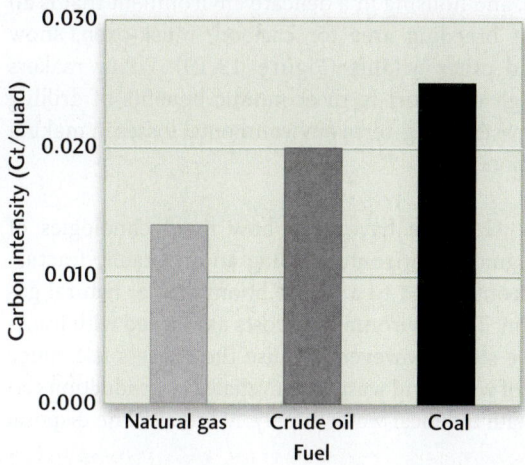

You can see that crude oil emits 40 percent more carbon per unit of energy than natural gas, and coal emits 70 percent more. Therefore, switching coal-fired electrical plants to gas-fired plants can substantially reduce the CO_2 emitted into the atmosphere.

Other measures of carbon intensity can be derived from these numbers. For example, according to Figure 13.5, the mix of fuels used to generate electrical power in the United States produces 38 quads of energy and emits 0.57 Gt of carbon for a combined carbon intensity of 0.015 Gt/quad. Note that this value is only slightly more than that of natural gas, the least carbon intensive of the fossil fuels, even though the U.S. energy system uses comparable amounts of crude oil and coal. Why? Because alternative sources of energy with very low carbon intensities—nuclear, hydroelectric, wind, and solar—also contribute substantially to energy production.

BONUS PROBLEM: Using the data in Figure 13.5 and assuming only the fossil fuels produce carbon dioxide, calculate the carbon intensity factor for the fossil-fuel component of the U.S. energy system.

mining—the removal of soil and surface sediments to expose coal beds—is safer for the miners, but it can ravage the countryside if the land is not restored. An especially destructive type of surface mining, now common in the Appalachian Mountains of the eastern United States, is "mountaintop removal," in which up to 300 vertical meters of a mountain crest is blasted away to expose underlying coal beds (**Figure 13.17**). The excess rock and soil are dumped into the surrounding valleys.

Coal is a notoriously dirty fuel. In addition to its high carbon intensity, most coal also contains appreciable amounts of pyrite, which is released into the atmosphere as noxious sulfur-containing gases when the coal is burned. Smog and acid rain, which forms when these gases combine with rainwater, has been a severe problem in many countries, especially China. U.S. government regulations now require industries that burn coal to adopt technologies for "clean" coal combustion, which have reduced emissions of sulfur and toxic chemicals. Federal laws also mandate the restoration of land disrupted by surface mining and the reduction of dangers to miners. These measures are expensive and add to the cost of coal, which is one reason why coal has declined substantially as a component of the U.S. energy mix (see Figure 13.2).

FIGURE 13.17 Mining of coal by mountaintop removal in the Appalachian Mountains of West Virginia. [Rob Perks, Natural Resources Defense Council.]

Oil On April 20, 2010, an explosion aboard the drilling platform *Deepwater Horizon* killed 11 men and injured 17 others. This blowout resulted in the largest marine oil spill in history, releasing 5 million barrels of crude oil into the Gulf of Mexico during the next 3 months (**Figure 13.18**). The oil spill caused significant environmental damage to ecosystems along the Gulf Coast.

This accident, like the infamous spills off the Yucatan coast in 1979 and Santa Barbara in 1969, renewed the long-running debate about whether to allow drilling for oil in fragile habitats such as the Arctic National Wildlife Refuge (ANWR) on the coastal plain of northern Alaska. The total petroleum resource in ANWR has not been fully evaluated, but it could be as much as 80 billion barrels of oil. The USGS estimates that if oil prices were high enough and environment restrictions relaxed, 6 billion to 16 billion barrels of this oil could be produced economically using current technologies. There is no doubt that these resources would contribute to the national economy. But oil and gas production would require the building of roads, pipelines, and housing in a delicate environment that is an important breeding area for caribou, musk-oxen, snow geese, and other wildlife (**Figure 13.19**). Policy makers must weigh the short-term economic benefits of drilling against possible long-term environmental losses in making this decision.

Natural Gas We have seen how the technologies of precisely guided horizontal drilling and hydraulic fracturing have contributed to a recent boom in U.S. natural gas production. The environmental costs associated with fracking can be steep, however, because the process uses huge amounts of water, and wastes from shale gas production can contaminate the local water supply. Moreover, the disposal

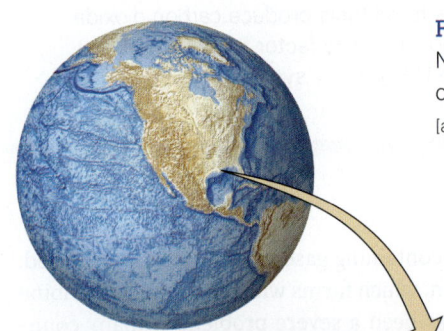

FIGURE 13.18 (a) Oil slick in the Gulf of Mexico imaged on May 24, 2010 by NASA's Terra satellite, 34 days after the explosion of the *Deepwater Horizon*. (b) The oil spilled by the *Deepwater Horizon* blowout harmed wildlife along the Gulf Coast.
[a: Michon Scott/NASA Earth Observatory/Goddard Space Flight Center; b: AP Photo/Bill Haber.]

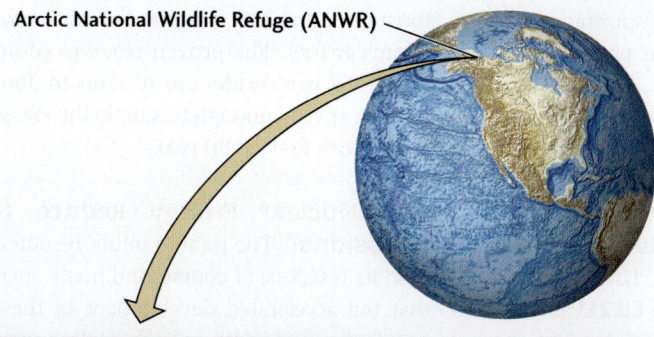

FIGURE 13.19 Herd of caribou in the Arctic National Wildlife Refuge (ANWR). An intense controversy surrounds proposals to drill for oil and natural gas in this pristine region. [Prisma by Dukas Presseagentur GmbH/Alamy.]

of waste water and chemicals used in fracking and other unconventional recovery methods is often done by injecting these fluids into deep wells, which lubricates old faults in Earth's crust, causing earthquakes. This practice has increased earthquake activity in many regions of the United States where the seismicity has been historically low, such as Oklahoma, Texas, and Ohio (see Earth Issues Box 10.2).

Alternative Energy Resources

Fossil fuels account for 81 percent of the world's primary energy production, about the same fraction as for the United States (Figure 13.20). The data from the previous section demonstrate that there are plenty of fossil-fuel resources to power civilization for decades, perhaps centuries, into the future. It is nevertheless clear that our society can ill afford to continue its dependence on fossil fuels, owing to the loading of the atmosphere with carbon dioxide. We will now explore the alternative sources of energy that have the greatest potential to replace fossil fuels, assessing the factors that are helping and hindering progress towards a "decarbonized" energy system.

Nuclear Energy

The first large-scale use of the radioactive isotope uranium-235 to produce energy was as an atomic bomb in 1944, but the nuclear physicists who first observed the

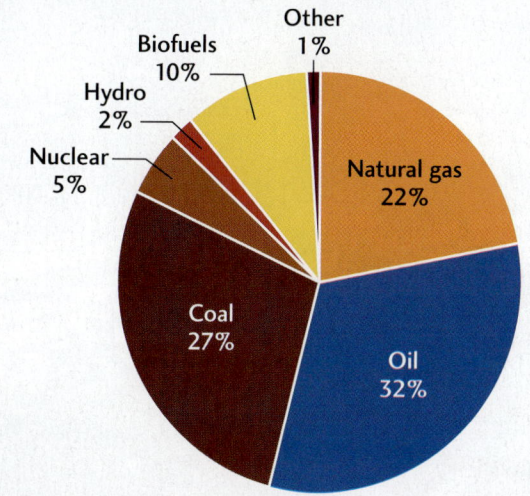

FIGURE 13.20 Distribution of world primary energy production by source. In 2017, the total energy production was 536 quads. [Data from the International Energy Agency, 2018.]

vast energy released when its nucleus split spontaneously (a phenomenon called *fission*) foresaw the possibility of peaceful applications of this new energy source. After World War II, countries around the world built nuclear reactors to produce **nuclear energy.** In these reactors, the fission of uranium-235 releases heat that is used to make steam, which then drives turbines to create electricity. A typical commercial reactor produces about 1000 megawatts of electricity (1 megawatt = 1 million watts). Large nuclear facilities may have multiple reactors (**Figure 13.21**).

Nuclear power supplies a substantial fraction of the electric energy used by some countries, such as France (72 percent in 2017), Slovakia (54 percent), and Sweden (40 percent), but this proportion is much smaller in the United States. Overall, the nation's 100 nuclear reactors produce 20 percent of U.S. electricity, accounting for about 8.6 percent of total U.S. energy demand (see Figure 13.5). Worldwide, this fraction is only 4.5 percent.

Uranium Reserves Uranium is an extremely rare element in the Earth, but the melting of mantle and crustal rocks tends to concentrate it in the upper crust, where it is 50 times more common than silver and 1800 times more common than gold. Minable concentrations are found as small quantities of uraninite, a uranium oxide mineral (UO_2, also called pitchblende), in veins in granites and other felsic igneous rocks. Uranium is highly soluble in groundwater and can be precipitated as uraninite in sedimentary rocks. High-grade uranium ores can contain several percent uranium. In its native form, only a small proportion (about 0.7 percent) is uranium-235. The much more abundant isotope, uranium-238, is not radioactive enough to be used as a nuclear fuel. The proven reserves of uranium, about 5.9 megatons worldwide, correspond to about 3200 quads of nuclear energy, enough to supply the existing nuclear power plants with fuel for 90 years.

Can Nuclear Energy Reduce Global Carbon Emissions? The total uranium resources are much larger than its reserves, of course, and many energy experts believe that the accelerated development of these resources could supply the world with enough nuclear energy to substantially reduce carbon emissions. Nuclear reactors release no greenhouse gases into the atmosphere—the carbon intensity of nuclear-fuel burning is essentially zero. It is not exactly zero if one takes into account the complete "energy life-cycle," because some fossil-fuel energy must be expended (e.g., in driving diesel-powered vehicles) to build, operate, and decommission nuclear plants and to the transport of nuclear energy from the plant to where it can be used. Taking these fossil-fuel expenditures into account, one finds that the carbon intensity of electricity generation by nuclear energy is only about 3 percent of that of natural gas, the cleanest fossil fuel.

As of January 2019, there were 450 nuclear power plants worldwide, accounting for 11 percent of global electricity generation, and another 60 or so are under construction, including two in the United States. A crash program to build hundreds more over the next few decades could reduce global carbon emissions by a billion tons per year—a prospect we will consider further in the next chapter. That is unlikely to happen, however, owing to societal anxieties about nuclear hazards.

FIGURE 13.21 Japan's Kashiwazaki-Kariwa facility is the world's largest nuclear power plant, with seven reactors and a total generating capacity exceeding 8200 megawatts. It was completely shut down after several reactors were damaged by a powerful earthquake (magnitude 6.6) that struck close by on July 16, 2007, and again after the great Tohuku earthquake (magnitude 9.0) badly damaged the Fukushima Daiichi nuclear plant on March 11, 2009. Two reactors are scheduled to restart in 2019. [STR/AFP/Getty Images.]

Hazards of Nuclear Energy The biggest drawbacks of nuclear energy are concerns regarding the safety of nuclear reactors, the risk of environmental contamination with radioactive material, and the potential use of radioactive fuels for making nuclear weapons.

In the United States, damage to a reactor at the Three Mile Island reactor in Pennsylvania in 1979 released radioactive debris. Although very little radioactive fuel left the containment building, and no one was harmed, it was a close call. Much more serious was the destruction of a nuclear reactor in the town of Chernobyl, Ukraine, in 1986. The reactor went out of control because of its poor design and human error. Thirty-one people were killed, and a plume of radioactive debris was carried by winds over Scandinavia and western Europe. Contamination of buildings and soil has made hundreds of square miles of land surrounding Chernobyl uninhabitable. Food supplies in many countries were contaminated by the radioactive fallout and had to be destroyed. Eventual deaths from cancer caused by exposure to the fallout may be in the thousands.

A second serious disaster occurred when the tsunami from the great Tohoku earthquake of March 11, 2011, inundated the Fukushima Daiichi nuclear power plant on the northeastern coast of Honshu, Japan (see Figure 10.31). The reactors shut down, as designed, but the tsunami destroyed the backup diesel generators, cutting power to the water pumps that were supposed to cool the still-hot reactors. Three of the six reactors suffered complete or partial meltdowns, and explosions of hydrogen gas generated during the meltdowns destroyed the reactor containment buildings, releasing radioactive debris into the atmosphere. Water sprayed to cool the damaged reactors carried radioactive material into the ocean. Radioactive materials from these reactors are still leaking into the environment.

The uranium consumed in nuclear reactors leaves behind dangerous radioactive wastes. A system of safe long-term waste disposal is not yet available, and reactor wastes are being held in temporary storage facilities at reactor sites. This is a dangerous practice; spent fuel rods stored on site contributed to the radioactive debris released at Fukushima. Many scientists believe that geologic containment—the burial of nuclear wastes in deep, stable, impermeable rock formations—would provide safe storage of the most dangerous wastes for the hundreds of thousands of years required before they decay. France and Sweden have built underground nuclear waste repositories. A national repository, the Yucca Mountain Nuclear Waste Repository, was being developed in the United States at its nuclear-weapon test site in Nevada (**Figure 13.22**), but local opposition caused the federal government to terminate funding for the site in 2011. At present the United States has no long-term plan for nuclear waste disposal.

Biofuels

Before the coal-fired Industrial Revolution of the mid-nineteenth century, the burning of wood and other biomass

FIGURE 13.22 Aerial view of the north entrance to the Yucca Mountain Nuclear Waste Repository developed at the Nevada Test Site, north of Las Vegas. Yucca Mountain is the high ridge to the right of the entrance. Federal funding for this project was terminated in 2011. [Department of Energy.]

derived from plants and animals satisfied most of society's energy needs. Even today, the burning of biomass contributes more than 10 percent to the global energy mix, exceeding the total derived from all other renewable resources (see Figure 13.20). Biomass is an attractive alternative to fossil fuels because, at least in principle, it is *carbon-neutral*; that is, the CO_2 produced by the combustion of biomass is eventually removed from the atmosphere by plant photosynthesis and used to produce new biomass.

Liquid **biofuels** derived from biomass, such as *ethanol* (ethyl alcohol: C_2H_6O), could replace gasoline as our main automobile fuel. The use of biofuels in transportation is hardly new. The first four-stroke internal combustion engine, invented by Nikolaus Otto in 1876, ran on ethanol, and the original diesel engine, patented by Rudolf Diesel in 1898, ran on vegetable oil. Henry Ford's Model T car, first produced in 1903, was designed to operate on ethanol. But soon thereafter, petroleum from the new reserves discovered in Pennsylvania and Texas became widely available, and cars and trucks were converted almost entirely to petroleum-based gasoline and diesel fuel.

Ethanol can be mixed with gasoline to run most car engines built today. It is produced mainly from corn in the United States and from sugarcane in Brazil. For the last 35 years, the Brazilian government has been pushing to replace imported oil with domestic ethanol; in 2017, about 90 percent of Brazil's automobile fuels were running on "flex-fuel," a combination of gasoline and ethanol, which is typically 70 percent cheaper than gasoline. Brazil and the United States account for 85 percent of the global biofuel production.

A promising biomass crop is switchgrass, a perennial plant native to the Great Plains (**Figure 13.23**). Switchgrass has the potential to produce up to 1000 gallons of ethanol per acre per year, compared with 665 gallons for sugarcane and 400 gallons for corn, and it can be cultivated

FIGURE 13.23 Switchgrass, a perennial plant native to the Great Plains, is an efficient source of ethanol, the most popular biofuel. Here, geneticist Michael Casler harvests switchgrass seed as part of a breeding program to increase the plant's ethanol yield. [Wolfgang Hoffmann/USDA.]

on grasslands of marginal utility for other types of agriculture. Nevertheless, biofuel production competes with food production, so increasing the former drives up the price of the latter, which reduces the economic benefits of biofuels.

How environmentally beneficial are biofuels? Are they really carbon-neutral? As we saw for nuclear power, the carbon intensity of biofuels is not zero if you consider the complete energy life cycle, because the energy used to fertilize plants, transform them into biofuels, and deliver the biofuels to market comes primarily from fossil fuels. Moreover, the basic assumption of carbon neutrality—that all of the carbon emitted into the atmosphere from biofuel burning will eventually be returned to the biosphere—is not true. About one-quarter of the carbon dioxide emitted into the atmosphere by fuel burning of any type is absorbed into the oceans, causing detrimental ocean acidification (see the Practicing Geology Exercise in Chapter 12). The widespread use of biofuels for transportation would no doubt reduce the pumping of carbon from the lithosphere to the atmosphere, but experts are still arguing about the magnitude of that reduction.

Hydroelectric Energy

About 2.4 percent of global energy production is **hydroelectric energy**, derived from water moving under the force of gravity to drive turbines that generate electricity. Artificial reservoirs behind dams usually provide the water. Hydroelectric energy is a renewable energy source that ultimately comes from the Sun, whose energy drives the climate system and creates rainfall. It is also relatively clean, risk-free, and cheap to produce.

The Three Gorges Dam on the Yangtze River in China (**Figure 13.24**) is the world's largest hydroelectric facility. It is capable of generating 22,500 megawatts—nearly 5 percent of China's total electricity demand. The project was controversial, however, because the damming of the Yangtze caused flooding that has displaced over a million people.

In the United States, hydroelectric dams deliver about 2.7 quads annually, a small but locally important share of the nation's annual energy consumption. The state of Washington is the largest U.S. producer, obtaining almost 90 percent of its electricity from hydroelectric power.

The U.S. Department of Energy has identified more than 5000 sites where new hydroelectric dams could be built and operated economically. Such expansion would be resisted, however, because the dams would drown farmlands and wilderness areas under artificial reservoirs while adding only a small amount of energy to the U.S. supply. For this reason, most energy experts expect that the proportion of the nation's energy produced by hydroelectric power will actually decline in the future.

Wind Energy

Wind energy is produced by windmills that drive electric generators. Today, the generation of electricity by high-efficiency wind turbines is a fast-growing source of renewable energy; global production is currently increasing by 17 percent per year. Wind farms containing hundreds of turbines can produce as much electric power as a mid-sized nuclear power plant (**Figure 13.25**). Denmark is a leading wind-power nation in the world, now producing over 40 percent of its electric power by wind. In the United States, electricity from wind sources tripled between 2008 and 2018, and it currently accounts for just over 6 percent of all U.S. electrical power production.

The U.S. Department of Energy estimates that winds sufficient for power generation blow across 6 percent of the land area of the continental United States, and that those winds have the potential to supply more than one and a half times the nation's current electricity demand. But harvesting

FIGURE 13.24 The Three Gorges Dam on China's Yangtze River is about 2335 m (7660 feet) long and 185 m (616 feet) high. Its 32 generators are capable of producing 22,500 megawatts of hydroelectric power. [AP photo/Xinhua Photo, Xia Lin.]

this energy would require placing millions of windmills, each 100 m tall, across hundreds of thousands of square kilometers of land. Changes to the landscape required for industrial wind farming, as well as the low-frequency noise generated by the turbines, have made the siting of new facilities a controversial environmental issue in some regions.

Solar Energy

Solar-energy enthusiasts are quick to remind you that "every hour Earth receives from sunlight more energy than civilization uses in one year." **Solar energy** is the prime example of a resource that cannot be depleted by usage: The Sun will continue to shine for at least the next several billion years. Although using solar energy to heat water for homes, industries, and agriculture is economically profitable with existing technology, the methods for the large-scale conversion of solar energy into electricity are still inefficient and expensive. Nevertheless, the solar generation of electricity is increasing rapidly as large power plants are being built in response to voter mandates and government subsidies. The Ivanpah solar electric–generating system in California's Mojave Desert, commissioned in 2013, is the world's largest, capable of producing up to 392 megawatts of electricity (**Figure 13.26**).

Solar energy is only a miniscule fraction of global energy production (0.2 percent), but its growth rate of 23 percent per year is higher than any other energy source. In Great Britain—not known for its sunny skies—solar generation of electricity exceeded that from coal-fired power plants for the first time in 2016. In the United States, solar-energy production rose from 0.2 quad in 2012 to almost 0.8 quad in 2017, a fourfold increase in 5 years. Optimistic projections are that, worldwide, solar conversion could increase to as

FIGURE 13.25 Photo taken on Feb. 20, 2012, shows the scene of windmills at Jinshan wind power plant in northwest China's Gansu Province. The installed capacity of grid-connected wind power in China has reached 53 million kilowatts so far. China has replaced America to be the number one wind power country in the world. [Ma Xiaowei/Xinhua News Agency/eyevine/Redux.]

FIGURE 13.26 The Ivanpah solar electric generating system in California's Mojave Desert, commissioned in 2013, is the world's largest. More than 170,000 mirrors focus sunlight on three towers filled with water, producing steam that spins turbines that can generate up to 392 megawatts of electricity. [Gilles Mingasson/Getty Images for Bechtel.]

much as 12 quads per year in a decade or so, which would amount to about 2 percent of total energy production.

Geothermal Energy

Earth's internal heat can be tapped as a source of *geothermal energy*, as we described in Chapter 12. According to one Icelandic estimate, as much as 40 quads of electricity could be generated each year from accessible geothermal energy sources, but so far only a tiny fraction of that amount, about 0.3 quad per year, is actually being generated. Another 0.3 quad of geothermal energy is used for direct heating. At least 46 countries now use some form of geothermal energy.

Geothermal energy is unlikely to replace petroleum as a major source of power, though it may help to meet the energy needs of a postcarbon economy. Like most of the other energy sources, geothermal power usage can cause environmental problems. Regional ground subsidence can occur if hot groundwater is withdrawn without being replaced. Hydrothermal waters can contain salts and toxic materials dissolved from the hot rock. As in the case of fracking, the disposal of these wastewaters by reinjection into the crust can trigger earthquakes.

Our Energy Future

As the human population continues to grow and standards of living improve, civilization's need for more energy will continue to rise. According to estimates by the U.S. Energy Information Administration, 70 percent of the increase over the next several decades will come from developing countries. The consumption of fossil fuels will continue to grow, though it will be outpaced by the expansion of alternative energy sources (**Figure 13.27**).

- The fossil-fuel fraction of the world energy supply will decline only slightly, from 81 percent in 2017 to 78 percent in 2040.
- Petroleum will remain the largest source of energy, its share decreasing from 33 percent in 2017 to 30 percent in 2040.
- Among the fossil fuels, coal consumption will grow the slowest, and natural gas the fastest.
- Nuclear energy use will increase more rapidly than fossil-fuel use.
- Renewable energy will continue to be the world's fastest-growing source of energy, at an average rate of 2.6 percent per year.

These projections are uncertain owing to unanticipated events (such as global military conflicts) and technological advances (new methods for generating energy), but they indicate that civilization will continue to pump carbon from the lithosphere into the atmosphere at very high rates. The potential effects on the climate system and the biosphere are the subjects of the next chapter.

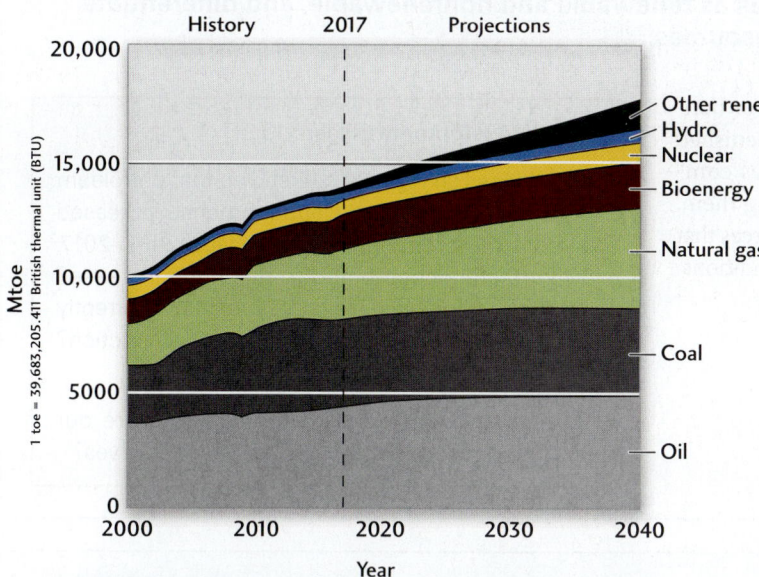

FIGURE 13.27 World energy consumption by source, 2000–2040. Values beyond 2017 are projections based on estimates of population and economic growth. [Data from the International Energy Agency, 2018.]

KEY TERMS AND CONCEPTS

- biofuel (p. 383)
- carbon intensity (p. 369)
- coal (p. 371)
- crude oil (p. 372)
- energy resources (p. 367)
- fossil fuels (p. 368)
- horizontal drilling (p. 373)
- Hubbert's peak (p. 375)
- hydraulic fracturing (fracking) (p. 373)
- hydrocarbon (p. 367)
- hydrocarbon economy (p. 369)
- hydroelectric energy (p. 384)
- natural gas (p. 372)
- nonrenewable resources (p. 368)
- nuclear energy (p. 382)
- oil trap (p. 372)
- oil window (p. 372)
- quad (p. 369)
- renewable resource (p. 367)
- reserves (p. 368)
- solar energy (p. 385)
- tight formation (p. 373)
- wind energy (p. 384)

REVIEW OF LEARNING OBJECTIVES

13.1 Explain how the impact of civilization on the Earth system qualifies it as a global geosystem.

Human society has harnessed the means of energy production on a global scale and now competes with the plate-tectonic and climate systems in modifying Earth's surface environment. Most of the energy used by civilization today comes from hydrocarbon fuels. The rise of this hydrocarbon economy has altered the natural carbon cycle by creating a huge new flux of carbon from the lithosphere to the atmosphere. If that flow continues unabated, CO_2 concentrations in the atmosphere will double by the mid-twenty-first century.

Study Assignment: Review the section *Growth and Impact of Civilization*.

Exercise: List three facts that illustrate how civilization competes with plate tectonics and the climate system.

Thought Questions: (a) How is civilization fundamentally different from the natural geosystems we have studied in this textbook? (b) What factors do you think will be most important in influencing human population growth during this century?

13.2 Categorize our natural resources as renewable and nonrenewable, and differentiate energy reserves from energy resources.

Natural resources can be classified as renewable or nonrenewable, depending on whether they are replenished by geological and biological processes at rates comparable to the rates at which we are consuming them. Reserves are the known supplies of natural resources that can be exploited economically under current conditions.

Study Assignment: Figure 13.3

Exercises: (a) Figure 13.12 shows that petroleum reserves of South and Central America increased almost threefold during the decade 2007–2017. What factors contributed to this huge increase? (b) Which nonrenewable energy resource currently contributes most to the U.S. energy production? Which renewable energy resource?

Thought Question: How much larger are our fossil-fuel resources than our fossil-fuel reserves?

13.3 Understand the geological processes that form fossil fuels, and the energy available from their reserves.

Coal is formed by the burial, compression, and diagenesis of wetland vegetation. More compression and heating of coal beds generally increase the energy content of the coal. Oil and natural gas form from organic matter deposited in oxygen-poor sedimentary basins, typically on continental margins. These organic materials are buried as the sedimentary layers grow in thickness. Under elevated temperatures and pressures, the buried organic matter is transformed into liquid and gaseous hydrocarbons. Oil and gas accumulate where geologic structures, called oil traps, create impermeable barriers to their upward migration. Hydrocarbon resources also include oil shales that are rich in organic material but never reached the oil window and tar sands that once contained oil but have since lost most of their volatile components. The energy available from proven reserves of fossil fuels is estimated to be about 53,000 quads; at current rates of energy consumption, this is enough to fuel civilization for at least another century.

Study Assignments: Figures 13.9 and 13.10

Exercises: (a) What are the prerequisites for oil traps to contain oil? (b) An aggressive drilling program in the Arctic National Wildlife Refuge could produce as much as 16 billion barrels of oil. At current consumption rates, for how many years would this resource supply U.S. oil demand? (c) Which three countries have the largest coal reserves?

Thought Question: Which of the following factors are important in estimating the future supply of oil and natural gas: (i) the rate of oil and gas accumulation in traps, (ii) the rate of depletion of known reserves, (iii) the rate of discovery of new reserves, (iv) the total amount of oil and gas now present on Earth.

13.4 Answer the question: When will civilization run out of oil?

The short answer is not for a very long time. At the current production rate, the world would consume all of today's oil reserves in about 55 years; however, new technologies such as precise horizontal drilling and hydraulic fracturing are increasing oil reserves at a rapid rate, by over 20 percent in just the last decade. Moreover, oil can be derived from huge, unconventional hydrocarbon resources including oil shales and tar sands. Pessimistic predictions that global oil production would reach Hubbert's peak in the early twenty-first century—that it would stop rising and begin to decline—have not come true. In fact, we will never really run out of oil; as resources diminish and prices rise, oil will become primarily a raw material for producing plastics, fertilizers, and other petrochemical products.

Study Assignment: Earth Issues 13.1

Exercises: (a) Global oil production in 2018 was about 30 billion barrels per year worldwide. What is the equivalent energy expressed in quads? (b) Is petroleum consumption increasing or decreasing in the United States? Is it increasing or decreasing worldwide?

Thought Questions: (a) Are you an "oil optimist" or an "oil pessimist"? Explain why. (b) Why do some experts believe our energy economy is transitioning from a "petroleum economy" to a "methane economy"?

13.5 Compute the carbon intensities of fossil fuels from the energy they produce and the carbon dioxide they emit, and use carbon intensities to compute the changes in carbon flux from changes in energy production.

The carbon intensity of a fuel is defined to be the mass of carbon emitted as CO_2 per unit of useful energy produced by burning the fuel. The carbon emitted by burning 1 gigaton (Gt) of methane is 0.75 Gt, and the energy produced is 52 quads; the ratio of these two quantities, 0.014 Gt/quad, is the carbon intensity of methane burning. The carbon intensity of oil burning is 40 percent higher than this value, and that of coal burning is 70 percent higher, showing that natural gas is the least carbon-intensive fossil fuel and coal is the most. The overall carbon intensity of the U.S. electrical system is 0.015 Gt/quad, which is only slightly higher than that of natural gas, because the higher carbon intensity of coal is offset by the near-zero carbon intensity of nuclear and renewable energy sources. Replacing coal-fired electrical plants with gas-fired electrical plants would reduce the carbon intensity of the U.S. electrical system to 0.012 Gt/quad [see Exercise (b)].

Study Assignment: Practicing Geology Exercise

Exercises: (a) Which fossil fuel produces the least amount of CO_2 per unit of energy: oil, natural gas, or coal? Which produces the most? (b) About one-third of U.S. electrical-energy generation comes from burning coal. Show that the carbon intensity of this system would be reduced from 0.015 to 0.012 by replacing coal-fired plants with gas-fired plants.

Thought Question: Is the overall carbon intensity of the U.S. energy system increasing or decreasing? Why?

13.6 Quantify the relative contributions of alternative energy resources to energy production, and estimate their potential to satisfy future energy needs.

Alternative energy sources include nuclear power, biofuels, and solar, hydroelectric, wind, and geothermal energy. Taken together, these energy sources currently supply only a small percentage of world energy demand. Nuclear energy produced by the fission of uranium, the world's most abundant minable energy resource, could be a major energy source, but only if the public can be assured of its safety and security. Nuclear fuels have the potential of providing a large, low-cost source of energy that produces essentially no carbon dioxide. This promise has not been realized, however, owing to problems with reactor safety, disposal of radioactive wastes, and nuclear security. With advances in technology and reductions in cost, renewable sources such as solar energy, wind energy, and biomass could become major contributors in the twenty-first century.

Study Assignment: Review the section *Alternative Energy Resources*, especially Figure 13.20.

Exercises: (a) Contrast the risks and benefits of nuclear fission and coal combustion as energy sources. (b) Why is it untrue to assert that biofuels are carbon-neutral; that is, that all of the carbon emitted into the atmosphere from biofuel burning will eventually be returned to the biosphere?

Thought Questions: (a) What issues related to the use of nuclear energy can be addressed by geologists? (b) Given that solar energy is so plentiful, why is its contribution to the global energy supply so small?

13.7 Project the worldwide growth of energy consumption by geographic region and fuel type.

Population and economic growth increases energy consumption. As countries develop and living standards improve, energy demand grows rapidly. Based on growth projections, 72 percent of the increase in world energy consumption between now and 2040 will come from the developing countries, mostly in Asia, and only 18 percent from developed countries in North America, Europe, and Australia. Owing to this demand, energy production from all major sources will continue rise. The consumption of nuclear energy will climb. The most rapid growth will be in renewable energy sources. Energy from coal will level off, replaced primarily by increases in energy from natural gas. Consequently, the overall fraction of the world energy supply from fossil-fuel burning will decline only slightly, from 81 percent today to about 78 percent in 2040.

Study Assignment: Figures 13.6 and 13.27

Exercises: In 2017, global carbon emissions were 9.9 Gt from the global energy production of 536 quads. (a) What was the carbon intensity of the global energy system in 2017? (b) Based on Figure 13.27, is this carbon intensity in the future expected to increase, stay the same, or decrease? Why? (c) From the data in this chapter, calculate a rough estimate of the carbon intensity of the global energy system in 2040.

Thought Question: What do you think will be the major sources of the world's energy in the year 2040? In the year 2100?

VISUAL LITERACY EXERCISE

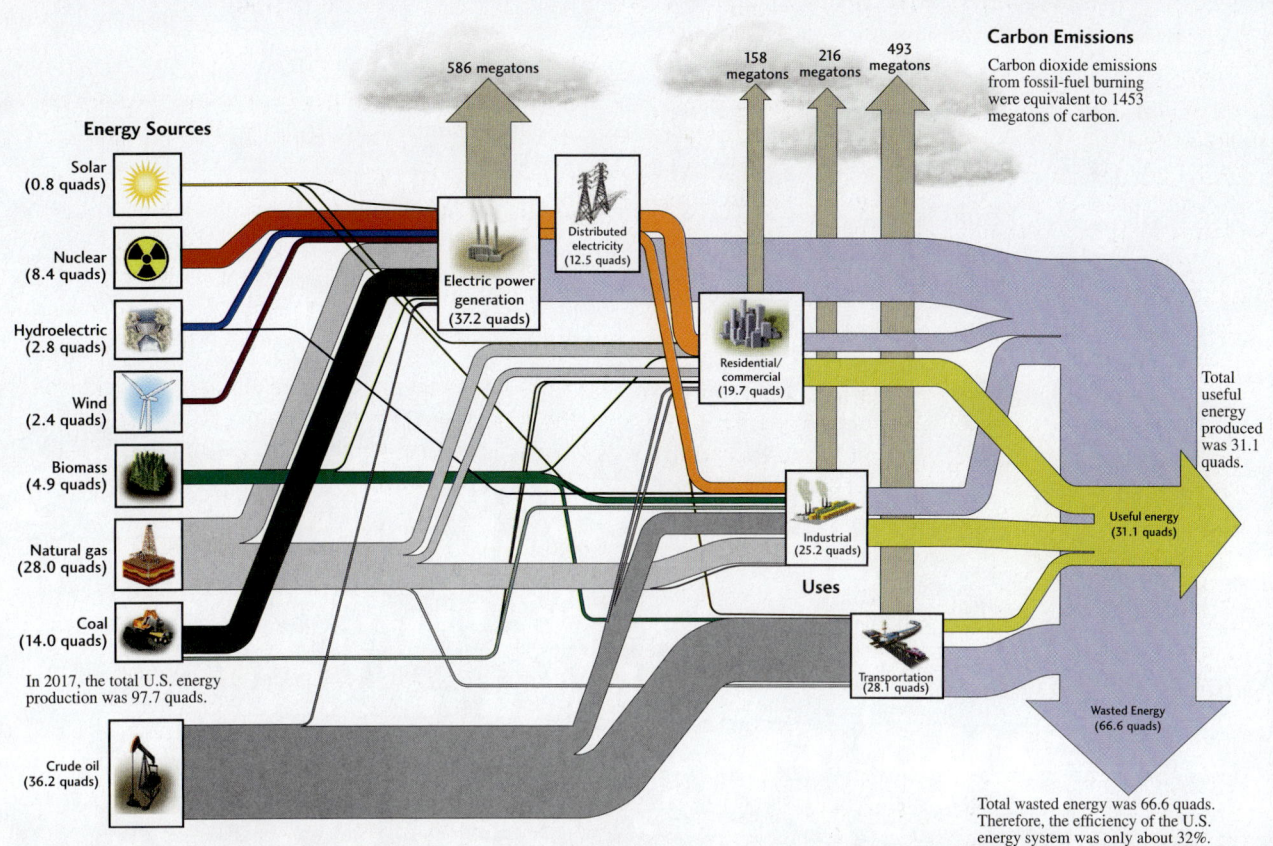

FIGURE 13.5 Energy consumption in the United States in 2017 (in quads). Energy from primary fuel sources (boxes on left side) is delivered to the residential, commercial, industrial, and transportation sectors (boxes in middle to right side). Not represented are small contributions to electric power generation from geothermal energy (0.2 quad). [Information from Lawrence Livermore National Laboratory, based on data from the Energy Information Administration.]

Figure 13.5 is a complicated diagram showing the energy consumption in the United States in 2017.

1. **The figure connects energy sources with human activities using a set of "pipes" of different widths and colors.**
   a. Does the color of a pipe on the left side of the figure represent an energy source or a human activity?
   b. Why is the width of the pipe connecting "coal" to "electric power generation" about twice as wide as the one connecting "nuclear power" to "electric power generation"?

2. **What determines the width of the yellow and blue arrows that point from the boxes representing human activities?**

3. Why is the arrow pointing upward from "electric power generation" wider than the arrows pointing upward from the other human-activity boxes?

4. Compute answers to the following questions from the numerical data on the figure:
   a. What fraction of U.S. energy production came from fossil fuels?
   b. Which activity consumed more energy, electrical power generation or transportation?
   c. What fraction of the total energy production was wasted?
   d. How much carbon did the U.S. energy system emit in 2017?

Anthropogenic Global Change 14

Malé City, capital of the Maldives, an island nation in the Indian Ocean, is one of the densest urban areas in the world. With a maximum elevation of less than 3 meters, the entire country is vulnerable to sea-level rise caused by anthropogenic climate change. [George Steinmetz/National Geographic Creative.]

Rise of Carbon Dioxide in the Atmosphere: The Keeling Curve	394
Types of Anthropogenic Global Change	395
Climate Change	400
Ocean Acidification	408
Loss of Biodiversity	410
Managing the Carbon Crisis	412

Learning Objectives

The goal of this chapter is to summarize what we know about anthropogenic global change, as well as the choices we have to manage this environmental change. From the information provided in this chapter, you should be able to:

14.1 Explain why scientists can assert with high confidence that fossil-fuel burning is increasing the atmospheric concentration of carbon dioxide.

14.2 Catalog the main types of anthropogenic global change and describe their main effects on the atmosphere, hydrosphere, cryosphere, and lithosphere.

14.3 Explain why scientists can assert with high confidence that fossil-fuel burning caused the twentieth-century warming and continues to cause the average surface temperature to increase.

14.4 Use scenarios developed by the Intergovernmental Panel on Climate Change (IPCC) to project how much greenhouse gas concentrations, the average surface temperature, and sea level will rise during this century.

14.5 Assess the potential effects of anthropogenic global change on the biosphere, and evaluate the possibility that the beginning of the Anthropocene epoch will be marked by a mass extinction.

14.6 Illustrate with specific examples of changes in global energy production and usage that could stabilize or reduce carbon emissions.

Throughout geologic history, Earth's climate has been highly variable, swinging between periods of tropical warmth and glacial cold. We learned in Chapter 12 that the glacial cycles of the Pleistocene epoch resulted from atmospheric variations in greenhouse gases driven by the Milankovitch cycles of solar forcing. We saw in Chapter 13 how fossil-fuel burning is transforming carbon in the lithosphere to carbon dioxide in the atmosphere at an unprecedented rate, and we examined alternative sources of energy that will eventually lower this rate.

We now focus on the global changes to the Earth system that are likely to occur because of continuing human activities. We will discuss three of the most serious forms of anthropogenic global change: (1) global climate change owing to increased concentrations of carbon dioxide and other greenhouse gases in the atmosphere, (2) ocean acidification owing to increased carbon dioxide dissolved in the hydrosphere, and (3) loss of species diversity owing to changes in the biosphere. We explore how geoscientists are observing these changes and combining data with models of the Earth system to predict future changes. We will see that reducing the impacts of anthropogenic global change, and adapting to its consequences, will require concerted worldwide actions on an unprecedented scale.

The dire consequences of anthropogenic global change are motivating governments to work together in new ways to avoid the "tragedy of the commons"—the spoiling of our commonly held environmental resources by unregulated overexploitation. New multinational treaties, such as the Paris Agreement adopted by the United Nations in 2015, are being formulated in attempts to reduce carbon pollution of the atmosphere and consequent damage to the global environment.

Rise of Carbon Dioxide in the Atmosphere: The Keeling Curve

The expression *global change* entered the world's vocabulary in the late twentieth century when it became clear that emissions from fossil-fuel burning and other human activities were beginning to alter the chemistry of the atmosphere. The most convincing evidence of global change was collected by the chemist Charles David Keeling, who began a program to measure the concentration of carbon dioxide in the atmosphere in 1958.

Keeling developed instruments for measuring atmospheric CO_2 more easily and accurately than had been previously possible. He installed one of his instruments at the Mauna Loa Observatory at an elevation 11,000 feet on the Big Island of Hawaii, where he could sample pristine air from the Pacific Ocean, uncontaminated by local fossil-fuel emissions, and he maintained those measurements until his death in 2005. Others, including his son Ralph Keeling, are continuing them at Mauna Loa and other observing stations around the world.

The product of these persistent efforts is the **Keeling curve**—the world's longest continuous record of carbon dioxide measurements in the atmosphere (**Figure 14.1**). The monthly values (red curve) show seasonal oscillations about yearly averages that are steadily rising (blue curve). The averages increase from 310 ppm in 1958, when Keeling first began his measurements, to 410 ppm in 2018, a 32 percent change in just 60 years. Owing to circulation and turbulence within the troposphere, the amount of CO_2 in the air sampled at Mauna Loa is representative of a global average. By direct observation, Keeling proved beyond a reasonable doubt that the atmospheric concentration of carbon dioxide, a powerful greenhouse gas, is increasing at the rapid clip of half a percent per year.

Scientists before Keeling, beginning with the chemist Svante Arrhenius in 1896, had speculated that human activities were increasing atmospheric carbon dioxide and that this rise might cause global warming. Arrhenius, a Swede, thought a little climate warming might be a good thing, at least for Sweden. Others argued that the natural variability in CO_2 concentration would be much larger and overwhelm any human contribution. Keeling confirmed that CO_2 measurements were very variable in dense forests and other biologically active zones, but not on the high, barren slopes of the Mauna Loa volcano.

There he observed the global activity of the biosphere from a completely new perspective: as small seasonal cycles imprinted on the anthropogenic rise of CO_2. The seasonal variation in atmospheric CO_2 ranges on average from 3 ppm higher than the annual mean value in May to 3 ppm lower in October, as shown in the inset diagram in Figure 14.1. These oscillations reflect cycles in global plant growth dominated by the temperate and boreal forests of the Northern Hemisphere. Net mass of land plants increases during the Northern Hemisphere summer, when photosynthesis draws CO_2 into plants, and decreases during the Northern Hemisphere winter, when plants respire CO_2 back into the atmosphere. This in-and-out flux of carbon dioxide, first observed by Keeling, is nothing less than the "global breathing of the biosphere."

But wait a minute. How can we be sure that the observed rise in the Keeling curve is not some sort of natural change unrelated to human activities? Keeling and fellow chemists answered this question by measuring the isotopes of carbon in his Mauna Loa air samples. These data demonstrated that the increase in atmospheric CO_2 could not be from natural sources, such as decaying vegetation; they instead matched the isotopic signature of fossil-fuel burning. Keeling's data imply that civilization is altering the chemistry of the atmosphere.

To get a geologic perspective on the Keeling curve, we can compare it with results from another heroic effort in

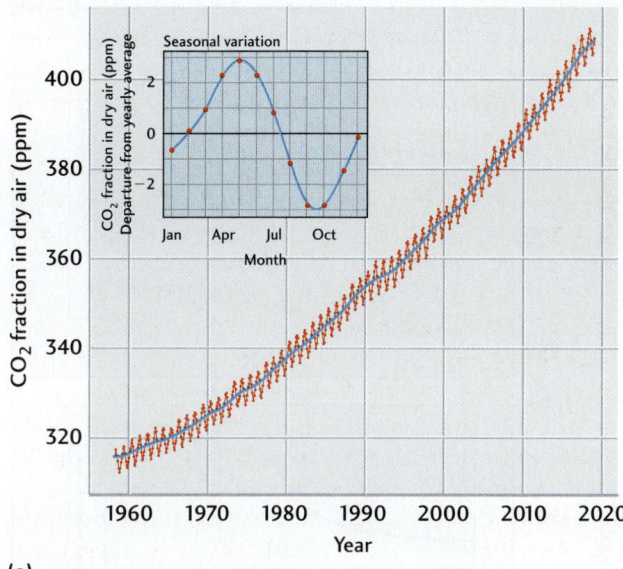

FIGURE 14.1 (a) The Keeling curve is a plot of a 60-year recording of the CO_2 content of the atmosphere, measured in parts per million (ppm) at the Mauna Loa Observatory. The blue curve shows the average yearly values, which have risen by 32 percent in the last 60 years, and the red curve is plotted month by month, oscillating in a season cycle about the yearly mean. The inset diagram shows the average seasonal cycle; the red dots are monthly averages. (b) Charles Keeling receiving the Medal of Science from President George W. Bush in 2002. [(b) National Science Foundation.]

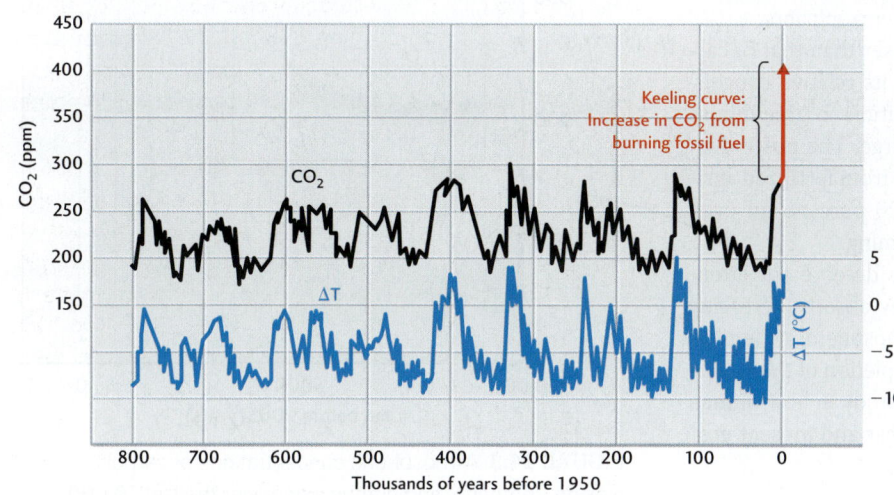

FIGURE 14.2 Atmospheric carbon dioxide and temperature data derived from Antarctic ice core measurements for the past 800,000 years. The Keeling curve is the nearly vertical segment appended to the ice core data at the upper right. [Data from D. C. Harris, *Anal. Chem.* 2010, 82, 7865–7870, and data from D. Lüthi et al., *Nature* 2008, 453, 379–382.]

environmental data collection—the CO_2 record derived from drilling into the Antarctic ice sheet (Chapter 12). What we see is alarming. In the two centuries since the Industrial Revolution, atmospheric CO_2 has spiked to levels way above those reached in the last 800,000 years of Earth history (**Figure 14.2**). At no time during the past 800,000 years have CO_2 concentrations been much higher than the preindustrial average of 280 ppm. In fact, you have to go far back into the geologic record to find a time when CO_2 concentrations were above 400 ppm. Studies of sediment cores indicate that those levels have not been reached since the Middle Miocene, more than 14 million years ago, when Earth's surface temperature was much warmer than it is today.

Types of Anthropogenic Global Change

The rising concentration of CO_2 documented by Keeling is a *chemical change* to the atmosphere. Along with this chemical change, there are *physical changes* to the climate system, and *biological changes* to the ecosystems that support life on Earth. One type of change can cause other types of change through interactions within the Earth system.

For instance, the atmospheric concentration of carbon dioxide is increasing by about half a percent per year. Some of this carbon dioxide is being absorbed by the oceans, causing acidification of seawater. The excess CO_2 not drawn into the oceans or continental forests is enhancing the greenhouse effect, leading to warming at the planetary surface and other physical changes to the climate system. Climate change and ocean acidification are in turn perturbing the biosphere, leading to the loss of species and potentially the collapse of entire ecosystems.

Chemical Change

Since the beginning of the industrial era, fossil-fuel burning, deforestation, land-use changes, and other human activities have caused a rapid rise in the concentrations of greenhouse gases in the atmosphere. **Figure 14.3** shows the atmospheric concentrations of three greenhouse gases—carbon dioxide, methane, and nitrous oxide—over the past 10,000 years. In all three cases, the concentrations remained relatively constant through most of the Holocene epoch, but shot upward after the Industrial Revolution.

The global atmospheric concentration of methane has increased by 150 percent from its preindustrial value, and that of carbon dioxide has increased almost 50 percent. In both cases, the observed increases can be explained by human activities. The energy sector is the predominant source of atmospheric methane—specifically from the production, processing, storage, and distribution of coal and natural gas. Another significant source of methane is from microbial activity in wetlands, rice paddies, and the guts of cows and other ruminants.

Methane's greenhouse effect is weaker than that of carbon dioxide, however, so even though its relative concentration has gone up more, its contribution to greenhouse warming is only about 30 percent as large. The postindustrial increase in nitrous oxide, primarily from fertilized agricultural soils, has contributed about 20 percent, an even smaller fraction of the greenhouse warming.

Two other global chemical changes deserve our attention: acidification of the oceans as they absorb increasing amounts of carbon dioxide from the atmosphere, a chemical process described in Chapter 12, and depletion of the stratospheric ozone layer. The latter is a case study in how human society can successfully respond to a clear and present ecological danger.

Reducing Ozone Depletion: An Environmental Success Story

Near Earth's surface, ozone is a major constituent of smog and a powerful greenhouse gas. Low-lying ozone forms when sunlight interacts with nitrogen oxides and other chemical wastes from industrial processes and automobile exhausts. Ozone in Earth's stratosphere, which is concentrated in a layer 20–30 km above the surface (see Figure 12.2), is another matter. There, solar radiation ionizes oxygen gas (O_2) into ozone (O_3^+), forming a protective layer that shields the planetary surface from ultra-violet (UV) radiation. Skin cancer, cataracts, impaired immune systems, and reduced crop yields are attributable to excessive UV exposure. Ozone acts as "Earth's sunscreen" to block much of this harmful radiation.

In 1995, the Nobel Prize in chemistry was awarded to Paul Crutzen, Mario Molina, and Sherwood Rowland for

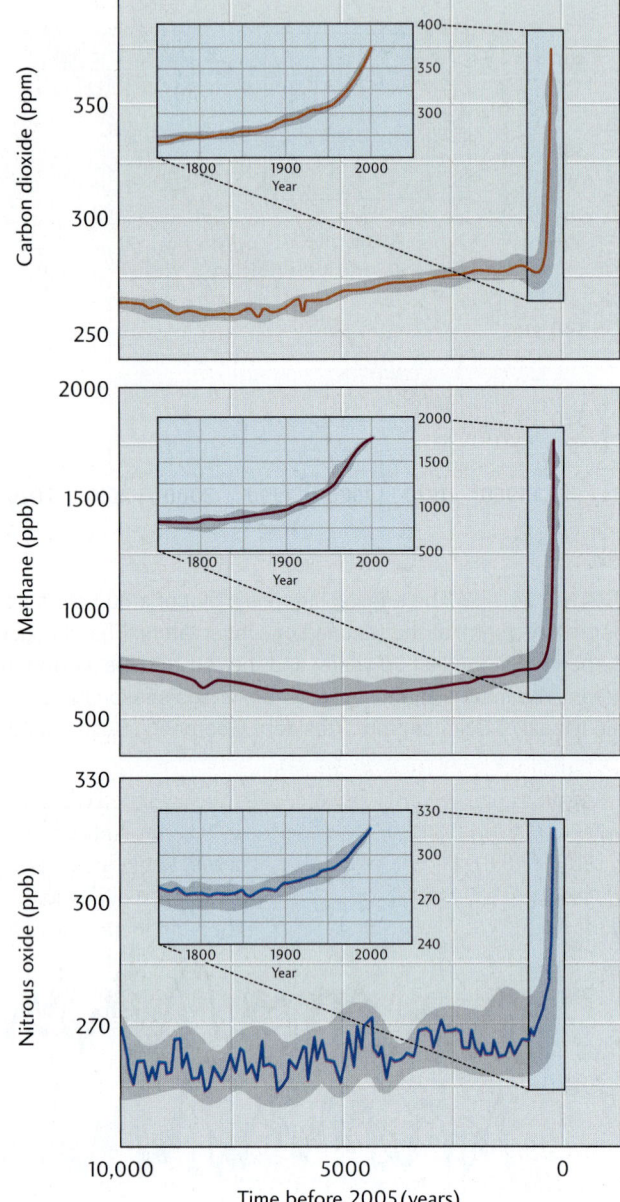

FIGURE 14.3 Atmospheric concentrations of carbon dioxide, methane, and nitrous oxide over the last 10,000 years (large panels) and for 1750–2000 (inset panels). These measurements, compiled by the Intergovernmental Panel on Climate Change, were derived from ice cores and atmospheric samples. Shaded bands show the uncertainties in the measurements. [Data from IPCC, *Climate Change 2007: The Physical Science Basis*. Figure SPM.1, Cambridge University Press.]

the hypotheses they had advanced more than 20 years earlier about how the protective ozone layer can be depleted by reactions involving human-made compounds. One class of compounds, chlorofluorocarbons (CFCs), were widely used as refrigerants, spray-can propellants, and cleaning solvents. CFCs are stable and harmless except when they migrate to the stratosphere. High above Earth, the intense sunlight breaks down these compounds,

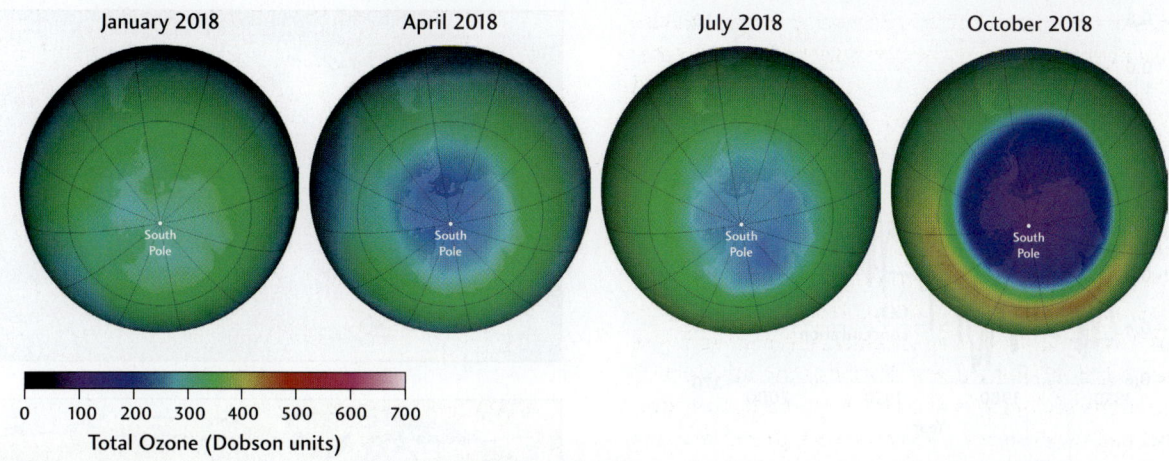

FIGURE 14.4 Total ozone in the Antarctic atmosphere, imaged by NASA satellites in January, April, July, and October 2018, measured in Dobson units. The Antarctic ozone hole is the dark blue/purple region seen on the October image (ozone values below 220 Dobson units). The ozone hole develops during the austral winters, when extreme cold precipitates ice clouds in the stratosphere that catalyze the destruction of ozone by the chlorine from CFCs. [NASA Ozone Watch.]

releasing their chlorine. Molina and Rowland proposed that chlorine reacts with the ozone molecules in the stratosphere and thins the protective ozone layer.

Molina and Rowland's hypothesis was confirmed when a large hole in the ozone layer was discovered over Antarctica in 1985 (**Figure 14.4**). Subsequently, **stratospheric ozone depletion** was found to be a global phenomenon. Elevated surface levels of UV radiation due to stratospheric ozone depletion have been observed since the early 1990s.

In the 1980s, when scientists were trying to convince government and industry officials that the ozone layer was possibly being depleted due to CFCs, a senior government official remarked that the solution was for people to wear hats, sunscreen, and dark glasses. Fortunately, environmental wisdom prevailed. In 1989, a group of nations entered into a global treaty to protect the ozone layer. This treaty, called the **Montreal Protocol** on Substances that Deplete the Ozone Layer, has been signed by all of the world's 195 countries. The Montreal Protocol phased out CFC production in 1996 and set up a fund, paid for by developed nations, to help developing nations switch to ozone-safe chemicals.

Continuing negotiations have produced amendments aimed at limiting emissions of other ozone-depleting chemicals. The most recent, called the Kigali Amendment, went into force on January 1, 2019; it commits countries to cut the production and consumption of hydrofluorocarbons (HFCs) by more than 80 percent over the next 30 years. Owing to these successful efforts, the long-term projections indicate diminished destruction of the ozone layer by anthropogenic chlorine. Monitoring of the ozone hole by satellites and other images confirms that chlorine levels in the stratosphere are declining and the ozone hole is gradually healing.

The Montreal Protocol has become a model for how scientists, industrial leaders, and government officials can work together to head off an environmental disaster.

Physical Change

Warming of Earth's surface is an example of global physical change. Humans have been tracking global temperatures for some time. The thermometer was invented in the early seventeenth century, and Daniel Fahrenheit set up the first standard temperature scale in 1724. By 1880, temperatures around the world were being reported by enough meteorological stations on land and on ships at sea to allow accurate estimation of Earth's average annual surface temperature. Although the average annual surface temperature fluctuates substantially from year to year and from decade to decade, the overall trend has been upward (**Figure 14.5**). Between the late nineteenth century and the early twenty-first, the average annual surface temperature rose by about 0.8°C (Figure 14.5a). This increase is referred to as the **twentieth-century warming.**

The twentieth-century warming was not uniform over the globe. **Figure 14.6** shows the geographic variation of the yearly average temperatures for 1912, 1962, and 2012, colored according to the temperature differences relative to the baseline period 1951–1980. Globally averaged, the difference between 1912 and 2012 is about 0.8°C, consistent with the twentieth-century warming. But some of the regional differences are larger, and some are smaller. In the Arctic region, for example, the temperature rise has been several times higher than the mean value, whereas in the central Pacific Ocean, there has been very little. In general, the land surfaces have warmed more than the oceans. Most of the warming has occurred during the last 50 years. In large regions of the northern continents, the temperature rise between 1962 and 2012 has exceeded 1°C.

We know that human activities are responsible for the increasing concentrations of CO_2 in the atmosphere because, as Keeling demonstrated, the carbon isotopes of fossil fuels have a distinctive signature that precisely

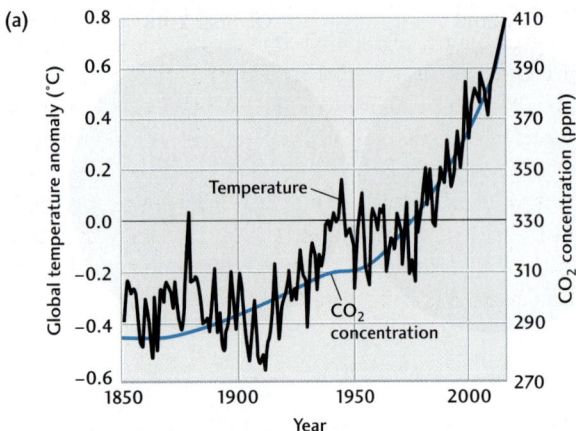

(a)

(b) 1000–2000

The twentieth-century warming is clearly anomalous when compared with climate variation over the last millennium.

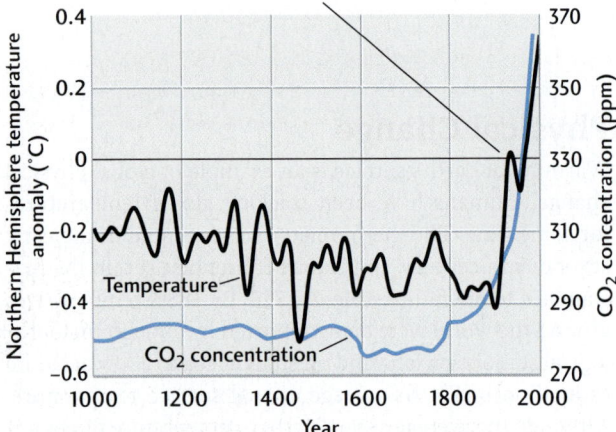

FIGURE 14.5 Comparison of average annual surface temperature anomalies (black lines) with atmospheric CO_2 concentrations (blue lines) shows a recent warming trend that is correlated with increases in atmospheric CO_2 concentrations. (a) Average global annual surface temperature anomalies, calculated from thermometer measurements, and CO_2 concentrations between 1850 and 2017. The small bump in the temperature rise during World War II, 1939–1945, is due to a wartime bias in the way temperature measurements were averaged by the British and U.S. fleets. (b) Average annual surface temperature anomalies for the Northern Hemisphere, estimated from tree rings, ice cores, and other climate indicators; and atmospheric CO_2 concentrations for the last millennium. In both of these figures, the temperature anomaly is defined as the difference between the observed temperature and the temperature average for the period 1961–1990. [Data from IPCC, *Climate Change 2001: The Scientific Basis*, and IPCC, *Climate Change 2013: The Physical Science Basis*.]

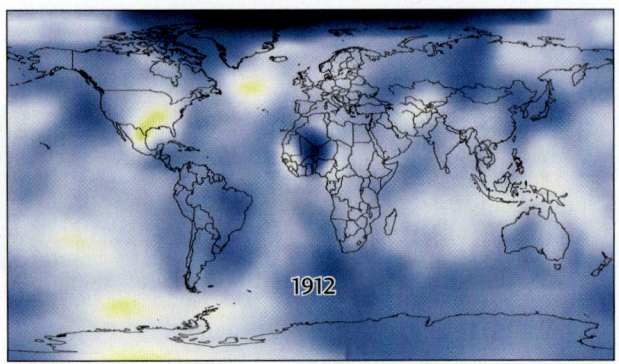
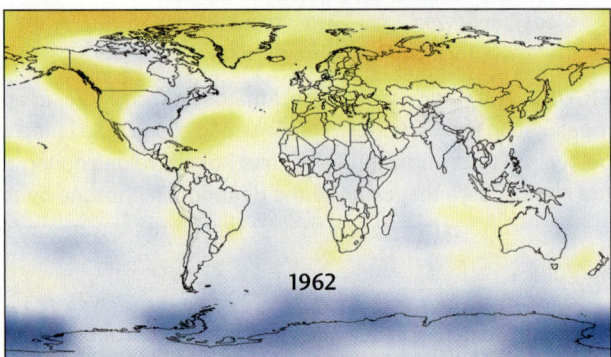
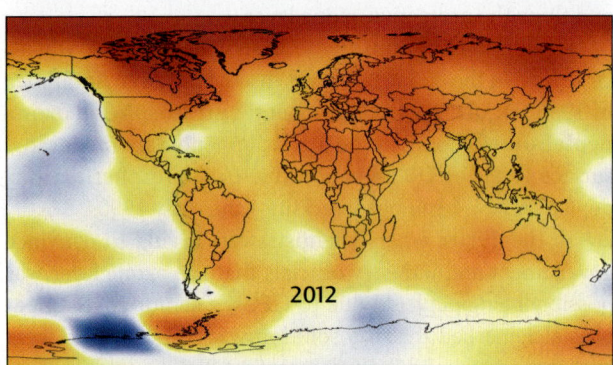
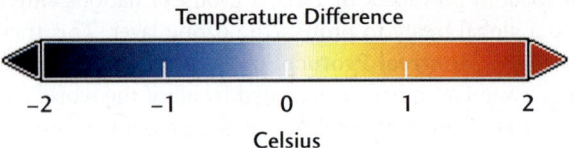

FIGURE 14.6 Surface temperature anomalies for the years 1912 (top), 1962 (middle), and 2012 (bottom) measured relative to the mean local temperatures for the baseline period 1951–1980. The globally averaged difference between 1912 and 2012 is about 0.8°C, consistent with the twentieth-century warming (see Figure 14.5). In the arctic region, the warming has been several times higher than this mean value, whereas in the central Pacific Ocean, it has been much less. [NASA's Goddard Space Flight Center Scientific Visualization Studio.]

matches the changing isotopic composition of atmospheric carbon. But how certain can we be that the twentieth-century warming was a direct consequence of the anthropogenic CO_2 increase—that is, a result of an enhanced greenhouse effect—and not some other kind of change associated with natural climate variation?

To answer this and other questions about how Earth's climate is changing, the United Nations established in 1988 the **Intergovernmental Panel on Climate Change (IPCC)**, an international scientific organization that reviews all research on climate and its variations. The IPCC is charged with developing a consensus, science-based view on how Earth's climate has changed in the past and what might happen in the future, including the potential environmental and socioeconomic impacts of anthropogenic climate change. This textbook portrays the scientific consensus on anthropogenic global change documented in the IPCC Assessment Reports (see Earth Issues 14.1) and in the more recent *Fourth National Climate Assessment,* released by U.S. Global Change Research Program in 2017.

Earth Issues 14.1 The Intergovernmental Panel on Climate Change

Earth's climate system is incredibly complex, so predicting its response to anthropogenic emissions of greenhouse gases is hardly a straightforward task. No one person can keep up with the vast amount of climate-change research that is being conducted worldwide by thousands of scientists. In 1988, the United Nations (UN) and the World Meteorological Organization (WMO) established the Intergovernmental Panel on Climate Change (IPCC) to provide government leaders and the public at large with a clear scientific view of current knowledge about climate change and its potential environmental and socioeconomic impacts.

The IPCC is open to all UN and WMO members, and all of the world's 195 countries are currently participating. The main product of the IPCC has been a series of Assessment Reports released every five to six years since 1990. Thousands of scientists from all over the world have contributed to the work of the IPCC on a voluntary basis as authors, contributors, and reviewers of these major reports. Each report in succession has laid out the most definitive scientific summaries of how climate has changed in the past and how it might change in the future.

IPCC's *First Assessment Report,* published in 1990, played a key role in the creation of the United Nations Framework Convention on Climate Change, the main international treaty to reduce global warming and deal with the consequences of climate change. The IPCC *Second Assessment Report* of 1995 provided important material for negotiators of the Kyoto Protocol in 1997. The *Third Assessment Report* was published in 2001, and the *Fourth* in 2007.

In 2007, the Nobel Peace Prize was awarded jointly to the IPCC and former U.S. Vice President Al Gore "for their efforts to build up and disseminate greater knowledge about man-made climate change, and to lay the foundations for the measures that are needed to counteract such change."

The *Fifth Assessment Report,* finalized in 2014, comprised sub-reports from the three IPCC working groups, entitled *The Physical Science Basis of Climate Change, Climate Change Impacts, Adaptation and Vulnerability,* and *Mitigation of Climate Change.* This report paved the way to the Paris Agreement, in which all 195 countries committed themselves to the goal of "holding the increase in the global average temperature to well below 2°C (3.6°F) above preindustrial levels and pursuing efforts to limit the temperature increase to 1.5°C (2.7°F)." The Paris Agreement was adopted on December 12, 2015, by the United Nations Framework Convention on Climate Change and unanimously ratified by all nations. As of this writing (December, 2018), only one country, the United States, has threatened to withdraw from the agreement.

The IPCC began the process of producing its sixth assessment in 2017. In October 2018, it released *Global Warming of 1.5°C: An IPCC Special Report,* which lays out in stark terms the manifestations and consequences of global warming to 1.5°C above preindustrial levels, which is 0.5°C above the current average surface temperature. The *New Yorker Magazine* (October 9, 2018) called this pessimistic report "a collective scream sieved through the stern, strained language of bureaucratese."

The material on climate change described in this chapter and elsewhere in this textbook draws heavily from the scientific consensus documented in the IPCC reports.

A 2018 meeting of the co-chairs and lead authors of the IPCC's *Fifth Assessment Report* (AR5). [Photo by IISD/Sean Wu (enb.iisd.org/climate/ipcc48/6oct.html).]

The twentieth-century warming lies within the range of temperature variations that have been inferred for the Holocene. In fact, average temperatures in many regions of the world were probably warmer 10,000 to 8000 years ago than they are today. The twentieth-century record is clearly anomalous, however, when compared with the pattern and rate of climate change documented during the last millennium. Although direct temperature measurements are not available from before the nineteenth century, climate indicators such as ice cores and tree rings have allowed climatologists to reconstruct a temperature record for the Northern Hemisphere during that period (Figure 14.5b). That record shows an irregular but steady global cooling of about 0.2°C in the nine centuries between 1000 and 1900. It also shows that fluctuations in average surface temperature during each of these centuries were less than a few tenths of a degree.

The second argument, and to many a more compelling one, comes from the agreement between the observed pattern of warming and the pattern predicted by the best climate models. Models that include changes in atmospheric greenhouse gas concentrations not only reproduce the twentieth-century warming, but also reproduce the observed patterns of temperature change both geographically and with altitude in the atmosphere—the fingerprints of an **enhanced greenhouse effect.** For example, these models predict that, as human-enhanced greenhouse warming occurs, nighttime low temperatures at Earth's surface should increase more rapidly than daytime high temperatures, thus reducing daily temperature variation. Climate data for the last century confirm this prediction.

Another fingerprint of global warming has been the changes seen in mountain glaciers at lower latitudes. Glaciers found above 5000 m in Africa, South America, and Tibet have been shrinking during the last hundred years, an observation that is also consistent with the predictions of climate models (Figure 14.7).

FIGURE 14.7 Glaciologist Lonnie Thompson at an altitude of 5300 m (17,390 ft) on Tibet's Dasuopu Glacier. Ice coring on this glacier provides evidence of abnormal global warming during the twentieth century. [Lonnie Thompson/Byrd Polar Research Center, Ohio State University.]

Biological Change

The biosphere has been evolving throughout its multi-billion-year history. The geologic record shows long periods of stability punctuated by brief intervals of rapid and radical changes to the biosphere. The most extreme events of global biological change are marked by *mass extinctions*—dramatic losses in the number of species over short periods. Some of these cataclysms were caused by extraterrestrial events, such as the meteorite impact that killed off the dinosaurs at the end of the Cretaceous period. Others may have been due to climate change and ocean acidification, including the biggest mass extinction ever, at the end of the Permian period, when 95 percent of all species vanished from the fossil record. The end-Permian extinction, which marks the boundary between the Paleozoic and Mesozoic eras, appears to have been stimulated by huge, gaseous outpourings of the basaltic lavas that formed the Siberian Traps (see Chapter 5).

Earth is now undergoing another period of global biological change, but this time not from the hammer blow of some extreme extraterrestrial or subterrestrial event, but because a single biological species—*homo sapiens*—has recently evolved into an active agent of global physical change. We are at the cusp of an extreme event that will leave its mark in the geologic record when seen millions of years from now. Many geoscientists believe that we are now leaving the equitable natural conditions of the Holocene epoch and entering a new epoch of human-dominated change—the *Anthropocene*. We do not know what features of this event will be most recognizable to some geologist far in the future, because that geologic record is just now being written. What it will say depends on how our species responds to its own success.

Climate Change

In its Fifth Assessment Report, finalized in 2014, the IPCC drew the following conclusions:

- From the beginning of the twentieth century until 2012, the average temperature of Earth's surface has risen, on average, by about 0.9°C.
- Most of this warming has been caused by anthropogenic increases in atmospheric greenhouse gas concentrations.
- Concentrations of atmospheric greenhouse gases will continue to increase throughout the twenty-first century, primarily because of human activities.
- The increase in atmospheric greenhouse gas concentrations will cause significant global warming during the twenty-first century.

The last prediction is strongly supported by temperature trends at the time of this writing (late 2018). The four years since the IPCC report was published have been the four warmest years since modern recordkeeping began in 1880. The warmest year on record, 2016, beat the previous record, set just the year before, by 0.13°C. The 10 warmest years on record have all occurred since 1998; the 20 warmest years have all occurred since 1995. The surface of our planet is definitely getting hotter (Figure 14.5b).

Projecting Future Climate Change

How much hotter will it get, and how will this global warming affect local climates and ecosystems? The projections are uncertain, first, because we do not completely understand how the climate system works—our models are uncertain—and, second, because the projections depend strongly on how human population and the global economy will evolve, including how energy resources will be exploited and what political decisions will be made to limit greenhouse gas emissions. The IPCC has forecast increases in atmospheric CO_2 concentrations under a series of scenarios that sample the possibilities. Each scenario is characterized by a **representative concentration pathway** or "RCP" that corresponds to a net concentration of greenhouse gases in Earth's atmosphere by the year 2100. Three of these scenarios are depicted in **Figure 14.8**:

- **Scenario A** (red line) assumes a continued reliance on fossil fuels as our major energy source and thus an increasing concentration of greenhouse gases. In this scenario, called "RPC8.5" by the IPCC, the carbon dioxide concentration would top 900 ppm by 2100, more than three times its preindustrial level, and the radiative forcing would be 8.5 W/m^2 in 2100. **Radiative forcing** by a climate variable such as greenhouse gas concentration is a change in Earth's energy balance between incoming solar energy and outgoing thermal infrared energy when the variable is changed while all other factors are held constant. A radiative forcing of 8.5 W/m^2, which specifies this particular RCP scenario, can be compared to the average solar forcing of 240 W/m^2.

- **Scenario B** (green line, "RCP6.0") assumes that carbon dioxide concentrations will begin to stabilize in the later part of the twenty-first century, reaching just over 600 ppm, more than twice the preindustrial level, by 2100. This scenario reduces the radiative forcing to 6.0 W/m^2, but achieving it would require a big shift toward nuclear energy and renewable energy sources, as well as fossil fuels with less carbon intensity, such as natural gas. As we saw in Chapter 13, the fossil-fuel burning still dominates the energy economy, although the transition to less carbon-intensive energy sources is under way.

- **Scenario C** (blue line, "RCP2.6") has a peak in carbon dioxide concentrations around 2050, followed by a modest decline to carbon dioxide concentrations near the present level (400 ppm) by the end of the century. To achieve a scenario with such little radiative forcing (2.6 W/m^2) would require a much more rapid conversion from fossil fuels to cleaner alternatives than scenario B.

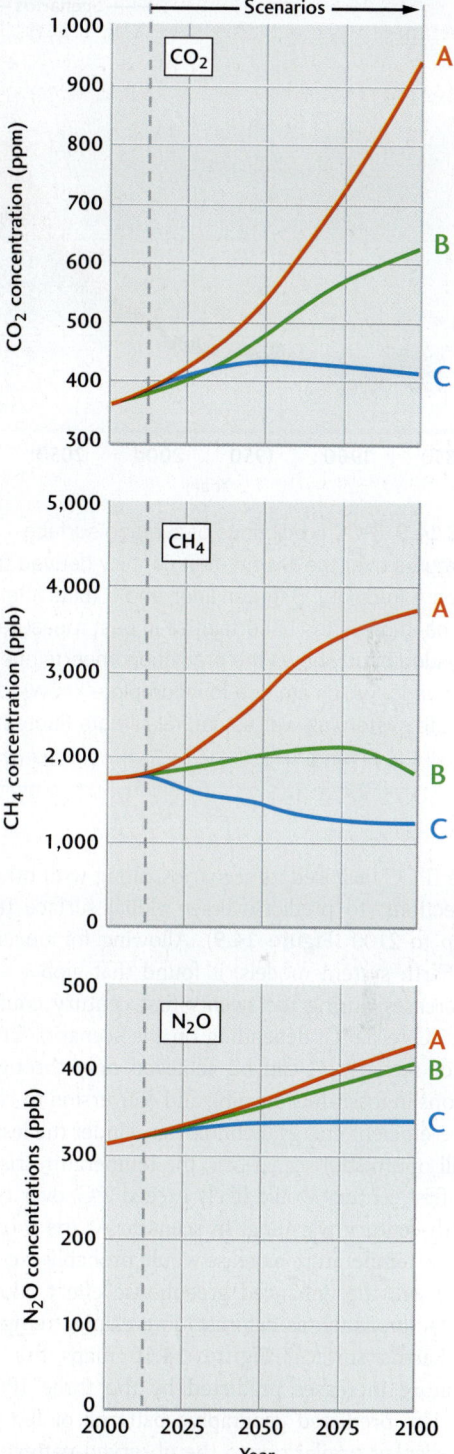

FIGURE 14.8 Three scenarios or "representative concentration pathways" (RCPs) projected by the IPCC for carbon dioxide, methane, and nitrous oxide during the twenty-first century. Scenario A (red line) implies continuing high rates of fossil-fuel burning (RCP8.5); scenario B (green line) implies stabilization of emission rates in the latter part of the twenty-first century (RCP6.0); scenario C (blue line) implies a rapid conversion to non-fossil fuels (RCP2.6). [Data from IPCC, *Climate Change 2013: The Physical Science Basis*.]

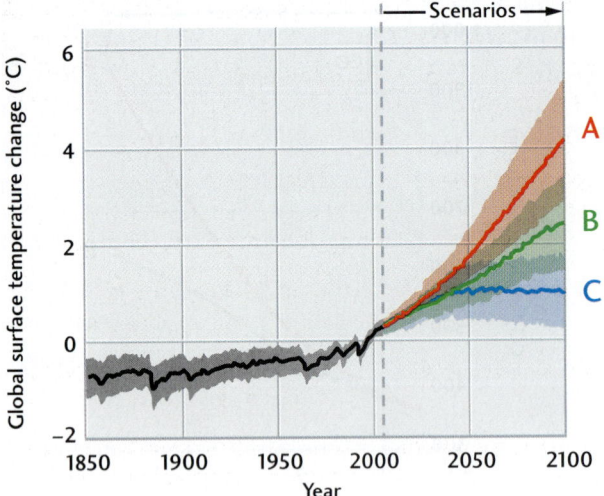

FIGURE 14.9 IPCC predictions of average surface temperatures over the twenty-first century derived from scenarios A (red line), B (green line), and C (blue line). Gray-shaded band gives the uncertainties in past measurements; color-shaded band shows the prediction uncertainties of each scenario, which are due to incomplete knowledge of the climate system as well as natural climate fluctuations. [Data from IPCC, *Climate Change 2013: The Physical Science Basis*.]

The IPCC has used its scenarios, along with other types of projections, to predict average global surface temperatures up to 2100 (**Figure 14.9**). Allowing for uncertainties in the Earth system models, it found that global temperature increases during the twenty-first century could range from 0.5°C to 5.5°C, depending on the scenario. The lower values of scenario C can be achieved only through rapid reductions in fossil-fuel burning and conversion to clean and resource-efficient energy technologies. Under the less radical (but still optimistic) scenario B, the temperature rise in the twenty-first century would likely exceed 2°C, over twice the twentieth-century warming. In scenario A, the most pessimistic, the temperature increase would probably exceed 4°C.

How will the enhanced greenhouse effect, along with related factors, such as deforestation, change temperatures across Earth's surface? **Figure 14.10** maps the regional temperature increases predicted by the three IPCC scenarios. The predicted geographic patterns of temperature change display similarities to the observed pattern of late-twentieth-century warming in Figure 14.5. In particular, the warming is greater over land than over the oceans, and the temperate and polar regions of the Northern Hemisphere show the most warming.

Human Population and Global Change

Anthropogenic global change is inextricably tied to human population size. This population exploded after the Industrial Revolution, increasing from about 1 billion people in 1800 to more than 7.6 billion today (**Figure 14.11**).

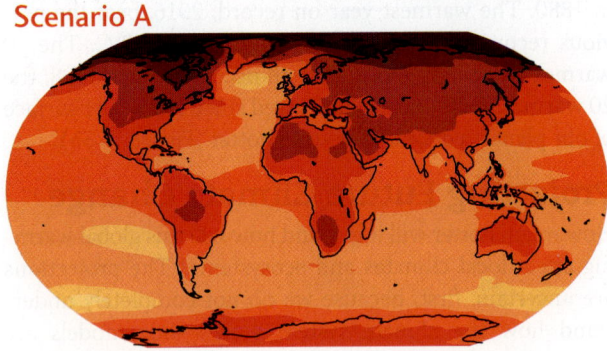
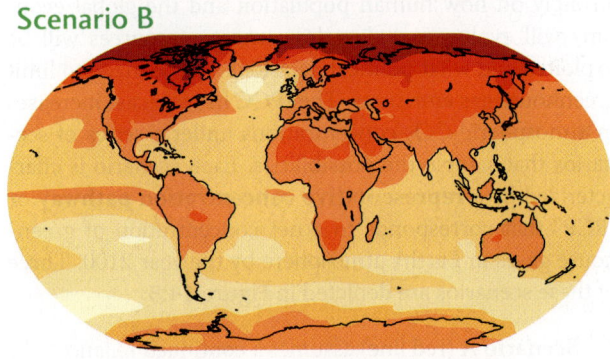
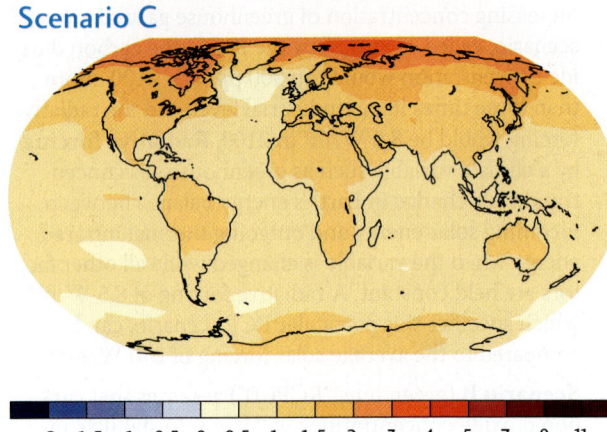

FIGURE 14.10 Average surface temperatures predicted for 2080–2100 by the three IPCC scenarios, expressed as differences from average surface temperatures measured at the same locations during the baseline period 1986–2005. [Data from IPCC, *Climate Change 2013: The Physical Science Basis*.]

Population is the dominant variable in any predictions of future anthropogenic change, because it so strongly controls anthropogenic emissions of greenhouse gases. Owing to their dependence on population, the representative concentration pathways used by the IPCC necessarily make implicit assumptions about population growth. Figure 14.11 shows the population growth curves consistent with these scenarios. Population projections to the year 2100 range from 12.4 billion for scenario A to 8.8 billion for scenario C.

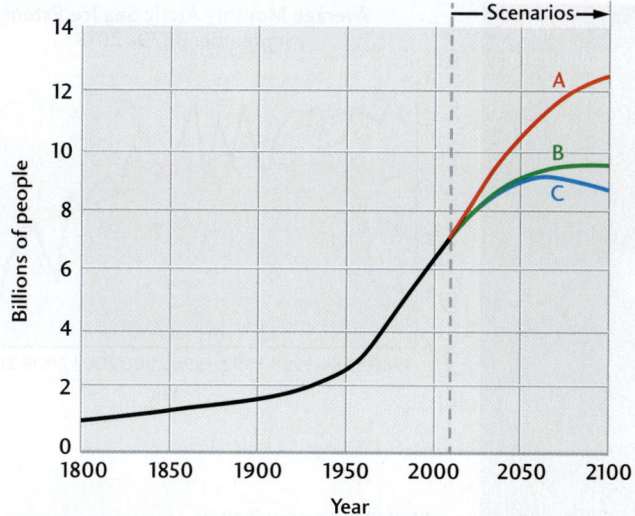

FIGURE 14.11 Black-line charts the global growth of human population since 1800. Colored lines show future population growth projected by the three IPCC scenarios in Figure 14.8. In scenario A (red line), world population continues to grow into the twenty-second century. In scenario B (green line), it levels off in the late twenty-first century, and in scenario C (blue line), it declines after 2070.
[Data from IPCC, *Climate Change 2013: The Physical Science Basis*.]

We can compare these scenario-based projections with demographic predictions by the United Nations. The UN's median projection, its "best guess," pegs the human population in 2100 at 11.3 billion, about halfway between scenarios A and B. The UN assigns uncertainties to its projections. When these uncertainties are taken into account, one finds that the chances are greater than 95 percent that the *actual* world population curve will fall somewhere within the band between scenarios A and C.

The probabilities given by the UN are not symmetrically distributed across this band, however. The chance that the population trajectory will lie between scenarios A and B is almost 80 percent, and there is less than a 5 percent chance that it will fall between B and C. In other words, scenario C is a much less likely representation of future global change than either A or B. In fact, over the last decade, human population and its greenhouse gas emissions are following scenario A rather closely. A climate future conforming to scenario B is becoming a harder goal to achieve.

Consequences of Climate Change

As documented by the IPCC, human emissions of greenhouse gases are certain to cause further global warming. These changes have the potential to affect civilization in both positive and negative ways. Some regional climates may become more hospitable to humans, while others may deteriorate. On the balance, however, the adaptation of civilization and the biosphere to a much warmer planet will not be easy, and it will certainly be costly. The authors' home city of Los Angeles, for example, is likely to become hotter and drier. Some potential effects of climate change are listed in **Table 14.1**.

TABLE 14.1 Potential Effects of Climate Change on Ecosystems and Resources

SYSTEM	POTENTIAL EFFECTS
Forests and other ecosystems	Migration of vegetation; reduction in ecosystem ranges; altered ecosystem composition
Species diversity	Loss of diversity; migration of species; invasion of new species
Coastal wetlands	Inundation of wetlands; migration of wetland vegetation
Aquatic ecosystems	Loss of habitat; migration to new habitats; invasion of new species
Coastal resources	Inundation of coastal structures; increased risk of flooding
Water resources	Changes in water supplies; changes in patterns of drought and flooding; changes in water quality
Agriculture	Changes in crop yields; shifts in relative productivity among regions
Human health	Shifts in ranges of infectious disease organisms; changes in patterns of heat-stress and cold-weather afflictions
Energy	Increase in cooling demand; decrease in heating demand; changes in hydroelectric energy resources

Source: Office of Technology Assessment, U.S. Congress.

Changes in Regional Weather Patterns The large-scale geographic patterns of climate change in the twenty-first century are likely to be similar to those observed over the past several decades. The IPCC has documented a number of current trends in regional weather patterns that are likely to continue:

- The relative humidity and frequency of heavy precipitation events have increased over many land areas, consistent with the observed temperature increases. Increased precipitation has been observed in eastern parts of North and South America, northern Europe, and northern and central Asia. For example, 2018 was wettest year on record in many cities in the eastern United States, including Washington, D.C., and Pittsburgh.

- Drying has been observed in the Sahel, the Mediterranean, southern Africa, and parts of southern Asia. More intense and longer droughts have been observed over wider areas since the 1970s, particularly in the tropics and subtropics.

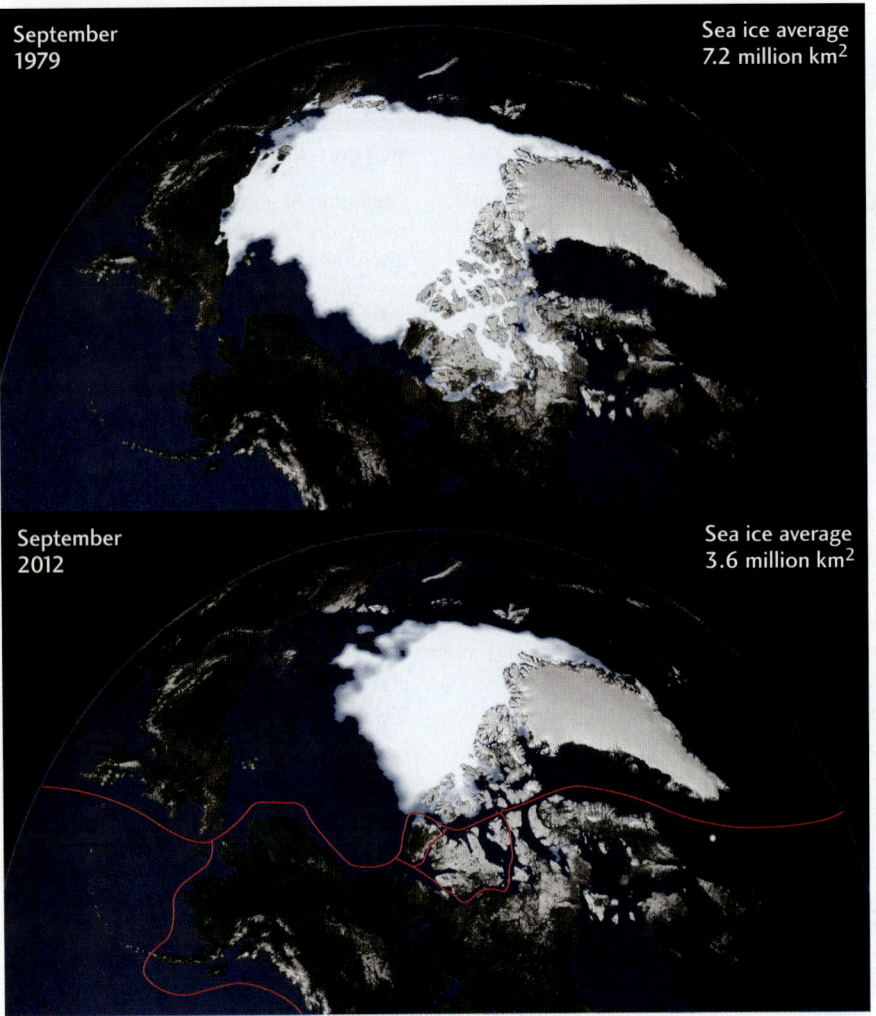
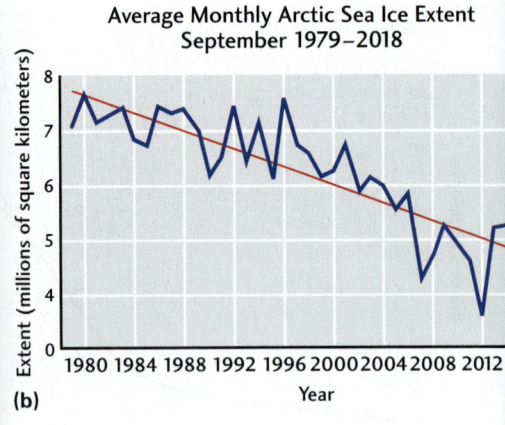

FIGURE 14.12 (a) Global warming is melting the Arctic ice cap. These images of the Arctic, derived from NASA satellite data, compare the minimum extent of the polar ice cap in September 1979 (top) with the minimum extent in 2012 (bottom). One near-term benefit to human society will be the opening of the Northwest Passage and other shorter sea routes between the Atlantic and Pacific Oceans (shown in the bottom panel as red lines). (b) Graph showing that the average sea ice extent during the month of September has decreased since 1979. [NASA/Goddard Scientific Visualization Studio.]

- Widespread changes in temperature extremes have been observed over the last 50 years. Cold days, cold nights, and frost have become less frequent, while hot days, hot nights, and heat waves have become more frequent.
- Intense hurricane activity in the North Atlantic has increased, consistent with increases in tropical sea surface temperatures. Although there is no clear trend in the annual number of hurricanes, the number of very strong hurricanes (category 4 and 5 storms) has almost doubled over the past three decades.

Changes in the Cryosphere Nowhere are the effects of global warming more evident than in polar regions. The amount of sea ice in the Arctic Ocean is decreasing with time. The sea ice cover in September 2012 was the lowest for that month since the keeping of satellite records began in 1978: 3.6 million square kilometers, down by a factor of two from the 1979 minimum of 7.2 million square kilometers (**Figure 14.12**). According to climate models, much of the Arctic Ocean will become ice-free within a few decades. The shrinkage of sea ice is already severely disrupting Arctic ecosystems (**Figure 14.13**).

Temperatures at the top of the permafrost layer—the exposed, permanently frozen ground of the Arctic continents (see Chapter 15)—have risen by 3°C since the 1980s, and the melting of permafrost is destabilizing structures such as the

FIGURE 14.13 Climate change is already disrupting Arctic ecosystems, adversely affecting the habitat of Arctic animals such as polar bears. [Ralph Lee Hopkins/Getty Images.]

FIGURE 14.14 Glacier National Park's Boulder Glacier in 1932 (left) and 2005 (right). [Photo courtesy T.J. Hileman, Glacier National Park Archives (*left*); photo by Greg Pederson, U.S. Geological Survey (*right*).]

Trans-Alaska oil pipeline. The maximum area covered by seasonally frozen ground has decreased by about 7 percent in the Northern Hemisphere since 1900, with a decrease in spring of up to 15 percent. Valley glaciers at lower latitudes retreated during the twentieth-century warming (**Figure 14.14**). The 37 glaciers remaining at Glacier National Park have lost about 40 percent of their mass in the last half century, and most of them will disappear in the next half century.

Sea-Level Rise The melting of sea ice does not affect the sea level, but the melting of continental glaciers causes the sea level to rise. The sea level also rises as the temperature of ocean water increases, expanding its overall volume. The sea level has risen more than 200 mm since the Industrial Revolution, and it is currently rising at about 3 mm per year. Most of this increase is due to ocean warming (see this chapter's Practicing Geology Exercise).

Climate models based on the IPCC scenarios indicate that the sea level could rise by as much as a meter during the twenty-first century (**Figure 14.15**), creating serious problems for low-lying countries such as Bangladesh (**Figure 14.16**). Island nations like the Maldives, where the highest ground is just a few meters above sea level, will be especially vulnerable (see the chapter-opening photo). On the Eastern Seaboard and Gulf Coast of the United States, flooding during coastal storm surges, such as those witnessed in Hurricanes Katrina (2005), Sandy (2012), and Michael (2018), could become much worse. Some parts of the southeastern United States are already experiencing the "sunny day" flooding of coastal towns during the highest (king) tides that occur once or twice per year (**Figure 14.17**).

Melting of the great continental ice sheets that cover Antarctica and Greenland has thus far contributed only a small amount to sea-level rise, but the rate of glacial thinning is increasing, primarily because of accelerating glacial flow. Satellite observations reveal that flow accelerations of 20 to 100 percent have occurred over the past decade (see

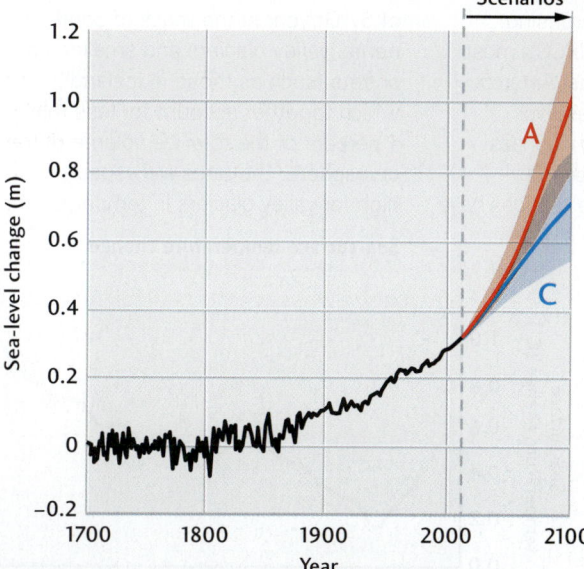

FIGURE 14.15 Sea-level rise 1700–2100. The black line shows the observed value through 2010. The red curve is the IPCC prediction of future sea-level rise during the remainder of the twenty-first century according to scenario A; the blue curve is the prediction according to scenario C.

FIGURE 14.16 Flooded village south of Dhaka, Bangladesh. [Yann Arthus-Bertrand/Getty Images.]

Chapter 15). Climate scientists are concerned that these accelerations will continue as the polar regions become warmer and the loss of ice leads to further warming through the albedo feedback described in Chapter 12.

Species and Ecosystem Migration As local and regional climates change, ecosystems will change with them. Many plant and animal species will have difficulty adjusting to rapid climate change or migrating to more suitable climates. The IPCC's most recent report on global warming estimates that, under conditions similar to scenario B, 18 percent of all insect species and 16 percent of all plants would lose more than half of their climatic habitat by 2100.

Ecosystem stress will be higher in arid zones than in humid zones. Deserts and arid vegetation will encroach on the Mediterranean climate zone, for example, causing changes not seen by humans in the last 10,000 years. Many

FIGURE 14.17 Seasonal high tides combined with rising sea levels caused street flooding on September 29, 2015, in Miami Beach, Florida. [Joe Raedle/Getty Images.]

PRACTICING GEOLOGY EXERCISE

Why Is Sea Level Rising?

Over the twentieth century, the sea level rose about 200 mm, and it is currently rising at a rate of about 3 mm/year, as shown in the figure below. Why is sea level rising?

We know that anthropogenic warming of the polar regions is reducing the amount of sea ice and causing the breakup of large ice shelves. Because of isostasy, however, this decrease in the volume of floating ice does not contribute to sea-level rise (see Chapter 15). Melting ice can cause the sea level to change only if the ice is on land, not floating in water (see Figure 15.13).

Most of the world's ice is locked up in the huge continental glaciers that cover Antarctica and Greenland. Is global warming causing these ice sheets to melt faster than they can be regenerated by new snowfall? Radar instruments mounted on Earth-orbiting satellites can directly measure changes in the ice volume of a region. The results have been surprising.

First, according to the IPCC's most recent assessment, the East Antarctic ice sheet, the largest ice reservoir on Earth, has been *gaining* ice mass at about 21 Gt/year in the period 1993–2010. Recent climate changes have evidently increased the amount of snowfall in East Antarctica. This net accumulation is good news because it subtracts from any sea-level rise. Unfortunately, the West Antarctic ice sheet is losing mass at a much higher rate, at about 118 Gt/year, and the smaller Greenland ice sheet is losing about 121 Gt/year.

Most surprising of all is a net loss of 57 Gt/year in the mass of continental valley glaciers and smaller ice sheets (such as those in Iceland), which together account for less than 1 percent of the total ice volume of the cryosphere. The rates are especially high for valley glaciers in temperate and

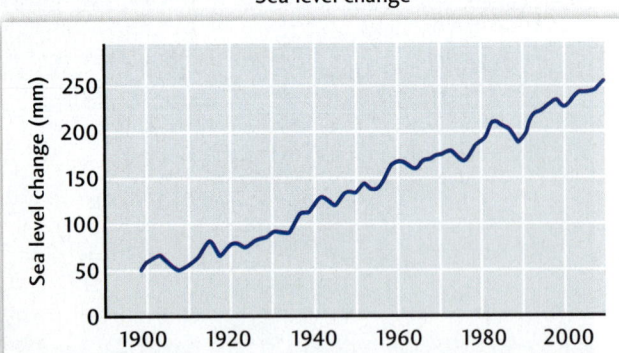

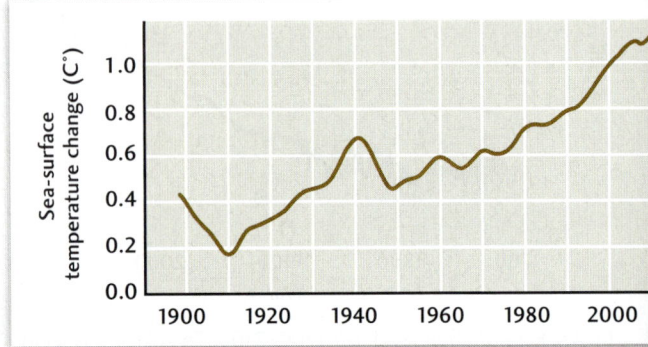

During the twentieth century, the sea level rose by about 200 mm (top panel), and the global average sea surface temperature increased by about 1°C (bottom panel). [Sea-level change data from B. C. Douglas; sea-surface temperature change data from British Meteorological Office.]

ecosystem impacts will be larger at higher latitudes owing to cold-season warming rates that are well above the global average. High-latitude tundra and boreal forest are particularly at risk; woody shrubs are already encroaching into tundra.

Marine organisms will face progressively lower oxygen levels and high rates of ocean acidification; ecological stress will be exacerbated by rising extremes in ocean temperature. Changes to water temperatures will drive some mobile species, such as plankton and fish, to relocate at higher latitudes. Novel ecosystems may appear. Especially vulnerable are coral reefs, which can move only slowly, and polar ecosystems, which have nowhere to go (**Figure 14.18**). Ecosystems that cannot migrate with climate change or adapt to its consequences may face disruption and even collapse. Under the most favorable climate scenarios, the majority of warm-water coral reefs that exist today will disappear by 2100; more realistic scenarios predict catastrophic losses—greater than 99 percent.

FIGURE 14.18 Bleached coral in the Indian Ocean in the Maldives. [the ocean agency / xl catlin seaview survey.]

tropical regions, which are vanishing very quickly (see Figure 14.14).

Summing up these numbers yields 275 Gt/year as the current rate of continental ice loss. Essentially all of this mass goes into the ocean. One gigaton of water occupies one cubic kilometer (its density is 1 g/cm^3), so the increase in ocean volume is about 275 km^3/year. We can convert this volume change into sea level change using the formula

sea-level rise = ocean volume increase ÷ ocean area

In Appendix 2, we find that the ocean area is 3.6×10^8 km^2, so

sea-level rise = 275 km^3/year
÷ 3.6×10^8 km^2 = 7.6×10^{-7} km/year

or about 0.8 mm/year.

This figure is only a fraction of the current rate of sea-level rise. The rest comes from the warming of the ocean itself. Over 90 percent of the heat generated by the enhanced greenhouse effect is taken up by the oceans. On average, surface waters have warmed by almost 1°C during the last century. This warming has caused the water in the upper portion of the ocean to expand a tiny fraction, about 0.01 percent. That small increase in volume can account for most of the 200 mm rise in sea level during that period. If Earth's climate changes according to scenario A, the average sea surface temperature could increase by 2.7°C during the twenty-first century, with even higher changes along some coastal regions (see the figure below).

BONUS PROBLEM: If seawater expands by 0.01 percent for each 1°C of temperature increase, how deep is the layer of the ocean that must be heated by 1°C to explain the twentieth-century sea-level rise of 200 mm?

Projected changes in sea surface temperature (°C) for the coastal United States under scenario A. Map shows the difference between the average sea-surface temperature for the period 2050–2099 period and the average sea-surface temperatures for 1956–2005. [NOAA, from *Climate Science Special Report: Fourth National Climate Assessment*, Volume I. Wuebbles, D.J., D.W. Fahey, K.A. Hibbard, D.J. Dokken, B.C. Stewart, and T.K. Maycock (eds.). U.S. Global Change Research Program.]

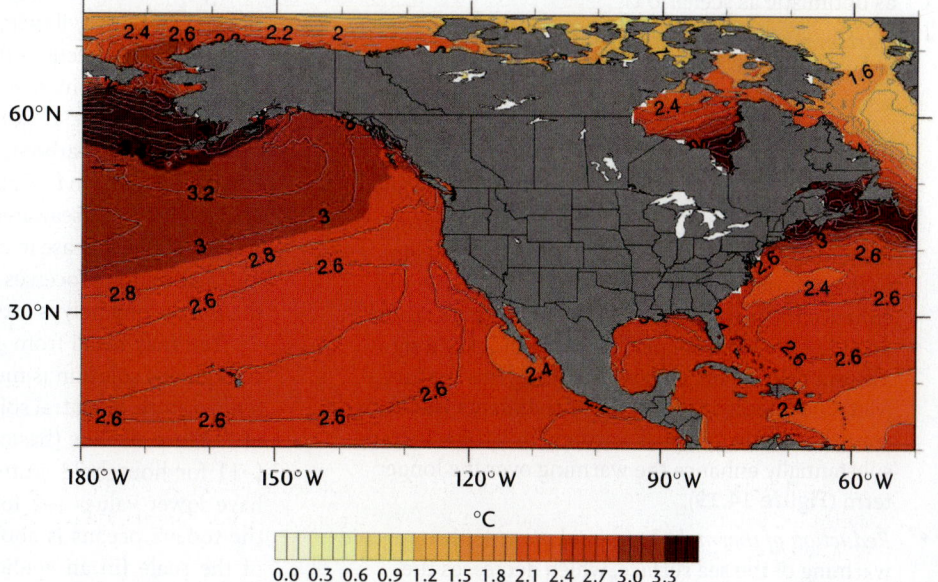

The Potential for Catastrophic Changes to the Climate System The current atmospheric concentrations of carbon dioxide and methane far exceed anything seen in the last 800,000 years (see Figure 14.2). Our climate system is entering unknown territory. While some criticize the IPCC climate change projections, believing they generate unnecessary anxiety, most scientists think that current projections are too conservative because they do not properly take into account some of the positive feedbacks that could greatly enhance global change. We briefly describe a few examples of feedbacks and tipping points that could create more catastrophic change than current observations predict.

- *Acceleration of climate change by albedo feedback.* Earth's *albedo* is the fraction of solar energy reflected by its surface. Albedo feedback occurs when a rise in temperature reduces the accumulation of ice and snow in the cryosphere, which decreases Earth's albedo and thus increases the heat energy Earth's surface absorbs; this increased warming enhances the temperature rise in a positive feedback. Climate models explicitly include this important feedback, and it has been accounted for in the IPCC projections, but aspects of the problem, such as short-term snow-albedo feedback and cloud-albedo feedback, remain uncertain.

- *Destabilization of continental glaciers.* The surface melting of the Greenland glacier in 2012 was the largest on record, and glacial streams within the ice sheet are accelerating much faster than expected. If the Greenland and Antarctic glaciers begin to shed ice much faster than snowfall can generate new glacial ice, the sea level could begin to rise much faster than the current IPCC predictions. The best estimate for Greenland puts this threshold only about 0.6°C above the current level of global warming, conditions that will be reached by mid-century, even under projections as optimistic as scenario C.

- *Carbon release from permafrost and seafloor sediments.* A massive release of methane from shallow seafloor sediments about 55 million years ago might have caused abrupt global warming and led to the mass extinction at the Paleocene-Eocene boundary. Today, there is far more organic carbon stored in permafrost and shallow seafloor sediments than was released at the end of the Paleocene. If global warming begins to thaw those carbon deposits, the feedback would enhance the warming. Models indicate that carbon dioxide released from permafrost could play a significant role in climate change during this century. Methane released from seafloor sediments will likely contribute little to global warming by 2100 but could substantially enhance the warming over the longer term (**Figure 14.19**).

- *Reduction of thermohaline circulation.* Anthropogenic warming of the sea surface, which decreases the

FIGURE 14.19 Bubbles of methane gas frozen into clear ice in Lake Baikal near the Mongolian border in Russia. [Streluk/Getty Images.]

density of surface waters, is causing the oceans to become more stratified. The stratification is strengthening because new precipitation and evaporation patterns are decreasing the salinity, and thus the density, of seawater at middle and high latitudes. The net effect is to inhibit surface water from sinking as part of oceans' thermohaline circulation (see Figure 12.5). Under scenario A, thermohaline circulation could weaken by as much as 10 to 50 percent, which would result in less absorption of heat and CO_2 by the ocean—a positive feedback to global warming. Major changes in the Gulf Stream, a strong current in the thermohaline circulation, could alter the climates of North America and Europe.

Ocean Acidification

Not only are the oceans warming, but their chemistry is changing as well. *Ocean acidification,* sometimes called global warming's evil twin, is also born from fossil-fuel burning. About 30 percent of the carbon dioxide emitted into the atmosphere by human activities is being absorbed into the oceans (see Figure 12.20). The carbon dioxide reacts with seawater to form carbonic acid, which is followed by a series of reactions (shown in Figure 12.18) that ultimately leach carbonate ions from seawater. The increase in seawater acidity and resulting decrease in carbonate ion concentration inhibit the calcification processes that sea creatures use to form their shells and coral polyps use to form a coral reef.

You may recall from chemistry class that the acidity of an aqueous solution is measured on the pH scale. At room temperature, a neutral solution such as pure water has a pH of 7. More alkaline (basic) solutions have higher pH values (~11 for household ammonia), and more acidic solutions have lower values (~2 for lemon juice). The mean pH of the today's oceans is about 8.1, so it's on the alkaline side of the scale (in an acidic ocean, no shells or coral could grow). The carbon dioxide content of the ocean is following the Keeling curve upward, and pH of seawater therefore is going downward (Figure 14.20). The total drop since the Industrial Revolution has been about 0.1 pH unit.

This pH difference may not seem like much, but remember that the pH scale, like the Richter scale for measuring earthquakes, is base-10 logarithmic. One unit represents a factor of ten in the concentration of H^+ ions, so a drop of 0.1 pH unit represents a 26 percent increase in seawater acidity ($10^{0.1} = 0.259$). The biochemical reactions needed to sustain life are very sensitive to small changes in pH. For instance, your body normally regulates your blood pH within a narrow range around 7.4. A sudden drop of 0.2–0.3 pH units can send you into a coma or even kill you. Similarly, a small change in the pH of seawater can harm marine life.

Ocean acidification is likely to affect many types of marine organisms, not just those with shells and skeletons. Anemones and jellyfish, for example, appear to be susceptible to even small changes in seawater acidity, and larger increases cause changes in seawater chemistry that can undermine the health of sea urchins and squid. The growing acidity of ocean surface waters is also likely to affect the concentrations of trace metals such as iron, an essential nutrient for the growth of many organisms. Oceanic ecosystems at higher latitudes typically have a lower buffering capacity against acidity influx, exhibiting seasonally corrosive conditions sooner than low-latitude ecosystems.

As human activities continue to pump more CO_2 into the atmosphere, the ocean will continue to acidify at a rate not seen at any time during the Cenozoic era. Under scenario A, the pH of surface water will likely drop from 8.1 to 7.8 or 7.7, corresponding to a 100 or 150 percent increase in acidity (Figure 14.21). In January 2009, more than 150 marine scientists from 26 countries convened under the auspices of the United Nations and issued the Monaco Declaration, which states, "We are deeply concerned by recent, rapid changes in ocean chemistry and their potential, within decades, to severely affect marine organisms. . . . Severe damages are imminent." The scientists pointed specifically to observations of acidification-related decreases in shellfish weights and slowed growth of coral reefs.

Whether marine organisms can adapt to the changes in store remains to be seen, but the effects on human society could be substantial. In the short term, damage to coral reef ecosystems and the fisheries and recreation industries that depend on them could result in economic losses amounting to many billions of dollars per year. In the longer term, changes in the stability of coastal reefs may reduce the protection they offer to coasts, and there may also be direct and indirect effects on commercially important fish and shellfish species.

Ocean acidification is essentially irreversible during our lifetimes. Even if we were magically able to reduce the atmospheric concentration of CO_2 to the level of 200 years ago, it would take tens of thousands of years for ocean chemistry to return to the conditions that existed at that time.

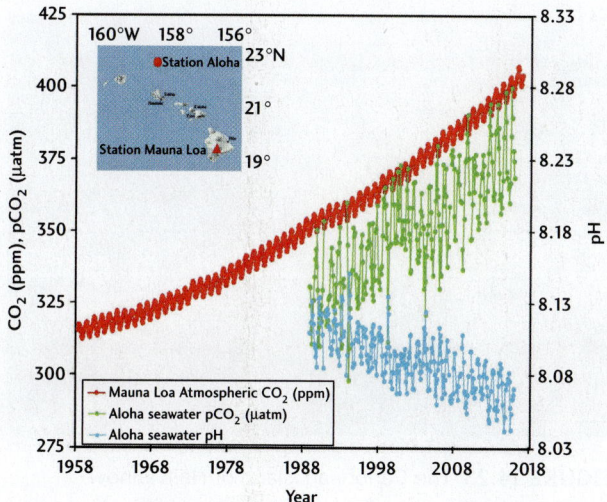

FIGURE 14.20 Comparison of the Keeling curve from the Mauna Loa Observatory (red line) with the CO_2 content (green points) and acidity (blue points) of seawater samples taken at Station Alpha, just north of Oahu (see inset map). CO_2 content of the atmosphere is measured in parts per million, and CO_2 content of seawater is measured as a partial pressure in millionths of an atmosphere (scale on left); acidity is measured in pH units (scale on right). [NOAA PMEL Carbon Program.]

FIGURE 14.21 Predicted change in sea-surface pH in 2090–2099 relative to 1990–1999 under scenario A. [U.S. Environmental Protection Agency.]

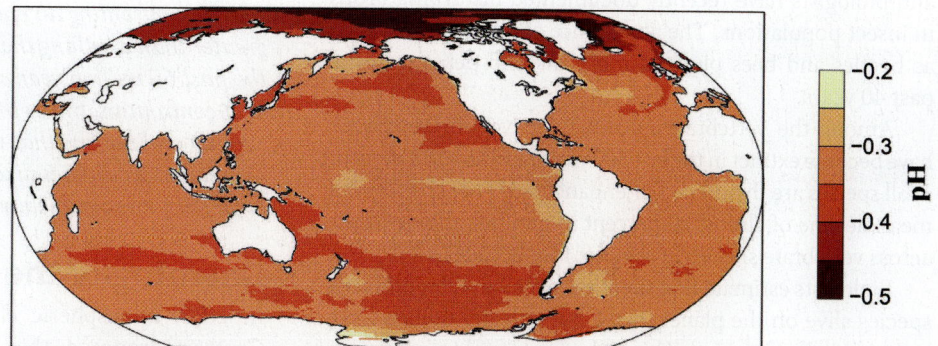

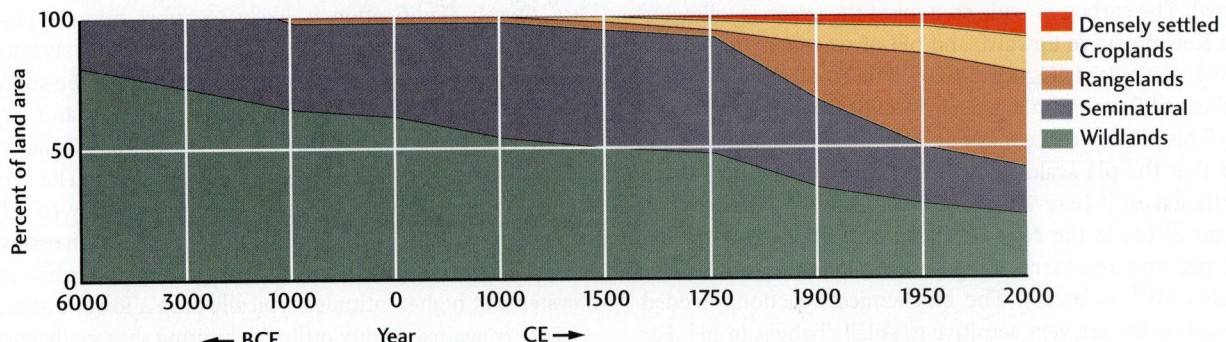

FIGURE 14.22 Anthropogenic transformation of the biosphere can be measured by land use, shown here for the last 8000 years as percentages of global ice-free land area. Land occupied by human settlements (red) or used by humans as croplands (yellow) or rangelands (orange) increased from less 10 percent in 1750 to almost 60 percent in 2000. The time scale is not linear. [Data from E. C. Ellis, 2011, *Proceedings of the Royal Society of London*, 369, 1015–1035.]

Loss of Biodiversity

Humans have been altering Earth's biosphere since before the development of agriculture around 8000 years ago, driving species to extinction and transforming entire ecosystems. Neolithic populations of a few million affected perhaps 20 percent of ice-free global lands in measurable ways, but they directly occupied only a tiny fraction of what was then a vast wilderness. Human land occupation remained a small fraction of the global total between the agricultural and industrial revolutions. In 1750, less than 10 percent of global land was occupied by settlements or directly used as croplands or rangelands (**Figure 14.22**). The subsequent population boom increased that fraction to almost 60 percent in 2000.

According to the United Nations, over 150,000 km² of tropical rain forests—about 1 percent of the total resource—are being converted each year to other land uses, mostly agricultural, and deforestation continues to rise. In 1950, forests covered approximately 25 percent of Haiti (a Caribbean island country the size of Maryland); its forested area now stands at less than 2 percent (**Figure 14.23**). Other developing nations face similar problems.

Given these rates of habitat loss, it is not surprising that the number of extant species—the most important measure of biodiversity—is declining. About one-quarter of the bird species on Earth have gone extinct in the last two millennia, and biologists have recently documented disturbing drops in insect populations. The abundance of invertebrates such as beetles and bees plunged more than 45 percent in the past 40 years.

Among the vertebrates, at least 322 vertebrate species have become extinct in the last 500 years, and 16 to 33 percent of all species are threatened or endangered. There has been a mean decline of almost 30 percent in number of individuals across vertebrate species in the past four decades.

Biologists estimate that there are over 9 million different species alive on the planet today, but only 1.5 million have been officially classified. This lack of data makes extinction rates difficult to quantify. Rough estimates suggest that 10,000 to 60,000 species are lost each year. Some scientists believe that up to one-fifth of all species could disappear during the next 30 years, and that as many as one-half may go extinct during the twenty-first century. One respected biologist, Peter Raven, has put the problem bluntly:

FIGURE 14.23 The Caribbean island of Haiti is now 98 percent deforested. This photo shows Haiti's brown landscape, which contrasts sharply with the rich forests of its neighbor, The Dominican Republic. [James P. Blair/National Geographic/Getty Images.]

> We are confronting an episode of species extinction greater than anything the world has experienced for the past 65 million years. Of all the global problems that confront us, this is the one that is moving the most rapidly and the one that will have the most serious consequences. And, unlike other global ecological problems, it is completely irreversible.

Dawning of the Anthropocene

In 2003, atmospheric chemist and Nobel laureate Paul Crutzen proposed the recognition of a new geologic epoch—the **Anthropocene,** or Age of Man—beginning

Loss of Biodiversity 411

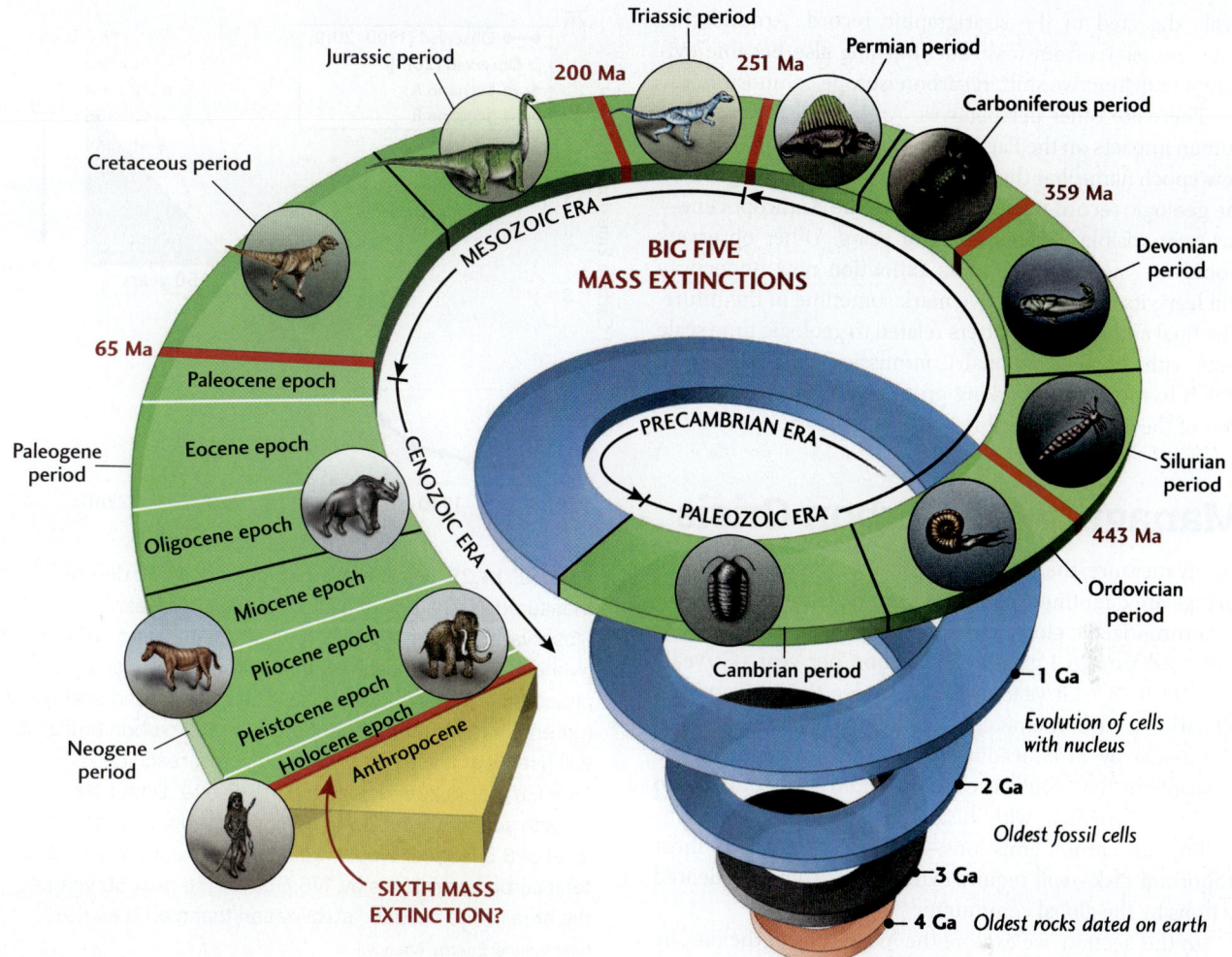

FIGURE 14.24 The Phanerozoic eon has seen many mass extinction events. The Big Five are marked as red lines on this spiral of geologic time. The current transition from the Holocene to the Anthropocene is likely to mark a sixth mass extinction.

about 1780, when James Watt's coal-powered steam engine launched the Industrial Revolution. The global changes that will characterize the Holocene-Anthropocene boundary are still happening, so future geologists may place that boundary at a somewhat different date. As with many previous geologic boundaries, the main marker will be a mass extinction.

Some observers, such as sociobiologist E. O. Wilson and science writer Elizabeth Kolbert, have gone so far as to call the current decline in biodiversity the "Sixth Extinction," placing it in the same rank as the "Big Five" mass extinctions of the Phanerozoic eon (see Figure 14.24 and Kolbert's piece in *The New Yorker*, May 25, 2009). The causes of the Big Five extinctions are different, but they are all marked in the geologic record by losses of more than three-quarters of all species. How many more years it will take for current extinction rates to produce species losses equivalent to the Big Five? The answer is not long at all.

If all "threatened" species became extinct within this century, and that rate continued, the extinction magnitude among land vertebrates would reach Big Five levels (75 percent) in surprisingly short times: approximately 240 years for terrestrial amphibians, 330 years for mammals, 540 years for birds. Limiting extinctions to "critically endangered" species over the next century and continuing those rates into the future increases the rise times by about a factor of four, which is still blindingly fast compared to the Big Five extinction rates. The current extinction rates of oceanic fauna are not as well known as those on land, but the situation is similarly dire. Under the most plausible climate-change scenarios (A and B), bleaching and die-offs will destroy 99 percent of all coral reefs before the end of this century, likely causing a collapse of reef-based ecosystems. The fish biomass that feeds much of humanity will also continue to decline.

Debate continues as to which key signatures of the Anthropocene should be used to define when that epoch actually started. A strong candidate is the mid-twentieth century, when the atomic explosions first rained down dateable, long-lasting radiogenic isotopes, such as plutonium-239, easily detected in the stratigraphic record. Around that time, emissions from fossil-fuel burning also became evident as a distinctive shift in carbon isotope compositions.

There are other perspectives. Archeologists agree that human impacts on the Earth are dramatic enough to merit a new epoch name, but they argue that the human imprint on the geologic record—the beginning of the Anthropocene—has been visible for thousands of years. Other observers would say "not yet"; the mass extinction now under way will leave its most distinctive mark sometime in the future. The final authority on matters related to geologic time scale rests with the International Commission on Stratigraphy, which has formed a working group to recommend definition of the Holocene-Anthropocene boundary.

Managing the Carbon Crisis

By any measure, the problems we face in confronting global change are daunting. Fossil-fuel burning is the main driver of anthropogenic global change. Fossil-fuel emissions have increased from 6.3 gigatons of carbon per year (Gt/year) in 2000 to 8.9 Gt/year in 2017. Under the high-growth scenario A, carbon emissions could increase to more than 22 Gt/year by 2100, leading to CO_2 concentration in the atmosphere that could exceed 900 ppm and continue to increase thereafter, with disastrous consequences. Controlling our carbon emissions—perhaps civilization's most important task—will require extraordinary, unprecedented actions by the global community.

In this section, we explore the magnitude of the task by considering the following problem. We suppose the human population and its per capita energy use continue to grow at their current rates. Then, fossil-fuel burning will cause the rate of carbon emissions to increase from 8.9 Gt/year in 2017 to about 15.9 Gt/year in 2067, an increase of 7 Gt/year in 50 years (**Figure 14.25**). What specific actions might be taken that could significantly reduce these increases?

Energy Policy

One set of questions policy makers must tackle is how much money we should spend to curb anthropogenic carbon emissions, and whether the benefits of doing so will justify the costs. Too much spending could depress the economy, yet preventing the most drastic effects of climate change might be less costly than coping with those disasters after they happen.

A partial solution—and certainly the most economical one—is to improve energy use efficiency and reduce waste. In a real sense, using energy more efficiently is like discovering a new source of fuel. We have seen that the U.S. energy system has an efficiency of only 32 percent; 68 percent of the total energy produced is wasted somewhere along the way (see Figure 13.5). Implementing efficiency measures costs relatively little—for example, insulating buildings,

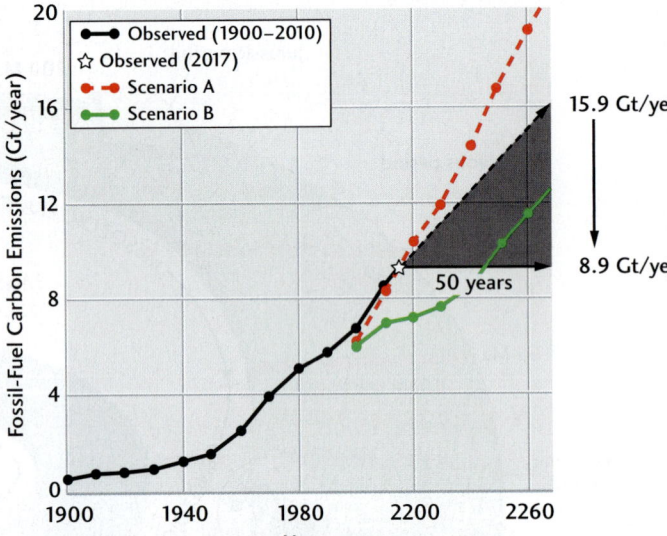

FIGURE 14.25 Fossil-fuel emissions of carbon dioxide measured in gigatons of carbon per year (Gt/year). Observations from 1900–2010 (black points) and 2017 (white star) are compared with fossil-fuel emissions projected according to scenario A (red points) and scenario B (green points). At the current growth rate, carbon emissions will rise by at least 7 Gt/year in 50 years, reaching 15.9 Gt/year in 2067 (dashed black arrow). Under this growth scenario, stabilizing carbon emissions at the 2017 level of 8.9 Gt/year would require special actions to reduce total carbon emissions by 175 Gt over the next 50 years—the area of the shaded "stabilization triangle." [Data from International Energy Agency.]

replacing incandescent light bulbs with fluorescent bulbs, increasing the fuel efficiency of motor vehicles, and making greater use of natural gas. The savings in energy costs could amount to hundreds of billions of dollars per year. These modest steps offer other fringe benefits, including improved air quality.

Another issue to consider is that fossil fuels are relatively cheap in the United States. At present, carbon emissions are not taxed, as they are in many other developed nations; creating incentive for energy conservation or conversion to new energy sources. The full economic costs of fossil fuels include the costs of cleaning up atmospheric pollution, oil spills, and other environmental damage; the costs of trade deficits; and the military costs of defending oil supplies, as well as the costs of global warming. If these costs were included in energy pricing, alternative energy sources would become much more competitive with fossil fuels. Such full-cost accounting has not been politically popular in the United States, however. In fact, the American Clean Energy and Security Act of 2009, which provided for several ways to curb carbon emissions, was approved by the House

FIGURE 14.26 A large coal-fired power plant near Ordos, a city in northern China. In 2007, China replaced the United States as the nation with the highest rate of greenhouse gas emissions. The carbon economies of China, India, and other developing countries will have a huge influence on future climates. [ZumaWire / Newscom.]

of Representatives in 2009, but was never made it to the Senate for a discussion or a vote.

We also face the issue of fairness in international politics. The United States, Canada, the European Union, and Japan—with less than one-quarter of the world's population—are responsible for about three-quarters of the global increase in atmospheric greenhouse gas concentrations. These rich industrial nations are better able to pay the costs of reducing their greenhouse gas emissions than the developing countries. China, for example, depends on its huge coal deposits for its rapid economic growth; it became the world's leader in greenhouse gas emissions in 2007 (Figure 14.26). Developing nations argue that they will need financial and technological support from the developed countries to help them reduce emissions. Policy makers have come to agree that the problems of global climate change cannot be solved on a national scale and will have to be addressed through international cooperation and investment.

Use of Alternative Energy Resources

As we have seen, no one alternative energy source will be able to replace fossil fuels quickly. However, some renewable energy resources, such as solar power, wind power, and biofuels, are becoming more important contributors to our energy system. If these technologies were aggressively implemented during the next 50 years, together they could reduce carbon emissions by gigatons per year.

Another step that could be taken is to increase the use of nuclear energy. The capacity of nuclear power plants, which today is approximately 400 gigawatts, could easily be tripled in the next 50 years, but this option is unattractive to many people for the environmental and security concerns discussed in Chapter 13. The potential exists for cleaner nuclear technologies, such as fusion power: the use of small, controlled thermonuclear explosions to generate energy. But scientific progress toward this goal has been slow, and conceptual breakthroughs will be required.

Engineering the Carbon Cycle

What about the possibility of engineering the carbon cycle to reduce the accumulation of greenhouse gases in the atmosphere? Several promising technologies aim to reduce greenhouse gas emissions by pumping the CO_2 generated by fossil-fuel combustion into reservoirs other than the atmosphere—a procedure known as **carbon sequestration** (Figure 14.27).

Carbon dioxide captured from oil and gas wells is already being pumped back into the ground as a means of moving oil toward the wells. If capture and underground storage of

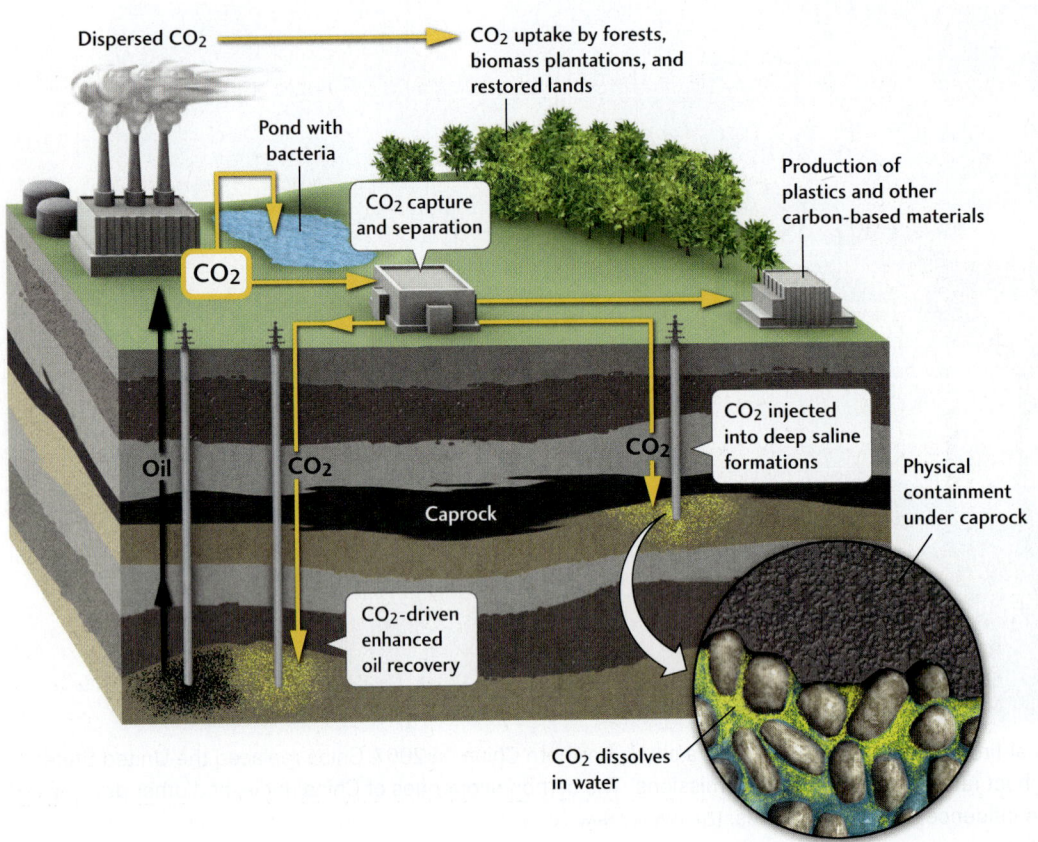

FIGURE 14.27 Carbon sequestration is the process of removing carbon from the atmosphere and depositing it into long-term reservoirs. A promising technology for carbon sequestration is capturing CO_2 at its source, such as a power plant, and storing it in underground reservoirs, such as depleted oil reservoirs and deep saline formations.

the CO_2 from coal-fired power plants were economically feasible, the world's abundant coal resources would become cleaner sources of fuel. So far, however, the technologies to remove and sequester carbon from the atmosphere are much too expensive to implement at a scale large enough to significantly reduce current carbon emissions.

The biosphere provides a natural mechanism for removing carbon already dispersed in the atmosphere. In Chapter 12, we saw that forests withdraw CO_2 from the atmosphere in surprisingly large amounts. Land-use policies that would not only slow the current high rates of deforestation but also encourage reforestation and other biomass production might help to mitigate anthropogenic climate change.

Another possibility is fertilization of the marine biosphere. We know that phytoplankton (small photosynthetic marine organisms) take up CO_2 from the atmosphere by photosynthesis. In most regions of the ocean, phytoplankton productivity is limited by the lack of nutrients, such as iron. Preliminary experiments in the 1990s suggested that the growth of phytoplankton could be stimulated by dumping modest amounts of iron into the ocean. Unfortunately, it appears that fertilizing the ocean in this manner also stimulates the growth of animals that eat the phytoplankton and quickly return the CO_2 to the atmosphere.

Stabilizing Carbon Emissions

At current growth rates, carbon emissions are expected to increase by at least 7 Gt/year during the next half century (see Figure 14.25). In the last two chapters, we have discussed ways that carbon emissions can be slowed. But there is no single "silver bullet."

In 2004, two scientists from Princeton University, Stephen Pacala and Robert Socolow, recognized that there will be a multitiered approach to lowering carbon emissions; they represented possible contributions to carbon reduction as **stabilization wedges,** each of which offsets the projected growth of carbon emissions by 1 Gt/year in the next 50 years (**Figure 14.28**). Therefore, one wedge roughly corresponds to one-seventh of the carbon reduction needed for stabilization.

Implementing each stabilization wedge will be a monumental task. To achieve wedge 1, for example, the average gasoline mileage of the world's entire fleet of passenger vehicles, which will grow to 2 billion by mid-century, will have to be steadily increased from 30 miles per gallon (mpg) to 60 mpg. This calculation assumes that a car is driven 10,000 miles per year, the current annual average. An alternative, not shown in Figure 14.29, would be to maintain gas mileage at 30 mpg but reduce the average amount of driving by half to 5000 miles per year. Yet

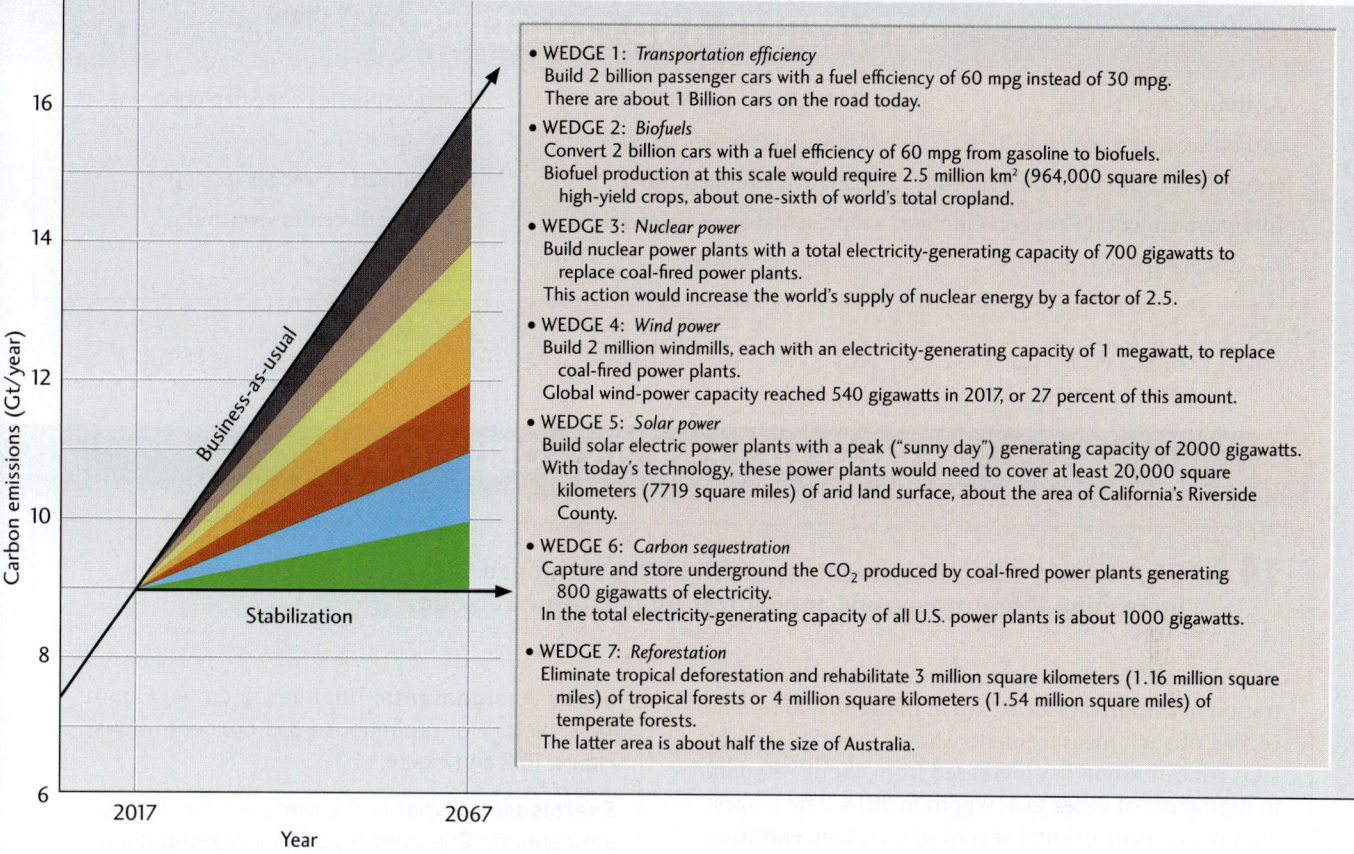

FIGURE 14.28 At current growth rates, carbon emissions are expected to increase by at least 7 Gt per year during the next 50 years. The problem of stabilizing carbon emissions at their 2017 level of about 9 Gt/year can be broken into seven stabilization wedges, each representing a reduction in emissions of 1 Gt per year by 2067. Possible actions that use existing technologies to achieve one-wedge reductions are listed next to each wedge. [Research from S. Pacala & R. Socolow. "Stabilization Wedges: Solving the Climate Problem for the Next 50 Years with Current Technologies." *Science, 305*: 968–972 (2004).]

another alternative (wedge 2) would be to convert all cars to biofuels. Growing that much biofuel would take up one-sixth of the world's total cropland, so this strategy could adversely affect agricultural productivity and food supplies.

Some of the stabilization wedges involve controversial or expensive technologies, such as expanding nuclear power by a factor of 2.5 (wedge 3), increasing the number of large windmills into the millions (wedge 4), or covering large desert areas with solar panels (wedge 5). As we have seen, at least one of the proposed wedges, the capture and storage of carbon emitted from coal-fired power plants (wedge 6), is at the margin of current technological feasibility. The last option, elimination of tropical deforestation and the reforestation of huge additional land areas (wedge 7), is favored by many people in principle, but would be difficult to achieve without imposing severe restrictions on developing countries such as Brazil.

The stabilization of carbon emissions at current emission rates would reduce, but not eliminate, the threat of global climate change. The 50-year stabilization scenario (which is intermediate to scenarios B and C in Figure 14.8) would still allow the atmospheric concentration of CO_2 to grow to 500 ppm, almost twice the preindustrial value. Further reductions in carbon emissions during the second half of the twenty-first century would be necessary to maintain atmospheric concentrations below that value. Climate models indicate that such a scenario would still increase the average global temperature by about 2°C, more than three times the total twentieth-century warming.

Nevertheless, the continued rise of atmospheric CO_2 concentrations is not inevitable. The available inventory of stabilization wedges constitutes a technological framework for concerted action by governments. Taking on the stabilization problem involves other difficulties such as developing broad public consensus and creating binding international agreements. Yet, as the Pacala-Socolow analysis demonstrates, there is still time for actions that can substantially reduce anthropogenic global change. Whether we can grasp this opportunity will depend on our understanding of the problem, its potential solutions, and the consequences of inaction.

KEY TERMS AND CONCEPTS

Anthropocene (p. 410)
carbon sequestration (p. 413)
enhanced greenhouse effect (p. 400)
Intergovernmental Panel on Climate Change (IPCC) (p. 399)
Keeling curve (p. 394)
Montreal Protocol (p. 397)
radiative forcing (p. 401)
representative concentration pathway (p. 401)
stratospheric ozone depletion (p. 397)
stabilization wedges (p. 414)
twentieth-century warming (p. 397)

REVIEW OF LEARNING OBJECTIVES

14.1 Explain why scientists can assert with high confidence that fossil-fuel burning is increasing the atmospheric concentration of carbon dioxide.

Atmospheric chemists can directly measure the concentration of carbon dioxide (CO_2) in the atmosphere. The record of such measurements shows that the average CO_2 concentration has increased from about 280 ppm in preindustrial times to 410 ppm in 2018. The longest continuous instrumental record of CO_2 concentration is the Keeling curve (Figure 14.1). Monthly values show seasonal oscillations about yearly averages that have risen from 310 ppm in 1958 to 410 ppm in 2018. Data on the changing isotope ratios of atmospheric carbon demonstrate that most of the CO_2 increase is being produced by fossil-fuel burning.

Using air samples from ice cores, climate scientists have extended the record of greenhouse gas concentrations back through the Holocene (Figure 14.3) and into the Pleistocene (Figure 14.2). At no time during the past 800,000 years have CO_2 concentrations been much higher than the preindustrial average of 280 ppm. Based on the sediment record, CO_2 concentrations above 400 ppm have not been reached since the Middle Miocene.

Study Assignments: The Keeling curve, Figure 14.1; history of greenhouse gas concentrations, Figure 14.2 and Figure 14.3.

Exercises: (a) What is the average rate at which atmospheric CO_2 concentration increased during the 60-year history of the Keeling curve? Express your answer in ppm per year. (b) From the Keeling curve, estimate average rate of CO_2 concentration increase during the 20-year intervals beginning in 1958, 1978, and 1998. (c) Do these 20-year averages indicate that the rate is either increasing or decreasing with time? (d) What do these results imply about the future rate of anthropogenic CO_2 emissions?

Thought Questions: (a) The average seasonal variation in atmospheric CO_2 ranges from 3 ppm higher than the annual mean in May to 3 ppm lower in October, as shown in the inset diagram in Figure 14.1. Why is the CO_2 concentration higher in May and lower in October? (b) If you look very closely at the Keeling curve (red line in Figure 14.1), you will find that this "global breathing of the biosphere" has actually increased in amplitude by about 50 percent, from ± 2.5 ppm in 1958 to ± 3.8 ppm in 2018. In qualitative terms, how can this 50 percent increase be explained as a feedback between atmospheric chemistry and the biosphere? (c) How might you use this "deeper breathing" to illustrate human impact on the biosphere?

14.2 Catalog the main types of anthropogenic global change and describe their main effects on the atmosphere, hydrosphere, cryosphere, and lithosphere.

Anthropogenic global change can be chemical, physical, or biological. Examples of anthropogenic chemical change are (1) the increase in atmospheric CO_2 and other greenhouse gases, (2) the acidification of the oceans, and (3) the depletion of stratospheric ozone catalyzed by human-made chlorine compounds.

Change (1) is responsible for change (2). Human emissions of carbon are enhancing the greenhouse effect by increasing the concentration of carbon dioxide in the atmosphere. Some of this carbon dioxide dissolves in the oceans, where it combines with water to form carbonic acid. This ocean acidification acts to increase the concentration of bicarbonate ions at the expense of carbonate ions, making it harder for marine organisms to form shells and skeletons of calcium carbonate.

Examples of anthropogenic physical change are global warming and related reductions in the mass of the cryosphere, which decreases Earth's albedo and transfers water to the hydrosphere, raising the sea level. The cryosphere-albedo feedback is positive, enhancing global warming. Climate models that account for these feedbacks indicate an increase in global temperatures ranging from 0.5°C to 5.5°C, depending on how humans cope with the carbon crisis.

Atmospheric concentrations of greenhouse gases are likely to rise throughout the twenty-first century, skewing the warming projections towards the higher values (Figure 14.9). Global warming of this magnitude will disrupt ecosystems and increase the rates of species extinction. The oceans will warm and expand, and continental glaciers will begin to melt, raising the sea level as much as a meter by 2100. The Arctic ice cap will continue to shrink, and much of the Arctic Ocean is expected to become ice-free during the summer months. Storms, floods, and droughts will intensify.

Study Assignment: Review the section *Types of Anthropogenic Global Change*.

Exercises: (a) What large-scale physical changes to the climate system might be caused by anthropogenic chemical changes to the atmospheric composition? (b) How much did the sea level rise during the twentieth century? (c) What was the primary physical cause of this sea-level rise?

Thought Questions: (a) Among the various anthropogenic changes to the climate system described in this chapter, which do you think will be the hardest for humans to adapt to, and why? (b) What actions can humans take to adapt to this type of global change?

14.3 Explain why scientists can assert with high confidence that fossil-fuel burning caused the twentieth-century warming and continues to cause the average surface temperature to increase.

The observed increase of about 0.8°C in Earth's average annual surface temperature during the twentieth century is correlated with a significant rise in atmospheric concentrations of CO_2 and other greenhouse gases relative to preindustrial times. Carbon isotopes demonstrate that the CO_2 rise is due to fossil-fuel burning. The geologic record documents that feedbacks within the climate system have maintained the tight correlation between CO_2 concentrations and average surface temperature throughout the Holocene (Figure 14.3) and back into the Pleistocene (Figure 14.2). Based on the available data and model-based predictions, essentially all experts on Earth's climate are now convinced that the twentieth-century warming was human-induced and that the warming will continue into the twenty-first century as atmospheric concentrations of greenhouse gases continue to rise. The projections of future global warming made by climate models depend primarily on which actions humans take to reduce fossil-fuel burning and how quickly they are implemented.

Study Assignments: Figures 14.5 and 14.6

Exercises: (a) How much did the CO_2 concentration of the atmosphere increase during the twentieth century? (b) How do climate scientists know that the twentieth-century increase in CO_2 concentration was due to fossil-fuel burning and not natural causes? (c) In which regions were the surface temperature increases during the twentieth century larger than the global average?

Thought Question: Is one justified in insisting that developing countries that have historically burned much less fossil fuel than developed countries agree to limit their future carbon emissions?

14.4 Use scenarios developed by the Intergovernmental Panel on Climate Change to project how much greenhouse gas concentrations, the average surface temperature, and sea level will rise during this century.

The United Nations has authorized the IPCC to report on anthropogenic global change and make recommendations about how to reduce global change and adapt to its probable effects. The climate projections made in the IPCC's *Fifth Assessment Report*, discussed extensively in this chapter, are based on "representative concentration pathways" specified by a level of radiative forcing, measured in watts per square meter (W/m^2). The IPCC's RCP8.5 projection (scenario A) assumes continued reliance on fossil fuels; the resulting rise in greenhouse gas concentrations, to over 900 ppm for CO_2, would increase the radiative forcing by 8.5 W/m^2. The RCP6.0 projection (scenario B) assumes a more rapid shift to alternative energy resources, reducing the radiative forcing to 6.0 W/m^2, and the most optimistic projection, RCP2.6 (scenario C), reduces it to only 2.6 W/m^2. The corresponding projections for average surface temperature and sea level are plotted in Figure 14.10 and Figure 14.15.

Study Assignment: Review the section *Projecting Future Climate Change*.

Exercises: (a) Scenario A is shorthand for what the IPCC calls scenario "RCP8.5." Explain the meaning of each letter in the acronym "RCP." What does "8.5" represent? In what units is it measured? (b) The description of scenario A compares the radiative forcing of anthropogenic greenhouse emissions with the average solar forcing of 240 W/m^2. How is the average solar forcing computed from the values of incoming solar radiation and Earth's albedo given in Figure 12.9? (c) In which regions will the surface temperature increase more quickly than the global average? (d) What is the best estimate of the surface temperature rise during the twenty-first century, and what is its uncertainty?

Thought Questions: (a) An economist once wrote, "The predicted change in global temperature due to human activity is less than the difference in winter temperature between New York and Florida, so why worry?" Should we worry, and if so, why? (b) Do you expect the per capita carbon intensity to go up or down during the twenty-first century? (c) Does your answer to (b) explain why future anthropogenic global change is so strongly dependent on the projected human population, as shown in Figure 14.11?

14.5 Assess the potential effects of anthropogenic global change on the biosphere, and evaluate the possibility that the beginning of the Anthropocene epoch will be marked by a mass extinction.

Global chemical and physical changes are affecting the biosphere, inevitably leading to global biological change. The biodiversity of ecosystems on land is declining through loss of habitat as well as the effects of climate change. The oceans are warming and acidifying, and many ecosystems are vulnerable, especially coral reefs and polar habitats. Even under the most moderate scenarios for global warming (scenario C), most coral reefs are expected to die by 2100. The current rapid rate of species extinction may eventually lead to a decline in biodiversity equal to the "Big Five" mass extinctions of the Phanerozoic eon.

Study Assignments: Review the sections *Ocean Acidification* and *Loss of Biodiversity*.

Exercises: (a) Review the geologic time scale in Figure 9.12, and note the boundaries between geologic periods identified as the Big Five mass extinctions. Which of the Big Five mark boundaries between geologic eras? (b) Suggest three anthropogenic signatures in the geologic record that could be used to identify the beginning of the Anthropocene. (c) In which geologic formations would these signatures be best expressed?

Thought Question: At projected extinction rates, how long will it be before extinctions among the vertebrates reach Big Five proportions?

14.6 Illustrate with specific examples changes in our energy production and usage that could stabilize carbon emissions to its current rate.

Stabilizing carbon emissions at their current levels of about 9 Gt/year will require major reductions in the carbon intensity of our energy sources. If civilization continues to rely on fossil fuels, anthropogenic carbon emissions will increase by at least 7 Gt/year during the next 50 years. The stabilization triangle (shaded region in Figure 14.25) can be broken down into stabilization wedges, defined by specific types of action that, if implemented in the next 50 years, could reduce the projected growth of carbon emissions by 1 Gt/year. Achieving seven such wedges would stabilize carbon emissions to the 2017 level (see Figure 14.28).

Study Assignment: Review the section *Managing the Carbon Crisis*.

Exercise: Visual Literacy Exercise.

Thought Questions: (a) Do you think we should act now to reduce carbon emissions or delay until the functioning of the climate system is better understood? (b) Do you think that future scientists and engineers will be able to modify the natural carbon cycle to prevent catastrophic changes in the climate system? (c) What technologies do you think would be most effective in reducing the amount of global warming?

VISUAL LITERACY EXERCISE

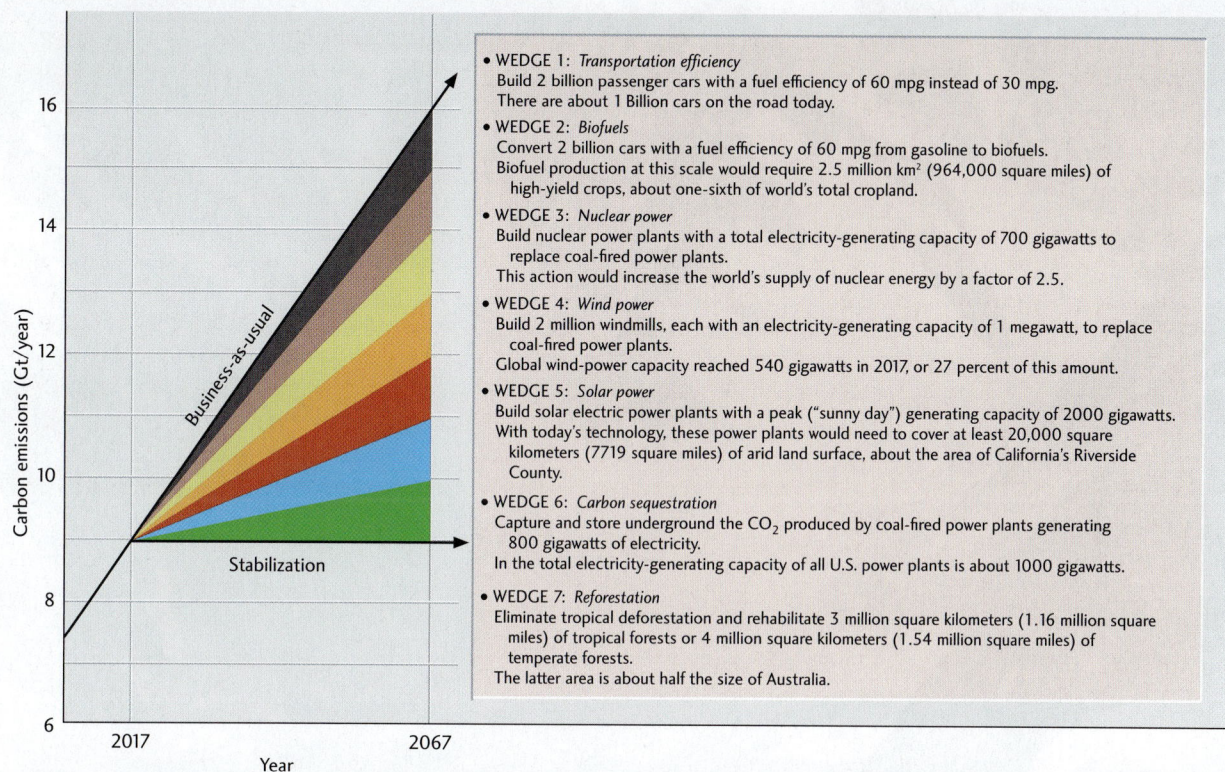

FIGURE 14.28 At current growth rates, carbon emissions are expected to increase by at least 7 Gt per year during the next 50 years. The problem of stabilizing carbon emissions at their 2017 level of about 9 Gt/year can be broken into seven stabilization wedges, each representing a reduction in emissions of 1 Gt per year by 2067. Possible actions that use existing technologies to achieve one-wedge reductions are listed next to each wedge. [Research from S. Pacala & R. Socolow. "Stabilization Wedges: Solving the Climate Problem for the Next 50 Years with Current Technologies." *Science, 305*: 968–972 (2004).]

Figure 14.28 shows the "stabilization triangle," the reduction in carbon emissions needed to stabilize them at the 2017 level.

1. **The vertical axis is carbon emissions measured in Gt/year.**
   a. How much carbon was emitted into atmosphere by fossil-fuel burning in 2017?
   b. If we project the "business-as-usual" scenario at current growth rates, how much carbon will be emitted by fossil-fuel burning in 2067?

2. **The area of the stabilization triangle can be measured in gigatons of carbon emission.**
   a. What is the area of the stabilization triangle?
   b. How much carbon is represented by each stabilization wedge?
   c. By what percentage would carbon-emission stabilization reduce the total carbon emitted in the next 50 years under the business-as-usual scenario of Figure 14.28?

3. **Among the seven stabilization wedges listed on the figure,**
   a. Which would you judge to be the easiest to attain?
   b. Which would you judge to be the most difficult to attain?

Glaciers: The Work of Ice 15

Types of Glaciers	424
How Glaciers Form	427
How Glaciers Move	429
Isostasy and Sea-Level Change	433
Glacial Landscapes	434
Glacial Cycles and Climate Change	442

Learning Objectives

Glaciers—massive accumulations of flowing ice—are powerful geologic agents that act to regulate Earth's climate system and shape its surface. After reading this chapter, you should be able to:

15.1 Identify the main types of glaciers.

15.2 Describe how glaciers grow and shrink, and how they move.

15.3 Explain why the melting of continental glaciers increases sea level but the melting of sea ice does not.

15.4 Recognize features of the landscape that have been formed and modified by glacial processes.

15.5 Understand what the geologic record tells us about past ice ages.

Several glaciers flow together near Mount Waddington in the Coast Mountains of British Columbia, Canada. [All Canada Photos/Alamy.]

The view of Earth from space is painted with the colors of water: vast blue oceans, swirling white clouds, and the frozen whites of solid ice and snow. The Earth system is constantly moving water across the planetary surface in ever-changing patterns. Among the main reservoirs of water, it is the system's icy component—the *cryosphere*—that waxes and wanes most visibly during climate cycles.

The ice sheets of Greenland and Antarctica, as huge as they are, now blanket only about one-tenth of Earth's land surface. But as recently as 20,000 years ago, continental glaciers covered almost three times more land than they do today, extending across all of Canada and deep into the midwestern United States. Within the next century or so, global warming could melt large parts of the existing ice sheets, with worldwide effects on human society. Sea level would rise, submerging low-lying cities. Climate zones would migrate, changing wet zones into deserts and vice versa. Given these threats, there is no doubt that understanding Earth's cryosphere—always an interesting scientific subject—has become an extremely practical goal.

The landscapes of many continents have been sculpted by glaciers now melted away. In mountainous regions, glaciers have eroded steep-walled valleys, scraped their bedrock surfaces smooth, and plucked huge blocks from their rocky floors. During the ice ages of the Pleistocene epoch, glaciers pushed across the northern continents, carving far more topography than water and wind. Glacial erosion creates enormous amounts of debris, and glaciers transport huge tonnages of sediment, depositing it at their edges, where it may be carried away by meltwater streams. Glacial processes affect the discharges and sediment loads of river systems, the erosion and sedimentation of coastal areas, and the quantity of sediment delivered to the oceans.

In this chapter, we take a close look at Earth's glaciers, how they form and change over time, and how they leave their marks on Earth's surface by eroding and depositing material as they advance and retreat. We examine the role glaciers play in the climate system and discover what the geologic record of glaciation can tell us about climate change over time.

Types of Glaciers

The distribution of glaciers on Earth's surface defines a special part of the polar climate zone—regions that have been both cold and snowy enough in the recent past to form and sustain glaciers. Glaciers cover 98 percent of Antarctica and 80 percent of Greenland, accounting for 99 percent of the ice on Earth's land surface. The remaining 1 percent is distributed across the high-latitude regions of continents, such as northern Canada and southern Patagonia and in high mountain ranges at lower latitudes, such as the Himalayas, Andes, and Rocky mountains.

Owing to global warming, the glacial climate zone is now shrinking, and the net transfer of water from the cryosphere to the oceans is increasing, causing sea level to rise (see Chapter 14). The ability to forecast sea level over the next century depends on our understanding of how fast glaciers will melt as the climate warms. Recent observations suggest that the melting of large ice sheets is occurring more quickly than scientists had previously forecast.

Ice Accumulations as Rock Formations

To a geologist, a block of ice is a rock, a mass of crystalline grains of the mineral *ice*, the frozen form of water (H_2O). Glacial ice, like sedimentary rock, initially forms from material deposited in layers at Earth's surface (**Figure 15.1**). As new deposits bury the old, the pressure increases, transforming loose-packed accumulations of individual particles into denser, intertwined crystalline structures. In other words, the rock *glacial ice* is formed by the burial and metamorphism of the sediment *snow*.

Ice has some unusual properties. Its melting temperature is very low (0°C), hundreds of degrees below the melting temperatures of silicate rocks. Most rocks are denser than their melts, which is why magma rises buoyantly through the lithosphere. But ice is less dense than its melt, which is why icebergs float on the ocean. And although very cold ice may seem hard, it is much weaker than most rocks.

FIGURE 15.1 Layering of snow and ice exposed in the wall of a large crevasse on Weissmies Glacier in the Swiss Alps. [Photo J. Alean.]

Because ice is so weak, it flows readily downhill like a viscous fluid. **Glaciers** are large masses of ice on land that are in motion, sliding under the force of gravity. Geologists divide glaciers, on the basis of their size and shape, into two basic types: *valley glaciers* and *continental glaciers.*

Valley Glaciers

Many skiers and mountain climbers are familiar with **valley glaciers,** sometimes called *alpine glaciers.* These rivers of ice form in the cold heights of mountain ranges, where snow accumulates. They then move downslope, sometimes coming together in flows massive enough to carve out deep, U-shaped valleys (see the chapter-opening photograph). A valley glacier usually occupies the complete width of the valley and may bury its floor under hundreds of meters of ice. In warmer, low-latitude climates, valley glaciers are found only at the heads of valleys on the highest mountain peaks. An example is the glacial ice that covers the Rwenzori Mountains at elevations over 5000 m, just north of the equator in east-central Africa (**Figure 15.2**).

In colder, high-latitude climates, valley glaciers may descend many kilometers, down the entire length of a valley. Broad lobes of ice may move into lower lands bordering mountain fronts. Valley glaciers that flow down coastal mountain ranges at high latitudes may terminate at the ocean's edge, where masses of ice break off and form icebergs—a process called **iceberg calving** (**Figure 15.3**).

FIGURE 15.3 Iceberg calving at Dawes Glacier, Alaska. Calving occurs when huge blocks of ice break off at the edge of a glacier that has moved to a shoreline. [Paul Souders/Getty Images.]

FIGURE 15.4 The Transantarctic Mountains rise to elevations of more than 4000 m, protruding through the thick ice of the Antarctic continental glacier. [Photo by Ed Stump.]

Continental Glaciers

A **continental glacier** is a thick *ice sheet* that covers a large part of a continent or other large landmass (**Figure 15.4**). Today, the world's largest continental glaciers overlie much of Greenland and Antarctica, covering about 10 percent of Earth's land surface and storing about 75 percent of the world's fresh water.

In Greenland, 2.6 million cubic kilometers of ice occupy 80 percent of the island's total area of 4.5 million square kilometers (**Figure 15.5**). The upper surface of the ice sheet resembles an extremely wide convex lens. At its highest point, in the middle of the island, the ice is more than 3200 m thick. From this central area, the ice surface slopes to the ocean on all sides. Near the mountain-rimmed coast, the ice sheet divides into narrow *ice streams* resembling

FIGURE 15.2. Tropical valley glaciers on Mount Stanley at altitudes close to 5000 m in the Rwenzori Range on the border between Uganda and the Democratic Republic of the Congo. This photograph was taken on an expedition by the Duke of Abruzzi in 1906. The glaciers have since retreated due to global warming and are likely to disappear by 2030. [DeAgostini/Getty Images.]

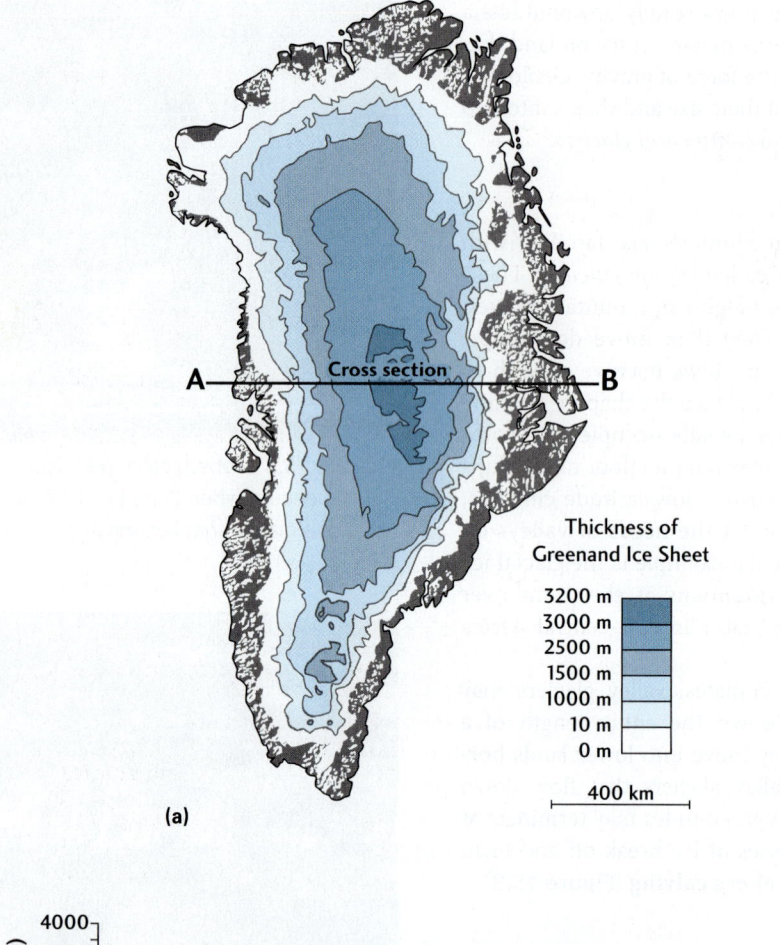
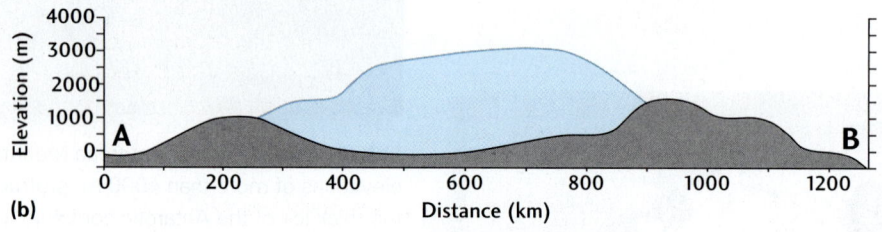

FIGURE 15.5 Ice-thickness map and cross section of the Greenland continental glacier. (a) The extent and elevation of the Greenland Ice Sheet. (b) Cross section A-B, showing the lenslike shape of the glacier. The ice flows outward toward the coast from the glacier's thick central dome.

valley glaciers that wind through the mountains to reach the ocean, where icebergs form by calving.

The bowl shape of the bedrock beneath the Greenland Ice Sheet, evident in the cross section at the bottom of Figure 15.5, is caused by the weight of the ice in the middle of the island. This depression of the bedrock, a consequence of isostasy, explains why mountains rim the Greenland coast.

Though very large, the Greenland glacier is dwarfed by the Antarctic Ice Sheet. Ice blankets 98 percent of Antarctica, covering an area of about 14 million square kilometers and reaching thicknesses of 4000 m (**Figure 15.6**). The total volume of Antarctic ice—about 30 million cubic kilometers—constitutes over 90 percent of the cryosphere.

As in Greenland, the ice forms domes near the center and slopes down to the margins of the continent.

Parts of Antarctica are rimmed by thinner sheets of ice—**ice shelves**—floating on the ocean and attached to the main glacier on land. The best known of these is the Ross Ice Shelf, a thick layer of ice about the size of Texas that floats on the Ross Sea.

Ice caps are the masses of ice that blanket Earth's North and South Poles. Most of the Arctic ice cap, located at the highest latitudes of the Northern Hemisphere, lies over water and is not a glacier. All of the Antarctic ice cap, except the ice shelves, lies on the continent and is therefore a continental glacier.

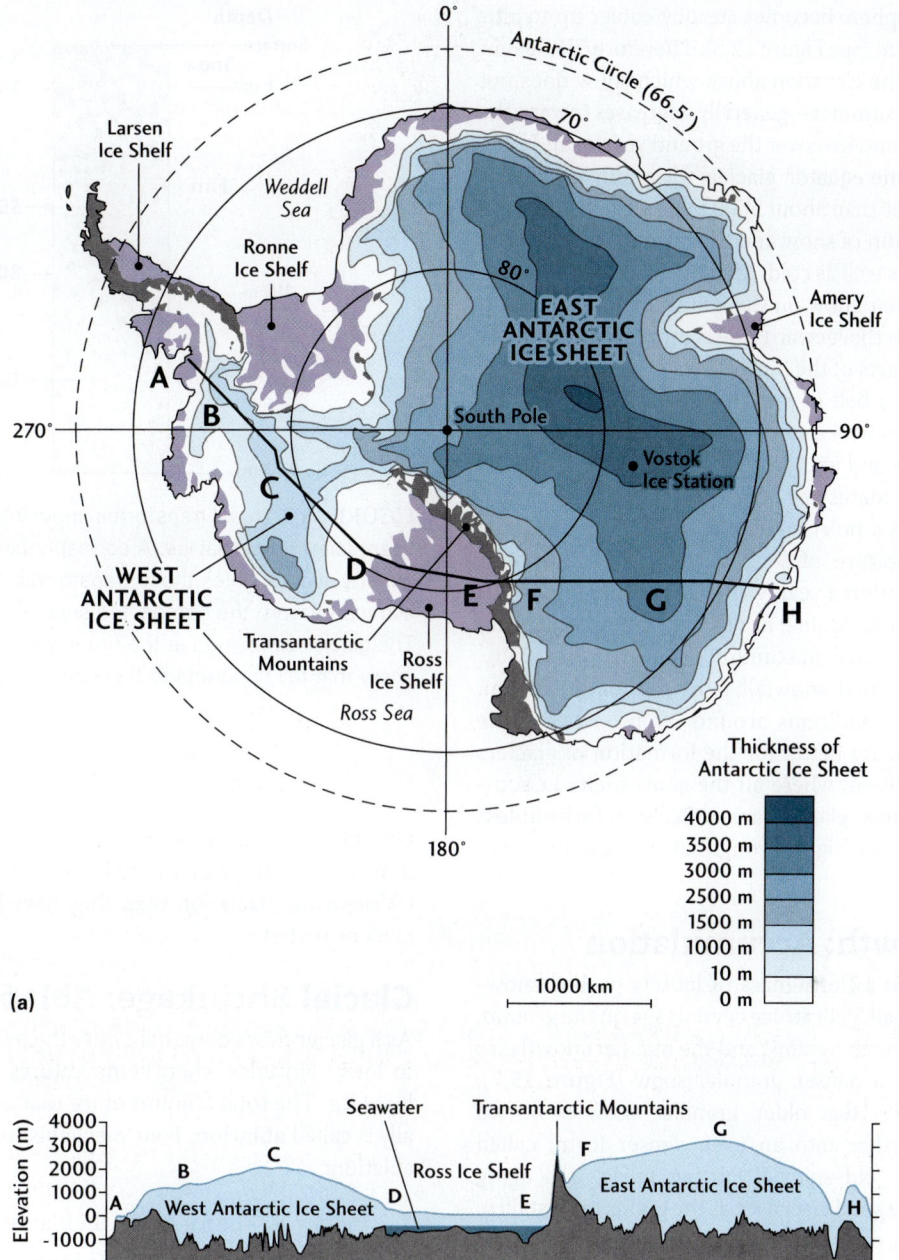

FIGURE 15.6 Map and cross section of the Antarctic Ice Sheet. (a) Colors contour the thickness of the ice sheet. Ice shelves are shown in purple. (b) Cross section A–H, showing the thick ice of the West Antarctic Ice Sheet (C) and East Antarctic Ice Sheet (G) and the thinner floating ice of the Ross Ice Shelf (D–E). [Data from British Antarctic Survey.]

How Glaciers Form

A glacier can form where abundant winter snowfall does not melt away in the summer. Old snow is buried by new snow, slowly squeezing and recrystallizing the particles into ice, and when the accumulating ice becomes massive enough, it begins to flow and becomes a glacier.

Basic Ingredients: Freezing Cold and Lots of Snow

For a glacier to form, temperatures must be low enough to keep snow on the ground year-round. These conditions occur at high latitudes, because the Sun's rays strike Earth at low angles there (see Figure 12.4), and at high elevations, because the atmosphere becomes steadily cooler up to altitudes of about 10 km (see Figure 12.3). Therefore, the height of the *snow line*—the elevation above which snow does not completely melt in summer—generally decreases toward the poles, where snow and ice cover the ground year-round even at sea level. Near the equator, glaciers form only on mountains that are higher than about 5000 m (see Figure 15.2).

The precipitation of snow and the formation of glaciers require moisture as well as cold. Moisture-laden winds tend to drop most of their snow on the windward side of a high mountain range, so the leeward side is more likely to be dry and unglaciated. Parts of the high Andes of South America, for instance, lie in a belt of prevailing easterly winds. Glaciers form on the moist eastern slopes, but the dry western side has little snow and ice.

The coldest climates are not necessarily the snowiest. Nome, Alaska, has a polar climate with an average annual maximum temperature of 9°C, but it gets only about 4.4 cm of precipitation a year, virtually all of it as snow. In comparison, Caribou, Maine, has a cool temperate climate with an average annual maximum temperature of 25°C, and its average annual snowfall is a whopping 310 cm. Nevertheless, the conditions around Nome, where little of the snow melts, are better for the formation of glaciers than those in Caribou, where all the snow melts in summer. In arid climates, glaciers are unlikely to form unless the temperature is so frigid throughout the year that very little snow melts.

Glacial Growth: Accumulation

A fresh snowfall is a fluffy mass of loosely packed snowflakes. As these small, delicate ice crystals age on the ground, they shrink and become grains, and the mass of snowflakes compacts to form a denser, granular snow (**Figure 15.7**). As new snow falls, that older, granular snow is buried and compacts further into an even denser form, called *firn* (German for "old snow"). Further burial and aging eventually produce solid glacial ice as the grains recrystallize and are cemented together. Transforming snow to glacial ice may take hundreds to thousands of years. A typical glacier adds a layer of ice each winter as snow falls on its surface. The amount of ice added to the glacier annually is its **accumulation**.

As glacial snow and ice accumulate, they entrap and preserve valuable relics of Earth's past. In 1991, mountaineers discovered the body of a prehistoric human preserved for more than 5000 years in Alpine ice on the border between Austria and Italy. In northern Siberia, extinct animals such as the woolly mammoth, a great elephantlike creature that once roamed icy terrains, have been found frozen and preserved in ancient ice. Ancient dust particles and bubbles of atmospheric gases are also preserved in glacial ice (see Figure 15.7). Chemical analyses of air bubbles found in very old, deeply buried Antarctic and

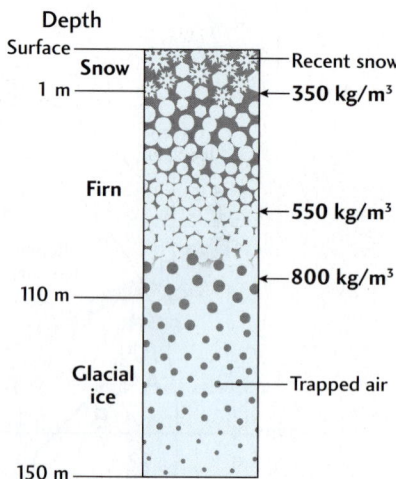

FIGURE 15.7 Burial transforms snow into firn and eventually into glacial ice. A corresponding increase in density accompanies these transformations as air is eliminated from the compacted snow (numbers on right). The glacial ice formed at the firn-ice transition is made of snow that fell hundreds to thousands of years ago.

Greenland ice have shown us that atmospheric carbon dioxide concentrations were lower during the most recent (Wisconsin) glaciation than they have been since the glaciers retreated.

Glacial Shrinkage: Ablation

As a glacier flows downhill under the pull of gravity, it flows to lower altitudes where temperatures are warmer, and it loses ice. The total amount of ice that a glacier loses annually is called **ablation**. Four mechanisms are responsible for ablation:

1. *Melting.* As the ice melts, the glacier loses material.
2. *Iceberg calving.* Pieces of ice break off and form icebergs when a glacier reaches a shoreline (see Figure 15.3).
3. *Sublimation.* In cold climates, water can be transformed directly from its solid state (ice) into its gaseous state (water vapor).
4. *Wind erosion.* Strong winds can erode glacial ice, primarily by melting and sublimation.

Most ablation takes place at the glacier's leading edge. Therefore, even when a glacier is advancing downward or outward from its center, its leading edge—the *ice front*—may be retreating. The two mechanisms by which glaciers lose the most ice are melting and iceberg calving.

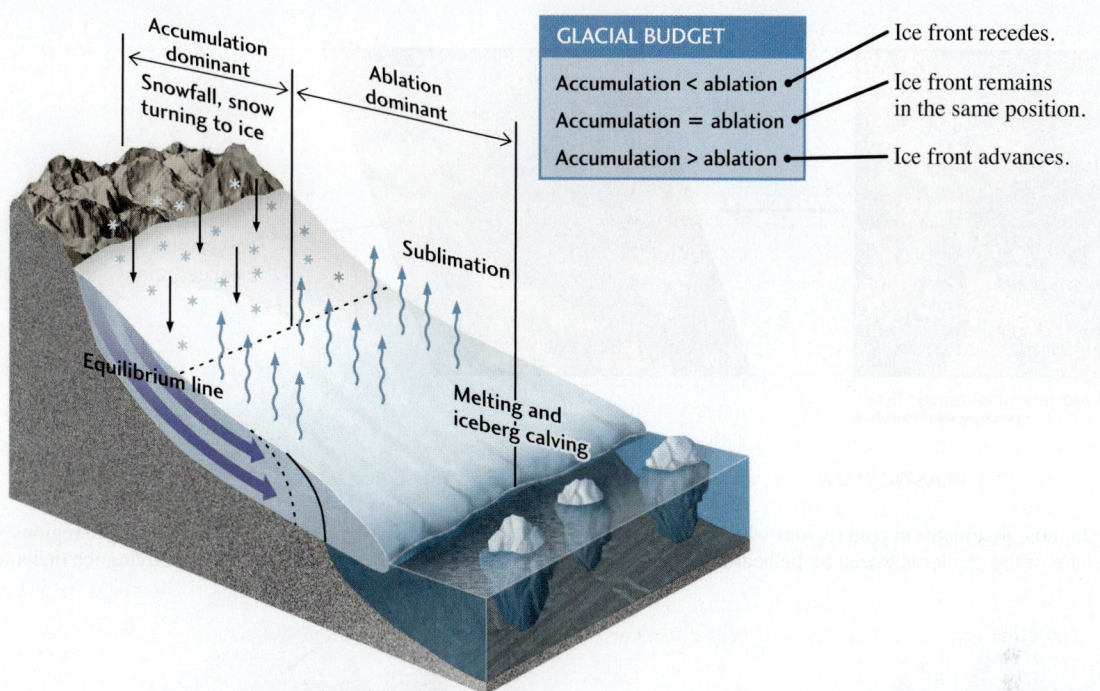

FIGURE 15.8 Accumulation takes place mainly by snowfall over a glacier's colder upper area. Ablation takes place by melting, iceberg calving, and sublimation in the warmer lower area. The two areas are separated by an equilibrium line (dashed) where the local accumulation equals local ablation. The difference between the total accumulation and total ablation is the glacial budget.

The Glacial Budget: Accumulation Minus Ablation

The relationship between accumulation and ablation, called the *glacial budget*, determines the growth or shrinkage of a glacier (Figure 15.8). Glaciers are divided into an upper accumulation zone and a lower ablation zone, separated by an equilibrium line. Above the equilibrium line the glacier is able to grow because some of last winter's snow survives the summer melt season. Below the equilibrium line the glacier loses more ice than can be replaced by the winter snowfall.

When total accumulation equals total ablation over a long period, the glacier remains a constant size, even as it continues to flow downslope from the area where it forms. Such a glacier accumulates snow and ice in its upper reaches and ablates an equal amount in its lower reaches. If accumulation exceeds ablation, the glacier grows; if ablation exceeds accumulation, the glacier shrinks.

Glacial budgets fluctuate from year to year. Over the past several thousand years, many glaciers have maintained a constant average size, though some show evidence of growth or shrinkage in response to short-term regional climate variations. In the last century, however, global warming has shifted the glacial budgets of many valley glaciers to net ablation, causing glaciers to shrink (see Figure 14.14). For example, the glacier cover of the Rwenzori Mountains shown in Figure 15.2 decreased from about 7.5 km^2 in 1906 to 1 km^2 today. All glaciers in this east-central African region are likely to disappear before 2030.

How Glaciers Move

When ice becomes thick enough for gravity to overcome its resistance to movement—normally at least several tens of meters—it starts to flow, and thus becomes a glacier. The ice in a glacier moves downhill in the same kind of laminar flow as a slowly flowing stream of water (see Figure 18.14). Unlike the readily observed flow of a stream, however, glacial movement is so slow that the ice seems not to move at all from day to day, giving rise to the expression "moving at a glacial pace."

Mechanisms of Glacial Flow

The flow of glaciers occurs in two ways: by plastic flow and by basal slip (Figure 15.9). In plastic flow, the movement occurs as deformation within the glacier. In basal slip, the glacier slides downslope as a single unit along its base, like a block of ice sliding down a ramp.

Movement by Plastic Flow The force of gravity exerted on a glacier causes individual crystals of ice to slip tiny distances relative to each other over short intervals

(a)

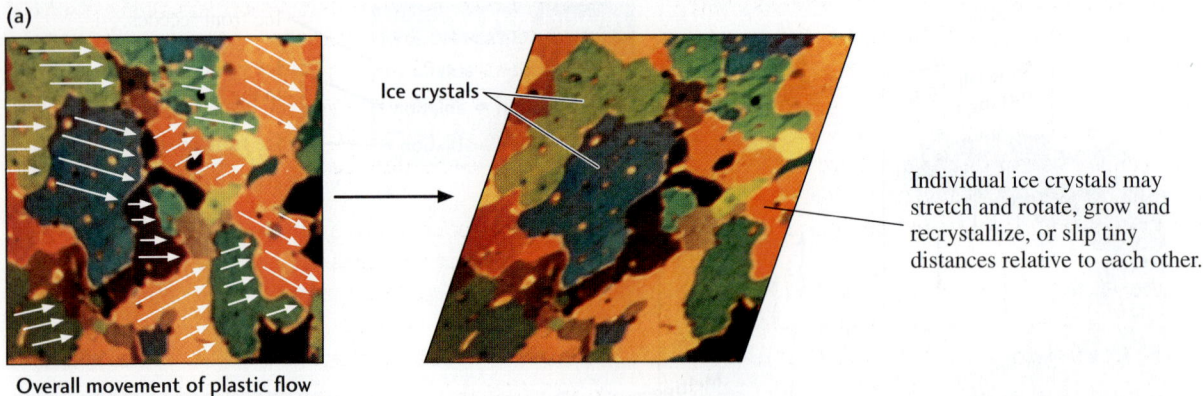

Ice crystals

Individual ice crystals may stretch and rotate, grow and recrystallize, or slip tiny distances relative to each other.

Overall movement of plastic flow

(b) **PLASTIC FLOW**

Plastic flow dominates in cold regions where the ice at the base of the glacier is frozen to the bedrock or soil.

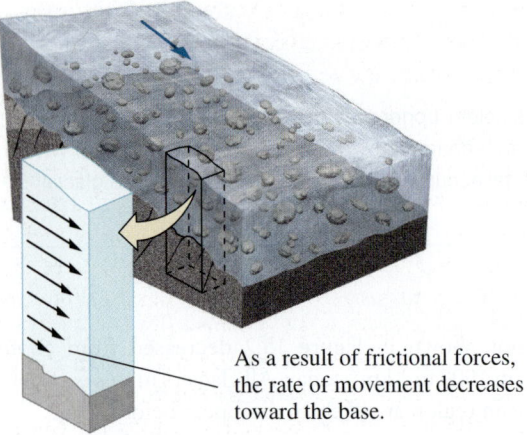

As a result of frictional forces, the rate of movement decreases toward the base.

(c) **BASAL SLIP**

Basal slip dominates in temperate regions where the pressure of overlying ice melts water at the glacier's base.

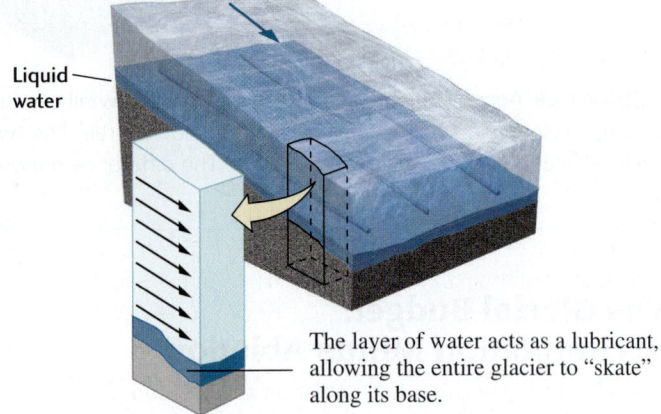

Liquid water

The layer of water acts as a lubricant, allowing the entire glacier to "skate" along its base.

(d) Valley glaciers in cold regions move mostly by plastic flow. If one drives a set of stakes deep into the glacier in a line across its flow,…

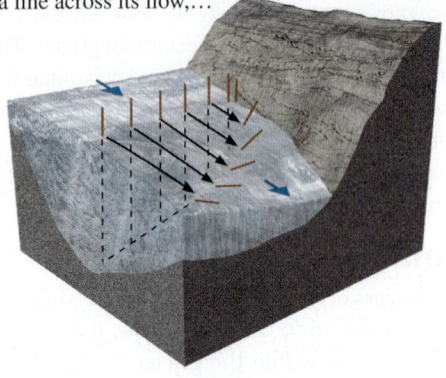

…after several years, the stakes in the center will have moved farther downhill and will slant forward, indicating faster movement in the center and at the top of the glacier.

(e) In continental glaciers, the ice moves down and out from the thickest section, like pancake batter poured on a griddle, as shown by the arrows.

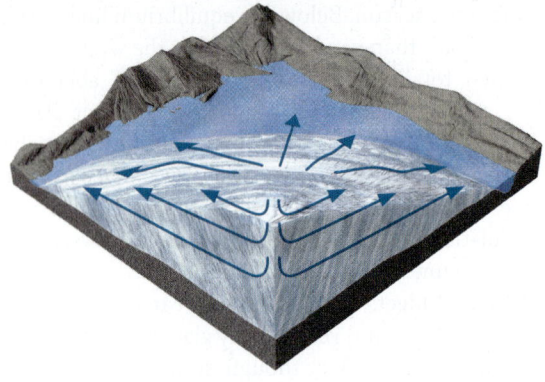

FIGURE 15.9 Glaciers move by two primary mechanisms: plastic flow and basal slip. (a) Deformation in plastic flow. (b) Plastic flow. (c) Basal slip. (d) Flow in valley glaciers. (e) Flow in continental glaciers.

(Figure 15.9a). The sum of many such movements among the enormous number of ice crystals that make up a glacier deforms the whole mass of ice in a process known as **plastic flow.** To visualize this process, think of a random pile of playing cards; the whole pile can be made to shift by inducing small slips on the many surfaces between cards. As ice crystals grow under pressure deeper in the glacier, their microscopic slip planes become more parallel, increasing plastic flow rates.

Plastic flow is most important in very cold regions, where the ice throughout the glacier, including its base, is well below the freezing point and the basal ice is frozen to the ground (Figure 15.9b). Most of the movement of these cold, dry glaciers takes place above the base by plastic flow. Movements near the frozen base detach and transport pieces of bedrock and soil. Because of this mixing of rock material with ice, the interface between the overlying ice and the underlying ground is usually not a sharp boundary, especially where the ground consists of sediments or weak sedimentary rocks. Instead, the interface becomes a transition between ice loaded with debris and deformed ground containing appreciable ice.

Movement by Basal Slip The other mechanism of glacial movement is **basal slip,** the sliding of a glacier along the boundary between the ice and the ground (Figure 15.9c). The melting point of ice decreases as pressure increases, so ice at the base of a glacier, where the weight of the overlying ice is greatest, melts at a lower temperature than ice within the glacier. The melted ice lubricates the base of the glacier, causing it to slip downslope. This same effect helps make ice skating possible: The weight of the body on the narrow skate blade provides enough pressure to melt just a little ice under the blade, which lubricates the blade so that it can slide easily along the surface.

In temperate regions, where the air temperature is above freezing during parts of the year, ice may be at the melting point within a glacier as well as at its base. Plastic flow contributes a small amount of internal heating from the friction generated by microscopic slips of ice crystals. In these glaciers, water occurs within the ice as small drops between crystals. Water seeping through cracks in the ice forms pools or streams of meltwater that carve tunnels in the ice. The water throughout the glacier eases internal slip between layers of ice.

Flow in Valley Glaciers

Louis Agassiz, a nineteenth-century Swiss zoologist and geologist, was the first to precisely measure how a valley glacier moves. As a young professor in the 1830s, he pounded stakes into a glacier in the Swiss Alps and measured their positions over several years. He observed that the stakes along the centerline of the glacier moved the fastest, at about 75 m/year, whereas the stakes nearer the valley walls traveled more slowly. Later, the deformation of long vertical tubes pounded deep into the glacier demonstrated that ice near the base flowed more slowly than ice in the center.

This type of deformation, in which the central part of a glacier moves faster than its sides or its base, is diagnostic of plastic flow (Figure 15.9d). Other valley glaciers have been observed to move at more uniform speeds, sliding as a single unit almost entirely by basal slip along a lubricating layer of meltwater next to the ground. Most often, however, valley glaciers flow by a combination of mechanisms—partly by plastic flow within the mass of ice and partly by basal slip.

A sudden period of fast movement of a valley glacier, called a **surge,** sometimes occurs after a long period of little movement. Surges may last several years, and during that time the ice may speed along at more than 6 km/year—a thousand times the normal velocity of a glacier. In many cases, surges follow the buildup of water pressure at or near the base of the glacier. This pressurized water greatly enhances basal slip.

The upper parts of a glacier have little pressure on them. At these low pressures, the ice at the surface of the glacier (shallower than about 50 m) behaves as a rigid, brittle solid, cracking as it is dragged along by the plastic flow of the ice below. The cracks, called **crevasses,** break up the surface ice into many small and large blocks (**Figure 15.10**). Crevasses are most likely to occur in places where glacial deformation is strong—such as where the ice drags against the valley walls at curves in the valley, at irregularities in the valley floor, and where the slope steepens sharply. The movement of brittle surface ice at these places is a "flow" resulting from slipping movements across the surfaces of these irregular blocks, similar in some ways to faulting in crustal rocks.

Antarctica in Motion

Antarctica may appear to be a land frozen in time, but the continental ice sheet that covers almost all of its surface certainly does not stand still. Geologists have used GPS satellites and airborne radar to map the movements of ice across Antarctica. These measurements reveal that the ice sheet is a composite structure, comprising hundreds of individual glaciers, each flowing along its own course from the interior of the continent towards the coast (**Figure 15.11**). The most rapid flows occur as **ice streams** 25 to 80 km wide and 300 to 500 km long. The ice stream of Lambert Glacier, shown in the blowup map of Figure 15.11, is the world's largest.

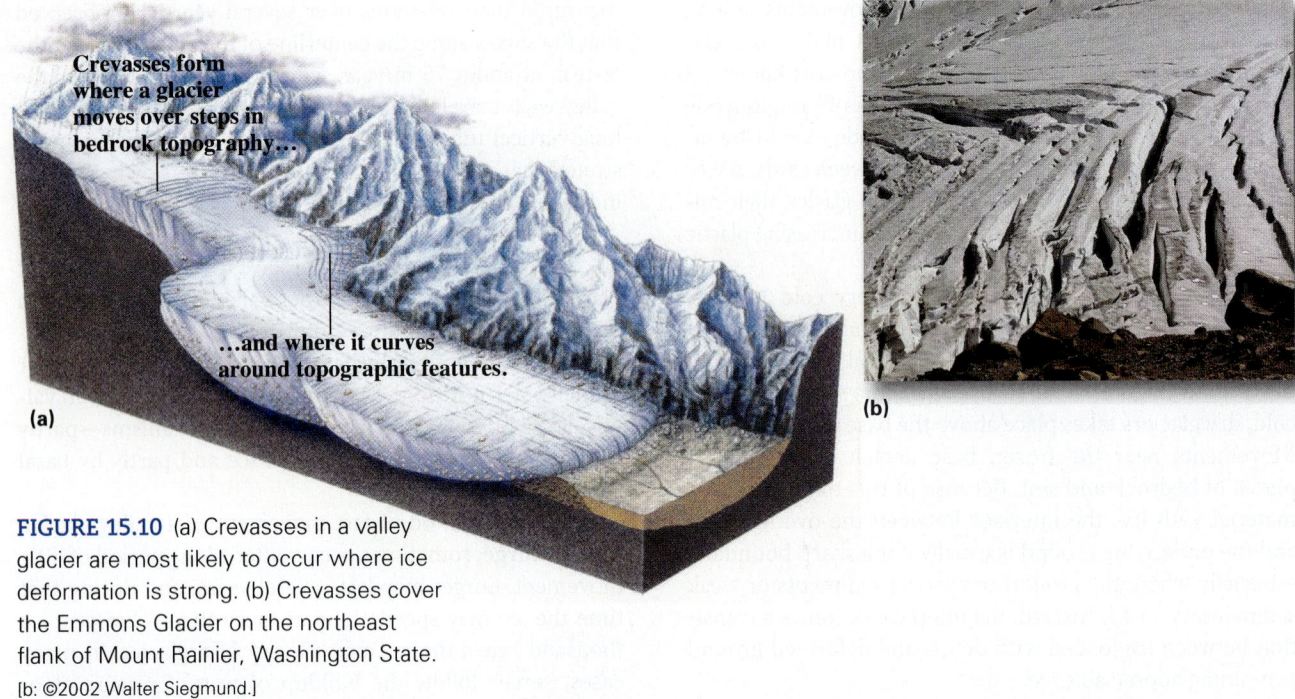

FIGURE 15.10 (a) Crevasses in a valley glacier are most likely to occur where ice deformation is strong. (b) Crevasses cover the Emmons Glacier on the northeast flank of Mount Rainier, Washington State. [b: ©2002 Walter Siegmund.]

The smaller tributary glaciers generally have low velocities of 100 to 300 m/year (green), which gradually increase as they flow down the sloping surface of the continent and intersect the upper reaches of Lambert Glacier. Most of Lambert Glacier has velocities of 400 to 800 m/year (blue). As it extends into and across the Amery Ice Shelf and the ice sheet spreads out and thins, velocities increase to 1000 to 1200 m/year (pink/red).

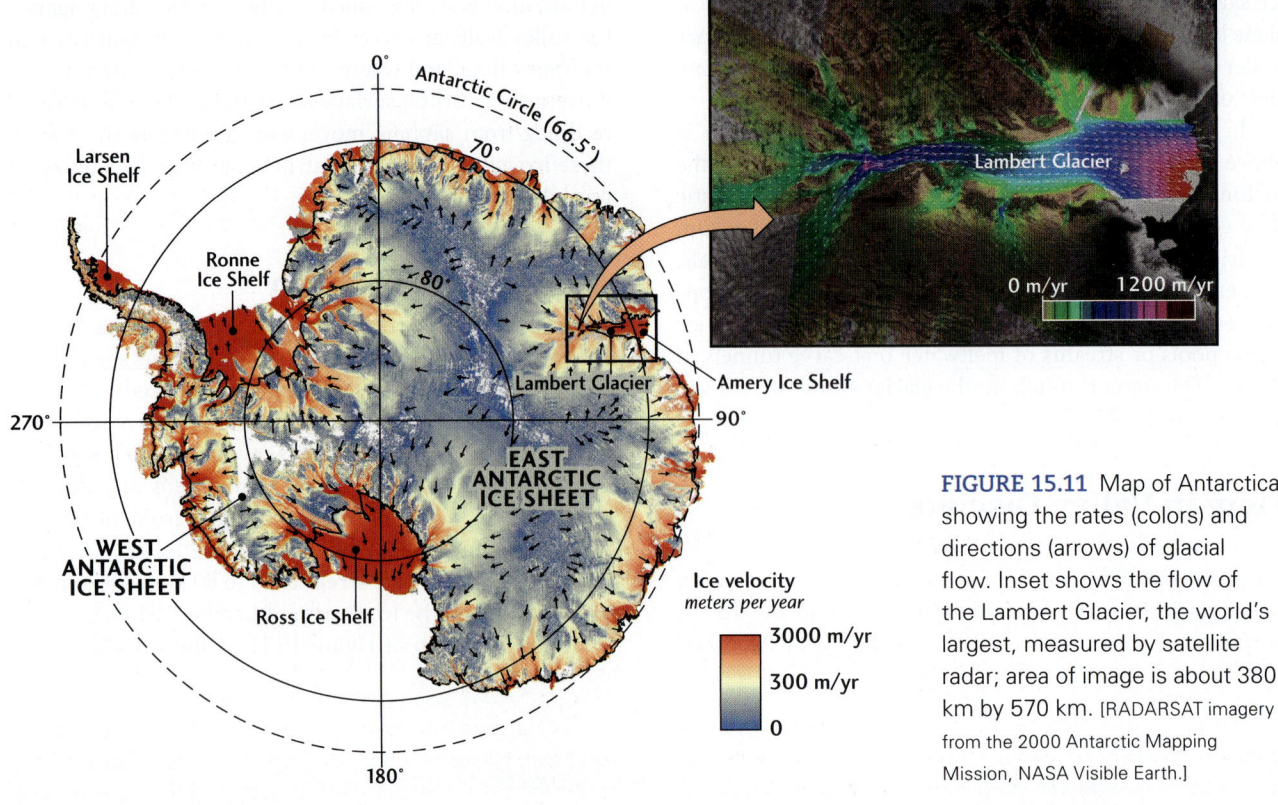

FIGURE 15.11 Map of Antarctica showing the rates (colors) and directions (arrows) of glacial flow. Inset shows the flow of the Lambert Glacier, the world's largest, measured by satellite radar; area of image is about 380 km by 570 km. [RADARSAT imagery from the 2000 Antarctic Mapping Mission, NASA Visible Earth.]

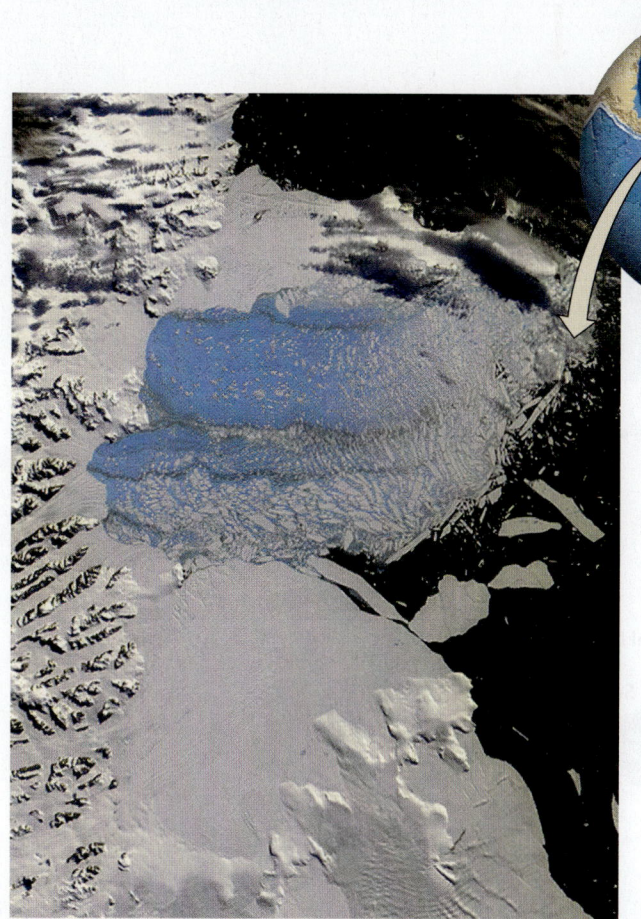

FIGURE 15.12 Collapse of the Larsen Ice Shelf. This satellite image was taken on March 7, 2002, toward the end of a 2-month period during which a huge piece of the ice shelf separated from land and splintered into thousands of icebergs. The darkest colors on the right-hand side of the image represent open seawater. The white parts are icebergs, the remaining parts of the ice shelf, and glaciers on land. The bright blue area is a mixture of seawater and highly fractured ice. The area of the image is about 150 km × 185 km.
[Image courtesy of the NASA/GSFC/LaRC/JPL, MISR Team.]

Using high-resolution satellite radar mapping, geologists have seen several Antarctic glaciers retreat more than 30 km in just 3 years. Over the past 20 years or so, enormous pieces of ice have snapped off Antarctic glaciers. In March 2000, an iceberg slightly less than 10,000 km^2 (larger than the state of Delaware) calved from the Ross Ice Shelf. In February and March of 2002, a portion of the Larsen Ice Shelf larger than Rhode Island (about 3250 km^2) shattered and separated from the northeastern side of the Antarctic Peninsula (Figure 15.12). The fracturing of this piece of the ice sheet produced thousands of icebergs.

Geologists who monitor the Larsen Ice Shelf were able to predict this episode of ice shelf collapse. Field and satellite-based observations showed that the flow rate of the ice stream leading to it had increased dramatically, which was interpreted as a sign of instability. After the collapse of the ice shelf, the flow rate of the ice stream increased further. In general, the destruction of ice shelves tends to destabilize the continental glaciers that feed them, causing the glaciers to flow more rapidly into the oceans.

Owing to warming of the surrounding oceans, instabilities of the Antarctica ice shelves, are increasing in size and frequency at an alarming rate. The Wilkins Ice Shelf, occupying an area of 14,000 km^2 on the southwestern side of the Antarctic Peninsula, began to break apart in early 2008 and appears to be on the verge of further collapse. In July 2017, another 5800-km^2 chunk of the Larsen Ice Shelf broke off, forming an iceberg more than 200 m thick that weighed more than a trillion tons.

Isostasy and Sea-Level Change

If the ice shelves around Antarctica continue to collapse, will sea level rise? It turns out that even if all of Earth's ice shelves were to break off into the ocean, sea level wouldn't change much at all. The reason is related to the principle of isostasy, which we discussed in Chapter 11. Ice shelves, like icebergs, float on ocean waters. When they melt, there is no change in sea level for the same reason that, when ice cubes in your drink melt, the level of the liquid in your glass doesn't change.

Icebergs float because the density of ice (0.92 g/cm^3) is less than that of seawater (1.03 g/cm^3). The volume of the ice below the sea surface thus weighs less than the volume

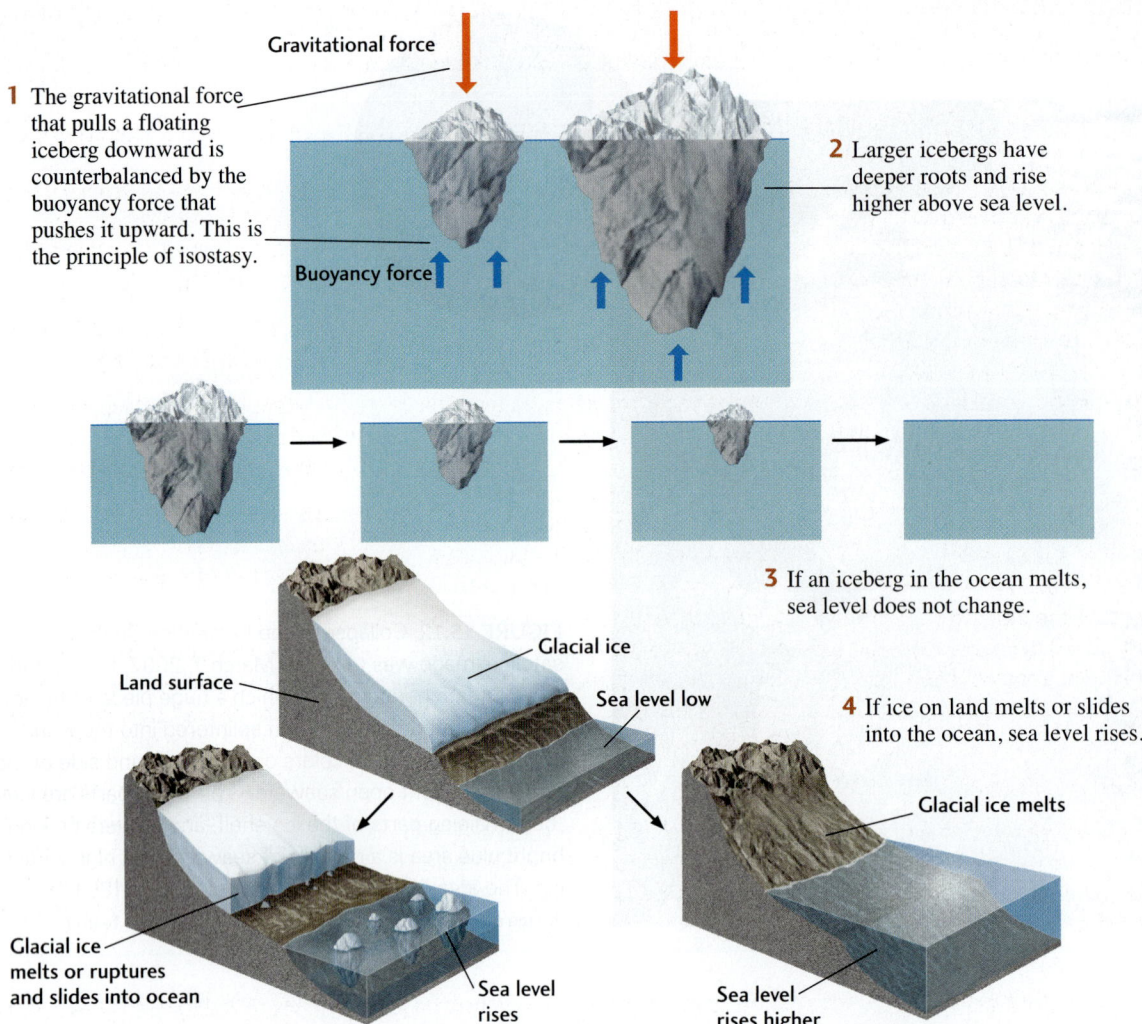

FIGURE 15.13 According to the principle of isostasy, floating ice shelves and icebergs displace a mass of water equal to their own mass; therefore, their melting does not change sea level. The melting of ice sheets on land, however, injects new water into the oceans, causing sea level to rise.

of seawater it displaces, and this weight difference causes a buoyancy force that must be balanced by the weight of the ice above the waterline (**Figure 15.13**). Bigger icebergs stand higher above the sea surface, because they have a larger volume below the surface that provides greater buoyancy. The total weight of an iceberg is thus equal to the weight of the seawater it displaces, so its melting has no effect on sea level.

In contrast, when glaciers on land melt, most of the water flows into the ocean, increasing the ocean volume and thereby raising the sea level. When glaciers on land flow directly into the ocean, the water in the oceans is displaced by the icebergs they shed, which also raises sea level. Global warming has reduced the glacial budget, causing the total ice volume of continental and valley glaciers to shrink. How much glacial melting can change sea level is investigated in the Practicing Geology Exercise that follows this section.

We conclude that the destruction of an ice shelf will raise sea level only if part of the shelf is grounded on land. In that case, when it slides into the ocean, the volume of the ice previously supported by the continent displaces seawater, causing a sea-level rise.

Glacial Landscapes

The movement of glaciers is responsible for the immense amount of geologic work done by ice: erosion, transportation, and sedimentation. Just as you cannot see your own footprint in the sand while your foot is covering it, you cannot see the effects of an active glacier at its base or sides. Only when the ice melts is its geologic work revealed.

PRACTICING GEOLOGY EXERCISE

How Much Would the Melting of Continental Ice Sheets Raise Sea Level?

In this time of global warming, there is great interest in knowing how much sea level will rise due to melting of the Greenland and Antarctic ice sheets. Let's consider the extreme case of *total* melting. The area of the oceans is 3.6×10^8 km² (see Appendix 2); hence, to raise sea level by one meter requires increasing the volume of the oceans by

$$3.6 \times 10^8 \text{ km}^2 \times 1 \text{ m} = 3.6 \times 10^{14} \text{ m}^3$$

The density of seawater is 1.03 gm/cm³ = 1030 kg/m³, which is greater than the density of pure water (1 gm/cm³) owing to seawater's salt content. The mass of seawater needed to raise sea level by one meter is therefore

$$3.6 \times 10^{14} \text{ m}^3 \times 1030 \text{ kg/m}^3 = 3.7 \times 10^{17} \text{ kg}$$

From careful surveys of ice thickness (see Figure 15.6), scientists have determined the total volume of Antarctic ice to be about 26.9 million cubic kilometers. However, were it to melt, not all this ice volume would contribute to sea level rise. In particular, melting of the Antarctic ice shelves would not change sea level (see Figure 15.13). This volume of floating ice is about 0.4 million cubic kilometers. Subtracting it off yields 26.5 million cubic kilometers for the "grounded ice volume."

There is another component of the ice volume that doesn't contribute to sea-level rise. The ice thickness measured in some parts of the continent exceeds its surface elevation, including much of the West Antarctic Ice Sheet (see profile in Figure 15.6). Some of the grounded ice lies below sea level. Melting this submerged ice, about 3 million cubic kilometers, would contribute little to sea-level rise. Reducing the ice volume by this amount, we find that only about 23.5 million cubic kilometers of the Antarctic ice volume, that is, 2.35×10^{16} m³, would contribute to sea-level rise upon melting. The average density of glacial ice is 0.92 gm/cm³ = 920 kg/m³, which yields a mass of

$$2.35 \times 10^{16} \text{ m}^3 \times 920 \text{ kg/m}^3 = 2.16 \times 10^{19} \text{ kg}$$

Hence, the sea-level rise that would be caused by the complete melting of the Antarctic Ice Sheet is

2.16×10^{19} kg $\div 3.7 \times 10^{17}$ kg per meter of sea-level rise
= 58 meters of sea-level rise

Of this amount, a 6-meter rise would be caused by the melting of the West Antarctic Ice Sheet, and 52 m would be caused by the much larger East Antarctic Ice Sheet.

BONUS PROBLEM: The total volume of the Greenland Ice Sheet above sea level is about 2.7×10^{18} kg. How much would the melting of this entire ice sheet contribute to sea-level rise?

We can infer the physical processes driven by moving ice from the topography of formerly glaciated areas and the distinctive landforms left behind.

Glacial Erosion and Erosional Landforms

Ice is a far more efficient agent of erosion than water or wind. A valley glacier only a few hundred meters wide can tear up and crush millions of tons of bedrock in a single year. The ice carries this heavy load of sediment to the ice front, where it is dropped as the ice melts. The total amount of sediment deposited in the world ocean per year has been several times larger during recent ice ages than during interglacial periods.

At its base and sides, a glacier engulfs jointed, cracked blocks of rock, breaks off pieces, and grinds them against the bedrock below. This grinding action fragments rocks into a great range of sizes, from boulders as big as houses to fine silt- and clay-sized material called *rock flour*. Rock flour is subject to rapid chemical weathering owing to its small size and correspondingly large surface area. Where glacial debris is still encased in ice and the ground is overlain by thick ice, chemical weathering is slower than on ice-free terrain. When rock flour is freed from the ice at the melting edge of a glacier, it dries out and is blown into the air as dust. Wind can carry this dust over great distances, ultimately depositing it as loosely compacted *loess* that is so common in formerly glaciated regions (see Chapter 19).

As a glacier drags rocks along its base, those rocks scratch or groove the bedrock beneath it. Such abrasions are termed **striations.** The orientation of striations shows us the direction of ice movement—an especially important observation in the study of continental glaciers. By mapping striations over wide areas formerly covered by a continental glacier, we can reconstruct the glacier's flow patterns (Figure 15.14).

Advancing glacial ice smooths small hills of bedrock—known as *roches moutonnées* ("sheep rocks") for their resemblance to a sheep's back—on their upstream side and plucks them to a rough, steep slope on their downstream

FIGURE 15.14 Glacial striations on bedrock in Quebec, Canada. Striations provide evidence of the direction of ice motion and are especially important clues for reconstructing the movement of continental glaciers. [MICHAEL GADOMSKI/Science Source.]

side (Figure 15.15). These contrasting slopes indicate the direction of ice movement.

A flowing valley glacier carves a series of erosional landforms as it flows from its origin to its lower edge (Figure 15.16). At the head of the valley, the plucking and tearing action of the ice tends to carve out an amphitheater-like hollow called a **cirque,** usually shaped like part of an inverted cone. With continued erosion, cirques at the heads of adjacent valleys gradually meet to form sharp mountaintops, or *horns,* and jagged crests called *arêtes* along the divide. As the glacier flows down from its cirque, it excavates a new valley or deepens an existing stream valley, creating a characteristic **U-shaped valley.** Glacial valley floors are wide and flat and have steep walls, in contrast to the V-shaped valleys of many mountain streams.

Glaciers and streams also differ in how their tributaries form junctions. Although the ice surface is level where a tributary glacier joins a main valley glacier, the floor of the main valley may be carved much more deeply than the tributary valley. When the ice melts, the tributary valley is left as a **hanging valley**—one whose floor lies high above the main valley floor (see Figure 15.16). After the ice is gone and streams occupy the valleys, the junction is marked by a waterfall as the stream in the hanging valley plunges over the steep cliff separating it from the main valley below.

Valley glaciers at coastlines may erode their valley floors far below sea level. When the ice retreats, these steep-walled valleys—which still maintain a U-shaped profile—are flooded with seawater (see Figure 15.16). These arms of the sea carved out by glaciers, called **fjords,** create the spectacular rugged scenery for which the coasts of Alaska, British Columbia, Norway, and New Zealand are renowned.

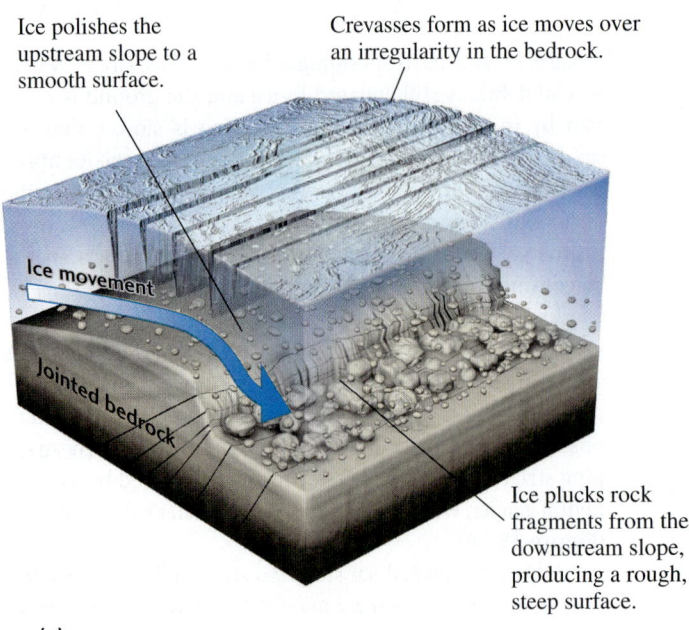
(a)

(b)

FIGURE 15.15 (a) A *roche moutonnée* is a small hill of bedrock, smoothed by glacial ice on the upstream side and plucked to a steep, rough face on the downstream side as the moving ice pulls rock fragments from joints and cracks. (b) A *roche moutonnée* known as The Beehive rises above Sand Beach at Acadia National Park in Maine. [mirceax/Getty Images.]

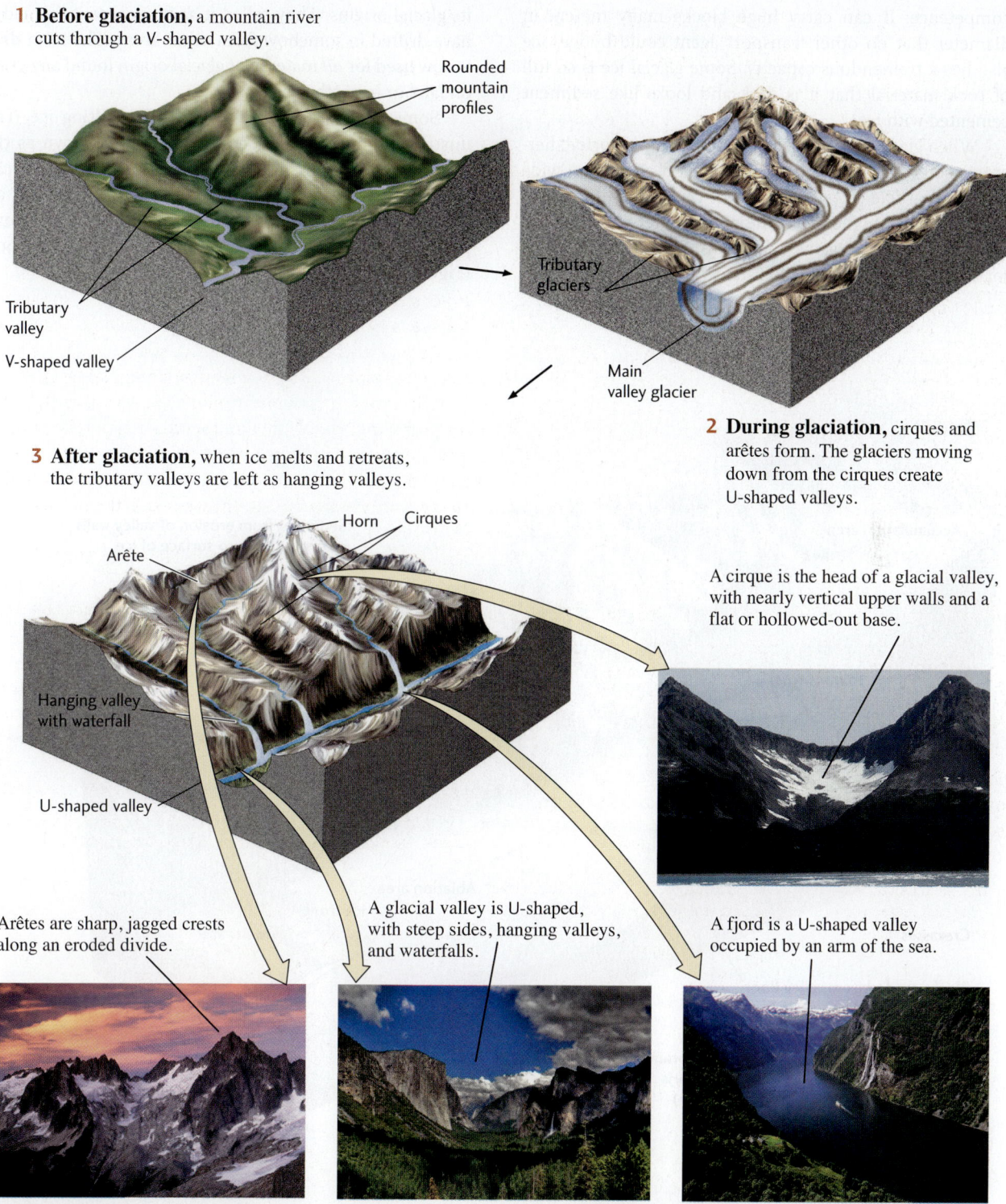

FIGURE 15.16 Erosion by valley glaciers creates distinctive landforms. [top right: Marli Miller; bottom, left to right: Stephen Matera/Alamy; Radomir Rezny/Alamy; Philippe Body/age fotostock/Robert Harding Picture Library.]

Glacial Sedimentation and Sedimentary Landforms

Glaciers transport eroded rock materials of all kinds and sizes downstream, eventually depositing them where the ice melts. Ice is a very effective transport agent because the material it picks up does not settle out like the load of sediment carried by a stream. Like water and wind currents, flowing ice has both a competence (an ability to carry particles of a certain size) and a capacity (a total amount of sediment that it can transport). Ice has extremely high competence: It can carry huge blocks many meters in diameter that no other transport agent could budge. Ice also has a tremendous capacity. Some glacial ice is so full of rock material that it is dark and looks like sediment cemented with ice.

When glacial ice melts, it deposits a poorly sorted, heterogeneous load of boulders, pebbles, sand, and clay. A wide range of particle sizes is the characteristic that differentiates glacial sediments from the much better sorted material deposited by streams and winds. This heterogeneous material puzzled early geologists, who were not aware of its glacial origins. They called it *drift* because it seemed to have drifted in somehow from other areas. The term **drift** is now used for *all* material of glacial origin found anywhere on land or beneath the ocean.

Some drift is deposited directly by melting ice. This unstratified and poorly sorted sediment is known as **till**, and it may contain all sizes of rock fragments from clay to boulders. The large boulders often contained in till are called *erratics* because of their seemingly random composition, often very different from that of local rocks (Figure 15.17).

FIGURE 15.17 A glacier deposits till as an end moraine at the ice front, as lateral moraines at the rocky valley walls, and as a ground moraine beneath the ice. Meltwater streams deposit glacial outwash downstream of the ice front. The inset photo shows glacial till deposited during the Pleistocene epoch on the eastern side of the Sierra Nevada in California. Note the wide range of particle sizes and the lack of stratification. The large boulder the person is standing on is an example of a glacial erratic. [Marli Miller.]

Other deposits of drift are laid down as the ice melts and releases water and sediment. Meltwater flowing in tunnels within and beneath the ice and in streams at the ice front may pick up, transport, and deposit some of the drift. Deposits of drift may trap some of the meltwater, causing it to pool and form lakes. Like any other waterborne sediment, the drift transported by meltwater is stratified and well sorted and may be cross-bedded. Drift that has been picked up and distributed by meltwater streams is called **outwash,** and it often forms broad sedimentary plains downstream of melting glaciers, known as *outwash plains*. Strong winds can transport fine-grained material from outwash plains over long distances and deposit it as loess.

Glacial sedimentary sequences can be identified by the distinctive textures of interbedded tills, outwash, and loess, as well as by striations and other erosional forms that may be preserved. Mapping of such sequences has allowed geologists to infer many glaciations of past geologic times.

Ice-Laid Deposits A **moraine** is an accumulation of rocky, sandy, and clayey material carried by glacial ice and deposited as till. There are many types of moraines, each named for its position with respect to the glacier that formed it (**Table 15.1**). One of the most prominent in size and appearance is an *end moraine*, formed at the ice front. As the ice flows steadily downstream, it brings more and more sediment to its melting edge. The unsorted material accumulates there as a hilly ridge of till. Regardless of their shape or location, moraines of all kinds consist of till. Figure 15.17 illustrates the various kinds of moraines that form as a glacier works its way through a valley.

Some continental glacial terrains display prominent landforms called **drumlins**: large, streamlined hills of till and bedrock that parallel the direction of ice movement (**Figure 15.18**). Drumlins, usually found in clusters, are shaped like long, inverted spoons with the gentlest slopes on the downstream sides. They may be 25 to 50 m high and a kilometer long. Drumlins form when the sediment-rich layer at the base of a glacier encounters a knob of bedrock or other obstacle and the excess pressure squeezes out water and drops the sediment.

Water-Laid Deposits Deposits of outwash by glacial meltwater take a variety of forms. *Kames* are small hills of sand and gravel created when drift fills a hole in a glacier and is left behind when the glacier recedes. Some kames are deltas built into a lake at the ice front. When the lake drains, the deltas are preserved as flat-topped hills. Kames are often exploited as commercial sand and gravel pits.

Eskers are long, narrow, winding ridges of sand and gravel found in the middle of ground moraines (see Figure 15.18). They may run for kilometers in a direction roughly parallel to the direction of ice movement. The origin of eskers is indicated by the well-sorted, water-laid character of their materials and the sinuous, channellike courses of the ridges themselves: Eskers are deposited by meltwater streams flowing in tunnels along the bottom of a melting glacier.

Kettles are hollows or undrained depressions that often have steep sides and may be occupied by ponds or lakes. Modern glaciers, which may leave behind huge isolated blocks of ice in outwash plains as they melt, offer the clue to the origin of kettles. A block of ice a kilometer in diameter may take 30 years or more to melt. During that time, the melting block may be partly buried by outwash sand and gravel carried by meltwater streams—usually braided—coursing around it. By the time the block has melted completely, the ice front has retreated so far that little outwash reaches the area. The sand and gravel that formerly surrounded the block of ice now surround a depression. If the bottom of the kettle lies below the groundwater table, a lake will form.

Varves are formed when valley glaciers deposit silts and clays on the bottom of a lake in a series of alternating coarse and fine layers (see Figure 15.18). A **varve** is a pair of layers

TABLE 15.1 Types of Glacial Moraines

TYPE OF MORAINE	LOCATION WITH RESPECT TO ICE FRONT	COMMENTS
End moraine	At ice front	After glacier melts, seen as ridge parallel to former ice front
Terminal moraine	At ice front marking farthest advance of ice	Type of end moraine
Lateral moraine	Along edge of glacier where it scrapes side walls of valley	Heavy sediment load eroded from valley walls; when ice melts, seen as ridge parallel to valley walls
Medial moraine	Formed as two joined glaciers merge their lateral moraines below junction	Inherits its sediment load from lateral moraines that formed it; forms ridge parallel to valley walls
Ground moraine	Beneath the ice as a layer of glacial debris	Ranges from thin and patchy to a thick blanket of tills

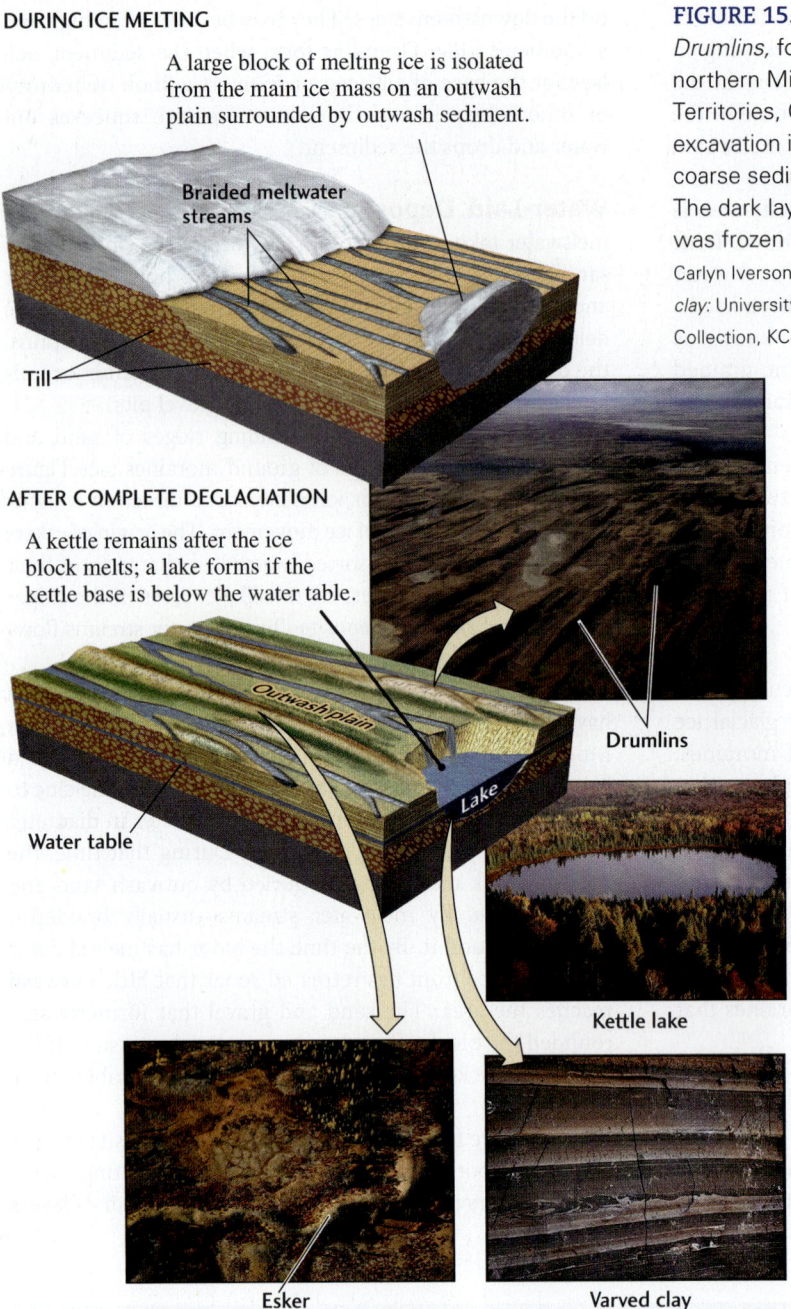

FIGURE 15.18 Ice-laid and water-laid glacial deposits. *Drumlins*, found in Patagonia, Argentina. Kettle lake, northern Minnesota. *Esker* near Whitefish Lake, Northwest Territories, Canada. Pleistocene *varved clay*, from an excavation in Stockholm, Sweden. The light layers are the coarse sediments deposited in a lake during warm seasons. The dark layers are the fine clays deposited when the lake was frozen in winter. [drumlins: © Hauke Steinberg; kettle lake: Carlyn Iverson/Getty Images; esker: All Canada Photos/Alamy; varved clay: University of Washington Libraries, Special Collections, John Shelton Collection, KC4536.]

deposited in one year by seasonal freezing of the lake surface. In summer, when the lake is free of ice, coarse silt is laid down by abundant meltwater streams flowing from the glacier into the lake. In winter, when the surface of the lake is frozen, the finest clays settle, depositing a thin layer on top of the coarse summer layer.

Some lakes formed by continental glaciers were huge, many thousands of square kilometers in area. The till dams that created these lakes were sometimes breached and carried away at a later time, causing the lakes to drain rapidly and creating huge floods. In eastern Washington State, an area called the Channeled Scablands (**Figure 15.19**) is covered by broad, dry stream channels, relics of torrential floodwaters draining from Lake Missoula, a large glacial lake, now completely emptied. From the giant ripples, sandbars, and coarse gravels found there, geologists have estimated that this flood discharged 21 million cubic meters of water per second, flowing as fast as 30 m/s.

FIGURE 15.19 The Channeled Scablands in eastern Washington State contain unique erosional features formed by catastrophic flooding that resulted from the draining of Lake Missoula, a huge glacial lake. This aerial photograph shows Dry Falls, a 350-foot-high, 3-mile-wide group of scalloped cliffs created by the flood. [Bruce Bjornstad.]

For comparison, the velocities of ordinary river flows are measured in fractions of a meter per second, and the discharge rate of the Mississippi River in full flood is less than 50,000 m³/s.

Permafrost

The ground is always frozen in very cold regions where the summer temperature never gets high enough to melt more than a thin surface layer. Perennially frozen soil, or **permafrost,** today covers as much as 25 percent of Earth's total land area. In addition to the soil itself, permafrost includes aggregates of ice crystals in layers, wedges, and irregular masses. The proportion of ice to soil and the thickness of the permafrost vary from region to region. Permafrost is defined solely by temperature, not by soil moisture content, overlying snow cover, or location: Any rock or soil remaining at or below 0°C for 20 or more years is considered permafrost.

In Alaska and northern Canada, permafrost may be as thick as 300 to 500 m. The ground below the permafrost layer, insulated from the bitterly cold temperatures at the surface, remains unfrozen, warmed from below by Earth's internal heat.

Permafrost is a difficult material to handle in engineering projects—such as roads, building foundations, and pipelines—because it melts as it is excavated. The meltwater cannot infiltrate the still-frozen soil below the excavation, so it stays at the surface, waterlogging the soil and causing it to creep, slide, and slump. Engineers decided to build part of the Trans-Alaska pipeline above ground when an analysis showed that the pipeline would thaw the permafrost around it in some places and lead to unstable soil conditions (**Figure 15.20**).

Glacial Cycles and Climate Change

Louis Agassiz, the same geologist who first measured the speed of a Swiss valley glacier, was the first to propose (in 1837) that the Alpine glaciers must have been much larger and thicker in the geologically recent past. He suggested that during a past ice age, Switzerland was covered by an extensive continental glacier almost as thick as its mountains were tall, similar to the one in Greenland today. Among the evidence he cited was the obvious glacial sculpting of high Alpine peaks such as the mighty Matterhorn (**Figure 15.21**). Agassiz's hypothesis was controversial and was not immediately accepted.

Agassiz immigrated to the United States in 1846 and became a professor at Harvard University, where he continued his studies in geology and other sciences. His

FIGURE 15.20 Permafrost melting could destabilize structures at high latitudes, such as the Trans-Alaska oil pipeline, whose 1300 km (800 mile) route from Prudhoe Bay to Valdez crosses 675 km of permafrost. Where the pipeline crosses permafrost, it is perched on specially designed vertical supports. Because thawing of the permafrost would cause the supports to become unstable, they are outfitted with heat pumps designed to keep the ground around them frozen. The heat pumps contain anhydrous ammonia that vaporizes below ground and rises and condenses above ground, discharging heat through the two aluminum radiators atop each of the vertical supports. [Ron Niebrugge/Alamy.]

Permafrost covers about 82 percent of Alaska and 50 percent of Canada, as well as a large proportion of Siberia (see Figure 12.6). Outside the polar regions, it is present in high mountainous areas such as the Tibetan Plateau. Permafrost extends downward several hundred meters in shallow marine areas off the Arctic coasts, presenting difficult engineering problems for offshore oil drillers. The IPCC estimates that, by the mid-twenty-first century, the area of permafrost in the northern hemisphere will have declined by 20–35% relative to its preindustrial coverage.

FIGURE 15.21 The high mountains of the Alps, such as the famous Matterhorn, shown here, were carved by a continental glacier nearly as thick as the peaks are tall. These ice-sculpted peaks provided Louis Agassiz with compelling evidence for an ice age in the recent geologic past. [Hubert Stadler/Getty Images.]

FIGURE 15.22 Irregular hills alternate with lakes in a terrain of glacial till in Coteau des Prairies, South Dakota. Such landscapes were formed by the great continental glaciations of the Pleistocene ice ages. [University of Washington Libraries, Special Collections, John Shelton Collection, KC10367.]

research took him to many places in the northern parts of Europe and North America, from the mountains of Scandinavia and New England to the rolling hills of the American Midwest. In all these diverse regions, Agassiz saw signs of glacial erosion and sedimentation. In the flat country of the American Great Plains, he observed deposits of glacial drift that reminded him of the end moraines of Swiss valley glaciers (Figure 15.22). The heterogeneous material of the drift, including erratic boulders, convinced him of its glacial origin, and the freshness of the soft sediments indicated that they were deposited in the recent past.

The areas covered by this drift were so vast that the ice that deposited them must have been a continental glacier much larger than the ones that now cover Greenland and Antarctica. Agassiz expanded his ice age hypothesis, proposing that a great continental glaciation had extended the polar ice caps far into regions that now enjoy more temperate climates. For the first time, people began to talk seriously about ice ages.

The Wisconsin Glaciation

Geologists have determined the ages of the glacial sediments Agassiz studied by isotopic dating, using carbon-14 in logs buried in the drift. The most recent drift was deposited by ice during the latter part of the Pleistocene epoch. Along the east coast of the United States, the southernmost advance of this ice is recorded by the enormous terminal moraines that form Long Island and Cape Cod. North American geologists named this glaciation after Wisconsin because its effects are particularly well manifested in the glacial terrains of that state. The Wisconsin glaciation reached its maximum 21,000 to 18,000 years ago. Figure 15.23 shows the distribution of ice near the end of the glacial maximum.

The Wisconsin glaciation was a global event, though geologists in various parts of the world have therefore given it their own local names (calling it the Würm glaciation in the Alps, for example). Ice sheets with thicknesses of 2 to 3 km built up over the northern parts of North America, Europe, and Asia. In the Southern Hemisphere, the Antarctic Ice Sheet expanded, and the southern tips of South America and Africa were covered with ice.

Glaciation and Sea-Level Change

At the Wisconsin glacial maximum, the continents were slightly larger than they are today because the continental shelves surrounding them—some more than 100 km wide—were exposed by a drop in sea level of about 130 m. This decrease of sea level was due to the enormous amount of water transferred from the hydrosphere to the cryosphere during the expansion of the continental ice sheets. Rivers extended across the newly emergent continental shelves and began to erode channels in the former seafloor. Early cultures, such as those of prehistoric Egypt, were evolving in the lands beyond the ice sheets, and humans lived in these low, coastal plains.

FIGURE 15.23 The extent of continental glaciers (white area) and sea ice (gray area) in the Northern Hemisphere near the end of the Wisconsin glacial maximum, around 18,000 years ago. [Mark McCaffrey, National Oceanic and Atmospheric Administration Paleoclimatology Program.]

The relationship between sea-level change and glacial cycles illustrates the interaction of the hydrosphere and the cryosphere, as we discussed in Chapter 12. As Earth warms or cools, the volume of the cryosphere shrinks or grows. However, as a result of isostasy, only changes in the ice volume on continents directly affect sea level (see Figure 15.13). As continental glaciers grow, the volume of the ocean decreases, and sea level falls; as continental glaciers melt, their volume decreases, and sea level rises. Thus, sea-level change is closely linked to climate change through changes in polar temperatures and ice volume. Were global warming to melt parts of the ice sheets in Greenland and Antarctica, sea level could rise by tens of meters, posing serious problems for human civilization (see the Practicing Geology Exercise).

The Geologic Record of Pleistocene Glaciations

Soon after Agassiz's ice age hypothesis became widely accepted in the mid-nineteenth century, geologists discovered that there had been multiple ice ages during the Pleistocene epoch, with warmer interglacial periods between them. As they mapped glacial deposits in more detail, they became aware of several distinct layers of drift, the lower ones corresponding to earlier ice ages. Between these older layers of glacial material were well-developed soils containing fossils of warm-climate plants. These fossils provided evidence that the glaciers had retreated as the climate warmed. By the early part of the twentieth century, scientists were convinced that at least four major glaciations had affected North America and Europe during the Pleistocene epoch. In North America, these ice ages, from youngest to oldest, are named after the U.S. states where the evidence of glacial advance is best preserved: Wisconsin, Illinois, Kansas, and Nebraska.

In the late twentieth century, geologists and oceanographers examined marine sediments for evidence of past glaciations, as described in Chapter 12. These sediments, which had accumulated continuously in undisturbed ocean basins, contained a much more complete geologic record of the Pleistocene than did continental glacial deposits, and they showed a much more complex history of glacial advance and retreat. By analyzing oxygen isotope ratios in marine sediments from around the world, geologists have constructed a record of climate history millions of years into the past (see Figure 12.15).

Variations During the Most Recent Glacial Cycle

Within glacial cycles, temperatures do not vary smoothly over time (see Figure 12.16). Superimposed on the 100,000-year glacial cycles are climate fluctuations of shorter duration, some nearly as large as the changes from glacial to interglacial periods. Geologists have combined information from cores in continental and valley glaciers, lake sediments, and deep-sea sediments to reconstruct a decade-by-decade—and in some cases, a year-by-year—history of short-term climate variations during the most recent glacial cycle.

Temperatures began to drop after the last interglacial warm period about 120,000 years ago, and they reached their lowest values during the Wisconsin glacial maximum 21,000 to 18,000 years ago. Temperatures then rebounded to warm interglacial levels 11,700 years ago, marking the end of the Pleistocene and the beginning of the Holocene. Here we summarize some of the basic features of this remarkable chronicle.

- During the Wisconsin glaciation, Earth's climate was highly variable, with shorter (1000-year) temperature oscillations occurring within longer (10,000-year) cycles. The most extreme variations appear to have been in the North Atlantic region, where average local temperatures rose and fell by as much as 15°C. Each 10,000-year cycle comprised a set of progressively cooler 1000-year oscillations and ended with an abrupt warming. Massive discharges of icebergs and fresh water resulting from these sudden warmings altered thermohaline circulation in the oceans and dumped large amounts of glacial material into deep-sea sediments.

- The transition from the Wisconsin glaciation to the current interglacial period, the Holocene, also involved rapid climate fluctuations. The climate abruptly warmed around 14,500 years ago. It then cooled back to glacial conditions in an ice age called the "Younger Dryas," and finally warmed to nearly present-day conditions about 11,700 years ago. Both warming periods were very rapid; broad regions of Earth experienced almost simultaneous changes from ice age to interglacial temperatures during intervals as short as 30 to 50 years. Evidently, the entire climate system can flip from one state (glacial cold) to another (interglacial warmth) in less than a human lifetime! This observation raises the possibility that anthropogenic changes could trigger abrupt shifts to a new—and unknown—climate state, rather than just a gradual warming.

- The Holocene has been unusually long and stable when compared with the previous interglacial periods of the Pleistocene epoch. Regional temperatures have fluctuated by about 5°C on time scales of 1000 years or so, but the global changes during this period have been much smaller, with a total range of only 2°C. These equable Holocene conditions were no doubt favorable for the rapid rise of agriculture and civilization that followed the end of the Wisconsin glaciation.

Some scientists think that if human civilization had not come along, Earth's climate might by now be plunging into another ice age, driven by decreasing amounts of solar energy due to Milankovitch cycles and accompanied by decreasing atmospheric concentrations of greenhouse gases. According to one hypothesis, the expansion of civilization began to release significant amounts of greenhouse gases into the atmosphere as early as 8000 years ago, primarily through deforestation and the rise of agriculture, extending the warm interglacial period beyond its natural limit.

Whatever the reason, measurements from ice cores indicate that, from the beginning of the Holocene until the dawn of the industrial age, atmospheric concentrations of the major greenhouse gases stayed relatively constant. The average CO_2 concentration, for example, fluctuated only between 260 and 280 ppm—less than a 10 percent variation over that entire period. But that situation ended early in the nineteenth century with the beginning of the Industrial Revolution, when human emissions of greenhouse gases shot upward (see Chapter 14).

The Geologic Record of Ancient Glaciations

The Pleistocene glacial cycles were not unique in Earth's history. Since the early part of the twentieth century, we have known from glacial striations and lithified ancient tills, called **tillites,** that glaciers covered parts of the continents several times in the distant geologic past, long before the Pleistocene. Tillites record major continental glaciations during Carboniferous-Permian glaciation time, during Ordovician time, and at least twice during Precambrian time (**Figure 15.24**). The Carboniferous-Permian glaciation covered much of southern Gondwana about 300 million years ago, leaving deposits that have been preserved as tillites across much of the Southern Hemisphere (see Figure 15.24a,b). The joining of the southern continents near the South Pole to form Gondwana may have triggered the cooling that led to this glaciation. The Ordovician glaciation was more limited in its distribution and is best preserved in northern Africa.

The oldest confirmed glaciation occurred during the Proterozoic eon, about 2.4 billion years ago. Its glacial deposits are preserved in Wyoming, along the Canadian portion of the Great Lakes, in northern Europe, and in South Africa. Some geologists argue for an even older glaciation in the Archean eon, almost 3 billion years ago, but that interpretation is disputed.

The youngest Proterozoic glaciation, which spanned a period between 750 million and about 600 million years ago, involved several ice ages separated by warm interglacial periods. Glacial deposits of this age have been found on every continent (Figure 15.24c). Curiously, the reconstruction of paleocontinents indicated that the ice sheets in the Northern Hemisphere had extended much farther south than during the Pleistocene glaciations, perhaps all the way to the equator. This evidence has provoked some geologists to speculate that Earth may have been completely covered by ice, from pole to pole—a bold hypothesis called *Snowball Earth* (Figure 15.24d).

According to the Snowball Earth hypothesis, there was ice everywhere—even the oceans were frozen. The average global temperature would have been about −40°C, like that of the Antarctic today. Except for a few warm spots near volcanoes, very little life would have survived. How could such an apocalyptic event have occurred? And how could it have ended, returning us to the climate we know today? The answers may lie in the feedbacks that occur within the climate system (see Chapter 12).

In one scenario, as Earth initially cooled, ice sheets at the poles spread outward, their white surfaces reflecting more and more sunlight away from Earth. The increase in Earth's albedo cooled the planet, which further expanded the ice sheets. This self-reinforcing process continued until it reached the tropics, encasing the planet in a layer of ice as much as 1 km thick. This scenario is an example of albedo feedback gone wild.

Earth remained buried in ice for millions of years, but the few volcanoes that poked above the surface slowly pumped carbon dioxide into the atmosphere. When the concentration of carbon dioxide reached a critical level, temperatures rose, the ice melted, and Earth again became a greenhouse.

The Snowball Earth hypothesis is controversial, and many geologists disagree with the idea that the oceans were completely frozen. Nevertheless, the evidence for glaciation at low latitudes is strong, and the hypothesis serves as an example of the potential of feedbacks in Earth's climate system to produce extreme change. Geologists have their work cut out for them in trying to understand the extremes of Earth's climate system!

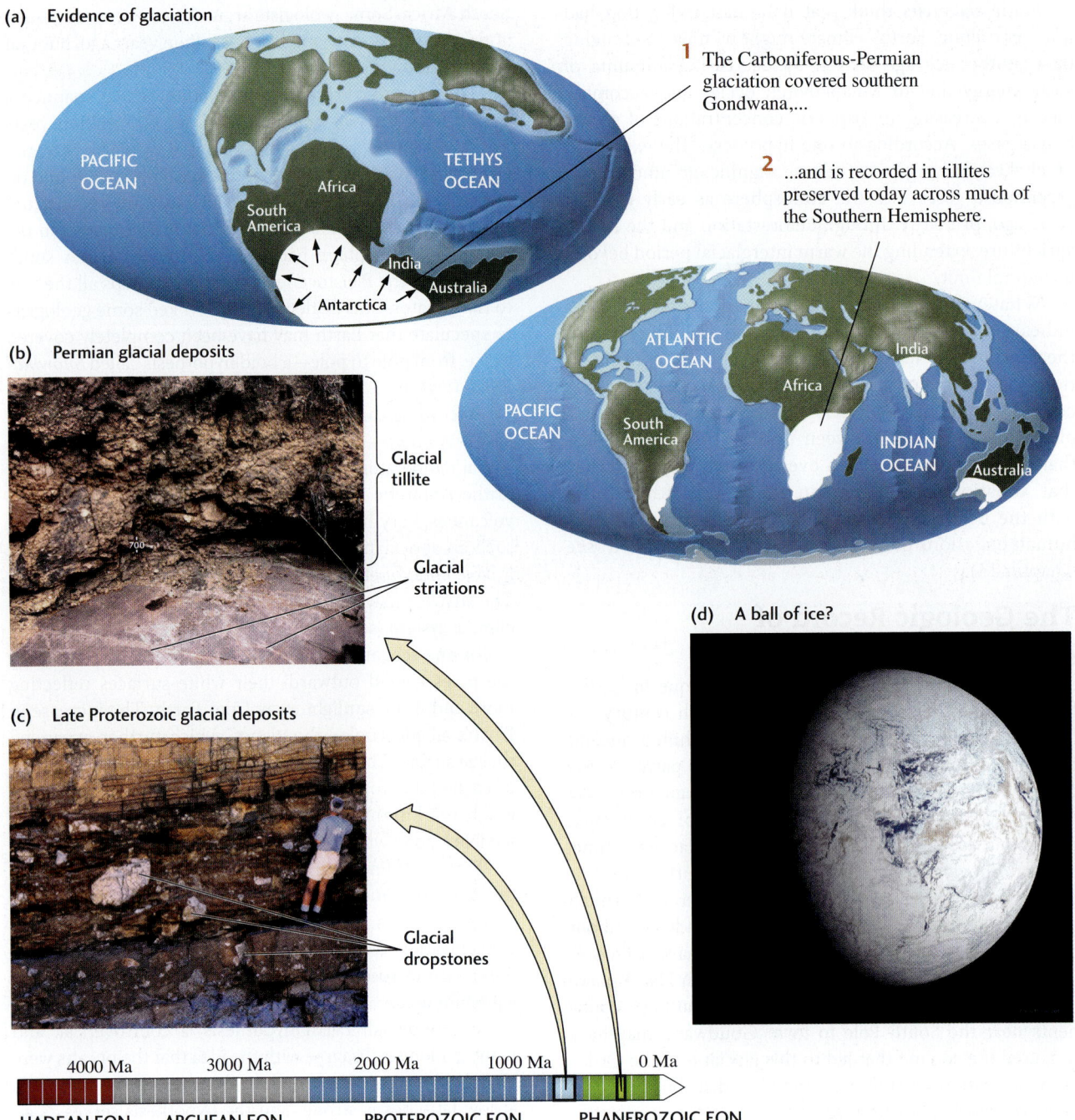

FIGURE 15.24 Ancient glaciations. (a) The first map shows the extent of the Carboniferous-Permian glaciation, which occurred more than 300 million years ago. At that time, the southern continents were assembled into the giant continent Gondwana, and the ice cap was situated in the Southern Hemisphere, centered over Antarctica, which is home to a continental glacier today. The second map shows the distribution of Permian-Carboniferous glacial deposits today. (b) Permian glacial deposits from South Africa. (c) Late Proterozoic glacial deposits. (d) Artist's depiction of a Snowball Earth, hypothesized to have formed in the late Proterozoic. Geologists debate the extent to which ice covered the globe, but some think even the oceans were frozen. [b and c: John Grotzinger; d: Chris Butler/Science Photo Library/Science Source.]

KEY TERMS AND CONCEPTS

ablation (p. 428)
accumulation (p. 428)
basal slip (p. 431)
cirque (p. 436)
continental glacier (p. 425)
crevasse (p. 431)
drift (p. 438)
drumlin (p. 439)
esker (p. 439)
fjord (p. 436)
glacier (p. 425)
hanging valley (p. 436)
ice shelf (p. 426)
ice stream (p. 431)
iceberg calving (p. 425)
kettle (p. 439)
moraine (p. 439)
outwash (p. 439)
permafrost (p. 441)
plastic flow (p. 431)
striation (p. 435)
surge (p. 431)
till (p. 438)
tillite (p. 445)
U-shaped valley (p. 436)
valley glacier (p. 425)
varve (p. 439)

REVIEW OF LEARNING OBJECTIVES

15.1 Identify the main types of glaciers.

Glaciers are divided into two main types. A valley glacier is a river of ice that forms in the cold heights of mountain ranges and flows downslope through a valley. A continental glacier is a thick ice sheet that covers part of a continent or other large landmass and flows towards the coastline in concentrated ice streams. Today, continental glaciers cover much of Greenland and Antarctica. Glaciers form where climates are cold enough that snow, instead of melting completely in summer, is buried and transformed into ice by recrystallization. As snow accumulates, either at the tops of valley glaciers or at the domed centers of continental glaciers, the ice thickens. Its thickness increases until it becomes so massive that gravity can pull it downhill.

Study Assignment: Review the section *Types of Glaciers*.

Exercises: (a) Do all glaciers move as gravitational flows? (b) Give an example of an ice cap that is not a continental glacier. (c) The flow of which type of glacier, continental or valley, is most constrained by topography?

Thought Questions: (a) How are the ice accumulations that constitute glaciers analogous to sedimentary rocks? Metamorphic rocks? (b) Should ice streams that flow as parts of continental glaciers be called glaciers?

15.2 Describe how glaciers grow and shrink, and how they move.

Glaciers lose ice by melting, sublimation, iceberg calving, and wind erosion. The glacial budget is the relationship between ablation (the amount of ice a glacier loses annually) and accumulation (the amount of new snow and ice added to the glacier annually). If ablation is balanced by accumulation, the size of the glacier remains constant. If ablation is greater than accumulation, the glacier shrinks; conversely, if accumulation exceeds ablation, the glacier grows. Glaciers move by a combination of plastic flow and basal slip. Plastic flow dominates in very cold regions, where the glacier's base is frozen to the ground. Basal slip is more important in warmer climates and where meltwater at the glacier's base lubricates the ice.

Study Assignment: Review the different types of glacial flow in Figure 15.9.

Exercises: (a) Consider a glacier for which the accumulation equals the ablation. (i) Does the size of the glacier stay the same? (ii) Is the glacier stationary? (b) For the tropical glaciers of the Rwenzori Mountains, which is currently greater, accumulation or ablation? (c) Which of the two mechanisms of glacial flow described in Figure 15.9 causes more internal deformation within a glacier? (d) Which of the two mechanisms is more prevalent in very cold climates?

Thought Questions: (a) Some parts of a glacier contain a lot of sediment; others contain very little. What accounts for the difference? (b) One of the dangers of exploring glaciers is the possibility of falling into a crevasse. What topographic features of a valley glacier or its surroundings might you use to predict whether that part of the glacier was badly crevassed?

15.3 Explain why the melting of continental glaciers increases sea level but the melting of sea ice does not.

Ice floats because its density (0.92 g/cm^3) is less than that of seawater (1.03 g/cm^3). The volume of an iceberg below the sea surface thus weighs less than the volume of seawater it displaces, causing a buoyancy force that must be balanced by the weight of the ice above the waterline. When an iceberg melts, there is no change in sea level, because the water volume resulting from the melting exactly equals the water volume that had been displaced by the iceberg. In contrast, when glaciers on land melt, most of the water flows into the ocean, increasing the ocean volume and thereby raising the sea level. When glaciers on land flow directly into the ocean, the water in the oceans is displaced by the icebergs they shed, which also raises sea level.

Study Assignment: Review Figure 15.13 and the Practicing Geology Exercise.

Exercises: (a) Combining the densities of ice and seawater with the principle of isostasy, compute what fraction of an iceberg's mass floats above the sea surface. (b) Why does sea level drop during ice ages? (c) Explain whether or not the melting of ice shelves by global warming of the oceans will increase sea level.

Thought Question: One result of continued global warming could be the shrinkage and collapse of the West Antarctic Ice Sheet. How much would sea level rise because of this melting and what impact would this rise have on coastal cities?

15.4 Recognize features of the landscape that have been formed and modified by glacial processes.

Glaciers erode bedrock by scraping, plucking, and grinding it into sizes ranging from boulders to fine rock flour. Valley glaciers erode cirques, horns, and arêtes at their heads; excavate U-shaped and hanging valleys; and create fjords by eroding their valleys below sea level at the coast. Glacial ice has both high competence and high capacity, which enable it to carry abundant sediment particles of all sizes. Glaciers transport huge quantities of sediments to the ice front, where melting releases them. The sediments may be deposited directly by the melting ice as till or picked up by meltwater streams and laid down as outwash. Moraines and drumlins are characteristic landforms deposited by ice. Eskers and kettles are formed by meltwater. Permafrost forms where summer temperatures never rise high enough to melt more than a thin surface layer of soil.

Study Assignment: Review the section *Glacial Landscapes*.

Exercises: (a) Describe three distinctive landforms made by glaciers. (b) What type of sedimentary deposit marks the farthest advance of a glacier? (b) Explain why kettles are described as water-laid features rather than ice-laid features.

Thought Questions: (a) How would the till from a glaciated terrain of granitic and metamorphic rocks differ from a till from a terrain of soft shales and loosely cemented sandstones? (b) Walking along a winding ridge of glacial drift, what evidence would you look for to determine whether the ridge was an esker or an end moraine? (c) What geologic evidence would you map to describe the directions that continental glaciers moved across Canada?

15.5 Understand what the geologic record tells us about past ice ages.

Glacial drift of Pleistocene age is widespread over high-latitude regions that now enjoy temperate climates. This widespread drift is evidence that continental glaciers once expanded far beyond the polar regions. Studies of the geologic ages of glacial deposits on land and in marine sediments show that continental ice sheets advanced and retreated many times during the Pleistocene epoch. The most recent glacial advance, known as the Wisconsin glaciation, covered the northern parts of North America, Europe, and Asia with ice and exposed large areas of continental shelves. Tillites record major continental glaciations during Carboniferous-Permian glaciation time, during Ordovician time, and at least twice during Precambrian time. The oldest confirmed glaciation occurred early in the Proterozoic eon, about 2.4 billion years ago. A late-Proterozoic glaciation left deposits on every continent, motivating some geologists to speculate that Earth was covered by ice from pole to pole—the Snowball Earth hypothesis.

Study Assignment: Review the section *Glacial Cycles and Climate Change*.

Exercises: (a) How much lower than now was sea level during the Wisconsin glacial maximum? (b) On which continents do we find geological evidence for the Carboniferous-Permian glaciation? (c) How is the distribution of the Carboniferous-Permian glaciation explained by plate tectonics?

Thought Questions: (a) Do you think humans played a role in extending the Holocene to be longer than previous Pleistocene interglacial periods? (b) Why is it unlikely that Earth's climate will return to a state of glacial cold in the next few millennia?

VISUAL LITERACY EXERCISE

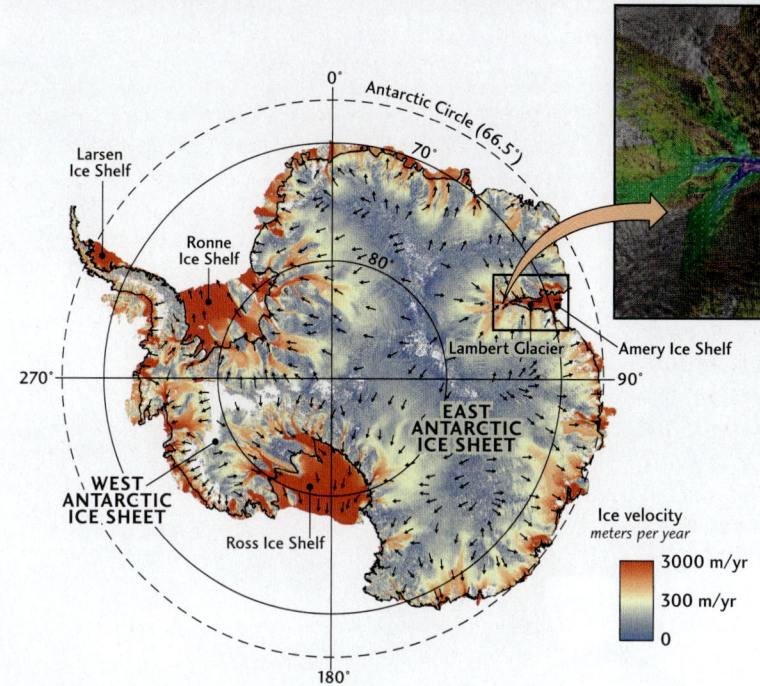

FIGURE 15.11 Map of Antarctica showing the rates (colors) and directions (arrows) of glacial flow. (b) Inset shows the flow of the Lambert Glacier, the world's largest, measured by satellite radar; area of image is about 380 km by 570 km. [RADARSAT imagery from the 2000 Antarctic Mapping Mission, NASA Visible Earth.]

1. Which of the Antarctic ice shelves labeled on the map lies at the lowest latitudes?
   a. Ross Ice Shelf
   b. Amery Ice Shelf
   c. Ronne Ice Shelf
   d. Larsen Ice Shelf

2. Where are the velocities of ice flows smallest?
   a. Near the center of the East Antarctic Ice Sheet
   b. Along the Antarctic coastline
   c. Near the center of the West Antarctic Ice Sheet
   d. In the ice shelves

3. The general direction of ice flow is towards
   a. the South Pole
   b. the thickest parts of the ice sheets
   c. the regions of lowest ice velocity
   d. the Antarctic coastline

4. Where are the velocities of ice flow the largest?
   a. Near the center of the East Antarctic Ice Sheet
   b. In the ice shelves along the Antarctic coastline
   c. Near the center of the West Antarctic Ice Sheet
   d. At the South Pole

Earth Surface Processes and Landscape Development 16

Controls on Weathering	454
Chemical Weathering	456
Physical Weathering	461
Soils: The Residue of Weathering	462
Erosion and Formation of Stream Valleys	464
Mass Wasting	467
Classification of Mass Movements	472
Geomorphology and Landscape Development	482

Learning Objectives

Chapter 16 describes the geologic surface processes that break down rocks and transport the products over short distances: weathering, erosion, and mass wasting. These three processes result from interactions between the climate and plate tectonic systems, and give rise to Earth's diverse range of landscapes.

16.1 Define the controls on weathering.
16.2 Describe the processes of chemical weathering.
16.3 Describe the processes of physical weathering.
16.4 Summarize the types of soil and what processes they record.
16.5 Describe how erosion creates stream valleys.
16.6 Evaluate the mechanisms that cause mass movements.
16.7 Distinguish the different kinds of mass wasting processes.
16.8 Summarize the processes that control landscape development.

A landslide in the Bluebird Canyon area of Laguna Beach saturated by unusually heavy winter rains gave way, destroying 7 homes and damaging 11 others. Photo taken June 1, 2005. [Mark Boster/Los Angeles Times via Getty Images.]

Have you ever gazed out over the horizon and wondered why Earth's surface has the shape it does, or what forces created that shape? From high snowcapped peaks to broad rolling plains, Earth's landscapes comprise a diverse array of landforms—large and small, rough and smooth. These landscapes develop through slow changes as the processes of tectonic uplift, weathering, erosion, transportation, and sedimentation combine to sculpt the land surface. As solid as the hardest rocks may seem, all rocks—like rusting old automobiles and yellowed old newspapers—eventually weaken and crumble when exposed to water and the gases of the atmosphere. Unlike cars and newspapers, however, rocks may take thousands of years to disintegrate.

Weathering—driven by climate—is the general process by which rocks are broken down at Earth's surface. Weathering is the first step in flattening the mountains that have been uplifted by plate tectonic processes. Even as mountains are being uplifted, chemical decay and physical fragmentation join with rainfall, wind, ice, and snow to wear them away. *Chemical weathering* occurs when the minerals in a rock are chemically altered or dissolved. *Physical weathering* takes place when solid rock is fragmented by mechanical processes that do not change its chemical composition. Chemical and physical weathering reinforce each other. Chemical weathering weakens rocks and makes them more susceptible to physical weathering. The smaller the pieces produced by physical weathering, the greater the surface area available for chemical weathering.

Once weathering reduces rocks to particles, they may accumulate as soil, or they may be removed by erosion, transported, and deposited elsewhere as sediments. **Erosion** is the process by which particles produced by weathering are dislodged and removed from their source, usually by means of currents of water or air. Erosion moves particles from hillslopes to the starting points of stream channels, and the movement of those particles abrades and deepens stream channels cut into bedrock. **Mass wasting** includes all the processes by which weathered and unweathered Earth materials move downslope in larger amounts and in large single events, usually by means of gravity. The products of mass wasting—particles released by weathering as well as large masses of unweathered rock—are also transported to the starting points of stream channels. Once these materials reach stream channels, streams and rivers can efficiently transport them farther downslope, perhaps across continents and all the way to the ocean. The transport of sediments by streams from their source areas in mountains to their sinks in the world ocean will be covered in more detail in Chapter 18.

Weathering is one of the major processes of the rock cycle. It shapes Earth's surface topography and alters rock materials, converting all kinds of rocks into sediments and soils. The early sections of this chapter emphasize chemical weathering because it is in some ways the fundamental driving force of the weathering process. For example, the effects of physical weathering, which we will examine next, depend largely on the chemical decay of minerals. Before we look at either type of weathering in detail, however, let's examine the factors that control weathering.

Controls on Weathering

All rocks weather, but the manner and rates of their weathering vary. The four key factors that control rates of weathering are the properties of the parent rock, the climate, the presence or absence of soil, and the length of time the rocks are exposed to the atmosphere. These four factors are summarized in Table 16.1.

TABLE 16.1 Major Factors Controlling Rates of Weathering

	WEATHERING RATE		
	SLOW ────────	────────	────► FAST
PROPERTIES OF PARENT ROCK			
Mineral solubility in water	Low (e.g., quartz)	Moderate (e.g., pyroxene, feldspar)	High (e.g., calcite)
Rock structure	Massive	Some zones of weakness	Very fractured or thinly bedded
CLIMATE			
Rainfall	Low	Moderate	High
Temperature	Cold	Temperate	Hot
PRESENCE OR ABSENCE OF SOIL AND VEGETATION			
Thickness of soil layer	None—bare rock	Thin to moderate	Thick
Organic content	Low	Moderate	High
LENGTH OF EXPOSURE	Short	Moderate	Long

The Properties of Parent Rock

The mineralogy and crystal structure of parent rock affect weathering because different minerals weather at different rates and because a rock's crystal structure affects its susceptibility to cracking and fragmentation. Old inscriptions on gravestones are evidence of the varying rates at which rocks weather. The carved letters on a recently erected gravestone stand out clearly from the stone's polished surface. After a hundred years in a moderately rainy climate, however, the surface of a limestone monument will be dull, and the letters inscribed on it will have almost melted away, much as the brand name on a bar of soap disappears after a few washes (**Figure 16.1**). Slate or granite, on the other hand, will show only minor changes. The differences in the weathering of limestone and slate result from their different mineral compositions. Given enough time, however, even a resistant rock will ultimately decay. After several hundred years, a granite monument will also have weathered appreciably, and its surface and letters will be somewhat dulled and blurred.

Climate: Rainfall and Temperature

The rates of both chemical and physical weathering vary not only with the properties of parent rock, but also with the climate—especially the temperature and amount of rainfall—where that parent rock is located. High temperatures and heavy rainfall promote faster chemical weathering; cold and dry climates slow the process. Water in cold climates cannot dissolve minerals when it is frozen. In arid regions, water is relatively unavailable.

On the other hand, climates that minimize chemical weathering may enhance physical weathering. For example, freezing water may act as a wedge, widening cracks and pushing a rock apart. In temperate climates, the alternating freezing and thawing that accompany changes in temperature cause rocks to contract and expand, helping to break them apart.

The Presence or Absence of Soil

Although soil is itself a product of weathering, its presence or absence affects the chemical and physical weathering of other materials. Soil production is a *positive feedback process*—that is, soil formation advances more soil formation. Once soil starts to form, it works as a geologic agent to weather rock more rapidly. The soil retains rainwater, and it hosts a variety of vegetation, bacteria, and other organisms. The metabolism of those organisms creates an acidic environment that, in combination with moisture, promotes chemical weathering. Plant roots and organisms tunneling through the soil promote physical weathering by helping to create fractures in rock. Chemical and physical weathering, in turn, lead to the production of more soil.

The Length of Exposure

The longer a rock weathers, the greater its chemical alteration, dissolution, and physical fragmentation. Rocks that have been exposed at Earth's surface for many thousands

FIGURE 16.1 These early-nineteenth-century gravestones at Wellfleet, Massachusetts, show the results of chemical weathering. The stone on the right, which is limestone, has become so weathered that it is unreadable. The stone on the left, which is slate, has remained legible under the same conditions. [Courtesy of Raymond Siever.]

of years form a *rind*—an external layer of weathered material ranging from several millimeters to several centimeters thick—that surrounds fresh, unaltered rock. In dry climates, some rinds have grown as slowly as 0.006 mm per 1000 years.

Now that we have examined the factors that control rates of weathering, we can consider the two types of weathering—chemical and physical—in more detail.

Chemical Weathering

Chemical weathering occurs when minerals react with air and water. In these chemical reactions, some minerals dissolve. Others combine with water and atmospheric components such as oxygen and carbon dioxide to form new minerals. We begin our investigation by examining the chemical weathering of feldspar, the most abundant mineral in Earth's crust.

The Role of Water: Feldspar and Other Silicates

Feldspar is one of many silicates that are altered by chemical reactions to form clay minerals. Feldspar's behavior during weathering helps us understand the weathering process in general, for two reasons:

1. Feldspar is a key mineral in a great many igneous, sedimentary, and metamorphic rocks and is one of the most abundant minerals in Earth's crust.
2. The chemical processes that characterize feldspar weathering also characterize weathering in many other kinds of minerals.

Feldspar is one component of granite, which, as you will recall, is made up of several different minerals, all of which decay at different rates. A sample of unweathered granite is hard and solid because an interlocking network of feldspar, quartz, and other crystals holds it tightly together. When the feldspar is transformed by weathering into a loosely adhering clay, the network is weakened and the mineral grains are separated (**Figure 16.2**). In this instance, chemical weathering, by producing the clay, also promotes physical weathering because the rock now fragments easily along widening cracks at the boundaries between grains.

The white- to cream-colored clay produced by the weathering of feldspar is called *kaolinite*, named for Gaoling, a hill in southwestern China where it was first obtained. Chinese artisans had used pure kaolinite as the raw material of pottery and porcelain for centuries before Europeans borrowed the idea in the eighteenth century.

Only in the arid climates of some deserts and polar regions does feldspar remain relatively unweathered. This observation points to water as an essential component of the chemical reaction by which feldspar becomes kaolinite. Kaolinite is a hydrous aluminum silicate. In the reaction that produces it, the solid feldspar undergoes *hydrolysis* (a decomposition reaction involving water; from *hydro*, "water," and *lysis*, "to loosen"). The feldspar is broken down and also loses several chemical components, while kaolinite gains water.

The only part of a solid that reacts with a fluid is the solid's surface, so as we increase the surface area of the solid, we speed up the reaction. For example, as we grind coffee beans into finer and finer particles, we increase the ratio of their surface area to their volume. The finer the coffee beans are ground, the faster their reaction with water, and the stronger the brew becomes. Similarly, the smaller the fragments of minerals and rocks, the greater their surface area. The ratio of surface area to volume increases greatly as the average particle size decreases, as shown in **Figure 16.3**.

FIGURE 16.2 Diagrammatic microscopic views of stages in the disintegration of granite. [John Grotzinger/Ramón Rivera-Moret/Harvard Mineralogical Museum.]

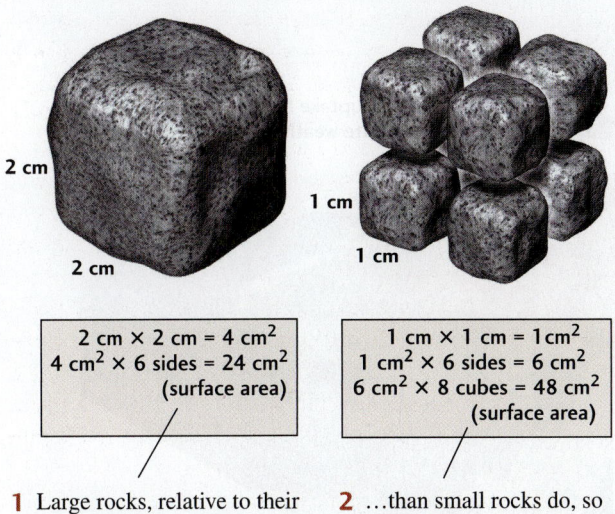

1. Large rocks, relative to their mass, have less surface area for chemical weathering…

2. …than small rocks do, so smaller rocks weather more quickly.

FIGURE 16.3 As a rock mass breaks into smaller pieces, more surface area becomes exposed to the chemical reactions of weathering.

Carbon Dioxide, Weathering, and the Climate System

Carbon dioxide, like water, is involved in the chemical reactions of weathering. Thus, variation in the atmospheric concentration of CO_2 leads to corresponding variation in the rate of weathering (**Figure 16.4**). Higher concentrations of CO_2 in the atmosphere lead to higher concentrations in the soil, which accelerate the weathering of rocks. As we saw in Chapter 14, increases in atmospheric CO_2, a greenhouse gas, make Earth's climate warmer and thus promote weathering. The weathering of calcium-rich rock, in turn, removes CO_2 from the atmosphere, making global climates cooler. In this way, chemical weathering links the plate tectonic system to the climate system. As more and more CO_2 is used up through weathering, the climate cools, and weathering decreases. As weathering decreases, the concentration of CO_2 in the atmosphere builds up again, and the climate warms, thus continuing the cycle.

The Role of Carbon Dioxide in Weathering
The reaction of feldspar with pure water in a laboratory is an extremely slow process that would take thousands of years to break down even a small amount of feldspar completely. We can speed weathering by adding a strong acid (such as hydrochloric acid) to the water, in which case the feldspar will break down in a few days. An *acid* is a substance that releases hydrogen ions (H^+) to a solution. A strong acid produces abundant hydrogen ions; a weak one, relatively few. The strong tendency of hydrogen ions to combine chemically with other substances makes acids excellent solvents.

On Earth's surface, the most common acid—and the one most responsible for increasing weathering rates—is carbonic acid (H_2CO_3). This weak acid forms when carbon dioxide (CO_2) gas from the atmosphere dissolves in rainwater:

$$\text{carbon dioxide} + \text{water} \rightarrow \text{carbonic acid}$$
$$CO_2 \qquad H_2O \qquad H_2CO_3$$

The amount of carbon dioxide dissolved in rainwater is small because the amount of CO_2 gas in the atmosphere is small. About 0.03 percent of the molecules in Earth's atmosphere are carbon dioxide. Thus, the amount of carbonic acid formed in rainwater is also very small, only about 0.0006 g/L.

As human activities increase the concentration of carbon dioxide in the atmosphere, the amount of carbonic acid in rainwater is increasing slightly. Acid rain accelerates weathering, but most of the acidity of acid rain comes not from carbon dioxide, but from sulfur dioxide and nitrogen gases, which react with water to form strong sulfuric and nitric acids, respectively. These acids promote weathering to a greater degree than carbonic acid does. Volcanoes and coastal wetlands emit gaseous forms of carbon, sulfur, and nitrogen into the atmosphere, but by far the largest source of these gases is industrial pollution.

Although rainwater contains only a relatively small amount of carbonic acid, that amount is enough to dissolve great quantities of rock over long periods. The chemical reaction for the weathering of feldspar is

$$\text{feldspar} + \text{carbonic acid} + \text{water} \rightarrow$$
$$2KAlSi_3O_8 \qquad 2H_2CO_3 \qquad H_2O$$

$$\text{dissolved} + \text{dissolved} + \begin{array}{c}\text{dissolved}\\\text{potassium}\\\text{ions}\end{array} + \begin{array}{c}\text{dissolved}\\\text{bicarbonate}\\\text{ions}\end{array}$$
$$\text{kaolinite} \qquad \text{silica}$$
$$Al_2Si_2O_5(OH)_4 \quad 4SiO_2 \qquad 2K^+ \qquad 2HCO_3^-$$

This simple weathering reaction illustrates the three main effects of chemical weathering on silicates:

1. It *leaches*, or dissolves away, cations and silica.
2. It *hydrates*, or adds water to, the minerals.
3. It makes solutions less acidic.

Specifically, the carbonic acid in rainwater helps to weather feldspar in the following way (see Figure 16.4):

- A small proportion of the carbonic acid molecules in rainwater ionize, forming hydrogen ions (H^+) and bicarbonate ions (HCO_3^-), thus making the water slightly acidic.

- The slightly acidic water dissolves potassium ions and silica from feldspar, leaving a residue of kaolinite, a solid clay. The hydrogen ions from the acidic water combine with the oxygen atoms of the feldspar to form the water in the kaolinite structure. The kaolinite becomes part of the soil or is carried away as the dissolved silica, potassium ions, and bicarbonate ions are carried away by rain and stream waters and are ultimately transported to the ocean.

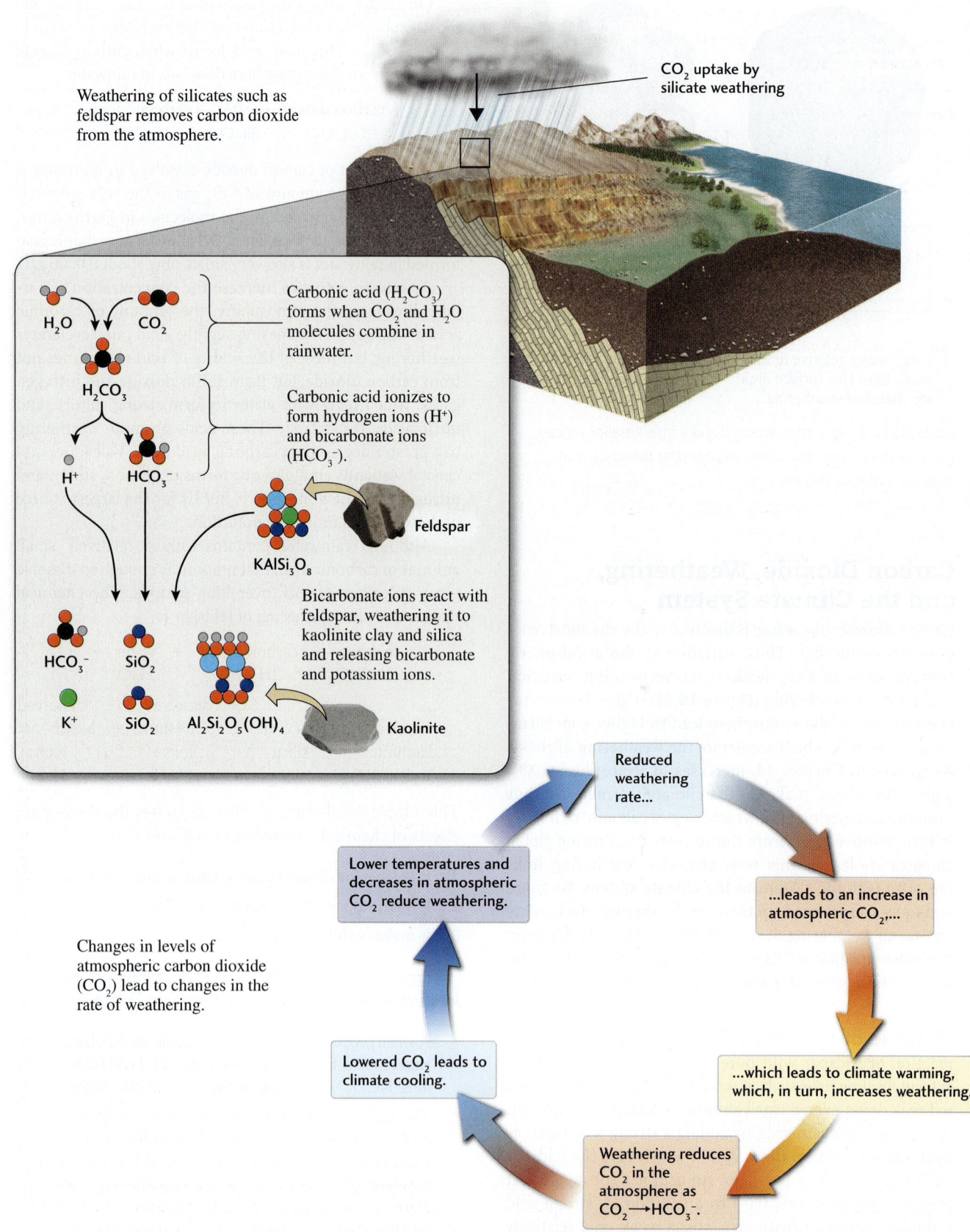

FIGURE 16.4 Variation in atmospheric carbon dioxide concentrations leads to corresponding variation in rates of weathering as well as in global temperatures, which also influence weathering. In this way, the lithosphere and the climate system are linked.

The Role of Soil in Weathering Now that we understand how acidic water weathers feldspar, we can better understand why feldspars on bare rock outcrops are better preserved than those buried in damp soils. The chemical reactions of weathering give us two separate but related clues to the factors responsible: the amount of water and the amount of acid available for those reactions. The exposed feldspar weathers only while the rock is moist with rainwater. During dry periods, the only moisture that touches bare rock is dew. The feldspar in damp soil, however, is constantly in contact with the small amounts of water retained in the pores between soil particles. Thus, feldspar weathers continuously in moist soil.

There is more acid in soil moisture than there is in falling rain. Rainwater carries its original carbonic acid into the soil. As water filters through the soil, it picks up additional carbonic acid and other acids produced by the roots of plants, by the many insects and other animals that live in the soil, and by the bacteria that degrade plant and animal remains. Recently, it was discovered that some bacteria release organic acids, even in waters hundreds of meters deep in the ground. These organic acids weather feldspar and other minerals in rocks below the surface. Bacterial respiration in soil may increase the soil's carbon dioxide concentration to as much as 100 times the atmospheric concentration!

Rock weathers more rapidly in the tropics than in temperate and cold climates, mainly because plants and bacteria grow more quickly in warm, humid climates, contributing more of the carbonic acid and other acids that promote weathering. Additionally, most chemical reactions, weathering included, speed up with an increase in temperature.

The Role of Oxygen: From Iron Silicates to Iron Oxides

Iron is one of the eight most abundant chemical elements in Earth's crust, but metallic iron, the element in its pure form, is rarely found in nature. It is present only in certain kinds of meteorites that fall to Earth from other places in the solar system. Most of the iron ores used for the production of iron and steel are formed by weathering. These ores are composed of iron oxide minerals originally produced by the weathering of iron-rich silicate minerals, such as pyroxene and olivine. The iron released by dissolution of the silicate minerals combines with oxygen from the atmosphere and hydrosphere to form iron oxide minerals.

The iron in minerals may be present in one of three forms: metallic iron, ferrous iron, or ferric iron. In the metallic iron found in meteorites (and in manufactured items), the iron atoms are uncharged: They have neither gained nor lost electrons by reacting with another element. In the *ferrous iron* (Fe^{2+}) found in silicate minerals, the iron atoms have lost two of the electrons they have in the metallic form and have thus become ions. In the *ferric iron* (Fe^{3+}) found in iron oxide minerals, the iron atoms have lost three electrons. The electrons lost by the iron are gained by oxygen atoms in a process called *oxidation*. Oxygen atoms in air and water oxidize ferrous iron to form ferric iron. Thus, all the iron oxides formed at Earth's surface, the most abundant of which is *hematite* (Fe_2O_3), are ferric. Oxidation, like hydrolysis, is one of the important reactions of chemical weathering.

When an iron-rich silicate mineral such as pyroxene is exposed to water, its silicate structure breaks down, releasing silica and ferrous iron into the water. In solution, the ferrous iron is oxidized to the ferric form (**Figure 16.5**). The strong chemical bonds that form between ferric iron and oxygen make ferric iron insoluble in most natural surface waters. It therefore precipitates from solution, forming a solid ferric iron oxide. We are familiar with ferric iron oxide in another

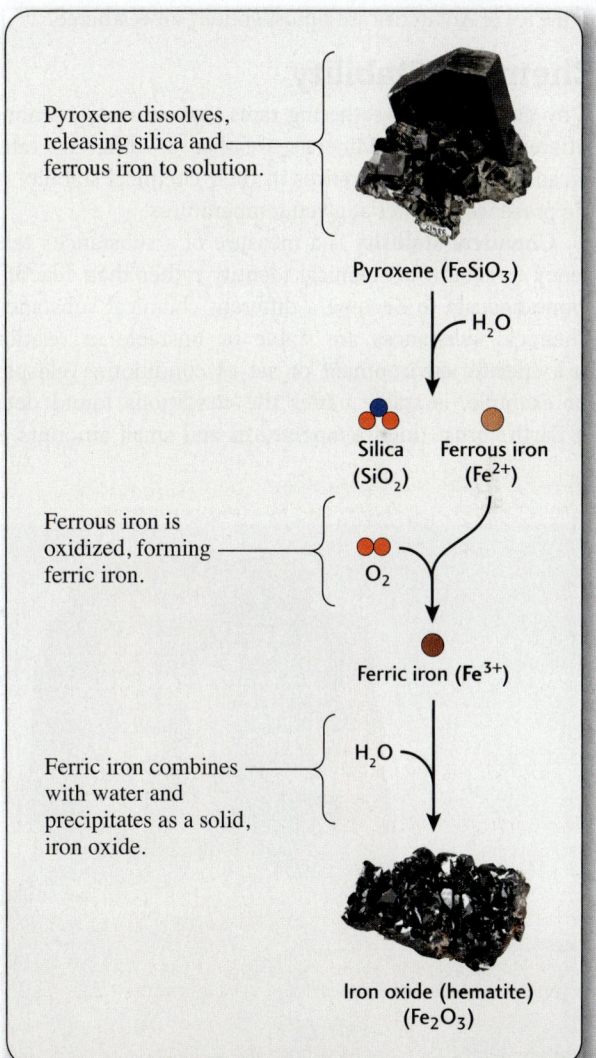

FIGURE 16.5 The general course of the chemical reactions by which an iron-rich silicate mineral, such as pyroxene, weathers in the presence of oxygen and water. [John Grotzinger/Ramón Rivera-Moret/Harvard Mineralogical Museum.]

form: rust, which is produced when metallic iron in manufactured items is exposed to the atmosphere.

We can show this overall weathering reaction with the following example:

iron-rich pyroxene + oxygen → hematite + dissolved silica

$4FeSiO_3 \quad O_2 \quad 2Fe_2O_3 \quad 4SiO_2$

Although the equation does not show it explicitly, water is required for this reaction to proceed.

Iron-containing minerals, which are widespread, weather to the characteristic red and brown colors of oxidized iron (Figure 16.6). Iron oxides are found as coatings and encrustations that color soils and weathered surfaces of iron-containing rocks. The red soils of Georgia and other warm, humid regions are colored by iron oxides. In contrast, iron-rich minerals weather so slowly in polar regions that iron meteorites frozen in the ice of Antarctica are almost entirely unweathered.

Chemical Stability

Why do chemical weathering rates vary so widely among different minerals? Minerals weather at different rates because there are differences in their chemical stability in the presence of water at given temperatures.

Chemical stability is a measure of a substance's tendency to retain its chemical identity rather than reacting spontaneously to become a different chemical substance. Chemical substances are stable or unstable in relation to a specific environment or set of conditions. Feldspar, for example, is stable under the conditions found deep in Earth's crust (high temperatures and small amounts of water), but unstable under the conditions at Earth's surface (lower temperatures and abundant water). Two characteristics of a mineral—its solubility and its rate of dissolution—help determine its chemical stability.

Solubility The *solubility* of a mineral is measured by the amount of that mineral dissolved in water when the solution is saturated. *Saturation* is the point at which the water cannot hold any more of the dissolved substance. The higher a mineral's solubility, the lower its chemical stability under most weathering conditions. Halite-containing evaporites, for example, are unstable under most weathering conditions. Their solubility is high (about 350 g/L), and they are leached from soil by even small amounts of water. Quartz, in contrast, is stable under most weathering conditions. Its solubility in water is very low (only about 0.008 g/L), and it is not easily leached from soil.

Rate of Dissolution A mineral's rate of dissolution is measured by the amount of that mineral that dissolves in an unsaturated solution in a given time. The faster the mineral dissolves, the less stable it is. Feldspar dissolves at a much faster rate than quartz, and primarily for that reason, it is less stable than quartz under weathering conditions.

Relative Stabilities of Common Rock-Forming Minerals The relative chemical stabilities of various minerals can be used to determine the intensity of weathering in a given area. In a tropical rain forest, only the most stable minerals will be left on an outcrop or in the soil, so we know that the weathering in that environment is intense. In an arid region such as the desert of North Africa,

FIGURE 16.6 Red and brown iron oxides color weathering rocks in Monument Valley, Arizona. [William Boyce/EyeEm/Getty Images.]

TABLE 16.2 Relative Stabilities of Common Minerals

STABILITY OF MINERALS	RATE OF WEATHERING
MOST STABLE	Slowest
Iron oxides (hematite)	↓
Aluminum hydroxides (gibbsite)	
Quartz	
Clay minerals	
Muscovite mica	
Orthoclase feldspar	
Biotite mica	
Sodium-rich plagioclase Feldspar (albite)	
Amphiboles	
Pyroxenes	
Calcium-rich plagioclase Feldspar (anorthite)	
Olivine	
Calcite	
Halite	
LEAST STABLE	Fastest

where weathering is minimal, alabaster (gypsum) monuments remain intact, as do many other unstable minerals. Table 16.2 shows the relative stabilities of all the common rock-forming minerals. Salt and carbonate minerals are the least stable, iron oxides the most stable.

Physical Weathering

Now that we have surveyed chemical weathering, we can turn to its partner process, physical weathering. We can see the workings of physical weathering most clearly by examining its role in arid regions, where chemical weathering is minimal.

How Do Rocks Break?

Rocks can break for a variety of reasons, including stress along natural zones of weakness and biological and chemical activity.

Natural Zones of Weakness Rocks have natural zones of weakness along which they tend to crack. In sedimentary rocks such as sandstone and shale, these zones are the bedding planes formed by successive layers of solidified sediment. Foliated metamorphic rocks such as slate form parallel cleavage planes, which enable them to be split easily. Granites and other nonfoliated rocks are often referred to as *massive*, which in this case means that they contain no preexisting planes of weakness. Massive rocks tend to crack along regular fractures spaced one to several meters apart,

FIGURE 16.7 Weathered, enlarged joint patterns have developed in two directions in these rocks at Point Lobos State Reserve, California. [Jeff Foott/Getty Images.]

called *joints* (**Figure 16.7**). As we saw in Chapter 8, joints and less regular fractures result from deformation and from cooling and contraction while rocks are still deeply buried in Earth's crust. Through uplift and erosion, the rocks eventually rise to Earth's surface. There, freed from the weight of overlying rock, the fractures open slightly. Once the fractures open a little, both chemical and physical weathering work to widen them.

Activities of Organisms The activities of organisms contribute to physical as well as chemical weathering. Bacteria and algae may invade cracks in rock, producing microfractures. These organisms, both those in cracks and those that may encrust the rock, produce acids, which then promote chemical weathering. In some regions, acid-producing fungi are active in soils, contributing to chemical weathering. Many of us have seen a crack in a rock that has been widened by a tree root (see Figure 6.2). Animals burrowing or moving through cracks can also break rock.

Frost Wedging One of the most efficient mechanisms for widening cracks in rock is **frost wedging**: breakage resulting from the expansion of freezing water. As water freezes, it expands, exerting an outward force strong enough to wedge open a crack and split a rock (**Figure 16.8**). This same process can crack the engine block of a car that is not protected by antifreeze. Frost wedging is most important where water episodically freezes and thaws, as in temperate climates and in mountainous regions.

Exfoliation One form of rock breakage is not directly related to preexisting zones of weakness. **Exfoliation** is a physical weathering process in which large flat or curved sheets of rock are detached from an outcrop. These

FIGURE 16.8 Granite boulder split by frost in the Sierra Mountains, California. [Susan Rayfield/Science Source.]

sheets may look like the layers peeled from a large onion (Figure 16.9). Even though exfoliation is common, no generally accepted explanation for it has yet emerged. Some geologists have suggested that exfoliation results from an uneven distribution of expansion and contraction caused by chemical weathering and temperature changes.

Soils: The Residue of Weathering

On moderate and gentle slopes, plains, and lowlands, where erosion is less intense, a layer of loose, heterogeneous weathered material remains overlying the bedrock. It may include particles of weathered and unweathered parent rock, clay minerals, iron and other metal oxides, and other products of weathering. Geologists use the term **soil** to describe layers of material, initially created by fragmentation of rock during weathering, that experience additions of new materials, losses of original materials, and modification through physical mixing and chemical reactions. Organic matter, called *humus,* is an important component of most of Earth's soils; it consists of the remains and waste products of the many organisms that live in soil. Leaf litter contributes significantly to the soil of forests. In addition, most soils have the ability to support rooted plants. Not all soils support life, however, and soils occur in places, such as Antarctica and Mars, where life is limited or possibly absent altogether.

Soils vary in color, from the brilliant reds and browns of iron-rich soils to the black of soils rich in organic matter. Soils also vary in texture. Some are full of pebbles and sand; others are composed entirely of clay. Soils are easily eroded, so they do not form on very steep slopes or where high altitude or frigid climate prevents the growth of plants that would hold them in place and contribute organic matter.

FIGURE 16.9 Exfoliation on Half Dome, Yosemite National Park, California. [Tony Waltham.]

Soil scientists, agronomists, and engineers, as well as geologists, study the composition and origin of soils, their suitability for agriculture and construction, and their value as a guide to climate conditions in the past.

Soils form at the interface between the climate and plate tectonic systems. They are crucial to life on Earth's continents, and they are one of human society's most valuable natural resources. Soils are the primary reservoir of nutrients for agriculture and the ecological systems that produce renewable natural resources. They filter our water and recycle our wastes, and they provide the necessary substratum for our buildings and infrastructure. In addition, they help regulate the global climate by storing and releasing carbon dioxide. Soils contain twice as much carbon as the atmosphere and three times more than all of the world's vegetation.

Soils as Geosystems

As we have seen, the concept of Earth as a set of interacting geosystems is of great value in understanding geologic processes. Soils, like many other components of the Earth system, can be described as a geosystem with inputs, processes, and outputs (**Figure 16.10**).

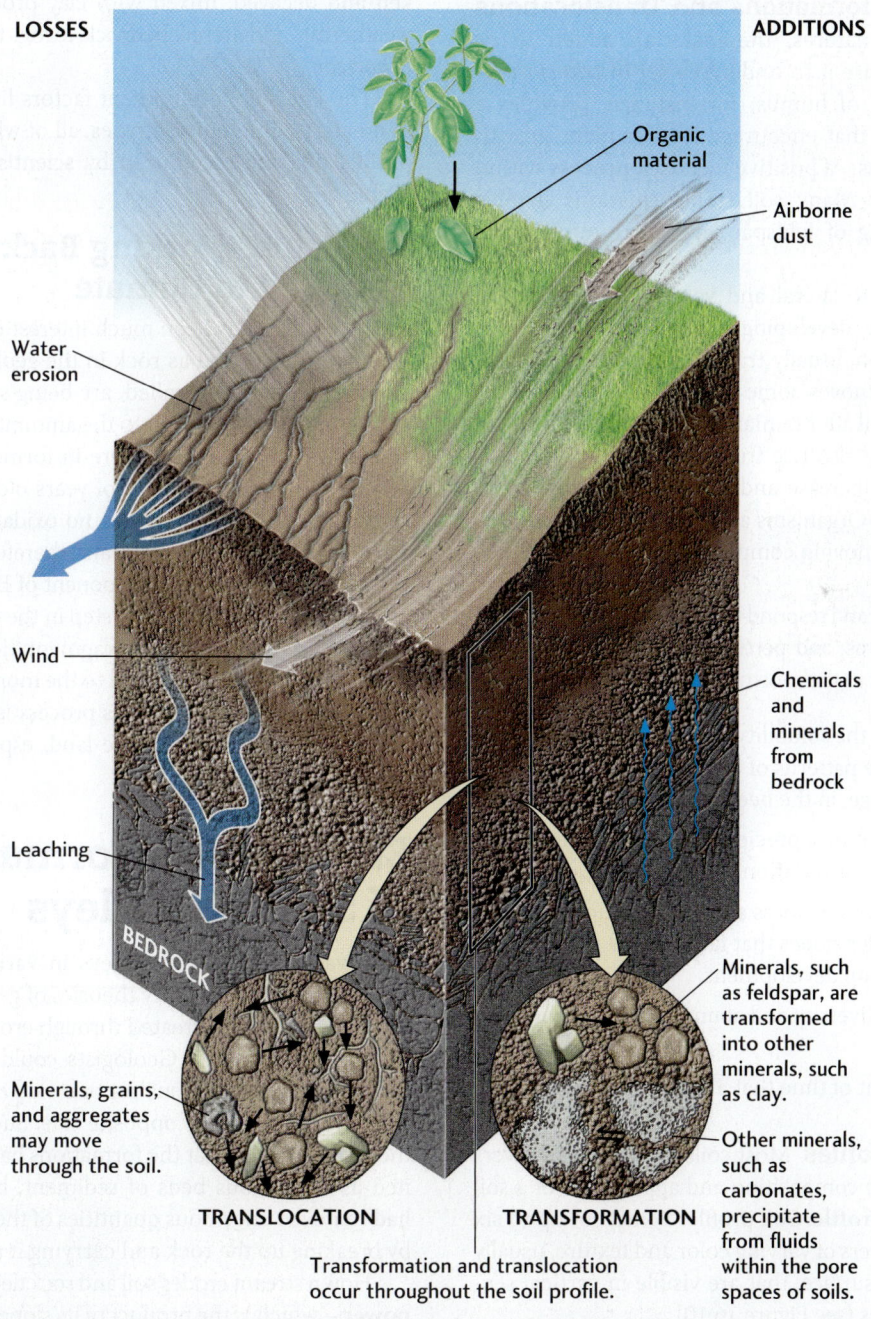

FIGURE 16.10 Soils are geosystems that develop through inputs of new materials, losses of original materials, and modification through physical mixing and chemical reactions. Soil modification processes can be divided into two basic types: translocations and transformations. The distinct soil horizons that make up the soil profile are also visible in this diagram.

Inputs: Weathered Rock, Organisms, and Dust

Soils develop from weathered rock, with additional inputs of organic matter from the biosphere and dust from the atmosphere. As discussed earlier, physical weathering breaks down rock into smaller pieces, and chemical weathering transforms minerals in that rock (such as feldspar) into other minerals (such as clays). Plants and other organisms may colonize the soil, and when they die, their tissues decompose to form humus. The atmosphere also contributes matter to the soil, but this material is predominantly inorganic dust.

Processes: Transformations and Translocations

As soil ages and matures, the materials added to or removed from it cause it to undergo a set of *transformations*. The addition of humus, for instance, provides a source of nutrients that encourage further plant growth and add more humus—a positive feedback process within the soil geosystem. Many soil transformations involve chemical weathering of feldspar and other minerals to form clay minerals.

Translocations are lateral and vertical movements of materials within the developing soil. Water is the main agent of translocation, usually transporting dissolved salts. Water selectively removes some materials as it percolates down through the soil after rainfall in a process called *leaching*. However, it may also rise from below the soil surface when temperatures increase and evaporation draws more water to the surface. Organisms also play an important role in translocation by moving components of the soil as they burrow through it.

Soils are dynamic and respond to changes in climate, interactions with organisms, and perturbations by humans. Five factors are important in their formation and development:

1. *Parent material:* the solubility of minerals, the sizes of grains, and the patterns of fragmentation, such as joints and cleavage, in the bedrock
2. *Climate:* temperatures, precipitation levels, and their seasonal patterns of variation
3. *Topography:* slope steepness and the direction the slope faces: gentler slopes that face toward the Sun promote better soil development
4. *Organisms:* the diversity and abundance of organisms living in the soil
5. *Time:* the amount of time that a soil has to form

Outputs: Soil Profiles Most soils form distinct layers as they develop. The composition and appearance of a soil is known as a **soil profile.** Soil profiles consist of up to six *horizons:* distinct layers of varying color and texture, usually parallel to the land surface, that are visible in vertical sections of exposed soils (see Figure 16.10).

The soil's topmost layer, called the *O-horizon,* is usually thin and consists of loose leaves and organic detritus. Beneath this topmost layer is the *A-horizon,* typically not much more than a meter or two thick and usually the darkest layer because it contains the highest concentration of humus. Next down is the *E-horizon,* which consists mostly of clay and insoluble minerals such as quartz, as soluble minerals will have been leached from this layer. Beneath the E-horizon is the *B-horizon,* in which organic matter is sparse. Soluble minerals and iron oxides accumulate in this layer. The climate influences the specific types of minerals that accumulate in the B-horizon; carbonate minerals and gypsum, for example, are found there in arid climates. The next layer, the *C-horizon,* is slightly altered bedrock, broken and decayed, mixed with clay produced by chemical weathering. Unaltered bedrock forms the lowest level, or *R-horizon.*

The five soil development factors listed above interact to create 12 different soil types, all of which have a distinct profile, that are recognized by scientists who study soils (**Table 16.3**).

Paleosols: Working Backward from Soil to Climate

Recently, there has been much interest in ancient soils that have been preserved as rock in the geologic record. These *paleosols,* as they are called, are being studied as guides to ancient climates and even to the amounts of carbon dioxide and oxygen in the atmosphere in former times. The mineralogy of paleosols billions of years old, for example, provides evidence that there was no oxidation of soils at that early stage of Earth's history, and therefore that oxygen had not yet become a major component of Earth's atmosphere.

Soil formation is just one step in the evolution of a landscape. Weathering and rock fragmentation often destabilize topographic features and lead to the more dramatic changes caused by mass wasting. This process is an important part of the general erosion of the land, especially in hilly and mountainous regions.

Erosion and Formation of Stream Valleys

Observations of stream valleys in various regions led to one of the important early theories of geology: the idea that stream valleys were created through erosion by the streams that flowed in them. Geologists could see that the sedimentary rock formations on one side of a valley matched the formations on the opposite side. Such observations led them to conclude that the formations had once been deposited as continuous beds of sediment, but that the stream had removed enormous quantities of the original formation by breaking up the rock and carrying it away.

How a stream erodes soil and rock depends on its **stream power**—which is the product of its slope and its discharge—balanced by the streambed's ability to resist erosion—which is the product of the volume and the particle size of the

TABLE 16.3 Twelve Recognized Soil Types

SOIL TYPE	DESCRIPTION	MOST IMPORTANT FORMATION FACTORS[a]
Alfisols	Soils of humid and subhumid climates with a subsurface horizon of clay accumulation, not strongly leached, common in forested areas	Climate, organisms
Andisols	Soils that formed in volcanic ash and contain compounds rich in organic matter and aluminum	Parent material
Aridisols	Soils formed in dry climates, low in organic matter and often having subsurface horizons with salt accumulation	Climate
Entisols	Soils lacking subsurface horizons because the parent material accumulated recently or because of constant erosion; common on floodplains, mountains, and badlands (highly eroded, rocky areas)	Time, topography
Gelisols	Weakly weathered soils formed in areas that contain permafrost (frozen soil) within the soil profile	Climate
Histosols	Soils with a thick upper layer very rich in organic matter (0.25%) and containing relatively little mineral material	Topography
Inceptisols	Soils with weakly developed subsurface horizons and little or no subsoil clay accumulation because the soil is young or the climate does not promote rapid weathering	Time, climate
Mollisols	Mineral soils of semiarid and subhumid midlatitude grasslands that have a dark, organic-rich A-horizon and are not strongly leached	Climate, organisms
Oxisols	Very old, highly leached soils with subsurface accumulations of iron and aluminum oxides, commonly found in humid tropical environments	Climate, time
Spodosols	Soils formed in cold, moist climates that have a well-developed B-horizon with accumulation of aluminum and iron oxides, formed under pine vegetation in sandy parent material	Parent material, organisms, climate
Ultisols	Soils with a subsurface horizon of clay accumulation, highly leached (but not as highly as oxisols), commonly found in humid tropical and subtropical climates	Climate, time, organisms
Vertisols	Soils that develop deep, wide cracks when dry (shrink and swell) due to high clay content (0.35%) and are not highly leached	Parent material

[a]All five soil formation factors (climate, organisms, parent material, topography, time) combine to create these soils, but only the most important factors are listed for each soil type.
Data from E. C. Brevik, *Journal of Geoscience Education* 50 (2002): 541.

sediment in the stream channel (**Figure 16.11**). If stream power is high enough to wash away the sediment, resistance to erosion is mainly a function of bedrock hardness.

As it turns out, rates of bedrock erosion increase dramatically as stream power increases. On most days, a flowing stream accomplishes little erosion because discharge, and thus stream power, is low. However, on the rare days when discharge (and thus stream power) is very high, erosion rates can be dramatically high. This relationship illustrates a fundamental characteristic shared by many of Earth's geosystems: Large, rare events often create much more change than small, frequent ones.

Three principal processes erode bedrock in mountainous terrain. The first is abrasion of the bedrock by suspended and saltating sediment particles moving along the bottom and sides of the channel (see Chapter 18). Second, the drag force of the current itself abrades the bedrock as it plucks rock fragments from the channel. Third, at higher elevations, glacial erosion forms valleys that can then be occupied by streams. Determining the relative importance of these three processes in mountainous terrain is one way geologists can distinguish between the influences of climate and of plate tectonic processes on landscape development.

Stream valleys have many names—canyons, gulches, arroyos, gullies—but all have the same general geometry. A vertical cross section through a young mountain stream valley with little or no floodplain has a simple V-shaped profile (Figure 16.11b). A broad, low stream valley with a wide floodplain has a cross section that is more open, but is still distinct from the U-shaped profile of a glacial valley. Regions with different topographies and types of bedrock

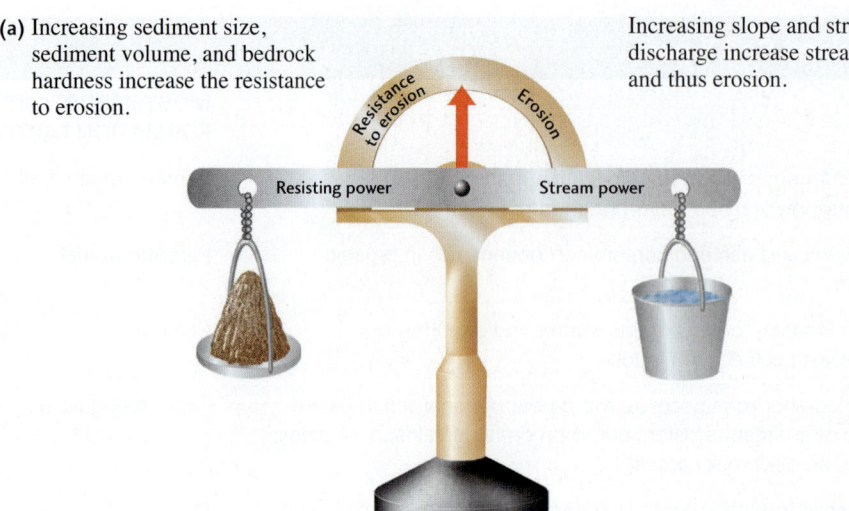

(a) Increasing sediment size, sediment volume, and bedrock hardness increase the resistance to erosion.

Increasing slope and stream discharge increase stream power and thus erosion.

(b) In steep, wet terrain, stream power overcomes resistance to erosion. Sediment particles are transported away, and bedrock hardness becomes the principal factor in resistance to erosion.

(c) Where slopes are gentler, stream discharge is lower, and therefore stream power is lower. Thus, sediment begins to be deposited, armoring the streambed and stopping erosion. At this point, stream power and resistance to erosion are in balance.

(d) Where slopes are much flatter, stream power is so decreased that much sediment is deposited and the streambed builds up and fills the valley with sediment.

FIGURE 16.11 (a) Erosion is controlled by a balance between stream power and resistance to erosion. (b) Yellowstone River, Yellowstone National Park; (c) Snake River, Suicide Point, Idaho; (d) Denali National Park, Alaska. [(a) Karl Weatherly/Getty Images; (b) Dave G. Houser/Getty Images; (c) Dennis Macdonald/Getty Images.]

FIGURE 16.12 Rapidly uplifting mountains form terraces, marking the former position of the river as it cuts through solid bedrock. The terrace in the middle foreground was formed along the Indus River, which cuts through the middle of the Himalaya mountains. [D. W. Burbank.]

produce stream valleys of varying shapes and widths (Figure 16.11b–d). Valleys range from the narrow gorges of erosion-resistant mountain belts to the wide, shallow valleys of easily eroded plains. Between these extremes, the width of a valley generally corresponds to the erosional state of the region. Valleys are somewhat broader in mountains that have begun to be lowered and rounded by erosion and are much broader in low-lying hilly topography. Valleys in young mountain ranges can be very steep and as they cut down through solid rock, they can leave behind terraces marking the former positions of the river (Figure 16.12).

Interactions Between Weathering and Erosion

Weathering and erosion are closely related, interacting processes. Physical weathering and erosion are tied by the processes by which wind, water, and ice work to dislodge particles of rock and move them away from their source. Physical weathering breaks a large mass of rock into smaller pieces, which are more easily eroded and transported.

The steepness of slopes affects both physical and chemical weathering, which in turn affect erosion. Weathering and erosion are more intense on steep slopes, and their action makes slopes gentler. Flows of rainwater are the primary agent of erosion, but wind may blow away the finest particles, and glacial ice can carry away large blocks torn from bedrock.

Chemical weathering rates are low at high elevations, where temperatures are generally low, soil is thin or absent, and vegetation is sparse. Physical weathering is greater at high elevations and in glacial terrains, where ice tears apart rock. We can see that erosional processes are closely related to the sizes of the fragments formed by physical weathering. As weathered material is eroded and transported, its size and shape may change further, and its composition may change as a result of chemical weathering. When transportation stops, deposition of the sediment formed by weathering begins. Figure 16.13 summarizes the factors that influence weathering and erosion.

Mass Wasting

On the morning of June 1, 2005, as residents of Laguna Beach, California, were waking up and enjoying their morning coffee, the hillside broke loose beneath their feet and collapsed. Seven multi-million-dollar homes were destroyed as a large mass of soil and weathered bedrock gave way and slid downhill. Twelve other homes were badly damaged, and hundreds more were evacuated as residents waited anxiously for geologists to evaluate their home sites and determine whether it was safe to return. Some houses completely collapsed; others literally broke in half; and still others were left stranded at the top of the hill, where they jutted out over a large gash formed where the sliding mass of earth broke away (see the chapter opening photo).

This mass wasting event was triggered by very high seasonal rainfall—the second highest ever recorded for that part of Southern California—which saturated the soil and bedrock and created the necessary conditions in an already unstable geologic environment to tip the scales in favor of disaster. Earlier in the same year, high rainfall had triggered similar events, including one that killed 10 people when homes were buried in La Conchita, California. In 2018, a similar set of events repeated itself, this time further north in Montecito, California (Figure 16.14).

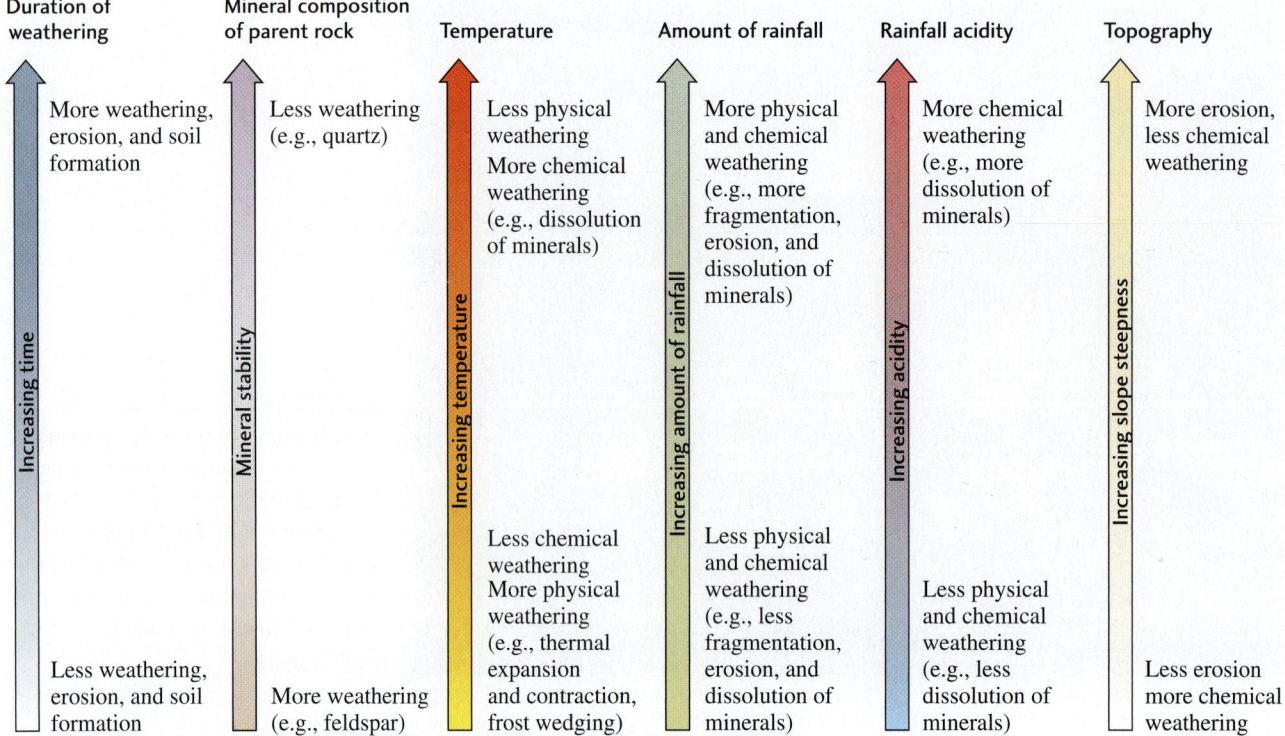

FIGURE 16.13 Summary of the factors that influence weathering and erosion.

FIGURE 16.14 Debris from a mudslide covers a home on January 10, 2018, in Montecito, California. [Justin Sullivan/Getty Images.]

These events in Southern California represent just one of many kinds of downslope movements of masses of soil, rock, mud, or other materials under the force of gravity, known collectively as **mass movements.** These masses are not pulled downslope primarily by the action of an erosional agent, such as wind, flowing water, or glacial ice. Instead, mass movements occur when the force of gravity exceeds the strength of the slope materials. The materials then move down the slope, sometimes very slowly, sometimes as a large, sudden, catastrophic movement. Mass movements can displace small, almost imperceptible amounts of soil down a gentle hillside, or they can be huge landslides that dump tons of earth and rock on valley floors below steep mountain slopes.

Every year, mass movements take a toll of lives and property throughout the world. In late October and early November of 1998, for example, one of the most catastrophic hurricanes of the twentieth century, Hurricane Mitch, dropped torrential rains on Central America, saturating the ground and causing raging floods and landslides. At least 9000 people were killed, and billions of dollars in damage was done as the floods and slides laid waste to once-fertile land and crops of corn, beans, coffee, and peanuts. One of the hardest-hit areas was near the Nicaragua-Honduras border, where a series of landslides and mudflows buried at least 1500 people. Dozens of villages were simply obliterated, engulfed by a sea of mud. The walls of a crater on Casita volcano collapsed and started a series of slides and flows that were described as a moving wall of mud more than 7 m high. Those in the direct path of the avalanche could not escape, and many were buried alive as they tried to outrun the fast-moving mud.

Because mass movements are responsible for so much destruction, we want to be able to predict them, and we certainly want to refrain from provoking them by unwise interference with natural processes. We cannot prevent most natural mass movements, but we can plan construction and land development so as to minimize losses. Mass movements change the landscape by scarring mountainsides as great masses of material fall or slide away from the slopes. The material that moves ends up as tongues or wedges of debris on the valley floor, sometimes piling up and damming a stream running through the valley. Such scars and debris deposits, mapped in the field or from aerial photographs, are clues to past mass movements. By reading these clues, geologists may be able to predict and issue timely warnings about new movements likely to occur in the future.

Mass wasting is influenced by three primary factors (**Table 16.4**):

1. *The nature of the slope materials.* Slopes may be made up of *unconsolidated materials,* which are loose and uncemented, or *consolidated materials,* which are compacted or bound together by mineral cements.

2. *The amount of water in the materials.* This factor depends on how porous the materials are and on how much rain or other water they have been exposed to.

3. *The steepness of the slope.* This factor contributes to the tendency of materials to fall, slide, or flow under various conditions.

All three factors operate in nature, but slope steepness and water content are most strongly influenced by human activity, such as excavation for building and highway construction. All three factors produce the same result: They lower resistance to movement. The force of gravity takes over, and the slope materials begin to fall, slide, or flow.

Slope Materials

Slope materials vary greatly because they are so dependent on the physical properties of the local terrain. Thus, the foliated bedrock of one hillside may be prone to extensive fracturing, whereas another slope only a few hundred meters away may be composed of massive granite. Slopes of unconsolidated material are the least stable of all.

TABLE 16.4 Factors that Influence Mass Movements

NATURE OF SLOPE MATERIAL	WATER CONTENT	STEEPNESS OF SLOPE	STABILITY OF SLOPE
UNCONSOLIDATED			
Loose sand or sandy silt	Dry	Angle of repose	High
	Wet		Moderate
Unconsolidated mixture of sand, silt, soil, and rock fragments	Dry	Moderate	High
	Wet		Low
	Dry	Steep	High
	Wet		Low
CONSOLIDATED			
Rock, jointed and deformed	Dry or wet	Moderate to steep	Moderate
Rock, massive	Dry or wet	Moderate	High
	Dry or wet	Steep	Moderate

Unconsolidated Sand and Silt The behavior of loose, dry sand and silt illustrates how the steepness of slopes influences mass wasting. Children's sandboxes have made nearly everyone familiar with the characteristic slope of a pile of dry sand. The angle between the slope of any pile of sand or silt and the horizontal is the same, whether the pile is a few centimeters or several meters high. For most sands and silts, that angle is about 35°. If you scoop some sand from the base of the pile very slowly and carefully, you can increase the slope angle a little, and it will hold temporarily. If you then jump on the ground near the sand pile, however, the sand will cascade down the side of the pile, which will again assume its original slope of 35°.

The slope angle of the sand pile is its **angle of repose**, the maximum angle at which a slope of loose material will lie without cascading down. A slope that is steeper than the angle of repose is unstable and will tend to collapse to the stable angle of repose. Sand or silt grains will form piles with slopes at and below the angle of repose because friction between the individual grains holds them in place. However, as more and more grains are placed on the pile and the slope steepens, the ability of the frictional forces to prevent sliding decreases, and the pile suddenly collapses. The angle of repose varies with a number of factors, one of which is the size and shape of the particles (**Figure 16.15a**). Larger, flatter, and more angular pieces of loose material

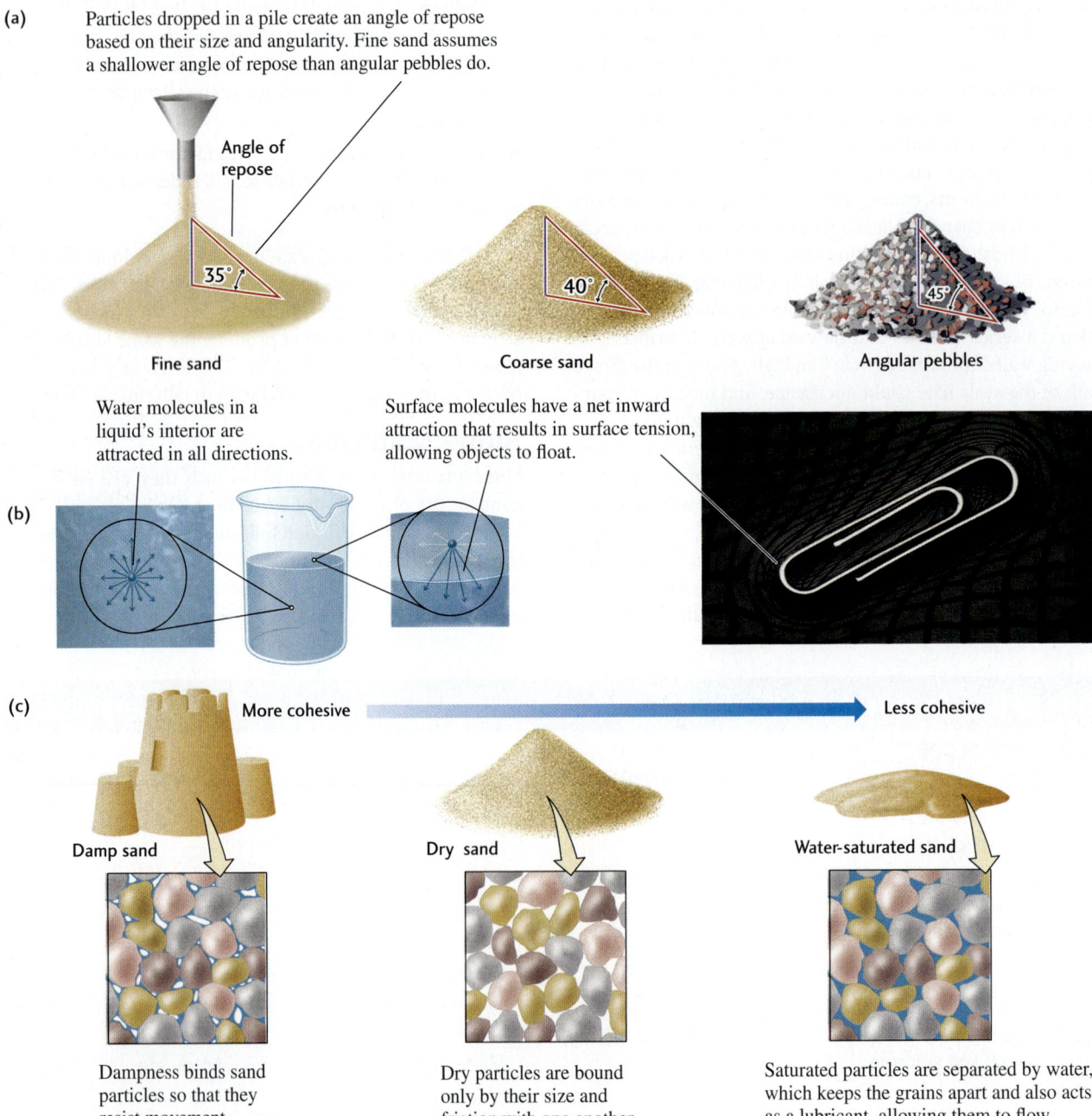

FIGURE 16.15 The angle of repose of a pile of unconsolidated material depends on the shape of the particles and the water content of the pile. [Photo: © 1990 Chip Clark–Fundamental Photographs.]

remain stable on steeper slopes. The angle of repose also varies with the amount of moisture between particles. The angle of repose of wet sand is higher than that of dry sand because the small amount of moisture between the grains tends to bind them together so that they resist movement. The source of this binding tendency is *surface tension:* the attractive force between molecules at a surface (Figure 16.15b). Surface tension is what makes water drops spherical and allows a razor blade or paper clip to float on a smooth water surface. Too much water between particles, on the other hand, separates the particles and allows them to move freely over one another. Saturated sand, in which all the pore space is occupied by water, runs like a fluid and collapses to a flat pancake shape (Figure 16.15c). The surface tension that binds damp sand allows beach sculptors to create elaborate sand castles (**Figure 16.16**), but when the tide comes in and saturates the sand, the structures collapse. Similarly, but at a much larger scale, landslides on hillslopes depend on water abundances in soils. Heavy rainfall may saturate pore spaces on a slope, causing catastrophic ground failure.

Unconsolidated Mixtures Slope materials composed of mixtures of unconsolidated sand, silt, clay, soil, and fragments of rock (often called *debris*) will form slopes with moderate to steep angles (see Table 16.4). The platy shape of clay minerals, the organic content of soils, and the rigidity of rock fragments are all key factors that affect the ability of the material to form slopes with a specific angle.

Consolidated Materials Slopes of consolidated dry materials—such as rock, compacted or cemented sediments, and vegetated soils bound together by plant roots—may be steeper and less regular than slopes consisting of unconsolidated materials. They can become unstable when they are oversteepened or denuded of vegetation. The particles of some consolidated sediments, such as dense clays, are bound together by cohesive forces associated with tightly packed particles. *Cohesion* is an attractive force between particles of a solid material that are close together. The greater the cohesive forces in a material, the greater the resistance to movement.

Water Content

The effect of water on consolidated materials is similar to its effect on unconsolidated materials. Mass movements of consolidated materials can usually be traced to the effects of moisture, often in combination with other factors, such as the loss of vegetation or the steepening of a slope. When the ground becomes saturated with water, the planes of weakness within the solid material are lubricated, the friction between particles is lowered, and the particles or larger aggregates can move past one another more easily, so that the material may start to flow like a fluid. This process is called **liquefaction.**

Slope Steepness

Rock slopes range from the relatively gentle inclines of easily weathered shales and volcanic ash beds to vertical cliffs of massive granite. The stability of rock slopes depends on the weathering and fragmentation of the rock. Shales, for example, tend to weather easily and fragment into small pieces that form a thin layer of loose, angular rock fragments (often called *rubble*) covering the bedrock (**Figure 16.17a**). The resulting angle of repose is similar to that of loose, coarse sand. The weathered rubble gradually builds up beyond the angle of repose and becomes unstable, and eventually some of it slides downhill.

In contrast, limestones and hard, cemented sandstones in arid environments resist erosion and break into large blocks, forming steep, bare bedrock slopes above and gentler slopes covered with broken rock (often called **talus**) below (Figure 16.17b). The bedrock cliffs are fairly stable, except for the occasional mass of rock that falls and rolls down to the rock-covered slope below. Where such limestones or sandstones are interbedded with shale, slopes may be stepped (see Figure 6.17a). As shale slides from under the harder beds, those beds are undercut, become less stable, and eventually fall as large blocks.

FIGURE 16.16 Sand castles hold their shape because they are made of damp sand. The steepness of their slopes is maintained by surface tension in the moisture between grains. [Kelly/Mooney Photography/Getty Images.]

FIGURE 16.17 The stability of a rock slope depends on the weathering and fragmentation pattern of the rock that forms it. (a) This small outcrop is being weathered to form broken blocks of rock known as rubble. (b) Talus accumulates on slopes where large blocks of rock fall or roll downhill to form a cone-shaped pile. [(a) John Grotzinger; (b) Phil Stoffer/USGS.]

The steepness of individual sedimentary beds also has an influence on slope stability. Mass movements are most likely when the dip of the beds close to the surface is parallel to the slope.

Triggers of Mass Movements

When the right combination of materials, moisture, and steepness makes a slope unstable, a mass movement is inevitable. All that is needed is a trigger. Sometimes a landslide, like the one at Laguna Beach, is provoked by a heavy rainstorm. Many mass movements are set off by vibrations, such as those that occur in earthquakes. Others may be precipitated by gradual steepening due to erosion that eventually results in sudden collapse of the slope.

Geologic reports can help to minimize the human cost of mass movements (see the Practicing Geology Exercise), but only if municipal planners and individual home buyers heed those reports and avoid building or buying in unstable areas. The devastating mass movements in Southern California during 2005 were clearly related to the unusually high seasonal rainfall during the winter of 2004–2005. That rainfall, however, was related to El Niño conditions (described in Chapter 12), which Earth scientists now understand to recur on a regular basis.

Similarly, most of the damage in the great Alaska earthquake of March 27, 1964, was caused by the slides it triggered. Mass movements of rock, earth, and snow wreaked havoc in residential areas of Anchorage, and there were major submarine slides along lakeshores and the seacoast. Huge landslides took place along the flat plains below the 30–35-m-high bluffs along the coast. The bluffs were composed of interbedded clays and silts. During the earthquake, the ground shook so hard that unstable, water-saturated sandy layers in the clay were transformed into fluid slurries. Enormous blocks of clay and silt were shaken down from the bluffs and slid along the flat ground with the liquefied sediments, leaving a completely disrupted terrain of jumbled blocks and broken buildings (**Figure 16.18**). Houses and roads were carried along by the slides and destroyed. The whole process took only 5 minutes, beginning about 2 minutes after the first shock of the earthquake. At one locality, three people were killed and 75 homes were destroyed. Studies of slope stability in both California and Alaska, and of the likelihood of repeated high rainfall or earthquakes in those regions, had indicated that both areas were prime candidates for landslides. A geologic report issued more than a decade before the quake had warned of the hazards of development in the part of Alaska that suffered the most damage, but the scenic beauty of the area overwhelmed people's judgment. The same is true of Southern California. In Alaska, people paid the price with their lives. Fortunately, at Laguna Beach, the cost was only the value of people's homes, but even that was steep in an area where the average price of a house was well over a million dollars.

Classification of Mass Movements

Although the popular press often refers to any mass movement as a "landslide," there are many different kinds of mass movements, each with its own characteristics. In this textbook, we use the term *landslide* only in its popular sense, to refer to mass movements in general.

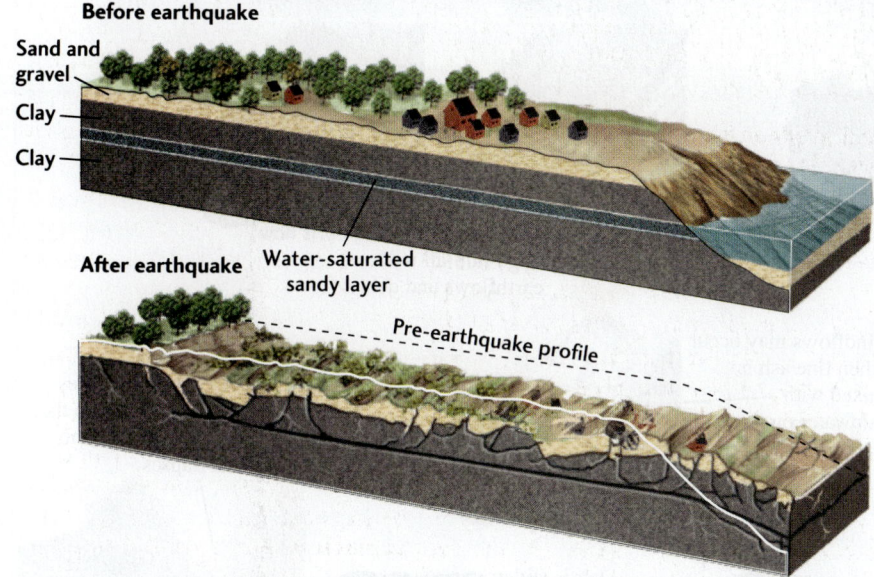

FIGURE 16.18 (a) A landslide triggered by the great Alaska earthquake of 1964. (b) Cross sections of the bluffs at Anchorage, Alaska, before and after the earthquake. [(a) USGS.]

Geologists classify mass movements in accordance with three characteristics, as summarized in **Figure 16.19**:

1. The nature of the moving material (for example, whether it is rock or unconsolidated material)
2. The velocity of the movement (from a few centimeters per year to many kilometers per hour)
3. The nature of the movement: whether it is sliding (the bulk of the material moves more or less as a unit) or flowing (the material moves as if it were a fluid)

The nature and velocity of the movement are greatly influenced by the water or air content of the moving material.

Some movements have characteristics that are intermediate between sliding and flowing. Most of the mass may move by sliding, for example, but parts of it along the base may move as a fluid. A movement is called a *flow* if that is the main type of motion. It is not always easy to tell the exact nature of a mass movement, however; the only evidence may be the debris deposited after the movement is over.

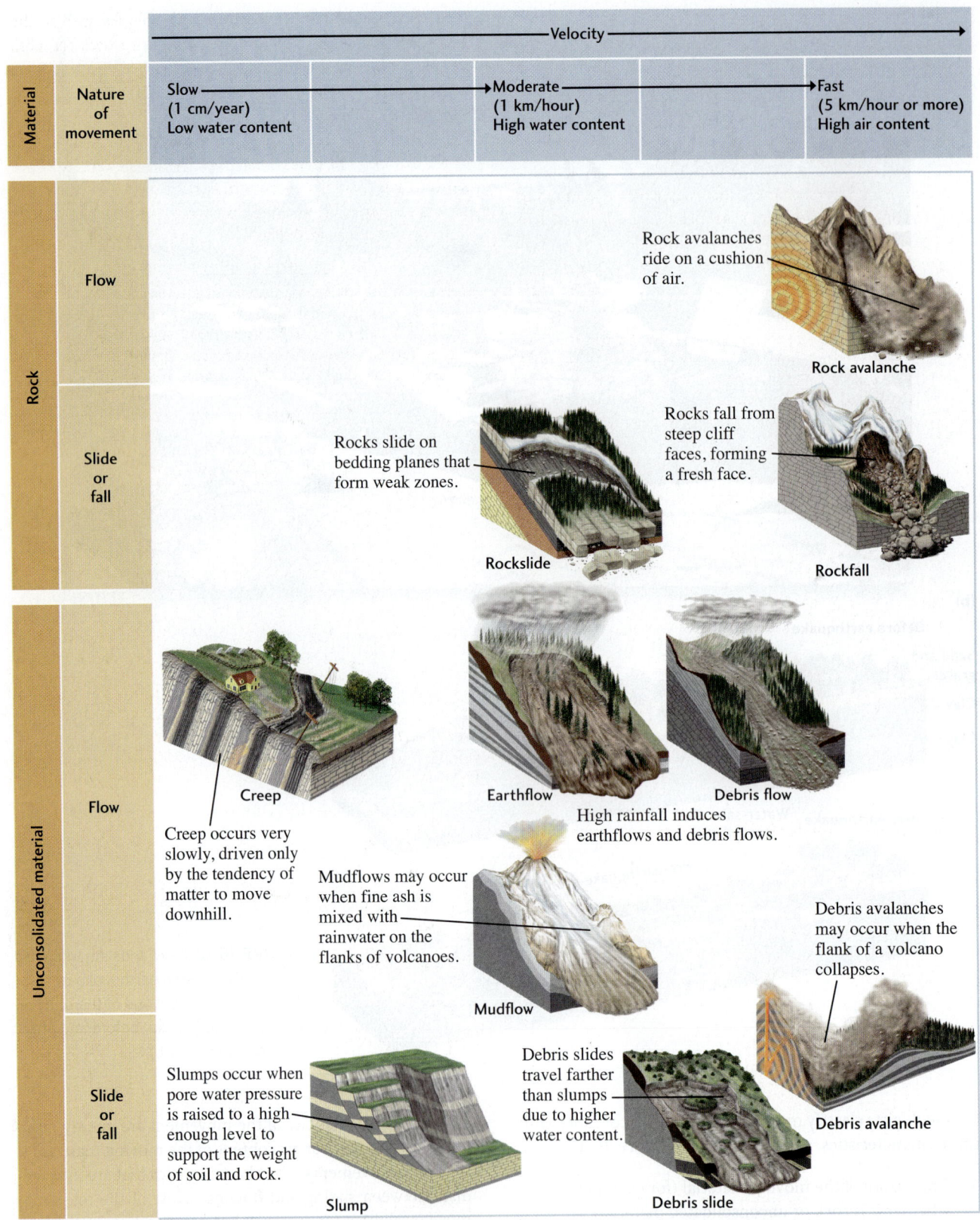

FIGURE 16.19 Mass movements are classified according to the nature of the moving material, the velocity of the movement, and the nature of the movement.

Mass Movements of Rock

Rock movements include rockfalls, rockslides, and rock avalanches. These movements may involve small blocks or larger masses of bedrock. During a *rockfall*, newly detached individual blocks of rock plummet suddenly in free fall from a cliff or steep mountainside (**Figure 16.20**). Weathering weakens bedrock along joints until the slightest pressure—often exerted by frost wedging—is enough to trigger a rockfall. The velocities of rockfalls are the fastest of all rock movements, but the travel distances are the shortest, generally only meters to hundreds of meters. Evidence for the origin of rockfalls is clear in the blocks seen in the accumulation of talus at the foot of a steep bedrock cliff, which can be matched with rock outcrops on the cliff. Talus accumulates slowly, building up into blocky slopes along the base of a cliff over long periods.

In *rockslides*, rocks do not fall freely, but rather slide down a slope. Although these movements are fast, they are slower than rockfalls because masses of bedrock slide more or less as a unit, often along downward-sloping bedding or joint planes (**Figure 16.21**).

The Palisades sill is composed of columnar basalt that fractures along columns leading to collapse to form a rock wall — the iconic western margin of the Hudson River in the New York City area. Areas where basalt columns collapsed recently show a light tone, which exposes fresher rock.

FIGURE 16.20 (a) Rockfall off the Hudson Palisades Cliffs at State Line Lookout, Alpine New Jersey. (b) In a rockfall, individual blocks plummet in free fall from a cliff or steep mountainside. [(a) Bill Menke.]

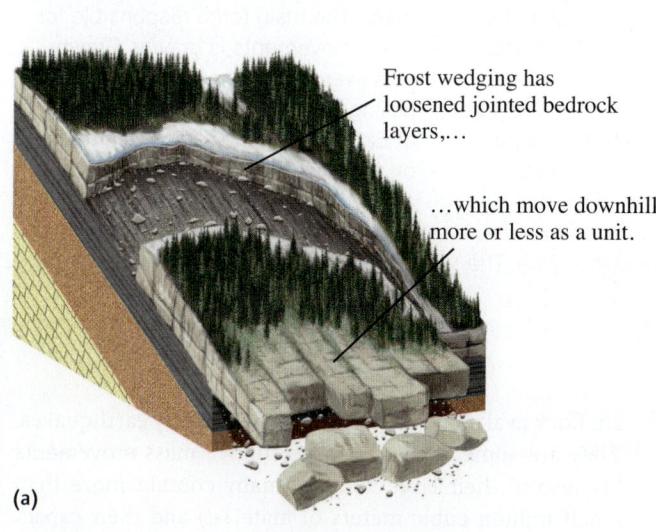

Frost wedging has loosened jointed bedrock layers,...

...which move downhill more or less as a unit.

FIGURE 16.21 (a) In a rockslide, large masses of bedrock move more or less as a unit in a fast, downward slide. (b) Rockslide in Northeastern Spain. [(b) CABALAR/EPA/Newscom.]

PRACTICING GEOLOGY EXERCISE

What Makes a Slope Too Unstable to Build On?

On hills with low slope, the ground is stable because the perpendicular component of gravity is large and tends to keep rocks and soils firmly in place.

On hills with moderate slope, the ground may become unstable because the component of gravity parallel to the hill slope is increased. This tends to push rocks and soils downslope and failure may occur.

On hills with steep slope, the ground is unstable because the parallel component of gravity is greatly increased and so the chance of failure is also increased.

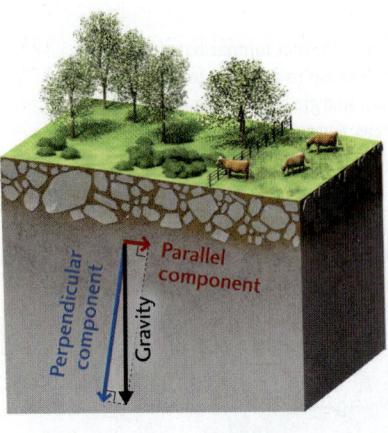
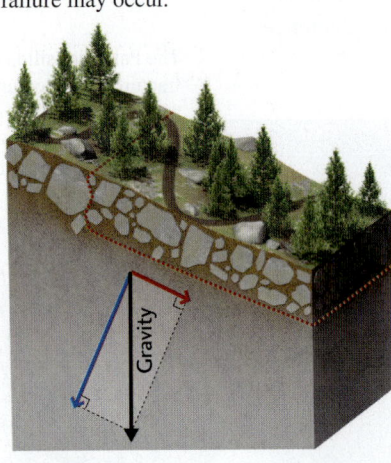
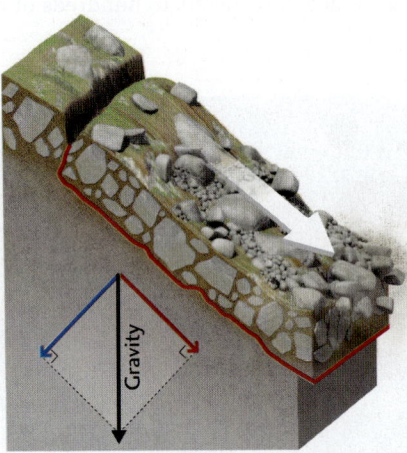

Forces acting on a block of soil or rock at different slopes.

How can the destruction of homes and other buildings by landslides be avoided? Landslides are most likely in areas where steep topography coincides with other key factors, such as episodic heavy rainfall or earthquakes. Understanding the risk associated with buying or building a home in such areas begins with an assessment of the terrain and its likelihood to undergo mass wasting. Geologists play an important role in making such assessments and in advising potential homeowners and local planners about what kinds of real estate are more likely than others to experience a landslide.

Intuition tells us that if we build a structure on a slope that is too steep, it will slide downhill. Recall the pile of sand we described in the text: As the pile gets steeper, less of it stays in place, and at some point, no matter how much sand we put on the pile, it will just keep slip-sliding away. The same thing is true of masses of soil and rock. They, too, will slide downhill if the slope gets too steep. The important question is, how steep is too steep? We use the three primary factors involved in mass movements to determine which slopes are too steep to build on.

The most important factor is the steepness of the slope. Other things being equal, a structure on a steeper slope will slide sooner than a structure of the same size on a gentler slope. The second most important factor is the nature of the slope materials. The better these materials stick together, the more stable the slope will be. The third factor is the presence of water in the slope materials. During heavy rainfall, soil and rock absorb water, their cohesiveness is reduced, and a landslide may be triggered, as some homeowners in Southern California discovered in 2005 (see the chapter-opening photo and Figure 16.14).

The accompanying diagram illustrates the forces acting on a mass of soil or rock—a potential slide mass. The main force responsible for mass movements is gravity. Gravity acts everywhere on Earth's surface, pulling everything toward the center of Earth. When the slide mass rests on a horizontal surface, gravity exerts a downward force on it, holding it firmly in place. On a slope, however, the force of gravity is directed across

Rock avalanches differ from rockslides in their much greater velocities and travel distances (**Figure 16.22**). They are composed of large masses of rocky material that break up into smaller pieces when they fall or slide. The pieces then flow farther downhill at velocities of tens to hundreds of kilometers per hour, riding on a cushion of air. Rock avalanches are typically triggered by earthquakes. They are some of the most destructive mass movements because of their large volume (many contain more than a half million cubic meters of material) and their capacity to transport materials for thousands of meters at high velocities.

the base of the slide mass at an angle because gravity pulls toward the center of Earth, no matter what the position of the slide mass is. In this case, the force of gravity can be divided into two components: a force perpendicular to the base of the slide mass and a force parallel to its base.

What do these forces tell us about the likelihood of a landslide? The parallel component of gravity creates *shear stress* parallel to the base of the slide mass, which pulls it in the downslope direction. The perpendicular component of gravity, known as *shear strength,* acts to resist the mass's downward slide. Friction at the base of the slide mass and cohesion between the particles of the slide mass contribute to shear strength.

Sliding tends to occur on steeper slopes because shear stress increases and shear strength decreases. When shear stress becomes greater than shear strength, the mass will slide downslope. Thus, landslides are most likely where shear stress is high (on steeper slopes) and shear strength is low (as on a slope saturated by high rainfall).

A simple equation known as the *safety factor*, F_s, can be used to predict where and when mass movement will occur:

$$F_s = \frac{\text{shear strength}}{\text{shear stress}}$$

If the safety factor for a slope is less than 1, then mass movement is expected to occur there.

We can consult Tables A and B for values of shear strength and shear stress to determine which combinations of slope steepness and slope material would provide safe building sites.

Table A

Slope	Shear Stress
5°	1
20°	5
30°	25

Table B

Material	Shear Strength
Loose soil	3
Slate	10
Granite	50

Let's calculate the safety factor for a building site on slate on a moderate slope of 20°.

$$F_s = \frac{\text{shear strength for sale}}{\text{shear stress for 20° slope}}$$
$$= 10/5 = 2$$

The slope is expected to be stable, but not by much. In many municipalities, this slope would be considered so marginally stable that it might not be permitted for building without meeting extremely expensive engineering standards.

BONUS PROBLEM: Fill in the blanks in Table C for the remaining combinations of slope and material (e.g., starting with a 5° slope on loose soil). Which slopes would be stable enough to build on?

Table C

Loose	Safety Factor (F_s)		
	Soil	Slate	Granite
5°	___	___	___
20°	___	___	___
30°	___	___	___

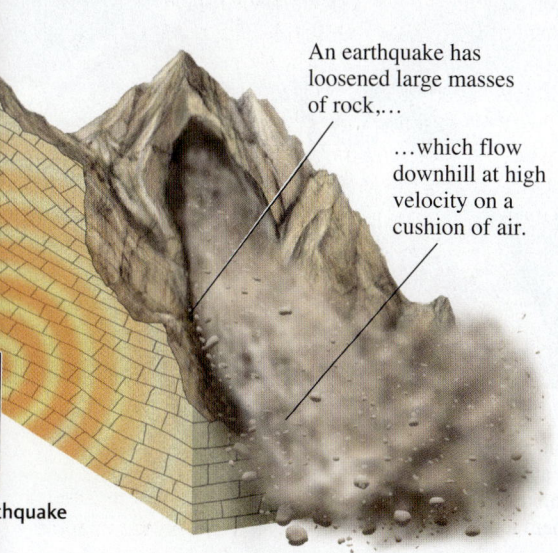

An earthquake has loosened large masses of rock,...

...which flow downhill at high velocity on a cushion of air.

(b)

FIGURE 16.22 (a) In a rock avalanche, large masses of broken rocky material flow, rather than slide, downslope at high velocity. (b) West Salt Creek rock avalanche occurred on the evening of May 25, 2014 near Collbran, Colorado, along the north side of the Grand Mesa, about 30 miles (48 km) east of Grand Junction. It was the largest landslide in Colorado's history. [(b) Mesa County Sheriff's Office.]

Most rock mass movements occur in high mountainous regions; they are rare in low hilly areas. Rock masses tend to move where weathering fragments rocks already predisposed to breakage by deformation features such as faults and joints, relatively weak bedding planes, or foliation. In many such regions, extensive talus accumulations have been built by infrequent but large-scale rockfalls and rockslides.

Mass Movements of Unconsolidated Material

Mass movements of unconsolidated material include various mixtures of sand, silt, clay, soil, and fragmented bedrock, as well as trees and shrubs and materials of human construction, from fences to cars and houses. Most mass movements of unconsolidated materials are slower than most rock movements, largely because of the lower slope angles at which these materials become unstable. Although some unconsolidated materials move as coherent units, many flow like very viscous fluids. (Viscosity, as you will recall from Chapter 4, is a measure of a fluid's resistance to flow.)

The slowest type of unconsolidated mass movement is **creep**: the gradual downhill movement of soil or other debris (**Figure 16.23**). Rates of creep range from about 1 to 10 mm/year, depending on the soil type, the climate, the steepness of the slope, and the density of the vegetation.

The movement is a very slow deformation of the soil, with the upper layers of soil moving down the slope faster than the lower layers. Such slow movements may cause trees, telephone poles, and fences to lean or move slightly downslope. The weight of the masses of soil creeping downhill can break poorly supported retaining walls and crack the walls and foundations of buildings. In icy regions where the deeper layers of soil are permanently frozen, a type of creep called *solifluction* occurs when water in the surface layers of soil alternately freezes and thaws, causing the soil to ooze downhill, carrying broken rocks and other debris with it.

Earthflows and debris flows are fluid mass movements that occur when rainfall soaks and loosens permeable material overlying a layer of less permeable rock. They usually travel faster than creep, as much as a few kilometers per hour, primarily because the moving materials are saturated with water and thus have little resistance to flow. An *earthflow* is a fluid mass movement of relatively fine-grained materials, such as soils, weathered shales, and clay (**Figure 16.24**). A *debris flow* is a fluid mass movement of rock fragments supported by a muddy matrix (**Figure 16.25**). Debris flows are made up mostly of material coarser than sand and tend to move more rapidly than earthflows. The slide at Laguna Beach, California, described above, was classified as a debris flow. In some cases, debris flows may reach velocities of 100 km/hour.

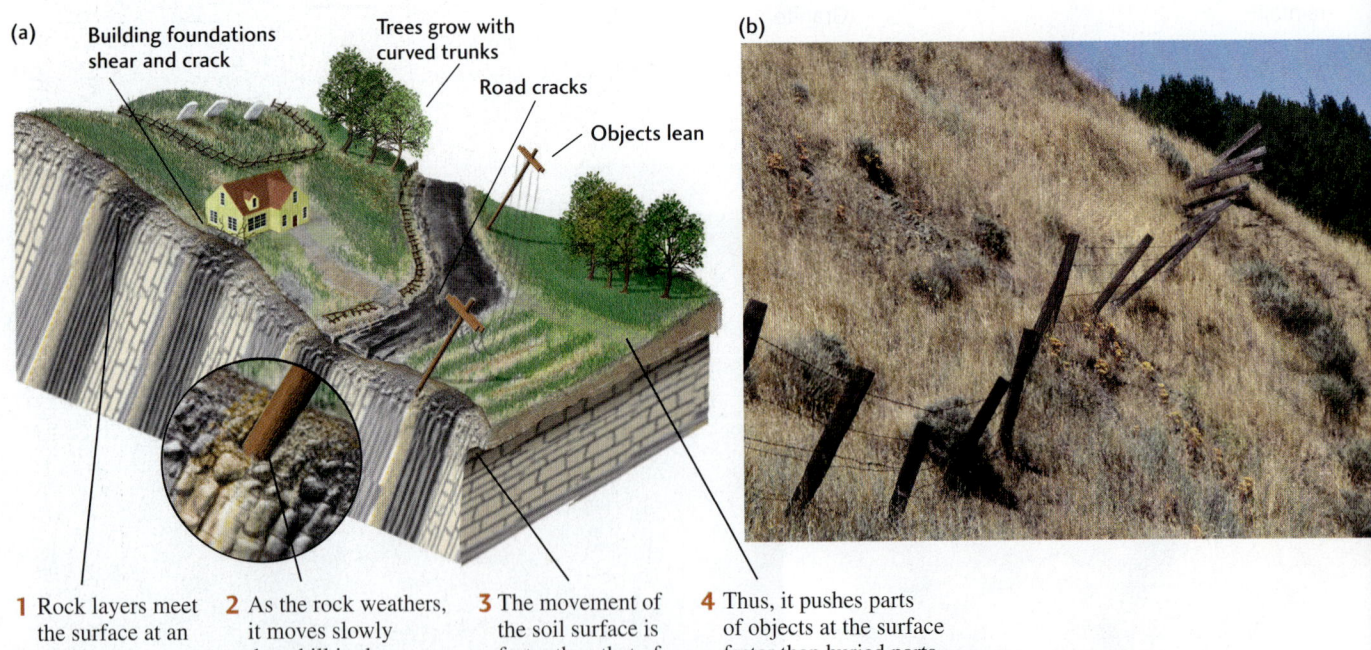

FIGURE 16.23 (a) Creep is the downhill movement of soil or other debris at a rate of about 1 to 10 mm/year. (b) A fence offset by creep in Marin County, California. [(b) Travis Amos.]

Classification of Mass Movements 479

FIGURE 16.24 (a) An earthflow is a fluid movement of relatively fine-grained material that may travel as fast as a few kilometers per hour. (b) Earthflow in the Buller Valley on the South Island of New Zealand. [(b) G. R. Dick Roberts/Science Source.]

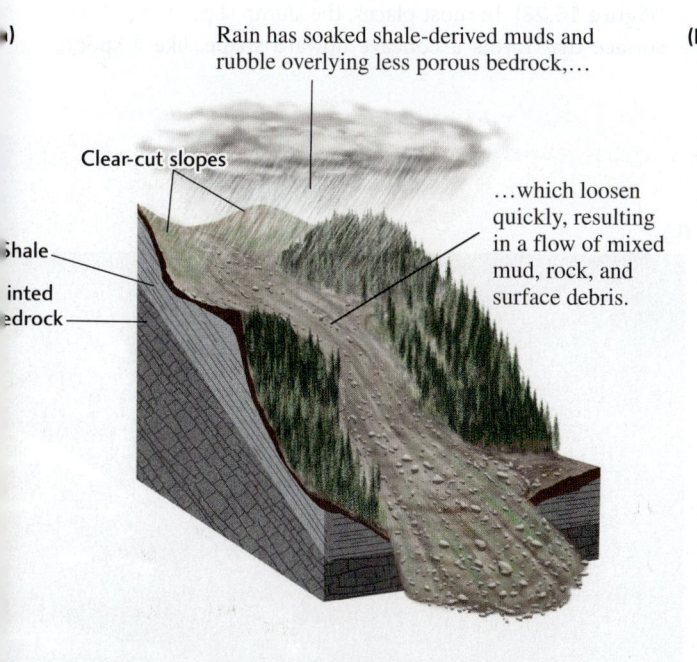

FIGURE 16.25 (a) A debris flow contains material that is coarser than sand and travels at rates from a few kilometers per hour to many tens of kilometers per hour. (b) Debris flow in Bear Canyon, Tucson, Arizona. [(b) J. P. Cook, Arizona Geological Survey.]

Mudflows are flowing masses made up mostly of material finer than sand, along with some rock debris, and containing large amounts of water (Figure 16.26). The mud offers little resistance to flow because of its high water content and thus tends to move faster than earth or debris. Many mudflows move at several kilometers per hour. Most common in hilly and semiarid regions, mudflows occur when fine-grained material becomes saturated. Mudflows of wet pyroclastic material, called *lahars,* may be triggered by volcanic eruptions, as when a lava flow melts snow and ice (see Chapter 5). Similarly, mudflows may start when dry, cracked mud on a slope is subjected to infrequent, sometimes prolonged, rains. If the mud keeps absorbing water as the rain continues, its physical properties change: Its internal friction decreases, and the mass becomes much less resistant to movement. The slope, which is stable when dry, becomes unstable, and any disturbance, such as an earthquake, triggers movement of waterlogged masses of mud. Mudflows may travel down tributary valleys on upper slopes and merge on the main valley floor. Where mudflows exit from confined upper valleys into broader, lower valley slopes and flats, they may splay out to cover large areas with wet debris. Mudflows can carry huge boulders, trees, and even houses.

Debris avalanches (Figure 16.27) are fast downhill movements of soil and rock that usually occur in humid mountainous regions. Their speed results from the combination of high water content and steep slopes. Water-saturated debris may move as fast as 70 km/hour, a speed comparable to that of water flowing down a moderately steep slope. A debris avalanche carries with it everything in its path.

In 1962, a debris avalanche on Nevado de Huascarán, Peru, one of the highest mountains in the Andes, traveled almost 15 km in about 7 minutes, engulfing most of eight towns and killing 3500 people. Eight years later, on May 31, 1970, an earthquake toppled a large mass of glacial ice at the top of the same mountain. As the ice broke up, it mixed with the debris of the high slopes and became an ice-debris avalanche. The avalanche picked up additional debris as it raced downhill, increasing its speed to an almost unbelievable 280 km/hour. Up to 50 million cubic meters of muddy debris roared down into the valleys, killing 18,000 people and wiping out scores of villages (Figure 16.27b). On May 30, 1990, an earthquake shook another mountainous area in northern Peru, in the same active subduction zone, again setting off mudflows and debris avalanches. It was the day before a memorial ceremony scheduled to commemorate the disaster that had occurred 20 years earlier. In regions close to convergent plate boundaries, where uplift and volcanism build up unstable slopes and earthquakes are frequent, there can be no doubt about the necessity of learning how to predict both earthquakes and the dangerous mass movements that follow.

In a **slump,** a mass of unconsolidated material slides slowly downslope as a unit, leaving a scar at its source (Figure 16.28). In most places, the slump slips along a basal surface that forms a concave-upward shape, like a spoon.

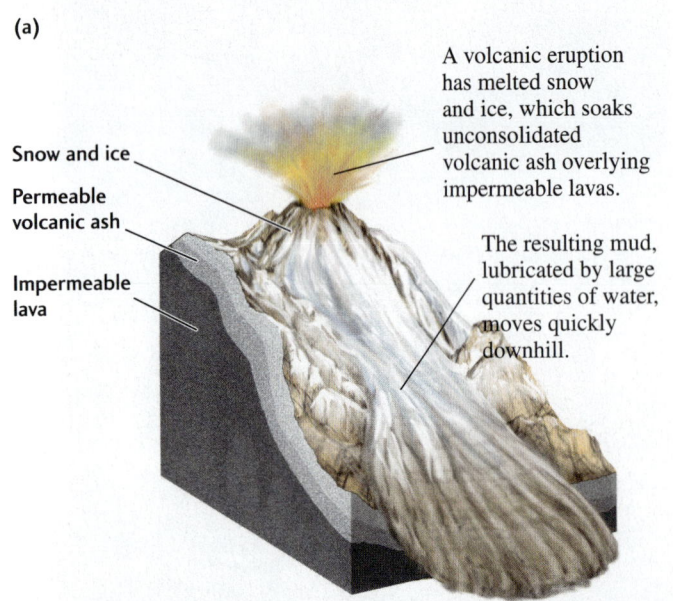

FIGURE 16.26 (a) Mudflows tend to move faster than earthflows or debris flows because they contain large quantities of water. (b) A mudflow from a clear-cut hillside spills across Highway 6 near in Lewis County, Washington, after heavy rains.
[(b) Washington State Department of Transportation/Seattle Times/MCT/Newscom.]

Classification of Mass Movements 481

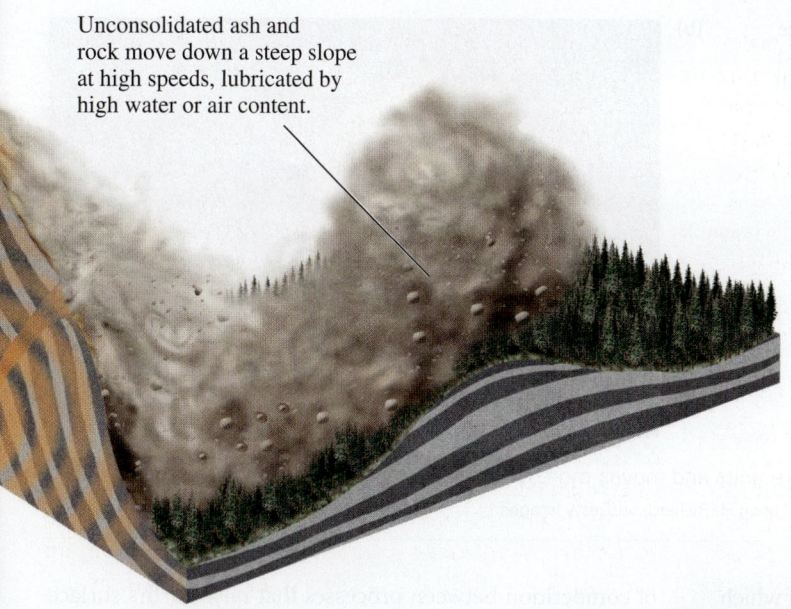

Unconsolidated ash and rock move down a steep slope at high speeds, lubricated by high water or air content.

FIGURE 16.27 (a) A debris avalanche is the fastest type of unconsolidated flow because of its high water content and movement down steep slopes. (b, c) In 1970, an earthquake-induced debris avalanche on Mount Huascarán, Peru, buried the towns of Yungay and Ranrahirca, killing some 18,000 people. The avalanche traveled 17 km at a speed of up to 280 km/hour and is estimated to have consisted of up to 50 million cubic meters of water, mud, and rocks. [(b,c) Lloyd S. Cluff/Steinbrugge Collection, NISEE-PEER, University of California, Berkeley.]

(c)

of Yungay and Ranrahirca before an earthquake-induced avalanche on Mount Huascarán, Peru, buried these towns.

Aftermath of the avalanche.

(b)

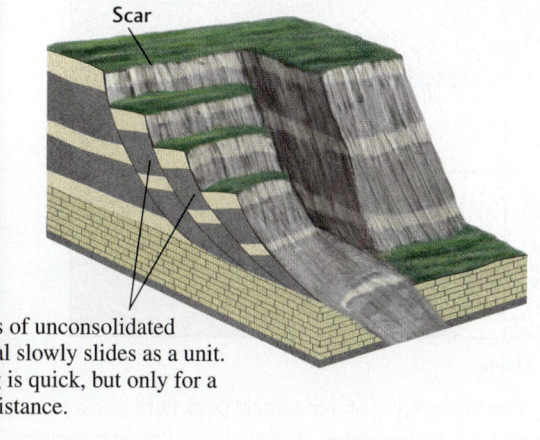

Scar

s of unconsolidated al slowly slides as a unit. g is quick, but only for a listance.

E 16.28 (a) A slump is a slow slide of unconsolidated al that travels as a unit. (b) Soil slump, Northern nia. [(b) Marli Bryant Miller.]

Scar

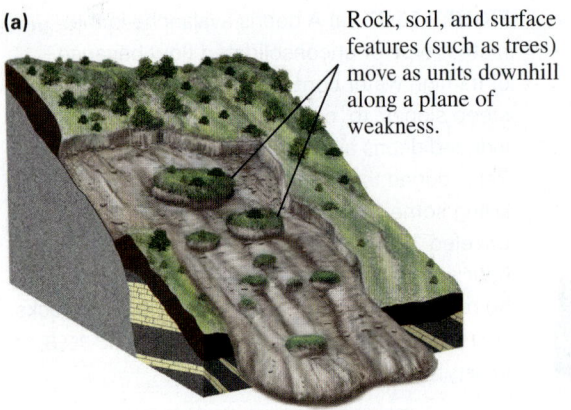 

FIGURE 16.29 (a) A debris slide travels as one or more units and moves more quickly than a slump. (b) A massive tree slide is seen along the Peak to Peak highway in Colorado. [(b) Helen H. Richardson/Getty Images.]

Faster than slumps are *debris slides* (Figure 16.29), in which rock material and soil move largely as one or more units along planes of weakness, such as a waterlogged clay zone either within or at the base of the debris. During the slide, some of the debris may behave as a chaotic, jumbled flow. Such a slide may become predominantly a flow as it moves rapidly downhill and most of the material mixes as if it were a fluid.

Geomorphology and Landscape Development

The term **geomorphology** refers both to the shape of a landscape and to the branch of Earth science concerned with that shape and how it develops. Understanding the slow processes by which landscapes develop helps us not only to manage our land resources, but also to appreciate the linkages between the plate tectonic and climate systems. In the most basic sense, landscapes can be viewed as the result of competition between processes that raise Earth's surface and those that lower it. Processes driven by plate tectonics raise the elevation of Earth's crust to form mountain ranges and high plateaus. The uplifted rocks are exposed to weathering and erosion—processes driven by climate.

We begin our study of geomorphology here with the elementary observations of any landscape that are obvious as one examines Earth's surface: the height and ruggedness, or roughness, of mountains and lowlands. **Topography** is the general configuration of varying heights that gives shape to Earth's surface (see Figure 1.8). We compare the heights of landscape features with sea level: the average height of the ocean surface throughout the world. We then express the vertical distance above or below sea level as **elevation.** A topographic map shows the distribution of elevations in an area. This distribution is most often represented by *contours*, lines that connect points of equal elevation (Figure 16.30). The more closely spaced the contours, the steeper the slope.

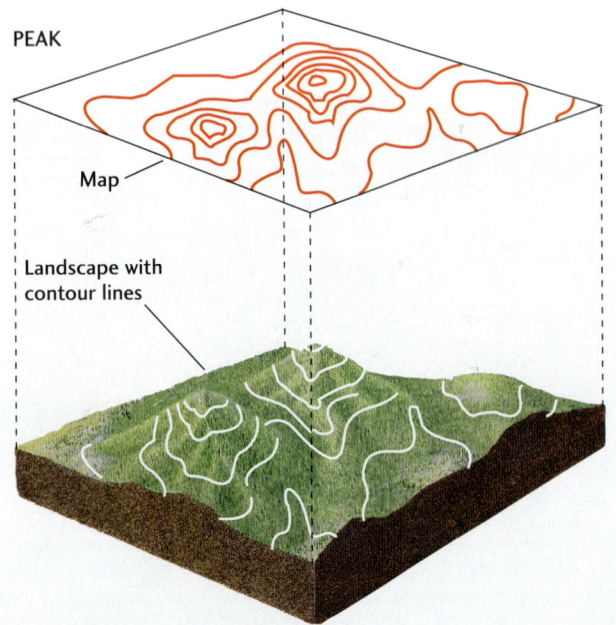 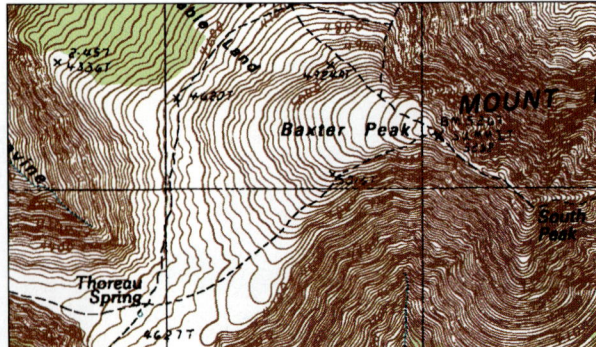

FIGURE 16.30 The topography of a mountain peak (left) and a stream valley (right) can be accurately depicted on a flat topographic map by contours, lines that connect points of equal elevation. The more closely spaced the contours, the steeper the slope. [Information from A. Maltman, *Geological Maps: An Introduction*. New York: Van Nostrand Reinhold, 1990, p. 17. Topographic maps from USGS/DRG.]

Geomorphology and Landscape Development

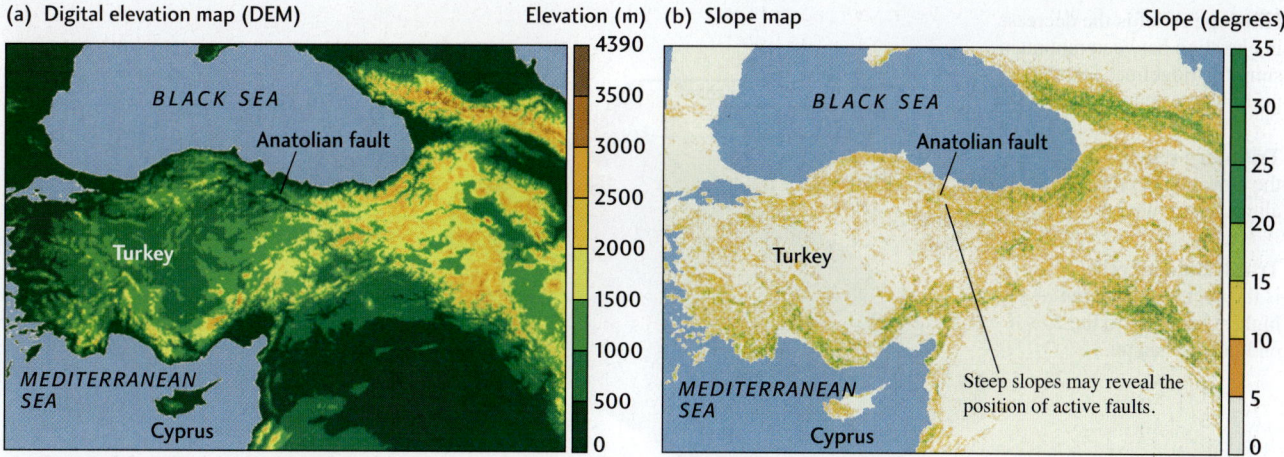

FIGURE 16.31 Topographic maps for Turkey and the adjacent region. (a) A digital elevation model, or DEM. The elevation values are depicted digitally, with each pixel representing one elevation value. (b) To produce this slope map, elevation values from the DEM were used to calculate the slopes between adjacent pixels. The slopes are represented by angles measured in degrees from the horizontal. A slope map is useful for identifying places where changes in topography are particularly abrupt, such as along mountain fronts or active fault scarps. [Information from Marin Clark.]

Centuries ago, geologists learned to survey topography and construct maps to plot and record geologic information. Although land surveying based on age-old methods is still used for some purposes, modern mapmakers rely on satellite photographs, radar imaging, airborne laser range finders, and other technologies that enable them to discern elevation and other topographic properties (**Figure 16.31**).

One of the properties of topography is **relief:** the difference between the highest and lowest elevations in a particular area (**Figure 16.32**). As this definition implies, relief varies with the scale of the area over which it is measured. In studies of geomorphology, it is useful to define three fundamental components of relief: hillslope relief (the decrease in elevation between mountain summits or ridgelines and the point at which stream channels begin), tributary relief (the decrease in elevation along tributary stream channels from their beginning to the main, or *trunk*, stream with which they merge), and trunk channel relief (the decrease in elevation from the highest tributary to the end of the trunk stream channel).

To estimate relief in an area of interest from contours on a topographic map, we subtract the elevation of the lowest contour, usually at the bottom of a stream valley, from that of the highest contour at the top of the highest hill or mountain. Relief is a measure of the roughness of a terrain: The greater the relief, the more rugged the topography. Mount Everest in the Himalaya—the highest mountain in the world, with an elevation of 8850 m—is located in an area of extremely

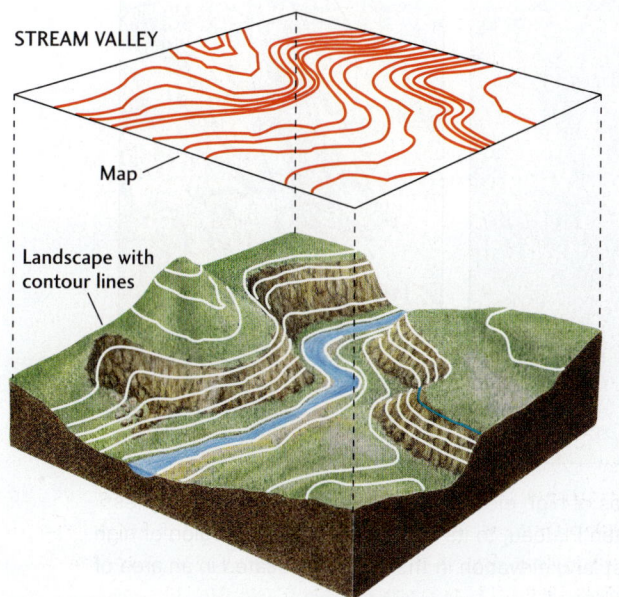

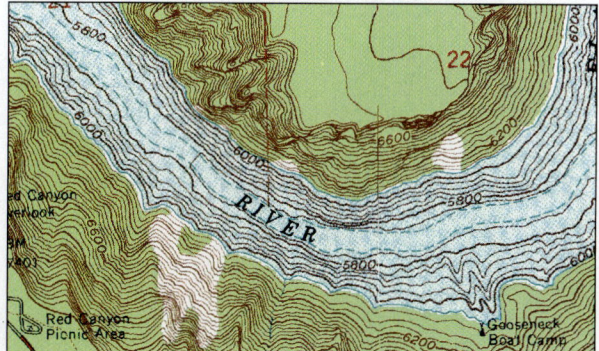

Flaming Gorge, Wyoming

FIGURE 16.30 (*Continued*)

Hillslope relief is the decrease in elevation between mountain summits/ridgelines and the point where channels begin.

Tributary relief is the elevation decrease along the tributaries.

Trunk channel relief is the decrease from the highest tributary to the end of the channel.

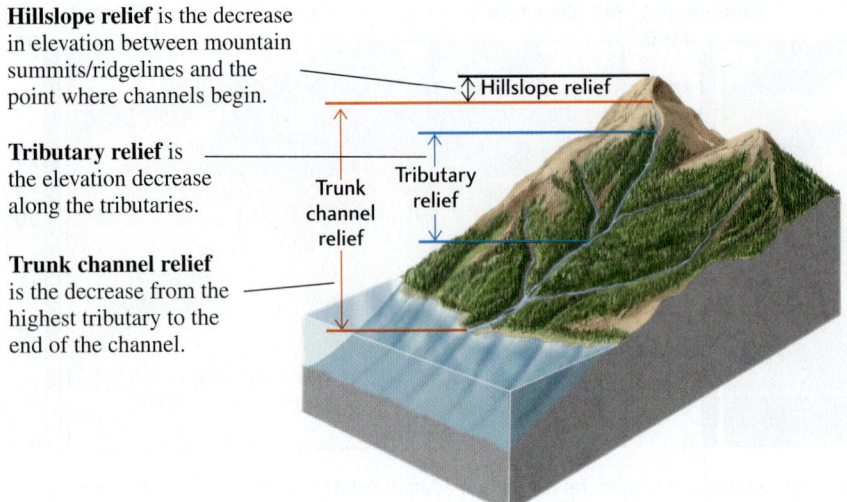

FIGURE 16.32 Relief is the difference between the highest and lowest elevations in a particular area.

high relief (Figure 16.33a). In general, most regions of high elevation are also areas of high relief, and most areas of low elevation are areas of low relief. There are exceptions, however. For example, the Dead Sea, between Israel and Jordan, has the lowest land elevation in the world at 392 m below sea level, but it is flanked by impressive mountains that provide significant relief in this small region (Figure 16.33b). Other regions, such as the Tibetan Plateau, lie at high elevations but have relatively low relief (see Figure 16.33a).

Types of Landforms

Streams, glaciers, and wind leave their marks on Earth's surface in a variety of **landforms:** rugged mountain slopes, broad valleys, floodplains, dunes, and many others. The scale of landforms ranges from regional to very local. At the largest scale (tens of thousands of kilometers), mountain belts form topographic walls along the boundaries of lithospheric plates. At the smallest scale (meters), the topography of an individual outcrop may be shaped by differential weathering due the differences in hardness of the rocks that compose the outcrop. This section focuses mostly on the regional-scale features that define the overall topography of Earth's surface.

Mountains and Hills We have used the word **mountain** many times in this textbook, yet we can define it no more precisely than to say that a mountain is a large mass of rock that projects well above its surroundings. Most mountains are found with others in ranges, where peaks of various heights

(a) Tibetan Plateau

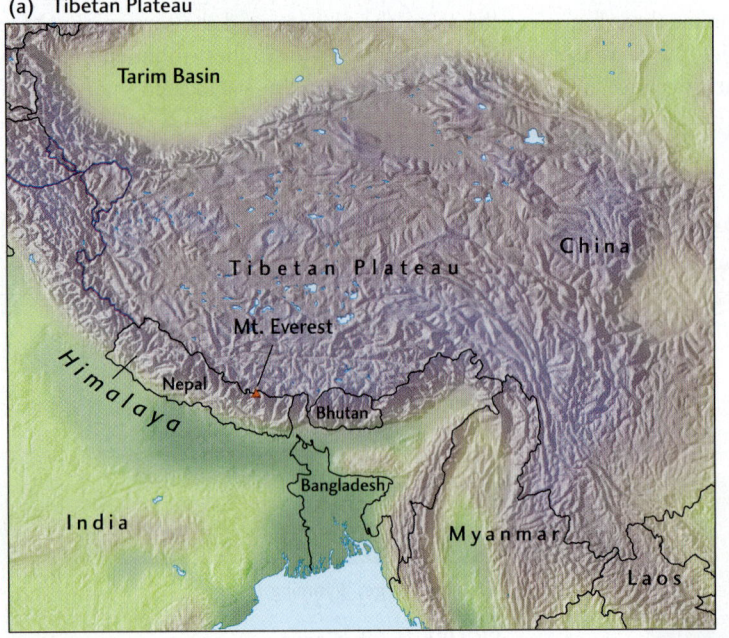

(b) Dead Sea

FIGURE 16.33 Areas of high relief are usually, but not always, areas of high elevation. (a) Mount Everest, the highest mountain in the world, is located in an area of high relief. The Tibetan Plateau, to its north, however, is a region of high elevation but relatively low relief. (b) The Dead Sea, with the lowest land elevation in the world, is located in an area of relatively high relief. [Information from Marin Clark and Nathan Niemi.]

are easier to distinguish than distinct separate mountains (**Figure 16.34**). Mountains that rise as single peaks above the surrounding lowlands are usually isolated volcanoes or erosional remnants of former mountain ranges.

We distinguish between mountains and hills only by size and custom. Landforms that would be called mountains in regions of overall lower elevation are called hills in regions of high elevation. In general, however, landforms more than several hundred meters above their surroundings are called mountains.

Mountains are direct or indirect manifestations of plate tectonic activity. The more recent the activity, the more likely the mountains are to be high. The Himalaya, the highest mountains in the world, are also among the youngest. The steepness of the slopes in mountainous and hilly areas generally correlates with elevation and relief. The steepest slopes are usually found on high mountains in areas of high relief. The slopes of mountains in areas of lower elevation and relief are less steep and rugged. As we will see later in this chapter, the relief of a mountain range depends greatly on how much the bedrock has been incised by glaciers and streams relative to the amount of tectonic uplift.

FIGURE 16.34 Most mountains are found in ranges, not as individual peaks. In this glacially sculpted terrain in southern Argentina, all the peaks are sharp arêtes. [Renato Granieri/Alamy.]

Plateaus A **plateau** is a large, broad, flat area of appreciable elevation above the neighboring terrain. Most plateaus have elevations of less than 3000 m, but the Altiplano of Bolivia lies at an elevation of 3600 m, and the extraordinarily high Tibetan Plateau, which extends over an area of 1000 km by 5000 km (well over half the size of the United States), has an average elevation of almost 5000 m (see Figure 16.33a). Plateaus form where tectonic activity produces regional uplift.

Smaller plateau-like features may be called *tablelands*. In the western United States, a small, flat, elevated landform with steep slopes on all sides is called a *mesa* (from the Spanish word for "table"). Mesas result from differential weathering of bedrock of varying hardness (see Figure 16.6).

Ridges and Valleys In young mountain belts, during the early stages of tectonic folding and uplift, the upward folds (anticlines) form ridges and the downward folds (synclines) form valleys (**Figure 16.35**). As weathering and erosion begin to predominate and gullies and valleys bite deeper into the

FIGURE 16.35 Valley and ridge topography formed on folded sedimentary rock in the Zagros Mountains of Iran. The deformation is so recent (Pliocene) that erosion has not yet significantly modified the original geologic structure of anticlines (ridges) and synclines (valleys). [Image provided by the USGS EROS Data Center Satellite Systems Branch.]

underlying geologic structures, the topography may become inverted, so that anticlines form valleys and synclines form ridges. This happens where the rocks—typically sedimentary rocks such as limestones, sandstones, and shales—exert strong control on the topography by their variable resistance to erosion. If the rocks beneath an anticline are easily erodible, as shales are, the core of the anticline may be eroded to form an anticlinal valley (**Figure 16.36**). In a region that has been eroded for many millions of years, a pattern of linear anticlines and synclines produces a series of ridges and valleys such as those of the Valley and Ridge Province of the Appalachian Mountains (**Figure 16.37**).

Interactions Between Climate, Tectonics, and Topography Control Landscapes

Broadly speaking, the interaction of Earth's internal and external heat engines controls the development of landscapes. Earth's internal heat engine drives plate tectonic processes, which elevate mountain belts and give rise to volcanoes. Earth's external heat engine, powered by the Sun, drives climate, and thus controls the processes at Earth's surface that wear away mountains and fill basins with sediment. Solar radiation powers the atmospheric circulation that produces Earth's climates, including its different temperature and precipitation regimes. Thus, landscapes are controlled by the interaction of the global tectonic and climate systems (**Figure 16.38**).

Many of the forces of weathering and erosion operate at different rates at different elevations. Thus, climate, which changes with elevation, modulates weathering and erosion, and therefore also modulates the uplift of mountain ranges.

This chapter describes some of the effects of climate on weathering, erosion, and mass wasting. Climate influences rates of freezing and thawing and of expansion and contraction of rock due to heating and cooling. Climate also affects the rate at which water dissolves minerals. Rainfall and temperature—the principal components of climate—affect weathering and erosion through infiltration and runoff, streamflow, and the formation of glaciers, all of which help to break up rock and mineral particles and carry them downslope.

High elevation and relief enhance the fragmentation and mechanical breakup of rock, partly by promoting freezing and thawing. At high elevations, where the climate is cool, mountain glaciers scour bedrock and carve out deep valleys. Rainfall lubricates rock on mountain slopes, which moves downhill quickly in landslides and other mass movements, exposing fresh rock to attack by weathering. Streams run faster in mountains than in lowlands and therefore erode and transport sediment more rapidly. Chemical weathering plays an important role in the erosion of high mountains, but the mechanical breakup of rocks is so rapid that most of the debris appears to be almost unweathered. The products of chemical weathering—dissolved materials and clay minerals—are carried down from steep mountain slopes as soon as they form. The intense erosion that occurs at high elevations produces a topography of steep slopes; deep, narrow stream valleys and narrow floodplains and drainage divides (see Figure 16.11b).

TIME 1
Harder, erosion-resistant rocks lie over softer, more erodible layers. Ridges are over anticlines and streams flow in valleys formed by synclines. Tributary streams on anticline slopes flow faster and with more power than valley streams. They erode the slopes faster than main streams erode the valleys.

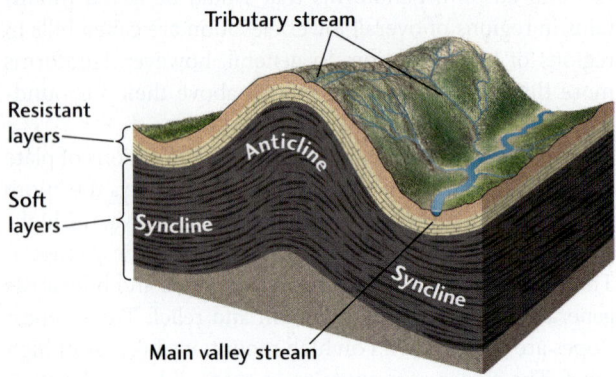

TIME 2
Tributaries over the synclines cut through resistant rock layers and start to quickly carve the softer underlying rock into steep valleys over the anticlines.

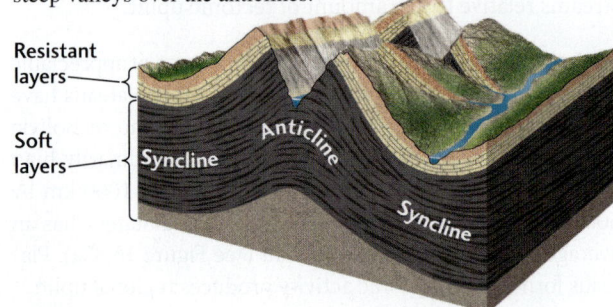

TIME 3
As the process continues, valleys form over the anticlines and ridges capped by resistant strata are left over the synclines.

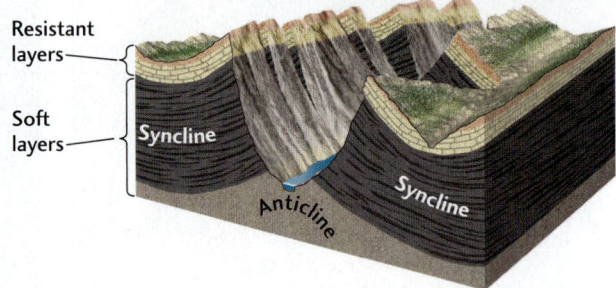

FIGURE 16.36 Stages in the development of ridges and valleys in folded sedimentary rock. In early stages (time 1), ridges are formed by anticlines and valleys by synclines. In later stages (times 2 and 3), the anticlines may be breached. Ridges held up by caps of resistant rock remain as erosion forms valleys in less resistant rock.

Geomorphology and Landscape Development 487

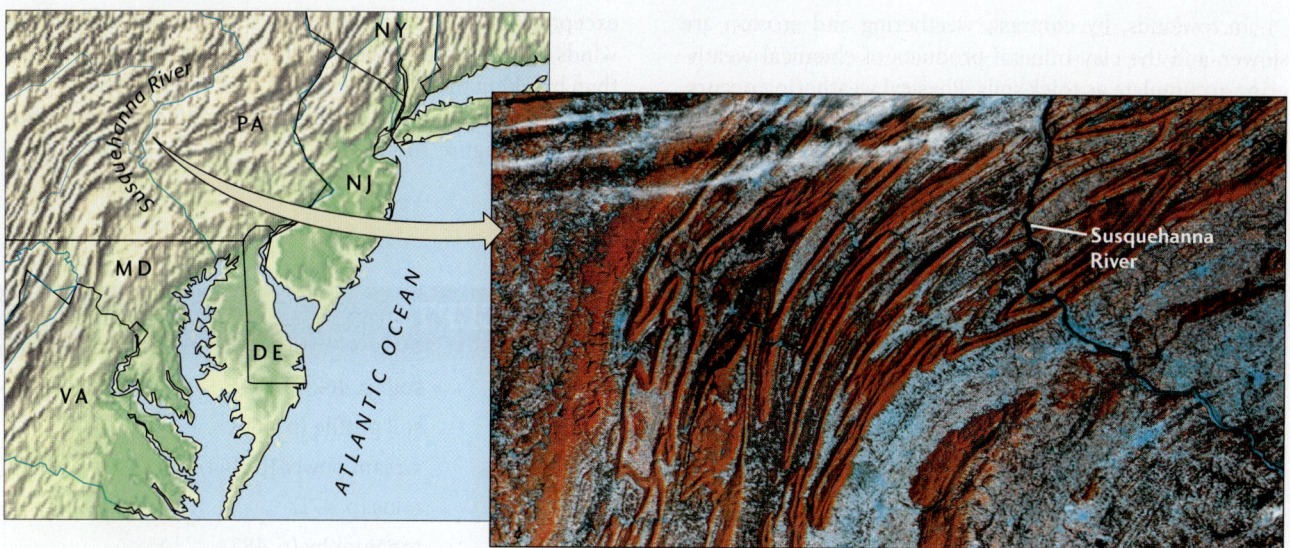

FIGURE 16.37 The Valley and Ridge Province of the Appalachian Mountains has the structurally controlled topography of linear anticlines and synclines exposed to millions of years of erosion. The prominent ridges, shown in reddish orange, are composed of erosion-resistant sedimentary rock. [Courtesy MDA Information Systems LLC.]

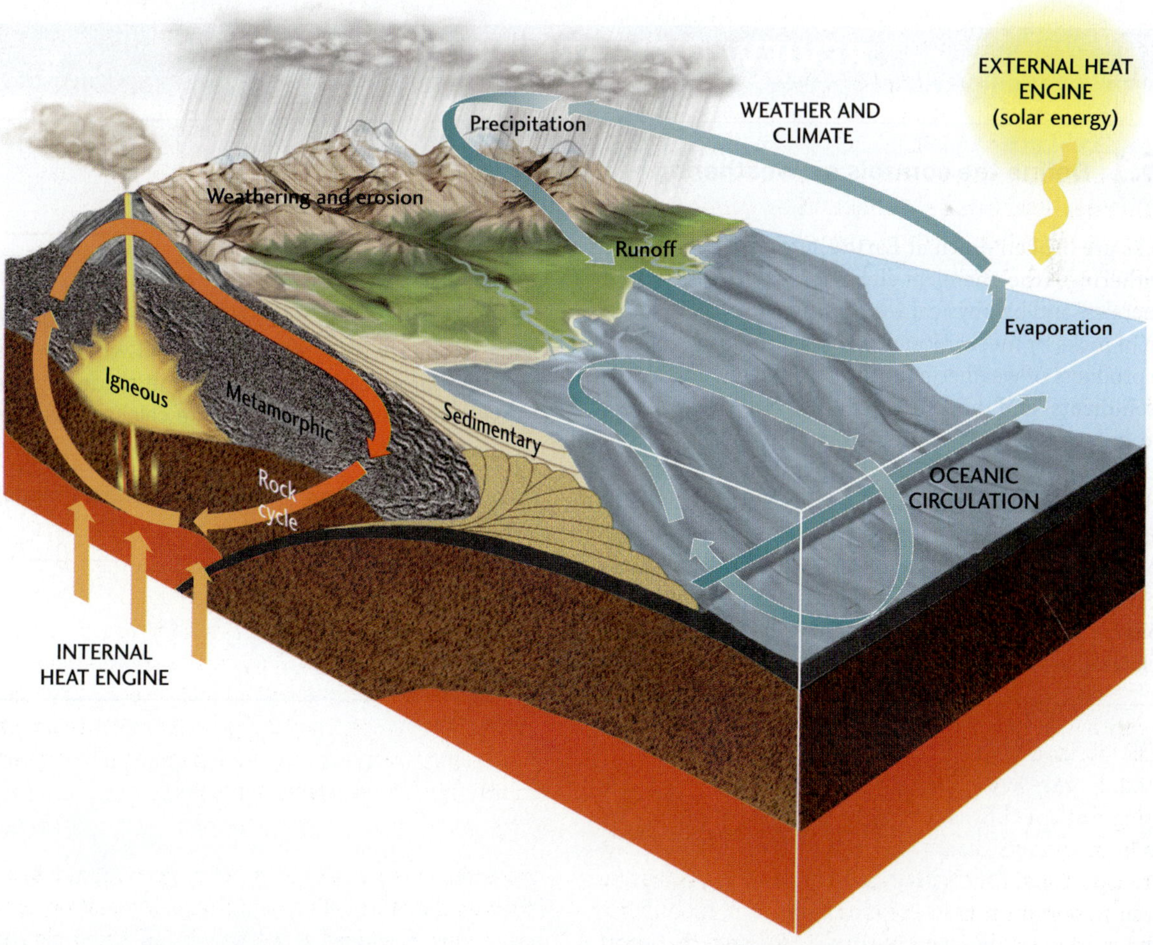

FIGURE 16.38 Landscape development is controlled by interactions between the plate tectonic system and the climate system.

In lowlands, by contrast, weathering and erosion are slower, and the clay mineral products of chemical weathering accumulate as thick soils. Physical weathering occurs, but its effects are small compared with those of chemical weathering. Most streams run over broad floodplains and do little mechanical cutting of bedrock. Glaciers are absent, except in polar regions. Even in lowland deserts, strong winds merely abrade rock fragments and outcrops rather than breaking them up. A lowland thus tends to have a gentle topography with rounded slopes, rolling hills, and flat plains (see Figure 16.11d).

KEY TERMS AND CONCEPTS

angle of repose (p. 470)
chemical stability (p. 460)
creep (p. 478)
elevation (p. 482)
erosion (p. 454)
exfoliation (p. 461)
frost wedging (p. 461)
geomorphology (p. 482)

landform (p. 484)
liquefaction (p. 471)
mass movement (p. 469)
mass wasting (p. 454)
mountain (p. 484)
plateau (p. 485)
relief (p. 483)
slump (p. 480)

soil (p. 462)
soil profile (p. 464)
stream power (p. 464)
talus (p. 471)
topography (p. 482)
weathering (p. 454)

REVIEW OF LEARNING OBJECTIVES

16.1 Define the controls on weathering.

Rocks are broken down at Earth's surface by chemical weathering—the chemical alteration or dissolution of minerals—and by physical weathering—the fragmentation of rocks by mechanical processes. Erosion dislodges the products of weathering, which are the raw materials of sediments, and moves them away from their source. The properties of the parent rock affect weathering because different minerals weather at different rates and have differing susceptibilities to fracturing. Climate strongly affects weathering: Warmth and heavy rainfall speed weathering; cold and dryness slow it down. The presence of soil accelerates weathering by providing moisture and acids secreted by organisms. The longer a rock weathers, the more completely it breaks down.

Study Assignment: Table 16.1

Exercise: Rank the following rocks in the order of the rapidity with which they would weather in a warm, humid climate: a sandstone made of pure quartz, a limestone made of pure calcite, a granite, an evaporite deposit of halite.

Thought Question: How does abundant rainfall affect weathering?

16.2 Describe the processes of chemical weathering.

The weathering of feldspar, the most abundant silicate mineral, serves as an example of the processes that weather most silicate minerals. In the presence of water, feldspar undergoes hydrolysis to form kaolinite. Carbon dioxide (CO_2) dissolved in water promotes chemical weathering by reacting with the water to form carbonic acid (H_2CO_3). The slightly acidic water dissolves away potassium ions and silica, leaving kaolinite. Iron (Fe), which is found in ferrous form in many silicate minerals, weathers by oxidation, producing ferric iron oxides. These processes operate at varying rates, depending on the chemical stability of the minerals involved under various weathering conditions.

Study Assignment: Figure 16.4

Exercise: What rock-forming minerals found in igneous rocks weather to clay minerals?

Thought Question: How does carbon dioxide in the atmosphere react with water to form acid, and then react with igneous minerals to form alteration minerals?

16.3 Describe the processes of physical weathering.

Physical weathering breaks rocks into fragments along preexisting zones of weakness or along joints and other fractures in massive rock. Physical weathering is promoted by frost wedging and by burrowing and tunneling by animals and tree roots, all of which expand cracks. Microorganisms contribute to both physical and chemical weathering. Patterns of breakage such as exfoliation probably result from interactions between chemical weathering and temperature changes.

Study Assignment: Figure 16.8

Exercise: Assume that a granite with grains about 4 mm across and a rectangular system of joints spaced about 0.5 to 1 m apart is weathering in a cold climate at Earth's surface. What size would you ordinarily expect the largest weathered particle to be?

Thought Question: What is the role of water in frost wedging?

16.4 Summarize the types of soil and what processes they record.

Soil is a mixture of rock particles, clay minerals, and other products of weathering, as well as humus. It develops through inputs of new materials, losses of original materials, and modification through physical mixing and chemical reactions. The five key factors that affect soil development are parent material, climate, topography, organisms, and time.

Study Assignment: Figure 16.10

Exercise: Summarize the inputs, outputs, and key processes that control soil development.

Thought Question: Assuming that life is absent on Mars, how might soils look different than on Earth?

16.5 Describe how erosion creates stream valleys.

Three principal processes erode bedrock in mountainous terrain. The first is abrasion of the bedrock by suspended and saltating sediment particles moving along the bottom and sides of the channel. Second, the drag force of the current itself abrades the bedrock as it plucks rock fragments from the channel. Third, at higher elevations, glacial erosion forms valleys that can then be occupied by streams.

Study Assignment: Figure 16.11

Exercise: Summarize the three processes that erode bedrock in mountainous terrain.

Thought Question: How do slope and discharge affect stream power?

16.6 Evaluate the mechanisms that cause mass movements.

The three factors that have the greatest bearing on the predisposition of material to move down a slope are the nature of the slope material, the water content of the material, and the steepness of the slope. Slopes made up of unconsolidated material become unstable when they are steeper than the angle of repose, the maximum slope angle that the material will assume without cascading downslope. Slopes made up of consolidated material may also become unstable when they are steepened or denuded of vegetation. Water absorbed by slope material contributes to instability by lowering internal friction and by lubricating planes of weakness in the material. Mass movements may be triggered by earthquakes, heavy rainfall, or gradual steepening of a slope due to erosion.

Study Assignments: Table 16.4, Figure 16.15

Exercise: Summarize the role of water in controlling the angle of repose.

Thought Question: How does steepness of a hillslope influence its stability?

16.7 Distinguish the different kinds of mass wasting processes.

Mass movements are slides, flows, or falls of large masses of material downslope in response to the force of gravity. The movements may be imperceptibly slow or too fast for a human to outrun. The masses may consist of consolidated material, including rock and compacted or cemented sediments; or unconsolidated material. Mass movements of rock include rockfalls, rockslides, and rock avalanches. Mass movements of unconsolidated material include creep, slumps, debris slides, debris avalanches, earthflows, mudflows, and debris flows.

Study Assignment: Figure 16.19

Exercise: Distinguish the relative velocities of a debris flow versus a debris avalanche.

Thought Question: What kind(s) of mass movements would you expect on a steep hillside with a thick layer of soil overlying unconsolidated sands and muds after a prolonged period of heavy rain?

16.8 Summarize the processes that control landscape development.

A landscape is described in terms of topography, which includes elevation, the vertical distance above or below sea level, and relief, the difference between the highest and the lowest elevations in a region. A landscape comprises the varied landforms produced through erosion and sedimentation by streams, glaciers, mass wasting, and the wind. The most common landforms are mountains and hills, plateaus, and structurally controlled cliffs and ridges—all of which are produced by tectonic activity modified by erosion. Plate tectonic processes lift up mountains and expose rock. Erosion carves rock into valleys and slopes. Climate, in turn, affects rates of weathering and erosion. Variations in climate and bedrock type strongly modify landscape development, making desert and glacial landscapes very different.

Study Assignments: Figures 16.36 and 16.38

Exercise: What changes in the landscape of the Rocky Mountains of Colorado might result from a change in the present temperate but somewhat dry climate to a warmer climate with a large increase in rainfall?

Thought Question: What is relief, and how is it related to elevation?

VISUAL LITERACY EXERCISE

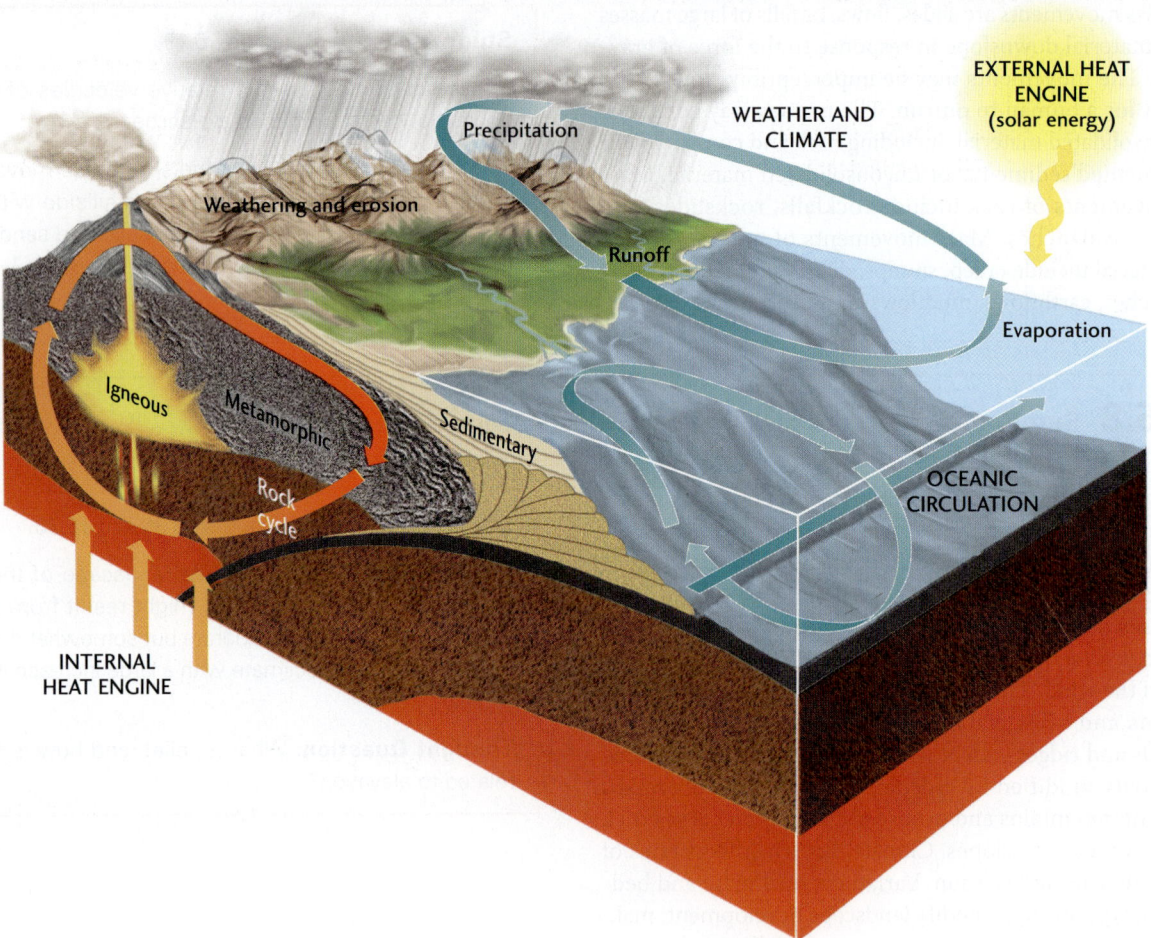

FIGURE 16.38 Landscape development is controlled by interactions between the plate tectonic system and the climate system.

1. **Which factor most directly influences weathering?**
   a. Oceanic circulation
   b. Melting to form igneous rocks
   c. Precipitation
   d. Generation of heat in Earth's interior

2. **What processes does solar energy control?**
   a. Evaporation of seawater
   b. Precipitation of rainwater
   c. Metamorphism
   d. Melting to form igneous rocks

3. **What process controls erosion?**
   a. Evaporation of seawater
   b. Oceanic circulation
   c. Metamorphism
   d. Runoff of rainwater

4. **What are the key components of the rock cycle?**
   a. Solar energy, evaporation, oceanic circulation, precipitation
   b. Weathering, erosion, burial, metamorphism and rock melting, uplift
   c. Weathering, soil, runoff, deposition
   d. Burial, subduction, melting, metamorphism

5. **How does precipitation control weathering and erosion?**
   a. Rainfall reacts with minerals and converts them to clays, thus making bedrock softer
   b. Rainfall flows over bedrock and erodes it due to abrasion
   c. Precipitation balances evaporation of seawater
   d. Runoff of water and sediment particles leads to deposition in delta

The Hydrologic Cycle and Groundwater

The Geologic Cycling of Water	496
Hydrology and Climate	498
The Hydrology of Groundwater	503
Erosion by Groundwater	515
Water Quality	516
Water Deep in the Crust	520

Learning Objectives

Chapter 17 studies the distribution, movements, and characteristics of water on, over, and under Earth's surface. We will follow the path of water in more detail as it sinks into the ground and flows through underground reservoirs. As we do so, we will see what makes groundwater a limited resource that must be carefully managed. After you have studied this chapter, you should be able to:

17.1 Describe how water moves through the hydrologic cycle.

17.2 Summarize how local hydrology is influenced by the local climate.

17.3 Explain the process of how water moves below the ground.

17.4 Classify the geologic processes that are affected by groundwater.

17.5 Describe the factors that govern human use of groundwater resources.

17.6 Illustrate how water deep in the crust affects hydrothermal processes.

Angel Falls, Venezuela. Known in Venezuela as Salto Angel, the falls cascade 3185 feet from Auyun Tepui, a towering sandstone plateau. The falls were named for pilot Jimmy Angel, who crash-landed atop the tepui in the 1930s. [Miquel Gonzalez/laif/Redux.]

CHAPTER 17   The Hydrologic Cycle and Groundwater

Most of us have heard the lines from Samuel Taylor Coleridge's "Rime of the Ancient Mariner": "Water, water everywhere, Nor any drop to drink." About 71 percent of Earth's surface is covered in water, but only a fraction of that water is available for human consumption. Humans cannot survive more than a few days without water. The amount of water consumed by modern society, however, far exceeds what we need for simple physical survival. We use water in immense quantities for industry, agriculture, and urban needs such as sewage systems. The study of the cycling of water is becoming ever more important as the demand on our limited water supplies increases.

In the previous two chapters, we have seen that water is essential to a wide variety of geologic processes. We saw in Chapter 12 that water exchanged between the oceans and the atmosphere forms a critical link in Earth's climate system; climate scientists now recognize that understanding the cycling of water is one of the most important steps in climate prediction. We saw in Chapter 16 that water is also important in weathering and erosion, both as a solvent of minerals in rock and soil and as a transport agent that carries away dissolved and weathered materials. The cycling of water links all of these processes. In Chapter 17 we will see how the streams and rivers formed by runoff help shape the landscapes of the continents. This chapter focuses on the water that sinks into Earth's crust to forms large reservoirs of groundwater.

The Geologic Cycling of Water

At what rates can we pump water from underground reservoirs without depleting them? What will be the effects of climate change on water supplies? Informed decision

(a)

(b)

(c)

(d)

(e)

FIGURE 17.1 The distribution of water on Earth. [(b) John Grotzinger; (c) Ron Niebrugge/wildnatureimages; (d) Viktor Lyagushkin/National Geographic Creative; (e) Charlie Munsey/Getty Images.]

making about the conservation and management of water resources requires knowledge of how water moves on, over, and under Earth's surface and how this flow responds to natural changes and human modifications. This field of study is known as **hydrology**.

Flow and Reservoirs

We can see water in Earth's lakes, oceans, and polar ice caps, and we can see water moving over Earth's surface in streams and glaciers. It is harder to see the massive amounts of water stored in the atmosphere and underground, or the flows of water into and out of these storage places. As water evaporates, it moves into the atmosphere as water vapor. As it falls from the sky as rain and sinks into the ground, it becomes **groundwater**—the mass of water that flows beneath Earth's surface. Because organisms use water, small amounts of water are also stored in the biosphere.

Each place that stores water is referred to as a *reservoir*. Earth's largest natural reservoirs, in the order of their size, are the oceans and seas, 95.96% (1.40×10^9 km^3). Freshwater accounts for the remainder: glacier and polar ice, 2.97% (4.34×10^7 km^3); groundwater, 1.05% (1.54×10^7 km^3); lakes and rivers, 0.009% (1.27×10^5 km^3); the atmosphere, 0.001% (1.5×10^4 km^3); and the biosphere, 0.0001% (2×10^3 km^3). Figure 17.1 shows the distribution of water among these reservoirs. Although the total amount of water in rivers and lakes is very small compared with the amounts in the oceans and even in groundwater, these reservoirs of water are important to human populations because they do not contain salt or high concentrations of other dissolved materials.

Reservoirs gain water from inflows, such as rain and streams running in, and lose water from outflows, such as evaporation and streams running out. If inflow equals outflow, the size of the reservoir stays the same, even though water is constantly entering and leaving it. These flows mean that any given quantity of water spends a certain average time, called the *residence time*, in a reservoir.

How Much Water Is There?

Earth's total water supply is enormous—about 1.4 billion cubic kilometers distributed among the various reservoirs. If all of that water covered the land area of the United States, it would submerge the 50 states under a layer of water about 145 km deep. This total is constant, even though the flows from one reservoir to another may vary from day to day, year to year, and century to century. Over these geologically short time intervals, there is neither a net gain nor a net loss of water to or from Earth's interior, nor any significant loss from the atmosphere to outer space.

The Hydrologic Cycle

All water on Earth cycles among the various reservoirs in the oceans, atmosphere, and on and under the land surface. The cyclical movement of water—from the oceans to the atmosphere by evaporation, to Earth's surface by precipitation, to streams through runoff over and under the ground, and back to the oceans—is called the **hydrologic cycle** (Figure 17.2).

Within the range of temperatures found at Earth's surface, water shifts among the three states of matter: liquid

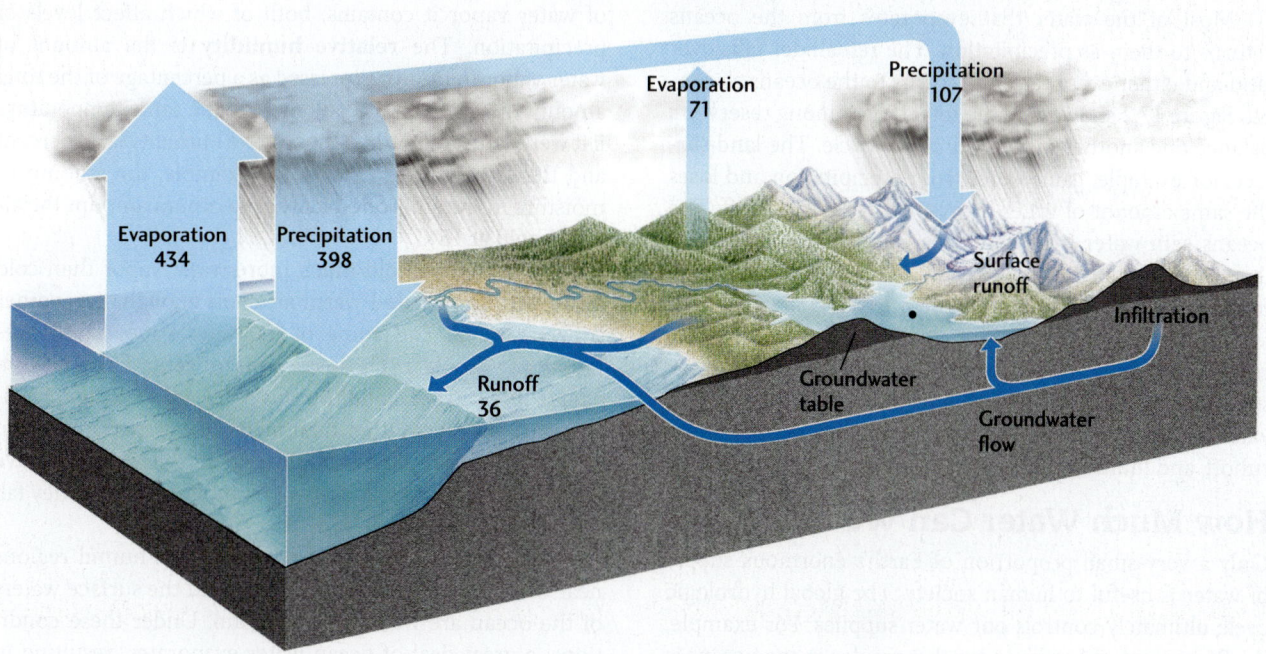

FIGURE 17.2 The hydrologic cycle is the movement of water through Earth's crust, atmosphere, oceans, lakes, and streams. The numbers indicate the amounts of water (in thousands of cubic kilometers per year) that flow between these reservoirs annually.

(water), gas (water vapor), and solid (ice). These transformations power some of the main fluxes from one reservoir to another. Earth's external heat engine, powered by the Sun, drives the hydrologic cycle, mainly by evaporating water from the oceans and transporting it as water vapor in the atmosphere.

Under the right temperature and humidity conditions, water vapor condenses into the tiny droplets of water that form clouds and eventually fall as rain or snow—together known as **precipitation.** Some of the precipitation that falls on land soaks into the ground by **infiltration,** a process by which water enters rock or soil through cracks or small pores between particles. Part of this groundwater evaporates through the soil surface and returns to the atmosphere as water vapor. Another part moves through the biosphere as it is absorbed by plant roots, carried up to the leaves, and returned to the atmosphere by a process called *transpiration*. Most of this groundwater, however, flows slowly underground. The residence time of water in groundwater reservoirs is long, but it eventually returns to the surface in springs that empty into rivers and lakes, and thus returns to the oceans.

The precipitation that does not infiltrate the ground runs off the land surface, gradually collecting into streams. The sum of all precipitation that flows over the land surface, including not only streams but also the fraction that temporarily infiltrates near-surface soil and rock and then flows back to the surface, is called **runoff.** Some runoff may later seep into the ground or evaporate from rivers and lakes, but most of it eventually flows into the oceans.

Snowfall may be converted into ice in glaciers, which return water to the oceans by melting and runoff, and to the atmosphere by *sublimation,* the transformation of water from a solid (ice) directly into a gas (water vapor).

Most of the water that evaporates from the oceans returns to them as precipitation. The remainder falls over land and either evaporates or returns to the oceans as runoff. Figure 17.2 shows how the total flows among reservoirs balance one another in the hydrologic cycle. The land surface, for example, gains water from precipitation and loses the same amount of water by evaporation and runoff. The oceans gain water from runoff and precipitation and lose the same amount by evaporation. More water evaporates from the oceans than falls on them as precipitation. This loss is balanced by the water returned as runoff from the continents. Thus, on a global scale, the size of each reservoir stays constant. Variations in climate, however, produce local variations in the balance among evaporation, precipitation, runoff, and infiltration.

How Much Water Can We Use?

Only a very small proportion of Earth's enormous supply of water is useful to human society. The global hydrologic cycle ultimately controls our water supplies. For example, the 96 percent of Earth's water that resides in the oceans is essentially off-limits to us. Almost all the water we use is *fresh water*—water that is not salty. Artificial desalination (the removal of salt) is now producing small but steadily growing amounts of fresh water from seawater in areas such as the arid Middle East. In the natural world, however, fresh water is supplied only by rain, rivers, lakes, groundwater, and water melted from snow or ice on land. All these waters are ultimately supplied by precipitation. Therefore, the practical limit to the amount of natural fresh water that we can ever envision using is the amount steadily supplied to the continents by precipitation.

Hydrology and Climate

For most practical purposes, geologists focus on local hydrology—the amount of water in the reservoirs of a region and how it flows from one reservoir to another—rather than global hydrology. The strongest influence on local hydrology is the local climate, especially temperature and precipitation levels. In warm areas where rain falls frequently throughout the year, water supplies—both at the surface and underground—are abundant. In warm arid or semiarid regions, it rarely rains, and water is a precious resource. People who live in icy climates rely on meltwaters from snow and ice. In some parts of the world, seasons of heavy rain, called *monsoons,* alternate with long dry seasons during which water supplies shrink, the ground dries out, and vegetation shrivels.

Humidity, Rainfall, and Landscape

Many geographic variations in climate are related to the average temperature of the air and the average amount of water vapor it contains, both of which affect levels of precipitation. The **relative humidity** is the amount of water vapor in the air, expressed as a percentage of the total amount of water the air could hold at the same temperature if it were saturated. When the relative humidity is 50 percent and the temperature is 15°C, for example, the amount of moisture in the air is one-half the maximum amount the air could hold at 15°C.

Warm air can hold much more water vapor than cold air. When unsaturated warm air cools enough, it becomes supersaturated, and some of its water vapor condenses into water droplets. Those condensed water droplets form clouds. We can see clouds because they are made up of visible water droplets rather than invisible water vapor. When enough moisture has condensed and the droplets have grown too heavy to stay suspended by air currents, they fall as rain.

Most of the world's rain falls in warm, humid regions near the equator, where both the air and the surface waters of the ocean are warmed by the Sun. Under these conditions, a great deal of ocean water evaporates, resulting in high relative humidity. When air is warmed, it expands, becomes less dense, and rises. When the humid air over tropical oceans rises to high altitudes and blows over nearby continents, it cools, condenses, and becomes supersaturated. The result is heavy rainfall over the land, even at great distances from the coast.

At about 30°N and 30°S latitude, the air that has dropped its precipitation in the tropics begins to sink back toward Earth's surface. This cold, dry air warms and absorbs moisture as it sinks, producing clear skies and arid climates. Many of the world's deserts are located at these latitudes.

Polar climates also tend to be very dry. The polar oceans and the air above them are cold, so the air can hold little moisture, and little ocean water evaporates. Between the tropical and polar extremes are the temperate climates, where rainfall and temperatures are moderate.

The climate patterns we have just described are driven by patterns of air circulation in the atmosphere, as we'll see in Chapter 12. Plate tectonic processes also influence climate processes. The uplifting of mountain ranges, for example, forms **rain shadows,** areas of low precipitation on their leeward (downwind) slopes. Humid winds rising over high mountains cool and precipitate on the windward slopes, losing much of their moisture by the time they reach the leeward slopes (**Figure 17.3**). The air warms again as it drops to lower elevations on the other side of the mountain range. Because the warmer air can hold more moisture, relative humidity declines, decreasing the likelihood of precipitation even more. The Cascade Range of Oregon, uplifted by the subduction of the Pacific Plate under the North American Plate, creates a rain shadow. The prevailing winds that blow inland from the Pacific Ocean release heavy rainfall on the mountains' western slopes, supporting a lush forest ecosystem. The eastern slopes, on the other side of the range, are dry and barren.

Just as landscape features can alter precipitation patterns, the resulting variations in precipitation patterns control rates of weathering and erosion, which shape the landscape.

Droughts

Droughts—periods of months or years when precipitation is much lower than normal—can occur in all climates, but arid regions are especially vulnerable to their effects. Lacking replenishment from precipitation, streams may shrink and dry up, ponds and lakes may evaporate, and the soil may dry and crack while vegetation dies. As human populations grow, demands on water supplies increase, so a drought can deplete already inadequate supplies.

The severest drought of the past few decades has affected a region of Africa known as the Sahel, along the southern border of the Sahara (**Figure 17.4**). This long drought has expanded the desert, as we'll see in Chapter 19, and has effectively destroyed farming and grazing. Hundreds of thousands of lives have been lost to famine in the area.

Another prolonged but less severe drought affected most of California from 1987 until February 1993, when torrential rains arrived. During the drought, groundwater and surface reservoirs dropped to their lowest levels in 15 years. Some restrictions on water use were instituted, but a move to reduce the extensive use of water supplies for irrigation encountered strong political resistance from farmers and the agricultural industry. As threats of water shortages loom, the use of water enters the arena of public policy debate (see Earth Policy 17.1).

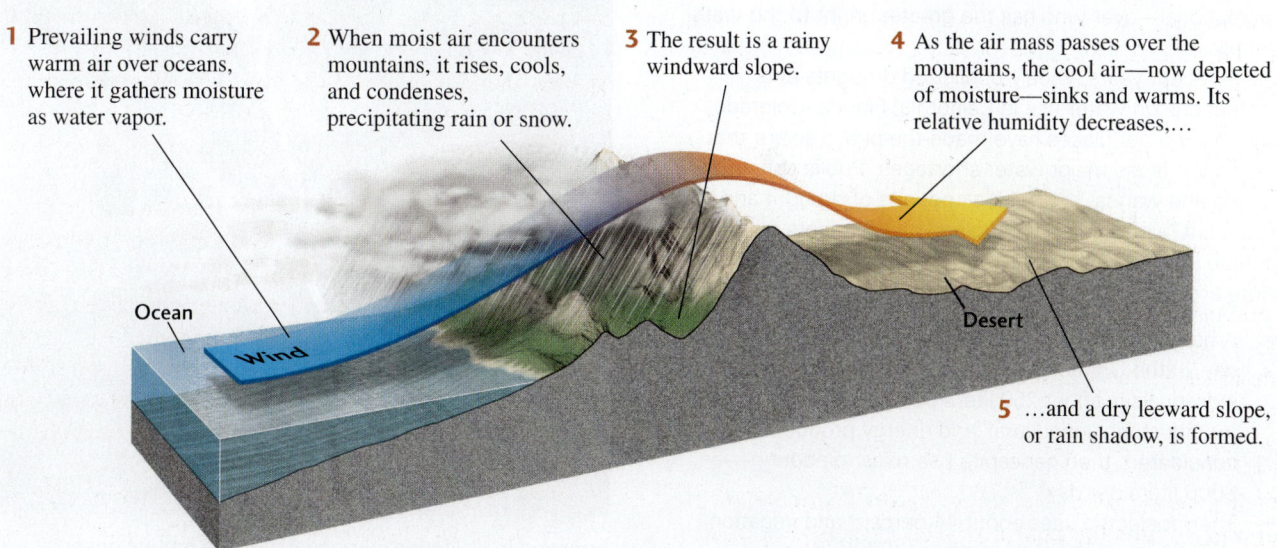

1 Prevailing winds carry warm air over oceans, where it gathers moisture as water vapor.

2 When moist air encounters mountains, it rises, cools, and condenses, precipitating rain or snow.

3 The result is a rainy windward slope.

4 As the air mass passes over the mountains, the cool air—now depleted of moisture—sinks and warms. Its relative humidity decreases,...

5 ...and a dry leeward slope, or rain shadow, is formed.

FIGURE 17.3 Rain shadows are areas of low rainfall on the leeward (downwind) slopes of a mountain range.

FIGURE 17.4 This millet field in a small village near Zinder, Niger, shows the effects of a long drought on soils and crops. This photo was taken on September 27, 2010, but the drought continues today. [Tomas van Houtryve/VII/Redux.]

Earth Policy 17.1 Water Is a Precious Resource: Who Should Get It?

Until recently, most people in the United States have taken their water supply for granted. In the near future, however, because of climate change and population growth, particularly in arid regions, many areas of the country will experience water shortages more and more frequently. These shortages will create conflict among several sectors of society—residential, industrial, agricultural, and recreational—over who has the greatest right to the water supply.

In recent years, widely publicized droughts and restrictions on water use in California, Florida, Colorado, and many other places have made the public aware that the nation faces major water shortages. Public concern waxes and wanes, however, as periods of drought and abundant rainfall come and go and governments fail to pursue long-term solutions with the urgency they deserve. Here are some facts to ponder:

- A human can survive with about 2 liters of water per day. In the United States, per capita water use by individuals is about 250 liters per day. If uses of water for industrial, agricultural, and energy production are considered, then per capita use rises to about 6000 liters per day.
- Thermoelectric uses about 44 percent and irrigation about 37 percent of the water withdrawn from U.S. reservoirs.

Irrigation in California's Imperial Valley, a natural desert. [David McNew/Getty Images.]

An example of a shorter-term, but high-impact, event is the 2013 drought in New Zealand. That country experienced a severe, widespread drought from late 2012 until April 2013. The extent of this drought made it unusual—it simultaneously occurred across the entire North Island and parts of the South Island. Many of the pastures are not irrigated and depend on rainfall. The lack of rain caused crop failures and affected pastures in a country where agriculture is a major industry.

Our climate history can give us perspective on the severity of droughts. The southwestern United States, for example, has been experiencing a recent drought. During the 400-year period from 1500 to 1900, however, the Southwest was drier, on average, than it has been during the last century. Moreover, the geologic record shows droughts that were more severe and of longer duration than the present drought has been (at least so far). Are the recent droughts just short-term fluctuations in climate, or do they signal a return to an extended dry period? How will global climate change affect rainfall in the Southwest? By exploring the past, geologists and climate scientists may find information that will help them predict the future.

The Hydrology of Runoff

How much of the precipitation that falls on a land area ends up as runoff? A dramatic short-term example of how precipitation levels affect local stream and river runoff can be seen when flash flooding occurs after torrential rains. When levels of precipitation and runoff are measured over a larger area (such as all the states drained by a major river) and over a longer period (such as a year), the relationship between them is less direct, but still strong. The maps in **Figure 17.5** illustrate this relationship. When we compare them, we see that in areas of low precipitation—such as Southern California, Arizona, and New Mexico—only a small fraction of precipitation ends up as runoff. In these dry regions, much of the precipitation leaves the land surface by evaporation and infiltration. In more humid areas, such as the southeastern United States, a much higher proportion of the precipitation runs off in streams. A large river may carry large amounts of water from an area with high rainfall to an area with low rainfall. The Colorado River, for example, begins in an area of moderate rainfall in Colorado and then carries its water through arid western Arizona and Southern California.

▶ Per capita domestic water use in the United States is two to four times greater than in western Europe, where consumers pay as much as 350 percent more for their water.

▶ Although the western United States receives one-fourth of the country's rainfall, per capita water use in the western states (mostly for irrigation) is 10 times greater than that in the eastern states, and water prices are much lower there. In California, for example, which imports most of its water, 85 percent of water use is for irrigation, 10 percent for municipalities and personal consumption, and 5 percent for industry. A 15 percent reduction in irrigation use would almost double the amounts of water available for use by cities and industries.

▶ The fresh water used in the United States eventually returns to the hydrologic cycle, but it may return to a reservoir that is not well located for human use, and its quality may be degraded. Recycled irrigation water is often saltier than natural fresh water and is loaded with pesticides. Polluted urban wastewater ends up in the oceans.

▶ The traditional ways of increasing water supplies, such as building dams and artificial reservoirs and drilling wells, have become extremely costly because most of the best (and therefore cheapest) sites have already been used. Furthermore, the building of more dams to hold larger reservoirs carries environmental costs, such as the flooding of inhabited areas, detrimental changes in river flows above and below the dams, and the disturbance of fish and other wildlife habitats. Factoring in these costs has led to delays in dam projects and rejection of proposals for new dams.

Thermoelectric power (43%)
Irrigation (36%)
Aquaculture (2%)
Industry (5%)
Domestic use (2%)
Public supply (12%)

U.S. water use by category in 2015.
[Data from U.S. Geological Survey.]

502 CHAPTER 17 The Hydrologic Cycle and Groundwater

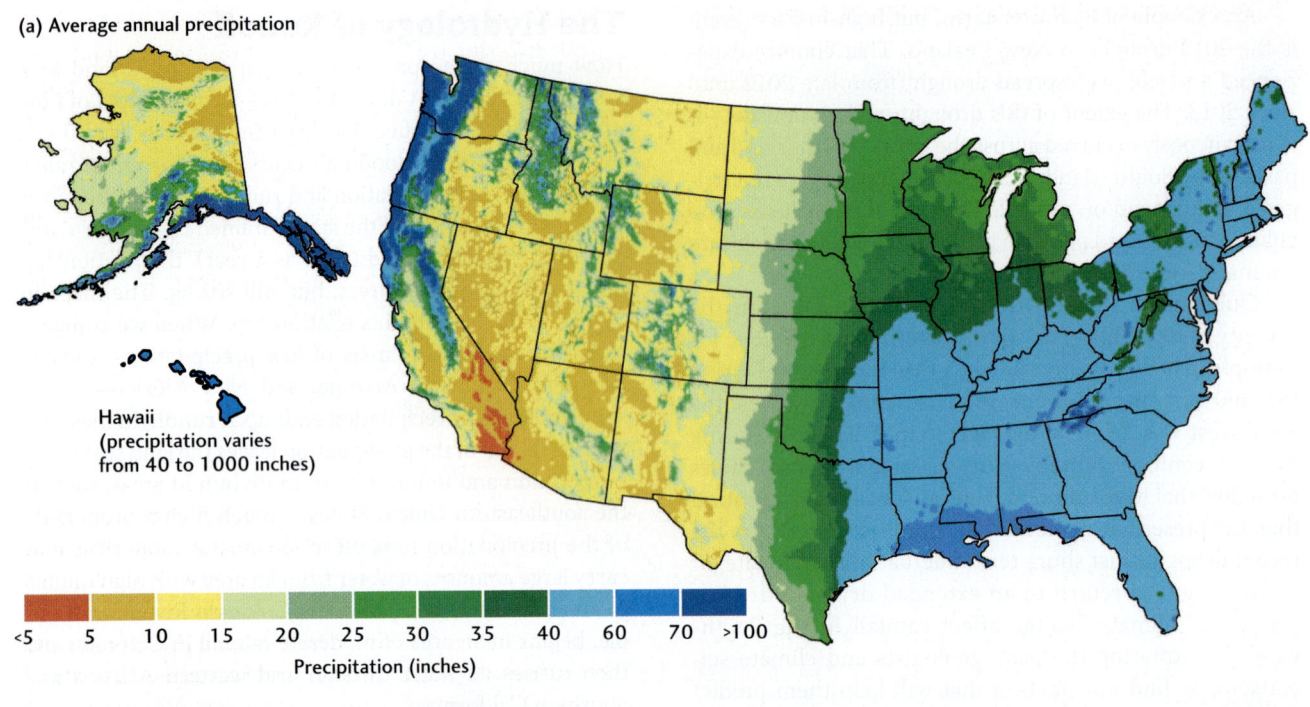

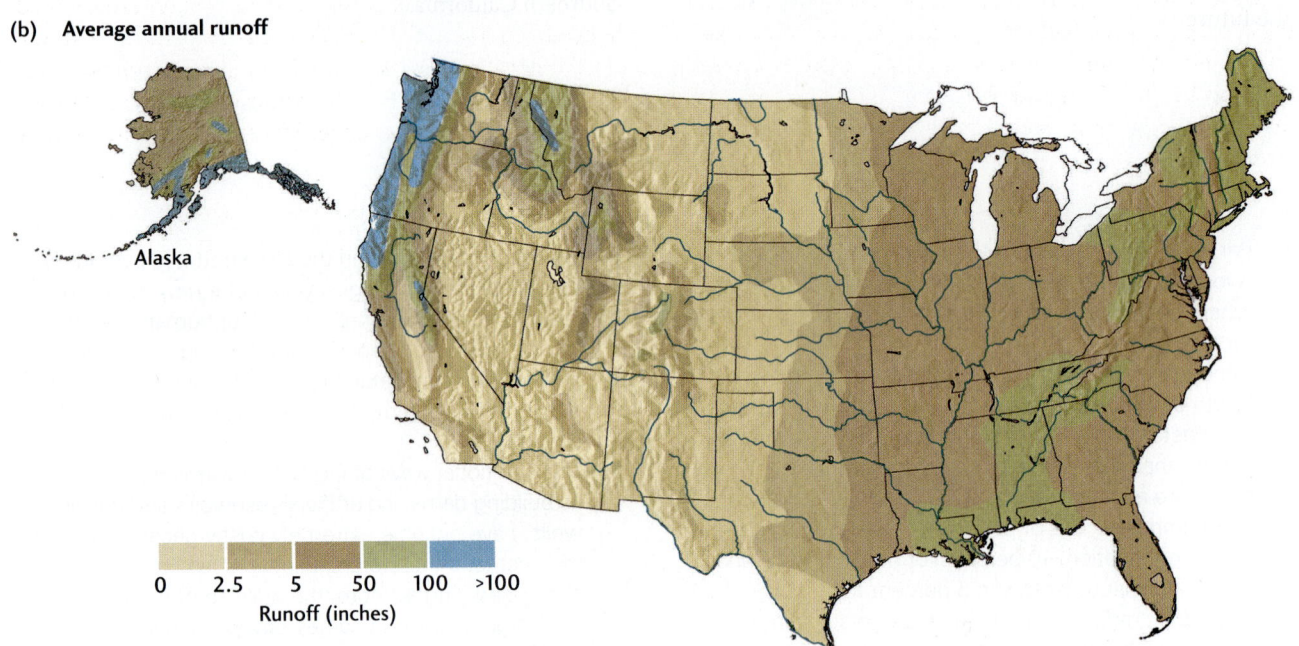

FIGURE 17.5 (a) Map of the average annual precipitation in the United States from 1981 to 2010. (b) Average annual runoff in the United States. [Data from USGS.]

Rivers and streams carry most of the world's runoff. The millions of small and medium-sized streams carry about half the world's runoff. About 70 major rivers carry the other half, and the Amazon River of South America carries almost half of that. The Amazon carries about 10 times more water than the Mississippi, the largest river of North America (Table 17.1). The major rivers transport great volumes of water because they collect it from large networks of streams and rivers that cover very large areas. The Mississippi, for example, collects its water from a network of streams that covers about two-thirds of the United States (Figure 17.6).

Runoff collects and is stored in natural lakes as well as in artificial reservoirs created by the damming of streams.

TABLE 17.1 Water Flows of Some Major Rivers

RIVER	WATER FLOW (m³/s)
Amazon, South America	175,000
La Plata, South America	79,300
Congo, Africa	39,600
Yangtze, Asia	21,800
Brahmaputra, Asia	19,800
Ganges, Asia	18,700
Mississippi, North America	17,500

Wetlands, such as swamps and marshes, also act as reservoirs for runoff (**Figure 17.7**). If these reservoirs are large enough, they can absorb short-term inflows from major rainfall events, holding some of the water that would otherwise spill over riverbanks. During dry seasons or droughts, these reservoirs release water to streams or to water systems built for human use. Thus, they help to control flooding by smoothing out seasonal or yearly variations in runoff and releasing steady flows of water downstream.

In addition to these roles, wetlands are important to biological diversity because they are breeding grounds for a great many types of plants and animals. For all these reasons, many governments have laws that regulate the artificial draining of wetlands for real estate development. Nevertheless, wetlands are disappearing rapidly as land development continues. In the United States, more than half the wetlands that existed before European settlement are now gone. California and Ohio have kept only 10 percent of their original wetlands.

The Hydrology of Groundwater

Groundwater forms as raindrops and melting snow infiltrate soil and other unconsolidated surface materials and even sink into the cracks and crevices of bedrock. This groundwater, formed from recent atmospheric precipitation, is known as **meteoric water** (from the Greek *meteoron*, "phenomenon in the sky," which also gives us the word *meteorology*). The enormous reservoir of groundwater stored beneath Earth's surface equals about 29 percent of all the fresh water stored in lakes and rivers, glaciers and polar ice, and the atmosphere. For thousands of years, people have drawn on this resource, either by digging shallow wells or

FIGURE 17.6 The Mississippi River and its tributaries form the largest drainage network in the United States.

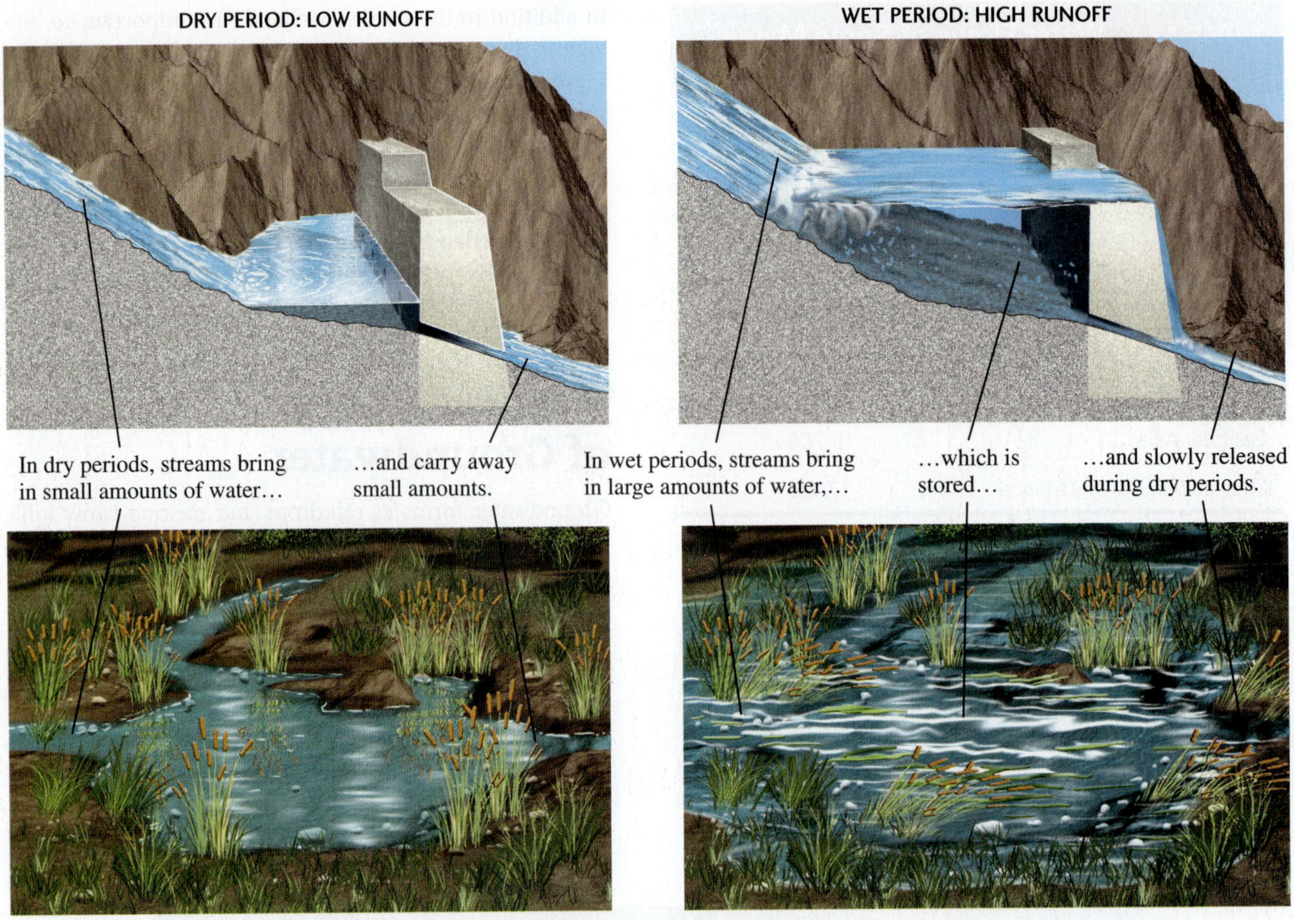

DRY PERIOD: LOW RUNOFF

In dry periods, streams bring in small amounts of water… …and carry away small amounts.

WET PERIOD: HIGH RUNOFF

In wet periods, streams bring in large amounts of water,… …which is stored… …and slowly released during dry periods.

FIGURE 17.7 Like a natural lake or an artificial reservoir behind a dam, a wetland stores water during times of rapid runoff and slowly releases it during periods of little runoff.

---

by storing water that flows out onto the surface at natural springs. These springs are direct evidence of water moving below the surface (Figure 17.8).

Porosity and Permeability

When water moves into and through the ground, what determines where and how fast it flows? With the exception of caves, there are no large open spaces underground for pools or rivers of water. The only spaces available for water are pores and cracks in soil and bedrock. Some pores, however small and few, are found in every kind of rock and soil, but large amounts of pore space are most often found in sandstones and limestones.

Recall from Chapter 6 that the amount of pore space in rock, soil, or sediment determines its *porosity*: the percentage of its total volume that is taken up by pores. This pore space consists mainly of space between grains and in cracks (Figure 17.9). It can vary from a small percentage of the total volume of the material to as much as 50 percent where rock has been dissolved by chemical weathering. Sedimentary rocks typically have porosities of 5 to 15 percent. Most metamorphic and igneous rocks have little pore space, except where fracturing has occurred.

There are three types of pores: spaces between grains (*intergranular porosity*), spaces in fractures (*fracture porosity*), and spaces created by dissolution (*vuggy porosity*). Intergranular porosity, which characterizes soils, sediments, and sedimentary rocks, depends on the size and shape of the grains that make up those materials and on how they are packed together. The more loosely packed the grains, the greater the pore space between them. The smaller the particles, and the more they vary in shape and size, the more tightly they fit together. Minerals that cement grains together reduce intergranular porosity. Intergranular porosity varies from 10 to 40 percent.

Porosity is lower in igneous and metamorphic rocks, in which pore space is created mostly by fractures, including joints and cleavage at natural zones of weakness. Fracture porosity values are commonly as low as 1 to 2 percent, though some fractured rocks contain appreciable pore space—as much as 10 percent of the rock volume—in their many cracks.

Pore space in limestones and other highly soluble rocks such as evaporites may be created when groundwater interacts with the rock and partly dissolves it, leaving irregular voids known as *vugs*. Vuggy porosity can be

The Hydrology of Groundwater

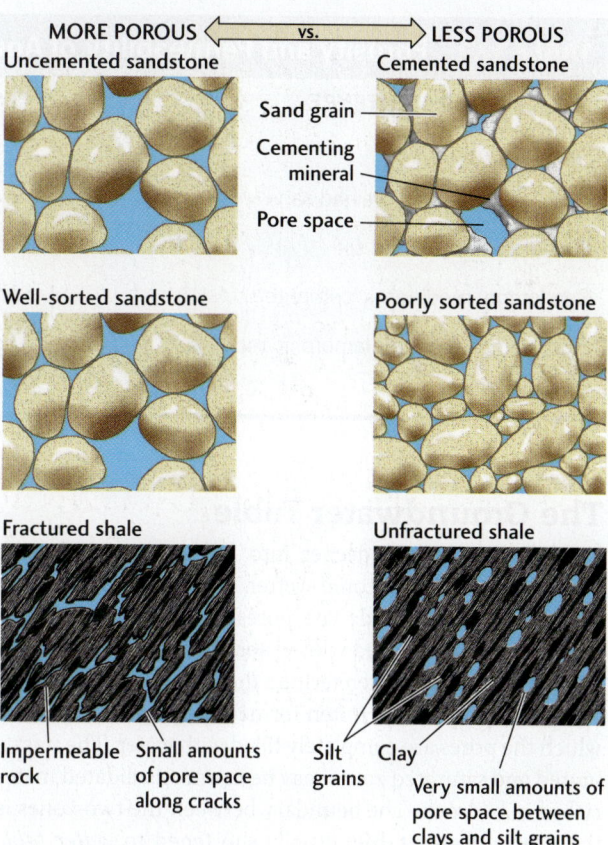

FIGURE 17.9 Porosity in rocks depends on several factors. In sandstones, the extent of cementation and the degree of grain sorting are both important. In shales, porosity is limited due to the small spaces between the tiny grains, but can be enhanced by fracturing.

FIGURE 17.8 Groundwater flows from a cliff in Vasey's Paradise, Marble Canyon, Grand Canyon National Park, Arizona, where hilly topography allows it to flow out onto the surface in natural springs. [Inge Johnsson/Alamy.]

very high (over 50 percent); caves are examples of extremely large vugs.

Although a rock's porosity tells us how much water it can hold if all its pores are filled, it gives us no information about how rapidly water can flow through those pores. Water travels through a porous material by winding between grains and through cracks. The smaller the pore spaces, and the more complex the path, the more slowly the water travels. The capacity of a solid to allow fluids to pass through it is its **permeability**. Generally, permeability increases as porosity increases, but permeability also depends on the sizes of the pores, how well they are connected, and how tortuous a path the fluid must travel to pass through the material. Vuggy pore networks in carbonate rocks may have extremely high permeabilities. Cave systems are so permeable that they allow people as well as water to move through them!

Both porosity and permeability are important considerations for geologists searching for groundwater supplies. In general, a good groundwater reservoir is a body of rock, sediment, or soil that has both high porosity (so that it can hold large amounts of water) and high permeability (so that the water can be pumped from it easily). Well drillers in temperate climates, for example, know that they are most likely to find a good supply of water if they drill into porous sand or sandstone beds not far below the surface. A rock with high porosity but low permeability may contain a great deal of water, but because the water flows so slowly, it is hard to pump it out of the rock. Table 17.2 summarizes the porosities and permeabilities of various rock types.

TABLE 17.2 Porosity and Permeability of Aquifer Rock and Sediment Types

ROCK OR SEDIMENT TYPE	POROSITY	PERMEABILITY
Gravel	Very high	Very high
Coarse- to medium-grained sand	High	High
Fine-grained sand and silt	Moderate	Moderate to low
Sandstone, moderately cemented	Moderate to low	Low
Fractured shale or metamorphic rock	Low	Very low
Unfractured shale	Very low	Very low

The Groundwater Table

As well drillers bore deeper into soil or rock, the samples they bring up become wetter. At shallow depths, the material is unsaturated: The pores contain some air and are not completely filled with water. This level is called the **unsaturated zone** (often termed the *vadose zone*). Below it is the **saturated zone** (often termed the *phreatic zone*), in which the pores are completely filled with water. The unsaturated and saturated zones may be in unconsolidated material or in bedrock. The boundary between the two zones is the **groundwater table,** usually shortened to *water table* (Figure 17.10). When a hole is dug below the water table, water from the saturated zone flows into the hole and fills it to the level of the water table.

Groundwater moves under the force of gravity, so some of the water in the unsaturated zone may be on its way down to the water table. A fraction of that water, however, remains in the unsaturated zone, held in small pore spaces by surface tension. Surface tension is what keeps the sand on a beach moist. Evaporation of this water into pore spaces in the unsaturated zone is slowed both by the effect of surface tension and by the relative humidity of the air in the pore spaces, which can be close to 100 percent.

If we were to drill wells at several sites and measure the elevations of the water levels in those wells, we could construct a map of the water table. A cross section of the landscape might look like the one shown in **Figure 17.11a**. The water table follows the general shape of the surface topography, but its slopes are gentler. It is exposed at the land surface in river and lake beds and at springs. Under the influence of gravity, groundwater moves downhill from places where the water table elevation is high—under a hill, for example—to places where the water table elevation is low—such as a spring where groundwater flows out onto the surface.

Water enters and leaves the saturated zone through recharge and discharge (Figure 17.11b). **Recharge** is the infiltration of water into any subsurface formation. Rain and melting snow are the most common sources of recharge. **Discharge,** the movement of groundwater to the surface, is the opposite of recharge. Groundwater is discharged by evaporation, through springs, and by pumping from artificial wells. Water may also enter and leave the saturated zone through streams. Recharge may take place through the bottom of a stream whose stream channel lies at an elevation above that of the water table. Streams that recharge groundwater in this way are called *influent streams*, and they are most characteristic of arid conditions, in which the water table is deep. Conversely, when a stream channel lies at an elevation below that of the water table, water is discharged from the groundwater into the stream. Such an *effluent stream* is typical of humid conditions. Effluent streams continue to flow long after runoff has stopped because they are fed by groundwater. Thus, the reservoir of groundwater may be increased by influent streams and depleted by effluent streams.

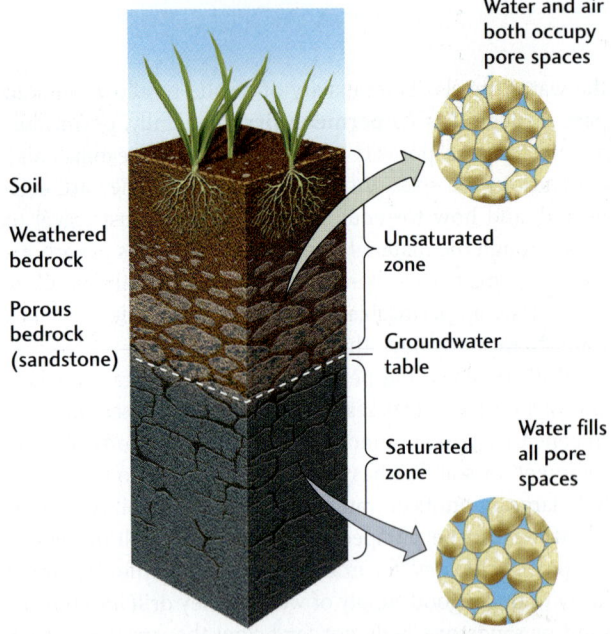

FIGURE 17.10 The groundwater table is the boundary between the unsaturated zone and the saturated zone. The saturated and unsaturated zones may be in unconsolidated material or in bedrock.

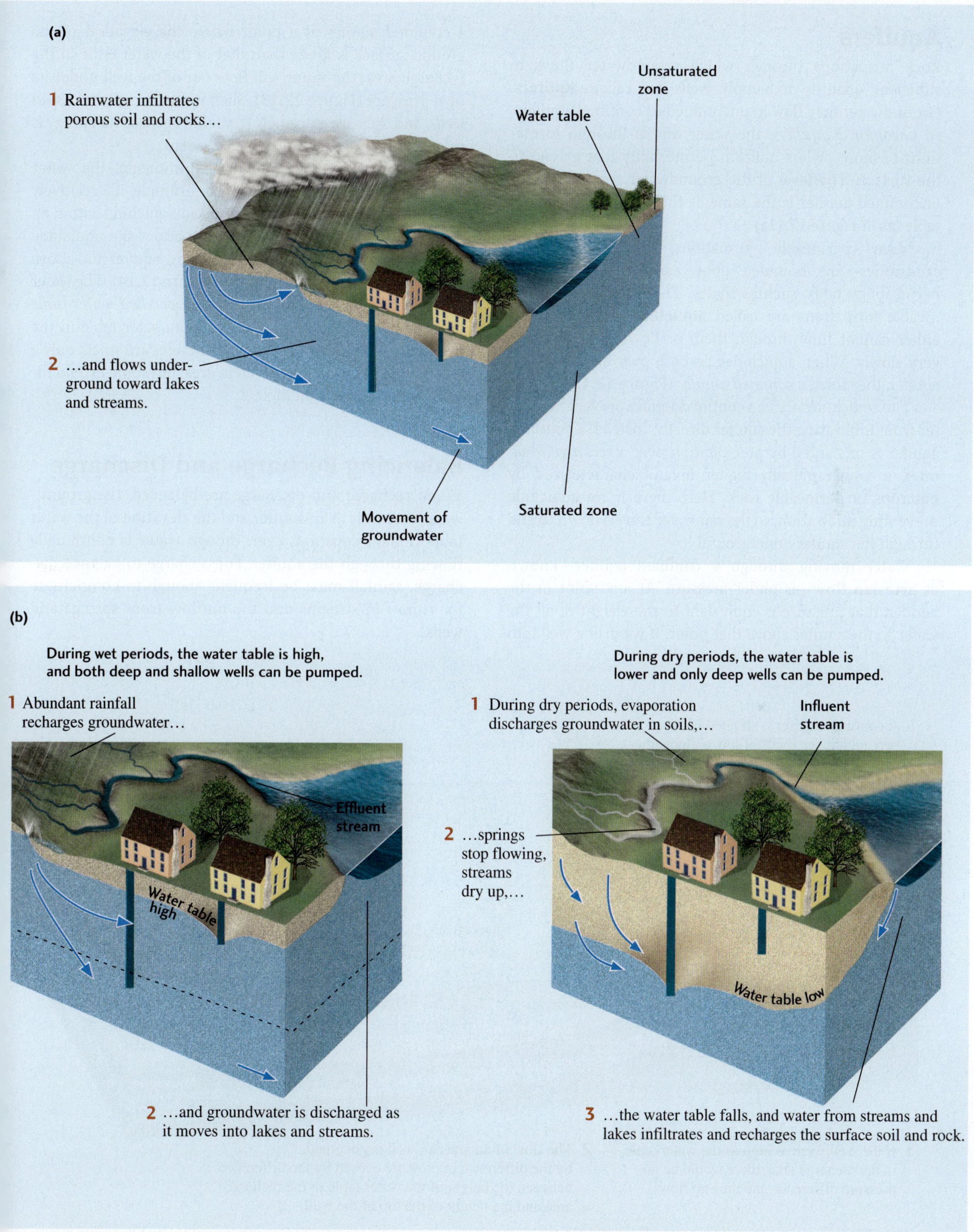

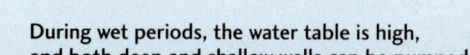
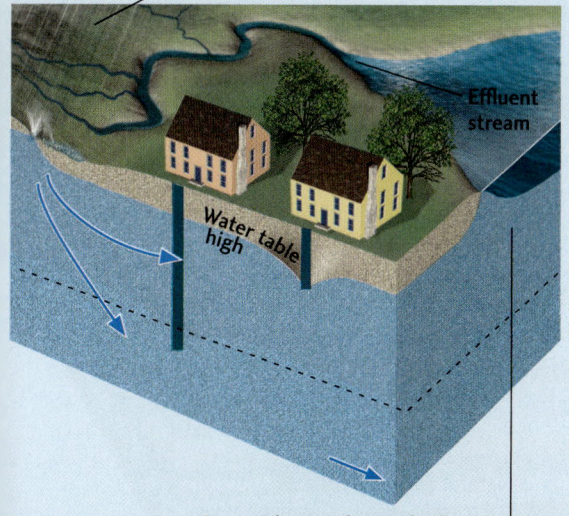

FIGURE 17.11 Dynamics of the groundwater table in a permeable shallow formation in a temperate climate. (a) The groundwater table follows the general shape of the surface topography, but its slopes are gentler. (b) The elevation of the water table fluctuates in response to the balance between water added by precipitation (recharge) and water lost by evaporation and from springs, streams, and wells (discharge).

Aquifers

Rock formations through which groundwater flows in sufficient quantity to supply wells are called **aquifers.** Groundwater may flow in unconfined or confined aquifers. In *unconfined aquifers*, the water travels through formations of more or less uniform permeability that extend to the surface. The level of the groundwater reservoir in an unconfined aquifer is the same as the height of the water table (as in Figure 17.11a).

Many permeable formations, however—typically sandstones—are bounded above and below by low-permeability beds, such as shales. These relatively impermeable formations are called **aquicludes.** Groundwater either cannot flow through them or flows through them very slowly. When aquicludes lie both over and under an aquifer, they form a *confined aquifer* (**Figure 17.12**).

The aquicludes above a confined aquifer prevent rainwater from infiltrating the aquifer directly. Instead, a confined aquifer is recharged by precipitation over a *recharge area*, often a topographically higher upland characterized by outcrops of permeable rock. Here there is no aquiclude preventing infiltration, so the rainwater travels down to and through the aquifer underground.

Water moving through a confined aquifer—known as **artesian flow**—is under pressure. At any point in the aquifer, that pressure is equivalent to the weight of all the water in the aquifer above that point. If we drill a well into a confined aquifer at a point where the elevation of the ground surface is lower than that of the water table in the recharge area, the water will flow out of the well under its own pressure (**Figure 17.13**). Such wells are called *artesian wells*, and they are extremely desirable because no energy is required to pump the water to the surface.

In more complex geologic environments, the water table may be more complicated. For example, if a relatively impermeable mudstone layer forms an aquiclude within an otherwise permeable sandstone formation, the aquiclude may lie below the water table of a shallow aquifer and above the water table of a deeper aquifer (**Figure 17.14**). The water table in the shallow aquifer is called a *perched water table* because it is "perched" above the main water table in the deeper aquifer. Many perched water tables are small, only a few meters thick and restricted in area, but some extend for hundreds of square kilometers.

Balancing Recharge and Discharge

When recharge and discharge are balanced, the groundwater reservoir in an aquifer and the elevation of the water table remain constant, even though water is continually flowing through the aquifer. For recharge to balance discharge, rainfall must be frequent enough to compensate for runoff in streams and the outflow from springs and wells.

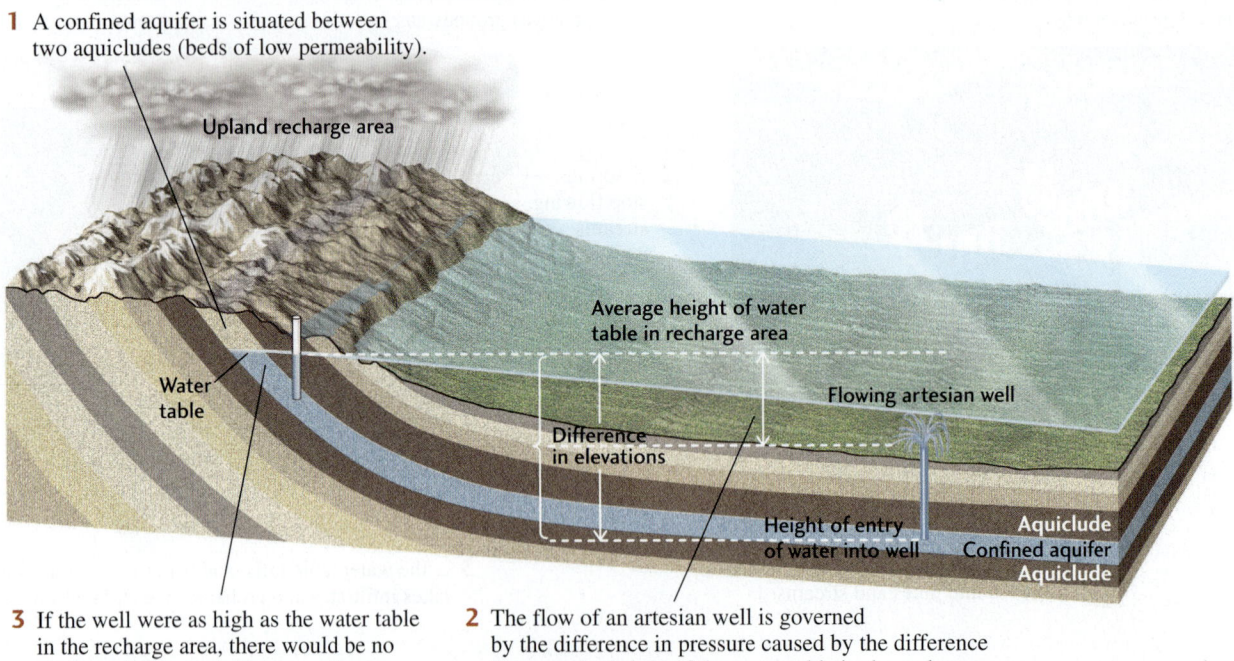

FIGURE 17.12 A permeable formation situated between two aquicludes forms a confined aquifer, through which water flows under pressure.

The Hydrology of Groundwater

FIGURE 17.13 Water flows from an artesian well under its own pressure. This artesian well supplies water in the southern Gulf Savannah region of Queensland, Australia. [ENVIRONMENTAL IMAGES/Science Source.]

But recharge and discharge are rarely equal because recharge varies with rainfall from season to season. Typically, the water table drops in dry seasons and rises in wet seasons (see Figure 17.11b). A longer period of low recharge, such as a prolonged drought, will be followed by a longer-term imbalance and further lowering of the water table.

An increase in discharge, usually due to increased pumping from wells, can also produce a long-term imbalance and

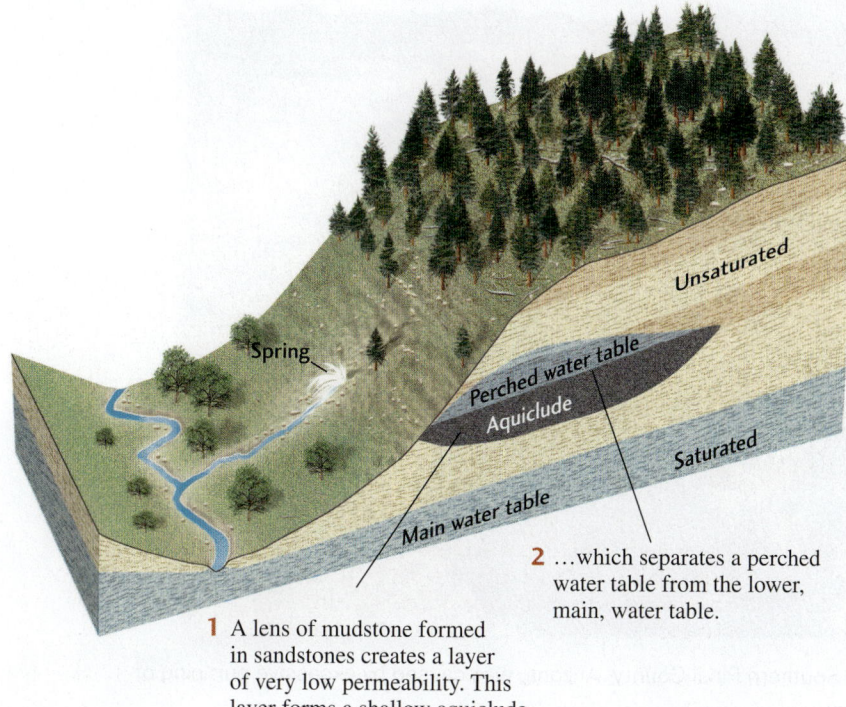

1. A lens of mudstone formed in sandstones creates a layer of very low permeability. This layer forms a shallow aquiclude,…

2. …which separates a perched water table from the lower, main, water table.

FIGURE 17.14 A perched water table forms in some geologically complex situations—in this case, where a mudstone aquiclude is located above the main water table in a sandstone aquifer. The dynamics of the perched water table's recharge and discharge may be different from those of the main water table.

a lowering of the water table. Shallow wells may end up in the unsaturated zone and go dry. When a well pumps water from an aquifer faster than recharge can replenish it, the water table is lowered in a cone-shaped area around the well, called a *cone of depression* (Figure 17.15). The water level in the well is lowered to the depressed level of the water table. If the cone of depression extends below the bottom of the well, the well goes dry. If the bottom of the well is above the base of the aquifer, extending the well deeper into the aquifer may allow more water to be withdrawn, even at continued high pumping rates. If the rate of pumping is maintained and the well is deepened so much that the full thickness of the aquifer is tapped, however, the cone of depression can reach the bottom of the aquifer and deplete it. The aquifer will recover only if the pumping rate is reduced enough to give it time to recharge.

Excessive withdrawals of water may not only deplete the aquifer, but may also cause another undesirable environmental effect. As water pressure in the pore spaces falls, the ground surface overlying the aquifer may subside, creating sinklike depressions (Figure 17.16). As water in some types of sediments is removed, those sediments compact, and the loss of volume lowers the ground surface, a phenomenon known as *subsidence*. Subsidence caused by excessive pumping has occurred in Mexico City and in Venice, Italy, as well as in many other regions of heavy pumping, such as the San Joaquin Valley in California. In these places, the rate of subsidence has reached almost 1 m every 3 years. Although there have been a few attempts to reverse the subsidence by pumping water back into the ground, they have not been very successful because most compacted materials do not expand to their former state. The best way to halt further subsidence is to restrict pumping.

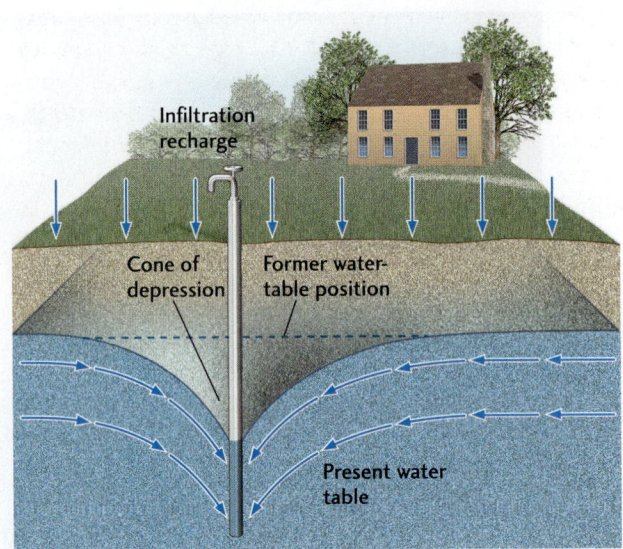

FIGURE 17.15 When discharge from a well exceeds recharge, the water table is lowered in a cone of depression. The water level in the well is lowered to the depressed level of the water table.

FIGURE 17.16 An earth fissure exposed in Southern Pinal County, Arizona, was caused by excessive pumping of groundwater. [J. P. Cook, Arizona Geological Survey.]

People who live near the ocean's edge may face a different problem when rates of discharge from an aquifer are high in relation to recharge: a flow of salt water into the aquifer. Near shorelines, or a little offshore, an underground boundary separates salty groundwater under the sea from fresh groundwater under the land. This *saltwater margin* slopes downward and inland from the shoreline in such a way that salt water underlies the fresh water of the aquifer (Figure 17.17a). Under many oceanic islands, a lens of fresh groundwater (shaped like a simple double-convex lens) floats on a base of seawater. The fresh water floats because it is less dense than the seawater (1.00 g/cm^3 versus 1.02 g/cm^3, a small but significant difference). Normally, the pressure of the fresh water keeps the saltwater margin slightly offshore. The balance between recharge and discharge in the freshwater aquifer maintains this freshwater–seawater boundary.

As long as recharge is at least equal to discharge, the aquifer will provide fresh water. If water is withdrawn from a well faster than it can be recharged, however, a cone of depression develops at the top of the aquifer, mirrored by an inverted cone rising from the saltwater margin below. The cone of depression makes it more difficult to pump fresh water, and the inverted cone leads to an intake of salt water at the bottom of the well (Figure 17.17b). People living closest to the coast are the first affected. Some towns on Cape Cod, Massachusetts, on Long Island, New York, and in many other coastal areas have had to post notices that town drinking water contains more salt than is considered healthful by environmental agencies. There is no ready solution to this problem other than to slow the pumping or, in some places, to recharge the aquifer artificially by funneling runoff into the ground.

One of the predicted effects of global warming is a rise in sea level. We can see that as sea level rises, the saltwater margins of coastal aquifers will also rise. Seawater will then invade coastal aquifers and turn fresh groundwater into salt water.

The Speed of Groundwater Flows

The speed at which water moves underground strongly affects the balance between discharge and recharge. Most groundwaters flow slowly—a fact of nature that is responsible for our groundwater supplies. If groundwater flowed as rapidly as streams, aquifers would run dry after a period without rain, just as many small streams do. But the slow flow of groundwater also makes rapid recharge impossible if groundwater levels are lowered by excessive pumping.

Although all groundwaters flow through aquifers slowly, some flow more slowly than others. In the middle of the nineteenth century, Henri Darcy, town engineer of Dijon, France, proposed an explanation for the difference in flow rates. While studying the town's water supply, Darcy measured the elevations of water in various wells

(a)

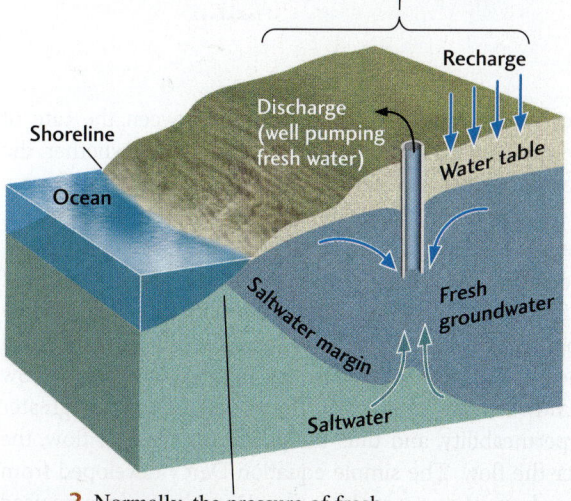

1. The boundary between fresh and salty groundwater along shorelines is determined by the balance between recharge and discharge in the freshwater aquifers.

2. Normally, the pressure of fresh water keeps the saltwater margin slightly offshore.

(b)

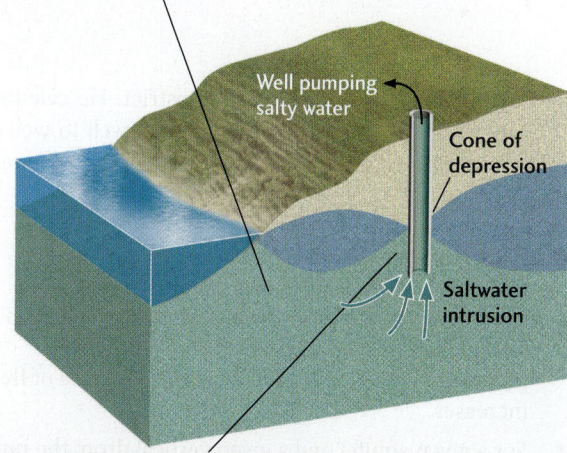

1. Extensive pumping lowers the pressure of the fresh water, allowing the saltwater margin to move inland.

2. This movement creates both a cone of depression and an inverted cone of depression that brings salty water into the well. A well that formerly pumped fresh water now pumps salty water.

FIGURE 17.17 The balance between recharge and discharge maintains the saltwater margin of a coastal aquifer. Maintaining the balance prevents salt water from entering wells (a), but over-pumping causes water to enter wells (b).

PRACTICING GEOLOGY EXERCISE

How Much Water Can Our Wells Produce?

The most important question anyone who is considering drilling a well can ask is whether that well can produce enough water to satisfy their needs. A well drilled in one kind of formation might produce plenty of water, while another not far away, but in a different formation, might not. How can we predict how much water a well in a certain location will produce?

Henri Darcy turned his conceptual understanding of the principles of groundwater flow into a simple and very useful mathematical equation. This equation—*Darcy's law*—shows how geologic factors control the rate of water flow through an aquifer:

$$Q = A\left[\frac{K(h_a - h_b)}{l}\right]$$

This equation relates the following values:

- The volume of water flowing in a certain time (Q);
- The cross-sectional area of the aquifer through which the volume of water flows (A);
- The permeability of the rock or the soil through which the water flows, called the *hydraulic conductivity* of the aquifer (K);
- The hydraulic gradient, which is determined by installing test wells at two points, a and b, measuring the difference in height between them ($h_a - h_b$);
- The distance between the test wells (l).

The equation tells us that water flow will increase if the cross-sectional area of the aquifer increases, if the hydraulic gradient increases, or if the hydraulic conductivity increases.

In rural and many suburban parts of the United States, it is still common practice to drill wells for the family water supply. When choosing a home site, a family must be careful to take into account the geology of the site and whether it is suitable for sufficient water flow. Will the water flowing through the well at point B in the accompanying diagram produce enough water for a family's needs? It depends on a number of factors, including the type of formation in which the well is drilled. We can use Darcy's law to evaluate the effects of hydraulic conductivity on the amount of water that will flow through the well.

Using the measurements provided in the diagram, we find the following values:

Cross-sectional area of well pipe
$$= A = 0.25 \text{ m}^2$$

$$\text{Hydraulic gradient} = \left[\frac{h_a - h_b}{l}\right]$$

$$= \left[\frac{440 \text{ m} - 415 \text{ m}}{1250 \text{ m}}\right]$$

$$= \left[\frac{25 \text{ m}}{1250 \text{ m}}\right]$$

$$= 0.02$$

---

and mapped the water table in the district. He calculated the distances that the water traveled from well to well and measured the permeabilities of the aquifers. Here are his findings:

- For a given aquifer and a given distance of travel, the rate at which water flows from one point to another is directly proportional to the vertical drop in elevation of the water table between the two points: As the vertical drop increases, the rate of flow increases.
- For a given aquifer and a given vertical drop, the rate of flow is inversely proportional to the distance the water travels: As the distance increases, the rate of flow decreases. The ratio of the vertical drop to the flow distance is known as the **hydraulic gradient.**

Darcy reasoned that the relationship between the rate of flow and the hydraulic gradient should hold whether the water is moving through a well-sorted gravel aquifer or a less permeable silty sandstone aquifer. As you might guess, water moves faster through the large pore spaces of the well-sorted gravel than through the torturous twists and turns of the finer-grained and less permeable silty sandstone. Darcy recognized the importance of permeability and included a measure of permeability in his final explanation of how groundwater flows. So, other things being equal, the greater the permeability, and thus the greater the ease of flow, the faster the flow. The simple equation Darcy developed from these observations, now known as **Darcy's law,** can be used to predict the behavior of groundwater and thus has important applications in the management of water resources, as discussed in the Practicing Geology Exercise.

Next, using the values for K shown below, we solve for Q for well-sorted sand:

Material	Hydraulic conductivity (K)
Clay	0.001 m/day
Silty sand	0.3 m/day
Well-sorted sand	40 m/day
Well-sorted gravel	3750 m/day

$$Q = A\left[\frac{K(h_a - h_b)}{l}\right]$$
$$= 0.25 \text{ m}^2 \times 40 \text{ m/day} \times 0.02$$
$$= 0.2 \text{ m}^3/\text{day (about 50 gallons/day)}$$

and for clay:

$$Q = A\left[\frac{K(h_a - h_b)}{l}\right]$$
$$= 0.25 \text{ m}^2 \times 0.001 \text{ m/day} \times 0.02$$
$$= 0.000005 \text{ m}^3/\text{day (about 1 teaspoon/day)}$$

It is clear that if the well is drilled in well-sorted sand, it might provide just enough water for a family of four, assuming each person uses 10 gallons per day for drinking, showering, toilet use, cooking, cleaning, and yard maintenance. In contrast, if the well is drilled in clay, it is likely to be a severe disappointment.

BONUS PROBLEM: Use Darcy's law to find the volume of water that could flow through the well in one day if it were drilled in silty sand or in well-sorted gravel.

Flow velocities calculated by Darcy's law have been confirmed experimentally by measuring how long it takes a harmless dye introduced into one well to reach another well. In most aquifers, groundwater moves at a rate of a few centimeters per day. In very permeable gravel beds near the surface, groundwater may travel as much as 15 cm/day. (This speed is still much slower than the speeds of 20 to 50 cm/s typical of river flows.)

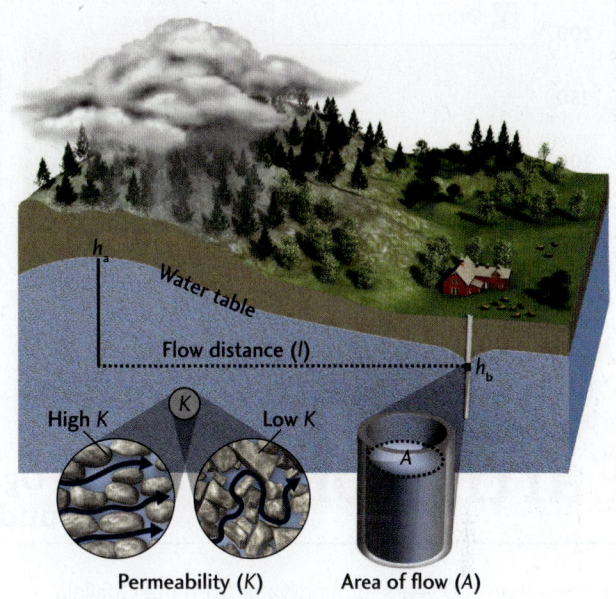

Groundwater Resources and Their Management

Large parts of North America rely solely on groundwater for all their water needs. The demand for groundwater resources has grown as populations have increased and uses such as irrigation have expanded (Figure 17.18). Many areas of the Great Plains and other parts of the Midwest rest on sandstone formations, most of which are confined aquifers that function like the one shown in Figure 17.12. These aquifers are recharged from outcrops in the western high plains, some very close to the foothills of the Rocky Mountains. From there, the water flows downhill in an easterly direction over hundreds of kilometers. Thousands of wells have been drilled into these aquifers, which constitute a major water resource.

Darcy's law tells us that water flows at a rate proportional to the slope of an aquifer between its recharge area and a given well. In the Great Plains, the slopes are gentle, and water moves through the aquifers slowly, recharging them at low rates. At first, many of the wells drilled in these aquifers were artesian, and the water flowed freely. As more wells were drilled, however, the water levels dropped, and the water had to be pumped to the surface. Today, water is being withdrawn from these aquifers faster than the slow flow from distant recharge areas can fill them, so the reservoirs of groundwater they contain are being depleted (see Earth Policy 17.2).

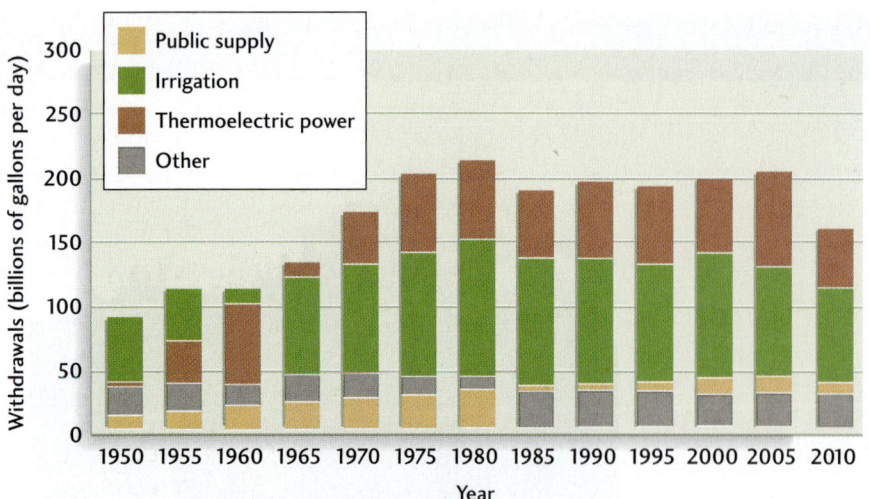

FIGURE 17.18 Groundwater withdrawals, United States, 1950–2010. [Data from U.S. Geological Survey.]

Earth Policy 17.2 The Ogallala Aquifer: An Endangered Groundwater Resource

For more than 100 years, groundwater from the Ogallala aquifer, a formation of sand and gravel, has supplied fresh water to the cities, towns, ranches, and farms of much of the southern Great Plains. The population of the region has climbed from a few thousand people late in the nineteenth century to about a million today. Pumping of water from the aquifer, primarily for irrigation, has been so extensive—about 6 billion cubic meters of water per year from 170,000 wells—that recharge from rainfall cannot keep up. Water pressure in the wells has declined steadily, and the water table has dropped by 30 m or more.

Natural recharge of the Ogallala aquifer is very slow because rainfall on the southern Great Plains is sparse, the degree of evaporation is high, and the recharge area is small. The waters in the Ogallala aquifer today may have been supplied as much as 10,000 years ago, during the Wisconsin glaciation, when the climate of the Great Plains was wetter. At current rates of recharge, if all pumping were to stop now, it would take several thousand years for the water table to recover its original elevation and for well pressure to be restored. Some scientists have attempted to recharge the aquifer artificially by injecting water from shallow lakes that form in wet seasons on the high plains. These experiments have managed to increase recharge, but the aquifer is still in danger over the long term.

It is estimated that the remaining supplies of groundwater in the Ogallala aquifer will last only into the early decades of the twenty-first century. If the water it produces cannot be replaced, about 5.1 million acres of irrigated land in western Texas and eastern New Mexico will dry up—and so will 12 percent of the country's supply of cotton, corn, sorghum, and wheat and a significant fraction of the feedlots for the nation's cattle.

Other aquifers in the northern Great Plains and elsewhere in North America are in a similar condition. In three

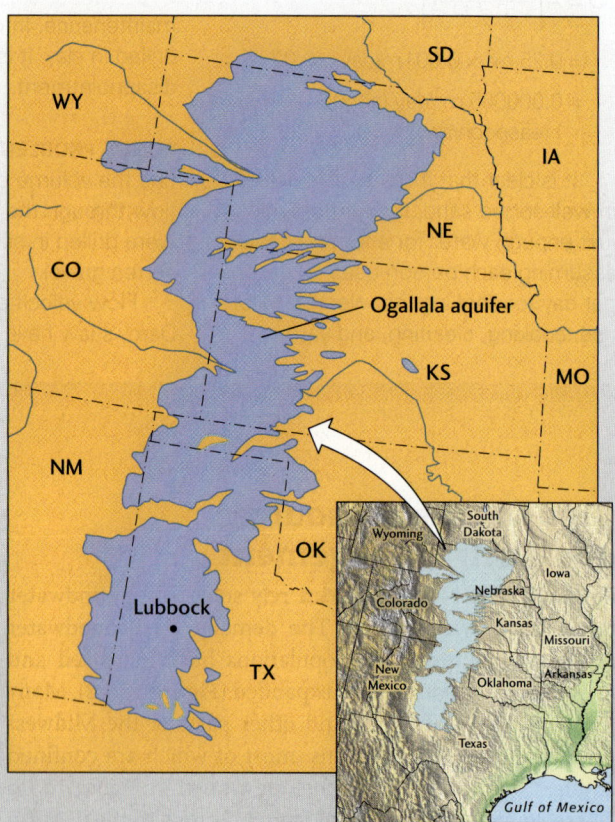

Much of the southern Great Plains region is underlain by the Ogallala aquifer. The blue area represents the aquifer. The general recharge area is located along the western margin of the aquifer. [Data from U.S. Geological Survey.]

Erosion by Groundwater

Every year, thousands of people visit caves, either on tours of popular attractions such as Mammoth Cave, Kentucky, or in adventurous explorations of little-known caves. These underground open spaces are actually enormous vugs produced by the dissolution of limestone—or, rarely, of other soluble rocks such as evaporites—by groundwater. Huge amounts of limestone have been dissolved to make some caves. Mammoth Cave, for example, has tens of kilometers of large and small, interconnected chambers. The Big Room at Carlsbad Caverns, New Mexico, is more than 1200 m long, 200 m wide, and 100 m high.

Limestone is widespread in the upper parts of Earth's crust, but caves form only where limestone is located at or near the ground surface and where enough carbon dioxide–rich or sulfur dioxide–rich water infiltrates the surface to dissolve extensive areas of this relatively soluble rock. As we saw in Chapter 16, atmospheric carbon dioxide dissolved in rainwater forms carbonic acid, which enhances the dissolution of limestone. Water that infiltrates soil may pick up even more carbon dioxide from plant roots, microorganisms, and other soil-dwelling organisms that give off this gas. As this carbon dioxide–rich water moves through the unsaturated zone to the saturated zone, it creates openings as it dissolves carbonate minerals. These openings are enlarged as the limestone dissolves along joints and fractures, forming a network of rooms and passages. Extensive cave networks form in the saturated zone, where—because the caves are filled with water—dissolution takes place over all surfaces, including floors, walls, and ceilings.

We can explore caves that were once below the water table but are now in the unsaturated zone because the water table has dropped. In these caves, now air-filled,

major areas of the United States—Arizona, the high plains, and California—groundwater supplies have been significantly depleted.

Ancient Water

In one of the deepest mines on Earth geologists recently discovered pockets of ancient water trapped in rock that is more than 1.5 billion years old. The discovery was based on a novel dating technique that exploits isotopes of xenon. Xenon and other noble gases accurately record when fluid masses last were in contact with the atmosphere.

The only waters older than this are minute, pinhead-size inclusions within minerals found in rocks that are over 3 billion years old. But water this abundant that actually flows from the rock has never been known before. The water occurs within open fractures, formed billions of years ago when tectonic forces related to continental formation created extensive fracture systems within metamorphic rocks. Some of these fractures became filled with economically valuable minerals but others were just filled with water that has never been in contact with the atmosphere.

This discovery has implications for the habitability of deep crustal environments. The water contains hydrogen and methane that could be used by microorganisms adapted to live in extreme environments (see Chapter 22, "Geobiology: Life Interacts with Earth"). If scientists were able to prove that microbes also live in this environment, then it would show that they too, along with the water, have been evolving in isolation for potentially billions of years. And, as we turn our attention increasingly to wonder about the potential habitability of Mars, it allows for the possibility that similar microbes could also occupy similar subterranean fracture systems that exist on planetary timescales.

A variety of innovative approaches are being used to enhance the sustainability of groundwater resources. In some areas, efforts to reduce excessive discharge have been supplemented by attempts to increase the recharge of aquifers artificially. On Long Island, for example, the water authority drilled a large system of recharge wells to pump treated wastewater into the ground. The water authority also constructed large, shallow basins over natural recharge areas to increase infiltration by catching and diverting runoff, including stormwater and industrial waste drainage. The officials in charge of the program knew that urban development can decrease recharge by interfering with infiltration. As urbanization progresses, the impermeable materials used to pave large areas for streets, sidewalks, and parking lots increase runoff and prevent water from infiltrating the ground. Such decreases in natural infiltration may deprive aquifers of much of their recharge. One remedy is to catch and use stormwater runoff in a systematic program of artificial recharge, as the Long Island water authority did. The multiple efforts of the water authority have helped to rebuild the Long Island aquifer, though not to its original levels.

Orange County, near Los Angeles, California, receives only about 15 inches of rainfall per year, yet this water must supply a population of 2.5 million people. Groundwater pumped from beneath the western part of the county meets about 75 percent of its requirements. The water table is gradually dropping, however, threatening to diminish this supply. To help replenish the supply, the Orange County Water District operates 23 wells that inject treated wastewater, mixed with groundwater from a second aquifer that is located beneath the county's main aquifer. The recycled water meets drinking water standards with additional treatment, but most of the contaminants are filtered out by the aquifer's pore network.

FIGURE 17.19 Luray Caverns, Virginia. Stalactites from the ceiling and stalagmites from the floor have joined to form a column. [Ivan Vdovin/Alamy.]

FIGURE 17.20 This aerial view shows the large sinkhole that continues to grow as workers try unsuccessfully to retrieve sunken sports cars from the depression in Winter Park, Florida, May 11, 1981. [AP Photo.]

water saturated with calcium carbonate may seep through the ceiling. As each drop of water drips from the cave's ceiling, some of its dissolved carbon dioxide evaporates, escaping to the cave's atmosphere. Its evaporation makes the calcium carbonate in the groundwater solution less soluble, so each water droplet precipitates a small amount of calcium carbonate on the ceiling. These deposits accumulate, just as an icicle grows, in a long, narrow spike of carbonate, called a *stalactite*, suspended from the ceiling. When the drop of water falls to the cave floor, more carbon dioxide escapes, and another small amount of calcium carbonate is precipitated on the floor below the stalactite. These deposits also accumulate, forming a *stalagmite*. Eventually, a stalactite and a stalagmite may grow together to form a column (**Figure 17.19**).

Microbial extremophiles (see Chapter 22) have been discovered living in caves, despite the lack of sunlight and highly acidic conditions that prevent most organisms from living in these environments. Some geologists think these microorganisms contributed to the formation of Carlsbad Caverns by using sulfates dissolved from gypsum ($CaSO_4$) evaporites as an energy source and releasing sulfuric acid as a by-product. The sulfuric acid then helped to dissolve limestone to form the caves.

In some places, dissolution may thin the roof of a limestone cave so much that it collapses suddenly, producing a **sinkhole**: a small, steep depression in the land surface above the cave (**Figure 17.20**). Sinkholes are characteristic of a distinct type of topography known as *karst*, named for a region in the northern part of Slovenia. **Karst topography** is an irregular, hilly type of terrain characterized by sinkholes, caves, and a lack of surface streams (**Figure 17.21**). Underground drainage channels replace the normal surface drainage system of small and large streams. The short,

scarce streams often end in sinkholes, detouring underground and sometimes, reappearing miles away.

Karst topography is found in regions with three characteristics:

1. A humid climate with abundant vegetation (providing carbon dioxide–rich waters)
2. Extensively jointed limestone formations
3. Appreciable hydraulic gradients

In North and Central America, karst topography is found in limestone terrains of Indiana, Kentucky, and Florida and on the Yucatán Peninsula of Mexico. It is also well developed on uplifted coral limestone terrains formed from tropical island arcs in the late Cenozoic era.

Karst terrains often have environmental problems, including the potential for catastrophic cave-ins and surface subsidence due to the collapse of underground spaces. The spectacular tower karst of southeastern China formed when cave networks collapsed to form sinkholes, which then expanded and merged, leaving "towers" behind (**Figure 17.22**).

Water Quality

Unlike people in many other parts of the world, North Americans are fortunate in that almost all of their public water supplies are free of bacterial contamination, and the vast majority are free enough of chemical contaminants to drink safely. Yet as more rivers become polluted, and more aquifers are contaminated by toxic wastes, North Americans are likely to see changes in water quality. Most residents of the United States are beginning to see their supply of fresh, pure water as a limited resource. Many people now travel

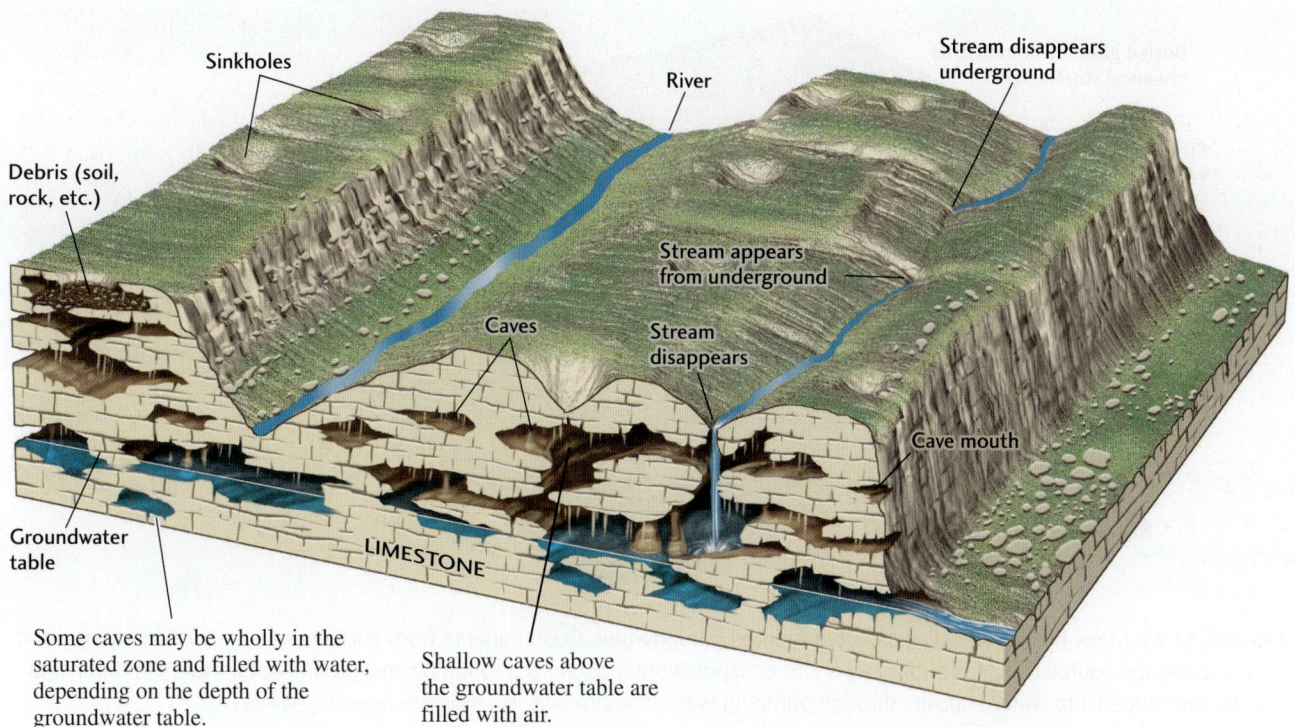

FIGURE 17.21 Some major features of karst topography are caves, sinkholes, and disappearing streams.

FIGURE 17.22 The tower karst of southeastern China is a spectacular terrain that features isolated hills with nearly vertical slopes. [Reimar/Alamy.]

with their own supply of bottled water, supplied by either home-installed purification systems or commercially available spring water.

Contamination of the Water Supply

The quality of groundwater is often threatened by a variety of contaminants. Most of these contaminants are chemicals, though microorganisms in water can also have negative effects on human health under certain conditions.

Lead Pollution Lead is a well-known pollutant produced by industrial processes that inject contaminants into the atmosphere. When water vapor condenses in the atmosphere, lead is incorporated into precipitation, which then transports it to Earth's surface. Lead is routinely eliminated from public water supplies by chemical treatment before the water is distributed through the water mains. In older homes with lead pipes, however, lead can leach into the water. Even in newer construction, lead solder used to connect copper pipes and metals used in faucets may be sources of contamination. Replacing old lead pipes with durable plastic pipes can reduce lead contamination. Even letting the water run for a few minutes to clear the pipes can help.

Other Chemical Contaminants A number of human activities produce chemicals that can contaminate groundwater (**Figure 17.23**). Some decades ago, when we knew much less about the health and environmental effects of toxic wastes, industrial, mining, and military wastes now known to be hazardous were dumped on the ground,

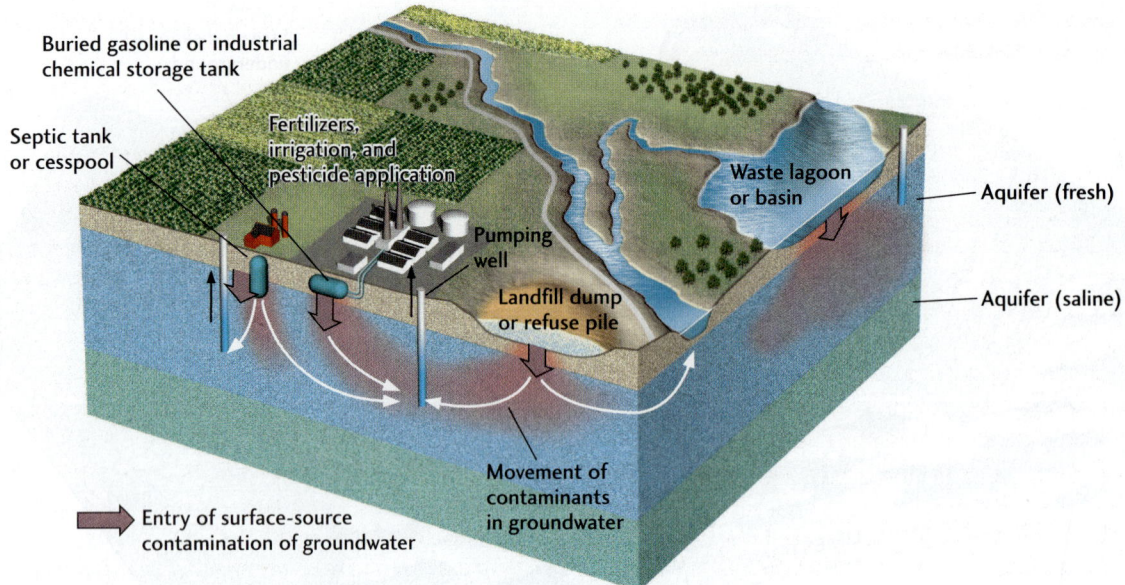

FIGURE 17.23 Many human activities can contaminate groundwater. Contaminants from surface sources such as dumps and from subsurface sources such as septic tanks and cesspools enter aquifers through normal groundwater flow. Contaminants may be introduced into water supplies through pumping wells. [Data from U.S. Environmental Protection Agency.]

disposed of in lakes and streams, or discharged underground. Even though many of these sources of pollution are being addressed, the contaminants are still making their way through aquifers by the slow flow of groundwater, and toxic chemicals are still entering groundwater from a number of other sources.

The disposal of chlorinated solvents—such as trichloroethylene (TCE), widely used as a cleaner in industrial processes—poses a formidable problem. These solvents persist in the environment because they are difficult to remove from contaminated waters. The burning of coal and the incineration of municipal and medical waste emit mercury into the atmosphere, which then contaminates water supplies. Buried gasoline storage tanks can leak, and road salt inevitably drains into the soil and ultimately into aquifers. Rain can wash agricultural pesticides, herbicides, and fertilizers into the soil, from which they percolate downward into aquifers. In some agricultural areas where nitrate fertilizers are heavily used, groundwaters contain high concentrations of nitrate. In one recent study, 21 percent of the shallow wells sampled exceeded the maximum amounts of nitrate (10 ppm) allowed in drinking water in the United States. Such high nitrate levels pose a danger of "blue baby" syndrome (an inability to maintain healthy oxygen levels) to infants 6 months old and younger.

Radioactive Wastes There is no easy solution to the problem of groundwater contamination by radioactive wastes. When radioactive wastes are buried underground, they may be leached by groundwater and find their way into aquifers. Storage tanks and burial sites at the atomic weaponry plants in Oak Ridge, Tennessee, and Hanford, Washington, have already leaked radioactive wastes into shallow groundwaters.

Microorganisms The most widespread causes of groundwater contamination by microorganisms are leaky residential septic tanks and cesspools. These containers, widely used in neighborhoods that lack full sewer networks, are buried settling tanks in which bacteria decompose the solid wastes from household sewage. To prevent contamination of drinking water, cesspools should be replaced by septic tanks, which must be installed at a sufficient distance from water wells in shallow aquifers.

Reversing Contamination

Can we reverse the contamination of groundwater supplies? Yes, but the process is costly and very slow. The faster an aquifer recharges, the easier it is to decontaminate. If the recharge rate is rapid, fresh water moves into the aquifer as soon as we close off the sources of contamination, and in a relatively short time, the water quality is restored. Even a fast recovery, however, can take a few years.

The contamination of slowly recharging aquifers is more difficult to reverse. The rate of groundwater movement may be so slow that contamination from a distant source takes a long time to appear. By the time it does, it is too late for rapid remediation. Even after the recharge area has been cleaned up, some contaminated deep aquifers extending hundreds of kilometers from the recharge area may not be free of contaminants for many decades.

When public water supplies are polluted, we can pump the water and then treat it chemically to make it safe, but that is an expensive procedure. Alternatively, we can try to treat the water while it remains underground. In one moderately successful experimental procedure, contaminated water was funneled into a buried bunker full of iron filings, which detoxified the water by reacting with the contaminants. The reactions produced new, nontoxic compounds that attached themselves to the iron filings.

Is the Water Drinkable?

Much of the water in groundwater reserves is unusable not because it has been contaminated by human activities, but because it naturally contains large quantities of dissolved materials. Water that tastes agreeable and is not dangerous to human health is called **potable** water. The amounts of dissolved materials in potable waters are very small, usually measured by weight in parts per million (ppm). Potable groundwaters of good quality typically contain about 150 ppm total dissolved materials because even the purest natural waters contain some dissolved substances derived from weathering. Only distilled water contains less than 1 ppm dissolved materials.

Geologic studies of streams and aquifers allow us to improve the quality of our water resources as well as their quantity. The many cases of groundwater contamination caused by human activity have led to the establishment of water quality standards based on medical studies. These studies have concentrated on the effects of ingesting average amounts of water containing various quantities of contaminants, both natural and anthropogenic. For example, the U.S. Environmental Protection Agency has set the maximum allowable concentration of arsenic, a well-known poison, at 0.05 ppm (Figure 17.24). Natural contamination of groundwater by arsenic is particularly acute in Bangladesh, where groundwater provides 97 percent of the drinking water supply. Geologists are helping to guide the placement of new wells that draw water with acceptable concentrations of arsenic.

Groundwater is almost always free of solid particles when it seeps into a well from a sand or sandstone aquifer. The pore networks in the rock or sand act as a fine filter, removing small particles of clay and other solids, and even straining out microorganisms and some large viruses. Limestone aquifers may have larger pores and so may filter water less efficiently. Any microbial contamination found at the bottom of a well is usually introduced from nearby underground sewage disposal systems, often when septic tanks leak or are located too close to the well.

Some groundwaters, although perfectly safe to drink, simply taste bad. Some have a disagreeable taste of "iron" or are slightly sour. Groundwaters passing through limestone dissolve carbonate minerals and carry away calcium, magnesium, and bicarbonate ions, making the water "hard." Hard water may taste fine, but it does not lather readily when used with soap. Water passing through waterlogged forests or swampy soils may contain dissolved organic compounds and hydrogen sulfide, which give the water a disagreeable smell similar to rotten eggs.

How do these differences in taste and quality arise in safe drinking waters? Some of the highest-quality, best-tasting public water supplies come from lakes and artificial surface reservoirs, many of which are simply collecting places

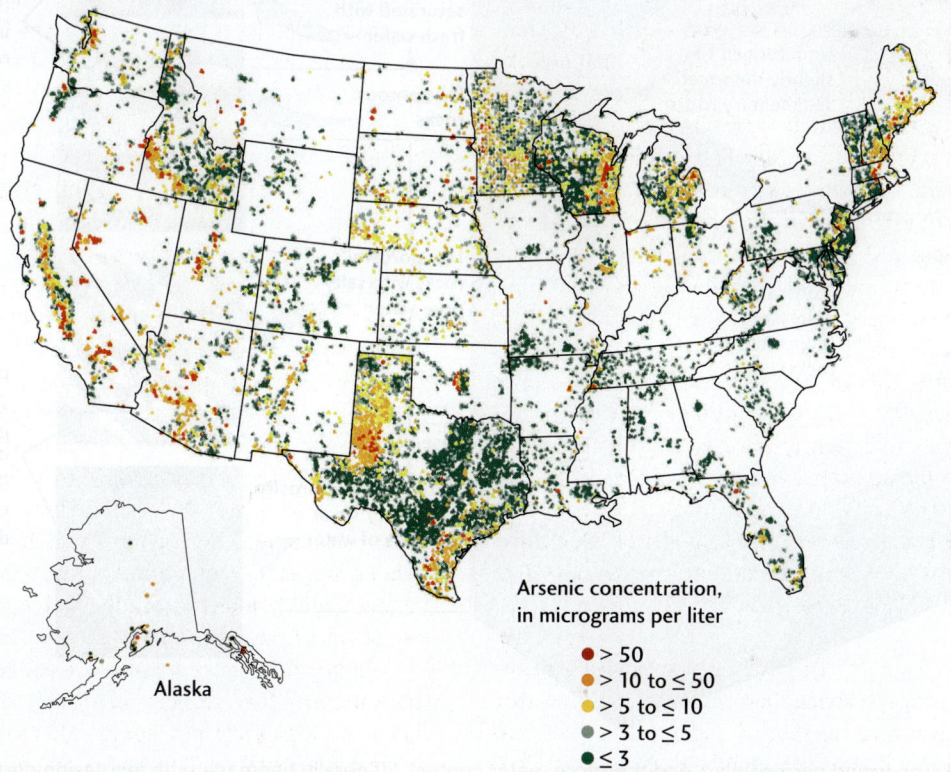

FIGURE 17.24 The National Water Quality Assessment Program measures arsenic, radon, and uranium levels in groundwater samples throughout the United States. This map shows arsenic concentrations measured in micrograms per liter (mg/L) in 2012. [Data from U.S. Geological Survey.]

for rainwater. Some groundwaters taste just as good; these tend to be waters that pass through rocks that weather only slightly. Sandstones made up largely of quartz, for example, contribute little in dissolved materials, and thus waters passing through them have a pleasant taste.

As we have seen, the contamination of groundwater in relatively shallow aquifers is a serious problem, and remediation is difficult. But are there deeper groundwaters that we can use?

Water Deep in the Crust

Most crustal rocks below the groundwater table are saturated with water. Even in the deepest wells drilled for oil, some 8 or 9 km deep, geologists find water in permeable formations. At these depths, groundwaters move so slowly—probably less than a centimeter per year—that they have plenty of time to dissolve minerals from the rocks through which they pass. Thus, dissolved materials become more concentrated in these waters than in near-surface waters, making them unpotable. For example, deep groundwaters that pass through salt beds, which dissolve quickly, tend to contain large concentrations of sodium chloride.

At depths greater than 12 to 15 km, deep in the basement igneous and metamorphic rocks that underlie the sedimentary formations of the upper crust, porosities and permeabilities are very low due to the tremendous weight of the overlying rocks. Although these rocks contain very little water, they are saturated (Figure 17.25). Even some mantle rocks are presumed to contain water, although in minute quantities.

Hydrothermal Waters

In some regions of the crust, such as along subduction zones, hot waters containing dissolved carbon dioxide play an important role in the chemical reactions of metamorphism, as we saw in Chapter 6. These *hydrothermal waters* dissolve some minerals and precipitate others.

Most hydrothermal waters of the continents come from meteoric waters that percolate downward to deeper regions

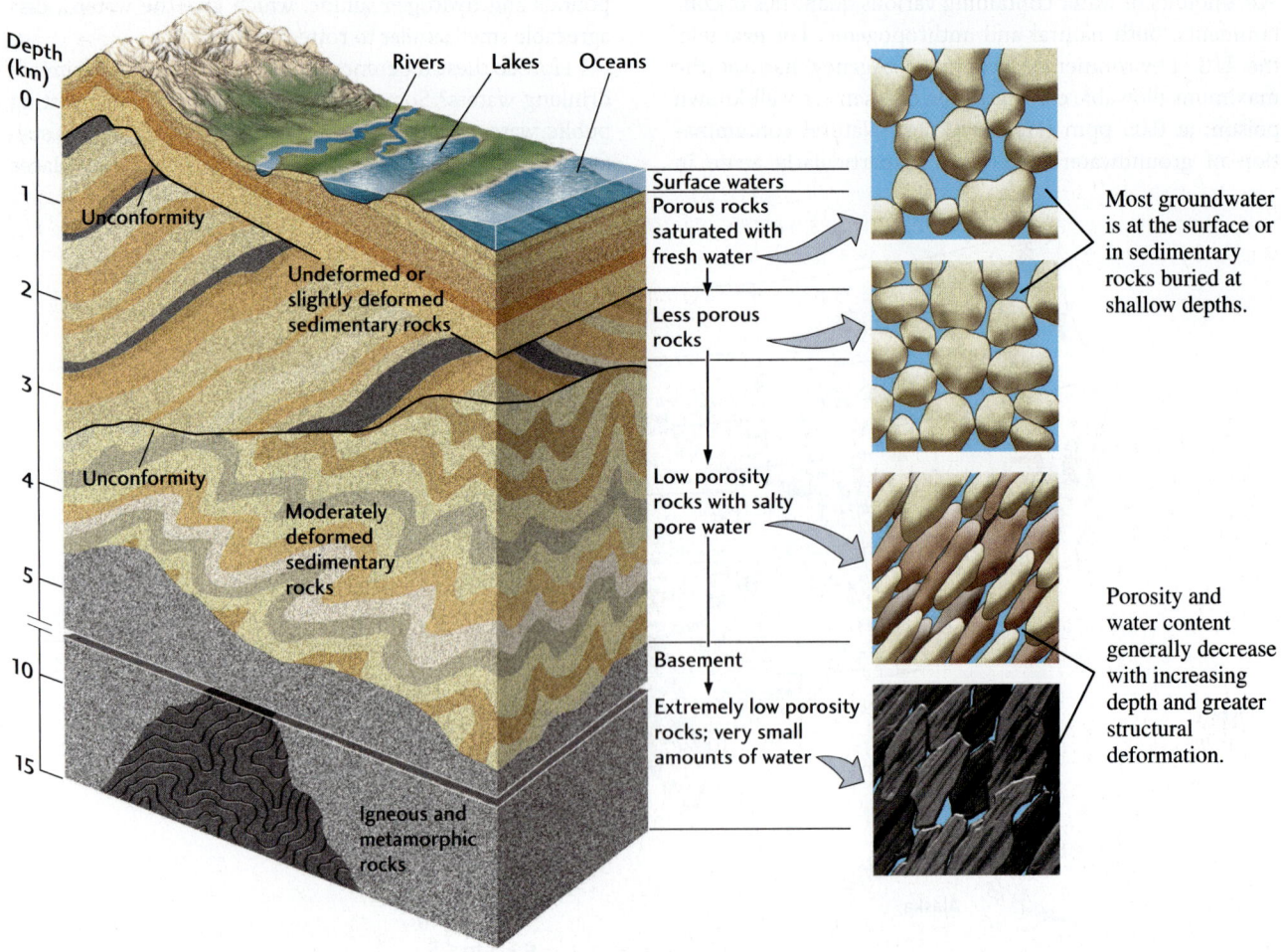

FIGURE 17.25 Porosity and permeability, and therefore water content, generally decrease with increasing depth in Earth's crust.

of the crust. The percolation rates for meteoric waters deep in the crust are very low, and thus the water may be very old. It has been determined that the water at Hot Springs, Arkansas, derives from rain and snow that fell more than 4000 years ago and slowly infiltrated the ground. Water that escapes from magma can also contribute to hydrothermal waters. In areas of igneous activity, sinking meteoric waters are heated as they encounter hot masses of rock. The hot meteoric waters then mix with water released from the nearby magma.

Hydrothermal waters are loaded with chemical substances dissolved from rocks at high temperatures. As long as the water remains hot, the dissolved material stays in solution. However, as hydrothermal waters reach the surface, where they cool quickly, they may precipitate various minerals, such as opal (a form of silica) and calcite or aragonite (forms of calcium carbonate). Crusts of calcium carbonate produced at some hot springs build up to form the rock travertine, which can form impressive deposits such as those seen at Mammoth Hot Springs in Yellowstone National Park (**Figure 17.26**). Amazingly, microbial extremophiles that can withstand temperatures above the boiling point of water have been discovered in these environments, where they may contribute to the formation of calcium carbonate crusts. Hydrothermal waters that cool slowly below the surface deposit some of the world's richest metallic ores, as we learned in Chapter 3. Hot springs and geysers exist where hydrothermal waters migrate rapidly upward without losing much heat and emerge at the surface, sometimes at boiling temperatures. Hot springs flow steadily; geysers erupt hot water and steam intermittently.

The theory explaining the intermittent eruption of geysers is an example of geologic deduction. We cannot observe the process directly because the dynamics of underground hydrothermal systems are hidden from sight hundreds of meters below the surface. Geologists hypothesized that geysers are connected to the surface by a system of very irregular and crooked fractures, recesses, and openings, in contrast to the more regular and direct plumbing of hot springs (**Figure 17.27**). The irregular fractures sequester some water in recesses, thus helping to prevent the deepest waters from mixing with shallower waters and cooling. The deepest waters are heated by contact with hot rock. When they reach the boiling point, steam starts to ascend and heats the shallower waters, increasing the pressure and triggering an eruption. After the pressure is released, the geyser becomes quiet as the fractures slowly refill with water.

In 1997, geologists reported the results of a novel technique used to study geysers. They lowered a miniature video camera to about 7 m below the surface of a geyser. They found that the geyser shaft was constricted at that point. Farther down, the shaft widened to a large chamber containing a wildly boiling mixture of steam, water, and what appeared to be carbon dioxide bubbles. These direct observations dramatically confirmed the previous theory of how geysers work.

Although hydrothermal waters are useful to human society as sources of geothermal energy and metallic ores, these waters do not contribute to surface water supplies, primarily because they contain so much dissolved material.

FIGURE 17.26 Travertine deposits at Mammoth Hot Springs, Yellowstone National Park, form large lobelike masses made of aragonite and calcite. [John Grotzinger.]

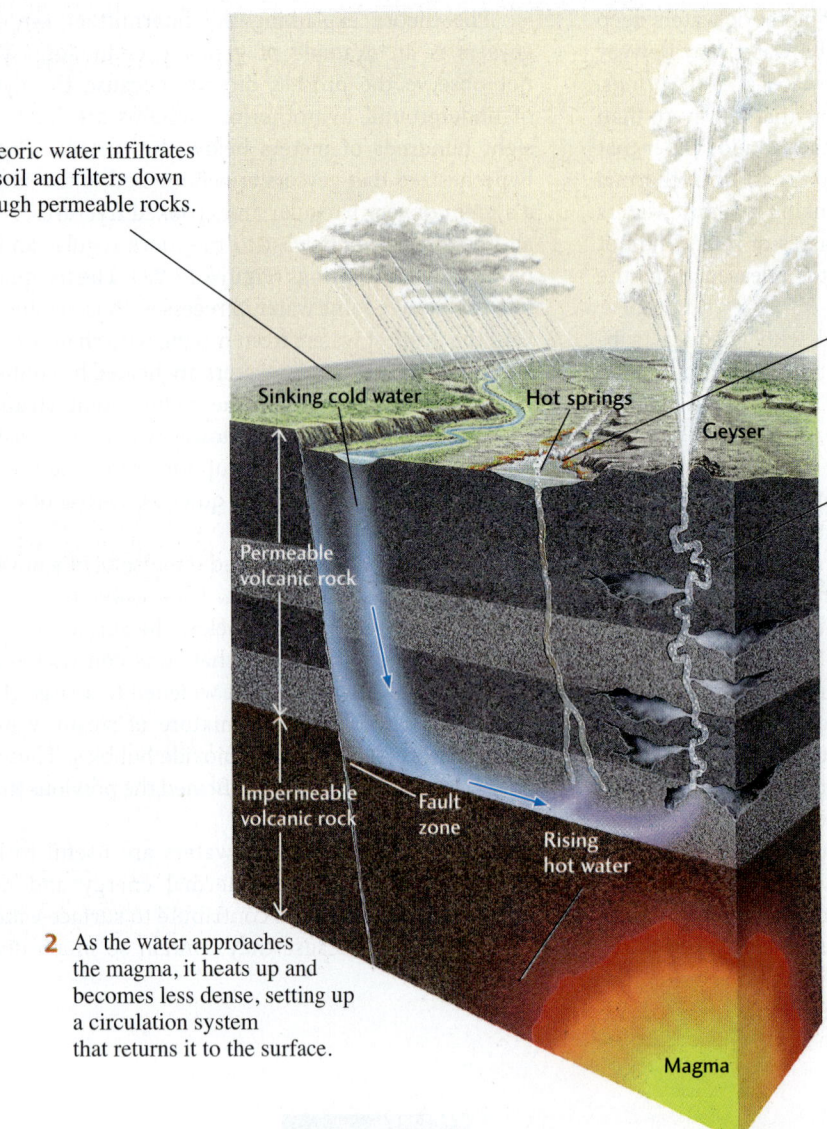

FIGURE 17.27 Circulation of water over magma or hot rock deep in the crust produces geysers and hot springs.

Ancient Microorganisms in Deep Aquifers

In recent years, geologists have explored aquifers deep underground (as much as several thousand meters) in search of potable groundwater. They failed to find it, but they did unveil a remarkable interaction between the biosphere and the lithosphere. They found microorganisms living in the groundwater in huge numbers. These chemoautotrophic microorganisms, well out of the reach of sunlight, derive their energy by dissolving and metabolizing minerals in rocks. These metabolic reactions, aside from serving as a source of energy for the microorganisms, continue the weathering process underground. The chemicals released by these reactions make the water unpotable.

Geobiologists think that the ancestors of these microorganisms were enclosed within the pores of sediments, which were then buried at great depths, where they became sealed off from the surface. In some cases, these deep aquifers may not have been in contact with Earth's surface for hundreds of millions of years. Yet the microorganisms persisted, living solely on chemicals provided by the dissolution of minerals and evolving new generations of descendants without interference from any other organisms. These ecosystems, involving only microorganisms, are probably the most ancient on Earth and testify to the remarkable balance that can be achieved between life and environment.

KEY TERMS AND CONCEPTS

- aquiclude (p. 508)
- aquifer (p. 508)
- artesian flow (p. 508)
- Darcy's law (p. 512)
- discharge (p. 506)
- drought (p. 499)
- groundwater (p. 497)
- groundwater table (p. 506)
- hydraulic gradient (p. 512)
- hydrologic cycle (p. 497)
- hydrology (p. 497)
- infiltration (p. 498)
- karst topography (p. 516)
- meteoric water (p. 503)
- permeability (p. 505)
- potable (p. 519)
- precipitation (p. 498)
- rain shadow (p. 499)
- recharge (p. 506)
- relative humidity (p. 498)
- runoff (p. 498)
- saturated zone (p. 506)
- sinkhole (p. 516)
- unsaturated zone (p. 506)

REVIEW OF LEARNING OBJECTIVES

17.1 Describe how water moves through the hydrologic cycle.

The water movements of the hydrologic cycle maintain a balance among the major reservoirs of water on Earth. Evaporation from the oceans, evaporation and transpiration from the continents, and sublimation from glaciers transfer water to the atmosphere. Precipitation returns water from the atmosphere to the oceans and the land surface. Runoff returns part of the precipitation that falls on land to the ocean. The remainder infiltrates the ground and forms groundwater. Differences in climate produce local variations in the balance among evaporation, precipitation, runoff, and infiltration.

Study Assignment: Figure 17.1

Exercise: What are the main reservoirs of water at and near Earth's surface?

Thought Question: Global warming causes evaporation from the oceans to increase greatly. How does this phenomenon affect the hydrologic cycle?

17.2 Summarize how local hydrology is influenced by the local climate.

The strongest influence on local hydrology is the local climate, especially temperature and precipitation levels. Many variations in climate are related to the average temperature of the air and the average amount of water vapor it contains, both of which affect precipitation. Humidity is the amount of water vapor in the air, expressed as a percentage of the total amount of water the air could hold at the same temperature if it were saturated. Droughts occur when there are periods of months or years with lower than normal precipitation, which affects vegetation, drinking water, farming, and grazing. Precipitation levels also affect runoff. In dry areas, the precipitation leaves the land by evaporation and infiltration. In more humid areas, a much higher proportion of the precipitation runs off in streams. Runoff collects and is stored in natural lakes, artificial reservoirs, and wetlands.

Study Assignment: Figure 17.3

Exercise: How do mountains form rain shadows?

Thought Question: Thinking about temperature and humidity, why do the tropics get lots of rain and the deserts get very little?

17.3 Explain the process of how water moves through the ground.

Groundwater forms as precipitation infiltrates the ground and travels through porous and permeable formations. The groundwater table is the boundary between the unsaturated and saturated zones. Groundwater moves downhill under the influence of gravity, eventually emerging at springs where the water table intersects the ground surface. Groundwater may flow through unconfined aquifers in formations of uniform permeability or in confined aquifers, which are bounded by aquicludes. Confined aquifers produce artesian flows and spontaneously flowing artesian wells. Darcy's law describes the rate of groundwater flow in relation to the hydraulic gradient and the permeability of the aquifer.

Study Assignment: Figure 17.11

Exercises: What is the difference between the saturated and unsaturated zones of an aquifer? How do aquicludes form a confined aquifer?

Thought Question: Your new house is built on soil-covered granitic bedrock. Although you think that prospects for drilling a water well are poor, a well driller who is familiar with the area says he has drilled many good water wells in this granite. What arguments might each of you offer to convince the other?

17.4 Classify the geologic processes that are affected by groundwater.

Erosion by groundwater in limestone terrains produces karst topography, characterized by caves, sinkholes, and disappearing streams. Limestone is widespread in the upper parts of the Earth's crust, but caves form only where limestone is located near the ground surface where enough carbon dioxide–rich or sulfur dioxide–rich water infiltrates the surface to dissolve extensive areas of this relatively soluble rock. If dissolution thins the roof of a limestone cave so much that it collapses suddenly, it produces a sinkhole. Sinkholes are characteristic of karst topography, which is an irregular, hilly terrain characterized by sinkholes, caves, and a lack of surface streams.

Study Assignment: Figure 17.21

Exercise: How does groundwater create karst topography?

Thought Question: You are exploring a cave and notice a small stream flowing on the cave floor. Where could the water be coming from?

17.5 Describe the factors that govern human use of groundwater resources.

As the human population grows, the demand for groundwater increases greatly, particularly where irrigation is widespread. As discharge exceeds recharge, many aquifers, such as those of the Great Plains of North America, are being depleted, and there is no prospect of their renewal for many years. Artificial recharge may help to renew some aquifers. The contamination of groundwater by industrial wastes, radioactive wastes, and sewage further reduces supplies of potable groundwater.

Study Assignment: Figure 17.23

Exercise: What are some common contaminants in groundwater?

Thought Question: Why would you recommend against extensive development and urbanization of the recharge area of an aquifer that serves your community?

17.6 Illustrate how water deep in the crust affects hydrothermal processes.

At great depths in the crust, rocks contain extremely small quantities of water because their porosities are very low. This water moves slowly and therefore dissolves minerals from the rocks through which they pass. These dissolved minerals become concentrated, making them unpotable. In some regions of the crust, such as along subduction zones, the heating of water forms hydrothermal waters, which may return to the surface as geysers and hot springs. Microorganisms also live in groundwater, out of the reach of sunlight, and derive their energy by dissolving and metabolizing minerals in rocks. These metabolic reactions continue the weathering process underground.

Study Assignment: Figure 17.27

Exercise: How do microorganisms survive deep in Earth's crust?

Thought Question: What geologic processes would you infer are taking place below the surface at Yellowstone National Park, which has many hot springs and geysers?

VISUAL LITERACY EXERCISE

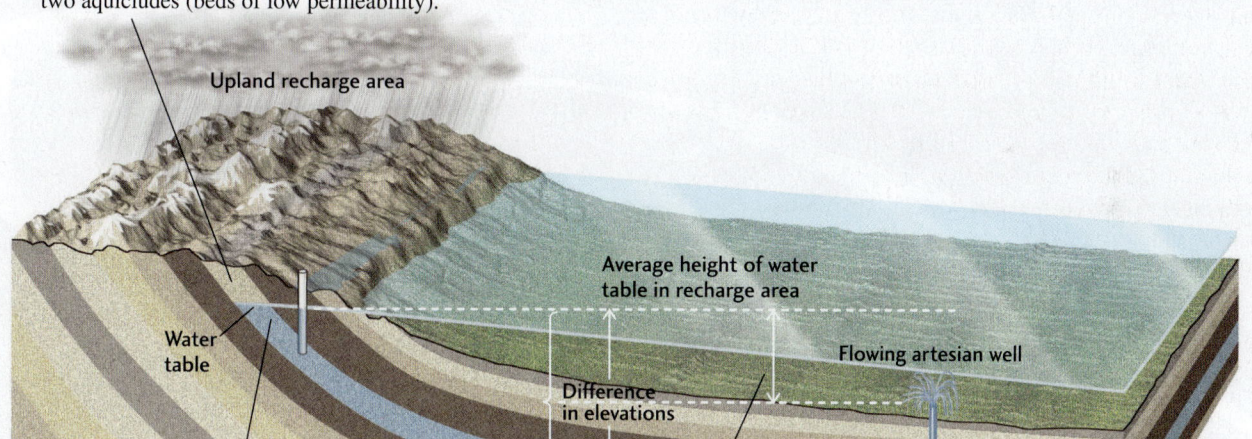

1. A confined aquifer is situated between two aquicludes (beds of low permeability).

2. The flow of an artesian well is governed by the difference in pressure caused by the difference between the height of the water table in the recharge area and the height of the top of the well.

3. If the well were as high as the water table in the recharge area, there would be no pressure difference and thus no flow.

FIGURE 17.12 A permeable formation situated between two aquicludes forms a confined aquifer, through which water flows under pressure.

1. How do uplands help with recharge?
2. What is an aquifer?
3. What is an aquiclude?
4. Why does an artesian well flow water?

Stream Transport: From Mountains to Oceans

18

The Form of Streams	528
Where Do Channels Begin? How Running Water Erodes Soil and Rock	535
How Currents Flow and Transport Sediment	537
Deltas: The Mouths of Rivers	541
Streams as Geosystems	543

Learning Objectives

Chapter 18 describes how streams form and how they accomplish their geologic work—how, on a large scale, streams carve valleys and develop vast networks of channels; and how, on a smaller scale, streams break up and erode solid rock. We examine how water flows in currents and how currents carry sediment. Then we return to a larger scale to look at streams as geosystems shaped by interactions between the plate tectonic and climate systems. After studying this chapter, you should be able to:

- **18.1** Identify how stream valleys and their channels and floodplains develop.
- **18.2** Summarize how running water erodes soil and rock.
- **18.3** Explain how flowing water in streams transports and deposits sediment.
- **18.4** Describe how drainage networks function as collection systems and deltas as distribution systems for water and sediment.
- **18.5** Define how a stream's longitudinal profile represents an equilibrium between erosion and sedimentation.

Remote river in Gates of the Arctic Park and National Preserve. The treeless slopes of the Arctic Divide near the Nunamuit village of Anaktuvuk Pass. [NPS Photo/Sean Tevebaugh.]

Before cars and airplanes existed, people traveled on rivers. In 1803, the United States purchased the Louisiana Territory from France. It was a huge tract of land, over 2 million square kilometers, taking in portions of what today is Texas and Louisiana and extending up to Montana and North Dakota. In 1804, President Thomas Jefferson asked Meriwether Lewis and William Clark to lead an expedition across this new territory and into western North America. One of their most important goals was to map the western rivers, which provided the key to opening up this uncharted frontier. Lewis and Clark decided to follow the Missouri River and its headwaters to their source. They then crossed the Rocky Mountains and followed the Columbia River westward to the Pacific Ocean. The total trip was 6000 km—with the section along the Missouri River alone extending over 3200 km—and upstream all the way.

The writings and maps produced by Lewis and Clark created a body of knowledge that could have been obtained only by following one of the great rivers that drain the interior of North America. On other continents and in other countries, other big rivers evoke a similar sense of adventure: in South America, the Amazon; in Asia, the Yangtze and Indus; and in Africa, the fabled Nile. Yet streams and rivers are not only the access routes for legendary explorations, but also the places where people settle and make their homes. A body of water flows through almost every town and city in most parts of the world. These streams have served as commercial waterways for barges and steamers and as water resources for resident populations and industries. The sediments they have deposited during floods have built fertile lands for agriculture. Living near a river also entails risks, however. When rivers flood, they destroy lives and property, sometimes on a huge scale.

In this chapter you will learn that streams are the lifelines of the continents. Their appearance is a record of the interaction of climate and plate tectonic processes. Tectonic processes lift up the land, producing the steep topography and slopes of mountainous regions. Climate determines where rain and snow will fall. Rainwater runs downhill, eroding the rocks and soils of the mountains, forming channels and carving out valleys as it gathers into streams. Streams carry back to the sea the bulk of the precipitation that falls on land and much of the sediment produced by erosion of the land surface. Streams are so important to understanding the role of climate and water on Earth that their discovery on Mars has fueled a generation of missions to search for evidence of water—and a different climate in the planet's ancient past.

The Form of Streams

We use the word **stream** for any body of water, large or small, that flows over the land surface, and **river** for the major branches of a large stream system. Most streams run through well-defined troughs called *channels*, which allow water to flow over long distances. As streams move across Earth's surface—in some places over bedrock, in others over unconsolidated sediments—they erode these materials and create *valleys*.

Identifying and mapping stream valleys were essential tasks for Lewis and Clark during their mission 200 years ago. As they traveled upstream and the river branched, they had to choose which branch was the larger of the two. They used two observations to help them make this choice: the width of the stream valley and the depth of the stream channel. Was the valley wide enough, and the channel deep enough, for their boats? Narrow valleys and shallow channels would mean that the branch led into a much shorter, and therefore less desirable, route; wider valleys and deeper channels, on the other hand, promised a longer passage up the main branch of the river.

Stream Valleys

A stream **valley** encompasses the entire area between the tops of the slopes on both sides of the stream. The cross-sectional profile of many stream valleys is V-shaped, but many other stream valleys have a broad, low profile like that shown in **Figure 18.1**. At the bottom of the valley

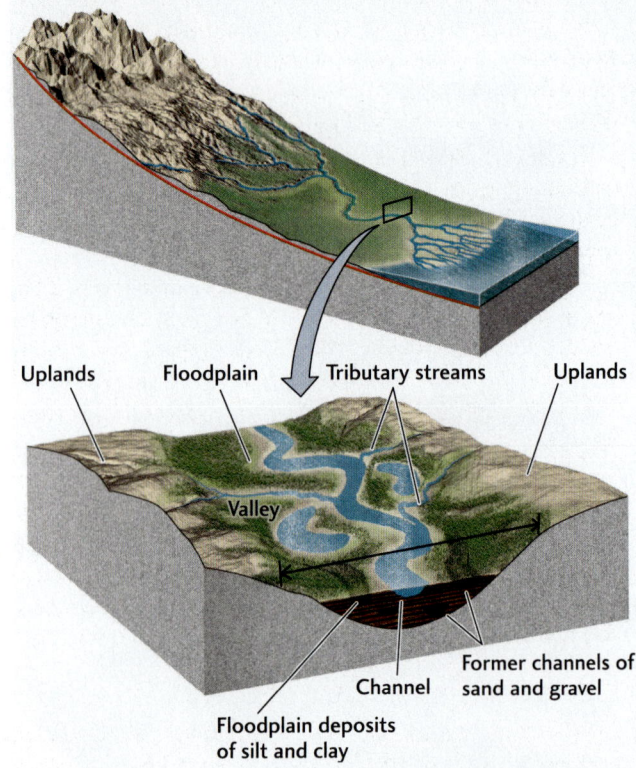

FIGURE 18.1 A stream flows in a channel that moves over a broad, flat floodplain in a wide valley. Floodplains may be narrow or absent in steep-walled valleys.

The Form of Streams

FIGURE 18.2 This section of the San Juan River in Utah is a good example of an incised meander belt, a deeply eroded, meandering, V-shaped valley with virtually no floodplain. [DEA/PUBBLI AER FOTO/Getty Images.]

is the **channel**, the trough through which the water runs. The channel carries all the water during normal, non-flood times. At low water levels, the stream may run only along the bottom of the channel. At high water levels, the stream occupies most of the channel. In broad valleys, a **floodplain**—a flat area about level with the top of the channel—lies on either side of the channel. It is this part of the valley that is flooded when the stream spills over its banks, carrying with it silt and sand from the channel.

In high mountains, stream valleys are narrow and have steep walls, and the channel may occupy most, or all, of the valley bottom (**Figure 18.2**). A small floodplain may be visible only at low water levels. In such valleys, the stream is actively cutting into the bedrock, a process that is characteristic of newly uplifted highlands in tectonically active areas. Its erosion of the valley walls is helped by chemical weathering and mass wasting. In lowlands, where tectonic uplift has long since ceased, the stream shapes its valley by eroding sediment particles and transporting them downstream. With a long time to operate, these processes produce gentle slopes and floodplains many kilometers wide.

Channel Patterns

As a stream channel makes its way along the bottom of a valley, it may run straight in some stretches and take a snaking, irregular path in others, sometimes splitting into multiple channels. The channel may run along the center of the floodplain or hug one edge of the valley.

Meanders On a great many floodplains, stream channels follow curves and bends called **meanders**, named for the Maiandros (now Menderes) River in Turkey, known in ancient times for its winding, twisting course. Meanders are the normal pattern for low-velocity streams flowing through gently sloping or nearly flat plains or lowlands, where their channels typically cut through unconsolidated sediments—fine sand, silt, or mud—or easily eroded bedrock. Meanders are less pronounced, but still common, in streams flowing down slightly steeper slopes over harder bedrock. In such terrain, meandering stretches may alternate with long, relatively straight ones.

A stream that has cut deeply into the curves and bends of its channel may produce incised meanders (see Figure 18.2). Other streams meander on somewhat wider floodplains bounded by steep, rocky valley walls. We are not sure why these two different patterns appear. We do know that meandering is widespread not only in streams, but also in many other kinds of flows. For example, the Gulf Stream, a powerful current in the western North Atlantic Ocean, meanders. Lava flows on Earth meander, and, as you will see in Chapter 20, planetary geologists have found meanders in former water channels on Mars as well as in lava flows on Mars and Venus.

Meanders on a floodplain migrate over periods of many years as the stream erodes the outside banks of bends, where the current is strongest (**Figure 18.3a**). As the outside banks are eroded, sediments are deposited to form curved sandbars, called **point bars**, along the inside banks, where the current is weaker (Figure 18.3b). In this way, meanders slowly shift position from side to side, as well as downstream, in a snaking motion something like that of a long rope being snapped. This migration may be quite rapid: Some meanders on the Mississippi River shift as much as 20 m/year. As meanders move, so do the point bars, building up an accumulation of sand and silt over the part of the floodplain across which the channel migrated.

As meanders migrate, sometimes unevenly, the bends may grow closer and closer together, until finally the stream bypasses one of them, often during a major flood. The stream then takes a new, shorter course. In its abandoned path, it leaves behind an **oxbow lake:** a crescent-shaped, water-filled loop (Figure 18.3c).

Engineers sometimes artificially straighten and confine a meandering river, channelizing it along a straight path with the aid of artificial levees made of concrete. The Army Corps of Engineers has been channelizing the Mississippi River since 1878. In 13 years, it decreased the length of the lower Mississippi by 243 km. Part of the severity of the disastrous Mississippi flood of 1993 was ascribed to channelization. Without channelization, floods are more frequent, but less damaging. With it, damage may be catastrophic when a flood breaches the artificial levees, as it did in 1993. Channelization has also been criticized for destroying wetlands and much of the natural life of the floodplain by cutting off the supply of sediments deposited by small, frequent floods.

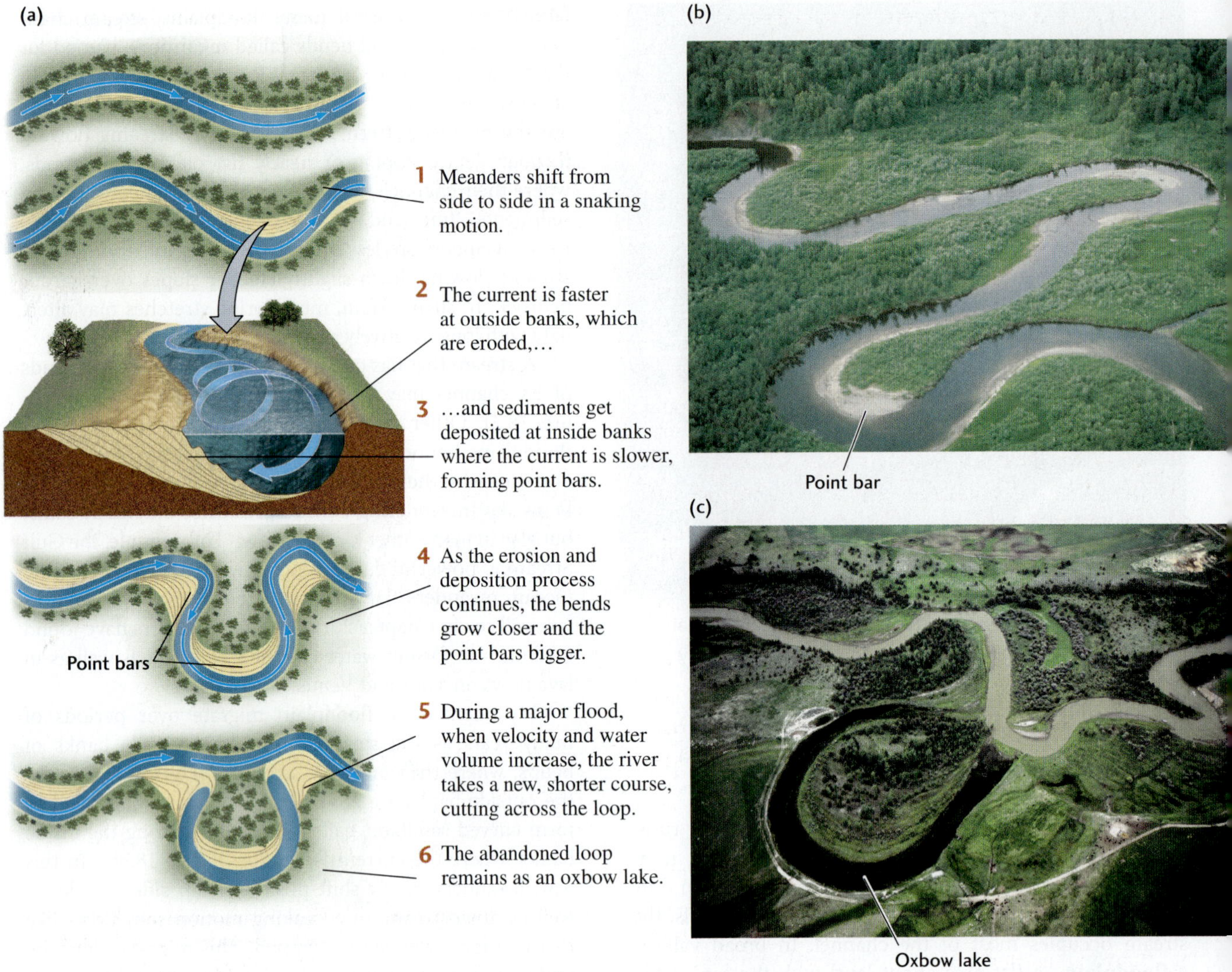

FIGURE 18.3 Meanders migrate over a period of many years. (a) How meanders move. (b) Meanders in an Alaskan river. (c) Oxbow lake in Blackfoot River Valley, Montana. [b: Peter Kresan; c: James Steinberg/Science Source.]

Such environmental concerns stimulated action to restore one channelized river, the Kissimmee in central Florida, to its original meandering course. Today, this restoration project is well under way. If left to its own natural processes, the Kissimmee might have taken many decades or hundreds of years to restore itself.

Braided Streams Some streams have many channels instead of a single one. A **braided stream** is a stream whose channel divides into an interlacing network of channels, which then rejoin in a pattern resembling braids of hair (Figure 18.4). Braided streams are found in many settings, from broad lowland valleys to wide, sediment-filled rift valleys adjacent to mountain ranges. Braids tend to form in streams with large variations in volume of flow combined with a high sediment load and banks that are easily eroded. They are well developed, for example, in the

FIGURE 18.4 This section of the Joekulgilkvisl River in Iceland is a braided stream. [Dirk Bleyer/imageBROKER/SuperStock.]

sediment-choked streams formed at the edges of melting glaciers. Currents in braided streams are usually swiftly flowing, in contrast to those in meandering streams.

Stream Floodplains

A stream channel migrating over the floor of a valley creates a floodplain. Point bars formed during migration build up the surface of the floodplain, as does sediment deposited by floodwaters. Erosional floodplains, covered with a thin layer of sediment, can form when a stream erodes bedrock or unconsolidated sediment as it migrates.

When a stream overflows its banks and floodwaters spread out over the floodplain, the velocity of the water slows, and the current loses its ability to carry sediment. The floodwater velocity drops most quickly along the immediate borders of the channel. As a result, the current deposits large amounts of coarse sediment, typically sand and gravel, along a narrow strip at the edge of the channel. Successive floods build up **natural levees**, ridges of coarse material that confine the stream within its banks between floods, even when water levels are high (**Figure 18.5**). Where natural levees have reached a height of several meters and the stream almost fills the channel, the floodplain level is below the stream level. You can walk the streets of an old river town built on a floodplain, such as Vicksburg, Mississippi, and look up at the levee, knowing that the river waters are rushing by above your head.

During floods, fine sediments—silts and muds—are carried well beyond the stream's banks, often over the entire floodplain, and are deposited there as floodwaters continue to lose velocity. Receding floodwaters leave behind standing ponds and pools of water. The finest clays are deposited there as the standing water slowly disappears by evaporation and infiltration. These fine-grained floodplain deposits, which are rich in mineral and organic nutrients, have been a major resource for agriculture since ancient times. The fertility of the floodplains of the Nile and other rivers of the Middle East, for example, contributed to the evolution of the cultures that flourished there thousands of years ago. Today, the great, broad floodplain of the Ganges in northern India continues to play an important role in India's life and agriculture. Many ancient and modern cities are sited on floodplains (see Earth Issues 18.1).

Drainage Basins

Every topographic rise between two streams, whether it measures a few meters or a thousand, forms a **divide:** a ridge of high ground along which all rainfall runs off down one side or the other. A **drainage basin** is an area of land, bounded by divides, that funnels all its water into the network of streams draining that area (**Figure 18.6**). Drainage basins occur at many scales, from a ravine surrounding a small stream to a great region drained by a major river and its tributaries (**Figure 18.7**).

FIGURE 18.5 (a) Floods form natural levees along the banks of a stream. (b) These natural levees lie along the main channel of the Mississippi River near South Pass, Louisiana. [b: U.S. Geological Survey National Wetlands Research Center.]

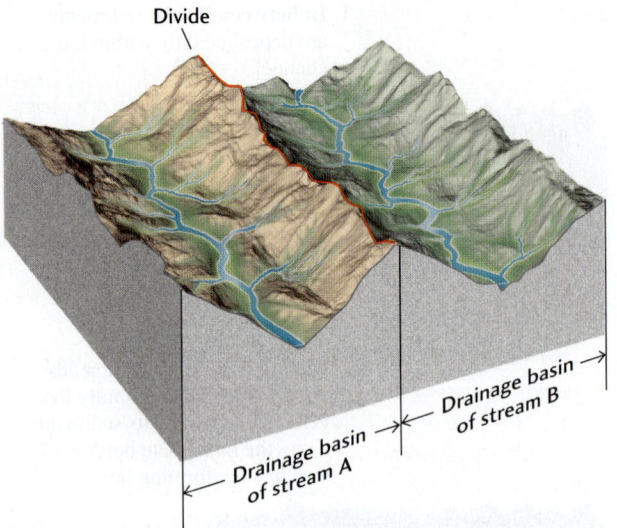

FIGURE 18.6 Drainage basins are separated by divides.

A continent has several major drainage basins separated by major divides. In North America, the continental divide formed by the Rocky Mountains separates all waters flowing into the Pacific Ocean from all those entering the Atlantic. Lewis and Clark followed the Missouri River upstream to its headwaters at the continental divide in western Montana. After crossing over the divide, they found the headwaters of the Columbia River, which they followed to the Pacific Ocean.

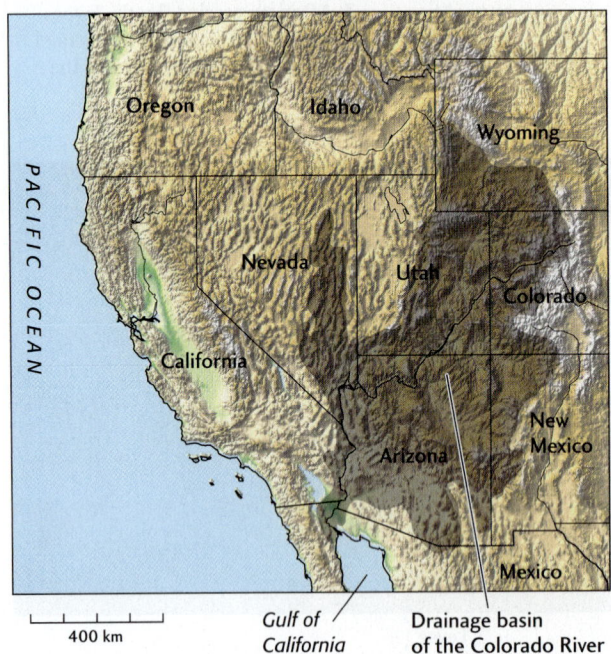

FIGURE 18.7 The drainage basin of the Colorado River covers about 630,000 km², constituting a large part of the southwestern United States. The basin is surrounded by divides that separate it from the neighboring drainage basins.

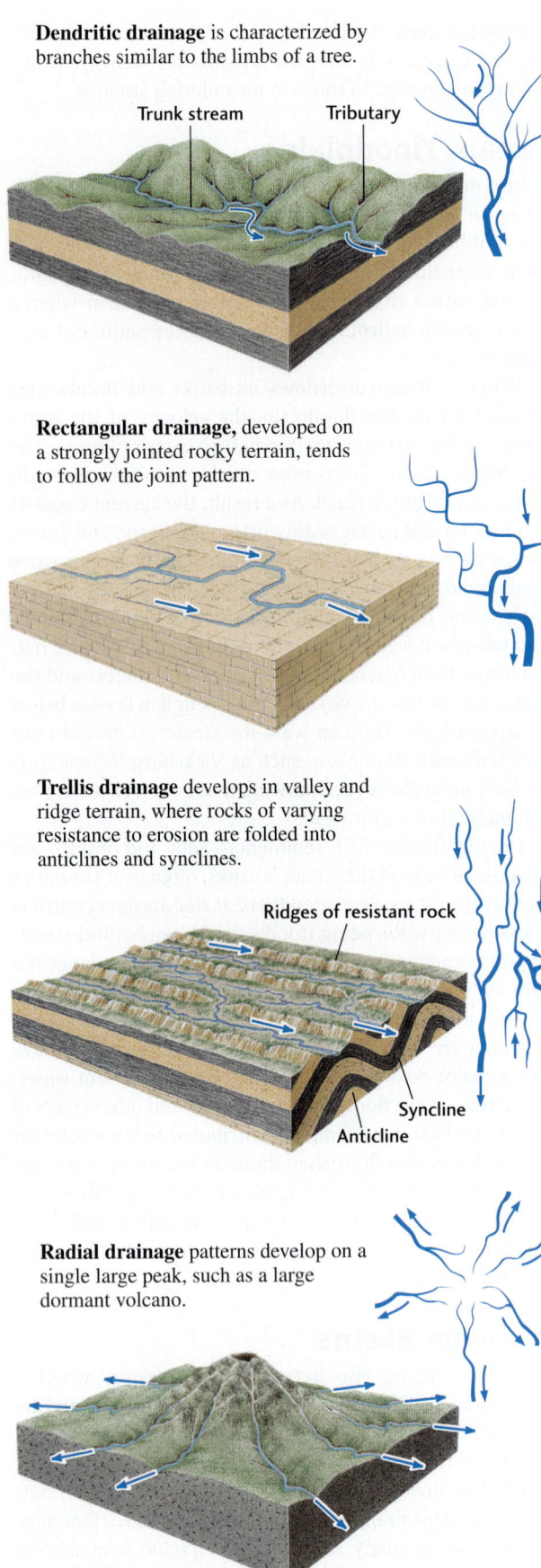

Dendritic drainage is characterized by branches similar to the limbs of a tree.

Rectangular drainage, developed on a strongly jointed rocky terrain, tends to follow the joint pattern.

Trellis drainage develops in valley and ridge terrain, where rocks of varying resistance to erosion are folded into anticlines and synclines.

Radial drainage patterns develop on a single large peak, such as a large dormant volcano.

FIGURE 18.8 Some typical drainage network patterns.

Drainage Networks

A map showing the courses of all the large and small streams in a drainage basin reveals a pattern of connections called a **drainage network.** If you followed a stream from its mouth (where it ends) to its headwaters (where it begins), you would see that it steadily divides into smaller and smaller **tributaries,** forming a drainage network that shows a characteristic branching pattern (**Figure 18.8**).

Branching is a general property of many kinds of networks in which material is collected and distributed. Perhaps the most familiar branching networks are those of tree branches and roots. Most rivers follow the same kind of irregular branching pattern, called **dendritic drainage** (from the Greek *dendron*, meaning "tree"). This fairly random drainage pattern is typical of terrains where the bedrock is of a uniform type, such as horizontally bedded sedimentary rock or massive igneous or metamorphic rock. Other drainage patterns are described as *rectangular, trellis,* and *radial* (see Figure 18.8).

Drainage Patterns and Geologic History

We can observe directly, or determine from historical or geologic records, how most stream drainage patterns developed. Some streams, for example, cut through erosion-resistant bedrock ridges to form steep-walled notches or gorges. What could cause a stream to cut a narrow valley directly through a ridge rather than running along the lowland on either side of it? The geologic history of the region provides the answers.

If a ridge is formed by deformation while a preexisting stream is flowing over it, the stream may erode the rising ridge to form a steep-walled gorge, as in **Figure 18.9**. Such a stream is called an **antecedent stream** because it existed before the present topography was created. The stream maintains its original course despite changes in the underlying rocks and in the topography.

(a)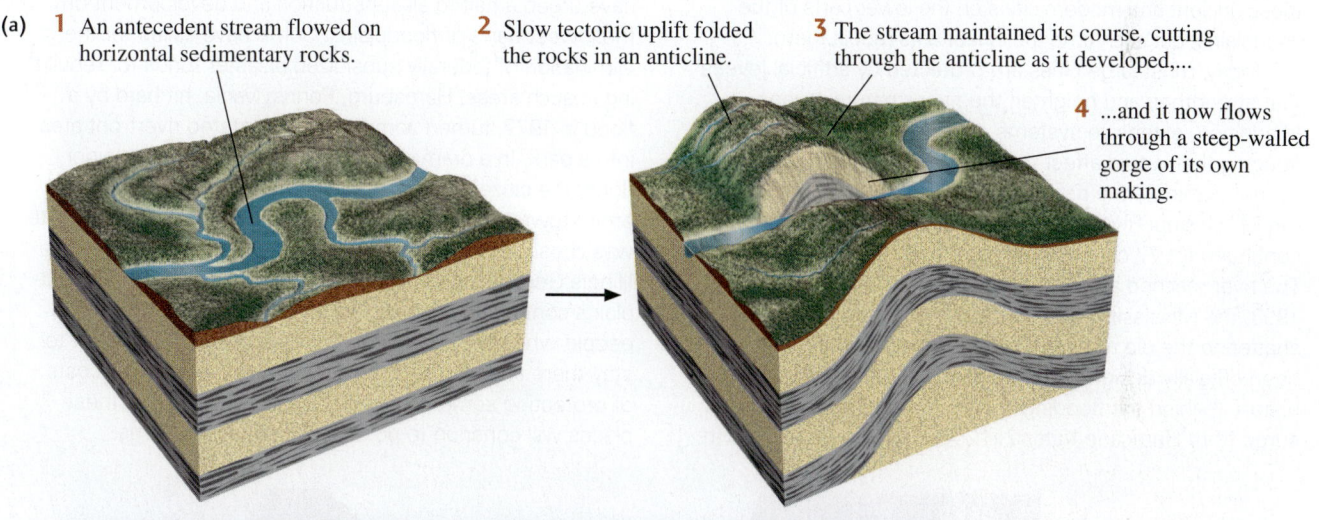

1. An antecedent stream flowed on horizontal sedimentary rocks.
2. Slow tectonic uplift folded the rocks in an anticline.
3. The stream maintained its course, cutting through the anticline as it developed,...
4. ...and it now flows through a steep-walled gorge of its own making.

(b)

FIGURE 18.9 (a) How an antecedent stream cuts a steep-walled gorge. (b) The Delaware Water Gap, located between Pennsylvania and New Jersey. At this point, the Delaware River is an antecedent stream. [b: Michael P. Gadomski/Science Source.]

Earth Issues   18.1 The Development of Cities on Floodplains

Floodplains have attracted human settlement since the beginning of civilization. Floodplains are natural sites for urban settlements because they combine easy transportation along a river with access to fertile agricultural lands. Such sites, however, remain subject to the floods that formed the floodplains. Small floods are common and usually cause little damage, but the larger episodes that happen every few decades or so can be quite destructive.

About 4000 years ago, cities began to dot floodplains in Egypt along the Nile, in the ancient land of Mesopotamia along the Tigris and Euphrates rivers, and in Asia along the Indus River of India and the Yangtze and Huang Ho of China. Later, many of the capital cities of Europe were built on floodplains: Rome on the Tiber, London on the Thames, Paris on the Seine. Floodplain cities in North America include St. Louis on the Mississippi, Cincinnati on the Ohio, and Montreal on the St. Lawrence. Floods periodically destroyed sections of these ancient and modern cities on the lower parts of the floodplains, but each time the inhabitants rebuilt them.

Today, most large cities are protected by artificial levees that strengthen and heighten the river's natural levees. In addition, extensive systems of dams help control the flooding that would affect these cities. But these structures cannot eliminate the risk entirely. In 1973, for example, the Mississippi River went on a rampage with a flood that continued for 77 consecutive days at St. Louis, Missouri. The river reached a record 4.03 m above flood stage. In 1993, the Mississippi and its tributaries broke loose again, shattering the old record in a disastrous flood that has been officially designated the second worst flood in U.S. history (behind the flooding of New Orleans by the storm surge from Hurricane Katrina in 2005). The flood resulted in 487 deaths and more than $15 billion in property damage. At St. Louis, the Mississippi stayed above flood stage for 144 of the 183 days between April and September. In an unexpected secondary effect, the floodwaters leached agricultural chemicals from farmlands and deposited them in the flooded areas, causing widespread pollution.

Figuring out how to protect society from floods presents some knotty problems. Some geologists believe that the construction of artificial levees to confine the Mississippi contributed to its record floods. A river hemmed in by artificial levees can no longer erode its banks and widen its channel to accommodate additional water during periods of high flow. In addition, the floodplain no longer receives deposits of sediment. In the case of New Orleans, the floodplain has sunk below the level of the Mississippi River, making future flooding more likely.

What are cities and towns in this position to do? Some have urged a halt to all construction and development on the lowest parts of floodplains. Some have called for the elimination of federally subsidized disaster funds for rebuilding in such areas. Harrisburg, Pennsylvania, hit hard by a flood in 1972, turned some of its devastated riverfront area into a park. In a dramatic step after the 1993 Mississippi flood, the citizens of Valmeyer, Illinois, voted to move the entire town to high ground several miles away. The new site was chosen with the help of a team of geologists from the Illinois Geological Survey. Yet the benefits of living on floodplains continue to attract people to those sites, and some people who have lived on floodplains all their lives want to stay there and are prepared to live with the risk. The costs of protecting some floodplains are prohibitive, and these places will continue to pose public policy problems.

Like many cities built on floodplains, Davenport, Iowa, is subject to flooding. This photo shows Peterson Paper Company employees going to work by canoe on July 2, 1993, across the flooded downtown area as the Mississippi River continues to rise in the background. [Chris Wilkins/AFP/Getty Images.]

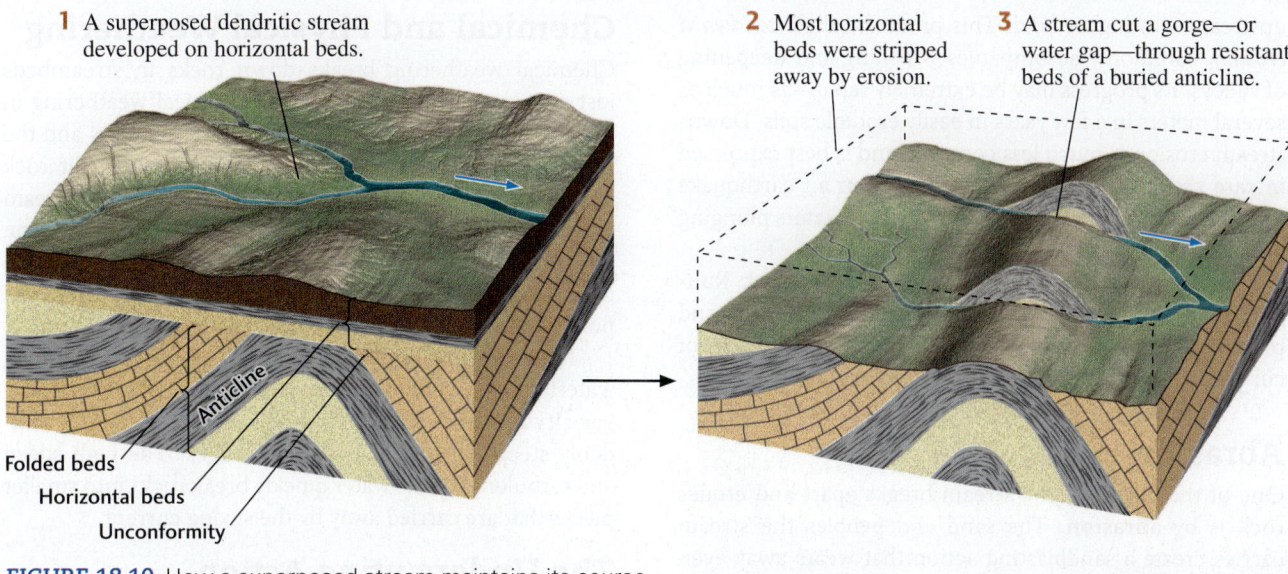

FIGURE 18.10 How a superposed stream maintains its course.

In another geologic situation, a stream may flow in a dendritic drainage pattern over horizontal beds of sedimentary rock that overlie folded and faulted rocks with varying resistance to erosion. Over time, as the softer beds are stripped away by erosion, the stream cuts into a harder bed of underlying rock and erodes a gorge in that resistant bed (Figure 18.10). Such a **superposed stream** flows through resistant formations because its course was established at a higher level, on uniform rock, before downcutting began. A superposed stream tends to continue the pattern that it developed earlier rather than adjusting to its new conditions.

Where Do Channels Begin? How Running Water Erodes Soil and Rock

Stream channels begin where rainwater, running off the surface of the land, flows so fast that it abrades the soil and bedrock and carves into it to form a *gully* (essentially a small valley). Once a gully forms, it captures more of the runoff, and thus the tendency of the stream to cut downward increases. As the gully progressively deepens, the rate of downcutting increases as more water is captured (Figure 18.11).

The erosion of unconsolidated material is relatively easy to observe. We can easily see a stream picking up loose sand from its bed and carrying it away. At high water levels and during floods, we can even see a stream scouring and cutting into its banks, which then slump into the flow and are carried away. Streams progressively cut their channels

FIGURE 18.11 Streams create gullies when the action of water flowing across Earth's surface causes erosion. The smallest gullies converge to form larger stream channels, and farther downslope these become river channels. These gullies were formed in the desert of Oman by occasional rainstorms that inundate the surface with rapidly flowing water, which erodes the bedrock. [Courtesy of Petroleum Development Oman.]

upstream into higher land. This process, called *headward erosion,* commonly accompanies widening and deepening of valleys. Its progress may be extremely rapid—as much as several meters in a few years in easily erodible soils. Downstream erosion is much less common and is best expressed in rare catastrophic events, such as when an earthquake collapses a natural dam and sends scouring waters plunging downstream.

We cannot so easily see the erosion of solid rock. Running water erodes solid rock by abrasion, by chemical and physical weathering, and by the undercutting action of currents.

Abrasion

One of the major ways a stream breaks apart and erodes rock is by **abrasion.** The sand and pebbles the stream carries create a sandblasting action that wears away even the hardest rock. In some streambeds, pebbles and cobbles rotating inside swirling eddies grind deep **potholes** into the bedrock (Figure 18.12). At low water, pebbles and sand can be seen lying quietly at the bottom of exposed potholes.

Chemical and Physical Weathering

Chemical weathering breaks down rocks in streambeds just as it does on the land surface. Physical weathering in streams can be violent, as the crashes of boulders and the constant smaller impacts of pebbles and sand split rock along natural zones of weakness. Such impacts in a stream channel break up rock much faster than slow weathering on a gently sloping hillside does. When these processes have loosened large blocks of bedrock, strong upward eddies can pull them up and out in a sudden violent plucking action.

Physical weathering is particularly strong at rapids and waterfalls. *Rapids* are places in a stream where the flow velocity increases because the slope of the streambed suddenly steepens, typically at rocky ledges. The high speed and turbulence of the water quickly break rocks into smaller pieces that are carried away by the strong current.

The Undercutting Action of Waterfalls

Waterfalls develop where hard rocks resist erosion or where faulting offsets the streambed. The tremendous impact of huge volumes of plunging water and tumbling boulders

FIGURE 18.12 Bourke's Luck Potholes in river rock in Blyde River Canyon Nature Reserve, South Africa. Flowing water rotates the pebbles inside the potholes, grinding deep holes in the bedrock. [Walter G. Allgöwer/imageBROKER/AGE Fotostock.]

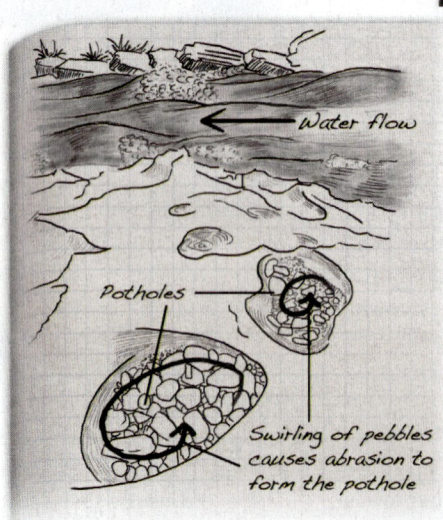

FIGURE 18.13 This waterfall on the Iguaçú River, Brazil, is retreating upstream as falling water and sediment pound the cliff's base and undercut it. Looking downstream, from the center to the upper left of the photo, one can see steep walls, the remnants of the waterfall's retreat upstream.
[Donald Nausbaum/Getty Images.]

quickly erodes streambeds below waterfalls. Waterfalls also erode the underlying rock of the cliffs that form the falls. As erosion undercuts these cliffs, the upper streambed collapses, and the falls recede upstream (**Figure 18.13**). Erosion by falls is fastest where the rock layers are horizontal, with erosion-resistant rocks at the top and softer rocks, such as shales, making up the lower layers. Historical records show that the main section of Niagara Falls, perhaps the best-known falls in North America, has been moving upstream in this way at a rate of a meter per year.

How Currents Flow and Transport Sediment

All currents, whether in water or in air, share the basic characteristics of fluid flow. We can illustrate two kinds of fluid flow by using lines of motion called *streamlines* (**Figure 18.14**). In **laminar flow,** the simplest kind of fluid movement, straight or gently curved streamlines run parallel to one another without mixing or crossing between layers. The slow movement of thick syrup over a pancake, with strands of unmixed melted butter flowing in parallel but separate paths, is a laminar flow. **Turbulent flow** has a more complex pattern of movement, in which streamlines mix, cross, and form swirls and eddies. Fast-moving stream waters typically show this kind of motion. Turbulence—the degree to which there are irregularities and eddies in the flow—may be low or high.

Whether a fluid flow is laminar or turbulent depends primarily on three factors:

1. Its velocity (rate of movement)
2. Its geometry (primarily its depth)
3. Its viscosity (resistance to flow)

Viscosity arises from the attractive forces between the molecules of a fluid. These forces tend to impede the slipping and sliding of molecules past one another. The greater the attractive forces, the greater the resistance to mixing with neighboring molecules, and the higher the viscosity. For example, when cold syrup or viscous cooking oil is poured, its flow is sluggish and laminar. The viscosity of most fluids, including water, decreases as their temperature increases. Given enough heat, a fluid's viscosity may decrease sufficiently to change a laminar flow into a turbulent one.

Water has low viscosity in the range of temperatures found at Earth's surface. For this reason alone, most streams in nature tend toward turbulent flow. In addition, the

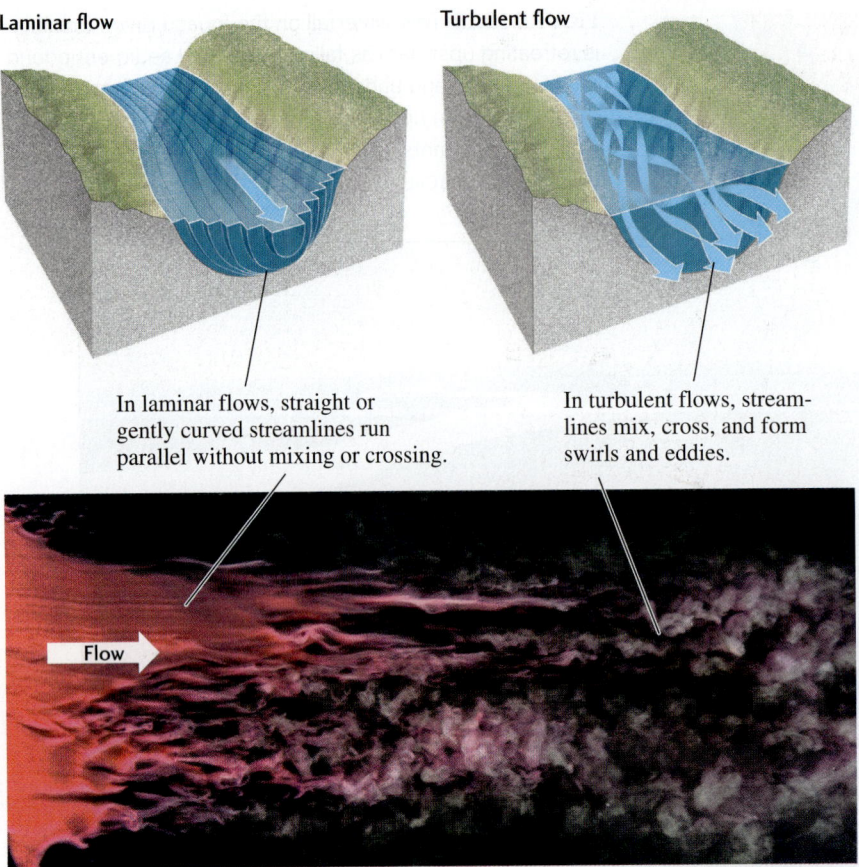

FIGURE 18.14 The two basic patterns of fluid flow: laminar flow and turbulent flow. The photograph shows the transition from laminar to turbulent flow in water along a flat plate, revealed by the injection of a dye. Flow is from left to right. [Henri Werlé, Onera, The French Aerospace Lab.]

In laminar flows, straight or gently curved streamlines run parallel without mixing or crossing.

In turbulent flows, streamlines mix, cross, and form swirls and eddies.

velocity and geometry of most natural streams make them turbulent. In nature, we are likely to see laminar flows of water only in thin sheets of rain runoff flowing slowly down nearly level slopes. In cities, we may see small laminar flows in street gutters.

Because most streams and rivers are broad and deep and flow quickly, their flows are almost always turbulent. A stream may show turbulent flow over much of its width and laminar flow along its edges, where the water is shallower and is moving more slowly. In general, the flow velocity is highest near the center of a stream; where a stream meanders, the flow velocity is highest at the outsides of the bends. We commonly refer to a rapid flow of water as a strong current.

Erosion and Sediment Transport

Currents vary in their ability to erode and carry sand grains and other sediments. Laminar flows of water can lift and carry only the smallest, lightest clay-sized particles. Turbulent flows, depending on their velocity, can move particles ranging in size from clay to pebbles and cobbles. As turbulence lifts particles into a flow, the flow carries them downstream. Turbulence also rolls and slides larger particles along the bottom of the channel. A current's **suspended load** includes all the material temporarily or permanently suspended in the flow. Its **bed load** is the material the current carries along the bed by sliding and rolling (**Figure 18.15**). The *bed*, in this context, is the layer of unconsolidated material in the channel that interacts with the current.

The greater the velocity of a current, the larger the particles it can carry as suspended load and bed load. A flow's ability to carry material of a given particle size is its **competence.** As current strength increases and coarser particles are suspended, the suspended load grows. At the same time, more of the bed material is in motion, and the bed load also increases. As we would expect, the larger the volume of a flow, the greater the sediment load (suspended load plus bed load) it can carry. The total sediment load carried by a flow is its **capacity.**

The velocity and the volume of a flow affect both the competence and the capacity of a stream. The Mississippi River, for example, flows at moderate speeds along most of its length and carries only fine to medium-sized particles (clay to sand), but it carries huge quantities of them. A small, steep, fast-flowing mountain stream, in contrast, may carry boulders, but only a few of them.

A current's suspended load depends on a balance between turbulence, which lifts particles, and the competing downward pull of gravity, which makes them settle out of the current and become part of the bed. The speed with which suspended particles of various weights settle to the bed is called their **settling velocity.** Small grains of silt and

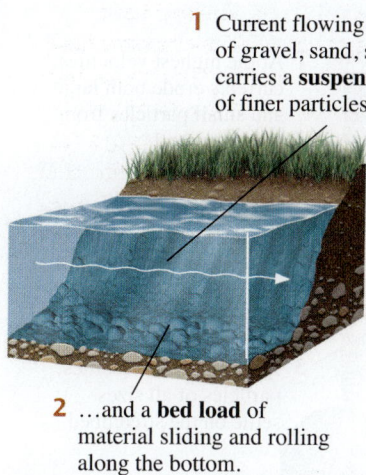
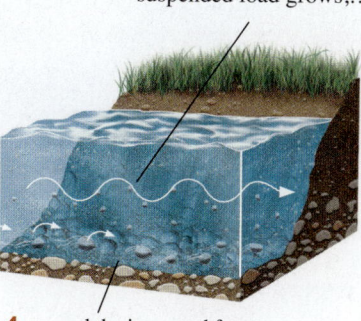
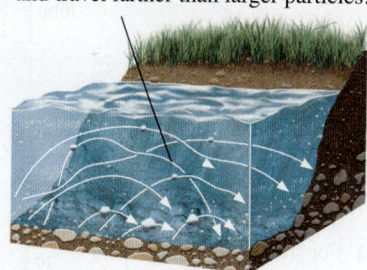

1. Current flowing over a bed of gravel, sand, silt, and clay carries a **suspended load** of finer particles…

2. …and a **bed load** of material sliding and rolling along the bottom.

3. As current velocity increases, the suspended load grows,…

4. …and the increased force of the flow generates an increase in the bed load.

5. Particles move by saltation, jumping along a bed. At a given current velocity, smaller particles jump higher and travel farther than larger particles.

FIGURE 18.15 A current flowing over a bed of unconsolidated material can transport particles in three ways.

clay are easily lifted into the stream and settle slowly, so they tend to stay in suspension. The settling velocities of larger particles, such as medium- and coarse-grained sand, are much higher. Most larger grains therefore stay suspended in the current only a short time before they settle.

As current velocity increases, sediment particles in the bed load begin to move by a third process, known as **saltation:** an intermittent jumping motion along the bed. Sand grains are most likely to move by saltation because they are light enough to be picked up from the bed, yet heavy enough not to be transported in suspension. The grains are sucked up into the flow by turbulent eddies, move with the current for a short distance, and then fall back to the bed (see Figure 18.15). If you were to stand in a rapidly flowing sandy stream, you might see a cloud of saltating sand grains moving around your ankles. The bigger the grain, the longer it will tend to remain on the bed before it is picked up. Once a large grain is in the current, it will settle quickly. The smaller the grain, the more frequently it will be picked up, the higher it will "jump," and the longer it will take to settle.

Worldwide, streams carry about 25 billion tons of siliciclastic sediments and an additional 2 billion to 4 billion tons of dissolved matter each year. Humans are responsible for much of the present sediment load. According to some estimates, prehuman sediment transport was about 9 billion tons per year, less than half the present value. In some places, we increase the sediment load of streams through agriculture and accelerated erosion. In other places, we decrease the sediment load by constructing dams, which trap sediment behind them.

To study how a particular stream carries sediments, geologists and hydraulic engineers measure the relationship between particle size and the force the flow exerts on the particles in both the suspended and bed loads. This relationship shows them how much sediment a particular flow can move and how rapidly it can move it. That information allows them to design dams and bridges or to estimate how quickly artificial reservoirs behind dams will fill with sediments. As we saw in Chapter 5, geologists can infer the velocities of ancient currents from the sizes of grains in sedimentary rocks.

Figure 18.16 graphs the relationship between the sizes of sediment particles on the streambed and the flow velocities required to erode them. You will notice that in this graph, contrary to our earlier discussion of competence, the current velocity required to erode some kinds of particles from the bed actually increases as particle size decreases. This relationship exists because it is easier for the current to lift noncohesive particles (particles that do not stick together) than cohesive particles (particles that stick together, as many clay minerals do). The finer the cohesive particles, the greater the velocity of the flow required to erode them. Settling velocities for these small particles, however, are so slow that even a gentle current, about 20 cm/s, can keep them in suspension and transport them over long distances.

Sediment Bed Forms: Dunes and Ripples

When a current transports sand grains by saltation, the sand tends to form dunes and ripples (see Chapter 5). **Dunes** are elongated ridges of sand up to many meters high that form in flows of wind or water over a sandy bed. **Ripples** are very small dunes—with heights ranging from less than a centimeter to several centimeters—whose long dimension is formed at right angles to the current. Although underwater ripples and dunes produced by water currents are harder to observe than those produced on land by air currents, they form in the same way and are just as common.

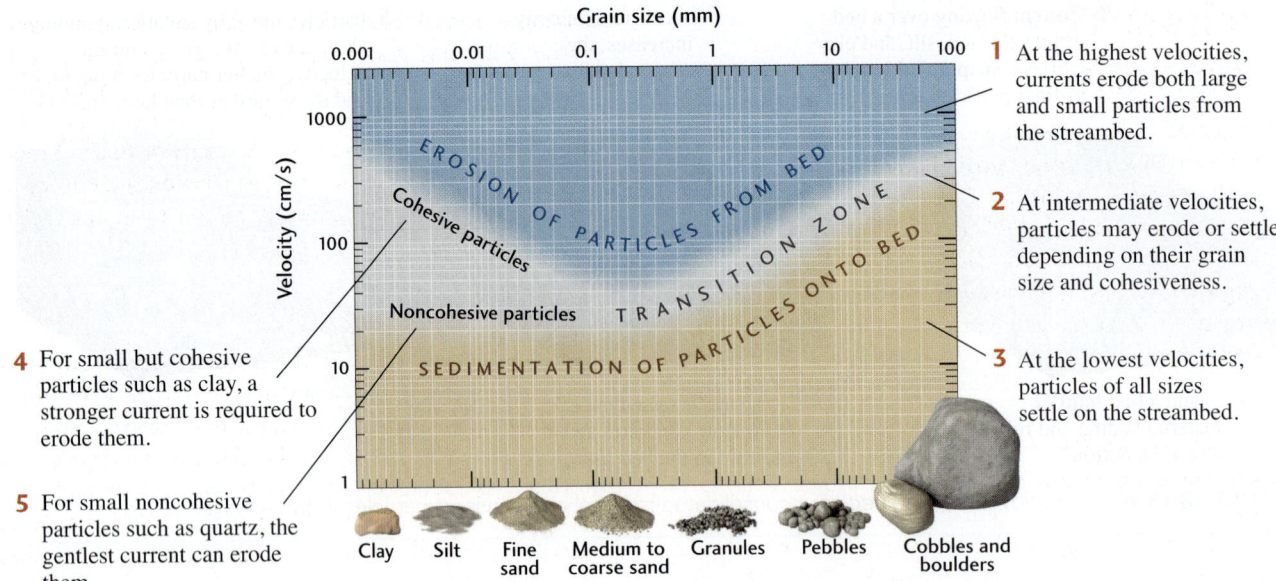

FIGURE 18.16 Relationship of current velocity to the erosion and settling of particles of different sizes. The blue area represents velocities at which particles are eroded from the streambed; the gray area, velocities at which particles may either be eroded or settle; and the brown area, velocities at which particles settle onto the bed.

As a current moves sand grains by saltation, they are eroded from the upstream side of ripples and dunes and deposited on the downstream side. The steady downstream transfer of grains across the ridges causes the ripple and dune forms to migrate downstream. The speed of this migration is much slower than the movement of individual grains and very much slower than the current. (We will look at ripple and dune migration in more detail in Chapter 19.)

The shapes of ripples and dunes and their migration speeds change as the velocity of the current increases. At the lowest current velocities, few grains are saltating, and the sediment bed is flat. At slightly higher velocities, the number of saltating grains increases. A rippled bed forms, and the ripples migrate downstream (**Figure 18.17**). As the velocity increases further, the ripples grow larger and migrate faster until, at a certain point, dunes replace the ripples. Both ripples and dunes have a cross-bedded structure (see Figure 5.11). As the current flows over their tops, it can actually reverse and flow backward along their downstream side. As the dunes grow larger, small ripples form on them. These ripples tend to climb over the backs of the dunes because they migrate more quickly than the dunes. Very high current velocities will erase the dunes and form a flat bed below a dense cloud of rapidly saltating sand grains. Most of these grains hardly settle to the bottom before they are picked up again. Some are in permanent suspension.

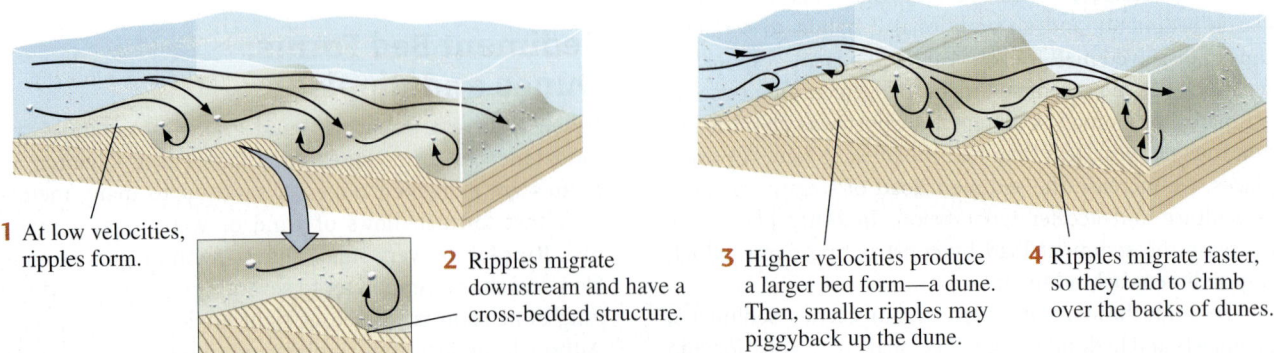

FIGURE 18.17 The form of a sediment bed changes with current velocity.

Deltas: The Mouths of Rivers

Sooner or later, all rivers end as they flow into a lake or an ocean, mix with the surrounding water, and—no longer able to travel downslope—gradually lose their forward momentum. The largest rivers, such as the Amazon and the Mississippi, can maintain some current many kilometers out to sea. Where smaller rivers enter a turbulent, wave-swept sea, the current disappears almost immediately beyond the river's mouth.

Delta Sedimentation

As its current gradually dies out, a river progressively loses its power to transport sediments. The coarsest material, typically sand, is dropped first, right at the mouths of most rivers. Finer sands are dropped farther out, followed by silt and, still farther out, by clay. As the floor of the lake or ocean slopes to deeper water away from the shore, the deposited materials build up a large, flat-topped deposit called a **delta**. (We owe the name *delta* to the Greek historian Herodotus, who traveled through Egypt about 450 B.C. The roughly triangular shape of the sediment deposit at the mouth of the Nile prompted him to name it after the Greek letter Δ, delta.)

As a river approaches its delta, where its slope is almost level with the ocean surface, it reverses its upstream-branching drainage pattern. Instead of collecting more water from tributaries, it discharges water into **distributaries**: smaller streams that receive water and sediments from the main channel, branch off *downstream*, and thus distribute the water and sediment into many channels. Materials deposited on top of the delta, typically sand, make up horizontal **topset beds**. Downstream, on the outer front of the delta, fine-grained sand and silt are deposited to form gently inclined **foreset beds**, which resemble large-scale cross-beds. Spread out on the seafloor seaward of the foreset beds are thin, horizontal **bottomset beds** of mud, which are eventually buried as the delta continues to grow. **Figure 18.18** shows how these structures form in a typical large marine delta.

The Growth of Deltas

As a delta builds outward into the ocean, the mouth of its river advances seaward, leaving new land in its wake. Much of this land is a *delta plain* just a few meters above sea level. At the seaward edge of the plain, broad depressions between distributary channels lie below sea level and form shallow bays that fill with fine-grained sediments. With time, they fill further and ultimately become salt marshes (see Figure 18.18).

As a delta grows, the river flow shifts from some distributaries to others that provide shorter routes to the sea. As a result of such shifts, the delta may grow in one direction for some hundreds or thousands of years, then the main stream may break out into a new distributary, sending sediments into the ocean in another direction. In this way, a major river may form a large delta thousands of square kilometers in area. The delta of the Mississippi has been growing for millions of years. About 150 million years ago, it started out around what is now the junction of the Ohio and the Mississippi rivers, at the southern tip of Illinois. It has advanced about 1600 km since then, creating almost the entire states of Louisiana and Mississippi as well as major parts of adjacent states. **Figure 18.19** shows the growth of the Mississippi Delta over the past 6000 years, as well as the direction its growth is likely to take in the future.

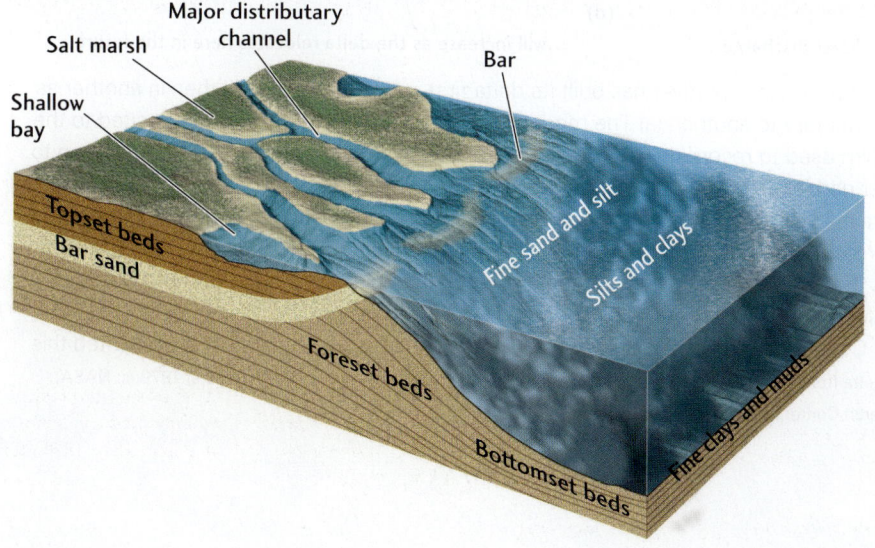

FIGURE 18.18 A typical large marine delta, many kilometers in extent, in which the fine-grained foreset beds are deposited at a very low angle, typically only 4° to 5° or less. Sandbars form at the mouths of the distributaries, where the current velocity suddenly decreases. The delta builds forward by the advance of these bars and the topset, foreset, and bottomset beds. Between distributary channels, shallow bays fill with fine-grained sediments and become salt marshes. This general structure is found on the Mississippi Delta.

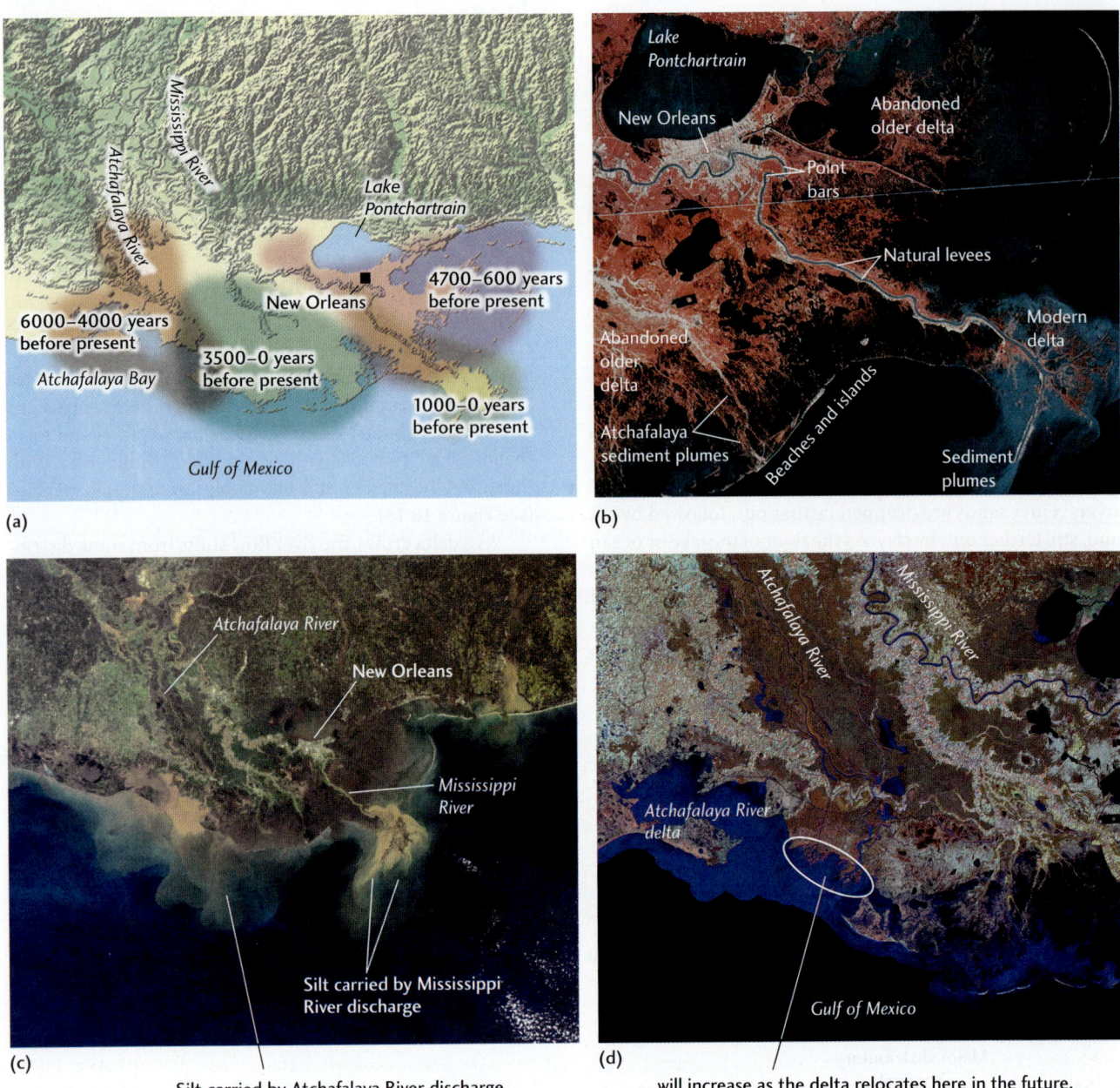

FIGURE 18.19 Over the past 6000 years, the Mississippi River has built its delta first in one direction and then in another as water flow has shifted from one major distributary to another. (a) The modern delta was preceded by deltas deposited to the east and west. (b) The infrared-sensitive film used to record this satellite image of the Mississippi Delta causes vegetation to appear red, relatively clear water to appear dark blue, and water with suspended sediment to appear light blue. At the upper left are New Orleans and Lake Pontchartrain. Well-defined natural levees and point bars can be seen at the center. At the lower left are beaches and islands that formed as ocean waves and currents transported river-deposited sand from the delta. (c) Satellite photograph of the Mississippi Delta. (d) This image shows the discharge of sediment into the Gulf of Mexico from the Mississippi River Delta and the Atchafalaya River. A major flood could divert the main flow of the Mississippi into the Atchafalaya, causing a new delta to form. Construction of artificial levees by the Army Corps of Engineers has prevented this so far. [b: From G. T. Moore, "Mississippi River Delta from Landsat 2," Bulletin of the American Association of Petroleum Geologists, 1979; c: NASA; d: U.S. Geological Survey National Wetlands Research Center.]

Deltas grow by the addition of sediment, and they sink as the sediment becomes compacted and Earth's crust subsides under the weight of the sediment load. Venice, built on part of the Po River Delta in northern Italy, has been sinking steadily for many years. Both crustal subsidence and depression of the ground attributable to the pumping of water from aquifers beneath the city are responsible for its sinking.

Human Effects on Deltas

The extensive wetlands found in delta plains are valuable natural resources because, like all wetlands, they store floodwaters and provide habitat for many diverse species of plants and animals, as noted in Chapter 17. The wetlands of the Mississippi River Delta, like delta wetlands in many other areas, have suffered a two-pronged attack. First, the extensive flood-control dams built on the river since the 1930s have decreased the volume of sediment brought to the delta, thereby reducing the sediment supply to the wetlands. Second, massive artificial levees have prevented the small but frequent floods that nourish the delta wetlands with sediments. At New Orleans, the Mississippi River floodplain has sunk below the level of the river, making future catastrophic flooding more likely.

The Effects of Ocean Currents, Tides, and Plate Tectonic Processes

Strong waves, ocean currents, and tides affect the growth and shape of marine deltas. Waves and ocean currents may move sediment along the shore almost as rapidly as it is dropped there by a river. The delta front then becomes a long beach with only a slight seaward bulge at the river mouth. Where tidal currents move in and out, they redistribute delta sediments into elongated bars parallel to the direction of the current, which in most places is at approximately right angles to the shore (see Figure 18.19b).

In some places, waves and tides are strong enough to prevent a delta from forming. Instead, the sediments a river transports to the ocean are dispersed along the shoreline as beaches and bars and transported into deeper waters offshore. The east coast of North America lacks deltas for this reason. The Mississippi has been able to build its delta because neither waves nor tides are very strong in the Gulf of Mexico.

Plate tectonic processes also exert some control over where deltas form because delta formation has two other preconditions: uplift in the drainage basin, which provides abundant sediments, and crustal subsidence in the delta region to accommodate the great weight and volume of those sediments. Two of the world's large deltas—those of the Mississippi and the Rhône (in France)—derive their sediments primarily from distant mountain ranges: the Rockies for the Mississippi and the Alps for the Rhône. Both are in the same type of plate tectonic setting: a passive margin originally formed by continental rifting. The active continent-continent convergence that is elevating the Himalaya has also formed the great deltas of the Indus and Ganges rivers.

Few large deltas are associated with active subduction zones. The reason may be that it is unusual for a large river (such as the Columbia River of the Pacific Northwest) to carry abundant sediment through a volcanic mountain belt (such as the Cascade Range) to the sea. Furthermore, these rapidly uplifting areas are too unstable for large deltas to develop. Oceanic island arcs are too small in land area to provide much siliciclastic sediment to their streams.

Streams as Geosystems

Streams are dynamic geosystems that are continually changing in response to the influences of climate and plate tectonic processes, and those changes, in turn, influence the transport of water and sediments. The flow of a stream may appear steady when you view it from a bridge for a few minutes or canoe along it for a few hours, but its volume and velocity may change appreciably from month to month and season to season. At any one location, the stream is constantly changing, shifting from low water to flood stage over a few hours or days and reshaping its valley over longer periods (Figure 18.20). The flow and channel dimensions of a stream also change as it moves downslope, from narrow valleys in its upland headwaters to broader floodplains in its middle and lower courses. Most of these longer-term changes are adjustments in the normal (nonflood) volume and velocity of the flow as well as in the depth and width of the channel.

From their headwaters to their mouths, all streams react to changes in climate (such as changes in precipitation) and to plate tectonic processes (such as uplift or subsidence of Earth's crust). As we have seen, streams gather into ever-larger streams, eventually forming a single large stream, as in the case of the Mississippi River. Precipitation in the headwaters may affect streamflow far downriver, where a river's volume may exceed the volume of the channel and then spill over the banks to create a flood. In this way, processes and events in one part of the stream network are propagated through the system to affect the behavior of a different part of the system.

Sediment transport changes in a similar way, though over a longer time scale. If precipitation in the headwaters increases over a long period—say, because the climate becomes rainier—or tectonic uplift rates increase, erosion rates and sediment yield increases. The stream network propagates a "wave" of sediment that eventually reaches the delta, where it may be preserved in the rock record as an interval of unusually high sediment accumulation. We will further explore these relationships and their effects on landscape development in Chapter 22.

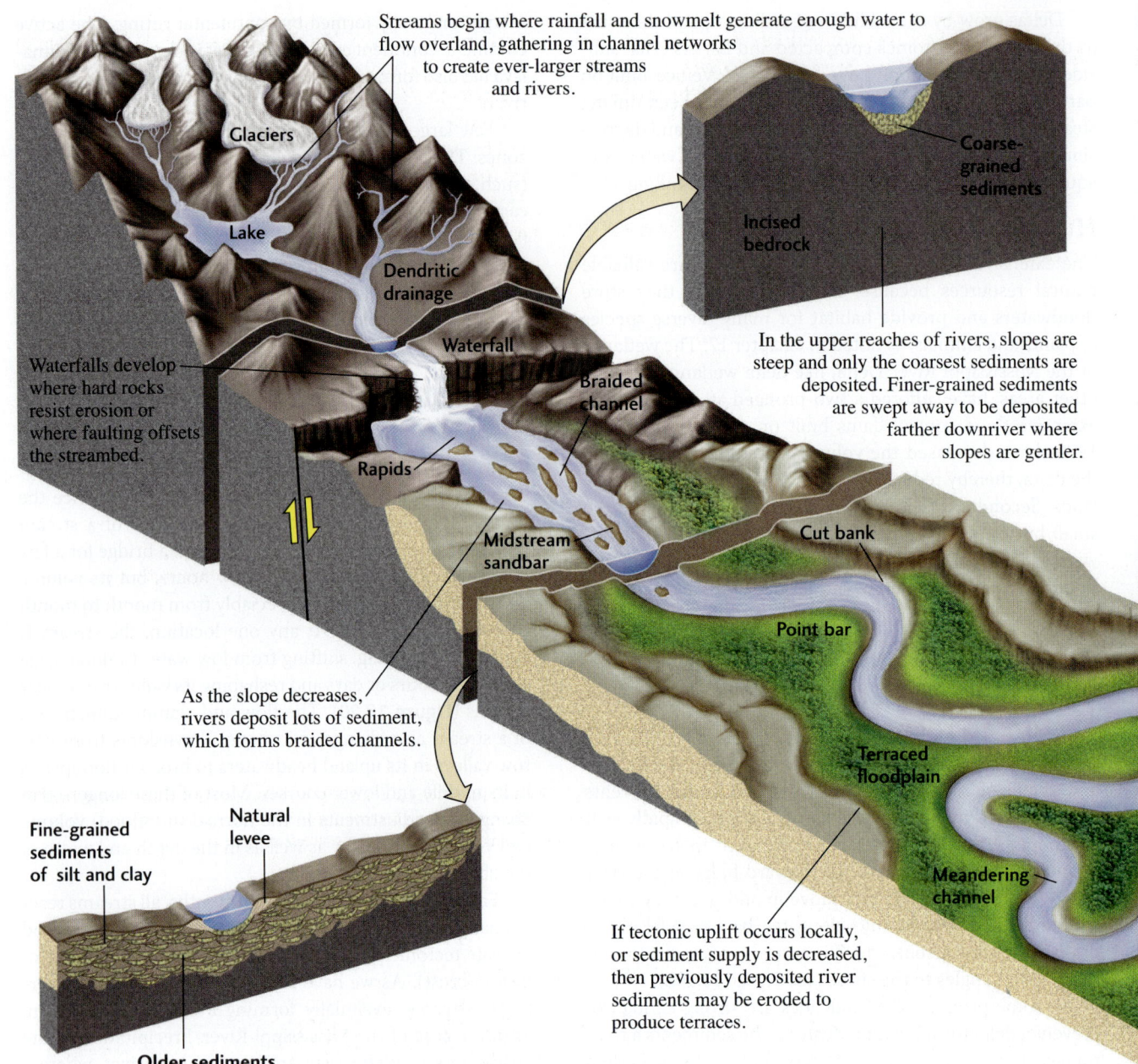

FIGURE 18.20 Stream networks transport water and sediments from their headwaters to the ocean.

Several factors are important in controlling how water and sediments move through stream geosystems. These factors include the stream's discharge, its longitudinal profile, and changes in its base level.

Discharge

We can measure the size of a stream's flow by its **discharge**: the volume of water that passes a given point in a given time as it flows through a channel of a certain width and depth. (In Chapter 17, we defined *discharge* as the volume of water leaving an aquifer in a given time. These definitions are consistent because they both describe volume of flow per unit of time.) Stream discharge is usually measured in cubic meters per second (m^3/s) or cubic feet per second. Discharge in small streams may vary from about 0.25 to 300 m^3/s. The discharge of a well-studied medium-sized river in Sweden, the Klarälven, varies from 500 m^3/s at low water levels to 1320 m^3/s at high water levels. The discharge of the Mississippi River can vary from as little as 1400 m^3/s at low water levels to more than 57,000 m^3/s during floods.

Discharge is matched by recharge at any location where rainfall or discharging groundwater contributes to the stream. When recharge is less than discharge, such as during a drought, stream levels may fall dramatically. When recharge is greater than discharge, stream levels will rise, and flooding will occur if the imbalance between discharge and recharge becomes too great.

To calculate discharge, we multiply the cross-sectional area (the width multiplied by the depth of the part of the channel occupied by water) by the velocity of the flow (distance traveled per second):

discharge = cross section × velocity
 (width × depth) (distance traveled/second)

Figure 18.21 illustrates this relationship. If discharge is to increase, then either velocity or cross-sectional area, or both, must increase. Think of increasing the discharge of a garden hose by opening the valve more, which increases the velocity of the water coming out the end of the hose. The cross-sectional area of the hose cannot change, so the discharge must increase. In a stream, as discharge at a particular point increases, both the velocity and the cross-sectional area of the flow tend to increase. (The velocity is also affected by the slope of the channel and the roughness of the bottom and sides, which we can neglect for the

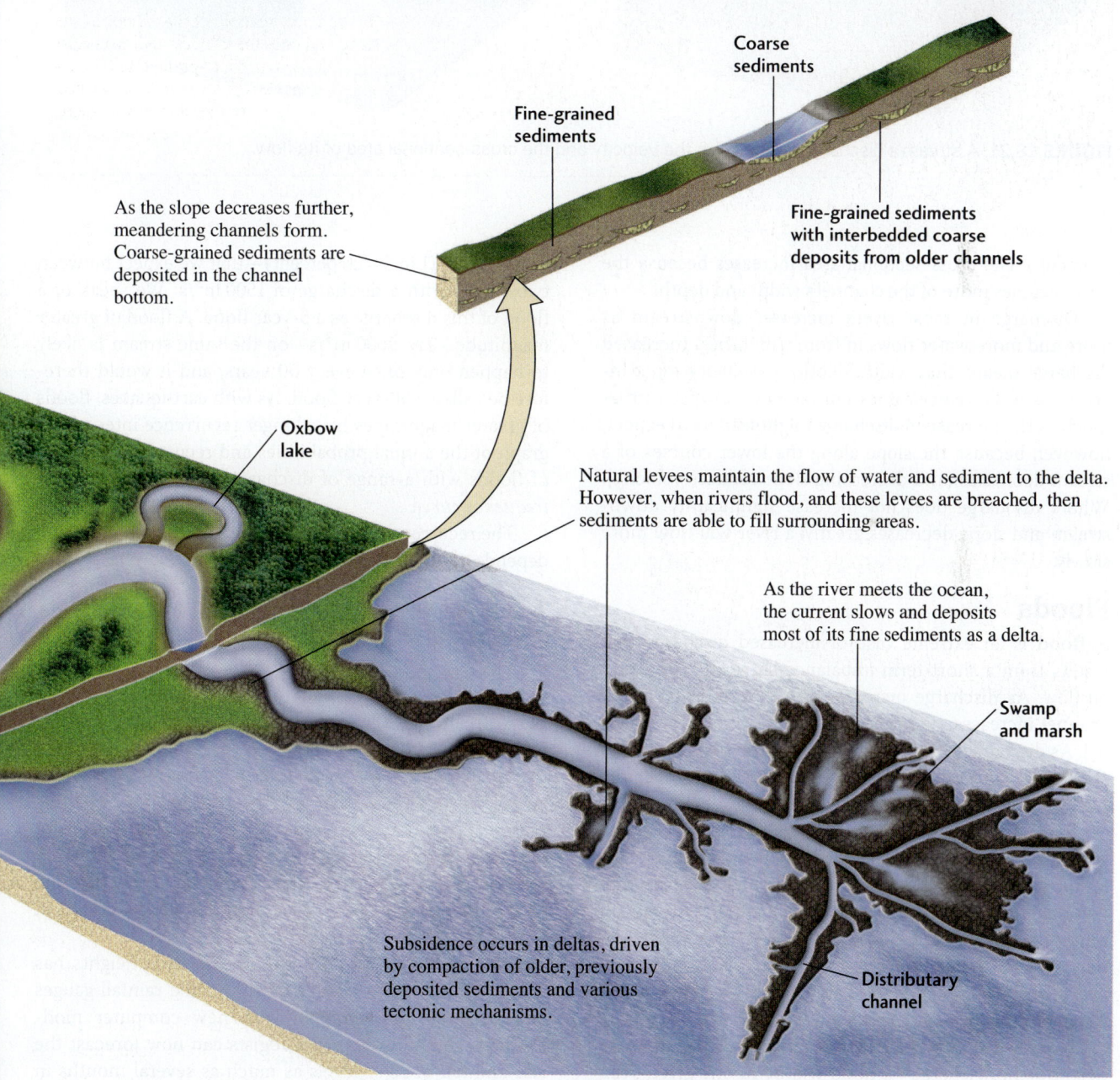

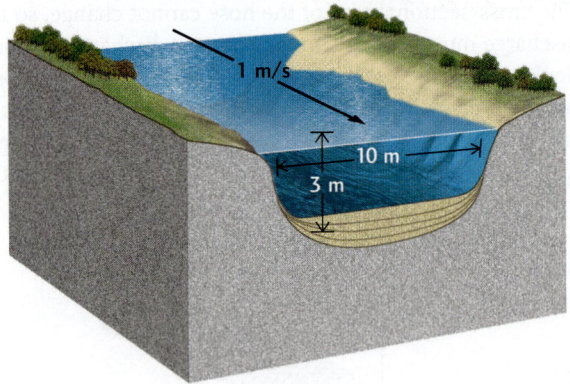
A stream with a smaller cross-sectional area and a lower velocity has a lower discharge (3 m × 10 m = 30 m² × 1 m/s = 30 m³/s discharge)…

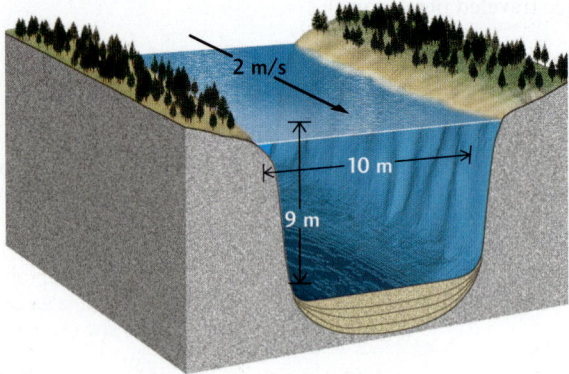
…than a stream with a larger cross-sectional area and a higher velocity (9 m × 10 m = 90 m² × 2 m/s = 180 m³/s).

FIGURE 18.21 A stream's discharge depends on the velocity and the cross-sectional area of its flow.

moment.) The cross-sectional area increases because the flow occupies more of the channel's width and depth.

Discharge in most rivers increases downstream as more and more water flows in from tributaries. Increased discharge means that width, depth, or velocity must increase as well. Velocity does not increase downstream as much as the increase in discharge might lead us to expect, however, because the slope along the lower courses of a stream decreases (and decreasing slope reduces velocity). Where discharge does not increase significantly downstream and slope decreases greatly, a river will flow more slowly.

Floods

A **flood** is an extreme case of increased discharge that results from a short-term imbalance between inflow and outflow. As discharge increases, the flow velocity in the channel increases, and the water gradually fills the channel. As discharge continues to increase, the stream reaches *flood stage* (the point at which water first spills over the banks).

Some streams overflow their banks almost every year when the snows melt or rains arrive; others flood at irregular intervals. Some floods bring very high water levels that inundate the floodplain for days. At the other extreme are minor floods that barely break out from the channel before they recede. Small floods are more frequent, occurring every 2 or 3 years on average. Large floods are generally less frequent, usually occurring every 10, 20, or 30 years.

No one can know exactly how severe—either in water height or in discharge—a flood will be in any given year, so predictions are stated as *probabilities*, not certainties. For a particular stream, we might say that there is a 20 percent probability that a flood of a given discharge—say, 1500 m³/s—will occur in any one year. That probability corresponds to an average recurrence interval—in this case, 5 years (1 in 5 = 20 percent)—that we expect between two floods with a discharge of 1500 m³/s. We speak of a flood of this discharge as a 5-year flood. A flood of greater magnitude—say, 2600 m³/s—on the same stream is likely to happen only once every 50 years, and it would therefore be called a 50-year flood. As with earthquakes, floods of greater magnitudes have longer recurrence intervals. A graph of the annual probabilities and recurrence intervals of floods with a range of discharges is known as a *flood frequency curve*.

The recurrence interval of floods of a certain discharge depends on three factors:

1. The climate of the region
2. The width of the floodplain
3. The size of the channel

For a stream in a dry climate, for example, the recurrence interval of 2600 m³/s floods may be much longer than the recurrence interval of 2600 m³/s floods for a similar stream in an area that gets intermittent rain. For this reason, flood frequency curves for individual rivers are necessary if towns along those rivers are to be prepared to cope with flooding. A flood frequency curve for one river—the Skykomish, in Washington State—is shown in **Figure 18.22**.

The prediction of river floods and their heights has become much more reliable as automated rainfall gauges and streamgauges, coupled with new computer models, have come into use. Geologists can now forecast the rise and fall of river levels as much as several months in advance, and they can issue reliable flood warnings days in advance. This information is useful for many other purposes as well, from water resource management to the planning of recreational river trips (see the Practicing Geology Exercise).

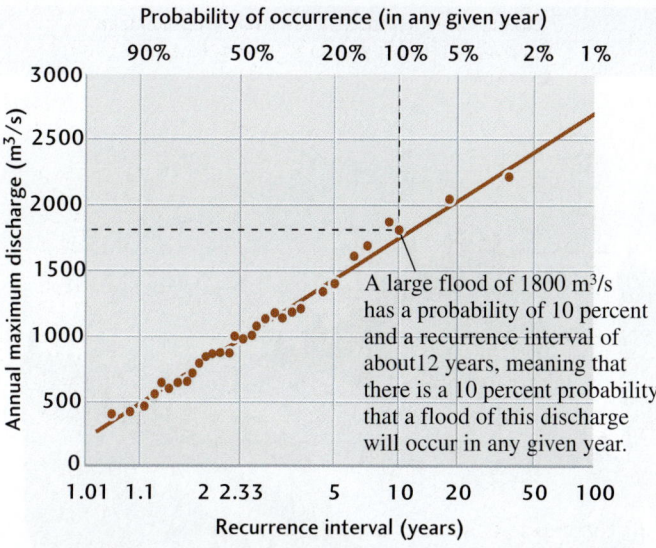

FIGURE 18.22 The flood frequency curve for the Skykomish River at Gold Bar, Washington. This curve predicts the probability that a flood of a certain discharge will occur in any given year.

PRACTICING GEOLOGY EXERCISE

Can We Paddle Today? Using Streamgauge Data to Plan a Safe and Enjoyable River Trip

How do geologists measure and record the flow of water in streams and rivers? In the United States, the U.S. Geological Survey (USGS) has been tracking water flow information for over a hundred years. The USGS uses *streamgauges* to measure and record the height of the water surface at repeated intervals (hourly, daily, weekly, or longer). In 2007, it operated and maintained more than 7400 streamgauges on rivers and streams across the nation. Geologists use USGS streamgauge data to manage the nation's water resources in a number of ways: to forecast floods and droughts, to manage and operate dams and artificial reservoirs, and to protect water quality, among others.

The ready availability of USGS streamgauge data can also make for safer and more enjoyable outings for anglers, kayakers, canoeists, and rafters. Many USGS-maintained streamgauges transmit near-real-time data through a satellite or telephone network directly to a Web site. Streamgauge data are updated at intervals of 4 hours or less and are available to the public on the Internet at http://water.usgs.gov.

Checking streamgauge data before a river trip can prevent the disappointment and loss of time associated with a long drive to a favorite paddling spot, only to find the water too low or too fast to paddle. On the other hand, paddlers may be willing to travel farther for a river run when they know that flow conditions on their favorite river are optimal. The USGS data allow paddlers to match the conditions of the water to their own abilities.

Streamgauges record the height of the water surface, often known as *stage*. The use of stage alone, however, can be misleading. "Stage" refers to the water surface elevation above a fixed reference point near the streamgauge and may not correspond directly to water depth. Never assume a stage reading is the equivalent of the distance between the water's surface and the streambed. Discharge is a more reliable indicator of the conditions that river enthusiasts will encounter.

As discussed in the chapter text, the discharge of a stream is determined by measuring the cross-sectional area (width × depth) and velocity of the streamflow. The depth of the stream below the fixed reference point and its width at each streamgauge are known, so discharge can be estimated if velocity and stage are known. The discharge values from each measurement can be plotted against the stage recorded at the same time to develop a rating curve for each streamgauge. Paddlers can find the streamgauge reading for their favorite river run, then read the rating curve to get an estimate of the discharge.

Suppose you consult the USGS Web site and see that the most recent streamgauge reading for the river you want to paddle is 3 feet.

▶ Find 3 feet on the vertical axis of the accompanying graph and read across to the rating curve.
▶ Then read down to find the discharge at a stage of 3 feet: 500 cubic feet per second (CFS).

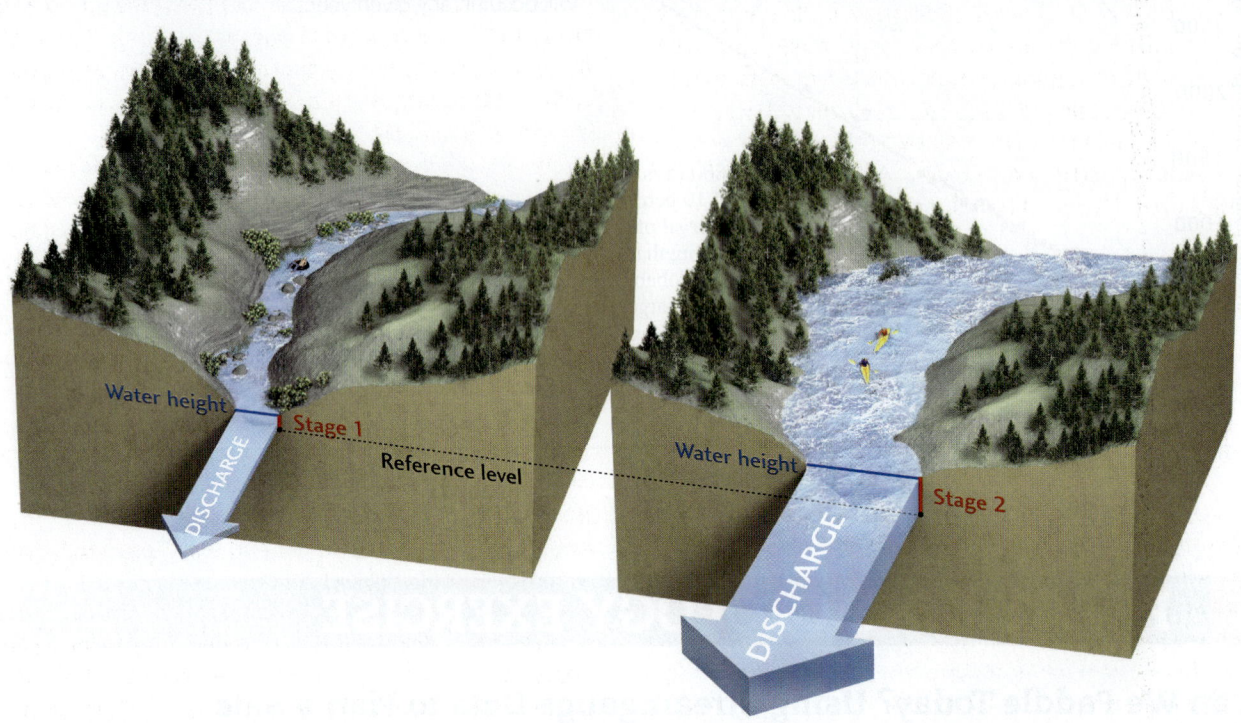

A river's discharge typically increases as rainfall or snowmelt in the watershed increases. The way people keep track of this is by measuring the stage. Stage is the height for the river relative to an arbitrary reference point. As the stage and discharge increase, the river will become increasingly turbulent, and potentially dangerous.

As the discharge increases at a particular spot, a river will become increasingly challenging and eventually dangerous to recreationists. How high the discharge must become before extra caution is warranted can be known only through experience with the stretch of river that is to be used. Paddlers should consider keeping notes or a logbook on the conditions they encounter at various discharges on a particular stretch of the river to learn the river and to plan future trips.

Streamgauge and discharge data available on the USGS Web site also allow river recreationists to project likely conditions on the river over several days. For example, anglers may be interested in knowing when it will be safe to wade a river as flows decline following a heavy rainstorm or snowmelt. By monitoring near-real-time hydrographs (graphs of discharge over time) on the site, recreationists who are interested in river flows can monitor changing conditions to determine when the water is ideal for their sport and skill level.

BONUS PROBLEM: Try reading the rating curve yourself. What is the discharge that corresponds to a stage of 10 feet? A stage of 25 feet?

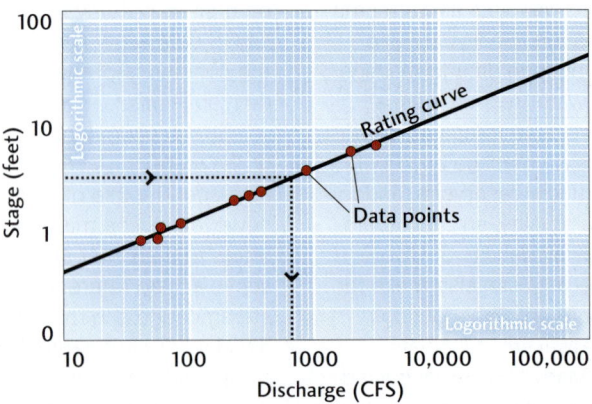

A rating curve records the relationship between stage and discharge at a particular streamgauge.

Longitudinal Profiles

We have seen that streamflow at any locality balances inflow and outflow, which become temporarily out of balance during floods. Studies of changes in discharge, flow velocity, channel dimensions, and topography along the entire length of a stream, from its headwaters to its mouth, reveal a larger-scale and longer-term balance. A stream is in dynamic equilibrium when erosion of the streambed balances sedimentation in the channel and floodplain over its entire length. This equilibrium is controlled by five factors:

1. Topography (especially slope)
2. Climate
3. Streamflow (including both discharge and velocity)
4. The resistance of rock in the streambed to weathering and erosion
5. Sediment load

A particular combination of these factors—such as steep slope, humid climate, high discharge and velocity, hard rock, and low sediment load—would mean that the stream is eroding a steep valley into bedrock and carrying downstream all sediment derived from that erosion. Downstream, where the slope is lower and the stream flows over easily erodible sediments, it would deposit sandbars and floodplain sediments, building up the elevation of the streambed by sedimentation.

We can describe the **longitudinal profile** of a stream from headwaters to mouth by plotting the elevation of its streambed against distance from its headwaters. **Figure 18.23** plots the slope of the Platte and South Platte rivers from the headwaters of the South Platte in central Colorado to the mouth of the Platte where it enters the Missouri River in Nebraska. The longitudinal profiles of all streams, from small rills to large rivers, form a similar smooth, concave-upward curve, from notably steep near the stream's headwaters to low, almost level, near its mouth.

Why do all streams follow this profile? The answer lies in the combination of factors that control erosion and sedimentation. All streams run downslope from their headwaters to their mouth. Erosion is greater in the higher parts of a stream's course than in the lower parts because slopes are steeper and flow velocities are higher, and both of these factors have an important influence on the erosion of bedrock (as we'll see in Chapter 22). In a stream's lower courses, where it carries sediments derived from erosion of the upper courses, erosion decreases and sedimentation increases. Differences in topography and the other factors listed above may make the longitudinal profile of an individual stream steeper or shallower, but the general shape remains a concave-upward curve.

Base Levels The longitudinal profile of a stream is controlled at its lower end by the stream's **base level:** the elevation at which it ends by entering a large standing body of water, such as a lake or ocean, or another stream. Streams cannot cut below their base level because the base level is the "bottom of the hill"—the lower limit of the longitudinal profile.

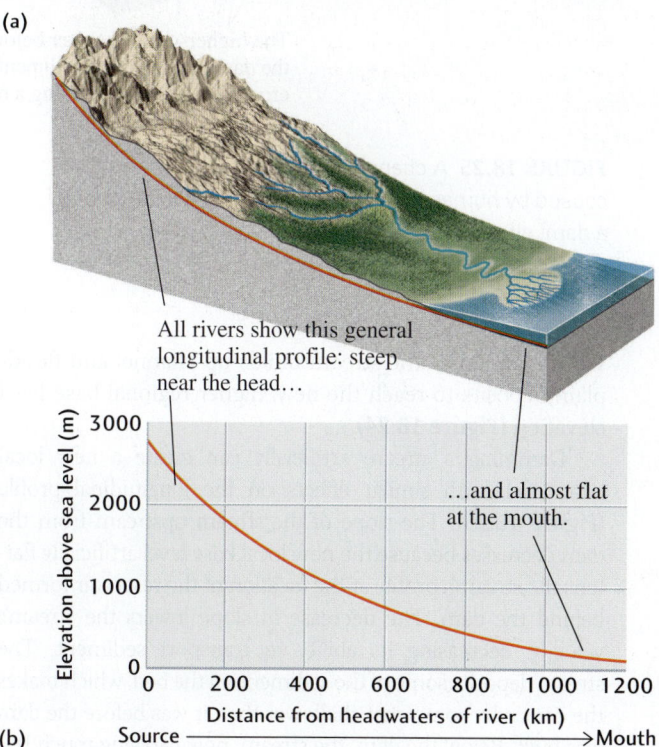

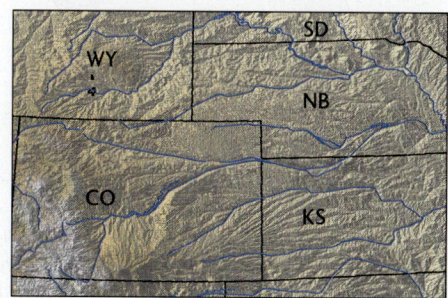

FIGURE 18.23 (a) A generalized longitudinal profile of a typical stream. (b) The longitudinal profile of the Platte and South Platte rivers from the headwaters of the South Platte in central Colorado to the mouth of the Platte at the Missouri River in Nebraska. [Data from H. Gannett, in *Profiles of Rivers in the United States*. USGS Water Supply Paper 44, 1901.]

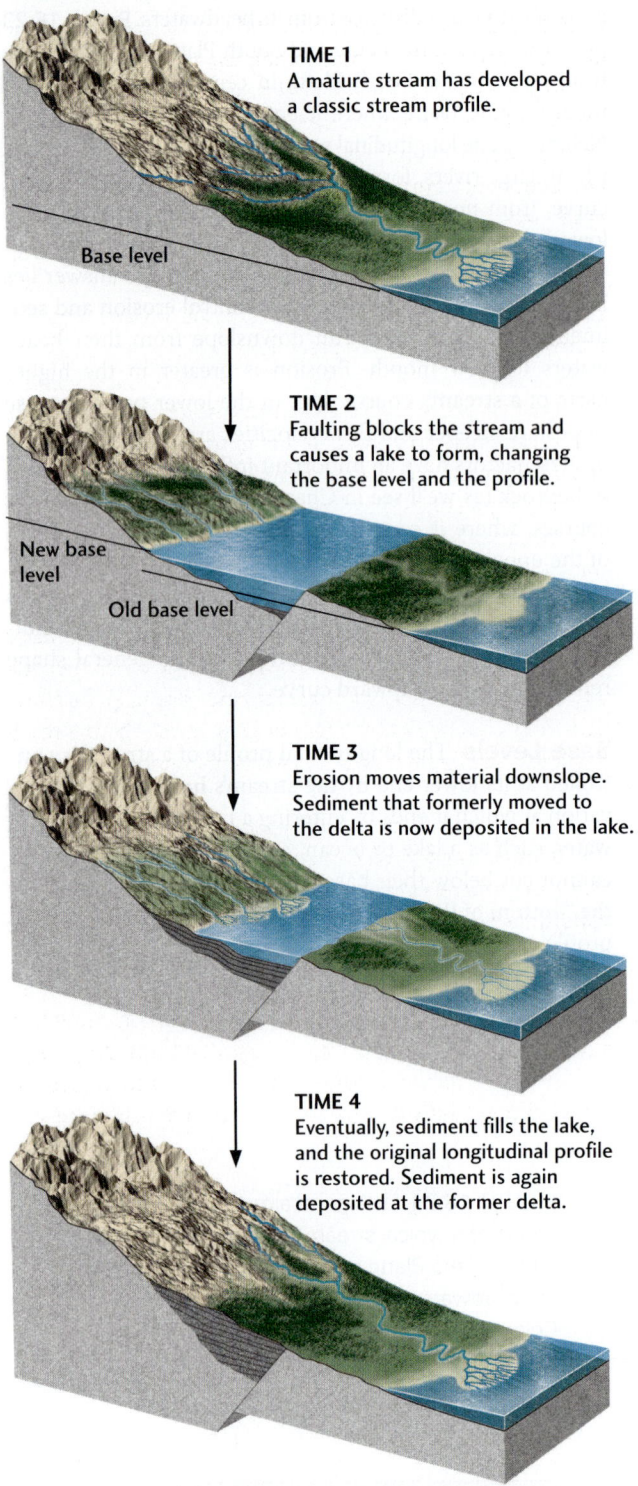

FIGURE 18.24 The base level of a stream controls the lower end of its longitudinal profile. If the base level of a stream changes, the longitudinal profile adjusts to the new base level over time.

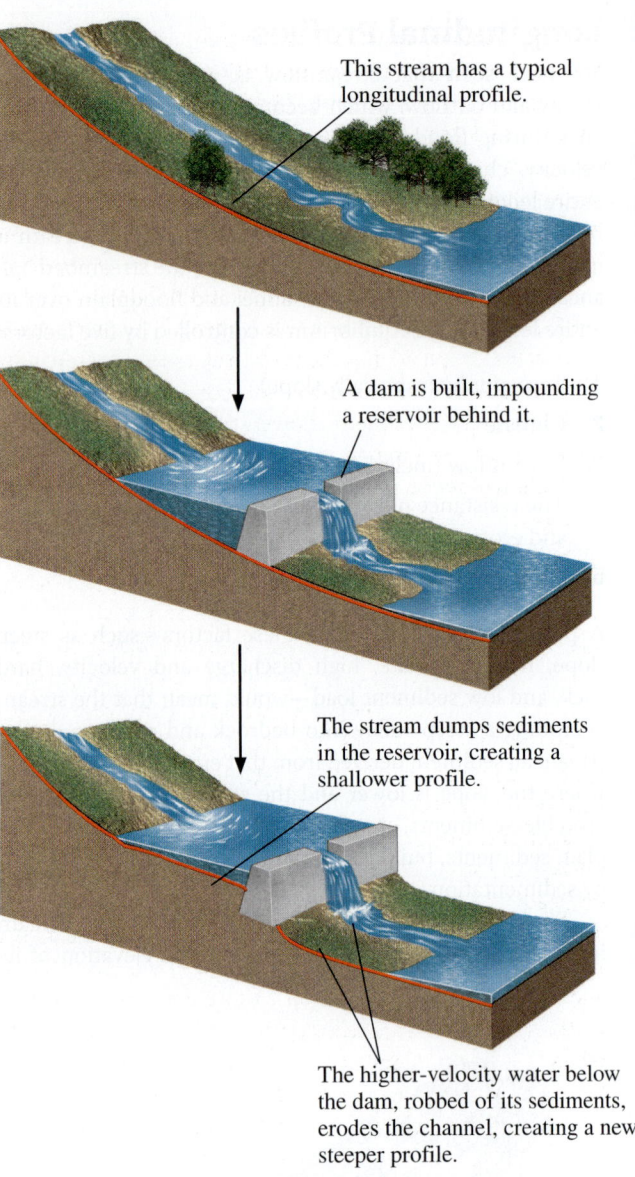

FIGURE 18.25 A change in the base level of a stream caused by human activities, such as the construction of a dam, alters the stream's longitudinal profile.

sedimentation as the stream builds up channel and floodplain deposits to reach the new, higher regional base-level elevation (**Figure 18.24**).

Damming a stream artificially can create a new local base level, with similar effects on the longitudinal profile (**Figure 18.25**). The slope of the stream upstream from the dam decreases because the new local base level artificially flattens the stream's profile at the location of the reservoir formed behind the dam. The decrease in slope lowers the stream's velocity, decreasing its ability to transport sediment. The stream deposits some of the sediment on the bed, which makes the concavity somewhat shallower than it was before the dam was built. Below the dam, the stream, now carrying much less sediment, adjusts its profile to the new conditions and typically erodes its channel in the section just below the dam.

When tectonic processes change the base level of a stream, that change affects its longitudinal profile in predictable ways. If the regional base level rises—perhaps because of faulting—the profile will show the effects of

This kind of erosion has severely affected sandbars and beaches in Grand Canyon National Park downstream from the Glen Canyon Dam. The erosion threatens animal habitats and archaeological sites as well as beaches used for recreation. River specialists calculated that if discharge during floods were increased by a certain amount, enough sand would be deposited to prevent depletion by erosion. This calculation was confirmed by an experiment in which a controlled flood was staged at the Glen Canyon Dam in 1996. As the gates of the dam were opened, about 38 billion liters of water spilled into the canyon at a rate fast enough to fill Chicago's 100-story Willis (formerly Sears) Tower in 17 minutes. This experiment showed that eroded areas could be restored by sedimentation during floods.

Falling sea level also alters regional base levels and longitudinal profiles. The regional base levels of all streams flowing into the ocean are lowered, and their valleys are cut into former stream deposits. When the drop in sea level is large, as it was during the last ice age, rivers erode steep valleys into coastal plains and continental shelves.

Graded Streams Over the years, a stream's longitudinal profile becomes stable as the stream gradually fills in low spots and erodes high spots, thereby producing the smooth, concave-upward curve that represents the balance between erosion and sedimentation. That balance is governed not only by the stream's base level, but also by the elevation of its headwaters and by all the other factors controlling the dynamic equilibrium of the stream. At equilibrium, the stream is a **graded stream:** one in which the slope, velocity, and discharge combine to transport its sediment load, with neither net sedimentation nor net erosion in the stream or its floodplain. If the conditions that produce a particular graded stream change, the stream's longitudinal profile will change to reach a new equilibrium. Such changes may include changes in depositional and erosional patterns and alterations in the shape of the channel.

In places where the regional base level is constant over geologic time, the longitudinal profile represents the balance between tectonic uplift and erosion on the one hand and sediment transport and deposition on the other—in other words, the stream is graded. If uplift is dominant, typically in the upper courses of a stream, the longitudinal profile is steep and expresses the dominance of erosion and transport. As uplift slows and the headwater region is eroded, the profile becomes shallower.

Effects of Climate Climate affects the longitudinal profile of a stream, primarily through the influences of temperature and precipitation on weathering and erosion (see Chapter 16). Warm temperatures and high rainfall promote weathering and erosion of soils and hillslopes and thus enhance sediment transport by streams. High rainfall also leads to greater discharge, which results in more erosion of the streambed. An analysis of sediment transport over the entire United States provides evidence that global climate change over the past 50 years is responsible for a general increase in streamflow. Short-term buildup of sediment or erosion may be the result of climate change, primarily variations in temperature.

Alluvial Fans Plate tectonic processes can force changes in the longitudinal profile of a stream in a number of ways. One place where a stream must adjust suddenly to changed conditions is at a *mountain front*, where a mountain range abruptly meets gently sloping plains. Here, streams leave narrow mountain valleys and enter broad, relatively flat valleys at lower elevations. Along sharply defined mountain fronts, typically at steep fault scarps, streams drop large amounts of sediment in cone- or fan-shaped accumulations called **alluvial fans** (Figure 18.26). This deposition results

FIGURE 18.26 An alluvial fan (Tucki Wash) in Death Valley, California. Alluvial fans are large cone- or fan-shaped accumulations of sediment deposited when stream velocity slows, as at a mountain front. [Marli Miller.]

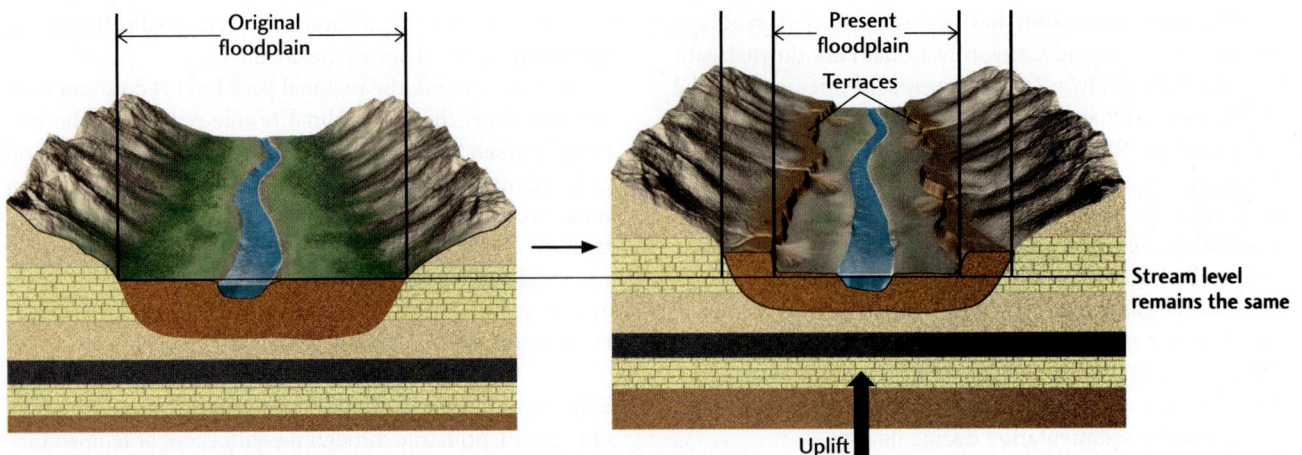

FIGURE 18.27 Terraces form when the land surface is uplifted, causing a stream to erode into its floodplain and establish a new floodplain at a lower level. The terraces are remnants of the former floodplain.

from the sudden decrease in current velocity that occurs as the channel widens abruptly. To a minor extent, the lowering of the slope below the front also slows the stream. The surface of the alluvial fan typically has a concave-upward shape connecting the steeper mountain profile with the gentler valley profile. Coarse materials, from boulders to sand, dominate on the steep upper slopes of the fan. Lower down, finer sands, silts, and muds are deposited. Fans from many adjacent streams along a mountain front may merge to form a long wedge of sediment whose appearance may mask the outlines of the individual fans that make it up.

Terraces Tectonic uplift can result in flat, steplike surfaces in a stream valley that line the stream above the floodplain. These **terraces** mark former floodplains that existed at a higher level before regional uplift or an increase in discharge caused the stream to erode into its former floodplain. Terraces are made up of floodplain deposits and are often paired, one on each side of the stream, at the same elevation (Figure 18.27). Terrace formation starts when a stream creates a floodplain. Rapid uplift then changes the stream's equilibrium, causing it to cut down into the floodplain. In time, the stream reestablishes a new equilibrium at a lower level. It may then build another floodplain, which will also undergo uplift and be sculpted into another, lower pair of terraces.

Lakes

Lakes are accidents of a stream's longitudinal profile, as we can see easily where a lake has formed behind a dam (see Figure 18.25). Lakes form wherever the flow of a stream is obstructed. They range in size from ponds only 100 m across to the world's largest and deepest lake, Lake Baikal in southwestern Siberia, which holds approximately 20 percent of the total fresh water in the world's lakes and rivers. It is located in a continental rift zone, a typical plate tectonic setting for lakes. The damming that takes place in a rift valley results from faulting that blocks a normal exit of water (see Figure 18.24). Streams can flow into a rift valley easily, but cannot flow out until water builds up to a high enough level to allow them to exit. Similarly, a great many lakes formed in the northern United States and Canada when glacial ice and glacial sediments disrupted normal stream drainage. Sooner or later, if plate tectonics and climate remain stable, such lakes will drain away as new outlets form and the longitudinal profile of the streams becomes smoother.

Because lakes are so much smaller than oceans, they are more likely than oceans to be affected by water pollution. Chemical and other industries have polluted Lake Baikal. In North America, Lake Erie has been polluted for many years, although there has been some improvement recently.

KEY TERMS AND CONCEPTS

abrasion (p. 536)	braided stream (p. 530)	discharge (p. 544)
alluvial fan (p. 551)	capacity (p. 538)	distributary (p. 541)
antecedent stream (p. 533)	channel (p. 529)	divide (p. 531)
base level (p. 549)	competence (p. 538)	drainage basin (p. 531)
bed load (p. 538)	delta (p. 541)	drainage network (p. 533)
bottomset bed (p. 541)	dendritic drainage (p. 533)	dune (p. 539)

Review of Learning Objectives 553

flood (p. 546)
floodplain (p. 529)
foreset bed (p. 541)
graded stream (p. 551)
laminar flow (p. 537)
longitudinal profile (p. 549)
meander (p. 529)
natural levee (p. 531)

oxbow lake (p. 529)
point bar (p. 529)
pothole (p. 536)
ripple (p. 539)
river (p. 528)
saltation (p. 539)
settling velocity (p. 538)
stream (p. 528)

superposed stream (p. 535)
suspended load (p. 538)
terrace (p. 552)
topset bed (p. 541)
tributary (p. 533)
turbulent flow (p. 537)
valley (p. 528)

REVIEW OF LEARNING OBJECTIVES

18.1 Identify how stream valleys and their channels and floodplains develop.

As a stream flows, it carves a valley and creates a floodplain on either side of its channel. The valley may have steep to gently sloping walls. The channel may be straight, meandering, or braided. During normal, non-flood periods, the channel carries the flow of water and sediments. During floods, the sediment-laden water overflows the banks of the channel and inundates the floodplain. The velocity of the floodwater decreases as it spreads over the floodplain. The water drops sediments, which build up natural levees and floodplain deposits.

Study Assignments: Figures 18.1 and 18.3

Exercise: How do braided and meandering stream channels differ?

Thought Question: You live in a town on a meander bend of a major river. An engineer proposes that your town invest in new, higher artificial levees to prevent the meander from being cut off. Give the arguments, pro and con, for and against this investment.

18.2 Summarize how running water erodes soil and rock.

Running water erodes solid rock by abrasion; by chemical weathering; by physical weathering as sand, pebbles, and boulders crash against rock; and by the plucking and undercutting actions of currents. The sand and pebbles the stream carries creates a sandblasting action that wears away rock, and if caught inside swirling eddies, they can form deep potholes in the bedrock. The large volume of water and tumbling boulders in waterfalls can quickly erode streambeds below waterfalls, as well as the underlying rock of the cliffs that form the falls.

Study Assignments: Figures 18.12 and 18.13

Exercise: Describe how a gully forms and how the rate of downcutting is affected by gully formation.

Thought Question: Explain the difference between headward erosion and downstream erosion and the scenarios that would cause each of them.

18.3 Explain how flowing water in streams transports and deposits sediment.

Any fluid can move in either laminar or turbulent flow, depending on its velocity, viscosity, and flow geometry. The flows of natural streams are almost always turbulent. These flows are responsible for transporting sediment in suspension, by rolling and sliding along the bed, and by saltation. When a current transports sand grains by saltation, cross-bedded dunes and ripples may form on the streambed. The settling velocity measures the speed with which suspended particles settle to the streambed.

Study Assignments: Figures 18.14 and 18.15

Exercise: What kind of bedding characterizes a ripple or a dune?

Thought Question: Why might the flow of a very small, shallow streambed be laminar in winter and turbulent in summer?

18.4 Describe how drainage networks work as collection systems and deltas as distribution systems for water and sediment.

A stream and its tributaries constitute an upstream-branching drainage network that collects water and sediments from a drainage basin. Each drainage basin is separated from other drainage basins by a divide. Drainage networks show various branching patterns—dendritic, rectangular, trellis, or radial. Where a river enters a lake or ocean, it may drop its sediments to form a delta. At the delta, the river tends to branch downstream to form distributaries, which drop the river's sediment load in topset, foreset, and bottomset beds. Deltas are modified or absent where waves, tides, and shoreline currents are strong. Plate tectonic processes influence delta formation by providing uplift in the drainage basin and subsidence in the delta region.

Study Assignment: 18.18

Exercise: What is the most common type of drainage network developed over horizontally bedded sedimentary rocks?

Thought Question: A major river, which carries a heavy sediment load, has no delta where it enters the ocean. What conditions might be responsible for the lack of a delta?

18.5 Define how a stream's longitudinal profile represents an equilibrium between erosion and sedimentation.

A stream is in dynamic equilibrium when erosion balances sedimentation over its entire length. Topography, climate, discharge and velocity, resistance to erosion, and sediment load affect this equilibrium. A stream's longitudinal profile is a plot of the stream's elevation from its headwaters to its base level.

Study Assignment: Figure 18.23

Exercise: How is a stream's longitudinal profile defined?

Thought Question: If global warming produces a significant rise in sea level as polar ice caps melt, how will the longitudinal profiles of the world's rivers be affected?

CHAPTER 18 Stream Transport: From Mountains to Oceans

VISUAL LITERACY EXERCISE

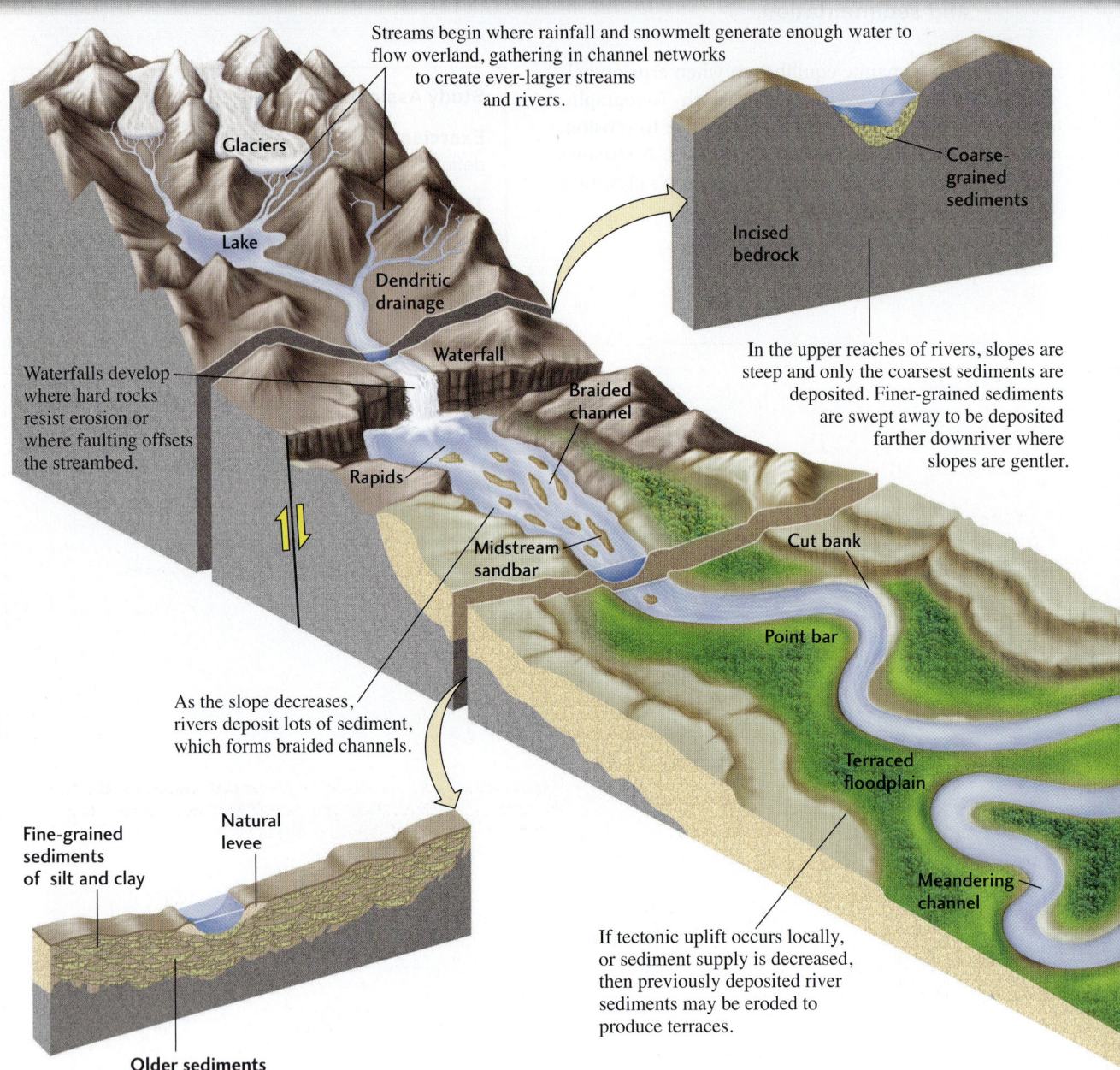

FIGURE 18.20 Stream networks transport water and sediments from their headwaters to the ocean.

1. Where do streams begin?
2. Where do streams end?
3. How are stream terraces produced?
4. How might changes in grain size of stream sediment reflect proximity to the upper versus lower reaches of a stream?

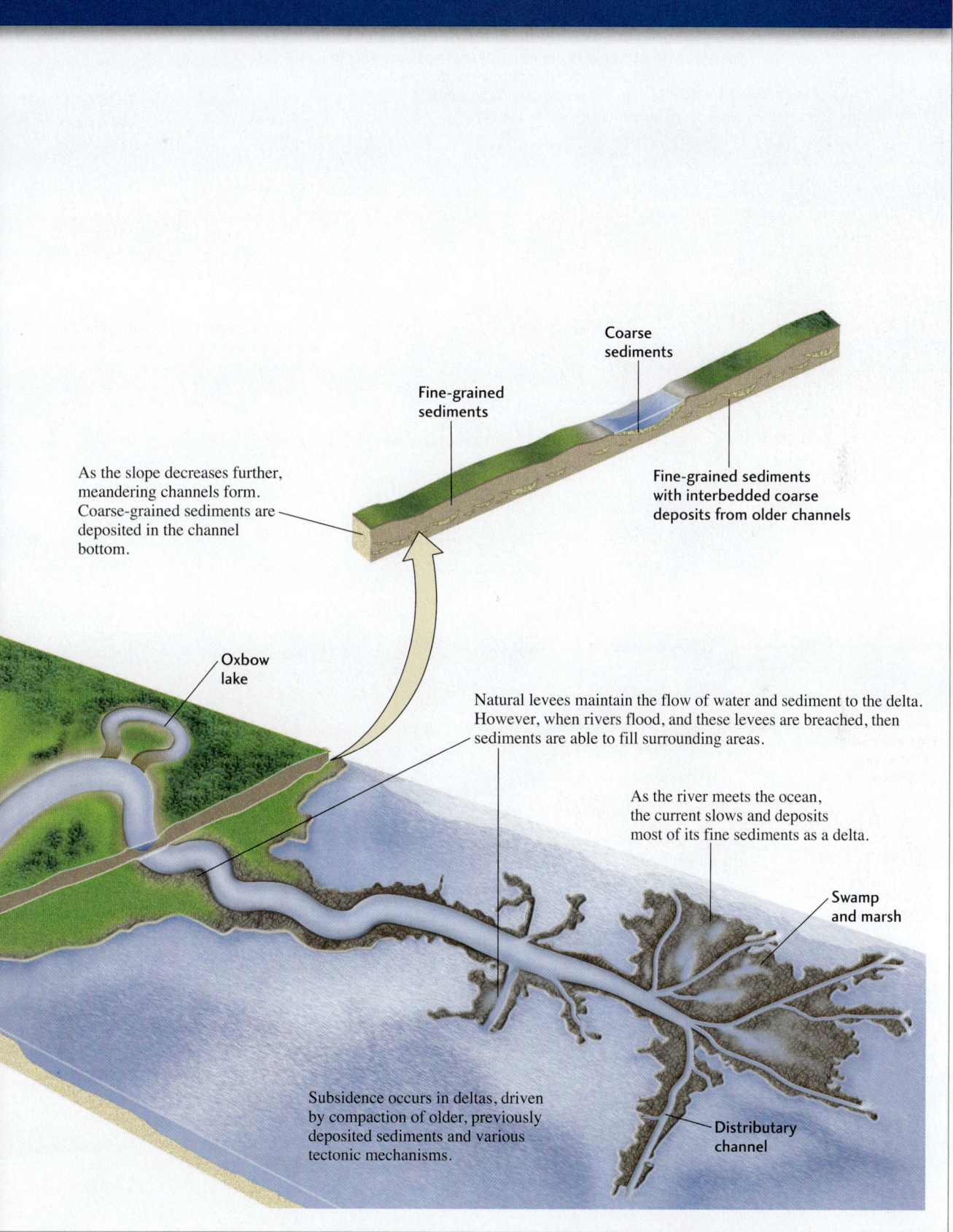

Coastlines and Deserts 19

Coastal Processes	**560**
The Shaping of Shorelines	**565**
Hurricanes and Coastal Storm Surges	**573**
Desert Processes	**579**
Windblown Sand and Dust	**581**
The Desert Environment	**588**
Tectonic, Climatic, and Human Controls on Deserts	**591**

Learning Objectives

After studying this chapter, you should be able to:

- **19.1** Identify the coastal processes that act on shorelines.
- **19.2** Summarize how processes shape shorelines.
- **19.3** Explain how hurricanes affect coastal areas.
- **19.4** Describe how winds transport and erode sand and finer-grained sediments.
- **19.5** Discuss how winds deposit sand dunes and dust.
- **19.6** Explain how wind and water combine to shape the desert environment and its landscape.
- **19.7** Summarize the tectonic, climatic, and human controls on deserts.

The desert meets the coastline at Sandwich Harbor, Namibia. These dunes are among the tallest in the world, and the drift of sand into the Atlantic Ocean is sculpted into sandy beaches. [Courtesy of Roger Swart.]

Coastlines and deserts are critically important components of Earth's surface environment. They also are among the most sensitive environments to global change. The large populations who live at the edge of the sea know well the impact of waves, the rise and fall of the tides, and the devastating effects of powerful storms. The forces that control these coastal processes result from interactions within the climate system and the solar system. Tides are caused by gravitational interactions between Earth and the Sun and Moon, and coastal surf and storms result from interactions between the atmosphere and the hydrosphere. Deserts are similarly governed by interactions with the climate systems, but here the force of wind rather than water tends to dominate landscape development. Deserts are important because so many of the geologic processes that shape those arid environments are related to the work of wind. Currents of water shape our shorelines and in a similar way currents of air shape the landforms of deserts. Both environments also involve the transport of large quantities of sand—one by water, the other by wind.

In Chapter 19, we focus on the geologic processes that form coastlines and deserts, and how they change over time in response to tectonic, climatic, and human influences. We describe and interpret the processes that affect shorelines and coastal areas and consider the effects of waves, tides, and damaging storms. We also examine how erosion, transportation, and deposition by wind shape the surface of the land. Finally, we discuss the elements that make up desert landscapes, and how those landscapes are spreading across the globe.

Coastal Processes

Coastlines are the broad regions where land and streams meet the ocean. Environmental problems such as coastal erosion and pollution of shallow waters make the geology of coastlines an active area of research. The landscapes of coastlines, even within a single continent, present striking contrasts (**Figure 19.1**). On the coast of North Carolina, for example, long, straight, sandy beaches stretch for miles along low coastal plains (Figure 19.1a). Here, tectonic activity is limited, and it is the currents produced by breaking waves that mold the coastline. The Maine coastline, on the other hand, is dominated by rocky cliffs (Figure 19.1b). Even though the effect of waves is considerable, it is postglacial rebound related to the retreat of ancient ice sheets that shapes this landscape. Many of the seaward edges of islands in the tropics are coral reefs, shaped by biological sedimentation (Figure 19.1d). As we will see, plate-tectonic processes, erosion, and sedimentation (Figure 19.1c) work together to create this great variety of coastline shapes and materials.

The five major oceans (Atlantic, Pacific, Indian, Arctic, and Southern) form a single connected body of water sometimes referred to as the *world ocean.* The term *sea* is often used to refer to smaller bodies of water set off somewhat from the oceans. The Mediterranean Sea, for example, is narrowly connected with the Atlantic Ocean by the Strait of Gibraltar and with the Indian Ocean by the Suez Canal. Other seas are more broadly connected, as is the North Sea with the Atlantic Ocean. Seawater—the salty water of the oceans and seas—is remarkably constant in its general chemical composition from year to year and from place to place. The chemical equilibrium maintained by the oceans is determined by the composition of river waters entering the oceans, the composition of the sediments they transport to the oceans, and the formation of new sediments in the oceans.

The major geologic forces operating at the **shoreline**—the line where the water surface meets the land surface—are ocean currents created by waves and tides. These currents eventually erode even the most resistant rocky shores. They also transport the sediments produced by erosion and deposit them on beaches and in shallow waters along the shore.

As we have seen in earlier chapters, currents are the key to understanding geologic processes at Earth's surface, and coastal processes are no exception. Let's examine the various types of currents that shape our shorelines.

Wave Motion: The Key to Shoreline Dynamics

Centuries of observation have taught us that waves are constantly changing. In quiet weather, waves with calm troughs between them roll regularly into shore. In the high winds of a storm, however, waves move in a confusion of shapes and sizes. They may be low and gentle far from shore, yet become high and steep as they approach land. High waves can break on the shore with fearful violence, shattering concrete seawalls and tearing apart houses built along the beach. To understand the dynamics of shorelines, and to make sensible decisions about coastal development, we need to understand how waves work.

Wind blowing over the surface of the ocean creates waves by transferring its energy of motion from air to water. As a gentle breeze of 5 to 20 km/hour starts to blow over a calm sea surface, ripples—little waves less than a centimeter high—take shape. As the speed of the wind increases to about 30 km/hour, the ripples grow to full-sized waves. Stronger winds create larger waves and blow off their tops to make whitecaps. The height of waves depends on three factors:

- The wind speed
- The length of time over which the wind blows
- The distance the wind travels over water

FIGURE 19.1 Coastlines exhibit a variety of geologic forms. (a) Long, straight, sandy beach, Pea Island, North Carolina. (b) Rocky coastline, Mount Desert Island, Maine. This formerly glaciated coastline has rebounded since the end of the last ice age, about 11,000 years ago. (c) The Twelve Apostles, Port Campbell, Australia, a group of stacks that developed from cliffs of sedimentary rock. These remnants of shoreline erosion are left as the shoreline retreats under the action of waves. (d) Coral reef along the Florida coastline. [a: Westend61/Getty Images; b: Neil Rabinowitz/Getty Images; c: Christopher Groenhout/Getty Images; d: Dr. Hays Cummins, Interdisciplinary Studies, Miami University.]

Storms blow up large, irregular waves that radiate outward from the storm center, like the ripples moving outward from a pebble dropped into a pond. As the waves travel outward in ever-widening circles, they become more regular, changing into low, broad, rounded waves called *swell*, which can travel hundreds of kilometers. Several storms at different distances from a shoreline, each producing its own pattern of swell, may account for the often irregular intervals between waves approaching the shore.

If you have seen waves in an ocean or a large lake, you have probably noticed how a piece of wood floating on the water moves a little forward as the crest of a wave passes and then a little backward as the trough between waves passes. Although it moves back and forth, the wood stays in roughly the same place—and so does the water around it. The water molecules move in a circle, even though the waves are moving toward the shore.

We can describe a wave form in terms of the following three characteristics (**Figure 19.2**):

1. *Wavelength*, the distance between wave crests
2. *Wave height*, the vertical distance between the crest and the trough
3. *Period*, the time it takes for two successive wave crests to pass a fixed point

We can measure the velocity at which a wave moves forward by using a simple equation:

$$V = \frac{L}{T}$$

where V is the velocity, L is the wavelength, and T is the period. Thus, a typical wave with a length of 24 m and a

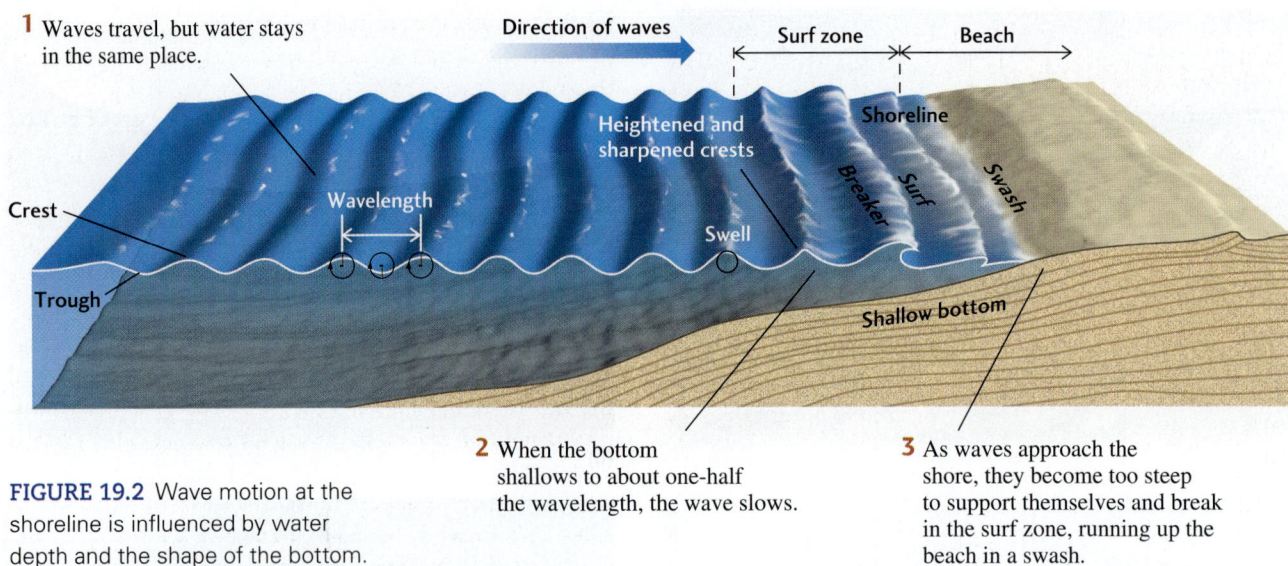

FIGURE 19.2 Wave motion at the shoreline is influenced by water depth and the shape of the bottom.

period of 8 s would have a velocity of 3 m/s. The periods of waves range from just a few seconds to as long as 15 or 20 s, and their wavelengths vary from about 6 m to as much as 600 m. Consequently, wave velocities vary from 3 to 30 m/s. Wave motion becomes very small below a depth equal to about one-half the wavelength. That is why deep divers and submarines are unaffected by the waves at the surface.

The Surf Zone

Swell becomes higher as it approaches the shore, where it assumes the familiar sharp-crested wave shape. These waves are called *breakers* because, as they come closer to shore, they break and form surf—a foamy, bubbly surface. The *surf zone* is the belt along which breaking waves collapse as they approach the shore. The transformation from swell to breakers starts where water depth decreases to less than one-half the wavelength of the swell. At that point, the wave motion just above the bottom becomes restricted because the water can only move back and forth horizontally. Above that, the water can move vertically just a little (see Figure 19.2). The restricted motion of the water molecules slows the whole wave. Its period remains the same, however, because the swell keeps coming in from deeper water at the same rate. From the wave equation, we know that if the period remains constant but the wavelength decreases, then the velocity must also decrease. The typical wave that we used as our example earlier might keep the same period of 8 s while its length decreased to 16 m, in which case it would have a velocity of 2 m/s. Thus, as waves approach the shore, they become more closely spaced, higher, and steeper, and their wave crests become sharper.

As a wave rolls toward the shore, it becomes so steep that the water can no longer support itself, and the wave breaks with a crash in the surf zone (see Figure 19.2). Gently sloping bottoms cause waves to break farther from shore; steeply sloping bottoms make waves break closer to shore. Where rocky shores are bordered by deep water, the waves break directly on the rocks with a force amounting to tons per square meter, throwing water high into the air. It is not surprising that concrete seawalls built to protect structures along the shore quickly start to crack and must be repaired constantly.

After breaking in the surf zone, the waves, now reduced in height, continue to move in, breaking again at the shoreline. They run up onto the sloping front of the beach, forming an uprush of water called *swash*. The water then runs back down again as *backwash*. Swash can carry sand and even large pebbles and cobbles onto the beach if the waves are high enough. The backwash carries the particles seaward again.

The back-and-forth motion of the water just offshore is strong enough to carry sand grains and even gravel. Wave action in water as deep as 20 m can move fine sand. Large waves caused by intense storms can scour the bottom at much greater depths, down to 50 m or more. At shallower depths, storms transport sediments in an offshore direction, often depleting beaches of their fine sand.

Wave Refraction

Far from shore, the lines of wave crests are parallel to one another, but are usually at some angle to the shoreline. As the waves approach the shore over a shallowing bottom, they gradually bend to a direction more parallel to the shore (**Figure 19.3a**). This bending is called *wave refraction*. It is similar to the bending of light rays in optical refraction, which makes a pencil half in and half out of water appear to bend at the water surface. Wave refraction begins as the part of a wave closest to the shore encounters the shallowing bottom first, and the front of the wave slows. Then the next part of the wave meets the bottom and it, too, slows. Meanwhile, the parts closest to shore have moved into even shallower water and have slowed even more. Thus, in a continuous transition along the wave crest, the line of waves bends toward the shore as it slows (Figure 19.3b).

Wave refraction results in more intense wave action on projecting headlands (Figure 19.3c) and less intense action in indented bays. The bottom becomes shallow more quickly around headlands than in the surrounding deeper water on either side. Thus, waves are refracted around headlands—that is, they are bent toward the projecting part of the shore on both sides. The waves converge around the headland and expend proportionally more of their energy breaking there than at other places along the shore. Because of this concentration of wave energy at headlands, they tend to be eroded more quickly than straight sections of shoreline.

The opposite happens as a result of wave refraction in a bay. The waters in the center of the bay are deeper, so the waves are refracted on either side into shallower water. The energy of wave motion is diminished at the center of the bay, which makes bays good harbors for ships.

Although refraction makes waves more parallel to the shore, most waves still approach it at some small angle. As the waves break on the shore, the swash moves up the beach slope in a direction perpendicular to that small angle. The backwash runs down the slope in the opposite direction at a similar small angle. The combination of these two motions moves the water a short way down the beach (Figure 19.3d). Sand grains carried by swash and backwash are thus moved along the beach in a zigzag motion known as *longshore drift*.

Waves approaching the shoreline at an angle can also cause a **longshore current,** a shallow-water current that flows parallel to the shore. The movement of swash and backwash creates a zigzag path of water molecules that transports sediments along the shallow bottom in the same direction as the longshore drift. Much of the transport of sand along many beaches results from longshore currents. Longshore currents are prime determiners of the shape and extent of sandbars and other depositional shoreline features. At the same time, because of their ability to erode loose sand, longshore currents may remove large amounts of sand from a beach. Longshore drift and longshore currents, working together, are potent agents of sand transport on beaches and in very shallow waters. In slightly deeper waters (less than 50 m), longshore currents—especially those running during intense storms—strongly affect the bottom.

Some types of flows related to longshore currents can pose a threat to unwary swimmers. A *rip current*, for

(a)

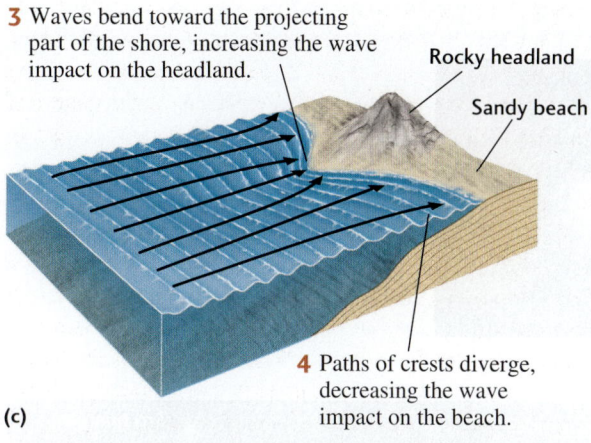

1 A fast-traveling wave approaches from deep water.

2 The part of the wave closest to the beach slows, causing the line of waves to refract toward the beach.

(b)

3 Waves bend toward the projecting part of the shore, increasing the wave impact on the headland.

4 Paths of crests diverge, decreasing the wave impact on the beach.

(c)

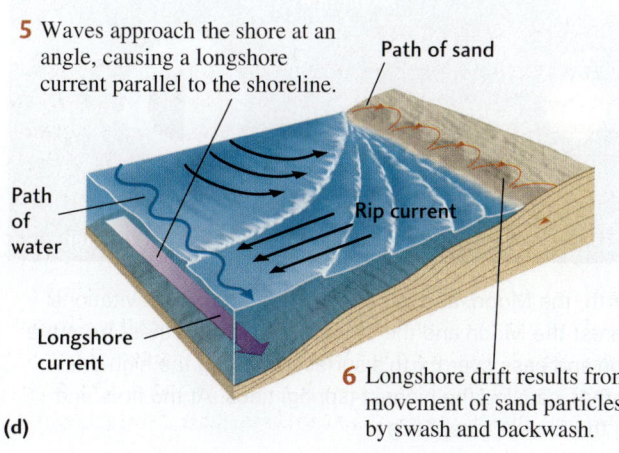

5 Waves approach the shore at an angle, causing a longshore current parallel to the shoreline.

6 Longshore drift results from movement of sand particles by swash and backwash.

(d)

FIGURE 19.3 Wave refraction. (a) Waves approach the shore at an angle. (b) As waves move closer to the shore, the angle of the wave crests becomes more parallel to the shoreline. (c) Wave refraction increases the erosion of projecting headlands. (d) Wave refraction gives rise to longshore drift and longshore currents. [a: Rob Crandall/Alamy.]

example, is a strong flow of water moving seaward at right angles to the shore (see Figure 19.3d). It occurs when a longshore current builds up along the shore and the water piles up imperceptibly until a critical point is reached. At that point, the water breaks out to sea, flowing through the oncoming waves in a fast current. Swimmers can avoid being carried out to sea by swimming parallel to the shore to get out of the rip.

Tides

For thousands of years, mariners and coastal dwellers have observed the twice-daily rise and fall of the ocean that we call **tides.** Many observers noticed a relationship between the position and phases of the Moon, the heights of the tides, and the times of day at which the water reaches high tide. Not until the seventeenth century, however, when Isaac Newton formulated the laws of gravitation, did we begin to understand that tides result from the gravitational pull of the Moon and the Sun on the water of the oceans.

The gravitational attraction between any two bodies decreases as they get farther apart. Thus, the strength of this attraction varies across Earth's surface. On the side of Earth closest to the Moon, the ocean water experiences a greater gravitational attraction than the average for the whole of the solid Earth. This pull produces a bulge in the water. On the side of Earth farthest from the Moon, the solid Earth, being closer to the Moon than the water, is pulled toward the Moon more than the water is, and the water therefore appears to be pulled away from Earth as another bulge. Thus, two bulges of water are formed in Earth's oceans: one on the side nearest the Moon, and the other on the side farthest from the Moon (Figure 19.4a). As Earth rotates, these bulges of water stay approximately aligned: One always faces the Moon, and the other is always directly opposite the Moon. These bulges of water passing over the rotating Earth are the high tides.

The Sun, although much farther away, has so much mass (and thus so much gravitational force) that it, too, causes tides. Sun tides are a little less than half the height of Moon tides, and they are not synchronous with Moon tides. Sun tides occur as Earth rotates once every 24 hours, the length of a solar day. The rotation of Earth with respect to the Moon is a little longer because the Moon is moving around the Earth, resulting in a lunar day of 24 hours and 50 minutes. In that lunar day, there are two high tides, with two low tides between them.

When the Moon, Earth, and Sun line up, the gravitational forces of the Sun and the Moon reinforce each other. This alignment produces the *spring tides*, which are the highest tides; their name is not related to the season, but to the German verb *springen*, meaning "to leap up."

(a)
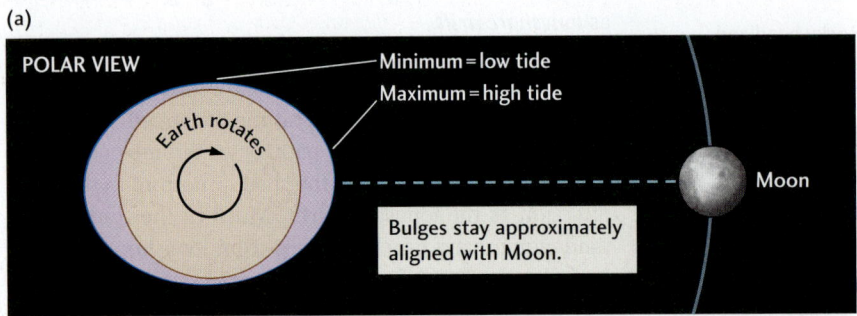

(b)
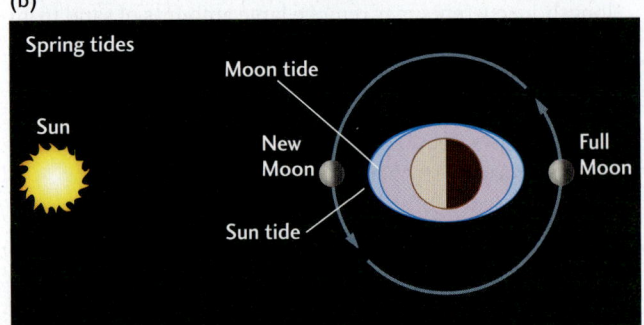
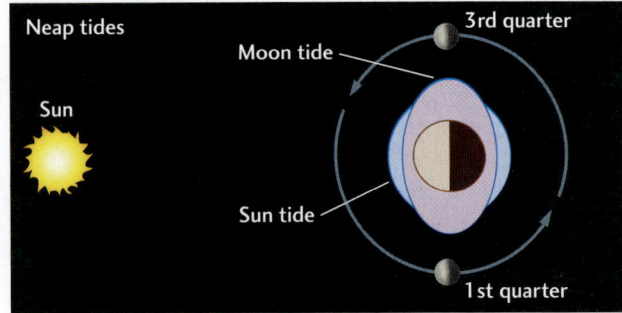

FIGURE 19.4 Tides are caused by the gravitational attraction of Earth, the Moon, and the Sun. (a) The Moon's gravitational pull forms two bulges of ocean water, one on the side of Earth nearest the Moon and the other on the side farthest from the Moon. As Earth rotates, these bulges remain aligned with the Moon and pass over Earth's surface, creating the high tides. (b) At the new and full Moon, Sun and Moon tides reinforce each other, causing the highest (spring) tides. At the first- and third-quarter Moon, Sun and Moon tides are in opposition, causing the lowest (neap) tides.

They appear every 2 weeks, at the full and new Moon. The lowest tides, the *neap tides,* come in between, at the first- and third-quarter Moon, when the Sun and Moon are at right angles to each other with respect to Earth (Figure 19.4b).

Although tides occur regularly everywhere, the difference between high and low tides varies in different parts of the ocean. As Earth rotates, the tidal bulges of water move along the surface of the ocean, encountering obstacles, such as continents and islands, which hinder the flow of the water. In the middle of the Pacific Ocean—in the Hawaiian Islands, for example, where there is little to obstruct the flow of the tides—the difference between low and high tides is only 0.5 m. Near Seattle, where the shape of the shoreline along Puget Sound is very irregular and the tidal flow must move through narrow passageways, the difference between low and high tides is about 3 m. Extraordinary tides occur in a few places, such as the Bay of Fundy in eastern Canada, where the tidal range can be more than 12 m. Many coastal residents need to know when tides will occur, so governments publish tide tables showing predicted tide heights and times. These tables combine local knowledge of water flow patterns with knowledge of the astronomical movements of Earth and the Moon with respect to the Sun.

Tides moving near shorelines cause currents that can reach speeds of a few kilometers per hour. As the tide rises, water flows in toward the shore as a *flood tide,* moving through narrow passages into inlets and bays, into shallow coastal wetlands, and up small streams. As the tide passes the high stage and starts to fall, the water flows out as an *ebb tide,* and low-lying coastal areas are exposed. Tidal currents meander across **tidal flats,** muddy or sandy areas that are exposed at low tide but flooded at high tide (**Figure 19.5**). Where obstacles restrict tidal flow and increase tidal range, current velocities may become very high. Large sand ridges many meters high may be formed in these tidal channels.

The Shaping of Shorelines

The effects of the coastal processes we have just described are best observed at shorelines. Waves, longshore currents, tidal currents, and storm surges interact with plate-tectonic processes and with the geologic structures of the coast to shape shorelines into a multitude of forms. We can see these factors at work in the most popular of shoreline environments: beaches.

Beaches

A **beach** is a shoreline environment made up of sand and pebbles. The shape of a beach may change from day to day, week to week, season to season, and year to year. Waves and tides sometimes broaden and extend a beach by depositing sand and sometimes narrow it by carrying sand away.

Many beaches are straight stretches of sand ranging from 1 km to more than 100 km long; others are smaller crescents of sand between rocky headlands. Belts of dunes border the landward edge of many beaches; bluffs or cliffs of sediment or rock border others. A beach may have a *tide terrace*—a flat, shallow area between the upper beach and an outer bar of sand—on its seaward side (**Figure 19.6**).

FIGURE 19.5 Tidal flats, such as this one at Mont-Saint-Michel, France, may be extensive areas covering many square kilometers, but most often are narrow strips seaward of the beach. When a very high tide advances on a broad tidal flat, it may move so rapidly that some areas are flooded faster than a person can run. The beachcomber is well advised to learn the local tides before wandering. [Thierry Prat/Sygma via Getty Images.]

FIGURE 19.6 A tide terrace exposed at low tide. This shallow depression between an outer ridge (a sandbar at high tide) and the upper beach is rippled by the tidal flow in many places. [David Hall/Alamy.]

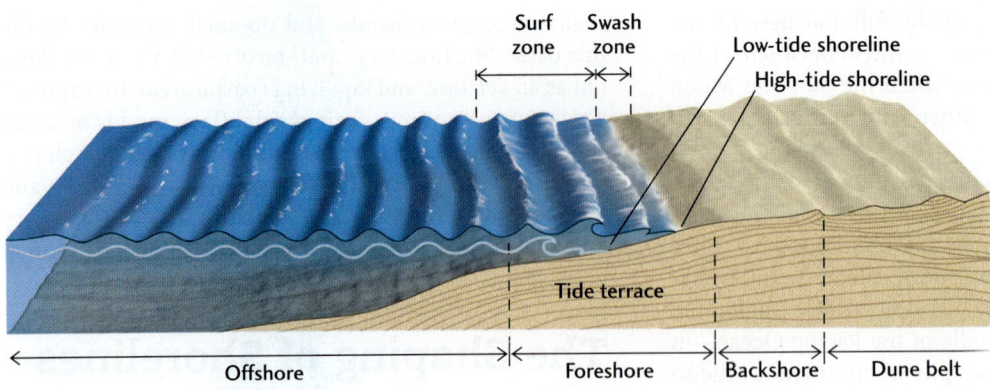

FIGURE 19.7 A profile of a beach, showing its major parts.

The Structure of a Beach Figure 19.7 shows the major parts of a beach. These parts may not all be present at all times on any particular beach. Farthest out is the *offshore*, which is bounded by the surf zone, where the bottom begins to become shallow enough for waves to break. The *foreshore* includes the surf zone; the tide terrace; and, right at the shoreline, the *swash zone*, a slope dominated by the swash and backwash of the waves. The *backshore* extends from the swash zone up to the highest level of the beach.

The Sand Budget of a Beach A beach is a scene of constant movement. Each wave moves sand back and forth with its swash and backwash. Both longshore drift and longshore currents move sand down the beach. At the end of the beach, and to some extent along it, sand is removed and deposited in deep water. In the backshore or along sea cliffs, sand and pebbles are freed by erosion and replenish the beach. Winds that blow over the beach transport sand, sometimes offshore into the water and sometimes onshore onto the land.

All these processes together maintain a balance between addition and removal of sand, resulting in a beach that may appear to be stable but is actually exchanging its material with the environments on all sides. Figure 19.8 illustrates the sand budget of a beach: the inputs and outputs caused by erosion, sedimentation, and transport. At any point along a beach, the beach gains sand from a number of sources: material eroded from the backshore; sand brought to the beach by longshore drift and longshore currents; and sediments carried to the shoreline by rivers. The beach also loses sand in a number of ways: Winds carry sand to backshore dunes, longshore drift and longshore currents carry it downcurrent, and deep-water currents and waves transport it during storms.

If the total sand input balances the total sand output, the beach is in dynamic equilibrium, and it keeps the same general form. If input and output are not balanced, the beach grows or shrinks. Temporary imbalances are natural over weeks, months, or even years. A series of intense storms, for example, might move large amounts of sand from the beach to deeper waters offshore, narrowing the beach. Then, weeks of mild weather and low waves might move sand onto the shore and rebuild a wide beach. Without this constant shifting of sand, beaches might be unable to recover from the effects of trash, litter, and other kinds of pollution. Within a year or two, even oil from spills is transported or buried out of sight, although the tarry residue may later be uncovered in spots.

Some Common Forms of Beaches Long, wide, sandy beaches grow where sand inputs are abundant, often where soft sediments make up the coast. Where the backshore is low and winds blow onshore, wide dune belts border the beach. Where the shoreline is tectonically elevated and the coast is made up of hard rock, cliffs line the shore, and any small beaches that form are composed of material eroded from those cliffs. Where the coast is low-lying, sand is abundant, and tidal currents are strong, extensive tidal flats are laid down and are exposed at low tide.

Preservation of Beaches What happens if one of the inputs to a beach is cut off—for example, by a concrete seawall built along a beach to prevent erosion? Because erosion supplies sand to the beach, preventing erosion cuts the sand supply and so shrinks the beach. Such attempts to save a beach, undertaken without an understanding of its dynamic equilibrium, may actually destroy it.

Humans are altering the dynamic equilibrium of more and more beaches by placing buildings on them and erecting structures to protect them from erosion. We build cottages and resort hotels on the shore; pave beach parking lots; erect seawalls; and construct groins, piers, and breakwaters. The consequence of such poorly planned development is shrinkage of the beach in one place and its growth in another. As landowners and developers bring suit against one another and against state governments, trial lawyers take the issue of "sand rights"—the beach's right to the sand that it naturally contains—into the courts.

To use a classic example, let's examine what happens when a narrow groin or jetty—a structure built out from the shore at right angles to it—is installed. In the subsequent

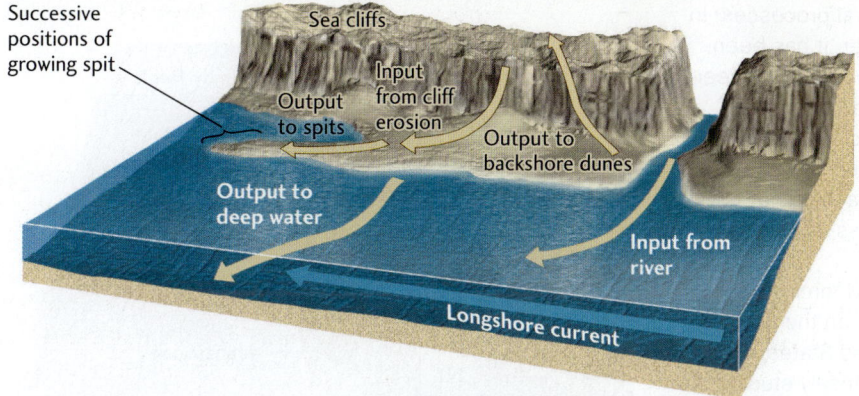

FIGURE 19.8 The sand budget of a beach is a balance between inputs and outputs of sand due to erosion, sedimentation, and transport.

months and years, the sand disappears from the beach on one side of the groin and greatly enlarges the beach on the other side (Figure 19.9). These changes are the predictable result of normal coastal processes. The waves, longshore current, and longshore drift bring sand toward the groin from the upcurrent direction (usually the prevailing wind direction). Stopped at the groin, they dump the sand there. On the downcurrent side of the groin, the current and drift pick up again and erode the beach. On this side, however, replenishment of sand is sparse because the groin blocks inputs of sand. As a result, the sand budget is out of balance, and the beach shrinks. If the groin is removed, the beach returns to its former state.

The only way to preserve a beach is to leave it alone. Groins and seawalls are only temporary solutions to the problem of beach erosion, and even if they are kept in repair with large expenditures of money—many times at public expense—the beach itself will suffer. Beach restoration projects, which involve pumping large volumes of sand from offshore, have had some success (see the Practicing Geology Exercise), but they, too, are extremely costly. Sooner or later, we must learn to let beaches remain in their natural state.

FIGURE 19.9 Construction of a groin to control beach erosion may result in erosion downcurrent of the groin and loss of parts of the beach there, while sand piles up on the other side of the groin. [Airphoto—Jim Wark.]

PRACTICING GEOLOGY EXERCISE

Does Beach Restoration Work?

Beach erosion is a problem facing many communities that have come to enjoy the scenic beauty of their beaches and depend on them to support tourism and economic development. Erosion of beaches is often driven by natural processes; in some cases, however, it has been greatly enhanced by failed engineering practices intended to prevent it. In recent years, scientists and engineers have pooled their efforts to create new approaches that have led to greater success in protecting beaches.

The beaches of Monmouth County, New Jersey, on the Atlantic coast of the United States, are among the most intensely studied on Earth. Human modification of these beaches commenced in 1870 with the construction of the New York and Long Branch Railroad. Access by railroad allowed tourism to develop and eventually permitted commuting to New York City by full-time residents, who began to alter the coastline. Concrete seawalls replaced beaches, and sand dunes and rock jetties were built about every quarter mile along the county's 12-mile shoreline. Little by little over the next 100 years, the Monmouth County beaches became very narrow, until miles of shoreline were without a sand beach of any kind. The only bathing beaches were found in tiny pockets tucked into the corners made by the seawall and a jetty. Winter storms in 1991 and 1992 did substantial damage to the entire Monmouth County shoreline, driving the boardwalk back onto streets as splintered debris. Damage to homes occurred as the ocean easily overtopped the almost nonexistent beaches and insufficient seawalls.

By 1994, the state of New Jersey became serious about finding a solution to the beach erosion problem and appealed to the federal government for help. Congress subsequently authorized funding for the nation's

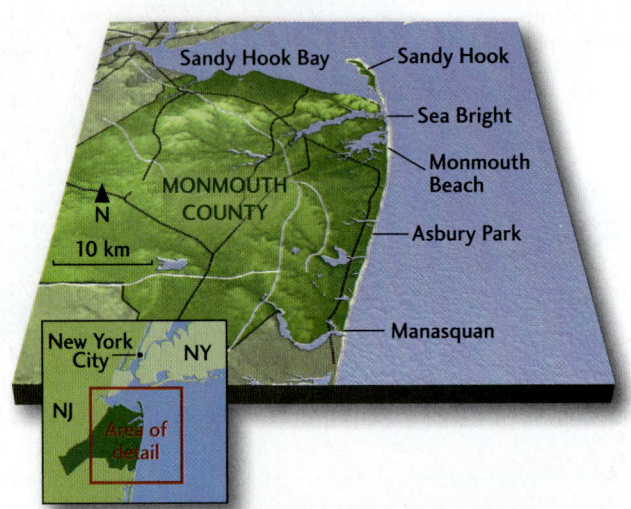

Sand placement at the southern end of Monmouth Beach, Monmouth County, New Jersey. This erosion control project by the U.S. Army Corps of Engineers includes periodic nourishment of the restored beaches on a 6-year cycle for a period of 50 years. [Courtesy of Army Corps of Engineers, New York District.]

Erosion and Deposition at Shorelines

The topography of shorelines is a product of the same forces that shape the continental interior: plate tectonic processes that elevate or depress Earth's crust, erosional processes that wear it down, and sedimentation that fills in the low spots. Thus, several factors are directly at work:

- Tectonic uplift of the coastal region, which leads to erosional coastal forms
- Tectonic subsidence of the coastal region, which leads to depositional coastal forms
- The nature of the rocks or sediments at the shoreline
- Changes in sea level, which affect the submergence or emergence of a shoreline
- The average and storm wave heights, which affect erosion
- The heights of the tides, which affect both erosion and sedimentation

largest beach restoration project ever attempted, covering 20 miles of shoreline in Monmouth County, from the township of Sea Bright to Manasquan Inlet. The restoration project involved pumping enough sand from offshore areas to construct a restored beach 100 feet wide with an elevation of 10 feet above mean low water. The project includes periodic nourishment of the restored beaches on a 6-year cycle for 50 years from the start of the initial beach construction in 1994.

Beginning in 1994 and ending in 1997, 57 million cubic meters of sand were pumped from about a mile offshore, at a cost of $210,000,000. This initial placement volume provided a vast supply of new sand to the beaches of 9 out of 12 oceanfront municipalities. The earliest sites restored have responded well, requiring little sand augmentation since the project started.

At the outset, it was not obvious that the Monmouth County beach restoration project would succeed. Some people predicted total loss of the sand within a year or two. Nevertheless, the project has performed far better than all expectations. The results have been tracked by monitoring of changes in sand volume along a 13-km-long segment of the restoration zone.

The accompanying table provides a more quantitative sense of seasonal changes in sand volume along the shoreline due to erosion and deposition by natural processes. Monitoring of sand erosion and deposition on a seasonal basis between 1998 and 2004 has yielded an average value for cubic meters of sand lost (or gained) per meter of shoreline (m^3/m) in each season. When this seasonal value is multiplied by the 13-km length of the shoreline, the change in the volume of the shoreline (m^3) can be calculated. Note that in fall 2002, the shoreline was augmented by an artificially supplied volume of sand. This maintenance fill was designed to offset the expected removal of sand by natural processes.

Changes in Sand Volume for a 13-km Length of Shoreline, Monmouth County, New Jersey, Fall 1998–Fall 2004

Loss (−) or Gain (+) per Meter of Shoreline (m^3)	Total Loss or Gain over Shoreline (m^3/m)	Period
+1.41	+18,330	Fall 1998
+0.16	+2080	Spring 1999
−22.97	−298,610	Fall 1999
−42.09	−547,170	Spring 2000
−24.7	−321,100	Fall 2000
−29.82	−2387,660	Spring 2001
−43.44	−564,720	Fall 2001
−1.02	−13,260	Spring 2002
+522.47	+6,792,110	Fall 2002*
−101.64	−1,321,320	Spring 2003
−77.00	−1,001,000	Fall 2003
−38.84	−504,920	Spring 2004
−79.53	−1,033,890	Fall 2004

*This gain represents the maintenance fill in fall 2002.

From these data we can draw the following conclusions:

1. The shoreline lost an average of 20 m^3/m of the initial placement volume per season from the time of initial placement through spring 2002. (This figure is the average of the first column of numbers up until the maintenance fill of fall 2002.)

2. The average seasonal shoreline loss rate increased to 74 m^3/m following the maintenance fill of fall 2002. (This figure is the average of the first column of numbers after the maintenance fill of fall 2002.)

3. The shoreline experienced a net loss of 162 m^3/m from the time of initial placement through spring 2002. (This figure is the average of the second column of numbers up until the maintenance fill of fall 2002.)

4. The shoreline experienced a net loss of 297 m^3/m following the maintenance fill of fall 2002. (This figure is the average of the second column of numbers after the maintenance fill of fall 2002.)

It is not known what factors contributed to the increase in sand loss after 2002, but scientists would want to investigate processes such as increased frequency of storms or increased intensity of storms over that period.

Did the volume of sand provided by the maintenance fill of fall 2002 make up for the losses between 1998 and 2004? We can answer this question by summing the numbers in the second column of the table (5,973,240 m^3) and comparing that sum with the volume of sand added in the maintenance fill of fall 2002 (6,792,110 m^3). These numbers are close enough that we can conclude that the losses due to natural causes were balanced by the artificial maintenance fill.

BONUS PROBLEM: Given the total cost of the initial restoration project that was started in 1994, and the volume of sand that was pumped to the shoreline at that time, calculate the average cost per cubic meter of sand. Then use this value to estimate the cost of the maintenance fill that was provided in fall 2002. Do you think this continuing cost—every 6 years—is worth it?

Erosional Coastal Forms Erosion is an important process along uplifted rocky coasts. Along these coasts, prominent cliffs and headlands jut into the sea, alternating with narrow inlets and irregular bays with small beaches. Waves crash against the rocky shorelines, undercutting cliffs and causing huge blocks of rock to fall into the water, where they are gradually eroded away. As the sea cliffs retreat, isolated remnants called *stacks* may be left standing in the sea, far from the shore (see Figure 19.1c). Erosion by waves also planes the rocky surface beneath the surf zone and creates a **wave-cut terrace,** which is sometimes visible at low tide (**Figure 19.10**). Wave erosion over long periods may straighten shorelines as headlands are eroded faster than recesses and bays.

Where relatively soft sediments or sedimentary rocks make up the coastal region, slopes are gentler and the heights of shoreline bluffs are lower. Waves erode these softer materials efficiently, and erosion of bluffs on such shores may be extraordinarily rapid. The high sea cliffs of soft glacial sediments at the Cape Cod National Seashore in Massachusetts, for instance, are retreating about a meter each year. Since Henry David Thoreau walked the entire length of the beach below those cliffs in the mid-nineteenth century and wrote of his travels in *Cape Cod*, about 6 km^2 of coastal land have been eaten away by the ocean, equivalent to about 150 m of beach retreat.

Our discussion of beaches illustrates the importance of erosional processes in those soft-sediment environments. In recent decades, more than 70 percent of the total length of the world's sand beaches has retreated at a rate of at least 10 cm/year, and 20 percent has retreated at a rate of more than 1 m/year. Much of this loss can be traced to the damming of rivers, which decreases sediment input to the shoreline.

Depositional Coastal Forms Sediments build up in areas where subsidence depresses Earth's crust along a coastline. Such coastlines are characterized by wide, low-lying coastal plains of sedimentary rock and by long, wide beaches. Shoreline forms along these coastlines include sandbars, low-lying sandy islands, and extensive tidal flats. Long beaches grow longer as longshore currents carry sand to the downcurrent end of the beach. There it builds up, first forming a submerged sandbar, then rising above the surface and extending the beach by a narrow addition called a **spit**.

Offshore, long sandbars may build up into **barrier islands** that form a barricade between open-ocean waves and the main shoreline. Barrier islands are common, especially along low-lying coasts composed of small sediment particles that are easily eroded and transported, or of poorly cemented sedimentary rocks where longshore currents are strong. As the sand builds up above the waves, vegetation takes hold, stabilizing the islands and helping them resist wave erosion during storms. Barrier islands are separated from the coast by tidal flats or shallow lagoons. Like beaches on the main shore, barrier islands are maintained at a dynamic equilibrium by the forces shaping them. That equilibrium can be disturbed by natural changes in climate or in wave and current patterns, as well as by human activities. Disruption or devegetation can lead to increased erosion, and barrier islands may even disappear beneath the sea surface. Barrier islands may also grow larger and more stable if sedimentation increases.

Over hundreds of years, sandy shorelines may undergo significant changes. Hurricanes and other intense storms may form new inlets, elongate spits, or breach existing spits and barrier islands. Such changes

FIGURE 19.10 Multiple wave-cut terraces on the coastline of San Clemente Island, California. Each terrace records a distinctly different sea level elevation. Sea level is controlled in turn by glacial ice volumes (see Chapter 15); when ice volumes are stable, sea level is fixed, and waves erode bedrock. [Dan Muhs, USGS Geosciences and Environmental Change Science Center.]

have been documented by aerial photographs taken at various time intervals. The shoreline of Chatham, Massachusetts, at the elbow of Cape Cod, has changed enough in the past 160 years or so that a lighthouse has had to be moved. **Figure 19.11** illustrates the many changes that have taken place in the configuration of the barrier islands to the north and to the long spit of Monomoy Island, including several breaches of the barrier islands.

(a) Beach near Chatham Light

The 1987 breach in the barrier spit, shown at the right below, closed again before this photo was taken.

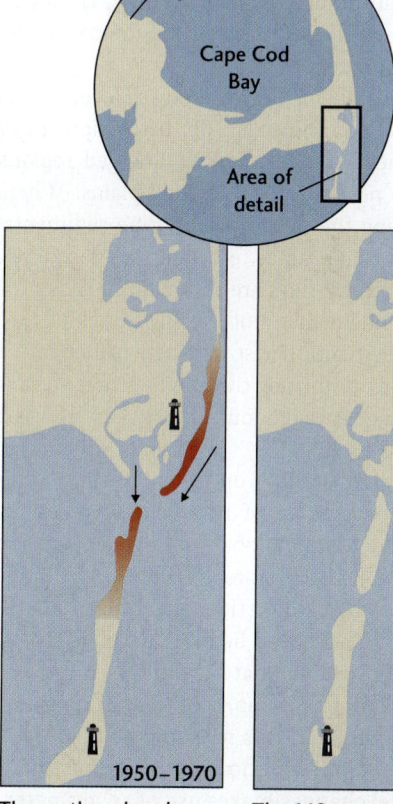

(b)

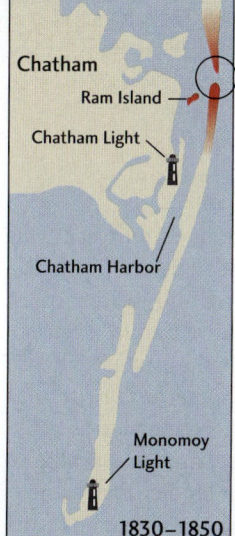
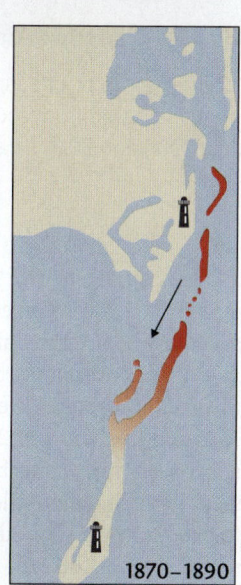

1830–1850	1870–1890	1910–1930	1950–1970	1987
The circle shows the approximate location of the 1846 breach in barrier island. Ram Island later disappears.	The beach south of the inlet breaks up and migrates southwest toward the mainland and Monomoy.	The southern beach has disappeared, and its remnants soon will connect Monomoy to the mainland.	The northern beach steadily grows with cliff sediment; Monomoy breaks from the mainland.	The 140-year cycle begins again with the Jan. 2 breach in the barrier spit across from the Chatham Light (circle).

FIGURE 19.11 Migrating barrier islands at Chatham, Massachusetts, at the southern tip of Cape Cod. (a) Aerial view of Monomoy Point. This spit has advanced into deep water to the south (foreground) from barrier islands along the main body of the Cape to the north (background). (b) Changes in the shoreline at Chatham over the past 160 years. [a: Steve Dunwell/Getty Images; b: Information from Cindy Daniels, *Boston Globe* (February 23, 1987).]

Many homes in Chatham are now at risk, but there is little that the residents or the state can do to prevent coastal processes from taking their natural course.

Effects of Sea-Level Change

The shorelines of the world serve as barometers for impending changes caused by many types of human activities. The pollution of our inland waterways sooner or later arrives at our beaches, just as sewage from city dumping and tar originating from leaking from ocean-going oil tankers wash up on the shore. As real estate development and construction along shorelines expands, we will see the continuing contraction, and even the disappearance, of some of our finest beaches. As global warming and glacial melting cause sea level to rise, we will see the effects of that change on our shorelines as well.

Shorelines are particularly sensitive to changes in sea level, which can alter tidal heights, change the approach patterns of waves, and affect the paths of longshore currents. The rise and fall of sea level at a shoreline can be local—the result of tectonic subsidence or uplift—or global—the result of glacial melting or growth. One of the primary concerns about human-induced global warming is its potential for causing sea-level rises that will flood coastal cities, as we learned in Chapter 14.

In periods of globally lowered sea level, areas that were offshore are exposed to agents of erosion. Rivers extend their channels over formerly submerged regions and cut valleys into newly exposed coastal plains. When sea level rises, flooding the backshore, marine sediments build up along former land areas, erosion is replaced by sedimentation, and river valleys are submerged. Today, long fingers of the sea indent many of the shorelines of the northern and central Atlantic coast of North America. These long indentations are former river valleys that were flooded as the last ice age ended about 11,000 years ago and sea level rose (**Figure 19.12**).

Sea-level variations on geologic time scales can be measured by studies of wave-cut terraces (see Figure 19.10), but detecting global sea-level changes on shorter (human) time scales is more difficult. Local changes can be measured by using a tide gauge that records sea level relative to a land-based benchmark. The major problem with this approach is that the land itself moves vertically as a result of deformation, sedimentation, and other geologic processes, and this movement is incorporated into the tide-gauge observations. A newer method of tracking sea-level changes makes use of an altimeter mounted on a satellite. The altimeter sends pulses of radar beams that are reflected off the ocean surface, providing measurements of the elevation of the ocean surface relative to the orbit of the satellite with a precision of a few centimeters.

Using these methods, oceanographers have determined that the global sea level has risen by 17 cm over the last

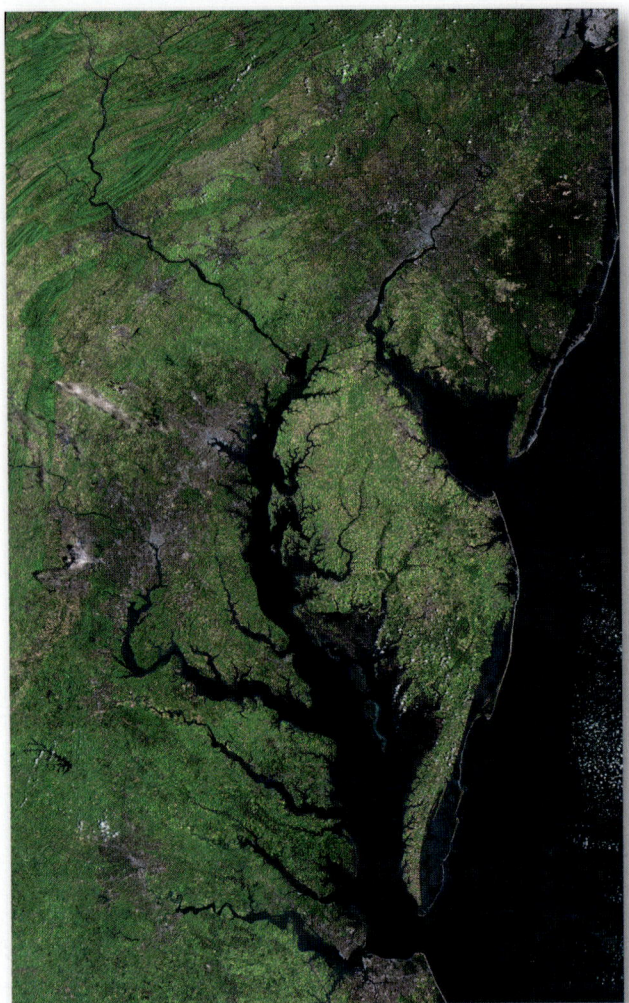

FIGURE 19.12 This photo of Chesapeake Bay was created by the USGS Landsat team from six Landsat 5 images collected in July 2009 and 2011. [NASA/USGS/Landsat 5.]

century, and is continuing to rise about 3 mm/year. This recent increase in sea level correlates with a worldwide increase in average global temperatures and melting of ice sheets, which most scientists now believe has been caused, at least in part, by anthropogenic emissions of greenhouse gases as we learned in Chapter 12. Some of the rise may result from short-term variations, but the magnitude of the rise is consistent with climate models that take greenhouse warming into account. These models predict that without significant worldwide efforts to reduce greenhouse gas emissions, the sea level will rise by another 3100 cm during the twenty-first century.

One of the major impacts associated with sea-level rise is the growing vulnerability of coastal communities that can expect increased flooding during major storms. This is because increased long-term sea level adds to the temporary rise in sea level as storms, including hurricanes, pass by.

Hurricanes and Coastal Storm Surges

Hurricanes are the greatest storms on Earth, swirling masses of dense clouds hundreds of kilometers across that suck their energy from the warm surface waters of tropical oceans. The term *hurricane* originates from the name *Huracan*, a god of storms to the Mayan people of Central America. In the western Pacific and China Sea, hurricanes are known as *typhoons*, from the Cantonese word *tai-fung*, meaning "great wind." In Australia, Bangladesh, Pakistan, and India, they are known as *cyclones*; in the Philippines, they are called *baguios*.

Whatever you call them, these intense tropical storms can wreak havoc. For example, a catastrophic cyclone struck the coastal lowlands of Bangladesh in 1970, drowning as many as 500,000 people—perhaps the deadliest natural disaster of modern times. Another cyclone hit the same region in 1991, drowning at least 140,000 (**Figure 19.13**). The 1991 storm was more intense, but its death toll was lower due to better disaster preparations; 2 million people were evacuated.

FIGURE 19.13 Devastation caused by a cyclone in Chittagong, Bangladesh, in 1991. [Peter Charlesworth/Getty Images.]

Earth Policy 19.1 The Great New Orleans Flood

On August 25, 2005, Hurricane Katrina struck southern Florida as a Category 1 storm, killing 11 people. Three days later, in the Gulf of Mexico, the hurricane grew to a monster Category 5 storm, with maximum sustained winds of up to 280 km/hour (175 miles/hour) and gusts up to 360 km/hour (225 miles/hour). On August 28, the National Weather Service issued a bulletin predicting "devastating" damage to the Gulf Coast, and the mayor of New Orleans ordered an unprecedented mandatory evacuation of the city.

When Katrina made landfall just south of New Orleans on August 29, it was a nearly Category 4 storm, with sustained winds of 204 km/hour (127 miles/hour). It had a minimum atmospheric pressure of 918 millibars (27.108 inches), making it the third strongest hurricane on record to make landfall in the United States. More than 100 people lost their lives during the early morning hours of August 29 as a result of the direct impact of the storm.

A 5- to 9-m storm surge came ashore over virtually the entire coastline of Louisiana, Mississippi, Alabama, and the Florida panhandle. The 9-m storm surge at Biloxi, Mississippi, was the highest ever recorded in the United States. The effects of this storm surge on New Orleans were devastating and unprecedented. Lake Pontchartrain, which is really a coastal embayment that is easily influenced by ocean conditions, was inundated by the storm surge. By midday on August 29, several sections of the levee system that held back the waters of Lake Pontchartrain from New Orleans had collapsed.

Water spills over a levee along the Inner Harbor Navigational Canal and floods the inner city of New Orleans. [POOL/AFP/Getty Images.]

Subsequent flooding of the city to depths of up to 7 or 8 m left 80 percent of New Orleans under water. The effects of flooding claimed at least another 300 lives, and by September 21, the death toll exceeded 1500 as disease and malnourishment indirectly caused by the flooding took effect.

Residents wade through a flooded street in New Orleans in the aftermath of Hurricane Katrina. [James Nielsen/AFP/Getty Images.]

Hurricane Katrina has surpassed Hurricane Andrew as the costliest natural disaster in U.S. history, with damages reaching almost $200 billion. In addition to the thousands of lives lost, over 150,000 homes were destroyed, and over a million people were displaced.

As with most natural disasters, the outcome was a result of rare but powerful geologic forces coupled with a lack of human preparation. No one had anticipated and planned for the worst-case scenario. Earth scientists had predicted for decades that a Category 4 or 5 hurricane would strike New Orleans eventually. The historical record of hurricanes made it clear that such an event was almost certain to happen. As Figure 19.18 shows, New Orleans is about in the middle of the "catcher's mitt" for hurricane landfalls in the United States. But the city was prepared to resist the damaging effects of only a Category 3 or smaller hurricane. Federal budget cuts had left only token support available to maintain and reinforce the east bank of hurricane levees that held back Lake Pontchartrain. This complex network of concrete walls, metal gates, and giant earthen berms was never completed, leaving the city vulnerable. Furthermore, it is not easy to protect a city from hurricane storm surges when its sidewalks and houses are, on average, 4 m below sea level. New Orleans is equally vulnerable to unusually large floods of the Mississippi River, which is also held back by an artificial levee system.

The damaging effects of a hurricane's extremely high, sustained winds and torrential rains are intuitively easy to understand. However, the associated storm surge, which may flood major regions of the coastline, is potentially the most destructive effect of a hurricane. When Hurricane Katrina struck New Orleans, Louisiana, on August 29, 2005, the disaster that followed was not so much the result of the direct impact of the hurricane itself as of the storm surge, which ultimately caused several sections of the artificial levee system protecting New Orleans to collapse (see Earth Policy 19.1, *The Great New Orleans Flood*). Subsequent flooding of parts of the city claimed hundreds of lives and left the city submerged and abandoned for almost a month. Hurricane Sandy was the second costliest hurricane in United States history (second to Hurricane Katrina) and hit the east coast of the United States in late October 2012. While Sandy caused water levels to rise along the entire East Coast, it caused a catastrophic storm surge along the New Jersey, New York, and Connecticut coastlines.

Hurricane Formation

Hurricanes form over tropical parts of Earth's oceans, between 8° and 20° latitude, in areas of high humidity, light winds, and warm sea surface temperatures (typically 26°C or greater). These conditions usually occur in the summer and early fall in the tropical North Atlantic and North Pacific. For this reason, hurricane "season" in the Northern Hemisphere runs from June through November (**Figure 19.14**).

The first sign of hurricane development is the appearance of a cluster of thunderstorms over the tropical ocean in a region where the trade winds converge. Occasionally, one of these clusters breaks out from this convergence zone and becomes better organized. Most hurricanes that affect the Atlantic Ocean and Gulf of Mexico originate in a convergence zone just off the coast of West Africa and intensify as they break out and move westward across the tropical Atlantic.

As the hurricane develops, water vapor condenses to form rain, which releases heat energy. In response to this atmospheric heating, the surrounding air becomes less dense and begins to rise, and the atmospheric pressure at sea level drops in the region of heating. As the warm air rises, it triggers more condensation and rainfall, which in turn releases more heat. At this point, a positive feedback process is set in motion, as the rising temperatures in the center of the storm cause surface

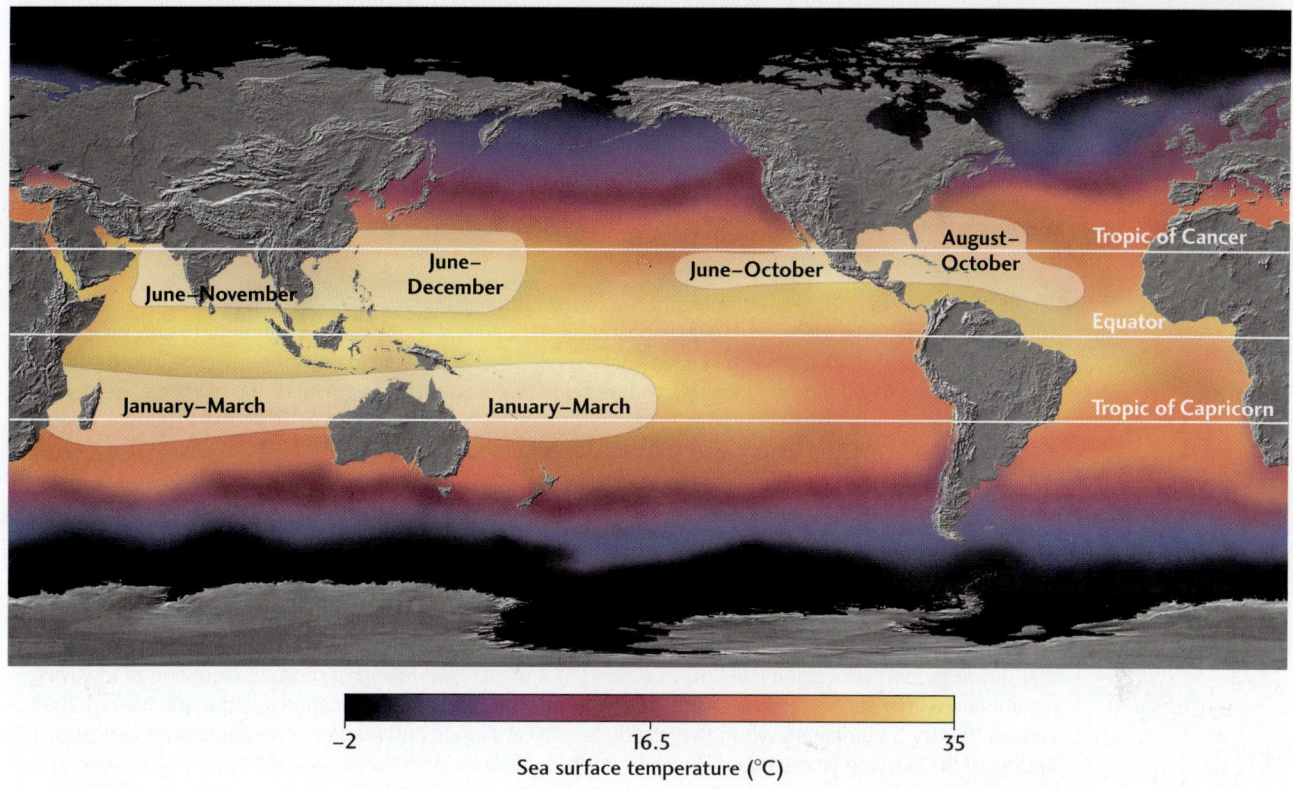

FIGURE 19.14 Hurricanes arise in summer and early fall when ocean temperatures are warmest. The light-shaded areas indicate places where hurricanes are most common. The times of year when they are most frequent are also shown. [NASA/GSFC.]

pressures to fall to progressively lower levels. In the Northern Hemisphere, because of the Coriolis effect, the increasing winds begin to circulate in a counterclockwise pattern around the storm's area of lowest pressure, which ultimately becomes the "eye" of the hurricane (**Figure 19.15**).

Once sustained wind speeds reach 37 km/hour (23 miles/hour), the storm system is called a *tropical depression*. As winds increase to 63 km/hour (39 miles/hour), the system is called a *tropical storm* and receives a name. This naming tradition started with the use of World War II code names, such as Andrew, Bonnie, Charlie, and so forth. Finally, when wind speeds reach 119 km/hour (74 miles/hour), the storm is classified as a hurricane. Once it becomes a hurricane, the storm is assigned a 1–5 rating based on the *Saffir-Simpson hurricane intensity scale*

FIGURE 19.15 Hurricane Katrina on August 28, 2005, a few hours before it struck New Orleans. In the Northern Hemisphere, winds circulate in a counterclockwise direction around the "eye" of the hurricane, which is the location of lowest atmospheric pressure. [NASA/Jeff Schmaltz, MODIS Land Rapid Response Team.]

TABLE 19-1	The Saffir-Simpson Hurricane Intensity Scale
STORM CLASSIFICATION	DESCRIPTION
Category 1	Winds 119–153 km/hour (74–95 miles/hour). Storm surge generally 1–1.5 m (4–5 feet) above normal. No real damage to building structures. Damage primarily to unanchored mobile homes, shrubbery, and trees. Some damage to poorly constructed signs. Some coastal road flooding and minor pier damage.
Category 2	Winds 154–177 km/hour (96–110 miles/hour). Storm surge generally 2–2.5 m (6–8 feet) above normal. Some roofing material, door, and window damage to buildings. Considerable damage to shrubbery and trees, with some trees blown down. Considerable damage to mobile homes, poorly constructed signs, and piers. Coastal and low-lying escape routes flood 2–4 hours before arrival of the hurricane center. Small craft in unprotected anchorages break moorings. Hurricane Frances of 2004 made landfall over the southern end of Hutchinson Island, Florida, as a Category 2 hurricane.
Category 3	Winds 178–209 km/hour (111–130 miles/hour). Storm surge generally 2.5–3.5 m (9–12 feet) above normal. Some structural damage to small residences and utility buildings with a minor amount of curtain-wall failure. Damage to shrubbery and trees with foliage blown off trees and large trees blown down. Mobile homes and poorly constructed signs are destroyed. Low-lying escape routes are cut off by rising water 3–5 hours before arrival of the hurricane center. Flooding near the coast destroys smaller structures, with larger structures damaged by battering from floating debris. Terrain continuously lower than 1.5 m (5 feet) above sea level may be flooded 3 m (10 feet) inland or more. Evacuation of low-lying residences within several blocks of the shoreline may be required. Hurricanes Jeanne and Ivan of 2004 were Category 3 hurricanes when they made landfall in Florida and Alabama, respectively. Hurricane Katrina of 2005 made landfall near Buras-Triumph, Louisiana, with winds of 204 km/hour (127 miles/hour). Katrina will prove to be the costliest hurricane on record, with estimates of more than $200 billion in losses.
Category 4	Winds 210–250 km/hour (131–155 miles/hour). Storm surge generally 3.5–5 m (13–18 feet) above normal. More extensive curtain-wall failures with some complete roof structure failures on small residences. Shrubs, trees, and all signs are blown down. Complete destruction of mobile homes. Extensive damage to doors and windows. Low-lying escape routes may be cut off by rising water 3–5 hours before arrival of the hurricane center. Major damage to lower floors of structures near the shore. Terrain lower than 3 m (10 feet) above sea level may be flooded, requiring massive evacuation of residential areas as far inland as 15 km (9 miles).
Category 5	Winds greater than 250 km/hour (155 miles/hour). Storm surge generally greater than 5 m (18 feet) above normal. Complete roof failure on many residences and industrial buildings. Some complete building failures with small utility buildings blown over or away. All shrubs, trees, and signs blown down. Complete destruction of mobile homes. Severe and extensive window and door damage. Low-lying escape routes are cut off by rising water 3–5 hours before arrival of the hurricane center. Major damage to lower floors of all structures located less than 4.5 m (15 feet) above sea level and within 500 m (1650 feet) of the shoreline. Massive evacuation of residential areas on low ground within 15–20 km (9–12 miles) of the shoreline may be required. Only three Category 5 hurricanes have made landfall in the United States since records began. Hurricane Andrew of 1992 made landfall over southern Miami-Dade County, Florida, causing $26.5 billion in losses—the third costliest hurricane on record. Hurricane Sandy, which hit the East Coast in 2012, was the second costliest at $68 billion.

(Table 19.1). This scale is used to estimate the potential property damage and flooding expected along the coast from hurricane landfall. It is analogous to the Mercalli intensity scale for earthquakes (see Table 10.1).

Storm Surges

As a hurricane intensifies, a dome of seawater—known as a **storm surge**—rises above the level of the surrounding ocean surface. The height of the storm surge is directly related to the atmospheric pressure in the eye of the hurricane and the strength of the winds that encircle it. Large swells, high surf, and wind-driven waves ride atop the surge. As the hurricane nears land, the surge moves ashore and floods coastal land areas, causing extensive damage to structures and the shoreline environment (Figure 19.16). Any landmass in the path of a storm surge will be affected to a greater or lesser extent, depending on a number of factors. The stronger the storm and the shallower the offshore

Hurricanes and Coastal Storm Surges

FIGURE 19.16 Hurricane storm surges along coastlines may result in the complete destruction of residential buildings, which pile up as lines of debris well inland of the shoreline. The damage seen here was caused by Hurricane Katrina in 2005. [U.S. Navy/Getty Images.]

waters, the higher the storm surge. When the effects of a storm surge coincide with a normal high tide, the result is known as a *storm tide* (**Figure 19.17**).

The storm surge is the deadliest of a hurricane's associated hazards, as underscored by Hurricane Katrina in 2005 and Hurricane Sandy in 2012. The magnitude of a hurricane is usually described in terms of its wind speed (see Table 19.1), but coastal flooding causes many more deaths than high wind. Boats ripped from their moorings, utility poles, and other debris floating atop a storm surge often demolish those buildings that are not destroyed by the winds. Even without the weight of floating debris, a storm surge can severely erode beaches and highways and undermine bridges. Because much of the United States' densely populated Atlantic and

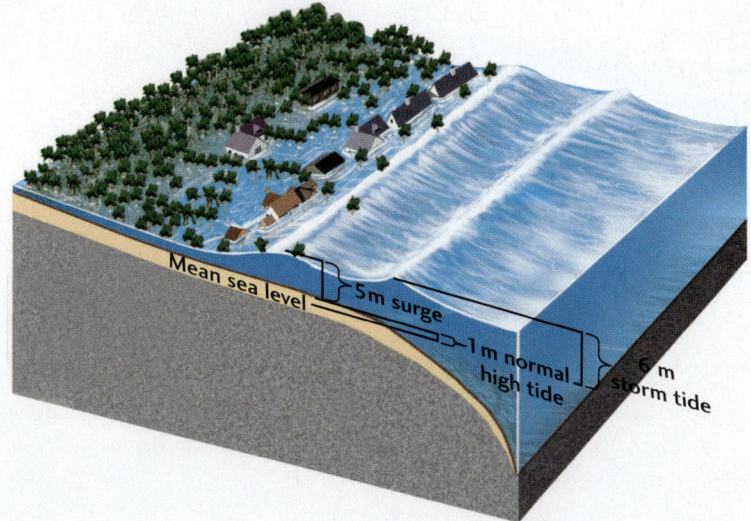

FIGURE 19.17 A storm tide is the combination of a storm surge and a normal high tide. If a storm surge arrives at the same time as a high tide, the water height will be increased. For example, if a normal high tide 1 m above sea level is combined with a storm surge of 5 m, the resulting storm tide will be 6 m in height.

Gulf Coast shorelines lie less than 3 m above sea level, the danger from storm surges there is tremendous.

As Hurricane Sandy formed, it affected much of the Caribbean, including Jamaica, Haiti, Dominican Republic, Puerto Rico, Cuba, and the Bahamas, then worked its way up the eastern coastline of the United States, affecting 24 states. The storm made landfall near Brigantine, New Jersey. The surge flooded streets, tunnels, and subway lines and cut power to much of the region. The path was correctly predicted nearly 8 days before it hit the eastern seaboard, so preparations were underway days before. Subway entrances and grates were covered in New York City, although flooding still occurred. There were mandatory evacuations, schools were closed, mass transit was shut down, airports were closed, and bus and rail services were suspended. Despite the preparation, communities were flooded with water and sand, houses were washed from their foundations, the New Jersey boardwalk and pier were destroyed, and cars and boats were tossed about. The storm also caused devastating fires in New York. Trees fell on power lines, transformers exploded, and wires fell into water, creating a dangerous fire that was hard to access and fight due to flooded communities.

Hurricane Landfall

Because hurricanes form over and move across tropical waters, most make landfall at low latitudes. Most North Atlantic hurricanes make landfall in Florida and the northern Gulf of Mexico (**Figure 19.18**). However, because there is a tendency for winds to be deflected northward (due to the Coriolis effect), hurricanes sometimes make landfall farther up the Atlantic coast. In rare cases, they may reach New England, but are always of lower intensity there because of the lower ocean surface temperatures. The most powerful Category 4 and 5 hurricanes are restricted to lower latitudes.

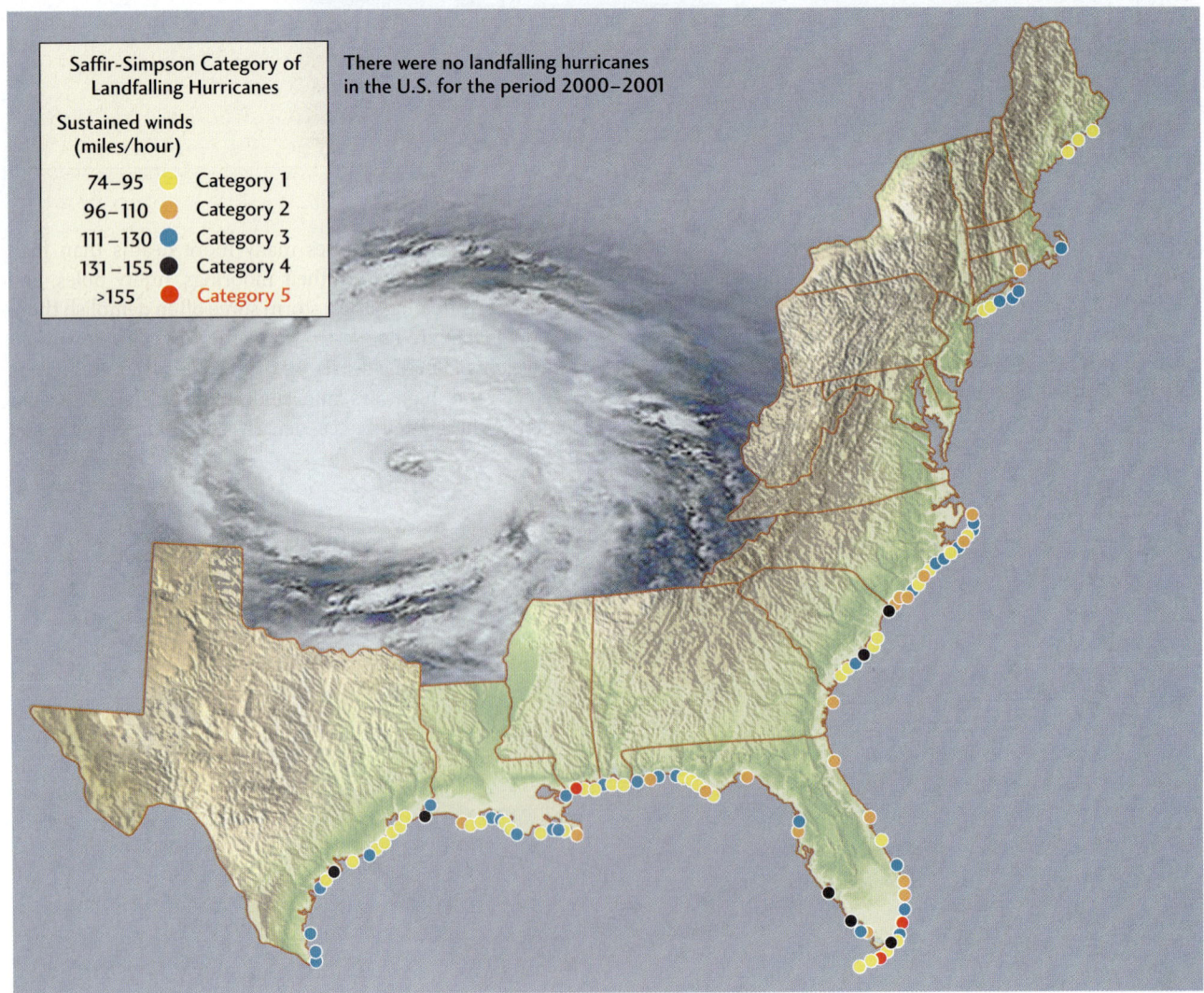

FIGURE 19.18 Hurricanes originating in the North Atlantic Ocean usually make landfall in the coastal areas of the southeastern United States, including the Gulf Coast states. Hurricanes lose energy as they move across cold water, so the number of hurricanes that make landfall drops dramatically for the central and northeastern states. [Information from NOAA.]

The tropical storms that grow into hurricanes can be tracked by satellites, and weather conditions inside the storms can be probed by airplanes. By feeding many types of data into computer models, meteorologists can predict a storm's track and changes in its intensity up to several days in advance of landfall with reasonable accuracy. The National Hurricane Center accurately predicted that Katrina would hit New Orleans as a severe hurricane 3 days before it actually did.

Hurricanes are notable for the damage that occurs during sustained strong winds. Deserts are another environment where strong winds can also occur, resulting in sandstorms that may transport sedimentary particles across entire oceans.

Desert Processes

At one time or another, we've all been caught in a wind strong enough to have blown us over, had we not leaned into it or held onto something solid. London, England, which rarely gets strong winds, experienced a major windstorm on January 25, 1990. Winds blowing at more than 175 km/hour ripped roofs off buildings, blew trucks over, and made it virtually impossible to walk on the streets. In deserts, strong winds are much more common, often howling for days on end. Dust storms are frequent, and many winds are strong enough to blow sand grains into the air, creating sandstorms.

The locations of the world's great deserts are determined by rainfall, which in turn is determined by a number of factors (**Figure 19.19**). The Sahara and Kalahari deserts of Africa and the Great Australian Desert get extremely low amounts of rainfall, normally less than 25 mm/year and in some places less than 5 mm/year. These subtropical deserts are found at about 30° N and 30° S, where prevailing wind patterns cause dry air to sink to ground level (see Figure 12.4). Because the relative humidity is extremely low in these zones of sinking air, clouds are rare, and the chance of precipitation is very small. The Sun beats down week after week.

Deserts also exist at temperate latitudes—between 30° N and 50° N and between 30° S and 50° S—in regions where rainfall is low because moisture-laden winds either are blocked by mountain ranges or must travel great distances from the ocean, their source of moisture. The Great Basin and Mojave deserts of the western United States, for example, lie in rain shadows created by the western coastal mountains. The Gobi and other deserts of central Asia are so deep in the continental interior that the winds reaching them have precipitated all their ocean-derived moisture long before they arrive there.

Another kind of desert is found in polar regions. There is little precipitation in these cold, dry areas because the frigid air can hold little moisture. The dry valley region of southern Victoria Land in Antarctica is so dry and cold that its environment resembles that of Mars.

As the global climate changes, these belts of sinking dry air may also change, expanding and shifting their margins in some places and contracting them in others. In this way, a region bordering a desert—perhaps already suffering from a shortage of rain—may begin to emerge as a persistent desert-like environment. Eventually, the region may become part of the desert. Conditions in southern

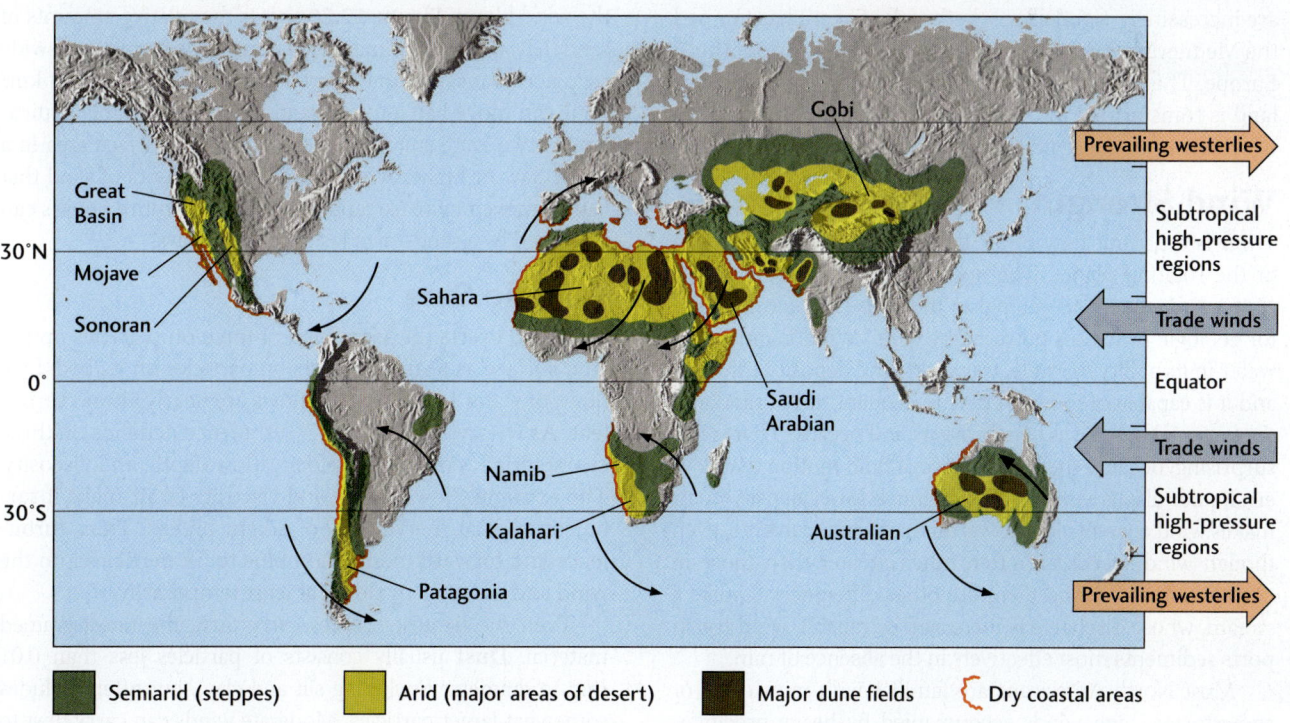

FIGURE 19.19 Major desert areas of the world (exclusive of polar deserts). Notice the relationships of their locations to prevailing wind belts and major mountain ranges. Notice, too, that sand dunes make up only a small proportion of the total desert area. [Information from K. W. Glennie, *Desert Sedimentary Environments.* New York: Elsevier, 1970.]

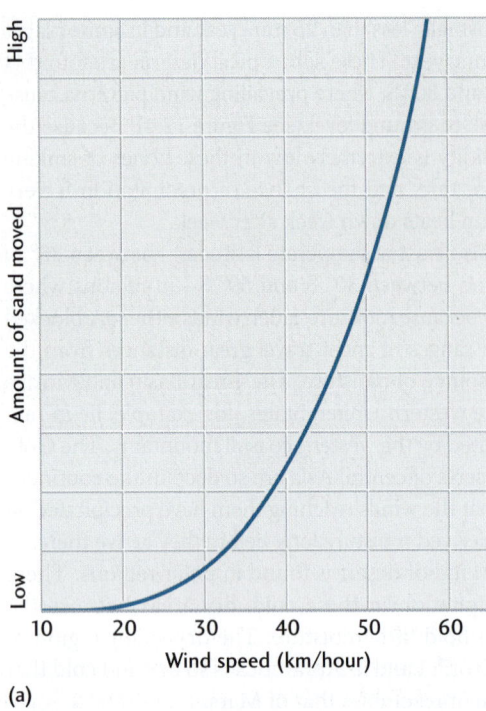

FIGURE 19.20 (a) The amount of sand moved daily across each meter of width of a dune's surface varies with wind speed. High-speed winds blowing for several days can move enormous quantities of sand. (b) A massive dust storm cloud (haboob) is close to enveloping a military camp as it rolls over Al Asad, Iraq, just before nightfall on April 27, 2005. [a: Data from R. A. Bagnold, *The Physics of Blown Sand and Desert Dunes*. London: Methuen, 1941; b: DOD photo by Corporal Alicia M. Garcia, U.S. Marine Corps.]

Spain, for example, have become so dry that people there are increasingly wondering whether the Sahara has jumped the Mediterranean Sea and is now encroaching on southern Europe. The process of *desertification*, in which non-desert land is transformed into desert, has become a major focus of scientists trying to understand Earth's climate system.

Wind Strength

Wind is a natural flow of air that is parallel to the surface of the rotating planet. The ancient Greeks called the god of winds Aeolus, and geologists today use the term **eolian** for geologic processes powered by wind. Wind is much like water in its ability to erode, transport, and deposit sediment, and it is capable of moving enormous quantities of sand and dust over large regions of continents and oceans. That is not surprising, because the general laws of fluid motion that govern liquids also govern gases. The much lower density of air makes wind currents less powerful than water currents, even though wind speeds are often much greater than those in currents of water. And there are other differences. Unlike a stream, whose discharge is increased by rainfall, wind transports sediments most effectively in the absence of rain.

Most North Americans are familiar with rainstorms or snowstorms: high winds accompanied by heavy precipitation. We may have less experience with dry storms, during which high winds blowing for days on end carry enormous amounts of sand and dust. The amount of material the wind can carry depends on the strength of the wind, the sizes of the particles, and the surface materials of the area over which the wind blows. **Figure 19.20** shows the relative amounts of sand that winds of various speeds can erode from a 1-m-wide strip across a sand dune's surface. A strong wind of 48 km/hour can move half a ton of sand (a volume roughly equivalent to two large suitcases) from this small surface area in a single day. At higher wind speeds, the amounts of sand that can be moved increase rapidly. No wonder entire houses can be buried by a sandstorm lasting several days!

Particle Size

The wind exerts the same kind of force on particles on the land surface as a stream exerts on particles on its bed. Like flows of water in streams, air flows are nearly always turbulent. As we saw in Chapter 17, turbulence depends on three characteristics of a fluid: velocity, flow depth, and viscosity. The extremely low density and viscosity of air make it turbulent even at the velocity of a light breeze. Thus, turbulence and forward motion combine to lift particles into the wind and carry them along, at least temporarily.

Even the lightest breezes carry dust, the finest-grained material. **Dust** usually consists of particles less than 0.01 mm in diameter (including silt and clay), but often includes somewhat larger particles. Moderate winds can carry dust to heights of many kilometers, but only strong winds can carry particles larger than 0.06 mm in diameter, such as sand grains. Moderate breezes can roll and slide these grains along a sandy bed, but it takes a stronger wind to lift them into the air current. Wind usually cannot transport particles larger than sand, however, because air has such low viscosity and density. Even though winds can be very strong, only rarely can they move pebbles in the way that rapidly flowing streams do.

Windblown Sand and Dust

Wind can lift sand and dust only from dry surface materials, such as dry soil, sediment, or bedrock. It cannot erode and transport wet soils because they are too cohesive. Wind can carry sand grains weathered from a loosely cemented sandstone, but it cannot erode grains from granite or basalt. As air moves, it picks up loose particles and transports them over surprisingly long distances. As we have seen, most of this material is dust, though sand can also be transported by wind.

The sand transported by wind may consist of almost any kind of mineral grain produced by weathering. Quartz grains are by far the most common because quartz is such an abundant constituent of many surface rocks, especially sandstones. Many windblown quartz grains have a frosted or matte (roughened and dull) surface (**Figure 19.21**) like the inside of a frosted light bulb. Some of the frosting is produced by wind-driven impacts, but most of it results from slow dissolution by dew. Even the tiny amounts of dew found in arid climates are enough to etch microscopic pits and hollows into sand grains, creating the frosted appearance. Frosting is found only in eolian environments, so it is good evidence that a sand grain has been blown by the wind.

Sandblasting

Windblown sand is an effective natural **sandblasting** agent. The common method of cleaning buildings and monuments with compressed air and sand works on exactly the same principle: The high-speed impact of sand particles wears away the solid surface. Natural sandblasting mainly works close to the ground, where most sand grains are carried.

FIGURE 19.21 Photomicrograph of frosted and rounded sand grains from Wahiba Desert, Sultanate of Oman. [John Grotzinger.]

FIGURE 19.22 These windblown ventifacts from the Taylor Valley, Antarctica, have been shaped by windblown sand in a frigid environment. [Ronald Sletten.]

Sandblasting rounds and erodes rock outcrops, boulders, and pebbles and frosts the occasional glass bottle.

Ventifacts are wind-faceted pebbles that have several curved or almost flat surfaces that meet at sharp ridges (**Figure 19.22**). Each surface or facet is formed by sandblasting of the pebble's windward side. Occasional storms roll or rotate the pebbles, exposing a new windward side to be sandblasted. Many ventifacts are found in deserts and in glacial gravel deposits, where the necessary combination of gravel, sand, and strong winds is present.

Deflation

As particles of clay, silt, and sand become loose and dry, blowing winds can lift and carry them away, gradually eroding the ground surface in a process called **deflation** (**Figure 19.23**). Deflation, which can scoop out shallow depressions or hollows, occurs on dry plains and deserts and on temporarily dry river floodplains and lake beds. Firmly established vegetation—even the sparse vegetation of arid and semiarid regions—retards it. Deflation occurs slowly in areas with plants because their roots bind the soil and their stems and leaves disrupt air flows and shelter the ground surface. Deflation is rapid where the vegetation cover is broken, either naturally by killing drought or artificially by cultivation, construction, or motor vehicle tracks.

When deflation removes the finer-grained particles from a mixture of pebbles, sand, and silt, it produces a remnant surface of pebbles too large for the wind to transport. Over thousands of years, as deflation removes the finer-grained particles, the pebbles accumulate as a layer of **desert pavement**—a coarse, gravelly ground surface that protects the soil or sediments below from further erosion.

This theory of desert pavement formation is not completely accepted because a number of pavements seem not to have formed in this way. A new hypothesis proposes that some of them are formed by the deposition of windblown dust. The coarse pebble pavement stays at the surface, while the windblown dust infiltrates below the surface layer of pebbles, is modified by soil-forming processes, and accumulates there (**Figure 19.24**).

FIGURE 19.23 Wind erosion of soil at the foot of Chimborazo, Ecuador. Wind has scoured the surface and eroded it to a slightly lower elevation. Deflation occurs in dry areas where vegetation cover is absent or broken. [Patricio Mena Vásconez.]

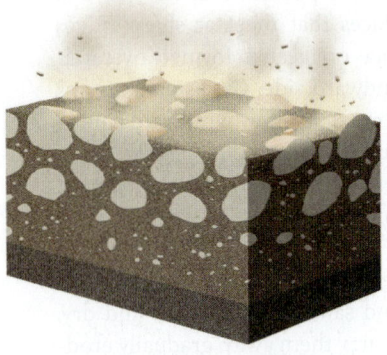

1 Desert pavement formation begins when the wind blows fine-grained materials into heterogeneous soil or sediment.

2 During rainstorms, the fine, windblown sediments infiltrate beneath the coarse layer of pebbles.

3 Microbes living beneath the pebbles produce bubbles that help raise the pebbles and maintain their position at the surface.

Stays consistent thickness

Gets thicker as more windblown dust is supplied over the millennia

4 Over time, these processes lead to thickening of the dust accumulating beneath the pebble layer.

5 A continued supply of windblown dust makes the deposit thicker.

(a) (b)

FIGURE 19.24 (a) According to a recent hypothesis, desert pavement is formed by the interaction of climate and microorganisms with windblown sediment and soil. (b) Example of desert pavement formed in Namib Desert, Namibia. [John Grotzinger.]

FIGURE 19.25 Linear dunes in the Qaidam Basin, China. [David Rubin.]

Sand Dunes

When the wind dies down, it can no longer transport the sand and dust it carries. The coarser material is deposited in **sand dunes** of various shapes, ranging in size from low knolls to huge hills more than 100 m high (see the chapter opening photo). The finer dust falls to the ground as a more or less uniform blanket of silt and clay. By observing these depositional processes working today, geologists have been able to link them with observed characteristics, such as bedding and texture, in sandstones and dust deposits to deduce past climates and wind patterns.

Where Sand Dunes Form Sand dunes occur in relatively few environmental settings. Many North Americans have seen the dunes that form behind ocean beaches or along large lakes. Some dunes are found on the sandy floodplains of large rivers in semiarid and arid regions. Most spectacular are the fields of dunes that cover large expanses of some deserts (**Figure 19.25**). Such dunes may reach heights of 250 m, truly mountains of sand.

Dunes form only in settings that have a ready supply of loose sand: sandy beaches along coasts, sandy river bars or floodplain deposits in river valleys, and sandy bedrock formations in deserts. Another common factor in dune formation is wind power. On oceans and lakes, strong winds blow onshore from the water. Strong winds, sometimes of long duration, are common in deserts.

As we have seen, wind cannot pick up wet materials easily, so most dunes are found in dry climates. The exception is beaches along a coast, where sand is so abundant and dries so quickly in the wind that dunes can form even in humid climates. In such climates, soil and vegetation begin to cover the dunes only a little way inland from the beaches, and the winds no longer pick up the sand there.

Dunes may stabilize and become vegetated when the climate grows more humid, then start moving again when an arid climate returns. There is geologic evidence that during droughts two to three centuries ago and earlier, sand dunes in the western high plains of North America were reactivated and migrated over the plains.

How Sand Dunes Form and Move When wind moves sand along a bed, it almost inevitably produces ripples and dunes much like those formed by water (**Figure 19.26**). Ripples in dry sand, like those under water, are *transverse*; that is, at right angles to the current. At low to moderate wind speeds, small ripples form. As the wind speed increases, the ripples become larger. Ripples migrate in the direction of the wind over the backs of larger dunes. Some wind is almost always blowing, so a sand bed is almost always rippled to some extent.

Given enough sand and wind, any obstacle—such as a large rock or a clump of vegetation—can start a dune. Streamlines of wind, like those of water, separate around the obstacle and rejoin downwind, creating a wind shadow downstream of the obstacle. Wind velocity is much lower in the wind shadow than in the main flow around the obstacle. In fact, it is low enough to allow sand grains blown into the wind shadow to settle there. The wind there is moving so slowly that it can no longer pick up these grains, and they accumulate as a *sand drift*, a small pile of sand in the lee of the obstacle (**Figure 19.27**). As the process continues, the sand drift itself

FIGURE 19.26 Wind ripples in White Sands National Monument, New Mexico. Although complex in form, these ripples are always transverse (at right angles) to the wind direction. [John Grotzinger.]

becomes an obstacle. If there is enough sand, and the wind continues to blow in the same direction long enough, the sand drift grows into a dune. Dunes may also grow by the enlargement of ripples, just as underwater dunes do.

As a dune grows, it starts to migrate downwind through the combined movements of a host of individual grains. The wind causes the sand grains to slide and roll along the surface by *saltation*, the jumping motion that temporarily suspends grains in a current of water or air (Figure 19.28). Saltation in air flows works the same way as it does in streams, except that the jumps in air flows are higher and longer. Sand grains suspended in an air current may rise to heights of 50 cm over a sand bed and 2 m over a pebbly surface—much higher than grains of the same size can jump in water. The difference arises partly from the fact that air is less viscous than water and therefore does not inhibit the bouncing of the grains as much as water does. In addition, the impact of grains falling in air induces higher jumps as they hit other grains on the surface. These collisions, which the air barely cushions, kick surface grains into the air in a sort of splashing effect. As saltating grains strike a sand bed, they can push forward grains too large to be thrown into the air, causing the bed to creep in the direction of the wind. A sand grain striking the surface at high speed can propel another grain as far as six times its own diameter.

Sand grains constantly saltate to the top of the low-angled windward slope and then fall over into the wind shadow on the leeward slope (see Figure 19.28). These grains gradually build up a steep, unstable accumulation on the upper part of

(a)

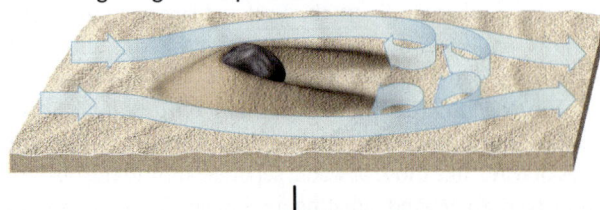
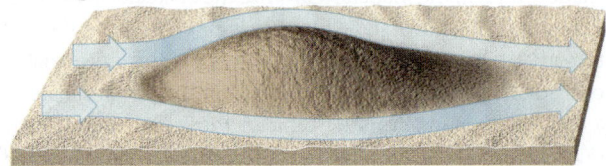

(b)

FIGURE 19.27 Sand dunes may form in the lee of a rock or other obstacle. (a) By separating the wind streamlines, the obstacle creates a wind shadow in which the eddies are weaker than the main flow. Windborne sand grains are thus able to settle in the wind shadow, where they pile up in drifts that eventually coalesce into a dune. (b) Sand drifts, Owens Lake, California. [b: Marli Miller.]

1. A ripple or dune advances by the movements of individual grains of sand. The whole form moves forward slowly as sand erodes from the windward slope and is deposited on the leeward slope.

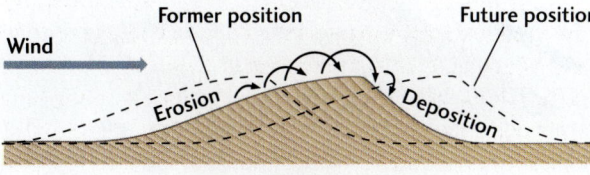

2. Particles of sand arriving on the windward slope of the dune move by saltation over the crest,...

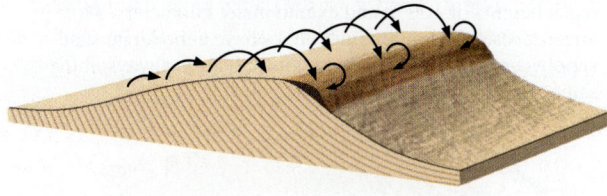

3. ...where the wind velocity decreases and the sand deposited slips down the leeward slope.

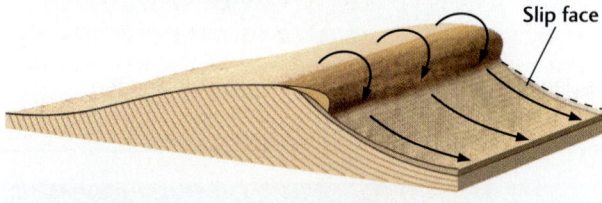

4. This process acts like a conveyor belt that moves the dune forward.

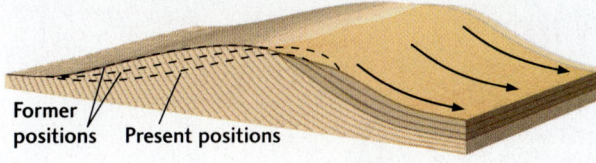

5. The dune stops growing vertically when it reaches a height at which the wind is so fast that it blows the sand grains off the dune as quickly as they are brought up.

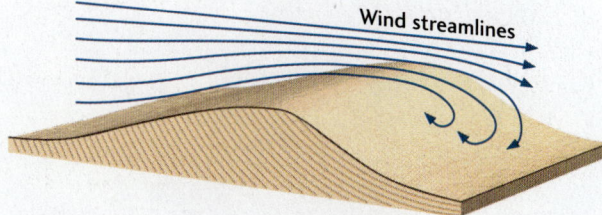

FIGURE 19.28 Sand dunes grow and move as wind transports sand particles by saltation.

the leeward slope. Periodically, the accumulation gives way and spontaneously slips or cascades down this *slip face,* as it is called, to a new slope at a lower angle. If we overlook these short-term, unstable steepenings of the slope, the slip face maintains a stable, constant slope angle—its angle of repose. As we saw in Chapter 17, the angle of repose increases with the size and angularity of the particles.

Successive slip face deposits at the angle of repose create the cross-bedding that is the hallmark of windblown dunes (see Figure 6.11). As dunes accumulate, interfere with one another, and become buried in a sedimentary sequence, the cross-bedding is preserved even though the original shapes of the dunes are lost. Sets of sandstone cross-beds many meters thick are evidence of high windblown dunes. From the directions of these eolian cross-beds, geologists can reconstruct wind directions of the past. Cross-bedding preserved on Mars (see Figure 20.28b) provides evidence of ancient windblown dunes there.

As more sand accumulates on the windward slope of a dune than blows off onto the slip face, the dune grows in height. Most dunes are meters to tens of meters in height, but the huge dunes of Saudi Arabia may reach 250 m, which seems to be the limit. The limit on dune height results from the relationship between wind streamline behavior, wind velocity, and topography. Wind streamlines advancing over the back of a dune become more compressed as the dune grows higher (see Figure 19.28). As more air rushes through a smaller space, the wind velocity increases. Ultimately, the air speed at the top of the dune becomes so great that sand grains blow off the top of the dune as quickly as they are brought up the windward slope. When this equilibrium is reached, the height of the dune remains constant.

Sand Dune Types A person standing in the middle of a large expanse of dunes might be bewildered by the seemingly orderless array of undulating slopes. It takes a practiced eye to see the dominant pattern, and it may even require observation from the air. The general shapes and arrangements of sand dunes depend on the amount of sand available and the direction, duration, and strength of the wind. Geologists recognize four main types of dunes: barchans, blowout dunes, transverse dunes, and longitudinal dunes (**Figure 19.29**).

Windblown Dust

Air has a staggering capacity to hold dust in suspension. Dust includes microscopic rock and mineral fragments of all kinds, especially silicates, as might be expected from their abundance as rock-forming minerals. Two of the most important sources of silicate minerals in dust are clays from soils on dry plains and volcanic ash from eruptions. Organic materials, such as pollen and bacteria, are also common components of dust. Charcoal dust is abundant downwind of forest fires; when found in buried sediments, it is evidence of forest fires in earlier geologic times. Since the beginning of the Industrial Revolution, humans have been

Barchans are crescent-shaped dunes, usually but not always found in groups. The horns of the crescent point downwind. Barchans are the products of limited sand supply and unidirectional winds.

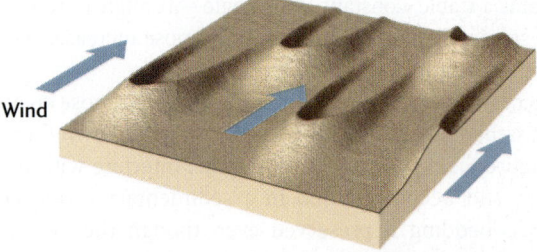

Wind

Transverse dunes are long ridges oriented at right angles to the wind direction. These dunes form in arid regions where there is abundant sand and vegetation is absent. Typically, sand-dune belts behind beaches are transverse dunes formed by strong onshore winds.

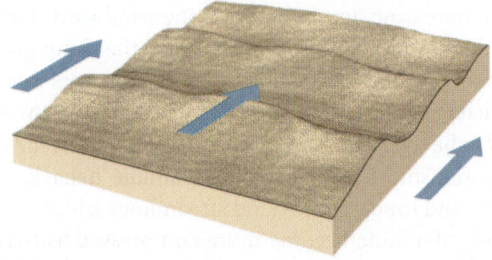

Blowout dunes are almost the reverse of barchans. The slip face of a blowout dune is convex downwind, whereas the barchan's is concave downwind.

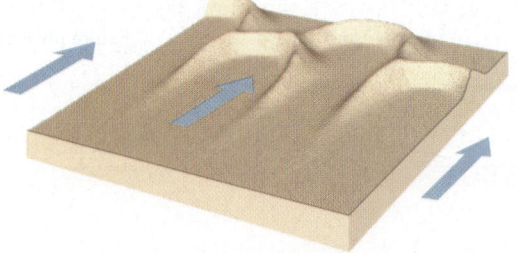

Longitudinal dunes are long ridges of sand whose orientation is parallel to the wind direction. These dunes may reach heights of 100 m and extend many kilometers. Most areas covered by longitudinal dunes have a moderate sand supply, a rough pavement, and winds that are always in the same general direction.

FIGURE 19.29 The general shapes and arrangements of sand dunes depend on the amount of sand available and the direction, duration, and velocity of the wind.

pumping new kinds of synthetic dust into the air—from ash produced by burning coal to the many solid chemical compounds produced by manufacturing processes, incineration of wastes, and motor vehicle exhausts.

In large dust storms, 1 km³ of air may carry as much as a thousand tons of dust, equivalent to the volume of a small house. When such storms cover hundreds of square kilometers, they may carry more than 100 million tons of dust and deposit it in layers several meters thick. Fine-grained particles from the Sahara have been found as far away as England and have been traced across the Atlantic Ocean to Florida. Wind annually transports about 260 million tons of material, mostly dust, from the Sahara to the Atlantic Ocean. Scientists on oceanographic research vessels have measured airborne dust far out to sea, and today it can be observed directly from space (Figure 19.30). Comparison of the composition of this dust with that of deep-sea sediments in the same region indicates that windblown dust is an important contributor to marine sediments, supplying

FIGURE 19.30 Satellite photograph of a dust storm originating in the Namib Desert in September 2002. Dust and sand are being transported from right (east) to left (west) by strong winds blowing out to sea. These sediments can be transported for hundreds to thousands of kilometers across the ocean. [NASA.]

FIGURE 19.31 Stacked layers of loess in Elba, Nebraska. [Daniel R. Muhs, U.S. Geological Survey.]

up to a billion tons of material each year. A large part of this dust comes from volcanoes, and there are individual ash beds on the seafloor marking very large eruptions.

Volcanic ash is an abundant component of dust because much of it is very fine grained and is erupted high into the atmosphere, where it can travel farther than nonvolcanic dust blown by winds closer to Earth's surface. Volcanic explosions inject huge quantities of dust into the atmosphere. The volcanic dust from the 1991 eruption of Mount Pinatubo in the Philippines encircled Earth, and most of the finest-grained particles did not settle until 1994 or 1995.

Mineral dust in the atmosphere increases when agriculture, deforestation, erosion, or other land-use changes disrupt soils. A large amount of the mineral dust in the atmosphere today may be coming from the Sahel, a semiarid region on the southern border of the Sahara where drought and overgrazing are responsible for a heavy load of dust.

Windblown dust has complex effects on the global climate. Mineral dust in the atmosphere scatters incoming visible light from the Sun and absorbs infrared energy radiated outward by Earth's surface. Thus, mineral dust has a net cooling effect in the visible portion of the spectrum and a net warming effect in the infrared portion.

Dust Falls and Loess

As the velocity of the wind decreases, the dust it carries in suspension settles to form **loess,** a blanket of sediment composed of fine-grained particles. Beds of loess lack internal stratification. In compacted deposits more than a meter thick, loess tends to form vertical cracks and to break off along sheer walls (**Figure 19.31**). Geologists theorize that the vertical cracking may be caused by a combination of root penetration and uniform downward percolation of groundwater, but the exact mechanisms are still unknown.

Loess covers as much as 10 percent of Earth's land surface. The largest loess deposits are found in China and North America. China has more than a million square kilometers of loess deposits (**Figure 19.32**). Its greatest deposits extend

FIGURE 19.32 Ancient cave houses dug into loess deposits in Shanxi Province, northern China. [Ashley Cooper/AGE Fotostock.]

over wide areas in the northwest; most are 30 to 100 m thick, although some exceed 300 m. The winds blowing over the Gobi Desert and the arid regions of central Asia provided the dust, which still blows over eastern Asia and the Chinese interior. Some of the loess deposits in China are over 2 million years old. They formed after an increase in the elevation of the Himalaya and related mountain belts in western China introduced rain shadows and dry climates to the continental interior. The uplift of these mountain belts was responsible for the cold, dry climates of the Pleistocene epoch in much of Asia. These climates inhibited vegetation and dried out soils, causing extensive wind erosion and transportation.

The best-known loess deposit in North America is in the upper Mississippi River valley. It originated as silt and clay deposited on the extensive floodplains of streams draining the edges of melting glaciers in the Pleistocene epoch. Strong winds dried the floodplains, whose frigid climate and rapid rates of sedimentation inhibited vegetation, and blew up tremendous amounts of dust, which then settled to the east. Geologists recognize that this loess deposit is distributed as a blanket of more or less uniform thickness on both hills and valleys, all in or near formerly glaciated areas. Changes in the regional thickness of the loess in relation to the prevailing westerly winds confirm its eolian origin. Its thickness on the eastern sides of major river floodplains is 8 to 30 m, greater than on the western sides, and decreases rapidly downwind to 1 to 2 m farther east of the floodplains.

Soils formed on loess are fertile and highly productive. Their cultivation poses environmental problems, however, because they are easily eroded into gullies by small streams and deflated by wind when they are poorly managed.

The Desert Environment

Of all Earth's environments, the desert is where wind is best able to do its work of erosion, transportation, and deposition. The deserts of the world are among the most hostile environments for humans. Yet many of us are fascinated by these hot, dry, apparently lifeless zones, full of bare rocks and sand dunes. The dry climate of deserts creates harsh yet fragile conditions, where human impacts last for decades.

All told, arid regions amount to one-fifth of Earth's land area, about 27.5 million square kilometers. Semiarid plains account for an additional one-seventh. Given the reasons for the existence of large areas of desert in the modern world—the effects of Earth's wind belts on climates, mountain building, and continental drift—we can be confident that, according to the principle of uniformitarianism, extensive deserts have existed throughout geologic time. Conversely, today's deserts may have been wet regions in the past, but may have dried out as a result of long-term climate change.

Desert Weathering and Erosion

As unique as deserts are, the same geologic processes operate there as elsewhere. Physical and chemical weathering work the same way in deserts as they do everywhere, but the balance between the two processes is different: In deserts, physical weathering predominates over chemical weathering. Chemical weathering of feldspars and other silicates into clay minerals proceeds slowly because the water required for those reactions is scarce. The little clay that does form is usually blown away by strong winds before it can accumulate. Slow chemical weathering and rapid wind transportation combine to prevent the buildup of any significant thickness of soil, even where sparse vegetation binds some of the weathered particles. Thus, desert soils are thin and patchy. Sand, gravel, rock fragments of many sizes, and bare bedrock are characteristic of much of the desert surface.

The Colors of the Desert The rusty, orange-brown colors of many weathered surfaces in the desert come from the ferric iron oxide minerals hematite and limonite. These minerals are produced by the slow chemical weathering of iron silicate minerals such as pyroxene. Even when present in only small amounts, they stain the surfaces of sands, gravels, and clays.

Desert varnish is a distinctive dark brown, sometimes shiny, coating found on many rock surfaces in the desert. It is a mixture of clay minerals with smaller amounts of manganese and iron oxides. Desert varnish probably forms when dew causes chemical weathering of primary minerals on an exposed rock surface to form clay minerals and iron and manganese oxides. In addition, tiny quantities of windblown dust may adhere to the rock surface. The process is so slow that Native American inscriptions scratched in desert varnish hundreds of years ago still appear fresh, with a stark contrast between the dark varnish and the light unweathered rock beneath (**Figure 19.33**). Desert varnish requires thousands of years to form, and some particularly ancient varnishes in North America are of Miocene age. However, recognizing desert varnish as such on ancient sandstones is difficult.

FIGURE 19.33 Petroglyphs scratched in desert varnish by Native Americans at Newspaper Rock, Canyonlands, Utah. The scratches are several hundred years old, but appear fresh on the varnish, which has accumulated over thousands of years. [Peter L. Kresan Photography.]

 (a)
 (b)

FIGURE 19.34 A large proportion of the streamflows in deserts occur as floods. (a) A desert valley during a summer thunderstorm at Saguaro National Park, Arizona. (b) The same valley a day after the storm. The coarse sediment deposited by such sudden desert floods may cover the entire valley floor. [Peter L. Kresan Photography.]

Streams: the Primary Agents of Erosion Wind plays a larger role in erosion in deserts than it does elsewhere, but it cannot compete with the erosive power of streams. Even though it rains so seldom that most streams flow only intermittently, streams do most of the erosional work in the desert when they do flow.

Even the driest desert gets occasional rain. In sandy and gravelly areas of deserts, rainfall infiltrates soil and permeable bedrock and temporarily replenishes groundwater in the unsaturated zone. There, some of it evaporates very slowly into pore spaces between particles. A smaller amount eventually reaches the groundwater table far below—in some places, as much as hundreds of meters below the surface. Desert oases form where the groundwater table comes close enough to the surface that the roots of palms and other plants can reach it.

When rain occurs in heavy cloudbursts, so much water falls in such a short time that infiltration cannot keep pace, and the bulk of the water runs off into streams. Unhindered by vegetation, the runoff is rapid and may cause flash floods along valley floors that have been dry for years. Thus, a large proportion of streamflows in deserts consist of floods (**Figure 19.34a**). When floods occur in deserts, they have great erosive power because little of the loose sediment is held in place by vegetation. Streams may become so choked with sediment that they look more like fast-moving mudflows. The abrasiveness of this sediment load moving at flood velocities makes desert streams efficient eroders of bedrock valleys.

Desert Sediments and Sedimentation

Deserts are composed of a diverse set of sedimentary environments. These environments may change dramatically when rain suddenly forms raging rivers and widespread lakes. Prolonged dry periods intervene, during which sediments are blown into sand dunes.

Alluvial Sediments As sediment-laden flash floods dry up, they leave distinctive alluvial deposits on the floors of desert valleys. In many cases, a flat fill of coarse sediment covers the entire valley floor, and the ordinary differentiation of the stream into channel, natural levees, and floodplain is absent (see Figure 19.34b). The sediments of many other desert valleys clearly show the intermixing of stream-deposited channel and floodplain sediments with eolian sediments. This combination of alluvial and eolian processes in the past formed extensive layers of eolian sandstones separated by channel sediments and ancient floodplain sandstones.

Large alluvial fans are prominent features at mountain fronts in deserts because desert streams deposit much of their sediment load on the fans. The rapid infiltration of stream water into the permeable sediments that make up the fan deprives the stream of the water required to carry the sediment load any farther downstream. Debris flows and mudflows make up large parts of the alluvial fans of arid, mountainous regions.

Eolian Sediments By far the most dramatic sedimentary accumulations in deserts are the sand dunes we have described above. Dune fields range in size from a few square kilometers to the "seas of sand" found on the Arabian Peninsula. These sand seas—or *ergs*—may cover as much as 500,000 km^2, twice the area of the state of Nevada.

Although film and television portrayals might lead one to think that deserts are mostly sand, only one-fifth of the world's desert area is actually covered by sand (see Figure 19.19). The other four-fifths are rocky or covered with desert pavement. Sand covers only a little more than one-tenth of the Sahara, and sand dunes are even less common in the deserts of the southwestern United States.

FIGURE 19.35 A desert playa lake in Death Valley, California. [Universal Images Group/Getty Images.]

Evaporite Sediments Desert streams carry large amounts of dissolved minerals, and those minerals accumulate in playa lakes. As the lake water evaporates, the minerals are concentrated and gradually precipitated. If evaporation is complete, the lakes become **playas,** flat beds of clay that are sometimes encrusted with precipitated salts. *Playa lakes* are permanent or temporary lakes that form in arid mountain valleys or basins where water is trapped after rainstorms (Figure 19.35). Playa lakes are sources of evaporite minerals such as sodium carbonate, borax (sodium borate), and other unusual salts.

Desert Landscapes

Desert landscapes are some of the most varied on Earth. Large low, flat areas are covered by playas, desert pavements, and dune fields. Uplands are rocky, cut in many places by steep stream valleys and gorges. The lack of vegetation and soil makes everything seem sharper and harsher than it would in a landscape in a more humid climate. In contrast to the rounded, soil-covered, vegetated slopes found in most humid regions, the coarse fragments of varying size produced by desert weathering form steep cliffs with masses of angular talus at their bases (Figure 19.36).

Much of the landscape of deserts is shaped by streams, but their valleys—called **dry washes** in the western United States and *wadis* in the Middle East—are dry most of the time. Stream valleys in deserts have the same range of profiles as valleys elsewhere. Far more of them, however, have steep walls because of the rapid erosion caused by stream flooding, combined with the lack of rainfall that might soften the slopes of the valley walls between flood events.

FIGURE 19.36 This desert landscape at Kofa Butte, Kofa National Wildlife Refuge, Arizona, shows the steep cliffs and masses of talus produced by desert weathering. [Peter L. Kresan Photography.]

Desert streams are widely spaced because of the relatively infrequent rainfall. Drainage patterns in deserts are generally similar to those in other terrains, with one important difference: Many desert streams die out before they can reach across the desert to join larger rivers flowing to the oceans. Most terminate at the base of alluvial fans. Damming by dunes or confinement within closed valleys with no outlet may lead to the development of playa lakes.

A special type of eroded bedrock surface, called a **pediment,** is a characteristic landform of the desert. Pediments are broad, gently sloping platforms of bedrock left behind as a mountain front erodes and retreats from its valley (**Figure 19.37**). The pediment forms like an apron around the base of the mountains as thin alluvial deposits of sand and gravel accumulate. Long-continued erosion eventually forms an extensive pediment below a few mountain remnants (**Figure 19.38**). A cross section of a typical pediment and its mountains would reveal a fairly steep mountain slope abruptly leveling into the gentle pediment slope. Alluvial fans deposited at the lower edge of the pediment merge with the sedimentary fill of the valley below the pediment.

There is much evidence that pediments are formed by running water, which cuts and forms the pediment surface as well as transporting and depositing sediments to create an apron of alluvial fans. At the same time, the mountain slopes at the head of the pediment maintain their steepness as they retreat, instead of becoming the rounded, gentler slopes found in humid regions. We do not know how specific rock types and erosional processes interact in an arid environment to keep the slopes steep as the pediment is enlarged.

Tectonic, Climatic, and Human Controls on Deserts

The Role of Plate Tectonics

In a sense, deserts are a result of plate-tectonic processes. The mountains that create rain shadows are raised at convergent plate boundaries. The great distance separating central Asia from the oceans is a consequence of the size of the Asian continent, a huge landmass assembled from smaller landmasses by continental drift. Large deserts are found at low latitudes because continental drift moved continents there from higher latitudes. If, in some future plate-tectonic scenario, the North American continent were to move south by 2000 km or so, the northern Great Plains of the United States and Canada would become a hot, dry desert. Something like that happened to Australia. About 20 million years ago, Australia was far to the south of its present position, and its interior had a warm, humid climate. Since then, Australia has moved northward into an arid subtropical zone, and its interior has become a desert.

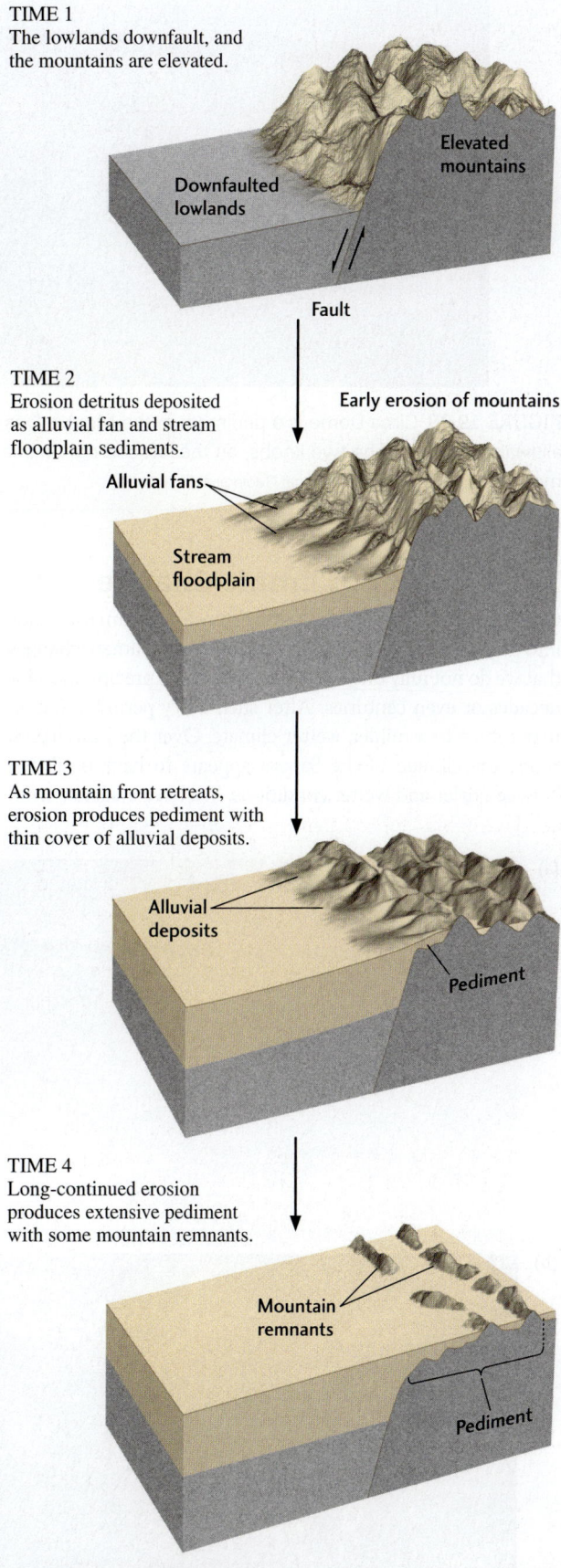

FIGURE 19.37 Pediments form as mountain fronts erode and retreat.

FIGURE 19.38 Cima Dome is a pediment in the Mojave Desert. The surface of the dome is covered by a thin veneer of alluvial sediments. The two knobs, on the left and right sides of the dome, are regarded as the final remnants of the former mountain. [Russ White, Lynn Canal Geological Services.]

The Impact of Climate Change

Changes in a region's climate may transform semiarid lands into deserts, a process called **desertification.** Climate changes that we do not fully understand may decrease precipitation for decades or even centuries. After such a dry period, a region may return to a milder, wetter climate. Over the past 10,000 years, the climate of the Sahara appears to have oscillated between drier and wetter conditions. We have evidence from orbiting satellites that an extensive system of river channels existed there a few thousand years ago (**Figure 19.39**). Now dry and buried by more recent sand deposits, these ancient drainage systems carried abundant running water across the northern Sahara during wetter periods.

The Sahara may now be expanding northward. The Desert Watch project, led by the European Space Agency, reports that over 300,000 km² of Europe's Mediterranean coast—an

(a)

(b)

FIGURE 19.39 The climate of the Sahara was not always as arid as it is today. (a) Remote sensing techniques that look only at Earth's surface see nothing but sand in the Sahara. (b) Remote sensing techniques that penetrate a few meters below the surface, however, see a dense network of buried riverbeds. [NASA/JPL Imaging Radar Team.]

area almost as large as the state of New York, with a population of 16 million—has been enduring the longest drought in recorded history. During 2005 and 2012, fires raged along the southern Spanish coast, and temperatures set new record highs for weeks on end. Were these merely long, hot summers, or are these the initial symptoms of desertification, made worse by overpopulation and overdevelopment within the fragile ecosystems of dry landscapes? Evidence to support the latter scenario is building. Soils have been loosened by the prolonged dryness, making them more susceptible to wind transportation and deflation. Groundwater levels have reached new lows. And there is little question that Europe is getting warmer: During the twentieth century, its average temperatures increased about 0.7°C. The 1990s marked the hottest decade since record keeping began in the mid-1800s, registering two of the five hottest years ever recorded.

The Influence of Humans

Climate oscillations occur naturally in the Sahara and in other deserts, but human activities are responsible for some of the desertification occurring today. The growth of human populations in semiarid regions, along with increased agriculture and animal grazing, may result in the expansion of deserts. When population growth and periods of drought coincide, the results can be disastrous. In Spain, the greatest urban and agricultural expansion is taking place on the Mediterranean coast—the nation's driest region. Former farmlands have been stripped of vegetation due to overfarming (up to four crops per year), which depletes water and strips soils. A tourism boom and its accompanying development are literally paving over the dry lands and desiccating the countryside that is left. In 2004, more than 350,000 new homes were built on the Mediterranean coast, many with backyard swimming pools and nearby golf courses requiring large amounts of water. In isolation, any one of these human activities might not have a negative effect. Together, however, they add up to desertification.

"Making the desert bloom," the opposite of desertification, has been a slogan of some countries with desert lands. They irrigate on a massive scale to convert semiarid or arid areas into productive farmlands. The Central Valley of California, where many of North America's fruits and vegetables are grown, is one example. If the waters used for irrigation contain dissolved substances (as almost all natural waters do), then, with time, these waters evaporate and deposit the dissolved substances as salts. Thus, ironically, irrigation in an arid or semiarid climate can eventually cause desertification through the slow accumulation of salts.

KEY TERMS AND CONCEPTS

barrier island (p. 570)	eolian (p. 580)	spit (p. 570)
beach (p. 565)	hurricane (p. 573)	storm surge (p. 576)
deflation (p. 581)	loess (p. 587)	tidal flat (p. 565)
desert pavement (p. 581)	longshore current (p. 563)	tide (p. 564)
desert varnish (p. 588)	pediment (p. 591)	wave-cut terrace (p. 570)
desertification (p. 592)	playa (p. 590)	ventifact (p. 581)
dry wash (p. 590)	sandblasting (p. 581)	
dust (p. 580)	shoreline (p. 560)	

REVIEW OF LEARNING OBJECTIVES

19.1 Identify the coastal processes that act on shorelines.

Winds blowing over the ocean generate swell; as those waves approach the shore, they are transformed into breakers in the surf zone. Wave refraction results in longshore drift and longshore currents, which transport sand along beaches. Tides, generated by the gravitational pull of the Moon and the Sun on ocean water, can also generate currents that transport sediments.

Study Assignment: Figures 19.2 and 19.4

Thought Question: In a 100-year period, the southern tip of a long, narrow, north-south beach has become extended about 200 m to the south by natural processes. What shoreline processes could have caused this extension?

Exercise: How are ocean waves formed?

19.2 Summarize how processes shape shorelines.

Waves and tides, interacting with plate tectonic processes, shape the topography of coastlines, which vary from beaches and tidal flats to uplifted rocky coasts.

Study Assignment: Figures 19.8, 19.10, and 19.11

Thought Question: How does the sand budget of a beach determine its shape? Consider inputs and outputs.

Exercise: How does wave refraction concentrate erosion at headlands?

19.3 Explain how hurricanes affect coastal areas.

Hurricanes are intense tropical storms with extremely high winds and very low atmospheric pressures. The low pressure results in the formation of a dome of seawater, known as a storm surge. As the hurricane moves ashore, the storm surge floods low-lying areas, often causing more extensive damage than the storm's high winds.

Study Assignment: Figures 19.16 and 19.17

Thought Question: After a period of calm along a section of the Gulf Coast of North America, a severe storm with hurricane force winds passes over the shore and up the Mississippi Valley. Describe the state of the surf zone, as well as the elevation of sea level, before the storm, during it, and after the storm has passed. What would happen to inland rivers?

Exercise: What is a storm surge?

19.4 Describe how winds transport and erode sand and finer-grained sediments.

Winds can pick up and transport dry particles in a manner similar to flowing water. Air flows are limited, however, in the size of particles they can carry (rarely larger than sand grains) and in their ability to keep particles in suspension. These limitations result from air's low viscosity and density. Windblown materials include volcanic ash, quartz grains, and other mineral fragments such as clay, as well as organic materials such as pollen and bacteria. Wind can carry great amounts of sand and dust. It moves sand grains primarily by saltation and carries finer-grained dust particles in suspension.

Sandblasting and deflation are the primary ways in which winds erode Earth's surface.

> **Study Assignment:** Figure 19.24
>
> **Thought Question:** What are the key processes involved in desert pavement formation?
>
> **Exercise:** What types of materials and sizes of particles can the wind move?

19.5 Discuss how winds deposit sand dunes and dust.

When winds die down, they deposit sand in dunes of various shapes and sizes. Dunes form in sandy desert regions, behind beaches, and along sandy floodplains, all of which are places with a ready supply of loose sand and moderate to strong winds. Dunes start as sand drifts in the lee of obstacles and may grow to heights of up to 250 meters, though most are tens of meters in height. Dunes migrate downwind as sand grains saltate up their gentler windward slopes and fall over onto their steeper downwind slip faces. The shapes and arrangements of sand dunes are determined by the direction, duration, and strength of the wind and by the abundance of sand. As the velocity of dust-laden winds decreases, the dust settles to form loess, a thick blanket of fine particles.

Loess layers have been deposited in many formerly glaciated areas by winds blowing over the floodplains of streams formed by glacial meltwater. Loess can accumulate to great thicknesses downwind of dusty desert regions.

> **Study Assignment:** Figures 19.29 and 19.31
>
> **Thought Question:** There are large areas of sand dunes on Mars. From this fact alone, what can you infer about conditions on the Martian surface?
>
> **Exercise:** Name three types of sand dunes and show the relationship of each to wind direction.

19.6 Explain how wind and water combine to shape the desert environment and its landscape.

When winds die down, they deposit sand in dunes of various shapes and sizes. Dunes form in sandy desert regions, behind beaches, and along sandy floodplains, all of which are places with a ready supply of loose sand and moderate to strong winds. Dunes start as sand drifts in the lee of obstacles and may grow to heights of up to 250 meters, though most are tens of meters in height. Dunes migrate downwind as sand grains saltate up their gentler windward slopes and fall over onto their steeper downwind slip faces. The shapes and arrangements of sand dunes are determined by the direction, duration, and strength of the wind and by the abundance of sand. As the velocity of dust-laden winds decreases, the dust settles to form loess, a thick blanket of fine particles.

Loess layers have been deposited in many formerly glaciated areas by winds blowing over the floodplains of streams formed by glacial meltwater. Loess can accumulate to great thicknesses downwind of dusty desert regions.

> **Study Assignment:** Figures 19.34, 19.35, and 19.36
>
> **Thought Question:** What kinds of landscape features would you ascribe to the work of the wind, to the work of streams, or to both?
>
> **Exercise:** What are the geologic processes that form playa lakes?

19.7 Summarize the tectonic, climatic, and human controls on deserts.

At a global scale, plate tectonics influences desert formation by moving continents, and associated land masses, into subtropical latitudes where climates are arid or semiarid. Continent size also matters and Asia, being so large, prevents ocean-derived moisture from reaching the interior of the continent. Climate also changes for reasons we do not fully understand. What we do know is that wet regions of Earth may become dry, and then transition back to wet again—and vice versa. When wet regions become dry due to climate change, this is called desertification. Human influences, such as overgrazing by farm animals, may accelerate this process.

> **Study Assignment:** Figures 19.37, 19.38, and 19.39
>
> **Thought Question:** What kinds of landscape features would you ascribe to the work of the wind, to the work of streams, or to both?
>
> **Exercise:** What is desertification?

VISUAL LITERACY EXERCISE

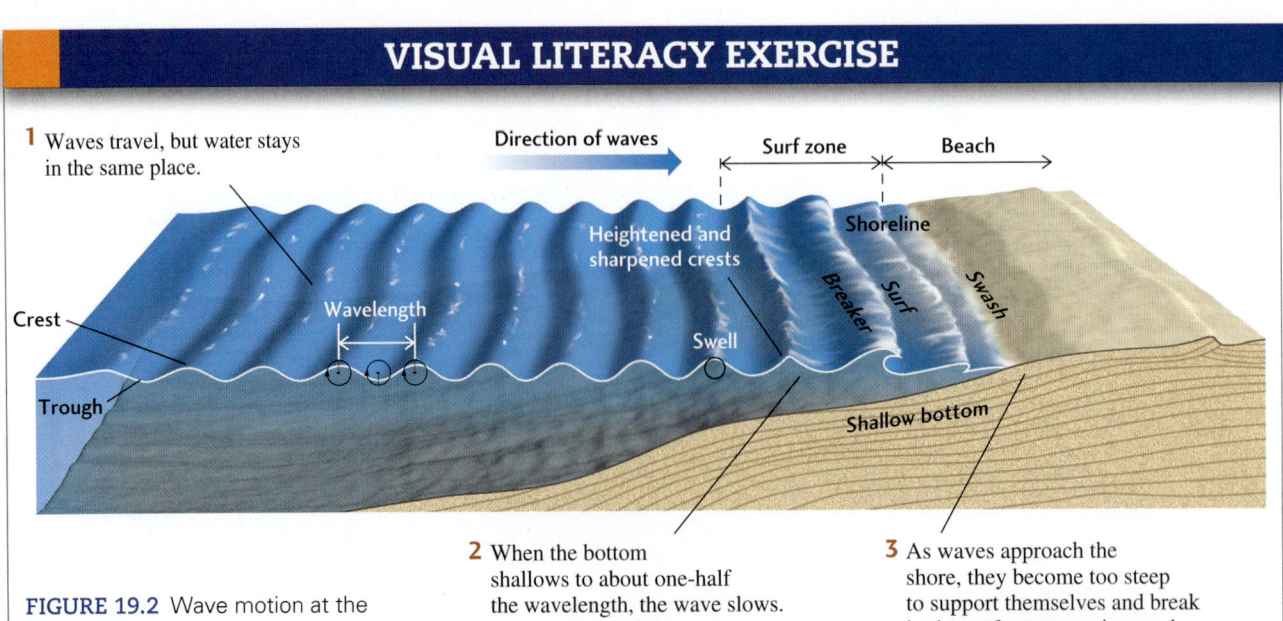

1. Waves travel, but water stays in the same place.
2. When the bottom shallows to about one-half the wavelength, the wave slows.
3. As waves approach the shore, they become too steep to support themselves and break in the surf zone, running up the beach in a swash.

FIGURE 19.2 Wave motion at the shoreline is influenced by water depth and the shape of the bottom.

Visual Literacy Exercise

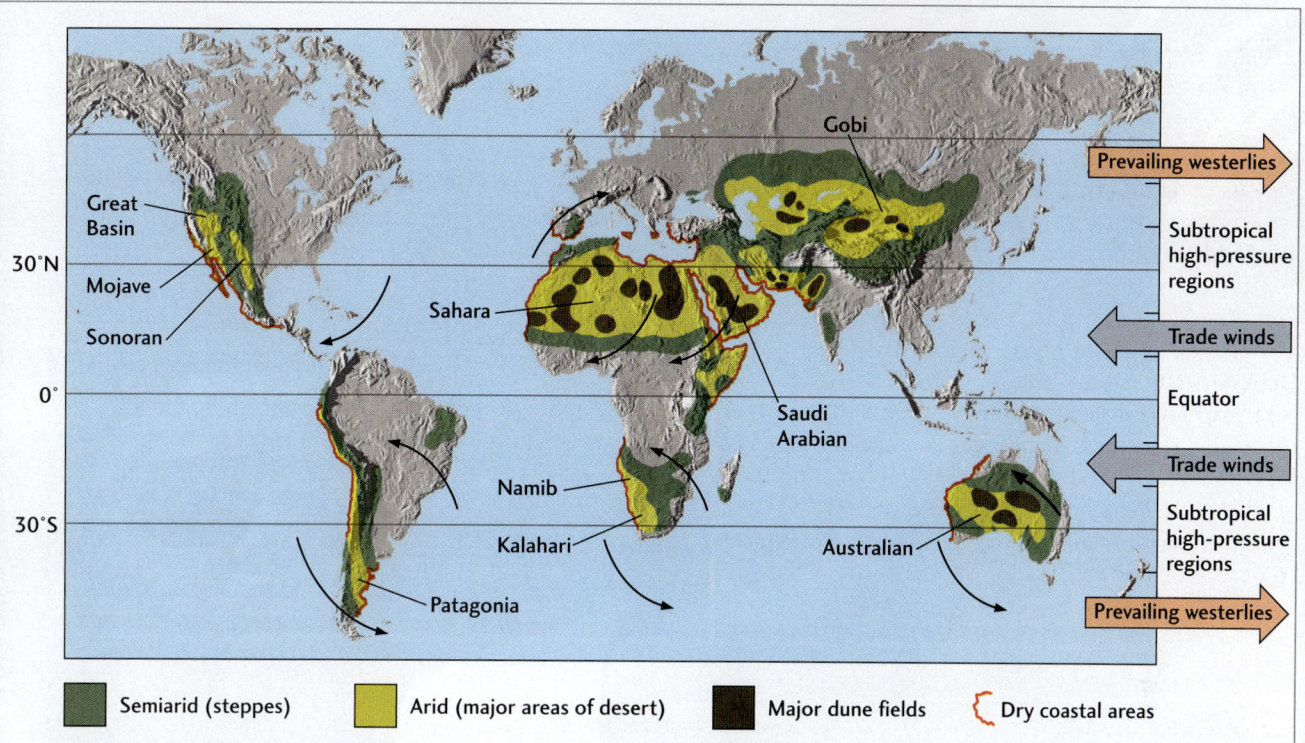

FIGURE 19.19 Major desert areas of the world (exclusive of polar deserts). Notice the relationships of their locations to prevailing wind belts and major mountain ranges. Notice, too, that sand dunes make up only a small proportion of the total desert area. [Information from K. W. Glennie, *Desert Sedimentary Environments.* New York: Elsevier, 1970.]

For the first three questions, refer to Figure 19.2.

1. How do waves affect water motion in deeper water?
2. At what point do waves slow down?
3. What causes waves to break?

For the next two questions, refer to Figure 19.19.

4. What latitudes do major deserts occur in?
   a. Polar regions
   b. Equatorial tropical regions
   c. Subtropical high pressure regions
   d. Antarctica

5. What are the two largest deserts on Earth?
   a. Mojave and Sonoran
   b. Sahara and Gobi
   c. Kalahari and Namib
   d. Patagonia and central Australia

Early History of the Terrestrial Planets

20

Origin of the Solar System	600
Early Earth: Formation of a Layered Planet	603
Diversity of the Planets	606
What's in a Face? The Age and Complexion of Planetary Surfaces	608
Mars Rocks!	616
Exploring the Solar System and Beyond	626

Learning Objectives

After studying this chapter, you should be able to:

20.1 Describe how the solar system originated.

20.2 Explain how the Earth formed and developed over time.

20.3 Compare and contrast the diversity of the planets and what makes them similar and different.

20.4 Summarize some of the major events in the early history of the solar system and describe how planetary surfaces can be dated.

20.5 Discuss how Mars and the other planets have been explored. Does spacecraft data show evidence for water on Mars?

20.6 Describe how we use light in exploring stars and the solar system. Is our solar system unique?

NASA astronaut Jessica Watkins in the field at Rio Grande del Norte National Monument, New Mexico. Watkins is a Ph.D. geologist who has studied mass wasting processes on both Earth and Mars. She was inducted into the NASA astronaut corps in 2017 and hopes to apply her skills someday on the Moon or Mars. [NASA.]

In a series of six landings from 1969 through 1972, the astronauts of the Apollo missions explored the surface of the Moon. These astronauts, trained in geology, took photographs, mapped outcrops, conducted experiments, and collected dust and rock samples for analysis back on Earth. This unprecedented achievement was possible only through the close collaboration of scientists, engineers, and the funding agencies that recognized the importance of basic research in developing new technologies. Perhaps the most important ingredient of all was the irrepressible drive, inherent in all human beings, to explore the unknown. The desire to explore our universe has existed for as long as humans have been able to think. Edwin Powell Hubble best captured the spirit of space exploration when he modestly noted, "Equipped with his five senses, man explores the universe around him and calls the adventure science."

The modern era of space exploration began in the early 1900s when a handful of scientists with a yearning to escape the confines of Earth's gravity began to develop the first generation of rockets. By the late 1920s, these backyard rockets, powered by liquid propellants, were ready for use. Developments occurred rapidly over the next few decades, culminating in the fevered cold-war race between the United States and the Soviet Union to put the first rocket into space, the first satellite into Earth orbit, the first human on the Moon, and the first robot on Mars. By the mid-1970s—50 years after the first liquid-fueled rockets were invented—all of these goals had been achieved.

The scientific dividends of space exploration have been tremendous. The age of the solar system, evidence for liquid water on early Mars, and the thick atmosphere of Venus were all revealed by the mid-1970s. Since that time, we have carried on our exploration of the solar system and beyond. Using instruments in Earth orbit and on spacecraft sent to the far limits of our solar system, we have obtained a much better view of what, literally, is *way* out there! Of all these instruments, none has produced such visually spectacular images of deep space as the Hubble Space Telescope, named for Edwin Powell Hubble. Not since Galileo turned his telescope toward the heavens in 1609 has any instrument so changed our understanding of the universe.

The crater-marked surfaces of the Moon and our neighboring planets, and the occasional meteorite that crashes through Earth's atmosphere, remind us of a disorganized, chaotic time when the solar system was young and Earth's environment was much less hospitable. How did the solar system become the well-ordered place it is today, with planets moving in stately orbits around the Sun? How did Earth's rocky mass come together and differentiate into a core, mantle, and crust? Why does Earth's surface, with its blue oceans and wandering continents, look so different from those of its planetary neighbors? Geologists can draw from many lines of evidence to answer these questions. The rocks of continents preserve a record of geologic processes more than 4 billion years old, and even more ancient materials have been collected from meteorites. And now we can reach beyond Earth for the answers.

In Chapter 20, we will explore the solar system not only outward through the vast reaches of interplanetary space, but also backward in time to its earliest history. We will see how Earth and the other planets formed around the Sun and differentiated into layered bodies. We will compare the geologic processes that have shaped Earth with those that formed Mercury, Mars, Venus, and the Moon, and we will see how exploration of the solar system by spacecraft might answer fundamental questions about the development of our planet and the life it contains.

Origin of the Solar System

Our search for the origins of the universe—and of our own small part of it—goes back to the earliest recorded mythologies. Today, the generally accepted scientific explanation is the Big Bang theory, which holds that our universe began about 13.7 billion years ago with a cosmic "explosion." Before that moment, all matter and energy were compacted into a single, inconceivably dense point. Although we know little of what happened in the first fraction of a second after time began, astronomers have a general understanding of the billions of years that followed: In a process that still continues, the universe has expanded and thinned out to form galaxies and stars. Geology explores the latter third of that time: the past 4.5 billion years, during which our *solar system*—the star we call the Sun and the planets that orbit it—formed and evolved. In particular, geologists look to the solar system to understand the formation of Earth and the Earthlike planets.

The Nebular Hypothesis

In 1755, the German philosopher Immanuel Kant suggested that the origin of the solar system could be traced to a rotating cloud of gases and fine dust, an idea called the **nebular hypothesis.** We now know that outer space beyond the solar system is not as empty as we once thought. Astronomers have recorded many clouds of the type that Kant surmised, which they have named *nebulae* (plural of the Latin word for "fog" or "cloud") (Figure 20.1). They have also identified the materials that form these clouds. The gases are mostly hydrogen and helium, the two elements that make up all but a small fraction of our Sun. The dust-sized particles are chemically similar to materials found on Earth.

FIGURE 20.1 Space exploration has progressed from its modest beginnings to address fundamental questions such as the origin of the solar system. (a) Robert H. Goddard, one of the fathers of rocketry, fired this liquid oxygen–gasoline rocket on March 16, 1926, at Auburn, Massachusetts. (b) Seventy years later, on November 2, 1995, the Hubble Space Telescope (in orbit around Earth) took this stunning photograph of the Eagle Nebula. The dark, pillarlike structures are columns of cool hydrogen gas and dust that give birth to new stars. [a: NASA; b: NASA/ESA/STSci.]

How could the solar system take shape from such a cloud? The diffuse, slowly rotating nebula contracted under the force of gravity (Figure 20.2). Its contraction, in turn, accelerated the rotation of the particles (just as ice skaters spin more rapidly when they pull in their arms), and the faster rotation flattened the cloud into a disk.

The Sun Forms

Under the pull of gravity, matter began to drift toward the center of the nebula, accumulating into a protostar, the precursor of our present Sun. Compressed under its own weight, the material in the proto-Sun became dense and hot. Its internal temperature rose to millions of degrees, at which point nuclear fusion began. The Sun's nuclear fusion, which continues today, is the same nuclear reaction that occurs in a hydrogen bomb. In both cases, hydrogen atoms, under intense pressures and at high temperatures, combine (fuse) to form helium. Some mass is converted into energy in the process. The Sun releases some of that energy as sunshine; an H-bomb releases it as an explosion.

The Planets Form

Although most of the matter in the original nebula was concentrated in the proto-Sun, a disk of gases and dust, called the **solar nebula,** remained to envelop it. The temperature of the solar nebula rose as it flattened into a disk. It became hotter in the inner region, where more of the matter accumulated, than in the less dense outer regions. Once formed, the disk began to cool, and many of the gases condensed— that is, they were transformed to their liquid or solid state, just as water vapor condenses into droplets on the outside of a cold glass and water solidifies into ice when it cools below the freezing point.

Gravitational attraction caused the dust and condensing material to clump together (accrete) into small, kilometer-sized chunks, or **planetesimals.** These planetesimals, in turn, collided and stuck together, forming larger, Moon-sized bodies (see Figure 20.2). In a final stage of cataclysmic impacts, a few of these larger bodies—with their stronger gravitational attraction—swept up the others to form the planets in their present orbits. Planetary formation happened rapidly, probably within 10 million years after the condensation of the nebula.

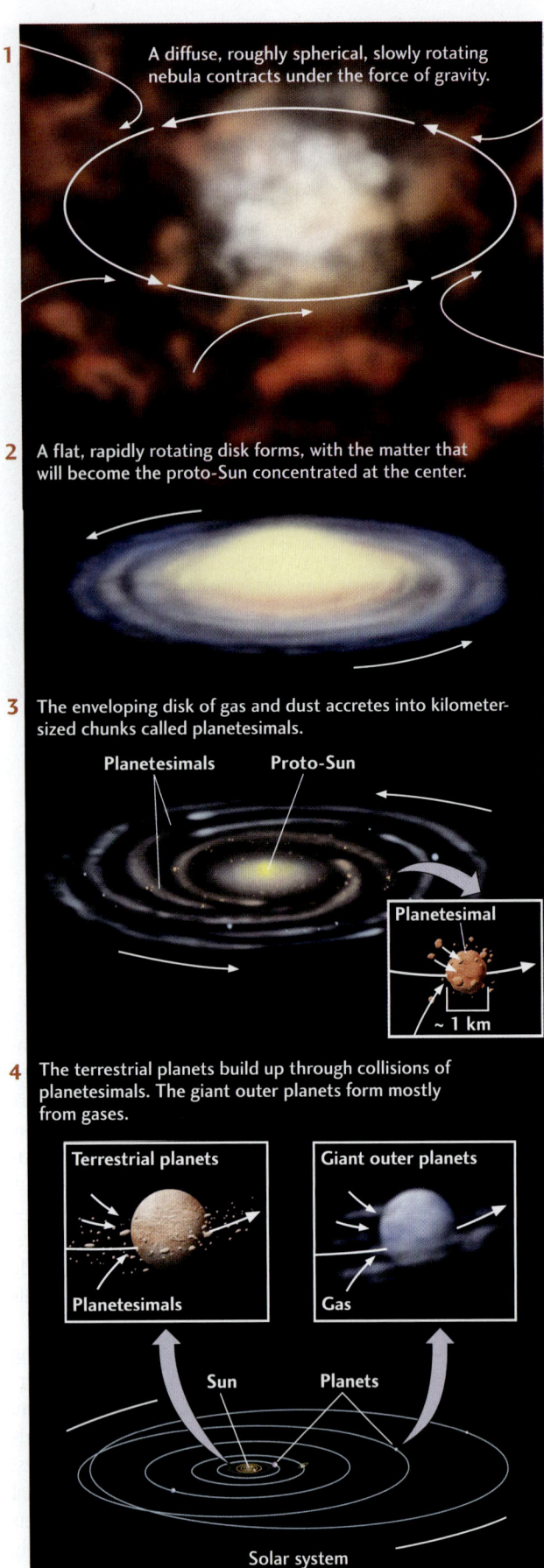

FIGURE 20.2 The nebular hypothesis explains the formation of the solar system.

As the planets formed, those in orbits close to the Sun and those in orbits farther from the Sun developed in markedly different ways. Thus, the composition of the inner planets is quite different from that of the outer planets.

Terrestrial Planets The four inner planets, in order of closeness to the Sun, are Mercury, Venus, Earth, and Mars (Figure 20.3). They are also known as the Earthlike planets, or **terrestrial planets.** They formed close to the Sun, where conditions were so hot that most of their volatile materials (those that most easily become gases) boiled away. Radiation and matter streaming from the Sun—the solar wind—blew away most of the hydrogen, helium, water, and other light gases and liquids on these planets. Thus, the inner planets were formed mainly from the dense matter that was left behind, which included the rock-forming silicates as well as metals such as iron and nickel. From isotopic dating of the meteorites that occasionally strike Earth, which are believed to be remnants of this pre-planetary process, we know that the inner planets began to accrete about 4.56 billion years ago (see Chapter 9). Computer simulations indicate that they would have grown to planetary size in a remarkably short time—perhaps as quickly as 10 million years or less.

Giant Outer Planets Most of the volatile materials swept from the region of the terrestrial planets were carried to the cold outer reaches of the solar system to form the giant outer planets—Jupiter, Saturn, Uranus, and Neptune—and their satellites. These giant planets were big enough, and their gravitational attraction strong enough, to enable them to hold onto the lighter nebular materials. Thus, although they have rocky and metal-rich cores, they, like the Sun, are composed mostly of hydrogen and helium and the other light materials of the original nebula.

Small Bodies of the Solar System

Not all the material from the solar nebula ended up in planets. Some planetesimals collected between the orbits of Mars and Jupiter to form the *asteroid belt* (see Figure 20.3). This region now contains more than 10,000 **asteroids** with diameters larger than 10 km and about 300 larger than 100 km. The largest is Ceres, which has a diameter of 930 km. Most **meteorites**—chunks of material from outer space that strike Earth—are tiny pieces of asteroids ejected from the asteroid belt during collisions with one another. Astronomers originally thought the asteroids were the remains of a large planet that had broken apart early in the history of the solar system, but it now appears they are

FIGURE 20.3 The solar system. This diagram shows the relative sizes of the planets as well as the asteroid belt separating the inner and outer planets. Although considered one of the nine planets since its discovery in 1930, Pluto was demoted from that status by the International Astronomical Union in 2006. With this revision, there are only eight true planets, not nine.

pieces that never coalesced into a planet, probably due to the gravitational influence of Jupiter.

Another important group of small, solid bodies is the *comets*, aggregations of dust and ice that condensed in the cooler outer reaches of the solar nebula. There are probably many millions of comets with diameters larger than 10 km. Most comets orbit the Sun far beyond the outer planets, forming concentric "halos" around the solar system. Occasionally, collisions or near misses throw a comet into an orbit that penetrates the inner solar system. We can then observe it as a bright object with a tail of gases blown away from the Sun by the solar wind. Perhaps the most famous of these is Halley's Comet, which has an orbital period of 76 years and was last seen in 1986. Comets are intriguing to geologists because they provide clues about the more volatile components of the solar nebula, including water and carbon-rich compounds, which they contain in abundance.

Early Earth: Formation of a Layered Planet

We know that Earth is a layered planet with a core, mantle, and crust surrounded by a fluid ocean and a gaseous atmosphere (see Chapter 1). How was Earth transformed from a hot, rocky mass into a living planet with continents, oceans, and a pleasant climate? The answer lies in **gravitational differentiation:** the transformation of random chunks of primordial matter into a body whose interior is divided into concentric layers that differ from one another both physically and chemically. Gravitational differentiation occurred early in Earth's history, as soon as the planet got hot enough to melt.

Earth Heats Up and Melts

Although Earth probably started out as an accretion of planetesimals and other remnants of the solar nebula, it did not retain this form for long. To understand Earth's present layered structure, we must return to the time when Earth was still subject to violent impacts by planetesimals and larger bodies. As these objects crashed into the primitive planet, most of their energy of motion (kinetic energy) was converted into heat—another form of energy—and that heat caused melting. A planetesimal colliding with Earth at a typical velocity of 15 to 20 km/s would deliver as much kinetic energy as 100 times its weight in TNT. The impact energy of a body the size of Mars colliding with Earth would be equivalent to exploding several trillion 1-megaton hydrogen bombs (a single one of which would destroy a large city)—enough to eject a vast amount of debris into space and to melt most of what remained of Earth.

Many scientists now think that such a cataclysm did occur during the middle to late stages of Earth's accretion. A giant impact by a Mars-sized body created a shower of debris from both Earth and the impacting body and propelled it into space. The Moon aggregated from that debris (**Figure 20.4**). According to this theory, Earth reformed as a body with an outer molten layer hundreds of kilometers thick—a *magma ocean*. The huge impact sped up Earth's rotation and changed the angle of its axis, knocking it from vertical with respect to Earth's orbital plane to its present 23° inclination. All this occurred about 4.51 billion years ago, between the beginning of Earth's accretion (4.56 billion years ago) and the formation of the oldest Moon rocks brought back by the Apollo astronauts (4.47 billion years ago).

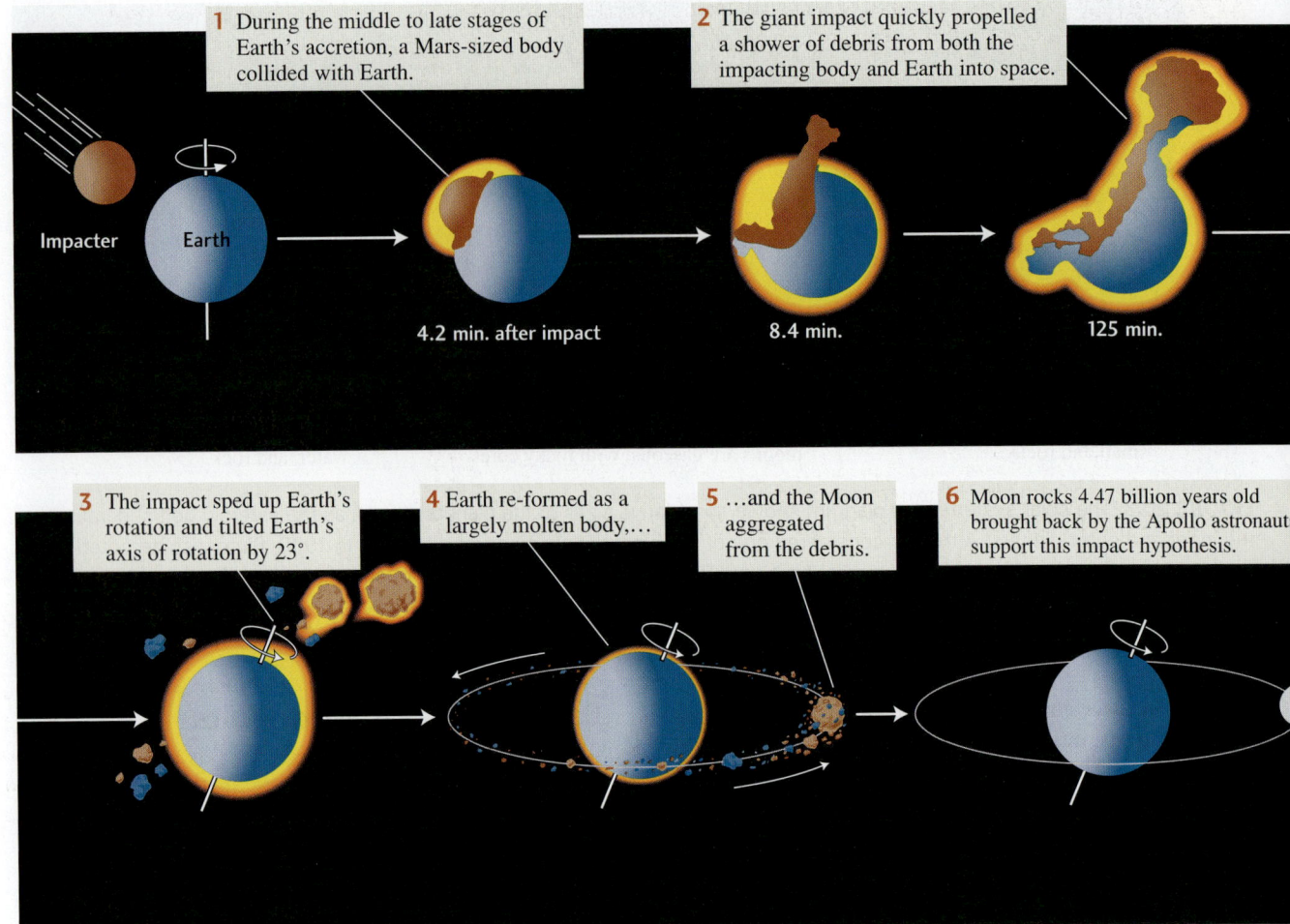

FIGURE 20.4 Computer simulation of the impact of a Mars-sized body on Earth. [Data from *Solid-Earth Sciences and Society*. National Research Council, 1993.]

Another source of heat that contributed to melting early in Earth's history was radioactivity. When radioactive elements decay, they emit heat. Although present in only small amounts, radioactive isotopes of uranium, thorium, and potassium have continued to keep Earth's interior hot.

Differentiation of Earth's Core, Mantle, and Crust

As a result of the tremendous impact energy absorbed during Earth's formation, its entire interior was heated to a "soft" state in which its components could move around. Heavy material sank to become the core, releasing gravitational energy and causing more melting, and lighter material floated to the surface and formed the crust. The rising lighter matter brought heat from the interior to the surface, where it could radiate into space. In this way, Earth differentiated into a layered planet with a central core, a mantle, and an outer crust (**Figure 20.5**).

Earth's Core Iron, which is denser than most of the other elements, accounted for about a third of the primitive planet's material (see Figure 1.11). This iron and other heavy elements, such as nickel, sank to form a central *core*, which begins at a depth of about 2890 km. By probing the core with seismic waves, scientists have found that it is molten on the outside but solid in a region called the *inner core*, which extends from a depth of about 5150 km to Earth's center at about 6370 km. Today the inner core is solid because the pressures deep in Earth's interior are too high for iron to melt.

Earth's Crust Other molten materials that are less dense than iron and nickel floated toward the surface of the magma ocean. There they cooled to form Earth's solid *crust*, which today ranges in thickness from about 7 km on the seafloor to about 40 km on the continents. We know that oceanic crust is constantly generated by seafloor spreading and recycled into the mantle by subduction. In contrast, continental crust began to accumulate early in Earth's history from silicates of relatively low density with a felsic composition and low melting temperatures. This contrast between dense oceanic crust and less dense continental crust is what helps drive oceanic crust into subduction zones, while continental crust resists subduction.

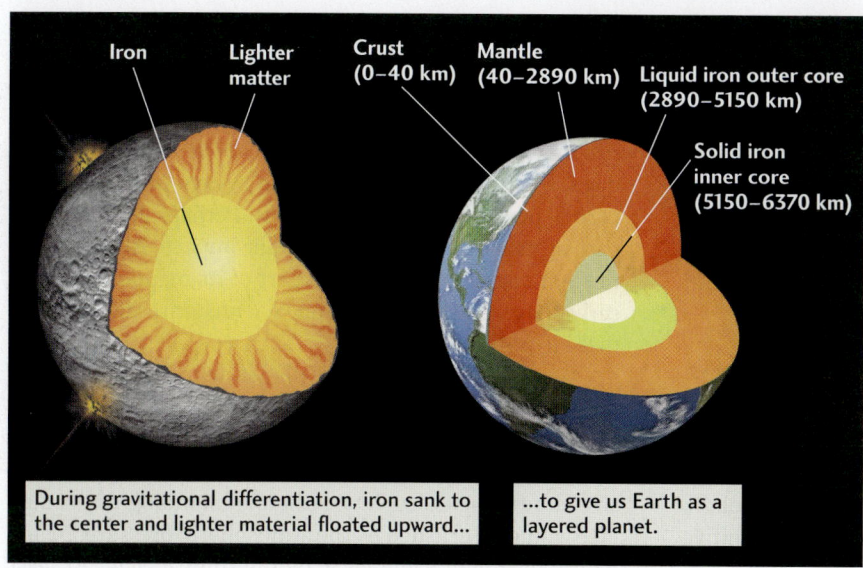

FIGURE 20.5 Gravitational differentiation of early Earth resulted in a planet with three main layers.

The 4.4-billion-year-old zircon grains recently found in Western Australia (see Chapter 9) are the oldest terrestrial material yet discovered. Chemical analysis indicates that they formed near Earth's surface under relatively cool conditions and in the presence of water. This finding suggests that Earth had cooled enough for a crust to exist only 100 million years after the planet re-formed following the giant impact that produced the Moon.

Earth's Mantle Between the core and the crust lies the *mantle*, the layer that forms the bulk of the solid Earth. The mantle is made up of the material left in the middle zone after most of the denser material sank and the less dense material rose toward the surface. It is about 2850 km thick and consists of ultramafic silicate rocks containing more magnesium and iron than crustal silicates do. Convection in the mantle removes heat from Earth's interior (see Chapter 2).

Because the mantle was hotter early in Earth's history, it was probably convecting more vigorously than it does today. Some form of plate tectonics may have been operating even then, although the "plates" were probably much smaller and thinner, and the tectonic features were probably very different from the linear mountain belts and long mid-ocean ridges we now see on Earth's surface. Some scientists think that Venus today provides an analog for these long-vanished processes on Earth. We will compare tectonic processes on Earth and Venus shortly.

Earth's Oceans and Atmosphere Form

The oceans and atmosphere can be traced back to the "wet birth" of Earth itself. The planetesimals that aggregated into our planet contained ice, water, and other volatiles, such as nitrogen and carbon, locked up in minerals. As Earth differentiated, water vapor and other gases were freed from these minerals, carried to the surface by magmas, and released through volcanic activity.

The enormous volumes of gases spewed from volcanoes 4 billion years ago probably consisted of the same substances that are expelled from present-day volcanoes (though not necessarily in the same relative abundances): primarily hydrogen, carbon dioxide, nitrogen, water vapor, and a few others (**Figure 20.6**). Almost all of the hydrogen escaped into space, while the heavier gases enveloped the

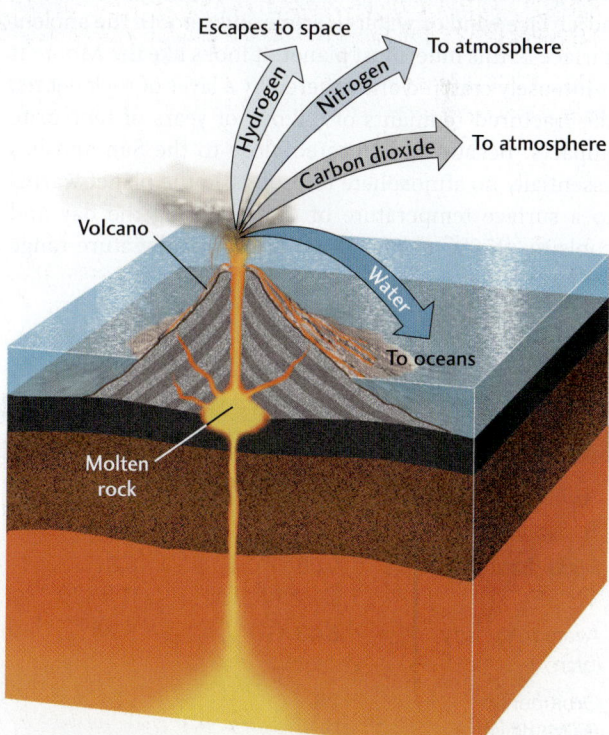

FIGURE 20.6 Early volcanic activity contributed enormous amounts of water vapor, carbon dioxide, and nitrogen to the atmosphere and oceans. Hydrogen, because it is lighter, escaped into space.

planet. Some of the air and water may also have come from volatile-rich bodies from the outer solar system, such as comets, that struck the planet after it had formed. Countless comets may have bombarded Earth early in its history, contributing water, carbon dioxide, and gases to the early oceans and atmosphere. The early atmosphere lacked the oxygen that makes up 21 percent of the atmosphere today. Oxygen did not enter the atmosphere until oxygen-producing organisms evolved, as we will see in Chapter 22.

Diversity of the Planets

By about 4.4 billion years ago, in less than 200 million years since its origin, Earth had become a fully differentiated planet. The core was still hot and mostly molten, but the mantle was fairly well solidified, and a primitive crust and continents had begun to develop. Oceans and atmosphere had formed, and the geologic processes that we observe today had been set in motion. But what about the other terrestrial planets? Did they experience a similar early history? Information transmitted from space probes indicates that the four terrestrial planets have all undergone gravitational differentiation into layered structures with an iron-nickel core, a silicate mantle, and an outer crust (**Table 20.1**).

Mercury has a thin atmosphere consisting mostly of helium. The atmospheric pressure at its surface is less than a trillionth of Earth's atmospheric pressure. There is no surface wind or water to erode and smooth the ancient surface of this innermost planet. It looks like the Moon: It is intensely cratered and covered by a layer of rock debris, the fractured remnants of billions of years of meteorite impacts. Because it is located close to the Sun and has essentially no atmosphere to protect it, the planet warms to a surface temperature of 470°C during the day and cools to −170°C at night—the largest temperature range for any planet.

Mercury's average density is nearly as great as Earth's, even though it is a much smaller planet. Accounting for differences in interior pressure (remember, higher pressures increase density), scientists have surmised that Mercury's iron-nickel core must make up about 70 percent of its mass, a record proportion for solar system planets (Earth's core is only one-third of its mass). Perhaps Mercury lost part of its silicate mantle in a giant impact. Alternatively, the Sun could have vaporized part of its mantle during an early phase of intense radiation. Scientists are still debating these hypotheses.

Venus developed into a planet with surface conditions surpassing most descriptions of hell. It is wrapped in a heavy, poisonous, incredibly hot (475°C) atmosphere composed mostly of carbon dioxide and clouds of corrosive sulfuric acid droplets. A human standing on its surface would be crushed by the atmospheric pressure, boiled by the heat, and eaten away by the sulfuric acid. At least 85 percent of Venus is covered by lava flows. The remaining surface is mostly mountainous—evidence that the planet has been tectonically active (**Figure 20.7**). Venus is close to Earth in mass and size, and its core seems to be about the same size as Earth's, with both liquid and solid portions. How it could develop into a planet so different from Earth is a question that intrigues planetary geologists.

Mars has undergone many of the same geologic processes that have shaped Earth (see Figure 20.7). The Red Planet is considerably smaller than Earth, with only about one-tenth of Earth's mass. However, the Martian core, like the cores of Earth and Venus, appears to have a radius of about half the planet's radius, and, like Earth's, it may have a liquid outer portion and a solid inner portion.

Mars has a thin atmosphere composed almost entirely of carbon dioxide. No liquid water is present on its surface today; the planet is too cold, and its atmosphere is too thin, so any water on its surface would either freeze or

TABLE 20.1 Characteristics of the Terrestrial Planets and Earth's Moon

	MERCURY	VENUS	EARTH	MARS	EARTH'S MOON
Radius (km)	2440	6052	6370	3388	1737
Mass (Earth = 1)	0.06 (3.3×10^{23} kg)	0.81 (4.9×10^{24} kg)	1.00 (6.0×10^{24} kg)	0.11 (6.4×10^{23} kg)	0.01 (7.2×10^{22} kg)
Average density (g/cm³)	5.43	5.24	5.52	3.94	3.34
Orbit period (Earth days)	88	224	365	687	27
Distance from Sun ($\times 10^6$ km)	57	108	148	228	—
Moons	0	0	1	2	0

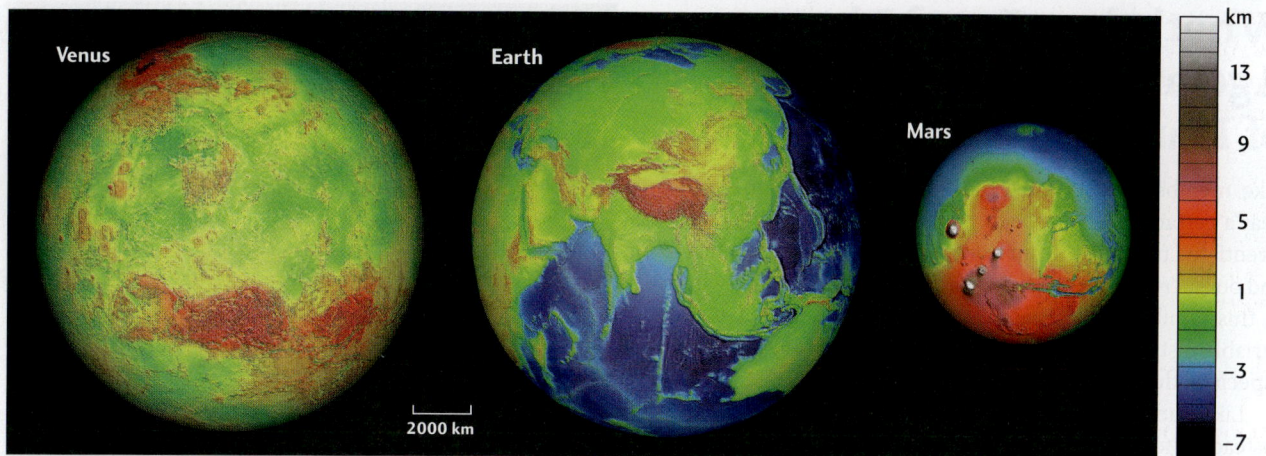

FIGURE 20.7 A comparison of the surfaces of Earth, Mars, and Venus, all at the same scale. The topography of Mars, which shows the greatest range, was measured in 1998 and 1999 by a laser altimeter aboard the orbiting *Mars Global Surveyor* spacecraft. That of Venus, which shows the smallest range, was measured from 1990 to 1993 by a radar altimeter aboard the orbiting *Magellan* spacecraft. Earth's topography, which is intermediate in range and dominated by continents and oceans, has been synthesized from altimeter measurements of the land surface, ship-based measurements of ocean depth, and gravity-field measurements of the seafloor surface from Earth-orbiting spacecraft. [Courtesy of Greg Neumann/MIT/GSFC/NASA.]

evaporate. Several lines of evidence, however, indicate that liquid water was abundant on the surface of Mars before 3.5 billion years ago, and that large amounts of water ice may be stored below the surface and in polar ice caps today. Life might have formed on the wet Mars of billions of years ago and could exist today as microorganisms below the surface.

Most of the surface of Mars is older than 3 billion years. On Earth, in contrast, most surfaces older than about 500 million years have been obliterated through the combined activities of the plate-tectonic and climate systems. Later in this chapter, we will compare surface processes on Earth and Mars in more detail.

Other than Earth itself, the *Moon* is the best-known body in the solar system because of its proximity to Earth and the manned and unmanned programs that have been designed to explore it. In bulk, its materials are lighter than Earth's, probably because the heavier matter of the giant impacting body remained embedded in Earth after the collision that formed it. The lunar core is therefore small, constituting only about 20 percent of the lunar mass.

The Moon has no atmosphere, and is mostly bone dry, having lost most of its water in the heat generated by the giant impact. There is some new evidence from spacecraft observations that water ice may be present in small amounts deep within sunless craters at the Moon's north and south poles. The heavily cratered lunar surface we see today is that of a very old, geologically dead body, dating back to a period early in the history of the solar system when crater-forming impacts were very frequent. Once topography is created on any planetary body, plate-tectonic and climate processes will work to "resurface" it, as they have on Venus and Mars. However, in the absence of these processes, the planet will remain pretty much the way it was just after its formation. Thus, the heavily cratered terrains of little-studied planetary bodies, such as Mercury, indicate that they lack both a convecting mantle and an atmosphere.

The giant gaseous outer planets—*Jupiter, Saturn, Uranus,* and *Neptune*—are likely to remain a puzzle for a long time. These huge gas balls are so chemically distinct and so large that their formation must have followed a course entirely different from that of the much smaller terrestrial planets. All four of the giant planets are thought to have rocky, silica-rich and iron-rich cores surrounded by thick shells of liquid hydrogen and helium. Inside Jupiter and Saturn, the pressures become so high that scientists believe the hydrogen turns into a metal.

Exactly what lies beyond the orbit of Neptune, the most distant giant planet, remains a mystery. Tiny *Pluto*, once regarded as the ninth planet, is a strange frozen mixture of gases, ice, and rock with an unusual orbit that sometimes brings it closer to the Sun than Neptune. Pluto, along with "2003 UB313" and two other bodies that share its attributes—tiny size, unusual orbit, rock-ice-gas composition—is now known as a **dwarf planet.** The dwarf planets lie within a belt of icy bodies that is the source region for the comets that periodically pass through the inner solar system. Other dwarf planet–sized objects are likely to be found as we explore the outer regions of the solar system. A spacecraft called *New Horizons* visited Pluto in 2015.

What's in a Face? The Age and Complexion of Planetary Surfaces

Like members of a family, the four terrestrial planets all bear a certain resemblance to one another. They are all differentiated planets, with an iron-nickel core, silicate mantle, and outer crust. But, as we have just seen, there are no twins in this family. Their different sizes and masses, and their variable distances from the Sun, make all four planets, and especially their surfaces, distinct.

Like human faces, the faces of planets reveal their ages. Instead of forming wrinkles as they get older, the surfaces of terrestrial planets are marked by craters. The surfaces of Mercury, Mars, and the Moon are heavily cratered and therefore obviously old. In contrast, Venus and Earth have very few craters because their surfaces are much younger. In this section, we will study planetary surfaces to learn about the tectonic and climate processes that have shaped them. Earth is excluded here because it is the subject of this textbook, and Mars will be mentioned only briefly because its surface is more thoroughly described in the following section.

FIGURE 20.8 The Moon has two types of terrain: the lunar highlands, with many craters, and the lunar lowlands, or maria, with few craters. The maria appear darker due to the presence of widespread basalts that flowed across their surfaces over 3 billion years ago. [Larry Keller, Lititz Pa./Getty Images.]

The Man in the Moon: A Planetary Time Scale

If you look at the face of the Moon through binoculars on a clear night, you will see two distinct types of terrain: rough areas that appear light-colored with lots of big craters, and smooth, dark areas, usually circular in shape, where craters are small or nearly absent (**Figure 20.8**). The light-colored regions are the mountainous *lunar highlands,* which cover about 80 percent of the surface. The dark regions are low-lying plains called *lunar maria,* from the Latin for "seas," because they looked like seas to early Earth-bound observers. It is the contrast between highlands and maria that forms the pattern we can see from Earth as the "Man in the Moon."

In preparation for the Apollo missions to the Moon, geologists such as Gene Shoemaker (**Figure 20.9**) developed

FIGURE 20.9 Astrogeologist Eugene Shoemaker leads an astronaut training trip on the rim of Meteor Crater, Arizona, in May 1967. (An aerial view of this crater is shown in Figure 1.6b.) Shoemaker and other geologists used their observations of craters to develop a relative time scale for dating lunar surfaces. [USGS.]

a relative time scale for the formation of lunar surfaces based on the following simple principles:

- Craters are absent on a new geologic surface; older surfaces have more craters than younger surfaces.
- Impacts by small bodies are more frequent than impacts by large bodies; thus, older surfaces have larger craters.
- More recent impact craters cross-cut or cover older craters.

By applying these principles, and by mapping the numbers and sizes of craters—a procedure known as *crater counting*—geologists showed that the lunar highlands are older than the maria. They interpreted the maria to be basins formed by the impacts of asteroids or comets that were subsequently flooded with basalts, which "repaved" the basins. They were able to assign different parts of the Moon's surface to geologic intervals analogous to those in the relative time scale worked out by nineteenth-century geologists for Earth.

In the pre-Apollo days, no one knew the absolute ages of either the maria or the highlands, but the smart money held that both were very old. The intense cratering evident in the highlands and the big impacts that formed the maria were consistent with the results of theoretical models of the early solar system. These models predicted a period called the **Heavy Bombardment,** during which the planets collided frequently with the residual materials that still cluttered the solar system after they had been assembled (**Figure 20.10**). According to the models, the numbers and sizes of impacting objects would have been greatest just after the planets formed and would have quickly decreased as the materials were swept up by the planets.

By applying the isotopic dating methods described in Chapter 22 to rock samples brought back by the Apollo astronauts, geologists were able to calibrate the absolute time scale for the Moon that they had developed by crater counting. Sure enough, the highlands turned out to be very ancient (4.4 billion to 4.0 billion years old) and the maria younger (4.0 billion to 3.2 billion years old). **Figure 20.11** plots these ages on the ribbon of geologic time.

The relatively young ages of the maria turned out to be a puzzle, however. The best computer simulations of the Heavy Bombardment indicated that it should have been over rather quickly, perhaps in a few hundred million years or even less. Why, then, did some of the biggest impacts observed on the Moon—those that formed the maria—occur so late in lunar history?

The simulations missed an important event. The rate at which large objects struck the Moon did decrease quickly, as the simulations predicted, but may have increased again in a period known as the *Late* Heavy Bombardment, which occurred between about 4.0 billion and 3.8 billion years ago (see Figure 20.10). The explanation of this event is controversial, and is based on hypothesized small changes in the orbits of Jupiter and Saturn about 4 billion years ago (caused by their gravitational interactions as they settled into their present orbits) that perturbed the orbits of the asteroids. The hypothesis predicts that some of the asteroids would have been thrown into the inner solar system, where they collided with the Moon and the terrestrial planets, including Earth. The Late Heavy Bombardment hypothesis may explain why so few rocks on Earth have ages greater than 3.9 billion years, thus marking the end of the Hadean eon and the beginning of the Archean eon (see Figure 20.11).

The time scale first developed for the Moon by crater counting has been extended to other planets by taking into account the differences in impact rates resulting from each planet's mass and position in the solar system.

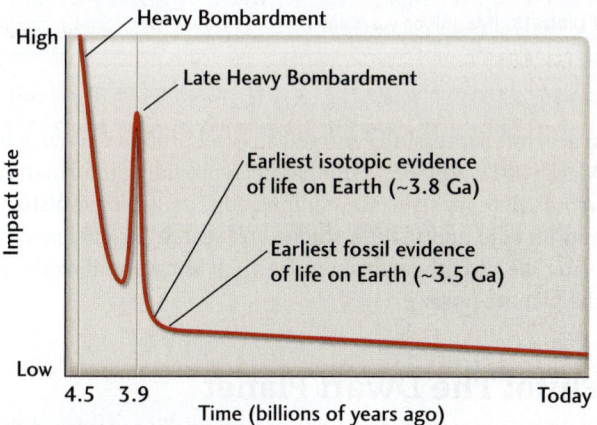

FIGURE 20.10 The number of planetary impacts has varied over the history of the solar system. After the planets formed, they continued to collide with the residual matter that still cluttered the solar system. These collisions tapered off over the first 500 million years of planetary development. However, there was another period of frequent impacts, known as the Late Heavy Bombardment, which peaked about 3.9 billion years ago. (Ga: billion years ago.)

Mercury: The Ancient Planet

The topography of Mercury is poorly understood. *Mariner 10* was the first and only spacecraft to visit Mercury when it flew by the planet in March 1974. It mapped less than half the planet, and we have little idea of what is on the other side.

Mariner 10 confirmed that Mercury has a geologically dormant, heavily cratered surface. It has the oldest surface

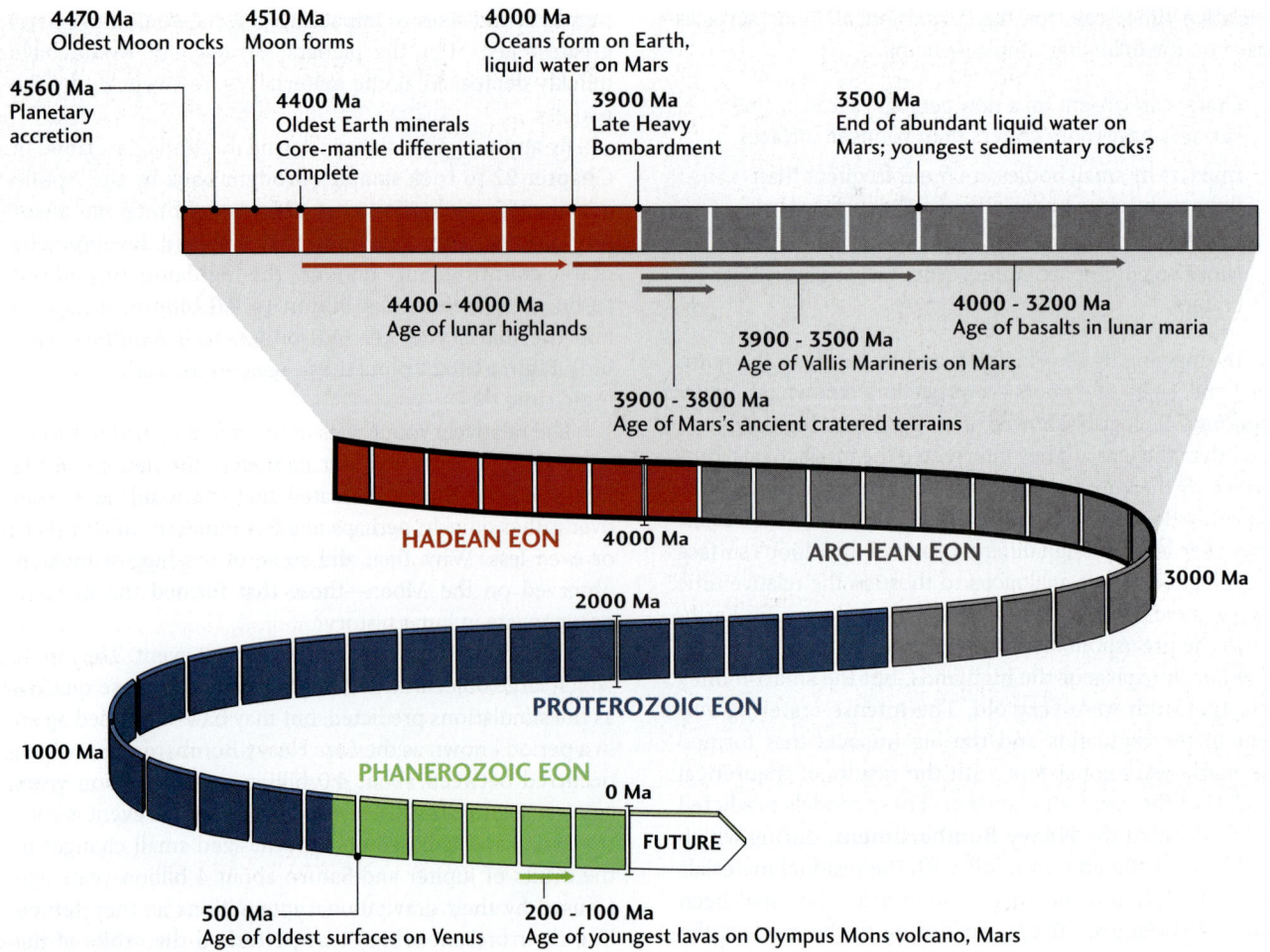

FIGURE 20.11 By calibrating the relative time scale developed by crater counting with the absolute ages of lunar rocks, geologists have constructed a geologic time scale for the terrestrial planets. (Ma: million years ago.)

of all the terrestrial planets (Figure 20.12). Between its large old craters lie younger plains, which are probably volcanic, like the lunar maria. The *Mariner 10* images show a difference in color between the craters and the plains, which supports this hypothesis. Unlike Earth and Venus, Mercury shows very few features that are clearly due to tectonic forces having reshaped its surface.

In many respects, the face of Mercury seems very similar to that of Earth's Moon. The two bodies are similar in size and mass, and most of their tectonic activity took place within the first billion years of their histories. There is one interesting difference, however. Mercury's face has several scars marked by scarps nearly 2 km high and up to 500 km long (Figure 20.13). Such features are common on Mercury, but rare on Mars and absent on the Moon. These cliffs appear to have resulted from horizontal compression of Mercury's brittle crust, which formed enormous thrust faults (see Chapter 8). Some geologists think they formed during the cooling of the planet's crust immediately after its formation.

On August 3, 2004, the first new mission to Mercury in 30 years was launched successfully. The *MESSENGER* spacecraft successfully entered an orbit around Mercury in March 2011. *MESSENGER* has been providing information about Mercury's surface composition, its geologic history, and its core and mantle, and it will search for evidence of water ice and other frozen gases, such as carbon dioxide, at the planet's poles.

Pluto: The Dwarf Planet

Pluto was discovered in 1930 and originally considered the ninth planet. However, in recent years it was re-designated as a dwarf planet because of its small size, icy-rock composition, and highly eccentric orbit. These attributes make it compare more favorably with a class of objects that also orbit the Sun, but are known as Kuiper Belt objects. Although Pluto is small, about one-sixth the mass of the Moon, it has five known moons—Charon, Styx, Nix, Kerberos, and Hydra.

On January 19, 2006, the *New Horizons* spacecraft was launched from Earth to fly by Pluto, study it, and then carry on to study other Kuiper Belt objects. On July 14, 2015, it

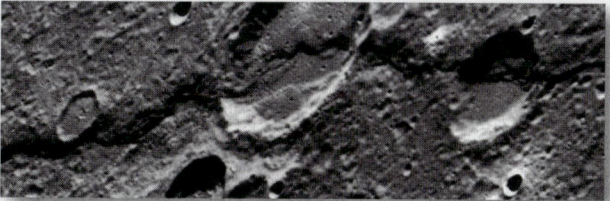

FIGURE 20.13 The prominent scarp that snakes across this image of Mercury is thought to have formed as the planet's crust was compressed, possibly as it cooled following its formation. Note that the scarp must be younger than the craters it offsets. [NASA/JPL/Northwestern University.]

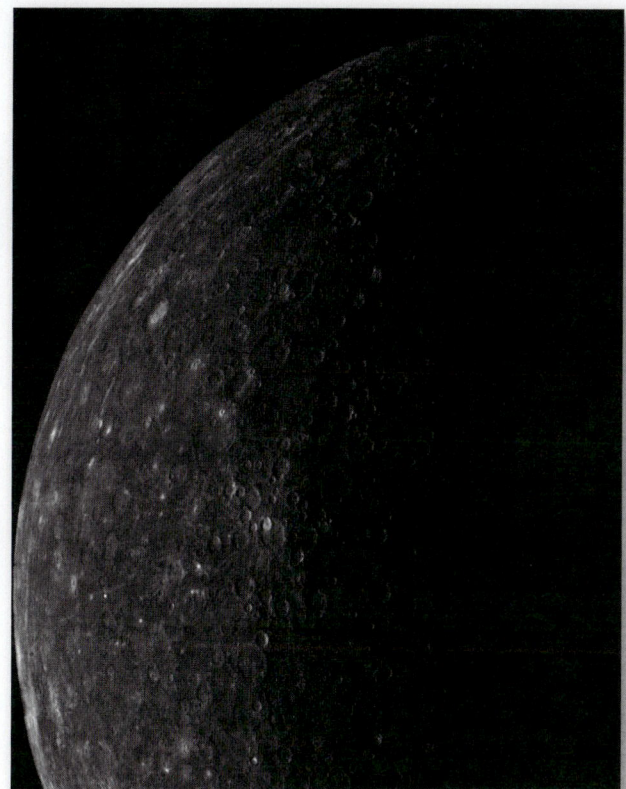

FIGURE 20.12 Mercury has a heavily cratered surface similar to that of Earth's Moon. [NASA/JPL/USGS.]

flew 12,500 km above the surface of Pluto, making it the first spacecraft to explore the dwarf planet. A number of significant discoveries were made, perhaps most interestingly that the surface of Pluto is "active," in the sense that its surface ices are mobile enough to erase the record of meteorite impacts that dimple other solid planets such as the Moon and Mercury. The surface of Pluto is marked by plains and mountains, such as the Sputnik Planitia region (**Figure 20.14**). The plains are composed of 98 percent nitrogen ice with traces of methane and carbon dioxide ices. In contrast, the mountains are made of water ice. Sputnik Planitia seems to have formed by upwelling of nitrogen liquid from Pluto's interior, which became frozen in a 1000 km-wide basin. The basin is divided into polygonal cells, which are regarded as evidence of former convection of the fluid precursor to the ice. Embedded within the nitrogen ice are broken blocks of what might have been a water ice crust (**Figure 20.15**).

Venus: The Volcanic Planet

Venus is our closest planetary neighbor, often brightly visible in the sky just before sunset. Yet in the early decades of space exploration, Venus frustrated scientists. The entire planet is shrouded in a dense fog of carbon dioxide, water vapor, and sulfuric acid, which prevents scientists from studying its surface with ordinary telescopes and cameras.

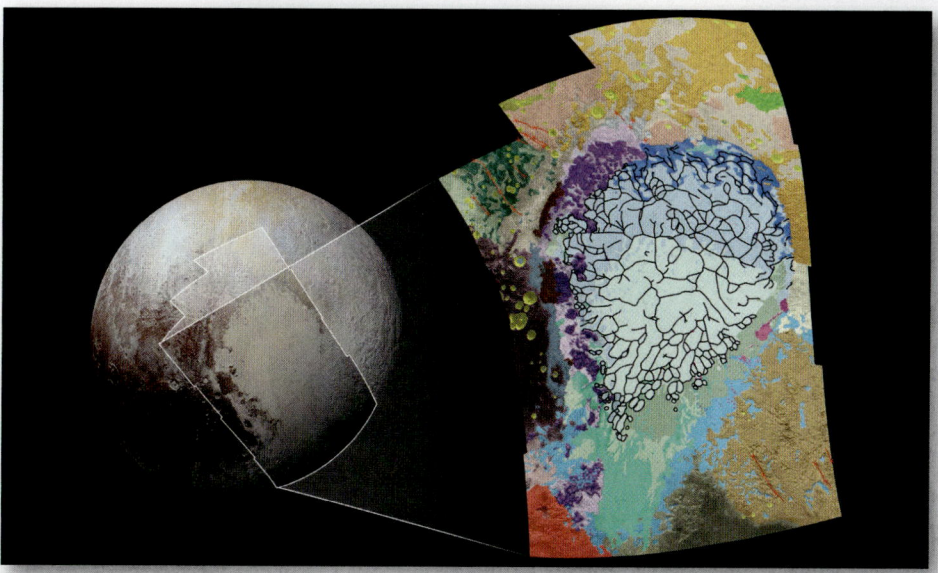

FIGURE 20.14 Pluto, a dwarf planet, has recently been explored by the *New Horizons* spacecraft, which discovered surprising complexity of its surface. A geologic map indicates that the plains are formed of nitrogen ice, such as at Sputnik Planitia, where polygonal cells (shown by black lines in map) may indicate frozen convection cells. [NASA/JHUAPL/SwRI.]

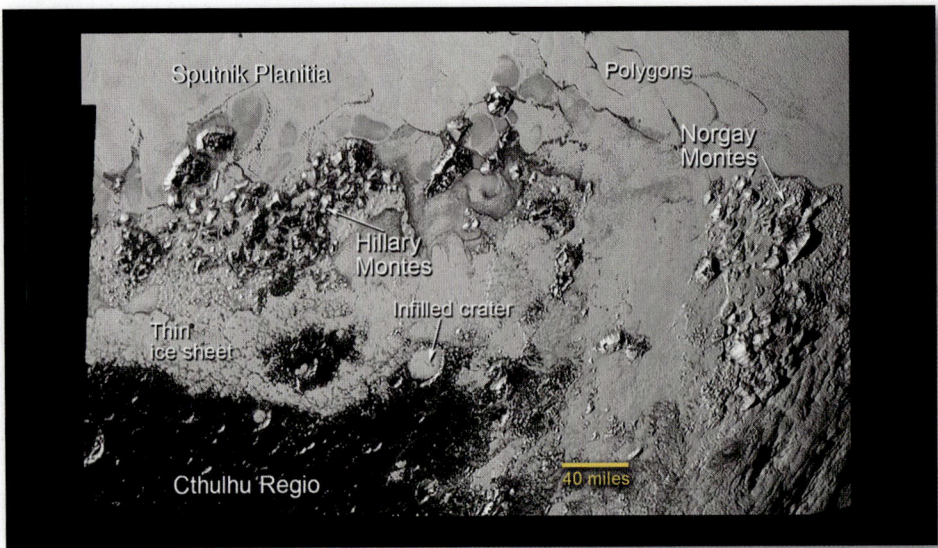

FIGURE 20.15 Close-up image mosaic of Sputnik Planitia reveals a flat surface, formed of nitrogen ice, separated from mountainous highlands, formed of water ice. These mountains may represent a former crust that was broken up and surrounded by liquid nitrogen that flowed up from deeper within Pluto's interior. The liquid nitrogen then froze to form Sputnik Planitia. [NASA/JHUAPL/SwRI.]

Although many spacecraft were sent to Venus, only a few were able to penetrate this acid fog, and the first ones that tried to land on its surface were crushed under the tremendous weight of its atmosphere.

It was not until August 10, 1990, after traveling 1.3 billion kilometers, that the *Magellan* spacecraft arrived at Venus and took the first high-resolution pictures of its surface (Figure 20.16). *Magellan* did this using *radar* (shorthand for *radio*

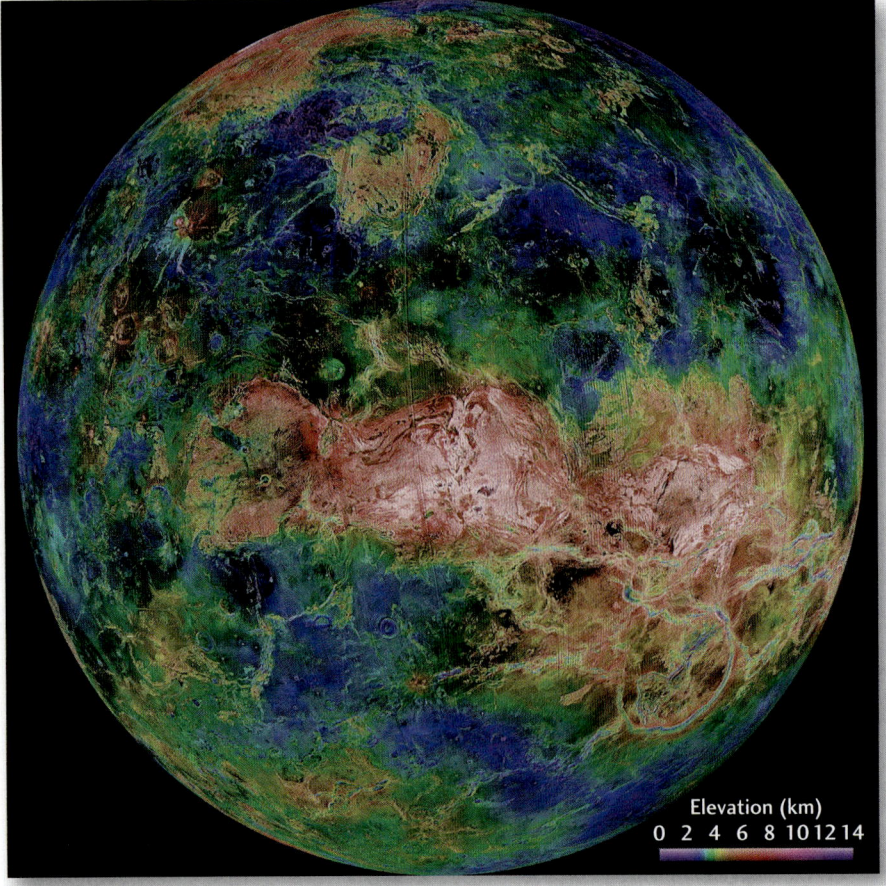

FIGURE 20.16 This topographic map of Venus is based on more than a decade of mapping, culminating in the 1990–1994 Magellan mission. Regional variations in elevation are illustrated by the highlands (tan colors), the uplands (green colors), and the lowlands (blue colors). Vast lava plains are found in the lowlands. [NASA/USGS.]

 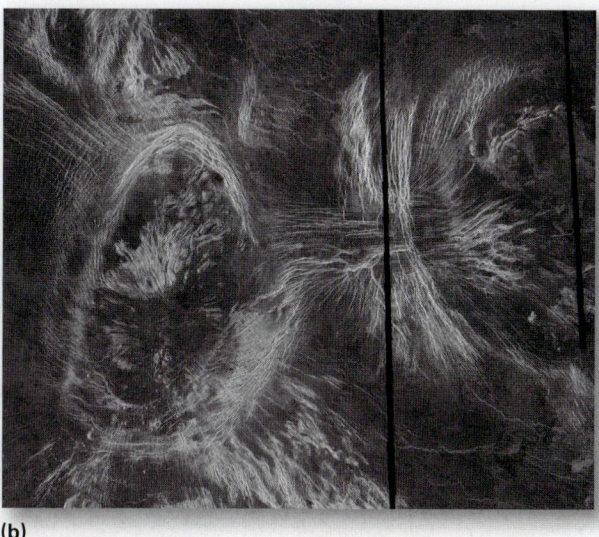

(a) (b)

FIGURE 20.17 Venus is a tectonically active planet with many surface features. (a) Maat Mons, a volcanic mountain that may be up to 3 km high and 500 km across. (b) Volcanic features called coronae are not observed on any other planet except Venus. The visible lines that define the coronae are fractures, faults, and folds produced when a large blob of hot lava collapsed like a fallen soufflé. Each corona is a few hundred kilometers across. [a: NASA/JPL; b: NASA/USGS.]

*d*etection *a*nd *r*anging) devices similar to the cameras that police officers use to enforce speed limits (they "see" at night, and through the fog, to clock your speed). Radar cameras form images by bouncing radio waves off stationary surfaces (like those of planets) or moving surfaces (like those of cars).

The images that *Magellan* returned to Earth show clearly that beneath its fog, Venus is a surprisingly diverse and tectonically active planet with mountains, plains, volcanoes, and rift valleys. The lowland plains of Venus—the blue regions in Figure 20.16—have far fewer craters than the Moon's youngest maria, indicating that they must be younger still. Estimates of their age range between 1600 million and 300 million years. Because there is no rain on Venus, there is very little erosion, and so the features we see today have been "locked in" for all that time. The relatively low number of craters suggests that many craters must have been covered by lava, and therefore that Venus must have been tectonically active relatively recently.

The young plains are dotted with hundreds of thousands of small volcanic domes 2 to 3 km across and perhaps 100 m or so high, which formed over places where Venus's crust got very hot. There are larger, isolated volcanoes as well, up to 3 km high and 500 km across, similar to the shield volcanoes of the Hawaiian Islands (**Figure 20.17a**). *Magellan* also observed unusual circular features called *coronae* that appear to result from blobs of hot lava that rose, created a large bulge or dome in the surface, and then sank, collapsing the dome and leaving a wide ring that looks like a fallen soufflé (Figure 20.17b).

Because Venus has so much evidence of widespread volcanism, it has been called the Volcanic Planet. Venus has a convecting mantle like Earth's, in which hot material rises and cooler material sinks (**Figure 20.18a**), but unlike Earth, it does not appear to have thick plates of rigid lithosphere. Instead, only a thin crust of frozen lava overlies the convecting mantle. As the vigorous convection currents push and stretch the surface, the crust breaks up into flakes or crumples like a rug, and blobs of hot magma bubble up to form large landmasses and volcanic deposits (Figure 20.18b). Scientists have called this process **flake tectonics.** When Earth was younger and hotter, it is possible that flakes, rather than plates, were the main expression of its tectonic activity.

Mars: The Red Planet

Of all the planets, Mars has a surface most similar to Earth's. Mars has features suggesting that liquid water once flowed across its surface, and liquid water may still exist in its deep subsurface. And where there is water, there may be living organisms. No other planet in the solar system has as much chance of harboring extraterrestrial life as Mars, though a mission to Europa, a moon of Jupiter, in 2024 will begin the search for habitable environments in the outer solar system.

The abundance of iron oxide minerals on the surface of Mars gives the Red Planet its name. Iron oxide minerals are common on Earth and tend to form where weathering of iron-bearing silicates occurs. We now know that many other minerals common on Earth, such as olivine and pyroxene, which form in basalt, are also present on Mars. But there are other relatively unusual minerals on Mars, such as sulfates, that record an earlier, wetter phase when liquid water may have been stable.

The topography of Mars shows a greater range of elevation than that of Earth or Venus (see Figure 21.7). Olympus Mons, at 25 km high, is a giant, recently active volcano—the tallest mountain in the solar system (**Figure 20.19a**).

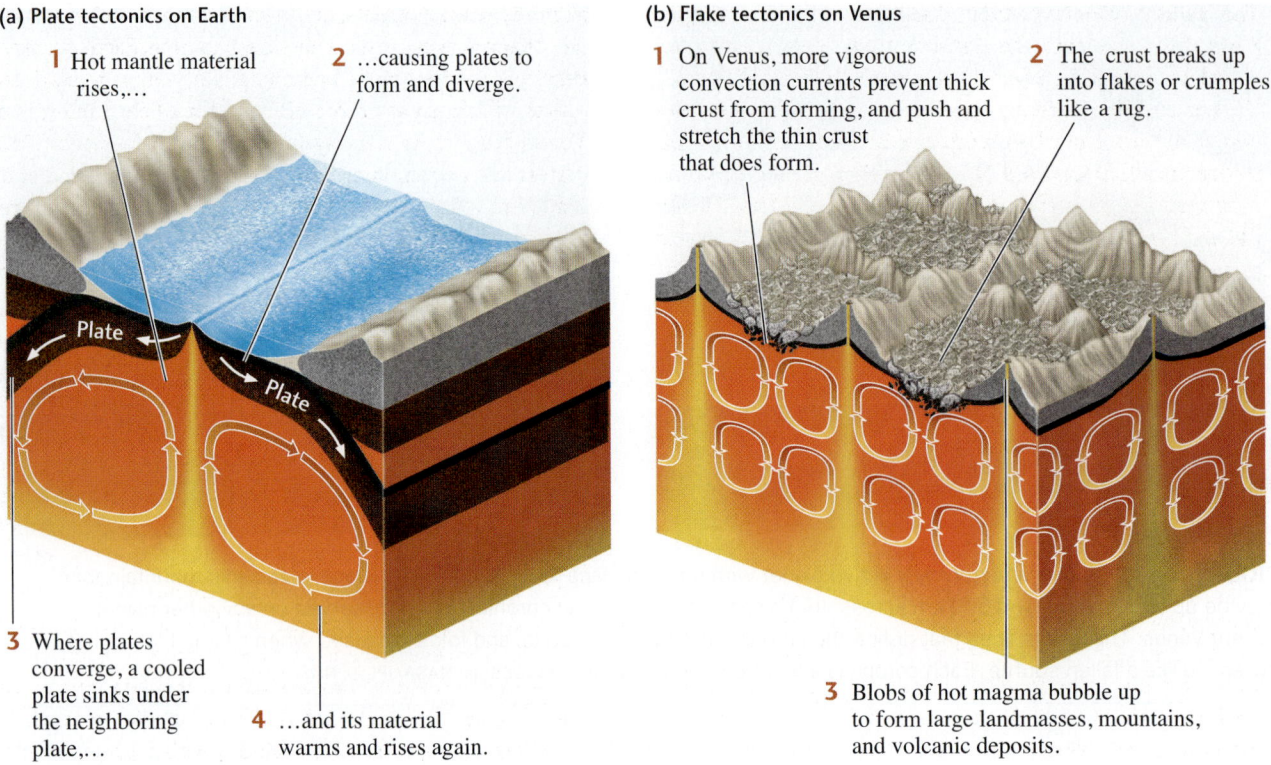

FIGURE 20.18 Flake tectonics on Venus (b) are very different from plate tectonics on Earth (a), but could be similar to tectonic processes on early Earth.

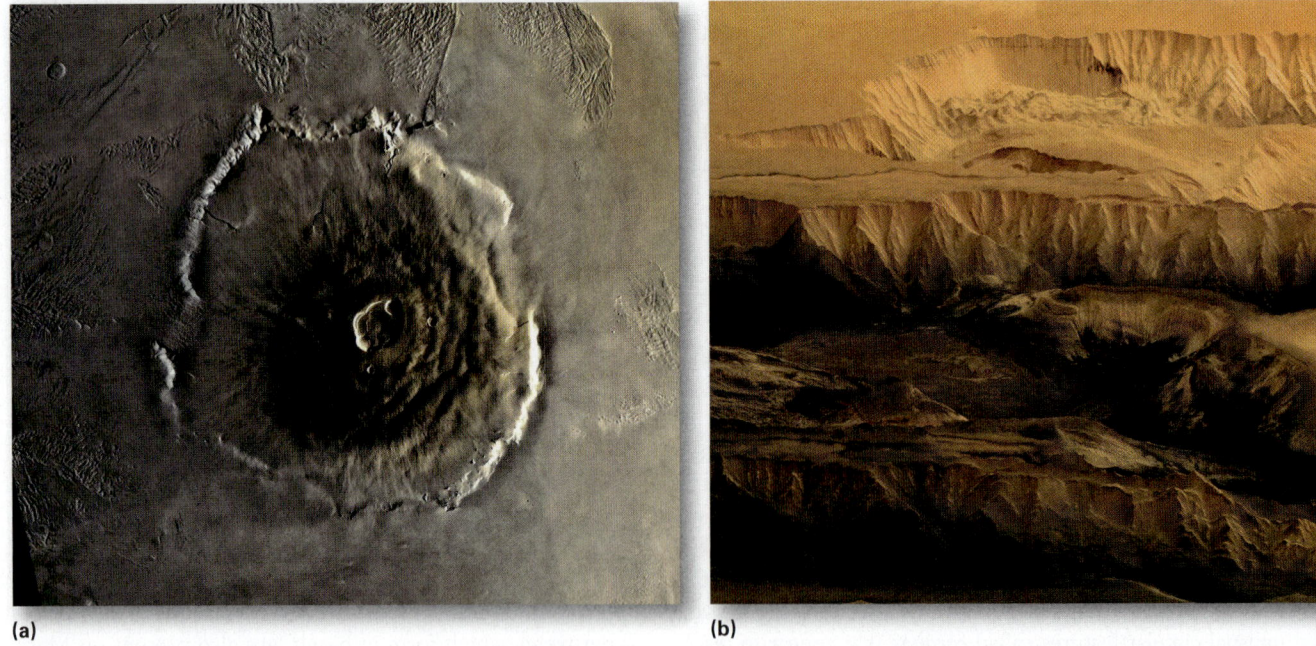

FIGURE 20.19 The topography of Mars shows a large range of elevation. (a) Olympus Mons is the tallest volcano in the solar system, with a summit almost 25 km above the surrounding plains. Encircling the volcano is an outward-facing scarp 550 km in diameter and several kilometers high. Beyond the scarp is a moat filled with lava, most likely derived from Olympus Mons. (b) Vallis Marineris is the longest (4000 km) and deepest (up to 10 km) canyon in the solar system. It is five times deeper than the Grand Canyon. In this image, the canyon is exposed as a series of fault-bounded basins whose sides have partly collapsed (as at upper left), leaving piles of rock debris. The walls of the canyon are 6 km high here. The layering of the canyon walls suggests deposition of sedimentary or volcanic rocks prior to faulting. [a: NASA/USGS; b: ESA/DLR/FU Berlin.]

The Vallis Marineris canyon, 4000 km long and averaging 8 km deep, stretches the distance from New York to Los Angeles and is five times deeper than the Grand Canyon (Figure 20.19b). Recently, geologists have discovered evidence of past glacial processes, when ice sheets similar to the ones that covered North America during the most recent ice age flowed across the surface of Mars. Finally, like the Moon, Mercury, and Venus, Mars has both heavily cratered ancient highlands and younger lowlands. However, unlike those of Mercury, Venus, and the Moon, the lowlands of Mars are created not only by lava flows, but also by sediments, sedimentary rocks, and accumulations of wind-blown dust.

The face of Mars may be sophisticated, but it has not always been easy to read, despite being visited and viewed more than any other planet except Earth. But, as we will see shortly, Mars's secrets are finally being revealed.

Earth: No Place Like Home

Every view of Earth underscores the unique beauty created by the overwhelming influences of plate tectonics, liquid water, and life. From its blue skies and oceans and its green vegetation to its rugged ice-covered mountains and moving continents, there truly is no place like home. Earth's remarkable appearance is maintained by the delicate balance of conditions necessary to support and sustain life.

The features that define the face of our planet are discussed throughout this textbook, but one process that is appropriate to review here is cratering. The impacts of meteorites and asteroids are preserved in the geologic record of all the terrestrial planets, but in contrast to the other planets, whose surfaces are essentially frozen in time, Earth preserves very few vestiges of its beginning. Recycling by plate tectonics, which is even more efficient than flake tectonics on Venus, has almost completely resurfaced our planet. Those craters that do remain are much younger than the end of the Late Heavy Bombardment, and they are preserved entirely on continents, which resist subduction (**Figure 20.20**).

Nevertheless, Earth accumulates a lot of junk from space. At present, some 40,000 tons of extraterrestrial material fall on Earth each year, mostly as dust and unnoticed small objects. Although the rate of impacts is now orders of magnitude lower than it was during the Heavy Bombardment, a large chunk of matter 1 to 2 km in size still collides with Earth every few million years or so. Although such collisions have become rare, telescopes are being assigned to search space and warn us in advance of sizable bodies that might slam into Earth. NASA astronomers recently predicted "with non-negligible probability" (1 chance in 300) that an asteroid 1 km in diameter will collide with Earth in March 2880. Such an event would threaten human civilization.

We already know that impacts with large objects can greatly upset the conditions that support life on Earth. As we will see in Chapter 23, a collision with a 10-km asteroid 65 million years ago caused the extinction of 75 percent of Earth's species, including all the dinosaurs. This event may have made it possible for mammals to become the dominant species and paved the way for humankind's emergence. **Table 20.2** describes the potential effects of impacts by objects of various sizes on our planet and its life-forms.

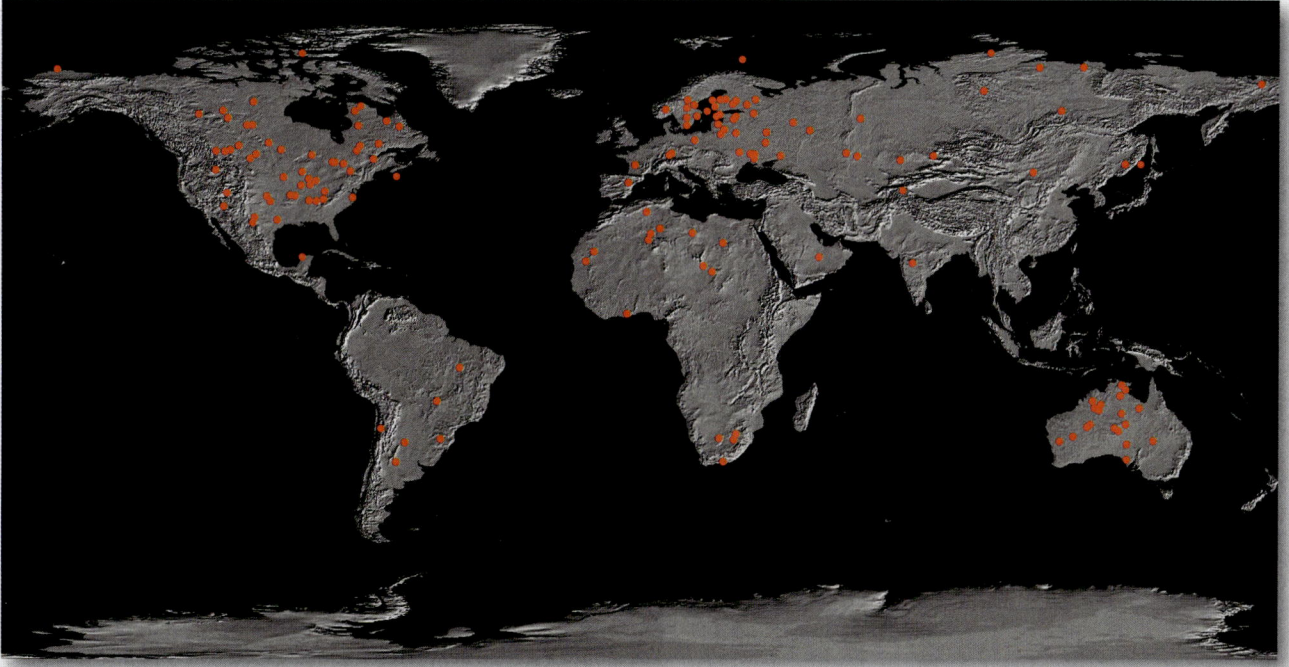

FIGURE 20.20 Impact craters formed by meteorites and asteroids are rare on Earth compared with the other terrestrial planets. Recycling of Earth's crust by plate tectonics has erased almost all evidence of impacts. Those craters that remain (red dots) are preserved only on the continents. [NASA/JPL/ASU.]

TABLE 20.2 Impacts by Asteroids and Meteorites and Their Effects on Life on Earth

	EXAMPLE OR SIZE EQUIVALENT	MOST RECENT	PLANETARY EFFECTS	EFFECTS ON LIFE
Supercolossal: radius $(R) > 2000$ km	Moon-forming event	4.51×10^9 years ago	Melts planet	Drives off volatiles; wipes out life on Earth
Colossal: $R > 700$ km	Pluto	More than 4.3×10^9 years ago	Melts crust	Wipes out life on Earth
Huge: $R > 200$ km	4 Vesta (large asteroid)	About 4.0×10^9 years ago	Vaporizes oceans	Life may survive below surface
Extra large: $R > 70$ km	Chiron (largest active comet)	3.8×10^9 years ago	Vaporizes upper 100 m of oceans	May wipe out photosynthesis
Large: $R > 30$ km	Comet Hale-Bopp	About 2×10^9 years ago	Heats atmosphere and surface to about 1000 K	Burns continents
Medium: $R > 10$ km	Cretaceous-Tertiary impactor; 433 Eros (largest near-Earth asteroid)	65×10^6 years ago	Causes fires, dust, darkness; chemical changes in ocean and atmosphere; large temperature swings	Cretaceous-Tertiary impact caused extinction of 75 percent of species and all dinosaurs
Small: $R > 1$ km	About size of near-Earth asteroids	About 300,000 years ago	Causes global dusty atmosphere for months	Interrupts photosynthesis; individuals die but few species extinct; threatens civilization
Very small: $R > 100$ m	Tunguska event (Siberia)	1908	Knocked over trees of kilometers away; caused minor hemispheric effects; dusty atmosphere	Newspaper headlines; romantic sunsets; increased birth rate

Data from J. D. Lissauer, *Nature* 402 (1998): C11–C14.

Mars Rocks!

We live in the golden age of Mars exploration. At the time of this writing, one robotic rover and one robotic lander are operating on the surface of the planet, and three orbiters are circling it. These five spacecraft are returning a seemingly endless stream of new data that are leading to significant new discoveries. Our understanding of the history of past surface environments on Mars is changing dramatically. And the fun won't end soon: Both NASA and the European Space Agency have promised to deliver an additional rover in the next few years, and plans are being made to collect and return rock samples from Mars to Earth. All the scientists involved in these missions are grateful to live in a time of such adventure.

Our understanding of surface processes on Mars was dramatically improved when two golf-cart-sized robots landed on Mars in January 2004. The Mars Exploration Rovers, named *Spirit* and *Opportunity*, began their 300-million-kilometer journey from Cape Canaveral, Florida, to the Red Planet in June 2003, accompanied by the *Mars Express*, an orbiter equipped with geologic remote sensing tools. These missions succeeded beyond anyone's expectations, making 2004 and 2005 two of the greatest years in the history of space exploration. Another new orbiter, *Mars Reconnaissance Orbiter*, which started its mission in 2006, has collected a vast set of observations that show evidence of aqueous processes over broad regions of the planet. Its camera is taking stunning pictures of the surface of Mars at unprecedented resolution (25 cm/pixel). The *Phoenix* lander conducted operations in the polar region of Mars from June to November 2008 and confirmed the presence of water ice just a few centimeters below the dusty surface. In August 2012, the Mars Science Laboratory *Curiosity* rover made its spectacular landing in Gale Crater on Mars (a video of the "Seven Minutes of Terror" can be found at http://www.jpl.nasa.gov/video/index.php?id=1090). At the time of writing, it is currently in its extended mission, driving up the base of the ~5-kilometer-high Mt. Sharp in the center of the crater that contains a rich record of the early environmental history of Mars when the planet may have been habitable. (One of the authors of this textbook, John Grotzinger, was the chief scientist for the *Curiosity* rover team during the primary mission.)

Missions to Mars: Flybys, Orbiters, Landers, and Rovers

Earlier missions to Mars helped lay the groundwork for the success of the current missions. All spacecraft sent to Mars since the early 1960s have worked in one of four ways. First, the pioneering Mars exploration spacecraft, such as *Mariner 4*, flew by Mars while quickly acquiring all the data they could before disappearing into deep space.

The second, and most common, mode of operation is to orbit Mars in the same way satellites orbit Earth. *Mariner 9*, launched in May 1971, was the first spacecraft to orbit another planet. Since that time, eight other orbiters have helped map the surface of Mars. *Mars Odyssey*, *Mars Express*, and *Mars Reconnaissance Orbiter* are still active today. After a very successful mission spanning over 10 years, *Mars Global Surveyor* ceased operations in 2006.

The third method of observing Mars involves landing a spacecraft on the Martian surface. The Viking mission deployed two spacecraft, each of which consisted of an orbiter and a lander. The *Viking 1* lander touched down on the surface of Mars on July 20, 1976, and became the first spacecraft to land on another planet and transmit useful data back to Earth. The Viking mission gave us our first look at the surface of another planet from the ground. It also provided our first chemical analyses of Martian rocks and performed the first life-detection experiments.

The fourth method of exploring Mars involves the use of a rover: a robotic vehicle that can move about on the surface of the planet. As exciting as the Viking mission was, it was two decades until another spacecraft landed safely on the surface of Mars. This time it was *Pathfinder*, which arrived on July 4, 1997. However, the *Pathfinder* lander also included a shoebox-sized rover—called *Sojourner*—that was able to ramble around on the surface, analyzing rocks and soils, within a few meters of the *Pathfinder* lander. *Sojourner* was the first mobile vehicle to operate successfully on another planet and became the prototype for the much larger and more capable Mars Exploration Rovers that landed in 2004.

Early Missions: Mariner (1965–1971) and Viking (1976–1980) The Mariner and Viking missions returned the first detailed images of Mars to Earth. We saw a cratered Moon-like terrain over part of its surface. In other areas, we saw spectacular features, including enormous volcanoes and canyons, vast dune fields, ice caps at both poles,

and the Martian moons Phobos and Deimos. Early images also confirmed global dust storms that had previously been observed from Earth. Orbiting spacecraft continue to monitor these dust storms (**Figure 20.21**).

In addition, extensive networks of stream channels were discovered, providing the first evidence that liquids—possibly water—may once have flowed across the surface of Mars (**Figure 20.22**). Collectively, these data also revealed something that had not been appreciated before: The planet is divisible into two main regions, northern low plains and southern cratered highlands.

The two *Viking* landers provided high-resolution views of the Martian terrain. Both landing sites were strewn with

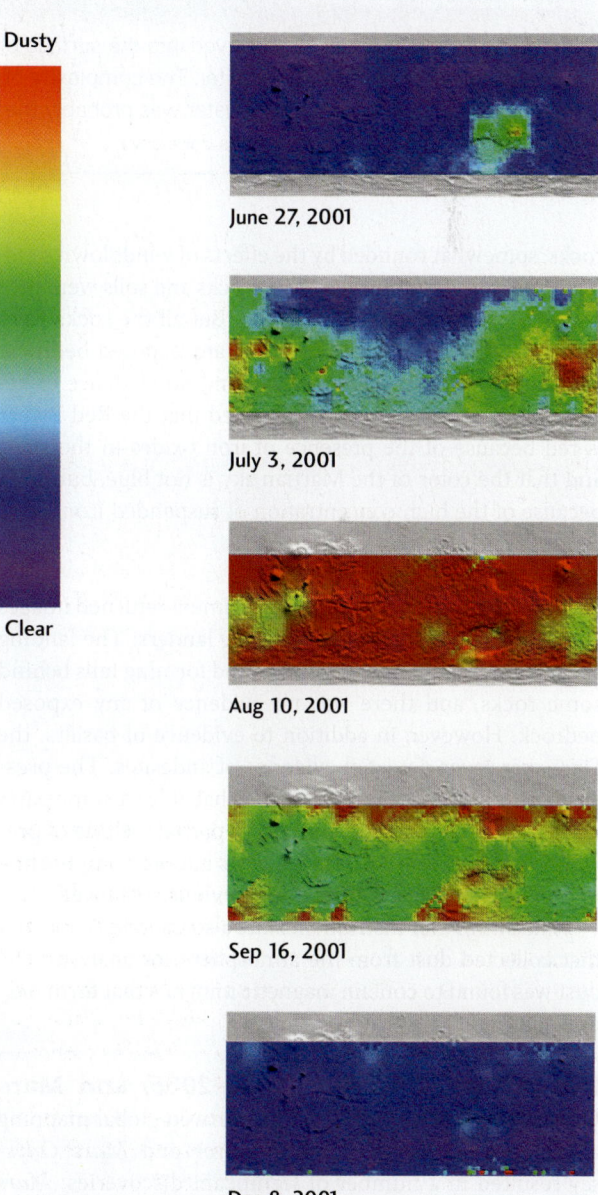

FIGURE 20.21 Global dust storms occur on Mars. The storms begin locally and gradually expand to envelop the entire planet, as seen in these images. [NASA/JPL/ASU]

FIGURE 20.22 Channel networks carved into the surface of Mars were revealed by the *Viking* orbiter. The complexity of these channels suggests that liquid water was probably the main force of erosion. [NASA/Washington University.]

deposits of loose sediment (**Figure 20.23**). The channels carved into bedrock observed by the *Viking* orbiters suggested flowing water; however, the presence of meandering stream deposits (see Chapter 18) is even stronger evidence for flowing water on the surface of Mars. But it was not until 2004 that the Mars Exploration Rovers first confirmed the presence of minerals that *require* liquid water to have been present.

Mars Global Surveyor and *Mars Odyssey* also showed that permafrost (soil rich in water ice; see Chapter 16) underlies the Martian soil from the mid-latitudes all the way to the poles. Widespread glaciers were also shown to have been present in the relatively recent past, suggesting that Mars—like Earth—may have experienced ice ages driven by changes in global climate. Finally, *Mars Global Surveyor* discovered rare patches of hematite (Fe_2O_3)—a mineral that often forms in water on Earth—scattered

rocks, somewhat rounded by the effects of windblown sand. Chemical sensors showed that the rocks and soils were predominantly basaltic in composition. But all the rocks were loose, and there was no evidence of any exposed bedrock. An onboard biology experiment found no evidence of life at either site. These missions revealed that the Red Planet is red because of the presence of iron oxides in the soils, and that the color of the Martian sky is not blue, but pink, because of the high concentration of suspended iron oxide dust particles.

Pathfinder (1997) *Pathfinder*'s camera returned images very similar to those from the *Viking* landers: The landing sites were rocky, with windblown sand forming tails behind some rocks, and there was no evidence of any exposed bedrock. However, in addition to evidence of basalts, the *Sojourner* rover detected evidence of andesites. The presence of andesites on Mars indicates that at least some parts of the Martian crust were formed by partial melting of previously formed basalts. This suggests a more complex history of crustal development than previously believed.

The *Pathfinder* instrument suite also included a magnet that collected dust from the atmosphere for analysis. The dust was found to contain magnetic minerals that form only in environments low in oxygen.

Mars Global Surveyor (1996–2006) and Mars Odyssey (2001–) The vastly improved global mapping capabilities of *Mars Global Surveyor* and *Mars Odyssey* resulted in a number of significant discoveries. *Mars Global Surveyor* carried a laser-based altimeter that surveyed Martian topography with unprecedented resolution. The new images provided the strongest evidence yet for liquid water, this time expressed as meandering stream

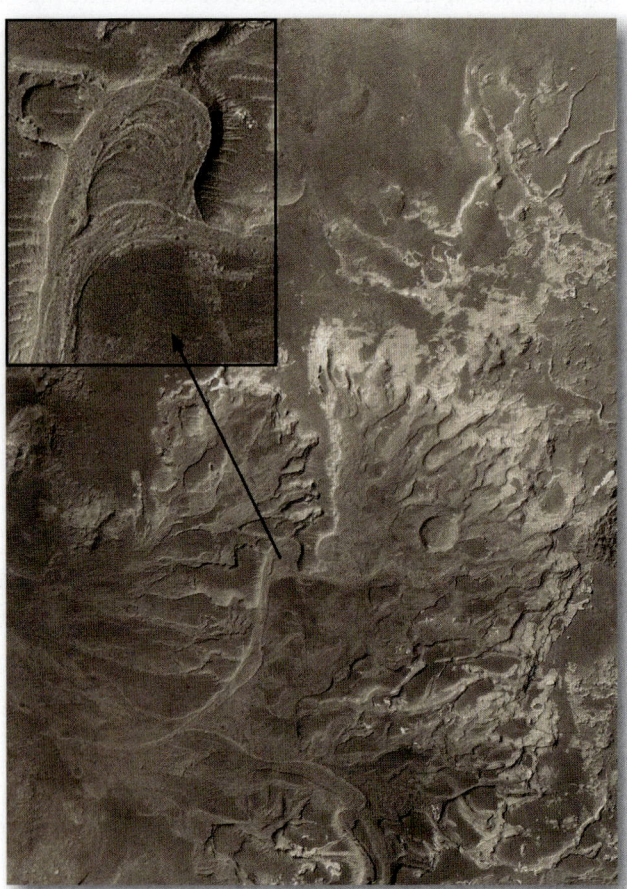

FIGURE 20.23 This image acquired by *Mars Global Surveyor* shows clear evidence of meandering patterns within sediments deposited inside Eberswalde Crater. Liquid water appears to have flowed across the Martian surface and entered the crater, where it deposited sediments in meandering channels similar to those seen in the Mississippi River on Earth today (see Chapter 18). [NASA/JPL/MSSS.]

over the surface of Mars. As we will see, this discovery contributed to the success of the Mars Exploration Rovers mission.

Mars Exploration Rovers (MER): Spirit and Opportunity

The Mars Exploration Rovers—*Spirit* and *Opportunity* (**Figure 20.24**)—were the first spacecraft sent to Mars that function almost as well as a human geologist would. Unlike orbiters, which look from afar, and landers, which cannot move from their landing site, *Spirit* and *Opportunity* can move around from rock to rock, picking and choosing which rocks to study in more detail. And when the right rock is found, the rover can look at it with a hand lens—just as geologists do here on Earth in the classroom and in the field. But unlike geologists on Earth, these rovers carry a mobile laboratory, so that the rocks can be analyzed on the spot without having to pay the enormous costs of flying them back to Earth. Because of this remarkable capability, *Spirit* and *Opportunity* have been dubbed the first *robotic geologists* on Mars.

The Mars Exploration Rovers were designed to survive 3 months under the hostile Martian surface conditions and to drive no farther than 300 m. They have since traveled more than 50 km in total across the Martian surface and *Opportunity* only just stopped operating at the time of this writing in 2018 (*Spirit* stopped operating in 2010). They have had to survive nighttime temperatures below −90°C, dust devils that could have tipped them over, global dust storms that diminished their solar power, and drives along rocky slopes of almost 30° and through piles of treacherous windblown dust. Despite all these obstacles, the rovers have discovered a treasure trove of geologic wonders.

Mars Science Laboratory (MSL): Curiosity

Curiosity launched in 2011 and landed in Gale Crater on Mars in August 2012. The overall scientific goals of the mission are to assess whether the field site where the rover landed ever had environmental conditions favorable to microbial life. *Curiosity* is similar to *Spirit* and *Opportunity*, but is about twice as long (3 meters or 10 feet) and five times as heavy—it literally weighs a ton (**Figure 20.25**). It also has the most sophisticated suite of instruments ever sent to another planet. While the MER rovers are solar powered, *Curiosity* is powered by a radioisotope power system, which gets its energy from the natural decay of plutonium.

Gale Crater has a ~5-kilometer-high mountain in its center—Mt. Sharp—which is made up of sedimentary strata with a diverse suite of hydrated minerals, which indicate that these strata at least in part formed the presence of water. *Curiosity* spent much of the first year of its primary

FIGURE 20.25 The Mars Science Laboratory *Curiosity* rover took this self-portrait (composed of dozens of exposures) with its Mars Hand Lens Imager (MAHLI) during the 177th Martian day, or sol, of *Curiosity*'s work on Mars. The lower left quadrant of the image shows gray powder and two holes where *Curiosity* used its drill on the rock target "John Klein." [NASA/JPL-Caltech/MSSS.]

FIGURE 20.24 *Spirit* (*left*), one of the Mars Exploration Rovers, is about the size of a golf cart. *Spirit* is standing next to a twin of *Sojourner*, a rover that was sent to Mars in 1997. Mars Science Laboratory (*right*) is about the size of a small car and was sent to Mars in 2011. [NASA/JPL-Caltech.]

mission in an area called Yellowknife Bay, where it found evidence of a potentially habitable environment, preserved in rocks over 3 billion years old. Streams once flowed from the crater rim toward the base of Mt. Sharp, where water pooled to form a lake that had low salinity and neutral pH—both favorable for life. *Curiosity* is currently investigating materials at the base of Mt. Sharp, where it will continue the search for additional habitable environments that may have characterized this part of Mars over 3 billion years ago.

What's Under the Hood? *Spirit* and *Opportunity* both come equipped with six-wheel drive, a color stereo camera with human vision, front-and-back hazard avoidance cameras, a magnifying "hand lens" for close-up inspection of rocks and soils, and instruments to detect the chemical and mineral composition of rocks and soils. A sample of *Curiosity*'s instruments include HD-resolution color cameras with video capability, detectors that measure radiation that might be harmful to humans, and a weather station (wind direction and velocity, atmospheric pressure, humidity, etc.). Buried inside the rover, *Curiosity* carries two laboratory instruments that provide information on the mineral composition of drilled samples (**Figure 20.26**), as well as their elemental and isotopic composition, and the presence of any organic compounds. The MER rovers are powered by solar energy, while MSL is powered by a nuclear power source, and all are controlled by scientists on Earth, who send daily command sequences to each rover via radio signals. Because it takes 10 minutes for these signals to travel between Earth and Mars, the rovers have some self-controlled navigation and hazard avoidance capabilities. However, almost every other decision is made by a team of humans back on Earth. This arrangement ensures that the rovers "think" as geologists do. Onboard computers receive the command sequences from Earth that control each rover activity, including driving; taking pictures of the terrain, rocks, and soils; analyzing rocks and minerals; and studying the atmosphere and moons of Mars.

Rover Landing Sites The Mars Exploration Rovers mission was motivated by the search for evidence of liquid water on Mars. The rovers were built with this goal in mind and sent to two locations where data from *Mars Global Surveyor* and *Mars Odyssey* suggested that the chances of finding geologic evidence for water would be high. (Some of the best places, however, were eliminated from consideration because of the extreme risk of landing in a rocky terrain; see the Practicing Geology Exercise at the end of the chapter.) Two of several hundred possible sites were chosen, both near the Martian equator but on opposite sides of the planet. The equatorial positions provide the rovers' solar panels with maximum energy throughout the year. By the time that *Curiosity* landed 10 years later, engineers at the Jet Propulsion Laboratory in Pasadena, CA, had learned how to develop a landing system that could deliver the rover to a very specific site with very high science value. Four final sites were selected by the science team, and all four were accessible by the landing system. The finalist—Gale Crater—was chosen because it had the greatest diversity of science targets. This turned out to be critically important and the first target explored—Yellowknife Bay—was successful in achieving the mission goal of a habitable ancient environment. Though *Curiosity* is extremely capable, the landing system was also critically important in helping to enable this discovery.

Spirit was sent to Gusev Crater, a large crater about 160 km in diameter that is thought to have once filled up with water to form a large lake (**Figure 20.27a**). *Opportunity* was sent to Meridiani Planum ("Plains of Meridiani"), where hematite had been detected by *Mars Global Surveyor* (Figure 20.27b). After landing, *Spirit* trekked across a volcanic plain, ascended the Columbia Hills, and crawled down the other side to arrive at an outcrop whose distinctive shape won it the name of Home Plate. After this long and difficult trek, one of *Spirit*'s left front wheels locked up. But, by turning around and driving backward so it could drag rather than push the broken wheel, *Spirit* finally made it to a part of Home Plate where it made an important discovery: mineral deposits made up of more than 90 percent silica. These deposits indicate that heated waters that once flowed at or near the surface of Mars carried high concentrations of dissolved silica, which precipitated to form hard crusts, similar to what occurs in the hot springs at Yellowstone National Park on Earth today—a place where microorganisms are known to thrive (see the chapter opening photo in Chapter 22). Thus, *Spirit*'s discovery of high-silica rocks

FIGURE 20.26 *Curiosity* drilled into this rock target, "Cumberland," during the 279th Martian day, or sol, of the rover's work on Mars and collected a powdered sample of material from the rock's interior. *Curiosity* used the Mars Hand Lens Imager (MAHLI) camera on the rover's arm to capture this view of the hole in Cumberland on the same sol that the hole was drilled. The diameter of the hole is about 0.6 inch (1.6 cm). The depth of the hole is about 2.6 inches (6.6 cm). [NASA/JPL-Caltech/MSSS.]

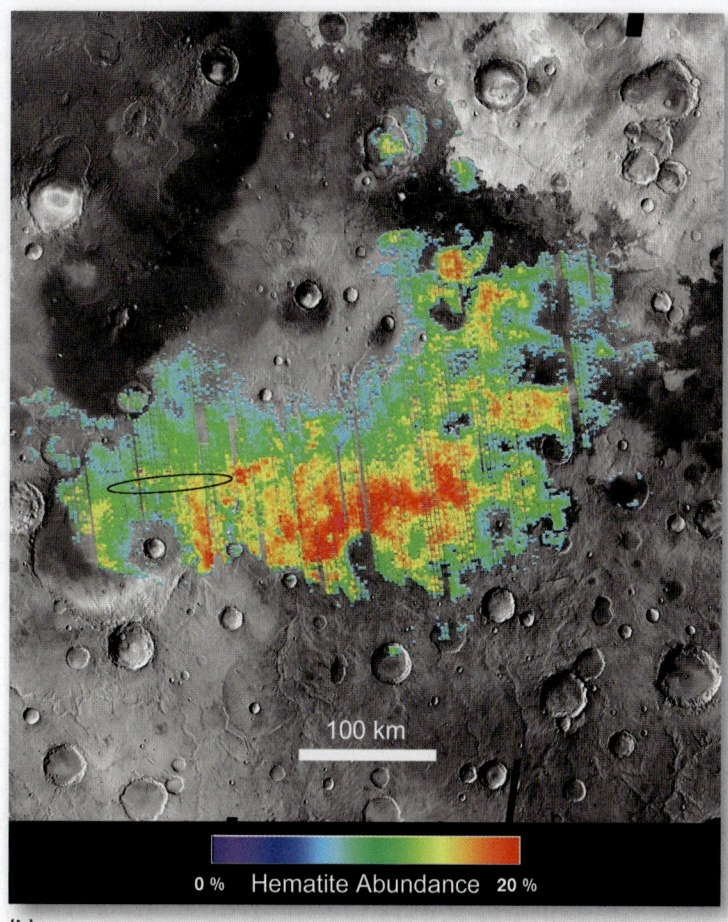

FIGURE 20.27 Mars Exploration Rover landing sites. (a) *Spirit* explored Gusev Crater, about 160 km in diameter, which is thought to have been filled with water, forming an ancient lake. A channel that might have supplied water to the crater is visible at the lower right. (b) *Opportunity* was sent to an area of Meridiani Planum where hematite—a mineral that often forms in water on Earth—is abundant. The image shows concentrations of hematite; the ellipse outlines the permissible landing area. [a: NASA/JPL/ASU/MSSS; b: NASA/ASU.]

suggests the potential for a once-habitable environment that could be confirmed by a future mission with a set of instruments similar to *Curiosity*'s.

Opportunity landed in Eagle Crater (a small crater about 20 m in diameter), where it spent 60 days studying the first sedimentary rocks ever found on another planet and gathering evidence that they must have formed in water. *Opportunity* then moved on to another, larger crater (Endurance Crater, about 180 m in diameter), where it spent the next 6 months putting those sedimentary rocks into a broader context of environmental evolution. *Opportunity* then traveled 5 km to a much larger crater (Victoria Crater, about 1 km in diameter), where it has explored even more expansive outcrops of sedimentary rock. *Opportunity* discovered an ancient sandy desert where shallow pools of water once filled depressions between the sand dunes. These pools of water are thought to have been very acidic and also extremely saline. Microorganisms can survive in extremely acidic waters, as we will see in Chapter 23; however, if salinity becomes too high, the availability of water to the microorganisms becomes limited, and they cannot survive. (In a similar way, but substituting sugar for salt, that is why honey does not spoil, even without refrigeration or addition of preservatives.) Thus, *Opportunity* has also discovered evidence of a potentially habitable environment, albeit one that would have required microorganisms to live in extreme conditions. *Opportunity* then drove over 10 kilometers to arrive at Endeavor Crater, where it discovered very ancient basaltic rocks that were altered to form clay-bearing deposits that indicate more neutral pH. This more ancient setting would have been more favorable for microorganisms had they ever originated on Mars.

Curiosity landed at the foot of Mt. Sharp—Gale's central mountain—near the end of an ancient alluvial fan that formed by sediments transported by streams from the crater rim. After landing it then drove ~600 meters to the east where it discovered sedimentary rocks that preserve evidence of ancient lake and stream environments characterized by low salinity and neutral pH. At the time of this writing *Curiosity* has driven over 20 km up into the lower reaches of the mountain, where more sedimentary rocks were discovered, deposited in a long-lived ancient lake, and that contain clay, sulfate, and iron-bearing minerals that required persistent water both as a lake and as groundwater in pores and fractures. Gale is special because it contains a wide diversity of ancient aqueous environments.

By exploring this diversity of geologic materials, scientists have discovered which kinds of rocks are favorable for preserving evidence of ancient habitable environments, as well as organic compounds that could be returned to Earth by future missions and analyzed for evidence of life.

InSight Lander Mission

InSight was launched on May 5, 2018 and landed on Mars at Elysium Planitia on November 26, 2018 (**Figure 20.28**). *InSight*'s objectives are to place a seismometer on the surface of Mars to measure seismic activity and provide information on its deep interior. It will also place a heat probe to study the thermal evolution of Mars, and its early geologic evolution including processes of accretion and differentiation. *InSight*'s design is based on the earlier *Phoenix* lander and is powered by solar panels. At the time of this writing, *InSight* had just reported a "Marsquake" in April of 2019. More data are required to assess the planetary interior.

Recent Missions: Mars Reconnaissance Orbiter (2006–) and Phoenix (May–November 2008)

Mars Reconnaissance Orbiter has been mapping the rocks and minerals of Mars at an unprecedented level of detail. Whereas the rovers are limited to a few kilometers of the Martian surface, the orbiter can map anywhere on the planet. It is equipped with several instruments, including a high-resolution stereo color camera capable of resolving objects on the surface of Mars as small as 1 m across. Another important device looks at the sunlight reflected from the Martian surface to reveal the presence of minerals that formed in water. One of the orbiter's most remarkable observations is the discovery of sedimentary layers that are so evenly bedded that they may preserve a record of periodic changes in the Martian climate that occurred billions of years ago (**Figure 20.29**).

In May 2008, a new lander touched down on the surface of Mars. It was named *Phoenix* because it was the twin of another lander (*Mars Polar Lander*) that crashed on the surface of Mars in 1999. NASA scientists studied the causes of the malfunction and became confident that they could get it right with the remaining twin. The name *Phoenix* seemed appropriate for a project that was resurrected from the ashes of a former ruin. In Egyptian and Greek mythology, the phoenix is a bird that can periodically burn and regenerate itself.

Phoenix was sent to search for ice in the polar region of Mars. Equipped with solar panels to generate energy from the Sun, it was never designed to survive the dark Martian winter; it had a planned life span of only a few months. Its mission focused on analyzing the composition of several

FIGURE 20.28 This image shows *InSight*'s domed Wind and Thermal Shield, which covers its seismometer. The image was taken on the 110th Martian day, or sol, of the mission. The seismometer is called Seismic Experiment for Interior Structure, or SEIS. On April 23, 2019, NASA reported that this instrument reported its first "Marsquake." [NASA/JPL-Caltech.]

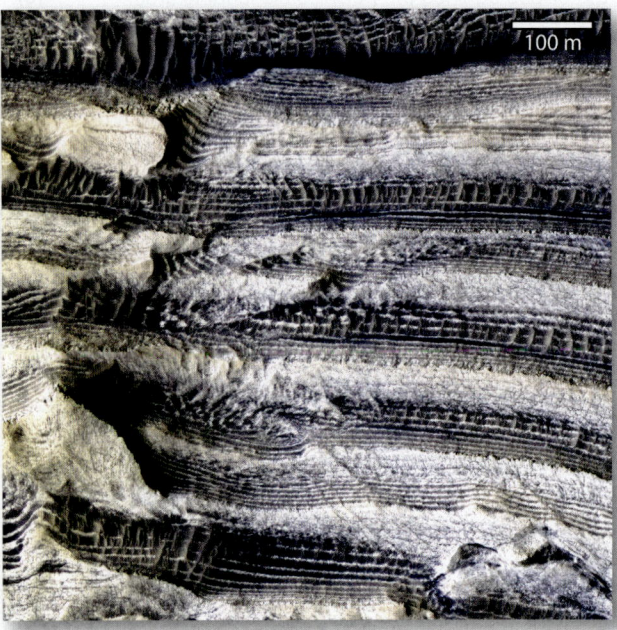

FIGURE 20.29 These sedimentary strata exposed in Becquerel Crater have a regular, almost periodic, appearance. Each bed is a few meters thick, and the beds are grouped in sequences a few tens of meters thick. These beds are thought to be composed of wind-deposited dust. The supply of sediment may have been regulated by periodic changes in climate. [NASA/JPL/University of Arizona.]

soil samples at the landing site. Within just a month of landing, it had accomplished its primary goal of demonstrating the presence of water ice within the soil. The presence of ice had been predicted by the *Mars Odyssey* orbiter, but it was important to confirm it on the ground.

In addition, *Phoenix* made its own surprising discovery concerning the surface environment of Mars. Based on data from the recent rovers and orbiters, a consensus had been developing that the global surface environment of Mars was likely to be very acidic. When *Phoenix* analyzed its first sample of polar soil, however, it found a neutral pH. This finding is another indicator of habitability, since most microorganisms prefer a neutral pH.

Recent Discoveries: The Environmental Evolution of Mars

The recent rover and orbiter missions to Mars have transformed our understanding of its early evolution. Like the Moon and the other terrestrial planets, Mars has ancient cratered terrains that preserve the record of the Late Heavy Bombardment. Therefore, these ancient terrains must be made of rocks older than 3.8 billion to 3.9 billion years (see Figure 20.11). Younger surfaces, which formed after the time of the Late Heavy Bombardment, are also widespread on Mars. Until recently, these younger surfaces were thought to be largely volcanic, as on Venus. However, data from the Mars Exploration Rovers, Mars Science Laboratory, and *Mars Express* show us that at least some—and perhaps many—of these younger surfaces are underlain by sedimentary rocks.

Some of these sedimentary rocks are composed of silicate minerals derived from erosion of old basaltic lavas and the pulverized basaltic rocks of the ancient cratered terrains. For example, the meandering stream deposits visible in Figure 20.23 may have formed largely by the accumulation of basaltic sediments. However, in most, if not all, of the sedimentary rocks beneath Meridiani Planum, where *Opportunity* has been exploring, sulfate minerals—which are chemical sediments—are mixed with silicate minerals. The sulfate minerals must have been precipitated when water evaporated, probably in shallow lakes or seas. The water must have been very salty to precipitate these minerals, and it must have contained common sulfate minerals such as gypsum ($CaSO_4$). In addition, the presence of unusual sulfate minerals such as *jarosite* (**Figure 20.30**)—an iron-rich sulfate mineral—tells us that the water must have been very acidic. On Mars, sulfuric acid probably formed when the abundant basaltic rocks interacted with water and were weathered, releasing their sulfur. The acid-rich water then flowed through rocks, heavily fractured from impacts, and over the surface to accumulate in lakes or shallow seas, where jarosite precipitated as chemical sediment.

As we have seen in Chapters 6 and 9, sedimentary rocks are valuable records of Earth's history. The vertical succession of sedimentary rocks—their *stratigraphy*—tells us how

FIGURE 20.30 The first outcrop studied on another planet (Mars). This outcrop is made of sedimentary rocks formed partly from sulfate minerals, including jarosite. Jarosite can form only in water—and only in acid-rich water. The area shown in the photograph is about 50 cm in width. [NASA/JPL/Cornell.]

environments change over time. One of the most exciting findings of the Mars Exploration Rovers mission so far has been the discovery of a stratigraphic record at Endurance Crater. Because the crater is so large, there is a lot of outcrop to observe, and it is mostly unaffected by the crater-forming impact. **Figure 20.31** shows the outcrop that contains all the stratigraphic clues. By using *Opportunity* to measure each layer, geologists were able to create a high-resolution stratigraphy (Figure 21.31b), the first of its kind generated for another planet. Remarkably, this interpretive drawing—from a planet 300 million miles away—provides the same level of understanding that is typically obtained here on Earth (as, for example, in Figure 6.15). *Curiosity* is now doing similar stratigraphic work at Gale Crater and also hoping to learn about the time-ordered sequence of events that characterized the early environmental evolution of Mars.

Perhaps one day we will have enough understanding of the stratigraphy of Mars to be able to correlate sedimentary and volcanic rocks from one part of the planet with those from another. To do this, we will need to link observations by the rovers on the ground with observations provided by orbiters overhead. The recent orbiters have shown that sulfates as well as clay minerals are abundant in several places on Mars—particularly in the Vallis Marineris, where they may form deposits up to several kilometers thick. This observation leads us to believe that their formation was related to a process that occurred globally, possibly over a long time.

PRACTICING GEOLOGY EXERCISE

How Do We Land a Spacecraft on Mars? Seven Minutes of Terror

When we send a lander to Mars, how do we decide where to land it? The riskiest part of such a mission comes when the spacecraft enters the Martian atmosphere, descends through it, and lands on the planet's surface. This step, called *Entry, Descent, and Landing,* or "EDL," takes about 7 minutes. During that time, the lander decelerates from 12,000 to 0 miles per hour, and its heat shield becomes as hot as the surface of the Sun (about 1500°C) due to friction caused by the atmosphere. A lot can go wrong here, so EDL has been called "seven minutes of terror."

The shape and elevation of the Martian surface play key roles in lander design. The lander holds a limited amount of fuel to power its engines, so if the land surface varies too much in its elevation, the lander has to spend time (and fuel) maneuvering. Your task as the EDL team geologist is to choose a safe landing site—one that does not vary too much in elevation. At the same time, you will want to choose a site that provides interesting outcrops for the lander or rover to study. So your problem is to determine how much variation in elevation is "too much."

To solve this problem, we need the following information: The lander's engines start up when the radar determines that the lander is 1000 m above the Martian surface. The engines slow the lander's descent, allowing it to descend at a rate of 50 meters per second (m/s) until it is 10 m above the ground. At that point, the lander descends at 2 m/s until touchdown. The engine's fuel consumption rate is 5 liters per second (L/s). The fuel tank holds 150 L.

First, how long does it take for the lander to descend to the surface? Note that two rates must be used here: one for the first 990 m of descent, and the other for the last 10 m of descent.

time = distance ÷ lander descent rate
= 900 m ÷ 50 m/s
= 20 s

time = distance ÷ lander descent rate
= 10 m ÷ 2 m/s
= 5 s

total time = 20 s + 5 s = 25 s

Next, how much fuel is consumed during landing?

fuel consumption = time × fuel consumption rate
= 25 s × 5 L/s
= 125 L

Given that 150 L of fuel are available, but only 125 L are used, there would be a reserve of 25 L remaining after landing. This calculation is for the

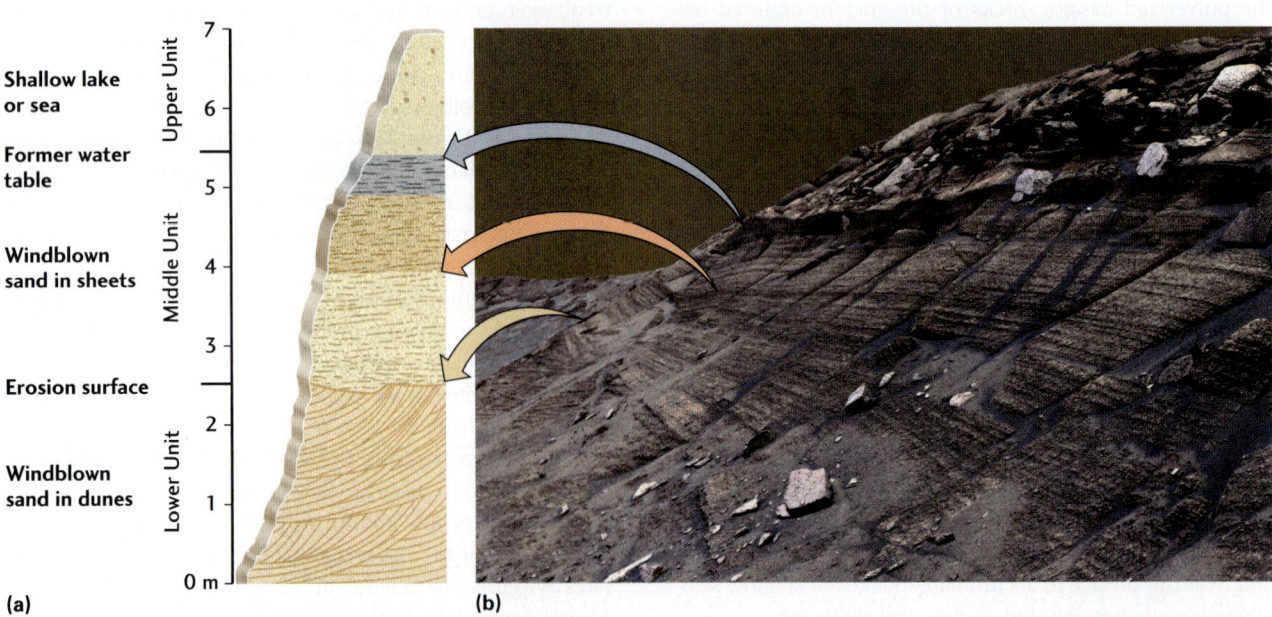

FIGURE 20.31 A sedimentary sequence exposed along the flank of Endurance Crater, photographed by the rover *Opportunity*. (a) An interpretive drawing showing each stage in the history of the outcrop. (b) The vertical succession of layers in the outcrop preserves an excellent record of early Martian environments. [NASA/JPL/Cornell.]

"perfect" landing condition in which the total descent distance is 1000 m (see "Case 1" in the accompanying figure).

Now let's consider what would happen if the lander drifted sideways while descending because the wind was blowing and moved over a low spot on the Martian surface (see "Case 2" in the accompanying figure). In this case, the total descent distance would be greater than 1000 m. If the low spot were too low, the lander would be at risk of depleting all its fuel reserves before it ever landed and crashing to the surface. Therefore, we need to determine how much elevation change would use up this fuel reserve.

First, we need to determine how much reserve time is provided by the 25 L of reserve fuel:

time = reserve fuel volume
= 25 L ÷ 5 L/s = 5 s

Now we can determine the additional descent distance that could be safely traveled before the fuel reserve was used up:

descent distance = reserve time × lander descent rate
= 5 s × 50 m/s
= 250 m

The solution tells us how much elevation change is "too much" to tolerate for a safe landing site: Anything more than 250 m is too much. The team geologist must find a landing site where elevation varies less than 250 m, yet which also provides interesting geologic features. In practice, there is a real trade-off between geologic interest and landing site safety.

BONUS PROBLEM: Determine the maximum variation in elevation at a landing site that could be tolerated by a lander with a fuel tank volume of 200 L. How much variation could be accepted if the final descent rate were 1 m/s rather than 2 m/s?

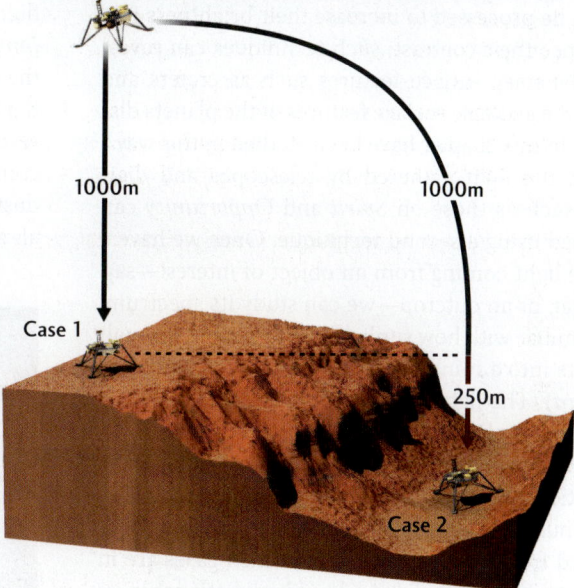

(*top*) Engineers building the *Phoenix* lander, which arrived at the surface of Mars in 2008. (*bottom*) Successful landing on the surface of Mars requires careful planning and consideration of the geologic environment of the surface, including variations in topography. [photo: NASA/JPL/UA/Lockheed Martin.]

There is some evidence suggesting that the clay minerals may have formed before the sulfate minerals. However, we do not yet know whether these deposits were formed all at the same time, signaling a global environmental event that may have been unique in Mars's history, or whether they were formed at multiple times in different places. *Curiosity*'s initial discoveries at Yellowknife Bay, Gale Crater, hints at this latter possibility. The conclusion would point to a more common process that operated throughout Mars's history wherever and whenever local conditions allowed.

The evidence is now compelling that at some point in Mars's history, there was liquid water on its surface and underground. The planet must have been warmer than it is today, unless the water was very short-lived, gushing to the surface briefly, then evaporating quickly or sinking back underground before being frozen, as would happen today. There are many questions left to answer. How much water was there? How long did it last? Did it ever rain, or was it all groundwater leaking to the surface? Did the water last long enough, and have the right composition, to allow life to get started? Only one thing is certain at this point: More missions are required to answer these questions.

Exploring the Solar System and Beyond

An astronomer staring through a telescope is the first image that comes to mind when most people think of exploring the solar system. But most modern telescopes have no eyepiece at all, and instead record their images with digital cameras. Many telescopes, such as the Hubble Space Telescope, are not even located on Earth, but are positioned in space.

Regardless of how a telescope takes its photographs or where it is stationed, its purpose is the same: to gather more light than we can with the naked human eye. Its photographs can be processed to increase their brightness further or enhance their contrast; such techniques can reveal important planetary surface features such as craters and canyons. All the geologic surface features of the planets discussed so far in this chapter have been studied in this way.

However, the light gathered by telescopes and digital cameras such as those on *Spirit* and *Opportunity* can also be studied using a second technique. Once we have a record of the light coming from an object of interest—say, a planet, a star, or an outcrop—we can study its spectrum. We are all familiar with how sunlight, when passed through a prism, splits into a rainbow of colors called its *spectrum* (plural *spectra*). The light generated by a star, or reflected off the surface of a planet or outcrop, also produces a spectrum. The colors of that spectrum can reveal the chemical composition of the light-producing or light-reflecting materials. Thus, geologists can look at the spectrum of light reflected from a planet and know which gases are in its atmosphere and which chemicals and minerals are in its rocks and soils.

Astronomers use this same principle to look at the light coming from faraway stars and galaxies. The spectra they see tell them the ages of those stars and galaxies, reveal how they evolved, and even provide mind-boggling insights into the origin and evolution of the universe.

Space Missions

Most observations of our solar system and beyond are still made from Earth. Over the past 50 years, however, we have sent all kinds of machines, robots, and even humans into space in our quest to explore the unknown. Space missions are a costly business, requiring a tremendous effort by hundreds and sometimes thousands of people at a cost of hundreds of millions to billions of dollars. The Mars Exploration Rovers mission cost on the order of $800 million for both rovers, and *Mars Science Laboratory*—about the size of a car—cost over $2 billion. And space missions are a risky business: Fewer than half the missions sent to Mars have succeeded. As the space shuttle program reminds us, space exploration is risky for humans, too.

Are these efforts, costs, and risks worth it? For thousands of years, people have looked up at the skies and pondered the universe. What are the stars and planets made of? How did the universe form? Is there any life out there? To answer these questions, we have to look for clues, and most of those clues will be provided only by missions to space. The issue is not so much whether or not to explore space, but how. Most debates focus on whether it is essential to send humans into space or whether the Mars rovers missions have demonstrated the adequacy of robots.

We are actively exploring space in many different ways. Spacecraft have been sent to orbit planets, moons, and asteroids and to fly by planets and comets in the outer solar system and beyond. On other occasions, we have instructed landers and other probes to descend to planetary surfaces and make direct measurements of rocks, minerals, gases, and fluids. On July 3, 2005, a probe was released from the *Deep Impact* spacecraft and instructed to collide deliberately with the comet Tempel 1. The depth of the resulting crater and the light emitted at the time of the collision (**Figure 20.32**) revealed what the interior of the comet is made of. The comet was found to consist of a mixture of dust and ice; the dust component included clay, carbonate, and silicate minerals and was enriched in sodium, which is rare in space.

FIGURE 20.32 The first moments after *Deep Impact*'s probe collided with the comet Tempel 1. Debris from the interior of the comet is expanding from the impact site. [NASA/JPL-Caltech/UMD.]

The Cassini-Huygens Mission to Saturn

An even more remarkable story of deep space exploration involves the Cassini-Huygens mission. In 2005, the *Huygens* lander became the farthest-traveled spacecraft to reach another planet and "live" to tell about it.

Cassini-Huygens is one of the most ambitious missions ever launched into space. The *Cassini-Huygens* spacecraft includes two components: the *Cassini* orbiter and the *Huygens* lander. The spacecraft was launched from Earth on October 15, 1997. After traveling over a billion kilometers across deep space in almost 7 years, *Cassini-Huygens* sailed through the rings of Saturn on July 1, 2004. Saturn's beautiful rings are what set it apart from the other planets (**Figure 20.33a**). It is the most extensive and complex ring system in the solar system, extending hundreds of thousands of kilometers from the planet. Made up of billions of particles of ice and rock—ranging in size from grains of sand to houses—the rings orbit Saturn at varying speeds. Understanding the nature and origin of these rings is a major goal of the Cassini-Huygens scientists. The Huygens lander touched down on the surface of Titan and discovered a planet whose surface is made up largely of hydrocarbons—liquid, solid, and gas—and which has lakes and rivers formed of these materials (Figure 20.33b).

On December 24, 2004, the *Huygens* lander was released from the orbiter and traveled over 5 million kilometers to reach Titan, one of Saturn's 18 moons. On January 14, 2005, it reached Titan's upper atmosphere, where it deployed a parachute and then plunged to the surface, where it landed successfully. Its cameras showed surface features that appear to be drainage networks, similar to those seen on Earth and Mars. The landing site was strewn with rocks up to 10–15 cm in diameter (Figure 20.33b). However, these "rocks" are probably ices made of methane (CH_4) and other organic compounds.

Bigger than the planet Mercury, Titan is of particular interest to scientists because it is one of the few moons in the solar system with its own atmosphere. It is cloaked in a thick, smoglike haze that scientists believe may be similar to Earth's atmosphere before life began more than 3.8 billion years ago (see Chapter 22). Organic compounds, including gases composed of methane, are plentiful on Titan. Further study of this moon promises to reveal much about planetary formation and, perhaps, about the early days of Earth.

(a)

(b)

FIGURE 20.33 (a) Saturn and its rings completely fill the field of view in this natural color image taken by the *Cassini-Huygens* spacecraft on March 27, 2004. Color variations in the rings reflect differences in the composition of the materials that make them up, such as ice and rock. The *Cassini-Huygens* scientists will investigate the nature and origin of the rings as the mission progresses. (b) The surface of Titan is strewn with "rocks" of ice composed of frozen methane and other carbon-containing compounds. [a: NASA/JPL/SSI/ESA/University of Arizona; b: ESA/NASA/University of Arizona.]

Other Solar Systems

For ages, scientists and philosophers have speculated that there may be planets around stars other than our Sun. In the 1990s, astronomers discovered planets orbiting nearby Sunlike stars. In 1999, the first family of **exoplanets**—planets that lie outside the solar system—was found. These planets are too dim to be seen directly by telescopes, but their existence can be inferred from their slight gravitational pull on the stars they orbit, which causes to-and-fro movements of the star that can be measured. We see these movements recorded in the spectra of the starlight. Spacecraft above Earth's atmosphere are able to search for the dimming of the parent star's light as an orbiting planet passes in front of it along the line of sight to Earth. Most planets found in this way are Jupiter-sized or larger and are close to their parent stars—many within scorching distance. Earth-sized planets have been discovered in recent years using other methods. By 2009, astronomers had discovered over 300 new planets, organized in 249 solar systems. At the time of this writing, in late 2018, there are now 3545 new planets that occur in 2660 solar systems; the rate of discovery is astounding.

We are fascinated by planetary systems around other stars because of what they might teach us about our own origins. Our overriding interest, however, is in the profound scientific and philosophical implications posed by the question, "Is anyone else out there?" NASA is currently building spacecraft that will carry instruments to analyze the atmospheres of exoplanets in our galaxy for signs of the presence of some kind of life. Based on what we know about biological processes, life on an exoplanet would probably be carbon-based and require liquid water. The benign temperatures we enjoy on Earth—not too far outside the range between the freezing and boiling points of water—appear to be essential for life (see Chapter 22). An atmosphere is needed to filter harmful radiation from the parent star, so the planet must be large enough for its gravitational field to keep the atmosphere from escaping into space. For a habitable planet with complex life *as we know it* to exist would require conditions even more limiting. For example, if the planet were too massive, delicate organisms such as humans would be too weak to withstand its gravitational force. Are these requirements too restrictive for life to exist elsewhere? Many scientists think not, considering the billions of Sunlike stars in our own galaxy.

KEY TERMS AND CONCEPTS

asteroid (p. 602)
dwarf planet (p. 607)
exoplanet (p. 628)
flake tectonics (p. 613)
gravitational differentiation (p. 603)
Heavy Bombardment (p. 609)
meteorite (p. 602)
nebular hypothesis (p. 600)
planetesimal (p. 601)
solar nebula (p. 601)
terrestrial planet (p. 602)

REVIEW OF LEARNING OBJECTIVES

20.1 Describe how the solar system originated.

According to the nebular hypothesis, the Sun and its planets formed when a cloud of gases and dust, known as the solar nebula, condensed about 4.5 billion years ago. Gravitational attraction caused the dust and condensing material to clump together into planetesimals, which in turn collided and stuck together, forming larger bodies. As the planets formed, the four planets closest to the Sun (terrestrial planets) developed very different compositions than the outer planets (giant outer planets).

Study Assignment: Figure 20.2

Thought Question: What is the origin of the asteroid belt between Mars and Jupiter?

Exercise: How and why do the inner planets of the solar system differ from its outer planets?

20.2 Explain how the Earth formed and developed over time.

Earth probably grew by the accretion of colliding planetesimals. Soon after it formed, it was struck by a large body about the size of Mars. Matter ejected into space from both Earth and the impacting body reassembled to form the Moon. The impact generated enough heat to melt most of what remained of Earth. Radioactivity and gravitational energy also contributed to this early heating and melting. Heavy matter, rich in iron, sank toward Earth's center to form the core, and lighter matter floated upward to form the crust. Still lighter gases formed Earth's atmosphere and oceans. In this way, Earth was transformed into a differentiated planet with distinct layers.

> **Study Assignment:** Figure 20.4
>
> **Thought Question:** Knowing how the Moon formed, what might you expect as a result if you were told that a large meteorite had collided with a planet twice its size? How would the result of the impact differ if the meteorite were significantly smaller than the planet?
>
> **Exercise:** What caused Earth to differentiate into a layered planet, and what was the result?

20.3 Compare and contrast the diversity of the planets and what makes them similar and different.

The four terrestrial planets have all undergone gravitational differentiation into layered structures with an iron-nickel core, a silicate mantle, and an outer crust. The gaseous outer planets are so chemically distinct and so large that their formation must have been drastically different than the terrestrial planets. All four of the outer gas planets are thought to have rocky silica-rich and iron-rich cores surrounding by thick shells of liquid hydrogen and helium. Pluto, which was once considered a planet, is now a dwarf planet—distinct with its tiny size, unusual orbit, and rock-ice-gas composition.

> **Study Assignment:** Table 20.1
>
> **Thought Questions:** Explain why Mercury's large temperature swings and Venus's incredibly hot temperature are dependent on the type of atmosphere they have. Why do you think Earth has a temperature range sustainable for life?
>
> **Exercise:** Mercury's average density is less than Earth's, but the relative size of its core is larger. How can you explain this?

20.4 Summarize some of the major events in the early history of the solar system and describe how planetary surfaces can be dated.

The age of the solar system, as determined from isotopic dating of meteorites, is about 4.56 billion years. Earth and the other terrestrial planets had formed within about 10 million years. The impact that formed the Moon occurred about 4.51 billion years ago. Minerals as old as 4.4 billion years have survived in Earth's crust. The Late Heavy Bombardment, which peaked about 3.9 billion years ago, marks the end of the Hadean eon on Earth. Rocks returned from the surface of the Moon by the Apollo missions have been dated using isotopic methods. The lunar highlands show ages from 4.4 billion to about 4.0 billion years. The lunar maria show ages from 4.0 billion to 3.2 billion years. These isotopic ages allowed geologists to calibrate the relative time scale they had developed by crater counting.

> **Study Assignment:** Figure 20.11
>
> **Thought Question:** If you were an astronaut landing on an unexplored planet, how would you decide whether the planet was differentiated and whether it was tectonically active?
>
> **Exercise:** If you counted all of the impact craters on the moon and kept track of their size, and also were given information on their age, how do you think size would relate to age?

20.5 Discuss how Mars and the other planets have been explored. Does spacecraft data show evidence for water on Mars?

Four types of spacecraft have been used to explore Mars and other planets. During a flyby, a spacecraft comes close to a planet only once. An orbiter circles a planet, making remote observations of its surface and interior. A lander can actually touch down on the surface of a planet to make local observations. A rover can leave the landing site and travel up to several kilometers to investigate new terrains. Currently, there are two rovers and three orbiters exploring Mars sending back new data every day, leading to significant new discoveries regarding evidence for water and whether the environment was habitable in the past. Today, water is present on Mars only as ice caps at the Martian poles and as permafrost. In the past, it may have been present as a liquid, as geologic evidence shows that it ran across the surface to carve stream channels and deposit sediments in meandering streams. It also accumulated in shallow lakes or seas, where it evaporated and precipitated a variety of chemical sediments, including sulfate minerals.

Study Assignment: Figures 20.21 and 20.22

Thought Question: During a dust storm on Mars, sediments fill the atmosphere with dust. But Mars has an atmosphere much thinner than Earth's. To move sand, would the wind have to blow faster on Mars to compensate for this difference?

Exercise: What surface features would you look for on Mars if you were searching for evidence of liquid water in its geologic past?

20.6 Describe how we use light in exploring stars and the solar system. Is our solar system unique?

In some cases, we can use enhanced photographs from telescopes, which may reveal surface features of distant objects. In other cases, we can use information from the spectrum of light, which varies depending on the composition of the object that produces or reflects that light. We have evidence of more than 300 planets that circle other stars. In several cases, there is more than one planet in these solar systems. Because these new planets lie outside our solar system, they are called exoplanets.

Study Assignment: Figure 20.32

Thought Question: How does the discovery of planets orbiting other stars contribute to the debate about the possibility of life elsewhere in the cosmos? What are the scientific and philosophical implications of the existence of life on the planets of other stars?

Exercise: Based on life as we know it here on Earth, discuss what characteristics another planet would need to be habitable.

VISUAL LITERACY EXERCISE

The inner planets are small and rocky.

The four giant outer planets and their moons are gaseous, with rocky cores.

Pluto is a snowball of methane, water, and rock.

FIGURE 20.3 The solar system. This diagram shows the relative sizes of the planets as well as the asteroid belt separating the inner and outer planets. Although considered one of the nine planets since its discovery in 1930, Pluto was demoted from that status by the International Astronomical Union in 2006. With this revision, there are only eight true planets, not nine.

1. How many planets are there in the solar system?
   a. 9
   b. 8
   c. 3
   d. 5

2. Where is the asteroid belt located?
   a. Between the Sun and Mercury
   b. Beyond Pluto
   c. Between Mars and Jupiter
   d. Asteroids are scattered throughout the solar system

3. What is the dominant composition of the inner planets?
   a. Gaseous
   b. Rocky
   c. Hydrogen
   d. Ice

4. Which planets have rings?
   a. Mercury, Pluto, and Earth
   b. Venus, Mars, and the Sun
   c. Neptune and Mercury
   d. Jupiter, Uranus, and Saturn

5. Which planets are largest and smallest?
   a. Jupiter and Mars
   b. Uranus and Pluto
   c. Jupiter and Mercury
   d. Saturn and Pluto

History of the Continents

21

The Structure of North America	634
Tectonic Provinces Around the World	640
How Continents Grow	643
How Continents Are Modified	646
The Origin of Cratons	655
Deep Structure of Continents	657

Learning Objectives

Continents are amalgamations of low-density, silica-rich rocks that float on the denser, convecting mantle. Using the information in this chapter, you will be able to:

21.1 Identify the major tectonic provinces on a map of North America.

21.2 Recognize the major types of tectonic provinces found on continents worldwide.

21.3 Describe the tectonic processes that add new crust to the continents.

21.4 Explain how orogenies modify continents.

21.5 Order the sequence of plate-tectonic events that constitute the Wilson cycle.

21.6 Categorize the main mechanisms of epeirogeny.

21.7 Explain how continental cratons have survived billions of years of plate-tectonic processes.

Shaded relief map of the North American continent. [USGS.]

Nearly two-thirds of Earth's surface—all of its oceanic crust—was created by seafloor spreading over the past 200 million years, an interval that spans a mere 4 percent of Earth's history. To understand how Earth evolved in earlier times, we must look at the evidence preserved in continental crust, which contains rocks over 4 billion years old.

The geologic record of the continental crust is very complex, but our ability to read that record has improved immensely in the last few decades. Our knowledge of plate tectonics can be used to interpret eroded mountain belts and ancient rock assemblages in terms of closing ocean basins and colliding continents. Precise isotopic dating and other geochemical tools help us decipher the history of continental rocks. Networks of seismographs and other sensors allow us to image the structure of the continents far below Earth's surface.

In this chapter, we will describe the structure of Earth's continents, and we will look back through their 4-billion-year history to understand the processes that formed them and continue to modify them today. We will see how plate-tectonic processes have added new material to the continental crust, how plate convergence has thickened that crust into mountain belts, and how those mountains have been eroded to expose the metamorphic basement rocks found in many older regions of the continents. We will reach back into the earliest period of continental evolution, the Archean eon (3.9 billion to 2.5 billion years ago), to ponder two of the great puzzles of Earth's history: How did continents form, and how have they survived through billions of years of plate tectonics and continental drift?

Continents, like people, show a great variety of surface features that reflect their origins and experience over time. Yet, also like people, continents share many similarities in their basic structure and growth patterns. Before considering continents in general, let's begin by outlining the major features of one particular continent: North America.

The Structure of North America

The long-term tectonic history of North America is reflected in its **tectonic provinces**—large-scale regions formed by distinctive tectonic processes (Figure 21.1).

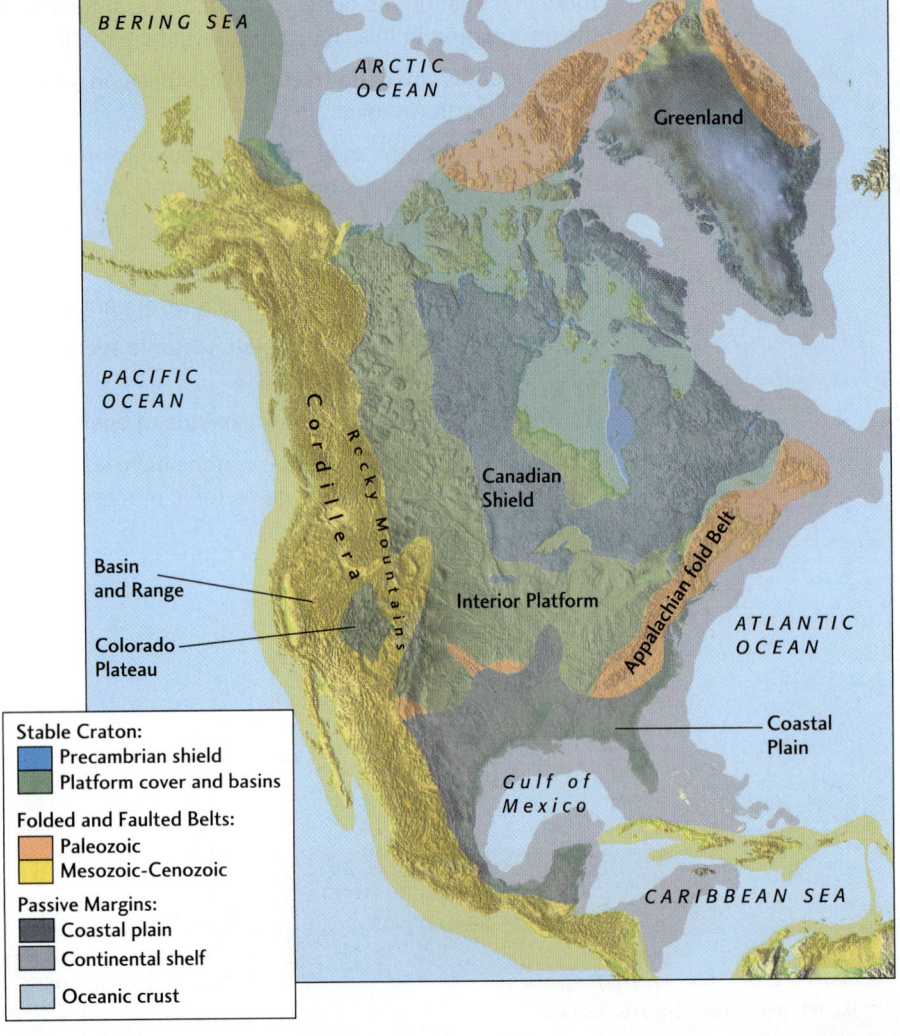

FIGURE 21.1 The major tectonic provinces of North America reflect the processes that formed the continent.

The oldest parts of North America's crust, built during the most ancient episodes of deformation, tend to be found in the northern interior of the continent. These regions, which include most of Canada and the closely connected landmass of Greenland, are *tectonically stable*. In other words, they have remained largely undisturbed by recent episodes of continental rifting, drift, and collision, and they have been eroded nearly flat. On the edges of these older tectonic provinces are younger metamorphic belts where most of the present-day mountain chains are found. These mountain chains form elongated topographic features near the margins of the continent. The two main examples are the *North American Cordillera*, which runs down the western side of North America and includes the Rocky Mountains, and the *Appalachian Fold Belt*, which runs southwest to northeast on the continent's eastern margin. In our description of tectonic provinces, we will often refer to the geologic time scale shown in Figure 9.16, so you might want to bookmark this figure for reference.

The Stable Interior

Much of central and eastern Canada is a landscape of very old crystalline basement rock—a huge tectonic province (8 million km^2) called the *Canadian Shield* (**Figure 21.2**). It consists primarily of Precambrian granitic and metamorphic rocks, such as gneisses, together with highly deformed and metamorphosed sedimentary and volcanic rocks, and it contains major deposits of iron, gold, copper, diamond, and nickel. Large portions of the shield were formed during the Archean eon, so it represents one of the oldest records of Earth's history. The nineteenth-century Austrian geologist Eduard Suess named these areas continental **shields** because they emerge from the surrounding sediments like a shield partly buried in the dirt of a battlefield.

In North America, extensive flat-lying (*platform*) sediments have been deposited on stable continental crust around the periphery of the Canadian Shield and near its center, beneath Hudson Bay (see Figure 21.1). The vast low-lying, sediment-covered region south and west of the Canadian Shield, which includes the Great Plains of Canada and the United States, is called the *Interior Platform*. The Precambrian basement rocks of the interior platform are a continuation of the Canadian Shield, although here they lie under nearly flat layers of Paleozoic sedimentary rocks, typically less than 2 km thick.

FIGURE 21.2 An aerial view of ancient, eroded metamorphic basement rocks in Nunavut, Canada, exposed on the surface of the Canadian Shield in the Northwest Territories. [Paolo Koch/Science Source.]

The North American platform sediments were laid down on the deformed and eroded Precambrian basement under a variety of conditions. Some rock formations (marine sandstones, limestones, shales, deltaic deposits, evaporites) indicate sedimentation in extensive shallow inland seas. Others (non-marine sediments, coal deposits) indicate deposition on floodplains or in lakes or wetlands.

Within the interior platform are a number of circular structures: broad *sedimentary basins*, roughly circular or oval depressions where the sediments are thicker than in the surrounding areas, and *domes*, areas where the platform sediments have been uplifted and eroded to expose the basement rock (Figure 21.3). Most of the basins are thermal subsidence basins; that is, they are regions that subsided when heated portions of the lithosphere cooled and contracted (see Chapter 6). An example is the Michigan Basin, a circular area of about 200,000 km² that covers most of the Lower Peninsula of Michigan (see Figure 8.15). This basin subsided throughout much of the Paleozoic era and received sediments more than 5 km thick in its central, deepest part. The sandstones and other sedimentary rocks of these basins, laid down under tectonically quiet conditions, have remained unmetamorphosed and only slightly deformed to this day. The interior platform basins contain important deposits of uranium, coal, oil, and natural gas. Rich mineral deposits in the basement rocks lie close to the surface in the domes, and they may also become traps for oil and gas.

The Appalachian Fold Belt

Along the eastern side of North America's stable interior are the old, eroded Appalachian Mountains. This classic fold and thrust belt, which we first examined in Chapter 8, extends along eastern North America from Newfoundland to Alabama. The rock assemblages and structures of the Appalachians resulted from the continent-continent collisions that formed the supercontinent Pangaea 470 million to 270 million years ago. The western side of the Appalachian belt is bounded by the *Allegheny Plateau*, a region of slightly uplifted, mildly deformed sediments that is rich in coal and oil. Moving eastward, we encounter regions of increasing deformation (Figure 21.4):

- *Valley and Ridge Province.* Thick Paleozoic sedimentary rocks laid down on an ancient continental shelf were folded and thrust to the northwest by compressive forces from the southeast. The rocks show that the deformation occurred in three mountain-building episodes, one beginning in the middle Ordovician period (about 470 million years ago), one in the middle to late Devonian period (380 million to 360 million years ago), and one in the late Carboniferous and early Permian periods (320 million to 270 million years ago).

- *Blue Ridge Province.* These eroded mountains are composed largely of highly metamorphosed Precambrian and Cambrian crystalline rocks. The

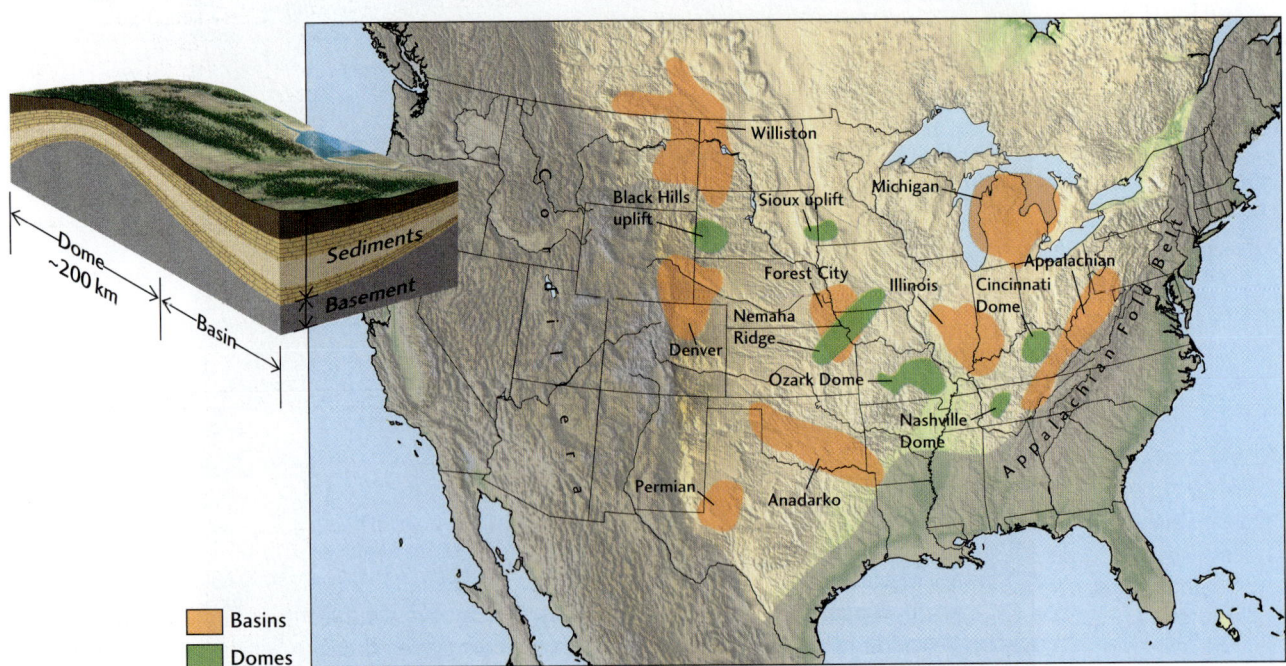

FIGURE 21.3 A map of the interior platform of North America, showing its basin and dome structure. The basins are nearly circular regions of thick sediments. The domes are regions where the sediments are anomalously thin. Basement rocks are exposed on the tops of some domes, such as the Black Hills uplift and the Ozark Dome.

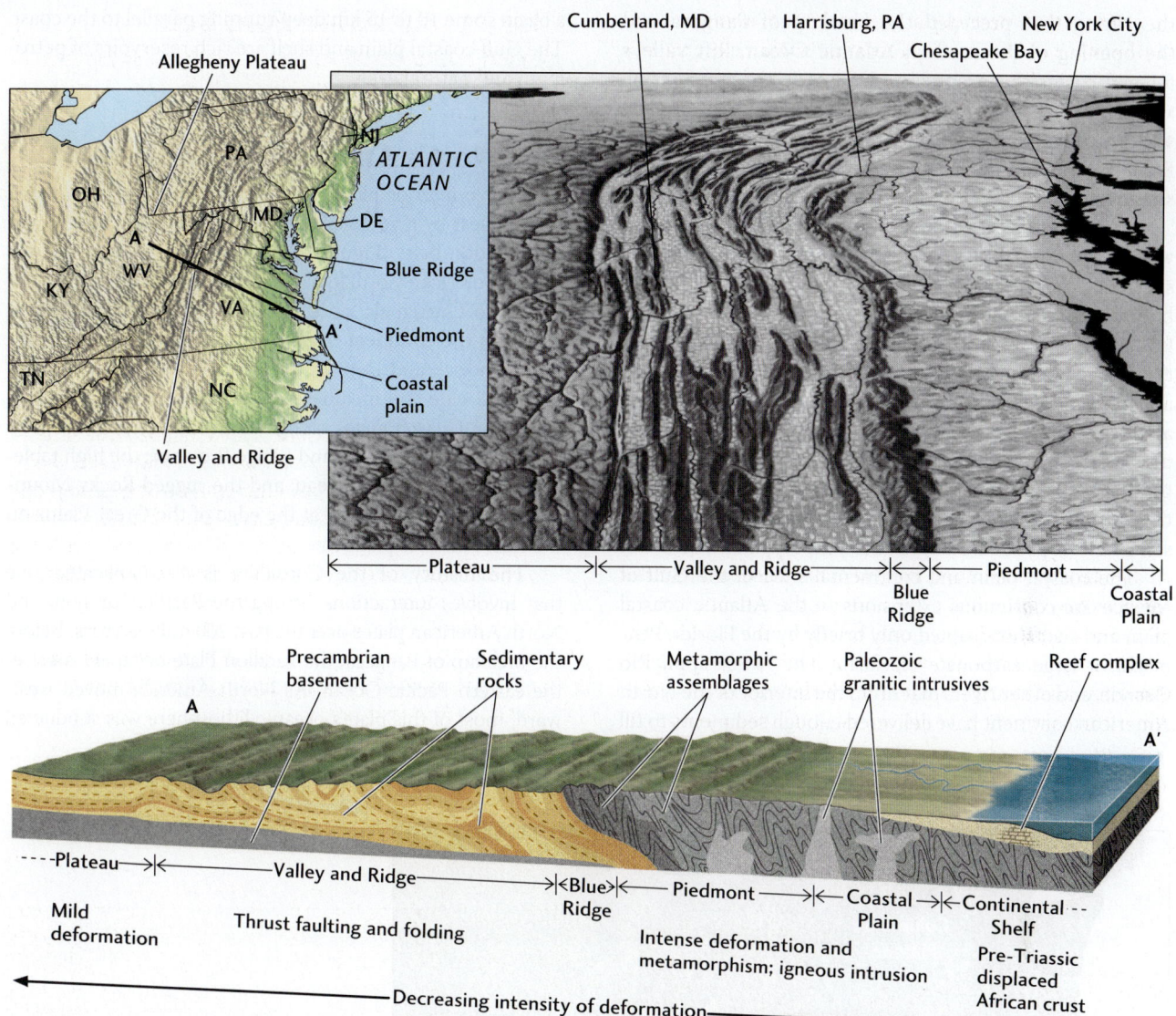

FIGURE 21.4 The Appalachian Fold Belt Province, shown in an aerial view to the northeast and an idealized cross section. The intensity of deformation increases from west to east. [Information from S. M. Stanley, *Earth System History*. New York: W. H. Freeman, 2005. Aerial view from NASA.]

Blue Ridge rocks were not intruded and metamorphosed in place, but rather thrust as sheets over the sedimentary rocks of the Valley and Ridge Province near the end of the Paleozoic era, about 300 million years ago.

- *Piedmont.* This hilly region contains metamorphosed Precambrian and Paleozoic sedimentary and volcanic rocks intruded by granite, all now eroded to low relief. Volcanism began in late Precambrian time and continued into the Cambrian. The Piedmont rocks were thrust over Blue Ridge rocks along a major thrust fault, overriding them to the northwest. At least two episodes of deformation are evident, coinciding with the last two mountain-building episodes in the Valley and Ridge Province.

The Coastal Plain and Continental Shelf

On the Coastal Plain, east and south of the Appalachian Fold Belt, relatively undisturbed sediments of Jurassic age and younger are underlain by rocks similar to those of the Piedmont. The Coastal Plain and its offshore extension, the continental shelf (see Figure 21.1), began to develop in the Triassic period, about 180 million years ago, with the rifting that preceded the breakup of Pangaea and the opening of the modern Atlantic Ocean. Rift valleys formed basins that trapped a thick series of non-marine sediments. As these deposits were accumulating, they were intruded by basaltic sills and dikes. The Connecticut River Valley and the Bay of Fundy are such sediment-filled rift valleys.

In the early Cretaceous period, as seafloor spreading widened the Atlantic Ocean, the deeply eroded, sloping surface of the Atlantic coastal plain and continental shelf began to cool, subside, and receive sediments from the continent. Cretaceous and Tertiary sediments as much as 5 km thick filled this slowly developing thermal subsidence basin, and even more material was dumped into the deeper water at the continental margin. This still-active basin continues to receive sediments. If the present stage of opening of the Atlantic is reversed some millions of years from now, the sediments in this basin will be folded and faulted in the same kind of process that produced the Appalachians.

The coastal plain and continental shelf of the Gulf of Mexico are continuous extensions of the Atlantic coastal plain and shelf, interrupted only briefly by the Florida Peninsula, a large carbonate platform. The Mississippi, Rio Grande, and other rivers that drain the interior of the North American continent have delivered enough sediments to fill a basin some 10 to 15 km deep running parallel to the coast. The Gulf coastal plain and shelf are rich reservoirs of petroleum and natural gas.

The North American Cordillera

The stable interior platform of North America is bounded on the west by a younger complex of mountain ranges and deformation belts (**Figure 21.5**). This region is part of the North American Cordillera, a mountain belt extending from Alaska to Guatemala, and it contains some of the highest peaks on the continent. Across its middle section, between San Francisco and Denver, the Cordilleran system is about 1600 km wide and includes several different tectonic provinces: the Coast Ranges along the Pacific Ocean; the lofty Sierra Nevada; the Basin and Range Province; the high tableland of the Colorado Plateau; and the rugged Rocky Mountains, which end abruptly at the edge of the Great Plains on the stable interior platform.

The history of the Cordillera is a complicated one that involves interactions among the Pacific, Farallon, and North American plates over the past 200 million years. Before the breakup of Pangaea, the Farallon Plate occupied most of the eastern Pacific Ocean. As North America moved westward, most of this plate's oceanic lithosphere was subducted

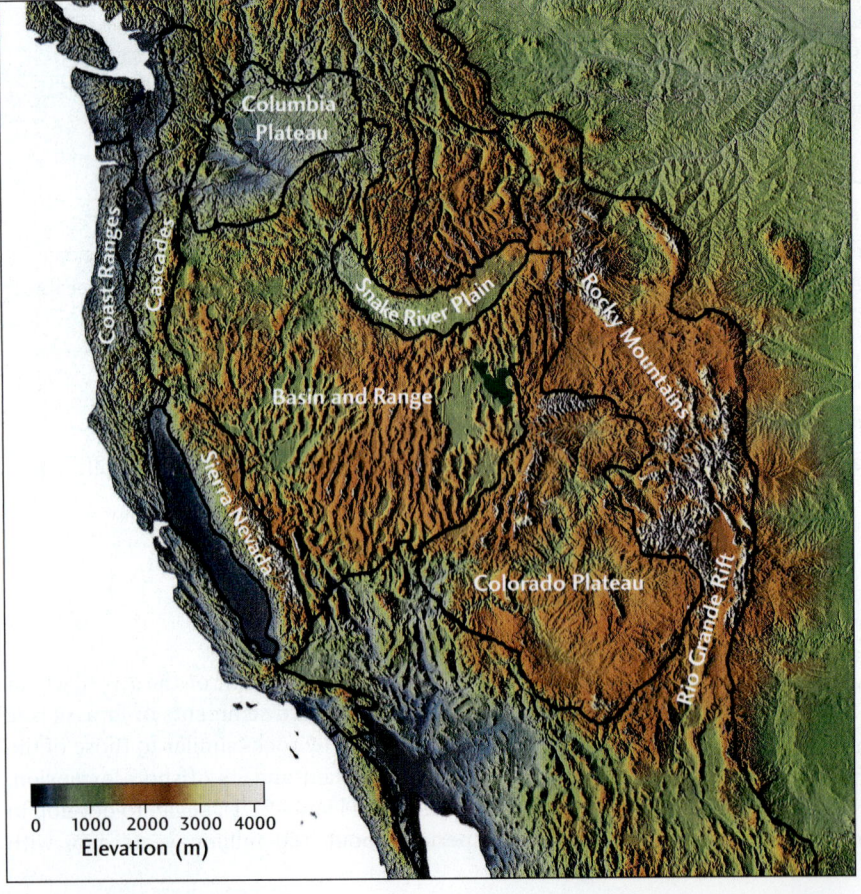

FIGURE 21.5 Topography of the North American Cordillera in the western United States. Computer manipulation of digitized elevation data produced this color shaded relief map. The major tectonic provinces of the area are clearly visible, as if illuminated by a light source low in the west.

eastward under the continent. The westward margin of the continent swept up island arcs and continental fragments, and the subduction zone eventually swallowed portions of the Pacific-Farallon spreading center, which converted the convergent boundary into the modern San Andreas transform-fault system (**Figure 21.6**). Today, all that is left of the Farallon Plate are small remnants, including the Juan de Fuca and Cocos plates, which are still subducting beneath North America.

The main phase of Cordilleran mountain building occurred in the last half of the Mesozoic era and in the early Paleogene period (150 million to 50 million years ago). The Cordilleran system is topographically higher than the Appalachians, which is not surprising, as there has been less time for erosion to wear it down. The form and height of the Cordillera that we see today resulted from even more recent events in the Neogene period, over the past 15 million or 20 million years, when the Pacific plate first encountered North America (see Figure 21.6). During these periods, the mountains underwent **rejuvenation**; that is, they were raised again and brought back to a more youthful stage. At that time, the central and southern Rockies attained much of their present height as a result of broad regional uplift. The Rockies were raised 1500 to 2000 m as Precambrian basement rocks and their veneer of later-deformed sediments were pushed above the level of their surroundings. Stream erosion accelerated, the mountain topography sharpened, and the canyons deepened. As you learned in Chapter 16, rejuvenation is driven not only by plate-tectonic processes, but also by interactions between the plate-tectonic and climate systems. For example, some of the increase in the relief of the Cordilleran mountain chains may have occurred as a result of the onset of glacial cycles in the Pleistocene.

The *Basin and Range Province* developed through the uplift and stretching of the crust in a northwest-southeast direction. This extension began with the heating of the lithosphere by upwelling convection currents in the mantle about 15 million years ago and continues to the present day (see Chapter 8). It has resulted in a wide zone of normal faulting extending from southern Oregon to Mexico and from eastern California to western Texas. The Basin and Range Province is volcanically active and contains extensive hydrothermal deposits of gold, silver, copper, and other valuable metals. Thousands of steeply dipping normal faults have sliced the crust into a pattern of upheaved and down-dropped blocks, forming scores of rugged and nearly parallel mountain ranges separated by sediment-filled rift valleys. The Wasatch Range of Utah and the Teton Range of Wyoming (**Figure 21.7**) are being uplifted on the eastern edge of the Basin and Range Province, while the Sierra Nevada of California is being uplifted and tilted on the province's western edge.

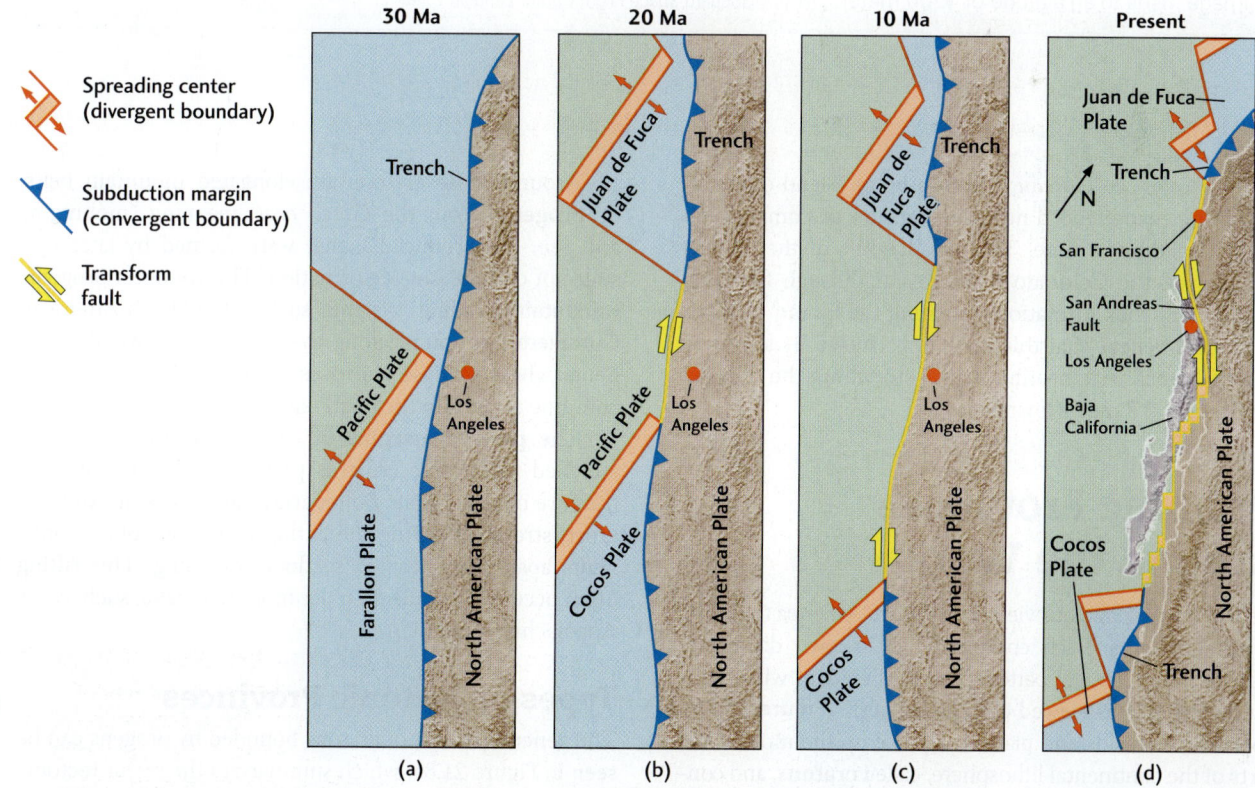

FIGURE 21.6 The interaction of the west coast of North America with the shrinking Farallon Plate as it was progressively subducted beneath the North American Plate, leaving the present-day Juan de Fuca and Cocos plates as small remnants. (Ma: million years ago.)

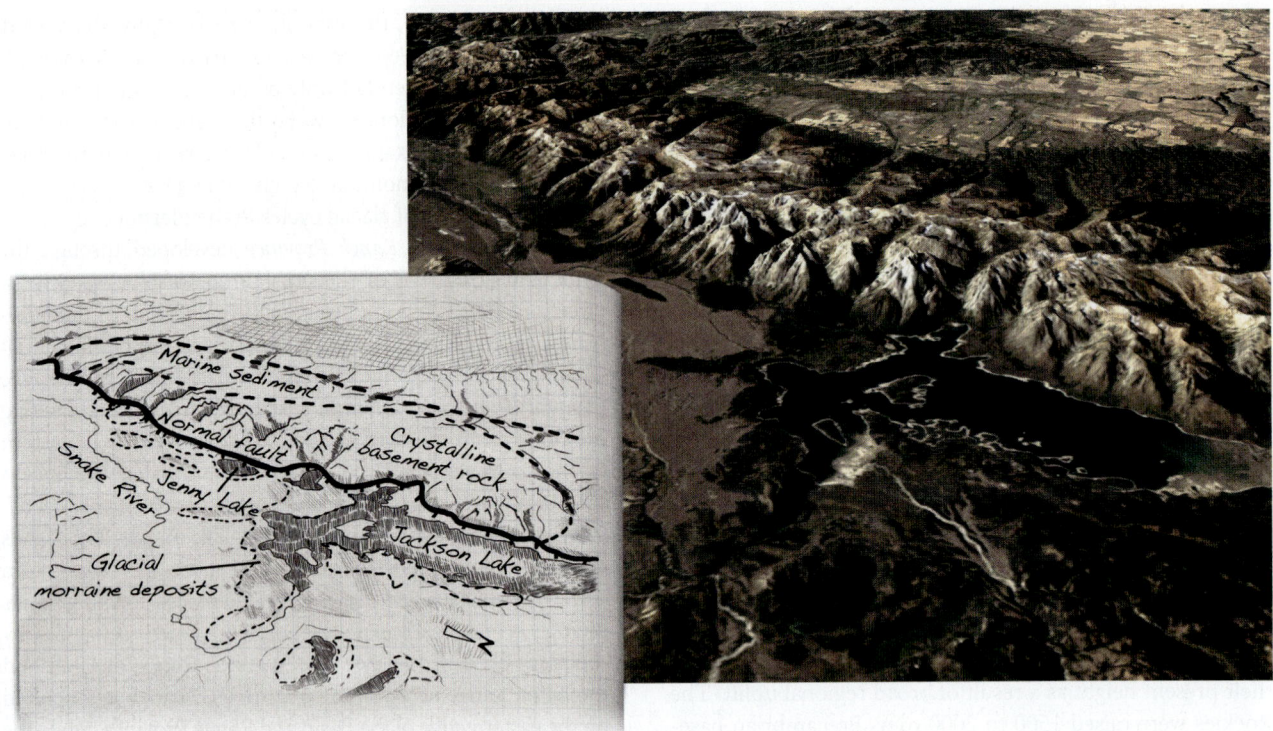

FIGURE 21.7 Image synthesized from satellite data of the Teton Range, Wyoming. The sharp eastern face of the mountain range, which has a vertical relief of more than 2000 m, is the result of normal faulting along the northeastern edge of the Basin and Range Province. The view is from the northeast looking to the southwest. Grand Teton Mountain, near the center of the image, rises to an altitude of 4200 meters. [NASA/Goddard Space Flight Center Landsat 7 Team.]

The *Colorado Plateau* seems to be an island of stability that has experienced no major tension or compression since Precambrian time. The broad uplift of the plateau has allowed the Colorado River to cut through flat-lying sedimentary rock formations, creating the Grand Canyon. Geologists believe that this uplift was caused by the same type of lithospheric heating that is stretching the crust in the Basin and Range Province.

Tectonic Provinces Around the World

We will now expand our view from North America to Earth's other continents. Each continent has its own distinctive features, but a general pattern becomes evident when continental geology is viewed on a global scale (**Figure 21.8a**). Continental shields and platforms make up the most stable parts of the continental lithosphere, called **cratons**, and contain the eroded remnants of ancient deformed rocks. The North American craton comprises the Canadian Shield and the interior platform (see Figure 21.1).

Around these cratons are elongated mountain belts, or **orogens** (from the Greek *oros*, meaning "mountain," and *gen*, "be produced"), that were formed by later episodes of compressive deformation. The youngest orogenic (mountain-building) systems, such as the North American Cordillera, are found along the **active margins** of continents, where tectonic processes caused by plate movements continue to deform the continental crust.

The **passive margins** of continents—those that are attached to oceanic crust as part of the same plate and thus are not near plate boundaries—are zones of extended crust, stretched during the rifting that broke older continents apart and initiated seafloor spreading. This rifting often occurred parallel to older mountain belts, such as the Appalachian Fold Belt.

Types of Tectonic Provinces

The general pattern of cratons bounded by orogens can be seen in Figure 21.8a, which summarizes the major tectonic provinces of the continents worldwide. The classifications portrayed on this map are closely related to those we used to describe the tectonic provinces of North America:

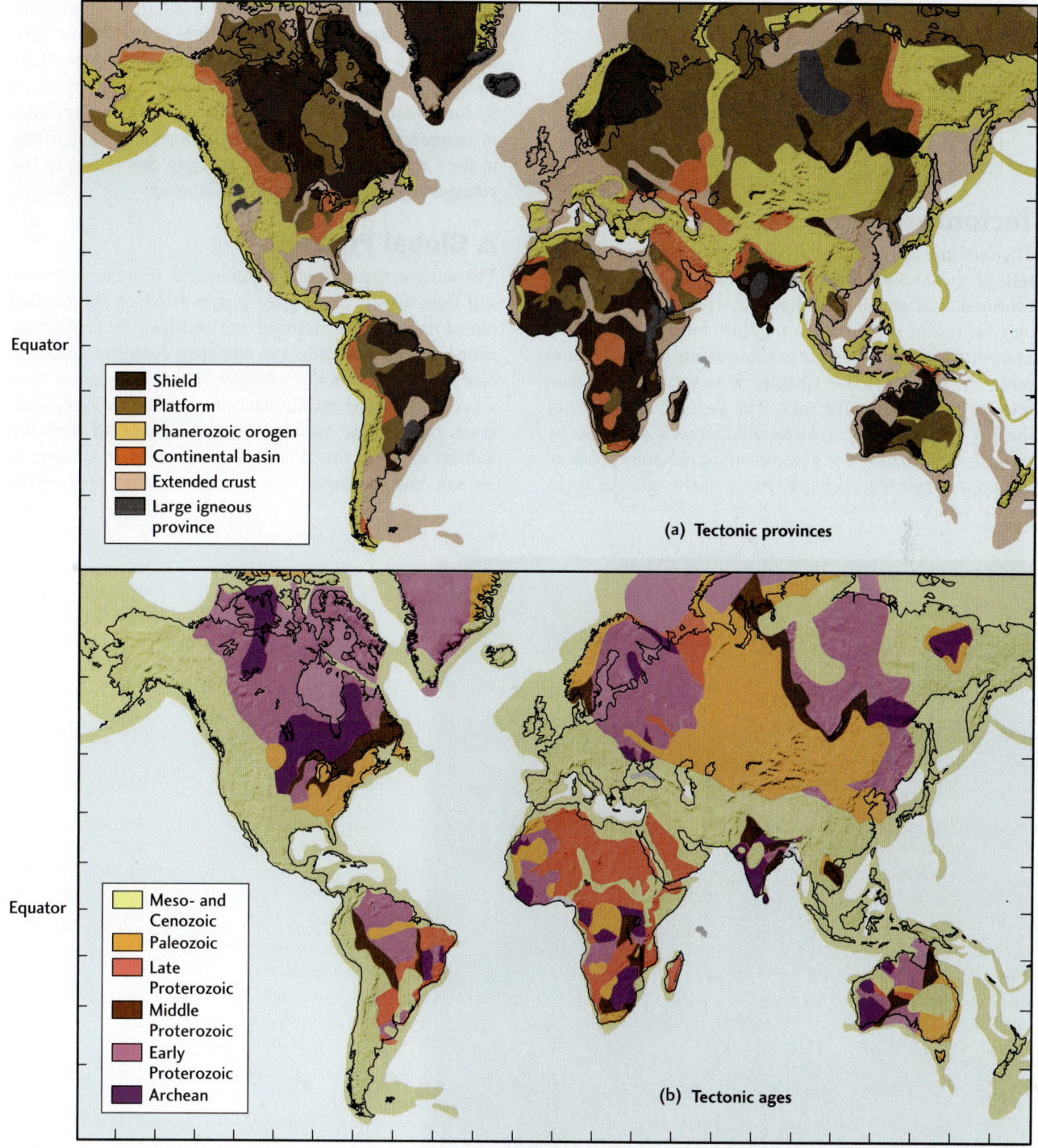

FIGURE 21.8 A global view of the continents, showing (a) their major tectonic provinces and (b) their tectonic ages.

- *Shield.* A region of uplifted and exposed crystalline basement rocks of Precambrian age, which have remained undeformed throughout the Phanerozoic eon (542 million years ago to the present). Example: Canadian Shield.
- *Platform.* A region where Precambrian basement rocks are overlain by less than a few kilometers of relatively flat-lying sediments. Examples: interior platform of central North America, Hudson Bay.
- *Continental basin.* A region of prolonged subsidence where thick sediments have accumulated during the Phanerozoic, with beds dipping into the center of the basin. Example: Michigan Basin.

- *Phanerozoic orogen.* A region where mountain building has occurred during the Phanerozoic. Examples: Appalachian Fold Belt, North American Cordillera.
- *Extended crust.* A region where the most recent deformation has involved large-scale crustal extension. Examples: Basin and Range Province, Atlantic Coastal Plain.

Tectonic Ages

The **tectonic age** of a rock is the time of the last major episode of crustal deformation of that rock (Figure 21.8b). Most continental basement rocks have survived a long and complex history of repeated deformation, melting, and metamorphism. We can often use isotopic dating techniques and other age indicators (see Chapter 9) to extract more than one age to any particular rock. The tectonic age indicates the *last* time the isotopic clocks within a rock were reset by tectonic deformation and accompanying metamorphism of the upper crust. For example, many of the igneous rocks in the southwestern United States were originally derived from the melting of crust and mantle in the middle Proterozoic (1.9 billion to 1.6 billion years ago) (**Figure 21.9**). However, those rocks were substantially metamorphosed by subsequent tectonic activity, including several episodes of compressive deformation in the Mesozoic and rifting in the Cenozoic. Geologists thus assign this region to the youngest age category, Mesozoic-Cenozoic.

A Global Puzzle

The current distribution of continental tectonic provinces and their ages is like a giant puzzle in which the original pieces have been rearranged and reshaped by continental rifting, continental drift, and continent-continent collisions over billions of years. Only the past 200 million years of plate movements can be reliably determined from existing oceanic crust. Earlier plate movements must be inferred from the indirect evidence found in continental rocks. In Chapter 2, we saw that geologists have made amazing progress in

FIGURE 21.9 The Vishnu Schist, part of the middle Proterozoic basement (1.8 billion years old), found at the bottom of the Grand Canyon.
[NPS photo by Michael Quinn.]

reconstructing earlier configurations of the continents from paleomagnetic and paleoclimate data and from the signatures of deformation exposed in ancient mountain belts.

In the next section, we trace the history of the continents even further back into geologic time. Once again, we use the history of North America as our prime example, starting with its youngest provinces along its west coast and working backward in time to the Canadian Shield. We focus on three key questions about continental evolution: What geologic processes built the continents we see today? How do these processes fit into the theory of plate tectonics? Can plate tectonics explain the original formation of the cratons? As we will see, these questions have been only partially answered by geological research.

How Continents Grow

Over the continents' 4-billion-year history, new crust has been added at an average rate of about 2 km^3/year. Earth scientists continue to debate whether the growth of continental crust has occurred gradually over geologic time or was concentrated early in Earth's history. In the modern plate-tectonic system, two basic processes work together to form new continental crust: magmatic addition and accretion.

Magmatic Addition

The process of magmatic differentiation of low-density, silica-rich rock in Earth's mantle and *vertical* transport of this buoyant, felsic material from the mantle to the crust is called **magmatic addition**.

Most new continental crust is born in subduction zones from magmas formed by fluid-induced melting of the subducting lithospheric slab and the mantle wedge above the slab (see Chapter 4). These magmas, which are of basaltic to andesitic composition, migrate toward the surface, pooling in magma chambers near the base of the crust. Here they incorporate crustal materials and differentiate further to form the felsic magmas that migrate into the upper crust, forming dioritic and granodioritic plutons capped by andesitic volcanoes.

Magmatic addition can emplace new crustal material directly at active continental margins. Subduction of the Farallon Plate beneath North America during the Cretaceous period, for example, created the batholiths along the western edge of the continent, including the rocks now exposed in Baja California and the Sierra Nevada. Subduction of the remnant Juan de Fuca Plate continues to add new material to the crust in the volcanically active Cascade Range of the Pacific Northwest, just as subduction of the Nazca Plate is building up the crust in the Andes of South America.

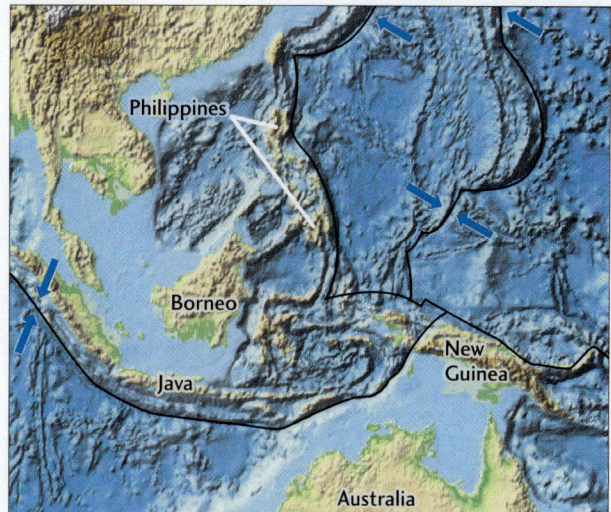

FIGURE 21.10 The Philippines and other island groups in the southwestern Pacific illustrate how island arcs can merge into thick sections of protocontinental crust at ocean-ocean convergence zones.

Buoyant felsic crust is also produced far away from continents, in volcanic island arcs at ocean-ocean convergence zones. Over time, these island arcs can merge into thick sections of silica-rich crust, such as those found today in the Philippines and other island groups of the southwestern Pacific (**Figure 21.10**). Plate movements transport these fragments of crust horizontally across the globe and eventually attach them to active continental margins by accretion.

Accretion

The integration of crustal material previously differentiated from mantle material into existing continental masses by *horizontal* transport during plate movements is called **accretion**.

Geologic evidence for accretion can be found on the active margins of North America. In the Pacific Northwest and Alaska, the crust consists of a mix of odd pieces—island arcs, seamounts (extinct underwater volcanoes), and remnants of basalt plateaus, old mountain ranges, and other slivers of continental crust—that were plastered onto the leading edge of the continent as it moved across Earth's surface. These pieces are sometimes referred to as **accreted terrains**. Geologists use this term (which they often spell as "accreted terranes") to define a large piece of crust, tens to hundreds of kilometers in geographic extent, with common characteristics and a distinct origin, usually transported great distances by plate movements.

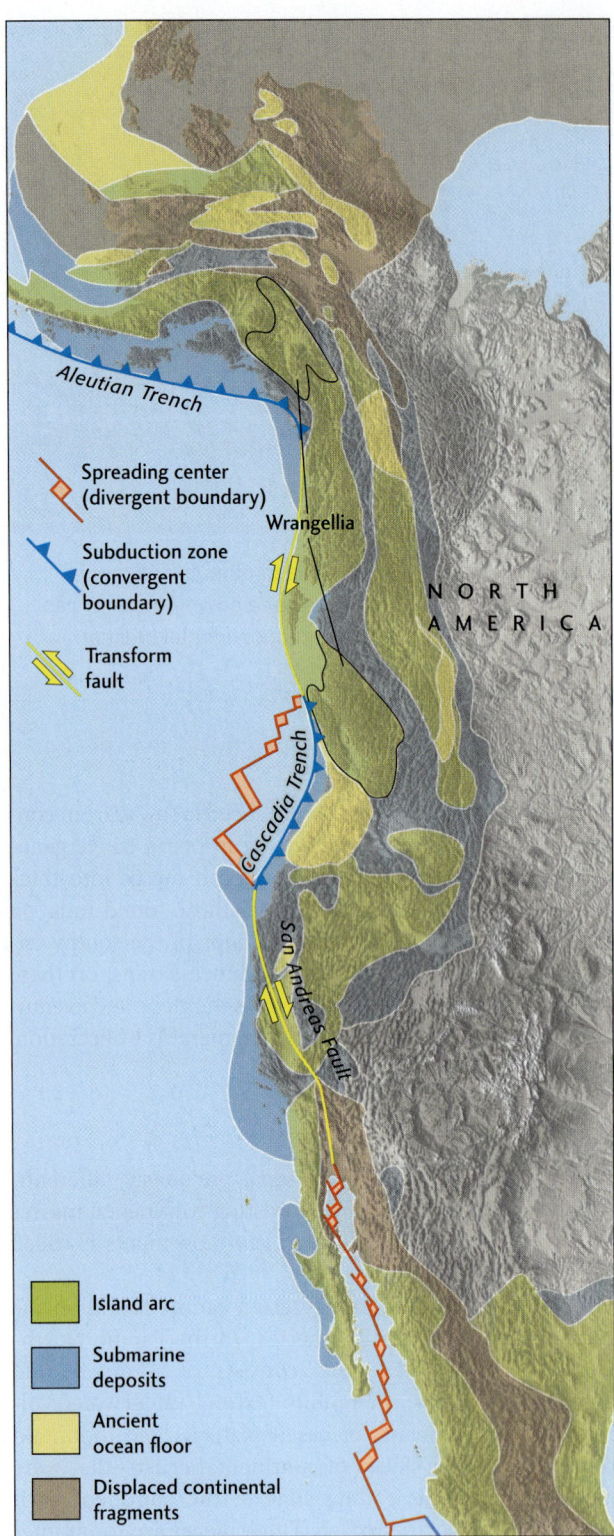

FIGURE 21.11 Much of the North American Cordillera has been formed by terrain accretion over the past 200 million years. Wrangellia, for example, is a former basalt plateau that was transported to its present location from 5000 km away. Other accreted terrains are made up of island arcs, ancient seafloor, and continental fragments.

The geologic arrangement of accreted terrains can be chaotic (Figure 21.11). Adjacent blocks of crust can contrast sharply in their rock types, the nature of their folding and faulting, and their history of magmatic activity and metamorphism. Geologists often find fossils indicating that these blocks originated in different environments, and at different times, than the rocks of the surrounding area. For example, an accreted terrain comprising ophiolite suites (pieces of seafloor) that contain deep-water fossils might be surrounded by remnants of island arcs and continental fragments containing shallow-water fossils of a completely different age. The boundaries between accreted terrains are almost always faults that have undergone substantial slippage, although the nature of the faulting is often difficult to discern. Blocks of crust that seem completely out of place are called *exotic terrains*.

Before the discovery of plate tectonics, exotic terrains were a subject of fierce debate among geologists, who had difficulty coming up with reasonable explanations for their origins. Accreted terrain analysis is a specialized field within plate-tectonic research. Well over a hundred areas of the North American Cordillera have been identified as exotic terrains accreted during the last 200 million years (many more than depicted in Figure 21.11). One such terrain, called Wrangellia, originally formed as a large basalt plateau (a region of oceanic crust thickened by a large outpouring of basaltic lava) and was then transported over 5000 km from the Southern Hemisphere to its current location in Alaska and western Canada. Extensive accreted terrains have also been mapped in Japan, Southeast Asia, China, and Siberia.

In only a few cases do we know precisely where these accreted terrains originated. We can begin to decipher how the others came together by considering four distinct tectonic processes that can result in accretion (Figure 21.12):

1. A crustal fragment that is too buoyant to be subducted may be transferred from a subducting plate to a continent on the overriding plate. Such fragments can be small pieces of continental crust ("microcontinents") or thickened sections of oceanic crust (large seamounts, basalt plateaus).

2. A sea that separates an island arc from a continent may be closed as the thickened island arc crust collides with and becomes attached to the advancing edge of the continent.

3. Two plates that slide past each other along a transform fault may result in strike-slip faulting and the movement of a crustal fragment from one plate to the other. Today, the southwestern part of California, which is attached to the Pacific Plate, is moving northwestward relative to the North American Plate along the San Andreas transform fault (see

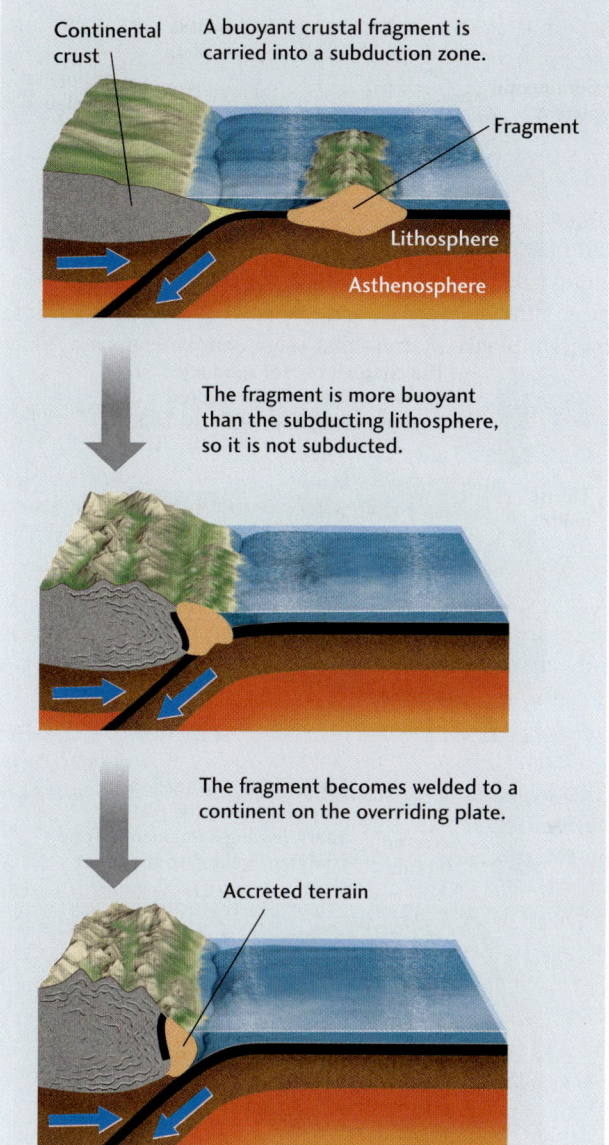
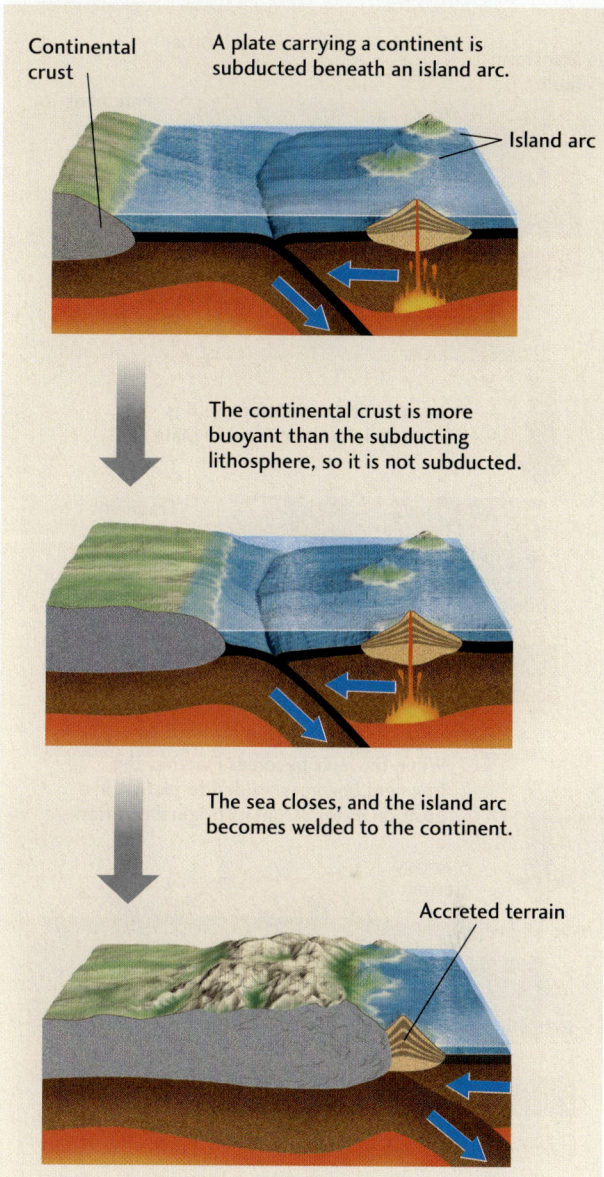

FIGURE 21.12 Four distinct processes explain the accretion of exotic terrains.

Figure 21.6). Strike-slip faulting landward of the deep-sea trench in oblique subduction zones can also transport terrains hundreds of kilometers.

4. Two continents may collide and be sutured together, then break apart later at a different location.

The fourth process explains some of the accreted terrains found on the passive eastern margin of North America. The Appalachian Fold Belt contains slices of ancient Europe and Africa as well as a variety of exotic terrains. Florida's oldest rocks and fossils are more like those in Africa than like those found in the rest of the United States, indicating that most of this peninsula was probably transported to North America when Pangaea was assembled and then left behind when North America and Africa split apart about 200 million years ago.

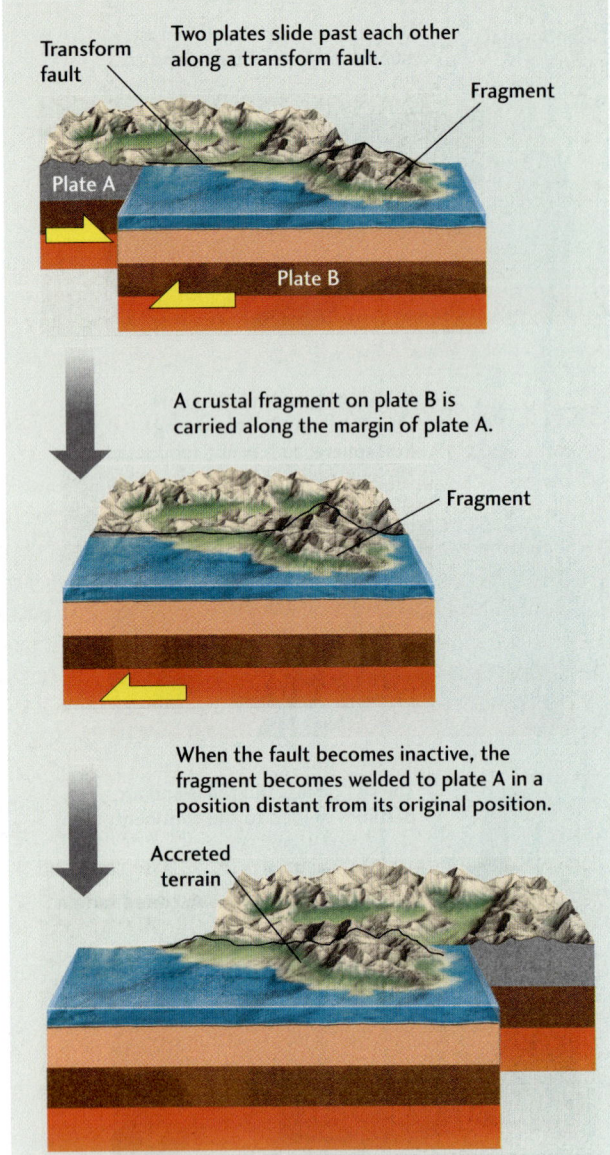
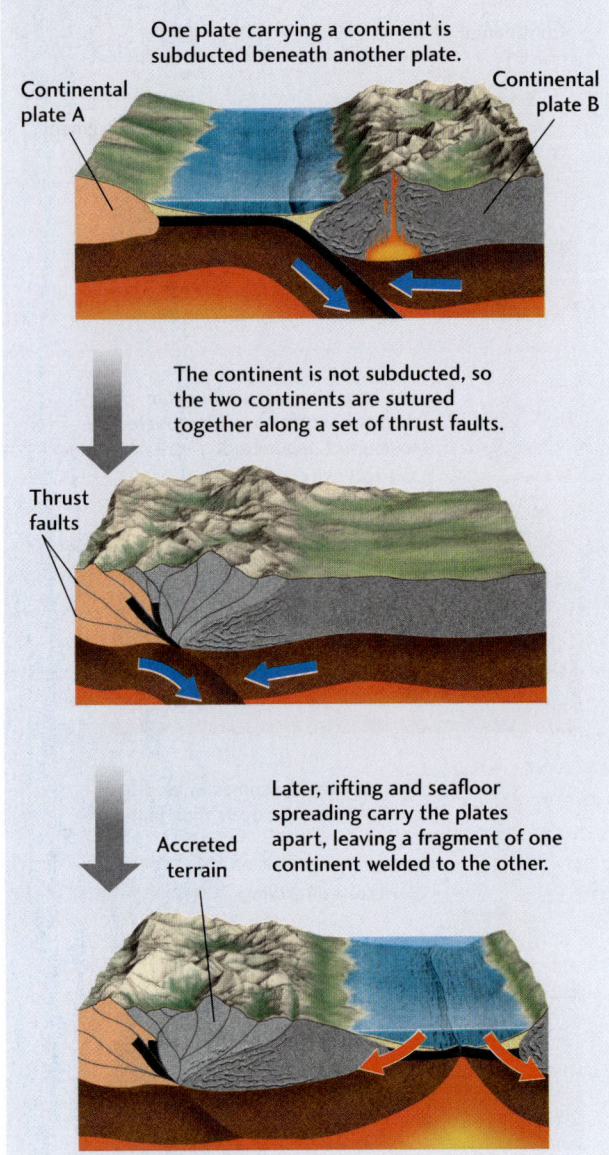

FIGURE 21.12 Four distinct processes explain the accretion of exotic terrains. (*Continued*)

How Continents Are Modified

The geology of the North American Cordillera, with its many exotic terrains, looks nothing like that of the ancient Canadian Shield, which lies directly east of the Cordillera. In particular, the accreted terrains of the youthful Cordilleran system do not show the high degree of melting or the high-grade metamorphism that characterize the Precambrian crust of the shield. Why such a difference? The answer lies in the tectonic processes that have repeatedly modified the older parts of the continent throughout its long history.

Orogeny: Modification by Plate Collision

Continental crust is profoundly altered by **orogeny**—the mountain-building processes of folding, faulting, magmatic addition, and metamorphism. Orogenic processes have repeatedly modified the edges of cratons. Most orogenies (episodes of mountain building) result from plate convergence. When one or both plates are made of oceanic lithosphere, their convergence usually results in subduction rather than orogeny. Orogenies can result when a continent rides forcibly over subducting oceanic crust, as in the Andean orogeny now under way in South America, but the most intense orogenies are caused by the collision of two

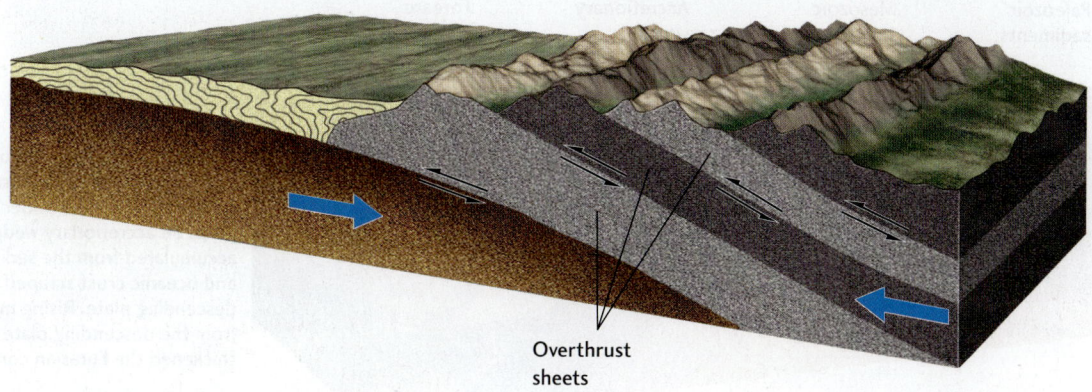

FIGURE 21.13 When continents collide, the continental crust can break into overthrust sheets stacked one above the other.

or more continents. As we saw in Chapter 2, when continents collide, a basic tenet of plate tectonics—the rigidity of plates—must be modified.

Continental crust is much more buoyant than mantle material, so colliding continents resist being subducted with the plates that carry them. Instead, the continental crust deforms and breaks in a combination of intense folding and faulting that can extend hundreds of kilometers from the collision zone, as described in Chapter 8. Thrust faulting caused by the convergence can stack the upper part of the crust into overthrust sheets tens of kilometers thick, deforming and metamorphosing the rocks they contain (**Figure 21.13**). Continental shelf sediments can be scraped off the basement rock on which they were deposited and thrust inland. Horizontal compression throughout the crust can double its thickness, causing the rocks in the lower crust to melt.

This melting can generate huge amounts of granitic magma, which rises to form extensive batholiths in the upper crust.

The Alpine-Himalayan Orogeny To see orogeny in action today, we look to the great chains of high mountains that stretch from Europe through the Middle East and across Asia, known collectively as the *Alpine-Himalayan belt* (**Figure 21.14**). The breakup of Pangaea sent Africa, Arabia, and India northward, causing the Tethys Ocean to close as its lithosphere was subducted beneath Eurasia (see Figure 2.16). These former pieces of Gondwana collided with Eurasia in a complex sequence, beginning in the western part of Eurasia during the Cretaceous period and continuing eastward through the Tertiary, raising the Alps in central Europe, the Caucasus and Zagros mountains in the Middle East, and the Himalaya and other high mountain chains across central Asia.

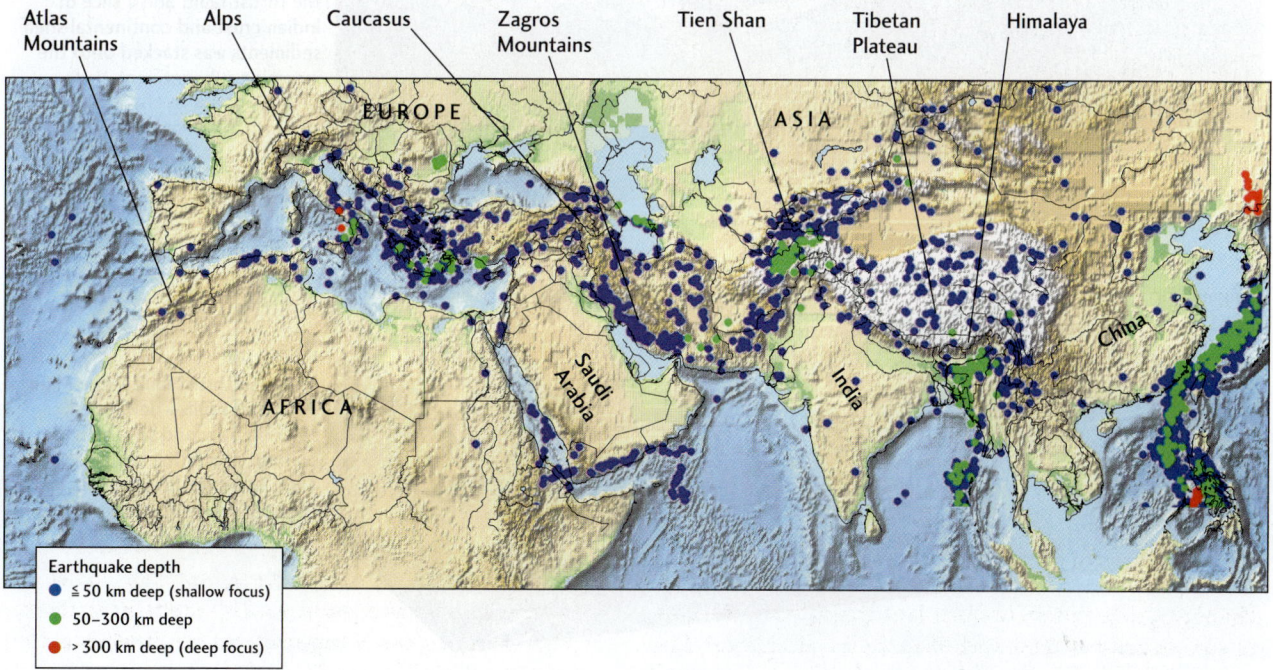

FIGURE 21.14 The Alpine-Himalayan belt, showing the chains of high mountains built by the ongoing collision of the African, Arabian, and Indian plates with the Eurasian Plate. This orogeny is marked by intense earthquake activity.

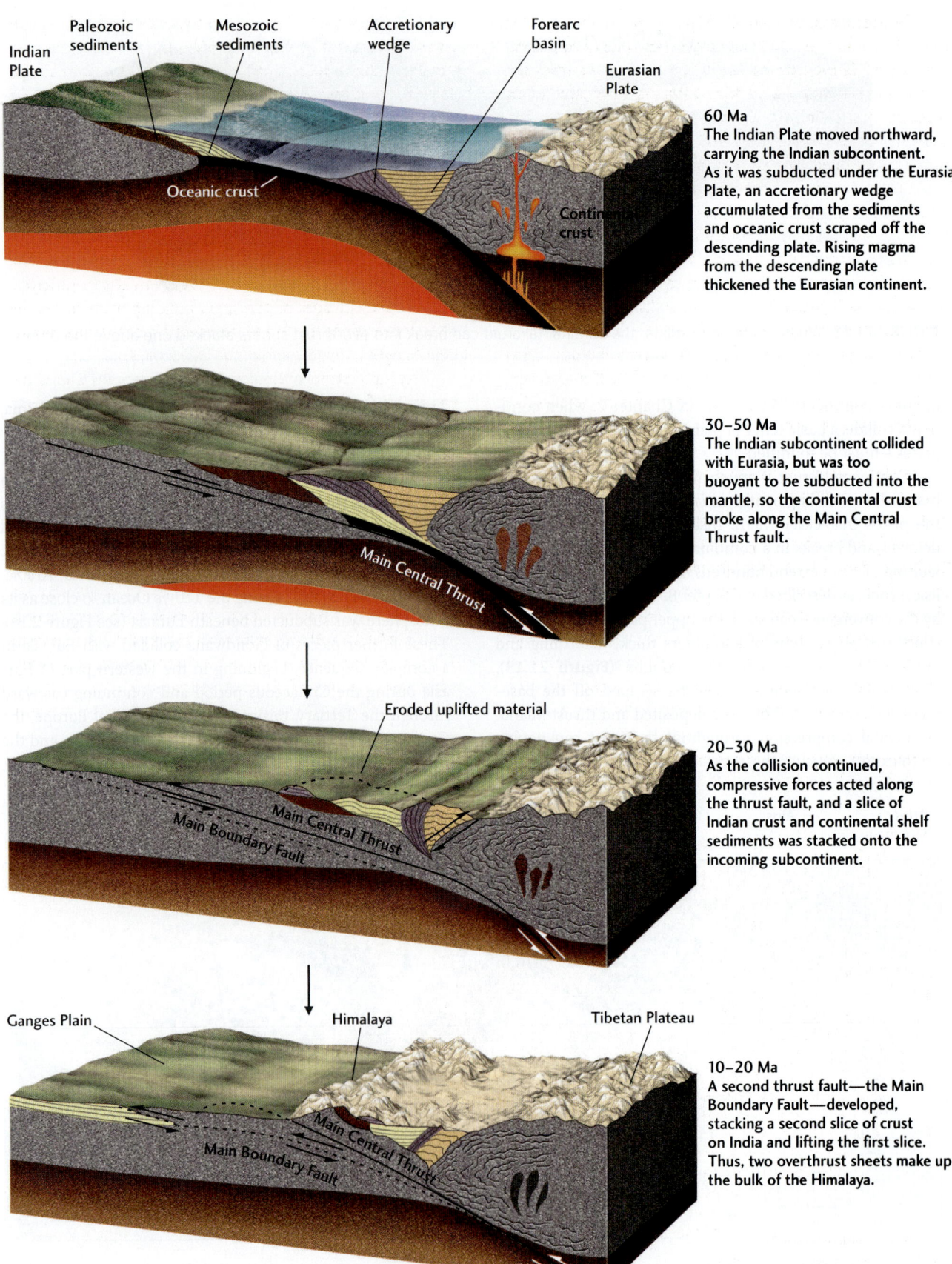

FIGURE 21.15 Cross sections showing the sequence of events that have caused the Himalayan orogeny, simplified and vertically exaggerated. (Ma: million years ago.)

The Himalaya, the world's highest mountains, are the most spectacular result of this modern episode of continent-continent collision (see the Practicing Geology Exercise following this section). About 50 million years ago, the Indian subcontinent, riding on the subducting Indian Plate, first encountered the island arcs and volcanic mountain belts that then bounded the Eurasian Plate (**Figure 21.15**). As the landmasses of India and Eurasia merged, the Tethys Ocean disappeared through subduction. Pieces of the oceanic crust were trapped along the suture zone between the converging continents and can be seen today as ophiolite suites along the Indus and Tsangpo river valleys that separate the high Himalaya from the Tibetan Plateau. The collision slowed India's advance, but the Indian Plate continued to drive northward. So far, India has penetrated over 2000 km into Eurasia, causing the largest and most intense orogeny of the Cenozoic era.

The Himalaya were formed from overthrust slices of the old northern portion of India, stacked one atop the other (see Figure 21.15). This process took up some of the compression. Horizontal compression and the formation of fold and thrust belts also thickened the crust north of India, causing the uplift of the huge Tibetan Plateau, which now has a crustal thickness of 60 to 70 km (almost twice the thickness of most continental crust) and stands nearly 5 km above sea level. These and other compression zones account for perhaps half of India's penetration into Eurasia. Further compression has pushed China and Mongolia eastward, out of India's way, like toothpaste squeezed from a tube. Most of this sideways movement has taken place along the Altyn Tagh Fault and other major strike-slip faults shown on the map in **Figure 21.16**. The mountains, plateaus, faults, and great earthquakes of Asia, extending thousands of kilometers from the Indian-Eurasian suture, are all results of the Alpine-Himalayan orogeny, which continues today as India plows into Asia at a rate of 40 to 50 mm/year.

Paleozoic Orogenies During the Assembly of Pangaea If we go further back in geologic time, we find abundant evidence of older orogenies. We have already mentioned, for example, that at least three distinct orogenies were responsible for the Paleozoic deformation now exposed in the eroded Appalachian Fold Belt of eastern North America. These three episodes of mountain building were caused by plate convergence that led to the assembly of the supercontinent Pangaea near the end of the Paleozoic era.

The supercontinent Rodinia began to break up toward the end of the Proterozoic eon, forming several paleocontinents (see Figure 2.16). One was the large continent of *Gondwana*. Two of the others were *Laurentia*, which included the North American craton and Greenland, and *Baltica*, comprising what are now the lands around the Baltic Sea (Scandinavia, Finland, and the European part of Russia). In the Cambrian period, Laurentia was rotated almost 90° from its present orientation and straddled the equator; its southern (today, eastern) side was a passive continental margin. To its immediate south was the proto-Atlantic, or *Iapetus*, Ocean (in Greek mythology, Iapetus was the father of Atlantis), which was being subducted beneath a distant island arc. Baltica lay off to the southeast, and Gondwana was thousands of kilometers to the south. **Figure 21.17** shows the sequence of events as the three continents converged.

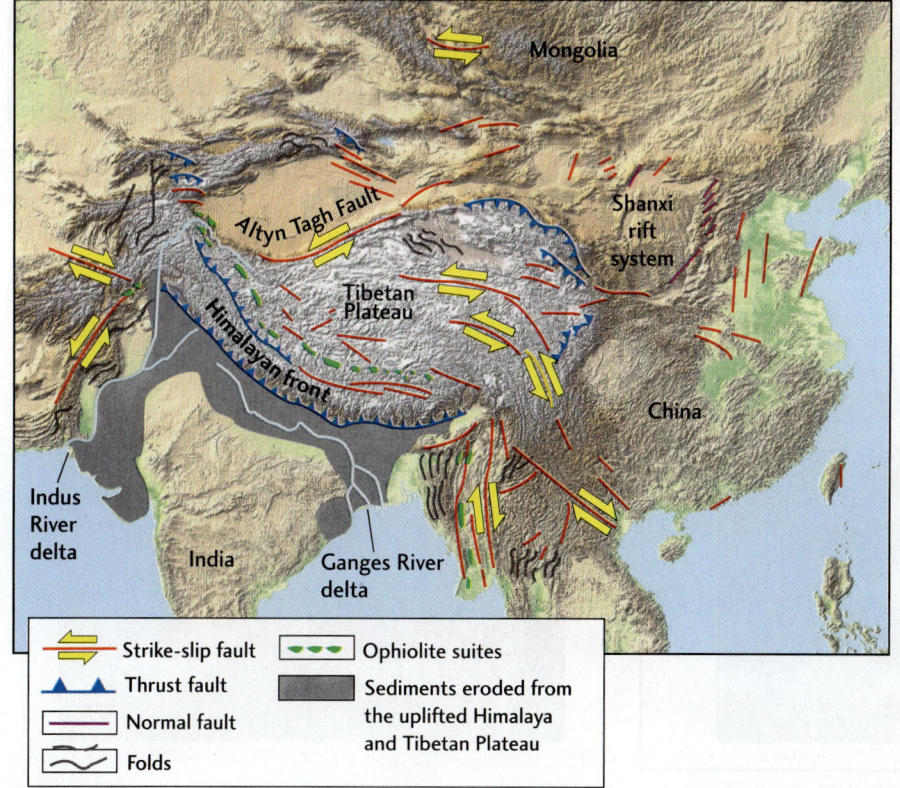

FIGURE 21.16 The collision between India and Eurasia has produced many spectacular tectonic features, including large-scale faulting and uplift.

Middle Cambrian (510 Ma)
After the breakup of Rodinia, Laurentia straddled the equator. Its southern side was a passive continental margin, bounded on the south by the Iapetus Ocean.

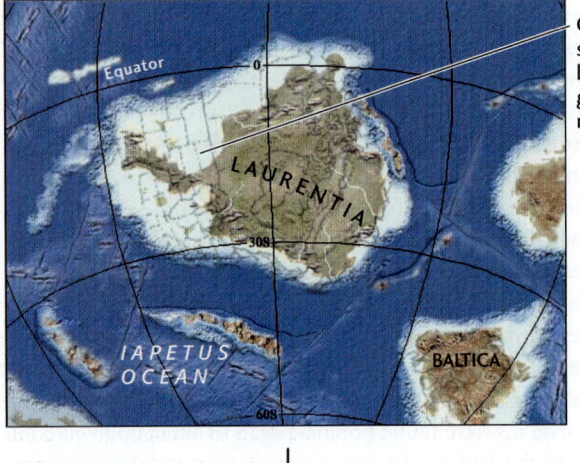

Outlines show U.S. state boundaries for geographic reference

Shelf and submerged continent

Early Carboniferous (340 Ma)
The collision of Gondwana with Laurussia began with the Variscan orogeny in what is now central Europe…

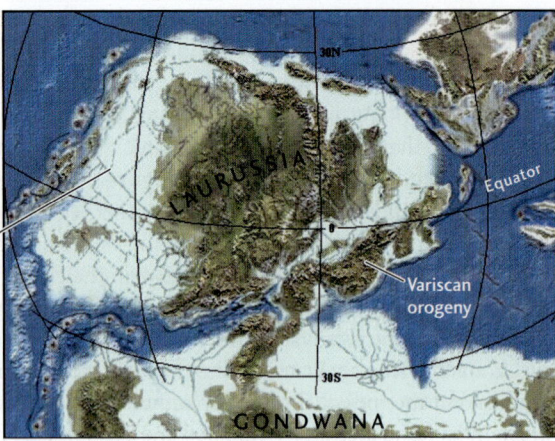

Late Ordovician (450 Ma)
The island arc built up by the southward-directed subduction of Iapetus lithosphere collided with Laurentia in the middle to late Ordovician, causing the Taconic orogeny.

Late Carboniferous (300 Ma)
… and continued along the margin of the North American craton with the Appalachian orogeny. At the same time, Siberia converged with Laurussia in the Ural orogeny to form Laurasia, while the Hercynian orogeny created new mountain belts across Europe and northern Africa.

Early Devonian (400 Ma)
The collision of Laurentia with Baltica caused the Caledonian orogeny and formed Laurussia. The southward continuation of the convergence caused the Acadian orogeny.

Early Permian (270 Ma)
The end product of these episodes of continental convergence was the supercontinent of Pangaea.

FIGURE 21.17 Paleogeographic reconstructions of the present North Atlantic region, showing the sequence of orogenic episodes that resulted from the assembly of Pangaea. (Ma: million years ago.) [Information from R. C. Blakey, Northern Arizona University, Flagstaff.]

The island arc built up by the southward-directed subduction of Iapetus lithosphere collided with Laurentia in the middle to late Ordovician (470 million to 440 million years ago), causing the first episode of mountain building: the *Taconic orogeny*. (You can see some of the rocks accreted and deformed during this period if you drive the Taconic State Parkway, which runs east of the Hudson River for about 160 km north of New York City.) The second orogeny began when Baltica and a connected set of island arcs began to collide with Laurentia in the early Devonian (about 400 million years ago). The collision deformed southeastern Greenland, northwestern Norway, and Scotland in what European geologists refer to as the *Caledonian orogeny*. The deformation continued into present-day North America as the *Acadian orogeny*, as island arcs that would become the terrains of maritime Canada and New England accreted to Laurentia in the middle to late Devonian (380 million to 360 million years ago).

The grand finale in the assembly of Pangaea was the collision of the behemoth landmass of Gondwana with Laurasia and Baltica, by then joined into a continent named *Laurussia*. The collision began about 340 million years ago with the *Variscan orogeny* in what is now central Europe and continued along the margin of the North American craton with the *Appalachian orogeny* (320 million to 270 million years ago). This latter phase of assembly pushed Gondwanan crust over Laurentia, lifting the Blue Ridge into a mountain chain that may have been as high as the modern Himalaya and causing much of the deformation now seen in the Appalachian Fold Belt. Also during this phase, Siberia and other Asian terrains converged with Laurussia in the *Ural orogeny*, forming the continent of *Laurasia* and pushing up the Ural Mountains. At the same time, extensive deformation created new mountain belts across Europe and northern Africa (the *Hercynian orogeny*).

The crunching together of all these continental masses profoundly altered the structure of the crust. The rigid cratons were little affected, but the younger accreted terrains caught in between were consolidated, thickened, and metamorphosed. The lower parts of this younger crust were partially melted, producing granitic magmas that rose to form batholiths in the upper crust and volcanoes at the surface. Uplifted mountains and plateaus were eroded, exposing high-grade metamorphic rocks that were once many kilometers deep and depositing thick sedimentary sequences. Sediments laid down following the first orogeny were deformed and metamorphosed by later mountain-building episodes.

Earlier Orogenies So far, we have investigated two major periods of mountain building: the Paleozoic orogenies associated with the assembly of Pangaea, and the Cenozoic Alpine-Himalayan orogeny. In Chapter 2, we discussed the assembly of the supercontinent Rodinia in the late Proterozoic eon. By now, it should not surprise you to learn that major orogenies accompanied the formation of that earlier supercontinent.

Some of the best evidence of these orogenies is found at the eastern and southern margins of the Canadian Shield in a broad belt known as the Grenville Province, where new crustal material was added to the continent in the middle Proterozoic, about 1.1 billion to 1.0 billion years ago (see Figure 21.8b). Geologists believe that these rocks, which are now highly metamorphosed, originally consisted of volcanic mountain belt and island arc terrains that were accreted and compressed by the collision of Laurentia with the western part of Gondwana. They have drawn analogies between what happened during this *Grenville orogeny* and what is happening today in the Himalayan orogeny. A Tibet-like plateau was formed by compressive thickening of the crust through folding and thrust faulting, which metamorphosed the upper crust and partially melted large parts of the lower crust. Once the orogeny ceased, erosion of the plateau thinned the crust and exposed crystalline rocks of high metamorphic grade. Geologists have found orogenic belts of similar age on continents worldwide. Although many of the details remain uncertain, they have reconstructed from this geologic record (which includes paleomagnetic data) a general picture of how Rodinia came together between 1.3 billion and 0.9 billion years ago.

The Wilson Cycle

From our brief look at the history of eastern North America, we can infer that the edges of many cratons have experienced multiple episodes of deformation in a general plate-tectonic cycle that comprises four main phases (**Figure 21.18**):

1. Rifting during the breakup of a supercontinent
2. Passive margin cooling and sediment accumulation during seafloor spreading and ocean opening
3. Active margin volcanism and terrain accretion during subduction and ocean closure
4. Orogeny during the continent-continent collision that forms the next supercontinent

This idealized sequence of events has been named the **Wilson cycle** after the Canadian pioneer of plate tectonics, J. Tuzo Wilson, who first recognized its importance in the evolution of continents.

The geologic record suggests that the Wilson cycle has operated throughout the Proterozoic and Phanerozoic eons (**Figure 21.19**), resulting in the formation of at least two supercontinents prior to Rodinia. One of these supercontinents (named *Columbia*) formed about 1.9 billion to 1.7 billion years ago. An even earlier one, whose assembly marks the transition from the Archean eon to the Proterozoic eon, formed about 2.7 billion to 2.5 billion years ago. Did the Wilson cycle also operate in the Archean eon? We will return to that question shortly.

Epeirogeny: Modification by Vertical Movements

So far, our consideration of continental evolution has emphasized accretion and orogeny, processes that involve horizontal plate movements and are usually accompanied by deformation in the form of folding and faulting. Throughout

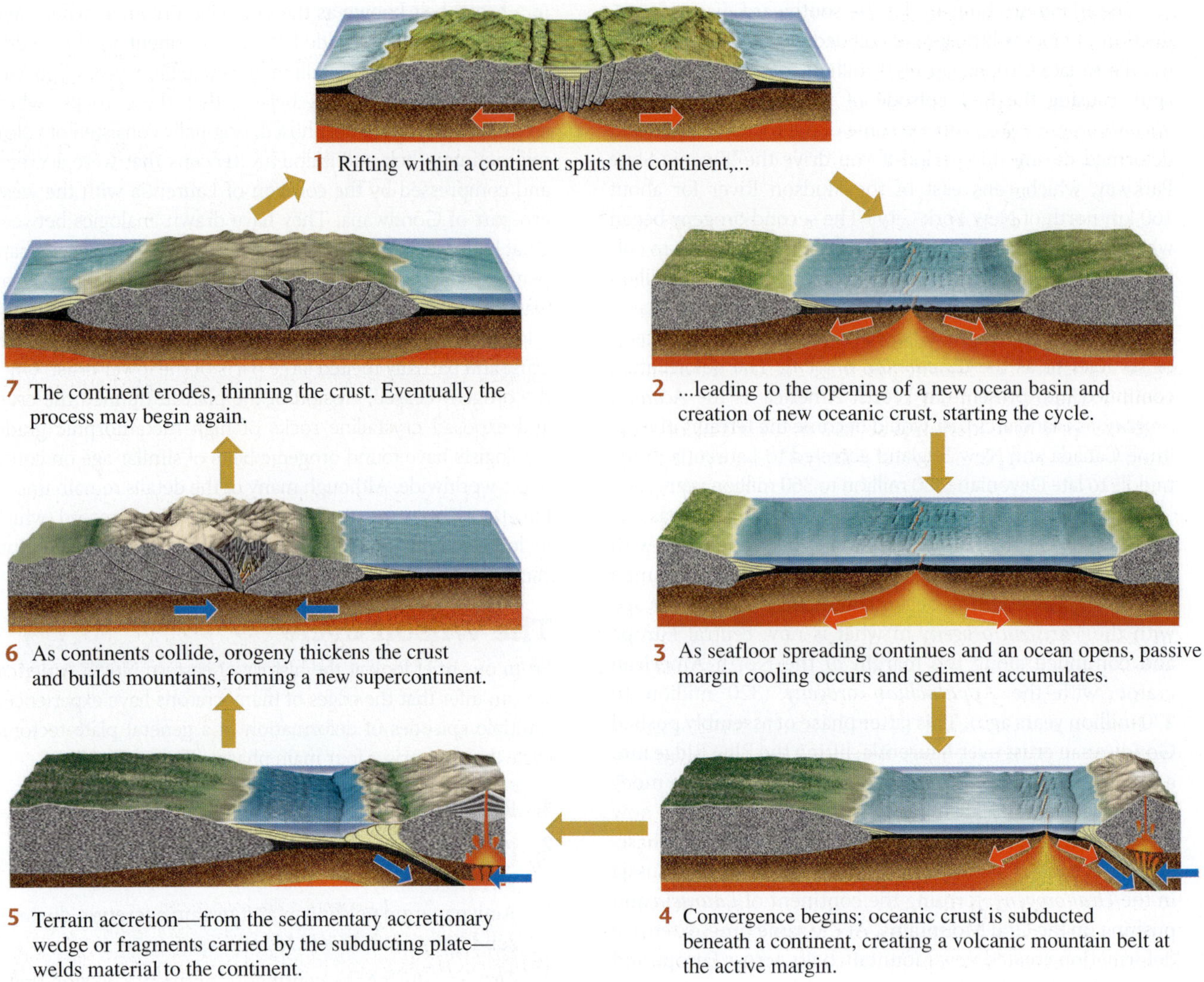

1 Rifting within a continent splits the continent,...
2 ...leading to the opening of a new ocean basin and creation of new oceanic crust, starting the cycle.
3 As seafloor spreading continues and an ocean opens, passive margin cooling occurs and sediment accumulates.
4 Convergence begins; oceanic crust is subducted beneath a continent, creating a volcanic mountain belt at the active margin.
5 Terrain accretion—from the sedimentary accretionary wedge or fragments carried by the subducting plate—welds material to the continent.
6 As continents collide, orogeny thickens the crust and builds mountains, forming a new supercontinent.
7 The continent erodes, thinning the crust. Eventually the process may begin again.

FIGURE 21.18 The Wilson cycle comprises the plate-tectonic processes responsible for the formation and breakup of supercontinents and the opening and closing of ocean basins.

Earth accretion
Oldest Earth minerals
Oldest continental rocks
Earliest cratons form
ARCHEAN EON
HADEAN EON
4000 Ma
3000 Ma
Assembly of early supercontinent?
Assembly of Columbia supercontinent
Assembly of Rodinia supercontinent
2000 Ma
PROTEROZOIC EON
1000 Ma
Assembly of Pangaea supercontinent
Alpine-Himalayan orogen
FUTURE
PHANEROZOIC EON
0 Ma

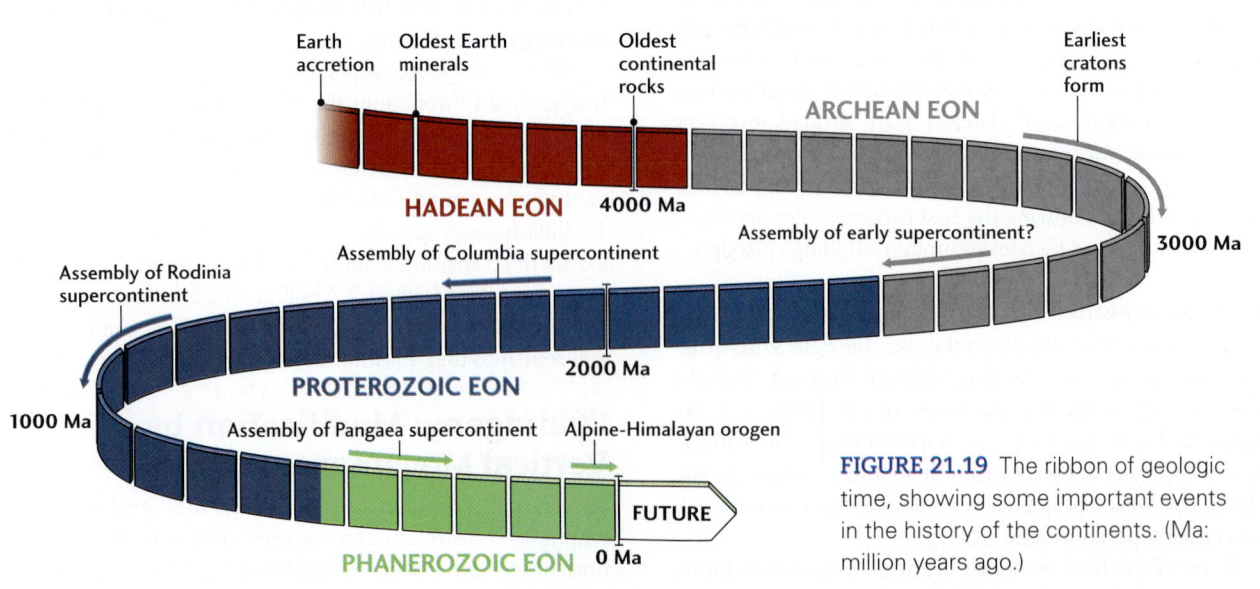

FIGURE 21.19 The ribbon of geologic time, showing some important events in the history of the continents. (Ma: million years ago.)

the world, however, sedimentary rock sequences record another kind of movement that has modified continents: gradual downward and upward movements of broad regions of crust without significant folding or faulting. These vertical movements are referred to as **epeirogeny**, a term coined in 1890 by the American geologist Clarence Dutton (from the Greek *epeiros*, meaning "mainland").

Epeirogenic downward movements usually result in a sequence of relatively flat-lying sediments, such as those found in the stable interior platform of North America. Upward movements cause erosion and gaps in the sedimentary record seen as unconformities. Erosion can lead to the exposure of crystalline basement rocks, such as those found on the Canadian Shield.

Geologists have identified several mechanisms of epeirogeny. One example is **glacial rebound** (Figure 21.20a). When large glaciers form, their weight depresses the continental crust. When they melt, the crust rebounds upward for tens of millennia. Glacial rebound explains the uplift of Finland and Scandinavia following the most recent glaciation, which ended about 17,000 years ago, as well as the raised beaches of northern Canada (Figure 21.21). Although glacial rebound seems slow by human standards, it is a rapid process, geologically speaking.

Heating and cooling of the continental lithosphere are important epeirogenic processes on longer time scales. Heating causes rocks to expand, decreasing their density and thus raising the continental surface (Figure 21.20b). A good example is the Colorado Plateau, which has been uplifted to about 2 km above sea level during the last 10 million years or so. Geologists think this heating results from active mantle upwelling, which is also stretching the crust in the Basin and Range Province on the western and southern sides of the plateau.

Conversely, the cooling of lithosphere increases its density, making it sink under its own weight and creating a thermal subsidence basin (Figure 21.20c). Cooling of once-hot areas in the continental interior may explain the Michigan Basin and other deep basins in central North America (see Figure 21.3). When a new episode of seafloor spreading splits a continent apart, the uplifted edges are eroded and eventually subside as they cool, forming basins in which sediments are deposited and carbonate platforms accumulate (Figure 21.20d). This process has led to the formation of a thick continental shelf along the east coast of the United States.

One intriguing puzzle is the South African Plateau, where a craton has been uplifted during the Cenozoic to almost 2 km above sea level—more than twice the elevation of most cratons. However, the lithosphere in this part of the continent does not appear to be unusually hot. One possible explanation is that the southern African craton may be uplifted by a hot, buoyant region of the lower mantle (see Chapter 11). This "superplume" could apply upward forces at the base of the lithosphere sufficient to raise the surface by about a kilometer (Figure 21.20e).

(a) GLACIAL REBOUND

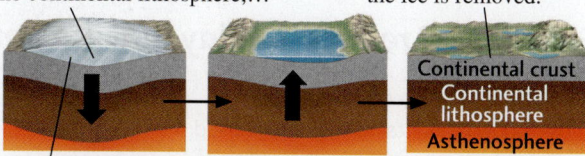

The weight of glacial ice downwarps the continental lithosphere,… …which rebounds once the ice is removed.

(b) HEATING OF LITHOSPHERE

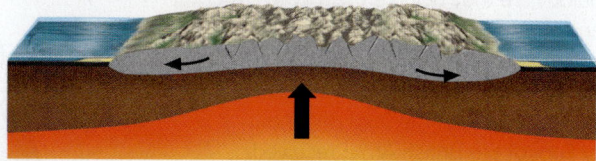

Upwelling of mantle material causes uplift and thinning of the continental lithosphere.

(c) COOLING OF LITHOSPHERE IN CONTINENTAL INTERIOR

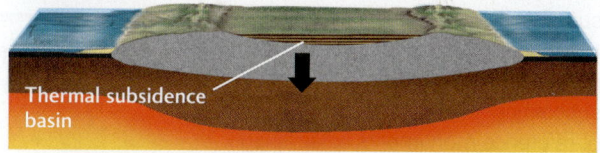

As the lithosphere cools and contracts, it subsides to form a basin within the continent.

(d) COOLING OF LITHOSPHERE ON CONTINENTAL MARGIN

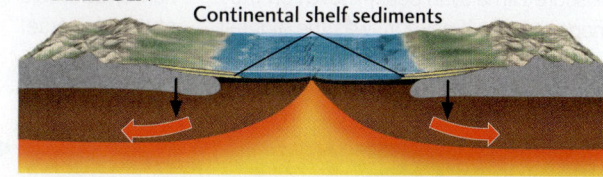

When seafloor spreading splits a continent apart, the edges subside as they cool, accumulating thick sediments.

(e) HEATING OF DEEP MANTLE

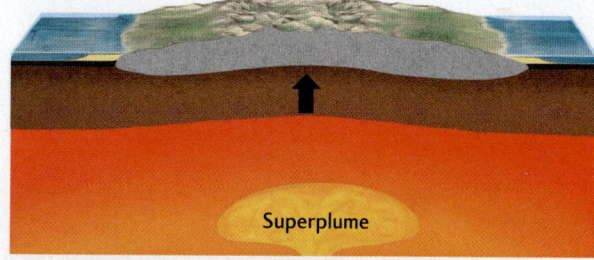

A superplume rising from the deep mantle heats the lithosphere and raises the base of the continent, upwarping the surface over a broad area.

FIGURE 21.20 Five major mechanisms of epeirogeny.

PRACTICING GEOLOGY EXERCISE

How Fast Are the Himalaya Rising, and How Quickly Are They Eroding?

The Himalaya, the world's highest and most rugged mountains, are being raised by thrust faulting caused by the collision of India with Asia (see Figure 21.15). How rapidly are they rising, and how quickly are they being eroded away? The answers to these questions depend on accurate topographic mapping.

On February 6, 1800, Colonel William Lambton, of the 33rd Regiment of Foot of the British Army, received orders to begin the Great Trigonometrical Survey of India, the most ambitious scientific project of the nineteenth century. Over the next several decades, intrepid British explorers led by Lambton and his successor, George Everest, hauled bulky telescopes and heavy surveying equipment through the jungles of the Indian subcontinent, triangulating the positions of reference monuments established at high points in the terrain, from which they could accurately establish Earth's size and shape. Along the way, in 1852, the surveyors discovered that an obscure Himalayan peak, known on their maps only as "Peak XV," was the highest mountain on Earth. They promptly named it Mount Everest, in honor of their former boss. Its official Tibetan name, Chomolungma, means "Mother of the Universe."

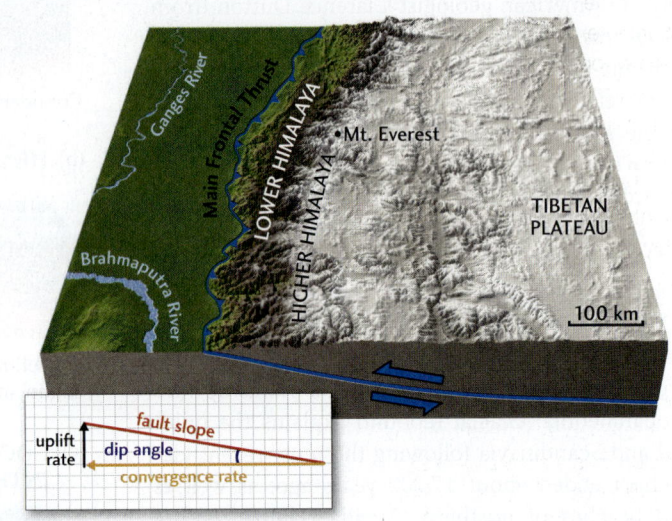

Cross section of the Himalaya, showing the approximate location of the thrust fault that is uplifting the mountains. The dip angle is about 10°.

On February 11, 2000, almost exactly 200 years after Lambton commenced his exploration, NASA launched another great survey, the Shuttle Radar Topography Mission (SRTM). The space

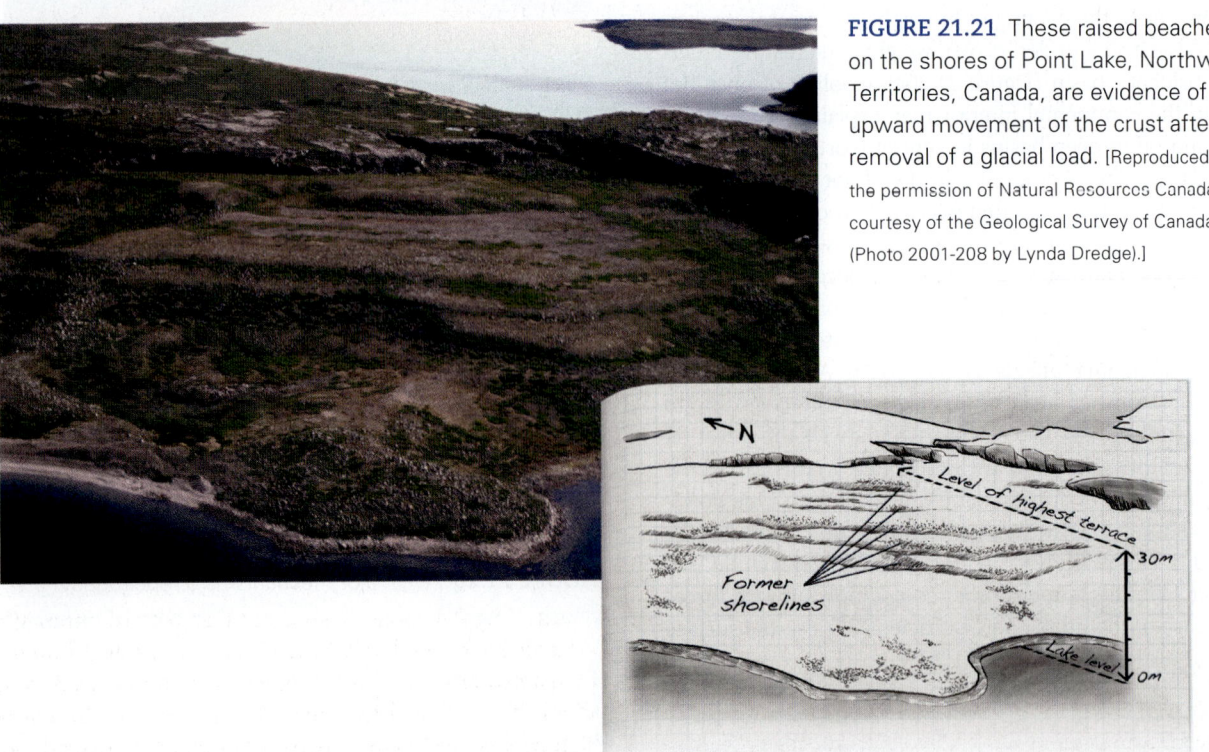

FIGURE 21.21 These raised beaches on the shores of Point Lake, Northwest Territories, Canada, are evidence of upward movement of the crust after removal of a glacial load. [Reproduced with the permission of Natural Resources Canada 2009, courtesy of the Geological Survey of Canada (Photo 2001-208 by Lynda Dredge).]

shuttle *Endeavour* carried two large radar antennas into low Earth orbit, one in the cargo bay and the second mounted on a mast that could extend up to 60 m outward. Working together like a pair of eyes, these antennas mapped the height of the land surface below the shuttle on a grid of very dense geographic points, rendering the terrain in unprecedented three-dimensional detail. Remarkably, the height of Mount Everest as confirmed by the SRTM (8850 m, or 29,035 feet) turned out to be only 10 m more than the original 1852 estimate.

Although the accuracy of the Great Trigonometrical Survey was impressive, data collection was a slow process. It took the British over 70 years to measure the positions of 2700 stations across the Indian subcontinent, an average of about one position every 3 months. In comparison, the SRTM collected about 3000 position measurements *each second*. In just 11 days, the SRTM mapped 2.6 billion points covering 80 percent of Earth's land surface, including many remote areas of the continents that had not been previously surveyed. And, unlike the British surveyors, the shuttle crew did not have to contend with malaria or tigers!

The SRTM position measurements have been used to create a *digital elevation model*, or DEM, of the Himalaya, shown here as a topographic map. An analysis of the features on this map, which includes Earth's highest peaks and deepest gorges, indicates that the average

The digital elevation model for the Mount Everest region of the Himalaya is derived from SRTM positions with a horizontal spacing of 90 m. [NASA images by Robert Simmon, based on SRTM data.]

height of the mountain range is staying approximately constant in time. In other words, the rate at which the Himalaya are rising is almost exactly balanced by the rate at which they are eroding:

uplift rate = erosion rate

As shown in the cross section, the geometry of the main thrust fault implies that

thrust fault slope = uplift rate ÷ convergence rate

Using GPS data, geologists have measured the convergence rate across the Himalaya to be about 20 mm/year. From earthquake locations, we know that the main thrust fault dips at an angle of about 10° below the mountain range. The slope of the fault is the tangent of its dip angle. Using a scientific calculator, we find tangent (10°) = 0.18. Therefore, the erosion rate is

erosion rate = thrust fault slope × convergence rate
= 0.18 × 20 mm/year
= 3.6 mm/year

This estimate is consistent with the erosion rate of 3–4 mm/year obtained from the pressure-temperature paths of metamorphic rocks in the Himalaya exhumed by erosion, using the techniques described in Chapter 7.

BONUS PROBLEM: Given that the convergence rate between the Indian and Eurasian plates is about 54 mm/year (see Figure 2.7), what fraction of the relative plate movement is taken up by thrust faulting in the Himalaya? How is the remaining plate movement accommodated by deformation in Eurasia?

None of these proposed epeirogenic mechanisms, however, explains a central feature of continental cratons: the existence of raised continental shields and subsided platforms. These regions are too vast, and have persisted too long, to be explained by the plate-tectonic processes we have discussed so far.

The Origin of Cratons

Every continental craton contains regions of ancient lithosphere that have been stable (i.e., undeformed) since the Archean eon (3.9 to 2.5 billion years ago). As we have seen, deformation has occurred at the edges of these stable landmasses, and new crust has accreted around them, during subsequent Wilson cycles. But how were these central parts of the cratons created in the first place?

We know that Earth was a hotter planet 4 billion years ago due to the heat generated by the decay of radioactive elements, which were more abundant then, as well as the energy released by differentiation and by impacts during the Heavy Bombardment (see Chapter 20). Evidence for a hotter mantle comes from a peculiar type of ultramafic volcanic rock found only in Archean crust, called *komatiite* (named after the Komati River in southeastern Africa, where it was first discovered). Komatiites contain a very high percentage (up to 33 percent) of magnesium oxide, so

FIGURE 21.22 Newly discovered rocks show that continental crust existed on Earth's surface during the Hadean eon. (a) The Acasta Gneiss from the Slave Craton has been dated at 4.0 billion years old. (b) Amphibole-bearing rocks from the Nuvvuagittuq greenstone belt, northern Quebec, Canada, have been dated at 4.28 billion years ago, making them the oldest rock formation yet discovered. [a: Courtesy of Sam Bowring, Massachusetts Institute of Technology; b: Jonathan O'Neil.]

their formation would have required a much higher melting temperature than is found anywhere in the mantle today.

A silica-rich continental crust existed at this early stage in Earth's history. Formations as much as 3.8 billion years old have been found on many continents; most are metamorphic rocks evidently derived from even older continental crust. In a few places, small pieces of this early crust survive. The Acasta Gneiss, in the northwestern part of the Canadian Shield, looks very similar to modern gneisses, although it has been dated at 4.0 billion years old (**Figure 21.22a**). Geologists recently discovered an even older rock formation, nearly 4.3 billion years old, in northern Quebec (Figure 21.22b). In Australia, single grains of zircon (a very hard mineral that survives erosion) have been dated as old as 4.4 billion years (see Chapter 9).

In the early part of the Archean, the continental crust that had differentiated from the mantle was very mobile. It may have been organized in small rafts that were rapidly pushed together and torn apart by intense tectonic activity—a version of the flake tectonic process that appears to be happening on Venus today. The first continental crust with long-term stability began to form about 3.3 billion to 3.0 billion years ago. In North America, the oldest surviving example is the central Slave Province in northwestern Canada (where the Acasta Gneiss is found), which stabilized about 3 billion years ago. Geologists have been able to show that this stabilization process involved not only the continental crust, but also chemical changes in the mantle portion of the continental lithosphere, as we will see shortly.

The rock formations in this Archean crust fall into two major groups (**Figure 21.23**):

1. **Granite-greenstone terrains** are areas of massive granitic intrusions that surround smaller pockets of greenstones, which in turn are capped with sediments. Greenstones, as we saw in Chapter 7, are low-grade metamorphic rocks derived from volcanic rocks, primarily of mafic composition. The origin of these greenstones is controversial, but many geologists think they were once pieces of oceanic crust formed at small spreading centers landward of island arcs accreted to the continents and later engulfed by the granitic intrusions.

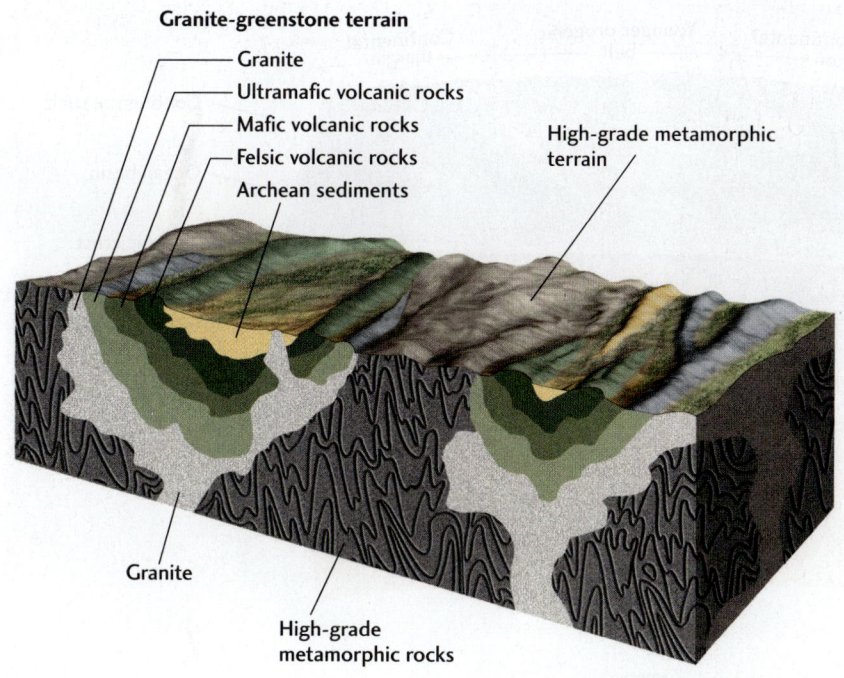

FIGURE 21.23 Two major types of rock formations are found in Archean regions of continental cratons: granite-greenstone terrains and high-grade metamorphic terrains.

2. **High-grade metamorphic terrains** are areas of high-grade (granulite facies) metamorphic rocks derived primarily from the compression, burial, and subsequent erosion of granitic crust. These areas look similar to the deeply eroded parts of modern orogenic belts, but the geometry of the deformation is different. Modern orogenies typically produce linear mountain belts where the edges of large cratons converge. Areas deformed in the Archean are more circular or S-shaped, reflecting the fact that the cratons were much smaller, with boundaries that were more curved.

By the end of the Archean, 2.5 billion years ago, enough continental lithosphere had been stabilized in cratons to allow the formation of larger and larger continents by magmatic addition and accretion. The plate-tectonic system was probably operating much as it does today. It is at about this time that we see the first evidence of major continent-continent collisions and the assembly of supercontinents. From this point onward in Earth's history, the history of the continents was governed by the plate-tectonic processes of the Wilson cycle.

Deep Structure of Continents

In this chapter, we have surveyed the most important processes in the development of Earth's continental crust. However, we have not yet explained one very basic aspect of continental behavior: the long-term stability of the cratons. How have the cratons survived being knocked around by plate-tectonic processes for billions of years? The answer to this question lies not in the crust, but in the lithospheric mantle below it.

Cratonic Keels

By using seismic waves to image Earth's interior, geologists have discovered a remarkable fact: The continental cratons are underlain by a thick layer of mechanically strong mantle material that moves with the cratons as the continents drift. These thickened sections of lithosphere extend to depths of more than 200 km—more than twice the thickness of the oldest oceanic lithosphere.

At 100 to 200 km beneath oceanic crust (as well as beneath most younger regions of the continents), the mantle rocks are hot and weak. They are part of the ductile asthenosphere, which flows relatively easily, allowing the plates to slide across Earth's surface. The lithosphere beneath the cratons extends into this region like the hull of a boat into water, so we refer to these mantle structures as **cratonic keels**, as shown in Figure 21.24. All cratons on every continent appear to have such keels.

Cratonic keels present many puzzles that scientists are still trying to solve. Less heat is emitted from the mantle beneath the cratons than from the mantle beneath oceanic crust. This observation indicates that the keels are several hundred degrees cooler than the surrounding asthenosphere, which explains their strength. If the rocks of the mantle beneath the cratons are so cool, however, why don't they sink into the mantle under their own weight, as cold, heavy slabs of oceanic lithosphere do in subduction zones?

Composition of the Keels

The cratonic keels would indeed sink into the mantle if their chemical composition were the same as that of ordinary mantle peridotites. To get around this problem, geologists have hypothesized that the cratonic keels are made of

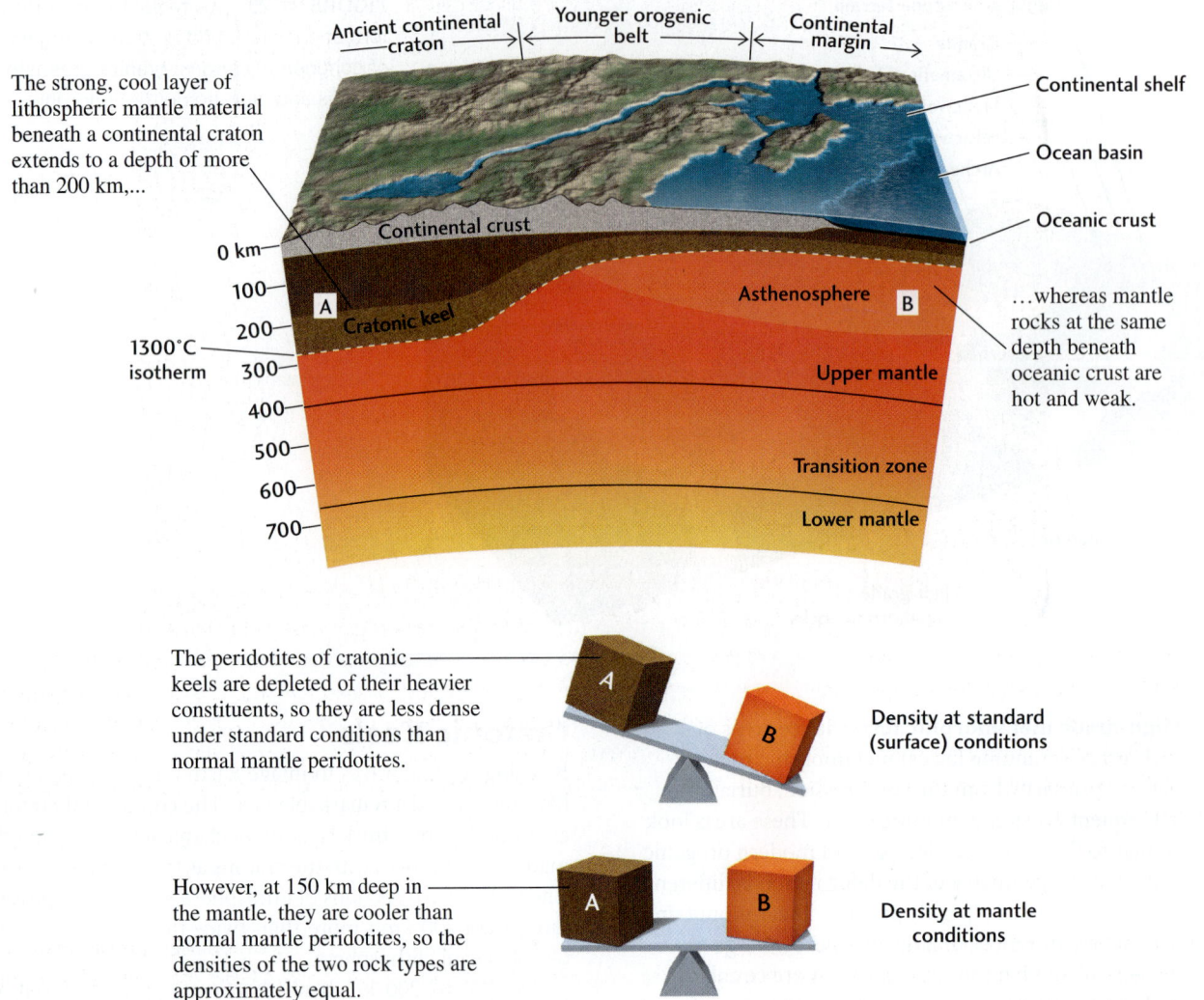

FIGURE 21.24 The chemical composition of cratonic keels counterbalances the effects of temperature to stabilize them against disruption by plate-tectonic processes. [Data from: T. H. Jordan, "The Deep Structure of Continents," *Scientific American* (January 1979).]

rock with a different, less dense chemical composition (see Figure 21.24). Their lower density counteracts the increase in density resulting from their cooler temperature.

Strong evidence in support of this hypothesis has come from mantle samples found in kimberlite pipes—the same types of volcanic deposits that produce diamonds, as we saw in Chapter 12. Kimberlite pipes are the eroded necks of volcanoes that have erupted explosively from tremendous depths (**Figure 21.25**). Almost all kimberlites that contain diamonds are located within the Archean regions of cratons. A diamond will revert to graphite at depths shallower than 150 km unless its temperature drops quickly. Therefore, the presence of diamonds in these pipes shows that the kimberlite magmas came from deeper than 150 km, and that they erupted through the keels when magma fractured the lithosphere very rapidly.

During a violent kimberlite eruption, fragments of the cratonic keel, some containing diamonds, are ripped off and brought to the surface in the magma as mantle xenoliths. The majority of these xenoliths turn out to be peridotites containing less iron (a dense element) and less garnet (a dense mineral) than ordinary mantle peridotites. Such rocks can be produced by extraction of a basaltic (or komatiitic) magma from the asthenosphere by partial melting. In other words, mantle rock beneath the cratons are the depleted residue left over after melting sometime earlier in Earth's history. A cratonic keel made of such depleted rocks can still float atop the mantle, despite its cooler temperature (see Figure 21.24).

Age of the Keels

By analyzing xenoliths from kimberlites and the diamonds they contain, we have learned that the cratonic keels are about the same age as the Archean crust above them.

FIGURE 21.25 Excavation of a kimberlite pipe at the Jwaneng diamond mine in Botswana. Diamonds are found in the dark-colored kimberlite rock in the center of the pit, which outlines the neck of an ancient, eroded volcano. The diamonds and other fragments of the African continental keel found at Jwaneng were erupted from depths of more than 150 km, and the analysis of these fragments supports the chemical stabilization hypothesis illustrated in Figure 10.24. Jwaneng is the world's richest diamond mine, expected to yield 100 million carats worth approximately $15 billion over the life of the mine. [Peter Essick/Robert Harding Picture Library.]

(The diamond in your ring or necklace is likely to be several billion years old!) Therefore, the rocks now present in the cratonic keels must have been depleted by the extraction of a basaltic melt very early in Earth's history, and they must have been positioned beneath the Archean crust about the time that crust was stabilized.

In fact, keel formation was probably responsible for the tectonic stabilization of the cratons. The existence of a cool, mechanically strong keel explains why the cratons have managed to survive through many continental collisions, including at least four episodes of supercontinent formation, without much internal deformation.

Many aspects of this process are still not understood. How did the keels cool down? How did they achieve the density balance illustrated in Figure 21.24? Why are the regions of the cratons with the thickest keels of Archean age? Some scientists believe that the continents play a major role in the mantle convection that drives the plate-tectonic system, but how the keels affect convection in the mantle is not completely understood. Indeed, many of the ideas presented in this chapter are hypotheses that have not yet been integrated into a fully accepted theory of continental evolution and deep structure. The search for such a theory remains a central focus of geologic research.

KEY TERMS AND CONCEPTS

accreted terrain (p. 643)	glacial rebound (p. 653)	shield (p. 635)
accretion (p. 643)	magmatic addition (p. 643)	tectonic age (p. 642)
active margin (p. 640)	orogen (p. 640)	tectonic province (p. 634)
craton (p. 640)	orogeny (p. 646)	Wilson cycle (p. 651)
cratonic keel (p. 657)	passive margin (p. 640)	
epeirogeny (p. 653)	rejuvenation (p. 639)	

REVIEW OF LEARNING OBJECTIVES

21.1 Identify the major tectonic provinces on a map of North America.

The continent's most ancient crust is exposed in the Canadian Shield. South of the Canadian Shield is the interior platform, where Precambrian basement rocks are covered by layers of Paleozoic sedimentary rocks. Around the edges of these provinces are elongated mountain chains. The Appalachian Fold Belt trends southwest to northeast on the eastern margin of the continent. The coastal plain and continental shelf of the Atlantic Ocean and Gulf of Mexico are parts of a passive continental margin that subsided after rifting during the breakup of Pangaea. The North American Cordillera is a mountainous region running down the western side of North America that contains several distinct tectonic provinces, including the Rocky Mountains, the Basin and Range Province, the Colorado Plateau, and the Cascade Range.

Study Assignment: Tectonic provinces of Figures 21.1 and 21.5.

Exercises: (a) Draw a rough topographic profile of the United States from San Francisco to Washington, D.C., and label the major tectonic provinces along this line. (b) Which coastlines of North America are active continental margins? (c) Which coastlines of North America are passive continental margins?

Thought Questions: (a) In which geologic era were the Appalachian Mountains at their highest elevation? (b) Why is the topography of the North American Cordillera higher than that of the Appalachian Mountains?

21.2 Recognize the major types of tectonic provinces found on continents worldwide.

The types of tectonic provinces seen in North America are found on other continents as well. Continental shields and platforms make up continental cratons, the oldest and most stable parts of continents. Around these cratons are orogens, the youngest of which are found at the active margins of continents, where tectonic deformation continues. The passive margins of continents are zones of crustal extension and sedimentation.

Study Assignment: Tectonic provinces and tectonic ages of Figure 21.8.

Exercises: (a) Which part of the Australian continent is tectonically older and more stable, its eastern half or its western half? (b) Drawing from the information in Figures 21.1 and 21.8, describe the tectonic province in which you live.

Thought Question: California is classified in Figure 21.8 as a Phanerozoic orogenic zone with a Mesozoic-Cenozoic tectonic age, yet Precambrian rocks are found in some parts of the state. Explain why this is not a contradiction.

21.3 Describe the tectonic processes that add new crust to the continents.

Two plate-tectonic processes, magmatic addition and accretion, add crust to continents. Buoyant silica-rich rocks are produced by magmatic differentiation, primarily in subduction zones, and added to the continental crust by vertical transport. Accretion occurs when preexisting crustal material is attached to existing continental masses by horizontal plate movement in one of four ways: the transfer of buoyant crustal fragments from a subducting plate to a continent on an overriding plate; the closure of a sea separating an island arc from a continent; the transport of crust laterally along continental margins by strike-slip faulting; or the collision and suturing of two continents and their subsequent rifting apart.

Study Assignment: Four processes of continental accretion described in Figure 21.12.

Exercises: (a) Illustrate two processes of continental accretion with examples of accreted terrains in North America. (b) Figure 21.8b shows more continental crust of Mesozoic-Cenozoic age than of any other tectonic age. Does this observation contradict the hypothesis that most of the continental crust was differentiated from the mantle in the first half of Earth's history? Explain your answer.

Thought Questions: (a) What criteria would you use to recognize an accreted terrain? (b) How could you tell whether it originated far away or nearby?

21.4 Explain how orogenies modify continents.

Horizontal tectonic forces, arising mainly from plate convergence, can produce mountains by folding and faulting. Thrust faulting can stack the upper part of the crust into overthrust sheets tens of kilometers thick, pushing up high mountains and metamorphosing buried rock formations. Compression can double the thickness of continental crust, causing the rocks in the lower crust to melt. This melting generates granitic magma, which rises to form extensive batholiths in the upper crust.

Study Assignment: Review the section *Orogeny: Modification by Plate Collision*.

Exercises: Two continents collide, thickening the crust from 35 km to 70 km and forming a high plateau. After hundreds of millions of years, the plateau is eroded down to sea level. (a) What types of rocks are exposed at the surface by this erosion? (b) What is the approximate crustal thickness after the erosion has occurred? (c) Where in North America has this sequence of events been recorded in surface geology?

Thought Questions: (a) How would you identify a region where orogeny is taking place today? (b) Give an example of such a region.

21.5 Order the sequence of plate-tectonic events that constitute the Wilson cycle.

The Wilson cycle is a sequence of tectonic events that occur during the assembly and breakup of supercontinents and the opening and closing of ocean basins. It has four main phases: rifting during the breakup of a supercontinent; passive margin cooling and sediment accumulation during seafloor spreading and ocean opening; active magmatic addition and accretion during subduction and ocean closure; and orogeny with crustal thickening during continent-continent collision. Orogeny is followed by erosion, which thins the crust.

Study Assignment: Wilson cycle, depicted in Figure 21.18.

Exercises: (a) How many times have the continents been joined in a supercontinent since the end of the Archean eon? (b) Use this number to estimate the typical duration of a Wilson cycle and the average speed at which plate-tectonic processes move continents.

Thought Question: Are the interiors of continents usually younger or older than their margins? Explain your answer using the concept of the Wilson cycle.

21.6 Categorize the main mechanisms of epeirogeny.

Epeirogeny is a downward or upward movement of a broad region of crust without folding or faulting. Epeirogenic upward movements can result from glacial rebound, heating of the lithosphere by upwelling mantle material, or uplifting of the lithosphere by a "superplume" in the deep mantle. The cooling of previously heated lithosphere can cause epeirogenic downward movements in the interior of a continent or at the margins of two continents separated by rifting. The densification of lithosphere caused by the cooling can form thermal subsidence basins, such as the Michigan Basin, that become filled with sediments.

Study Assignment: Epeirogenic processes illustrated in Figure 21.20.

Exercises: (a) What epeirogenic process explains the roughly circular basins shown in Figure 21.3? (b) Over what time span does glacial rebound occur? (c) Over what time span does thermal subsidence from cooling of the lithosphere occur?

Thought Question: The ocean basins are just about the right size to contain all the water on Earth's surface. What processes, acting together, maintain this approximate equality?

21.7 Explain how continental cratons have survived billions of years of plate-tectonic processes.

The oldest regions of the cratons, formed in the Archean eon, are underlain by a layer of cool, strong mantle material more than 200 km thick that moves with the continents as they drift. These cratonic keels are made up of mantle peridotites that have been depleted of their denser chemical constituents by the extraction of basaltic magmas through partial melting. The lower densities and lower temperatures strengthen the keels and stabilize them against disruption by plate-tectonic processes.

Study Assignment: Review the section *The Deep Structure of Continents*.

Exercises: (a) How was orogeny in the Archean eon different from orogeny during the Proterozoic and Phanerozoic eons? (b) What factors might explain these differences?

Thought Questions: (a) What would happen at Earth's surface if the cold keel beneath a craton were suddenly heated up? (b) How might this effect be related to the formation of the Colorado Plateau?

VISUAL LITERACY EXERCISE

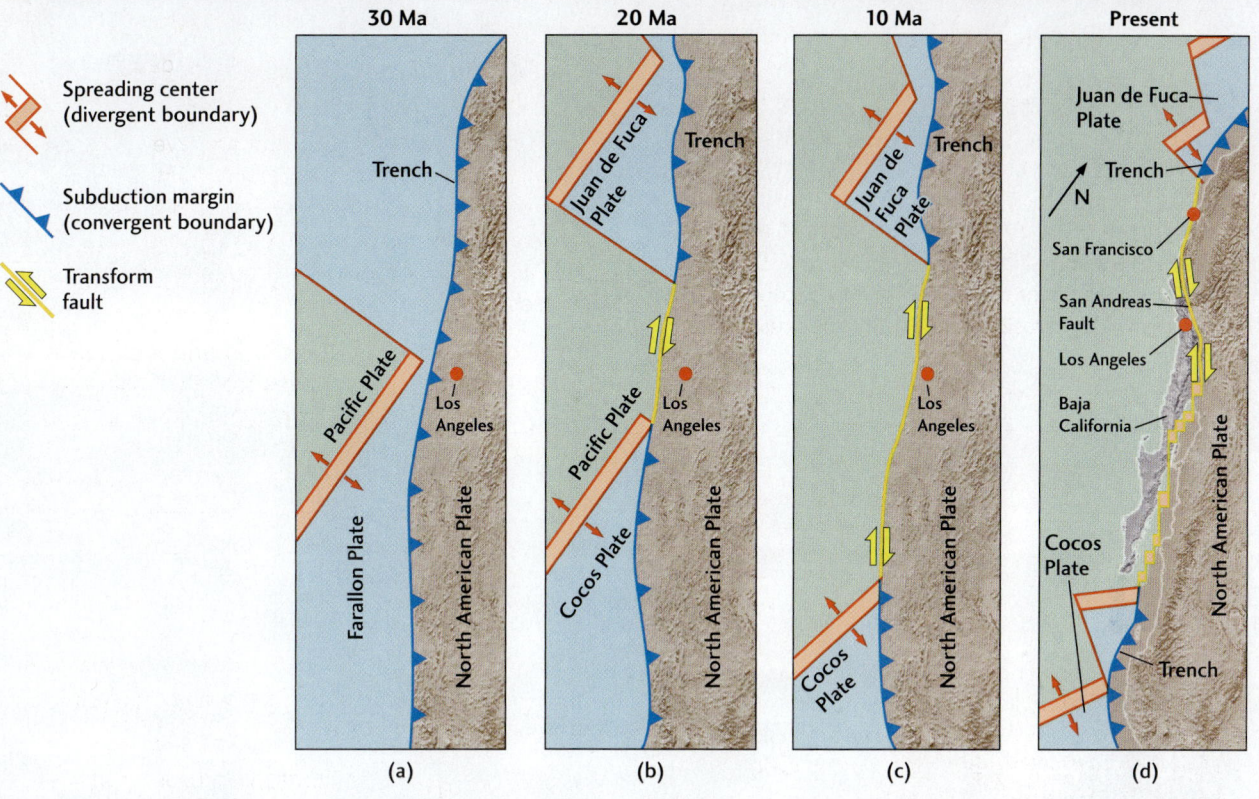

FIGURE 21.6 The interaction of the west coast of North America with the shrinking Farallon Plate as it was progressively subducted beneath the North American Plate, leaving the present-day Juan de Fuca and Cocos plates as small remnants. (Ma: million years ago.)

1. The plate-tectonic interactions shown in this figure occurred in which geologic epoch?
   a. Jurassic
   b. Cretaceous
   c. Paleogene
   d. Neogene

2. During this interval, in what direction was the motion of the Pacific Plate relative to the North American Plate?
   a. north
   b. northwest
   c. east
   d. southeast

3. Today, the continental crust of Baja California is part of which plate?
   a. North American Plate
   b. Pacific Plate
   c. Cocos Plate
   d. Farallon Plate

4. Twenty million years ago, the continental crust of Baja California was part of which plate?
   a. North American Plate
   b. Pacific Plate
   c. Cocos Plate
   d. Farallon Plate

Geobiology: Life Interacts with Earth

22

The Biosphere as a System	666
Microorganisms: Nature's Tiny Chemists	671
Geobiologic Events in Earth's History	679
Evolutionary Radiations and Mass Extinctions	684
Astrobiology: The Search for Extraterrestrial Life	690

Learning Objectives

After studying this chapter, you should be able to

22.1 Define geobiology and the biosphere.

22.2 Compare some of the processes of metabolism.

22.3 Summarize the biogeochemical cycle and discuss some ways in which metabolism affects the physical environment.

22.4 Discuss how microorganisms interact with the physical environment.

22.5 Discuss how life originated and led to the rise of atmospheric oxygen.

22.6 Compare the evolutionary processes of radiation and extinction.

22.7 Formulate some ways we can search for life on other worlds.

Grand Prismatic Hot Spring, Yellowstone National Park, Wyoming. The striking array of colors reflects different communities of microorganisms that are very sensitive to water temperature. Water flowing away from the center of the spring (blue) cools down, causing a given community of microorganisms to be replaced by a different community that grows best at the new, lower temperature. The boardwalk visible in the lower part of the photo allows tourists to peer into its depths and provides a sense of scale. [Luis Castañeda/AGE Fotostock.]

CHAPTER 22 Geobiology: Life Interacts with Earth

Geology is the study of the physical and chemical processes that control the Earth system, today and in the past. Biology is the study of life and living organisms, including their structure, function, origin, and evolution. As separate as geology and biology may seem, organisms and their physical environment interact in many ways. We have long recognized that biology and geology are intimately related, but until recently, we have not known exactly how. Fortunately, technological advances in both Earth and life sciences now allow us to ask and answer questions that were previously beyond our scope. Over the past decade, scientists working at the frontiers of both fields have begun to understand how several important geobiological processes work.

We know that organisms can change Earth. For example, Earth's atmosphere is distinct from that of every other planet in having a significant concentration of oxygen—the result of the evolution of oxygen-producing microorganisms billions of years ago. Organisms also contribute to the weathering of rocks by releasing chemicals that help break down minerals; through this process, they obtain nutrients essential to their growth. Similarly, geologic processes can change life, as when an asteroid struck Earth 65 million years ago, causing a mass extinction that killed off the dinosaurs.

In Chapter 22, we will explore the link between organisms and Earth's physical environment. We will see how the biosphere works as a system and what gives Earth its ability to support life. Next, we will explore the remarkable roles microorganisms play in geologic processes, and discuss some of the major geobiological events that have changed our planet. Finally, we will consider the key ingredients for sustaining life and ponder the eternal question posed by astrobiologists: Is there life out there?

The Biosphere as a System

Life is everywhere on Earth. The **biosphere** is that part of our planet that contains all of its living organisms. It includes the plants and animals with which we are most familiar as well as the nearly invisible microorganisms that live in some of the most extreme environments on Earth. These organisms live on Earth's surface, in its atmosphere and ocean, and within its upper crust, and they interact continuously with all of these environments. Because the biosphere intersects with the lithosphere, hydrosphere, and atmosphere, it can influence or even control basic geologic and climate processes. **Geobiology** is the study of these interactions between the biosphere and Earth's physical environment.

The biosphere is a system of interacting components that exchanges energy and matter with its surroundings. Inputs into the biosphere include energy (usually in the form of sunlight) and matter (such as carbon, nutrients, and water). Organisms use these inputs to function and grow. In the growing process, they create an amazing variety of outputs, some of which have important influences on geologic processes. At a local scale—such as that of a water-filled pore within loose sediment particles—a small group of organisms may have a geologic effect that is limited to a particular sedimentary environment. At larger scales, the activities of organisms may influence the concentrations of gases in the atmosphere or the cycling of certain elements through Earth's crust.

Ecosystems

Think of a class project in which each member of a team has special skills that allow the team as a whole to exceed the capabilities of individuals working alone. Groups of organisms act in similar ways: Individual organisms play roles that contribute to the survival of other organisms as well as their own. In the case of human groups, we accomplish this teamwork as a result of conscious decisions. For the organisms living together in a particular environment—referred to as a *community*—it happens through trial and error and involves feedbacks between the community and individuals. These feedbacks determine the structure and functioning of the community.

Whether at local, regional, or global scales, the interactions of biological communities with their environments define organizational units known as **ecosystems**. Ecosystems are composed of biological and physical components that function in a balanced, interrelated fashion. Ecosystems occur at many different scales (**Figure 22.1**). They may be separated by geologic barriers such as mountains, deserts, or oceans at the largest scale, or by barriers such as different water temperatures within a single hot spring at a much smaller scale (see the chapter opening photo). But no matter how large or small they are, all ecosystems are characterized by a flow, or *flux*, of energy and matter between organisms and their environment.

A typical ecosystem might involve, say, a river and its surroundings, where different groups of organisms are adapted to live in the water (fish), in the sediment (worms, snails), on the banks (grass, trees, muskrats), and in the sky above (birds, insects). In one sense, the river controls where the organisms live by supplying the ecosystem with water, sediment, and dissolved mineral nutrients. Conversely, the organisms influence how the river behaves; for example, grass and trees stabilize the riverbanks against the destructive effects of floods. The balance between such biologically controlled and geologically controlled processes ensures the long-term stability of the ecosystem.

Ecosystems respond sensitively to biological changes, such as the introduction of new groups of organisms. When severe imbalances in ecosystems occur, responses are often dramatic. Consider the effects of introducing a new organism into your neighborhood environment, such as a pretty new plant for your garden. In all too many cases, the new organism is better suited for its new environment than the

The Biosphere as a System

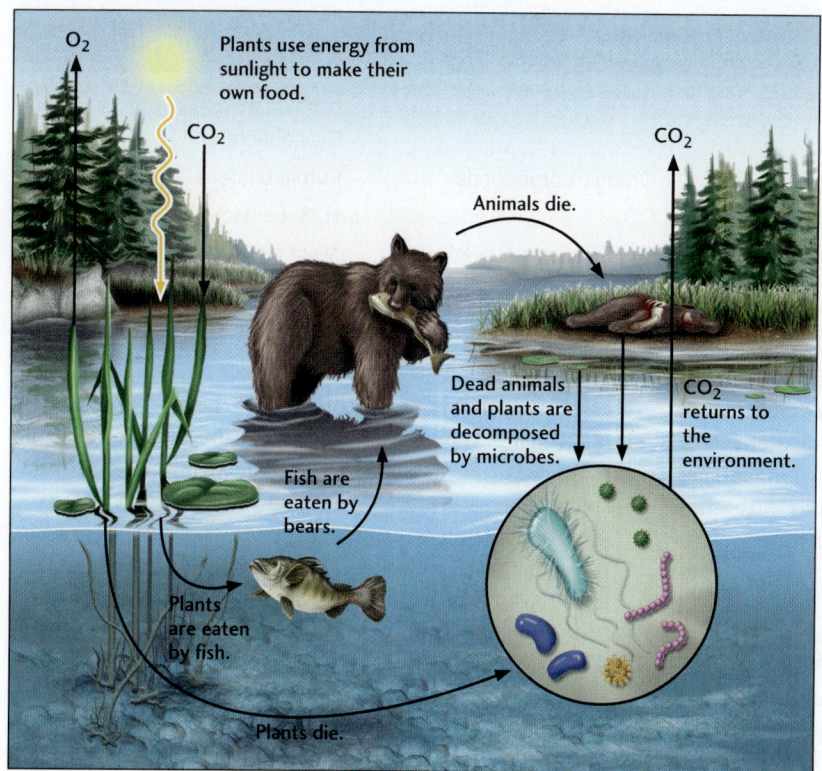

FIGURE 22.1 Ecosystems are characterized by a flux of energy and matter between organisms and their environment. In this example, sunlight is used as an energy source by plants, which are eaten by fish, which are eaten by bears. The plants, fish, and bears eventually die and are decomposed by microorganisms. In this way, the matter that made up those organisms is returned to the physical environment, where it can be used again.

current inhabitants and becomes *invasive*, multiplying rapidly and squeezing out the previous inhabitants, which may then become extinct (**Figure 22.2**). A successful invader often comes from a place where the physical environment is similar but where biological competition is less intense, so it is likely to win the battle for nutrients and space in its new home.

(a)

(c)

(b)

FIGURE 22.2 Invasive organisms create problems by dominating their local ecosystems. (a) Kudzu, introduced into North America to stop highway erosion, rapidly overgrows other plants. (b) Purple loosestrife, introduced from Europe as a garden flower, has invaded many North American wetlands. (c) Zebra mussels aggressively colonize and overwhelm ordinary mussels. In 2002, the U.S. Fish and Wildlife Service estimated that $5 billion was spent by electric utility companies just to unclog water intake pipes blocked by zebra mussels. [a: Kerry Britton/U.S. Department of Agriculture, Forest Service, Athens, Georgia, USA; b: Gaertner/Alamy Stock Photo; c: U.S. Fish and Wildlife.]

TABLE 22.1　Organisms as Producers and Consumers

TYPE	ENERGY SOURCE	CARBON SOURCE	EXAMPLE
Photoautotroph	Sun	CO_2	Cyanobacteria
Photoheterotroph	Sun	Organic compounds	Purple bacteria
Chemoautotroph	Chemicals	CO_2	H, S, Fe bacteria
Chemoheterotroph	Chemicals	Organic compounds	Most bacteria, fungi, and animals, including humans

Earth's history shows us that ecosystems respond sensitively to geologic processes as well. Impacts by meteorites, huge volcanic eruptions, and rapid global warming are just a few of the processes that have contributed to the extinction of major groups of organisms. We will explore some of their effects later in this chapter.

Inputs: The Stuff Life Is Made Of

The organisms of any ecosystem can be subdivided into producers and consumers according to the way they obtain their *food*, which is their source of energy and nutrients (Table 22.1). Producers, or **autotrophs,** are organisms that make their own food. They use energy from sunlight or, in some cases, energy derived from chemicals in their environment to manufacture organic compounds such as carbohydrates. Examples of these chemicals include hydrogen, sulfur, and iron. Consumers, or **heterotrophs,** get their food by feeding directly or indirectly on producers.

It is often said that you are what you eat, and that statement is true not only for humans, but for all organisms. Our foods are all made up of more or less the same materials: molecules composed of carbon, hydrogen, oxygen, nitrogen, phosphorus, and sulfur. So it doesn't matter whether the organism is an autotroph or a heterotroph, it still utilizes the same six elements as food. What differs is the form (that is, the molecular structure) their food comes in. When heterotrophs such as humans eat bread, we are feeding ourselves on *carbohydrates:* large molecules formed of carbon, hydrogen, and oxygen. Even the lowliest microorganisms dine on carbon compounds such as carbon dioxide (CO_2) or methane (CH_4). The difference is that what is food for a microorganism may not seem like food to us!

Carbon The fundamental building block of all life on Earth is carbon. If water is removed, the composition of all organisms, including humans, is dominated by carbon. No other chemical element can match carbon in the variety and complexity of the compounds it can form. Part of the reason for its versatility is that it can form four covalent bonds with itself and other elements (see Figure 3.3), which allows for a wide variety of structures. Carbon acts as the template around which all organic molecules, such as carbohydrates and proteins, are built. Thus, carbon is critically important to organisms because it goes into manufacturing every part of them, from genes to body structures.

The biosphere largely controls the flux of carbon through the Earth system. Marine organisms extract carbon— which is present in seawater as dissolved CO_2—to form carbonate shells and skeletons. When the organisms die, their skeletons settle to the seafloor, where they accumulate as sediments, effectively moving carbon from the biosphere to the lithosphere. The accumulation of the organic remains of organisms in freshwater wetlands and on the seafloor also moves carbon from the biosphere to the lithosphere. Over geologic time, these organic remains are transformed into oil, natural gas, and coal deposits. Today, when we extract and burn these deposits, we are moving carbon from the lithosphere to the atmosphere in the form of CO_2 emissions.

Nutrients Nutrients are chemical elements or compounds that organisms require to live and grow. Common plant nutrients include the elements phosphorus, nitrogen, and potassium—the ones most commonly found in garden fertilizer. Other organisms also depend on iron and calcium. Some organisms can manufacture their own nutrients, but others must obtain them in their diets from materials in their environment. Some specialized microorganisms can obtain nutrients by dissolving minerals.

Water Life, as we know it, requires water (H_2O). All organisms on Earth, including humans, are composed primarily of water, typically 50 to 95 percent. It is well known that humans can live for weeks without food, but most will perish in a few days without water. Even microorganisms that live in the atmosphere must obtain water from tiny droplets that condense around dust particles, and viruses must obtain water from their hosts.

Water is the habitat in which life first emerged and in which much of it still thrives. Water's chemical properties and the way it responds to changes in temperature make it an ideal medium for biological activity. The cells of all organisms are made up primarily of an aqueous solution that promotes the chemical reactions of life. Water also helps moderate Earth's climate, which has supported life for at least 3.5 billion years (see Chapter 12). Water is such an important ingredient for life that the search for extraterrestrial life must begin with the search for water, as we will see at the end of this chapter.

Energy All organisms need energy to live and grow. Some of the simplest organisms, such as single-celled algae, obtain energy from sunlight. Others acquire energy by breaking down chemicals in their environment. Heterotrophs get energy by feeding on other organisms. Energy is important because it fuels the conversion of simple molecules such as carbon dioxide and water into larger molecules, such as carbohydrates and proteins, which are essential for life.

Processes and Outputs: How Organisms Live and Grow

Metabolism encompasses all the processes organisms use to convert inputs into outputs. In one type of metabolic process, organisms use small molecules, such as CO_2, H_2O, and CH_4, and energy to create larger molecules, such as proteins and certain types of carbohydrates, that enable them to function and grow. Other carbohydrates—for example, a sugar called glucose—are stored for later use as an energy source—that is, food. In another type of metabolic process, organisms break down food to release the energy it contains.

One particularly well-known metabolic process is **photosynthesis** (Figure 22.3). Through this process,

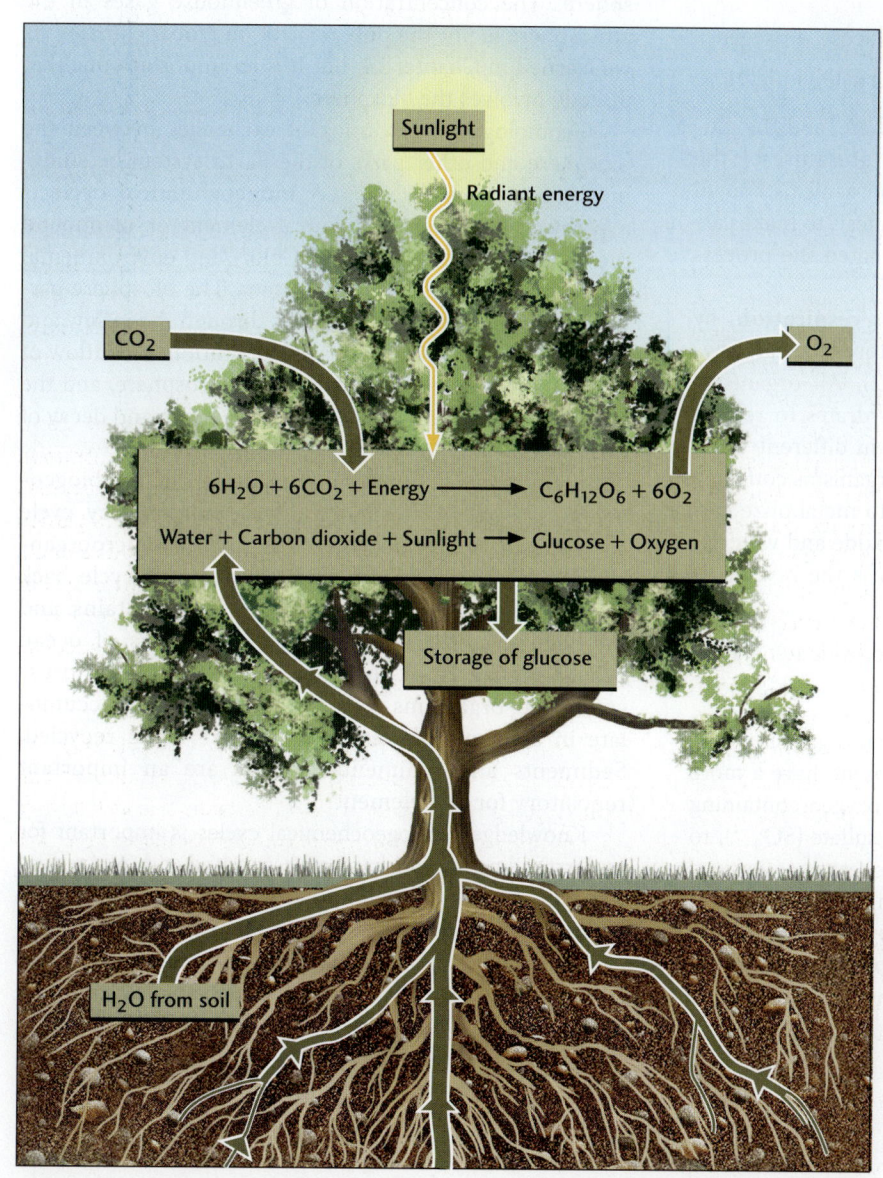

FIGURE 22.3 During the metabolic process of photosynthesis, organisms use carbon dioxide, water from the environment, and the energy of sunlight to make carbohydrates such as glucose.

TABLE 22.2	Comparison of Photosynthesis and Respiration
PHOTOSYNTHESIS	**RESPIRATION**
Stores energy as carbohydrates	Releases energy from carbohydrates
Uses CO_2 and H_2O	Releases CO_2 and H_2O
Increases mass	Decreases mass
Produces oxygen	Consumes oxygen

organisms such as plants and algae use energy from sunlight to convert water and carbon dioxide into carbohydrates (such as glucose) and oxygen (Table 22.2). This reaction proceeds as follows:

Water + carbon dioxide + sunlight → glucose + oxygen

$$6\,H_2O + 6\,CO_2 + energy \rightarrow C_6H_{12}O_6 + 6\,O_2$$

The oxygen is released into the atmosphere, and the glucose is stored as an energy source for future use by the organism. An important group of microorganisms known as the **cyanobacteria** also use photosynthesis to make carbohydrates; in fact, they probably originated the process early in life's history.

The other key metabolic process is **respiration,** by which organisms release the energy stored in carbohydrates such as glucose (see Table 22.2). All organisms use oxygen to burn, or *respire*, carbohydrates to release energy, but different organisms respire in different ways. For example, humans and many other organisms consume oxygen gas (O_2) from the atmosphere to metabolize carbohydrates, and they release carbon dioxide and water as by-products. In this case, the reaction is the reverse of photosynthesis:

glucose + oxygen → water + carbon dioxide + energy

$$C_6H_{12}O_6 + 6\,O_2 \rightarrow 6\,H_2O + 6\,CO_2 + energy$$

But other organisms, such as microorganisms that live in environments where oxygen is absent, have a more difficult task. They must break down oxygen-containing compounds dissolved in water, such as sulfate (SO_4^{-2}), to obtain oxygen. During the course of these reactions, various gases—such as hydrogen (H_2), hydrogen sulfide (H_2S), and methane (CH_4)—may be produced as by-products. These gases escape to the atmosphere and contribute to global warming. Conversely, when organisms consume these gases, they contribute to global cooling.

The metabolism of organisms affects the geologic components of their environment. For example, the oxygen released by photosynthesis reacts with iron-bearing silicate minerals such as pyroxene and amphibole to form iron-bearing oxide minerals such as hematite (see Chapter 16).

Biogeochemical Cycles

In the course of living and dying, organisms continuously exchange energy and matter with their environment. This exchange occurs at the scale of the individual organism, the ecosystem of which it is a part, and the global biosphere. The metabolic consumption and production of gases such as CO_2 and CH_4 is a good example of how organisms may exert global controls on Earth's climate. Carbon dioxide and methane are *greenhouse gases:* gases that absorb heat emitted by Earth and trap it in the atmosphere. The concentration of greenhouse gases in the atmosphere is not the only control on global climates, as we learned in Chapter 14, but it is an important one that directly involves the biosphere.

Geobiologists keep track of exchanges between the biosphere and other parts of the Earth system by studying biogeochemical cycles. A **biogeochemical cycle** is a pathway by which a chemical element or compound moves between the biological ("bio") and environmental ("geo") components of an ecosystem. The biosphere participates in biogeochemical cycles through the inflow and outflow of atmospheric gases by respiration, the inflow of nutrients from the lithosphere and hydrosphere, and the outflow of those nutrients through the death and decay of organisms.

Because ecosystems vary in scale, so do biogeochemical cycles. Phosphorus, for example, may cycle back and forth between the water and the microorganisms in the pores of sediments, or it may cycle back and forth between uplifted rocks in mountains and the sediments deposited along the margins of ocean basins (Figure 22.4). In either case, when phosphorus-containing organisms die, the phosphorus may accumulate in a temporary repository before being recycled. Sediments and sedimentary rocks are an important repository for this element.

Knowledge of biogeochemical cycles is important for understanding the mechanisms associated with major geobiological events throughout Earth's history, as we will see later in this chapter. It is also critical for understanding how elements and compounds that humans emit into the atmosphere and ocean are interacting with the biosphere, as we saw in Chapters 13 and 14.

Microorganisms: Nature's Tiny Chemists 671

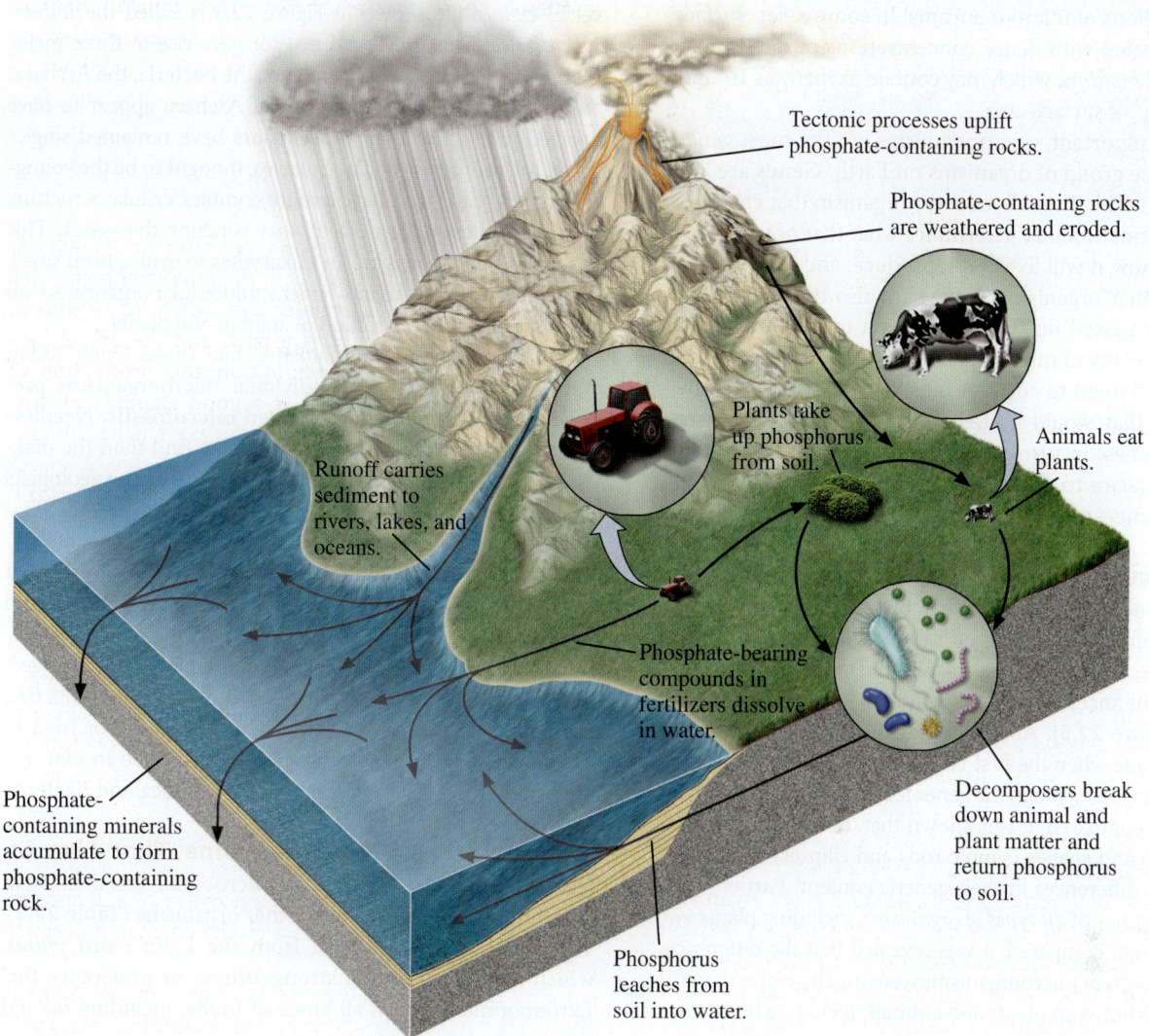

FIGURE 22.4 The biosphere plays key roles in the biogeochemical cycle of phosphorus.

Microorganisms: Nature's Tiny Chemists

Single-celled organisms, which include bacteria, archaea, some fungi, some algae, and most protists, are known as **microorganisms,** or *microbes*. Wherever there is water, there are microorganisms. Microorganisms, like other organisms, need water to live and reproduce. Microorganisms can be as small as a few microns in size (1 micron = 10^{-6} m) and can inhabit almost any nook or cranny you can think of, from at least 5 km beneath Earth's surface to more than 10 km high in the atmosphere. They live in air, in soil, on and in rocks, inside roots, in piles of toxic waste, on frozen snowfields, and in water bodies of every type, including boiling hot springs. They live at temperatures that range from lower than −20°C to higher than the boiling point of water (100°C).

People have exploited the useful effects of microbial metabolism for thousands of years to produce bread, wine, and cheese. Today, people also use microorganisms to produce antibiotics and other valuable drugs. Geobiologists study microorganisms to understand their roles in biogeochemical cycles and to understand the early evolution of the biosphere before the advent of more complex organisms.

Abundance and Diversity of Microorganisms

Microorganisms dominate Earth in terms of numbers of individuals. Concentrations ranging from 10^3 to 10^9 microorganisms/cm^3 have been reported from soils, sediments, and natural waters. Every time you walk on the ground, you step on billions of microorganisms! In some cases, surfaces become coated with dense concentrations of microorganisms called *biofilms*, which may contain as many as 10^8 individuals/cm^2 of surface area.

More important, microorganisms are the most genetically diverse group of organisms on Earth. **Genes** are large molecules within the cells of every organism that encode all of the information that determines what that organism will look like, how it will live and reproduce, and how it differs from all other organisms. Genes are also the basic hereditary units passed on from generation to generation. The genetic diversity of microorganisms is important because it has allowed them to colonize, adapt to, and thrive in environments that would be lethal to most other organisms. These abilities, in turn, are important because they allow microorganisms to recycle important materials in a broad—even extreme—range of geologic environments.

The Universal Tree of Life Biologists have learned how to use the genetic information contained in living organisms to understand which forms of life are most closely related to one another. This knowledge has allowed them to organize the hierarchy of ancestors and descendants into a universal tree of life (**Figure 22.5**). About 30 years ago, a startling discovery was made when the first family trees for microorganisms were constructed. When the genes for *all types of microorganisms* were compared, it was shown that, despite their similar sizes (tiny) and shapes (simple rods and ellipses), there were enormous differences in their genetic content. Furthermore, when the genes of *all types of organisms*, including plants and animals, were compared, it was revealed that the differences among groups of microorganisms were much greater than the differences between plants and animals, including humans.

The Three Domains of Life The single root of the universal tree of life shown in Figure 22.5 is called the *universal ancestor*. This universal ancestor gave rise to three major groups, or domains, of descendants: the Bacteria, the Archaea, and the Eukarya. The Bacteria and Archaea appear to have evolved first; all of their descendants have remained single-celled microorganisms. The Eukarya, thought to be the youngest branch of the tree, have a more complex cellular structure, which includes a cell nucleus that contains the genes. This structure made it possible for eukaryotes to evolve from small, single-celled organisms into larger, multicellular organisms—an essential step in the evolution of animals and plants.

Precambrian microorganisms, like those living today, were tiny. The traces of individual microorganisms preserved in rocks are therefore called **microfossils**. Needless to say, such features are much harder to find than the macroscopic fossils of shells, bones, and twigs used by geologists to study the evolution of animals and plants during the Phanerozoic eon (recall that *Phanerozoic* means "visible life").

For geobiologists, the universal tree of life is a map that reveals how microorganisms relate to one another and interact with Earth. The names of microorganisms, such as *Halobacterium*, *Thermococcus*, and *Methanopyrus*, suggest that these organisms can live in extreme environments that are very salty (*halo*, "halite"), or hot (*thermo*), or high in methane (*methano*). Microorganisms that live in extreme environments are almost exclusively Archaea and Bacteria.

Extremophiles: Microorganisms That Live on the Edge **Extremophiles** are microorganisms that live in environments that would kill other organisms (**Table 22.3**). The suffix *phile* is derived from the Latin word *philus*, which means "to have a strong affinity or preference for." Extremophiles live on all kinds of foods, including oil and

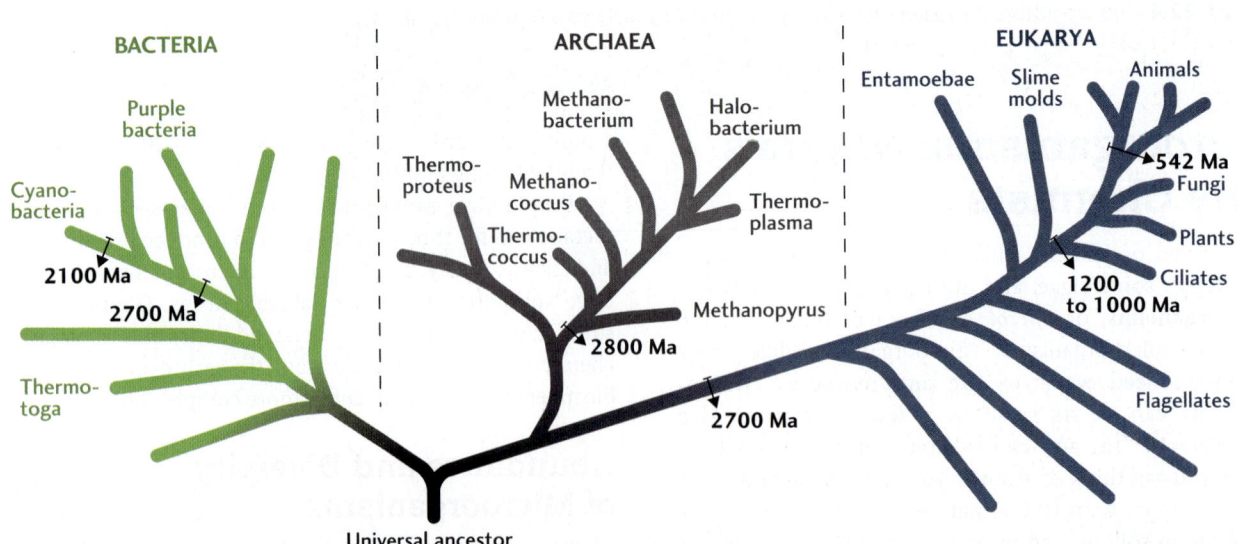

FIGURE 22.5 The universal tree of life shows how all organisms are related to one another. Organisms are subdivided into three great domains: the Bacteria, Archaea, and Eukarya. These domains are all descended from a universal common ancestor. All three domains are dominated by microorganisms. Note that animals appear at the tip of the eukaryote branch. (Ma: million years ago.)

Microorganisms: Nature's Tiny Chemists

TABLE 22.3 Characteristics of Extremophiles

TYPE	TOLERANCE	ENVIRONMENT	EXAMPLE
Halophile	High salinity	Playa lakes Marine evaporates	Great Salt Lake, Utah
Acidophile	High acidity	Mine drainage Water near volcanoes	Rio Tinto, Spain
Thermophile	High temperature	Hot springs Mid-ocean ridge vents	Yellowstone National Park
Anaerobe	No oxygen	Pores of wet sediments Groundwater Microbial mats Mid-ocean ridge vents	Cape Cod Bay sediments

toxic wastes. Some use substances other than oxygen, such as nitric acid, sulfuric acid, iron, arsenic, or uranium, for respiration.

Acidophiles are microorganisms that thrive in acidic environments. Acidophiles can tolerate pH levels low enough to kill other organisms. These microorganisms live by eating sulfide! They are able to survive in such acidic habitats because they have developed a way to pump out the acid that accumulates inside their cells. Such extremely acidic habitats occur naturally (see Earth Issues 22.1), but are more commonly associated with mining.

Thermophiles are microorganisms that live and grow in extremely hot environments. They grow best in temperatures that are between 50°C and 70°C, and can tolerate temperatures up to 120°C. They will not grow if the temperature drops to 20°C. Thermophiles live in geothermal habitats, such as hot springs and hydrothermal vents at mid-ocean ridges, and in environments that create their own heat, such as compost piles and garbage landfills. The microorganisms that cover the bottom of Grand Prismatic Hot Spring (see the chapter opening photo) are dominated by thermophiles. Of the three domains of life, the Eukarya (which include humans) are generally the least tolerant of high temperatures (60°C seems to be their upper limit). The Bacteria are more tolerant (with an upper limit close to 90°C), and the Archaea are the most tolerant, able to withstand temperatures of up to 120°C. Microorganisms that can stand temperatures above 80°C are called *hyperthermophiles*.

Halophiles are microorganisms that live and grow in highly saline environments. They can tolerate salt concentrations up to 10 times that of normal ocean water. Halophiles live in naturally hypersaline playa lakes such as the Great Salt Lake and the Dead Sea and in some parts of the ocean, such as the southern end of San Francisco Bay, where seawater is commercially evaporated to extract salt (**Figure 22.6**). These microorganisms can control the salt concentration inside their cells by expelling extra salt from their cells into the environment.

FIGURE 22.6 Humans have dammed off parts of the ocean to create ponds where seawater can evaporate to precipitate halite for table salt and other uses. The halophilic bacteria that thrive in these hypersaline environments produce a distinctive pigment that turns the ponds pink.
[NNehring/Getty Images.]

Earth Issues

22.1 Sulfide Minerals React to Form Acidic Waters on Earth and Mars

Many economically significant mineral deposits are associated with high concentrations of sulfide minerals. When water comes into contact with sulfide minerals, the sulfide they contain reacts with oxygen to form sulfuric acid. Thus, during the course of mining and afterward, rainwater and groundwater may interact with these minerals to produce highly acidic surface water and groundwater. Unfortunately, these acidic waters are lethal to most organisms. As they spread throughout the environment, extensive devastation may result. In some cases, the only organisms that survive are acidophilic extremophiles.

In a few places on Earth where sulfide minerals occur in high enough concentrations, acidic waters are produced naturally. One of these places is the Rio Tinto in Spain. Here, geologists have been able to study a system in which a naturally occurring ore deposit, almost 400 million years old, interacts with groundwater that flows through it by hydrothermal circulation. With the help of mineral-dissolving acidophilic microorganisms, sulfide minerals such as pyrite (FeS_2) in the ore deposit react with oxygen in the groundwater to produce sulfuric acid, sulfate ions (SO_4^{-2}), and iron ions (Fe^{3+}). The warm spring water that flows out of the deposit as a river (*rio* in Spanish) is extremely acidic. Your skin would dissolve if you went swimming in that water.

The river is red (*tinto* in Spanish) because of the dissolved Fe^{3+} ions. The Fe^{3+} ions combine with oxygen to produce the iron oxide minerals goethite and hematite, which may be reddish or brownish in color. In addition, unusual iron sulfate minerals such as jarosite (yellow-brown in color) form abundantly in the Rio Tinto. When geologists encounter this mineral on Earth, they know that the water from which it precipitated must have been extremely acidic.

What is a rare—and environmentally damaging—geologic setting on Earth may once have been widespread on Mars. As we saw in Chapter 20, past exploration of Mars has revealed abundant sulfate minerals similar to those found in the Rio Tinto, including jarosite. Understanding how this unusual mineral forms on Earth allows geologists to make inferences about past environments on Mars. In this case, the presence of jarosite indicates that some ancient waters on Mars were very acidic, perhaps because of the interaction of groundwater with igneous rocks composed of basalt with trace amounts of sulfide.

This scenario has implications for how we think about the possibility of life—past or present—on other planets. Environments such as the Rio Tinto on Earth show that microorganisms have learned to adapt to highly acidic conditions, and they help motivate the search for ancient life on Mars. Some scientists, however, think that although life may have learned to adapt to such harsh conditions, it may not have been able to originate under those conditions. In any event, the search for life on other planets will be strongly guided by our understanding of rocks, minerals, and extreme environments on Earth.

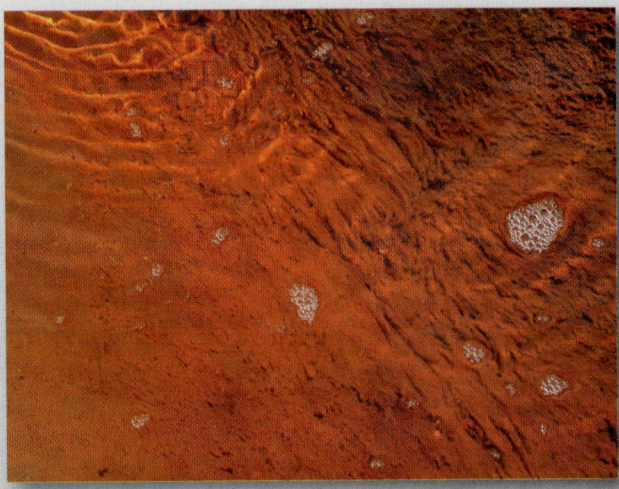

Microorganisms thrive in the acidic water of the Rio Tinto, Spain. [John Grotzinger]

Anaerobes are microorganisms that live in environments completely devoid of oxygen. At the bottoms of most lakes, streams, and oceans, the fluids in the pores of sediments just a few millimeters or centimeters below the sediment-water interface are starved of oxygen. Microorganisms that live at the sediment-water interface use up all the oxygen during respiration, creating an *anaerobic* (oxygen-free) *zone* beneath them in the sediment, where only anaerobes thrive. The oxygen-rich upper sediment layer is known as the *aerobic zone*. Many microorganisms that live in the aerobic zone could not survive in the anaerobic zone, and vice versa. The boundary between these zones is often very sharp, as shown in **Figure 22.7**.

Microorganism-Mineral Interactions

Microorganisms play a critical role in many geologic processes, including mineral precipitation, mineral dissolution, and the flux of elements through Earth's crust in biogeochemical cycles. As we will learn later in this chapter, they have also been crucial factors in the evolutionary history of larger, more complex organisms.

Mineral Precipitation Microorganisms precipitate minerals in two distinct ways: *indirectly* by influencing the composition of the water surrounding them and *directly* in

FIGURE 22.7 Microorganisms can form layered deposits called microbial mats. The top part of the mat, which is exposed to the Sun, contains photosynthetic autotrophic microorganisms, as revealed by the green color. Farther down in the mat, but still within the aerobic zone, are nonphotosynthetic autotrophs, as revealed by the purple color. Deeper in the mat, the color turns gray, revealing the anaerobic zone where heterotrophs live. [John Grotzinger.]

their cells as a result of their metabolism. Indirect precipitation occurs when dissolved minerals in an oversaturated solution precipitate on the surfaces of individual microorganisms. This happens because the surface of a microorganism has sites that bind dissolved mineral-forming elements. Mineral precipitation often leads to the complete encrustation of the microorganisms, which are effectively buried alive. Microbial precipitation of carbonate minerals and silica in hot springs are good examples of this type of microbial biomineralization (**Figure 22.8a**). Thermophiles may become completely overgrown by the mineral deposits they help to precipitate.

Minerals are directly precipitated by the metabolic activities of some microorganisms. For example, microbial respiration causes precipitation of pyrite (**Figure 22.9**) in the anaerobic zone of sediments that contain iron-bearing minerals and water in which sulfate is dissolved. As we have learned, all organisms—including microorganisms—need oxygen for respiration. In the anaerobic zone, however, O_2 is not available. Some microbial decomposers have adapted

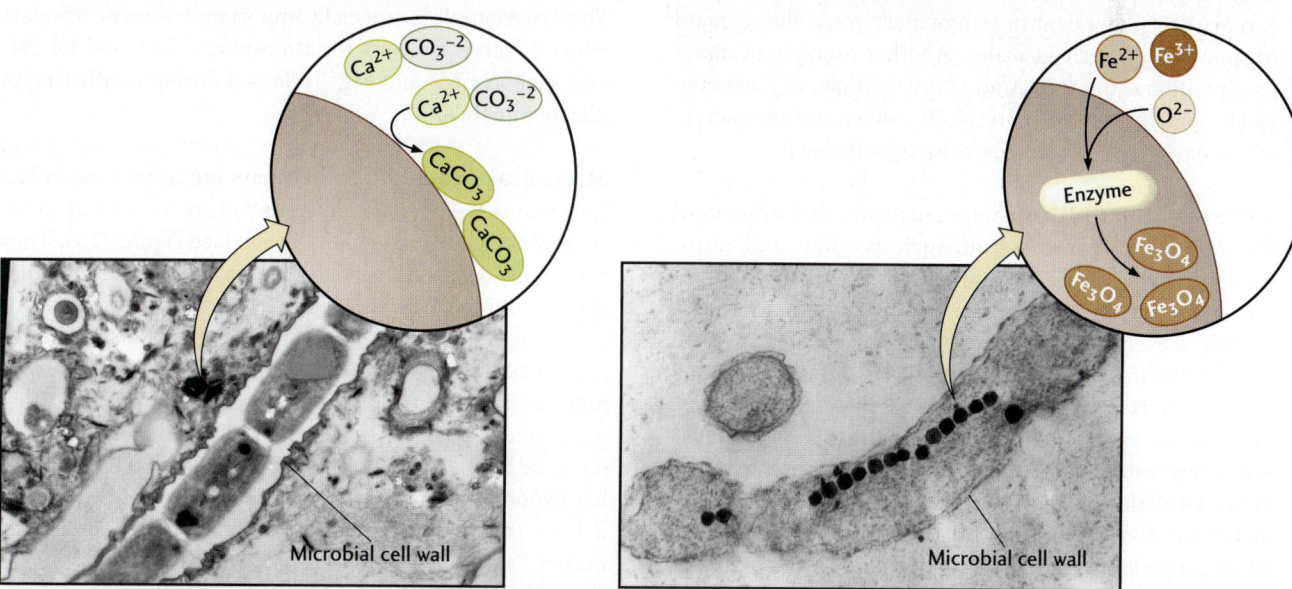

(a) Indirect precipitation of calcium carbonate

(b) Direct precipitation of magnetite

FIGURE 22.8 Microorganisms can precipitate minerals indirectly or directly. (a) The precipitation of calcium carbonate on the surfaces of bacteria is an example of indirect precipitation. (b) Intracellular production of magnetite (Fe_3O_4) crystals by some bacteria is an example of direct precipitation. Some organisms use these crystals to find their way by sensing Earth's magnetic field. [a: Grant Ferris, University of Toronto; b: Richard B. Frankel, Ph.D., California Polytechnic State University.]

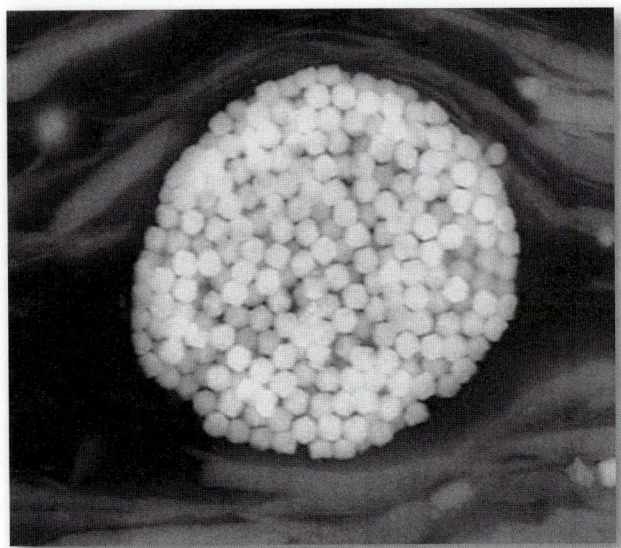

FIGURE 22.9 Pyrite commonly forms small globules in the pore fluids of anaerobic sediments. [Courtesy of Dr. Jüergen Schieber.]

to this harsh, but very common, environment by evolving ways to obtain oxygen from other sources. These microorganisms can remove the oxygen contained in sulfate (SO_4), which is abundant in most sediment pore fluids. In the process, they make hydrogen sulfide gas (H_2S), which produces the unpleasant odor of rotten eggs that is released when you dig into sandy or muddy sediments at low tide. In the final step of the process, hydrogen sulfide reacts with iron, which replaces the hydrogen to form pyrite (FeS_2). Pyrite is remarkably abundant in sedimentary rocks that contain organic matter, such as shales. Another example of direct precipitation is the formation of tiny particles of magnetite inside some bacteria (Figure 22.8b), which use these crystals to navigate by sensing Earth's magnetic field.

Mineral Dissolution Some elements that are essential for microbial metabolism, such as sulfur and nitrogen, are readily available from natural waters in dissolved form, but others, such as iron and phosphorus, must be actively scavenged from minerals by microorganisms. All microorganisms need iron, but iron concentrations in near-surface waters are generally so low that the microorganisms must obtain it by dissolving nearby minerals. In a similar way, some microorganisms obtain phosphorus—required for construction of biologically important molecules—by dissolving minerals such as apatite (calcium phosphate). Some autotrophs derive their energy not from sunlight, but from the chemicals produced when minerals are dissolved. These organisms are known as **chemoautotrophs** (see Table 22.1). For example, manganese (Mn^{2+}), iron (Fe^{2+}), sulfur (S), ammonium (NH_4^+), and hydrogen (H_2) supply microorganisms with energy when they are released from minerals.

Microorganisms dissolve minerals by producing organic molecules that react with those minerals to liberate ions from mineral surfaces. Rates of mineral dissolution are normally slow, but may be enhanced where minerals containing nutrient elements are coated by microbial biofilms. Mineral-dissolving acidophiles thrive in waters where mineral dissolution results in prolific acid formation.

Microorganisms and Biogeochemical Cycles

Pyrite precipitation by microorganisms plays an important role in the global biogeochemical cycling of sulfur (**Figure 22.10**). As we have seen, iron and sulfur are precipitated as pyrite, which accumulates abundantly within sediments. As layers of sediment are deposited, the pyrite becomes buried and encapsulated in sedimentary rocks. The pyrite remains buried until the rocks are returned to Earth's surface by tectonic uplift. When the rocks are weathered, the iron and sulfur in the pyrite are dissolved as ions in water or become incorporated into new minerals that accumulate in sediments, starting the biogeochemical cycle over again.

On a global scale, microorganisms play roles in several other biogeochemical cycles. Microbial precipitation of phosphate minerals contributes to the flow of phosphorus into sediments, particularly along the west coasts of South America and Africa, where phosphorus-rich deep ocean water that rises to the surface is available to microorganisms that live in shallower water, as we saw in Chapter 6. The chemical weathering of continental rocks is influenced by microorganisms that can increase the acidity of soils, leading to faster weathering rates. And finally, as we also saw in Chapter 6, the precipitation of carbonate minerals in marine environments is stimulated by microbial processes. This last example is especially important because carbonate minerals serve as a sink for atmospheric CO_2 and for cations such as Ca^{2+} and Mg^{2+} released during weathering of silicate minerals.

Microbial Mats **Microbial mats** are layered microbial communities. The microbial mats you are most likely to see are those that are exposed to the Sun (see Figure 22.7). They commonly occur in tidal flats, hypersaline lagoons, and hot springs. On the top, you will usually find a layer of oxygen-producing cyanobacteria that use energy from sunlight for photosynthesis. This uppermost layer is green because cyanobacteria contain the same light-absorbing pigment that plants and algae have. This layer may be as thin as 1 mm, yet it can be as effective in producing energy from the Sun as a hardwood forest or grassland. This uppermost green layer defines the aerobic zone of the mat. The anaerobic zone occurs below the cyanobacterial layer and is often a dark gray color. Although this anaerobic part of the mat contains no oxygen, it still can be very active. The anaerobic heterotrophs in this layer derive their food from the organic matter produced by the cyanobacteria. Their respiration often results in the precipitation of pyrite, as described earlier in this chapter.

Microorganisms: Nature's Tiny Chemists

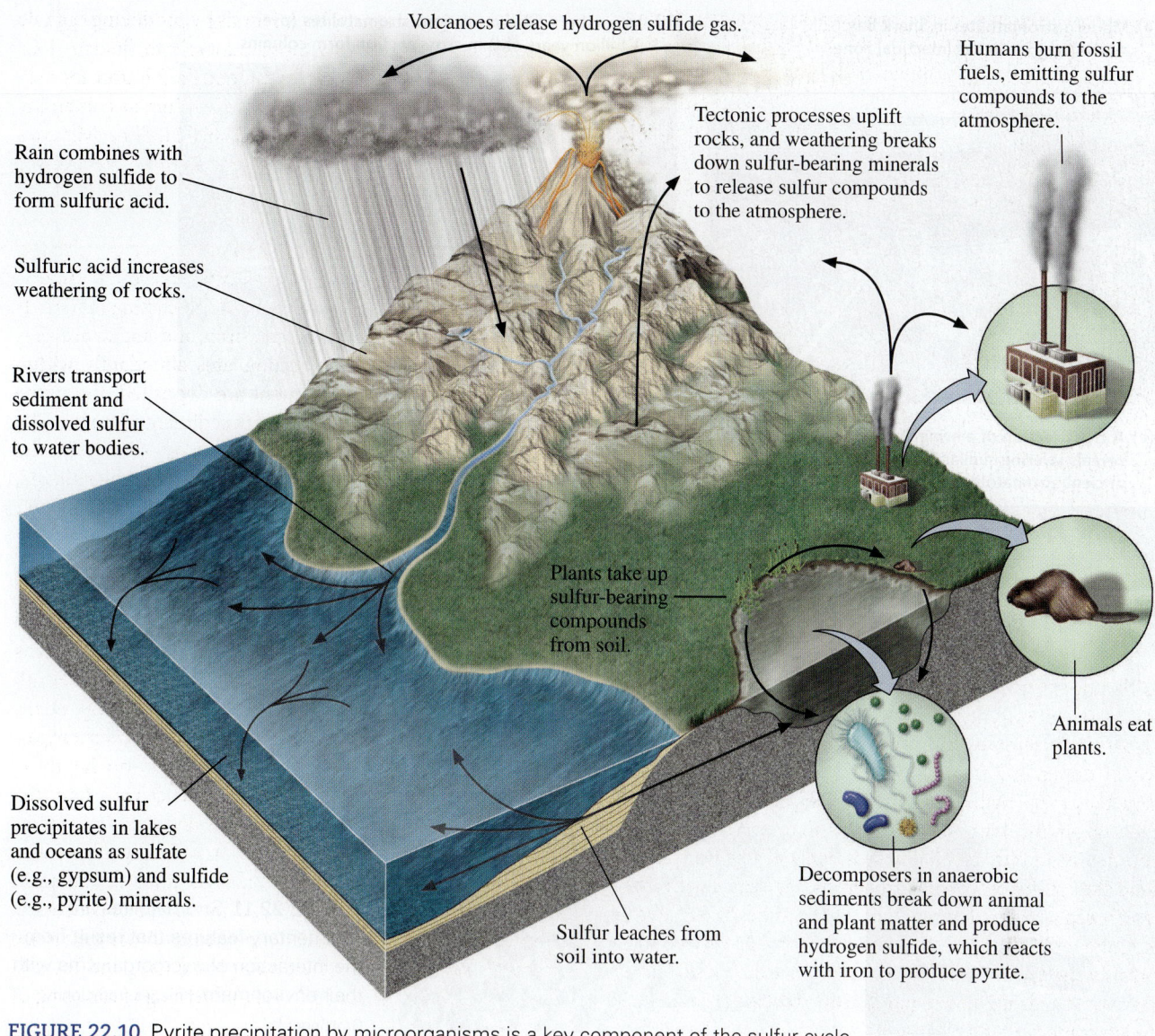

FIGURE 22.10 Pyrite precipitation by microorganisms is a key component of the sulfur cycle.

Microbial mats are miniature models of the same biogeochemical cycles that occur at regional or even global scales. In a microbial mat, photosynthetic autotrophs use energy from sunlight to convert carbon in atmospheric CO_2 into carbon in larger molecules such as carbohydrates. After the photoautotrophs die, the heterotrophs use the carbon in their bodies as an energy source. In the process, the heterotrophs convert some of this carbon into CO_2, which is returned to the atmosphere, where it can be used by the next generation of photoautotrophs, and so on. In the case of microorganisms, this cycle is confined to the very small scale of a layer of sediment, but it is directly analogous to the process by which rain forests—at a global scale—extract CO_2 from the atmosphere during photosynthesis. Although individual trees do the actual work, one can think of a rain forest as a giant photosynthesis machine that removes enormous quantities of CO_2 from the atmosphere and produces enormous quantities of carbohydrates. When the trees die, their organic matter is used by heterotrophs on the forest floor to produce energy. This process returns enormous amounts of carbon—in the familiar form of CO_2—to the atmosphere.

Stromatolites Today, microbial mats are restricted to places on Earth where plants and animals cannot interfere with their growth. Before the evolution of plants and animals, however, microbial mats were widespread, and they are one of the most common features preserved in Precambrian sedimentary rocks formed in aquatic environments. **Stromatolites**—rocks with distinctive thin layers—are believed to have been formed from ancient microbial mats.

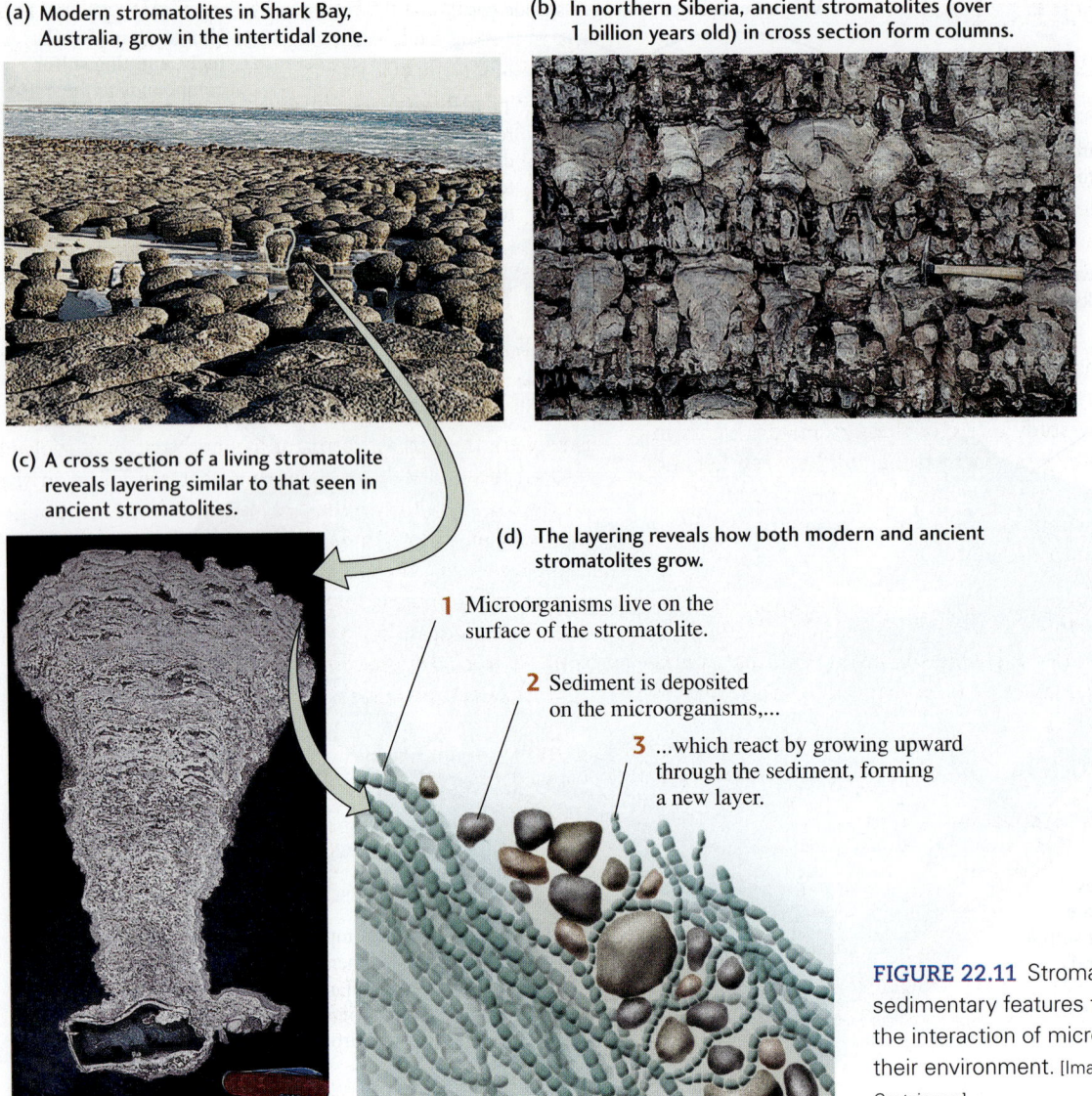

FIGURE 22.11 Stromatolites are sedimentary features that result from the interaction of microorganisms with their environment. [Images from John Grotzinger.]

Stromatolites range in shape from flat sheets to dome-shaped structures with complex branching patterns (Figure 22.11). They are one of the most ancient types of fossils on Earth and give us a glimpse of a world once ruled by microorganisms.

Most stromatolites probably formed when sediment raining down on microbial mats was trapped and bound by microorganisms living on the surfaces of the mats (Figure 22.11d). Once covered with sediment, the microorganisms grew upward between the sediment particles and spread laterally to bind the particles in place. Each stromatolite layer corresponds to the deposition of a sediment layer followed by the trapping and binding of that layer. Microbial communities can be observed building such structures today in intertidal environments such as Shark Bay, Western Australia (Figure 22.11a).

In other cases, however, stromatolites form by mineral precipitation, rather than by trapping and binding of sediment by microorganisms. That mineral precipitation may be indirectly controlled by microorganisms, or it may simply be the result of oversaturation of the surrounding water. As we saw in Chapter 6, the ocean contains abundant calcium and carbonate, which react to form the minerals calcite and aragonite. These minerals are important for the growth of stromatolites formed by mineral precipitation.

The potential role of microorganisms in stromatolite formation is important to understand because these layered, dome-shaped structures have been used as evidence for life on early Earth. But if stromatolites can be built by nonmicrobial mineral precipitation, their use as evidence for early life is uncertain. Only by carefully studying the processes by which microorganisms interact with minerals and sediments, and the chemical and textural fingerprints of these interactions, will we be able to determine whether the formation of stromatolites on early Earth required the presence of microorganisms.

Geobiologic Events in Earth's History

The geologic time scale divides time based on the comings and goings of fossil assemblages (see Chapter 9). These biological patterns provide a convenient ruler for subdividing Earth's history, but they were almost always associated with global environmental changes. At many of the major boundaries in the geologic time scale, Earth experienced a one-time event that caused dramatic changes in conditions for life. Some of these changes were triggered by organisms themselves, others by geologic events, and still others by forces from outside the Earth system.

We will now study a few of these dramatic events in Earth's history—events in which the link between life and the physical environment is clearly visible. **Figure 22.12** shows the great antiquity of life on Earth and the timing of several of these major events.

Origin of Life and the Oldest Fossils

When Earth first formed some 4.5 billion years ago, it was lifeless and inhospitable. A billion years later, it was teeming with microorganisms. How did life begin? Along with other grand puzzles such as the origin of the universe, this question remains one of science's greatest mysteries.

The question of *how* life may have originated is very different from the question of *why* life originated. Science offers an approach only to understanding the "how" part of this mystery because, as you may recall from Chapter 1, it uses observations and experiments to create testable hypotheses. These hypotheses may explain the series of steps involved in the origin and evolution of life, and they can be tested by searching for evidence in the fossil and geologic records. However, observations and experiments do not provide a testable approach to the question of why life evolved.

The fossil record tells us that single-celled microorganisms were the earliest forms of life, and that they evolved into all the multicellular organisms that are found in the younger parts of the geologic record. The fossil record also shows us that most of life's history involved the evolution of microorganisms. We can find microfossils in rocks 3.5 billion years old, yet we can conclusively identify fossils of multicellular organisms only in rocks younger than 1 billion years. It therefore appears that microorganisms were the only organisms on Earth for at least 2.5 billion years!

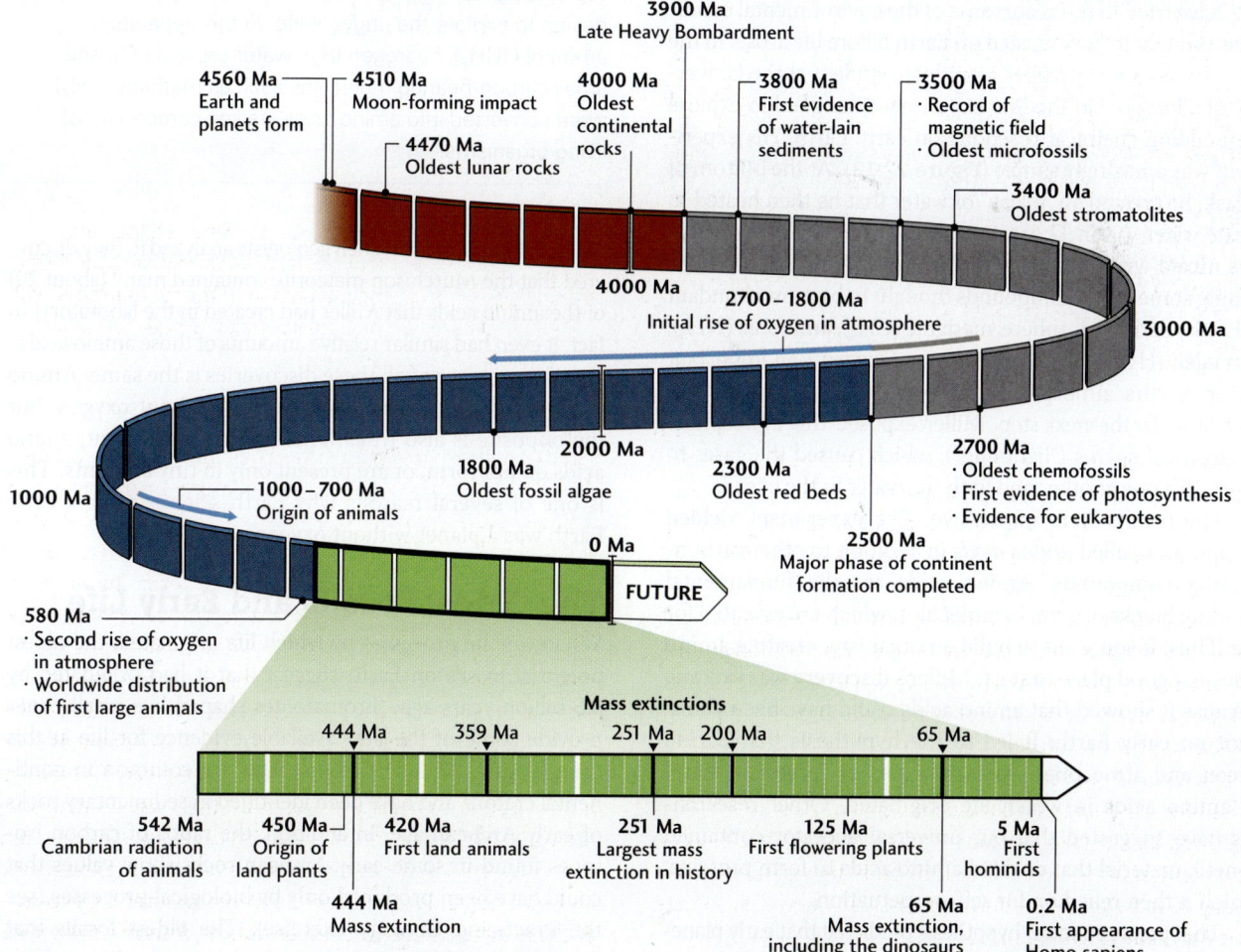

FIGURE 22.12 The geologic time scale, showing major events in the history of life. (Ma: million years ago.)

The theory of evolution predicts that these first microorganisms—and all life that came after them—evolved from a universal ancestor (see Figure 22.5). What did this universal ancestor look like? We really don't know, but most geobiologists agree that it must have had several important characteristics. The most crucial of these would be genetic information: instructions for growth and reproduction. Otherwise, it would have had no descendants. The universal ancestor must also have been composed of carbon-rich compounds. As we have seen, all organic substances, including organisms, are made principally of carbon.

How did the universal ancestor arise? One approach to answering this question would be to search for clues in rocks. However, well-preserved fossils are found only in sedimentary rocks that have not been significantly affected by metamorphism or deformation. There are no well-preserved sedimentary rocks from the time when life first evolved, so scientists must use other approaches. Laboratory chemists have played an important role here.

Prebiotic Soup: The Original Experiment on the Origin of Life

In laboratory experiments that probe the origin of life, scientists have tried to re-create some of the environmental conditions thought to have existed on Earth before life arose. In the early 1950s, Stanley Miller, a graduate student at the University of Chicago, did the first experiment designed to explore life-building chemical reactions on early Earth. His experiment was amazingly simple (Figure 22.13). At the bottom of a flask, he created an "ocean" of water that he then heated to create water vapor. The water vapor emitted from the ocean was mixed with other gases to create an "atmosphere" containing some of the compounds thought to be most abundant in Earth's early atmosphere: methane (CH_4), ammonia (NH_3), hydrogen (H_2), and the water vapor. Oxygen—an important gas in Earth's atmosphere today—was probably absent at that time. In the next step, Miller exposed this atmosphere to electrical sparks ("lightning"), which caused the gases to react with one another and with the water in the ocean.

The results were impressive. The experiment yielded compounds called *amino acids* in addition to other carbon-bearing compounds. Amino acids are the fundamental building blocks of protein molecules, which are essential for life. Thus, if you want to build an organism, creating amino acids is a good place to start. Miller's discovery was exciting because it showed that amino acids could have been abundant on early Earth. It led to the hypothesis that Earth's ocean and atmosphere formed a sort of "prebiotic soup" of amino acids in which life originated. Other researchers have suggested that our universal ancestor contained genetic material that enabled amino acids to form proteins, which it then relied on for self-perpetuation.

The "prebiotic soup" hypothesis predicted that early planetary materials might contain amino acids. That prediction was borne out years later when, in 1969, a meteorite hit Earth near

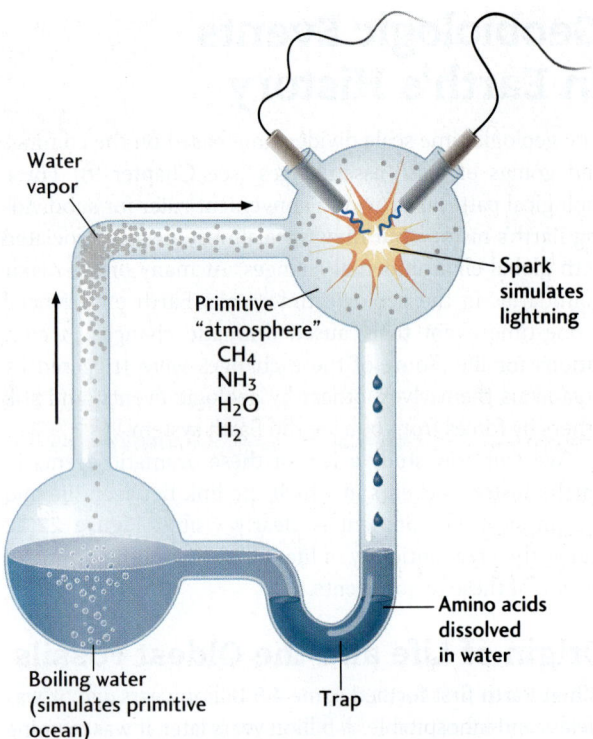

FIGURE 22.13 Stanley Miller used this simple experimental design to explore the origin of life. In this apparatus, ammonia (NH_3), hydrogen (H_2), water vapor (H_2O), and small carbon-bearing molecules such as methane (CH_4) were converted into amino acids—a key component of living organisms.

Murchison, Australia. When geologists analyzed it, they discovered that the Murchison meteorite contained many (about 20) of the amino acids that Miller had created in the laboratory! In fact, it even had similar relative amounts of those amino acids.

The message of all these discoveries is the same: Amino acids could have formed on a planet without oxygen. But the opposite is also true: Where oxygen is present, amino acids do not form, or are present only in tiny amounts. This is one of several reasons why Earth scientists think early Earth was a planet without oxygen.

The Oldest Fossils and Early Life

Whatever the processes by which life originated, the oldest potential fossils on Earth suggest that it had originated by 3.5 billion years ago. Stromatolites shaped like small cones provide some of the best available evidence for life at this time (Figure 22.14a). Stromatolites are common in continental cratons and have been identified in sedimentary rocks of early Archean age. In addition, the ratios of carbon isotopes found in some early Archean rocks show values that could have been produced only by biological processes (see the Practicing Geology Exercise). The oldest fossils that preserve possible morphological evidence for life are tiny threads that are similar in size and appearance to modern

 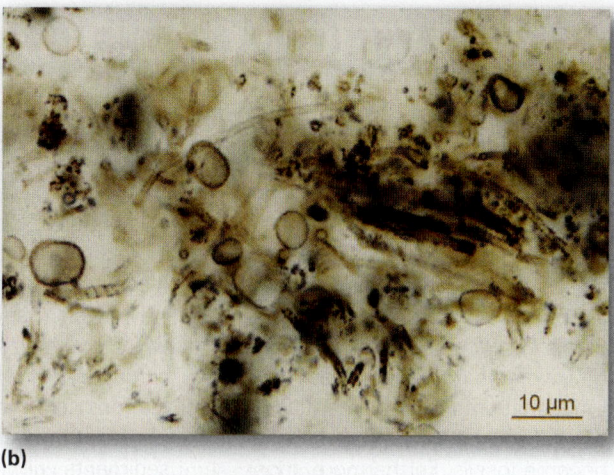

FIGURE 22.14 (a) Early Archean (3.4-billion-year-old) stromatolites in the Warrawoona Formation, Western Australia. The conical shapes suggest that the microbial mats that formed these rocks grew toward the sunlight. (b) Abundant microfossils are well preserved in the 2.1-billion-year-old Gunflint Formation of southern Ontario, Canada. [Images courtesy of H. J. Hofmann.]

microorganisms, encased in chert. These features were found in formations in Western Australia that may be as old as 3.4 billion years, although their interpretation as microfossils remains controversial. Younger, better-preserved microfossils occur in the 3.2-billion-year-old Fig Tree Formation of South Africa and in the 2.1-billion-year-old Gunflint Formation of southern Canada (Figure 22.14b). The Gunflint fossils, discovered in 1954, were the first ever discovered in Precambrian rock, and they set off a tidal wave of research that continues to this day. In the past 50 years, we have seen, in many new localities, just how ancient life on Earth is and how well it can be preserved under the right geologic circumstances.

Most geobiologists agree that there was life on Earth 3.5 billion years ago, but are uncertain about how those early organisms functioned or obtained energy and nutrients. Some scientists argue that the oldest organisms on the universal tree of life were chemoautotrophic, obtaining their energy directly from chemicals in the environment. Furthermore, those oldest organisms may have been hyperthermophilic. This possibility suggests that life may have originated in very hot water, such as that in hot springs or hydrothermal vents on mid-ocean ridges, where sunlight was unavailable as an energy source, but chemicals were abundant (Figure 22.15).

FIGURE 22.15 Hot water released from hydrothermal vents along mid-ocean ridges (visible here as a plume of what looks like black smoke) is full of mineral nutrients from which chemoautotrophic microorganisms obtain their energy. It is possible that life originated in such environments. [NOAA PMEL EOI Program.]

PRACTICING GEOLOGY EXERCISE

How Do Geobiologists Find Evidence of Early Life in Rocks?

Perhaps the most important question a geobiologist can ask is, "What evidence of life is preserved in rocks?" If fossils of animal shells and skeletons are present in rocks, then this question is easy to answer. In many cases, however, the geologic processes that turn sediments into sedimentary rocks destroy the materials that could have become fossils. Furthermore, most organisms do not have hard shells or skeletons that are easily preserved, so we would not expect them to become fossils. And on early Earth, before the advent of animals with hard shells and skeletons, most organisms were microscopic in size. Simply put, how can the former presence of life on Earth be detected in situations in which fossils are not preserved?

One approach that geobiologists often depend on is the use of chemical signatures of former life. Carbon provides the most obvious example of an element that might have been concentrated by biological processes. Not all concentrations of carbon are of biological origin, however, so additional tests must be applied.

One of these tests asks whether the carbon present has a distinctive *isotopic* composition. Recall from Chapter 3 that isotopes are atoms of the same element with different numbers of neutrons. Many elements of low atomic weight have two or more stable (nonradioactive) isotopes. A carbon atom has six protons, but may have six, seven, or eight neutrons, which give it atomic masses of 12, 13, or 14, respectively. Carbon-12 is by far the most common isotope, so samples of carbon from ancient rocks or modern sediments will yield mostly carbon-12.

Fortunately, it turns out that metabolic processes such as photosynthesis use carbon-12 and carbon-13 differently. The difference in atomic mass between carbon-12 (often denoted ^{12}C) and carbon-13 (^{13}C) results in differences in their uptake by organisms. Photosynthetic organisms, for example, use carbon dioxide and water to form carbohydrates. They use carbon dioxide molecules that contain ^{12}C preferentially over those that contain ^{13}C. As a result, photosynthetic organisms become enriched in ^{12}C relative to the environment from which they draw carbon dioxide.

We can therefore use carbon isotopes as a tool to detect ancient life by measuring the amounts of ^{12}C and ^{13}C present in sedimentary rocks. If sediments are formed in the presence of organic matter that is enriched in ^{12}C (or in any particular isotope), this enrichment may be passed along to the sediments, and then on to the resulting rock. Thus, a shale that might be billions of years old can preserve a "signature" of life recorded by its carbon isotope composition.

We begin by measuring the amounts of ^{12}C and ^{13}C in a rock sample and calculating the ratio between them ($^{12}C/^{13}C$). We then compare that ratio with the $^{12}C/^{13}C$ ratio in a *standard*. The standard is a material (often a pure mineral, such as calcite) whose $^{12}C/^{13}C$ ratio is precisely known and varies very little. The standard can be compared over and over again with samples of other carbon-bearing rocks and sediments, as well as with living organisms and other natural substances. By comparing both rock samples and organisms with the standard, we can search for similarities that link rock samples to particular biological processes.

The following table gives the $^{12}C/^{13}C$ ratios for a standard, three rock samples, and two natural substances—plant material and methane gas:

Standard	Rock A	Rock B	Rock C	Plant Material	Methane Gas
1000	995	1020	1050	1025	1060

Chemofossils and Eukaryotes Form and size alone are not enough to allow us to deduce the function of microorganisms, so microfossils are ultimately limited in the information they can provide. Additional information can be gleaned from **chemofossils**, the chemical remains of organic compounds made by ancient microorganisms while they were alive. When an organism dies, most of the organic compounds in its body are quickly broken down into much smaller molecules, usually by heterotrophs. Some of these molecules, however, are very stable and resist recycling. *Cholestane*, for example, is a remarkably durable substance, made only by eukaryotes, that is very similar to the well-known compound cholesterol. Cholestane chemofossils have been identified in 2.7-billion-year-old rocks from Western Australia. The presence of these chemofossils tells us that single-celled eukaryotic microorganisms must have emerged by that time. It is the eukaryotes that would eventually evolve into multicellular organisms, including animals, but not until much later.

Origin of Earth's Oxygenated Atmosphere

The rise of oxygen—the stuff we breathe—is another important milepost in the history of interactions between life and its environment. As we learned in Chapter 9,

The following equation* allows us to compare these data:

$R_{(sample)} = [^{12}C/^{13}C \text{ of the standard}] - [^{12}C/^{13}C \text{ of the sample}]$

where R stands for the value of the difference between the two ratios.

For most rocks, the value of R will be close to zero, but could be slightly positive or negative. In contrast, if photosynthesis was involved in the formation of organic matter incorporated into a sedimentary rock—for example, a shale—the value of R for a sample of that shale could be very negative—close to –20. Some chemoautotrophic microorganisms that consume methane gas produce carbonate rocks with an extremely negative R value, on the order of –50.

Using the data and equation above, let's try to identify which of our rock samples formed in the presence of biological processes. Let's begin with rock B:

$R_{(rock\ B)} = [^{12}C/^{13}C \text{ of the standard}] - [^{12}C/^{13}C \text{ of the rock B}]$
$= 1000 - 1020$
$= -20$

This result shows that rock B has an R value that is substantially different from zero, suggesting that ancient biological processes may have been at work when the rock formed. We can confirm that its value of –20 is a close match for the R value predicted for photosynthesis by calculating the R value for plant material:

$R_{(plant)} = [^{12}C/^{13}C \text{ of the standard}] - [^{12}C/^{13}C \text{ of the plant material}]$
$= 1000 - 1025$
$= -25$

The R value for the plant material, at –25, is close enough to the R value for rock B, at –20, that our hypothesis of ancient photosynthesis is strengthened.

BONUS PROBLEM: Try calculating R values for rocks A and C. Which rock does not record a distinctive signature of biological processes? Is there a rock among our samples that might have formed in the presence of methane-consuming microorganisms? If so, how can you check this result?

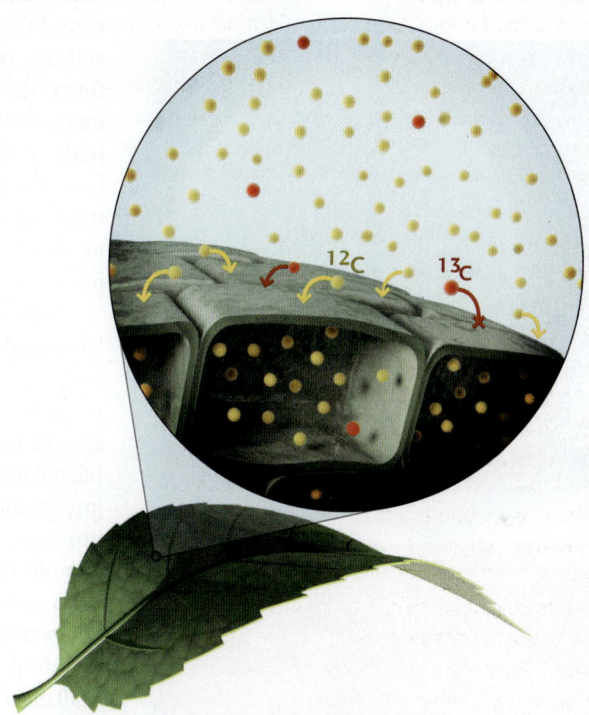

Plants take up carbon dioxide during photosynthesis. Because they take up CO_2 molecules containing ^{12}C more easily than those containing ^{13}C, plants become enriched in ^{12}C relative to their environment.

*This equation is simplified from what is normally used in practice. It neglects the standard "delta" notation, which normalizes the actual abundances of the isotopes in the sample to those in the standard.

Earth's early atmosphere contained little oxygen. Our current oxygen-rich atmosphere was produced by early life through photosynthesis. Remarkably, the same Australian rocks that preserve chemofossil evidence of eukaryotes also preserve chemofossil evidence of cyanobacteria. Because of this evidence, geologists believe that photosynthesis had become an important metabolic process by 2.7 billion years ago. Thus, one group of organisms (cyanobacteria) permanently altered Earth's environment by changing the composition of its atmosphere, while another group of organisms (eukaryotes) was influenced by that change to evolve in new directions.

The oxygenation of Earth's atmosphere probably occurred in two main steps, separated by more than a billion years. The first major increase began with the evolution of the cyanobacteria. The oxygen they produced reacted with iron dissolved in seawater, causing iron oxide minerals, such as magnetite and hematite, and silica-rich minerals, such as chert and iron silicates, to precipitate and sink to the seafloor. These minerals accumulated in thin,

FIGURE 22.16 Unusual sedimentary rocks and new, larger eukaryotes mark the rise of oxygen concentrations in the atmosphere between 2.7 billion and 2.1 billion years ago. (a) A banded iron formation. (b) These fossils of *Grypania*, a type of eukaryotic algae, are visible with the naked eye. (c) Red beds are made up of sandstones and shales cemented together by iron oxide minerals. [a: Francois Gohier/Science Source; b: Courtesy of H. J. Hofmann; c: John Grotzinger.]

alternating layers of sediments called **banded iron formations** (Figure 22.16a). Iron is soluble in water when oxygen concentrations are low, as would have been the case on Earth before cyanobacteria evolved. When oxygen concentrations are high, however, iron reacts with oxygen to form highly insoluble compounds. Therefore, the oxygen produced by cyanobacteria would have immediately caused iron to precipitate from seawater and sink to the seafloor. This process would have continued until most of the dissolved iron was used up, allowing oxygen to accumulate in the ocean and atmosphere.

Atmospheric oxygen concentrations began to build about 2.4 billion years ago and reached an initial plateau about 2.1 billion to 1.8 billion years ago, when the first eukaryotic fossils, of a type of algae, entered the geologic record (Figure 22.16b). The large size of these organisms—at least 10 times larger than anything that came before them—is thought to be a consequence of the oxygen increase. This time also marks the first appearance of **red beds,** unusual stream deposits of sandstones and shales bound together by iron oxide cement, which gives them their red color (Figure 22.16c). The presence of iron oxides in these deposits indicates that oxygen must have been present in the atmosphere to precipitate them.

After eukaryotic algae came on the scene, not much happened for over a billion years. Then, about 580 million years ago, atmospheric oxygen concentrations rose dramatically, almost to their modern level. The reason for this second increase is still not understood, though it may be related to an increase in the burial of organic carbon by sedimentation. In a process somewhat similar to the one that produces banded iron formations, oxygen reacts easily with organic matter, usually with the help of microorganisms. Thus, as long as there is organic matter around, oxygen will be used up. If organic matter is removed from the system by burial in sediments, however, it cannot react with the available oxygen. Thus, the second step in the rise of atmospheric oxygen might have been related to an increase in sediment production. Such an increase may have occurred when mountains were built—and then eroded—during global tectonic events, such as the assembly of supercontinents. In any case, the consequences were dramatic: The first large multicellular animals suddenly appeared, and all the modern groups of animals evolved shortly thereafter, ushering in the Phanerozoic eon with its wonderfully complex and diverse organisms.

Evolutionary Radiations and Mass Extinctions

In most cases, the boundaries of the eras and periods within the Phanerozoic eon are marked by the demise, or *extinction*, of a particular group of organisms, followed

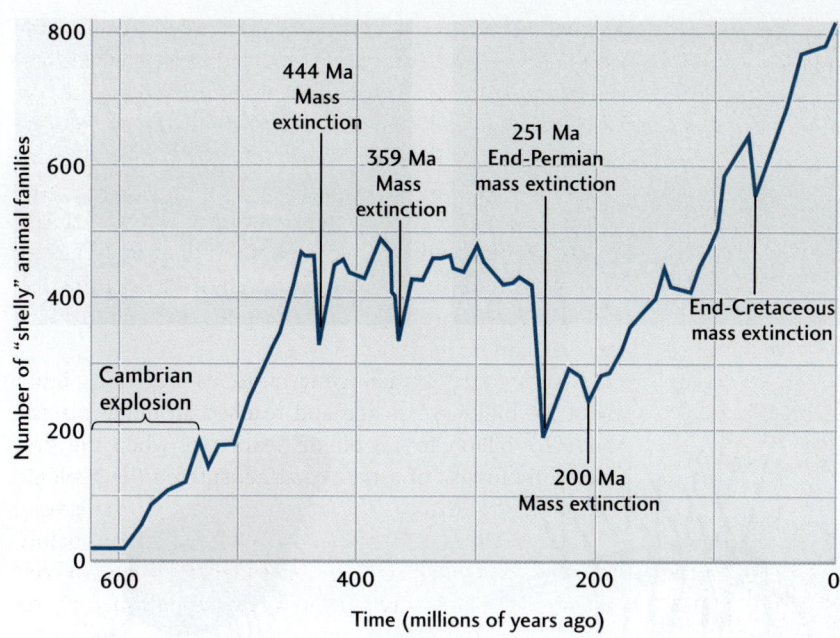

FIGURE 22.17 The diversity of animal fossils reveals both mass extinctions and radiations. This graph shows the number of "shelly" animal families found in the fossil record during the last 600 million years; each family comprises many species. During a radiation, such as the Cambrian explosion, the number of new families increases. During a mass extinction, such as the one at the end of the Cretaceous period, the number of families decreases. (Ma: million years ago.)

by the rise, or *radiation*, of a new group of organisms. When groups of organisms are no longer able to adapt to changing environmental conditions or compete with more successful groups of organisms, they become extinct. An interval when many groups of organisms become extinct at the same time is called a *mass extinction* (**Figure 22.17**) (see Chapter 9). In a few cases, the boundaries of the geologic time scale are marked by environmental catastrophes of truly global magnitude. Radiations are stimulated by the availability of new habitats when a mass extinction eliminates highly competitive and established groups of organisms.

Radiation of Life: The Cambrian Explosion

Perhaps the most remarkable geobiological event in Earth's history, aside from the origin of life itself, was the sudden appearance of large animals with shells and skeletons at the end of Precambrian time (**Figure 22.18**). This rapid development of new types of organisms from a common ancestor—what biologists call an **evolutionary radiation**—had such an extraordinary effect on the fossil record that its culmination 542 million years ago is used to mark the most profound boundary of the geologic time scale: the beginning of the Phanerozoic eon. This boundary also coincides with the start of the Paleozoic era and the Cambrian period (see Chapter 9 and Figure 22.12).

Evolutionary radiations are rapid by nature; if they were not, they would not be noticed in the fossil record. However, the radiation of animals during the early Cambrian, after almost 3 billion years of very slow evolution, was so fast that it is often called the **Cambrian explosion,** or biology's Big Bang. Every major animal group that exists on Earth today, as well as a few more that have since become extinct, appeared within less than 10 million years. All the major branches (*phyla*) on the animal tree of life (**Figure 22.19**) originated during the Cambrian explosion. Note, however, that as impressive as this tree of animals seems, it is a single, short branch of the universal tree of life (see Figure 22.5).

Geobiologists have raised two major questions about the Cambrian explosion. First, what allowed these early animals to develop such complex body forms so rapidly, and therefore to become so diverse? Systematic change in organisms over many generations is referred to as **evolution.** Evolution is driven by **natural selection,** the process by which populations of organisms adapt to changes in their environment. The theory of *evolution by natural selection* states that, over many generations, individuals with the most favorable traits are most likely to survive and reproduce, passing those traits on to their offspring. If environmental conditions change over time, the traits that are favored change as well. This process can lead eventually to the emergence of new species.

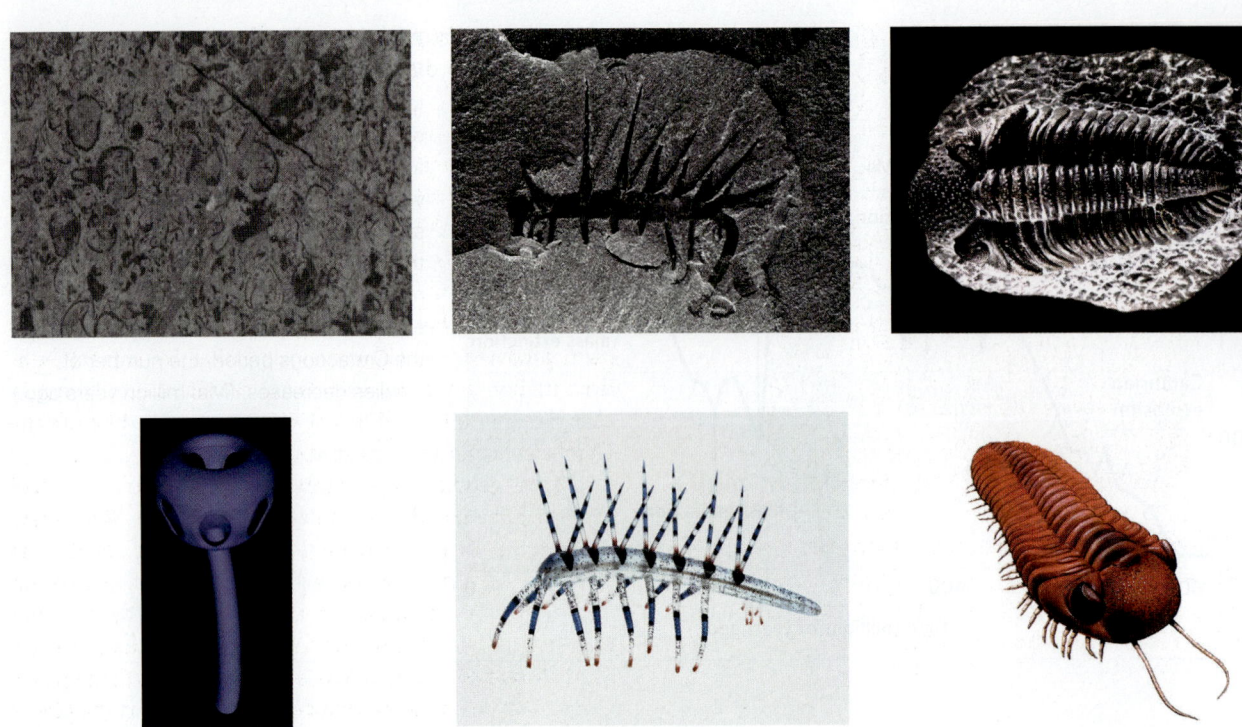

FIGURE 22.18 Fossils that record the Cambrian explosion. Precambrian organisms such as *Namacalathus* (*left*) were the first organisms to use calcite in making shells. These organisms became extinct at the Precambrian-Cambrian boundary. Their extinction paved the way for a strange new group of organisms, including *Hallucigenia* (*center*) and the more familiar trilobites (*right*), that formed weak shells made of organic material similar to fingernails. In each example, the fossils are shown on top and the reconstructed organism is shown on the bottom. [*left, top:* John Grotzinger; *left, bottom:* W. A. Watters; *center, top:* Burgess Shale *Hallucigenia* 18-5 by Chip Clark, Smithsonian; *center, bottom:* Chase Studio/Science Source; *right, top:* Courtesy of Musée cantonal de géologie, Lausanne. Photo by Stéphane Ansermat; *right, bottom:* Chase Studio/Science Source.]

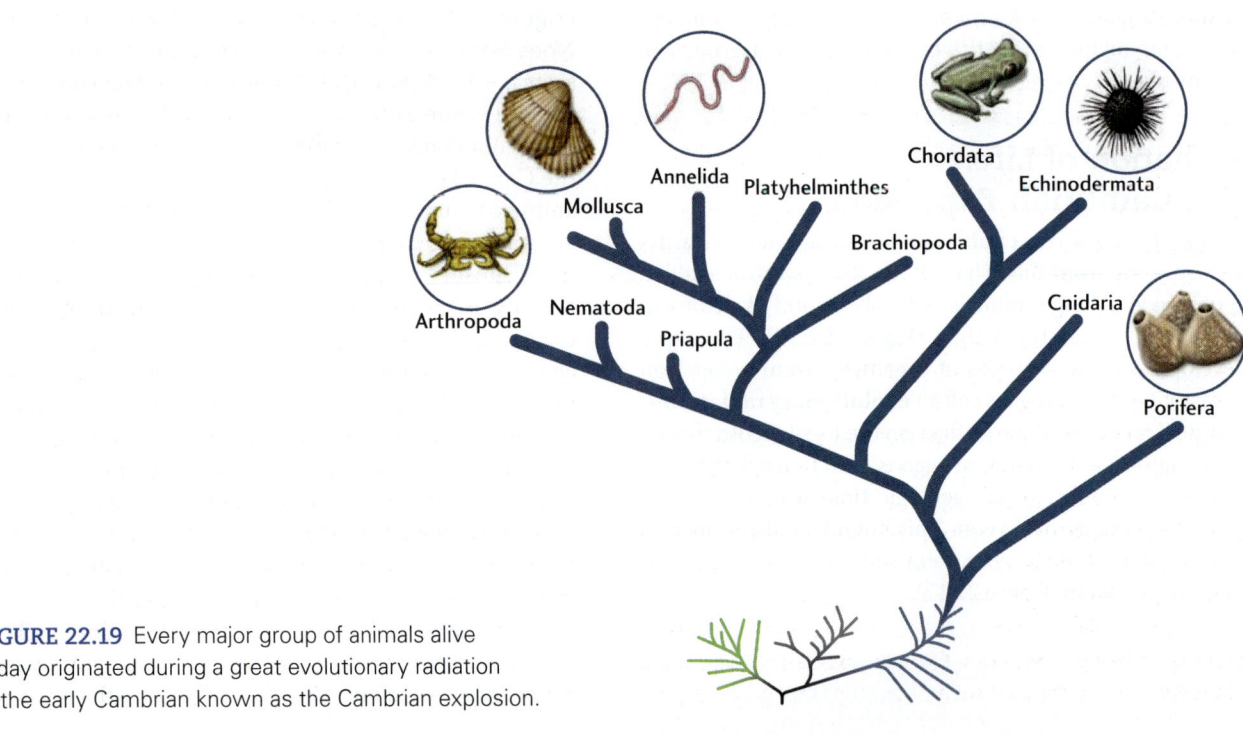

FIGURE 22.19 Every major group of animals alive today originated during a great evolutionary radiation in the early Cambrian known as the Cambrian explosion.

FIGURE 22.20 A fossilized animal embryo from the latest part of Precambrian time. Such fossils show that multicellular animals had evolved before the Cambrian period and are the ancestors of the animals that evolved during the Cambrian explosion. [Courtesy Shuhai Xiao, Virginia Tech.]

One hypothesis for the cause of the Cambrian explosion is that the genes of these early animals changed in some way that made it possible for them to exceed some sort of evolutionary barrier. The stage was set by the development of multicellularity in late Precambrian time (**Figure 22.20**), which opened up new evolutionary possibilities. It is also possible that the ancestral animals had to reach a certain size before they could diversify. Some Precambrian animals, such as the fossil animal embryo shown in Figure 22.20, are so small they can be seen only with a microscope. The development of shells and skeletons might have been an important trigger of further diversification: Once one group of animals had evolved hard parts, the others had to as well, or they would have been eliminated through competition.

The second riddle of the Cambrian explosion is why these animals differentiated *when* they did. Geobiologists have puzzled over the timing of the Cambrian explosion for more than 150 years. Back in the days of Charles Darwin, it wasn't clear if the Cambrian explosion represented the origin of life itself. But the sudden appearance of complex and diverse animal fossils in the geologic record presented a challenge to Darwin's theory of natural selection. His theory predicted slow changes in the form and function of organisms; hence, it predicted that less-complex life-forms should have occurred before the first animals, and it could not easily accommodate these complex creatures that apparently had no simpler ancestors. Therefore, Darwin hypothesized that the expected ancestors must be absent from the record because the rocks containing the Cambrian fossils must lie above an unconformity. He predicted that rocks from the time of the proposed unconformity would eventually be discovered, and that those rocks would contain the "missing" ancestors. Darwin turned out to be right, but it was only in the past several decades that geobiologists discovered the fossils described earlier in this chapter, proving that animals did indeed originate before the Cambrian explosion.

So it seems clear that the Cambrian animals did have ancestors, perhaps lurking between tiny grains of sand at the bottom of shallow seas. However, isotopic dating techniques show that these tiny animals were probably less than 100 million years older than their Cambrian descendants. Other dating techniques, based on studies of the genes of modern organisms, suggest that the origin of animals may have predated the Cambrian explosion by several hundred million years. But even these estimates hardly matter compared with the billions of years that passed before the Cambrian explosion occurred.

Most geobiologists agree that once animals had evolved, they could have radiated at any time. Why, then, did they radiate about 542 million years ago and not at some other time? Perhaps the timing of the Cambrian explosion was driven by the dramatic environmental changes that occurred near the end of Precambrian time. To human eyes, Earth at that time would have seemed a very strange place: Long chains of great mountains were forming as the pieces of the giant continent Gondwana were being fused together, and the climate was in turmoil, flipping between frigid periods when the entire Earth may have been covered in ice and extremely warm, ice-free periods (see Chapter 15). Oxygen concentrations in the oceans and atmosphere were increasing as erosion of the rising mountains produced sediments, which buried the organic matter whose decomposition would otherwise have consumed that oxygen. This last change may have been the most important. Without sufficient oxygen, animals simply cannot grow large.

Whatever the ultimate cause of the Cambrian explosion, one point stands clear: Evolutionary radiations are the result of genetic possibility combined with environmental opportunity. The radiation of organisms is not just the result of having the right genes, and it is not just the result of living in the right environment. Organisms must take advantage of both to evolve.

Tail of the Devil: The Demise of Dinosaurs

The mass extinction that marks the Cretaceous-Tertiary boundary and the end of the Mesozoic era (about 65 million years ago; see Figures 9.11 and 9.15) represents one of the greatest such events in Earth's history. Entire global ecosystems were obliterated, and about 75 percent of all species on Earth, both on land and in the ocean, were extinguished forever. The dinosaurs are only one of several groups that became extinct at the end of the Cretaceous period, but they are certainly the most prominent. Other groups, such as ammonites, marine reptiles, certain types of clams, and many types of plants and plankton, also perished.

In contrast to the Cambrian explosion, almost all scientists agree on the cause of the Cretaceous-Tertiary mass extinction. We are now virtually certain that the cause was

FIGURE 22.21 The pocketknife marks a light-colored layer of clay, containing both extraterrestrial materials and materials from local rocks at the Chicxulub impact site, which accumulated in the Raton Basin of the southwestern United States. Such deposits have been found worldwide. [Dr. David A. Kring.]

a gigantic asteroid impact. In 1980, geologists discovered a thin layer of dust containing *iridium*—an element that is typical of extraterrestrial materials—in sediments deposited at the end of the Cretaceous in Italy (**Figure 22.21**).

This extraterrestrial dust was subsequently found at many other locations around the world, on every continent and in every ocean, but always exactly at the Cretaceous-Tertiary boundary. The geologists argued that the accumulation of this much iridium-bearing dust would require an asteroid about 10 km in diameter to hit Earth, explode, and send its cosmic detritus across the globe. Publication of this hypothesis spurred a search for the impact crater. That search was bound to be difficult for two reasons. First, most of Earth's surface is covered by oceans, so the crater could easily have been under water. Second, since the crater would be 65 million years old, it could have been eroded or filled in with sediments and sedimentary rock. In the early 1990s, however, geologists found a huge crater, almost 200 km in diameter and 1.5 km deep, buried under sediments near a town on Mexico's Yucatán Peninsula, called Chicxulub.

Geologic evidence from Chicxulub, as well as from the surrounding region and around the world, has allowed geologists to paint a picture of what happened there. The name *Chicxulub* means "tail of the devil" in the local Mayan language, and the immediate aftermath of the impact would have been hellish indeed. The asteroid struck Chicxulub at a speed of Mach 40, coming in from the south at an angle of about 20° to 30° from the horizontal. Its explosion would have produced a blast 6 million times more powerful than the 1980 eruption of Mount St. Helens. It would have created winds of unimaginable fury and a tsunami as high as 1 km (100 times higher than the great Indian Ocean tsunami of 2004). The sky would have turned black with massive amounts of dust and vapor. A global firestorm may have resulted as the flaming fragments from the blast fell back to Earth (**Figure 22.22**).

Materials from the impact crater spread out in a radial kill zone focused toward western and central North America. Creatures around at that time, assuming they weren't in the kill zone, might have witnessed the following events: a brilliant flash as the asteroid rammed into Chicxulub, vaporizing Earth's upper crust at temperatures up to 10,000°C; an arc of flaming hot rocks that bolted across the sky at speeds of up to 40,000 km/hour, then crashed into North America; and a plume of debris, gas, and molten material that heated part of the atmosphere to several hundred degrees, punched into space, and then collapsed back to Earth. Over the next several days or weeks, the finer materials in this plume would have settled across Earth's entire surface.

The direct effects of the impact would have been devastating for many organisms. But worse yet would have been the aftermath for months and years to come, which scientists think led to the actual mass extinction. The high concentration of debris in the atmosphere would have blocked out the Sun, vastly reducing the light available for photosynthesis. In addition to solid particles of debris, poisonous sulfur- and nitrogen-bearing gases would have been injected into the atmosphere, where they would have reacted with water vapor to form toxic sulfuric and nitric acids that would have rained down on Earth. The combination of these two effects, and others, would have been devastating to plants and other photosynthetic autotrophs, and thus to both marine and terrestrial ecosystems that depended on them as the base of the food chain. Heterotrophs, including the dinosaurs, would have been next; once their food sources died off, they would have died off as well. A cascading series of such effects leading to the collapse of ecosystems was probably the ultimate cause of the mass extinction.

FIGURE 22.22 The artist's rendition of the Cretaceous-Tertiary scene after the asteroid impact shows Cryolophosaurus dinosaurs fleeing from a forest fire. [Mark Stevenson/Stocktrek Images/Getty Images.]

Global Warming Disaster: The Paleocene-Eocene Mass Extinction

The mass extinction at the Paleocene-Eocene boundary (about 55 million years ago; see Figure 9.11) was not one of the largest such events. It was an important event in the evolution of life, however, because it paved the way for the mammals, including the primates, to radiate into an important group. Unlike the mass extinction that wiped out the dinosaurs, it had no extraterrestrial cause. Instead, it was caused by abrupt global warming. Earth scientists are very interested in the details of what happened because global warming—this time produced by human activities—will likely threaten ecosystems in the coming decades (see Chapter 14).

We now believe that the global warming at the end of the Paleocene epoch occurred when the oceans suddenly belched an enormous amount of methane—a potent greenhouse gas—into the atmosphere. The resulting global warming was the primary cause of the mass extinction. But where did all that methane come from? To unravel this mystery, we must weave together many of the processes we have learned about in this chapter, including microbial metabolism, biogeochemical cycles, and the global behavior of the biosphere.

Microorganisms Sow the Seeds of Disaster
The story begins with the biogeochemical cycle of carbon, which was described in detail in Chapter 12. Normally, carbon is removed from the atmosphere by photoautotrophs, including algae and cyanobacteria in the oceans. After these marine organisms die, they slowly settle to the seafloor, where they accumulate as organic debris. Some of this carbon-rich debris is buried in sediments, but some is consumed by heterotrophic microorganisms as food. As you may recall, some heterotrophic microorganisms that live in anaerobic environments produce methane as a by-product of respiration. The methane produced by these anaerobes accumulates in the pores of seafloor sediments. If the seafloor is as cold as it is in our present climate (about 3°C), the methane combines with water to form a frozen solid (a methane-water ice), which remains within the sediments. Geologists searching for oil and natural gas have found layers with abundant methane-water ices in the upper 1500 m of sediments along many continental margins. If temperatures rise by even a few degrees, however, the methane-water ices melt, and the methane is quickly transformed into a gas.

The Oceans Bubble Methane
At the end of the Paleocene epoch, average temperatures in the deep sea may have risen by as much as 6°C. Once the first methane-water ices thawed and were transformed back into gases, they bubbled up through the oceans and entered the atmosphere, where they reinforced the greenhouse effect. This effect raised temperatures on the seafloor even further, which accelerated the rate of thawing. These positive feedbacks eventually resulted in a sudden—and catastrophic—release of methane that caused average global temperatures to rise dramatically. As much as 2 *trillion* tons of carbon, in the form of methane, may have escaped to the atmosphere over a period as short as 10,000 years or less!

Because methane reacts easily with oxygen to produce carbon dioxide, the release of methane also caused oxygen concentrations in the oceans to plummet. Marine organisms were essentially suffocated when oxygen concentrations dropped below a critical level. The oxygen decrease and temperature rise were devastating to seafloor ecosystems, and up to 80 percent of bottom feeders, such as clams, became extinct.

Recovery and the Evolution of Modern Mammals
Following the catastrophe, it took about 100,000 years for Earth to return to its previous state. During this time, temperatures remained unusually high until Earth was able to absorb all the extra carbon that had been released into the

atmosphere. The warmer temperatures allowed rapid expansion of forests into higher latitudes. Redwoods—related to the giant sequoias of California—grew as far north as 80°, rain forests were widespread in Montana and the Dakotas, and tropical palms flourished near London, England. Primitive mammals rapidly evolved into the ancestors of today's modern mammals, which adapted to cope with the high temperatures of that time. One particular group of mammals—the *primates*—eventually gave rise to humans.

Methane Deposits Today: A Ticking Time Bomb?

Could we see a repeat of the Paleocene-Eocene global warming disaster today? In the frozen tundra of northern Canada and other Arctic regions of the world, there may be as much as half a trillion tons of frozen methane, and deep-sea sediments around the world contain much more. The global inventory of methane deposits is estimated to be 10 trillion to 20 trillion tons of carbon present as methane, far more than what was released to cause the Paleocene-Eocene mass extinction. Human activities are adding greenhouse gases to the atmosphere at an unprecedented rate, causing the climate to warm significantly. If this trend continues and the oceans warm up, it is possible that those methane deposits could thaw. We would be wise to pay attention to the lessons of our geologic history.

The Mother of All Mass Extinctions: Whodunit?

The Cretaceous-Tertiary and Paleocene-Eocene extinctions are clear-cut examples of dramatic changes in Earth's environment that caused the catastrophic collapse of ecosystems and led to mass extinction. Those events were big, but not the biggest. In the mass extinction that marked the end of the Permian period and the Paleozoic era (see Figure 22.17), 95 percent of all species on Earth became extinct.

In this case, it seems unlikely that something as straightforward as an asteroid impact could explain how almost every species on Earth was killed. Not surprisingly, the absence of clear-cut evidence for any single cause has resulted in a long list of hypotheses, as we saw in Chapter 1. Some scientists point to extraterrestrial events, such as a comet impact or an increase in the solar wind. Others argue for events generated by Earth itself, such as an increase in volcanism, depletion of oxygen in the oceans, or a sudden release of carbon dioxide from the oceans. As in the Paleocene-Eocene extinction, a sudden release of methane from the oceans has also been proposed.

Recently, it has been shown that the mass extinction at the end of the Permian occurred exactly 251 million years ago. Perhaps it is no coincidence that the age of an enormous deposit of flood basalts in Siberia is also 251 million years. *Flood basalts*, as we learned in Chapter 5, are extrusive igneous rocks formed from huge volumes of lava that pour out across the surface of Earth in a relatively short time. In Siberia, volcanic fissures spewed out some 3 million cubic kilometers of basaltic lava, covering an area of 4 million square kilometers, almost twice the size of Alaska. Isotopic dating of the basalt shows that all of it was formed within 1 million years or less. It is hard to escape the conclusion that the Permian mass extinction was somehow related to this catastrophic eruption, which would have injected enormous amounts of carbon dioxide and sulfur dioxide gases into the atmosphere. Carbon dioxide contributes to global warming, and sulfur dioxide is the principal source of acid rain. Both are harmful to life if atmospheric concentrations get too high.

More work is required to test all these hypotheses. For example, the Deccan basalts of India are about 65 million years old, and it is possible that the massive outpouring of lava that formed them enhanced the Cretaceous-Tertiary mass extinction. However, equally large outpourings have occurred at other times in Earth's history without such apparently devastating effects.

Whatever the cause of the Permian mass extinction, one point is clear: Just as in the Cretaceous-Tertiary and Paleocene-Eocene mass extinctions, the ultimate cause was the collapse of ecosystems. We know that this collapse occurred, although we don't know exactly how. The message that we should take away from this history lesson is that history may repeat itself. The environmental changes that humans are making today will inevitably influence ecosystems—we just don't know exactly how, at least not yet.

Astrobiology: The Search for Extraterrestrial Life

Looking up at the stars on a clear night, it's hard not to wonder whether we are alone in the universe. As we have learned, the activities of life on our planet create distinctive biogeochemical signatures. Some of these signatures of life could be detected remotely, such as the presence of oxygen in the atmosphere of a planet in another solar system. In other cases, we might be able to land a spacecraft equipped with instruments to detect chemofossils or morphological fossils preserved in rocks.

In the past few decades, **astrobiologists** have begun to search systematically for evidence of life on other worlds. Although no organisms have yet been discovered beyond Earth, we should be encouraged to pursue this quest. Life may have gotten started somewhere, even if it failed to flourish. In our own solar system, Mars and Europa (a moon of Jupiter) are tantalizing targets because they are similar to Earth in several important ways. In addition, new discoveries of planets orbiting other stars have allowed us to extend this search to other solar systems.

The search for life on other worlds requires a patient, systematic, scientific approach. The most widely accepted approach has been to recognize that life, as we know it here on Earth, is based on liquid water and carbon-bearing organic compounds. Therefore, a sensible strategy might begin with a search for these two principal components of life. Compounds made of carbon are common throughout the universe; astronomers find evidence for them

Astrobiology: The Search for Extraterrestrial Life

FIGURE 22.23 The Allende meteorite, which fell to Earth near Allende, Mexico, in 1969, is full of carbon compounds. Such findings provide evidence that these compounds, one of the two key components of life, are common throughout the universe. [John Grotzinger/Ramón Rivera-Moret/Harvard Mineralogical Museum.]

was designed to look for habitable environments. If the two rovers, *Spirit* and *Opportunity*, had failed to detect any evidence of water on Mars, any future plans to search for life or habitable environments on Mars (such as the Mars Science Laboratory mission) might well have been abandoned.

Of course, there is some risk in this "life as we know it" approach to searching for extraterrestrial life. We might miss forms of life we know nothing about. One could imagine a whole host of other elements and chemical compounds that life could be based on. In general, however, these alternative schemes mainly provide fuel for science fiction writers. At least for the time being, carbon and water are regarded as the key components of all life in the universe.

Habitable Zones Around Stars

At the broadest scale, we assume that life is restricted to surfaces of planets and moons that orbit stars (**Figure 22.24**). The trick is to identify planets where water could remain stable as a liquid for a long enough time that life could originate. That could take hundreds of millions of years, based on our experience on Earth. If the surface of a planet is too close to its star, water will boil off and become a gas. That is what happened on Venus, which is 30 percent closer to the Sun than Earth and whose surface temperature is 475°C. If the surface of a planet is too far from its star, the water will freeze and become a solid. That is the case on Mars today, which is 50 percent farther from the Sun than Earth and whose surface temperature may fall below −150°C. Earth is in the middle zone, where water is stable as a liquid and surface temperatures are just right for life. For every star, there is a **habitable zone,** marked by the distances from the

everywhere, from interstellar gases and dust particles to meteorites that land on Earth (**Figure 22.23**). Therefore, astrobiologists have focused on searching for liquid water. The Mars Exploration Rover mission, described in Chapter 20, was designed to search for evidence of water on the surface of Mars, while the Mars Science Laboratory mission

FIGURE 22.24 Stars have habitable zones where life on an orbiting planet could exist. The habitable zone is determined by distance from the star; it extends from the point at which water would boil away (too close to the star) to the point at which water would freeze solid (too far from the star).

star at which water is stable as a liquid. If a planet is within the habitable zone, there a chance that life might have originated there.

Greenhouse gases such as carbon dioxide and methane also play an important role in determining the habitable zone. The Martian atmosphere may have had high concentrations of greenhouse gases early in its history. Thus, even though Mars is farther from the Sun than Earth, it might have been warmed through the greenhouse effect, as Earth is today. Indeed, new discoveries suggest that liquid water was once present on the surface of Mars, although we don't know how long it might have been stable. Thus, it is possible that Mars was habitable at some time in the past. But once the greenhouse gases were lost, Mars was transformed into the icy desert it is today.

Habitable Environments on Mars

People have long wondered about life on Mars. Mars is the planet most closely resembling Earth and is therefore the most likely planet in our solar system to host, or to have hosted, life. As we saw in Chapter 20, the Mars Exploration Rovers and Mars Science Laboratory found clear evidence of liquid water on the Martian surface at some point in the past. Based on their estimates of the ages of surface features, geologists estimate that water on Mars was stable 3 billion years ago, when it carved deep canyons across the planet's surface, dissolved rocks and minerals, and then precipitated them in a variety of basins where the water evaporated.

Water is present on Mars today, but only as ice. Any life that evolved early on would have had to seek refuge deep beneath the surface from the frigid modern climate. Any organisms that had remained on the surface would now be thoroughly frozen. However, the interior of Mars, like that of Earth, is warmed by radioactive decay, so at some depth within Mars, the ice that is present at or just below its surface must turn into liquid water. It is therefore possible that organisms—perhaps microbial extremophiles—live within a watery zone located a few hundred meters to a few kilometers below the surface of Mars.

Unfortunately, the lack of liquid water is not the only challenge that modern or ancient life would have to face on Mars. As we saw in Chapter 20, the sedimentary rocks discovered by the Mars Exploration Rover *Opportunity* are full of jarosite, an unusual iron sulfate mineral that precipitates from highly acidic water (**Figure 22.25**). On Earth, jarosite accumulates in some of the most acidic waters ever observed in natural environments.

Thus, it seems that life on Mars would have to cope not only with limited water, but possibly with very acidic water. The encouraging news is that extremophiles on Earth can live under such conditions (see Earth Issues 22.1). But the more important question is whether life can originate in such environments. Experiments on the origin of life

FIGURE 22.25 Sedimentary rocks recently discovered on Mars contain a variety of sulfate minerals that form by precipitation from water. The presence of jarosite shows that the waters from which they precipitated were extremely acidic. Extremophiles can live in these conditions, but it is not yet clear whether they could originate in such acidic waters. The holes in the rocks were drilled in 2004 by *Opportunity,* one of the Mars Exploration Rovers, to analyze their composition. [NASA/JPL/Cornell.]

suggest that it might be difficult. Some of the simple reactions that Miller observed in the 1950s would not be possible in an ocean of highly acidic water.

Not all environments on Mars may be highly acidic, however. The *Curiosity* rover has recently discovered a habitable lake environment represented by rocks formed over 3 billion years ago whose chemistry indicates the presence of more neutral to alkaline conditions. Furthermore, this ancient environment was not very salty, in strong contrast to the extremely salty environment discovered by *Opportunity*—the same environment that was also very acidic. The findings of the *Curiosity* mission are encouraging, as they suggest that Mars has environments that might be favorable for life as well as environments that might challenge life. The stunning discoveries made by *Opportunity, Spirit, Phoenix,* and *Curiosity* confirm that Mars may well have been habitable at some time. But only continued exploration will show whether life ever originated there.

KEY TERMS AND CONCEPTS

astrobiologists (p. 690)
autotrophs (p. 668)
banded iron formations (p. 684)
biogeochemical cycle (p. 670)
biosphere (p. 666)
Cambrian explosion (p. 685)
chemoautotrophs (p. 676)
chemofossils (p. 682)
cyanobacteria (p. 670)
ecosystems (p. 666)
evolution (p. 685)
evolutionary radiation (p. 685)
extremophiles (p. 672)
genes (p. 672)
geobiology (p. 666)
habitable zone (p. 691)
heterotrophs (p. 668)
metabolism (p. 669)
microbial mats (p. 676)
microfossils (p. 672)
microorganisms (p. 671)
natural selection (p. 685)
photosynthesis (p. 669)
red beds (p. 684)
respiration (p. 670)
stromatolites (p. 677)

REVIEW OF LEARNING OBJECTIVES

22.1 Define geobiology and the biosphere.

Geobiology is the study of how organisms have influenced and been influenced by Earth's physical environment. The biosphere is the part of our planet that contains all of its living organisms. Because the biosphere intersects with the lithosphere, hydrosphere, and atmosphere, it can influence or even control basic geologic and climate processes. The biosphere is a system of interacting components that exchanges energy and matter with its surroundings. Organisms use inputs of energy and matter to function and grow. In the process, they generate outputs such as oxygen and certain sedimentary minerals.

Study Assignment: Figure 22.1

Thought Question: Can the biosphere be considered an Earth system? How would that system be described?

Exercise: How do autotrophs differ from heterotrophs?

22.2 Compare some of the processes of metabolism.

Metabolism is a process that organisms use to convert inputs to outputs. Photosynthesis is a metabolic process in which organisms use energy from sunlight to convert water and carbon dioxide into carbohydrates, releasing oxygen as a by-product. Respiration is a metabolic process in which organisms use oxygen to release the stored energy of carbohydrates. Many organisms take up oxygen from the atmosphere and release carbon dioxide and water as by-products of respiration. Others, such as microorganisms that live in environments where oxygen is absent, must obtain oxygen by breaking down oxygen-containing compounds, producing substances such as hydrogen, hydrogen sulfide, or methane as by-products of respiration.

Study Assignment: Figure 22.3

Thought Question: How does the metabolism of organisms affect the geologic components of their environment?

Exercise: What is the difference between photosynthesis and respiration?

22.3 Summarize the biogeochemical cycle and discuss some ways in which metabolism affects the physical environment.

The biogeochemical cycle is a pathway by which a chemical element or compound moves between the biological and environmental components of an ecosystem. When organisms produce oxygen, it is released into the atmosphere, where it can react with other elements and compounds. When organisms release carbon dioxide or methane, which are both greenhouse gases, they contribute to global warming. Conversely, when organisms consume these gases, they contribute to global cooling.

Study Assignment: Figure 22.4

Thought Question: How does the biogeochemical cycle of carbon affect global climates?

Exercise: Draw a diagram of a biogeochemical cycle. What are the inputs and outputs? What are the processes that power the cycle?

22.4 Discuss how microorganisms interact with the physical environment.

Microorganisms are the most abundant and the most diverse organisms on Earth. Some microorganisms, called extremophiles, can live in extremely hot, acidic, salty, oxygen-depleted, or otherwise inhospitable environments. Microorganisms are involved in many geologic processes, such as weathering, mineral precipitation, mineral dissolution, and the release of gases into the atmosphere. In these ways, they play critical roles in the flux of elements through the Earth system in biogeochemical cycles.

Study Assignment: Figure 22.5 and Table 22-3

Thought Question: How does the precipitation of pyrite by microorganisms play an important role in the global biogeochemical cycling of sulfur?

Exercise: In what environments might you find extremophiles? Can humans adapt to live in extreme conditions of temperature and salinity?

22.5 Discuss how life originated and led to the rise of atmospheric oxygen.

Experiments show that compounds thought to be abundant on early Earth, such as methane, ammonia, hydrogen, and water, could have combined to form amino acids, which could then have combined to form proteins and genetic materials. These results have been supported by the finding of meteorites that are rich in amino acids and other carbon-bearing compounds. The oldest potential fossils on Earth are 3.5 billion years old and appear to be the remnants of microorganisms, based on their shape and size. Chemofossils from about 2.7 billion years ago suggest that photosynthetic bacteria and eukaryotes were both present at that time. Banded iron formations, red beds, and the appearance of eukaryotic algae testify to an initial rise in atmospheric oxygen by about 2.1 billion years ago. A second, more dramatic rise in oxygen occurred near the end of Precambrian time and may have triggered the evolution of animals.

Study Assignment: Figure 22.12

Thought Question: How are banded iron formations indicative of increased oxygen in the atmosphere and seawater?

Exercise: Carbon is regarded as the starting point for all life. What else is important?

22.6 Compare the evolutionary processes of radiation and extinction.

When groups of organisms are no longer able to adapt to changing environmental conditions or to compete with more successful groups, they become extinct. In a mass extinction, many groups of organisms become extinct at the same time. An evolutionary radiation is the relatively rapid evolution of new types of organisms from a common ancestor. Radiations may be stimulated by the availability of new habitats when a mass extinction eliminates highly competitive and established groups. The greatest radiation of animals in Earth's history occurred during the early Cambrian period, when all the animal phyla living today evolved. Several mass extinctions have occurred throughout the Phanerozoic eon. A major mass extinction occurred at the end of the Cretaceous period, when an asteroid hit Earth and 75 percent of all species were wiped out. Global warming resulting from a release of methane caused a mass extinction at the Paleocene-Eocene boundary. The cause of the greatest mass extinction of all time, which wiped out 95 percent of all species at the end of the Permian period, is unknown.

Study Assignment: Figure 22.17

Thought Question: What is the leading hypothesis for the extinction of dinosaurs?

Exercise: In the diagram shown in Figure 22.12, how many period boundaries of the geologic time scale are marked by mass extinctions? How many era boundaries are marked by mass extinctions?

22.7 Formulate some ways we can search for life on other worlds.

Astrobiologists searching for extraterrestrial life recognize that life, as we know it on Earth, is based on carbon-containing compounds and liquid water. There is ample evidence that carbon compounds are common throughout the universe, so astrobiologists search for evidence of the presence of liquid water, today or in the past. There is a habitable zone at a certain distance from every star where liquid water could be stable. If a planet is within the habitable zone, there is a chance that life might have originated there. There is unambiguous evidence that Mars had liquid water on its surface, and thus may have been habitable, at some time in the past.

Study Assignment: Figure 22.24

Thought Question: Carbon and water are the basis for all life, as we know it. If a "silicon giraffe," whose biomolecules contain the element silicon instead of carbon, walked past one of the Mars Exploration Rovers, how would we know it was alive?

Exercise: What controls the habitable zone around stars? Is Neptune in the habitable zone of our solar system?

696 CHAPTER 22 Geobiology: Life Interacts with Earth

VISUAL LITERACY EXERCISE

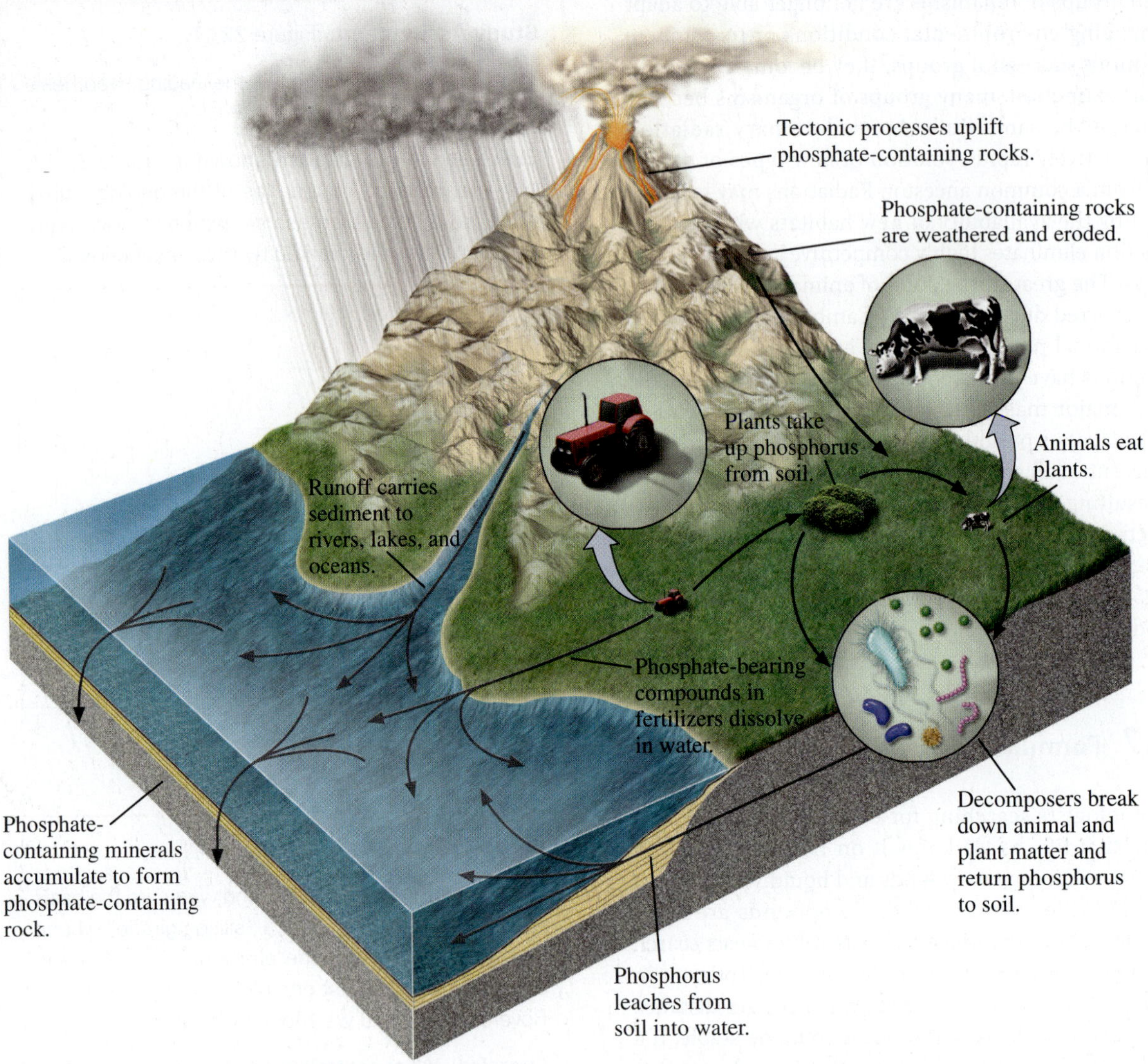

FIGURE 22.4 The biosphere plays key roles in the biogeochemical cycle of phosphorus.

Labels in figure:
- Tectonic processes uplift phosphate-containing rocks.
- Phosphate-containing rocks are weathered and eroded.
- Plants take up phosphorus from soil.
- Animals eat plants.
- Runoff carries sediment to rivers, lakes, and oceans.
- Phosphate-bearing compounds in fertilizers dissolve in water.
- Decomposers break down animal and plant matter and return phosphorus to soil.
- Phosphate-containing minerals accumulate to form phosphate-containing rock.
- Phosphorus leaches from soil into water.

1. Are organisms essential in the biogeochemical cycle of phosphorus?
   a. No
   b. Yes
   c. Possibly
   d. Probably not

2. What is the ultimate source of phosphorus?
   a. Oceans
   b. Rocks
   c. Tractors
   d. Soil

3. How is phosphorus derived from rocks?
   a. Circulation of water in oceans
   b. Animals eating plants
   c. Weathering and erosion
   d. Dissolution of fertilizers

4. Phosphorus is an essential nutrient for plant growth. How do they obtain it?
   a. From volcanoes
   b. From animals
   c. From soil
   d. From the oceans

5. How are phosphate-containing minerals formed?
   a. Decomposition of plant matter
   b. Decomposition of animal matter
   c. Uplift of rock
   d. Precipitation of dissolved phosphorus in seawater to form sedimentary deposits

APPENDIX 1 Conversion Factors

LENGTH

1 centimeter	0.3937 inch
1 inch	2.5400 centimeters
1 meter	3.2808 feet; 1.0936 yards
1 foot	0.3048 meter
1 yard	0.9144 meter
1 kilometer	0.6214 mile (statute); 3281 feet

LENGTH

1 mile (statute)	1.6093 kilometers
1 mile (nautical)	1.8531 kilometers
1 fathom	6 feet; 1.8288 meters
1 angstrom	10^{-8} centimeter
1 micrometer	0.0001 centimeter

VELOCITY

1 kilometer/hour	27.78 centimeters/second
1 mile/hour	17.60 inches/second

AREA

1 square centimeter	0.1550 square inch
1 square inch	6.452 square centimeters
1 square meter	10.764 square feet; 1.1960 square yards
1 square foot	0.0929 square meter
1 square kilometer	0.3861 square mile
1 square mile	2.590 square kilometers
1 acre (U.S.)	4840 square yards

VOLUME

1 cubic centimeter	0.0610 cubic inch
1 cubic inch	16.3872 cubic centimeters
1 cubic meter	35.314 cubic feet
1 cubic foot	0.02832 cubic meter
1 cubic meter	1.3079 cubic yards
1 cubic yard	0.7646 cubic meter
1 liter	1000 cubic centimeters; 1.0567 quarts (U.S. liquid)
1 gallon (U.S. liquid)	3.7853 liters

MASS

1 gram	0.03527 ounce
1 ounce	28.3495 grams
1 kilogram	2.20462 pounds
1 pound	0.45359 kilogram

PRESSURE

1 kilogram/square centimeter	0.96784 atmosphere; 0.98067 bar; 14.2233 pounds/square inch
1 bar	0.98692 atmosphere; 10^5 pascals

ENERGY

1 joule	0.239 calorie; 9.479×10^{-4} Btu
1 British thermal unit (Btu)	251.9 calories; 1054 joules
1 quad	10^{15} Btu

POWER

1 watt	0.001341 horsepower (U.S.); 3.413 Btu/hour

APPENDIX 2 Numerical Data Pertaining to Earth

Equatorial radius	6378 kilometers
Polar radius	6357 kilometers
Radius of sphere with Earth's volume	6371 kilometers
Volume	1.083×10^{12} cubic kilometers
Surface area	5.1×10^8 square meters
Percent surface area of oceans	71
Percent surface area of land	29
Average elevation of land	623 meters
Average depth of oceans	3600 meters
Total mass	5.976×10^{24} kilograms
Mean density	5.517 grams/cubic centimeters
Gravity at equator	9.78 meters/second/second
Mass of atmosphere	5.1×10^{18} kilograms
Mass of ice	$25-30 \times 10^{18}$ kilograms
Mass of oceans	1.4×10^{21} kilograms
Mass of crust	2.5×10^{22} kilograms
Mass of mantle	4.05×10^{24} kilograms
Mass of core	1.90×10^{24} kilograms
Mean distance to Sun	1.496×10^8 kilometers
Mean distance to Moon	3.844×10^5 kilometers
Ratio: Mass of Sun/mass of Earth	3.329×10^5
Ratio: Mass of Earth/mass of Moon	81.303
Total geothermal energy reaching Earth's surface each year	10^{21} joules; 949 quads
Earth's daily receipt of solar energy	1.49×10^{22} joules; 14,137 quads
U.S. energy consumption, 2017	97.7 quads

APPENDIX 3 Chemical Reactions

Electron Shells and Ion Stability

Electrons surround the nucleus of an atom of each element in a unique set of concentric spheres called electron shells. Each shell can hold a certain maximum number of electrons. In the chemical reactions of most elements, only the electrons in the outermost shells interact. In the reaction between sodium (Na) and chlorine (Cl) that forms sodium chloride (NaCl), for example, the sodium atom loses an electron from its outer shell of electrons, and the chlorine atom gains an electron in its outer shell (see Figure 3.4).

Before reacting with chlorine, the sodium atom has one electron in its outer shell. When it loses that electron, its outer shell is eliminated and the next shell inward, which has eight electrons (the maximum that shell can hold), becomes the outer shell. The original chlorine atom had seven electrons in its outer shell, with room for a total of eight. By gaining an electron, it fills its outer shell. Many elements have a strong tendency to acquire a full outer electron shell, some by gaining electrons and some by losing them in the course of a chemical reaction.

Many chemical reactions entail gains and losses of several electrons as two or more elements combine. The element calcium (Ca), for example, becomes a doubly charged cation, Ca^{2+}, as it reacts with two chlorine atoms to form calcium chloride. In the chemical formula for calcium chloride, $CaCl_2$, the presence of two chloride ions is symbolized by the subscript 2. Chemical formulas thus show the relative proportions of atoms or ions in a compound. Common practice is to omit the subscript 1 next to single ions in a formula.

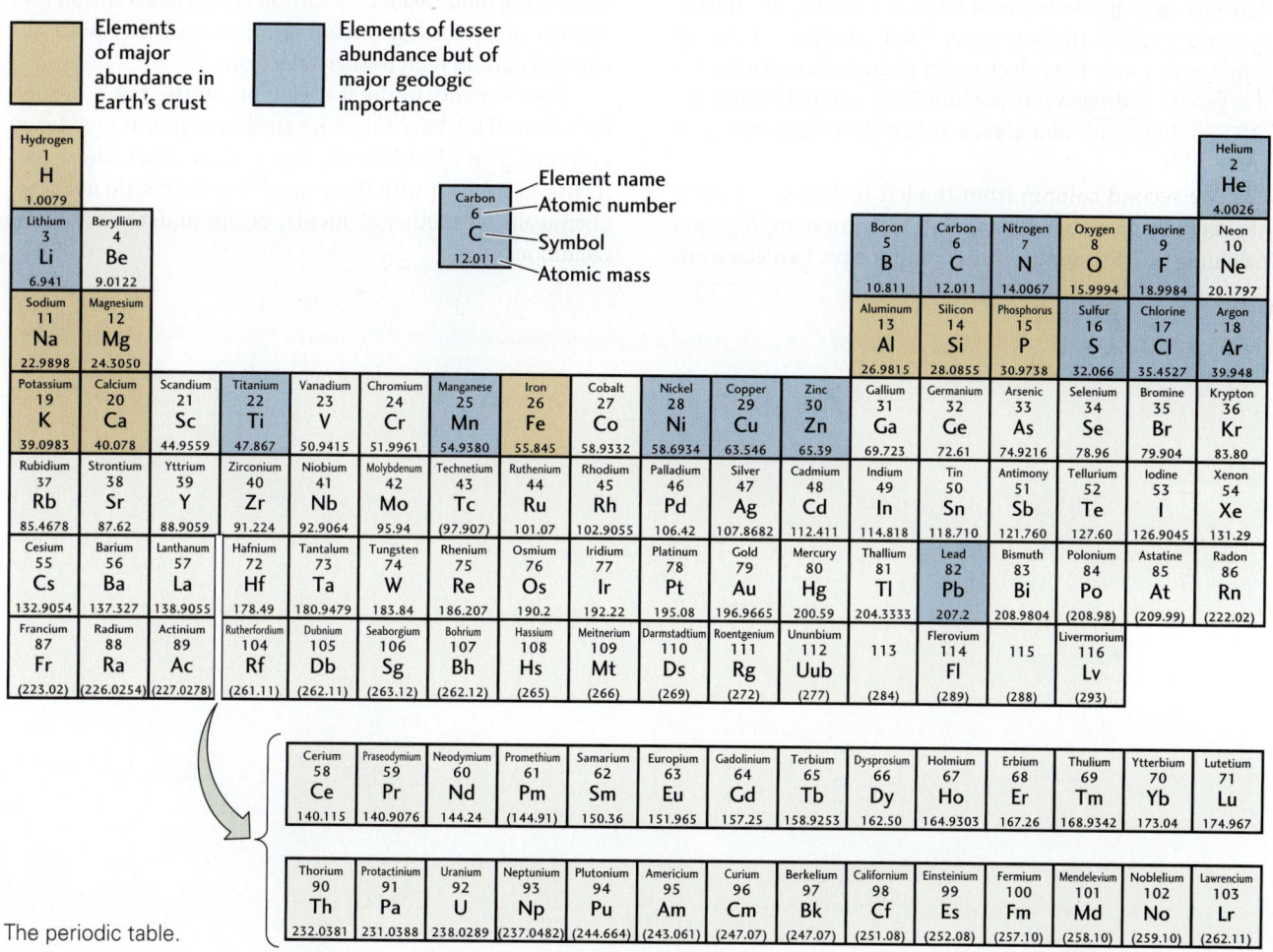

The periodic table.

The periodic table organizes the elements (from left to right along a row) in order of atomic number (the number of protons), which also means increasing the numbers of electrons in the outer shell. The third row from the top, for example, starts at the left with sodium (atomic number 11), which has one electron in its outer shell. The next is magnesium (atomic number 12), which has two electrons in its outer shell, followed by aluminum (atomic number 13), with three, and silicon (atomic number 14), with four. Then come phosphorus (atomic number 15), with five; sulfur (atomic number 16), with six; and chlorine (atomic number 17), with seven. The last element in this row is argon (atomic number 18), with eight electrons, the maximum possible, in its outer shell. Each column in the table forms a vertical grouping of elements with similar electron-shell patterns.

Elements That Tend to Lose Electrons

The elements in the leftmost column of the table all have a single electron in their outer shells and have a strong tendency to lose that electron in chemical reactions. Of this group, hydrogen (H), sodium (Na), and potassium (K) are found in major abundance at Earth's surface and in its crust.

The second column from the left includes two more elements that are abundant on Earth: magnesium (Mg) and calcium (Ca). Elements in this column have two electrons in their outer shells and a strong tendency to lose both of them in chemical reactions.

Elements That Tend to Gain Electrons

Toward the right side of the table, the two columns headed by oxygen (O), the most abundant element on Earth, and fluorine (F), a highly reactive toxic gas, contain elements that tend to gain electrons in their outer shells. The elements in the column headed by oxygen have six of the possible eight electrons in their outer shells and tend to gain two electrons. Those in the column headed by fluorine have seven electrons in their outer shells and tend to gain one.

Other Elements

The elements in the columns between those farthest to the left and those farthest to the right have varying tendencies to gain, lose, or share electrons. The column toward the right of the table headed by carbon (C) includes silicon (Si), another of the most abundant elements on Earth. Both silicon and carbon tend to share electrons.

The elements in the last column on the right, headed by helium (He), have full outer shells and thus no tendency either to gain or to lose electrons. As a result, these elements, in contrast with those in other columns, do not react chemically with other elements, except under very special conditions.

APPENDIX 4 Properties of the Most Common Minerals of Earth's Crust

	Mineral or Group Name	Structure or Composition	Varieties and Chemical Composition	Form, Diagnostic Characteristics	Cleavage, Fracture	Color	Hardness
LIGHT-COLORED MINERALS, VERY ABUNDANT IN EARTH'S CRUST IN ALL MAJOR ROCK TYPES	FELDSPAR	FRAMEWORK SILICATES	ORTHOCLASE FELDSPARS $KAlSi_3O_8$ Sanidine, Orthoclase, Microcline	Cleavable coarsely crystalline or finely granular masses; isolated crystals or grains in rocks, most commonly not showing crystal faces	Two at right angles, one perfect and one good; pearly luster on perfect cleavage	White to gray, frequently pink or yellowish; some green	6
			PLAGIOCLASE FELDSPARS $NaAlSi_3O_8$ Albite, $CaAl_2Si_2O_8$ Anorthite		Two at nearly right angles, one perfect and one good; fine parallel striations on perfect cleavage	White to gray, less commonly greenish or yellowish	
	QUARTZ		SiO_2	Single crystals or masses of 6-sided prismatic crystals; also formless crystals and grains or finely granular or massive	Very poor or nondetectable; conchoidal fracture	Colorless, usually transparent; also slightly colored smoky gray, pink, yellow	7
	MICA	SHEET SILICATES	MUSCOVITE $KAl_2(AlSi_3O_{10})(OH)_2$	Thin, disc-shaped crystals, some with hexagonal outlines; dispersed or aggregates	One perfect; splittable into very thin, flexible, transparent sheets	Colorless; slight gray or green to brown in thick pieces	$2-2\frac{1}{2}$
DARK-COLORED MINERALS, ABUNDANT IN MANY KINDS OF IGNEOUS AND METAMORPHIC ROCKS			BIOTITE $K(Mg,Fe)_3AlSi_3(OH)_2$	Irregular, foliated masses; scaly aggregates	One perfect; splittable into thin, flexible sheets	Black to dark brown; translucent to opaque	$2\frac{1}{2}-3$
			CHLORITE $(Mg,Fe)_5(Al,Fe)_2Si_3O_{10}(OH)_8$	Foliated masses or aggregates of small scales	One perfect; thin sheets flexible but not elastic	Various shades of green	$2-2\frac{1}{2}$
	AMPHIBOLE	DOUBLE CHAINS	TREMOLITE-ACTINOLITE $Ca_2(Mg,Fe)_5Si_8O_{22}(OH)_2$	Long, prismatic crystals, usually 6-sided; commonly in fibrous masses or irregular aggregates	Two perfect cleavage directions at 56° and 124° angles	Pale to deep green Pure tremolite white	5–6
			HORNBLENDE Complex Ca, Na, Mg, Fe, Al silicate			Most commonly black, but varies from pale gray to black	
	PYROXENE	SINGLE CHAINS	ENSTATITE-HYPERSTHENE $(Mg,Fe)_2Si_2O_6$	Prismatic crystals, either 4- or 8-sided; granular masses and scattered grains	Two good cleavage directions at about 90°	Green and brown to grayish or greenish white	5–6
			DIOPSIDE $(Ca,Mg)_2Si_2O_6$			Light to dark green	
			AUGITE Complex Ca, Na, Mg, Fe, Al silicate			Very dark green to black	

(Continued)

APPENDIX 4 Properties of the Most Common Minerals of Earth's Crust

	Mineral or Group Name	Structure or Composition	Varieties and Chemical Composition	Form, Diagnostic Characteristics	Cleavage, Fracture	Color	Hardness
LIGHT-COLORED MINERALS, TYPICALLY AS ABUNDANT CONSTITUENTS OF SEDIMENTS AND SEDIMENTARY ROCKS	OLIVINE	ISOLATED TETRAHEDRA	$(Mg,Fe)_2SiO_4$	Granular masses and disseminated small grains	Conchoidal fracture	Olive to grayish green and brown	$6\frac{1}{2}-7$
	GARNET		Ca, Mg, Fe, Al silicate	Isometric crystals, well formed or rounded; high specific gravity, 3.5–4.3	Conchoidal and irregular fracture	Red and brown, less commonly pale colors	$6\frac{1}{2}-7$
	CALCITE	CARBONATES	$CaCO_3$	Coarsely to finely crystalline in beds, veins, and other aggregates; cleavage faces may show in coarser masses; calcite effervesces rapidly in acid, but dolomite effervesces slowly and only if crushed into powder	Three perfect cleavages, at oblique angles; splits to rhombohedral cleavage pieces	Colorless, transparent to translucent; variously colored by impurities	3
	DOLOMITE		$CaMg(CO_3)_2$				$3\frac{1}{2}-4$
	CLAY MINERALS	HYDROUS ALUMINO-SILICATES	KAOLINITE $Al_2Si_2O_5(OH)_4$ ILLITE Similar to muscovite $+Mg, Fe$ SMECTITE Complex Ca, Na, Mg, Fe, Al silicate $+ H_2O$	Earthy masses in soils; bedded; in association with other clays, iron oxides, or carbonates; plastic when wet; montmorillonite swells when wet	Earthy, irregular	White to light gray and buff; also gray to dark gray, greenish gray, and brownish depending on impurities and associated minerals	$1\frac{1}{2}-2\frac{1}{2}$
	GYPSUM	SULFATES	$CaSO_4 \cdot 2H_2O$	Granular, earthy, or finely crystalline masses; tabular crystals	One perfect, splitting to fairly thin slabs or sheets; two other good cleavages	Colorless to white; transparent to translucent	2
	ANHYDRITE		$CaSO_4$	Massive or crystalline aggregates in beds and veins	One perfect, one nearly perfect, one good; at right angles	Colorless, some tinged with blue	$3-3\frac{1}{2}$
	HALITE	HALIDES	NaCl	Granular masses in beds; some cubic crystals; salty taste	Three perfect cleavages at right angles	Colorless, transparent to translucent	$2\frac{1}{2}$
	OPAL-CHALCEDONY	SILICA	SiO_2 [Opal is an amorphous variety; chalcedony is a formless microcrystalline quartz.]	Beds in siliceous sediments and chert; in veins or banded aggregates	Conchoidal fracture	Colorless or white when pure, but tinged with various colors by impurities in bands, especially in agates	$5-6\frac{1}{2}$
DARK-COLORED MINERALS, COMMON IN MANY ROCK	MAGNETITE	IRON OXIDES	Fe_3O_4	Magnetic; disseminated grains, granular masses; occasional octahedral isometric crystals; high	Conchoidal or irregular fracture	Black, metallic luster	6

Properties of the Most Common Minerals of Earth's Crust   AP-7

Mineral or Group Name	Structure or Composition	Varieties and Chemical Composition	Form, Diagnostic Characteristics	Cleavage, Fracture	Color	Hardness
HEMATITE	IRON OXIDES	Fe_2O_3	Earthy to dense masses, some with rounded forms, some granular or foliated; high specific gravity, 4.9–5.3	None; uneven, sometimes splintery fracture	Reddish brown to black	5–6
"LIMONITE"		GOETHITE [the major mineral of the mixture called "limonite," a field term] $FeO(OH)$	Earthy masses, massive bodies or encrustations, irregular layers; high specific gravity, 3.3–4.3	One excellent in the rare crystals; usually an early fracture	Yellowish brown to dark brown and black	5–5½
KYANITE	ALUMINO-SILICATES	Al_2SiO_5	Long, bladed or tabular crystals or aggregates	One perfect and one poor, parallel to length of crystals	White to light-colored or pale blue	5 parallel to crystal length; 7 across crystals
SILLIMANITE		Al_2SiO_5	Long, slender crystals or fibrous, felted masses	One perfect parallel to length, not usually seen	Colorless, gray to white	6–7
ANDALUSITE		Al_2SiO_5	Coarse, nearly square prismatic crystals, some with symmetrically arranged impurities	One distinct; irregular fracture	Red, reddish brown olive-green	7½
FELDSPATHOIDS		NEPHELINE $(Na,K)AlSiO_4$	Compact masses or as embedded grains, rarely as small prismatic crystals	One distinct; irregular fracture	Colorless, white, light gray; gray-greenish in masses, with greasy luster	5½–6
		LEUCITE $KAlSi_2O_6$	Trapezohedral crystals embedded in volcanic rocks	One very imperfect	White to gray	5½–6
SERPENTINE		$Mg_6Si_4O_{10}(OH)_8$	Fibrous (asbestos) or platy masses	Splintery fracture	Green; some yellowish brownish or gray; waxy or greasy luster in massive habit; silky luster in fibrous habit	4–6
TALC		$Mg_3Si_4O_{10}(OH)_2$ masses or aggregates	Foliated or compact masses or aggregates	One perfect, making thin flakes or scales; soapy feel	White to pale green; pearly or greasy luster	1
CORUNDUM		Al_2O_3	Some rounded, barrel-shaped crystals; most often as disseminated grains or granular masses (emery)	Irregular fracture	Usually brown, pink, or blue; emery black. Gemstone varieties: ruby, sapphire	9

LIGHT-COLORED MINERALS, MAINLY IN METAMORPHIC AND IGNEOUS ROCKS AS COMMON OR MINOR CONSTITUENTS

APPENDIX 4 Properties of the Most Common Minerals of Earth's Crust

	Mineral or Group Name	Structure or Composition	Varieties and Chemical Composition	Form, Diagnostic Characteristics	Cleavage, Fracture	Color	Hardness
DARK-COLORED MINERALS, COMMON IN METAMORPHIC ROCKS	EPIDOTE	SILICATES	$Ca_2(Al,Fe)Al_2Si_3O_{12}(OH)$	Aggregates of long prismatic crystals, granular or compact masses, embedded grains	One good, one poor at greater than right angles; conchoidal and irregular fracture	Green, yellow-green, gray, some varieties dark brown to black	6–7
	STAUROLITE		$Fe_2Al_9Si_4O_{22}(O,OH)_2$	Short prismatic crystals, some cross-shaped, usually coarser than matrix of rock	One poor	Brown, reddish, or dark brown to black	7–7½
METALLIC LUSTER, COMMON IN MANY ROCK TYPES, ABUNDANT IN VEINS	PYRITE	SULFIDES	FeS_2	Granular masses or well-formed cubic crystals in veins and beds or disseminated; high specific gravity, 4.9–5.2	Uneven fracture	Pale brass-yellow	6–6½
	GALENA		PbS	Granular masses in veins and disseminated; some cubic crystals; very high specific gravity, 7.3–7.6	Three perfect cleavages at mutual right angles, giving cubic cleavage fragments	Silver-gray	2½
	SPHALERITE		ZnS	Granular masses or compact crystalline aggregates; high specific gravity, 3.9–4.1	Six perfect cleavages at 60° to one another	White to green, brown and black; resinous to submetallic luster	3½–4
	CHALCOPYRITE		$CuFeS_2$	Granular or compact masses; disseminated crystals; high specific gravity, 4.1–4.3	Uneven fracture	Brassy to golden-yellow	3½–4
	CHALCOCITE		Cu_2S	Fine-grained masses; high specific gravity, 5.5–5.8	Conchoidal fracture	Lead-gray to black; may tarnish green or blue	2½–3
MINERALS FOUND IN MINOR AMOUNTS IN A VARIETY OF ROCK TYPES AND IN VEINS OR PLACERS	RUTILE	TITANIUM OXIDES	TiO_2	Slender prismatic crystals; granular masses; high specific gravity, 4.25	One distinct, one less distinct; conchoidal fracture	Reddish brown, some yellowish, violet, or black	6–6½
	ILMENITE		$FeTiO_3$	Compact masses, embedded grains, detrital grains in sand; high specific gravity, 4.79	Conchoidal fracture	Iron-black; metallic to submetallic luster	5–6
	ZEOLITES	SILICATES	Complex hydrous silicates; many varieties of minerals, including analcime, natrolite, phillipsite, heulandite, and chabazite	Well-formed radiating crystals in cavities in volcanics, veins, and hot springs; also as fine-grained and earthy bedded deposits	One perfect for most	Colorless, white, some pinkish	4–5

APPENDIX 5 Practicing Geology Exercises: Answers to Bonus Problems

Chapter 1 $1.083 \times 10^{12} \text{ km}^3$

Chapter 2 The distance between the North American and Charleston, South Carolina, continental margin and the African continental margin near Dakar, Senegal, measured using the Google Earth Ruler tool, is about 6300 km. From the isochron map in Figure 2.15, you can estimate that the two continents began to rift apart about 200 million to 180 million years ago (see also Figure 2.16). Assuming that the continents rifted apart 200 million years ago gives

6300 km ÷ 200 million years = 31.5 km/year = 31.5 mm/year

Assuming that the rifting age is 180 million years gives

6300 km ÷ 180 million years = 35 km/year = 35 mm/year

Chapter 3 $\$163{,}200{,}000 - \$120{,}000{,}000 = \$43{,}200{,}000$ profit. So yes, it is worth it.

Chapter 4 Plagioclase feldspar will settle at a rate of 1.18 cm/hour, which is slower than olivine.

Chapter 5 The production rate of the Hawaiian basalts is

$$100{,}000 \text{ km}^3 \div 1 \text{ million years} = 0.1 \text{ km}^3/\text{year}$$

The length of the Nazca–Pacific plate boundary needed to produce this amount is given by the following equation:

$$1.4 \times 10^{-4} \text{ km/year} \times 7 \text{ km} \times length = 0.1 \text{ km}^3/\text{year}$$

or

$$length = 0.1 \text{ km}^3/\text{year} \div (1.4 \times 10^{-4} \text{ km/year} \times 7 \text{ km}) = 102 \text{ km}$$

Chapter 6 125°C

Chapter 7 A change in pressure at constant temperature could indicate that the rocks were moved upward or downward in a subduction zone. Movements in subduction zones can be so fast that temperatures don't have time to change even though pressures may be changing quickly.

Chapter 8 The chances of encountering oil in the reservoir rock at point C are poor. From the geologic structure exposed at the surface, you can see that the anticline is plunging to the northeast with a dip of about 30°. Therefore, the depth to the sandstone reservoir rock is increasing to the northeast, and drilling at point C is likely to encounter water, not oil.

Chapter 9 The rock age is given by

$$T = \log(1.0143)/\log(2)$$
$$= 0.00617/0.301$$
$$= 0.0205 \text{ half-lives}$$

Multiplying this result by the half-life of rubidium-87 yields an age of

$$0.0205 \times 49 \text{ billion years} = 1.00 \text{ billion years}$$

Chapter 10 A magnitude 6 rupture has 100 times the area of a magnitude 4 rupture (because $10^{(6-4)} = 10^2$) and 10 times the slip (because $10^{(6-4)/2} = 10^1$); therefore, it takes $100 \times 10 = 1000$ magnitude 4 earthquakes to equal a magnitude 6 earthquake.

Chapter 11 The appropriate isostatic equation is

Tibetan Plateau elevation = $(0.15 \times$ Tibetan Plateau crustal thickness$) - (0.12 \times 7.0 \text{ km}) - (0.70 \times 4.5 \text{ km})$

Solving for crustal thickness, we obtain

$$\text{Tibetan Plateau crustal thickness} = \frac{(\text{Tibetan Plateau elevation}) + (0.12 \times 7.0 \text{ km}) + (0.70 \times 4.5 \text{ km})}{0.15}$$

For an elevation of 5 km, this formula gives a crustal thickness of 60 km, which agrees with seismic data collected in the Tibetan Plateau.

Chapter 12 The net carbon flux from the land surface is given by the difference between the positive flux into the atmosphere due to land-use changes and the negative flux from the atmosphere into the land sink due to net plant growth.

1980–1989: 1.4 Gt − 1.5 Gt = −0.1 Gt
1990–1999: 1.6 Gt − 2.7 Gt = −1.1 Gt
2000–2009: 1.1 Gt − 2.6 Gt = −1.5 Gt

The net carbon flux from the land surface was negative (net absorption) for all three decades. One explanation for why the net negative flux is growing with time is *plant growth feedback* (p. 347), the negative feedback due to the stimulation of plant growth by increasing CO_2 concentrations in the atmosphere.

Chapter 13 The carbon intensity for the fossil-fuel component of the U.S. energy system is calculated by dividing the total carbon flux by the total rate of energy production by fossil fuels.

Total carbon flux
$= (0.586 + 0.158 + 0.216 + 0.493) \text{ Gt/year} = 1.453 \text{ Gt/year}$

Total energy production
$= (36.2 + 14.0 + 28.0) \text{ quad/year} = 78.2 \text{ quad/year}$

Carbon intensity of U.S. fossil-fuel energy production
$= 1.453 \text{ Gt/year}/78.2 \text{ quad/year} = 0.0186 \text{ Gt/quad}$

Chapter 14 If D is the depth of the heated surface layer, then 200 mm = 0.0001 D. Solving for D gives D = 200 mm/0.0001 = 0.2 m × 10,000 = 2000 m = 2 km.

Chapter 15 200 mm ÷ 0.0001 = 2×10^6 mm = 2 km

Chapter 16

Table C

	SAFETY FACTOR (F_S)		
	Loose Soil	Slate	Granite
5°	3	10	50
20°	0.6	2	10
30°	0.12	0.4	2

A safety factor value greater than one indicates that a slope is stable enough to build on.

Chapter 17 Silty sand: 0.0015 m^3/day
Well-sorted gravel: 18.75 m^3/day

Chapter 18 10 feet = about 6000 cubic feet per second
25 feet = about 40,000 cubic feet per second

Chapter 19 The 2002 maintenance fill cost about $25 million.

Chapter 20 The maximum variation in elevation at a landing site that could be tolerated by a lander with a fuel tank volume of 200 L is 750 m. The maximum acceptable variation if the final descent rate were 1 m/s instead of 2 m/s is 500 m.

Chapter 21 The fraction of relative plate movement taken up by thrust faulting is 20 mm/year ÷ 54 mm/year = 0.37. The remaining movement, about 60 percent of the total, is accommodated by faulting and folding north of the Himalaya, primarily by movement on the Altyn Tagh and other major strike-slip faults as China and Mongolia are pushed eastward (see Figure 21.16).

Chapter 22 Rock A: $R = 5$; Rock A does not record a distinctive signature of biological processes.

Rock C: $R = -50$; this large negative ratio, which is similar to that of the ratio for methane, does show a distinctive signature of biological processes.

APPENDIX 6 Answers to Visual Literacy Exercises

Chapter 1

1. The plate tectonic system and the geodynamo
2. The geodynamo
3. The climate system
4. All three
5. The lithosphere interacts strongly with the other components of the climate system—atmosphere, hydrosphere, cryosphere, and biosphere—and it is also a principal component of the plate tectonic system.

Chapter 2

1. c
2. d
3. b
4. c
5. a

Chapter 3

1. b
2. c
3. d
4. a
5. c

Chapter 4

1. d
2. b and d
3. c
4. a
5. c

Chapter 5

1. a
2. c
3. a
4. All of the above
5. d

Chapter 6

1. c
2. c
3. b
4. d
5. d

Chapter 7

1. b
2. b
3. b
4. d
5. a

Chapter 8

1. a
2. a
3. c
4. c
5. c

Chapter 9

1. a. There are four eons: Hadean, Archean, Proterozoic, and Phanerozoic.
   b. The longest eon is the Proterozoic, which began at 2500 Ma and ended at 542 Ma.
   c. The oldest eon is the Hadean, which began at 4560 Ma and ended at 3900 Ma.
2. a. The Paleozoic era comprises six periods, and the Mesozoic era comprises three periods.
   b. The Mesozoic–Cenozoic boundary corresponds to the boundary between the Cretaceous period and the Paleogene period.
3. Cenozoic era, 65 Ma–0 Ma
4. The Neogene period spans the last 23 Ma and includes four epochs.

Chapter 10

1. d
2. c
3. c
4. b

Chapter 11

1. a. White
   b. Yellow
   c. Blue
2. a. The outer core
   b. The inner core
   c. Because it is colder and denser than average mantle

AP-11

3. a. Panel (b)
   b. Cold lithosphere of stable continental cratons
   c. Subduction of cold oceanic lithosphere into the lowermost mantle

Chapter 12

1. c
2. b
3. Because the carbon flux is in both directions
4. True

Chapter 13

1. a. An energy source
   b. Because the electrical energy produced from coal burning is twice the amount produced from nuclear fission
2. The width of the yellow arrow is proportional to the amount of useful energy; the width of the blue arrow is proportional to the amount of wasted energy.
3. Because electric power generation produces more carbon waste than the other human activities
4. a. $(36.2 + 14.0 + 28.0)/97.7 = 0.80$
   b. Electric power generation
   c. $66.6/97.7 = 0.68$
   d. 1453 Mt

Chapter 14

1. a. About 9 Gt
   b. About 16 Gt
2. a. 350 Gt
   b. 50 Gt
   c. $350 \text{ Gt}/(350 \text{ Gt} + 450 \text{ Gt}) = 44\%$
3. a. subjective answer
   b. subjective answer

Chapter 15

1. a
2. a
3. d
4. b

Chapter 16

1. c
2. a and b
3. d
4. b
5. a

Chapter 17

1. Rainfall over hills and mountains injects water into rocks that can flow to lower elevations.
2. A permeable rock layer that extends underground
3. A rock layer of lower permeability that helps guide the flow within an aquifer
4. If a recharge area is at higher elevation than a well, then the well will naturally flow water because of the difference in pressure created by the difference in elevation of the recharge area versus the well.

Chapter 18

1. Where snowmelt and rainfall generate enough water to flow across the land surface
2. In deltas, formed at the margins of lakes and oceans where currents slow and deposit sediment
3. If the stream is subjected to tectonic uplift, or a decrease in sediment flux, it will cut down into previously deposited sediments.
4. Stream sediments are coarser in the upper reaches, where slopes are steeper, and finer in the lower reaches, where slopes are less steep.

Chapter 19

1. Waves move by, but the water stays in place.
2. When the water depth gets to be about one-half the wavelength
3. When water depths get very shallow, waves steepen and then collapse.
4. c
5. b

Chapter 20

1. b
2. c
3. b
4. d
5. c

Chapter 21

1. d
2. b
3. b
4. a

Chapter 22

1. b
2. b
3. c
4. c
5. d

Glossary

Words in *italics* have separate entries in the Glossary. Specific minerals are defined and described in Appendix 4.

ablation The total amount of ice that a *glacier* loses each year. (Compare *accumulation*.)

abrasion The erosive action that occurs when suspended and saltating *sediment* particles move along the bottom and sides of a stream channel.

absolute age The actual number of years elapsed from a geologic event until now. (Compare *relative age*.)

abyssal hill A hill on the slope of a *mid-ocean ridge*, typically 100 m or so high and parallel to the ridge crest, formed primarily by normal faulting of newly formed oceanic *crust* as it moves out of a rift valley.

abyssal plain A wide, flat plain that covers large areas of the ocean floor at depths of about 4000 to 6000 m.

accreted terrain A piece of continental *crust*, tens to hundreds of kilometers in extent, with common characteristics and a distinct origin, usually transported great distances by plate movements and plastered onto the edge of a continent.

accretion A process of continental growth in which buoyant fragments of *crust* are attached (accreted) to existing continental masses by horizontal transport during plate movements. (Compare *magmatic addition*.)

accumulation The amount of snow added to a *glacier* annually. (Compare *ablation*.)

active margin A *continental margin* where tectonic forces caused by plate movements are actively deforming the continental crust. (Compare *passive margin*.)

aftershock An *earthquake* that occurs as a consequence of a previous earthquake of larger magnitude. (Compare *foreshock*.)

albedo The fraction of *solar energy* reflected by a surface. (From the Latin *albus*, meaning "white.")

alluvial fan A cone- or fan-shaped accumulation of *sediment* deposited where a *stream* widens abruptly as it leaves a mountain front and enters a broad, relatively flat *valley*.

amphibolite (1) A usually *granoblastic rock* made up mainly of amphibole and plagioclase feldspar, typically formed by medium- to high-grade metamorphism of mafic volcanic rock. Foliated amphibolites can be produced by *deformation*. (2) The metamorphic grade above *greenschist*.

andesite An *intermediate igneous rock* with a composition between that of *dacite* and that of *basalt*; the extrusive equivalent of *diorite*.

andesitic lava A lava type of intermediate composition that has a higher silica content than *basalt*, erupts at lower temperatures, and is more viscous.

angle of repose The maximum angle at which a slope of loose material can lie without sliding downhill.

anion A negatively charged ion. (Compare *cation*.)

antecedent stream A *stream* that existed before the present *topography* was created and so maintained its original course despite changes in the structure of the underlying *rocks* and in the topography. (Compare *superposed stream*.)

Anthropocene The "Age of Man," a geologic epoch beginning about 1780, when the coal-powered steam engine launched the Industrial Revolution; proposed by atmospheric chemist Paul Crutzen to recognize the speed and magnitude of the changes industrial society is causing in the *Earth system*.

anthropogenic global change Modification of the global environment by humans, including climate change, ocean acidification, and loss of biodiversity.

anticline An archlike fold of layered *rocks* that contains older rock layers in the core of the fold. (Compare *syncline*.)

aquiclude A relatively impermeable *formation* that bounds an *aquifer* above or below and acts as a barrier to the flow of *groundwater*.

aquifer A porous *formation* that stores and transmits *groundwater* in sufficient quantity to supply wells.

arkose A sandstone containing more than 25 percent feldspar.

artesian flow A spontaneous flow of *groundwater* through a confined *aquifer* to a point where the elevation of the ground surface is lower than that of the *groundwater table*.

ash-flow deposit An extensive sheet of hard volcanic *tuff* produced by a continental eruption of *pyroclasts*.

asteroid One of the more than 10,000 small celestial bodies orbiting the Sun, most of them between the orbits of Mars and Jupiter.

asthenosphere The weak, *ductile* layer of *rock* that constitutes the lower part of the upper *mantle* (below the *lithosphere*) and over which the lithospheric plates slide. (From the Greek *asthenes*, meaning "weak.")

astrobiologist A scientist who searches for the chemical building blocks of life, environments that may have been habitable for life, or even life itself, on other worlds.

atomic mass The sum of an atom's protons and neutrons.

atomic number The number of protons in the nucleus of an atom.

autotroph A producer; an organism that makes its own food by manufacturing organic compounds, such as carbohydrates, that it uses as sources of energy. (Compare *heterotroph*.)

GL-1

badland A deeply gullied landscape resulting from the rapid *erosion* of easily erodible *shales* and *clays*.

banded iron formation A *sedimentary rock* formation composed of alternating thin layers of iron oxide minerals and silica-rich minerals, precipitated from seawater when oxygen was first produced by *cyanobacteria* and reacted with iron dissolved in seawater.

barrier island A long, offshore sandbar that builds up to form a barricade between open ocean waves and the main *shoreline*.

basal slip The sliding of a *glacier* along the boundary between the ice and the ground. (Compare *plastic flow*.)

basalt A dark, fine-grained, *mafic igneous rock* composed largely of plagioclase feldspar and pyroxene; the *extrusive* equivalent of *gabbro*.

basaltic lava A lava type of mafic composition that has a low silica content, erupts at high temperatures, and flows readily.

base level The *elevation* at which a *stream* ends by entering a large standing body of water.

basin A synclinal structure consisting of a bowl-shaped depression of *rock* layers in which the beds *dip* radially toward a central point. (Compare *dome*.)

batholith A great irregular mass of *intrusive igneous rock* that covers at least 100 km^2; the largest type of *pluton*.

beach A *shoreline* environment made up of *sand* and pebbles.

bed load The material a *stream* carries along its bed by sliding and rolling. (Compare *suspended load*.)

bedding The formation of parallel layers, or beds, by deposition of *sediment* particles.

bedding sequence A sequence of interbedded and vertically stacked layers of different *sedimentary rock* types.

bioclastic sediment A shallow-water *sediment* made up of fragments of shells or skeletons directly precipitated by marine organisms and consisting primarily of two calcium carbonate *minerals*—calcite and aragonite—in variable proportions.

biofuel A fuel, such as ethanol, derived from biomass.

biogeochemical cycle The pattern of flux of a chemical between the biological ("bio") and environmental ("geo") components of an *ecosystem*.

biological sediment A *sediment* formed near its place of deposition as a result of direct or indirect mineral *precipitation* by organisms. (Compare *chemical sediment*.)

biosphere The component of the *Earth system* that contains all its living organisms.

bioturbation The process by which organisms rework existing *sediments* by burrowing through them.

blueschist A *metamorphic rock* formed under high pressures and moderate temperatures, often containing glaucophane, a blue amphibole.

bomb A *pyroclast* 2 mm or larger, usually consisting of a blob of *lava* that cools in flight and becomes rounded, or a chunk torn loose from previously solidified volcanic rock. (Compare *volcanic ash*.)

bottomset bed A thin, horizontal bed of *mud* deposited seaward of a *delta* and then buried by continued delta growth.

braided stream A *stream* whose *channel* divides into an interlacing network of channels, which then rejoin in a pattern resembling braids of hair.

breccia A volcanic *rock* formed by the *lithification* of large *pyroclasts*. (Compare *tuff*.)

brittle Pertaining to a material that undergoes little *deformation* under increasing stress until it breaks suddenly. (Compare *ductile*.)

building code A set of standards for the design and construction of new buildings that specifies the intensity of shaking a structure must be able to withstand during an *earthquake*.

burial metamorphism Low-grade metamorphism in which buried *sedimentary rocks* are altered by a progressive increase in pressure exerted by growing layers of overlying *sediments* and by the increase in heat associated with increased depth of burial.

caldera A large, steep-walled, basin-shaped depression formed by a violent volcanic eruption in which large volumes of *magma* are discharged rapidly from a large *magma chamber*, causing the overlying volcanic structure to collapse catastrophically through the roof of the emptied chamber.

Cambrian explosion The rapid *evolutionary radiation* of animals during the early Cambrian period, after almost 3 billion years of very slow evolution, in which all the major branches of the animal tree of life originated within about 10 million years.

capacity The total *sediment* load carried by a current. (Compare *competence*.)

carbon cycle The continual movement of carbon among different components of the *Earth system*.

carbon sequestration The process of removing carbon from the atmosphere and depositing it in long-term reservoirs.

carbonates A class of minerals composed of carbon and oxygen—in the form of the carbonate anion (CO_3^{2-})—in combination with calcium and magnesium.

carbonate compensation depth The ocean depth below which the seawater is sufficiently undersaturated with calcium carbonate that calcium carbonate shells and skeletons dissolve.

carbonate rock A *sedimentary rock* formed from *carbonate sediment*.

carbonate sediment A *sediment* formed from the accumulation of carbonate minerals directly or indirectly precipitated by marine organisms.

carbon intensity The amount of carbon released into the atmosphere per amount of energy produced by burning of a fossil fuel. For example, burning methane releases 14.5 Gt of carbon per quad of energy produced, so its carbon intensity is 14.5 Gt/quad.

cation A positively charged ion. (Compare *anion*.)

cementation A diagenetic change in which *minerals* are precipitated in the pores between *sediment* particles and bind them together.

channel A well-defined trough through which the water in a *stream* flows.

chemical sediment A *sediment* formed at or near its place of deposition from dissolved materials that *precipitate* from water. (Compare *biological sediment*.)

chemical stability A measure of a substance's tendency to retain its chemical identity rather than reacting spontaneously to become a different chemical substance.

chemical weathering Weathering in which the *minerals* in a rock are chemically altered or dissolved. (Compare *physical weathering*.)

chemoautotroph An *autotroph* that derives its energy not from sunlight but from the chemicals produced when minerals are dissolved.

chemofossil The chemical remains of an organic compound made by an organism while it was alive.

chert A *sedimentary rock* made up of chemically or biologically precipitated silica.

cirque An amphitheater-like hollow carved at the head of a glacial valley by the plucking and tearing action of ice.

clay A *siliciclastic sediment* in which most of the particles are less than 0.0039 mm in diameter and which consists largely of clay minerals; the most abundant component of fine-grained *sedimentary rocks*.

claystone A *sedimentary rock* made up exclusively of *clay*-sized particles.

cleavage (1) The tendency of a *crystal* to break along planar surfaces. (2) The geometric pattern produced by such breakage.

climate The average conditions of Earth's surface environment and their variation.

climate model Any representation of the *climate system* constructed to reproduce one or more aspects of its behavior.

climate system The global *geosystem* that includes all the components of the *Earth system* and all the interactions among these components needed to determine climate on a global scale and how it changes over time.

coal A *biological sedimentary rock* composed almost entirely of organic carbon formed by the *diagenesis* of wetland vegetation.

color A property of a *mineral* imparted by transmitted or reflected light.

compaction A diagenetic decrease in the volume and *porosity* of a *sediment* as its particles are squeezed closer together by the weight of overlying sediments.

competence The ability of a current to carry material of a given size. (Compare *capacity*.)

compressional wave A *seismic wave* that propagates by expanding and compressing the material it moves through. (Compare *shear wave*.)

compressive force A force that squeezes or shortens a body. (Compare *shearing force; tensional force*.)

concordant intrusion An igneous intrusion whose boundaries lie parallel to layers of bedded *country rock*. (Compare *discordant intrusion*.)

conduction The mechanical transfer of heat energy by the jostling of thermally agitated atoms and molecules. (Compare *convection*.)

conglomerate A *sedimentary rock* composed of pebbles, cobbles, and boulders; the lithified equivalent of *gravel*.

consolidated material Sediment that is compacted and bound together by mineral cements. (Compare *unconsolidated material*.)

contact metamorphism Metamorphism resulting from heat and pressure in a small area, as in rocks in contact with and near an igneous intrusion.

continental drift The large-scale movements of continents across Earth's surface, driven by the *plate tectonic system*.

continental glacier A thick, slow-moving sheet of ice that covers a large part of a continent or other large landmass. (Compare *valley glacier*.)

continental margin The *shoreline*, shelf, and slope of a continent.

continental rise An apron of muddy and sandy *sediment* extending from the foot of the *continental slope* to the *abyssal plain*.

continental shelf A broad, flat, submerged platform, consisting of a thick layer of flat-lying shallow-water *sediment*, that extends from the *shoreline* to the edge of the *continental slope*.

continental slope A steep slope that descends from the edge of the *continental shelf* to the *continental rise*.

contour A line that connects points of equal *elevation* on a topographic map.

convection The mechanical transfer of heat energy that occurs as a heated material expands, rises, and displaces cooler material, which is itself heated and rises to continue the cycle.

convergent boundary A boundary between lithospheric plates where the plates move toward each other and one plate is recycled into the *mantle*. (Compare *divergent boundary; transform fault*.)

core The dense central part of Earth below the *core-mantle boundary*, composed principally of iron and nickel. (See also *inner core*; *outer core*.)

core-mantle boundary The boundary between Earth's *core* and its *mantle*, about 2890 km below Earth's surface.

country rock The rock surrounding an igneous intrusion.

covalent bond A bond between atoms in which electrons are shared. (Compare *ionic bond*.)

crater (1) A bowl-shaped pit found at the summit of most *volcanoes*, centered on the vent. (2) A depression caused by the impact of a *meteorite*.

craton A stable region of ancient continental *crust*, often made up of continental *shields* and platforms.

cratonic keel A mechanically stable and chemically distinct portion of the lithospheric *mantle* that extends some 200 to 300 km beneath a *craton* into the *asthenosphere* like the hull of a boat into water.

creep A slow downhill *mass movement* of *soil* or other debris at a rate ranging from about 1 to 10 mm/year.

crevasse A large vertical crack in the surface of a *glacier*, caused by the cracking of brittle surface ice as it is dragged along by the *plastic flow* of the ice below.

cross-bedding A *sedimentary structure* consisting of beds deposited by currents of wind or water and inclined at angles as much as 35° from the horizontal.

crude oil An organic sediment formed by diagenesis from organic material in the pores of sedimentary rocks; a diverse class of liquids composed of complex hydrocarbons. Also called petroleum.

crust The thin outer layer of Earth, averaging from about 8 km thick under the oceans to about 40 km thick under the continents, consisting of relatively low-density silicates that melt at relatively low temperatures.

crystal An ordered three-dimensional array of atoms in which the basic arrangement is repeated in all directions.

crystal habit The shape in which a mineral's individual *crystals* or aggregates of crystals grow.

crystallization The formation of a solid *mineral* from a gas or liquid whose constituent atoms come together in the proper chemical proportions and ordered three-dimensional arrangement.

cuesta An asymmetrical ridge formed from a tilted and eroded series of beds with alternating weak and strong resistance to *erosion*.

cyanobacteria A group of microorganisms that produce carbohydrates and release oxygen by *photosynthesis* and that probably originated the process early in life's history.

dacite A light-colored, fine-grained *intermediate igneous rock* with a composition between that of *rhyolite* and that of *andesite*; the *extrusive* equivalent of *granodiorite*.

Darcy's law A summary of the relationships among the volume of water flowing through an *aquifer* in a certain time, the vertical drop of the flow, the flow distance, and the *permeability* of the aquifer.

decompression melting The spontaneous melting of rising *mantle* material as it reaches a level where pressure decreases below a critical point, without the introduction of any additional heat. (Compare *fluid-induced melting*.)

deflation The removal of *clay*, *silt*, and *sand* from dry *soil* by strong winds, which gradually scoop out shallow depressions in the ground.

deformation The modification of rocks due to folding, faulting, shearing, compression, or extension by plate tectonic forces.

delta A large, flat-topped deposit of *sediments* formed where a *river* enters an ocean or lake and its current slows.

dendritic drainage An irregular *drainage network* that resembles the limbs of a branching tree. (From the Greek *dendron*, meaning "tree.")

density The mass per unit volume of a substance, commonly expressed in grams per cubic centimeter (g/cm^3). (Compare *specific gravity*.)

depositional remanent magnetization A weak magnetization of *sedimentary rock* created by the parallel alignment of magnetic *sediment* particles in the direction of Earth's *magnetic field* as they settle and are preserved when the sediments are lithified.

desert pavement A coarse, gravelly ground surface left when continued *deflation* removes the smaller *sand* and *silt* particles from desert *soils*.

desert varnish A distinctive dark brown, sometimes shiny coating found on many desert rock surfaces, consisting of a mixture of clay minerals with smaller amounts of manganese and iron oxides.

desertification The transformation of semiarid lands into deserts.

diagenesis The physical and chemical changes, caused by pressure, heat, and chemical reactions, by which buried *sediments* are lithified to form *sedimentary rocks*.

diatreme A structure formed when a volcanic vent and the feeder channel below it are left full of *breccia* as an explosive eruption wanes.

dike A sheetlike *discordant igneous intrusion* that cuts across layers of bedded *country rock*. (Compare *sill*.)

diorite A coarse-grained *intermediate igneous rock* with a composition between that of *granodiorite* and that of *gabbro*; the *intrusive* equivalent of *andesite*.

dip The amount of tilting of a *rock* layer; the angle at which a rock layer inclines from the horizontal, measured at right angles to the *strike*.

dipole Pertaining to two oppositely polarized magnetic poles.

dip-slip fault A *fault* on which the relative movement of opposing blocks of rock has been up or down the *dip* of the fault plane.

discharge (1) The volume of *groundwater* leaving an *aquifer* in a given time. (Compare *recharge*.) (2) The volume of water that passes a given point in a given time as it flows through a *channel* of a certain width and depth.

discordant intrusion An igneous intrusion that cuts across the layers of the *country rock* it intrudes. (Compare *concordant intrusion*.)

disseminated deposit A deposit of ore minerals that is scattered through volumes of rock much larger than a vein.

distributary A smaller *stream* that receives water and *sediment* from the main *channel* of a *river*, branches off downstream, and thus distributes the water and sediment into many channels; typically found on a *delta*.

divergent boundary A boundary between lithospheric plates where two plates move apart and new *lithosphere* is created. (Compare *convergent boundary*; *transform fault*.)

divide A ridge of high ground along which all rainfall runs off down one side or the other.

dolostone An abundant *carbonate rock* composed primarily of dolomite and formed by the *diagenesis* of *carbonate sediments* and *limestones*.

dome An anticlinal structure consisting of a broad circular or oval upward bulge of rock layers in which the beds dip radially away from a central point. (Compare *basin*.)

drainage basin An area of land, bounded by *divides*, that funnels all its water into the network of *streams* draining the area.

drainage network The pattern of connections of all the large and small *streams* in a *drainage basin*.

drift All material of glacial origin found anywhere on land or at sea.

drought A period of months or years when precipitation is much lower than normal.

drumlin A large streamlined hill of *till* and bedrock deposited by a *continental glacier* that parallels the direction of ice movement.

dry wash A desert *valley* that carries water only briefly after a rain. Called a *wadi* in the Middle East.

ductile Pertaining to a material that undergoes smooth and continuous *deformation* under increasing stress without fracturing and does not spring back to its original shape when the stress is released. (Compare *brittle*.)

dune An elongated mound or ridge of *sand* formed by a current of wind or water.

dust Windborne material usually consisting of particles less than 0.01 mm in diameter (including *silt* and *clay*), but often including somewhat larger particles.

dwarf planet Any of several tiny objects of the outer solar system (including Pluto) that are composed of a frozen mixture of gases, ice, and rock and that orbit the Sun in an unusual pattern that sometimes brings them closer to the Sun than Neptune.

Earth system The collection of Earth's open, interacting, and often overlapping *geosystems*.

earthquake The violent motion of the ground that occurs when brittle *rock* under stress suddenly breaks along a *fault*.

eclogite An ultra-high-pressure *metamorphic rock* formed at the base of the *crust* at moderate to high temperatures, typically containing *minerals* such as coesite (a very dense, high-pressure form of quartz).

ecosystem An organizational unit at any scale composed of biological and physical components that function in a balanced, interrelated fashion.

El Niño An anomalous warming of the eastern tropical Pacific Ocean that occurs every 3 to 7 years and lasts for a year or so.

elastic rebound theory A theory of faulting and *earthquake* generation holding that, as the crustal blocks on either side of a *fault* are deformed by tectonic forces, they remain locked in place by friction, accumulating elastic strain energy, until they fracture and rebound to their undeformed state.

electron sharing The mechanism by which a *covalent bond* is formed between the elements in a chemical reaction.

electron transfer The mechanism by which an *ionic bond* is formed between the elements in a chemical reaction.

elevation The vertical distance above or below sea level.

enhanced greenhouse effect Global warming of Earth's atmosphere due to anthropogenic increases in CO_2 and other greenhouse gases, demonstrated by the IPCC to be the primary cause of the twentieth-century warming.

ENSO (El Niño–Southern Oscillation) A natural cycle of variation in the exchange of heat between the atmosphere and the tropical Pacific Ocean, of which *El Niño* and a complementary cooling event, known as La Niña, are a part.

eolian Pertaining to wind.

eon The largest division of geologic time, including multiple *eras*.

epeirogeny Gradual downward and upward movements of broad regions of *crust* without significant folding or faulting. (From the Greek *epeiros*, meaning "mainland.")

epicenter The geographic point on Earth's surface directly above the *focus* of an *earthquake*.

epoch A division of geologic time representing one subdivision of a *period*.

era A division of geologic time representing one subdivision of an *eon* and including multiple *periods*.

erosion The set of processes that loosen *soil* and *rock* and move them downhill or downstream.

esker A long, narrow, winding ridge of *sand* and *gravel* found in the middle of a ground *moraine*, running roughly parallel to the direction of ice movement, deposited by meltwater streams flowing in tunnels along the bottom of a melting *glacier*.

evaporite rock A *sedimentary rock* formed from *evaporite sediment*.

evaporite sediment *Chemical sediment* that is precipitated from evaporating seawater or lake water.

evolution Systematic change in organisms over time, driven by the process of *natural selection*.

evolutionary radiation   The relatively rapid evolution of many new types of organisms from a common ancestor.

exfoliation   A *physical weathering* process in which large flat or curved sheets of *rock* are detached from an outcrop.

exhumation   The transportation of subducted *metamorphic rocks* back to Earth's surface.

exoplanet   A planet outside the solar system.

extremophile   A *microorganism* that lives in environments that would kill most other organisms.

extrusive igneous rock   A fine-grained or glassy *igneous rock* formed from *magma* that erupts at Earth's surface as lava and cools rapidly. (Compare *intrusive igneous rock*.)

fault   A fracture in *rock* that displaces the rock on either side of it.

fault mechanism   The orientation of the fault rupture and the slip direction of a fault that caused an *earthquake*.

fault slip   The distance of the displacement of the two blocks of *rock* on either side of a *fault* that occurs during an *earthquake*.

faunal succession, principle of   See *principle of faunal succession*.

felsic rock   Light-colored *igneous rock* that is poor in iron and magnesium and rich in high-silica *minerals* such as quartz, orthoclase feldspar, and plagioclase feldspar. (Compare *mafic rock*; *ultramafic rock*.)

fissure eruption   A volcanic eruption emanating from a large, nearly vertical crack in Earth's surface rather than a central vent.

fjord   A former glacial valley with steep walls and a U-shaped profile, now flooded with seawater.

flake tectonics   The tectonic process of a planet with a vigorously convecting mantle underlying a thin *crust*, which breaks up into flakes or crumples like a rug; thought to occur on Venus and possibly on early Earth.

flexural basin   A type of *sedimentary basin* that develops at a *convergent boundary* where one lithospheric plate pushes up over the other and the weight of the overriding plate causes the underlying plate to bend or flex downward.

flood   Inundation that occurs when increased *discharge*, resulting from a short-term imbalance between inflow and outflow, causes a *stream* to overflow its banks.

flood basalt   An immense basalt *plateau* formed by *fissure eruptions* of highly fluid *basaltic lava*.

floodplain   A flat area about level with the top of a *channel* that lies on either side of the channel; the part of a *valley* that is flooded when a *stream* overflows its banks.

fluid-induced melting   Melting of *rock* induced by the presence of water, which lowers its melting point. (Compare *decompression melting*.)

focus   The point along a *fault* at which slipping initiates an *earthquake*.

fold   A curved deformation structure formed when an originally planar structure, such as a sedimentary sequence, is bent by tectonic forces.

foliated rock   *Metamorphic rock* that displays *foliation*. Foliated rocks include *slate*, *phyllite*, *schist*, and *gneiss*. (Compare *granoblastic rock*.)

foliation   A set of flat or wavy parallel *cleavage* planes produced by *deformation* under directed pressure; typical of regionally metamorphosed rock.

foot wall   The block of rock below a dipping *fault* plane. (Compare *hanging wall*.)

foraminifera   A group of single-celled planktonic organisms that live in ocean surface waters and whose calcite shells account for most of the *carbonate sediments* of the deep seafloor.

foraminiferal ooze   A sandy and silty *sediment* composed of the shells of dead *foraminifera*.

foreset bed   A gently inclined deposit of fine-grained *sand* and *silt*, resembling large-scale cross-beds, on the outer front of a *delta*.

foreshock   A small *earthquake* that occurs in the vicinity of, but before, a main shock. (Compare *aftershock*.)

formation   A distinct set of *rock* layers that can be identified throughout a region by its physical properties and possibly by the assemblage of *fossils* it contains.

fossil   A trace of an organism that has been preserved in the *geologic record*.

fossil fuel   An energy *resource* formed by the burial and heating of dead organic matter, such as *coal*, *crude oil*, or *natural gas*.

fractional crystallization   The process by which the *crystals* formed in cooling *magma* are segregated from the remaining liquid *rock*, usually by settling to the floor of the *magma chamber*.

fracture   The tendency of a *crystal* to break along irregular surfaces other than *cleavage* planes.

frost wedging   A *physical weathering* process in which the expansion of freezing water in cracks in *rock* breaks the rock.

gabbro   A dark gray, coarse-grained *igneous rock* containing an abundance of mafic *minerals*, particularly pyroxene; the *intrusive* equivalent of *basalt*.

gas   See *greenhouse gas*; *natural gas*.

genes   Large molecules within the cells of every organism that encode all the information that determines what the organism will look like, how it will live and reproduce, and how it differs from all other organisms.

geobiology   The study of interactions between the *biosphere* and Earth's physical environment.

geochemical cycle   The pattern of flux of a chemical from one component of the *Earth system* to another.

geochemical reservoir A component of the *Earth system* where a chemical is stored at some point in its *geochemical cycle*.

geodesy The science of measuring the shape of Earth and locating points on its surface.

geodynamo The global *geosystem* that produces Earth's *magnetic field*, driven by *convection* in the *outer core*.

geologic cross section A diagram showing the geologic features that would be visible if vertical slices were made through part of the *crust*.

geologic map A two-dimensional map representing the rock formations exposed at Earth's surface.

geologic record Information about geologic events and processes that has been preserved in rocks as they have formed at various times throughout Earth's history.

geologic time scale A worldwide history of geologic events that divides Earth's history into intervals, many of which are marked by distinctive sets of *fossils* and bounded by times when those sets of fossils changed abruptly.

geology The branch of Earth science that studies all aspects of the planet: its history, its composition and internal structure, and its surface features.

geomorphology (1) The shape of a landscape. (2) The branch of Earth science concerned with the shapes of landscapes and how they develop.

geosystem A subsystem of the *Earth system* that produces specific types of geologic activity.

geotherm The curve that describes how Earth's temperature increases with depth.

geothermal energy Energy produced when underground water is heated as it passes through a subsurface region of hot *rock*.

glacial cycle A climate cycle alternating between cold glacial periods, or *ice ages*, during which temperatures decline, water is transferred from the hydrosphere to the cryosphere, ice sheets expand into lower latitudes, and sea level falls, and warm *interglacial periods*, during which temperatures rise abruptly, water is transferred from the cryosphere to the hydrosphere, and sea level rises.

glacial rebound A mechanism of epeirogeny in which continental lithosphere depressed by the weight of a large glacier rebounds upward for tens of millennia after the glacier melts.

glacier A large mass of ice on land that shows evidence of being in motion, or of once having moved, under the force of gravity. (See also *continental glacier*; *valley glacier*.)

global change Change in the *climate system* that has worldwide effects on the *biosphere*, atmosphere, and other components of the *Earth system*.

gneiss A light-colored, poorly *foliated*, high-grade *metamorphic rock* with coarse bands of segregated light and dark *minerals* throughout.

graded bedding A bed that shows progressive change in grain size from large sediment particles at the bottom to small particles at the top, usually indicating a weakening of the current that deposited the particles.

graded stream A *stream* in which the slope, velocity, and *discharge* combine to transport its *sediment* load, with neither net sedimentation nor net *erosion* in the stream or its floodplain.

grain A crystalline particle of a *mineral*.

granite A felsic, coarse-grained *igneous rock* composed of quartz, orthoclase feldspar, sodium-rich plagioclase feldspar, and micas; the *intrusive* equivalent of *rhyolite*.

granoblastic rock A nonfoliated *metamorphic rock* composed mainly of *crystals* that grow in equant shapes, such as cubes and spheres, rather than in platy or elongate shapes. Granoblastic rocks include *hornfels*, *quartzite*, *marble*, *greenstone*, *amphibolite*, and *granulite*. (Compare *foliated rock*.)

granodiorite A light-colored, coarse-grained *intermediate igneous rock* that is similar to *granite* in containing abundant quartz, but whose predominant feldspar is plagioclase, not orthoclase; the *intrusive* equivalent of *dacite*.

granulite (1) A high-grade, medium- to coarse-grained *granoblastic rock*. (2) The highest metamorphic grade.

gravel The coarsest *siliciclastic sediment*, consisting of particles larger than 2 mm in diameter and including pebbles, cobbles, and boulders.

gravitational differentiation The transformation of a planet by gravitational forces into a body whose interior is divided into concentric layers that differ from one another both physically and chemically.

graywacke A sandstone composed of a heterogeneous mixture of *rock* fragments and angular *grains* of quartz and feldspar in which the *sand* grains are surrounded by a fine-grained *clay* matrix.

greenhouse effect A global warming effect that results when a planet with an atmosphere containing *greenhouse gases* radiates *solar energy* back into space less efficiently than it would without such an atmosphere.

greenhouse gas A gas that absorbs and reradiates energy when it is present in a planet's atmosphere. Greenhouse gases in Earth's atmosphere include water vapor, carbon dioxide, and methane.

greenschist (1) A low-grade *metamorphic rock* formed from mafic volcanic rock and containing abundant chlorite. (2) The metamorphic grade above the zeolite grade.

greenstone A low-grade *granoblastic rock* produced by the metamorphism of mafic volcanic *rock* and containing abundant chlorite, which accounts for its greenish cast.

groundwater The volume of water that flows beneath Earth's surface.

groundwater table The boundary between the *unsaturated zone* and the *saturated zone*.

guyot A large, flat-topped *seamount* resulting from the *erosion* of a volcanic island when it was above sea level.

habitable zone The distance from a star at which water is stable as a liquid; if a planet's orbit is within this zone, there is a chance that life might have originated there.

half-life The time required for one-half the original number of parent atoms in a radioactive *isotope* to decay.

hanging valley A *valley* formed by a tributary glacier that enters a deeper glacial valley high above the main valley floor.

hanging wall The block of rock above a dipping *fault* plane. (Compare *foot wall*.)

hardness A measure of the ease with which the surface of a *mineral* can be scratched.

Heavy Bombardment A time early in the early history of the solar system when planets were subjected to very frequent crater-forming impacts.

hematite The principal iron *ore*; the most abundant iron oxide at Earth's surface.

heterotroph A consumer organism that gets its food by feeding directly or indirectly on *autotrophs*. (Compare *autotroph*.)

high-pressure metamorphism Metamorphism occurring at pressures of 8 to 12 kbar.

hogback A landscape feature similar to a *cuesta*, consisting of steep, narrow, more or less symmetrical ridges, formed by the *erosion* of steeply dipping or vertical beds of hard strata.

horizontal drilling A technique for navigating drill bits on horizontal paths through flat-lying sedimentary rocks, used extensively in the production of oil and gas by hydraulic fracturing of tight formations.

hornfels A *granoblastic rock* of uniform grain size that has undergone little or no *deformation*; usually formed by *contact metamorphism* at high temperatures.

hot spot A region of intense, localized volcanism found far from a plate boundary; hypothesized to be the surface expression of a *mantle plume*.

Hubbert's peak The high point of a bell-shaped curve representing the rate of oil production; the point at which oil production peaks and then begins to decline.

humus An organic component of *soil* consisting of the remains and waste products of the many organisms living in that soil.

hurricane A great storm that forms over the warm surface waters of tropical oceans (between 8° and 20° latitude) in areas of high humidity and light winds, producing winds of at least 119 km/hour (74 miles/hour) and large amounts of rainfall.

hydraulic fracturing (fracking) A technique for withdrawing oil and gas from shale and other tight formations by first pumping water and sand into a borehole at high pressures to create fractures through which the oil and gas can more readily flow.

hydraulic gradient The ratio between the difference in *elevation* between two points in the *groundwater table* and the flow distance that water travels between the two points.

hydrocarbon economy The economy of industrial civilization that uses fossil fuels as its primary energy source.

hydroelectric energy Energy derived from water moving under the force of gravity driving a turbine that generates electricity.

hydrologic cycle The cyclical movement of water from the ocean to the atmosphere by evaporation, to the surface by *precipitation*, to *streams* through *runoff* and *groundwater*, and back to the ocean.

hydrology The science that studies the movements and characteristics of water on and under Earth's surface.

hydrothermal activity The circulation of water through hot volcanic *rocks* and *magmas*.

hydrothermal solution A hot water solution formed when circulating *groundwater* or seawater comes into contact with a hot magmatic intrusion, reacts with it, and carries off significant quantities of elements and ions released by the reaction, which may be deposited later as ore minerals.

ice age The cold period of a *glacial cycle*, during which Earth cools, water is transferred from the hydrosphere to the cryosphere, ice sheets expand, and sea level drops. Also called glacial period. (Compare *interglacial period*.)

ice shelf A sheet of ice floating on the ocean that is attached to a *continental glacier* on land.

ice stream A current of ice within a *continental glacier* that flows faster than the surrounding ice.

iceberg calving The process by which pieces of ice break off a *valley glacier* and form icebergs when the glacier reaches a *shoreline*.

igneous rock A *rock* formed by the solidification of *magma*. (From the Latin *ignis*, meaning "fire.")

infiltration The movement of water into *rock* or *soil* through cracks or small pores between particles.

inner core The central part of Earth below a depth of 5150 km, consisting of a solid sphere, composed of iron and nickel, suspended within the liquid *outer core*.

intensity scale A scale for estimating the intensity of a destructive geologic event, such as an earthquake or a hurricane, directly from the event's destructive effects.

interglacial period The warm period of a *glacial cycle* during which ice sheets melt, water is transferred from the cryosphere to the hydrosphere, and sea level rises. (Compare *ice age*.)

Intergovernmental Panel on Climate Change (IPCC) An international scientific organization established by the United Nations in 1988 to review climate research, develop a scientific consensus on how climate has changed in the past, and make scientific projections of future climate change, including the potential environmental and socio-economic impacts of anthropogenic climate change.

intermediate igneous rock An *igneous rock* midway in composition between mafic and felsic, neither as rich in silica as *felsic rock* nor as poor in it as *mafic rock*.

intrusive igneous rock A coarse-grained *igneous rock* formed from *magma* that intrudes into *country rock* deep in Earth's *crust* and cools slowly. (Compare *extrusive igneous rock*.)

ion An atom or group of atoms that has an electrical charge, either positive or negative, because of the loss or gain of one or more electrons.

ionic bond A bond formed by electrostatic attraction between *ions* of opposite charge when electrons are transferred. (Compare *covalent bond*.)

iron formation A *sedimentary rock* that usually contains more than 15 percent iron in the form of iron oxides and some iron silicates and iron carbonates.

island arc A chain of volcanic islands formed on the overriding plate at a *convergent boundary* by *magma* that rises from the *mantle* as water released from the subducting lithospheric slab causes *fluid-induced melting*.

isochron A *contour* that connects rocks of equal age.

isostasy A principle stating that the buoyancy force that pushes upward a lower-density body (such as a continent or an iceberg) floating in a higher-density medium (such as the *asthenosphere* or seawater) must be balanced by the gravitational force that pulls it downward. (From the Greek for "equal standing.")

isotope One of two or more forms of atoms of the same element that have different numbers of neutrons and therefore different *atomic masses*.

isotopic dating The use of naturally occurring radioactive elements to determine the ages of *rocks*.

joint A crack in a *rock* along which there has been no appreciable movement.

kaolinite A white to cream-colored *clay* produced by the *weathering* of feldspar.

karst topography An irregular, hilly type of terrain characterized by *sinkholes*, caves, and a lack of surface *streams*; formed in regions with humid climates, abundant vegetation, extensively jointed limestone formations, and appreciable *hydraulic gradients*.

Keeling curve World's longest continuous record of carbon dioxide measurements in the atmosphere, initiated at the Mauna Loa Observatory by Charles Keeling in 1958. The Keeling curve documents an increase in average CO_2 concentration from 310 ppm in 1958 to 410 ppm in 2018, a 32 percent change in 60 years.

kettle A hollow or undrained depression that often has steep sides and may be occupied by a pond or lake; formed in glacial deposits when *outwash* is deposited around a residual block of ice that later melts.

lahar A torrential mudflow of wet volcanic debris.

laminar flow Fluid movement in which straight or gently curved streamlines run parallel to one another without mixing or crossing between layers. (Compare *turbulent flow*.)

landform A characteristic landscape feature on Earth's surface shaped by the processes of *erosion* and sedimentation.

large igneous province (LIP) A voluminous emplacement of predominantly *mafic extrusive* and *intrusive igneous rock* whose origins lie in processes other than normal *seafloor spreading*. LIPs include continental *flood basalts*, oceanic basalt plateaus, and aseismic ridges produced by hot spots.

lava *Magma* that flows out onto Earth's surface.

limestone A *carbonate rock* composed mainly of calcium carbonate in the form of the mineral calcite.

liquefaction The temporary transformation of solid material to a fluid state when it is saturated with water.

lithic sandstone A sandstone containing many particles derived from fine-grained rocks, mostly *shales*, volcanic rocks, and fine-grained *metamorphic rocks*.

lithification The conversion of *sediment* into solid *rock* by *compaction* and *cementation*.

lithosphere The strong, rigid outer shell of Earth that comprises the *crust* and the uppermost part of the *mantle* down to an average depth of about 100 km. (From the Greek *lithos*, meaning "stone.")

loess A blanket of unstratified, wind-deposited, fine-grained *sediment*.

longitudinal profile The smooth, concave-upward curve that represents a cross-sectional view of a *stream*, from notably steep near its head to almost level near its mouth.

longshore current A shallow-water current that runs parallel to the shore.

lower mantle A relatively homogeneous region of the *mantle* about 2200 km thick, extending from the *phase change* at about 660 km in depth to the *core-mantle boundary*.

low-velocity zone A layer near the base of the *lithosphere*, beginning at a depth of about 100 km, where *S-wave* speed abruptly decreases, marking the top part of the *asthenosphere*.

luster The way the surface of a *mineral* reflects light. (See Table 3.3.)

mafic rock Dark-colored *igneous rock* containing *minerals* such as pyroxenes and olivines that are rich in iron and magnesium and relatively poor in silica. (Compare *felsic rock*; *ultramafic rock*.)

magma Hot, molten *rock*.

magma chamber A large pool of *magma* that forms in the *lithosphere* as rising magmas melt and push aside surrounding solid rock.

magmatic addition A process of continental growth in which low-density, silica-rich *rock* differentiates in the *mantle* and is transported vertically to the *crust*. (Compare *accretion*.)

magmatic differentiation A process by which *rocks* of varying composition arise from a uniform parent *magma* as various *minerals* are withdrawn from it by *fractional crystallization* as it cools, changing its composition.

magnetic anomaly One in a pattern of long, narrow bands of high or low magnetic intensity on the seafloor that are parallel to and almost perfectly symmetrical with respect to the crest of a mid-ocean ridge.

magnetic field The region of influence of a magnetized body or an electric current.

magnetic time scale The detailed history of Earth's *magnetic field* reversals as determined by measuring the thermoremanent magnetization of *rock* samples whose ages are known.

magnitude scale A scale for estimating the size of an *earthquake* using the logarithm of the largest ground motion registered by a *seismograph* (Richter magnitude) or the logarithm of the area of the *fault* rupture (moment magnitude).

mantle The region that forms the main bulk of Earth, between the *crust* and the *core*, containing *rocks* of intermediate *density*, mostly compounds of oxygen with magnesium, iron, and silicon.

mantle plume A narrow, cylindrical jet of hot, solid material rising from deep within the *mantle*, thought to be responsible for intraplate volcanism.

marble A *granoblastic rock* produced by the metamorphism of *limestone* or *dolostone*.

mass extinction A short interval during which a large proportion of the species living at the time disappear from the *geologic record*.

mass movement A downslope movement of masses of *soil*, *rock*, *mud*, or other materials under the force of gravity.

mass wasting All the processes by which weathered and unweathered Earth materials move downslope in large amounts and in large single events, usually under the influence of gravity.

meander A curve or bend in a *stream* that develops as the stream erodes the outer bank of a bend and deposits *sediment* against the inner bank.

mélange A distinct metamorphic assemblage that forms where oceanic *lithosphere* is subducted beneath a plate carrying a continent on its leading edge.

mesa A small, flat, elevated landform with steep slopes on all sides, created by differential *weathering* of bedrock of varying hardness. (From the Spanish word for "table.")

metabolism All the processes organisms use to convert inputs (such as sunlight, water, and carbon dioxide) into outputs (such as oxygen and carbohydrates).

metallic bond A type of *covalent bond* in which freely mobile electrons are shared and dispersed among ions of metallic elements, which have the tendency to lose electrons and pack together as *cations*.

metamorphic facies Groupings of *metamorphic rocks* of various mineral compositions formed under different grades of metamorphism from different parent rocks.

metamorphic rock Rock formed by high temperatures and pressures that cause changes in the mineralogy, texture, or chemical composition of any kind of preexisting rock while maintaining its solid form. (From the Greek *meta*, meaning "change," and *morphe*, meaning "form.")

metasomatism Change in the composition of a *rock* by fluid transport of chemical substances into or out of the rock.

meteoric water Rain, snow, or other forms of water derived from the atmosphere.

meteorite A chunk of material from outer space that strikes Earth.

microbial mat A layered microbial community commonly occurring in *tidal flats*, hypersaline lagoons, and thermal springs.

microfossil A trace of an individual *microorganism* preserved in the *geologic record*.

microorganism A single-celled organism. Microorganisms include bacteria, some fungi and algae, and most protists.

mid-ocean ridge An undersea mountain chain at a *divergent boundary*, characterized by earthquakes, volcanism, and rifting, all caused by the tensional forces of mantle convection that are pulling the two plates apart.

migmatite A mixture of *igneous* and *metamorphic rock* produced by incomplete melting, typically badly deformed and contorted and penetrated by many *veins*, small pods, and lenses of melted rock.

Milankovitch cycle A pattern of periodic variations in Earth's movement around the Sun that affects the amount of solar energy received at Earth's surface. Milankovitch cycles include variations in the eccentricity of Earth's orbit, the tilt of Earth's axis of rotation, and precession—Earth's wobble about its axis of rotation.

mineral A naturally occurring, solid crystalline substance, generally inorganic, with a specific chemical composition.

mineralogy (1) The branch of geology that studies the composition, structure, appearance, stability, occurrence, and associations of *minerals*. (2) The relative proportions of a *rock*'s constituent minerals.

Mohorovičić discontinuity The boundary between the *crust* and the *mantle*, at a depth of 5 to 45 km, marked by an abrupt increase in *P-wave* velocity to more than 8 km/s. Also called Moho.

Mohs scale of hardness An ascending scale of mineral *hardness* based on the ability of one *mineral* to scratch another. (See Table 3.2.)

moraine An accumulation of rocky, sandy, and clayey material carried by glacial ice and deposited as *till*.

mud A fine-grained *siliciclastic sediment* mixed with water, in which most of the particles are less than 0.062 mm in diameter.

mudstone A blocky, poorly bedded, fine-grained *sedimentary rock* produced by the *lithification* of *mud*.

natural gas Methane gas (CH_4), the simplest hydrocarbon.

natural levee A ridge of coarse material built up by successive *floods* that confines a *stream* within its banks between floods, even when water levels are high.

natural resource A supply of energy, water, or raw material used by human civilization that is available from the natural environment. (See also *resource*.)

natural selection The process by which inherited traits within a population of organisms make it more likely for an organism to survive and successfully reproduce over successive generations.

nebular hypothesis The idea that the solar system originated from a diffuse, slowly rotating cloud of gas and fine dust (a "nebula") that contracted under the force of gravity and eventually evolved into the Sun and planets.

negative feedback A process in which one action produces an effect (the feedback) that tends to counteract the original action and stabilize the system against change. (Compare *positive feedback*.)

nonrenewable resource A *natural resource* that is produced at a rate much slower than the rate at which human civilization is using it up; for example, *fossil fuels*. (Compare *renewable resource*.)

normal fault A *dip-slip fault* in which the *hanging wall* moves downward relative to the *foot wall*, extending the structure horizontally.

nuclear energy Energy produced by the fission of the radioactive isotope uranium-235, which can be used to make steam and drive turbines to create electricity.

obsidian A dense, glassy volcanic *rock*, usually of felsic composition.

ocean acidification A process in which carbon dioxide from the atmosphere dissolves into the ocean and reacts with seawater to form carbonic acid (H_2CO_3), increasing the acidity of the ocean.

oil See *crude oil*.

oil shale A fine-grained, clay-rich *sedimentary rock* containing relatively large amounts of organic matter, from which combustible oil and gas can be extracted.

oil trap An impermeable barrier that blocks the upward migration of *crude oil* or *natural gas*, allowing them to collect beneath the barrier.

oil window The limited range of pressures and temperatures, usually found at depths between about 2 and 5 km, at which *crude oil* forms.

ophiolite suite An assemblage of *rocks*, characteristic of the seafloor but found on land, consisting of deep-sea *sediments*, submarine *basaltic lavas*, and *mafic igneous* intrusions.

ore A *mineral* deposit from which valuable metals can be recovered profitably.

organic sedimentary rock A *sedimentary rock* that consists entirely or partly of organic carbon-rich deposits formed by the burial and *diagenesis* of once-living material.

original horizontality, principle of See *principle of original horizontality*.

orogen An elongated mountain belt, usually formed by an episode of compressive *deformation*.

orogeny Mountain building by tectonic forces, particularly through the folding and faulting of *rock* layers, often with accompanying volcanism. (From the Greek *oros*, meaning "mountain," and *gen*, meaning "be produced.")

outer core The layer of Earth extending from the *core-mantle boundary* to the *inner core*, at depths of 2890 to 5150 km, composed of molten iron and nickel and minor amounts of lighter elements, such as oxygen or sulfur.

outwash Glacial *drift* that has been picked up and distributed by meltwater *streams*.

oxbow lake A crescent-shaped, water-filled loop created in the former path of a *stream* when it bypasses a *meander* and takes a new, shorter course.

oxides A class of minerals that are compounds of the oxygen anion (O^{2-}) and metallic cations.

P-T path The history of changing temperature (T) and pressure (P) conditions that is reflected in the texture and *mineralogy* of a *metamorphic rock*.

P wave The first type of *seismic wave* to arrive at a *seismograph* from the *focus* of an *earthquake*; a type of *compressional wave*.

paleomagnetism The *geologic record* of ancient magnetization.

Pangaea A supercontinent that coalesced in the late Paleozoic *era* and comprised all present continents, then began to break up in the Mesozoic era.

partial melting Incomplete melting of a *rock* that occurs because the *minerals* that compose it melt at different temperatures.

passive margin A *continental margin* far from a plate boundary. (Compare *active margin*.)

peat A rich organic material, made up of accumulated vegetation preserved from decay in a wetland environment, that contains more than 50 percent carbon.

pediment A broad, gently sloping platform of bedrock left behind as a mountain front erodes and retreats from its *valley*.

pegmatite A *vein* of extremely coarse-grained *granite*, crystallized from a water-rich *magma* in the late stages of solidification, that cuts across much finer grained *country rock* and may contain rich concentrations of rare *minerals*.

pelagic sediment An open-ocean *sediment* composed of small terrigenous and biologically precipitated particles that slowly settle out of suspension in seawater.

peridotite A coarse-grained, dark greenish gray, *ultramafic intrusive igneous rock* composed primarily of olivine with smaller amounts of pyroxene and other minerals such as spinel or garnet; the dominant rock in Earth's *mantle* and the source rock of basaltic magmas.

period A division of geologic time representing one subdivision of an *era*.

permafrost Perennially frozen soil containing aggregates of ice crystals; any *rock* or *soil* remaining at or below 0°C for 2 or more years.

permeability The ability of a solid to allow fluids to pass through it.

phase change A transformation of a rock's *crystal* structure (but probably not its chemical composition) by changing conditions of temperature and pressure, signaled by a change in *seismic wave* velocity.

phosphorite A chemical or biological *sedimentary rock* composed of calcium phosphate precipitated from phosphate-rich seawater and formed diagenetically by the interaction of calcium phosphate with muddy or *carbonate sediments*. Also called phosphate rock.

photosynthesis The process by which organisms such as plants and algae use energy from sunlight to convert water and carbon dioxide into carbohydrates and oxygen.

phyllite A *foliated* rock that is intermediate in metamorphic grade between *slate* and *schist*, containing small *crystals* of mica and chlorite that give it a more or less glossy sheen.

physical weathering *Weathering* in which solid *rock* is fragmented by mechanical processes that do not change its chemical composition. (Compare *chemical weathering*.)

planetesimal Any of the numerous kilometer-sized chunks of material that accreted by gravitational attraction early in the history of the solar system.

plastic flow The deformation of a *glacier* that results from the sum of all the small slips of the ice *crystals* within it. (Compare *basal slip*.)

plate tectonic system The global *geosystem* that includes the convecting *mantle* and its overlying mosaic of lithospheric plates.

plate tectonics The theory that describes and explains the creation and destruction of Earth's lithospheric plates and their movement over Earth's surface. (From the Greek *tekton*, meaning "builder.")

plateau A large, broad, flat area of appreciable *elevation* above the neighboring terrain.

playa A flat bed of *clay* and encrusting precipitated salts, formed by the complete evaporation of a *playa lake*.

playa lake A permanent or temporary lake in an arid mountain *valley* or *basin*, where dissolved *minerals* may be concentrated and precipitated as the water evaporates.

pluton A large igneous intrusion, ranging in size from a cubic kilometer to hundreds of cubic kilometers, formed deep in the *crust*.

point bar A curved sandbar deposited along the inside bank of a *stream*, where the current is weakest.

polymorph One of two or more alternative possible *crystal* structures for a single chemical compound; for example, the *minerals* quartz and cristobalite are polymorphs of silica (SiO_2).

porosity The percentage of a rock's volume consisting of open pores between particles.

porphyroblast A large *crystal*, surrounded by a much finer grained matrix of other *minerals*, formed in *metamorphic rock* from a mineral that is stable over a broad range of temperatures and pressures.

porphyry An *igneous rock* of mixed *texture* in which large *crystals* (phenocrysts) "float" in a predominantly fine-grained matrix.

positive feedback A process in which one action produces an effect (the feedback) that tends to enhance the original action and amplify change in the system. (Compare *negative feedback*.)

potable Pertaining to water that tastes agreeable and is not dangerous to human health.

pothole A hemispherical hole in the bedrock of a streambed, formed by *abrasion* by small pebbles and cobbles rotating in a swirling eddy.

precipitate (1) (verb) To drop out of a saturated solution as *crystals*. (2) (noun) The crystals that drop out of a saturated solution.

precipitation (1) A deposit on Earth's surface of condensed atmospheric water vapor in the form of rain, snow, sleet, hail, or mist. (2) The condensation of a solid from a solution during a chemical reaction.

pressure-temperature path See *P-T path*.

principle of faunal succession A stratigraphic principle stating that the *sedimentary rock* strata in an outcrop contain distinct *fossils* in a definite sequence.

principle of original horizontality A stratigraphic principle stating that *sediments* are deposited as essentially horizontal beds.

principle of superposition A stratigraphic principle stating that each *sedimentary rock* stratum in a tectonically undisturbed sequence is younger than the one beneath it and older than the one above it.

principle of uniformitarianism A principle stating that the processes we see in action on Earth today have worked in much the same way throughout the geologic past.

pumice A volcanic *rock*, usually *rhyolitic* in composition, containing numerous cavities (vesicles) that remain after trapped gas has escaped from solidifying *lava*.

pyroclast A *rock* fragment ejected into the air by a volcanic eruption. (See also *bomb; volcanic ash*.)

pyroclastic flow A glowing cloud of hot ash, dust, and gases ejected by a volcanic eruption that rolls downhill at high speeds.

quad A unit consisting of 1 quadrillion (10^{15}) British thermal units (Btu), used to measure large quantities of energy.

quartz arenite A sandstone made up almost entirely of quartz grains, usually well sorted and rounded.

quartzite A very hard, white *granoblastic rock* derived from quartz-rich *sandstone*.

radiation See *evolutionary radiation*.

radiative forcing Change in Earth's energy balance between incoming solar energy and outgoing thermal infrared energy when a climate variable, such as the concentration of a greenhouse gas in Earth's atmosphere, is changed instantaneously while all other variables are held constant. Measures in radiative flux units [W/m^2] the sensitivity of the greenhouse effect to the change in the climate variable. Numerical values can be directly compared to the solar radiative forcing of 240 W/m^2.

rain shadow An area of low rainfall on the leeward slope of a mountain range.

recharge The *infiltration* of water into any subsurface rock formation.

recurrence interval The average time between large *earthquakes* at a particular location; according to the elastic rebound theory, the time required to accumulate the strain that will be released by fault slipping in a future earthquake.

red bed An unusual stream deposit of *sandstones* and *shales* bound together by iron oxide cement, which gives the bed its red color.

reef A moundlike or ridgelike organic structure constructed of the carbonate skeletons and shells of marine organisms.

regional metamorphism Metamorphism caused by high pressures and temperatures that extend over large regions; typical of convergent boundaries where two continents collide. (Compare *contact metamorphism*.)

rejuvenation Renewed uplift in a previously existing mountain chain that returns it to a more youthful stage.

relative age The age of one geologic event in relation to another. (Compare *absolute age*.)

relative humidity The amount of water vapor in the air, expressed as a percentage of the total amount of water the air could hold at the same temperature if it were saturated.

relative plate velocity The velocity at which one lithospheric plate moves relative to another.

relief The difference between the highest and lowest *elevations* in a particular area.

renewable resource A *natural resource* that is produced at a rate rapid enough to match the rate at which human civilization is using it up; for example, wood. (Compare *nonrenewable resource*.)

representative concentration pathway (RCP) A scenario developed by the IPCC in its Fifth Assessment Report for how greenhouse-gas concentrations in Earth's atmosphere will change during the twenty-first century. Each scenario is characterized by a concentration of greenhouse gases in the year 2100, which is expressed as a net radiative forcing in radiative flux units [W/m^2] relative to the preindustrial atmosphere.

reserve The supply of a *natural resource* that has already been discovered and can be exploited economically and legally at the present time. (Compare *resource*.)

reservoir See *geochemical reservoir*.

residence time The average time an atom of a particular element spends in a *geochemical reservoir* before leaving it.

resource (1) The entire amount of a given material, including the amount that may become available for use in the future; includes *reserves* plus known but currently unrecoverable supplies plus undiscovered supplies that geologists think may eventually be found. (Compare *reserve*.) (2) A *natural resource*.

respiration The metabolic process by which organisms release the energy stored in carbohydrates; requires oxygen.

reverse fault A *dip-slip fault* in which the *hanging wall* moves upward relative to the *foot wall*, compressing the structure horizontally.

rhyolite A light brown to gray, fine-grained *felsic igneous rock*; the *extrusive* equivalent of *granite*.

rhyolitic lava The lava type that is the richest in silica, erupts at the lowest temperatures, and is the most viscous.

rift basin A *sedimentary basin* that develops at a *divergent boundary* at an early stage of plate separation as the stretching and thinning of the continental crust results in subsidence. (Compare *thermal subsidence basin*.)

ripple A very small ridge of *sand* or *silt* whose long dimension is at right angles to the current that formed it.

river A major branch of a *stream* system.

rock A naturally occurring solid aggregate of minerals or, in some cases, nonmineral solid matter.

rock cycle The set of geologic processes that convert *rocks* of each of the three major types—*igneous, sedimentary*, and *metamorphic*—into the other two types.

Rodinia A supercontinent older than *Pangaea* that formed about 1.1 billion years ago and began to break up about 750 million years ago.

runoff The sum of all *precipitation* that flows over the land surface, including not only streams but also the fraction that temporarily infiltrates near-surface *soil* and *rock* and then flows back to the surface.

S wave The second type of *seismic wave* to arrive at a *seismograph* from the *focus* of an *earthquake*; a type of *shear wave*. S waves cannot travel through liquids or gases.

salinity The total amount of dissolved substances in a given volume of water.

saltation The transportation of *sand* or smaller *sediment* particles by a current in such a manner that the particles move along in a series of short intermittent jumps.

sand A *siliciclastic sediment* consisting of medium-sized particles, ranging from 0.062 to 2 mm in diameter.

sandblasting *Erosion* of a solid surface by *abrasion* caused by the high-speed impact of *sand* grains carried by wind.

sandstone The lithified equivalent of *sand*.

saturated zone The level below the *groundwater table*, in which the pores of *soil* or *rock* are completely filled with water. Also called the phreatic zone. (Compare *unsaturated zone*.)

schist An intermediate-grade *metamorphic rock* characterized by pervasive coarse, wavy *foliation* known as schistosity.

scientific method A general procedure, based on systematic observations and experiments, by which scientists propose and test hypotheses that explain some aspect of how the physical universe works.

seafloor metamorphism A form of *metasomatism* associated with mid-ocean ridges, in which seawater infiltrates hot basaltic lava, is heated, circulates through the newly forming oceanic crust by convection, and reacts with and alters the chemical composition of the basalt.

seafloor spreading The mechanism by which new oceanic *crust* is formed at a *spreading center* on the crest of a mid-ocean ridge. As two plates move apart, *magma* wells up into the rift between them to form new crust, which spreads laterally away from the rift and is replaced continually by newer crust.

seamount A submerged *volcano*, usually extinct, found on the seafloor.

sediment Material deposited on Earth's surface by physical agents (wind, water, and ice), chemical agents (*precipitation* from oceans, lakes, and *rivers*), or biological agents (living and dead organisms).

sedimentary basin A region where the combination of sedimentation and *subsidence* has formed thick accumulations of *sediment* and *sedimentary rock*.

sedimentary environment A geographic location characterized by a particular combination of climate conditions and physical, chemical, and biological processes.

sedimentary rock A *rock* formed by the burial and *diagenesis* of layers of *sediment*.

sedimentary structure Any kind of *bedding* or other feature (such as *cross-bedding, graded bedding,* or *ripples*) formed at the time of *sediment* deposition.

seismic hazard The intensity of shaking and ground disruption by *earthquakes* that can be expected over the long term at some specified location.

seismic ray path The path along which seismic energy propagates. Ray paths are perpendicular to wave fronts.

seismic risk The earthquake damage that can be expected over the long term in a specified region, usually measured in average dollar losses per year.

seismic tomography A technique that uses differences in the travel times of *seismic waves* produced by *earthquakes* and recorded on *seismographs* to construct three-dimensional images of Earth's interior.

seismic wave A ground vibration produced by an *earthquake*. (See also *P wave; S wave; surface wave*.) (From the Greek *seismos*, meaning "earthquake.")

seismograph An instrument that records the *seismic waves* generated by earthquakes.

settling velocity The speed at which particles of various weights suspended in a current settle to the bed.

shadow zone (1) A zone beyond 105° from the *focus* of an *earthquake* where *S waves* are not recorded because they are not transmitted through Earth's liquid *outer core*. (2) A zone at angular distances of 105° to 142° from the focus of an earthquake where *P waves* are not recorded because they are refracted downward into the core and emerge at greater distances after the delay caused by their detour through the core.

shale A fine-grained *sedimentary rock* composed of *silt* plus a significant component of *clay*, which causes it to break readily along bedding planes.

shear wave A seismic wave that propagates by moving the material it travels through from side to side. Shear waves cannot propagate through any fluid—air, water, or the liquid iron in Earth's *outer core*. (Compare *compressional wave*.)

shearing force A force that pushes two sides of a body in opposite directions. (Compare *compressive force; tensional force*.)

shield A large *tectonic province* within a continent that is tectonically stable and where ancient crystalline basement rocks are exposed at the surface.

shield volcano A broad, shield-shaped *volcano* many tens of kilometers in circumference and more than 2 km high, built by successive flows of *basaltic lava* from a central vent.

shock metamorphism Metamorphism that occurs when *minerals* are subjected to high pressures and temperatures by heat and shock waves generated when a *meteorite* collides with Earth.

shoreline The line where the ocean surface meets the land surface.

silicates The most abundant class of minerals in Earth's crust, composed of oxygen (O) and silicon (Si), mostly in combination with cations of other elements.

siliceous ooze A biologically precipitated *pelagic sediment* produced by sedimentation of the silica shells of diatoms and radiolarians.

siliciclastic sediment Sediment formed from clastic particles produced by the *weathering* of *rocks* and physically deposited by running water, wind, or ice. (From the Greek *klastos*, meaning "broken.")

sill A sheetlike *concordant igneous intrusion* formed by the injection of *magma* between parallel layers of bedded *country rock*. (Compare *dike*.)

silt A *siliciclastic sediment* in which most of the particles are between 0.0039 and 0.062 mm in diameter.

siltstone A *sedimentary rock* that contains mostly *silt* and looks similar to *mudstone* or very fine grained *sandstone*; the lithified equivalent of silt.

sinkhole A small, steep depression in the land surface formed when the thin roof of a limestone cave collapses suddenly.

slate A fine-grained *foliated rock* that is easily split into thin sheets, formed primarily by low-grade metamorphism of *shale*.

slip face The steep leeward slope of a *dune* on which *sand* is deposited in cross-beds at the *angle of repose*.

slump A slow *mass movement* of *unconsolidated material* that travels as a unit.

soil An intricate combination of weathered *rock* and organic material.

soil profile The composition and appearance of a *soil*, usually characterized by distinct layers.

solar energy Energy derived from the Sun.

solar forcing Cyclical variation in the amount of solar energy received at Earth's surface.

solar nebula According to the *nebular hypothesis*, a disk of gas and dust that surrounded the proto-Sun from which the planets of the solar system formed.

sorting The tendency for variations in current velocity to segregate *sediments* according to size.

specific gravity The weight of a substance divided by the weight of an equal volume of pure water at 4°C. (Compare *density*.)

spit A narrow extension of a *beach* formed by *longshore currents* that carry *sand* to its downcurrent end.

spreading center A *divergent boundary*, marked by a rift at the crest of a mid-ocean ridge, where new oceanic *crust* is formed by *seafloor spreading*.

stabilization wedge A strategy for reducing carbon emissions by 1 gigaton per year in the next 50 years relative to a business-as-usual scenario. About seven stabilization wedges will be necessary to stabilize carbon emissions at current levels.

stock A *pluton* less than 100 km^2 in area.

storm surge A dome of seawater, formed by a *hurricane*, that rises above the level of the surrounding ocean surface.

stratigraphic succession A chronologically ordered set of *rock* strata.

stratigraphy The description, correlation, and classification of strata in *sedimentary rocks*.

stratosphere The cold, dry layer of the atmosphere above the *troposphere* that extends from about 11 to 50 km in altitude. (Compare *troposphere*.)

stratospheric ozone depletion A net decrease in stratospheric ozone concentration due to chemical changes in the stratosphere, most notably that caused by anthropogenic emissions of chlorofluorocarbons in the twentieth century, which have since been curtailed by the Montreal Protocol.

stratovolcano A concave-shaped *volcano* formed from alternating layers of lava flows and beds of *pyroclasts*.

streak The *color* of the fine deposit of *mineral* powder left on an abrasive surface when a mineral is scraped across it.

stream Any body of water, large or small, that flows over the land surface.

stream power The product of stream slope and stream *discharge*.

stress The force per unit area acting on any surface within a solid body.

striation A scratch or groove left on bedrock by a *glacier* dragging rocks along its base; may show the direction of glacial movement.

strike The compass direction of a line formed by the intersection of a rock layer's surface or a fault surface with a horizontal surface.

strike-slip fault A *fault* on which the relative movement of the opposing blocks of rock has been horizontal, parallel to the *strike* of the fault plane.

stromatolite A rock with distinctive thin layers, believed to have been formed by ancient *microbial mats*; one of the most ancient *fossil* types on Earth.

subduction The sinking of oceanic *lithosphere* beneath overriding oceanic or continental lithosphere at a convergent plate boundary.

subsidence Depression or sinking of a broad area of *crust* relative to the surrounding crust, induced partly by the weight of *sediments* on the crust but driven mainly by plate tectonic processes.

sulfates A class of *minerals* that are compounds of the sulfate anion (SO_4^{2-}) and metallic *cations*.

sulfides A class of minerals that are compounds of the sulfide anion (S^{2-}) and metallic *cations*.

superposed stream A *stream* that erodes a gorge in a resistant *formation* because its course was established at a higher level on uniform *rock* before downcutting began. (Compare *antecedent stream*.)

superposition, principle of See *principle of superposition*.

surface wave A type of *seismic wave* that travels around Earth's surface from the *focus* of an *earthquake* and arrives at a *seismograph* later than S waves.

surge A sudden period of fast movement of a *valley glacier*.

suspended load All the material temporarily or permanently suspended in the flow of a current. (Compare *bed load*.)

sustainable development Development that meets the needs of the present without compromising the ability of future generations to meet their own needs.

suture A narrow zone where two continental blocks have been juxtaposed by plate convergence and the ocean basin that once separated them has been entirely subducted. Suture zones are often marked by *ophiolite suites*.

syncline A troughlike fold of layered *rocks* that contains younger rock layers in the core of the fold. (Compare *anticline*.)

talus Large blocks of broken *rock* that fall from a steep cliff of *limestone* or hard, cemented sandstone and accumulate in a gentler slope at the foot of the cliff.

tar sands A deposit of *sand* or *sandstone* that once contained oil but has lost many of its volatile components, leaving a tarlike substance called natural bitumen.

tectonic age The time that a *rock* was last subjected to crustal *deformation* intense enough to reset the isotopic clocks within the rock by metamorphism.

tectonic province A large-scale region formed by particular tectonic processes.

tensional force A force that stretches a body and tends to pull it apart. (Compare *compressive force; shearing force*.)

terrace A flat, steplike surface in a stream valley that parallels a stream above its *floodplain*, often paired one on each side of the stream, marking a former floodplain that existed at a higher level before regional uplift or an increase in *discharge* caused the stream to erode into the former floodplain.

terrestrial planet Any of the four inner planets of the solar system (Mercury, Venus, Earth, and Mars) that formed from dense matter close to the Sun, where conditions were so hot that most of their volatile materials boiled away. Also called Earthlike planets.

terrigenous sediment *Sediment* eroded from the land surface.

texture The sizes and shapes of a rock's mineral crystals and the way they are put together.

thermal subsidence basin A *sedimentary basin* that develops in the later stages of plate separation as *lithosphere* that was thinned and heated during the earlier rifting stage cools, becomes more dense, and subsides below sea level. (Compare *rift basin*.)

thermohaline circulation A global three-dimensional oceanic circulation pattern driven by differences in the temperature and the salinity—and therefore in the density—of ocean waters.

thermoremanent magnetization Permanent magnetization of magnetizable materials in *igneous rocks* when groups of atoms of the material align themselves in the direction of the *magnetic field* that exists when the material is hot and are then locked into place when the material cools below about 500°C.

thrust fault A low-angled *reverse fault*—one with a dip of less than 45°.

tidal flat A muddy or sandy area that is exposed at low *tide* but is flooded at high tide.

tide The twice-daily rise and fall of the ocean caused by the gravitational attraction between Earth and the Moon.

tight formations Impermeable source beds that hold petroleum resources, including extensive gas-rich shale deposits.

till Unstratified and poorly sorted *drift* deposited directly by a melting *glacier*, containing particles of all sizes from *clay* to boulders.

tillite The lithified equivalent of *till*.

topography The general configuration of varying heights that gives shape to Earth's surface, which is measured with respect to sea level.

topset bed A horizontal bed of *sediment*—typically *sand*—deposited on top of a *delta*.

trace element An element that makes up less than 0.1 percent of a mineral.

transform fault A plate boundary at which the plates slide horizontally past each other and *lithosphere* is neither created nor destroyed.

transition zone The portion of the *mantle* bounded by two abrupt *phase changes* at depths of about 410 and 660 km.

tributary A *stream* that discharges water into a larger stream.

troposphere The lowest layer of the atmosphere, which has an average thickness of about 11 km, contains about three-fourths of the atmosphere's mass, and convects vigorously due to the uneven heating of Earth's surface by the Sun.

(From the Greek *tropos,* meaning "turn" or "mix.") (Compare *stratosphere.*)

tsunami A fast-moving sea wave, generated by an *earthquake* that lifts the seafloor, that propagates across the ocean and increases in size when it reaches the shore.

tuff A volcanic *rock* formed by the *lithification* of small *pyroclasts.* (Compare *breccia.*)

turbidity current A *turbulent flow* of water carrying a suspended load of *mud* that flows down the *continental slope* beneath the overlying clear water.

turbulent flow Fluid movement in which streamlines mix, cross, and form swirls and eddies. (Compare *laminar flow.*)

twentieth-century warming The rise in Earth's average surface temperature by about 0.6°C between the end of the nineteenth century and the beginning of the twenty-first.

U-shaped valley A deep *valley* with steep upper walls that grade into a flat floor; the typical shape of a valley eroded by a *glacier.*

ultra-high-pressure metamorphism Metamorphism occurring at pressures greater than 28 kbar.

ultramafic rock An *igneous rock* consisting primarily of mafic *minerals* and containing less than 10 percent feldspar. (Compare *felsic rock; mafic rock.*)

unconformity A surface between two *rock* layers in a *stratigraphic succession* that were laid down with a time gap between them.

unconsolidated material *Sediment* that is loose and uncemented. (Compare *consolidated material.*)

uniformitarianism, principle of See *principle of uniformitarianism.*

unsaturated zone The level above the *groundwater table,* in which the pores of the *soil* or *rock* are not completely filled with water. Also called the vadose zone. (Compare *saturated zone.*)

upper mantle The portion of the *mantle* that extends from the *Mohorovičić discontinuity* to the base of the *transition zone* at about 660 km in depth.

valley The entire area between the tops of the slopes on both sides of a *stream.*

valley glacier A river of ice that forms in the cold heights of a mountain range, where snow accumulates, then moves downslope, either flowing down an existing stream valley or carving out a new valley. (Compare *continental glacier.*)

varve One pair in a series of alternating coarse and fine *sediment* layers deposited on a lake bottom by a valley glacier, formed in one year by the seasonal freezing of the lake surface.

vein A sheetlike deposit of *minerals* precipitated in fractures or *joints* in *country rock,* often by a *hydrothermal solution.*

ventifact A pebble with several curved or almost flat surfaces that meet at sharp ridges, formed by *sandblasting* of the pebble's windward side.

viscosity A measure of a fluid's resistance to flow.

volcanic ash *Pyroclasts* less than 2 mm in diameter, usually glass, that form when escaping gases force a fine spray of *magma* from a *volcano.* (Compare *bomb.*)

volcanic geosystem The total system of *rocks, magmas,* and processes needed to describe the entire sequence of events from melting to the eruption of *lava* from a *volcano* at Earth's surface.

volcano A hill or mountain constructed from the accumulation of erupted *lava* and *pyroclasts.*

wadi A desert *valley* that carries water only briefly after a rain. Called a *dry wash* in the western United States.

wave-cut terrace A level surface formed by wave *erosion* of a rocky *shoreline* beneath the surf zone, which may be visible at low *tide.*

weathering The processes by which *rocks* are broken down at Earth's surface to produce *sediment* particles. (See also *chemical weathering; physical weathering.*)

Wilson cycle The sequence of tectonic events on continents caused by the formation and closure of ocean basins. The cycle comprises (1) rifting during the breakup of a continent, (2) *passive margin* cooling and *sediment* accumulation during *seafloor spreading* and ocean opening, (3) *magmatic addition* and *accretion* during *subduction* and ocean closure, and (4) orogeny during continent-continent collision.

wind energy Energy produced by wind-driven electric generators, amounting to 6 percent of total U.S. electrical energy production in 2017.

zeolite (1) A class of silicate *minerals* containing water within their *crystal* structure, formed by metamorphism at very low temperatures and pressures. (2) The lowest metamorphic grade.

Index

Page numbers in *italic* indicate figures; page numbers in **bold** indicate key terms; and *t* following a page number indicates a table.

Ablation, **428**, *429*
Abrasion, 159, *159*, 536, *536*
Absolute ages, **241**, 256–258, *257*
Absolute plate movements, 138
Absolute time, measuring with isotopic clocks, 249–250
Acadia National Park, *436*
Acadian orogeny, 651
Acasta Gneiss, 656, *656*
Accreted terrains, **643**
Accretion, 603, **643**–645, *644*, *645*
Accumulation, 428, **428**, *429*
Acidic water, sulfide minerals reactions in forming, on Earth and Mars, 674, *674*
Acidification, ocean, 408–409, *409*
Acidophiles, 673, 673*t*
  mineral-dissolving, 676
Acid rain, 376, 379, 457
Acids, 457
  carbonic, 457, 515
  strong, 457
  weak, 457
Acid test, 65
Active faults, 269
Active margins, **640**
Aerobic zone, 674
Aftershocks, **271**–273, *273*
Agassiz, Louis, 431, 442–443
  ice age hypothesis of, 443
Age(s)
  absolute, 241, 256–258, *257*
  ancient ice, and warm periods, 354–355
  of Earth materials, 254
  ice, 350
  of petroleum source rocks, 249
  relative, 241
Aggregate, 70
Agriculture
  effects of climate change on, *403*
  impact of human activities on, 367
A-horizon, 464
Air bubbles, chemical analyses of, 428
Alaska
  Dawes Glacier in, *425*
  permafrost in, 441
Alaska earthquake (1964), 280, 284, 290
Albedo feedback, **346**, 347, 406, 445
  acceleration of climate change by, 408
Alberta, Canada, tar sands of, 378
Alberts, Bruce, 3
Aleutian Islands, 134

Alfisols, 465*t*
Allegheny Plateau, 373, 636
Allende meteorite, *691*
Alloy, 8
Alluvial environments, 162
Alluvial fans, *551*, **551**–552, 589, 591, *591*
Alluvial sediments, 589
Alpine fault, *22–23*, 34
Alpine glaciers, 425, 442
Alpine-Himalayan belt, 647, *647*
Alpine-Himalayan orogeny, 647, *647*, *648*, 649, 651
Alpine ice, 428
Alps, *232*
Alta California, 40
Alternative energy resources, 413
Altiplano, 485
Altyn Tagh Fault, 649
Aluminum, 58, 64
Amazon River, 502, 503*t*, 528
American Clean Energy and Security Act (2009), 412–413
Amery Ice Shelf, 432, *432*
Amethyst, crystals of, *58–59*
Amino acids, 680
Ammonia, 680
Ammonites, 687
Ammonium, 676
Amphiboles, *62*, 680
  cleavage of, 67, *67*
  crystal structure of, 193
  metamorphism and, 74, 96, 97
  as silicate mineral, 95
Amphibolites, 112, 195, **198**, 200, *201*
Anaerobes, 673*t*, 674
Anaerobic heterotrophs, 676
Anaerobic zone, 674
Anatolian Plate, 27
Ancient water, 515
Andalusite, 190
Andesite, **97**
Andesitic lavas, 113, *121*, **121**–122
Andes Mountains, *31*, 33, 111, 428
  formation of, 193
Andisols, 465*t*
Angel Falls, Venezuela, *494–495*
Angle of repose, **470**, *470*, 585
Angular unconformity, 245, *246*
Anhydrite, 61, 65
Anions, **56**, 58
Antarctica
  ice-core drilling in, 352, *352*

  ice sheets of, 424, 425–426, 443
  melting of continental ice sheet covering, 405
Antarctic Circumpolar Current, 342
Antarctic ice, volume of, 425
Antarctic ice cap, 426, *427*
Antarctic ice sheet, 395
Antecedent streams, **533**, *533*
Anthracite, 371
Anthropocene, **416**
  dawning of the, 410–412
Anthropogenic emissions, 14
  of greenhouse gases, 572
Anthropogenic global change, 14, **345**, 393–416
  climate change and, 399, *399*, 400–408, *401*, *402*, *403*, *404*, *405*, *406*, *407*, *408*
  loss of biodiversity and, *410*, 410–412, *411*
  managing carbon crisis in, *412*, 412–415, *413*, *414*, *415*
  ocean acidification and, 408–409, *409*
  rise of carbon dioxide in the atmosphere and, 394–395, *395*
  rise of the sea level, *406*, 406–407, *407*
  types of, 395–400
Anticlinal theory, 225
Anticlinal trap, *372*, 373
Anticlines, **222**, 372
Apatite, 676
Apollo, 604
  image of, from Earth, *1*
  missions to the Moon, 600
Appalachian Fold Belt, 635, 636–637, *637*, 642, 645, 649, 651
Appalachian Mountains, 34, *487*, 639
  mountaintop removal of coal in, 379
  surface mining in, 379
Appalachian orogeny, 651
Aquatic ecosystems, effects of climate change on, *403*
Aquicludes, **508**
Aquifers, **508**
  ancient microorganisms in deep, 522
  confined, 508, *509*
  Ogallala, *514*, 514–515
  unconfined, 508
Arabian Plate, *33*
Archaea, 672
Archean crust, 655
  rock formations in, *656*, 656–657, *657*

I-1

INDEX

Archean eon, **257**, 609, *610*, 634
Arctic ice cap, 426
Arctic National Wildlife Refuge (ANWR), 380, *381*
Arctic Ocean, 343, 560
Area rule in controlling earthquakes, 282–283
Arêtes, 436, *437*
Aridisols, 465*t*
Arid zones, 338
Aristotle, 6
Arkoses, **173**, *174*
Arrhenius, Svante, 394
Arsenic, contamination of groundwater by, 519
Artesian flow, **508**
Artesian wells, 508, *509*
Artificial desalination, 198
Asbestos, 69–70
Aseismic ridge, 136
Ash, volcanic, **94**, *95*
Ash-flow deposits, **130**
Asteroid belt, 302
Asteroids, **602**
   impacts by on Earth, 616*t*
Asthenosphere, *13*, **14**, 24, 32, 118, 313
Astrobiologists, **690**
Astrobiology, 690–692
Astronauts, *4*
Astronomers, 302, 626
Astronomical positioning, 37–38
Asymmetrical folds, *223*, 223–224
Atlantic Coastal Plain, 642
Atlantic Ocean, 560
Atmosphere, *13*, 355
   evolution of, 16
   greenhouse gases in Earth's, 346–347
   origin of Earth's oxygenated, 682–683
   rise of carbon dioxide in, 394–395
   volcanism and, *132*, 132–133
Atmosphere-biosphere gas exchange, 357
Atmosphere-ocean gas exchange, 355, 357
Atom(s), 54
   structure of, *55*, 55
Atomic mass, **56**
Atomic number, **55**–56
Atomic structure of minerals, *57*, 58, *58*
Autotrophs, **668**
   photosynthetic, *677*
Avalanches, debris from, 480, *481*
Axial volcanoes, 134

Backshore, 566
Backwash, 562, *563*
Bacteria, 585, 672, 673
Bacterial respiration, 459
Baguios, 573
Baja California, *33*, 643
   geography of, *40*, 40–41
Baku region of Azerbaijan, 373
Baltica, 649
Baltic Sea, 649

Banded iron formations, 684, **684**
Barchans, *586*
Barrier islands, **570**–571
   migrating, *571*
Basal slip, **431**
   movement of glaciers by, *430*, 431
Basalt, *71*, 92, 95, *95*, **97**
   flood, *39*, 130, *130*, 138, 690
   granite from, 101–104, *104*
   Siberian, 141
Basaltic lavas, 119–121, *119*–121, *120*
Base level, **549**–551, *550*
Basement rocks, *636*
Basin and Range Province, 639, 640, 642, 653
Basins, **224**, 226
Batholiths, **105**, 107
Bay of Fundy, 638
Beaches, 162, 565, **565**–567
   erosion of, 568
   forms of, 566
   preservation of, 566–567
   profile of, 566, *566*
   restoration of, *568*, 568–569
   sand budget of a, 566, 567, *567*
Bear Canyon (Tucson), 479
Becquerel, Henri, 250
Bedding, **74**, 165
   cross-, 165, *166*
   graded, 165–166
   sequences in, 167–168, *168*
Bed load, **538**, *539*
Bedrock, 214
Beryllium, 82, 85
B-horizon, 464
Big Bang theory, 18, 83, 600
Big Bend compression, 230, *231*
Biochemical cycles, microorganisms and, 678
Bioclastics, 173
Bioclastic sediments, 157, **157**
Biodiversity, loss of, *410*, 410–412
Biofilms, 672
Biofuels, **383**, 383–384
   liquid, 383
Biogeochemical cycle, **670**
Biological changes, ecosystem response to, 666–667
Biological evolution, 18
Biological precipitation of carbonate sediments, 176–178, *177*, 178
Biological sediments, **74**, 156–157, *157*
   classification of, 175–178, 175*t*
Biology, 24
   defined, 666
Biomass, 367
Biomineralization, microbial, 675, *675*
Biosphere, 13, *13*, 338, 340, 344–345, *345*, 355, 400, 668
   atmosphere gas exchange and, 357
   evolution of, 16
   magnetic field and, *328*, 328–329

role of biogeochemical cycle of phosphorus, *671*
as a system, **666**–671, *667*, 671
Biotite (mica), *71*, 95, *96*, 97,197
Biotite mica, 95, 96
Bioturbation, **166**, *167*
Bituminous coal, 371
Black Hills, 226, *636*
Black lung, 378
Blowout dunes, 586
Blue Ridge Province, 636–637
Blueschist, **200**, *201*, 204, 205
Bombs, **94**, *94*
   volcanic, 124, *124*
Bonds
   chemical, 56–57
   covalent, 56, *56*, 66
   ionic, 56–57, *57*, 66
   metallic, 56, *57*
Boreal zones, 338
Boron, 82, 85
Bottomset beds, **541**
Bourke's Luck Potholes, *536*
Bowen's reaction series, 100, *100*
Bowknot Bend, *240*
Box Canyon, *246*
Brahmaputra River, 503*t*
Braided streams, *539*, 539–540
Brazil, ethanol production in, 383
Breakers, 562
Breccia, **124**, i124
   fault, 227, *227*
Bright Angel Shale, 251
Brittle behavior, **218**, *219*
   of rocks in the Earth's crust, 219
   rocks in the laboratory, 217–219, *218*
Bronze Age, 366
Bryce Canyon, 251, *252*
Building codes, earthquakes and, **296**
Built environment, 268
Buller Valley (New Zealand), *479*
Buoyancy factor, 315
Burial, *55*, 155, 168
Burial metamorphism, *194*, **194**–195
Bush, George W., 395
Bushveld deposits, 102

Calcite, 54, *55*, 61, *63*, 65, *67*, 74
Calcium, 668
Calcium carbonate, 160, 178, 355
   indirect precipitation of, *675*
Calcium sulfate, 65, 180
Caldera, *126*, 127, **127**
   collapse of, 143–144
   resurgent, 127
Caledonian progeny, 651
California Coast Ranges, Franciscan formation of the, 205
Cambrian explosion, **685**, *686*, 687
   fossils recording, *686*
Cambrian period, 250
Campbell, Colin, 377

Canada, permafrost in northern, 441
Canadian Shield, 635, *635*, 640, 641, 643, 646, 651
Capacity, **538**
Cape Ann earthquake (1755), 286
Cape Cod, Massachusetts, 443, 511, 571
Cape Cod National Seashore, 570
Carbohydrates, 668, 669
Carbon, 605, 668, 682
 atomic mass of, 56
 balancing carbon emission with accumulation, 358–359
 cycling of, 355–356, *356*
Carbon-12, 56, 682
Carbon-13, 682
Carbon-14, 253, *253*, 256, 260, 682
Carbonate environments, 165
Carbonate platforms, 178
Carbonate rocks, *176*, **176**–178, *177, 178*
Carbonates, **61**, *61*, 63–64, *63–64*, 74
Carbonate sediments, 176–**178**
 biological precipitation of, 176–178, *177*, 178
Carbon crisis, managing the, *412*, 412–415, *413, 414, 415*
Carbon cycle, **355**–359, *356*
 chemical reactions in, 355
 engineering, 413–414, *414*
 geochemical cycles in, 355, 357
 human perturbations of, 357
 residence time in, 355
Carbon dioxide, 65, 355, 366, 457, 668
 anthropogenic increases in, 370
 as greenhouse gas, 339, 345, 366, 370, 396, 670, 692
 in regulating climate, 14
 in weathering, 457, *458*
Carbon emissions
 balancing with carbon accumulation, 358–359
 increases in, *415*
 nuclear energy in reducing, 382
 stabilizing, 414–415, *415*
Carbon flux, from energy production, 370
Carbonic acid, 457, 515
Carboniferous period, 247, 368, 636
Carboniferous-Permian glaciation, 445, *446*
Carbon intensity, **369**
 of fossil-fuel burning, 378–379
Carbon neutrality, 383, 384
Carbon release, from permafrost and seafloor sediments, 408
Carbon sequestration, **413**–414, *414*
Caribou, Maine, temperate climate in, 428
Carlsbad Caverns, *4*, 515, 516
Carlsbad, New Mexico, salt mines nears, 180
Carrara marble, 198
Carrizo Plate, *34*
Cascade Range, 33–34, 111, 499, 543, 643
Cascadia Fault, 284

Cascadia subduction zone, 33, 34
Casita volcano, 469
Cassini-Huygens mission to Saturn, 627, *627*
Cassini orbiter, 627
Catastrophic changes, potential to climate system, 408
Cations, **56**, 58, 63
 substitution, 58
Caucasus Mountains, 647
Cavendish, Henry, 8
Cementation, 74, **169**
Cenozoic era, 247, 249, 256, 642
Central eruptions, 118
Central Valley of California, 593
Ceres, 602
Cesium, 85
Chalcocite, *81*
Chalcopyrite, *81*
Challenger Deep, 6
Channeled Scablands, 440, *441*
Channelization, 530
Channels, 528, **529**
Charleston earthquake (1886), 286
Chatham, Massachusetts
 barrier islands at, 571
 shoreline of, 571
Chelyabinsk meteor, explosion of, *14*
Chemical bonds, 56–57
Chemical change, reduction of ozone depletion, 396–397
Chemical contaminants, 517–518, *518*
Chemical reactions, 56, *56*, 355
Chemical sediments, **74**, **156**, *156*, 162
 classification of, 175–178, 175t
Chemical stability, **460**–461
Chemical stratigraphy, 260
Chemical substitution, 193
Chemical wearing, glacial erosion and, 435
Chemical weathering, **154**, 155, 454, 455, 456–461, 536
 carbon dioxide in, 457, *458*
 chemical stability and, 460
 of feldspars, 588
 oxygen in, *459*, 459–460
 rates of, 467
 water and, 456, *457*
Chemoautotroph, 668t, **676**
Chemofossils, **682**
Chemoheterotroph, 668t
Chernobyl, 383
Chert, *176*, **181**, 683
 sources of, 181
Chicxulub, geologic evidence from, 688, *688*
Chile earthquake (1960), 280, 284
Chimborazo, Ecuador, wind erosion of sand at, *582*
China
 coal deposits in, 413
 coal-fired plants in, *413*
 deaths of coal miners in, 378

 electricity demand in, 384
 energy use in, 370
 loess deposits in, 587–588
 Three Gorges Dam in, 384, *385*
Chlorine, 56
Chlorite, 200
 formation of, 195, 199
Chlorite isograd, 199
Chlorofluorocarbons (CFCs), 396
Cholestane, 682
Chondrites, 83, 85
C-horizon, 464
Christchurch earthquake (2010), 273, *273*, 280, 290
Chromite, *82*
Chromium, 64, 69, 82, 102
Chrons
 Gauss normal, *36*, 37
 Gilbert reversal, *36*, 37
 magnetic, 35, 328
Chrysotile, *70*
Cima Dome, *592*
Cincinnati, 534
Cinder cones, 125, *126*
Cinnabar, 80, *80*
Cirque, **436**, *437*
Cities, development of, on floodplains, 531, 534, *534*
Civilization, as a global geosystem, 365–387
Clastic particles, 74, 156
Clasts, 176
Clay, **175**
Clay minerals, 74
Claystone, **175**
Cleavage of minerals, **66**–68, *67, 70*
Climate, **14**
 defining, 338–339
 effect on streams, 551
 hydrology and, 498–499, *499*, 500, 501–503, *502, 503*
 interactions between tectonics, topography and, 486, *487*, 488
 long-term global variations in, 350–353
 rainfall and, 455
 temperature and, 455
 variations in, 348–355, *349, 350, 351, 352, 353*
 zones of, 338
Climate change, 400–408
 acceleration of, 408
 anthropogenic, 14
 consequences of, 403–408, *404, 405, 406, 407, 408*
 glacial cycles and, *442*, 442–443
 global, 579–580
 human population and global change and, 402–403
 Intergovernmental Panel on Climate Change (IPCC) on, 399
 projecting future, *401*, 401–402

Climate system(s), 13, *13*, 14, **14,** *336–337,*
  337, 337–360
 atmosphere in, 339–341
 balancing through feedbacks, 347–348
 biosphere in, 344–345, *345*
 clocking, 260, *260*
 components of the, *339,* 339–345, *340,*
  341
 cryosphere in, 342–343, *343*
 global carbon cycle and, 370
 global warming and, 14, 345, 357, 370,
  400, 405–406
 greenhouse effect in, 345–349, *346*
 hydrosphere in, 341–342
 lithosphere in, 343–344
 plate tectonics and, 75–77, *76,* 78,
  206–207
 potential for catastrophic changes to,
  395, 408
Climate zones, **338**
 arid, 338, *339*
 boreal, 338, *339*
 polar, 338, *339*
 temperate, 338, *339*
 tropical, 338, *339*
Coal, 54, 154, **182,** 368, **371**–372
 extraction and combustion of, 378–379,
  379
 future use of, 386
 geologic formation of, *371,* 371–372
 mountaintop removal of, *379*
 surface mining of, 379
 underground mining of, 378
Coal-fired plants in China, *413*
Coastal Plain, 637–638
Coastal processes, 560–565, *561*
 surf zone and, 562, *562*
 tides and, 560, *564,* 564–565
 wave motion and, 560–562, *562*
 wave refraction and, 562–564, *563*
Coastal storm surges, hurricanes and,
  573–579
Coastal wetlands, effects of climate change
 on, *403*
Coastlines, 560, *560*
 sensitivity to global change, 560
Coconino Sandstone, 251
Cocos Plate, 639, *639*
Coesite, 195
Cohesion, 471
Colorado National Monument, *5*
Colorado Plateau, 640, 653
 stratigraphy of, 130, **130,** *130,* 251,
  252
Colorado River, 501, 640
Color of minerals, **68**–69, *69, 70*
Columbia supercontinent, 651
Columbia River, 528, 532, *532,* 543
Columbia River basalts, 97
Columbus, Christopher, 8
Comets, 603

Communities, 666
 interactions of biological, 666
Compaction, 74, **169**–170
Competence, **538**
Compressional waves, 9, 10, 275, **308,**
  310–311
Compressive forces, **214**
Compressive tectonics, 229, *230*
Computerized axial tomography (CAT),
  320
Concordant intrusions, **105**
Conduction, **316**
 through the lithosphere, 316, *317*
Conductivity, hydraulic, 512
Cone of depression, 510, *510,* 511
Confining pressure, 192
Conglomerate, **172,** *173,* i173
Congo River, 503t
Connecticut River Valley, 638
Connemara marble, *189*
Consolidated materials, 471
Consumers, organisms as, 668t
Contact metamorphism, 74, 75, **193**–194,
  194, 197, 202
Contamination
 reversing, 518–519
 of water supply, 517–518
Continental basin, 641
Continental buoyancy, 10
Continental-continent collision, 192
Continental-continent convergence, *31,*
  34
Continental crust, 10, *10,* 27
 geologic record of, 634
Continental deformation, styles of, *228,*
  228–230
Continental drift, 14, **24**–25, 44
Continental glaciers, *425,* **425**–426, *430,*
  443, 444
 destabilization of, 408
Continental ice sheets, melting of, 435
Continental interiors, 202
Continental rifting, 32
Continental rift zone, 30
Continental sedimentary environments,
  162, 163
Continental shelf, **161,** 164, 637, 647
Continental transform fault, *31*
Continent-continent collision, 205–206
Continents, 6
 deep structure of, 657–659, *658, 659*
 growth of, 643–645
 modification of, 646–647, *648*
Contour lines, *482, 482*
Convection, **15,** *15,* 16, **316,** 605
 in the mantle and core, 316–317
Convection cells, 340–341
Convection currents, nature of rising,
  46–47, *47*
Convergence, **31, 33**–34
 ocean-continent, 204–205, *205*

Convergent boundaries, **27,** 30–31, 202,
  284, 286
 continent-continent convergence, *31,* 34
 ocean-continent convergence, *31,* 33–34
 ocean-ocean convergence, *30,* 32–33
Copper, 635
 chemical substitution in forming, 193
Copper sulfide, 77
Coral reefs, 178, 179, *179,* 357, 560
 vulnerability of, 407, *407*
Cordón Caulle, eruption of, *132*
Core (Earth's), **8**–10, 315–316, 604
 convection in the, 316–317
 inner, 7, 10, 11, *13,* 604
 outer, 7, 11, *13*
Core-mantle boundary, 10, *312,* **314**–315
Coriolis effect, 341, 575
Coronae, *613*
Corriolis effect, 578
Cortes, Hernando, 40
Coteau des Prairies, South Dakota, *443*
Country rock, **94**
Covalent bonds, 56, **57,** 66
Covellite, 77
Crater counting, 609
Craters, **125,** *126,* 127
Cratonic keels, 657, *657,* 658
Cratons, **640,** 653
 origin of, 655–657, *656*
Creep, 478, **478,** *478*
Cretaceous period, 249, 256, 638, 643
Cretaceous-Tertiary mass extinction, 687
Crevasses, **431,** 431–432, *432,* 438
Cristobalite, formation of, 60
Cross-bedding, *152*–153, *165, 166,* 585, *585*
Cross-cutting relationships, 246–247, *247,*
  248
Crude oil, **182,** 368, **372**
 geologic formation of, 372
Crust (Earth), 7, **10,** *10,* 312, 604–605
 continental, 10, *10,* 27
 felsic rocks in, 312
 mafic rock in, 312
 oceanic, 10, *10,* 27
 ultramafic rock in, 312
 water in, *520,* 520–521, *521, 522*
Crutzen, Paul, 396, 410
Cryosphere, 13, *13,* 340, 342–343, *343,*
  355, 424
 changes in, 405–406
Crystal habit, **69**
 of minerals, *59,* 69–70, *70*
Crystallization, **58**
 fractional, *100,* 100–101, *101*
 laboratory studies of, 93
 of minerals, 58–60, *59*
Crystals, **58**
 faces, 58, *59*
 magmatic differentiation through
  setting, *102,* 102–103
 reading geologic history in, 206–207

Current flow of streams, 537–540, *538*
Currents, 560
  longshore, 563
  moderately strong, 158
  rip, 563–564
  in shaping shorelines, 560
  strength, particle size, and sorting, 157–160
  strong, 158, *159*
  as transport agents, 157, *158*
  turbidity and, 164
  weak, 158–159
Cyanobacteria, **670,** *670,* 683

Dacite, **97**
Dalton, John, 54
Darcy, Henri, 511–512
Darcy's law, 512, **512,** 513
"Darkness" (Lord Byron), 133
Darwin, Charles, 240, 243, 687
  evolution theory of, 245, 247
  study of coral reefs by, 179
  theory of natural selection, 245, 687
Davenport, Iowa, *534*
Dawes Glacier, *425*
Dead Sea, 484, *484,* 673
Debris, 471
Debris avalanches, 480, *481*
Debris flow, 478, *479*
Debris slides, 482, *482*
Deccan flood basalts, 97
Decompression melting, **98,** 139
Deep Impact spacecraft, 626, *626*
Deep mantle, *13*
Deep-sea drilling, 37, *37,* 109
Deep-sea environments, 164
Deep time, 240
Deepwater Horizon, explosion aboard, 380, *380*
Deffeyes, Ken, 377
Deflation, **581,** *582*
Deforestation, impact of human activities on, 367
Deformation, **214**–232, 642, 680
  brittle and ductile behavior of rocks in the Earth's crust and, 219
  brittle and ductile behavior of rocks in the laboratory and, 217–219, *218*
  circular structures in, 224–226, *226*
  compressive forces in, 214
  compressive tectonics in, 229, *230*
  continental, *228,* 228–230
  faults and, *219,* 219–221, *220*
  folds and, *222,* 224–224
  geologic maps and, 216, *217*
  joints and, 226, *227*
Deimos, 617
Deltas, 162, 541, **541**
  effects of ocean currents, tides, and plate tectonics on, 543
  growth of, 541, 543

  human effects on, 543
Delta sedimentation, 541
Dendritic drainage, **532,** *532,* 533, **533**
Dendrochronology, 256
Density, **60**
  of the Earth, 7–8
  of minerals, 69, *70*
Deposition, 154, *155*
  of shorelines, 568, 570–572
Depositional coastal forms, 570–571
Depositional remanent magnetization, **327,** *327*
Deposits
  ash-flow, 130
  bushveld, 102
  coal, 413
  disseminated, 81–82
  hydrothermal, 79–82, *80, 81*
  igneous, 82, *82*
  loess, 587–588
  pyroclastic, *123,* 123–124
  sedimentary, 82–83, *83*
  Stillwater, 102
  water-laid, 439–440, *440*
Desertification, 580, **592**–593
Desert pavement, **581,** *582*
Deserts, 162, 560, *579,* 579–581, 588–591
  colors of, 588
  impact of climate change on, *592,* 592–593
  influence of humans on, 593
  landscapes in, *590,* 590–591
  oases in, 589
  particle sizes in, 580–581
  plate tectonics and, 591
  sediments and sedimentation in, 589–590, *590*
  weathering and erosion in, 588–589, *589*
Desert varnish, **588,** *588*
Devonian period, 636
Diagenesis, 155, *155,* **168**–170, *169,* 193, 194, 371
Diamonds, 56, 57, 58, 60, *61,* 66, 635, 658
  microscopic, 195
Diatremes, **127**–128, *128,* 195
Diesel, Rudolf, 383
Differential stress, 192
Differentiation, magmatic, 99–104
Digital elevation model (DEM), 655
Dikes, **106,** *106,* 246
  textures of, 106
Dinosaur National Monument, *241*
Dinosaurs, 244
  demise of, 687–688
Diorite, **97**
Dip, **214**
  measuring, 214, 216, *216*
Dipole field, 323
Dip-slip fault, **220,** *220,* 221
Directed pressure, 192
Discharge, **506,** **544**–546, *545, 546,* 548

  balancing with recharge, *507,* 508–511, *509, 510, 511*
Disconformity, 245, *245*
Discordant intrusion, 105
Disseminated deposits, **81**–82
Dissolution, rate of, 460
Distributaries, **541**
Divergent boundaries, **27,** *30,* 32, 202, 284
  continental rifting, *32*
  oceanic spreading centers, *32*
Divide, **531**
Dolomite, 63–64, 74, 176
Dolostone, **176, 178,** 193
Domes, 226, 636
  volcanic, 125, *126*
Downstream erosion, 536
Drainage basins, **531**–532
Drainage networks, **533,** *533*
Drainage patterns, geologic history and, 533, 535
Drake, Edwin, 224, *368,* 368–369
Drake's Folly, 368, *368*
Drift, **438,** *439,* 443
Drift hypothesis, 25
Droughts, **499,** *500,* 501
  severity of, 501
Drumlins, **439,** *440*
Dry washes, **590**
Ductile behavior, **219**
  in rocks in the Earth's crust, 218, 219
  in rocks in the laboratory and, 217–219, *218*
Dunes, **540,** *540*
Dust, 464, **580**–581
  mineral, 587
  volcanic, 587
  windblown, 585–587, *586, 587*
Dust falls, 587–588
Dust storms, 585–587, *586*
Dutton, Clarence, 653
Dwarf planet, **607**

Eagle Nebula, 601
Earth, 615
  accretion, 603
  atmosphere of, 600, 605–606, 666
  characteristics of, 606*t*
  chemical composition of layers of, 10–12
  circumference of, *8, 9*
  climate system on, *14*
  core of (*See* Core (Earth))
  crust of, *7,* 10, 604
  density of, 7–8
  distribution of water on, *496*
  evolution of life on, 17–18
  formation of, 603–604
  formation of acidic waters on, 674, *674*
  geodynamo and, 320, *321,* 323–329
  gravitational field of, 320
  gravity of, 600
  greenhouse atmosphere of, 346–347

INDEX

Earth (*Continued*)
  image of, from *Apollo*, *1*
  impact by asteroids and meteorites on, 616t
  impact craters on, 615, *615*
  impact of Mars-sized body on, 603, *604*
  interior of (See Interior of Earth)
  internal temperature of, 316–317, *317*
  layers of, 6–7, *7*
  magnetic field of, *15*, 15–16, *17*, 317
  mantle of, *7*, 8–10, 25, 605
  oceans on, 605
  origin of, 16–17
  origin of oxygenated atmosphere on, 682–683
  paths of seismic waves through, **308**–311, *309*, *310*
  as planet, 602, *603*
  plate tectonic system on, 14–15, *614*
  radius of, 9
  rotation of, 15, 341, 346, 564
  shape and surface of, 6, 8, 322, *322*
  as system of interacting components, 12–16
  temperature of, 10, 317, 319
  topography of, *607*
  visualizing the three-dimensional structure, 319–320
  water on, 497
Earth fissure, *510*
Earthflow, 478, *479*
Earthquakes, 267–300, **269**. See also specific earthquakes
  aftershocks following, 272–273
  controlling, 282–283, *283*
  determining fault mechanisms, *281*, 281–284
  early warning systems for, 297
  elastic rebound theory of, 269, *270*, 271, 272, 300
  faulting patterns in, 284–287, *285*
  fault rupture during, 271
  fires as secondary hazard of, 292
  foreshocks of, 271–273, *272*, 273
  generation of, 228
  GPS measurements and silent, 283–284
  hazards and risks of, 287–297, 293, *294*, *295*, *296*
  intraplate, 286
  land-use policies in reducing damage from, 295–296
  locating the focus, *276*, 277
  magnitude and frequency of, *278*, 278–279
  main shocks of, 273
  measuring the size of, 277, *278*, 279–280
  moment magnitude of, 279
  origin time for, 277
  plate tectonics and, 284, *285*, 286
  predicting, 297–300
  regional fault systems and, 286–287
  seismic waves in studying, 275, *276*, 277
  seismographs in studying, 274, *274*
  shaking intensity of, 280–281
  silent, 283–284
  steps for safety, 298–299
Earth system, 2, *12*, **12**–16
East Africa, rift valleys of, 32, 229
East Pacific Rise, 37
Ebb tides, 565
Eclogite, 112
Ecology, 3
Economic basement, 194
Economy
  as dependent on nonrenewable energy resources, 366
  energy production and, 371
  hydrocarbon, 369
  methane, 376
  petroleum, 376
Ecosystems, **666**–668, *667*
  responses to biological changes, 666–667
Effluent streams, 506
E-horizon, 464
Elastic rebound theory, 269, *270*, 271, 272, 300
Elba, Nebraska, stacked layers of loess in, *587*
El Capitan, *164*
El Chichón eruption (1982), 344
Electron sharing, **56**
Electron transfer, **56**
Elevation, **482**
El Niño, 172, 178, 349, 350
El Niño–Southern Oscillation (ENSO), *349*, 349–350, *350*
Emeralds, 69
Emergency preparedness and response, 297
Emmons Glacier, *432*
End moraine, *438*, 439, 439t
Energy, 669
  alternative resources of, *381*, 381–386, *382*, *383*, *384*, *385*, *386*
  carbon flux from production of, 370
  consumption of, *369*, 369–370
  economics of production of, 371
  geothermal, 145, 386
  hydroelectric, 384
  renewable, 386
  solar, 385–386, *386*
  wind, 384–385, *385*
  world consumption of, *387*
Energy policy in managing the carbon crisis, *412*, 412–413, *413*
Energy resources, *367*, **367**–368
Engineering
  of the carbon cycle, 413–414. *414*
  earthquake, 296
Enhanced greenhouse effect, **400**
Enstatite, 63
Entisols, 465t

Environment
  alluvial, 162
  built, 268
  carbonate, 165
  continental sedimentary, 162, *163*
  deep-sea, 164
  evaporite, 165
  glacial, 162
  habitable, on Mars, *692*, 692
  human impact on, 367
  lake, 162
  marine, *163*, 164
  sedimentary, 162–165, *163*, 165t
Environmental costs of extracting fossil fuels, 378–381
Eolian sediments, **580**, 589
Eons, **256**–258
Epeirogeny, 651, 653, **653**, 655
Epicenter, **271**, 272
Epidote, 191
Epochs, **247**
Equivalent energy, 369
Eras, **247**
Eratosthenes, 6, 8, 9
Ergs, 589
Erosion, **73**, 154, *155*, 155–157, **454**
  beach, 568
  desert weathering and, 588–589
  downstream, 536
  in formation of stream valleys, 464–465, *466*, 467
  glacial, 424
  by groundwater, 515–516
  interactions between weathering and, 467, *468*
  sediment transport and, 538–539, *539*
  of shorelines, 568, 570–572
  streams as primary agent of, 589
  by valley glaciers, *437*
Erosional coastal forms, 570
Erosional floodplains, 531
Erratics, 438, *438*
Eruption clouds, 144
Eruptions. See also Volcanoes; specific by name
  central, 118
  fissure, 128–130, *129*, *129*
  recurrence intervals, 145
Eskers, **439**, 440
Ethanol, 383
Eukarya, 672, 673
Eukaryotes, 682
Euphrates River, 534
Eurasian Plate, 27, 34, *648*, 649
Europe, volcanic ash clouds over, 135, *135*
European Space Agency, 616
Evaporite environments, 165
Evaporite rocks, **177**, 178
Evaporites
  marine, 178, *180*, 180–181
  nonmarine, 181, *181*

Evaporite sediments, **177,** 178, **178,** 590, *590*
Everest, George, 654
Evolution, **685**
  Darwin's theory of, 245, 247
  of life, *17,* 17–18
  of minerals, 83, *84,* 85
Evolutionary radiation, **685,** 687
  mass extinctions and, 684–690, *685, 686, 687, 688, 689*
Ewing, Maurice, 26
Exfoliation, **461**–462, *462*
Exhumation, 190, **202,** 206–207
Exoplanets, **628**
Exotic terrains, 644
Extended crust, 642
External heat engine, 12, *12*
Extraterrestrial life, search for, 690–692
Extreme events, study of, 5
Extremophiles, **672**–674, 692, *692*
  characteristics of, 673t
Extrusive igneous rocks, *72,* 73, **94,** *94*
Extrusive pyroclasts, *94*
Eyjafjallajökull volcano, eruption of, 135, *135*

Facies, metamorphic, 201, *202*
Fahrenheit, Daniel, 397
Farallon Plate, 205, 320, 639, *639,* 643
Fault(s), 214, *219,* **219**–221, *220,* 246
  dip-slip, 220, *220,* 221
  left-lateral, 221
  normal, 220
  oblique-slip, 220, *220,* 221
  reverse, 221
  right-lateral, 221
  strike-slip, 220, *220*
  thrust, 221
  transform, 202, 230
Fault breccia, 227, *227*
Fault mechanism, **281,** 281–284
Fault slip, **269,** 298
Fault traces, 221
Fault trap, *372,* 373
Faunal succession, *243,* **243**–244
  principle of, 243
Feedback
  albedo, 346, 347, 406, 408, 445
  balancing the climate system through, 347–348
  negative, 347
  plant growth, 347
  positive, 347, 455
  radiative damping, 347
  water vapor, 347
Feldspar, 60, *62,* 63, 70, 71, 74, 83, 200
  chemical weathering of, 588
  dissolution of, 460
  orthoclase, 62, *71,* 96
  plagioclase, 62, *71,* 96, 100
  in schists, 197
  as silicate mineral, 95
  weathering of, 456, 457, 459
Felsic rocks, *94,* 95, 95t, **96**
  in Earth's crust, 312
Ferric (ferrous) iron, 459
Field and thrust belt, *228,* 229
Fig Tree Formation of South Africa, 681
Finger Mountains, Antarctica, 105
Fires as hazard of earthquakes, 292
Firn, 548, *548*
Fission, 382
Fissure eruptions, 128–130, **129,** *129*
Fixed-hot-spot hypothesis, 138–139
Fjords, **436,** *437*
Flake tectonics, **613,** *614*
Flaming Gorge, Wyoming, *483*
Flank collapse, 142–143
Flash floods, 589, *589*
Flexfuel, 383
Flexural basin, **162**
Flood basalts, **130,** *130,* 138, *139,* 690
Floodplains, 162, **529,** *529,* 588
  cities on, 531, 534, *534*
Floods, 528, **546,** *547*
  flash, 589, *589*
Flood stage, 546
Flood tides, 565
Florida keys, reefs of, 178
Florida Peninsula, 638
Flows, 473, 497
  current, 537–540, *538*
  debris, 478, *479*
  glacial, 429, *430,* 431
  groundwater, *505,* 511–512
  laminar, 537
  lava, 530
  pyroclastic, 124, *125*
  rock, 435
Fluid-induced melting, **99,** *112,* **112**–113
Fluids as cause of metamorphism, 193
Fluorine, 82
Focus, **271**
Fold and thrust belt, *228,* 229
Fold axis, 223
Folds, 214, 222, **222**–224
  asymmetrical, *223,* 223–224
  overturned, *223*
  plunging, *223, 223*
  symmetrical, *223*
Foliated rocks, 195–197, *197*
Foliation, 74, **195,** *196*
Foliated metamorphic rocks, 461
Food, 668
Fool's gold, 64, *64*
Foot wall, **220**
Foraminifera, **177**
Forces
  compressive, 214
  shearing, 214
  tensional, 214
Foreset beds, **541**

Foreshocks, **273,** *273*
Foreshore, 566
Forests, effects of climate change on, *403*
Formations, **214**
Fossil(s), **17,** *241,* 241–242
  gunflint, 681
  oldest, and early life, 680–682
  plant, 244
  as recorders of geologic time, 242–245, *243, 244*
  recording the Cambrian explosion, *686*
  systematic study of, 244–245
Fossil assemblages, 679
Fossil fuels, 366, **368,** 370
  burning of, 367, 370, 401, 412
  carbon intensity of burning of, 378–379
  carbon isotopes of, 397
  costs of, in U.S., 412
  environmental costs of extracting, 378–381
  exhaustion of reserves, 368
  future of, 386
  resources of, *370,* 370–381
Fossil record, 679
Fracking, 174, **373,** *374,* 375
  economic costs associated with, 380–381
Fractional crystallization, *100,* **100**–101, *101*
Fractures, **73**
  joints as, 226, *227*
Franciscan formation, 205
Franklin, Benjamin, 24
Frost wedging, **461**
Fuels. See Fossil fuels
Fukushima-Daiichi nuclear power plant, 296, 383
Fumaroles, 131, *131*
Fusion power, 413

Gabbro, *94,* 95, **97,** 119
Galápagos Islands, 137
Galena, 68, 80, *80*
Galileo, 600
Ganges River, 503t, 531
Garnet porphyroblasts, *198,* 203
Garnets, 68, 74, 191, 192, 202–203
Garnierite, 77
Gas. See also Greenhouse gases; Natural gas
  production of, from tight formations, 373
  searching for, 170–171, *171*
Gas industry, burial metamorphism and, 194
Gates of the Arctic Park, *526*–527
Gauss normal chrons, *36,* 37
Geco Topaz, 311
Gelisols, 465t

General Map of Strata in England and Wales (Smith), 244, *244*
Genes, **672**
Geobarometers, 193
Geobiologists, 671, 682, 685
Geobiology, 3, *4*, 85
　defined, **666**
Geochemical cycles, **355**, 357
Geochemical reservoirs, **355**
Geochemical tools, 634
Geochemistry, 3
Geodesy, **6**, 8, **37**
　measurements of plate movements by, *3*, 37–38
Geodynamo, 13, *15*, 15–**16**, 320, *321*, 323–329, 338
　computer simulations of the, 325, *325*
Geoid, 322, *322*
Geologic cross sections, **216**–217, *217*, *218*
Geologic cycling of water, *496*, 496–498
Geologic deductions, 521
Geologic formation of hydrocarbons, *371*, 371–372
Geologic history
　reconstructing from the stratigraphic record, *241*, 241–247, *242*, *243*, *244*, *245*, *246*
Geologic maps, **216**, *217*
　in finding oil, 224–225, *225*
Geologic record, **4**–5, *5*, *6*
　ancient glaciations in, 445, *446*
　documentation of volcanoes in, 118
　of Pleistocene glaciations, 444
Geologic structure, mapping, 214, *215*, 216–217
Geologic time, 652
　changes in Earth system through, 2
　fossils as recorders of, 242–245, *243*, *244*
　overview of, 16–18
　perspectives of, 258, *258*
Geologic time scale, **247**–249, *249*, *250*, 256–258, *257*, 610, 635, 679, 685
Geologists, adoption of "Sherlock Holmes" approach by, 54
Geology, **2**
　defined, 666
　goals of, 2
　modern, 5
　as a science, 3–5, *4*
　subfields of, 3, *4*
Geomorphology, **482**
　landscape development and, 482–488
Geophysics, 3
Geosystems, **13**
　civilization as global, 365–387
　interactions among, in supporting life, 16
　soil as, *463*, 463–464
　streams as, 543–547, *544*, *545*, *546*, *547*
　volcanic, 118, *119*, 130–133, *131*

Geotherm, **317**, *317*
Geothermal energy, **145**, 386
Geothermal gradient, 191
Geothermometer, 191–192
Geysers, 118, **131**, 132, *522*
　studying, 521
The Geysers, 147, *147*
Ghawar field (Saudi Arabia), 374
Giant's Causeway, 97
Gilbert reversal chron, *36*, 37
Gilbert, William, 323
Giraud-Soulavie, Abbé J.L., 24
Glacial budget, 429, *429*
　global warming and, 434
Glacial cycles, **350**
　climate change and, *442*, 442–443, *443*
　variations during the most recent, 444–445
Glacial environments, 162
Glacial erosion, 424, 435–436
Glacial flow, movement by, 429, *430*, 431
Glacial ice, 424
Glacial landscapes, 434–442, *436*, *437*, *438*, *440*, *441*
　glacial erosion and erosional landforms, 435–436
　glacial sedimentation and sedimentary landforms, 437–441, *438*, *440*, *441*
　permafrost, 441–442, *442*
Glacial maximum, 351
Glacial melting in sea-level changes, 572
Glacial moraines, types of, *438*, 439t
Glacial rebound, **653**, *653*
　isostasy and, *318*, *318*
Glacial sedimentation and sedimentary landforms, 437–441, *438*, *440*, *441*
Glacial thinning, 405
　rate of, 405
Glacial till, *443*
Glacial valley, *437*
　U-shaped profile of, 465
Glaciation
　Carboniferous-Permian, 445, *446*
　geologic record of ancient, 445, *446*
　Ordovician, 445
　Pleistocene, 444, 445
　Proterozoic, 445
　sea-level change and, 443–444
　Wisconsin, 443, *443*, 444
　Würm, 443
Glacier National Park, 405, *405*
Glaciers, *422*–*423*, 423–447, **425**
　ablation in shrinkage of, 428
　accumulation in growth of, 428, *428*
　continental, 425, *425*–426, *430*, *443*, 444
　formation of, 427–429, *428*
　ice accumulation in, 424, *424*
　isostasy and sea-level change and, *434*, 434–434

　movement of, 429, *430*, 431–433, *432*
　types of, 424–426
　valley, 425, *425*, 429, *430*, 431, 436
Global carbon cycle, climate system and, 370
Global change, 394
　anthropogenic, 394
　human population and, 402–403
Global cooling, 670, 680
Global Earthquake Model Foundation, 295
Global geosystem, 326, 366. *See also* Civilization as a global geosystem
Global hydrologic cycle, 498
Global pattern of volcanism, *133*, 133–137, *134*
Global Positioning System (GPS), *4*, 38, *38*, 41, 240
　in mapping ice movement, 431
　silent earthquakes and, 283–284
　in studying earthquakes, 274
Global variations in the climate, 350–353
Global warming, 14, **345,** 357, 370, 400, 670
　changes in the cryosphere and, 405–406
　glacial budget and, 434
　human-induced, 572
　melting of ice sheets and, 424
　sea-level changes and, 435, 572
Global warming disaster, 689–690
Glucose, 669
Gneiss, 196, **197**, *197*, 635
Gobi Desert, 579, 588
Goddard, Robert H., 601, *601*
Gold, 68, 77, 79, 635
　panning for, *83*
Gondwana, 24, *25*, 445, 647, 649, 651, 687
Gore, Al, 399
Grade, 78
Graded bedding, **165**
Graded streams, **551**
Grains, **59**
Grand Canyon, *252*, 640
　rocks at bottom of, *6*
Grand Prismatic Hot Spring, 664–665, 673
Granite, 8, 54, *71*, 72, 74, *91*, 92, *92*, 95, **96**, 455
　from basalt, 101–104, *104*
　density of, 7
　disintegration of, 456, *456*
　as evidence of slow cooling, *93*, 93–94, *94*
Granite-greenstone terrains, 656, *657*
Granoblastic rocks, **197**–198, *198*
Granodiorite, **97**
Granofels, 198
Granulite, **198,** 200, *201*
Graphite, 60, **61**, *61*
Gravel, **172**
Gravitational attraction, 601
Gravitational differentiation, 315, **603**

Gravity
  of Earth, 320, 600
  specific, 69
Gray, Peter, 4
Graywacke, **173**–174, *174*
Great Australian Desert, 579
Great Basin, 99, 191, 229, 579
Great continental ice sheets, melting of, 405
Great New Orleans flood, 573, 573–574, *574*
Great Plains, 591, *591*, 635
Great Rift Valley, *30*, 32
Great Salt Lake, 162, *181*, 673
Great Synthesis (1963-1968), 26–27
Great Trigonometrical Survey of India, 654–655
Great Unconformity, *246*, 251
Great Victoria Desert, 340
Greenhouse atmosphere, 346–347
Greenhouse effect, 14, 345–349, **346**, *346*, 396, 689
  enhanced, 400
Greenhouse gases, 14
  anthropogenic increases in, 370
  carbon dioxide as, 339, 345, 366, 370, 396, 670, 692
  methane as, 345, 396, 670, 692
  ozone as, 339, 396
  planet without, 346
  rise in, 396
Greenland, 635, 649
  ice-core drilling in, 352, *352*
  ice sheets over, 424, 425–426, *426*, 443
  melting of continental ice sheet covering, 405
Green River, Bowknot Bend on the, *240*
Greenschists, **200**, *201*
Greenstones, **198**, 656
Grenville orogeny, 651
Grenville Province, 651
Groins, 567, *567*
Grotzinger, John, 617
Ground failure, 290–291
Ground moraine, *438*, 439t
Groundwater, **497**
  ancient microorganisms in, 522
  balancing recharge and discharge of, 506, *507*, 508–511, *509*, *510*, *511*
  drinkability of, *519*, 519–520
  erosion by, 515–516
  flow of, *505*, 511–512
  hydrology of, 503–514
  movement of, 506, *507*
Groundwater table, **506**
  dynamics of the, *507*
Grypania, 684
Gryposaurus, *241*
Gulf of California, 32
Gulf of Mexico, 638
  oil slick in, *380*

Gulf of Suez, *33*
Gulf Stream, 342
Gullies, 535, *535*
Gunflint Formation of southern Canada, 681, *681*
Gunflint fossils, 681
Gutenberg, Beno, 10, 310
Gypsum, 59, 65, 74, *176*, 180, 623

Habitable environments on Mars, 692, *692*
Habitable zones around stars, *691*, 691–692
Hadean eon, **256**–257
Hager, Brad, 322
Haiti
  deforestation in, *410*
  2010 earthquake in, 290, *290*
Half Dome, *462*
Half-life, 253, **253**, *253*
Halides, *61*
Halite, 57, 60, *60*, 74, 162, *176*, 180
Halley's Comet, 603
Hallucigenia, 686
Halobacterium, 672
Halophiles, 373t, 673
Hanging valley, **436**
Hanging wall, 220
Hardness of minerals, **66**, 66t, 70
Harrisburg, Pennsylvania, 534
Hawaii, 46
  beaches of, 162
  as volcanic island, 119–121, *120*, 124, *124*, *137*, 137–139, *138*
Hazards, volcanic, 141–144
Headlands, wave refraction in protecting, 563
Headwater erosion, 536
Heat engine, 12
  external, 12, *12*
  internal, 12, *12*
Heat flow, through the Earth's interior, 316–317, *317*
Heat reservoir, 145
Heavy Bombardment, **609**, 615, 655
Hector Mine earthquake (1999), 287
Heezen, Bruce, 26, *26*
Heimaey, Iceland, 145
Helium, 600, 601, 602
Hematite, 61, 64, *68*, 68–69, 77, 459, 670, 683
  discovery on Mars, 618
Hercynian orogeny, 651
Hermit Shale, 251
Heterotrophs, **668**, *677*, 682, 688
  anaerobic, 676
High-grade metamorphic terrains, 657, *657*
High-pressure metamorphism, **195**
Hills, 484–485
Himalaya Mountains, 6, *232*, 485, 647, 649
  formation of, 193
  rise and erosion of, *654*, 654–655, *655*

Himalayan orogeny, *648*
Histosols, 465t
Holmes, Arthur, 25–26, 44
Holocene epoch, 247, 338
Homo sapiens, 367, 400
Horizontal drilling, **373**
Horizontality, principle of original, 242
Hornblende, 63
Hornfels, 197, 198
Horns, 536
Hot spots, 46, 136–**137**
Hot-spot tracks, measuring plate movements with, 138–139
Hot springs, 118, 132, *522*, 676
Hubbert, M. King, 375, 377
Hubbert's peak, **375**, 377
Hubble, Edwin Powell, 600
Hubble Space Telescope, 600, 601, 626
Hudson Bay, 641
Hudson River, 651
Human observation, record of, 240
Humans
  effects of climate change on, *403*
  effects on deltas, 543
  global change and, 402–403
  impact on environment, 367
  influence on deserts, 593
  perturbations of carbon cycle and, 357
  population growth of, *366*, 366–367
  volcanism and, 140–145, *147*
Humidity, relative, 498–499
Humus, *462*
Hunt, T. Sterry, 225
Hurricane(s), **573**
  coastal storm surges and, 573–579
  eye of, *575*
  formation of, 574–576, *575*
  increase in intense activity of, 404
  landfall by, *578*, 578–579
  naming of, 575
  rating of, 575–576, 576t
Hurricane Andrew, 574
Hurricane Katrina, 534, 573–574, *574*, *575*, *577*, 579
Hurricane Mitch, 469
Hurricane Sandy, 574, 578
Hutton, James, 5, 93–94, 240
Huygens lander, 627
Hydraulic conductivity, 512
Hydraulic fracturing, 174–175, **373**, *374*
Hydraulic gradient, **512**
Hydrocarbon(s), **367**
  geologic formation of, *371*, 371–372
Hydrocarbon economy, **369**
  rise of, 368–369
Hydroelectric energy, 384, **384**
Hydrogen, 600, 602, 668, 670, 680
Hydrogen sulfide, 519, 670, 676, *676*
Hydrologic cycle, 497, **497**–498, 501
  global, 498

Hydrology, 13, **497**
    climate and, 498–499, *499, 500,*
        501–503, *502, 503*
    of groundwater, 503–514
    of runoff, 501–503, *502*
Hydrolysis, 456
Hydrosphere, 13, *13, 340,* 341–342, 355
    volcanism and, *131,* 131–132
Hydrothermal activity, **131,** 145
Hydrothermal deposits, 79–82, *80, 81*
Hydrothermal fluids in metamorphism, 193, 194
Hydrothermal solutions, **79**–80
Hydrothermal veins, 107
Hydrothermal vents, *681*
Hydrothermal water, 520–521, *521, 522*
Hypersaline lagoons, 676
Hypersaline playa lakes, 673, *673*
Hyperthermophiles, 673
    chemoautotrophic, 132
Hypotheses, 2–3. *See also* Theories
    drift, 25
    of mantle plumes, 47
    nebular, **600**–601, *602*
    prebiotic soup, 690
    seafloor spreading, 26–27, 35
    Snowball Earth, 445
    as sometimes controversial, 3
    testing, 8–9
Hypothesis, Cambrian explosion, 685, 687

Iapetus lithosphere, 651
Iapetus Ocean, 649
Ice
    accumulation of, as rock formations, 424, *424*
    glacial, 424
    properties of, 424
Ice age, **350**
    hypothesis of, 443
Icebergs
    calving of., 425, *425, 428*
    float of, 433–434
Ice caps, 426–427
Ice-core drilling, in Antarctica and Greenland, 352, *352*
Ice front, 428
Iceland, Mid-Atlantic Ridge in, 134
Ice sheets, 225
    melting of, in global warming, 424
    over Antarctica, 424, 425–426, 443
    over Greenland, 424, 425–426, *426,* 443
Ice shelves, **425**
    collapse of, *433*
Ice streams, 425–426, **431**
Igneous deposits, 82, *82*
Igneous rocks, **71**–73, *72, 73,* 91–113
    chemical and mineral composition of, *95,* 95–97, *96*
    classification model for, *96*
    common minerals of, 73, 73t, 95t

extrusive, *72,* 73, **94,** *94*
forms of igneous intrusions, *104,* 104–107, *105, 106, 107*
intermediate, 96–97
intrusive, 72, *72,* **94,** *94*
magma formation and, 97–99, *98*
magmatic differentiation in, 99–104, *100, 101, 102*
plate tectonics and, 107–113, *108, 109, 110, 111, 112*
porosity in, 504
texture of, *92,* 92–95
Iguaçú River, *537*
Imperial Valley, California, irrigation in, *500*
Inceptisols, 465t
Index minerals, 199
Indian Ocean, 560
    2004 tsunami in, 688
Indian Plate, 34, *648,* 649
Induced seismicity, **288**
Indus River, 528
Industrial Revolution, 367, 368, 369, 395, 405, 411
Infiltration, **498**
Influent streams, 506
Inner core, 7, **10,** 11, 13, 604
Inner planets, 602, *603*
Inputs, 668–669
Insight, 622, *622*
Intensity scale, **280**
Interglacial period, **350**
Intergovernmental Panel on Climate Change (IPCC), 399, *399,* 402, 403, 406
    climate change projections of, 408
    climate models based on scenarios, 405
    on permafrost, 442
    "representative concentration pathways" of, 401, *401*
Interior of Earth, 307–329
    exploring with seismic waves, 308–316, *309, 310, 311, 312, 313*
    heat flow through, 316–317, *317*
    layering and composition of, *312,* 312–316, *313, 314*
    magnetic field
        and the biosphere, 328–329
        and the geodynamo, 320, *321,* 323–329
    paleomagnetism and, *325,* 326–328
    temperature of, 316–317, *317*
    three-dimensional structure of, 319–320
    waves reflected by internal boundaries in, *310,* 310–311
Interior Platform, 635
Intermediate igneous rocks, **96**–97
Internal heat engine, 12, *12*
International Astronomical Union, 603
International Decade of Natural Disaster Reduction, 295, *295*

Interplanetary space, 600
Intertropical convergence zone, 340
Intraplate earthquakes, 286
Intraplate volcanism, 136–137, *137*
Intrusions, discordant, 105
Intrusive igneous rocks, 72, *72,* **94,** *94*
Ionic bonds, **56**–57, *57,* 66
Ions, **56**
    mineral color and, 69
Iridium, 688
Iron, 11, *11,* 64, 69, 102, 602, 604, 635, 668, 676
Iron formations, **182**
Iron oxides, *459,* 459–460
Iron oxide sediments, 182
Irrigation, water supplies for, 499
Island arc, **32**–33
Islands, barrier, 570
Isochrons, **38**
    seafloor, 38–39, *39*
Isograds, 199
Isolated tetrahedra, 63
Isostasy, 433–434, *434*
    glacial rebound and, 318, *318*
    principle of, **312,** *314,* 314–315
Isotopes, **56,** 682
    radioactive, 253, 253t, 256
Isotopic clocks, 642
Isotopic dating, 118, **250,** 254–255, 634, 642, 690
    measuring absolute time with, 249–250
    of meteorites, 602
    methods of, 256, 609
    radioactive elements used in, 253t
Ivanpah solar electric generating system, *386*

Jack Hills, 258, *258*
Japan
    Fukushima-Daiichi nuclear power plant in, 383
    Kashiwazaki-Kariwa nuclear facility in, *382*
    Tsunami offshore in 2011 in, 268
Jarosite, 623, *623,* 674
Jet streams, 341
Jetty, 566
Johnson, David A., 146
JOIDES Resolution, 37, 260
Joints, **226**–227, 461, *461*
Jökulhlaup, 142
Juan de Fuca Plate, 27, *28–29,* 33, 111, 639, *639,* 643
Jupiter, 602, 607
Jurassic period, 247
Juvinas meteorite, 254, *255*

Kaibab Limestone, 251
Kalahari Desert, 579
Kamchatka Peninsula, 286

INDEX   I-11

Kames, 439
Kant, Immanuel, 600–601, *602*
Kaolinite, 63, 456, 457
Karst topography, **516,** *517*
Keeling, Charles David, 394, 395, *395,* 397
Keeling curve, **394**–395, *395,* 409, *409*
Keeling, Ralph, 394
Keels
    age of the, 658–659, *659*
    composition of the, 657–658, *658*
    cratonic, 657, *658*
Kettles, **439**
Keystone thrust fault, 230, *230*
Kigali Amendment (2019), 397
Kilauea
    eruption of, *137,* 137–139, *138*
    flank collapse at, 142
Kimberlite pipes, 128, 658, *659*
Kimberlites, 128, 194, 195, 658
Kimberly diamond mines, 83
Kissimmee River, 530
Klarälven River, 544
Kofa Butte, Kofa National Wildlife Refuge, Arizona, *590*
Kolbert, Elizabeth, 411
Komatite, 655–656
Komati River, 655–656
Krakatau eruption (1883), 122, 292, 344
Kudzu, *667*
Kuiper Belt objects, 610
Kuroshio Current, 342
Kyanite, 74, 190

La Conchita, California, 467, *468*
Lagoons, 178
Laguna Beach, landslide at, *452*–453, 478
Lahars, **142,** 480
Lake(s), **552,** *552*
    oxbow, 529, *530*
    playa, 590
Lake Baikal, Siberia, 552
Lake environments, 162
Lake Missoula, 440
Lake Pontchartrain, 573
Laki eruption, 130
Lambert Glacier, 431–432, *432*
Lambton, William, 654
Laminar flow, **537**
Landers earthquake (1992), 287
Landforms, **484**
    types of, 484–486, *485*
Landscape development, geomorphology and, 482–488
Landslides, 172–473, 268, 290–291, *452*–453, *473,* 476
Land-use policies, *295,* 295–296, *296*
La Niña, 349
La Plata River, 503*t*
L'Aquila earthquake (2009), 300
Large igneous provinces, **138**–139
Large-scale cross-bedding, *153*

Lariat Loop Scenic Byway, *218*
Larochelle, A., 35
Larsen Ice Shelf, 433
    collapse of, *433*
Late Heavy Bombardment, 615, 623
Lateral moraine, *438,* 439*t*
Laurasia, 41
Laurentia, 651
Lava flows, 530
Lavas, **92**–93, 94, 118
    andesitic, *121,* 121–122
    basaltic, 119–121
    pillow, 111, 121
    rhyolitic, *122,* 122–123
Lead, 517
    chemical substitution in forming, 193
Lead pollution, 517, **517**
Left-lateral fault, 221
Lehmann, Inge, 10, 311, *311*
Life
    domains of, 672
    evidence of early, in rocks, 682–683, *683*
    evolution of, *17,* 17–18
    interactions among geosystems in supporting, 16
    origin of, 679–680
    search for extraterrestrial, 690–692
Light waves, 308
Lignite, 371
Limestone, 54, 63, 74, **176,** *176,* **178,** 455, 515
    porosity of, 504
    recrystallization in, 192–193
    transformation into marble, 190
Liquefaction, **471**
Liquefied natural gas (LNG), 376
Liquid biofuels, 383
Listric faults, 229
Lithic sandstones, **173**
Lithification, **74,** 170
Lithium, 82, 85
Lithosphere, *13,* **14**–15, 24, 27, *340,* 343–344, 355
    atmosphere gas exchange and, 357
Lithosphere-atmosphere gas exchange, 357
Lithospheric graveyards, *46, 46*
Loess, 435, *587,* **587**–588
Loma Prieta earthquake (1989), 287, 290
Long Island, New York, 443, 511
Longitudinal dunes, *586*
Longitudinal profile, **549**
Longshore current, **563,** 570
Long-term global variations, 354–355
Long Valley Caldera, 127
Lower mantle, 11, **314**
Low-velocity zone, **313**
Lunar highlands, 608, *608*
Lunar lowlands, 608, *608*
Lunar maria, 608, *608*
Luray Caverns, Virginia, *516*

Luster of minerals, **68,** 68*t, 70*
Lyell, Charles, 5, 240

Maat Mons, *613*
Mafic minerals, 95, 96
Mafic rocks, *94,* 95*t,* **97**
    in Earth's crust, 312
Magellan spacecraft, 607, 612, 613
Magma, 32, **60**
    asthenosphere as source of, 118
    composition of, 113
    formation of, 97–99, *98*
    subduction zones as factories of, 111–113, *112*
Magma chambers, **99**
    formation of, 99
Magma factories, spreading centers as, *108,* 108–111
Magma ocean, 604
Magmatic addition, **643**
Magmatic differentiation, **99**–104
    complexities of, 101–104, *104*
    through crystal settling, *102,* 102–103
Magmatic stopping, 105, *105*
Magma viscosity, 103
Magnesium, 11, *11,* 58, 64
Magnesium olivine, 58
Magnetic anomalies, **35,** *36*
    mapping on the seafloor, 37
    negative, 35
    positive, 35
Magnetic chrons, 35, 328
Magnetic compass, 323
Magnetic field, **15,** *15,* 328–329
    biosphere and, *328,* 328–329
    complexity of, *321,* 323–329
    depositional remanent magnetization, 327
    dipole field and, 323
    of Earth, 317, 320, *321,* 323–329
    geodynamo and, 320, *321,* 323–329
    magnetic reversals and, 325–326
    nondipole field and, 323–324, *324*
    paleomagnetism and, *324,* 326–328
    secular variation and, *314,* 314–315, *315*
    thermoremanent magnetization and, 326–328
Magnetic reversals, 16, 320, *325,* 325–326
    rock record of, 35
Magnetic subchrons, 35
Magnetic time scale, **35,** *36*
Magnetite, *69,* 683
    direct precipitation of, *675*
Magnetization
    depositional remanent, *327, 327*
    thermoremanent, 326–328
Magnitude scale, **277,** 278
Main Boundary Fault, *648*
Maine, coastline in, 560, *561*
Mainshocks, 271, 272, 273
Malachite, *81*

Maldives, 405, *407*
  bleached coral in, *407*
Mammals. *See also* Animals
  recovery and evolution of modern, 689–690
Mammoth Cave, 515
Mammoth Hot Springs, 521, *521*
Manganese, 676
Mantle, *7*, **8**, 312–314, *313*
  deep, *13*
  lower, 11, 314
  peridotite in the, 109–110
  upper, 11, 312
Mantle convection, 44–47, *45*, *46*, *47*, 316–317
Mantle plumes, **46**–47, *47*, 107, **137**
  hypothesis of, 47, *47*, 136–137
  as magma factories, 113
Mapping
  of geologic structure, 214, *215*, 216–217
  of magnetic anomalies, on the seafloor, 37
  seismic, 285, 373
Maps, geologic, 216, *217*
Marble, 197, 198
  Carrara, 198
  Connemara, *189*
  strength of, 192
  transformation of limestone into, 190
Marble Canyon, *243*
Marcellus Formation, 373
Marcellus Shale, 174–175, *175*
Mariana Islands, *30*, 33
Marianas Trench, 6, *7*, 30, 32
Marine biosphere, fertilization of, 414
Marine environments, *163*, *164*
Marine evaporites, 178, **178**, *180*, 180–181, *181*
Mariner missions, 617
  Mariner 4, 617
  Mariner 10, 609–610
Mars, 329, 600, 602, *603*, 606–607, *607*, 613, 615
  Becquerel Crater on, *622*
  channel networks carved into surface of, *618*
  characteristics of, 606t
  dust storms on, *618*
  Elysium Planitia on, 622
  Endeavor Crater on, 621
  Endurance Crater on, 623
  environmental evolution of, 623, 625–626
  exploration of, 616–623
  formation of acidic waters on, 674, *674*
  Gale Crater on, 616–617, 619–623, 625
  Gusev Crater on, 620, 621
  habitable environments on, 692, *692*
  Home Plate on, 620
  landing spacecraft on, 624–625, *625*
  liquid water on, 600

  Meridiani Planum on, 620, 623
  Mt. Sharp on, 619–620, 621
  sedimentary rocks on, 692, *692*
  topography of, *607*, 613, 615
  Vallis Mariner on, 623
  Victoria Crater on, 621
  Yellowknife Bay on, 620, 625
Mars exploration rovers (MER), 619, *619*, 620, 623, 691, 692
  Opportunity, 616, 691, 692
  Phoenix, 692
  Spirit, 616, 692
Mars Express, 616, 617, 623
Mars Global Surveyor, 617, *618*, 618–619
Mars Hand Lens Imager (MAHLI) camera, 619, *620*
Marshes, 503
Mars Odyssey, 617, 618–619
Mars Polar Lander, 622–623
Mars Reconnaissance Orbiter, 616, 617, 622–623
Mars Science Laboratory (MSL), 623, 626, 691, 692
  Curiosity, 599–600, 616, 617, *619*, 619–622, *620*
Mass, 78
Mass extinction, 18, **247**, 349
  evolutionary radiations and, 684–690, *685*
  marking of interval boundaries, 247
  Paleocene-Eocene, 689–690
  Siberian Traps and, *140*, 140–141
Mass movements
  classification of, 172–473, *473*, *474*
  factors that influence, **469**t
  of rock, *475*, 475–478, *477*
  triggers of, 172
  of unconsolidated material, 478, *479*, 480
Mass spectrometer, 256
Mass wasting, **454**, 467, 469
Materials, consolidated, 471
Mathews, D. H., 35
Matter, structure of, 54
Matterhorn, 442, *442*
Mauna Loa, 125, 394
Mauna Loa Observatory, 394
McPhee, John, 5, 16
Meanders, **529**, *529*
Medial moraine, 439t
Mediterranean Sea, 560
Megathrusts, 229, 284
  Cascadia, 284
Mélange, **204**
Melting
  in ablation, 428
  decompression, 98, 139
  fluid-induced, **99**, *112*, 112–113
  partial, 98
  pressure and, 98
  temperature and, 98, 971
  water and, 98–99

Mercalli, Giuseppe, 280
Mercalli intensity scale for earthquakes, 576
Mercury, 600, 602, *603*, 606, 609–610
  characteristics of, 606t
  cratered surface of, *611*
Mesa, 485
Mesopotamian Basin, 162
Mesosaurus, fossils of, *25*
Mesozoic-Cenozoic, 642
Mesozoic era, 247, 249, 256, **258**, 642
MESSENGER spacecraft, 610
Metabolism, **669**
Metallic bonds, 56, **57**
Metamorphic facies, **201**, 202
Metamorphic grade, 190
  parent rock composition and, 200–201, *201*
  regional metamorphism and, 199–201
Metamorphic pressure-temperature paths, 202–204, *203*
Metamorphic rocks, **71**, 72, *73*, 74, *75*, 77, 190
  common minerals of, *73*, 73t, 74
  foliated, 461
  high-grade, 190
  low-grade, 190, 194
  porosity of, 504
  silicate minerals in, 190–191
  texture of, 185–199, *196*, *197*, *198*, *199*
Metamorphic textures
  foliated rocks and, 195–197, *197*
  granoblastic rocks and, 197–198, *198*
  porphyroblasts and, *198*, 198–199
Metamorphism, 74, 189–207, 214, 680
  burial, *194*, 194–195
  causes of, *190*, 190–193
  contact, 74, **74**, *75*, *75*, 193–194, **193**–194, *194*, 197, 202
  ductile deformation and, 219
  exhumation and, 206–207
  fluids of 193
  high-pressure, 74, *75*, 195
  plate tectonics and, *194*, 202–207
  recrystallization during, 256
  regional, 74, *75*, 193, *194*, 199–201, *200*, *201*
  role of pressure in, *192*, 192–193
  role of temperature in, *191*, 191–192
  seafloor, *194*, 194, 202
  shock, 185
  subduction-related, *200*, 204, 204–205
Metasomatism, *193*, 194
Meteor Crater, Arizona, 6, 608, *608*
Meteoric water, **503**
Meteorites, **602**
  Allende, 691
  crater created by, 6
  impacts by on Earth, 616t
  isotopic dating of, 602
  types of, *9*

Meteorologica (Aristotle), 6
Meteorology, 3
Meteors, 12–13
Methane, 372, 668, 670, 680, 689, 690
    as greenhouse gas, 345, 396, 670, 692
    sources of, 396
    on Titan, 627
Methane economy, 376
Methanopyrus, 672
Mica, 62, 70, 71, 74, 192
    cleavage of, 67
    crystal structure of, 193
    formation of, 195, **195**
    as silicate mineral, 95
Michigan Basin, 224, *224*, 636, 641
Michigan, Lower Peninsula of, 636
Microbial biomineralization, 675, *675*
Microbial mats, *675*, 678–679, **678**–679
Microfossils, **672,** 679
Microorganism-mineral interactions, 674
    biochemical cycles and, 678
    microbial mats in, *675*, 678–679
    mineral dissolution in, 678
    stromatolites in, 677–678, *678*
Microorganisms, 518, **671**–678
    abundance and diversity of, 671–674
    ancient, in deep aquifers, 522
    biochemical cycles and, 678
    interactions with minerals and, 674–678, *677*, *678*
Microscopic diamonds, 195
Mid-Atlantic Ridge, 26, *26*, 30, 32, 34
Middle East
    oil fields in, 373–374
    oil reserves of, 229
Mid-ocean ridges, *31*, **32**
    rift valleys of, 229
Mid-Ocean Ridge Transform Fault, *31*
Migmatite, 196, **197,** *197*
Milankovitch cycles, 352, *353*, 353–354, 394
Miller, Stanley, 680
Mineral dissolution, 678
Mineral-dissolving acidophiles, 676
Mineral dust, 587
Mineral isograds, 199–200
Mineralogy, **54,** 70
Mineral precipitation, 674–676, *675*, 678
Minerals, 54, **54**. *See also specific minerals*
    atomic structure of, *57*, *58*, *58*
    carbonate, **61,** *61*, 63–64, *63–64*, 74
    characteristics of, 54
    chemical classes of, 61*t*
    chemical composition of, 54
    classes of rock-forming, 60–65, *61*
    cleavage of, 66–68, *67*, *70*
    color of, 68–69, *69*, *70*
    concentrations of valuable, 77, 79
    crystal habit of, *59*, 69–70, *70*
    crystallization of, 58–60, *59*
    defined, 54
    density of, 69, *70*
    evolution of, 83, *84*, 85
    exploration of, *76*, 76–77
    felsic, 95
    formation of, 57–60
    fracture of, 68, *70*
    hardness of, 66, *66*, 70
    luster of, 68, 68*t*, *70*
    malic, 95, 96
    physical properties of, *65*, 65–70, *66*, 70*t*
    relative stabilities of, 460–461, 464*t*
    silicate, 61, *62*, 63, 66, *67*
    solubility of, 459
    sulfate, **61,** *61*, 65, 180
    sulfide, **61,** *61*, 64, 77
    trace elements and color of, 69
Mining
    economics of, *76*, 76–77
    open-pit, *81*
    surface, 379
    underground, 378
Miocene epoch, 247
Mississippi River, 502, *503*, 503*t*, 534, 543, 544, 638
    channelizing of, 530
    delta of, 540, *542*
    floodplain of, 543
Mississippi River Valley, loess deposits in, 588
Missouri River, 528, 532
Modified Mercalli intensity scale, 280, 280*t*, *281*, 288
Moenkopi Formation, 251
Mohorovičić discontinuity, **312**
Mohs, Friedrich, 66
Mohs scale of hardness, **66,** 66*t*
Mojave Desert, 579
Molina, Mario, 396, 397
Mollisols, 465*t*
Moment magnitude, 278
Monaco Declaration, 409
Monmouth Beach, New Jersey, *568*, 568–569
Monomoy Island, 571
Monsoons, 498
Mont Pelée, eruption of, 124
Montreal, 534
Montreal Protocol, **397**
Mont-Saint-Michel, France, *565*
Moon, 600
    Apollo missions to the, 600, 608–609
    characteristics of, 606*t*, 607
    crater-marked surfaces of, 600
    gravitational pull of, and tides, *564*, 564–565
    lunar highlands on, 608, *608*
    lunar lowlands on, 608, *608*
    lunar maria on, 608, *608*
Moraines, **439,** 439*t*
    end, *438*, 439*t*
    ground, *438*, 439*t*
    lateral, *438*, 439*t*
    medial, *438*, 439*t*
Morgan, W. Jason, 47
Morley, L., 35
Mountain front, 551
Mountains, 484–485, **484**–485, *485*. *See also specific mountain ranges*
Mount Etna, eruption of, 92, 125, 127
Mount Everest, 6, *7*, 32, 154, *654*
    topography of, 483–484
Mount Fuji, eruption of, 125
Mount Katahdin, Maine, *482*
Mount Merapi, eruption of, 142
Mount Pinatubo, 124
    eruption of, *125*, 133, 145, 344, 587
Mount Rainier, *145*
    eruption of, 125
    volcanic risk of, 144
Mount Sinabung, 134, *135*
Mount Stanley, valley glaciers of, *425*
Mount St. Helens, *4*, 33–34
    eruption of, 121, *121*, 135, 137, 146, *146*, 688
    flank collapse at, 142
Mount Tambora eruption (1815), 133, 344
Mount Waddington, glaciers near, 422–423
Mount Whitney, *91*
Muav Limestone, 251
Mud, **174**
Mudflows, 480, *480*
Mudslide, debris from, *468*
Mudstone, **174**
Murchison meteorite, 680
Muscovite, *62*, 63, 66, *67*, 67–68, 197
    as silicate mineral, 95
Muscovite mica, 96
Mylomites, 228

Namacalathus, *686*
Namib Desert, *586*
NASA, 616, 628
National Earthquake Prediction Evaluation Council, 300
Native elements, 60, *61*
Natural bitumen, 377
Natural gas, 154, **182, 372,** 375–376
    consumption of, 376
    future of, 386
    geologic formation of, 372
    production of, *376*, 380–381
Natural levees, **531,** *531*
Natural resources
    growth in demand for, 367
    limitation of Earth's, 366
    from volcanoes, 145, 147
Natural selection, **685**
    Darwin's theory of, 245, 687
Navajo Sandstone, 251
Nazca Plate, 33, 34, 45, 111, 643

I-14 INDEX

Nazca Ridge, 137
Neap tides, 565
Nebulae, 600
Nebular hypothesis, **600**–601, *602*
Negative feedback, **347**
Neogene period, 247, 256, 639
Neptune, 602, 607
Neutrons, 55
Nevada, Grand Basin in, 191
Nevado de Huascarán, debris avalanche on, 480, *481*
Nevado del Ruiz, 33
New Horizons spacecraft, 607, 610–611
New Madrid earthquake (1811), 280, 286
New Orleans, 534, 543
　great flood of, *573*, 573–574, *574*
Newspaper Rock, Canyonlands, Utah, *588*
Newton, Isaac, 564
　laws of gravity, 2–3, 8
New York City, seasonal temperatures in, 338, 338t
New Zealand, drought in, 501
Niagara Falls, 537
Nickel, 602, 604, 635
Nile River, 528, 531
Niolium, 82
Nitrogen, 605, 668, 676
Nitrogen-14, 256
Nitrous oxide, as greenhouse gas, 396
Nome, Alaska, polar climate in, 428
Nonconformity, 245
Nondipole field, 323–324, *324*
Nonmarine evaporites, 181, *181*
Nonrenewable resources, **368**
Normal fault, **220**
North America
　Appalachian Fold Belt in, 636–637, *637*
　Coastal Plain and Continental Shelf in, 637–638
　stable interior of, 635–636, *636*
　structure of, *634*, 634–635
North American Cordillera, 635, 638, *638*, 638–640, 642, 644, *644*, 646
　topography of, *638*
North American craton, 640
North American Plate, 27, *33*, 34, 40, 205, 284, 644
　boundary between Pacific Plate and, 219, *219*, 228
　subduction of Pacific Plate under, 499
North Atlantic Gyre, 342
North Atlantic Oscillation, 350
North Carolina, coastline in, 560
Northern Hemisphere, 341, 342
North Pacific Gyre, 342
Northridge earthquake (1994), 230, 287, *287*
North Sea, 560

Nuclear energy, *381*, 381–383, **382**, *382*, 383
　future of, 386
　hazards of, 383, *383*
　increased use of, 413
　in reducing global carbon emissions, 382
Nuclear fusion, 601
Nuclear power plants, siting of, 296
Nucleus, 55
Nunavut, Canada, *635*
Nutrients, 668

Oases, desert, 589
Oblique-slip fault, 220, *220*, 221
Obsidian, **94,** 123
Ocean(s). *See also* Seafloor; Seafloor spreading; *specific oceans*
　as chemical mixing vats, 160
　coastal processes and, 560–565
　formation of, *605*, 605–606
　shorelines of, 565–572
Ocean acidification, 383, 396, 408–409, *409*
Ocean basins, 6
Ocean-continent convergence, *31*, 33–34, **204**–205, *205*
　evidence of, 205
Ocean crust, 27
Ocean currents, effects on deltas, 543
Oceanic spreading center, 30, *32*
Ocean-ocean convergence, *30*, 32–33, 107, *108*, 643
Oceanography, 3
Offshore, 566
Ogallala Aquifer, *514*, 514–515
Ohio River, 534
O-horizon, 464
Oil, 154. *See also* Petroleum
　crude, 182
　geologic maps in finding, 224–225, *225*
　production of, from tight formations, 373
　searching for, 170–171, *171*
Oil reserves, distribution of, 373–375, *374*, *375*
Oil shale, 174, 377
Oil spills, 380, *380*
Oil traps, **372,** 373
　seismic mapping of, 373
Oil window, **372**
Oklahoma, seismicity in, 288–289, *289*
Old Faithful, 131, *131*
Oldham, R. D., 310
Oldham, Robert, 10
Olivine, 54, 58, 60, *62*, 63, 69, 103, 312, 459
　magnesium, 58
　perovskite, 69
　as silicate mineral, 95
　spinel, 69

Olympus Mons, 613, *614*
Ontong Java Plateau, 138
Open-pit mining, *81*
Open system, 12
Ophiolite suites, **108**–109, **205**
Opportunity, 619, 620, *621*, 623, 626
Ordovician glaciation, 445
Ordovician period, 636
Ores, **77**
Organic reefs, 164
Organic sedimentary rock, **182**
Organic sediments, 182
Organisms
　activities of, in weathering, *461*, 464
　metabolic process of, *669*, 669–670
　as producers and consumers, 668t
Original horizontality, principle of, **242**
Orinoco Basin, 375
Orogens, **640**
Orogeny, **646**–647, 649, *650*, 651
　Alpine-Himalayan, 647, *647*, *648*, 649
Orthoclase feldspars, 62, 71, 96
Outcrops, 214, *258*
Outer core, *7*, **10,** 11, *13*
Outer planets, 602, *603*
Outwash, 438, **439**
Overthrust structures, 229
Overturned folds, *223*
Owens Valley earthquake (1872), 287
Oxbow lakes, **529,** *530*
Oxidation, 85, 459
Oxides, **61,** *61*, 64, *64*, 77
Oxisols, 465t
Oxygen, 11, *11*, 61, 668
Ozark Dome, *636*
Ozone, 396
　as greenhouse gas, 339, 396
　reducing depletion of, 396–397

Pacala, Stephen, 414
Pacific Coast Range, *232*
Pacific-Farallon spreading center, 639
Pacific Ocean, 560
Pacific Plate, 33, *33*, 34, *40*
　boundary between North American Plate and, 219, *219*, 228
　subduction of, under the North American Plate, 499
Pacific Tsunami Warning Center, 297
Pahoehoe lava, 120, *120*
Paleocene-Eocene mass extinction, 689–690
Paleocontinents, 41
Paleogene period, 247
Paleogeographic reconstructions, *650*
Paleomagnetic stratigraphy, 260, *327*, 327–328
Paleomagnetism, *325*, 326–328, 329
Paleontology, 243, 244
Paleosols, 464

Paleozoic era, 247, 256, **258,** 637
Paleozoic-Mesozoic boundary, 249
Paleozoic orogenies during the assembly of Pangaea, 649, *649, 650*
Palisades, 100, 101, 103, *475*
Palladium, 102
Pangaea, **24,** *34,* 38, 249, 636, 645, 651
   assembly of, 41, *42,* 44
   breakup of, 41, *43,* 44, 638, 647
Parent rock, 455
   metamorphic grade and, 200–201, *201*
   properties of, 455, *455*
Paris Agreement (2015), 394, 399
Partial melting, **97,** 98
Passive margins, **640**
Pathfinder, 617, 618
Patterson, Clair, 250
Peat, **182,** 371
Pediments, **591,** *591,* 592
Pegmatite, 82, *93,* **107**
Perched water table, 508, *509*
Peridotite, **97,** 109–**110**
   in the mantle, **109–110**
Periods, **247,** 561
Permafrost, **441**–442, *442,* 618
   carbon release from, 408
Permeability, **505,** 512
Permian Kupferschier, 82
Permian mass extinction, 690
Permian period, 636
Peru-Chile subduction zone, 34
Peru-Chile Trench, *31,* 45
Petrified Forest, Arizona, *241*
Petroleum, 370. *See also* Oil
   future of, 386
   search for, 373
Petroleum economy, 376
Petroleum reservoir, 13
Petroleum source rocks, 249
Phanerozoic eon, **257**–258, 672
Phanerozoic orogen, 641, 642
Phase change, **313**
Phenocrysts, *94*
Philippines, *643*
Phobos, 617
Phoenix, 616, 622–623
Phosphate rock, 181
Phosphorite, **181**
Phosphorus, 668, 670, 676
   role of biosphere in the biogeochemical cycle of, *671*
Photoautotrophs, 668*t, 677,* 689
Photoheterotroph, 668*t*
Photomicrographs, 92
Photosynthesis, *669,* **669**–670, 682, 683
   comparison of respiration and, 670*t*
Photosynthetic autotrophs, *677*
Phreatic zone, 506
Phyllites, 196, **197**
Physical laws, 2

Physical weathering, **154,** 454, 455, 456, 461–462, 536
   exfoliation in, 461–462, *462*
   frost wedging in, 461
Physics, 24
Piedmont, 637
Pillow lavas, *12,* 110, 111, 121
Placers, 82–83, *83*
Plagioclase feldspar, *62, 71,* 96, 100
Planetary science, 3–4
Planetary surfaces, age and complexion of, *6,* 608–615, *609, 610, 611, 612, 613, 614*
Planetesimals, **601,** 602, *602,* 605
Planets. *See also specific by name*
   diversity of, 606–607, 606*t*
   dwarf, 607
   formation of, 601–602
   inner, 602, *603*
   outer, 602, *603*
   terrestrial, 602, *602*
Plant growth feedback, 347
Plastic flow, **431,** *431*
   movement of glacier by, 429, *430,* 431
Plateaus, 485, **485**
Plate boundaries
   convergent, 284, 286
   divergent, 284
   transform-fault, 284
Plate recycling, 45–46
Plate tectonics, 2, 13, *13,* **15, 27,** 54, 192, 268, 338, 605, 634
   assembly of Pangaea by continental drift and, 41, *42*
   breakup of Pangaea and, 41, *43*
   climate system and, 206–207
   continental drift and, *24,* 24–25, *25*
   convective motions of, 16
   convergent boundaries and, 27, *28–29, 30–31*
   deep-sea drilling and, 37, *37*
   depth of plate recycling and, 45–46, *46*
   discovery of, 24–27
   divergent boundaries and, 27, *28–29, 30,* 32–64
   on Earth, *614*
   effects on deserts, 591
   effect on deltas, 543
   forces of, 214
   geologists reconstruction of plate movements, *40,* 40–41
   grand reconstruction and, 38–44, *39, 40, 42–43*
   Great Synthesis and, 26–27, *27*
   history of plate movements and, 35, *36,* 37
   igneous processes and, 107–113, *108, 109, 110, 111, 112*
   interactions between climate system and, 75–77, *76, 78*
   in landscape development, *487*

   mantle convection and, 44–47, *45, 46, 47*
   mantle plume hypothesis and, 47, *47*
   measuring plate movements by geodesy and, 37–38, *38*
   measuring plate movements with hot-spot tracks, 138–139
   metamorphism and, *194,* 202–207
   reconstruction of plate movements, *40,* 40–41
   rising convection currents and, 46–47, *47*
   seafloor spreading and, 25–26, *26*
   theory of, 14, 24, 27
   transform faults and, 27, *31,* 34, 39
Platform, 641
Platinum, 82, 102
Platte River, 549
Platy, 69
Playa lakes, 590
Playas, **590,** *590*
Pleistocene epoch, 247, 260, 444
Pleistocene glaciation, 424, 445
   geologic record of, 444
Pleistocene ice ages, 350–353
   timing the, *351,* 351–353
Pliocene epoch, 247, 260
Plumes, mantle, 46–47, *47,* 107, 137
Plunging folds, 223, *223*
Pluto, 607, 610–611, *611*
   moons of, 610
   New Horizons visit to, 607, 611
Plutons, **105,** *105*
Point bars, **529,** *529,* 531
Point Lake, Northwest Territories, Canada, *654*
Point Lobos State Reserve, *461*
Polar climates, 499
Polar region, desserts in, 579
Polar vortex, 341
Polar zones, 338
Pollen, 585
Pollution, lead, 517
Polymorphs, **60**
Population growth, human, *366,* 366–367
Po River Delta, 543
Porosity, **169,** 504–505
   fracture, 504
   intergranular, 504
   in rocks, *505*
   vuggy, 504–505
Porphyritic crystals, *94*
Porphyroblasts, **198–199,** *199*
   garnet, 198, 203
Porphyry, **94–95**
Positive feedback, **347,** 455
Potable water, *519,* **519**–520
Potassium, 604, 668
   radioactive isotopes of, 604
Potassium-40, 253, *253*

Potholes, 536, **536**
Prebiotic soup hypothesis, 680
Precambrian, **250,** 640
Precambrian-Cambrian boundary, 686
Precipitate, **60**
Precipitation, **498,** 501, 543
Pressure
    as cause of metamorphism, *192,* 192–193
    melting and, 98
Primates, 690
Producers, organisms as, 668*t*
Proteins, 668, 669
Proterozoic eon, **257,** 642
Proterozoic glaciation, 445
Proto-Atlantic Ocean, 649
Protons, 55
Proto-Sun, 601, *602*
P-T (pressure-temperature) path, 202–**204,** *203,* 206
Pumice, **94,** *94,* 123
Purple loosestrife, *667*
P (primary) waves, **275,** *276,* 277, 284, 308, *308,* 309, *309,* 316
Pyrite, 61, 64, *64,* 80, *80,* 157, 676, *676*
    precipitation of, 676, *677*
Pyroclastic deposits, *123,* 123–124
Pyroclastic flows, **124,** *125*
Pyroclasts, **94,** 123
    extrusive, *94*
Pyroxenes, *62,* 66–67, *67,* 74, 95, 97, 100, 200, 312, 459, 680

Quads, *369,* 376
Quartz, 70, *71,* 83
    crystal faces of, *52–53, 59, 69, 71*
    formation of, 60
    grains of, 581
    in schist, 197
    a silicate mineral, 63, 68, 74, 95
    structure of, *62*
Quartz arenites, **173,** *174*
Quartzite, 197, 198
Quasi-stellar radio sources, 37–38

Radar altimeters, 322
Radial drainage, **532,** *532*
Radiation, 685
Radiative damping, 347
Radiative damping feedback, 347
Radiative forcing, **401**
Radioactive isotopes, 253, 253*t,* 256
    and ages of Earth materials, 252, 254–255
Radioactive wastes, 518
Radioactivity, 604
    discovery of, 250
Radium, 250
Rain, 498–499
    acid, 376, 379, 457
    and climate, 455
Rain forests, *677*

Rain shadow, 344, **499,** *499*
Ramp morphology, 178
Rapids, 536
Raven, Peter, 410
Recharge, **506**
    balancing with discharge, *507,* 508–511, *509, 510, 511*
Record
    geologic, 4–5, *5, 6*
    reconstructing geologic history from the stratigraphic, *241,* 241–247, *242, 243, 244, 245, 246*
Rectangular drainage, **532,** *532*
Recurrence intervals, **271,** *272*
Red beds, **684,** *684*
Red Sea, 32
Redwall Limestone, 251
Reefs, *177,* **177**–178
    coral, 178, 179, *179,* 357
    evolutionary processes and, 178
    organic, 164
Refraction, wave, 562–564, *563*
Regional metamorphism, **74,** 75, **193,** *194*
    metamorphic grade and, 199–201
Reid, Henry Fielding, 269
Rejuvenation, **639**
Relative age, **241**
Relative dating, 251, *252*
Relative humidity, **498**–499
Relative plate velocity, 37
Relief, **483**–484, *484*
Renewable energy, future of, 386
Renewable resources, **367**
Representative concentration pathway, **401,** *401*
Reserves, **368**
Reservoirs, 497, 503
Residence time, **355,** 497
Resources
    energy, *367,* 367–368
    nonrenewable, 368
    renewable, 367
Respiration, **670**
    bacterial, 459
    comparison of photosynthesis and, 670*t*
Reverse fault, 221
Reykjavik, geothermal heating of, 147
Rhine Valley, 32, 229
Rhône River, 543
R-horizon, 464
Rhyolite, 95, **96**
Rhyolitic lavas, *122,* 122–123, **122**–123
Richards, Mark, 322
Richter, Charles, 277, 279
Richter scale, 409
Riff basins, **160**
Rift, 26
Rifting, 160–161
Rift valleys, 229, 552
Right-lateral fault, 221

"Rime of the Ancient Mariner" (Coleridge), 496
Rimmed shelf morphology, 178
Rinds, 456
Ring of Fire, 27, *27,* 133, 320
Rio Grande River, 638
Rio Tinto, 674, *674*
Rip currents, 563–564
Ripples, **166,** *166, 167,* **540,** *540*
River, **528**
Rivera Plate, 40
Roches moutonnuées, 435–436, *436*
Rock(s), 54, **70**–75
    age of Earth's, 17
    brittle and ductile behavior of
        in the Earth's crust, 219
        in the laboratory, 217–219, *218*
    coarse-grained, 92
    evaporite, 178
    evidence of early life in, 682–683, *683*
    felsic, *94,* 95*t*
    fine-grained, 92
    foliated, 195–197, *197*
    granoblastic, **197**–198, *198*
    identity of, 70–71
    igneous, 71–73, *72, 73* (*See* Igneous rocks)
    mafic, *94,* 95*t*
    mass movements of, *475,* 475–478, *477*
    melting of, 97–99, *98*
    metamorphic (*See* Metamorphic rocks)
    natural zones of weaknesses in, 461–462
    phosphate, 181
    porosity of, 505, *505*
    properties of, 70–71, *71*
    of seafloor, 35–36
    sedimentary (*See* Sedimentary rocks)
    ultramafic, 97
    volcanic, 92–93, *94,* 123, *123*
    weathered, 464
Rock avalanches, 476, *477*
Rock cycle, **75**–77, *76, 78*
    surface processes of the, 154–160, *155*
    weathering as process in, 454
Rockfalls, 475, *475*
Rock flow, 435
Rockslides, 475, *475, 476*
Rocky Mountains, *232,* 343, 635
Rodinia, **44,** 649, 651
Ross Ice Shelf, 426, 433
Ross Sea, 426
Rowland, Sherwood, 396, 397
Rubble, 471
Rubidium, 254
Rubidium-87, 253, *253*
Ruby, *69*
Runoff, **498**
    collection and storage of, 502–503
    hydrology of, 501–503, *502*

Rwenzori Mountains, *425*
　glacier cover of, 429

Saffir-Simpson hurricane intensity scale, 575–576, *576t*
Saguaro National Park, *589*
Sahara Desert, 340, 499, 579, 580, 592–593
　climate in, *592*, 593
Sahel, 499
St-Laurent, Louis S., *4*
St. Lawrence River, 534
St. Louis, 534
St. Pierre, 124
Saline lakes, 162
Salinity, **160**
Saltation, **539**, 540, 584, *585*
Salt dome trap, *372*, 373
Salton Sea, 104
Saltwater margin, 511
San Andreas fault, 644
　"Big Bend" in the, 230, *231*, 286, *286*–287
　deformation texture and, 228
　segments of, *269*
　space shuttle photo of, *231*
　as transform fault, *31*, 34, *34*, 40, 41, 219, *219*, 268
San Bernardino Mountains, 230
San Clemente Island, California, *570*
Sand, 172–173, **470**
　grains of, *581*
　mineralogy of, 173
　particle size, 580–581
　sizes and shapes of grains, 173
　unconsolidated, 470–471
Sandblasting, **581**, *581*
Sand budget, 566, 567, *567*
Sand drift, 583, *584*
Sand dunes, *583*, 583–585
　formation of, 583–585, *584*
　linear, *583*
　movement of, 583–585, *585*
　types of, 585, *586*
Sand rights, 566–567
Sandstone, 74, 173, **173**, *173*, 520
　cross-bedding in, *152–153*
　kinds of, 173–174, *174*
　mineralogy of, 173
Sandwich Harbor, Namibia, *558–559*
San Fernando earthquake (1971), 296
San Francisco earthquake (1906), *27*, *31*, 34, 268, *268*, 270, 271, 292
San Gabriel Mountains, 230
San Joaquin Valley, California, 510
San Juan Bautista, 269
San Juan River, *529*
Santorini, 141
Sapphires, 69, *69*
Saturated zone, **506**
Saturation, 460
Saturn, 602, 607, *627*

Scarp, 221
Scenic ocean, 10, *10*
Schist, 74, 190, 196
　feldspar in, 197
　quartz in, 197
　Vishnu, 251, *642*
Schistosity, 197
Science
　geology as a, 3–5, *4*
　goal of, 2
Scientific collaboration, importance of, 3
Scientific method, **2**, 9
　overview of, 2–3
Scientific research, 2
Scientific theory, 2–3
Sea, 560
Seafloor
　inferring age of, 35–36
　magnetic anomalies on, 35, *36*, *37*
　relative plate velocity and, 35–36
Seafloor isochrons, 38–40, *39*
Seafloor metamorphism, **194**, *194*, 202
Seafloor sediments, carbon release from, 408
Seafloor spreading, 25–26, *26*, 37, 41, 45, 46, 109, 110–111, *111*, 317, 320, 604, 640
　hypothesis of, 26–27, 35
Seafloor topography, theory of, 316
Sea ice, 342
Sea level, 6
Sea-level change
　effects of, 572, *572*
　glaciation and, 443–444
Sea-level rise, 405–406, 435, 572
　reasons for, *406*, 406–407
Seawalls, 567
Seawater, 560
Secular variation, *314*, 314–315, *315*
Sedimentary basins, 636
　formation of, 154, **160**–162, *161*
Sedimentary deposits, 82–83, *83*
Sedimentary environments, **162**–165, *163*, 165, 165t
　continental, 162, *163*
　shoreline, 162, *163*
　siliciclastic *versus* chemical and biological, 164–165, *165*, 165t
Sedimentary landforms, 434–441, *438*
Sedimentary rocks, **71**, *72*, **73**, *73*–74, 93, 670
　classification of, *172*, 172–178, *175*, 180
　as coal resource, 376
　common minerals of, 73, *73*, 74, 74t
　on Mars, 692, *692*
　organic, 182
　seismic imaging of, 259, *259*
　stratigraphy of, 623
Sedimentary sequence, 245

Sedimentary structures, **165**–168, *166*, *167*
　bioturbation in, 166, *167*
　cross-bedding in, 165, *166*
　graded bedding in, 165–166
　ripples in, 166, *166*, *167*
Sedimentation, 153–182, *154*, *155*
　burial in, 168
　chemical, 162
　classification of chemical and biological sediments and sedimentary rocks, 175–178, *175t*, *176*, *177*, *178*, *181*, 181–182
　classification of siliciclastic sediments and, 172–175, *172t*, *173*, *175*
　delta, 541
　diagenesis in, 168–170, *169*
　glacial, 434–441, *438*
　looking for oil and gas, 170–171, *171*
　oceans as chemical mixing vats in, 160
　surface processes of the rock cycle and, 154–160, *155*
　transportation and deposition in, 157–160, *158*, *159*
　weathering and erosion in, 155–157, *156*
Sediment bed forms, *540*, 540–541
Sediments, **73**, 77, 670
　alluvial, 589
　bioclastic, 157, *157*
　biological, 156–157, *157*
　chemical, 74, 156, *156*
　eolian, *579*, 589
　evaporite, 178, 590, *590*
　iconoclastic, 74
　iron oxide, 182
　layers of, 74
　minerals present in, 156t
　organic, 182
　phosphorite, 181
　siliceous, 181
　siliciclastic, 74, 156, *156*
　sources of, 155–157
Seismic Experiment for Interior Structure (SEIS), 622
Seismic exploration of near-surface layering, *311*, 311–312
Seismic hazards, **293**, *294*, *295*
Seismic imaging, 373, *373*
Seismic mapping, *285*
　of oil traps, 373
Seismic moment, 278
Seismic profiling, 311
Seismic ray paths, **308**–311, *309*, *310*
Seismic risk, **293**, *294*
Seismic tomography, three-dimensional image created by, 320, *321*
Seismic waves, **7**, 8–9, 306–307
　in studying earthquakes, 275, *276*, 277, 278
　in studying Earth's interior, 290, 308–311, *309*, *310*
Seismograms, 274, 309

Seismographs, **274**
    in studying earthquakes, 274, *274*
Seismology, 308, 315–316, 319
Seismometers, 7
Self-organized natural system, 326
Sequence stratigraphy, *259*, 259–260
Settling velocity, 103, **538**–539
Shadow zone, **310**
ShakeMaps, 297
Shale, *173*, **174**, 190, 200
Shale oil, 174, 377
Shearing forces, **214**
Shearing tectonics, *228*, 230, *231*
Shear waves, 9–10, 308, **308**, 310
Shields, **635**, 641
Shield volcanoes, 125, *126*
Shiprock, 127
Shock metamorphism, **195**, 202
Shoemaker, Eugene, 608, *608*
Shorelines, **560**
    deposition of, 568, 570–572
    sedimentary environments in, 162, *163*
    shaping of, 565–572
Shuttle Radar Topography Mission (SRTM), 654–655
Siberian basalts, 97, 141
Siberian traps, 400
    mass extinction and, *140*, 140–141
Sierra Mountains, *462*
Sierra Nevada Mountains, 205, 343, 639, 643
Silent earthquakes, 283–284
Silica, 8, 95
Silicates, 61, **61**, *61*, *62*, 63, 77
    chemical weathering and, 456
    composition of, 63
    frameworks of, 63
    in metamorphic rocks, 190–191
    structure of, 63
    weathering of, *458*
Silicate tetrahedra, 63
Silicate weathering, 357
Siliceous sediments, 181
Siliciclastic sedimentary environments, 164–165, *165*
Siliciclastic sediments, **74, 156**
    coarse-grained, 172, 172t, *173*
    fine-grained, 172t, *173*, 174–175, *175*
    medium-grained, 172–174, 172t, *173*
Silicon, 11, *11*, 58, 61
Sillimanite, 190
Sills, **105**–106, *106*, 246
    textures of, 106
Silt, **174**, **470**
    unconsolidated, 470–471
Siltstone, **174**
Sink area, 154
Sinkholes, **516**, *516*
Sklodowska-Curie, Marie, 250
Skykomish River, 546, *547*
Slate, 455
Slaty cleavage, 195
Slave Craton, *656*
Slave Province, 656
Slip direction, 219
Slip face, 585
Slip rule in controlling earthquakes, 282–283
Slope, building on, *476*, 476–477
Slope materials, 469–471
Slumps, **480**, *481*, 482
Smith, William, 242–243
Smog, 379
Snowball Earth hypothesis, 445
Snowfall, 498
Snow line, 428
Society Islands, 136–137
Socolow, Robert, 414
Sodium, 56
Sodium chloride, 56, *57*, 58, 160, 180, 520
Soil, **462**–464
    as geosystem, *463*, 463–464
    production of, 455
    types of, 465t
    weathering and, 459
    wind erosion of, *582*
Soil liquefaction, 290
Soil profile, **464**
Sojourner, 617, 618
Solar energy, 12, 385–386, **385**–386, *386*
Solar forcing, **338**, 394
Solar nebula, **601**
Solar power, 368
Solar radiation, 346–347, 486
Solar system(s), 600
    exploring, 626–627
    exploring other, 628
    formation of planets in, 601–602
    formation of Sun in, 301
    nebular hypothesis of formation of, 600–601, *602*
    origin of, 600–603
Solar wind, 16, *17*, 329
Solifluction, 478
Solubility of mineral, 459, **460**
Solutions, 60
    hydrothermal, 79–80
Somali Subplate, 32
Sorting, **159**, *159*
Soufriere Hills, eruption of, *117*, 142, *143*
Sound waves, 308
Source area, 154
South African Plateau, 653
South American Plate, 33, 284
Southern Hemisphere, 341
Southern Ocean, 44, 560
South Platte River, 549
Space exploration
    modern era of, 600
    scientific dividends of, 600
Space missions, 626
Species diversity, effects of climate change on, *403*
Specific gravity, **69**
Sphalerite, 80, *80*
Sphere, volume of, 9
Spinels, 64, *64*
Spirit, 619, *619*, 620–621, 626
Spit, **570**, 571
Spreading center, **32**
    as magma factory, *108*, 108–111
    volcanism at, *131*, 133–134
Spring tides, 564–565
Sputnik Planitia, 611, *612*
Stabilization wedges, **414**–415, *415*
Stacks, 570
Stalactites, 516
Stalagmite, 516
Stars, habitable zones around, *691*, 691–692
Staurolite, 74, 191
Steam technology, 368
Steam vents, 118
Steno, Nicholaus, 242, *242*, 244
Stillwater deposits, 102
Stishovite, 195
Stocks, **105**
Stokes' law, 103
Storm surge, **573**, 576–578, *577*
Storm tide, 577, *577*
Strabo, 92
Strain, 269
Strait of Gibraltar, 560
Stratification, 165
Stratified convection, *46*, *46*
Stratigraphic record, reconstructing geologic history from the, *241*, 241–247, *242*, *243*, *244*, *245*, *246*
Stratigraphic succession, **242**
Stratigraphic traps, *372*, 373
Stratigraphy, **242**
    chemical, 260
    of the Colorado Plateau, 251, *252*
    paleomagnetic, 260, *327*, 327–328
    sequence, *259*, 259–260
Stratosphere, **339**
Stratospheric ozone depletion, **397**
Stratovolcanoes, **125**, *126*
Streak, **68**, *68*
Streak plate, 68, *68*
Stream(s), **528**
    braided, *539*, 539–540
    channel beginnings and, *535*, 535–537
    channel patterns and, 529–531, *530*
    current flow of, 537–540, *538*
    damming of, 550
    discharge of, 544–546, *545*, *546*, *548*
    drainage basins and, *531*, 531–532, *532*
    drainage networks and, 533, *533*
    drainage patterns of, 533, *533*, 535
    effects of climate on, 551
    effluent, 506
    erosion and sediment transport and, 538–539, *539*

flooding by, 546, *547*
floodplains and, 531, *531*
form of, *528*, 528–529, *529*
as geosystems, 543–547, *544, 545, 546, 547*
graded, 551
influent, 506
longitudinal profiles of, *549*, 549–552, *551, 552*
as primary erosion agent, 589
sediment transport by, 538–539, *539*
superposed, 535, *535*
Streamgauge data in planning a river trip, 547–548, *548*
Streamlines, 537
Stream power, **464**–465
Stream valleys, *528*, 528–529
 erosion and formation of, 464–465, *466*, 467
Stress, **192**, 269
Striations, **435**, *436*
Strike, **214**
 measuring, 214, 216, *216*
Strike-slip faults, **220**, *220*, 269, 281, 282
Stromatolites, **677**–678, *678*, 680
Strontium, 254
Structural traps, 372–373
Subchrons, 328
 magnetic, 35
Subduction, **32**, 192, 604
Subduction-related metamorphism, *200, 204*, 204–205
Subduction zones, 604
 earthquakes in, 271
 as magma factories, 111–113, *112*
 volcanism in, *133*, 134, 136
Sublimation, 498
 in ablation, 428
Subpolar lows, 340
Subsidence, **160**, 510
Subsidence basins, thermal, 160–161, *161*
Subtropical highs, 340
Succession
 faunal, *243*, 243–244
 stratigraphic, 242
Suess, Eduard, 24, 635
Suez Canal, 560
Sulfates, **61**, *61*
Sulfide minerals, formation of acidic waters on Earth and Mars, 674, *674*
Sulfides, **61**, *61*, 64, 77
Sulfur, 668, 676
Sulfur dioxide, residence time of, 355
Sumatra earthquake (2004), 268, 280, 284, 291–292
Sun
 formation of, 601
 gravitational pull of, *564*, 564–565
 nuclear fusion of, 601
 tides and, *564*, 564–565
Supai Formation, 251

Supernovas, 83
Superplumes, 320, 653, *653*
Superposed streams, **535**, *535*
Superposition, principle of, **242**
Supervolcanoes, 127
Surface mining, 379
Surface temperature, 338, 338t
Surface tension, 471
Surface waves, 275, **275**, *276*, 277
Surf zone, 562, *562*
Surge, **431**
Suspended load, **538**, *539*
Suture, **205**
Swamps, 503
Swash, 562, 563
Swash zone, 566
S (secondary) waves, **275**, *276*, 277, 308, *308*, 309, *309*
Swell, 561
Swiss Alps, Weissmies Glacier in the, *424*
Switchgrass, 383–384, *384*
Sylvite, 60
Symmetrical folds, *223*
Synclines, **222**, 486, *486*

Tablelands, 485
Taconic orogeny, 651
Taconic State Parkway, 651
Talus, **471**
Tapeats sandstone, 246, 251
Tar sands, 378, *378*
Tectonic age, **642**
Tectonic deformation, 27
Tectonic provinces, **634**, *634*, 640–643
 types of, 640–642, *641*
Tectonics. *See also* Plate tectonics
 compressive, 229, *230*
 interactions between climate, topography and, 486, *487*, 488
 shearing, *228*, 230, *231*
 tensional, 229, *229*
Tektites, 195
Temperate climates, 499
Temperate zones, 328
Temperature
 as cause of metamorphism, *191*, 191–192
 and climate, 455
 inside the Earth, 317, 319
 melting rocks and, 98
 surface, 338, 338t
Temple Butte Limestone, 251
Tensional forces, **214**
Tensional tectonics, 229, *229*
Terminal moraine, 439t
Terraces, **552**, *552*
Terrestrial planets, **602**, *602*
 characteristics of, 606t
Terrigenous sediments, **165**
Tethys Ocean, 41, *42, 43,* 647, 649
Teton Range, 639

Tetrahedron, *56, 61, 62. See also* Silicates
 carbon atoms bonding to, 57
 explanation of, 58
 isolated, 63, 66, 68, 95t
 in sheet structures, 63, 66
 silicate, 61, *62*, 63, 66, *67*
 sulfate, 65
Texture(s), **71**
 deformation, 227–228, *228*
 of igneous rocks, *92*, 92–95
 of metamorphic rocks, 185–199, *196, 197, 198, 199*
 of volcanic rocks, 123, *123*
Thames River, 534
Tharp, Marie, 26, *26*
Theories, 3. *See also* Hypotheses
 Big Bang, 18, 83, 600
 elastic rebound, 269, *270*, 271, 272, 300
 of evolution, 245, 247
 of natural selection, 245
 of plate tectonics, 14, 24, 27
 scientific, 2–3
Thera, explosion of, 141
Thermal subsidence basins, 160–**161**, *161*
Thermococcus, 672
Thermohaline circulation, **342**
 reduction of, 408
Thermometer, invention of, 397
Thermophiles, 673, 673t
Thermoremanent magnetization, **326**–328
Thompson, Lonnie, *400*
Thoreau, Henry David, 570
Thorium, 604
 radioactive isotopes of, 604
Three Georges Dam, 384, *385*
Three Mile Island, 383
Thrust fault, **221**
Tiber River, 534
Tibetan Plateau, *31, 33, 34,* 442, 485, 649
Tidal flats, 162, **565**, *565*, 570, 676
Tidal waves, 291
Tides, 560, **564**, 564–565
 causes of, 560
 ebb, 565
 effects on deltas, 543
 flood, 565
 neap, 565
 spring, 564–565
 storm, 577, *577*
Tide terrace, 565, *565*
Tight formation, **373**
 production of oil and gas from, 373
Tight gas, 375
Tight oil, 377
Tigris River, 534
Till, **438**, *438*
Tillites, **445**
Time, measuring absolute, with osotopic clocks, 249–250
Tin, 102

Titan, 627
　methane on, 627
Titanium, 64, 69, 102
Tohoku earthquake (2011), 34, *267,* 273, 284, 288, 297
"Tongue stones," 242, *242*
Topography, **6,** *7,* **482**
　interactions between tectonics, climate, and, 486, *487,* 488
　karst, 516, *517*
Topset beds, **541**
Tourmaline, 82
Trace elements, **69**
　mineral color and, 69
Trade winds, 341
"Tragedy of the commons," 394
Trans-Alaska pipelines, 441, *442*
Transantarctic Mountains, *425*
Transformations, 464
Transform-fault boundaries, *31,* 39, 284
　continental transform fault, *31*
　mid-ocean ridge transform fault, *31*
Transform faults, **27,** *31,* 34, 202, 230
　continental rift zone, *30*
　ocean spreading center, *30*
Transition zone, 11, **313**–314
Translocations, 464
Transpiration, 498
Transportation, 154, *155*
　abrasion during, 159, *159*
Transverse dunes, *586*
Travel-time curves, 277
Tree River Folds, *212*–213
Trellis drainage, **532,** *532*
Triassic period, 637–638
Tribolites, 686
Tributaries, 533, **533**
Trichlorethylene (TCE), 518
Trilobites, *239,* 244
Tropical depression, 575
Tropical rain forests, conversion to other land uses, 410
Tropical storms, 575
Tropical zones, 338
Troposphere, **339**
Tsunami barrier, 297, *297*
Tsunamis, 268, 288, *291,* **291**–292, *292,* 293
　warning systems for, 291–292, 297, *297*
Tsunami stone of Aneyoshi, 275, *275*
Tuffs, **124**
Turbidity current, 164
Turbulence, 537
Turbulent flow, **537**
Twentieth-century warming, **397,** *398,* 400
Typhoons, 573

Ultisols, 465t
Ultra-high-pressure metamorphism, **195**
Ultramafic rocks, **97**
　in Earth's crust, 312

Unconformities, *245,* **245**–246
　angular, 245. *246*
　classification of, 245
Unconsolidated material, mass movements of, 478, *479,* 480
Unconsolidated mixtures, 471
Uniformitarianism, **5,** 240
United Nations Framework Convention on Climate Change, 399
United States
　energy consumption in, 369–370, *370*
　energy-efficient technologies in, 370
　energy production in, *369,* 369–370
　natural gas reserves in, 376
　oil production in, 375, *375*
　oil reserves in, *374,* 374–375
Universal ancestors, 672
Universal tree of life, 672, *672*
Unsaturated zone, **506**
Upper mantle, 11, **312**
Ural Mountains, 651
Ural orogeny, 651
Uranium, 82, 85, 604
　radioactive isotopes of, 604
　reserves, 382, 383
Uranium-235, 253, *253,* 256
Uranium-238, 253, *253*
Uranus, 602, 607
U-shaped valley, 425, **436**

Vadose zone, 506
Valles Caldera, 127
Valley and Ridge Province, 636, 637
Valley glaciers, **425,** *425,* 430, 436
　erosion by, *437*
　flow in, *430,* 431
Valleys, 528, **528**–529
　rift, 552
　stream, *528,* 528–529
　U-shaped, 425, 436
　V-shaped, *437*
Vallis Marineris canyon, *614,* 615
Variscan orogeny, 651
Varves, **439**–440, *440*
Vasey's Paradise, Marble Canyon, 505
Veins, **80, 107,** *107*
　hydrothermal, 107
Velocity, settling, 538–549
Venezuela, Orinoco Basin in, 375
Ventifacts, **581,** *581*
Vents
　hydrothermal, *681*
　steam, 118
Venus, 602, *603,* 605, 606, *607,* 611–613, *612, 613,* 691
　atmosphere of, 600
　characteristics of, 606t
　flake tectonics on, 613, *614*
　topography of, *607*
　volcanism on, 613
Vertebrates, endangered, 410

Vertical movements, modification by, 651, 653, 655
Vertisols, 465t
Viking missions, 617–618
　Viking I lander, 617
Vine, F. J., 35
Viscosity, **97,** 537–538
　magma, 103
　settling, 103
Vishnu schist, 251, *642*
Volcanic activity, 605, *605*
Volcanic ash, **94, 95,** 135, 587
Volcanic bombs, 124, *124*
Volcanic domes, 125, *126*
Volcanic dust, 587
Volcanic ejecta, 123–124, *124*
Volcanic eruptions
　predicting, *144,* 144–145
Volcanic geosystem, **118**
Volcanic glass, 95
Volcanic hazards, 118, 141–144
　reducing the risks of, *144,* 144–145
Volcanic rocks, 92–93, 94
　textures of, 123, *123*
Volcanism, 93, 118, 344, 637
　atmosphere and, *132,* 132–133
　global pattern of, *133,* 133–137, *134*
　human affairs and, 140–145, *147*
　hydrosphere and, *131,* 131–132
　interplate, 136–137, *137*
　at spreading centers, *131,* 133–134
　in subduction zones, *133,* 134, 136
Volcanoes, 13, 117–147, *118.* See also Eruptions; *specific by name*
　central eruptions of, 125–128, *126, 127*
　eruptive styles and landforms, 124–130, *127, 128, 129, 130*
　fissure eruptions of, 128–130, **129,** *129*
　as geosystems, 118, *119*
　interactions with other geosystems, 130–133, *131*
　natural resources from, 145, 147
　shield, 125, *126*
V-shaped valleys, *437*

Wadis, 590
Wapatai shale, *246*
Warming
　global, 14, 345, 400, 405–406
　twentieth-century, **397,** *398,* 400
Warrawoona Formation, *681*
Wasatch Formation, 251
Wasatch Range, 639
Washes, dry, 590
Wastes, radioactive, 518
Water, 56, 668–669
　ancient, 515
　in chemical weathering, 456, *457*
　in crust of the Earth, 520, 520–521, *521, 522*

fresh, 198
geologic cycling of, *496*, 496–498
hydrothermal, 520–521, *521, 522*
melting and, 98–99
potable, *519*, 519–520
quality of, 516–520
rights to, *500*, 500–501
total supply of, on Earth, 497
Waterfalls, **536**
undercutting action of, 536–537, *537*
Water-laid deposits, 439–440, *440*
Water resources, effects of climate change on, *403*
Water supplies
contamination of, 517–518
increasing, 501
Water table, 506
perched, 508, *509*
Water vapor, 497, 498
Water vapor feedback, 347
Watt, James, 368, 411
Wave-cut terrace, **570**
Wave refraction, 562–564, *563*
Waves
in coastal processes, 560–562, *562*
compressional, 9, 10, 275, 308, 310–311
height of, 561
length of, 561, 562
light, 308
P (primary), **275,** *276,* 277, 284, 308, *308,* 309, *309,* 316
periods of, 561, 562
S (secondary), **275,** *276,* 277, 308, *308,* 309, *309*

seismic, 7, 8–9, 275, *276,* 277, *278,* 290, 308–311, *309, 310*
shear, 9–10, 308, 310
sound, 308
speed of, 560
surface, 275, *276,* 277
tidal, 291
velocity of, 561–562
Weather, 14
Weathered rock, 464
Weathering, **73,** 85, 154, *155,* 155–157, **454**
chemical, 154, 454, 455, 456, 536
factors controlling rates of, 454–456, 454*t*
interactions between erosion and, 467, *468*
physical, 154, 454, 455, 456, 536
precipitation and, 499
of rock, 666
silicate, 357
Wedging, 105
Wegener, Alfred, *24,* 24–25
Weissmies Glacier, *424*
Wellfleet, Massachusetts, gravestones in, *455*
Wenchuan earthquake (2008), 229
Westeries, 341
Wetlands, 503
White Sands National Monument, New Mexico, wind ripples in, *584*
Whole-mantle convection, *46, 46*
Wiechert, Emil, 7–10
Wilkins Ice Shell, 433
Wilson cycle, **651,** *652,* 655
Wilson, E. O., 411
Wilson, J. Tuzo, 27, 651

Wind
in sand dune formation, 583
solar, 329
strength of, 580, *580*
Windblown dust, 585–587, *586, 587*
Wind energy, 384–385, **384**–385, *385*
Wind erosion in ablation, 428
Wisconsin glaciation, 443, *443,* 444
Wood, combustion of, 367
Wooly mammoth, 428
World Meteorological Organization (WMO), 399
World ocean, 560
World population, 366
Wrangellia, 644
Würm glaciation, 443

Xenoliths, 105, 658

Yangtze River, 503*t,* 528
Yellowstone Caldera, 137
Yellowstone National Park, 118, 131
Younger Dryas, 444
Yucatan coast, oil spill off of, 380, *380*
Yucca Mountain Nuclear Waste Repository, 383, *383*

Zagros Mountains, 485, 647
Zebra mussels, *667*
Zeolite, **200**
Zinc, chemical substitution in forming, 193
Zinder, Niger, drought in, *500*
Zion Canyon, 251, *251*
Zircon, 656

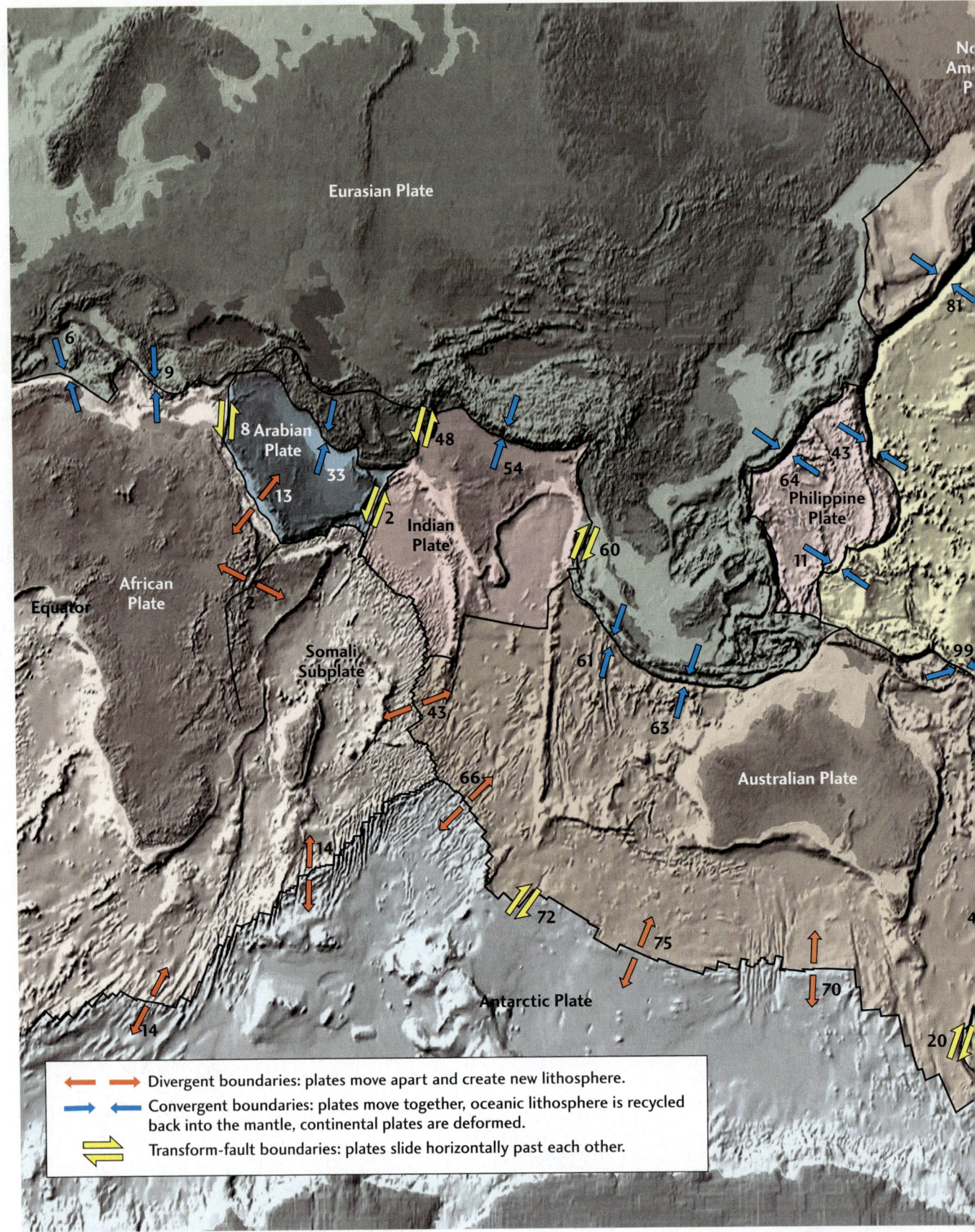

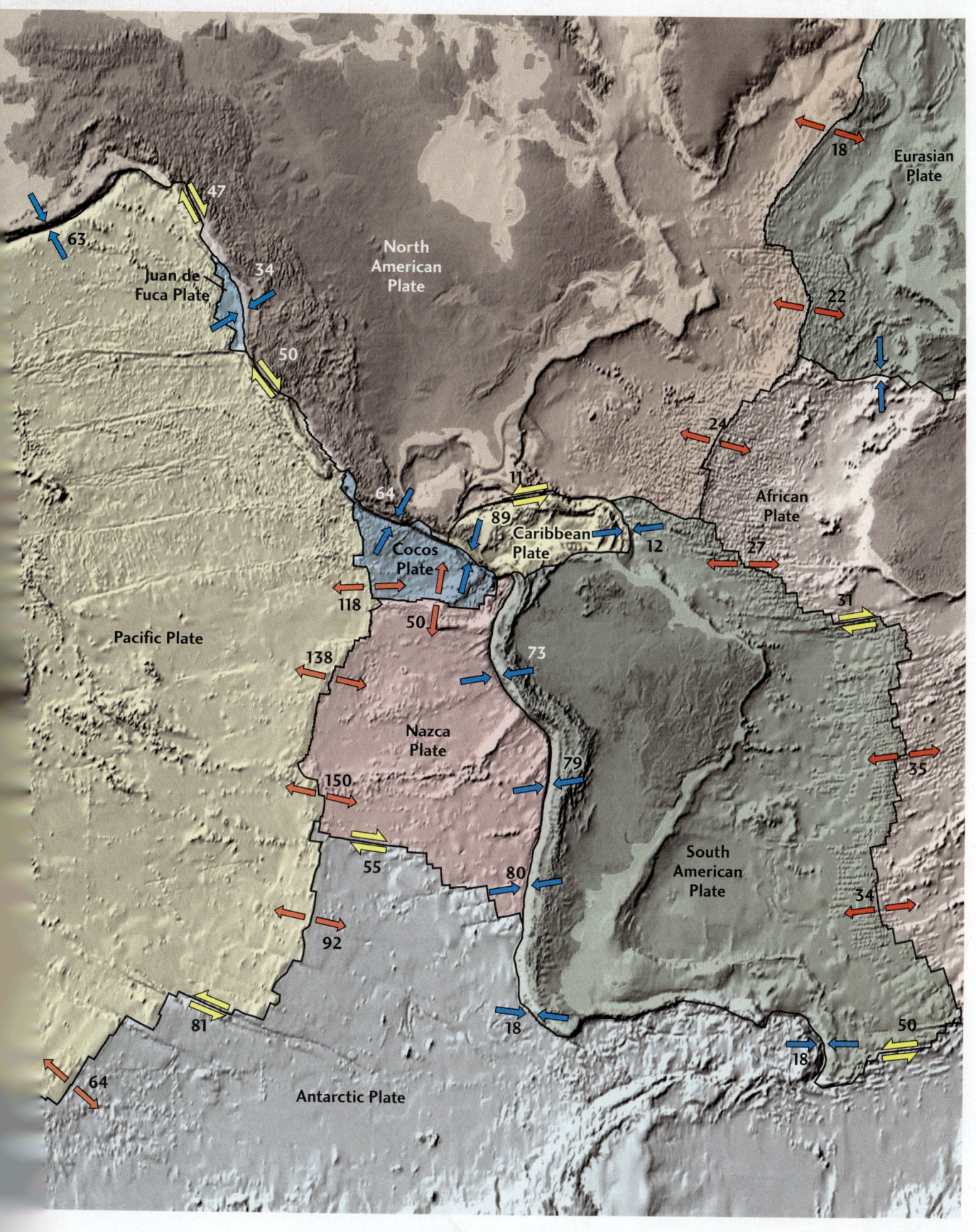